# Signaling in the Heart

# Signaling in the Heart

By

José Marín-García, M.D.
Director, The Molecular Cardiology and Neuromuscular Institute, Highland Park, NJ, USA

With contributions by

**Alexander Akhmedov, Ph.D. and Vitalyi Rybin, Ph.D.**
Senior Research Scientist, The Molecular Cardiology and Neuromuscular Institute, Highland Park, NJ, USA

Springer

José Marín-García, M.D.
Director, The Molecular Cardiology and Neuromuscular Institute
Raritan Avenue 75
08904 Highland Park
NJ, USA
tmci@att.net

ISBN 978-1-4419-9460-8 e-ISBN 978-1-4419-9461-5
DOI 10.1007/978-1-4419-9461-5
Springer New York Dordrecht Heidelberg London

Library of Congress Control Number: 2011928241



Printed on acid-free paper

Springer is part of Springer Science+Business Media (www.springer.com)

*To my wife Danièle and daughter Mélanie with love*

## ART WORKS for 'Signaling in the Heart'

Danièle M. Marin ' Signaling pathway #1' Oil painting

Danièle M. Marin 'Signaling pathway #2' Cast Acrylic Print

Danièle M. Marin 'Signaling pathway #3' Mixed Media Painting

Danièle M. Marin 'Signaling pathway #4' Oil painting

The molecular phenomena of signal transduction pathways are at the core of most biological processes and are critical regulators of heart physiology and pathophysiology. These four paintings, the work of the author's wife, Danièle M. Marin were inspired by some of these signaling pathways.

Artwork #1 Mitochondria: a receiver and integrator of signals.
Artwork #2 Cell cycle signaling.
Artwork #3 Rapid signaling pathways.
Artwork #4 Growth factors signaling pathways.

# Preface

Signal transduction is at the core of most biological processes and represents a critical area of research. Signal transduction is extremely important not only to molecular biology research, but also to clinical medicine in general since many diseases, such as diabetes, cardiovascular diseases, autoimmunity, and cancer, arise from defects in signal transduction pathways.

At the present time, the complex area of cardiovascular signal transduction is in its infancy, and much of the available information came to light as a by-product of extensive research effort to understand the mechanisms of hypertrophy, apoptosis/cell death, and myocardial remodeling. The heart acts as both transmitter and dynamic receiver of a variety of intracellular and extracellular stimuli, as well as an integrator of numerous interacting transducers, including protein kinases and effectors, the G proteins and small G protein activators which are profoundly influenced by their location in the cell. Given that the targeting and localization of signaling factors and enzymes to discrete subcellular compartments or substrates are important regulatory mechanisms, ensuring specificity of signaling events in response to local stimuli, these systems deserve examination from a subcellular/organellar and a functional standpoint under both physiological and pathophysiological conditions. Moreover, cardiovascular signaling includes a built-in specificity, reversibility and a redundancy of its components, which while making their analysis a very complex undertaking, provides the cardiac cells with great plasticity to respond to insult, as well as to growth stimuli. Understanding this plasticity is fundamental in the discovery of new cardiovascular signal transduction techniques and new therapies.

Gene transfer studies have shown promising results in re-engineering defective signal transduction pathways in animal models of cardiac dysfunction, including heart failure, as well as providing cardioprotection against insults such as myocardial ischemia. Similarly, transduction engineering approaches with vascular remodeling (e.g., angiogenesis) and dysfunction (e.g., hypertension) have been successful in clinical trials. In this volume, we present what is presently known in cell signal transduction pathways, genetics, and cellular biology in heart failure, development of novel therapies for to improve cardiac function, as well as where this field is heading in the future.

As the role of genetic screening in cardiology is strengthened and as research on the multiple signaling pathways involved in cardiac organogenesis and pathology progresses, the time seems appropriate for a book that comprehensively integrates known facts, what is developing and what will be known in the near future. In addition to providing a recount of past discoveries, this book deals with areas that are of emerging interest to medical students, cardiologists, and researchers in diverse fields, eyeing new therapeutic modalities that may improve currently available therapies and interventions in the management of human cardiac diseases. Furthermore, we are now witnessing the transition from the Cardiology of the past to the study of systems biology, the constructive cycle of computational model building, and the experimental verification capable of providing the input for exciting new discoveries and hope. The chapters in this book have been arranged in a way that the readers, who browse it, can to some degree recognize and appreciate the current thoughts and ideas on cardiovascular signaling pathways. We have tried to include original and creative scientific works as much as possible, although humbly we must say that this is a work still in progress.

Hopefully, this book will be a valuable guide to signaling of the heart from a post-genomic perspective, and also an important introduction to new ideas and future progress.

*Deciphering the mystery....*
*progress continues*
*Signals are received and sent......*
*for us to interpret*

José Marín-García, M.D.
Highland Park, NJ

# Contents

# Part I
# Methodologies

# Chapter 1
# Tools to Study Signaling

**Abstract** Recently, input from various genomic and transcriptomic projects have resulted in a rapid accumulation of information on sequence and transcription of proteins participating in different signal-transducing cascades. At the same time, several functional aspects of cell-signaling systems are beyond the scope of genomic approaches, including the presence of many genes of unknown function, the absence in many cases of good correlation between messenger RNA transcript level and protein expression level, several different functional protein products of one gene, and posttranslational modifications. Combining molecular biology and molecular genetics techniques with modern proteomics and imaging methodology have produced fruitful insights into our understanding of the mechanisms that underlie signal transduction pathways in the organism in general, and in the cardiovascular system in particular.

**Keywords** Signal transduction • Cardiovascular signaling • Genome • Proteomics • FRET

## Introduction

Signal transduction events involve the transmission and amplification of signals from transmembrane receptors to the nucleus. Large-scale analysis of genome structure and gene expression are new technologies available for studying cardiovascular-signaling systems under normal conditions and during the development of diseases known to modify cardiac function. Genome-wide transcriptomic microarray analysis can reveal genes involved in cell growth and signal transduction that are altered under pathological conditions. With the development in the last 15 years of technologies that allow manipulation of the mammalian genome, it is possible now to create transgenic animals carrying known mutations and determine the subsequent effects at the molecular, biochemical, cellular, and whole organ levels. In this chapter, we describe genetic manipulations which are commonly used today in developing models that test the effects of mutations associated with individual components of different signal-transducing pathways – gene targeting and insertional transgenesis. We also describe gene targeting Cre-loxP system technology utilizing DNA recombinases which is commonly used for tissue-specific gene knockouts, and point out current techniques of RNA interference posttranscriptional gene silencing by small, double-stranded RNAs (siRNAs). Moreover, using the siRNA approach, it is possible to carry out inducible, reversible gene knockdown at different times of development or during unique physiological stresses.

It is worth noting that important mechanisms underlying signal-transducing events are involved in a plurality of molecular responses, such as regulatory reversible posttranslational modifications (PTMs) of signaling molecules and protein–protein interactions (PPIs). PTMs and PPIs (from binary to protein interaction networks) are involved in the regulation of many cellular processes, e.g., modulation of protein folding, function, activity, targeting, and degradation, being the interaction between components of diverse-signaling cascades an important mechanism of the signaling machinery.

## Molecular Biology Methodology

### *PTMs/PPIs and Functional Proteomics*

At the outset, PTMs and PPIs cannot be readily predicted from the genome and therefore have to be analyzed at the protein level. A powerful method for monitoring molecular responses following the activation of signal transduction pathways is proteomics, and in particular functional proteomics. Functional proteomics aims to analyze the global changes of signaling pathways and related signaling molecules. Later, we present an overview of basic proteomics methodologies as well as several complementary techniques, such as PTM-protein enrichments, pull-down approaches, two-hybrid screening systems (Y2H) which are efficiently used to study specific cellular-signaling networks and to identify the critical proteins in the signal transduction pathways.

J. Marín-García, *Signaling in the Heart*,
DOI 10.1007/978-1-4419-9461-5_1, © Springer Science+Business Media, LLC 2011

Another major advance in modern cell biology and physiology has been the ability to perform real-time imaging of signaling pathways in living cells. Expression of fluorescent-labeled biosensors for imaging of fluorescent resonance energy transfer (FRET) (Fig. 1.1) demonstrates the value of methods that allow the preservation of cell integrity and enable visualization of PPIs and other molecular events in living cells as they happen.

## Molecular Cloning

It is not possible to overestimate the significance of molecular biology methodology in the discoveries related to the functioning of cellular signal transduction systems in general, and cardiovascular-signaling pathways, in particular. DNA/RNA-related manipulations underlie the vast majority of modern disease diagnostic tools and drug discovery approaches. Here, we only describe basic molecular biology methods concentrating in a number of molecular genetics/genetic engineering techniques, which have been used (or could be used) to identify and investigate mechanisms of signal transduction cascades in the cardiovascular system.

Major general molecular biology techniques include amplification of nucleic acids [cloning, PCR (Fig. 1.2)] and methods of separation and identification of nucleic acids.

One of the most basic techniques of molecular biology is the amplification of target DNA sequence – molecular cloning. With this approach, the DNA fragment of interest (transgene)

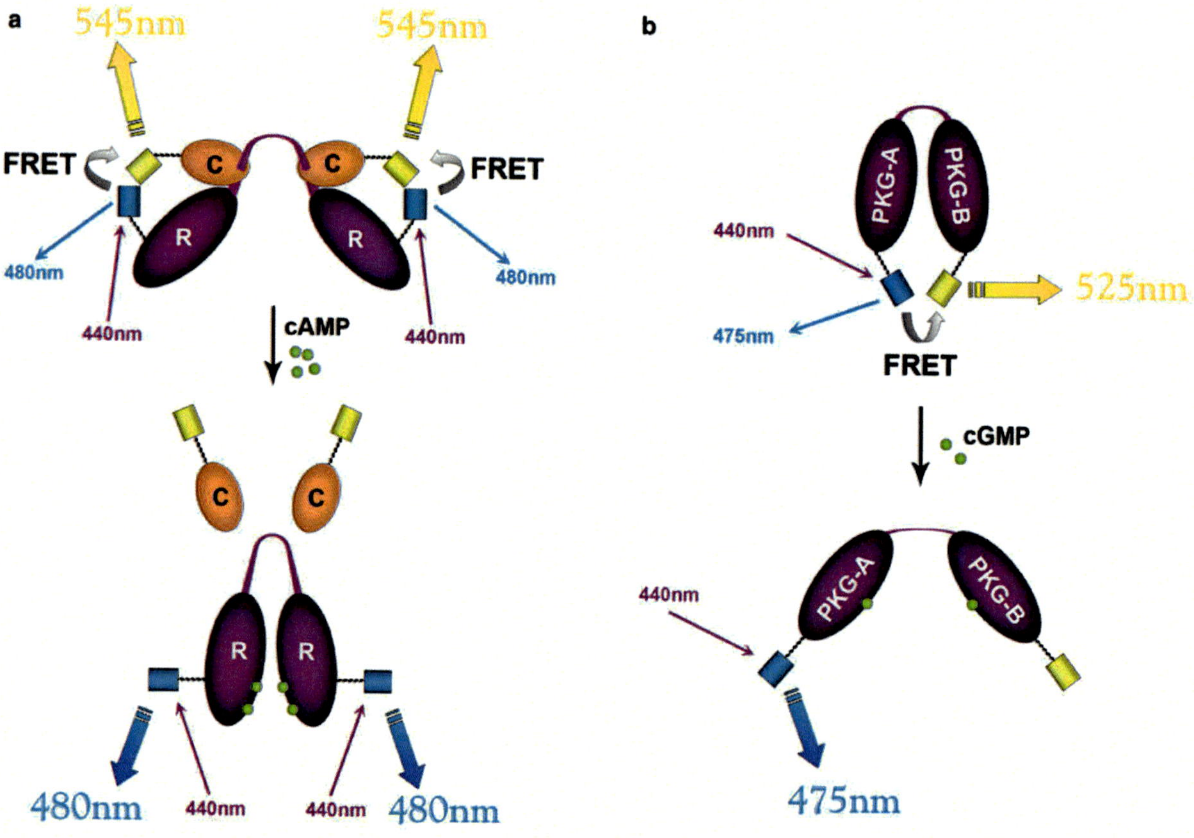


**Fig. 1.1** FRET-based biosensors. (**a**) cAMP-detecting sensor. Two fusion proteins are generated, CFP-tagged regulatory (R) subunit of PKA and YFP-tagged catalytic (C) subunit of PKA. At low cAMP, tagged PKA is a tetramer: $(CFP\text{-}R)_2(YFP\text{-}C)_2$. Upon excitation of the donor CFP (excitation wavelength 440 nm), part of the energy is transferred to the acceptor YFP (FRET), and YFP emits at wavelength 545 nm. When cAMP rises, it binds to R, induces conformational change and dissociation of C-subunits from R-dimer. Increased distance between CFP and YFP abolishes FRET, and only donor fluorescence at wavelength 480 nm takes place. Changes in FRET (measured as donor:acceptor fluorescence ratio, 480 nm/545 nm) directly correlate to changes in PKA holoenzyme concentration and thus to cAMP level. (**b**) cGMP-detecting biosensor. An indicator is designed as one molecule: a fragment of PKG (tandem of cGMP-binding domains PKG-A and PKG-B, respectively) fused N-terminally to CFP and C-terminally to YFP. At low cGMP, CFP-fluorophore locates in close proximity to YFP so FRET is possible when donor CFP is excited at 440 nm. As a result of FRET, YFP emits at 525 nm. cGMP binding induces conformational change of the indicator, and FRET is terminated as distance between CFP and YFP increases. Only donor fluorescence can be observed for cGMP-bound biosensor at wavelength 475 nm. *CFP* cyan fluorescent protein, *FRET* fluorescence resonance energy transfer, *PKA* protein kinase A, *PKG* protein kinase G, *YFP* yellow fluorescent protein. CFP and YFP are shown as *blue* and *yellow cylinders*, respectively

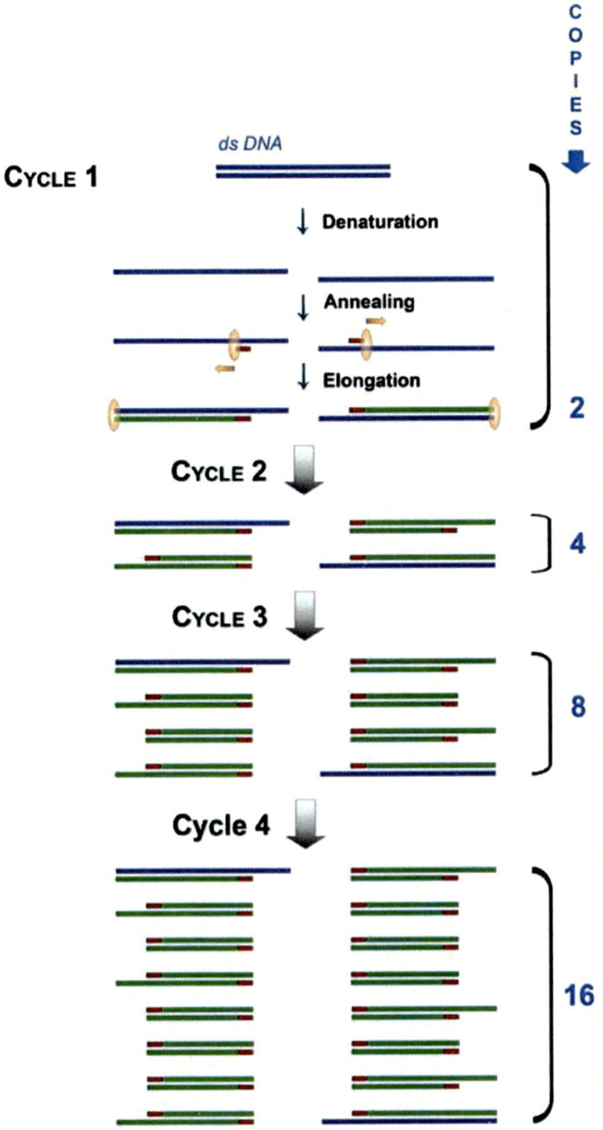


**Fig. 1.2** PCR cycle. Each cycle starts from denaturation of double-stranded DNA template (dsDNA, shown as a *blue lines*) at 94–96°C which leads to strand separation. Then temperature drops down to ~65°C to anneal primers (shown in *red*) and bind DNA polymerase (*brown oval*). Third step, elongation, is an extension of primers by the DNA polymerase at 72°C to give shorter DNA products (*green lines*). DNA products are used as templates as PCR progresses. The amount of targeted DNA is doubled in each cycle, so four cycles shown here produce 16 dsDNA products (*blue numbers at right* represent an amount of dsDNA copies at the end of each cycle)

is transferred into target bacterial cell by specialized vehicle DNA molecule, vector. Vector is able to transcribe in the target cell which leads to the amplification of inserted transgene. Finally, substantial amounts of genetic material can be obtained for further manipulations, including sequencing, restriction analysis, probes for hybridizations, and subsequent transfection.

Different types of DNA can be inserted in the vector: DNA fragments coding proteins, promoters, noncoding DNA fragments, or synthetic oligonucleotides. Very often transgene is a cDNA – DNA complementary to mRNA coding certain protein (see PCR paragraph below). There are many kinds of transgene delivery vectors, most common being bacterial plasmids and genetically engineered viruses. All plasmid vectors as minimal contain replication sequences that allow for semi-independent replication of the plasmid in the host. Besides transcription-only vectors, numerous types of plasmids and viruses contain additional regulatory elements (promoter, Kozak, and polyadenylation sequence) to express transgene-encoded protein in the target cell, to purify recombinant protein or to study functional activity of expressed/overexpressed protein.

To insert a transgene into a vector, both vector DNA and transgene are cut with the same restriction endonuclease, and then glued together with the assistance of DNA ligase. There are several types of introduction of recombinant vector into host organism: plasmids can be inserted into bacteria (transformation) or to eukaryotic cells (transfection); viral vectors can be delivered to bacteria (transduction) or to eukaryotic cells (infection). Genes delivered by vectors to eukaryotic cells either stay separate from host genome (transient transfection/infection) or become incorporated into host genome (stable transfection/infection: stable cell lines, transgenic animals – see below).

## *DNA Libraries*

DNA library is a collection of DNA fragments inserted into host cells (every individual DNA fragment in separate cell) for storage and further manipulations. Major types of DNA libraries include cDNA libraries and genomic DNA libraries. A cDNA library is a collection of complementary DNAs (cDNAs) produced from mRNAs isolated from certain cell, tissue, or entire organism. Thus, cDNA library represents all genes actively transcribed in the source at certain time under certain conditions. cDNA libraries usually are hosted by bacterial cells or yeasts and can be used for screening to find out whether protein of interest is expressed in certain cell/tissue, to compare cell/tissue/species specificity of expression of target protein. Cell carrying cDNA of interest can be retrieved from the library and used as a producent of recombinant protein or cDNA-containing plasmid.

Genomic DNA library is constructed on the basis of chromosomal DNA and represents entire genome of certain organism fragmented into pieces and stored within host cells (again, one individual DNA fragment per one host cell). In contrast to cDNA libraries, genomic DNA library includes nontranslated regions of protein-encoding genes (introns), regulatory elements (promoters, enhancers), DNA fragments encoding rRNAs, tRNAs, siRNAs, etc. Among many applications of genomic DNA libraries, we can mention to study regulatory mechanisms of transcription of target protein, and the use of genomic sequences for generation of transgenic animals.

To create cDNA library, mRNAs after purification from the source of interest must be converted back to DNA templates (cDNAs) by reverse transcriptase (synthesizes single-stranded DNA complementary to mRNA) and DNA polymerase (converts single-stranded DNA to double-stranded DNA), incorporated into the vector (plasmid or bacteriophage – see "Molecular cloning") and introduced into the host cell (bacterial cell or yeast). In the case of genomic DNA library, initial steps include isolation of genomic DNA and fragmentation to manageable sizes. The DNA molecules of an organism of interest are isolated. The DNA molecules are then partially digested by restriction endonuclease. The DNA fragments are ligated into vectors by recombination techniques (see "Molecular cloning").

To screen DNA libraries for DNA fragment of interest, host cells grown as individual colonies are tested with a probe designed to specifically recognize target DNA fragment. Usually, labeled DNA fragment of target gene or labeled synthetic oligonucleotide are used to hybridize with DNA from colonies (see below "identification of RNA/DNA fragments" below), although for cDNA libraries Western blotting with specific antibodies can be undertaken to visualize colonies that express target protein.

## Polymerase Chain Reaction

In many cases, it is possible to retrieve only small amounts of DNA fragments from the cell or tissue. For further analysis and/or other applications, DNA needs to be amplified. A technique to generate thousands to millions of copies of a particular DNA sequence from few copies of a piece of DNA is called polymerase chain reaction (PCR) (Fig. 1.2). Core element of PCR is a heat-stable DNA polymerase. This enzyme assembles a new DNA strand, by using single-stranded DNA as a template. Also DNA oligonucleotides (primers) complementary to the selective region of initial DNA are required for the initiation of DNA synthesis. PCR sample undergoes a series of temperature changes, heatings, and coolings. Heating causes physical separation of two strands in a DNA double helix (denaturation). During following cooling phase, primers anneal the target DNA, and the DNA polymerase synthesizes DNA using both strands of the target DNA (elongation). The amplification process is selective because of primers that are complementary to the flanking regions of DNA fragment chosen for amplification. During PCR, thermal cycles repeat up to 40 times, and as PCR progresses, the newly synthesized DNA is also used as a template for replication. This chain reaction results in exponential amplification of the DNA template.

Very important variant of PCR is reverse transcription polymerase chain reaction (RT-PCR). This approach allows to amplify RNA. Additional critical step in RT-PCR assay is the initial reverse transcription (RT): reverse transcriptase in the presence of primers converts RNA into cDNA. This is necessary for DNA polymerase to start PCR because it can use only DNA templates.

Both PCR and RT-PCR can be performed under specific conditions, when the amount of amplified DNA is detected after each thermal cycle, i.e., in real time. For the detection of DNA, real-time PCR assays contain either fluorescent dyes that intercalate with double-stranded DNA or sequence-specific fluorescent DNA probes. Fluorescence is measured in the real-time PCR thermocycler to determine the abundance of a particular DNA sequence in sample. There are numerous applications for quantitative real-time PCR methods: detection of biomarkers, they can be used for diseases diagnostics; in research, they can be used to detect changes in a particular gene expression over time in response to extracellular signal, to pharmacological agent or environmental changes.

## Separation of Nucleic Acids

Gel electrophoresis is one of the principal methods that allows the separation of complex mixtures of RNA, DNA fragments, or plasmids. Combined with blotting techniques, it can be used for analytical purposes (to identify nucleic acid by size, by interaction with specific probe, or by restriction map) or for purification purposes. In horizontal agarose gel electrophoresis, nucleic acids separate according to their size while running through an agarose gel with an electric field. By varying the concentration of agarose, fragments of DNA from about 200 to 50,000 bp can be separated using standard electrophoretic techniques. One modification of agarose gel electrophoresis is called pulsed field electrophoresis. In this technique, the direction of current flow in the electrophoresis chamber is periodically altered. This allows fractionation of pieces of DNA ranging from 50,000 to 5 million bp, which are too large to be resolved on standard gels.

Routinely, DNA fragments are visualized by staining with ethidium bromide (EtBr). This fluorescent dye intercalates between bases of DNA and RNA. It is often incorporated into the gel so that staining occurs during electrophoresis, but the gel can also be stained after electrophoresis by soaking it in a dilute solution of EtBr. To visualize DNA or RNA, the gel is placed on an ultraviolet (UV) transilluminator. Two disadvantages of EtBr include its mutagenicity and that UV light is required for excitation of fluorescence (UV light can cause damage to DNA and reduce the efficiency of subsequent manipulations). There are another DNA stains which are not toxic (SYBR Safe, GelRed) and they are excited by a blue light (SYBR Safe).

In the case of DNA, polyacrylamide is used for separating fragments of less than 500 bp. Polyacrylamide gels have a relatively narrow range of separation, but very high resolving power: fragments of DNA differing in length by a single base pair can be resolved.

There are several other techniques that allow to isolate/enrich certain categories of nucleic acids from complex mixtures. One of them is chromatographic isolation of mRNAs: when cellular total RNA is applied onto oligo(dT) cellulose matrix, only poly(A)-tailed mRNAs specifically bind to it, whereas the majority of RNA (rRNA, tRNA) washes out. Another technique involves the differential binding and elution of bacterial plasmid DNA and chromosomal DNA from crude lysates on a commercially available ion-exchange resins.

## *Identification of RNA/DNA Fragments*

Common but very important task in molecular biology-related studies and technologies is the identification of RNA/DNA fragments. A routine method for detection of a specific DNA sequence in DNA samples is Southern blot. As a first step, it involves separation of DNA fragments by gel electrophoresis (see above). After electrophoresis, nucleic acids are transferred onto a carrier membrane (blotting), and DNA fragment of interest is visualized by specific hybridization with probe. The probe is a fragment of DNA complementary to the DNA to be detected. The probe is also labeled with either radioactive isotope (usually $^{32}P$) or fluorescent molecule.

Northern blot is a technique to identify and estimate (in many cases) the relative abundance of a specific type of RNA among different samples of RNA. Similarly to Southern blot, it includes RNA gel electrophoresis and blotting; RNA is separated on the basis of size, transferred to a membrane, and is then probed with a labeled hybridization probe complementary of a sequence of interest. Visualization of the appropriate size band indicates expression of certain RNA, whereas the intensity of the band usually reflects the amount of the target RNA in the analyzed sample. Northern blot is a basic tool for investigation of regulation of certain genes expression *in vivo* (tissues) and *in vitro* (cultured cells). Using Northern blotting has several advantages compared to other methods of RNA analysis (microarrays, RT-PCR), including the ability to estimate RNA size and the alternative splice products.

Dot blot represents a simplification of the Southern blot or Northern blot: a mixture containing the molecule to be detected is applied directly on a membrane as a dot. This is then followed by detection using hybridization probe. This technique offers no information on the size of the target biomolecule. Furthermore, if two molecules of different sizes are present, they will still appear as a single dot. Dot blots, therefore, can only confirm the presence or absence of a target molecule which can be detected by the hybridization probe.

Alternatively (or in addition) to techniques that use hybridization approaches for detection, RT-PCR and PCR when used with primers specific to target molecule, can identify RNA in cell or tissue lysates (by RT-PCR) or cDNA in cDNA-library (by PCR). Moreover, the level of RNA expression can be determined and compared between several biological samples by real-time RT-PCR (see above). An advantage of methods based on PCR is that they are very sensitive, and that the starting material can be used directly (gel electrophoresis is not needed).

## *Microarrays*

Microarrays can be viewed as reverse Southern blots because they use isolated DNA fragments affixed to a substrate, and hybridization with a labeled DNA-probe made from cellular RNA. The advantage that microarray has over Northern blot is that thousands of genes can be visualized at a time with a microarray while Northern blot is usually looking at one or a small number of genes. Screening by DNA microarrays allows the simultaneous analysis of ~55,000 transcripts per assay and results in both qualitative (switched on/off genes) and quantitative data (transcriptional level of single genes). Combined with laser microdissection technique, microarray screening allows studies on a single cell population, which resolves the cellular heterogeneity problem when tissue is the object of investigation. At the same time, there are some advantages in using whole tissue samples for microarray screenings since it is possible to identify unexpected components expressed under certain conditions revealing novel biomarker genes.

In standard microarrays, hundreds-to-thousands of microscopic spots of the probes that correspond to cellular RNAs (cDNAs, oligonucleotides, or PCR products) are covalently attached to a solid surface of glass, quartz, plastic, or silicon. Before analysis, total RNA should be isolated from the cell or tissue, converted into cDNA via reverse transcription and labeled (using fluorescent cyanine dyes; biotin/streptavidin). Labeled sample is applied onto microarray chip for hybridization. During hybridization, cDNAs from samples that have complementary partners in individual spots of microarray chip, base pair with them and remain bound after washing out of nonbonding cDNAs. The labels on the bound cDNAs produce signals which can be detected by scanning devices. Scan analysis allows to identify cDNA and to estimate the amount of it.

Comparison of microarray profiles obtained at different developmental stages or from different cells/tissues can reveal developmental changes in proteins/pathways, and cell type-related differences in signaling systems. Microarray screening under different pathological conditions have the potentials for the discovery of biomarkers and proteins/pathways that change in relation to disease. For example, Kittleson et al. [1] developed an Affymetrix microarray platform containing 22,000 transcripts to identify genes with altered expression in two types of human cardiomyopathy: 41 genes were found differentially expressed, and they were mainly involved in cell growth and signal transduction. When a CardioChip containing 10,848 nonredundant human heart-related transcripts was used in human heart failure (HF), atrial natriuretic peptide (ANP) was found upregulated, whereas cell-signaling channels and several proteins involved in $Ca^{2+}$ pathways were downregulated [2]. Recently, several investigators reported data obtained with microarrays, used for genome-wide transcriptomic screening, for genes which are differentially regulated in experimental physiological and pathological hypertrophy, as well as in HF in rats. Analysis of these microarrays revealed several very important points. For example, in physiological and pathological hypertrophy the spectra of genes changing differ (genes involved in metabolism and cell growth vs. genes associated with oxidative stress responses, inflammation, and apoptosis) in HF, where several signaling genes become active (including *STAT3*, *STAT5B*, *FYN*, *PTEN*, and *AKT1*) [3, 4].

In addition to microarrays designed to screen mRNAs, there are others for tracking microRNAs in humans, mice, rats, and dogs.

## Molecular Genetics. Genetic Engineering Techniques

In this section, we briefly describe several gene manipulation techniques operating on the molecular level (DNA, RNA). These techniques are widely used in modern studies related to the investigation of natural variation in genes, RNA, and proteins over time (such as an organism's development) or space (such as its body regions), as well as in studies when natural or experimental functional disruptions affect genes, chromosomes, RNAs, or proteins.

### *Mutagenesis*

Using modern molecular biology techniques, it is possible to mutate exact region of DNA encoding protein of interest in order to investigate the role this region (or one particular amino acid) plays in functioning of the protein. Creation of mutation at a defined site in a DNA molecule is called site-directed mutagenesis. First, an oligonucleotide primer containing a base change must be synthesized. Next steps include hybridization of the primer with single-stranded DNA encoding gene of interest and building double-stranded DNA with DNA polymerase. Finally, mutated double-stranded DNA can be inserted into the vector (plasmid, virus), amplified/selected and used. Alternatively to the approach with DNA polymerase, cassette mutagenesis can be used. In the latter case, an oligonucleotide with mutation also contains flanking sites for a restriction enzyme. To introduce a mutation, both mutated oligonucleotide and vector carrying gene of interest are treated with the same restriction enzyme to produce sticky ends on the plasmid and the oligonucleotide. Finally, mutated oligonucleotide is incorporated into the gene by base pairing sticky ends followed by ligation. PCR can be used with the mutated oligonucleotide primers and gene of interest as a template. After amplification, the resulting mutated DNA can be inserted into the vector using recombinant molecular biology techniques.

Following methods described above, not only mutations but many protein modifications are possible, for instance, construction of tagged-proteins, fusion proteins, and proteins with targeting-sequences. Genetic/protein-engineering techniques help to obtain information about the localization and interaction of the desired protein. Thus, the addition of sequence recognizable by antibodies (tag) allows to track protein expression developmentally and/or in response to regulatory stimulae, purify protein, investigate protein translocation, and trafficking. Recently epitope-based and more complex tags fusion techniques were developed to purify signaling proteins and determine their binding partners [5]. By fusion with a green fluorescent protein (GFP), one can visualize the expression and localization of proteins in the living cell.

### *Genetically Engineered Animal Models*

Modern genetic engineering methodology has made possible to introduce known mutations into the mammalian genome and determine the subsequent effects at the molecular, biochemical, cellular, and whole organ levels. Using defined mutagenesis, one can determine the function of a particular protein by either eliminating expression or overexpression. By incorporating amino acids that mimic PTMs, it is possible to investigate even the role of PTMs. Hundreds of animal models have been made to study multiple aspects of cardiovascular function, including the regulation of particular signaling pathways and proteins critical for cardiac function (normal or pathologically changed).

## Transgenesis Techniques

Transgenesis techniques involve the random insertion of ectopic genetic material of interest into a "host" genome to examine systemic gain-of-function. Transgenesis includes the injection of transgene into the pronucleus of a fertilized egg. After a limited number of *ex vivo* divisions, the embryo is implanted into a pseudopregnant female and the resulting offspring are screened for the presence of the transgene. Limitations of this approach include: (1) No control of the site of transgene insertion, so insertion may be mutagenic; (2) No control of the number of transgene copies inserted into the host genome, so abnormally high transgene protein product expression may disturb other signaling pathways or cellular processes; (3) The presence of the endogenous gene and its protein may affect interpretation of the data. Despite these disadvantages, transgenesis remains a widely used technique for studying cardiovascular function and pathogenesis. Modern transgenesis techniques utilize organ or cell type-specific promoters to simplify interpretation of resulting organ-specific phenotype. For example, cardiomyocyte-specific expression of the α-myosin heavy chain promoter in mouse is very useful because its ability to drive high levels of expression. Usually, transgene encodes a dominant protein that prevails over an endogenous gene, which remains active in a transgenic animal. This leads to gain-of-function. Sometimes, with a transgenesis approach, it is possible to produce loss-of-function as well. For example, De Windt et al. [6] inhibited cardiac calcineurin by the expression of fragments of calcineurin-inhibiting proteins. Interestingly, Mochly-Rosen et al. [7] designed a transgenic mouse model of cardiac protein kinase Cε (PKCε) activation (gain-of-function) or inhibition (loss-of-function) in transgenic animals expressing peptides that stimulate or prevent PKCε translocation.

## Gene Targeting

Gene targeting techniques using homologous recombination at a preselected *locus* mostly produce null alleles (knockouts) to examine loss-of-function. Gene targeting is based on homologous recombination of transgene at a preselected *locus* in mice. To insert transgene, the targeting construct is transfected into mice totipotent embryonic stem (ES) cells with subsequent selections of cells with the correct homologous recombination. These cells are injected into an early blastocyst and the resulting embryo implanted into a pseudopregnant female. Homologous recombination specifically eliminates endogenous gene, thus resulting in loss of particular protein expression ("knockout") and loss-of-function. As a variation of this approach, genes with point mutations can be created ("knock-in") to determine structure/function relationships or a mutation's effect(s). Sometimes, point mutations result in a hyperallele, i.e., lead to the gain-of-function. The limitations of this technique include: (1) Lethal mutations; (2) Presence of an active selection marker gene in the targeted *locus* (to allow ES cell selection) that might affect the mutant phenotype in an unpredictable way.

Genetically engineered mice models can define the precise signaling pathways that are affected in some cardiac anomalies (in particular, Noonan syndrome, LEOPARD syndrome), as demonstrated by studies related to the cardiac phosphoprotein tyrosine phosphatase, SHP2. Both transgenesis [8] and gene targeting [8, 9] studies demonstrated that selective overexpression of gain-of-function mutant SHP2 in cardiomyocytes is sufficient to cause cardiac pathology, probably through inappropriate Erk1 or Erk2 hyperphosphorylation during cardiac development. Similarly, observations from genetic mice models, with global or conditional cell-restricted deletion of membrane guanylyl cyclase A (GC-A), revealed an important role for the natriuretic peptide/GC-A/cGMP system in the regulation of arterial blood pressure and cardiac remodeling [10].

Gene targeting technique was further improved with the so-called Cre-loxP recombination system approach. With this targeting strategy, selection of the marker gene in the initial targeting construct is flanked by 34 bp-motifs, called loxP. The loxP motif is a recognition site for Cre recombinase. Following insertion of the initial targeting construct and selection of ES cells, which incorporated the construct into the target *locus*, the selection marker gene can be removed from the ES cell genome. For this purpose, cells are transiently transfected with a vector-plasmid expressing Cre. Cre finds loxP sites and excises the DNA-segment between two loxP sites (selection marker gene). The significance of the Cre-loxP recombination system is that it helps to overcome developmentally lethal knockout problems and allows to achieve conditional, i.e., cell-type-specific gene targeting. Briefly, if a functional, but loxP-flanked gene is introduced to the ES cells, the mutant mice will develop normally. Breeding of these mice with transgenic animals expressing Cre in cell/tissue-specific manner (for example, driven by cardiac specific promoters, like α-MHC or myosin light chain 2 V sequence) results in offsprings that carry both loxP-flanked gene and cell-type-controlled Cre gene. These animals develop normally and lose the transgene only in the adult ventricular muscle, when Cre is expressed and eliminates loxP-flanked transgene. Cre-loxP technology has been successfully used to dissect the vascular effects of ANP *in vivo*; receptor for this peptide (GC-A) was selectively inactivated either in smooth muscle cells [11], in cardiomyocytes [12], or in endothelium [13] (see Chap. 3).

Genetic manipulation methods may advance our understanding of cellular signaling and disease mechanisms when combined with proteomics and metabolomics methodologies. Mayr et al. [14, 15] performed proteomic and metabolomic studies on vascular smooth muscle cells from knockout mice lacking PKCδ, and demonstrated that PKCδ is involved in the regulation of glucose and lipid metabolism, which is important for preconditioning-mediated myocardial protection.

## *RNA Interference Methods*

By these methods, it is possible to inhibit a gene or several genes (subgenomes) expression and assign possible functions of inhibited gene(s) based on the observed phenotype. One approach is to use antisense oligonucleotides (RNA or DNA) complementary to mRNA encoding protein of interest. Base pairing of antisense oligonucleotide to sense mRNA prevents translation of target protein. Similarly, cells can be transfected with a transgene coding for antisense mRNA to block the expression of a target gene through the same mechanism. The effects of antisense RNA should not be confused with the effects of RNA interference.

Living cells contain a system, RNA interference system, that posttranscriptionally controls genes' activity and/or the degree of activation via production of small RNAs. These RNAs specifically bind to messenger RNAs causing either increase or decrease of translation. RNA interference is a cell defense mechanism against pathogenic genes (viruses, transposons) as well as a mechanism of gene expression. Small RNAs are produced in the cell from dsRNAs. There are two major sources of small RNAs, exogenous and endogenous. The exogenous dsRNAs are introduced into the cytoplasm of the cell by RNA viruses (or by experimental manipulations). The endogenous dsRNAs are transcription products of certain cellular RNA-coding genes. Initial transcripts of such genes are enriched in palindromic sequences and have characteristic hairpin-loop ds-structure. They are processed in the nucleus to form premicroRNA (~70 nucleotides in length; also called small hairpin RNA; short hairpin RNA; shRNA) and then exported to the cytoplasm. Some transcripts named mirtrons (microRNAs that are located in the introns of the mRNA encoding host genes) do not undergo processing.

In the cytoplasm, dsRNAs (exogenous, shRNAs, mirtrons) are cleaved by endonuclease called dicer. Dicer-dependent cleavage produces short ds-fragments of RNA around 20–25 nucleotides in size. Usually, cleavage-derived products of exogenous dsRNA are called small interfering RNAs (short interfering RNAs; silencing RNAs; siRNAs), whereas the after-cleavage fragments of endogenous dsRNA are microRNAs (miRNAs). One strand of siRNA/miRNA (mature siRNA/miRNA) interacts with several proteins to form the RNA Induced Silencing (RISC) loading apparatus. RISC with mature siRNA/miRNA then targets mRNA in order to induce silencing. Silencing happens because siRNA/miRNA base pairs with the target mRNA thus preventing from protein translation of the target mRNA and accelerating target mRNA degradation.

The very selective and strong effect of RNA interference system on gene expression makes it a powerful research tool; it is possible to suppress target gene by the introduction of synthetic dsRNA into living organism or cultured cells. In general, any gene can be targeted based on sequence complementarity with an siRNA. Experimentally, in RNA interference-related studies the exogenous RNAs may be designed as a long ds-molecule (to be cleaved by dicer), or short siRNAs. They can be delivered to the cells directly, or by transfection with a plasmid encoding siRNAs, or by viral vector system encoding shRNAs. Recent studies have shown the successful application of RNAi technique in studying different components of the cardiac-signaling machinery, such as muscarinic $M_2$ acetylcholine receptor [16], adaptor protein Shc [17], $Na^+$–$Ca^{2+}$ exchanger [18]. Suckauet et al. [19] recently reported the efficacy of the RNAi strategy in the treatment of experimental HF. They developed a cardiotropic adeno-associated virus vector and used it to deliver phospholamban-shRNA to the rat myocardium. Long-term therapy with phospholamban-shRNA significantly reduced levels of this protein and effectively normalized cardiac function.

## Proteomics

There are numerous aspects of signaling in the cardiovascular system that still require further intensive investigation in order to understand the underlying mechanisms of myocardium functioning, as well as to find new specific and effective ways to fight cardiovascular diseases. Just representing the tip of the iceberg, we need further research efforts focused on the identification and characterization of components of different signaling pathways, as well as studies on the regulatory modifications made to the proteins participating in the cardiac signal transduction. Moreover, even for relatively well-investigated pathways, our knowledge about the quantitative and/or qualitative changes in the protein players during development, or in response to diverse requirements or stresses that the organism undergoes, is still far from complete.

The aforementioned and many other related questions are the subjects of proteomics, the large-scale study of proteome – entire pool of proteins, produced by an organism or system. Proteomics deals with protein structure and function. For the purposes of this section, we can narrow the term "proteome" to "cardiovascular signaling proteome," as a pool of proteins involved in the functioning of cardiovascular inter- and intracellular-signaling cascades that control and modulate development and functioning of cardiovascular tissues under physiologic and pathologic conditions.

In this section, we focus on the methodology of modern signaling proteomics. Initial step in all proteomics-related techniques is the preparation of cell or tissue extract containing proteins which later will be separated and analyzed. Biological extracts are extremely complex mixtures of thousands of different proteins. Thereby, in many cases sample complexity can be reduced by cell/tissue fractionation with subsequent more sensitive and specific analysis of proteins in subcellular compartments (e.g., cytosol, nucleus, and mitochondria).

The core technologies of proteomics for separation of proteins are one- or two-dimensional polyacrylamide gel electrophoresis (1DE or 2DE). While samples of low complexity can be subjected to 1DE, complex samples should be separated by high resolution 2DE. Due to its superior resolution, more than 10,000 proteins can be separated on a single 2D gel. In addition, with 2DE it is possible to study differences between varying states of a biological sample. Moreover, protein isoforms and different levels of PTM of a single protein can be separated by 2DE. However, 2DE hardly resolves hydrophobic proteins [20], so membrane receptors and ion channels may escape detection.

The nature of proteins separated by 1DE, 2DE or different types of chromatography can be further analyzed by mass spectrometry (MS). Two features of a mass spectrometer that are important for the identification of proteins, and especially for analysis of their PTMs, are mass accuracy and resolution. MS technique is based on the ionizing of compounds to generate charged molecules (whole proteins or peptide fragments), separation of the ions according to their mass-to-charge ratios, and detection of the relative abundance of ions. A variety of mass spectrometers, which differ in material ionization techniques, ion separation technologies, and ion detection approaches, are available. Proteins and/or peptides are ionized often either by electrospray ionization (ESI) or by matrix-assisted laser desorption/ionization (MALDI). To separate charged molecules, modern mass analyzers use electric and/or magnetic fields (static or dynamic) and include the following types: Sector field, time-of-flight (TOF), quadrupole (q), linear trap quadrupole (LTQ), Fourier transform ion cyclotron resonance (FTICR), LQT Orbitrap combination.

While protein identifications are often based on identifying multiple independent peptides from the same protein, there are situations when detection of a given protein is needed in the presence of other molecules, or PTM of protein must be indentified from a single spectrum. For these purposes, tandem mass spectrometry (MS/MS) is used, which involves several rounds of mass spectrometry and molecule fragmentation. Peptide of interest is isolated in the first mass analyzer from many entering a mass spectrometer; this peptide undergoes fragmentation in the second mass analyzer and subsequently a third mass analyzer sorts the fragments produced from the peptide. Information about the peptide obtained by MS/MS can be used to identify this peptide in a protein database. Several methods of peptide fragmentation include collision-induced dissociation (CID), electron capture dissociation (ECD), and electron transfer dissociation (ETD).

Many mass spectrometers available for investigators differ in the specific configuration of the ionization source, analyzer, and detector. For example, MALDI-TOF mass spectrometer combines MALDI source with TOF analyzer. Tandem mass spectrometers are usually equipped with fragmentation devices, including orbitrap, q-TOF, LTQ, MALDI-TOF/TOF, FTICR with CID and FTICR with ECD. In addition, mass spectrometers can include liquid chromatography which separates compounds chromatographically before they enter the mass spectrometer.

Each of MS approaches has advantages and disadvantages. Thus, ESI is the best ionization way to identify small acidic peptides, but requires a mass analyzer with high resolution and mass accuracy (like FTICR) to obtain good quality spectra. MALDI instruments can be applied to either peptides or whole proteins and are suitable for slightly basic peptides. CID-fragmentation has some limitations: basic residues, proline, and some PTMs can inhibit this type of fragmentation. The ECD-generated fragmentations are independent of peptide amino acid composition and the presence of PTMs, but require large amounts of sample. Recently, Farley and Link [21] have described in detail different types of MS.

In summary, the major applications of proteomics may be grouped into three categories: (1) Identification of proteins (including splice-variants, isoforms) with determination of level of expression, (2) PTMs, and (3) Establishing PPIs.

## *Identification of Proteins/Determination of Level of Expression*

Major approaches to identify proteins and to estimate levels of their expression include gel electrophoresis and mass spectrometry, as we have described above. Here, we highlight several aspects which should be taken into consideration when components of cell-signaling systems are the subject of investigation.

First, many proteins involved in signal transduction (receptors, ion channels, effector molecules) are low abundant and hydrophobic by nature. Apparently, this complicates their identification and characterization of isoform composition, and a significant effort of modern proteomics should be directed to the discovery of new effective solubilizing agents (detergents), improve 2DE and MS technologies, and design effective enrichment strategies.

Second, one should keep in mind that many of the signaling molecules can be expressed as multiple splice variants with different functional properties, and that the level of their expression may significantly change during pathological conditions [22].

## *Posttranslational Modifications*

One very powerful application of functional proteomics is the identification of novel proteins which respond to the activation of signal transduction system(s) by regulatory PTM, proteolytic processing and/or covalent modification(s). The detection of new signal transduction PTM-substrates might lead to the identification of novel-signaling pathways, which in turn may give new insights into the cellular processes.

Phosphorylations are among the most abundant PTMs in proteins. Identification of different phosphoproteins and their phosphorylation sites provide informational insights into signaling pathways triggered by many kinds of factors. A classical approach for the screening of changes in phosphorylation is labeling cells *in vivo* with $^{32}$P- or $^{33}$P-orthophosphate, activate or inhibit specific signaling pathway and quantitate the differences in phosphoproteins between two or more samples (nontreated vs. treated, treated in the presence of drugs, etc.). For quantification of differentially phosphorylated proteins, cell extracts are subjected to 1DE or 2DE with subsequent visualization of protein phosphorylation by autoradiography. Finally, proteins that change their phosphorylation status in response to a signal can be sequenced and identified by MS and MS/MS. Using this approach, Immler et al. [23] were able to characterize thrombin-dependent phosphorylation of different isoforms of the myosin light chain. In addition, there is a novel fluorescence-based Pro-Q Diamond staining technique for specific and sensitive tracking of the changes in protein phosphorylation directly in 1DE or 2DE gel. This technique allows protein phosphorylation level and expression level to be monitored in the same gel, and is compatible with MS techniques [24]. An alternative strategy to detect phosphorylated proteins blotted onto a membrane combines phospho-specific antibody immunoblotting with 2DE. For example, with a specific antiphosphotyrosine antibodies, it is possible to detect about 260 proteins phosphorylated in response to activation of PDGF β-receptor [25]. Although it is worth noting the limitations of this method, at present there are no high-quality specific antibodies to phosphoserine and phosphothreonine to study signaling pathways involving serine and/or threonine phosphorylations on target proteins.

In many cases, it is necessary to enrich phosphoproteins in order to reduce the background and increase the sensitivity of an assay. Tyrosine-phosphorylated proteins, for instance, can be specifically immuno-precipitated from cell extract with antiphosphotyrosine antibodies prior to separation and analysis [26]. Similarly, phosphoprotein-binding motifs (SH2, PTB) can be used to characterize differences in tyrosine phosphorylation patterns between samples [27]. Another approach, Phosphate Metal Affinity Chromatography (PMAC), has a broader spectrum of applications because it allows any phosphorylated protein to be enriched from cellular extracts [28].

Currently, a number of MS/MS-based technologies are available to identify phospho sites on proteins. Most of them require protease-dependent breaking down of protein into peptides to determine the sequence of phosphopeptide(s), whereas FTICR MS is able to identify PTMs at the protein level [29].

Identification of new phosphoprotein(s) as a target(s) of certain signaling cascade requires investigation of the molecular mechanism involved in the regulatory phosphorylation of these proteins. At this stage, traditional assays of signal transduction can complement phosphoproteomics. Following this, Yan et al. [30] combined PMAC with proteomics to identify annexin A2 as a new phosphotarget of the signaling cascade initiated by the Epstein-Barr virus-encoded latent membrane protein 1 (LMP1) in epithelial cells. They further analyzed the relationship between LMP1 and phospho-annexin A2 by traditional biochemical and immuno-cytochemical techniques (Western blotting, inhibitory analysis, cell subfractionation, and immunofluorescence staining) to conclude that LMP1-mediated activation of protein kinase C leads to phosphorylation of serine(s) on cytoplasmic annexin A2. Phosphorylated annexin A2 enters the nucleus to modulate DNA synthesis and mRNA transport. Combination of phosphoproteomics with other selective and sensitive analytical techniques makes it also possible to study dynamic changes of protein phosphorylation and determination of different phospho sites in the same protein.

Besides phosphorylation, hundreds of other PTMs are known which take part in enzyme regulation, signal transduction, mediation of protein localization, interactions, and stability. The type of PTM to be analyzed dictates the kind of sample preparation technique to be used, as well as the methods of separation and detection. Glycosylation is by far the most common PTM and plays key functions in biological processes, including recognition of hormones, cell-to-cell interactions, protease-dependent degradation, determination of localization, activity, and function of proteins. Proteomics of glycosylated proteins (glycoproteomics) is a difficult task because glycan chains are polydisperse and very often variably expressed on a protein. Nevertheless, glycoproteins can be separated by 1DE or 2DE, and glycans can be enzymatically released for subsequent analysis by MALDI-TOF or ESI MS. After deglycosylation, protein can be identified by MALDI-TOF MS [31, 32]. To characterize the site of glycosylation, glycoprotein can be excised from 2DE and analyzed by FTICR MS [33]. Prior to proteomic analysis, glycoproteins can be selectively enriched using lectin-affinity chromatography [34].

A very important type of PTM involved in the recruitment of modified proteins to membranes as well as in facilitating PPIs is lipidation. Actually, this is a generalized term covering a number of different regulatory modifications, such as farnesylation or geranylgeranylation (a form of prenylation) of cysteine residues close to the C-termini of proteins, palmitoylation, myristoylation, and attachment of glycosyl-phosphatidyl-inositol-(GPI)-anchors to the C-termini of proteins. Methods of detection of these modifications include metabolic labeling of lipid moieties with radioactive isotopes, separation of proteins with subsequent visualization of modified proteins by autoradiography (similar to phosphoproteomics approach described above) or tagging-via-substrate approach (metabolic incorporation of modified lipid moiety into proteins followed by biotinylation and affinity purification of PTMed-proteins). Several types of MS can be applied for identification of lipidated proteins [35].

Many intracellular processes are regulated through an enzymatic conjugation of cellular proteins' lysines to the C-terminal glycine residue of conserved 76-amino acid protein ubiquitin (Ub), process named ubiquitination. Ub contains seven lysines that can also be ubiquitinated, leading to the formation of polyubiquitin chains (polyUb). The internal Ub lysines involved into polyubiquitination can determine the biological consequence of this modification. K48-linked polyUb chains direct modified protein to degradation in proteasomes, monoUb, and K63-linked polyUb chains are involved in DNA damage responses, protein trafficking, and NF-κB-dependent signaling [36]. Proteins also can undergo modification with another small protein molecule, small ubiquitin-like modifier (SUMO). Sumoylation is shown to affect transcriptional regulation, nuclear pore complexes, and DNA repair. Recent studies of Ub-related proteome combine Ub affinity purification with high-resolution mass spectrometry [37, 38].

Under physiologic conditions, deamidation of asparagines and glutamines is mechanistically similar to an enzymatically driven proteolytic reaction. It was recently found that deamidation of antiapoptotic protein Bcl-$x_L$ is an important component of the signal transduction pathway that regulates the cellular response to DNA damage [39]. Deamidation of proteins can be studied by their fragmentation to peptides with subsequent separation and quantification of deamidated/nondeamidated peptides by reversed-phase chromatography, capillary electrophoresis, or isoelectric focusing [40].

Another PTM, nitrosylation, plays a significant role in nitric oxide (NO)-dependent signaling. Nitrosylation of cysteines (S-nitrosylation), for example, regulates G protein-coupled receptor signaling, death receptor-mediated apoptosis, and vesicular trafficking pathways. An approach to identify nitrosative protein PTMs is to resolve proteins in 2DE and to visualize nitrosylated proteins using the specific antibody. The immunopositive proteins can be excised from the gel, digested and subjected to MS identification [21, 35]. An alternative approach, biotin-switch technique (BST), was introduced by Jaffrey et al. [41] employing specific biotinylation of S-nitrosylated proteins. This technique allows to detect S-nitrosylated proteins in a complex mixture after 1DE/2DE with antibiotin antibodies, track S-nitrosylated protein of interest after pulldown with streptavidin or identify by MS-specific cysteine residues targeted by S-nitrosylation. BST was successfully applied to study S-nitrosylation of G protein-coupled receptor kinase 2 (GRK2) via GPCR/NOS-signaling pathway [42]. Recently, BST was improved by the use of fluorescent agents to label S-nitrosylated proteins [43, 44].

## Protein–Protein Interaction (Interactional Proteomics)

Multiple levels of regulation either within one particular signaling pathway or during cross talk between different signaling pathways involve interactions between proteins, such as oligomerization of receptors and other signaling proteins, targeting to different compartments and regulation of catalytic activity. Below we discuss several major methodologies used to establish PPIs.

### Immunoprecipitation and Subsequent Analysis of the Co-immunoprecipitated Proteins

Immunoprecipitation of the metabotropic glutamate receptor 5 (mGluR5) from rat brain lysates and subsequent analysis of the co-immunoprecipitated proteins could identify eight known and ten novel members of this signaling complex [45]. Similarly, Edmondson et al. [46] have analyzed proteins co-immunoprecipitated with cardiac PKCε. 2DE combined with liquid chromatography and high sensitivity MS/MS revealed functional PKCε-related subproteome of more than 50 proteins, including structural, signaling, stress-activated, metabolism-related, transcription/translation-related as well as proteins of unknown function.

### Pull-Down Approach

This is an approach where specific peptide sequences (tags) are genetically grafted onto either the C- or N-terminal end of the protein of interest. These tags are used to isolate proteins as a complex with binding partners from crude biological source (cell or tissue extract) using an affinity technique or immunoprecipitation with antitag-specific antibodies. Subsequently, 2DE and MS can be used to separate and identify associated proteins [47, 48]. In several cases, the structure of some PPI peptide domains is known, e.g., Src homology 2 (SH2), pleckstrin homology (SH3), Ena/VASP homology, phosphotyrosine-binding domain (PTB), and PSD-95/Disc-large/Zonula occludens-1 (PDZ). These peptides can be synthesized and used as a bait to isolate proteins interacting with these domains. For instance, 14-aa-peptides corresponding to C-terminal PDZ ligand domains of serotonin receptors (types 5-$HT_{2A}$, 5-$HT_{2C}$, 5-$HT_{4(a)}$, 5-$HT_{4(e)}$) recruit several PDZ proteins *in vitro* [49, 50]. Interestingly, two of these proteins specifically direct 5-$HT_{4(a)}$ receptor to different cell compartments; interaction with $Na^+$/$H^+$-exchanger regulatory factor (NHERF) targets the receptor to microvilli while interaction with sorting nexin 27 (SNX27) sequesters the receptor in early endosomes [50]. PDZ domain-related interaction of NHERF was also demonstrated for purinergic $P2Y_1$ receptor when PDZ domain protein array was subjected to screening with C-terminal tail of the $P2Y_1$ [51]. MacBeath et al. [52] undertook a very elegant proteomic approach to show the phosphorylation-specific interaction partners of epidermal growth factor receptors family, ErbB. They constructed a protein array containing almost all phosphotyrosine-binding domains (SH2, PTB) in human genome, synthesized the fluorescently labeled phosphotyrosine-containing peptides corresponding to phosphorylation sites on ErbB receptors, and used these peptides to probe the protein array to determine the binding between each phosphotyrosine-containing peptide and each protein on the array. These investigators identified known interaction partners of ErbB receptors as well as a number of previously unknown partners.

## Two-Hybrid Screening Approach

Two-hybrid screening technique (yeast two-hybrid system, Y2H) is an example of a fruitful contribution of gene engineering/molecular biology methods to proteomic needs. Y2H is based on the nature of yeast transcription factors, which consist of a binding domain (BD) and an activating domain (AD). Regular transcription of yeast reporter gene starts after the binding of BD to reporter gene upstream activating sequence (UAS), which allows AD to activate transcription. It was found that even when the transcription factor is split into BD- and AD-fragments, it can still activate transcription when the two fragments become indirectly connected. In a typical Y2H system, designed for proteomic screening of binding-partners of protein of interest, two types of plasmids are engineered. One type of plasmid encodes a protein in which the BD-fragment is fused to a protein of interest ("bait"). Another type of plasmid is actually a panel of plasmids derived from a cDNA library selected for screening; the cDNA library is converted to plasmids encoding proteins fused to the AD-fragment ("preys"). Then, separate bait and prey plasmids are simultaneously introduced into the yeast cells. In the cells, which express bait and its interacting prey-partner, AD- and BD-domains of the transcription factor become indirectly connected and trigger transcription of a reporter gene. In the cells, which express bait and noninteracting prey, transcription of reporter gene does not take place. Yeast cells with interacting proteins express an active reporter gene product that changes their phenotype, so they can be separated from cells with inactive reporter gene. For example, the product of *lacZ* reporter gene, β-galactosidase, converts 5-Br-4-Cl-3-indolyl-β-D-galactoside (X-gal) to a blue colorized product which allows investigators to choose cells with interacting proteins. Alternatively, mutant strains of yeast can be used that survive under certain conditions, only if they express the essential product of the reporter gene (usually, an enzyme required for synthesis of amino acid, nucleotide, etc.). The primary structure of the binding partner(s) can be determined by retrieving the protein-encoding sequence originally inserted into the prey-plasmid(s) of cells selected by Y2H.

One disadvantage of the "classic" Y2H system is that it is limited to the analysis of soluble proteins, and many integral membrane-signaling molecules cannot be studied. Stagljar et al. [53] designed an elegant version of Y2H (split-ubiquitin system) useful for screening and analysis of interactions between membrane proteins. As a bait they used the endoplasmic reticulum (ER) transmembrane protein Wbp1p fused to the C-terminal part of ubiquitin and the transcription factor of *lacZ* reporter gene. When expressed alone, this construct targeted ER and did not regulate the reporter gene. Several preys utilized in the study carried the N-terminal part of modified ubiquitin. Those preys which interacted with Wbp1p in ER membrane formed "split-ubiquitin." Split-ubiquitin was recognized by endogenous ubiquitin-specific protease, resulting in the transcription factor to cleave off the bait, translocate to the nucleus and activate the reporter gene transcription.

Another recent modification of Y2H screen system takes advantage of the mating process described for haploid yeast strains a and α. By building two cDNA-libraries (bait library in strain a and prey library in strain α) with subsequent combinatorial mating of these strains, it is possible to subject genome-wide proteomes to high-throughput scale screening for PPIs [54, 55].

One should be mindful of the high number of false identifications accompanying Y2H screenings, for example sometimes BD- or AD-fused proteins can activate transcription of the reporter gene in the absence of mutual interaction, and some PPIs are prevented *in vivo* because proteins are compartmentalized or are not expressed at the same time (false positives). In some cases, the interaction requires PTM of one or both partners, which is not possible in the yeast cell (false negatives). High probability of false positives dictates interaction that is revealed by Y2H screening approach to be confirmed by another high confidence assay (e.g., co-immunoprecipitation of the endogenous proteins).

## Microproteome Analysis

Monitoring signaling molecules abundances and PTMs in clinical specimens can prove to be useful in targeting therapies and validating biomarkers. Unfortunately, very often this is limited by the small amount of biopsy material available. Recently, a number of mini-proteomic technologies that make it possible to fractionate and analyze small amounts of tissue samples have emerged.

Several technologies (nanofluidic proteomic immunoassay, lysate microarray, flow cytometry-based Luminex xMAP assay) allow to detect as little as 2 pg of protein, track drug-related changes in PTM of selected proteins in clinical samples and quantify up to 100 different analytes in a single reaction [56]. They all include immunochemical techniques and are currently limited by the availability of selective antibodies. Kim and Lubman [57] have developed a microproteomic approach which involves microchromatofocusing technique (separation of complex protein mixtures in pH gradient generated on an anion exchange microcolumn) with further analysis by tandem MS. This method can detect 700–800 proteins from microgram amounts of cell lysate.

## Imaging

Imaging is a methodology that investigates compartmentalization and dynamics of cell signaling-related pathways (migration/translocation and PPIs of individual components of signaling machinery as well as distribution of second messenger molecules). It is possible to label proteins *in vivo* and study very different kinds of processes, from fast single molecule events to hours/days-lasting developmental or pathological changes.

Modern imaging techniques frequently rely on fluorescence microscopy. Recent studies reported the use of fluorescently tagged proteins to investigate the dynamic changes of second messenger molecules in living cell. Biosensors based on fluorescent proteins fused to IP3-binding domain of IP3 receptor, diacylglycerol-binding domain of protein kinase C, or cGMP binding site of cGMP-dependent protein kinase are very effective tools to track intracellular agonist-dependent changes of IP3, diacylglycerol and cGMP, respectively [58–61]. GFP-fused regulatory protein domains were also successfully used to study intracellular translocations of signaling proteins and receptor-dependent activation of signaling pathways [62, 63].

Advanced modification of fluorescence microscopy is called FRET technique (Fig. 1.1). Basically, two fluorescent proteins are tightly associated, and the excited energy state of a donor fluorophore can be transferred to an acceptor fluorophore, which then can emit its own characteristic fluorescence. Accordingly, uncoupling of proteins increases the distance between them and terminates FRET. FRET can be detected by imaging technique. FRET-based microscopy has much better resolution than standard light microscopy and allows real-time detection of changes in PPIs in specific compartments of the living cell. Frequently, one of the proteins of interest is fused to cyan fluorescent protein (CFP; donor fluorophore) while another is fused to yellow fluorescent protein (YFP; acceptor fluorophore). Fusion proteins are introduced into the cells by vectors (plasmids, adenoviruses) and analyzed by FRET-based microscopy. For example, FRET-analysis of CFP-fused endothelial isoform of nitric oxide synthase (eNOS) and YFP-fused calmodulin in vascular endothelial cells demonstrated local regulation between them by the $Ca^{2+}$-mobilizing agonist, vascular endothelial growth factor (VEGF) [64]. FRET-based approach has been used for investigation of the receptor-dependent dissociation of the Gα (CFP) and Gβ (YFP) subunits following the activation of G protein within cell plasma membrane [65]. Also, by the expression of donor–acceptor FRET sensor (CFP-fused catalytic subunit of PKA and YFP-fused regulatory subunit of PKA) it is possible to monitor cAMP signaling in the living cardiac cell [66].

Often fluorescence-related techniques are combined with the high-resolution Total Internal Reflection Fluorescence Microscopy (TIRFM). In TIRFM, fluorophores (fluorescent proteins) are selectively excited by evanescent wave of light, which penetrates to very short depth into the sample. Thus, the TIRFM enables a selective visualization of fluorescent proteins in the cell plasma membrane and the detection of signaling pathways that control the cytoskeleton and adhesion [60, 65, 67]. In addition, other powerful high-resolution methods of microscopy that have emerged in last several years include, multiphoton excitation microscopy, scanning light sheet microscopy, confocal linescan imaging, and force microscopy [67–70]. They allow visualization of 3D cellular imaging, changes in cell shape, cell movements during embryogenesis and many other aspects of cell physiology.

## Conclusions

With the development of new technologies and methodologies, modern molecular biology, biochemistry, and molecular genetics have successfully demonstrated a significant progress in the investigation of cell-signaling processes, regulatory mechanisms, expression of components, and the target of signal-transducing cascades. The combination of genetic and biochemical methods with proteomics and cell imaging can allow the discovery of new proteins and signal transduction pathways, and bring about new insights in the study of expression, PTMs and interactions of signaling proteins.

## Summary

- Major general molecular biology techniques include amplification of nucleic acids (cloning, PCR); methods of separation and identification of nucleic acids.
- Molecular cloning is a transfer of DNA fragment by vector into target host cell where it can transcribe. There are many kinds of transgene delivery vectors; the most common are bacterial plasmids and genetically engineered viruses.
- DNA library is a collection of DNA fragments inserted into host cells (every individual DNA fragment in separate cell) for storage and further manipulations. Major types of DNA libraries include cDNA libraries and genomic DNA libraries.
- PCR is an amplification of DNA fragment catalyzed by heat-stable DNA polymerase. This is a chain reaction resulting in an exponential amplification of the DNA template.
- RT-PCR is a variant of PCR which allows to amplify RNA.
- The principle method allowing to separate complex mixtures of RNA, DNA fragments, or plasmids is gel electrophoresis. There are several other techniques that allow to isolate/enrich certain categories of nucleic acids from complex mixtures.
- Major methods for detection of a specific nucleic acid sequences in samples include Southern and Northern blots, PCR and RT-PCR.
- Screening by DNA microarrays allows the simultaneous analysis of thousands of transcripts per assay and results in both qualitative and quantitative data.
- Site-directed mutagenesis is the creation of a mutation at a defined site in a DNA molecule. By a mutation in the exact region of DNA encoding protein of interest, it is possible to investigate the role this region (or one particular amino acid) plays in the function of the protein. Using modern molecular biology techniques, not only mutations but many other protein modifications are possible.
- Modern genetic engineering methodology makes it possible to create genetically engineered animals by introducing known mutations into the mammalian genome. Using defined mutagenesis, one can determine the function of a particular protein by either eliminating its expression (gene targeting technique) or overexpression (transgenesis technique). Cre-loxP recombination approach of gene targeting helps to overcome developmentally lethal knockout phenotypes and allows to achieve cell-type-specific gene targeting.
- The cellular RNA interference system controls the activity of genes posttranscriptionally via production of small RNAs, which specifically bind to messenger RNAs and change their translation. Very selective and strong effect of RNA interference system on gene expression makes it a powerful research tool. It is possible to suppress the target gene by the introduction of synthetic dsRNA or siRNA into living organism or cultured cells.
- Investigation of different aspects of signaling in the cardiovascular system requires the application of new technologies, such as modern proteomics. Proteomics is the large-scale study of large pools of proteins produced by an organism or system. The core technologies of proteomics include polyacrylamide gel electrophoresis and a variety of mass spectrometry methods.
- One very powerful application of functional proteomics is in the identification of proteins which respond to the activation of signal transduction system(s) by regulatory PTMs, such as phosphorylations, glycosylations, lipidations, ubiquitinations, nitrosylations, and many other modifications.
- An important aspect of regulation either within one particular signaling pathway or during cross talk between different signaling pathways involves the interactions between proteins. Several major methodologies of interactional proteomics can be used to establish PPIs, such as immunoprecipitation, pull-down approach and two-hybrid screening technique (Y2H).
- Recently, a number of mini-proteomic technologies emerged which make it possible to fractionate and analyze small amounts of biological samples. These technologies are very useful in targeted therapies because they allow monitoring of signaling molecules in clinical specimens and can validate biomarkers.
- Imaging is a methodology used to investigate compartmentalization and dynamics of cell signaling-related pathways. Modern imaging techniques very frequently rely on fluorescence microscopy. Advanced modification of fluorescence microscopy is called FRET technique. FRET-based microscopy allows real-time detection of changes in PPIs in specific compartments of the living cell.

# References


1. Kittleson MM, Minhas KM, Irizarry RA, et al. Gene expression analysis of ischemic and nonischemic cardiomyopathy: shared and distinct genes in the development of heart failure. Physiol Genomics. 2005;21:299–307.
2. Barrans JD, Allen PD, Stamatiou D, Dzau VJ, Liew CC, et al. Global gene expression profiling of end-stage dilated cardiomyopathy using a human cardiovascular-based cDNA microarray. Am J Pathol. 2002;160:2035–43.
3. Strom CC, Aplin M, Ploug T, Christoffersen TE, Langfort J, Viese M, et al. Expression profiling reveals differences in metabolic gene expression between exercise-induced cardiac effects and maladaptive cardiac hypertrophy. FEBS J. 2005;272:2684–95.
4. Kong SW, Bodyak N, Yue P, et al. Genetic expression profiles during physiological and pathological cardiac hypertrophy and heart failure in rats. Physiol Genomics. 2005;21:34–42.
5. Yang W, Steen H, Freeman MR. Proteomic approaches to the analysis of multiprotein signaling complexes. Proteomics. 2008;8:832–51.
6. De Windt LJ, Lim HW, Bueno OF, Liang Q, Delling U, Braz JC, et al. Targeted inhibition of calcineurin attenuates cardiac hypertrophy in vivo. Proc Natl Acad Sci USA. 2001;98:3322–7.
7. Mochly-Rosen D, Wu G, Hahn H, Osinska H, Liron T, Lorenz JN, et al. Cardiotrophic effects of protein kinase C ε: analysis by in vivo modulation of PKCε translocation. Circ Res. 2000;86:1173–9.
8. Nakamura T, Colbert M, Krenz M, Molkentin JD, Hahn HS, Dorn II GW, et al. Mediating ERK 1/2 signaling rescues congenital heart defects in a mouse model of Noonan syndrome. J Clin Invest. 2007;117:2123–32.
9. Araki T, Mohi MG, Ismat FA, Bronson RT, Williams IR, Kutok JL, et al. Mouse model of Noonan syndrome reveals cell type- and gene dosage-dependent effects of *Ptpn11* mutation. Nat Med. 2004;10:849–57.
10. Kuhn M. Function and dysfunction of mammalian membrane guanylyl cyclase receptors: lessons from genetic mouse models and implications for human diseases. In: Schmidt HHHW et al., editors. cGMP: generators, effectors and therapeutic implications, Handbook of experimental pharmacology, vol. 191. Berlin: Springer; 2009. p. 47–69.
11. Holtwick R, Gotthardt M, Skryabin B, Steinmetz M, Potthast R, Zetsche B, et al. Smooth muscle-selective deletion of guanylyl cyclase-A prevents the acute but not chronic effects of ANP on blood pressure. Proc Natl Acad Sci USA. 2002;99:7142–7.
12. Holtwick R, Van Eickels M, Skryabin BV, Baba HA, Bubikat A, Begrow F, et al. Pressure-independent cardiac hypertrophy in mice with cardiomyocyte restricted inactivation of the atrial natriuretic peptide receptor guanylyl cyclase-A. J Clin Invest. 2003;111:1399–407.
13. Sabrane K, Kruse MN, Fabritz L, Zetsche B, Mitko D, Skryabin BV, et al. Vascular endothelium is critically involved in the hypotensive and hypovolemic actions of atrial natriuretic peptide. J Clin Invest. 2005;115:1666–74.
14. Mayr M, Siow R, Chung YL, Mayr U, Griffiths JR, Xu Q. Proteomic and metabolomic analysis of vascular smooth muscle cells: role of PKCδ. Circ Res. 2004;94:e87–96.
15. Mayr M, Metzler B, Chung YL, Mayr U, Troy H, Hu Y, et al. Ischemic preconditioning exaggerates cardiac damage in PKC-δ null mice. Am J Physiol Heart Circ Physiol. 2004;287:H946–56.
16. Rinne A, Littwitz C, Bender K, Kienitz M-C, Pott L. Adenovirus-mediated delivery of short hairpin RNA (shRNA) mediates efficient gene silencing in terminally differentiated cardiac myocytes. Methods Mol Biol. 2009;515:107–23. Viral applications of green fluorescent protein.
17. Guo J, Gertsberg Z, Ozgen N, Steinberg SF. p66Shc links $\alpha_1$-adrenergic receptors to a reactive oxygen species-dependent AKT-FOXO3A phosphorylation pathway in cardiomyocytes. Circ Res. 2009;104:660–9.
18. Hurtado C, Ander BP, Maddaford TG, Lukas A, Hryshko LV, Pierce GN. Adenovirally delivered shRNA strongly inhibits $Na^+$-$Ca^{2+}$ exchanger expression but does not prevent contraction of neonatal cardiomyocytes. J Mol Cell Cardiol. 2005;38:647–54.
19. Suckau L, Fechner H, Chemaly E, et al. Long-term cardiac-targeted RNA interference for the treatment of heart failure restores cardiac function and reduces pathological hypertrophy. Circulation. 2009;119:1241–52.
20. Nothwang HG, Schindler J. Two-dimensional separation of membrane proteins by 16-BAC-SDS-PAGE. Methods Mol Biol. 2009;528:269–77.
21. Farley AR, Link AJ. Identification and quantification of protein posttranslational modifications. Methods Enzymol. 2009;463:725–63.
22. Jugdutt BI, Sawicki G. $AT_1$ receptor blockade alters metabolic, functional and structural proteins after reperfused myocardial infarction: detection using proteomics. Mol Cell Biochem. 2004;263:179–88.
23. Marcus K, Moebius J, Meyer HE. Differential analysis of phosphorylated proteins in resting and thrombin-stimulated human platelets. Anal Bioanal Chem. 2003;376:973–93.
24. Steinberg TH, Agnew BJ, Gee KR, Leung WY, Goodman T, Schulenberg B, et al. Global quantitative phosphoprotein analysis using multiplexed proteomics technology. Proteomics. 2003;3:1128–44.
25. Soskic V, Gorlach M, Poznanovic S, Boehmer FD, Godovac-Zimmermann J. Functional proteomics analysis of signal transduction pathways of the platelet-derived growth factor beta receptor. Biochemistry. 1999;38:1757–64.
26. Zhang Y, Wolf-Yadlin A, Ross PL, Pappin DJ, Rush J, Lauffenburger DA, et al. Time-resolved mass spectrometry of tyrosine phosphorylation sites in the epidermal growth factor receptor signaling network reveals dynamic modules. Mol Cell Proteomics. 2005;4:1240–50.
27. Yaoi T, Chamnongpol S, Jiang X, Li X. Src homology 2 domain-based high throughput assays for profiling downstream molecules in receptor tyrosine kinase pathways. Mol Cell Proteomics. 2006;5:959–68.
28. Yan G, Li L, Tao Y, Liu S, Liu Y, Luo W, et al. Identification of novel phosphoproteins in signaling pathways triggered by latent membrane protein 1 using functional proteomics technology. Proteomics. 2006;6:1810–21.
29. Zahedi RP, Begonja AJ, Gambaryan S, Sickmann A. Phosphoproteomics of human platelets: a quest for novel activation pathways. Biochim Biophys Acta. 2006;1764:1963–76.
30. Yan G, Luo W, Lu Z, Luo X, Li L, Liu S, et al. Epstein-Barr virus latent membrane protein 1 mediates phosphorylation and nuclear translocation of annexin A2 by activating PKC pathway. Cell Signal. 2007;19:341–8.
31. Kuster B, Wheeler SF, Hunter AP, Dwek RA, Harvey DJ. Sequencing of N-linked oligosaccharides directly from protein gels: in-gel deglycosylation followed by matrix-assisted laser desorption/ionization mass spectrometry and normal-phase high-performance liquid chromatography. Anal Biochem. 1997;250:82–101.
32. Packer NH, Lawson MA, Jardine DR, Sanchez JC, Gooley AA. Analyzing glycoproteins separated by two-dimensional gel electrophoresis. Electrophoresis. 1998;19:981–8.
33. Hakansson K, Emmett MR, Marshall AG, Davidsson P, Nilsson CL. Structural analysis of 2D-gel-separated glycoproteins from human cerebrospinal fluid by tandem high-resolution mass spectrometry. J Proteome Res. 2003;2:581–8.

34. Morelle W, Canis K, Chirat F, Faid V, Michalski J-C. The use of mass spectrometry for the proteomic analysis of glycosylation. Proteomics. 2006;6:3993–4015.
35. Reinders J, Sickmann A. Modificomics: posttranslational modifications beyond protein phosphorylation and glycosylation. Biomol Eng. 2007;24:169–77.
36. Pickart CM, Fushman D. Polyubiquitin chains: polymeric protein signals. Curr Opin Chem Biol. 2004;8:610–6.
37. Peng J, Schwartz D, Elias JE, Thoreen CC, Cheng D, Marsischky G, et al. A proteomics approach to understanding protein ubiquitination. Nat Biotechnol. 2003;21:921–6.
38. Denison C, Kirkpatrick DS, Gygi SP. Proteomic insights into ubiquitin and ubiquitin-like proteins. Curr Opin Chem Biol. 2005;9: 69–75.
39. Deverman BE, Cook BL, Manson SR, Niederhoff RA, Langer EM, Rosova I, et al. Bcl-$x_L$ deamidation is a critical switch in the regulation of the response to DNA damage. Cell. 2002;111:51–62.
40. Reubsaet JL, Beijnen JH, Bult A, van Maanen RJ, Marchal JA, Underberg WJ. Analytical techniques used to study the degradation of proteins and peptides: chemical instability. J Pharm Biomed Anal. 1998;17:955–78.
41. Jaffrey SR, Erdjument-Bromage H, Ferris CD, Tempst P, Snyder SH. Protein S-nitrosylation: a physiological signal for neuronal nitric oxide. Nat Cell Biol. 2001;3:193–7.
42. Whalen EJ, Foster MW, Matsumoto A, Ozawa K, Violin JD, Que LG, et al. Regulation of β-adrenergic receptor signaling by S-nitrosylation of G protein-coupled receptor kinase 2. Cell. 2007;129:511–22.
43. Kettenhofen NJ, Wang X, Gladwin MT, Hogg N. In-gel detection of S-nitrosated proteins using fluorescence methods. Methods Enzymol. 2008;441:53–71.
44. Han P, Zhou X, Huang B, Zhang X, Chen C. On-gel fluorescent visualization and the site identification of S-nitrosylated proteins. Anal Biochem. 2008;377:150–5.
45. Farr CD, Gafken PR, Norbeck AD, Doneanu CE, Stapels MD, Barofsky DF, et al. Proteomic analysis of native metabotropic glutamate receptor 5 protein complexes reveals novel molecular constituents. J Neurochem. 2004;91:438–50.
46. Edmondson RD, Vondriska TM, Biederman KJ, Zhang J, Jones RC, Zheng Y, et al. Protein kinase Cε signaling complexes include metabolism- and transcription/translation-related proteins: complimentary separation techniques with LC/MS/MS. Mol Cell Proteomics. 2002;1:421–33.
47. Becamel C, Alonso G, Galeotti N, Demey E, Jouin P, Ullmer C, et al. Synaptic multiprotein complexes associated with 5-$HT_{2C}$ receptors: a proteomic approach. EMBO J. 2002;21:2332–42.
48. Becamel C, Galeotti N, Poncet J, Jouin P, Dumuis A, Bockaert J, et al. A proteomic approach based on peptide affinity chromatography, 2-dimensional electrophoresis and mass spectrometry to identify multiprotein complexes interacting with membrane-bound receptors. Biol Proced Online. 2002;4:94–104.
49. Becamel C, Gavarini S, Chanrion B, Alonso G, Galeotti N, Dumuis A, et al. The serotonin 5-$HT_{2A}$ and 5-$HT_{2C}$ receptors interact with specific sets of PDZ proteins. J Biol Chem. 2004;279:20257–66.
50. Joubert L, Hanson B, Barthet G, Sebben M, Claeysen S, Hong W, et al. New sorting nexin (SNX27) and NHERF specifically interact with the 5-$HT_{4(a)}$ receptor splice variant: roles in receptor targeting. J Cell Sci. 2004;117:5367–579.
51. Fam SR, Paquet M, Castleberry AM, Oller H, Lee CJ, Traynelis SF, et al. $P2Y_1$ receptor signaling is controlled by interaction with the PDZ scaffold NHERF-2. Proc Natl Acad Sci USA. 2005;102: 8042–7.
52. Kaushansky A, Gordus A, Budnik BA, Lane WS, Rush J, MacBeath G. System-wide investigation of ErbB4 reveals 19 sites of Tyr phosphorylation that are unusually selective in their recruitment properties. Chem Biol. 2008;15:808–17.
53. Stagljar I, Korostensky C, Johnsson N, Te HS. A genetic system based on splitubiquitin for the analysis of interactions between membrane proteins *in vivo*. Proc Natl Acad Sci USA. 1998;95: 5187–92.
54. Rual JF, Venkatesan K, Hao T, Hirozane-Kishikawa T, Dricot A, Li N, et al. Towards a proteome-scale map of the human protein-protein interaction network. Nature. 2005;437:1173–8.
55. Stelzl U, Worm U, Lalowski M, Haenig C, Brembeck FH, Goehler H, et al. A human protein-protein interaction network: a resource for annotating the proteome. Cell. 2005;122:957–68.
56. Gujral TS, MacBeath G. Emerging miniaturized proteomic technologies to study cell signaling in clinical samples. Sci Signal. 2009;2:1–3.
57. Kim H, Lubman DM. Micro - proteome analysis using microchromatofocusing in intact protein separations. J Chromatogr A. 2008;1194:3–10.
58. Tanimura A, Nezu A, Morita T, Turner RJ, Tojyo Y. Fluorescent biosensor for quantitative real-time measurements of inositol 1,4,5-trisphosphate in single living cells. J Biol Chem. 2004;279: 38095–8.
59. Remus TP, Zima AV, Bossuyt J, et al. Biosensors to measure inositol 1,4,5-trisphosphate concentration in living cells with spatiotemporal resolution. J Biol Chem. 2006;281:608–16.
60. Codazzi F, Teruel MN, Meyer T. Control of astrocyte $Ca^{2+}$ oscillations and waves by oscillating translocation and activation of protein kinase C. Curr Biol. 2001;11:1089–97.
61. Nausch LW, Ledoux J, Bonev AD, Nelson MT, Dostmann WR. Differential patterning of cGMP in vascular smooth muscle cells revealed by single GFP-linked biosensors. Proc Natl Acad Sci USA. 2008;105:365–70.
62. Matsuoka S, Iijima M, Watanabe TM, et al. Single-molecule analysis of chemoattractant-stimulated membrane recruitment of a PH-domain-containing protein. J Cell Sci. 2006;119:1071–9.
63. Sasaki AT, Firtel RA. Regulation of chemotaxis by the orchestrated activation of Ras, PI3K, and TOR. Eur J Cell Biol. 2006;85: 873–95.
64. Jobin CM, Chen H, Lin AJ, et al. Receptor-regulated dynamic interaction between endothelial nitric oxide synthase and calmodulin revealed by fluorescence resonance energy transfer in living cells. Biochemistry. 2003;42:11716–25.
65. Elzie CA, Colby J, Sammons MA, Janetopoulos C. Dynamic localization of G proteins in Dictyostelium discoideum. J Cell Sci. 2009;122:2597–603.
66. Warrier S, Belevych AE, Ruse M, et al. β-Adrenergic and muscarinic receptor induced changes in cAMP activity in adult cardiac myocytes detected using a FRET-based biosensor. Am J Physiol Cell Physiol. 2005;289:C455–61.
67. Iwadate Y, Yumura S. Molecular dynamics and forces of a motile cell simultaneously visualized by TIRF and force microscopies. Biotechniques. 2008;44:739–50.
68. Helmchen F, Denk W. Deep tissue two-photon microscopy. Nat Methods. 2005;2:932–40.
69. Verveer PJ, Swoger J, Pampaloni F, et al. High-resolution three-dimensional imaging of large specimens with light sheet-based microscopy. Nat Methods. 2007;4:311–3.
70. Zima AV, Blatter LA. Inositol-1,4,5-trisphosphate-dependent $Ca^{2+}$ signalling in cat atrial excitation–contraction coupling and arrhythmias. J Physiol. 2004;555:607–15,

# Part II
# Normal Signaling Processes

# Chapter 2
# Cell-Cycle Signaling, Epigenetics, and Nuclear Function

**Abstract** Like many other cells, mammalian cardiac cells are able to divide and proliferate during development and during a short time after birth. However, the ability to divide decreases dramatically in the neonatal period, and adult cardiomyocytes are unable to proliferate. In contrast, postnatal cardiomyocytes from some lower vertebrates (e.g., zebrafish, newt) maintain the capability to divide, although the mechanisms underlying these species differences are unknown.

As a result of the inability of adult cardiac cells to proliferate, the heart is unable to regenerate new functional tissue following injury, which can cause myocardium dysfunction and even death. Recently, the dogma that the heart is a terminally differentiated nonproliferating organ has been challenged as cardiac stem cells capable of converting to cardiomyocyte-like cells were identified. Thus, in the last several years, two strategies have been used for cardiac repair: induction of endogenous cardiomyocyte proliferation and cell replacement therapy.

Our understanding of the mechanisms that control proliferation of cardiovascular cells has increased significantly in recent years. Studies in proliferating cells and animal models have identified groups of genes and proteins that control cell division; cyclins, cyclin-dependent kinases, and their inhibitors are essential for cell cycle progression, and the retinoblastoma protein and transcription factors (i.e., E2F) modulate the activities of cell cycle-regulators.

In this chapter, we review the cell cycle machinery and discuss how this controls the proliferation of cardiomyocytes. In addition, we analyze the role of sirtuin-depended deacylation in cell cycle progression and proliferation, the functioning and regulation of telomere/telomerase system, and the integration of reactive oxygen species into cell proliferation. New insights into the epigenetic components of cell inheritance, the stable transmission of cellular information beyond just DNA, highlighting DNA methylation, and chromatin organization as major candidates for carriers of epigenetic information are also presented.

**Keywords** Cell cycle • Epigenetics • Telomerase • IGF-1 • Redox signaling

## Introduction

Cell proliferation is a sequence of events in which cells duplicate their contents and then divide. The cell cycle encompasses series of events that takes place in a cell leading to its division and duplication (replication). Before a cell can enter cell division, it needs to take up nutrients. All of the preparations are done during the interphase. Interphase proceeds in three stages (phases), $G_1$, S, and $G_2$. Cell division operates in a cycle. Therefore, interphase is preceded by the previous cycle of mitosis and cytokinesis. Progression of the cell through the different phases is highly regulated.

Proliferation of cardiac cells mediates mammalian heart growth during fetal development, and in some species heart regeneration as well. During the first weeks after birth, mammalian cardiomyocytes *in vivo* lose their ability to proliferate and exit the cell cycle. Thus, the vast majority of adult cardiomyocytes that have stopped proliferating enter a state of quiescence called $G_0$. Recently, it has been reported that the mammalian heart contains minor populations of resident cardiac stem cells (CSCs) retaining the ability to proliferate.

The natural compensatory processes of the injured heart are limited to hypertrophy of the remaining cardiomyocytes and replacement of necrotic regions with fibrotic scar tissue. To correct the loss of cardiomyocytes, two strategies have been employed in recent years: cell replacement through cell therapy and induction of endogenous cardiomyocyte proliferation.

In this chapter, we discuss how the cell cycle machinery controls the growth of cardiac myocytes, and other cellular components of the cardiovascular system.

## Regulators of Cell Cycle: Cyclin-Related Mechanism

Quiescent cells are usually found in the $G_0$ phase, state where mRNA and protein syntheses are minimal. They reenter the cycle at the $G_1$ phase following binding of a growth factor to its extracellular receptor. For instance, mitotically competent

J. Marín-García, *Signaling in the Heart*,
DOI 10.1007/978-1-4419-9461-5_2, 

adult mouse CSCs express stem cell antigen-1 (Sca-1). Sca-1 is the phosphatidylinositol (PtdIns)-anchored protein that stimulates CSC cell cycle progression via Akt. Knockdown of Sca-1 stops proliferation of CSCs *ex vivo* and during regeneration of the ischemic myocardium [1]. Similarly, EGF and bFGF stimulate the neural stem cell cycle progression through the phosphorylation of Akt [2]. A very unique mechanism of cell growth negative regulation was described by Kang et al. [3]. It involves G protein coupled receptor (GPCR) and β-arrestin1. In human neuroblastoma cell, activation of δ-opioid receptor (DOR) results in growth inhibition. This DOR-regulated epigenetic pathway is G protein-independent and begins with DOR-induced trafficking of β-arrestin1 to the nucleus. In the nucleus, β-arrestin1 targets the promoter region of p27 (see below), recruits histone acetyltransferase p300, resulting in enhanced local histone H4 acetylation and stimulated transcription of p27. p27 inhibits cyclin–cyclin-dependent kinase (CDK) complex, leading to cell cycle inhibition.

Regulation of mammalian cell cycling is achieved by the sequential formation, activation, and inactivation of a series of cell cycle regulatory molecules. Cyclins are a family of eight proteins (A–H), which are synthesized and destroyed during each cell cycle. All cyclins contain a homologous nine residue sequence near the N terminus ("cyclin box") that binds to specific CDKs. Several cyclins (C, D, E) are short-life proteins and function during the $G_1$ phase, being later ubiquitinated and degraded. Cyclins A and B remain stable in interphase, but degrade during mitosis also by a ubiquitin-dependent pathway (Fig. 2.1).

Cyclins are regulatory subunits for CDKs. The CDKs are a family of seven protein kinases (1–7) which bind to, and are activated by specific cyclins. CDKs are constitutively expressed in cells, but they have catalytic activity only in the complex with cyclins. As cyclins are synthesized at specific stages of the cell cycle, in response to various molecular signals, different cyclin–CDK combinations (active CDKs) operate at different phases of the cell cycle. For example, CDKs 4–6 complex with cyclin D and function during the $G_0$/$G_1$ phases of the cell cycle; CDK2 associates with cyclins A and E and functions during the $G_1$ phase and during the $G_1$/S transition; and CDK1 binds to cyclins A and B and functions in the S, $G_2$, and M phases. CDK-dependent phosphorylations activate or inactivate target proteins to orchestrate a coordinated entry into the next phase of the cell cycle.

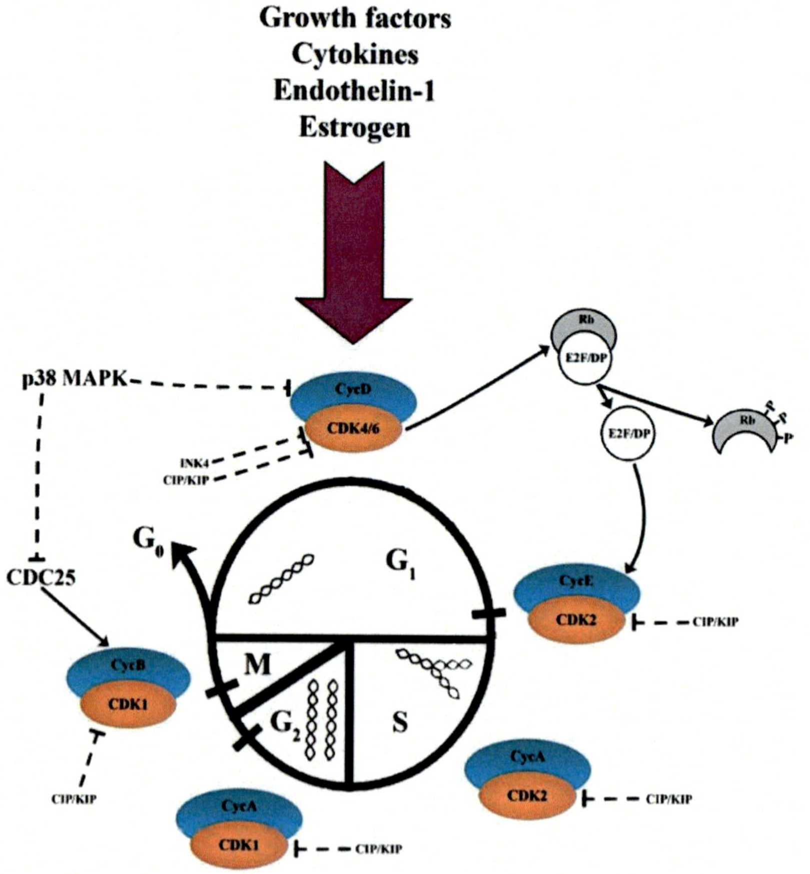


**Fig. 2.1** Mammalian cell cycle. Distinct cyclin (Cyc)/cyclin-dependent kinase (CDK) complexes regulate different phases of cell cycle. Activities of CDKs are, in turn, inhibited by CDK inhibitors, INK4 and CIP/KIP. Diagram also shows the relationship between Cyc/CDKs and several major regulators of cell cycle. Abbreviations: *CDC25* protein tyrosine phosphatase, *CIP/KIP* a family of CDK inhibitors (p21, p27, p57), *E2F/DP* a family of transcription factors regulating the transcription of the genes involved in the progression to the S phase from the $G_1$ ($G_0$) phase of cells, *INK4* inhibitor of kinase 4, *p38 MAPK* p38 mitogen-activated protein kinase, *Rb* retinoblastoma susceptibility protein

Upon receiving a promitotic extracellular signal (growth factor), $G_1$ cyclin–CDK complex is activated to prepare the cell for S phase; cyclin D binds to CDK4, forming the active cyclin D–CDK4 complex. It promotes the expression of transcription factors that stimulate the expression of S cyclins and of enzymes required for DNA replication. In particular, active cyclin D–CDK4 complex phosphorylates the retinoblastoma susceptibility protein (Rb). The hyperphosphorylated Rb dissociates from the E2F/DP1/Rb complex (which was bound to the E2F responsive genes, effectively "blocking" them from transcription), activating E2F. Activation of E2F results in transcription of various genes such as cyclin E, cyclin A, DNA polymerase, and thymidine kinase. Thus, the cyclin E produced binds to CDK2, forming the cyclin E–CDK2 complex, which pushes the cell from $G_1$ to S phase ($G_1$/S transition). Also, during the $G_1$ phase a prereplication complex (pre-RC) is formed on DNA replication origins by the assembly of several factors: origin recognition complex, Cdc6, Cdt1, and minichromosome maintenance (MCM) 2–7 helicase complex.

There are several S phase inhibitory proteins which halt the cell cycle in $G_1$ phase. Some of them belong to the *CIP/KIP* family (p21, p27, and p57). p21 is activated by p53, and p27 is activated by transforming growth factor β (TGF β). *CIP/KIP* proteins inhibit cyclin–CDK complex.

Another family of inhibitory proteins, INK4a/ARF family, includes p16INK4a protein (binds to CDK4) and p14arf protein. The $G_1$ cyclin–CDK complexes promote the degradation of S phase inhibitors.

Several processes are responsible for activation of pre-RC in S phase. First, active S cyclin (cyclin A–CDK2 complex) phosphorylates Cdc6. Phosphorylation is an inhibitory modification of Cdc6, and inactivated Cdc6 gets ubiquitinated and degrades in proteosome. Second, Cdt1 becomes inhibited by geminin. With Cdc6 and Cdt1 no longer bound, MCM-proteins can unwind the double-stranded DNA, and DNA replication begins. Also during S and $G_2$ phases, mitotic cyclin–CDK complex (cyclin B) is synthesized but stays inactivated.

Cyclin B along with CDK1 forms active cyclin B–CDK1 complex, which initiates the $G_2$/M transition. Initiation of mitosis by active mitotic cyclin B–CDK1 complex happens via stimulation of downstream proteins involved in chromosome condensation and mitotic spindle assembly. One of them is the anaphase-promoting complex (APC) with a ubiquitin ligase activity. APC promotes a cascade of events leading to degradation of structural proteins associated with chromosomes; initially, APC targets securin for degradation. In the absence of this inhibitory protein, protease separase cleaves cohesin allowing separation of sister chromatids and the beginning of anaphase. At the end of anaphase, APC triggers ubiquitination/degradation of mitotic cyclins, so telophase and cytokinesis can proceed (Fig. 2.1).

### *Phosphatases Cdc25*

Prior to prophase, CDK1 stays in the cytosol in a phosphorylated (phospho-threonine-14 and phosphor-tyrosine-15), inactive form. During prophase, CDK1/cyclin B1 complexes accumulate in the nucleus and are activated through the phosphatase Cdc25C by dephosphorylation of threonine 14 and tyrosine 15 of CDK1.

## Proliferation of Embryonic Myocytes

There is now extensive knowledge about the genetic and molecular regulation of myocardial cellular development and cardiac morphogenesis. Cardiac muscle is derived from cells in the anterior lateral plate mesoderm. Myoblasts, also known as the embryonic progenitor cells that gives rise to myocytes undergo a series of proliferation to form the primitive straight heart tube. Multiple agents are involved in the regulation of cardiac myocyte proliferation including, insulin-like growth factor (IGF-1) [4], fibroblast growth factors (FGF) [5], neuregulin and receptors (erbB3, erbB4) [6], erythropoietin and erythropoietin receptor [7], retinoic acid and retinoic acid receptor, and cytokines, including cardiotropin-1 and interleukin-6. Cardiac muscle cells remain mitotically active and proliferate through the fetal and early perinatal period, but shortly after birth, mitotic division of cardiac myocytes becomes routinely undetectable.

### *Nonproliferating Adult Cardiomyocytes*

During postnatal maturation of the mammalian heart, proliferation of cardiomyocytes essentially ceases. One interpretation is that cardiomyocytes of the maturing mammalian heart irreversibly exit from the cell cycle and become terminally differentiated. An alternative interpretation is that cardiomyocytes withdraw from the cell cycle and develop blocks at the $G_0$/$G_1$ and $G_2$/M transition phases of the cell cycle with the potential for reactivation.

Little is known about factors that induce cardiomyocytes to withdraw from the cell cycle. The majority of mature cardiomyocytes (85%) are growth-arrested at the $G_0$ or $G_1$ phase, suggesting that factors such as Rb protein, which blocks the $G_1$/S transition, become constitutively active in cardiomyocytes. Active Rb or related factors preclude E2F transcription factors from activating genes required for DNA synthesis. Rb is present in an active (hypophosphorylated) form because in adult cardiomyocytes there are no CDK-1 and CDK-2 which inactivate Rb by phosphorylation.

It is apparent that the absence of CDK-1 and CDK-2 is not enough to stop proliferation because (1) even under these conditions adult cardiomyocytes are able to synthesize DNA which leads to polyploidization of nuclei or multinucleation [8]. (2) Although IGF-1 induced upregulation of cyclins and cell cycle proteins, it was unable to cause proliferation of adult cardiomyocytes [9]. (3) Gene targeting studies show that overexpression of $G_1$ and S phase-promoting cell cycle regulators are not sufficient to override the mechanisms that limit cell division in mature cardiomyocytes [10]. There is some evidence that active mitotic cyclin B1/CDK1 complex enables adult cardiomyocytes to overcome the $G_2$/M block [11].

Persistent stress leads to ongoing remodeling in which cardiomyocyte death exceeds cardiomyocyte renewal, resulting in progressive heart failure. Cell-based therapies capable to reactivate the cell cycle in adult cardiomyocytes will promote myocardial regeneration. Recently, a number of molecular strategies showed new targets for such therapies. For example, Tseng et al. [12] reported that inhibition of glycogen synthase kinase-3 (GSK-3) reactivates the cell cycle in cardiomyocytes via increasing the expressions of $G_1$- and S-phase-acting cyclins D1 and A and decreasing the expression of the CDK inhibitor, p27. Other studies have shown that cell cycle reactivation in adult cardiomyocytes can be initiated by extracellular matrix protein, periostin. This effect, at least in part, is realized via integrin-mediated signal transduction pathways [13]. It seems important to note that induction of cardiomyocyte proliferation in the damaged areas of myocardium might be not enough for heart regeneration and has only a transient healing effect if other cells [vascular smooth muscle cells (VSMCs), fibroblasts, endothelial cells] are not restored.

Another possibility is that of cardiomyocyte renewal mediated by endogenous CSCs. In 2003, a seminal discovery documented that the myocardium contains endogenous CSCs and cardiomyocyte progenitor cells [14, 15].

Finally, it is worth to mention that in contrast to mammalian cells, adult myocardial cells from newt and zebrafish preserve the ability to proliferate [16, 17]. In adult newt, for instance, approximately one-third of the initial cardiomyocyte population grown in culture progress through mitosis and enter successive cell divisions [18].

## *Proliferating Vascular Cells*

Proliferation of VSMCs is a main contributor to vein bypass graft failure and restenosis, thus the targeting of VSMC proliferation is of great significance. One approach is to inhibit VSMC cycle. For example, sirolimus (rapamycin) arrests $G_1$ phase via the upregulation of the CDK inhibitor, p27 [19]. Another agent, paclitaxel, stabilizes microtubules and leads to a $G_2$/M phase arrest [19]. Also, it is possible to reduce VSMC proliferation by selective inhibition of transcription factor E2F3 (promotor of $G_1$-to-S phase transition) [20] or by overexpression of transcription factor FoxO3 (inductor of p27) [21].

## Regulators of Cell Cycle: Sirtuins

Histone acetylation is the main type of covalent histone modification. Histone acetyltransferases (HATs) acetylate histones on lysines, whereas histone deacetylation involves histone deacetylases (HDACs). HATs and HDACs act in an opposing manner to control the acetylation state of histones and other proteins. Three mammalian coactivators, p300/CBP, P/CAF, and TAFII 250, have been identified as HATs. HDACs are grouped into four classes: class I HDACs (1, 2, 3, and 8) are related to yeast RPD3, class II HDACs (4, 5, 6, 7, 9, and 10) are related to yeast HDA1, and class III HDACs (SIRT1–7) are related to yeast Sir2. HDAC11 falls into a fourth class. Class I and II HDACs are zinc-dependent enzymes.

Gathered observations have indicated that HDACs, which belong to class III (sirtuins), induce cell cycle progression and proliferation of many cell types. Sirtuins deacetylate substrates range from histones to transcriptional regulators. They are not inhibited by trichostatin and require $NAD^+$ as a cofactor for their activity. These enzymes have been found to regulate apoptosis and muscle differentiation. In the heart, SIRT2–6 are expressed at high levels, whereas SIRT7 have low abundance.

Some experimental data suggest that the regulatory effect of SIRT1 is realized via modification of certain major players in cell cycle $G_1$ to S transition control, such as Rb and E2F1. As described above, hypophosphorylated Rb in resting and early $G_1$ cells represses transcription of E2F-regulated genes necessary for S-phase entry. As an additional regulatory mechanism, Rb (and its family members p107 and p130) can be acetylated by HATs, like p300 or Tip 60 [22, 23]. Acetylation obstructs efficient phosphorylation of Rb, and cells are maintained in a growth-arrested state. SIRT1 deacetylates Rb-family proteins [24]. Deacetylated Rb can now be phosphorylated to relieve repression of E2F-dependent cycle genes. Another important mechanism leading to increased proliferation was described by Rathbone et al. [25]. In rat muscle precursor cells, SIRT1-dependent cell proliferation was accompanied by decreased level of p21, an inhibitor of cyclin D–CDK4/6 activity. In this case, it is likely that SIRT1 deacetylated p53 [26], which lead to suppression of p21 synthesis and activation of CDK4/6. Thus, SIRT1 in combination with CDK4/6 activities restarts the cell cycle.

SIRT1 is highly expressed in vascular endothelial cells during blood vessel growth and controls their angiogenic activity. Endothelial-restricted SIRT1-deficient mice develop normally in the absence of ischemic stress but are unable to produce ischemia-induced neovascularization, indicating that endothelial SIRT1 is a mediator of stress-induced sprouting angiogenesis signaling. Gain- and loss-of-function approaches undertaken by Potente et al. [27] showed that under ischemic stress conditions SIRT1 interacts, deacetylates, and represses the transcriptional activity of the forkhead transcription factor Foxo1, an essential negative regulator of blood vessel development. Cardiac defects observed in SIRT1-deficient mice indicate that SIRT1 likely plays an important role in the development of heart [28].

In contrast to SIRT1, SIRT2 participates in the cell cycle at the $G_2$ to M checkpoint and negatively regulates cellular proliferation. Also, SIRT2 causes extension of mitosis in the normal cell cycle and prevents cells progression into mitosis in response to mitotic stress (by blocking chromatin condensation) [29–31].

The levels, activity, and localization of SIRT2 are tightly regulated by phosphorylation/dephosphorylation. SIRT2 is phosphorylated both *in vitro* and *in vivo* on serine 368 by the cell-cycle regulator cyclin B1–CDK1, and phosphorylation at this site is required for SIRT2-mediated delay in cell cycle progression [29, 31]. Phosphorylation of SIRT2 can be reversed by the phosphatase Cdc14. Apparently, dephosphorylation provokes degradation of SIRT2 via ubiquitination/26S proteasome-dependent pathway [29]. Thus, Cdc14 controls the cell cycle-dependent abundance of SIRT2, and mitotic exit.

Many aspects of SIRT2 function remain unknown. So far, two major potential targets have been described for this deacetylase: chromatin and the microtubules. A direct influence of chromatin condensation during the $G_2$ to M transition is possible due to SIRT2-catalyzed histone H4 lysine 16 deacetylation and deacetylation of this lysine during mitosis ($G_2$/M transition) [32]. Notwithstanding, further research is needed to clarify the role of histone acetylation for mitosis progression. The other known SIRT2 substrate is α-tubulin [33]. Since acetylation of tubulin seems to stabilize microtubules, SIRT2-catalyzed deacetylation might disrupt them and inhibit cell proliferation.

## Regulators of Cell Cycle: Telomerase

Telomerase and its target, telomeres, play very important role as "mitotic clock" that regulates a number of cell divisions and the time when cell has to cease proliferation (replicative senescence). A number of signaling pathways are involved in the regulation of telomerase activity (see below). To define the functional importance of telomerase/telomeres "axle," we will start with a description of some aspects of DNA replication and explanation of telomere/telomerase function. Subsequently, we will highlight the function and regulation of telomerase/telomere system in the cardiovascular system.

Eukaryotic genomic DNAs are linear molecules and exist as highly organized complexes with nuclear proteins and RNAs, the chromosomes. At their ends, the chromosomal DNA is composed of an array of guanine-rich, 6–8 base-pair-long repeats that terminates with a 3′ single-stranded-DNA, and specifically binds a number of proteins. These DNA–protein structures at the ends of chromosome, the telomeres, are very important structures to control cell genomic stability and cell replicative senescence.

The major function of telomeres is to compensate for incomplete DNA replication at chromosomal ends. The problem of incomplete DNA replication arises from the nature of DNA polymerase which replicates DNA during the S phase of cell cycle. This enzyme can only synthesize DNA in a 5′–3′ direction. At the replication fork, there is no abnormalities for DNA polymerase to synthesize *de novo* strains on the basis of parental leading strand (strand oriented 3′–5′ direction), as the enzyme just follows moving the replication fork and keeps adding nucleotides complementary to the unzipped leading strand in a 5′–3′ direction.

A completely different mechanism takes place at the lagging parental template strand (strand oriented in the 5′–3′ direction). In this strand, DNA polymerase has to start the replication in the nearby fork, which requires a short RNA primer attached to lagging DNA strand – double-stranded initiation point. From this initiation point, the enzyme moves apart from the replication fork and synthesizes a new strand until it riches a double-stranded DNA, created previously (previous initiation point). As a result of every single action of DNA polymerase, a short DNA fragment (100–200 nucleotides) is formed on the basis of lagging parental template strand. While replication continues, the lagging parental template strand contains a series of these short fragments of DNA called Okazaki fragments that initially are separated by RNA-primers. Later RNA bases are replaced with DNA bases by the endonuclease/DNA polymerase, and DNA fragments are "glued" together by the DNA ligase to create a continuous strand of daughter DNA strand.

When the replication fork reaches the very end of the chromosome, the RNA primer binds to the 5′-end of the lagging template strand, and the replication of the most-5′-piece of template strand proceeds as described. However, after removal of RNA template bases from this fragment it is impossible for DNA polymerase to substitute them with DNA bases because there is no 5′-upstream double-stranded primer. Finally, the single-stranded 5′-ends of replicated chromosomes (approximately 20 bases initially pared with

RNA primers) get degraded. As a result, if there is no protective mechanism, chromosome would lose genetic material and would grow shorter and shorter in subsequent replications.

Fortunately, eukaryotic cells do have telomeres which prevent loses of genetic information and slow down replication-related chromosomal fraying. In addition, telomeres protect chromosomes from being fused. As mentioned previously, telomeres are 6–8 base-pair-long repeats (in vertebrates, the repeat is 5′-TTAGGG-3′), and the total length of these repeats reaches several kilobases in humans. In most somatic cells, the telomeres serve as disposable DNA sequences that shorten during replication instead of meaningful genetic information-carrying DNA sequences. According to some hypotheses, telomeres determine certain number of cell divisions; eventually after a number of divisions, telomeric sequences are running out, and this induces replicative senescence of the cell followed by blockage of cell division.

Most of somatic cells lose telomeric sequences as a result of a series of replications, because they are unable to restore them. At the same time, germ cells, stem cells, and certain leukocytes express active enzyme telomerase, responsible for the synthesis of telomeres. Telomerase is a "ribonucleoprotein complex" composed of a protein component, the telomerase reverse transcriptase (TERT) and an RNA primer sequence, the telomerase RNA component (TERC). TERC is several hundred bases in length noncoding RNA with a template region, 3′-CAAUCCCAAUC-5′ (telomere repeat – see above). Telomerase can base pair the first few nucleotides of the TERC template to the last telomere sequence on the chromosome, and then the TERT-component adds new telomere repeat using TERC as a template. This process of telomere elongation repeats several times.

As mentioned earlier, cardiomyocytes remain mitotically active and proliferate through the fetal period, and exit cell cycle during the early perinatal period. In mice genetic models, when TERT is transgenically expressed, it maintains telomere length and delays cardiomyocyte cell cycle exit. Proliferative signaling by TERT happens under mitogenic conditions, but is insufficient to control exiting cardiac cells from proliferation. This is why in TERT-overexpressing mice, cardiac proliferation is also subsided eventually (by 3 months of age) [34]. In a number of noncardiac cells (human fibroblasts, keratinocytes), forced expression of functional TERT also prevents the loss of telomeric DNA and prolongs the cellular life span. Conversely, the overexpression of a catalytically inactive form of TERT causes premature cell senescence and apoptosis [35].

The vast majority of adult cardiomyocytes are of the postmitotic nature, and the mitotic clock could not be applied to them. However, c-kit-positive undifferentiated resident cardiac progenitor cells (CPCs) have also been detected in the adult heart, and they give rise to new myocytes in the adult myocardium. When a progenitor cell divides, two daughter cells are formed; they may maintain parental cell properties or become amplifying cells. Amplifying cells divide rapidly and simultaneously differentiate. Amplifying cardiac cells have a limited number of population doublings. Such a restriction in cell division correlates with the progressive downregulation of telomerase during differentiation. It has been assumed that aging effects on CPCs lead to an imbalance between telomerase activity and length of telomeres, resulting in critical telomeric shortening, permanent withdrawal from the cell cycle, and CPC senescence [36]. There is evidence that in mouse cardiac CPCs, IGF-1 activates telomerase activity and induces CPC division. This mechanism involves PI3K-Akt pathway and leads to the phosphorylation/activation of TERT [37].

Another potential target for telomerase in myocardium is the endothelium. Upregulation of endogenous telomerase by fibroblast growth factor-2 (FGF-2) [38] or overexpression of TERT [39] promotes endothelial cell proliferative capacity and function. Conversely, atherogenic factors suppress telomerase activity and accelerate endothelial cell senescence [40]. Similar to IGF-1 on cardiomyocytes, estrogens can increase telomerase activity in endothelial cells via PI3K-Akt pathway and subsequent phosphorylation of TERT [41]. Furthermore, critically short telomeres in mice markedly impair angiogenesis [42].

In conclusion, the modulation of telomerase activity and the control of telomeric length may represent an important therapeutic tool in regenerative medicine, in particular, when regeneration of cardiac tissue after injury is required. Restoring telomerase activity may be of benefit to the native vasculature and to angiogenic therapies for ischemia.

## Regulators of Cell Cycle: Redox Signaling

Several lines of evidence suggest that redox signaling exists and plays a role in the regulation of cell cycle and proliferation. For instance, nuclear levels of the major redox indicator, thiol tripeptide glutathione (GSH), change during the cell cycle, with the highest levels found in the S and $G_2$/M phases, and depletion of GSH leads to reduced proliferation [43, 44]. Also, the majority of oxidant-sensitive proteins conduct transcription, nucleotide metabolism, (de)phosphorylation, and (de)ubiquitinylation during the $G_2$/M phase indicating that oscillations of the intracellular redox environment may regulate oxidant-sensitive proteins.

A number of proteins, which belong to the classes of transcription factors, chromatin-modifying enzymes, kinases, and phosphatases, are regulated by oxidative stress (OS). Thus, Src family kinases (Src, Yes) are activated under OS and tyrosine phosphorylate TERT. Phospho-$Tyr_{707}$-TERT forms a complex with export receptor CRM-1 and the nuclear GTPase Ran, which leads to translocation of TERT out of the nucleus [45, 46]. Tyrosine phosphorylation of TERT under OS conditions is further accelerated by oxidation/inhibition of protein tyrosine phosphatase Shp-2, so it is unable to maintain unphosphorylated TERT within the nucleus [46]. Therefore, an imbalance in the redox status seems to enhance the activity of the nuclear Src family of tyrosine kinases and to inhibit the nuclear tyrosine phosphatase, which contributes to diminished telomerase activity and accelerated replicative senescence.

Another phosphatase, T-cell protein tyrosine phosphatase (TC-PTP), seems to be responsible for OS-caused reduction of sprouting angiogenesis. In endothelial cell, OS induces translocation of the 45 kDa TC-PTP from the nucleus to the cytosol, where it dephosphorylates specific tyrosine residues on vascular endothelial growth factor receptor 2 (VEGFR2). Dephosphorylation delays vascular endothelial growth factor (VEGF)-induced VEGFR2 internalization and thus inhibits growth of endothelial cells and reduces angiogenesis [47]. Oxidation of another phosphatase, Cdc25C, also influences cell cycle progression. In the latter case, oxidation-enhanced formation of the disulfide bond between cysteine-330 and -377 in Cdc25C promotes phosphorylation of serine-216. This phosphorylation leads to binding of Cdc25C to 14-3-3 protein and to the nuclear export of Cdc25C [48]. As a result, Cdk1 cannot be activated during prophase, and cells do not proliferate staying arrested at $G_2$/M-checkpoint. Moreover, the association of OS with telomere shortening and senescence has been suggested by observations on vascular endothelial cells [38], VSMCs [49], and cardiomyocytes. For example, elevated production of ROS in mice with experimentally developed diabetes leads to a significant decrease of telomeric length in cardiomyocytes and CPCs. Deletion of $p66^{shc}$ (adaptor protein which enhances ROS-mediated cell injury) prevents OS in transgenic animals; in the diabetic heart, the CPC pool may be protected and, thereby, myocyte regeneration and vessel formation can occur [50].

## Regulators of Cell Cycle: MicroRNAs

Recently, a novel mechanism involving posttranscriptional regulation by small microRNAs (miRNAs) has emerged as a regulator of cell proliferation. In cultured myoblasts and cardiac progenitor cells, for example, *miR-133* stimulates proliferation by targeting serum response factor (SRF), which is important in muscle proliferation [51, 52]. Surprisingly, transgenic mice lacking *miR-133a-1* and *miR-133a-2* (double-mutant) have excessive cardiac proliferation [53]. It is possible that dysregulation of the cell cycle control observed in double-mutant mice, is related to the upregulation of the miR-133a mRNA target, cyclin D2.

As mentioned earlier, postnatal mouse cardiomyocytes terminally exit the cell cycle after the first 10 days of life. However, adult mice have an increase in mitotic cardiac myocytes along with cardiac hyperplasia when they lack another microRNA, *miR-1-2*. Genome-wide profiling of *miR-1-2*-deficient adult hearts reveals a broad upregulation of positive regulators of the cell cycle and downregulation of tumor suppressors, indicating cell reentering the cell cycle [54].

MicroRNA was shown to be involved also in angiogenic signaling in endothelial cells. Normally endothelial cells express *miR-126*. This microRNA represses Sprouty-related protein 1 (SPRED-1) and the p85-β regulatory subunit of phosphatidylinositol 3-kinase (PI3KR2), negative regulators of VEGF signaling. Endothelial cells from *miR-126*-deficient mice show diminished angiogenesis because they have elevated levels of inhibitors of VEGF/Erk/Akt-dependent angiogenic cascade [55, 56].

## Epigenetic Component of Cell Inheritance

Cell genomic DNA is organized into chromatin – complex combination of DNA, RNA, and proteins (mainly histones, but also many other proteins), which maintains DNA structure and orchestrates the pattern of active/silent genes unique for each cell type. During replication, chromatin undergoes genome-wide alterations in structure, which have to be restored in the daughter cells to maintain cell identity and the physiological status that existed in the former parental cell. At the same time, there must be some flexibility in the genome regulatory machinery to allow programmed changes in the identity of the daughter cells when needed (differentiation). Recently, several regulatory systems have been discovered that recover chromatin structure and function after DNA replication during phase S of cell cycle. They can be referred as mechanisms of epigenetic inheritance, stable transmission of information about regulation of genome function that occurs without alterations in the DNA sequence [57].

Epigenetic transfer of chromatin structure from parental to daughter cell through S phase/mitosis is possible due to a number of epigenetic marks – heritable instructions, that determine whether, when, and how particular genetic information will be read. These marks function during DNA replication and cell division, and include DNA methylation, histone modifications, histone variants, histone modifiers, other nonhistone chromatin proteins, and nuclear RNA. Thanks to epigenetic marks, the cell type specific-state of chromatin can survive through perturbations that occur during the replication fork in the S phase (Fig. 2.2).

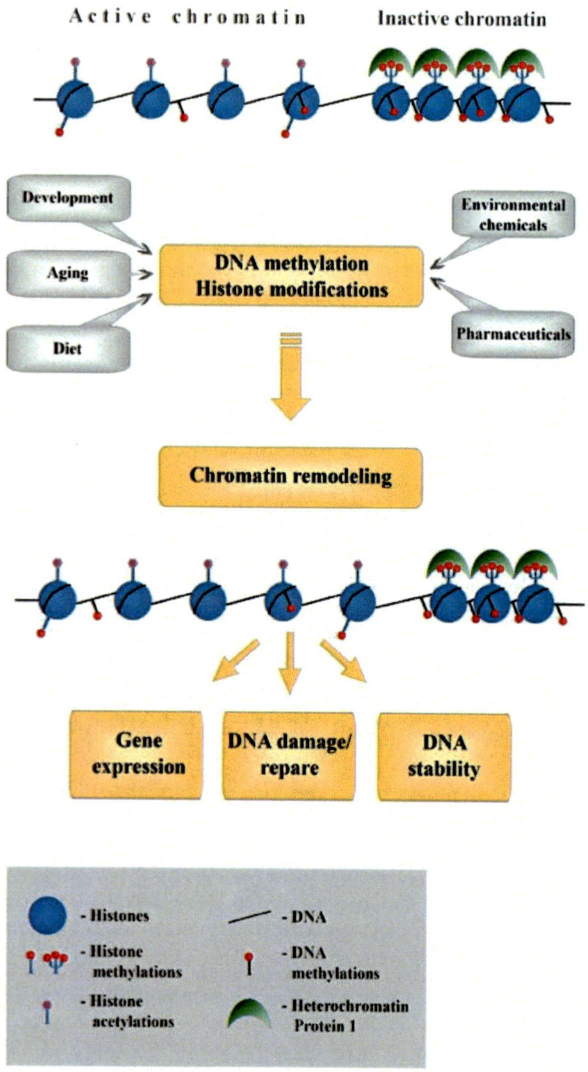


**Fig. 2.2** Epigenetic factors mediating gene expression, DNA damage/repair and DNA stability

When DNA replicates, chromatin undergoes a wave of disruption and subsequent epigenetic marks-regulated restoration in the wake of the passage of a replication fork. Also, DNA replication is a stage where epigenetic changes lead to cell differentiation and development. DNA methylation of newly replicated DNA at the fork is ensured by NP95/DNA methyltransferase 1 using hemimethylated DNA as a template. The resulting methylamines in the daughter strand represent parental DNA. This newly synthesized methylated DNA is a template for several histone modifiers (HDAC, Lys methyltransferases G9a and SETDB). Also, in the replication fork area one or several mechanisms are believed to be available to restore the histone-based nucleosomal pattern of newly synthesized strands, similar to parental DNA. Although the processes that underlie this phenomenon are not well understood, histones H3 and H4 are likely candidates to transmit information about chromatin organization from one cell cycle to the next. Several hypotheses on the mechanism of histone epigenetic inheritance have been recently proposed [58].

Some carriers of epigenetic information function independently from DNA replication. Examples include inheritance of transcriptionally active regions of genomic DNA and several types of condensed chromatin, known as heterochromatin (mainly centromere). The types of cell genomic material mentioned above are characterized by enrichment in special kind of histones, histone H3.3 and centromere-specific histone H3 (CenH3), respectively. These histones seem to serve as epigenetic marks via incorporation into particular regions of DNA in a replication-independent manner (late telophase–early $G_1$ phase) [33]. It is worth to note that many DNA- and histone-modifying and remodeling proteins contribute to the mitotic inheritance of nuclear DNA organization, such as HDACs, Lys methyltransferases, and chaperones.

## Conclusions

Scientific data accumulated in the last several years support the concept that under normal conditions there is a low level of cardiomyocyte death and turnover in the heart, based on the new cardiomyocytes originating from endogenous cardiac precursor (stem) cells. A challenge for research is to create strategies to optimize and enhance CSC response to myocardial stresses in order to rescue myocardial remodeling effectively.

Targeting the cell cycle has many important implications in cardiovascular medicine. For instance, changes in the proliferative activity of the vasculature may prevent vein bypass graft failure and restenosis of vessels after angioplasty interventions. Recent observations suggest that reprogramming of the cardiomyocyte cell cycle is also possible although the precise mechanism(s) by which cell cycle-regulating molecules withdraw cardiomyocytes from the cell cycle remains to be determined. The potential for re-initiation of cardiac cell division is to provide a powerful approach for repairing damaged areas of cardiac tissue following an injury (infarct).

Further studies are needed to understand the regulation of the cardiac cell cycle in order to therapeutically reactivate cell cycle in existing cardiomyocytes, and/or to design methodologies which use differentiated CSCs.

## Summary

- Quiescent cells are usually found in the $G_0$ phase. They reenter the cycle at the $G_1$ phase following binding of a growth factor to its extracellular receptor. Regulation of mammalian cell cycling is achieved by cyclins which are synthesized at specific stages of the cell cycle, in response to various molecular signals. Cyclins are regulatory subunits for protein kinases called CDKs.

- $G_1$ active cyclin D–CDK4 complex phosphorylates the retinoblastoma susceptibility protein, which leads to the activation of transcription factor E2F and transcription of various genes necessary for cell cycle.
- There are several proteins that regulate CDK activity: two families of protein inhibitors (CIP/KIP family and INK4a/ARF family) and the activatory Cdc25 phosphatases.
- Multiple agents are involved in the regulation of fetal cardiomyocyte proliferation (IGF-1, FGF, neuregulin, erbB, etc.). Shortly after birth, mitotic division of cardiac cells becomes routinely undetectable.
- The majority of mature cardiomyocytes (85%) are growth arrested at the $G_0$ or $G_1$ phase. Little is known about the factors that induce cardiomyocytes to withdraw from the cell cycle, although there is some evidence that active Rb protein, in the absence of several CDKs, is involved in cardiac cell cycle arrest.
- The seminal discovery that the myocardium contains endogenous CSCs and cardiomyocyte progenitor cells has been documented.
- In contrast to mammalian cells, adult myocardial cells from newt and zebrafish preserve the ability to proliferate.
- Proliferation of VSMCs is a main contributor to vein bypass graft failure and restenosis. Thus, the targeting of VSMC proliferation is of great interest.
- Sirtuins (SIRT) are class III HDACs which regulate cell cycle progression and proliferation of many cell types. The regulatory effect of SIRT1 is carried out via modification of Rb and E2F1 and leads to increased proliferation.
- SIRT1 is highly expressed in the vascular endothelial cells during blood vessel growth and controls their angiogenic activity.
- SIRT2 participates in the cell cycle at the $G_2$ to M checkpoint and negatively regulates cellular proliferation. Many aspects of SIRT2 function remain unknown. Two major potential targets described for SIRT2 are chromatin and the microtubules.
- Telomerase and its target telomeres play a very important role of "mitotic clock" that regulates the number of cell divisions and the time when cell has to cease proliferation (replicative senescence). Major function of telomeres is to compensate for incomplete DNA replication at chromosomal ends.
- Telomerase is an enzyme responsible for synthesis of telomeres. Most somatic cells do not express telomerase, and they lose telomeric sequences as a result of a series of replications because they are unable to restore them.
- In mouse CPCs, IGF-1 activates telomerase activity and induces CPC division. This mechanism involves the PI3K-Akt pathway and leads to the phosphorylation/activation of TERT.
- In endothelium, FGF-2 and estrogens activate telomerase and promote endothelial cell proliferative capacity.
- Several lines of evidence have established that redox signaling exists and plays a role in the regulation of the cell cycle and proliferation. Oscillations of the intracellular redox environment may regulate oxidant-sensitive proteins during the $G_2$/M phase.
- Src family tyrosine kinases are activated under oxidative stress and phosphorylate/inactivate telomerase.
- Oxidative stress affects several phosphatases (Shp-2, TC-PTP, Cdc25C), which results in accelerated replicative senescence, reduction of angiogenesis, and the arrest of proliferation.
- A novel mechanism of cell proliferation regulation involves posttranscriptional regulation by microRNAs.
- Chromatin orchestrates the pattern of active/silent genes unique for each cell type. Epigenetic transfer of chromatin structure from parental to daughter cell through S phase/mitosis is possible due to a number of epigenetic marks. Major marks include DNA methylation and histone modification patterns. Some carriers of epigenetic information function independently from DNA replication.

## References

1. Tateishi K, Ashihara E, Takehara N, et al. Clonally amplified cardiac stem cells are regulated by Sca-1 signaling for efficient cardiovascular regeneration. J Cell Sci. 2007;120:1791–800.
2. Groszer M, Erickson R, Scripture-Adams DD, et al. PTEN negatively regulates neural stem cell self-renewal by modulating G0–G1 cell cycle entry. Proc Natl Acad Sci USA. 2006;103:111–6.
3. Kang J, Shi Y, Xiang B, et al. A nuclear function of β-arrestin1 in GPCR signaling: regulation of histone acetylation and gene transcription. Cell. 2005;23:833–47.
4. Evans-Anderson HJ, Alfieri CM, Yutzey KE. Regulation of cardiomyocyte proliferation and myocardial growth during development by FOXO transcription factors. Circ Res. 2008;102:686–94.
5. Nakajima Y, Sakabe M, Matsui H, Sakata H, Yanagawa N, Yamagishi T. Heart development before beating. Anat Sci Int. 2009; 84:67–76.
6. Meyer D, Birchmeier C. Multiple essential functions of neuregulin in development. Nature. 1995;378:386–90.
7. Wu H, Lee SH, Gao J, Liu X, Iruela-Arispe ML. Inactivation of erythropoietin leads to defects in cardiac morphogenesis. Development. 1999;126:3597–605.
8. Buja LM, Vela D. Cardiomyocyte death and renewal in the normal and diseased heart. Cardiovasc Pathol. 2008;17:349–74.
9. Reiss K, Cheng W, Pierzchalski P, et al. Insulin-like growth factor-1 receptor and its ligand regulate the reentry of adult ventricular myocytes into the cell cycle. Exp Cell Res. 1997;235:198–209.
10. Bicknell KA, Brooks G. Reprogramming the cell cycle machinery to treat cardiovascular disease. Curr Opin Pharmacol. 2008;8: 193–201.
11. Bicknell KA, Coxon CH, Brooks G. Forced expression of the cyclin B1–CDC2 complex induces proliferation in adult rat cardiomyocytes. Biochem J. 2004;382:411–6.
12. Tseng AS, Engel FB, Keating MT. The GSK-3 inhibitor BIO promotes proliferation in mammalian cardiomyocytes. Chem Biol. 2006;13:957–63.

13. Kuhn B, del Monte F, Hajjar RJ, et al. Periostin induces proliferation of differentiated cardiomyocytes and promotes cardiac repair. Nat Med. 2007;13:962–9.
14. Beltrami AP, Barlucchi L, Torella D, et al. Adult cardiac stem cells are multipotent and support myocardial regeneration. Cell. 2003;114:763–76.
15. Oh H, Bradfute SB, Gallardo TD, Nakamura T, et al. Cardiac progenitor cells from adult myocardium: homing, differentiation, and fusion after infarction. Proc Natl Acad Sci USA. 2003;100: 12313–8.
16. Laube F, Heister M, Scholz C, Borchardt T, Braun T. Re-programming of newt cardiomyocytes is induced by tissue regeneration. J Cell Sci. 2006;119:4719–29.
17. Poss KD, Wilson LG, Keating MT. Heart regeneration in zebrafish. Science. 2002;298:2188–90.
18. Bettencourt-Dias M, Mittnacht S, Brockes JP. Heterogeneous proliferative potential in regenerative adult newt cardiomyocytes. J Cell Sci. 2003;116:4001–9.
19. Wessely R, Schomig A, Kastrati A. Sirolimus and paclitaxel on polymer-based drug-eluting stents: similar but different. J Am Coll Cardiol. 2006;47:708–14.
20. Giangrande PH, Zhang J, Tanner A, et al. Distinct roles of E2F proteins in vascular smooth muscle cell proliferation and intimal hyperplasia. Proc Natl Acad Sci USA. 2007;104:12988–93.
21. Park KW, Kim DH, You HJ, et al. Activated forkhead transcription factor inhibits neointimal hyperplasia after angioplasty through induction of p27. Arterioscler Thromb Vasc Biol. 2005;25:742–7.
22. Chan HM, Krstic-Demonacos M, Smith L, Demonacos C, La Thangue NB. Acetylation control of the retinoblastoma tumour-suppressor protein. Nat Cell Biol. 2001;3:667–74.
23. Leduc C, Claverie P, Eymin B, et al. p14ARF promotes RB accumulation through inhibition of its Tip60-dependent acetylation. Oncogene. 2006;25:4147–54.
24. Wong S, Weber JD. Deacetylation of the retinoblastoma tumour suppressor protein by SIRT1. Biochem J. 2007;407:451–60.
25. Rathbone CR, Booth FW, Lees SJ. Sirt1 increases skeletal muscle precursor cell proliferation. Eur J Cell Biol. 2009;88:35–44.
26. Vaziri H, Dessain SK, Ng EE, et al. hSIR2(SIRT1) functions as an NAD-dependent p53 deacetylase. Cell. 2001;107:149–59.
27. Potente M, Laleh Ghaeni L, Baldessari D, et al. SIRT1 controls endothelial angiogenic functions during vascular growth. Genes Dev. 2007;21:2644–58.
28. Cheng H-L, Mostoslavsky R, Saito S, et al. Developmental defects and p53 hyperacetylation in Sir2 homolog (SIRT1)-deficient mice. Proc Natl Acad Sci USA. 2003;100:10794–9.
29. Dryden SC, Nahhas FA, Nowak JE, Goustin AS, Tainsky MA. Role for human SIRT2 NAD-dependent deacetylase activity in control of mitotic exit in the cell cycle. Mol Cell Biol. 2003;23:3173–85.
30. Inoue T, Hiratsuka M, Osaki M, et al. SIRT2, a tubulin deacetylase, acts to block the entry to chromosome condensation in response to mitotic stress. Oncogene. 2007;26:945–57.
31. North BJ, Verdin E. Mitotic regulation of SIRT2 by cyclin-dependent kinase 1-dependent phosphorylation. J Biol Chem. 2007;282:19546–55.
32. Vaquero A, Scher MB, Lee DH, et al. SirT2 is a histone deacetylase with preference for histone H4 Lys 16 during mitosis. Genes Dev. 2006;20:1256–61.
33. North BJ, Marshall BL, Borra MT, Denu JM, Verdin E. The human Sir2 ortholog, SIRT2, is an NAD+-dependent tubulin deacetylase. Mol Cell. 2003;11:437–44.
34. Oh H, Taffet GE, Youker KA, et al. Telomerase reverse transcriptase promotes cardiac muscle cell proliferation, hypertrophy, and survival. Proc Natl Acad Sci USA. 2001;98:10308–13.
35. Kim M, Xu L, Blackburn EH. Catalytically active human telomerase mutants with allele-specific biological properties. Exp Cell Res. 2003;288:277–87.
36. Kajstura J, Rota M, Urbanek K, et al. The telomere–telomerase axis and the heart. Antioxid Redox Signal. 2006;8:2125–41.
37. Torella D, Rota M, Nurzynska D, et al. Cardiac stem cell and myocyte aging, heart failure and IGF-1 overexpression. Circ Res. 2004;94:514–24.
38. Erusalimsky JD, Skene C. Mechanisms of endothelial senescence. Exp Physiol. 2009;94.3:299–304.
39. Yang J, Nagavarapu U, Relloma K, et al. Telomerized human microvasculature is functional in vivo. Nat Biotechnol. 2001;19:219–24.
40. Breitschopf K, Zeiher AM, Dimmeler S. Proatherogenic factors induce telomerase inactivation in endothelial cells through an Akt-dependent mechanism. FEBS Lett. 2001;493:21–5.
41. Imanishi T, Hano T, Nishio I. Estrogen reduces endothelial progenitor cell senescence through augmentation of telomerase activity. J Hypertens. 2005;23:1699–706.
42. Franco S, Segura I, Riese HH, Blasco MA. Decreased B16F10 melanoma growth and impaired vascularization in telomerase-deficient mice with critically short telomeres. Cancer Res. 2002;62:552–9.
43. Markovic J, Borras C, Ortega A, Sastre J, Vina J, Pallardo FV. Glutathione is recruited into the nucleus in early phases of cell proliferation. J Biol Chem. 2007;282:20416–24.
44. Markovic J, Mora NJ, Broseta AM, et al. The depletion of nuclear glutathione impairs cell proliferation in 3t3 fibroblasts. PLoS One. 2009;4:e6413.
45. Haendeler J, Hoffmann J, Rahman S, Zeiher AM, Dimmeler S. Regulation of telomerase activity and antiapoptotic function by protein–protein interaction and phosphorylation. FEBS Lett. 2003;536:180–6.
46. Jakob S, Schroeder P, Lukosz M, et al. Nuclear protein tyrosine phosphatase shp-2 is one important negative regulator of nuclear export of telomerase reverse transcriptase. J Biol Chem. 2008;283:33155–61.
47. Mattila E, Auvinen K, Salmi M, Ivaska J. The protein tyrosine phosphatase TCPTP controls VEGFR2 signalling. J Cell Sci. 2008;121:3570–80.
48. Peng CY, Graves PR, Thoma RS, Wu Z, Shaw AS, Piwnica-Worms H. Mitotic and G2 checkpoint control: regulation of 14-3-3 protein binding by phosphorylation of Cdc25C on serine-216. Science. 2997;277:1501–5.
49. Cakir Y, Ballinger SW. Reactive species-mediated regulation of cell signaling and the cell cycle: the role of MAPK. Antioxid Redox Signal. 2005;7:726–40.
50. Rota M, LeCapitaine N, Hosoda T, et al. Diabetes promotes cardiac stem cell aging and heart failure, which are prevented by the deletion of the p66shc gene. Circ Res. 2006;99:42–52.
51. Chen JF, Mandel EM, Thomson JM, et al. The role of microRNA-1 and microRNA-133 in skeletal muscle proliferation and differentiation. Nat Genet. 2006;38:228–33.
52. Ivey KN, Muth A, Arnold J, et al. MicroRNA regulation of cell lineages in mouse and human embryonic stem cells. Cell Stem Cell. 2008;2:219–29.
53. Liu N, Bezprozvannaya S, Williams AH. microRNA-133a regulates cardiomyocyte proliferation and suppresses smooth muscle gene expression in the heart. Genes Dev. 2008;22:3242–54.
54. Zhao Y, Ransom JF, Li A, et al. Dysregulation of cardiogenesis, cardiac conduction, and cell cycle in mice lacking miRNA-1-2. Cell. 2007;129:303–17.
55. Wang S, Aurora AB, Johnson BA. The endothelial-specific microRNA miR-126 governs vascular integrity and angiogenesis. Dev Cell. 2008;15:261–71.
56. Fish JE, Santoro MM, Morton SU. miR-126 regulates angiogenic signaling and vascular integrity. Dev Cell. 2008;15:272–84.
57. Riggs AD, Martiennssen RA, Russo VEA. Epigenetic mechanisms of gene regulation 1–4. Cold Spring Harbor, NY: Cold Spring Harbor Laboratory Press; 1996.
58. Probst AV, Dunleavy E, Almouzni G. Epigenetic inheritance during the cell cycle. Nat Rev Mol Cell Biol. 2009;10:192–206.

# Chapter 3
# Signaling in the Endothelium

**Abstract** The endothelium regulates the tone of vascular smooth muscle cells at rest and during exercise, and the thrombotic and adhesive properties of the vascular wall. Endothelial cells are also responsible for vessel growth (angiogenesis) and inhibition of the positive inotropic and chronotropic responses of catecholamines in cardiomyocytes. Regulation occurs by releasing of relaxing factors and among them nitric oxide is the most important modulator of myocardial function. A comprehensive analysis of the signaling pathways network that regulates endothelium structure and function will be presented in this chapter.

**Keywords** Endothelium • Angiogenesis • Angiotensin • Cardiomyocytes • NOS • Prostanoids

## Introduction

The endothelium is a layer of thin, single layer protective epithelial cells that lines the inside of the heart and the lumina of the lymphatic and blood vessels. It interfaces between their walls and the circulating blood, separating the blood from the other layers of vessels.

Cardiac endothelium contains two types of endothelial cells: the vascular endothelial cells lining coronary blood vessels, and the endocardial endothelial cells lining the cardiac chambers.

Nitric oxide (NO) is, among other relaxing factors, the most important modulator of myocardial function. This highly reactive signaling gas is formed enzymatically by the oxidation of L-arginine by oxygen to produce NO and L-citrulline. The enzyme that catalyzes this reaction is called nitric oxide synthase (NOS). NOS is a family of enzymes encoded by separate genes. Three NOS isoforms are present in the heart: neuronal NOS (nNOS or NOS1), endothelial NOS (eNOS or NOS3), and inducible NOS (iNOS or NOS2).

## NO Production

### Distribution of NO Synthases

nNOS and eNOS are constitutively expressed in the heart. iNOS is absent in the healthy heart, but its expression is induced by pro-inflammatory cytokines. iNOS produces high unregulated NO concentrations, which are connected with the generation of peroxynitrites and NO cytotoxic action. nNOS is mostly expressed in neuron terminals and regulates the release of catecholamines in the heart by suppression of norepinephrine release and inhibition of norepinephrine uptake into sympathetic neurons [1, 2]. Nonetheless, there is some evidence showing that nNOS is also present in human and rodent cardiomyocytes [3].

The major isoform responsible for the production of endothelial NO is eNOS. This enzyme is activated by agonists such as histamine, acetylcholine, and vasoactive peptides (endothelin 1, angiotensin II, and bradykinin), which increase intracellular $Ca^{2+}$ levels followed by $Ca^{2+}$/calmodulin-dependent modulation of eNOS. NO released by the endothelium as a result of eNOS activation acts as an autocrine factor promoting endothelial cell sprouting and vessel growth as well as a paracrine factor affecting cardiac myocytes.

### Regulation of NOS Activity

NOS isoforms contain binding sites for heme, L-arginine, tetrahydrobiopterin (Fig. 3.1), and calmodulin. Binding of calmodulin is very important because it is required for the functioning of heme. nNOS and eNOS are $Ca^{2+}$-dependent because binding of calmodulin to the enzyme is regulated by physiological concentrations of $Ca^{2+}$. Many hormones (bradykinin, estradiol, serotonin, vascular endothelial growth factor, and histamine) mobilize intracellular $Ca^{2+}$ transients,

J. Marín-García, *Signaling in the Heart*,
DOI 10.1007/978-1-4419-9461-5_3, 

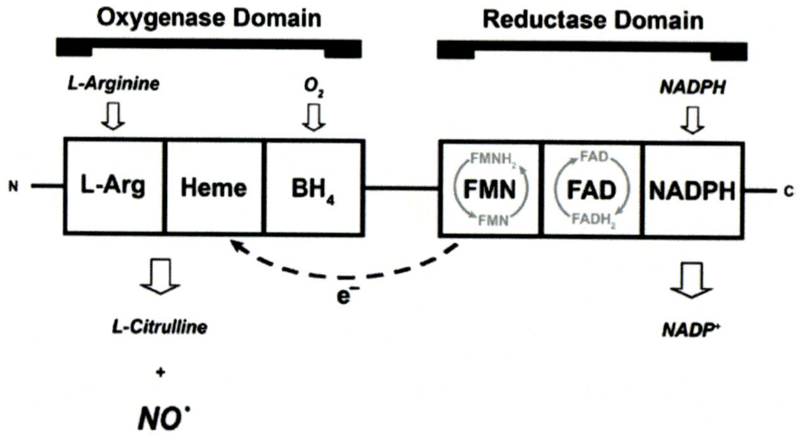


**Fig. 3.1** Domain structure of nitric oxide synthase. Nitric oxide synthase consists of two major domains: C-terminal reductase domain (RD) and N-terminal oxygenase domain (OD). In the RD, NADPH donates electrons, which travel through flavin adenine dinucleotide (FAD) and flavin mononucleotide (FMN). Then, they are shuttled through an intermediate $Ca^{2+}$/calmodulin-occupied region (not shown) to heme located in the OD. Heme catalyzes conversion of L-arginine (L-Arg) and molecular oxygen into L-citrulline and nitric oxide (NO). OD also contains a binding site for tetrahydrobiopterin ($BH_4$) – cofactor that is required for NO generation

and thus provide rapid mechanisms of eNOS activation via calmodulin. Agonists like bradykinin (BK) and acetylcholine through G protein-coupled receptors (GPCRs) activate phospholipase C (PLC) that cleaves the membrane component phosphatidylinositol 4,5-bisphosphate into diacylglycerol (an activator of protein kinase C) and inositol 1,4,5-trisphosphate (IP3). IP3 binds to IP3 receptors which regulate intra-cellular calcium through pleiotropic effects on ion channels.

It is interesting that iNOS is $Ca^{2+}$-independent because calmodulin is tightly bound to this isoform, even at low concentrations of intracellular $Ca^{2+}$.

In addition to "classical" $Ca^{2+}$-dependent regulation, other models of eNOS regulation have emerged during the last decade. They involve regulatory posttranslational modifications of eNOS (phosphorylation and nitrosylation) and protein–protein interactions.

## eNOS Phosphorylation

The most important regulatory phosphorylation site on eNOS is Serine (Ser) 1177 (primary sequence numbering corresponds to human eNOS). Phosphorylation of eNOS-Ser 1177 is associated with increased activity and NO production in response to a growing list of stimuli (bradykinin [4], β-adrenergic [5], vascular endothelial growth factor [6], and apelin [7]). eNOS-Ser 1177 can be phosphorylated by several protein kinases, including Akt, cAMP-dependent protein kinase (PKA), AMP-activated protein kinase, cGMP-dependent protein kinase, and $Ca^{2+}$/calmodulin-dependent protein kinase (CaM kinase). Phosphorylation of Ser 1177 stabilizes complex "calmodulin–eNOS" and enhances electron transfer, which is manifested as an increased catalytic activity.

In addition to Ser 1177, phosphorylation of two other serines, Ser 615 and Ser 633, was shown to stimulate eNOS. In ECs, these serines are phosphorylated in response to bradykinin, ATP, and vascular endothelial growth factor (VEGF). Data with specific inhibitors suggest agonist-dependent phosphorylation of Ser 615 and Ser 633 via Akt and PKA, respectively [8]. Phospho-Ser 615 increases sensitivity of eNOS to $Ca^{2+}$/calmodulin, whereas phospho-Ser 633 positively influences eNOS catalytic activity [8, 9].

In addition to stimulatory phosphorylations, two phosphor-sites have been described, which inhibit endothelial NO synthesis by eNOS. For example, phosphorylation at threonine (Thr) 495 (via protein kinase C, AMP-activated protein kinase) attenuates the binding of calmodulin by eNOS [4]. Bradykinin causes rapid dephosphorylation of Thr 495 (via protein phosphatase 1) and enhances the association of eNOS with calmodulin [4]. Similarly, protein kinase C (PKC)-dependent phosphorylation of Ser 114 is inhibitory [9, 10]. In bovine aortic endothelial cells, VEGF induces phosphoprotein phosphatase calcineurin, which leads to dephosphorylation of Ser 114 on eNOS and increases its catalytic activity [10].

## eNOS Nitrosylation

*S*-nitrosylation at cysteine residues is another reversible covalent modification that plays a role in posttranslational regulation of eNOS in endothelial cells. There are two cysteines, Cys 94, and Cys 99, that are tonically nitrosylated in caveolae-localized eNOS under resting conditions. Denitrosylation is associated with an increase in the enzymatic activity of eNOS. There is evidence that agonists such as VEGF and insulin promote rapid denitrosylation of eNOS associated with enzyme activation [11].

The mechanism of the inhibitory action of *S*-nitrosylation seems to involve impaired substrate and/or tetrahydro-L-biopterin binding to the oxygenase domain of eNOS, or electron transfer between reductase and oxygenase domains. eNOS is required to be targeted to the caveolae membrane for eNOS *S*-nitrosylation: nonmyristoylated mutant eNOS, which is targeted to the cytosol, does not undergo *S*-nitrosylation [11].

## eNOS and Protein–Protein Interactions

It has been well documented that eNOS (and likely nNOS) in endothelial cells and cardiomyocytes are associated with small invaginations of the plasmalemma called caveolae. Caveolae are flask-shaped membrane microdomains containing a scaffolding protein called caveolin (caveolin-1 isoform in endothelial cells and caveolin-3 isoform in cardiomyocytes). The caveolae are enriched in cholesterol and sphingolipids. In these plasma membrane subdomains numerous transmembrane signaling proteins and their effectors concentrate.

Localization of eNOS to caveolae is dependent on irreversible, cotranslational myristoylation of its N-terminal glycine. N-myristoylated eNOS targets to the cell membrane, where reversible posttranslational palmitoylation of the cysteine 15 and cysteine 26 residues occurs. Three acyls anchor eNOS to the caveolar lipid bilayer. Palmitoylation of eNOS is reversible, since prolonged bradykinin stimulation of eNOS induces depalmitoylation, which might serve as a mechanism for agonist-dependent modulation of eNOS activity.

In caveolae, eNOS interacts with caveolin. This direct binding is possible due to specific binding domains present on both eNOS (aa 350-358) and caveolin (aa 60-101). In the endothelium, caveolin not only serves as a scaffold to ensure coupling of eNOS to specific receptors in caveolae, but also maintains the enzyme inactive through direct allosteric blocking of the calmodulin binding site in eNOS [12]. Based on these findings, the current understanding of $Ca^{2+}$-dependent NOS regulation takes into account the initial disruption of the complex "NOS–caveolin" by $Ca^{2+}$/calmodulin. In addition, another possible role of caveolin is to serve as a "signalosome" to traffic eNOS to and from caveolae.

Recently, another biochemical partner of the eNOS has been identified that adds to our understanding of agonist-dependent eNOS activation. This is the 90-kDa heat shock protein, Hsp90. Hsp90 contains eNOS-binding site (aa 442-600), which directly interacts with specific region of eNOS (aa 300-400). Association between the two proteins is increased upon stimulation of endothelial cells with diverse eNOS agonists (VEGF, estrogen, histamine and shear stress) and correlates with enhanced NO production. Hsp90 binding stimulates eNOS activity directly and indirectly. Hsp90 enhances the affinity of eNOS for binding calmodulin; on the other hand it balances the output of NO versus superoxide and facilitates heme binding. In addition, Hsp90 plays an important role as a scaffold for active Akt, and this leads to activatory Akt-dependent phosphorylation of C-terminal serine 1177 on eNOS. It is possible that Hsp90-related regulation is more complicated, and it could involve other regulatory proteins (like co-chaperones cdc37 and CHIP), which bind eNOS–Hsp90–Akt ternary complex and regulate the interactions of both eNOS and Hsp90 with Akt.

Another protein identified to associate with eNOS is the 34 kDa-eNOS interacting protein (NOSIP). This protein was shown to compete with caveolin to bind to the oxygenase domain of eNOS. NOSIP decreases eNOS activity (by uncoupling eNOS from its interaction with caveolar co-localized effectors of upstream agonists) and redistributes eNOS from caveolae to the actin cytoskeleton.

The oxygenase domain of eNOS is also a binding site for the 58 kDa-eNOS trafficking inducer protein (NOSTRIN). By means of its SH3 domain NOSTRIN is capable to trimerize, so it may serve as a multivalent adaptor for the association of eNOS with other SH3-binding protein partners. Icking et al. [13] hypothesized that NOSTRIN plays a critical role in eNOS internalization because it binds GTPase dynamin-2 and neural Wiskott–Aldrich syndrome protein (N-WASP), which is necessary for caveolar endocytosis.

As previously noted, the caveolar localization appears to be essential for efficient agonist-mediated stimulation of eNOS. In quiescent ECs the myristoylated and palmitoylated eNOS is caveolae bound and is tonically inhibited by binding to caveolin-1, by inhibitory nitrosylation of cysteines 94 and 99, and by inhibitory phosphorylation of threonine 495 (Fig. 3.2).

Many agonists (bradykinin, acetylcholine, histamine, adenosine, ADP/ATP, VEGF, and sphingosine 1-phosphate) mobilize intracellular $Ca^{2+}$. Increased intracellular $Ca^{2+}$ transients lead to eNOS-$Ca^{2+}$-calmodulin binding. Significantly, the association of eNOS and calmodulin may be facilitated by dephosphorylation of eNOS at threonine 495.

$Ca^{2+}$ and calmodulin work together with Hsp90 to displace caveolin from eNOS to release its tonic inhibition. This results in the initial rapid activation of eNOS. Simultaneously, several agonists (e.g., VEGF and sphingosine 1-phosphate) activate phosphatidylinositol 3-kinase (PI3K)-dependent phosphorylation pathway resulting in membrane translocation/activation of Akt. eNOS/calmodulin/Hsp90 recruit active Akt to an oligomeric protein complex that phosphorylates serine 1177. This process is accompanied by rapid eNOS denitrosylation.

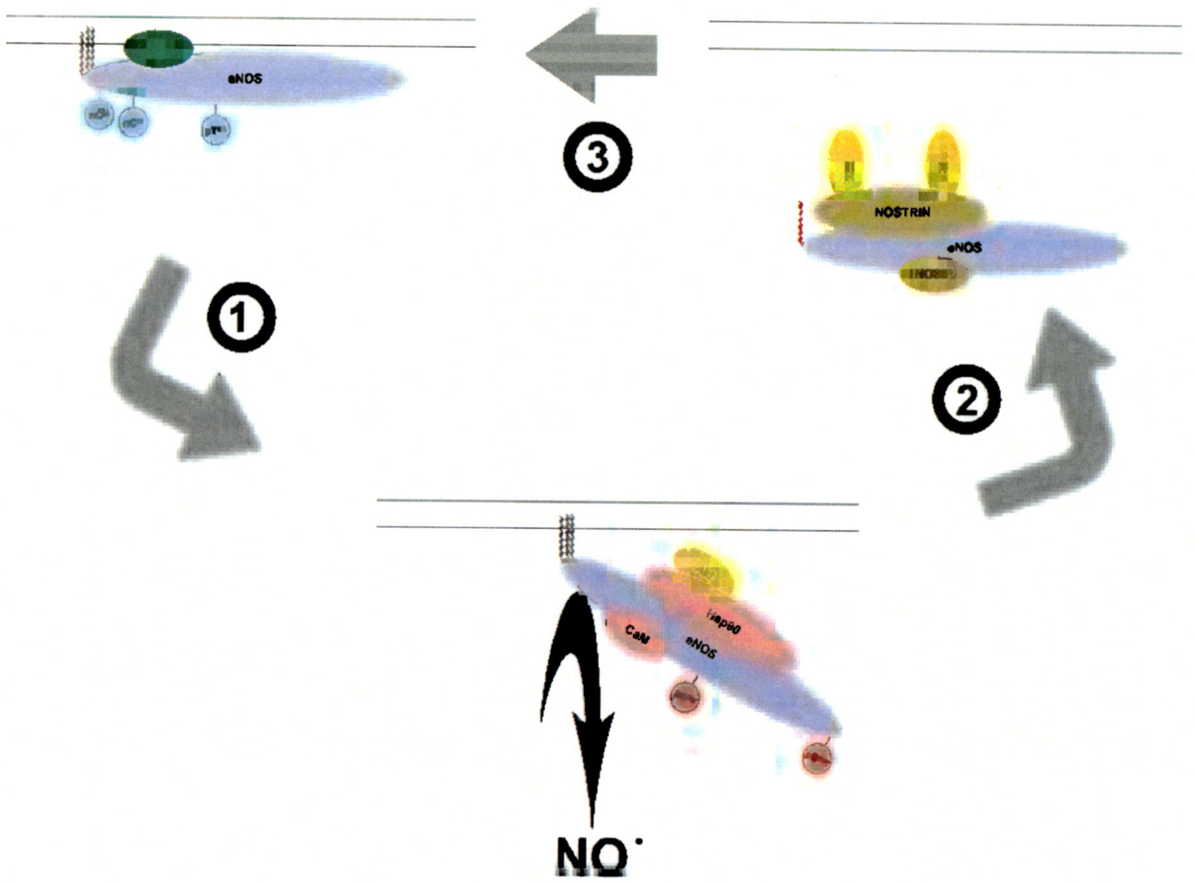


**Fig. 3.2** Regulation of eNOS activity and localization. Under quiescent conditions, myristoylated (*red* chain) and palmitoylated (*gray* chains) eNOS is anchored to caveolae. Enzyme in this state is inhibited by caveolin-1 (CAV-1) binding and several inhibitory covalent modifications (phosphorylation at threonine 495, nitrosylation at cysteines 94 and 99). (1) $Ca^{2+}$-mobilizing agonists (bradykinin, acetylcholine, histamine, etc.) increase intracellular $Ca^{2+}$. As a result, $Ca^{2+}$-calmodulin (CaM), together with Hsp90 protein, displaces CAV-1 on eNOS to increase catalytic activity. This process is accompanied by dephosphorylation of threonine 495, denitrosylation, and Akt-dependent activatory phosphorylation of serine 1179, which further enhance eNOS activity. (2) Inactivation/internalization of eNOS happens as a result of depalmitoylation, dephosphorylation, and dissociation of calmodulin. NOSIP, NOSTRIN, and several other proteins bind to eNOS at this step and navigate translocation to intracellular compartments. (3) Reassociation of eNOS with caveolae: renitrosylation and palmitoylation play role in relocation of eNOS to caveolae and rebinding of CAV-1. See text for details

Together with calmodulin binding, denitrosylation and phosphorylation of serine 1177 enhances the eNOS catalytic activity (Fig. 3.2).

A period of agonist stimulation and eNOS activation is followed by a slower mechanism of inactivation and eNOS internalization. One mechanism of inactivation is realized through eNOS depalmitoylation/translocation from the caveolae to intracellular compartments (e.g., Golgi, peri-nuclear region, mitochondria and cytoskeleton). Another mechanism is NOSTRIN- or NOSIP-mediated trafficking of eNOS to intracellular compartments (including NOSTRIN-regulated endocytosis of caveolae). It has been reported that different agonists can induce eNOS translocation to different cellular compartments; for example, in ECV-304 cells acetylcholine induces preferential movement of eNOS to the Golgi apparatus, while platelet-activating factor moves eNOS to the cytosol [14]. There are several deactivating events which accompany eNOS internalization, including eNOS uncoupling from upstream signaling pathways and activating molecules; also the loss of $Ca^{2+}$/calmodulin-dependent activation, and phosphorylation of inhibitory and dephosphorylation of stimulatory sites take place [15, 16] (Fig. 3.2).

We should be mindful that translocation of eNOS is probably a more complex and polyfunctional process than originally thought. Under certain conditions agonists can activate internalized eNOS [17], and activation of internal eNOS includes denitrosylation, Akt-dependent phosphorylation, and/or interaction with $Ca^{2+}$/calmodulin [17–19].

Probably, NO generated by active internalized eNOS plays a role in the nitrosylation of intracellular proteins involved in regulation of endothelial cell function(s), such as guanylyl cyclase, mitochondrial iron–sulfur cluster enzymes, transcription factors, and cytoskeletal proteins.

Trafficking of eNOS from caveolae to intracellular compartments is a reversible process. The enzyme must reenter caveolae for reassociation with caveolin and *S*-nitrosylation. That, palmitoylation plays a role in eNOS relocation appears to be an attractive hypothesis that needs to be confirmed experimentally.

## Mechanism of Action of NO (Targets for NO)

### Activation of Guanylyl Cyclases, cGMP-Regulated Cyclic Nucleotide Phosphodiesterases, and cGMP-Dependent Protein Kinases

There are several mechanisms by which NO affects the biology of endothelial cells. Most of the effects of NO involve the activation of cytoplasmic soluble guanylyl cyclases (sGC). sGC is a heterodimeric protein that consists of an α and a β subunit. Once generated, NO binds to the prosthetic heme group of sGC to induce a conformational change – breakage of the histidine-to-iron bond and activation of the enzyme. Activated sGC synthesizes cyclic guanosine monophosphate (cGMP) [20].

The main targets for cGMP in myocardium are cGMP-dependent protein kinases (PKGs), cGMP-stimulated and -inhibited cyclic nucleotide phosphodiesterases (PDEs), and cGMP-modulated cation channels. The former pathway potentially alters myocyte function through phosphorylation of various kinases, whereas the latter modulates intracellular levels of cyclic adenosine monophosphate.

In the heart, several effects have been attributed to cGMP, including modulation of sarcolemmal $Ca^{2+}$ influx, changes in the action potential characteristics, reduction in myofilament $Ca^{2+}$ sensitivity, and reduction in $O_2$ consumption [20].

### Nitrosylation of Proteins

Another regulatory effect of NO is a posttranslational modification of proteins via nitration of tyrosines and *S*-nitrosation of thiol-groups of cysteines. Upon modification, proteins may change their properties. Important targets of *S*-nitrosation that influence cardiac function include ion channels, such as calcium release channel ryanodine receptor [21], and regulator of G protein signaling 4 (RGS4) [22]. Recently, it was shown the nitrosation of sGC in the presence of low levels of NO. *S*-nitrosation led to a reduction in NO-stimulated activity of sGC [23]. Thus, this process may account for desensitization of sGC. Moreover, the nitrosation of adenylyl cyclase, which decreased the $V_{max}$ of enzyme (without altering the $K_m$ for ATP), has been reported [24].

## NO-Regulated Functions

### Effects on Cardiac Cells

The effects of endothelium-produced NO on myocardial function include modulation of contractile performance, metabolism, and cell growth and survival.

In general, NO effect on cardiomyocyte contraction is rather complex; namely low levels of NO have a positive inotropic effect, whereas higher levels have a negative inotropic effect. The positive inotropic effect is mediated by cyclic adenosine monophosphate, whereas the negative effect is induced by cGMP activation [25].

There is increasing evidence that NO is a paracrine factor regulating the growth of cardiomyocytes. Under certain conditions, NO demonstrates inhibitory effect on cardiomyocytes hypertrophy induced by epinephrin through α1-adrenoceptors [26]. This effect of NO is realized through targeting the calcineurin-nuclear factor of activated T-cells (NFAT) signaling pathway; moreover, NO via sGC/cGMP/PKGI suppresses the L-type $Ca^{2+}$-channels. This reduces cellular $Ca^{2+}$ and the activity of $Ca^{2+}$/calmodulin-activated serine/threonine phosphatase calcineurin. As a result, calcineurin cannot dephosphorylate transcription factor NFAT. Inactive phosphorylated NFAT does not translocate from the cytosol to the nucleus to induce the hypertrophic program of gene expression [20, 26]. Other possible targets of antihypertrophic action of NO include down-regulated expression of the cytoskeleton-associated muscle LIM protein [27], inhibition of *c-fos*, and β-MHC expression [28]. The findings from Ozaki et al. [29] are in agreement with a cardioprotective function of NO, and according to these authors, mice overexpressing eNOS in endothelium developed less myocardial hypertrophy after treatment with isoproterenol.

It is worthy of note that the role of NO in the regulation of cardiomyocyte growth is rather complex. Recently, it has been shown that increased endothelial cell mass promotes myocardial hypertrophy, and this effect was mostly prevented by treatment with an eNOS inhibitor. Thus, in this model, increased NO (as a result of increased endothelial cell mass) served as a myocardial prohypertrophic factor. A potential mechanism might be that NO at high levels destabilizes an important regulator of cardiac hypertrophy, the RGS, by nitrosylation of this molecule. Normally, RGS negatively modulates $G_q/G_i$ protein-mediated myocardial hypertrophy by deactivating the α-subunit of G protein. NO-dependent modification leads to RGS ubiquitination/degradation and derepression of the hypertrophic program [20, 30].

### Induction of Vascular Relaxation

Several agonists, acetylcholine, endothelin (ET-1), angiotensin II (Ang II), bradykinin, and histamine (via activation of $H_1$ receptors) cause relaxation of the vascular smooth muscle by acting on endothelial cells [31–33]. They stimulate the releasing of relaxing factors by endothelium. For example, low doses of ET-1 provoke dose-dependent vasodilatation in the rat isolated perfused mesentery artery, but endothelial removal abolishes the vasodilatation induced by ET-1 [34].

There is evidence that the major target for the relaxing hormones, in the endothelium, is eNOS, NO being the main relaxing factor produced. NO diffuses into the smooth muscle and activates guanylyl cyclase and the production of cGMP, which mediates relaxation of the vascular smooth muscle.

Inhibitors of eNOS blocked the relaxing effect of acetylcholine on the mouse aorta, carotid, and coronary arteries [32]. In the rat carotid artery, a NOS inhibitor and an inhibitor of guanylyl cyclase significantly reduced ET-1-mediated relaxation, confirming the participation of the NO-cGMP pathway in ET-1-mediated vasorelaxation [35]. The relaxation induced by ETs in the rat aorta [36] and in the rat mesenteric bed [34] was described to involve the activation of the guanylyl cyclase enzyme. Angiotensin II is also able to induce relaxation via activation of the endothelial NO-cGMP pathway. Interestingly, in the aorta and in several arteries this peptide interacts with traditional receptors (e.g., AT 2 type), whereas in the carotid artery high levels of Ang II are required to activate an atypical angiotensin receptor, Ang(1-7) [37]. The effect of bradykinin as a vasodilator is related, at least in part, to the activation of endothelial $B_2$ receptors, release of NO, and activation of sGC [37].

The role of eNOS/PKG in endothelium-induced relaxation of vascular smooth musculature has been further demonstrated in genetic mouse models; relaxation of aorta, carotid, and coronary arteries was abrogated in eNOS KO mice [31, 32]. In addition, endothelium-induced vasorelaxation has not been observed in sGC KO or PKGI KO mice [38, 39].

Endothelial cells also release contracting factors such as vasoconstrictor prostanoids. Under physiological conditions, there is a balanced release of relaxing and contracting factors, and this balance can be altered in cardiovascular diseases. In particular, in patients with essential hypertension dysfunction of endothelium leads to decreased bioavailability of NO and impairs endothelium-dependent vasodilation. Under these pathophysiological conditions the endothelium is transformed from an organ-protector to a source of vasoconstrictors (see below).

### Angiogenesis

As mentioned above, endothelial NO plays an important role in angiogenesis. NO has been identified as a downstream mediator of several growth factors that initiate the angiogenic signaling cascade in endothelial cells: vascular endothelial growth factor (VEGF), transforming growth factor β1 (TGF-β1), and substance P [40]. Autocrine NO production, following the activation of eNOS in endothelial cells, sets in motion the first steps of angiogenesis: cell migration and organization in tubes [12]. Transgenic endothelial NOS overexpression enhanced angiogenesis and blood flow recovery in an ischemic model [41], while modulation of endogenous NO bioactivity may promote angiogenesis in ischemic tissue [42].

The compartmentalization of eNOS to caveolae is very important for the role of eNOS in angiogenesis. For example, decreases in caveolin abundance in microvascular endothelial cells by statins derepressed eNOS activity, induced NO production and microvessel formation [43]. Paradoxically, studies on caveolin-1 KO mice gave opposite results. In this model, animals had a deficient angiogenesis [12]. An understanding of this phenomenon comes from the mechanism of action of angiogenic factor, VEGF. In the endothelial cell, receptors for VEGF are localized to caveolae. Binding of VEGF to its receptors recruits complexes “chaperone protein hsp90/active Akt” and “chaperone protein hsp90/calcineurin” to caveolae, where eNOS becomes activated through phosphorylation of Serine-1177 and dephosphorylation of Threonine-495 [12, 43, 44]. In the absence of caveolae in the caveolin-1 KO endothelial cell, the VEGF/eNOS pathway is uncoupled. Interestingly, re-expression of caveolin in caveolin-1 KO cells restored the VEGF/eNOS coupling [12, 44].

## Other Endothelial Pathways

### *Prostanoids (Prostaglandins and Prostacyclin)*

Prostanoids are local hormones that are synthesized as oxygenated derivatives of three different 20-carbon essential fatty acids. They are produced in the cell in response to extracellular signal (mechanical trauma, cytokines, growth factors, etc). Extracellular signal triggers activation of phospholipase $A_2$, which catalyzes release of 20-carbon essential fatty acid from phospholipid. Next, the fatty acid converts to prostanoids by oxygenation. This two-step process is catalyzed by a very important regulatable enzyme, cyclooxygenase (COX); initially, the short-lived, unstable prostaglandin G (PGG) forms, and subsequently converts into PGH. All other prostanoids are synthesized from PGH by several enzymes (PGE synthetase, PGE 9-reductase, PGD synthetase, prostacyclin synthase, and thromboxane synthase).

### *Function of Prostanoids*

Prostanoids have a short half-life, ranging from seconds to minutes. They act on the same cell or nearby cells (i.e., they are autocrine and paracrine mediators). A number of vasoactive agents were shown to influence vascular smooth muscle cell (SMC) contraction by stimulation of prostanoid release from endothelial cells.

Prostacyclin is the principal prostanoid synthesized by blood vessels. In most vessels, prostacyclin causes relaxation of the vascular smooth muscle through the activation of its receptors (IP receptors) leading to stimulation of adenylyl cyclase [45]. A subsequent increase of production of cAMP leads to the activation of PKA, phosphorylation/inhibition of myosin light-chain kinase (MLCK) and results in muscle relaxation. The secretion of prostacyclin and several other vasodilatory prostanoids is regulated by acetylcholine (production of prostacyclin in a higher amount than $PGE_2$ and $PGF_{2\alpha}$ and much higher thromboxane in rat endothelium) [46], bradykinin (production of prostacyclin in human endothelium) [47], and ET-1 (production of prostacyclin) [37]. Vasodilatory effects of acetylcholine, bradykinin, and ET-1 can be blocked by inhibitors of COX [35] by COX1 knocking out, or by prostanoid receptor antagonists [48].

Importantly, under several conditions (animal models: spontaneously hypertensive rats (SHR) and aged rats; in humans: aged subjects, essential hypertension) endothelium-dependent relaxations are impaired because of decreased bioavailability of NO [increased production of oxygen-derived free radicals (superoxide anions), under pathological conditions mentioned above, destroys NO]. In hypertension, with aging function of endothelium in aorta, the mesenteric arteries change to the opposite and this results in an endothelium-derived contraction [46, 49, 50]. Surprisingly, this endothelium-dependent contraction also involves COX-catalyzed metabolites different from well-known vasoconstrictor, thromboxane [46]. It turns out that in the SHR and aging rats, the endothelium-dependent contractions, elicited by acetylcholine most likely, also involve the release of prostacyclin with an associated contribution of $PGH_2$ [46] because: firstly, more prostacyclin is released from the aorta of SHR compared to normal animals [46]; secondly, IP receptor is less expressed in SHR than in wild type rats and the IP receptor gene expression decreases with age. Thus, under conditions where functional vessel-relaxing IP receptors disappear, massively produced prostacyclin activates thromboxane receptors (TP) on vascular smooth muscle cell. Stimulation of G protein-coupled TP leads to smooth muscle contraction [46, 50].

Wilson et al. [51] identified components of the signaling pathway involved in TP-mediated contraction of rat caudal artery. According to their model, activated TP couples to the $G_{12/13}$ protein and activates Rho guanine nucleotide exchange factor (Rho-GEF) via direct interaction of the G protein α-subunit with the Rho-GEF. Activated Rho-GEF, in turn, activates small GTPase, RhoA. The downstream target of activated RhoA is Rho-associated kinase (ROK). Downstream targets of ROK include L-type $Ca^{2+}$ channels and myosin light chain phosphatase (MLCP). ROK-mediated phosphorylation (directly or indirectly) of $Ca^{2+}$ channels results in the entry of extracellular $Ca^{2+}$. $Ca^{2+}$ diffuses to the contractile machinery, where it binds to calmodulin to activate MLCK, increases the level of myosin regulatory light-chain (LC) phosphorylation, and activates cross-bridge cycling and contraction. Phosphorylation of MLCP decreases it activity, shifts MLCK/MLCP-balance in favor of MLCK and results in a greater degree of LC phosphorylation and contraction [51].

ATP in high concentrations is also a vasoconstrictor, and this effect can be blocked by inhibitors of COX (indomethacin and iboprofen) [52, 53]. ATP activates purinergic P2Y receptors on endothelial cells, which results in synthesis of several prostanoids (prostacyclin and thromboxane) [52, 53]. Prostanoids paracrinally activate TP on vascular smooth muscle cells and induce contraction. This way ATP may cause arterial vasospasm under pathological conditions, such as ischemic or injured vessels, when it reaches high concentrations.

Since, prostanoids cause vasoconstriction by activation of the TPs on the vascular smooth muscle cells, selective antagonists at these receptors prevent endothelium-dependent contractions, and curtail the endothelial dysfunction in hypertension and diabetes.

Not all arteries are affected in hypertension. For example, coronary arteries in SHR maintain the endothelium-dependent relaxation and is not associated with the prostanoid-dependent vasoconstriction [54].

## Vasoactive Peptides (Endothelin 1, Angiotensin II, and Bradykinin)

Vasoactive peptides, such as endothelin 1, angiotensin II, and bradykinin, play a fundamental role in controlling the functional and structural integrity of blood vessel wall and may be important in physiological processes and in pathological mechanisms underlying vascular diseases. These peptides induce vasoconstriction or relaxation by the release of endothelium-derived relaxing factors, such as NO and prostanoids. The actions of ET-1, Ang II, and BK are mediated via specific intracellular signaling pathways that are stimulated following initial binding of peptides to their respective extracellular receptors. Intracellular signaling pathways transmit information from the receptors to intracellular proteins that regulate cell activities such as contraction and relaxation.

### Endothelin 1

ET-1 represents a potent vasoactive agent released by the endothelium and vascular smooth muscle. ET-1 is synthesized as 38-amino-acid precursor (big ET-1). The big ET-1 is processed to ET-1 by ET-converting endopeptidase-1 (ECE-1) in many blood vessels [37]. There are four isoforms of human

ECE-1: ECE-1a, -1b, -1c, and -1d, which are expressed via alternative promoters from the same gene. These isoforms perform as dimers and localize to different subcellular regions: ECE-1a, -1c, and -1d are located on the plasma membrane, whereas ECE-1b is intracellular. Interaction of plasma membrane isoforms with ECE-1b anchor them inside the cell and thus decreases extracellular ECE-1 activity [55]. Recently it was shown that expression of endothelial ECE-1b is regulated by transcription factor E2F2 and RNA-binding transcription cofactor Sam68 [56]. Interestingly, down-regulation of ECE-1b in transgenic E2F2 defective mice resulted in high blood pressure and elevated aorta contractility because in the absence of ECE-1b more of the other ECE-1 isoforms reached the plasma membrane with subsequent increase in extracellular ET-1 synthetase activity. The physiological importance of ECE-1b also revealed in clinical studies: C-338A polymorphism in regulatory region of the ECE-1b gene, which alters binding of E2F2, strongly correlated with increases in blood pressure in women [57, 58].

ET-1 binds to specific receptors, $ET_A$ and $ET_B$ (endothelial $ET_{B1}$ and muscular $ET_{B2}$ subtypes). Pharmacologically and immunohistochemically, ET receptors were identified in the endothelium as well as in the vascular smooth muscle. Interestingly, endothelial cells express only $ET_{B1}$ receptor, whereas vascular smooth muscle cells express both $ET_A$ and $ET_{B2}$ receptors.

Infusion of ET-1 into rats caused an initial drop in blood pressure that was followed by intense and prolonged hypertension. The initial decrease in blood pressure results from the activation of endothelial $ET_{B1}$ receptors, which are linked to production of vasodilatory agents, NO and prostacyclin (described above).

In addition to production of NO and prostacyclin, ET-1 was shown to inhibit activity of NAD(P)H oxidase Nox1 and superoxide production in human abdominal aortic endothelial cells. Thus, ET-1 through $ET_{B1}$ receptors may protect endothelial cells by attenuating intracellular reactive oxygen species (ROS) level. Inhibitory effect of ET-1 is mediated by Pyk2-Rac1 pathway: activation of $ET_{B1}$ receptors by ET-1 suppressed activatory phosphorylation of protein tyrosine kinase Pyk2 and small signaling G protein Rac1, while silencing of $ET_{B1}$ receptors increased phosphorylation of Pyk2 and Rac1, and abolished ET-1-induced inhibition in phosphorylation of Pyk2 and Rac1 [59].

The vasoconstrictor response of smooth muscle cells is due to the direct action of ET-1 on these cells via $ET_A$ and $ET_{B2}$ receptors. Functional significance of ET receptor subtypes in SMC depends on the species and blood vessel type: $ET_B$ receptor plays a minor role in ET-1 induced contraction of the rat carotid artery, rat mesenteric small arteries, and porcine coronary arteries [37], but contributes 80% of contraction in the rabbit pulmonary artery [60]. In humans, endogenous ET also increases basal vascular tone, which can be decreased by systemic administration of an ET receptor antagonist [61]. ET-1-induced SMC contraction in several arteries was found to be markedly inhibited by antagonists of L-type $Ca^{2+}$ channels or ryanodine. Thus, extracellular $Ca^{2+}$ influx and mobilization of $Ca^{2+}$ from the sarcoplasmic reticulum participate in ET-1-related signal transduction mechanisms. Increase in intracellular $Ca^{2+}$ causes phosphorylation of myosin kinase and, in turn, long-lasting smooth muscle cell contraction.

ET is activated in several animal models of hypertension as well as in humans with salt-sensitive essential hypertension. Possible role of ET in hypertension is discussed in Chap. 13.

Besides vasoactivity, endothelin modifies several cell-survival signaling pathways in the heart. Major pathways include MAPK, PI3-K/Akt, NF-κB, and calcineurin signaling, and will be described in separate chapters.

## Angiotensin II

Angiotensin II is another vasoactive peptide which plays an important role in controlling the functional and structural integrity of the vessel wall. In the classical renin–angiotensin system, the liver produces angiotensinogen, which is converted into decapeptide angiotensin I (Ang I) by circulating kidney-derived protease renin. In lungs dipeptidyl carboxypeptidase, angiotensin-converting enzyme (ACE), converts Ang I to the physiologically active Ang II.

A number of studies suggest that the heart and blood vessels express components of the renin–angiotensin system. Active ACE was found in rat carotid artery [37] and human and rodent arteries [62]. In addition, an alternative pathways to ACE for Ang II generation exist: First, chymase an Ang II-forming enzyme, which is expressed in the human heart and blood vessels, including endothelial cells [62]. Second, elastase-2 was shown to play a physiological role generating Ang II in rat carotid and mesenteric arteries [63], including endothelial cells [64]. These evidences indicate that Ang II, beyond being produced systemically, can be produced locally and can induce local vascular constriction.

Two Ang II receptor subtypes, $AT_1$ and $AT_2$, have been identified. They both have 7-transmembrane domains and belong to GPCR family. The constrictor effect of Ang II involves the activation of both $AT_1$ and $AT_2$ receptors in rat mesenteric venules and portal vein SMC, whereas only $AT_1$ receptors are found in the rat carotid artery [37]. AT-mediated increase in tension depends on the influx of extracellular $Ca^{2+}$ and the mobilization of intracellular $Ca^{2+}$ [65]. Evidence shows that when activated by an agonist, the $AT_1$ receptors couple to $G_{\alpha q/11}$, $G_{\alpha 12/13}$, and $G_{\beta\gamma}$ complexes, which activate downstream effectors including several phospholipases: phospholipase C (PLC), phospholipase $A_2$ ($PLA_2$), and phospholipase D (PLD). Activation of PLC produces inositol-1,4,5-trisphosphate and diacylglycerol (DAG) within seconds. IP3 binds to its receptor on sarcoplasmic reticulum, opening a channel that allows calcium efflux into the cytoplasm. $Ca^{2+}$ binds to calmodulin and activates MLCK, which phosphorylates the myosin light chain and enhances the interaction between actin and myosin, causing smooth muscle cell contraction. PLD activation results in hydrolysis of phosphatidylcholine (PC) to choline and phosphatidic acid (PA). PA is rapidly converted to DAG, leading to sustained PKC activation and sustained muscle contraction. Activation of $PLA_2$ leads to production of arachidonic acid (AA) and its metabolites [66].

Endothelial cells also express angiotensin receptors. Binding of Ang II to endothelial AT-receptors causes the release of endothelium-derived NO which is a potent vasodilator (see above). Thus, the contractile response of vessels to Ang II results from the combined activation of endothelial and smooth muscle AT-receptors. For example, contractile activity of rat carotid artery decreases the maximal response to the Ang II in the presence of the endothelium [37].

There are experimental evidences of vasorelaxant actions of Ang II under conditions where the levels of Ang II are high (0.03–3 mM). Several mechanisms are involved in this phenomenon. First, at high concentrations, a fraction of Ang II can be converted into Ang(1-7) by ACE-2, a novel ACE homologue integral membrane protease. Ang(1-7) is an endogenous ligand for the atypical angiotensin-type G protein-coupled receptor Mas. Binding of Ang(1-7) to Mas activates the signaling cascade resulting in NO release, which autocrinally produces SMC relaxation [67, 68]. It should be mentioned that at the high concentrations Ang II can directly activate Mas. Based on a number of observations ACE-2/Ang(1-7)/Mas axis is realized in mice aorta and rat carotid artery [37]. On the other hand, in a number of vessels (fowl aorta, dog renal and cerebral arteries, and rat aorta) Ang II exerts its vasodilator effect by traditional receptors, $AT_2$ [37]. In this case, $AT_2$ stimulation increases activity of the amiloride-sensitive $Na^+/H^+$ exchanger and produces intracellular acidification, resulting in activation of kininogenase. Kininogenase activation leads to production of bradykinin. Bradykinin through an autocrine–paracrine mechanism stimulates $B_2$ receptors on SMC, which in turn is the trigger for the release of vasodilators, NO, and cyclooxygenase products [69]. Ang II-induced vasodilatory effects can be endothelium-dependent. For instance, in rabbit selective $AT_2$ receptor stimulation in endocardial endothelium induces a negative inotropic and lusitropic effect on myocardium [70]. Similarly, Ang II produces a vasodilator effect through the $AT_2$ receptor in aorta of mice and requires intact endothelium [71]. In both cases, the action of Ang II involves bradykinin/NO and targets myocardial large-, intermediate-, and small-conductance $Ca^{2+}$-activated $K^+$ channels [70, 71].

One of the important effects of $AT_1$-receptor activation in the cardiovascular system is the production and release of ROS. $AT_1$-receptor-mediated ROS production is mainly linked to activation of the superoxide radical-producing NAD(P)H oxidase in vascular cells. In addition, Ang II may cause uncoupling of eNOS and activation of xanthine oxidase, leading to enhanced superoxide production by these enzymes [72]. An excessive production of ROS, overriding cellular antioxidant defense systems, leads to oxidative stress and pathological process(es) underlying hypertension (see Chap. 13).

In addition to vasoactivity, Ang II has intracellular signal transduction pathways that lead to long-term biological effects, such as cell growth, migration, extracellular matrix deposition, and inflammation.

## Bradykinin

Bradykinin is a potent vasoactive nonapeptide that is formed in the plasma by the proteolytic cleavage of a larger precursor termed kininogen. Kininogen is produced mainly by hepatocytes. However, in the vascular wall it is synthesized by endothelial and vascular smooth muscle cells. Cleavage of kininogen is catalyzed by proteolytic enzymes, kininogenases: kallikrein and a family of genetically related kallikrein-like serine proteases [73]. Kininogenases are expressed endogenously by endothelial and smooth muscle cells of various blood vessels.

Bradykinin is a very powerful vasodilator. Biological effect of bradykinin is mediated by two G protein-coupled receptors, the $B_1$ and $B_2$ receptors. Vascular endothelial and smooth muscle cells express $B_2$ receptors. Activation of $B_2$ receptors induces relaxation of SMC, and this role of BK is mediated at least in part by NO release in endothelial cells (canine and rat carotid artery, bovine middle cerebral arteries, and porcine coronary arteries). Specifically, $B_2$ receptor in the endothelial cell is functionally coupled to phospholipase C, which generates inositol trisphosphate arid diacylglycerol. Activation of phospholipase C leads to a stimulation of protein kinase C and the mobilization of calcium ions from both intracellular and extracellular pools. Increased $Ca^{2+}$ results in activation of eNOS with subsequent release of NO.

Constitutive NO synthase undergoes phosphorylation of a serine residue during stimulation of endothelial cells by shear stress or bradykinin, which leads to a translocation of the enzyme to the cytosol. Bradykinin treatment of cells results in activation of NO synthesis within seconds, and enzyme activity returns to baseline within 5 min. In contrast, phosphorylation of NO synthase in endothelial cells in response to bradykinin is maximal only after 5 min of agonist exposure and persists for at least 20 min, long after bradykinin-induced enzyme activation has decayed [74].

BK-induced relaxation also involves the release of vasorelaxant prostanoids (prostacyclin) [75].

Plasma levels of bradykinin are much lower than the threshold concentration for the systemic hypotensive effect of bradykinin. Thus, the molecular basis of the autocrine and paracrine functions of BK in blood vessels is the regulation of activity of enzymes responsible for synthesis and degradation of BK, i.e., kininogenases and kininases. In experimental animal models, specific up-regulation of angiotensin II receptors type $AT_2$, resulted in vasodilatory effect because Ang II through the receptors of this type activated kininogenase in endothelium and vascular SMCs (see above, "Angiotensin II"). Kininogenase generated bradykinin, which increased NO production, autocrinally/paracrinally by endothelial cells [71, 76] or autocrinally by SMCs [69]. Contribution of endothelial $B_2$ receptors to the Ang II-mediated vasodilation was also demonstrated in human heart coronary microarteries [77]. One of the kininases that catalyze the breakdown of bradykinin is the angiotensin-converting enzyme (ACE). In cultured endothelial cells, bradykinin accumulates during ACE inhibition and elicits the generation of NO and prostacyclin [78]. Also, ACE inhibition by perindopril improved hemodynamic characteristics in patients with coronary disease, and this effect was accompanied by increased systemic serum levels of bradykinin in patients. Moreover, effects of ACE inhibition were attenuated by $B_2$-receptor antagonist HOE-140, suggesting that bradykinin is importantly involved in the endothelial effects of ACE inhibition [79].

### Redox Signaling

ROS, particularly superoxide ($O_2^{\bullet-}$) and hydrogen peroxide ($H_2O_2$), are important signaling molecules in cardiovascular cells. Their production is regulated by hormone-sensitive enzymes such as the vascular NAD(P)H oxidases, and their metabolism is coordinated by antioxidant enzymes such as superoxide dismutase, catalase, and glutathione peroxidase. ROS function as second messengers activating multiple intracellular proteins and enzymes, including tyrosine kinases (epidermal growth factor receptor and c-Src), tyrosine phosphatases, MAP kinases (p38 mitogen-activated protein kinase), Ras, Akt/protein kinase B, and ion channels, primarily through oxidative modification of proteins and activation of transcription factors. In physiological conditions, low concentrations of intracellular ROS play an important role in normal redox signaling involved in maintaining vascular function and integrity. Thus, ROS participate in vascular smooth muscle cell growth and migration, modulation of endothelial function, including endothelium-dependent relaxation, and modification of the extracellular matrix. Under pathological conditions ROS contribute to vascular dysfunction and remodeling through oxidative damage. In hypertension, activation of prooxidant enzymes such as NAD(P)H oxidase, NOS, xanthine oxidase, and mitochondrial enzymes or altered thioredoxin and glutathione systems results in increased ROS formation, which have damaging effects on the vasculature.

The primary physiological producers of $O_2^{\bullet-}$ in vascular tissue are NAD(P)H oxidases. Production of superoxide by NAD(P)H oxidases is modulated by vasoactive hormones and the low-molecular weight G protein Rac [80–84]. Dismutation of $O_2^{\bullet-}$ by superoxide dismutase (SOD) produces $H_2O_2$. Removal of ROS is catalyzed by catalase and glutathione peroxidase, which convert $H_2O_2$ into water. Expression of antioxidant enzymes (SOD, catalase, and glutathione peroxidase) is regulated by Ang II, tumor necrosis factor α (TNF-α), and interleukin (IL)-1β.

$O_2^{\bullet-}$ generation is catalyzed by a membrane bound heterodimeric flavocytochrome (cytochrome *b*558), consisting of a 91-kDa glycoprotein (gp91*phox*) and a 22-kDa protein (p22*phox*). Electron transfer from NAD(P)H to oxygen requires interaction of the cytochrome *b*558 with two cytosolic proteins, p47*phox* and p67*phox*, and small GTPase (Rac1 or Rac2). Upon activation, the cytosolic components translocate to the plasma membrane to assemble functional NAD(P)H oxidase. All the above mentioned components of NAD(P)H oxidase are expressed in the endothelial cells, whereas NADPH oxidase-1 (nox-1) substitutes gp91*phox* in vascular SMCs [83, 85–87].

Ang II stimulates NAD(P)H oxidase-dependent $O_2^{\bullet-}$ and $H_2O_2$ production in SMCs and endothelium. This process is biphasic: the first rapid phase (peak at 30 s) involves phospholipase D-mediated phosphatidylcholine hydrolysis. This results in the production of phosphatidic acid. Phosphatidic acid is a source of DAG that activates PKC. Stimulation of PKC is important for this phase of NAD(P)H oxidase activation because PKC inhibitors attenuate Ang II-dependent ROS generation [88, 89]. Also, phosphatidic acid can directly activate NAD(P)H oxidase [90].

The second phase of ROS generation is of much greater magnitude than the first and continues for several hours. This phase is likely to mostly influence cell growth and differentiation. According to current views, transactivation of endothelial growth factor (EGF) receptor (EGFR) plays an important role in sustained Ang II-mediated activation of NAD(P)H oxidase: initial $H_2O_2$ (formed at phase one) activates Src, leading to EGFR transactivation and stimulation of phosphatidylinositol 3-kinase. PI3K produces phosphatidylinositol (3,4,5)-trisphosphate ($PIP_3$), which in turn activates Rac. Rac most likely binds to the NAD(P)H oxidase complex, leading to the generation of $O_2^{\bullet-}$ and $H_2O_2$. $H_2O_2$ stimulates Src, which then further amplifies NAD(P)H oxidase activity [89].

Besides Ang II, several other growth factors were shown to activate NAD(P)H oxidase in SMCs: platelet-derived growth factor (PDGF) (probably, via stimulation of phospholipase

Cγ and activation of PKC), thrombin, and TNF-α [82–84]. Mechanical forces also play an important role in ROS generation. In general, laminar flow is associated with an increased expression of antioxidants glutathione peroxidase and SOD, thus having protective effect on vascular wall against oxidative injury. On the other hand, oscillatory shear causes increased ROS production with consequent oxidative damage, as occurs in hypertension.

ROS through oxidation/inactivation of protein tyrosine phosphatases (PTP) and activation of transcription factors, such as NF-κB, activator protein-1, and hypoxia-inducible factor-1 (HIF-1), modify the activity of protein kinases that participate in the signaling pathways regulating vascular cell growth, apoptosis, migration (to be described elsewhere), protein tyrosine kinases (Src, JAK2, Pyk2, and EGFR), and mitogen-activated protein kinases (p38-MAPK, JNK, and ERK5) [82, 83, 91]. These pathways, if uncontrolled, could contribute to hypertension and atherosclerosis.

## Natriuretic Peptides

Natriuretic peptides is a family of three peptide hormones possessing a common core structure. Two of them, atrial natriuretic peptide (ANP) and B-type natriuretic peptide (BNP), are cardiac hormones that are produced mainly in the atrium and ventricle, respectively [92]. C-type natriuretic peptide (CNP) is generated not only in the vascular endothelium, but also in several other tissues [93].

Targets for natriuretic peptides are plasma membrane forms of guanylyl cyclase receptors, enzymes that synthesize the second-messenger cGMP. Several membrane GCs share a basic topology: they have an extracellular ligand binding domain, a transmembrane region, and an intracellular region that contain the catalytic guanylyl cyclase domain. To note, the intracellular region also contains the protein kinase-homology domain which binds ATP, but has no protein kinase activity. Kinase-homology domain is highly phosphorylated (at least six amino acid residues) and negatively modulates the cyclase activity. The kinase-homology domain also may be a docking site for association of membrane GC with other proteins. It has been shown that homodimerization is essential for the activation of the enzymatic activity of GC [94].

According to the current model of transmembrane signal transduction, natriuretic peptide reverses the inhibitory effect of protein kinase-homology domain on guanylyl cyclase domain and induces optimal orientation of two guanylyl cyclase domains within the GC dimmer. Activated membrane GC rises the synthesis of cGMP, which then modulates the activity of specific target proteins, phosphodiesterases (PDEs 2, 3, and 5), ion channels, cGMP dependent protein kinases (PKG I and II), and thereby modifies cellular functions (Fig. 3.3).

Other consequences of natriuretic peptide binding include decreasing the affinity of GC for ligand and dephosphorylation of protein kinase-homology domain leading to desensitization [94].

Membrane GC-A is the receptor for both ANP and BNP. GC-A regulates arterial blood pressure and volume homeostasis and has local antihypertrophic and antifibrotic effects in the heart. Within the vascular system, the GC-A receptor is expressed in smooth muscle cells, endothelial cells, fibroblasts, and also can be found in cardiomyocytes. Potentially, ANP and BNP can have paracrine (on smooth muscles, endothelial cells, and fibroblasts) and autocrine (on cardiomyocytes) functions. To dissect the vascular actions of ANP *in vivo*, Kuhn et al. [95–97], using the Cre-lox technology (in smooth muscle cells or in endothelial cells, or cardiomyocytes),

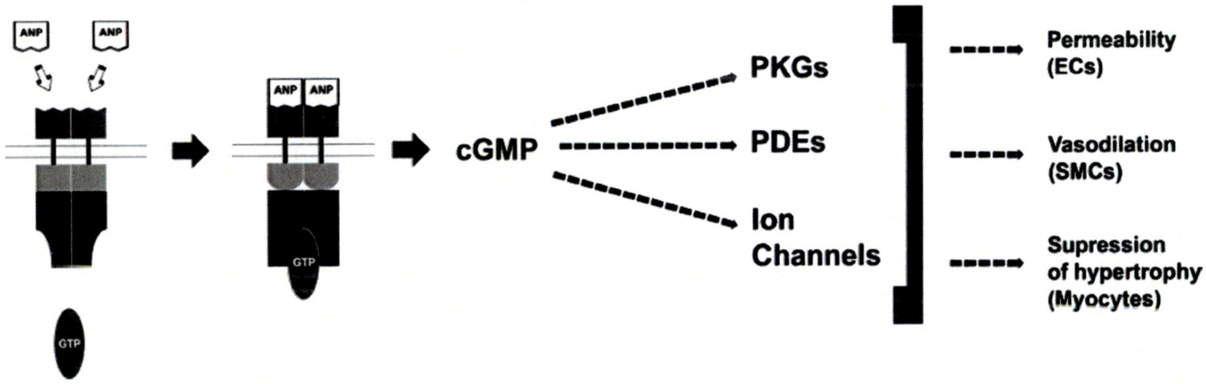


**Fig. 3.3** Regulation of natriuretic peptide-dependent guanylyl cyclase receptor. Guanylyl cyclase receptor (GCR) has ligand-binding domain, transmembrane region, inhibitory kinase-homology domain (*gray square*), and catalytic guanylyl cyclase domain. Binding of atrial natriuretic peptide (ANP) to dimeric GCR reverses inhibitory effect of protein kinase-homology domain on guanylyl cyclase domain. Activated GCR produces cGMP, which in turn modulates several target proteins: cGMP dependent protein kinases (PKGs), phosphodiesterases (PDEs), ion channels, and thereby modifies cellular functions. Other abbreviations: *ECs* endothelial cells, *SMCs* smooth muscle cells. See text for details

elegantly inactivated the GC-A gene in a tissue-specific manner. These studies made them to come to several very important conclusions: First, ANP regulates intravascular volume and blood pressure by activation of endothelial GC-A and modulation of endothelial permeability. EC GC-A KO mice preserved ANP vasodilatation, but were hypertensive as a result of chronic hypervolemia [97]. Second, the vasodilatory effect of ANP is attributed to GC-A receptors in vascular smooth muscle cells. In SMC GC-A KO animals the vasodilatory effect of ANP was completely abolished, whereas their arterial blood pressure was normal [95]. Third, ANP is an important local autocrine factor suppressing hypertrophic growth of cardiomyocytes: mice with selective deletion of GC-A in ventricular cardiomyocytes showed cardiomyocyte hypertrophy in the absence of any increase in arterial pressure [96].

As we mentioned above, both ANP and BNP interact with GC-A receptors. Under physiological conditions, plasma concentration of BNP is much lower than ANP. Moreover, BNP has a much lower affinity to GC-A when compared to ANP [94]. Most likely, BNP does not play the role of vasodilatant, and the defect in BNP expression (in BNP KO mice) does not lead to hypertension or cardiac hypertrophy [98]. It is possible that BNP serves as an antifibrotic factor in the heart by acting on the fibroblast GC-A receptors and suppressing profibrotic extracellular matrix genes [98, 99].

Another type of membrane guanylyl cyclase receptor is GC-B. This receptor is a target for the third member of the natriuretic peptide family, CNP [100]. CNP acts in a local, paracrine fashion [100]. The GC-B expresses in vascular endothelial and smooth muscle cells, at high density in fibroblasts, and other tissues [93]. Deletion of CNP or GC-B genes does not result in arterial hypertension suggesting that CNP does not play a role in vasodilatation or reduction of blood pressure [98]. Possible functions of CNP-triggered signaling pathway in cardiovascular system include a role in angiogenesis (stimulation of proliferation of endothelial cells, attenuation of proliferation of smooth muscle cells) [101, 102] and a role in cardiac hypertrophy and fibrosis (inhibition of cardiac fibroblast proliferation and myocyte hypertrophy) [103]. Moreover, GC-B plays a critical role during bone formation.

## Neuregulins

Neuregulins (NRG) is a family of signaling proteins that are expressed in the heart, as well as in the nervous system and the mammary glands. They are ligands for receptor tyrosine kinases of the ErbB family and activate intracellular signaling cascades, which result in the induction of various cellular responses. Four NRGs encoded by different genes have been discovered, NRG1–NRG4, although very little is known about the functions of NRG2, 3, and 4.

NRG1 is specifically expressed by ventricular endocardial cells in the developing mouse heart. Targeted disruption of the NRG1 gene in the endocardial endothelium prevented the development of the myocardium. Targeting of NRG1 to ErbB4 receptors on myocardial cells is essential for myocardial trabeculation, whereas NRG1-regulated ErbB3 receptors on mesenchymal cells play a role in the development of the endocardial cushion [104]. In contrast to NRG1, the highest expression of NRG2 was observed in atrial endocardial cells [105].

As many as 31 different NRG1 isoforms can be produced as a result of NRG1 gene transcription from multiple promoters and alternative splicing. Individual isoforms differ in domain structure (type of EGF-like domain, the N-terminal sequence) and have different biological properties. Many NRG1s are synthesized as transmembrane proteins. Then, their ectodomain is getting "shed" from the membrane by proteolytic processing by metalloprotease to produce the paracrine signaling molecule.

ErbB receptors are a family of receptor protein tyrosine kinase. Under physiological conditions, NRG1 binds to ErbB3 or ErbB4. ErbB2 does not interact directly with NRG1, but forms a heterodimer with ErbB3 or ErbB4: ErbB2/ErbB3 or ErbB2/ErbB4. Receptor heterodimerization stimulates the tyrosine phosphorylation. Phosphorylation of tyrosine residues in the cytoplasmic domain of the ErbB receptor creates docking sites for various adaptor proteins (Shc and Grb2) and for the regulatory subunit of PI3-kinase. These steps lead to the activation of the MAP kinase. MAP kinase- and PI3-kinase pathways modulate the transcriptional activity of the cardiac cell [106].

NRGs also have functions in the adult heart. In cell culture neuregulins promote survival and growth of cardiac myocytes [107]. Sources of the NRG1 in the adult heart may be the cardiac and coronary microvascular endothelium [107, 108]. Synthesis of NRG1 is under control of vasoactive agents: vasoconstrictors (angiotensin II and epinephrine) suppress the levels of NRG1 whereas vasodilators (endothelin-1 and mechanical stress) stimulate expression of NRG1. Binding of NRG1 to ErbB4 on cardiomyocytes leads to the activation of downstream signaling pathways including activation of Erk1/2 (hypertrophic response) and PI3-kinase/Akt (antiapoptotic effect).

In addition, there is evidence that NRG1/ErbB interact with the cardiac neurohormonal system: NRG1 inhibits the inotropic response of the rabbit papillary muscle to an adrenergic agonist, isoproterenol [109]. Thus, endothelium-derived NRG1, through an antiadrenergic effect, can decrease blood pressure. Possibly, the antiadrenergic effect of NRG1 is mediated by NO, produced by NOS, in cardiomyocytes [109].

It has been observed that the NRG1-knockout mice exhibit dilated cardiomyopathy (abnormal cell structure and sarcomeric organization), which indicates that NRG1 is involved in maintaining the normal sarcomeric structure of cardiac cells [110] and may play a role in the development of chronic heart failure. Moreover, this observation is consistent with reported data about decline in the expression of NRG1, ErbB2, and ErbB4 during the development of ventricular hypertrophy in rat [111, 112], and about the improvement of cardiac function and survival of chronic heart failure in rats by recombinant NRG1 [113].

### Neurotrophins

Neurotrophins (NTs) are a family of proteins that originally were considered to play a regulatory function in the nervous system. Recently, these proteins have been shown to exert important cardiovascular functions. During development, NTs play a role in the formation of the heart and the myocardial vascular system. Post-natally, neurotrophins control the survival of endothelial cells, vascular SMCs, and cardiomyocytes and regulate angiogenesis and vasculogenesis.

NT family includes β-nerve growth factor (NGF), brain-derived neurotrophic factor (BDNF), neurotrophin-3 (NT-3), and neurotrophin-4/5 (NT-4/5). NTs initially are produced in neurons, vascular cells, and cardiomyocytes as proproteins. Pro-NTs can be converted to active signaling proteins by intracellular proteolysis, and then be secreted as a mature homodimers. In some cases, cells secrete pro-NTs, which undergo proteolytic processing extracellularly.

Target cells express two types of receptors for NTs: the tropomyosin receptor kinase (TRK) and the neurotrophin receptor p75 ($p75^{NTR}$). TRKs (A, B, C) are tyrosine kinase receptors with domain structure: extracellular ligand-binding immunoglobulin-like C2 type domain, single transmembrane domain, and cytoplasmic tyrosine kinase domain. TRK A- and B-receptors are expressed in all cardiovascular cells, including endothelium. Binding of NT promotes dimerization of TRK and multiple tyrosine phosphorylation of cytoplasmic domain *in trans*. Phospho tyrosines are the binding sites for phospholipase C-γ (PLC-γ) and a number of adapter proteins (Shc, SH2B, Grb2, etc). As a result, NTs through TRK activate several signaling pathways: Ras/Rap-MAPK/Erk, PLC-γ, and PI3K-Akt. The latter cascade is critically involved in NGF-induced survival of endothelial cells, their migration, and in angiogenesis [114].

$p75^{NTR}$ belongs to the proapoptotic tumor necrosis factor receptor superfamily and contains extracellular domain, transmembrane domain, and a cytoplasmic death domain. This receptor is not expressed in endothelial and SM cells under normal conditions, but is induced in pathological conditions (ischemia, atherosclerosis, and diabetes). Pro-NTs bind to $p75^{NTR}$ receptor and promote apoptosis.

NTs and TRKs play a crucial role in the development of the heart and the coronary vessels. For example, Shmelkov et al. [115] were able to demonstrate that BDNF promotes differentiation of human fetal stem cells toward the endothelial cells and beating cardiomyocytes. Studies on genetically-modified mice revealed that BDNF and functional TRKB are essential for normal cardiac endothelial cell–cell contacts and their survival and for normal blood vessel density in the embryonic heart [116, 117].

Vascular endothelial cells are the cells responsible for blood vessel growth, which can occur in the adult organism (wound healing, post-ischemic reparative neovascularization, etc). NGF was shown to induce several EC activities which are important for angiogenesis: proliferation, survival, and migration/invasion. NGF regulates EC through TRKA, and this regulation is mediated by the activation of Erk and PI3K-Akt signaling pathways [114]. Similar to NGF, BDNF, via TRKBs, also supports EC survival. Interestingly, neovascularization of ischemic muscle is impaired in diabetic mice. It is known that Type 1 diabetes induces expression of $p75^{NTR}$ in capillary EC. Probably, overexpression of this proapoptotic receptor leads to the increased apoptosis of diabetic EC and impairs neovascularization. In fact, functional knocking down of $p75^{NTR}$ normalized post-ischemic angiogenesis in diabetic mice [114].

## Protein Kinases

The activation or inhibition of protein kinases is an important step in the intracellular signal transduction pathways, converting extracellular stimuli into cellular reactions. Since protein kinases are to be discussed in several chapters, here it will suffice to note that protein kinases are involved in the regulation of practically all biological functions of endothelial cells: production of NO, releasing of paracrinal factors, as wells as in reaction to the mechanical forces elicited by blood flow (shear stress). Major serine/threonine protein kinase families in the endothelium include cAMP-protein kinase, protein kinases C, CaM kinase, protein kinases G, Akt, AMP-activated protein kinase, MAPKs [Erk1/2, JNKs/stress-activated protein kinases (SAPKs), and p38], and many others.

Protein kinases participate in the regulation of eNOS activity. For example, a number of agonists that activate eNOS (bradykinin and acetylcholine) release IP3 and diacylglycerol. IP3 mobilizes intra-cellular calcium, which results in an activation of CaM kinase. CaM kinase-dependent phosphorylation of serine 1177 increases the catalytic activity of eNOS. Several other protein kinases also maybe involved in the activatory phosphorylation of eNOS, i.e., Akt, protein kinase A, AMP-activated protein kinase, and PKG.

Another second messenger molecule, diacylglycerol is increased by activators of eNOS. DAG is also a potent activator of protein kinase C. PKC can phosphorylate threonine 495 and serine 114 on eNOS, which inactivates the enzyme.

Another process regulated by protein kinases is the synthesis and release of prostacyclin. This process includes $Ca^{2+}$-dependent translocation of PLA2 from the cytosol to phospholipid membranes. Membrane-bound $PLA_2$ is a substrate for Erk1/2 activated by PKC. Phosphorylation of $cPLA_2$ is required for the release of arachidonic acid and further synthesis of prostacyclin [118].

Protein kinases appear to be involved in early mechanotransduction of shear stress in endothelial cells. Receptor tyrosine kinase Flk-1 and integrins (e.g., the vitronectin receptor $a_v b_3$) can serve as mechanoreceptors by associating with Shc. Subsequently, Ras is activated by the complex of Shc/Grb2/Sos. As a result, Erk and JNK pathways are activated, and ultimately lead to the shear stress-induced gene activation, stress fiber formation, and cell alignment [119, 120].

## Conclusions

Endothelial cells are involved in many aspects of vascular biology, including control of blood pressure, formation of new blood vessels, modulation of contractile performance, and metabolism of cardiomyocytes. Regulation occurs by releasing vasoactive, trophic factors that autocrinally/paracrinally regulate vascular smooth muscle cells, cardiomyocytes, and endothelial cells themselves.

Among endothelium-released factors, NO exerts multifactorial effects on various cell types in the cardiovascular system, including relaxation of the vascular smooth muscle, first steps of angiogenesis, and regulation of growth of cardiomyocytes. Endothelial cells also produce a number of other vasoactive agents that influence vascular smooth muscle cell contraction: prostanoids and vasoactive peptides (endothelin 1, angiotensin II, and bradykinin).

Endothelium is a source of neuregulin-1, which has an essential function in the development of the heart. Recent observations demonstrated an important role for neuregulin in the postnatal myocardium. It appears to regulate cell growth, myofilament organization, survival, and angiogenesis.

Functioning of the endothelium is under control of a number of factors such as responding to hormones, hemodynamic stimuli (pressure, shear stress, and wall strain), and local mediators (bradykinin, prostaglandins, angiotensin II, and NO). Secretion of vasoactive substances like NO and prostanoids is regulated by shear stress, acetylcholine, serotonin, VEGF, histamine, endothelin 1, angiotensin II, and bradykinin; angiotensin II activates the production of bradykinin. Production of ROS by endothelium is regulated by Ang II, PDGF, thrombin, TNF-α, and ET-1. ANP paracrinally produced by atrial cardiomyocytes modulates endothelial permeability, whereas the C-type natriuretic peptide plays a role in angiogenesis. Neurotrophins (NGF and BDNF) are important hormones that control endothelial cell-related blood vessel formation in the embryonic heart and regulate angiogenesis in the adult organism.

Finally, endothelial dysfunction is associated with a number of cardiovascular abnormalities. A full understanding of the signals, transducers and effectors in the signaling pathways present in the endothelium, under normal and pathological conditions, is necessary for the development of novel, efficient, and specific pharmaceutical agents to correct the endothelial dysfunction.

## Summary

- Cardiovascular endothelial cells produce an important relaxative modulator of myocardial function, nitric oxide (NO). The major enzyme that is responsible for the production of NO is endothelial nitric oxide synthase, eNOS. Endothelial cells produce NO as a result of eNOS activation by a number of hormones and vasoactive peptides. Regulation of eNOS activity involves posttranslational modifications (phosphorylation and nitrosylation) and protein–protein interactions. Compartmentalization of eNOS to caveolae is essential for efficient agonist-mediated stimulation of activity.
- Endothelium-produced NO paracrinally modulates myocardial contractile performance and growth. Anti hypertrophy effect of NO is realized through targeting the calcineurin-NFAT signaling pathway. Abnormal increase of endothelial cell mass promotes an opposite effect, myocardial hypertrophy, by NO-dependent destabilization of RGS.
- Endothelial eNOS-produced NO is the main vasodilatory factor that causes relaxation of the vascular smooth muscles. Endothelial cells also release contracting factors, such as vasoconstrictor prostanoids. Under physiological conditions, there is a balanced release of relaxing and contracting factors.
- Several growth factors (VEGF, TGF-β1, and substance P) initiate the angiogenic signaling cascade in endothelial cells by activation of eNOS. Released NO autocrinally regulates the first steps of angiogenesis: cell migration and organization in tubes.
- Endothelial cells secrete vasodilatory prostanoid, prostacyclin, in response to acetylcholine, bradykinin, and ET-1. Prostacyclin paracrinally activates IPs on vascular smooth muscle, which leads to the muscle relaxation. Under some conditions (essential hypertension, diabetes, and aged individuals) endothelium-dependent relaxations are impaired, and prostacyclin abnormally interacts with thromboxane receptors, which results in induced contraction, and may cause vasospasm.

- Cardiovascular endothelial cells participate in the regulation of the level of several vasoactive peptides: endothelin 1, angiotensin II, and bradykinin. Infusion of ET-1 into rats causes an initial decrease in blood pressure followed by prolonged hypertension. Initial effect results from the activation of endothelial ET-receptors, which are linked to production of vasodilatory agents (NO and prostacyclin). The vasoconstrictor response of smooth muscle cells is due to the direct action of ET-1 on these cells via ET-receptors.
- Endothelial cells express angiotensin II-forming proteases, and thus take part in local production of Ang II and induction of local Ang II-dependent vascular constriction. Ang II also binds to AT-receptors on endothelial cells and causes the release of vasodilator, NO. Thus, the contractile response of vessels to Ang II results from the combined activation of endothelial and smooth muscle AT-receptors.
- Endothelial cells also express kininogenases that can locally convert kininogen into active vasodilatory peptide, bradykinin. Relaxative effect of bradykinin is based on the stimulation of $B_2$ receptors on endothelial and SM cells with subsequent generation of NO and prostacyclin.
- Angiotensin II stimulates NAD(P)H oxidase in the vascular endothelial cells. This enzyme plays a key role in the production of ROS, superoxide and hydrogen peroxide. ROS, through oxidation/inactivation of protein tyrosine phosphatases and activation of transcription factors, are involved in maintaining vascular function and integrity. Mechanical forces also play an important role in ROS generation.
- Natriuretic peptides are important regulators of endothelial cells. Atrial natriuretic peptide produced by the atrial cardiomyocytes modulates endothelial permeability. C-type natriuretic peptide stimulates proliferation of endothelial cells.
- Endocardial endothelium specifically express signaling protein neuregulin 1. Neuregulin 1 targets to ErbB receptors on cardiomyocytes and plays essential role in the development of myocardium. Recent investigations revealed an importance of neuregulin for functioning of adult myocardium: promotion of survival and growth of cardiac cells and modulation of neurohormonal regulation of myocardium.
- Neurotrophins activate TRK on endothelial cells. This is the mechanism of neurotrophin-dependent differentiation of embryonic stem cells into endothelial cells and regulation of vasculature embryogenesis. NGF and BDNF are potent regulators of blood vessel growth in the adult organism. Under certain pathologic conditions neurotrophins stimulate degradation of endothelial cells via proapoptotic $p75^{NTR}$ receptors, which results in impaired blood vessel regeneration.

## References


1. Schwarz P, Diem R, Dun NJ, Forstermann U. Endogenous and exogenous nitric oxide inhibits norepinephrine release from rat heart sympathetic nerves. Circ Res. 1995;77:841–8.
2. Kaye DM, Wiviott SD, Balligand J-L, Smith TW. Nitric oxide inhibits norepinephrine uptake into cardiac sympathetic neurons. Circulation. 1995;92 Suppl 1:I. Abstract.
3. Okhotin VE, Shuklin AV. Significance of neuronal, endothelial and inducible NO-synthase isoforms in the cardiac muscle histophysiology. Morfologiia. 2006;129(1):7–17.
4. Fleming I, Fisslthaler B, Dimmeler S, Kemp BE, Busse R. Phosphorylation of $Thr^{495}$ regulates $Ca^{2+}$/calmodulin-dependent endothelial nitric oxide synthase activity. Circ Res. 2001;88: E68–75.
5. Ladage D, Brixius K, Hoyer H, Steingen C, Wesseling A, Malan D, et al. Mechanisms underlying nebivolol-induced endothelial nitric oxide synthase activation in human umbilical vein endothelial cells. Clin Exp Pharmacol Physiol. 2006;33:720–4.
6. Brouet A, Sonveaux P, Dessy C, Balligand JL, Feron O. Hsp90 ensures the transition from the early $Ca^{2+}$-dependent to the late phosphorylation-dependent activation of the endothelial nitric-oxide synthase in vascular endothelial growth factor-exposed endothelial cells. J Biol Chem. 2001;276:32663–9.
7. Zhong JC, Yu XY, Huang Y, Yung LM, Lau CW, Lin SG. Apelin modulates aortic vascular tone via endothelial nitric oxide synthase phosphorylation pathway in diabetic mice. Cardiovasc Res. 2007;74:388–95.
8. Michell BJ, Harris MB, Chen ZP, Ju H, Venema VJ, Blackstone MA, et al. Identification of regulatory sites of phosphorylation of the bovine endothelial nitric-oxide synthase at serine 617 and serine 635. J Biol Chem. 2002;277:42344–51.
9. Bauer PM, Fulton D, Boo YC, Sorescu GP, Kemp BE, Jo H, et al. Compensatory phosphorylation and protein–protein interactions revealed by loss of function and gain of function mutants of multiple serine phosphorylation sites in endothelial nitric oxide synthase. J Biol Chem. 2003;278:14841–9.
10. Kou R, Greif D, Michel T. Dephosphorylation of endothelial nitric-oxide synthase by vascular endothelial growth factor. Implications for the vascular responses to cyclosporin A. J Biol Chem. 2002;277:29669–73.
11. Erwin PA, Lin AJ, Golan DE, Michel T. Receptor-regulated dynamic S-nitrosylation of endothelial nitric-oxide synthase in vascular endothelial cells. J Biol Chem. 2005;280:19888–94.
12. Feron O, Balligand J-L. Caveolins and the regulation of endothelial nitric oxide synthase in the heart. Cardiovasc Res. 2006;69: 788–97.
13. Icking A, Matt S, Opitz N, Wiesenthal A, Müller-Esterl W, Schilling K. NOSTRIN functions as a homotrimeric adaptor protein facilitating internalization of eNOS. J Cell Sci. 2005;118:5059–69.
14. Sanchez FA, Savalia NB, Duran RG, Lal BK, Boric MP, Duran WN. Functional significance of differential eNOS translocation. Am J Physiol Heart Circ Physiol. 2006;291:H1058–64.
15. Jagnandan D, Sessa WC, Fulton D. Intracellular location regulates calcium–calmodulin-dependent activation of organelle-restricted eNOS. Am J Physiol Cell Physiol. 2005;289:C1024–33.
16. Church JE, Fulton D. Differences in eNOS activity because of subcellular localization are dictated by phosphorylation state rather than the local calcium environment. J Biol Chem. 2006; 281:1477–88.
17. Zhang Q, Church JE, Jagnandan D, Catravas JD, Sessa WC, Fulton D. Functional relevance of Golgi- and plasma membrane-localized endothelial NO synthase in reconstituted endothelial cells. Arterioscler Thromb Vasc Biol. 2006;26:1015–21.

18. Erwin PA, Mitchell DA, Sartoretto J, Marletta MA, Michel T. Subcellular targeting and differential *S*-nitrosylation of endothelial nitric-oxide synthase. J Biol Chem. 2006;281:151–7.
19. Jobin CM, Chen H, Lin AJ, Yacono PW, Igarashi J, Michel T, et al. Receptor-regulated dynamic interaction between endothelial nitric oxide synthase and calmodulin revealed by fluorescence resonance energy transfer in living cells. Biochemistry. 2003;42:11716–25.
20. Tirziu D, Simons M. Endothelium-driven myocardial growth or nitric oxide at the crossroads. Trends Cardiovasc Med. 2008;18:299–305.
21. Gonzalez DR, Treuer A, Sun QA, Stamler JS, Hare JM. S-nitrosylation of cardiac ion channels. J Cardiovasc Pharmacol. 2009;54(3):188–95.
22. Hu RG, Sheng J, Qi X, et al. The N-end rule pathway as a nitric oxide sensor controlling the levels of multiple regulators. Nature. 2005;437:981–6.
23. Sayed N, Baskaran P, Ma X, van den Akker F, Beuve A. Desensitization of soluble guanylyl cyclase, the NO receptor, by S-nitrosylation. Proc Natl Acad Sci USA. 2007;104:12312–7.
24. McVey M, Hill J, Howlett A, Klein C. Adenylyl cyclase, a coincidence detector for nitric oxide. J Biol Chem. 1999;274:18887–92.
25. Vila-Petroff MG, Younes A, Egan J, et al. Activation of distinct cAMP-dependent and cGMP-dependent pathways by nitric oxide in cardiac myocytes. Circ Res. 1999;84:1020–31.
26. Fiedler B, Lohmann SM, Smolenski A, Linnemuller AS, Pieske B, Schroder F, et al. Inhibition of calcineurin-NFAT hypertrophy signaling by cGMP-dependent protein kinase type I in cardiac myocytes. Proc Natl Acad Sci USA. 2002;99:11363–8.
27. Heineke J, Kempf T, Kraft T, et al. Downregulation of cytoskeletal muscle LIM protein by nitric oxide: impact on cardiac myocyte hypertrophy. Circulation. 2003;107:1424–32.
28. Cheng TH, Shih NL, Chen SY, et al. Nitric oxide inhibits endothelin-1-induced cardiomyocyte hypertrophy through cGMP mediated suppression of extracellular-signal regulated kinase phosphorylation. Mol Pharmacol. 2005;68:1183–92.
29. Ozaki M, Kawashima S, Yamashita T, et al. Overexpression of endothelial nitric oxide synthase attenuates cardiac hypertrophy induced by chronic isoproterenol infusion. Circ J. 2002;66:851–6.
30. Tirziu D, Chorianopoulos E, Moodie KL, Palac RT, Zhuang ZW, Tjwa M, et al. Myocardial hypertrophy in the absence of external stimuli is induced by angiogenesis in mice. J Clin Invest. 2007;117:3188–97.
31. Huang PL, Huang Z, Mashimo H, Bloch KD, Moskowitz MA, Bevan JA, et al. Hypertension in mice lacking the gene for endothelial nitric oxide synthase. Nature. 1995;377:239–42.
32. Chataigneau T, Félétou M, Huang PL, Fishman MC, Duhault J, Vanhoutte PM. Acetylcholine-induced relaxation in blood vessels from endothelial nitric oxide synthase knockout mice. Br J Pharmacol. 1999;126:219–26.
33. Jansen-Olesen I, Ottosson A, Cantera L, Strunk S, Lassen LH, Olesen J, et al. Role of endothelium and nitric oxide in histamine-induced responses in human cranial arteries and detection of mRNA encoding H1- and H2-receptors by RT-PCR. Br J Pharmacol. 1997;121:41–8.
34. De Nucci G, Thomas R, D'Orléans-Juste P, Antunes E, Walder C, Warner TD, et al. Pressor effects of circulating endothelin are limited by its removal in the pulmonary circulation and by the release of prostacyclin and endothelium-derived relaxing factor. Proc Natl Acad Sci USA. 1988;85:9797–800.
35. Tirapelli CR, Casolari DA, Yogi A, Monte-zano AC, Tostes RC, Legros E, et al. Functional characterization and expression of endothelin receptors in rat carotid artery: involvement of nitric oxide, a vasodilator prostanoid and the opening of K+ channels in ET B-induced relaxation. Br J Pharmacol. 2005;146:903–12.
36. Sakata K, Ozaki H, Kwon SC, Karaki H. Effects of endothelin on the mechanical activity and cytosolic calcium level of various types of smooth muscle. Br J Pharmacol. 1989;98:483–92.
37. Tirapelli CR, Bonaventura D, Tirapelli LF, de Oliveira AM. Mechanisms underlying the vascular actions of endothelin 1, angiotensin II and bradykinin in the rat carotid. Pharmacology. 2009;84:111–26.
38. Friebe A, Mergia E, Dangel O, Lange A, Koesling D. Fatal gastrointestinal obstruction and hypertension in mice lacking nitric oxide-sensitive guanylyl cyclase. Proc Natl Acad Sci USA. 2007;104:7699–704.
39. Pfeifer A, Klatt P, Massberg S, Ny L, Sausbier M, Hirneiss C, et al. Defective smooth muscle regulation in cGMP kinase I-deficient mice. EMBO J. 1998;17:3045–51.
40. Papapetropoulos A, Garcia-Cardena G, Madri JA, Sessa WC. Nitric oxide production contributes to the angiogenic properties of vascular endothelial growth factor in human endothelial cells. J Clin Invest. 1997;100:3131–9.
41. Amano K, Matsubara H, Iba O, et al. Enhancement of ischemia-induced angiogenesis by eNOS overexpression. Hypertension. 2003;41:156–62.
42. Simons M. Molecular multitasking: statins lead to more arteries, less plaque. Nat Med. 2000;6:965–6.
43. Brouet A, Sonveaux P, Dessy C, Moniotte S, Balligand J-L, Feron O. Hsp90 and caveolin are key targets for the proangiogenic nitric oxide-mediated effects of statins. Circ Res. 2001;89:866–73.
44. Sonveaux P, Martinive P, DeWever J, Batova Z, Daneau G, Pelat M, et al. Caveolin-1 expression is critical for vascular endothelial growth factor-induced ischemic Hindlimb collateralization and nitric oxide-mediated angiogenesis. Circ Res. 2004;95:154–61.
45. Wise H, Jones RL. Focus on prostacyclin and its novel mimetics. Trends Pharmacol Sci. 1996;17:17–21.
46. Gluais P, Lonchampt M, Morrow JD, Vanhoutte PM, Feletou M. Acetylcholine-induced endothelium-dependent contractions in the SHR aorta: the Janus face of prostacyclin. Br J Pharmacol. 2005; 146:834–45.
47. Barrow SE, Dollery CT, Heavey DJ, Hickling NE, Ritter JM, Vial J. Effect of vasoactive peptides on prostacyclin synthesis in man. Br J Pharmacol. 1986;87:243–7.
48. Tang EH, Ku DD, Tipoe GL, Feletou M, Man RY, Vanhoutte PM. Endothelium-dependent contractions occur in the aorta of wild-type and COX2–/– knockout but not COX1–/– knockout mice. J Cardiovasc Pharmacol. 2005;46:761–5.
49. Heymes C, Habib A, Yang D, Mathieu E, Marotte F, Samuel J-L, et al. Cyclo-oxygenase-1 and -2 contribution to endothelial dysfunction in ageing. Br J Pharmacol. 2000;131:804–10.
50. Vanhoutte PM, Feletou M, Taddei S. Endothelium-dependent contractions in hypertension. Br J Pharmacol. 2005;144:449–58.
51. Wilson DP, Susnjar M, Kiss E, Sutherland C, Walsh MP. Thromboxane $A_2$-induced contraction of rat caudal arterial smooth muscle involves activation of $Ca^{2+}$ entry and $Ca^{2+}$ sensitization: rho-associated kinase-mediated phosphorylation of MYPT1 at Thr-855, but not Thr-697. Biochem J. 2005;389:763–74.
52. Gluais P, Vanhoutte PM, Feletou M. Mechanisms underlying ATP-induced endothelium-dependent contractions in the SHR aorta. Eur J Pharmacol. 2007;556:107–14.
53. Dominiczak AF, Quilley J, Bohr DF. Contraction and relaxation of rat aorta in response to ATP. Am J Physiol. 1991;261:H243–51.
54. Bund SJ. Influence of mode of contraction on the mechanism of acetylcholine-mediated relaxation of coronary arteries from normotensive and spontaneously hypertensive rats. Clin Sci (Lond). 1998;94:231–8.
55. Muller L, Barret A, Etienne E, Meidan R, Valdenaire O, Corvol P, et al. Heterodimerization of endothelin-converting enzyme-1 isoforms regulates the subcellular distribution of this metalloprotease. J Biol Chem. 2003;278:545–55.
56. Zhou J, Zhu Y, Cheng M, Dinesh D, Thorne T, Poh KK, et al. Regulation of vascular contractility and blood pressure by the E2F2 transcription factor. Circulation. 2009;120:1213–21.

57. Funke-Kaiser H, Reichenberger F, Kopke K, Herrmann SM, Pfeifer J, Orzechowski HD, et al. Differential binding of transcription factor E2F-2 to the endothelin-converting enzyme-1b promoter affects blood pressure regulation. Hum Mol Genet. 2003;12: 423–33.
58. Funalot B, Courbon D, Brousseau T, Poirier O, Berr C, Cambien F, et al. Genes encoding endothelin-converting enzyme-1 and endothelin-1 interact to influence blood pressure in women: the EVA study. J Hypertens. 2004;22:739–43.
59. Dammanahalli JK, Sun Z. Endothelin (ET)-1 inhibits nicotinamide adenine dinucleotide phosphate oxidase activity in human abdominal aortic endothelial cells: a novel function of $ET_{B1}$ receptors. Endocrinology. 2008;149:4979–87.
60. Calo G, Gratton JP, Telemaque S, D'Orleans-Juste P, Regoli D. Pharmacology of endothelins: vascular preparations for studying $ET_A$ and $ET_B$ receptors. Mol Cell Biochem. 1996;154:31–7.
61. Verhaar MC, Strachan FE, Newby DE, Cruden NL, Koomans HA, Rabelink TJ, et al. Endothelin-A receptor antagonist-mediated vasodilatation is attenuated by inhibition of nitric oxide synthesis and by endothelin-B receptor blockade. Circulation. 1998;97: 752–6.
62. Hollenberg NK, Fisher NDL, Price DA. Pathways for angiotensin II generation in intact human tissue: evidence from comparative pharmacological interruption of the renin system. Hypertension. 1998;32:387–92.
63. Becari C, Sivieri Jr DO, Santos CF, Moyses MK, Oliveira EB, Salgado MC. Role of elastase-2 as an angiotensin II-forming enzyme in rat carotid artery. J Cardiovasc Pharmacol. 2005;46:498–504.
64. Santos CF, Caprio MA, Oliveira EB, Salgado MC, Schippers DN, Munzenmaier DH, et al. Functional role, cellular source, and tissue distribution of rat elastase-2, an angiotensin II-forming enzyme. Am J Physiol Heart Circ Physiol. 2003;285:H775–83.
65. Fukada SY, Iyomasa MM, Cunha FQ, Correa FM, de Oliveira AM. Mechanisms of impaired vascular response to Ang II in perivascular injured carotid arteries of ovariectomized rat. J Cardiovasc Pharmacol. 2004;44:393–400.
66. Mehta PK, Griendling KK. Angiotensin II cell signaling: physiological and pathological effects in the cardiovascular system. Am J Physiol Cell Physiol. 2007;292:C82–97.
67. Santos RAS, Simões e Silva AC, Maric C, Silva DMR, Machado RP, Buhr I, et al. Angiotensin-(1–7) is an endogenous ligand for the G protein-coupled receptor Mas. Proc Natl Acad Sci USA. 2003;100:8258–63.
68. Igase M, Strawn WB, Gallagher PE, Geary RL, Ferrario CM. Angiotensin II AT 1 receptors regulate ACE2 and angiotensin-(1–7) expression in the aorta of spontaneously hypertensive rats. Am J Physiol. 2005;289:H1013–9.
69. Fukada SY, Tirapelli CR, de Godoy MA, de Oliveira AM. Mechanisms underlying the endothelium-independent relaxation induced by angiotensin II in rat aorta. J Cardiovasc Pharmacol. 2005;45:136–43.
70. Castro-Chaves P, Soares S, Fontes-Carvalho R, Leite-Moreira AF. Negative inotropic effect of selective $AT_2$ receptor stimulation and its modulation by the endocardial endothelium. Eur J Pharmacol. 2008;578:261–9.
71. Yayama K, Okamoto H. Angiotensin II-induced vasodilation via type 2 receptor: role of bradykinin and nitric oxide. Int Immunopharmacol. 2008;8:312–8.
72. Wassmann S, Nickenig G. Pathophysiological regulation of the $AT_1$-receptor and implications for vascular disease. J Hypertens. 2006;24 Suppl 1:S15–21.
73. Bhoola KD, Figueroa CD, Worthy K. Bioregulation of kinins: kallikreins, kininogens, and kininases. Pharmacol Rev. 1992;44:1–80.
74. Michel T, Li GK, Busconi L. Phosphorylation and subcellular translocation of endothelial nitric oxide synthase. Proc Natl Acad Sci USA. 1993;90:6252–6.
75. Barrow SE, Dollery CT, Heavey DJ, Hickling NE, Ritter JM, Mial J. Effect of vasoactive peptides on prostacyclin synthesis in man. Br J Pharmacol. 1986;87:243–7.
76. Tsutsumi Y, Matsubara H, Masaki H, Kurihara H, Murasawa S, Takai S, et al. Angiotensin II type 2 receptor overexpression activates the vascular kinin system and causes vasodilation. J Clin Invest. 1999;104:925–35.
77. Batenburg WW, Garrelds IM, Bernasconi CC, Juillerat-Jeanneret L, van Kats JP, Saxena PR, et al. Angiotensin II type 2 receptor-mediated vasodilation in human coronary microarteries. Circulation. 2004;109:2296–301.
78. Busse R, Lamontagne D. Endothelium-derived bradykinin is responsible for the increase in calcium produced by angiotensin-converting enzyme inhibitors in human endothelial cells. Naunyn Schrniedebergs Arch Pharmacol. 1991;344:126–9.
79. Landmesser U, Drexler H. Effect of angiotensin II type 1 receptor antagonism on endothelial function: role of bradykinin and nitric oxide. J Hypertens. 2006;24 Suppl 1:S39–43.
80. Zafari AM, Ushio-Fukai M, Akers M, Yin Q, Shah A, Harrison DG, et al. Novel role of NADH/NADPH oxidase-derived hydrogen peroxide in angiotensin II-induced hypertrophy of rat vascular smooth muscle cells. Hypertension. 1998;32:488–95.
81. Irani K, Xia Y, Zweier JL, Sollott SJ, Der CJ, Fearon ER, et al. Mitogenic signaling mediated by oxidants in ras-transformed fibroblasts. Science. 1997;275:1649–52.
82. Marumo T, Schini-Kerth VB, Fisslthaler B, Busse R. Platelet-derived growth factor-stimulated superoxide anion production modulates activation of transcription factor NF-κB and expression of monocyte chemoattractant protein-1 in human aortic smooth muscle cells. Circulation. 1997;96:2361–7.
83. Patterson C, Ruef J, Madamanchi NR, Barry-Lane P, Hu Z, Horaist C, et al. Stimulation of a vascular smooth muscle cell NAD(P)H oxidase by thrombin: evidence that p47(phox) may participate in forming this oxidase in vitro and in vivo. J Biol Chem. 1999;274:19814–22.
84. De Keulenaer GW, Alexander RW, Ushio-Fukai M, Ishizaka N, Griendling KK. Tumor necrosis factor-α activates a p22phox-based NADH oxidase in vascular smooth muscle cells. Biochem J. 1998;329:653–7.
85. Gorlach A, Brandes RP, Nguyen K, Amidi M, Dehghani F, Busse RA. gp91phox containing NADPH oxidase selectively expressed in endothelial cells is a major source of oxygen radical generation in the arterial wall. Circ Res. 2000;87:26–32.
86. Suh Y, Arnold RS, Lassègue B, Shi J, Xu X, Sorescu D, et al. Cell transformation by the superoxide-generating oxidase mox1. Nature. 1999;401:79–82.
87. Jones SA, O'Donnell VB, Wood JD, Broughton JP, Hughes EJ, Jones OTG. Expression of phagocyte NADPH oxidase components in human endothelial cells. Am J Physiol. 1996;271:H1626–34.
88. Touyz RM, Schiffrin EL. Ang II-stimulated superoxide production is mediated via phospholipase D in human vascular smooth muscle cells. Hypertension. 1999;34:976–82.
89. Seshiah PN, Weber DS, Rocic P, Valppu L, Taniyama Y, Griendling KK. Angiotensin II stimulation of NAD(P)H oxidase activity: upstream mediators. Circ Res. 2002;91:406–13.
90. Gomez-Cambronero J, Keire P. Phospholipase D: a novel major player in signal transduction. Cell Signal. 1998;10:387–97.
91. Paravicini TM, Touyz RM. Redox signaling in hypertension. Cardiovasc Res. 2006;71:247–58.
92. de Bold AJ, Ma KK, Zhang Y, de Bold ML, Bensimon M, Khoshbaten A. The physiological and pathophysiological modulation of the endocrine function of the heart. Can J Physiol Pharmacol. 2001;79:705–14.
93. Olney RC. C-type natriuretic peptide in growth: a new paradigm. Growth Horm IGF Res. 2006;16:S6–14.
94. Kuhn M. Function and dysfunction of mammalian membrane guanylyl cyclase receptors: lessons from genetic mouse models

and implications for human diseases. In: Schmidt HHHW et al., editors. cGMP: generators, effectors and therapeutic implications, Handbook of experimental pharmacology, vol. 191. Berlin: Springer; 2009. p. 47–69.

95. Holtwick R, Gotthardt M, Skryabin B, Steinmetz M, Potthast R, Zetsche B, et al. Smooth muscle-selective deletion of guanylyl cyclase-A prevents the acute but not chronic effects of ANP on blood pressure. Proc Natl Acad Sci USA. 2002;99:7142–7.
96. Holtwick R, Van Eickels M, Skryabin BV, Baba HA, Bubikat A, Begrow F, et al. Pressure-independent cardiac hypertrophy in mice with cardiomyocyte-restricted inactivation of the atrial natriuretic peptide receptor guanylyl cyclase-A. J Clin Invest. 2003;111: 1399–407.
97. Sabrane K, Kruse MN, Fabritz L, Zetsche B, Mitko D, Skryabin BV, et al. Vascular endothelium is critically involved in the hypotensive and hypovolemic actions of atrial natriuretic peptide. J Clin Invest. 2005;115:1666–74.
98. Tamura N, Ogawa Y, Chusho H, Nakamura K, Nakao K, Suda M, et al. Cardiac fibrosis in mice lacking brain natriuretic peptide. Proc Natl Acad Sci USA. 2000;97:4239–44.
99. Kapoun AM, Liang F, O'Young G, Damm DL, Quon D, White RT, et al. B-type natriuretic peptide exerts broad functional opposition to transforming growth factor-beta in primary human cardiac fibroblasts: fibrosis, myofibroblast conversion, proliferation, and inflammation. Circ Res. 2004;94:453–61.
100. Kemp-Harper B, Schmidt HHHW. cGMP in the Vasculature. In: Schmidt HHHW et al., editors. cGMP: generators, effectors and therapeutic implications, Handbook of experimental pharmacology, vol. 191. Berlin: Springer; 2009. p. 447–67.
101. Yamahara K, Itoh H, Chun TH, Ogawa Y, Yamashita J, Sawada N, et al. Significance and therapeutic potential of the natriuretic peptides/cGMP/cGMP-dependent protein kinase pathway in vascular regeneration. Proc Natl Acad Sci USA. 2003;100:3404–9.
102. Komatsu Y, Ito H, Suga S, Ogawa Y, Hama N, Kishimoto I, et al. Regulation of endothelial production of C-type natriuretic peptide in coculture with vascular smooth muscle cells. Role of the vascular natriuretic peptide system in vascular growth inhibition. Circ Res. 1996;78:606–14.
103. Pagel-Langenickel I, Buttgereit J, Bader M, Langenickel TH. Natriuretic peptide receptor B signaling in the cardiovascular system: protection from cardiac hypertrophy. J Mol Med. 2007;85:797–810.
104. Meyer D, Birchmeier C. Multiple essential functions of neuregulin in development. Nature. 1995;378:386–90.
105. Carraway III KL, Weber JL, Unger MJ, Ledesma J, Yu N, Gassmann M, et al. Neuregulin-2, a new ligand of ErbB3/ErbB4-receptor tyrosine kinases. Nature. 1997;387:512–6.
106. Baliga RR, Pimental DR, Zhao YY, Simmons WW, Marchionni MA, Sawyer DB, et al. NRG-1-induced cardiomyocyte hypertrophy. Role of PI-3-kinase, p70(S6K), and MEK-MAPKRSK. Am J Physiol. 1999;277:H2026–37.
107. Zhao Y, Sawyer DR, Baliga RR, Opel DJ, Han X, Marchionni MA, et al. Neuregulins promote survival and growth of cardiac myocytes. Persistence of ErbB2 and ErbB4 expression in neonatal and adult ventricular myocytes. J Biol Chem. 1998;273:10261–9.
108. Lemmens K, Segers VF, Demolder M, De Keulenaer GW. Role of neuregulin-1/ErbB2 signaling in endothelium-cardiomyocyte cross-talk. J Biol Chem. 2006;281:19469–77.
109. Lemmens K, Fransen P, Sys SU, Brutsaert DL, De Keulenaer GW. Neuregulin-1 induces a negative inotropic effect in cardiac muscle: role of nitric oxide synthase. Circulation. 2004;109:324–6.
110. Crone SA, Zhao YY, Fan L, Gu Y, Minamisawa S, Liu Y, et al. ErbB2 is essential in the prevention of dilated cardiomyopathy. Nat Med. 2002;8:459–65.
111. Rohrbach S, Yan X, Weinberg EO, Hasan F, Bartunek J, Marchionni MA, et al. Neuregulin in cardiac hypertrophy in rats with aortic stenosis: differential expression of erbB2 and erbB4 receptors. Circulation. 1999;100:407–12.
112. Lemmens K, Doggen K, De Keulenaer GW. Role of neuregulin-1/erbB signaling in cardiovascular physiology and disease. Implications for therapy of heart failure. Circulation. 2007;116: 954–60.
113. Liu X, Gu X, Li Z, Li X, Li H, Chang J, et al. Neuregulin-1/erbB-activation improves cardiac function and survival in models of ischemic, dilated, and viral cardiomyopathy. J Am Coll Cardiol. 2006;48:1438–47.
114. Caporali A, Emanueli C. Cardiovascular actions of neurotrophins. Physiol Rev. 2009;89:279–308.
115. Shmelkov SV, Meeus S, Moussazadeh N, Kermani P, Rashbaum WK, Rabbany SY, et al. Cytokine preconditioning promotes codifferentiation of human fetal liver CD133+ stem cells into angiomyogenic tissue. Circulation. 2005;111:1175–83.
116. Donovan MJ, Lin MI, Wiegn P, Ringstedt T, Kraemer R, Hahn R, et al. Brain derived neurotrophic factor is an endothelial cell survival factor required for intramyocardial vessel stabilization. Development. 2000;127:4531–40.
117. Wagner N, Wagner KD, Theres H, Englert C, Schedl A, Scholz H. Coronary vessel development requires activation of the TrkB neurotrophin receptor by the Wilms' tumor transcription factor Wt1. Genes Dev. 2005;19:2631–42.
118. Anfuso CD, Lupo G, Romeo L, Giurdanella G, Motta C, Pascale A, et al. Endothelial cell-pericyte cocultures induce PLA2 protein expression through activation of PKCα and the MAPK/ERK cascade. J Lipid Res. 2007;48:782–93.
119. Chen KD, Li YS, Kim M, Li S, Yuan S, Chien S, et al. Mechanotransduction in response to shear stress. Roles of receptor tyrosine kinases, integrins, and Shc. J Biol Chem. 1999;274: 18393–400.
120. Li S, Chen BP, Azuma N, Hu YL, Wu SZ, Sumpio BE, et al. Distinct roles for the small GTPases Cdc42 and Rho in endothelial responses to shear stress. J Clin Invest. 1999;103:1141–50.

# Chapter 4
# Rapid Signaling Pathways

**Abstract** Recent evidence has shown that in the heart a plurality of receptor systems regulate heart rate and the force of contraction. A number of cardiac receptors are targets for neurotransmitters released from sympathetic and parasympathetic neurons innervating the myocardium. Other cardiac receptors transduce signals from circulating hormones as well as from several paracrine factors. In this chapter, we describe the structural and functional aspects of major signaling pathways participating in the contractile activity of normal healthy myocardium.

**Keywords** Rapid signaling • Neurotransmitters • G Protein • Adrenoceptors • Serotonin • Purinergic receptors

## Introduction

Signaling molecules that influence cardiac muscle differ in structure, originate from different places, and initiate different signaling cascades within the cardiac cell. Important group of signaling molecules are biogenic amines – amino-derivatives of several amino acids (norepinephrine, epinephrine, acetylcholine, histamine, serotonin). They are produced by and released from neurons to act as neurotransmitters, travel short distances, and target cardiac cells. Some of them also can function as hormones (epinephrine, histamine). Peptides are members of another large family of heart regulators. Some of them, neuropeptides, are neurotransmitters (neuropeptide Y, vasoactive intestinal peptide, calcitonin gene-related peptide, substance P, neurotensin, somatostatin), whereas other are hormones (angiotensin II, produced by the renin–angiotensin system) or paracrine factors being released locally on demand (endothelin-1, urocortins, natriuretic peptides). Again, several peptides can serve as neurotransmitter/paracrine factor (substance P) or hormone/paracrine factor (angiotensin II). Finally, adenosine is a nucleoside which has negative chronotropic and inotropic effects on the heart.

The vast majority of receptors that respond to the above-mentioned molecules belong to the superfamily of so-called "G protein-coupled receptors" (GPCRs). They are integral sarcolemmal proteins with seven transmembrane domains. The key element of GPCR functioning is their interaction with regulatory heterotrimeric G protein consisting of $\alpha$, $\beta$, and $\gamma$ subunits. Binding of ligand allows GPCR to act as a guanine nucleotide exchange factor: it promotes an exchange of guanyl nucleotides in associated G protein. Specifically, GDP bound to $G_\alpha$ exchanges for a GTP. Binding of GTP promotes dissociation of $G_\alpha$-GTP from $G_{\beta\gamma}$ and targeting of $G_\alpha$ (and $G_{\beta\gamma}$ in some cases) to effector(s). There are several isoforms for each of G protein subunit. In many cases transduction of the signal from one particular GPCR to specific effector system depends on the nature of $G_\alpha$: $G_{\alpha s}$, $G_{\alpha i/o}$, or $G_{\alpha q/11}$.

GPCRs/G proteins regulate the activity of two major effector enzymes synthesizing intracellular second messenger molecules: adenylate cyclase (AC; produces cAMP) and phospholipase C [PLC; produces inositol 1,4,5-trisphosphate ($IP_3$) and diacylglycerol (DAG)]. Second messengers in the cardiac cell affect (directly or indirectly, via protein kinases) the final targets responsible for contractile activity: ion channels that are involved in the contraction cycle and regulatory components of myocyte contractile machinery (e.g., troponin C, troponin I, myosin light chains).

In humans several receptor systems increase heart rate and contractility through accumulation of intracellular cAMP via $G_s$ protein/adenylate cyclase pathway: histamine, serotonin 5-HT4, vasoactive intestinal peptide, adenosine A2a, urocortin receptors. Most powerful are the $\beta$-adrenoceptors activated by catecholamines (norepinephrine, epinephrine). Positive effects of some agonists are based on receptor-dependent activation of phospholipase C/DAG/$IP_3$ pathway (e.g., $\alpha_1$-adrenoceptor, serotonin 5-HT2, somatostatin SSTR3, angiotensin II, and endothelin-1 $ET_B$ receptors). In addition, several receptor systems act through inhibition of cAMP synthesis and cause negative inotropic effects in the heart (muscarinic m2, serotonin 5-HT1, neuropeptide Y, somatostatin, adenosine A1, endothelin-1 $ET_A$ receptors). Furthermore, activation of acetylcholine m2 or adenosine A1 receptors leads to the release of $G_{\beta\gamma}$ subunits from $G_{i/o}$ protein followed by direct activation of $K^+$ channel GIRK in the atrial pacemaker cells.

J. Marín-García, *Signaling in the Heart*,
DOI 10.1007/978-1-4419-9461-5_4, © Springer Science+Business Media, LLC 2011

## Neurohormonal Signaling

Neurohormones are produced and released by neurons. Some of them are released and act as neurotransmitters traveling short distances to the target cells. In the cardiovascular system, norepinephrine, epinephrine, and acetylcholine are neurotransmitters that mediate the regulation of myocardial contractility by the autonomic nervous system. Histamine and serotonin also can function as neurotransmitters, being released by neurons. It is noteworthy that norepinephrine, epinephrine, and histamine also function as hormones: the first two are released from the adrenal glands under stress conditions, whereas histamine is released from basophils and mast cells as a pro-inflammatory agent in response to allergic reactions or tissue damage.

### *Biogenic Amines*

#### β-Adrenoceptors

Elliott [1] in 1904 hypothesized that sympathetic neurotransmission may be mediated by a catecholamine liberated by nerve impulses. Four decades later, in 1946, noradrenaline [or norepinephrine (NE)] was identified as a sympathetic transmitter [2].

NE is the primary neurotransmitter of the cardiac sympathetic nervous system. It is synthesized in the neuron, is stored in vesicles, and upon receiving a stimulus, is released into synaptic cleft. NE binds to postsynaptic adrenergic receptors. NE may be recycled into the neuron by the NE transporter 1 (NET-1).

Cardiomyocytes express a family of receptors for catecholamines which are called β-adrenoceptors (β-ARs) and represent a type of plasma transmembrane GPCR (Fig. 4.1).

Stimulation of β-AR by sympathetic neuronal activation or by circulating catecholamines increases heart rate (positive chronotropic effect), force of cardiac contraction (positive inotropic effect), rate of cardiac relaxation (positive lusitropic effect), and automaticity. To date, three β-AR subtypes have been cloned and pharmacologically characterized: $\beta_1$-AR, $\beta_2$-AR, and $\beta_3$-AR. Both the $\beta_1$- and $\beta_2$-ARs are found in the atria and ventricles of human and rat hearts [3], whereas the $\beta_2$-AR may be absent in the ventricles of cat and guinea pig [4]. It has been reported that $\beta_3$-AR is also present in the myocardium [5]. A major subtype expressed in the mammalian heart is $\beta_1$-AR (75–85%), whereas the remaining receptors belong to the $\beta_2$-AR subtype. The spectrum of β-AR subtypes changes under pathological conditions: the expression of the ventricular $\beta_1$-AR, but not the $\beta_2$-AR, is decreased in the animal model of congestive heart failure (HF) as well as in patients with end-stage dilated heart failure [6, 7].

Among the catecholamines, isoprenaline and epinephrine evoke their positive inotropic effects through stimulation of $\beta_1$- and $\beta_2$-adrenoceptors, whereas norepinephrine induces its positive inotropic effect predominantly (if not exclusively) via $\beta_1$-adrenoceptor stimulation [8]. It appears that in humans, under normal conditions, the force of contraction and heart rate is regulated only by cardiac $\beta_1$-ARs (by norepinephrine released from sympathetic neurons) whereas in stress situations (when epinephrine is released from the adrenal medulla), stimulation of cardiac $\beta_2$-ARs could contribute to the positive inotropic and chronotropic effects.

β-ARs have extracellular amino terminus, seven membrane-spanning domains connected by intracellular and extracellular loops, and an intracellular carboxyl terminus. β-ARs reside in the sarcolemma of cardiac cell. Binding of NE to β-AR initiates a cascade of events that result in myocyte contraction. Thus, activated NE-bound β-AR changes conformation which facilitates the stimulatory G protein ($G_s$) binding. Under resting conditions, $G_s$ is an inactive heterotrimer composed of three subunits: α, β, and γ. α-Subunit of inactive $G_s$ contains guanosine diphosphate (GDP). Interaction with specific domains of activated β-AR triggers the exchange of GDP for guanosine triphosphate (GTP) on the α-subunit of $G_s$ resulting in the dissociation of the heterotrimer into active $G_{\alpha s}$- and $G_{\beta\gamma}$-subunits; $G_{\alpha s}$-subunit activates adenylate cyclase.

Two closely related AC isoforms are predominant in the heart, type V and type VI. AC catalyzes the synthesis of cyclic adenosine monophosphate (cAMP), which activates cAMP-dependent protein kinase A (PKA). PKA-dependent phosphorylation of $Ca^{2+}$ channels, phospholamban, and contractile proteins leads to a functional response.

The effect of PKA on $Ca^{2+}$ channels plays a key role in β-adrenergic regulation of cardiac muscle contraction because $Ca^{2+}$ channels substantially contribute to the generation and regulation of cardiac automaticity (in pacemaker cells) and take part in the development of cardiac cell action potential. Two kinds of $Ca^{2+}$ channels have been found in heart cells: low-voltage activated T-type and high-voltage activated L-type $Ca^{2+}$ channels [9, 10]. β-Adrenergic stimulation does not seem to function on T-type $Ca^{2+}$ channels whereas β-adrenergic enhancement of cardiac L-type $Ca^{2+}$ channels is well documented [11, 12].

L-type $Ca^{2+}$ channel is a heterotetrameric polypeptide complex comprising the $\alpha_1$, $\alpha_2/\delta$, and β subunits. The $\alpha_1$ subunit contains the $Ca^{2+}$-selective pore and voltage sensor, whereas the accessory subunits $\alpha_2/\delta$ and β modulate the biophysical properties and trafficking of the $\alpha_1$ subunit to the membrane. To date, at least ten $\alpha_1$ subunits have been identified, and the $\alpha_{1C}$ subunit is expressed at high levels in cardiac muscle. PKA-dependent phosphorylation of the channel

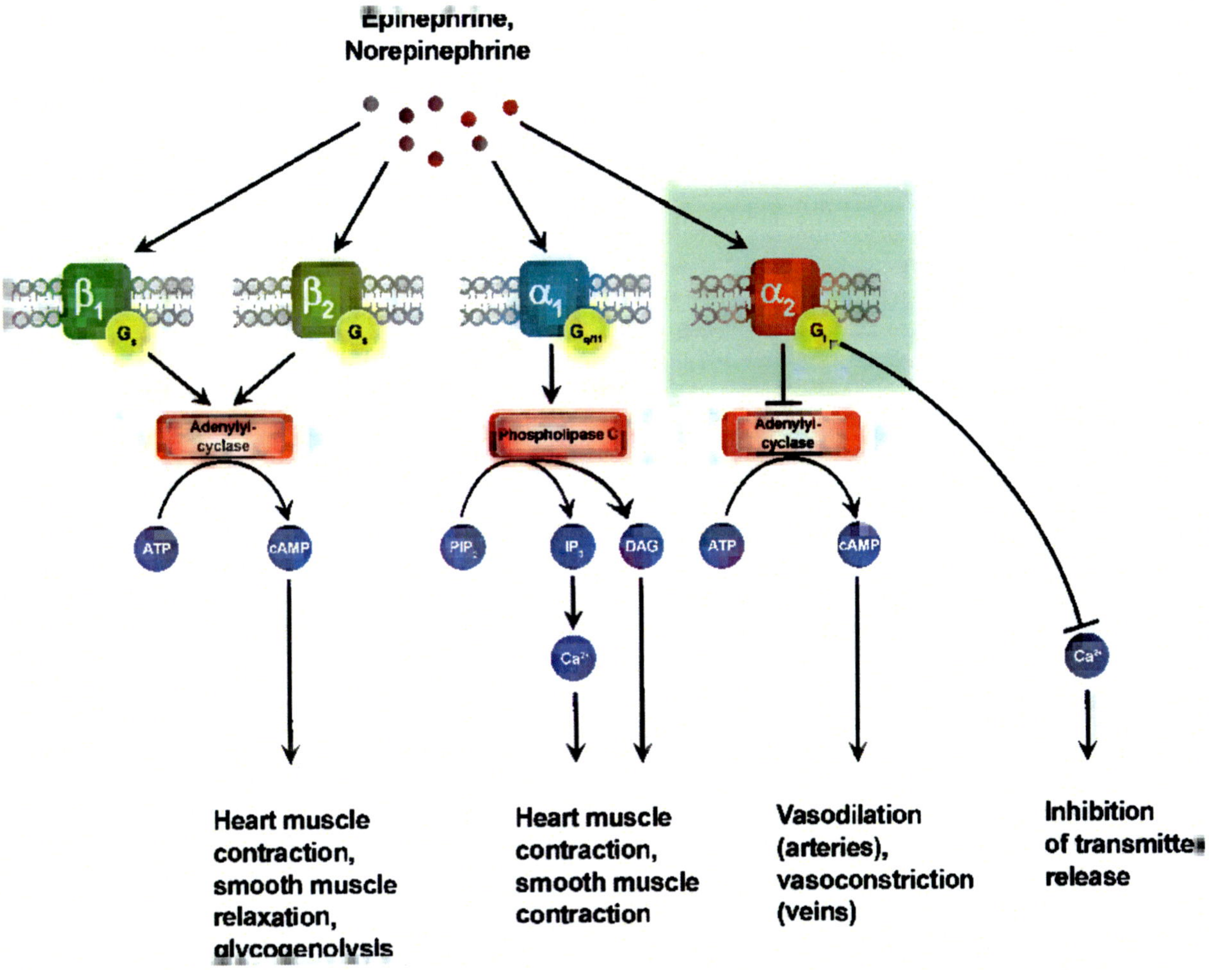


**Fig. 4.1** Signaling pathways of the major cardiac adrenoceptor subtypes. The three adrenergic receptors coupled to two positive inotropic effector pathways in cardiomyocytes are shown. Agonist occupancy of $\beta_1$- and $\beta_2$-adrenoceptors results in activation of the stimulatory G protein ($G_s$). Following activation, the α-subunit of $G_s$ interacts with adenylate cyclase to enhance formation of cAMP. cAMP-dependent activation of protein kinase A leads to increases in heart rate, contractility and energy metabolism by phosphorylation of a panel of proteins. The $\alpha_1$-adrenoceptor couples to $G_{q/11}$ protein and activates phospholipase C and the downstream mediators inositol trisphosphate ($IP_3$), diacylglycerol (DAG), and protein kinase C, which are involved in cardiac and in smooth muscle contraction. The $\alpha_2$-adrenoceptor couples to inhibitory G protein ($G_i$), which inhibits adenylate cyclase activity. The $\alpha_2$-adrenoceptor is not expressed in cardiomyocytes (*green background*) but causes arterial vasodilation, vasoconstriction of veins, and neurotransmitter release inhibition

forming $\alpha_{1C}$ subunit occurs at serine-1928 [13] and leads to a higher open-state probability of $Ca^{2+}$ channel in the phosphorylated state [14]. Also, it has been found that PKA phosphorylates the β subunit of the $Ca^{2+}$ channel at three sites (Ser-459, Ser-478, and Ser-479) *in vitro* [15]. Yatani et al. [16] have observed activation of cardiac L-type $Ca^{2+}$ channel when it was reconstituted with the $G_s$ protein in lipid bilayers. These results indicated that in addition to AC/cAMP/PKA-pathway, $G_s$ protein can directly stimulate L-type $Ca^{2+}$ channel.

Another target of PKA which is involved in the β-adrenergic regulation of cardiac cells is the transmembrane protein phospholamban (PLB). PLB regulates $Ca^{2+}$-ATPase (SERCA2a), an ATP-driven pump that translocates $Ca^{2+}$ from cytosol into the sarcoplasmic reticulum (SR), initiating muscle relaxation. PLB as an inhibitor of SERCA2a reduces enzyme affinity for $Ca^{2+}$ and thereby regulates cardiac contractility. Employing a transgenic mice deficient in PLB or reexpressing either wild-type or mutant PLBs in the heart of the PLB knockout background, Luo et al. [17] reported that ablation of PLB results in enhanced basal cardiac contractile parameters assessed at the cellular, organ, and intact animal levels. Reinsertion of wild-type PLB in the knockout background led to reversal of the hyperdynamic function of PLB-deficient heart and restored sensitivity of heart contraction to the β-adrenergic agonist. Moreover, the mutant form of PLB, which is unphosphorylatable by PKA at serine-16, cannot restore β-adrenergic-dependent contractility of PLB-deficient heart because nonphosphorylated PLB does not support the relief of SERCA2a inhibition.

One more mechanism for the β-adrenergic signaling cascade to regulate and modulate cardiac function is through cAMP/PKA-dependent phosphorylation of cardiac Troponin I (TnI). TnI is one of the subunits forming heterotrimeric troponin complex, which is involved in the $Ca^{2+}$-dependent functioning of cardiac actin–myosin–tropomyosin contractile machinery. Studies in transgenic animal models have provided strong evidence of the important role that TnI plays in cardiac function. More specifically, studies on mice expressing mutant forms of TnI tell us that phosphorylation of TnI is a significant factor in cardiac relaxation, similarly to that of PLB phosphorylation [18, 19]. According to a modern concept, β-AR-induced events lead to PKA-dependent phosphorylation of N-terminal serine-23 and serine-24 on cardiac TnI followed by altered intermolecular and intramolecular interactions: phosphorylation weakens the interaction of the N terminus of TnI with troponin C and facilitates interaction of N terminus with so-called "inhibitory region" of TnI. As a result, dissociation of TnI from the actin–tropomyosin complex allows strong, force generating reactions of actin with myosin heads [20].

Recent discoveries suggest that $\beta_1$- and $\beta_2$-adrenoceptor-regulated cAMP pools are compartmentalized so they differ in availability to effectors. In addition, these cAMP pools are negatively controlled by different cAMP-degrading phosphodiesterases (PDE). According to Christ et al. [21], PDE3 and PDE4 isoforms negatively regulate the level of cAMP which forms in response to activated $\beta_1$-AR, and activates L-type $Ca^{2+}$ channel. At the same time, only PDE4 blunts $\beta_1$-AR/cAMP/PKA-dependent increases in contractile force. Also, only PDE3 specifically reduces both $I_{Ca\text{-}L}$ and force of rat ventricular contraction, increased by epinephrine through $\beta_2$-AR.

Several mechanisms are involved in the deactivation of β-AR signaling including events that alter the function of the receptors themselves as well as postreceptor processes. One of them is hydrolysis of GTP to GDP by an intrinsic GTPase activity of G protein which returns the $G_{\alpha s}$-subunit to an inactive GDP-bound state.

There are several mechanisms that regulate β-AR functioning in order to prevent overstimulation, achieve signal termination, and provide the receptor responsive to following stimuli. Among these mechanisms there is the heterologous (agonist-independent) desensitization of receptor and it is realized through phosphorylation/inactivation of agonist-free β-AR by activated PKA and/or protein kinase C (PKC). Besides receptor, PKC can also phosphorylate the α-subunit of G protein, which decreases reassociation and availability of heterotrimeric G protein.

In contrast to PKA, G protein receptor kinases (GRKs) are able to phosphorylate only agonist-bound β-AR causing homologous (agonist-dependent) desensitization of receptor. Of six GRK isoforms, types 2 (also known as β-AR kinase 1 or βARK1) and 5 are abundantly expressed in the mammalian heart. They selectively phosphorylate agonist-occupied $\beta_1$- and $\beta_2$-ARs on serine residues located in the C-terminal tails. The specificity of GRK2 for agonist-bound β-AR is enhanced by $G_{\beta\gamma}$ subunits which are released from activated G protein and interact with the kinase [22]. Interestingly, $\beta_3$-AR lacks GRK-phosphorylation sites and thus, is not deactivated by the GRKs [23].

GRK-phosphorylated receptor is a target for a protein named β-arrestin. Association of receptor with β-arrestin prevents the interaction of the receptor with $G_s$ protein and therefore reduces the response to agonists. In addition, β-arrestin facilitates internalization of β-AR: binding of β-arrestin to phosphorylated β-AR causes formation of a clathrin-coated pit, which leads to endosomal internalization of the receptor [24]. Internalization seems to play an important role in the resensitization of $\beta_2$-AR via dephosphorylation in endosomal vesicles [25]. After internalization into endosomes, the receptors may return to the cell surface as active receptors or may traffic to lysosomes for degradation.

Gauthier et al. [26] recently found that ventricular biopsy samples of heart transplant recipients contain $\beta_3$-AR. These receptors mediate negative inotropic effects via $G_i$ protein/nitric oxide (NO) pathway.

## $\alpha_1$-Adrenoceptors

β-Adrenoceptors are not the only receptors for catecholamines. Cardiomyocytes are known to express two types of receptors that belong to the family of $\alpha_1$-adrenoceptors ($\alpha_1$-AR): $\alpha_{1A}$ and $\alpha_{1B}$. The $\alpha_{1A}$-adrenoceptor is the most abundant $\alpha_1$-adrenoceptor subtype in the human heart.

$\alpha_1$-ARs are coupled to $G_q/G_{11}$ proteins and mediate positive inotropic responses: the prolongation of myocyte contraction and sensitization of myofibrils to $Ca^{2+}$ [27]. The mechanism of $\alpha_1$-AR signal transduction is different from β-ARs. $\alpha_1$-ARs do not influence cAMP levels but activate phospholipase C. PL-C in turn catalyzes the breakdown of phosphatidylinositol 4,5-bisphosphate ($PIP_2$) into two products, $IP_3$ and DAG. $IP_3$ stimulates mobilization of intracellular $Ca^{2+}$ transient from intracellular stores by acting on its receptor on the sarcoplasmic reticulum. DAG is an activator of PKC. Increased $Ca^{2+}$ modulates a variety of calcium-sensitive regulatory proteins including multiple enzymes and kinases whereas activated PKC phosphorylates multiple targets (nuclear transcription factors, enzymes, transporters, receptors, including the $\beta_1$- and $\beta_2$-ARs). Two other phospholipases, PL-$A_2$ and PL-D are also activated by $\alpha_1$-AR [28].

The $\alpha_1$-AR may play an important role in the modulation of ventricular hypertrophy. For example, phenylephrine through $\alpha_1$-ARs can elevate levels of heparin-binding EGF-like growth factor (HB-EGF) in neonatal and adult cardiomyocytes. Increased HB-EGF autocrinally stimulates the growth of cardiac cells [29].

## $\alpha_2$-Adrenoceptors

A separate family of G protein coupled and catecholamine-regulated receptors, $\alpha_2$-adrenoceptors, consists of three members: $\alpha_{2A}$, $\alpha_{2B}$, and $\alpha_{2C}$. $\alpha_2$-ARs are not expressed in cardiomyocytes but are found in presynaptic membrane of sympathetic neurons in human atrium. They prejunctionally inhibit norepinephrine release by reducing $Ca^{2+}$ influx into the sympathetic nerve endings [30].

## Cholinergic Muscarinic m2 Receptors

The functioning myocardium is under a dual regulation by the nervous system. As we described before, the sympathetic system has positive inotropic effect on the heart. On the other hand, the parasympathetic nervous system modulates and buffers the sympathetic nervous system. The primary agonist released from the parasympathetic neurons is acetylcholine. The cardiac cell receptor for acetylcholine belongs to the family of muscarinic cholinergic receptors consisting of five subtypes (m1–m5), predominantly m2 receptor (although recent studies have reported the possible presence of m1 and m3 receptors in the heart) [31]. Muscarinic receptors are glycoproteins which belong to the superfamily of GPCRs.

m2 receptor decreases cAMP production either through coupling to the inhibitory G protein ($G_i$) and inhibition of the adenylate cyclase or via a still unknown $G_i$-independent mechanism. Stimulation of m2 receptors leads to bradycardia (in atria) and a negative inotropic response (in atria and ventricles). Moreover, in the ventricles m2-dependent negative inotropic effect is indirect: only force of contraction enhanced by other agonist (e.g., norepinephrin) is negatively regulated [32].

The difference between atrial and ventricular myocardium regulation by acetylcholine can be explained by the differences in effector systems activated by acetylcholine in atria and ventricles. As described above (see section on "β-Adrenoceptors"), in both tissues cAMP-elevating agents increase the $Ca^{2+}$ current, which in turn enhances the force of contraction. This is the result of cAMP-dependent phosphorylation of the high-voltage activated L-type $Ca^{2+}$ channel, which leads to an increase in maximal $Ca^{2+}$ conductance [33].

Activation of m2 receptor under these conditions has been shown to inhibit "pre-activated" L-type $Ca^{2+}$ current, contractile amplitude, and beating rate in atrial and ventricular myocytes [34, 35]. This process is mediated by *Pertussis toxin*-sensitive $G_{i/o}$ protein and requires inhibition of adenylate cyclase activity [36, 37]. m2-dependent overriding inhibition of adenylate cyclase decreases the level of cAMP [38], reduces the $Ca^{2+}$ current, and hence inhibits the force of contraction (indirect negative effect of acetylcholine).

The muscarinic regulation of L-type $Ca^{2+}$ channel also includes activation of the endothelial nitric oxide synthase (eNOS)/cGMP component: targeted disruption of the eNOS gene interrupts m2 cholinergic control of $I_{Ca\text{-}L}$ in mice ventricular myocytes. However, the responsiveness of eNOS-KO myocytes can be restored by transfection with wild-type eNOS [34]. It appears, that eNOS-produced NO stimulates the activity of guanylate cyclase which, by elevating the intracellular concentration of cGMP, promotes cAMP breakdown by PDE-2. In the functioning of this signaling cascade, compartmentalization of components plays an essential role. First, mutant eNOS, which could not target to caveolae, was unable to restore m2-dependent regulation of L-type $Ca^{2+}$ channel and contractility in eNOS-deficient cardiomyocytes [35]. Second, m2 receptor is not located to caveolae in the absence of agonist but moves into myocyte caveolar microdomains and interacts with caveolin-3 following exposure to muscarinic agonist [39]. Inhibitory effects of acetylcholine do not always correlate with changes in cAMP levels. To explain this phenomenon, some investigators have speculated that acetylcholine might antagonize cAMP-dependent responses by stimulating phosphatases and enhancing protein dephosphorylation [40, 41].

Another cardiac ion channel which depends on cAMP and therefore is sensitive to cAMP-regulating agonists is the hyperpolarization-activated cyclic nucleotide-gated (HCN) channel (see also Chap. 6). The mixed $Na^+/K^+$ current flowing through HCN channel (hyperpolarization-activated or "funny" current, $I_h$ or $I_f$) plays a key role in the control of cardiac rhythmicity ("pacemaker current"). HCN channels are encoded by four genes (HCN1–4). HCN4 is the major HCN channel isoform contributing to $I_h$ in rabbit, guinea pig, mouse, and dog cardiac conduction system (sinoatrial and atrioventricular nodes, Purkinje fibers), whereas remaining fraction of $I_h$ flows through HCN1 (rabbit) or HCN2 (mouse) [42, 43]. Interestingly, dominant isoform in atrial and ventricular myocytes is HCN2.

The voltage-dependent opening of HCN channels is directly regulated by the binding of cAMP to cyclic nucleotide-binding domain located in the cytosolic C-terminal part of the molecule [44]. Sympathetic stimulation, via β-adrenoceptor-triggered cAMP production, activates HCN and, hence, accelerates heart rate while vagal stimulation via m2 receptor inhibits cAMP synthesis, inhibits $I_h$, and consequently lowers the heart rate [44, 45].

The direct negative effect of acetylcholine, observed in atrial tissue, is related to atrial specific regulation by acetylcholine of $K^+$ efflux across the cardiac cell membrane. Hutter and Trautwein [46] in 1955 demonstrated an increase of $K^+$ efflux across the sinoatrial cell membrane with vagal stimulation. Increase of $I_k$ results in hyperpolarization, slows beating rate, and reduces the force of contraction. It has been well documented that atrial cells express $K^+$ channels which are functionally coupled to and regulated by m2 receptors [47]. These channels belong to the subfamily of inwardly rectifying $K^+$ channels (Kir). This subfamily is called "G protein-regulated $K^+$ channels" (GIRK) and includes five channels cloned during 1993–1995: GIRK1–GIRK5 [47]. Two of them, GIRK1 and GIRK4, are expressed in cardiac cells. GIRK in atrial cells functions as a heterotetramer of two of each GIRK1 and GIRK4 molecules [48, 49].

m2 receptor is coupled to GIRK channel through the *Pertussis toxin*-sensitive $G_{i/o}$ protein. Unique for the system that we are discussing is that the actual mediator of G protein-induced activation of GIRK is the βγ subunits ($G_{\beta\gamma}$) but not the α subunit ($G_\alpha$) of the G protein. The interaction between $G_{\beta\gamma}$ and $K^+$ channel has been shown with cloned $K^+$ channel and G protein subunits [50]. The current understanding of the role of G protein in the m2-dependent activation of GIRK is that in the absence of acetylcholine, G protein is a heterotrimer: $G_{\beta\gamma}$ and GDP-bound form of $G_\alpha$. Agonist-stimulated m2 receptor accelerates the GDP/GTP exchange on G protein. GTP-$G_\alpha$ dissociates from $G_{\beta\gamma}$. The dissociated $G_{\beta\gamma}$ interacts with the $G_{\beta\gamma}$-binding site located on the C terminus of GIRK. There are several proposals in the literature on the mechanism underlying $G_{\beta\gamma}$-dependent activation of GIRK. According to one opinion, the $K^+$ channel is intrinsically inhibited by the C- and/or N-terminal domains of GIRK1, and $G_{\beta\gamma}$ activates the channel by removing this inhibition [51, 52]. Another opinion is based on the observations by Huang et al. [53] that GIRK is maximally activated by $PIP_2$ which binds to the C terminus of the channel, although the channel has low sensitivity to $PIP_2$ in the absence of $G_{\beta\gamma}$. $G_{\beta\gamma}$ activates the $K^+$ channel by increasing the sensitivity of GIRK to $PIP_2$.

It seems that the nature of the $G_{\beta\gamma}$ does not play a role in the regulation of the $K^+$ channel. For example, *in vitro* $G_{\beta1\gamma1}$, $G_{\beta1\gamma2}$, $G_{\beta1\gamma5}$, $G_{\beta1\gamma7}$, $G_{\beta2\gamma5}$, and $G_{\beta2\gamma7}$ differ less than ten times in potency of GIRK activation [54]. The lack of specificity of $G_{\beta\gamma}$ rises the question of why in myocytes m2-dependent release of $G_{\beta\gamma}$ from $G_i$ protein results in GIRK activation, whereas $\beta_1$-adrenoceptor-dependent release of $G_{\beta\gamma}$ from $G_s$ protein has no effect. It is possible that the m2 receptor, the $G_i$ protein, and GIRK are compartmentalized into separate cell microdomains (caveolae?); however, to date there is no experimental evidence to confirm or disregard this hypothesis.

There is evidence that agonist-mediated activation of $K^+$ channel is accelerated by regulator of G protein signaling 3 protein (RGS3) [55], but the mechanism of action is not known.

## Histamine

Histamine has multiple direct and indirect effects on the cardiovascular system. Effects include changes in rate, rhythmicity, and conduction; the vasoactivity of histamine results from its effect on vascular smooth muscle. Mammalian myocardium contains a substantial pool of releasable histamine. Main sources of cardiac histamine are cardiac mast cells and terminals of certain cardiac histaminergic neurons. Outside of the heart, the major pool of histamine is linked to elements of the immune system. Thus, histamine increased in response to sympathetic nerve stimulation or during anaphylaxis can modify cardiac rhythm and may contribute to the development of several types of cardiac dysrhythmias.

H1 and H2 receptors are involved in the direct action of histamine on the cardiovascular system. They belong to the superfamily of G protein-coupled receptors. The type of histamine receptor varies between species and also in the various regions of the heart. Human heart contains H2 receptors while dog heart has H1 receptors. Rabbit atria contain both H1 and H2 receptors while the ventricles have only H1. In the guinea pig heart H1 receptors are expressed in the left atria and ventricles while H2 receptors are found in the right atria and are the predominant histamine receptors in the ventricles. Rat and cat heart do not seem to have histamine receptors.

In the heart, histamine increases sinus rate (positive chronotropic effect), the amplitude of ventricular contraction (positive inotropic effect), and impairs atrioventricular conduction (negative dromotropic effect). At high concentration histamine induces ventricular dysrhythmias.

Histamine increases the sinus rate of isolated hearts from the guinea pig and human fetus, and increases rabbit and monkey atrial contractility. Positive chronotropic effect in the guinea pig is the result of direct action of histamine on the atrial pacemaker cells via H2 receptors. Chronotropic effects of histamine are not seen in man.

In rabbit, the positive inotropic effect of histamine on ventricular and atrial contraction is also attributed to H2 receptors. Interestingly, in guinea pig the H2 receptors mediate the positive inotropic response of the right atrium to histamine, whereas the inotropic response of the left atrium is H1-mediated [56]. Histamine slows atrioventricular conduction, and in human this effect is H1 receptor-mediated.

The positive inotropic and chronotropic effects of histamine in cardiac preparations from rat and cat are related to the action of histamine as a neuromodulator: it stimulates the release of noradrenaline.

H2 receptors are coupled to $G_s$ protein. Activated H2 receptors stimulate adenylate cyclase and increase cyclic AMP levels, which leads to increase in free intracellular $Ca^{2+}$ concentration. H1 receptors are not associated with cyclic nucleotides in the heart, but also increase $Ca^{2+}$ in the cardiac cells. H1 receptors mobilize $Ca^{2+}$ via $G_{q/11}$-mediated activation of phospholipase C.

A third histamine receptor subtype (H3) was discovered in 1983 in the central histaminergic pathways as a receptor, which presynaptically autoinhibits the release of histamine neurotransmitter [57]. Later Endou et al. [58] demonstrated that histamine via H3 receptors presynaptically reduces the release of endogenous norepinephrine from sympathetic nerve endings in the guinea pig myocardium. Neuromodulatory effect of activated H3 receptors includes a *Pertussis toxin*-sensitive $G_{i/o}$ protein and leads to decrease in $Ca^{2+}$ influx through N-type channels. H3-dependent reduction of norepinephrine release results in the inhibition of adrenergic positive inotropic and chronotropic responses in guinea pig atria.

Histamine plays role in the genesis or modification of dysrhythmias. The mechanisms underlying dysrhythmogenic actions of histamine include alterations in normal automaticity, the induction of abnormal automaticity or triggered activity, and the induction of abnormal impulse conduction. These changes cause dysrhythmogenic events such as slowing of atrioventricular conduction, increases in sinus rate, and ventricular automaticity.

### Serotonin

Serotonin (5-hydroxytryptamine; 5-HT) is a biogenic monoamine. Physiologically, 5-HT is mainly a neurotransmitter in the central nervous system. In addition, 5-HT participates in gastrointestinal peristalsis, blood coagulation, and the maintenance of blood pressure. 5-HT is an important regulator of cardiovascular function stimulating chemosensitive nerves in the heart and acting directly on cardiomyocytes.

The major source of serotonin in humans are platelets whereas in rodents it is stored in mast cells. Also serotonin is synthesized in the central nervous system and can be found in serotonergic neurons.

The effects of 5-HT are mediated by at least seven main types of receptors: 5-HT1–5-HT7. One of them, 5-HT3, is integral to the $Na^+/K^+$ ion channel whereas other belong to the superfamily of GPCRs. The majority of 5-HT receptors (5-HT4, 5-HT6, 5-HT7) through $G_s$ proteins are positively coupled to adenylate cyclase [59]. In atrial cardiomyocytes, stimulation of 5-HT4 receptors causes an increase in contractile force and accelerates the onset of muscle relaxation via increased cyclic adenosine monophosphate levels and cAMP-dependent protein kinase A activity. Increased contractile force is a result of phosphorylation of membrane bound and sarcolemmal L-type $Ca^{2+}$ channels by PKA leading to increased atrial cytoplasmic $Ca^{2+}$. PKA also phosphorylates phospholamban and troponin I, followed by shortening of muscle relaxation [60]. Several 5-HT receptors (5-HT1, 5-HT5) inhibit AC via $G_{i/o}$ proteins, whereas 5-HT2 receptor is linked to $G_q$ protein and stimulates phospholipase C [59].

There are several levels of serotonin-dependent regulation of the myocardium. First, 5-HT is involved in the regulation of the cardiovascular system by the central nervous system. Serotonergic neurons in the raphe nuclei in the brain neurohormonally modulate the sympathetic and vagal innervation of cardiovascular structures, including the myocardium. Predominant targets of 5-HT are 5-HT1, 5-HT2, and 5-HT3 receptors in the presynaptic areas of brain neurons involved in cardiovascular regulation. Activation of 5-HT1 receptors causes central sympathoinhibition and vagal bradycardia, whereas activation of 5-HT2 receptors causes sympathoexcitation leading to tachycardia. Moreover, 5-HT3 receptors are involved in the regulation of blood pressure [60, 61].

Second level of regulation involves the peripheral effects of serotonin. A number of conditions (e.g., angina pectoris, myocardial infarction) cause release of serotonin from platelets. A local increase in circulating serotonin might stimulate the cardiac sympathetic afferent neurons in cardiomyocytes. For example, 5-HT is responsible for positive chronotropic, inotropic, and lusitropic effects on the heart via the stimulation of 5-HT4 receptors in cardiomyocytes. More specifically, 5-HT has been shown to increase the pacemaker current in atrial myocytes, and thus contributing to atrial dysrhythmias, including atrial fibrillation [60]. Normally, functional 5-HT4 receptors are absent in healthy ventricles, but have been found abnormally expressed after extensive myocardial infarction (human failing ventricles, rats with congestive heart failure). Stimulation of 5-HT4 receptors in the failed ventricles lead to a positive inotropic response [62, 63]. Another mechanism of the peripheral effect of serotonin in the rabbit atria has been detected: in this tissue 5-HT via 5-HT3 receptors stimulates release of noradrenaline from postganglionic cardiac sympathetic neurons, which results in atachycardic response [64].

## *Neuropeptides*

The control of heart rate by the sympathetic and parasympathetic systems is well established and has been described above. In addition, peptidic transmitters, neuropeptides, are present in the myocardium and have direct or indirect actions on cardiac function. Employing recent advances in immunohistochemical methods, several neuropeptide-containing neurons have been found in the heart, generating and conducting electrical impulses in the working myocardium (sinoatrial node): neuropeptide Y, vasoactive intestinal peptide, somatostatin, substance P, dynorphine B, calcitonin gene-related peptide, and neurotensin. Furthermore, it has been shown that some neuropeptides can be coexpressed in the same neuron or coexist with "classical" neurotransmitters (somatostatin and neuropeptide Y, somatostatin and dynorphine B, somatostatin and substance P, somatostatin and norepinephrine) [65, 66]. Neuropeptides released from neurons can regulate directly the sinoatrial node or interact with "classical" neurotransmitters to modulate heart rate.

### Neuropeptide Y

36-residue neuropeptide Y (NPY) is expressed by sympathetic neurons innervating the cardiovascular system. It can be colocalized with norepinephrine. Neurons containing NPY were found not only in the sinus node, but also in the atria and around the coronary vessels. Also, NPY can perform as a local paracrine factor released from endocardial endothelial cells [67]. NPY binds to a group of GPCRs: $Y_1$, $Y_2$, $Y_3$, $Y_4$, $Y_5$, $Y_6$. These receptors are coupled to the heterotrimeric $G_{i/o}$ protein. Interestingly, $Y_1$, $Y_2$, and $Y_5$ receptors are expressed in the heart.

NPY increases the frequency of spontaneous contraction on embryonic chick ventricular myocytes. This modulation occurs via activation of $Y_1$ receptor and involves stimulation of $Ca^{2+}$ influx through the L-type $Ca^{2+}$ channel [68]. Interestingly, NPY decreases the heart rate in rats [69]. In addition, NPY is able to induce hypertrophy of adult ventricular cardiomyocytes through $Y_1$ and $Y_5$ receptors.

### Vasoactive Intestinal Peptide

Vasoactive intestinal peptide (VIP) is a 28-residue peptide. VIP-containing nerves have been localized in the atrial and ventricular myocardium, sinoatrial and atrioventricular nodes, and in the coronary vessels [65]. VIP is released in the coronary vessels and in the heart during parasympathetic (vagal) nerve stimulation. Cholinergic agonists such as serotonin, dopaminergic agonists, prostaglandins (PGE, PGD) and nerve growth factor also stimulate VIP release in other parts of the body. Release of VIP in the heart increases atrial and ventricular contractility.

The VIP receptor is a member of a family of G protein-coupled receptors. Two subtypes of VIP receptors, VPAC1 and VPAC2, have been cloned from rat and sequenced. Both subtypes of the VIP receptor have amino-terminal VIP recognition sequence, several extracellular glycosylation sites, seven transmembrane-spanning domains and a number of intracellular sites for phosphorylation by protein kinase C. VPAC1 and VPAC2 receptors are expressed in the heart [70].

Binding of VIP to VPAC receptor promotes $G_s$ protein-mediated activation of AC followed by accumulation of intracellular cAMP [71, 72]. A VIP-induced increase in cAMP can increase the activity of protein kinase A, which enhances calcium channel phosphorylation, the L-type calcium current, and the release of calcium from the sarcoplasmic reticulum. As a consequence, intracellular calcium concentration increases, enhancing the rate and extent of cardiomyocyte contraction. Moreover, PKA phosphorylates troponin I and phospholamban, which decreases the affinity of troponin for calcium and enhances intracellular calcium sequestration. As a result, VIP has positive chronotropic and inotropic effects on cardiomyocytes. In the sinoatrial node, a VIP-induced increase in cAMP also activates hyperpolarization-activated pacemaker $I_f$ current and increases the heart rate [73].

After VIP binds to its receptor, the peptide is rapidly internalized, probably by receptor-mediated endocytosis. This internalization decreases the cell surface receptor density. Most of the VPAC receptors are recycled back to the cell surface, but some receptors are degraded in the lysosomes.

In animal models of heart failure and in patients with cardiomyopathy, the concentration of VIP can decrease in the myocardium by more than 50% as well as the density of the VPAC receptor. These changes suggests that the cardiovascular VIP signaling pathways may be of critical importance in the pathogenesis of HF as well as in hypertension.

### Calcitonin Gene-Related Peptide

Thirty-seven amino acids-calcitonin gene-related peptide (CGRP) is found in neurons innervating the atria. Within the atria, CGRP especially localizes in the sinoatrial and atrioventricular nodes; often CGRP colocalizes with substance P. In lesser amounts CGRP-immunoreactivity has been also observed in the ventricles.

In many animal species and in humans, CGRP has positive chronotropic and inotropic properties.

### Substance P

Substance P (SP) is an 11-residue peptide with some structural similarities to bradykinin. SP-containing nerves have been localized in the sinoatrial and atrioventricular nodes, and usually it colocalizes with CGRP (see above). However, SP does not exert a direct effect on heart rate, but it may play a modulatory role in sinoatrial function. Also, SP-immunoreactive fibers are potent vasodilators of coronary blood vessels [65].

### Neurotensin

Neurotensin is a 13-residue peptide neurohormone and/or neuromodulator, located in the synaptic vesicles. Neurotensin-containing nerves have been localized in the sinus node, and are involved in the regulation of sinus node blood flow, and in impulse generation as well as in extrinsic and intrinsic cardiac reflex mechanisms [65]. Neurotensin functions are mediated by neurotensin receptors, NTS1 and NTS2. These receptors belong to the family of the G protein-coupled receptors. NTS1 stimulation could lead to the activation of several G proteins. For example, NTS1 can activate phospholipase C through the $G_{q/11}$ protein coupled pathway. NTS1 was also found to stimulate AC through interaction with $G_s$ protein. Furthermore, neurotensin can inhibit AC and stimulate arachidonic acid production through interaction with $G_{i/o}$ proteins [74]. It seems that the expression of G protein polypotential capability of NTS1 depends on the host cell. NTS1 mediates the multiple functions of neurotensin, such as hypotension, hyperglycemia, and hypothermia.

NTS1 and NTS2 are mainly expressed in the brain. A large portion of endogenous NTS2 located in the rat spinal cord neurons is associated mostly with intracellular vesicles and the *trans*-Golgi network, and does not respond to neurotensin.

In the rat heart, neurotensin evokes marked concentration-dependent positive inotropic responses, and acts as a potent neuromodulator which presynaptically stimulates release of norepinephrine. The neurotensin-induced contractile effects can be abolished by propranolol, a β-adrenoceptor antagonist [75].

### Somatostatin

Somatostatin-14 (somatostatin) is a tetradecapeptide which is synthesized by posttranslational cleavage of the precursor peptide, somatostatin-28. Somatostatin colocalizes with acetylcholine in presynaptic endings of parasympathetic vagus nerve innervating the heart and may be released by high-frequency stimulation of the vagus nerve. The main cardiac effects of somatostatin include heart rate deceleration, decrease myocardial contractility, and slowing of propagation velocity along the conduction system.

The cardiovascular effects of somatostatin result from its action as a neurotransmitter, through interaction with its own receptors. Some of the effects of somatostatin may result from its neuromodulatory action on presynaptic release of acetylcholine or noradrenaline.

There are five different somatostatin receptors: SSTR1–SSTR5. All of them belong to the GPCR superfamily, are coupled with $G_{i/o}$ proteins and inhibit adenylate cyclase [76]. Excitation of somatostatin receptors causes a decrease of intracellular cAMP content, inhibition of L-type $Ca^{2+}$ channels [77], and activation of potassium channels [78]. In addition, SSTR2, 3, and 5 can couple to the phospholipase C-β via $G_{q/11}$ protein [79].

Cardiac expression of SSTRs varies between species, developmental stage, and cellular type. SSTR3, 4, and 5 are abundant in adult rat cardiomyocytes, and two others (SSTR1 and 2) are less abundant [80]. In contrast, SSTR4 is absent in adult mouse heart [81]. Only SSTR1 and 2 are present in human cardiomyocytes [82].

While somatostatin elicits negative inotropic and chronotropic effects in atrial myocardium, the effects of this neuropeptide on the ventricular myocardium are more complex. Somatostatin shows direct positive and negative effects in ventricular cardiomyocytes isolated from the adult rat heart. Interestingly, the inhibitory effect of somatostatin is observed at relatively high concentrations of the peptide ($IC_{50}$ = 13 nM) and negatively modulates β-adrenergic stimulation of cell contraction. This effect is mediated by SSTR4 receptors and requires activation of the transient outward potassium current ($I_{to}$). Most likely, activation of $I_{to}$ reduces the time available for transmembrane influx of $Ca^{2+}$ and thus attenuates cardiomyocyte contractile response. Surprisingly, inhibition of $I_{to}$ unmasks very potent activatory effect of somatostatin on the basal rate of cardiomyocyte contraction ($EC_{50}$ = 118 fM). The signaling cascade associated with the positive contractile effect of somatostatin involves different subtypes of receptors (SSTR3, with the additional contribution of SSTR2 and SSTR5), stimulation of the phospholipase C, mobilization of intracellular $Ca^{2+}$, and activation of L-type $Ca^{2+}$ channels [80, 83].

Somatostatin receptors are expressed in postganglionic sympathetic neurons. They are linked to $G_{i/o}$ proteins at presynaptic sites and block voltage-gated $Ca^{2+}$ channels. This leads to the inhibition of $Ca^{2+}$ entry which is the crucial link between depolarization and the release of neurotransmitter at presynaptic nerve terminals [84].

## Purinergic Receptors

Adenosine has the negative chronotropic, dromotropic, and inotropic effects on myocardium. Receptors to adenosine are heterogeneously distributed within the heart and are coupled to different effector systems. Under basal conditions interstitial adenosine concentration is below the value necessary for regulation, but in the heart exposed to ischemia or hypoxia concentration of circulating adenosine increases as a result of catabolism of extracellular nucleotides or released from within heart cells.

Four adenosine receptor subtypes have been cloned: A1, A2a, A2b, and A3. They are encoded by separate genes and belong to the superfamily of GPCRs. The adenosine A1 and A2a subtypes have been identified in cardiac myocytes. Activation of A1 receptor has no effect on basal adenylate cyclase activity but inhibits the stimulation of the enzyme by other agonists (e.g., by norepinephrin). Inhibitory effect of A1 receptor is mediated by regulatory *Pertussis toxin*-sensitive

$G_{i/o}$ proteins. A2a receptor is coupled to $G_s$ protein and stimulates AC activity.

Adenosine-dependent negative chronotropic and dromotropic effects are carried out via A1 receptors in the pacemaker cells, in the sinus and atrioventricular nodes. Signal from A1 receptors is transduced by $G_{i/o}$ protein to the potassium channel, and cyclic nucleotides do not mediate this process.

Adenosine also acts as a negative inotropic agent. It directly reduces the force of contraction of atrial muscle from several species (human, rat, guinea pig, and dog) [32, 85–87]. Again, this process is cAMP/cGMP-independent and influences outward potassium current (stimulation) and entry of $Ca^{2+}$ via L-type $Ca^{2+}$ channels (inhibition). Interestingly, adenosine has no direct effects on ventricular muscle [32]. Furthermore, the indirect "antiadrenergic" negative inotropic effect of adenosine in both atrial and ventricular tissues is well documented: adenosine via A1 receptors reverses catecholamine-stimulated force of contraction and accumulation of cAMP [32]. Indirect effect of adenosine depends on cAMP and changes on the phosphorylation of $Ca^{2+}$ and $K^+$ channels [88–90]. In rabbit atrioventricular nodal myocytes, Al receptor-mediated attenuation of catecholamine-stimulated inward $Ca^{2+}$ current ($I_{Ca}$) involves the activation of NO synthase [91]. It is likely that NO in nodal cells stimulates guanylate cyclase leading to an increase in cGMP. cGMP in turn activates cGMP-stimulated cAMP phosphodiesterase. The latter enzyme can reverse catecholamine-induced accumulation of cAMP and cAMP-dependent activation of $I_{Ca}$.

The antiadrenergic action of adenosine may be protective in ischemic myocardium, when released catecholamine could cause dysrhythmias.

Recently, Hove-Madsen et al. [92] have demonstrated that human atrial cardiomyocytes express A2a receptors, and that they colocalize with ryanodine receptors within the cell. This study showed that selective A2a agonist (CGS21680) increased $Ca^{2+}$ release from the sarcoplasmic reticulum, and that adenosine through A2a receptors stimulates protein kinase A via $G_s$/AC/cAMP pathway. PKA-dependent phosphorylation/activation of the ryanodine receptor leads to the elevation of intracellular $Ca^{2+}$ and contributes to the regulation of $I_{Ca}$. Ventricular cardiomyocytes also express A2a receptors. In rat, the functional activity of this receptor can be shown after specific inactivation of dominating A1 receptors. In the presence of 1,3-dipropyl-8-cyclopentylxanthine (selective A1 antagonist), specific A2a receptor agonist (CGS21680) increases the level of cAMP and the contractile amplitude of cardiac cells [93]. The role of ventricular A2a receptor under physiological conditions remains unclear because "inhibitory" A1 receptor masks the signaling cascades regulated by A2a.

Adenosine may influence cardiovascular function indirectly through presynaptic receptors that modulate norepinephrin release from sympathetic nerve terminals [94]. It is worth noting that neurosecretory processes depend strongly on extracellular $Ca^{2+}$ and that activation of prejunctional A1 receptor reduces $Ca^{2+}$ influx into sympathetic nerve endings [58], which inhibits the secretion of norepinephrin.

## Peptide Hormones

The contractile activity of the heart can be modulated by regulatory peptides originating from components of the cardiovascular system (vascular and endocardial endothelium, atrial and ventricular myocytes). Peptides could be released locally on demand and modulate the heart rate paracrinally or autocrinally.

### Angiotensin II

Angiotensin II (Ang II) is an octapeptide hormone produced by the renin–angiotensin system. Interestingly, the existence of local cardiac renin–angiotensin system is now well established.

Cardiac cells express AT1 receptors which bind Ang II. AT1 receptors couple to heterotrimeric G proteins and regulate several cell responses: vasoconstriction, hypertrophy, and growth. Two AT1 receptor subtypes are expressed in the myocardium, AT1a, and AT1b. AT1 receptors have positive inotropic effect acting independent of cAMP possibly, through the phospholipase C/DAG/$IP_3$ pathway.

Repeated stimulation causes rapid desensitization of AT1 that can return to its initial sensitivity within 1 h [95]. There is evidence that activation of AT1 receptor is accompanied by its phosphorylation but the role of phosphorylation in desensitization is unclear. Both subtypes of AT1 receptor undergo internalization upon stimulation by Ang II. AT1a and AT1b receptors are regulated differentially in myocardial infarction: AT1a subtype is increased whereas AT1b subtype stays unchanged [96].

Besides being a direct positive chronotropic agent, Ang II can serve as a neuromodulator by stimulation of release of catecholamines from cardiac sympathetic neurons and by reducing of vagal tone.

### Endothelin

The endothelin family contains three peptides: endothelin 1, 2, and 3. Endothelin 1 (ET-1) is produced by endothelial cells. This peptide has powerful positive chronotropic and inotropic effect on the myocardium.

ET-1 regulates cardiac function through EDNRA and EDNRB receptors expressed in cardiomyocytes. Positive inotropic and chronotropic effects of ET-1 are possibly mediated through an EDNRB receptor-mediated cAMP-independent mechanism, which involves the activation of the phospholipase C/DAG/IP$_3$ pathway. Also, there is an inhibitory component that is involved in the ET-1-induced positive chronotropic effect. This effect is observed at high concentrations of ET-1 (30–100 nM) and is mediated through the EDNRA receptor. This receptor is coupled to the *Pertussis toxin*-sensitive G$_i$ protein and causes a reduction in cAMP production [97]. The EDNRA receptor-mediated inhibition of heart rate raised by stimulating the β-adrenoceptor may play an important compensatory role in pathophysiological conditions like myocardial infarction and cardiogenic shock when sympathetic tone is increased while plasma levels of ET-1 are elevated. Interestingly, Humbert and Simmoneau [98] have reported on the development of endothelin-receptor antagonists (ERAs) for the treatment of pulmonary hypertension (PAH) associated with rheumatic diseases (Fig. 4.2).

Bosentan, an antagonist with dual specificity for the endothelin-receptor subtypes A and B, was shown to be effective and well tolerated in placebo-controlled clinical trials. Subsequently, other ERAs with specificity for the endothelin-receptor subtype A have been developed, including sitaxsentan and ambrisentan, the later may be an option for patients who have discontinued bosentan and/or sitaxsentan therapy due to abnormal liver function test [99].

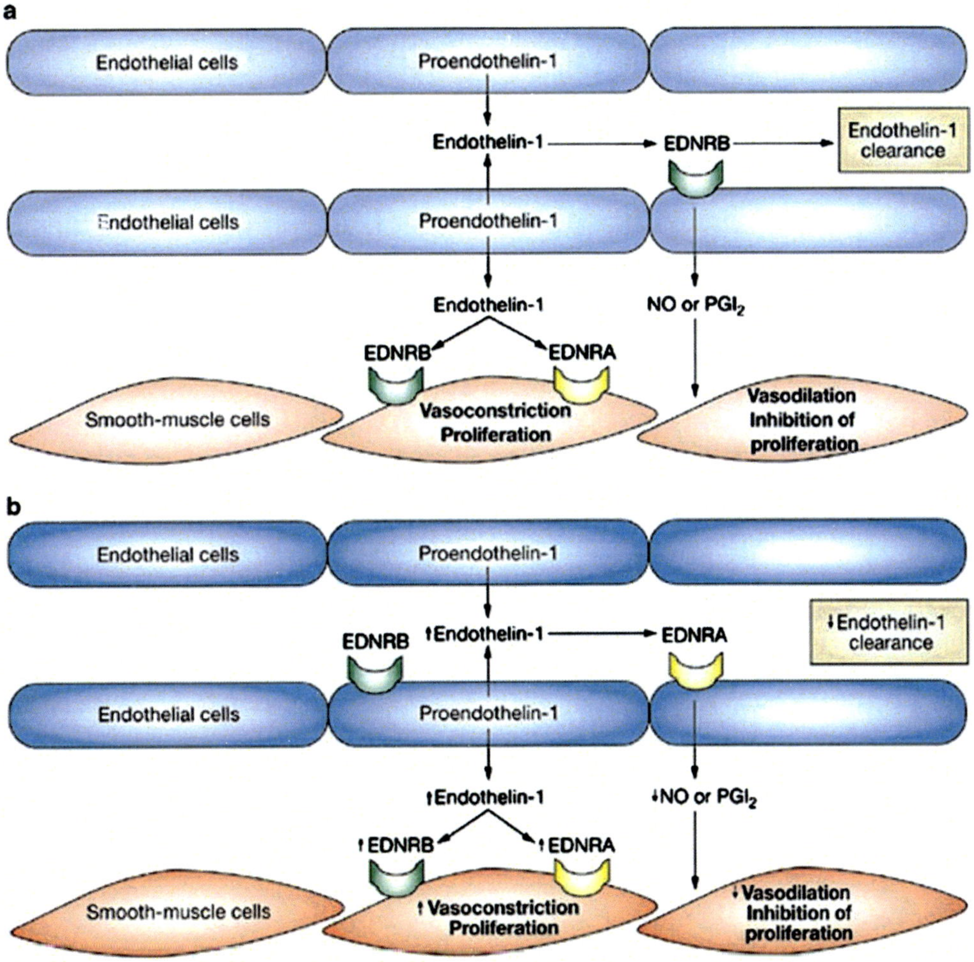


**Fig. 4.2** The endothelin signaling pathway. (**a**) Normal endothelial cells (*pale blue*) produce endothelin-1 from its precursor proendothelin-1. Released endothelin-1 binds to type A and/or type B endothelin receptors (EDNRA and EDNRB, respectively), on vascular smooth muscle cells (*pale red*), promoting vasoconstriction and smooth muscle cell (SMC) growth. When interacting with EDNRB on endothelial cells, endothelin-1 stimulates the generation of local vasorelaxants, such as nitric oxide (NO) and prostaglandin I$_2$ (PGI$_2$). Endothelial EDNRBs also contribute to the clearance of circulating endothelin-1. (**b**) In hypertension, dysfunctional endothelial cells (*dark blue*) produce increased amounts of endothelin-1 and decreased amounts of NO and PGI$_2$. Vasoconstrictor–vasodilator imbalance promotes vasoconstriction and proliferation of SMCs after interaction with EDNRAs and/or EDNRBs, which are overexpressed in SMCs (*dark red*). (Reprinted from Humbert and Simmoneau [98] with kind permission from Nature Publishing Group)

### Urocortin

Urocortin is a family of peptide hormones which increase contractility and cardiac output without causing changes in mean arterial blood pressure. Three urocortins are expressed locally in the heart and are detectable in plasma: urocortin (Ucn 1), urocortin 2 (Ucn 2), and urocortin 3 (Ucn 3). Ucn 1 can bind and activate two receptors which belong to the GPCR family: CRH-R1 and CRH-R2. Ucn 2 and Ucn 3 activate only CRH-R2. Both receptors are expressed in the brain, but only CRH-R2 is detected in the heart. Specifically, quantitative RT-PCR revealed expression of CRH-R2 in cardiomyocytes and at much higher level in vascular endothelium (aorta, coronary artery, microvessels).

Ucn-dependent activation of CRH-R2 leads to increase of cAMP, increased PKA activity, and phosphorylation of phospholamban. Ucn is also known to activate the phosphatidylinositol 3-kinase (PI3K)-Akt pathway.

Intravenously administered urocortins have inotropic and lusitropic effects on rat and sheep hearts. Similarly, Ucn 2 directly enhances contractility in rabbit ventricular myocytes via stimulation of PKA. Urocortins can also protect isolated cardiomyocytes from simulated ischemia and reperfusion injury through activation of acute cardioprotective pathways (PI3K-Akt, Erk1/2) and induce a hypertrophic response via activation of PI3K-Akt and PKA. Although the role that urocortins play in the heart is not clear, the aforementioned examples indicate a direct role for urocortin on cardiomyocytes function.

Plasma levels of urocortins increase more than twice in HF patients [100]. However, how circulating urocortins can cross the endothelial layer which lines the cardiac vasculature is not known. One possibility is that urocortin binds CRH-R2 receptors on endothelial cells and then is transported to the cardiomyocytes. Moreover, urocortins are powerful regulators of vascular and endothelial cells enhancing their survival, proliferation, and function.

### Natriuretic Peptides

The natriuretic peptide family consists of three peptides, atrial natriuretic peptide (ANP), brain natriuretic peptide (BNP), and C-type natriuretic peptide (CNP). Twenty-eight-amino-acid ANP is produced primarily in the atria, whereas 32-amino-acid BNP is present in the ventricles. The 22-amino-acid CNP is present in the vascular endothelial cells.

There is no direct effect of ANP on heart rate, but ANP through natriuretic peptide receptor A could act as a neuromodulator of heart rate by suppression of the release of catecholamines from autonomic nerves, and especially by suppression of sympathetic regulation from the central nervous system. CNP has positive chronotropic and inotropic paracrine effects in canine models acting through natriuretic peptide receptors B.

## $Ca^{2+}$ as a Signaling Molecule

In cardiac muscle $Ca^{2+}$ regulates cell division, growth, and cell death. Intracellular $Ca^{2+}$ is responsible for initiating contraction in cardiac myocytes. Furthermore, relaxation of cardiac cell also depends on the concentration of cytosolic $Ca^{2+}$. Concentration of intracellular $Ca^{2+}$ is controlled by a number of protein systems located in the sarcolemma (ion channels, $Ca^{2+}$ transporters, and $Na^+/Ca^{2+}$ exchanger), sarcoplasmic reticulum ($Ca^{2+}$ release channel, $Ca^{2+}$-ATPase), and mitochondria.

Under resting conditions, concentration of free intracellular $Ca^{2+}$ is around 100 nM. It has to reach approximately 10 μM during the action potential to activate contraction in cardiomyocytes, The majority of $Ca^{2+}$ is released from the intracellular store, in the sarcoplasmic reticulum via ryanodine-sensitive $Ca^{2+}$ release channel (ryanodine receptor, RyR). This release happens in response to triggering amounts of external $Ca^{2+}$ entering the cell (i.e., $Ca^{2+}$-induced $Ca^{2+}$-release, CICR) (Fig. 4.3).

The most important trigger of the systolic transient is the $Ca^{2+}$ that enters the cell through the sarcolemmal L-type $Ca^{2+}$ channel: inhibition of L-type $Ca^{2+}$ channel by nisoldipine significantly diminishes systolic $Ca^{2+}$ transient in the cardiomyocyte. Interestingly, myocardial rapid inotropic responses can be initiated by intracellular $Ca^{2+}$ entered into the cell during the action potential via the L-type $Ca^{2+}$ channel (without the release of $Ca^{2+}$ from SR) [101].

As mentioned above, blocking of L-type $Ca^{2+}$ channel results in significant but not complete inhibition of systolic $Ca^{2+}$. The L-type $Ca^{2+}$ channel-independent $Ca^{2+}$ entry responsible for CICR under these conditions is mediated by $Na^+/Ca^{2+}$ exchanger (NCX) working in the "reverse mode" [100]. NCX is a family of three proteins, NCX1, NCX2, and NCX3. The only isoform present in the mammalian heart is NCX1 [102]. NCX1 has nine transmembrane domains with a large intracellular loop between fifth and sixth transmembrane domains. Two conservative regions in the NCX1 ($\alpha_1$ and $\alpha_2$ regions) are responsible for the binding and transport of $Na^+$ and $Ca^{2+}$. The NCX transports three $Na^+$ and one $Ca^{2+}$ in opposite directions and the exchanger can either remove or bring $Ca^{2+}$ into the cells depending on the conditions. Major factors that determine the activity of NCX and the direction of the flow of ions are the electrochemical gradient of $Na^+$ across the membrane and the membrane potential. During diastole, NCX functions in the "forward mode" and plays the important role of $Ca^{2+}$ extrusion. Recently, a number of proteins were discovered to interact with NCX1. Those include protein kinases (PKA, PKC)

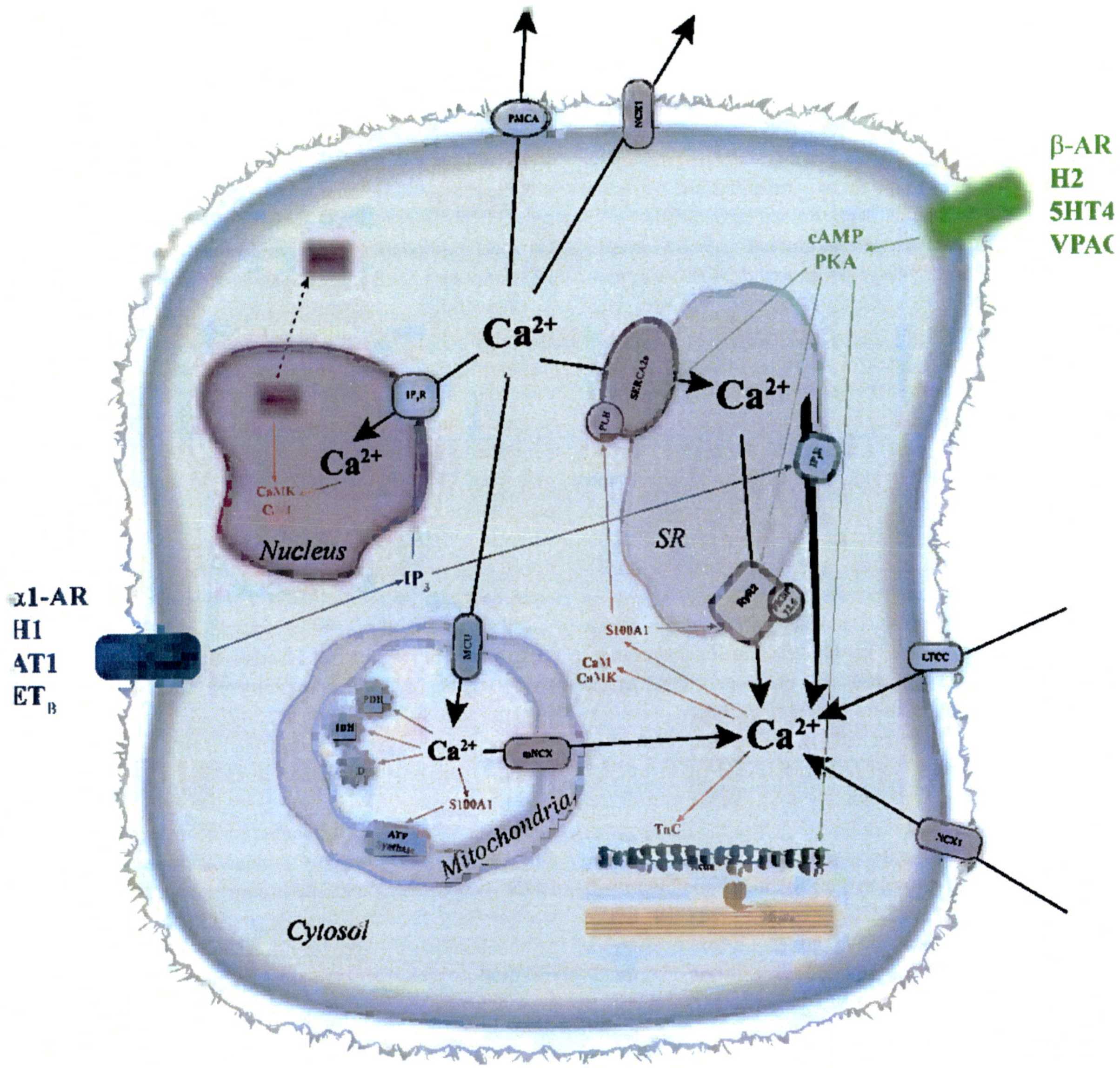


**Fig. 4.3** Major $Ca^{2+}$-related signaling pathways. Major pathways influencing $Ca^{2+}$ inflow/outflow cytosol as well as several intracellular compartments are shown by *black arrows*. Intracellular $Ca^{2+}$ regulates target enzymes and proteins directly or via $Ca^{2+}$-binding proteins (shown in *red*). Many receptor-initiated signaling cascades modulate $Ca^{2+}$-dependent processes via second messenger molecules: $IP_3$ (depicted in *blue*) or cAMP (depicted in *green*; only cAMP raising receptors are listed for simplification). *Abbreviations: SR* sarcoplasmic reticulum; *$\alpha_1$-AR, β AR* adrenoceptor ($\alpha_1$- and β-subtypes, respectively); *H1, H2* histamine receptors (type 1 and type 2, respectively); *5HT4* serotonin receptor, type 4; *VPAC* vasoactive intestinal peptide receptor; *AT1* angiotensin II receptor, type 1; *$ET_B$* endothelin receptor, type B; *NCX1* $Na^+/Ca^{2+}$ exchanger, isoform 1; *mNCX* mitochondrial $Na^+/Ca^{2+}$ exchanger; *LTCC* high-voltage activated L-type $Ca^{2+}$ channel; *PMCA* plasma membrane $Ca^{2+}$ ATPase; *RyR2* ryanodine receptor, subtype 2; *FKBP12.6* FK-506 binding protein; *$IP_3$* inositol 1,4,5-trisphosphate; *$IP_3R$* $IP_3$ receptor; *PLB* phospholamban; *SERCA2a* sarco(endo)plasmic reticulum $Ca^{2+}$ ATPase pump, isoform 2a; *MCU* mitochondrial $Ca^{2+}$ uniporter; *PDH* pyruvate dehydrogenase; *IDH* isocitrate dehydrogenase; *α-KDH* α-ketoglutarate dehydrogenase; *HDAC5* histone deacetylase 5; *PKA* protein kinase A; *CaM* calmodulin; *CaMK* $Ca^{2+}$/CaM-dependent protein kinase; *TnC* troponin C; *S100A1* S100 protein, isoform A1

and phosphoprotein phosphatases (PP1, PP2A, PP2B), phospholemman, sorcin, 14-3-3, ankyrin, caveolin-3, and annexin-5. These proteins can regulate the function of NCX [101]. In addition, there are several other potential triggers of $Ca^{2+}$ entries: stretch activated $Ca^{2+}$ channels, B type channels, store-operated and TRP channels [103–106]. $Ca^{2+}$ entry pathways alternative to L-type $Ca^{2+}$ channel are minor regulators of $Ca^{2+}$ release from SR.

Three subtypes of ryanodine receptor exist: RyR1, RyR2, and RyR3. RyR2 is the major subtype in the cardiac muscle. RyR is the largest ion channel known to exist (more than 5,000 amino acids: ~600 kDa). The complete receptor is a homotetramer. Ryanodine blocks RyR in the open state allowing $Ca^{2+}$ to leak out of SR. This practically empties the SR. RyRs are located in the membrane of the SR. At resting $Ca^{2+}$, RyR2 is inhibited by $Mg^{2+}$. Raised cytoplasmic $Ca^{2+}$ is sufficient to mitigate the $Mg^{2+}$ inhibition of RyR2. Moreover, under resting conditions RyR is inhibited by FK-506 binding protein, FKBP12.6, while RyR2 activity was stimulated by isoproterenol and blocked by propranolol. β-AR activates RyR2 via PKA-dependent phosphorylation: cAMP-activated PKA phosphorylates Ser-2809 on RyR2. Phosphorylation of RyR2 uncouples it from FKBP12.6, and free RyR2 becomes more sensitive to the $Ca^{2+}$. A kinase adaptor protein, mAKAP (aka AKAP100), colocalizes with RyR2 and participates in the interaction of PKA with RyR2. Phosphorylation of RyR2 can be rapidly reversed by phosphatase. There is evidence that RyRs can interact with voltage-gated $Ca^{2+}$ channels.

Cytosolic $Ca^{2+}$ levels, which increase during cardiac muscle contraction return back low levels during relaxation. Cytosolic $Ca^{2+}$ is partly sequestered back into the SR by the sarco(endo)plasmic reticulum $Ca^{2+}$ ATPase pump (SERCA) and partly released from the cell by the $Na^+/Ca^{2+}$ exchanger and plasma membrane $Ca^{2+}$ ATPase (PMCA). In human, 70% of the cytosolic $Ca^{2+}$ reloads the SR. Of the two different isoforms encoded by different genes, cardiomyocytes express SERCA2a. This transmembrane SR enzyme transports two $Ca^{2+}$ from cytosol to the SR lumen upon hydrolysis of one ATP molecule. SERCA2a controls both the rate of $Ca^{2+}$ removal from cytosol and SR $Ca^{2+}$ load. Activity of SERCA2a is negatively regulated by SR transmembrane protein phospholamban. At low $Ca^{2+}$ concentrations, PLB reversibly inhibits the affinity of SERCA2a for $Ca^{2+}$, while elevations in $Ca^{2+}$ concentration lead to uncoupling of PLB from SERCA2a. Dissociation of PLB is the result of $Ca^{2+}$-induced conformational changes of SERCA2a [107]. In addition, β-adrenergic agonists induce PKA-mediated phosphorylation of PLB (see above) which relieves inhibition of SERCA2a.

In addition to $Ca^{2+}$ reuptake to the SR, during relaxation there is $Ca^{2+}$ extrusion from the cell. The main pathway for $Ca^{2+}$ extrusion from myocytes is through NCX1 working in the "forward mode" (see above). Another system for $Ca^{2+}$ expulsion is plasmalemmal $Ca^{2+}$ ATPase. Three isoforms of this enzyme, 1, 2, and 4, are expressed in myocardium but contribute very little to $Ca^{2+}$ removal. Instead, recent evidence suggest that PMCA plays a significant role in signal transduction [108].

Besides RyR, cardiac cells express another type of $Ca^{2+}$ channel involved in the compartmentalization of $Ca^{2+}$ within the cell. This is the inositol 1,4,5-trisphosphate receptor ($IP_3R$) which localizes in the membrane fraction of cardiomyocytes including SR and nucleus. There are three $IP_3R$ isoforms, denoted type 1, type 2, and type 3. Type 1 $IP_3R$ is dominant in human atrial and rat Purkinje myocytes, whereas atrial and ventricular myocytes from most other animal species express predominantly type 2 $IP_3R$ and, to a lesser extent, type 3 $IP_3R$. $IP_3R$ isoforms differ in $IP_3$ affinity, regulation by $Ca^{2+}$, ATP, phosphorylation, and other modulatory factors. $IP_3R$ releases $Ca^{2+}$ from endogenous stores in response to $IP_3$, an intracellular messenger generated by membrane receptors that couple to phospholipase C.

In the SR, $IP_3R$ represents only a small fraction of $Ca^{2+}$ releasing channels (1/100 to 1/50 of RyR) and does not seem to play a major role in raising intracellular $Ca^{2+}$ during cell contraction. Rather, in atrial cardiomyocytes the extra component of $Ca^{2+}$ mobilization from $IP_3Rs$ adds to the action potential-induced RyR-mediated $Ca^{2+}$ release. In other words, $Ca^{2+}$ release through $IP_3R$ underlies the positive inotropic effect of receptors coupled to the $PIP_2$-PLC-$IP_3$ cascade in atrial myocardium [109]. There is evidence that $IP_3Rs$ play a fundamental role in the regulation of $Ca^{2+}$ oscillations in myocytes during the first few days of the developing heart, prior to the maturation of the excitation–contraction coupling system (up to 8–9 days mouse embryos) [110].

Recent studies have demonstrated the expression of $IP_3Rs$ within or around the nucleus in cardiac myocytes. In the nuclear envelope, $IP_3Rs$ face both the cytoplasm and the nucleoplasm [111]. G protein-coupled receptors (e.g., for angiotensin II, endothelin-1, phenylephrine), and growth factor receptors can regulate transcription via nuclear $IP_3R$/nuclear $Ca^{2+}$: they increase $IP_3$ concentration via activation of phospholipase C. $IP_3$ may activate nuclear $IP_3Rs$. Those receptors which face the nucleoplasm will increase the nuclear $Ca^{2+}$ concentration directly. In addition, $IP_3$ may also activate $IP_3Rs$ on the nuclear envelope facing the cytoplasm. This $Ca^{2+}$ may then increase nuclear $Ca^{2+}$ concentration indirectly via cytoplasmic $Ca^{2+}$ diffusing into the nucleus through nuclear pores. Actually, nuclear $Ca^{2+}$ in cardiac cells is not entirely independent from cytoplasmic $Ca^{2+}$ and follows systolic cytoplasmic $Ca^{2+}$ increase with some delay, but $IP_3$ can locally initiate the cascade that involves $Ca^{2+}$ release into the nucleus, activation of nuclear $Ca^{2+}$/calmodulin-dependent protein kinase II, phosphorylation/nuclear export of class II histone deacetylase 5 (HDAC5) and, thereby, derepressing of transcription [112].

The first reports describing $Ca^{2+}$ uptake by heart mitochondria appeared in 1964 and until the late 1970s these organelles were considered an important structure in the control of $Ca^{2+}$ homeostasis [113]. Several major $Ca^{2+}$ influx/efflux pathways have been discovered in mitochondria, namely the mitochondrial $Ca^{2+}$ uniporter (MCU), mitochondrial $Na^+/Ca^{2+}$-exchanger (MNCX), and mitochondrial permeability transition pore (MTP). MCU is the channel that permits $Ca^{2+}$ entry into the matrix of mitochondria whereas MNCX provides a mechanism for $Ca^{2+}$ extrusion. Neither MCU nor MNCX have been purified or cloned. MTP pore is not specific for $Ca^{2+}$ but can act as a $Ca^{2+}$ efflux mechanism. The MTP pore is probably closed under physiological conditions and does not seem to play any significant role in mitochondrial $Ca^{2+}$ movements.

Later detailed studies of mitochondrial $Ca^{2+}$ channels indicated that they are too slow to play a role in intracellular $Ca^{2+}$ signaling during excitation–contraction coupling. Intramitochondrial $Ca^{2+}$ plays an important role in the regulation of energy production and in the modulation of whole-cell $Ca^{2+}$ signaling in the heart.

The role of mitochondria is to translate the cytosolic $Ca^{2+}$ signal into the mitochondria, so that $Ca^{2+}$ could orchestrate mitochondrial oxidative phosphorylation (OXPHOS)/ATP synthesis with cytosolic ATP hydrolysis. Experimental observations support this concept: mitochondria $Ca^{2+}$ transients follow the kinetics of the cytosolic $Ca^{2+}$ transients very closely [114]. Also, there is evidence of the presence of RyR1 in the mitochondrial membrane which can take part in the mitochondrial physiological rapid $Ca^{2+}$ transients [115]. An increase in mitochondrial $Ca^{2+}$ resulting from a rise in cytosolic $Ca^{2+}$, leads within the mitochondria to $Ca^{2+}$-dependent activation of several dehydrogenases, increases in NADH supply and hence ATP synthesis [116]. While pyruvate dehydrogenase is activated by dephosphorylation via $Ca^{2+}$-sensitive phosphatase, isocitrate dehydrogenase and $\alpha$-ketoglutarate dehydrogenase are activated by $Ca^{2+}$ directly [117, 118]. $Ca^{2+}$ is also an activator of mitochondria ATP synthase (F1-FO-ATPase), although the mechanism of the activation of the F1-FO-ATPase by $Ca^{2+}$ is still unknown. Thus, there are data on $Ca^{2+}$-sensitive phosphorylation of F1-FO-ATPase $\gamma$-subunit, $Ca^{2+}$-induced de-repression of F1-FO-ATPase by protein inhibitor, and $Ca^{2+}$-dependent activation of F1-FO-ATPase by S100A1 protein [119–121]. The balanced activation of OXPHOS by $Ca^{2+}$ is one of the mechanisms to balance the rate of ATP production with the rate of ATP utilization in the working myocardium.

Mitochondria may modulate whole cardiac cell $Ca^{2+}$ oscillations during excitation–contraction coupling. Mitochondrial $Ca^{2+}$ transport pathways are in close contact with the cardiomyocyte SR or sarcolemma [122, 123]. $Ca^{2+}$ can enter mitochondria through MCU directly from the SR (via RyR2) or sarcolemma (via L-type $Ca^{2+}$ channels), and $Ca^{2+}$ efflux from mNCX contributes to SR refilling during relaxation.

As previously mentioned, $Ca^{2+}$ plays a central role in the physiology of cardiac muscle. $Ca^{2+}$ function as a second messenger occurs through a number of $Ca^{2+}$-sensor proteins which contain the specific $Ca^{2+}$ binding sites, and some of them are involved in the regulation of cardiac muscle contraction: troponin C, essential myosin light chain, and regulatory myosin light chain [124, 125]. Another $Ca^{2+}$-binding protein, calmodulin, is ubiquitously distributed in all tissues including the myocardium. Calmodulin interacts with more than 100 different proteins and enzymes and regulates numerous cell processes, including, glycogen metabolism, intracellular motility, $Ca^{2+}$ transport, cyclic nucleotide metabolism, protein phosphorylation/dephosphorylation, cell cycle, and gene expression. The largest subgroup within the superfamily of $Ca^{2+}$ binding proteins are the S100 proteins (more than 20 S100 proteins are known in human). The most abundant in human heart is S100A1 isoform, but S100A4, S100A6, and S100B proteins are also found in cardiomyocytes in much lower amounts [126]. S100A1 has been shown to interact in a $Ca^{2+}$-dependent manner with RyR2, SERCA2a, phospholamban, titin, and the mitochondrial ATP synthase. The main functions of this protein in cardiac tissue include positive inotropy and lusitropy as well as $Ca^{2+}$-dependent regulation of mitochondrial energy production [127].

## Conclusions

There are several cardiac receptor systems that regulate myocardial functional activity (i.e., force of contractility, heart rate). A number of cardiac receptors transduce and amplify signals from neurotransmitters released from sympathetic and parasympathetic neurons innervating the myocardium (including, norepinephrine, acetylcholine, histamine, serotonin, neuropeptide Y, vasoactive intestinal peptide, and somatostatin), respond to circulating hormones (epinephrine, histamine, angiotensin II) and also to paracrinal factors (neuropeptide Y, adenosine, angiotensin II, endothelin, urocortin, natriuretic peptides). In addition to direct regulation of cardiac cell, some signaling molecules (norepinephrine, histamine, serotonin, substance P, neurotensin, somatostatin, adenosine, angiotensin II, atrial natriuretic peptide) function as indirect modulators via release/secretion of other neurotransmitters or paracrine/autocrine factors.

Many of cardiac receptors are transmembrane cell-surface molecules which belong to the superfamily of G protein-coupled receptors: they contain seven membrane-spanning domains and link to heterotrimeric guanine nucleotide-binding proteins. Among these, there are receptors coupled to the $G_s$ protein-adenylate cyclase pathway ($\beta$-adrenoceptors, histamine receptors, serotonin receptors), receptors coupled to the $G_{i/o}$ protein-adenylate cyclase pathway (muscarinic receptors, adenosine receptors), and receptors that couple to the $G_{q/11}$ protein-phospholipase C/protein kinase C pathway ($\alpha_1$-adrenoceptors, endothelin receptors, angiotensin II receptors).

In humans the most powerful mechanism to increase heart rate and contractility is the $\beta$-adrenoceptor/$G_s$ protein/adenylate cyclase pathway. Several other receptor systems can also mediate positive inotropy through accumulation of intracellular cAMP (histamine, serotonin 5-HT4, vasoactive intestinal peptide, adenosine A2a, urocortin receptors) or through the phospholipase C/DAG/$IP_3$ pathway ($\alpha_1$-adrenoceptors, serotonin 5-HT2, somatostatin SSTR3, angiotensin II, endothelin-1 $ET_B$ receptors) but positive effects of these receptor systems are submaximal compared to β-ARs. In addition, several receptor systems act through inhibition of cAMP synthesis and cause negative inotropic effects in the heart (muscarinic m2, serotonin 5-HT1, neuropeptide Y, somatostatin, adenosine A1, endothelin-1 $ET_A$ receptors).

A critical signaling molecule that triggers and regulates contraction is $Ca^{2+}$. Intracellular $Ca^{2+}$ concentration is the result of a coordinated functioning of several systems that move $Ca^{2+}$ between separate intracellular compartments, enter $Ca^{2+}$ from the extracellular environment, and extrude $Ca^{2+}$ from the cell.

## Summary


- Contractile activity of myocardium is regulated by sympathetic and parasympathetic nervous systems. Neurohormones that modulate the functioning of the heart belong to two major categories, biogenic amines and neuropeptides.
- A very potent biogenic amine synthesized and released by sympathetic neurons is the catecholamine, norepinephrine. Another catecholamine, epinephrine, is a stress hormone released from the adrenal medulla.
- Catecholamines increase heart rate (positive chronotropic effect), force of cardiac contraction (positive inotropic effect), rate of cardiac relaxation (positive lusitropic effect), and automaticity.
- Catecholamines evoke their effects through stimulation of β- ($\beta_1$-, $\beta_2$-) and $\alpha_1$-adrenoceptors expressed in cardiomyocytes. Stimulation of β-adrenoceptor initiates a signaling cascade involving $G_s$ protein-dependent activation of adenylate cyclase, synthesis of cyclic adenosine monophosphate (cAMP), and activation of cAMP-dependent protein kinase A (PKA). PKA-dependent phosphorylation of L-type $Ca^{2+}$ channels, phospholamban, and contractile proteins leads to a functional response. $\alpha_1$-adrenoceptors also mediate catecholamine signaling in the cardiac tissue, but they are coupled to $G_{q/11}$ protein and activate phospholipase C-dependent synthesis of two second messangers, inositol trisphosphate ($IP_3$), and diacylglycerol (DAG). $IP_3$/DAG-induced mobilization of $Ca^{2+}$ and phosphorylation of multiple targets leads to the modulation of contraction and ventricular hypertrophy.
- The parasympathetic nervous system modulates and buffers the sympathetic nervous system. Primary biogenic amine released from the parasympathetic neurons is acetylcholine. Acetylcholine interacts with specific muscarinic m2 receptors expressed in the myocardium.
- Stimulation of m2 receptors leads to bradycardia (in atria) and a negative inotropic response (in atria and ventricles). Mechanism which underlies inhibitory actions of acetylcholine includes m2-dependent $G_{i/o}$ protein-dependent inhibition of adenylate cyclase and/or activation of cAMP phosphodiesterase 2, decrease of cAMP level and resulting deactivation of ion channels participating in cardiomyocyte contraction (L-type $Ca^{2+}$ channels, hyperpolarization-activated cyclic nucleotide-gated channels). In addition, m2 receptor uniquely activates atrial G protein-regulated $K^+$ channels (GIRK) via βγ subunit of $G_{i/o}$ protein. This leads to the increase of $I_k$, and hyperpolarization, slows beating rate of atrial cell, and reduces the force of contraction.
- Histamine and serotonin are two biogenic amines which have multiple direct and indirect effects on the heart. The cardiac distribution of two subtypes of histamine receptors, H1 and H2 varies between species.
- Whereas H2-inducible signaling pathway in cardiac cell is associated with $G_s$ protein and AC, H1 receptor mobilizes $Ca^{2+}$ via $G_{q/11}$-mediated activation of phospholipase C.
- Serotonin receptor 5-HT4 is expressed in atrial myocytes and is linked to $G_s$ protein/AC signaling pathway. Besides direct actions as neurotransmitter/hormone/paracrine factor, each histamine and serotonin have indirect effects on the myocardium: they can act presynaptically and influence secretion of neurotransmitters from neurons innervating heart tissue.
- Several peptidic transmitters, neuropeptides, are present in myocardium and have direct or indirect effect on cardiac contractility. Each of them functions via specific receptors coupled to regulatory heterotrimeric G proteins/effectors. For example, neuropeptide Y receptors are coupled to the heterotrimeric $G_{i/o}$ protein and inhibit AC activity; vasoactive intestinal peptide receptors promote $G_s$ protein-mediated activation of AC.
- Interestingly, different subtypes of somatostatin receptors (SSTR) are linked to different signaling pathways: while SSTR4 negatively modulates β-adrenergic stimulation of rat cardiomyocytes via $G_{i/o}$ protein, SSTR3 in the same cell activates basal rate of contraction via $G_{q/11}$ protein/phospholipase C mechanism.
- Some neuropeptides do not seem to have direct effects on heart rate, but rather play a neuromodulatory role (substance P, calcitonin gene-related peptide, neurotensin).
- Adenosine under basal conditions does not play a significant role as a regulator of cardiac function because normal interstitial levels of adenosine are below the value necessary for regulation. Under pathological conditions,

the concentration of circulating adenosine in the heart increases (as a result of catabolism of extracellular nucleotides or by it is released from within the cells) and it can bind to specific receptors expressed in cardiomyocytes. One of them, A1, is coupled to $G_{i/o}$ protein, influences several effector systems (GIRK, AC) and is responsible for the negative chronotropic, dromotropic (in atrial pacemaker cells), and inotropic (atrial and ventricular tissues) effects of adenosine.

- Cardiomyocytes have also been shown to express adenosine receptor A2a coupled to $G_s$/AC/cAMP pathway. The significance of A2a receptor under physiological conditions remains unclear because dominating "inhibitory" A1 receptor masks the signaling cascades regulated by A2a.
- The contractile activity of the heart can be modulated by circulating peptide hormones (angiotensin II) or peptides released locally on demand (endothelin-1, urocortins, natriuretic peptides, angiotensin II).
- Angiotensin II and endothelin-1 are positive chronotropic/inotropic agents that act via specific receptors stimulating phospholipase C/DAG/$IP_3$ pathway (although at high concentrations, an inhibitory effect of endothelin-1 can be observed).
- Urocortins also increase contractility and cardiac output but via the receptor activating $G_s$ protein/AC/cAMP cascade. In addition, urocortins have protective/hypertrophic effects on cardiomyocytes via activation of PI3K/Akt. Some peptide hormones function as neuromodulators (angiotensin II, atrial natriuretic peptide).
- Intracellular $Ca^{2+}$ is responsible for initiating the contraction and relaxation of cardiac myocytes. $Ca^{2+}$ also is involved in numerous processes, including cell metabolism, trafficking, growth, and death.
- During the action potential the majority of $Ca^{2+}$ releases from the sarcoplasmic reticulum (SR) in response to triggering amounts of external $Ca^{2+}$ entering the cell through L-type $Ca^{2+}$ channel. Some other sarcolemmal ion channels also contribute to the entering of "trigger $Ca^{2+}$." Major systolic $Ca^{2+}$ leaks out of SR through a highly regulated channel, the ryanodine receptor.
- During relaxation cytosolic $Ca^{2+}$ levels return back to low levels: $Ca^{2+}$ ATPase pump (SERCA) sequesters part of $Ca^{2+}$ back into the SR, whereas sarcolemmal $Na^+/Ca^{2+}$ exchanger extrudes part of $Ca^{2+}$ from the cell.
- Several cardiomyocyte organelles contain systems for $Ca^{2+}$ influx/efflux. The nuclear envelope has $IP_3$-activated $Ca^{2+}$ channel, so $IP_3$-releasing agonists can initiate $Ca^{2+}$ release into the nucleus, followed by $Ca^{2+}$-dependent modulation of gene transcription.
- In mitochondria, the $Ca^{2+}$ uniporter, $Na^+/Ca^{2+}$ exchanger, and ryanodine receptor synchronize oscillations of cytosolic $Ca^{2+}$ with those within the organelle. As a result, $Ca^{2+}$ via regulation of several mitochondrial dehydrogenases and ATP synthase balances the rate of ATP production with the rate of ATP utilization in the working myocardium.

## References


1. Elliott TR. On the action of adrenalin. J Physiol (London). 1904;31:20–1.
2. von Euler US. A specific sympathomimetic ergone in adrenergic nerve fibres (sympathin) and its relation to adrenaline and noradrenaline. Acta Physiol Scand. 1946;12:73–97.
3. Vago T, Bevilacqua M, Dagani R, Meroni R, Frigeni G, Santoliss C, et al. Comparison of rat and human left ventricle beta-adrenergic receptors: subtype heterogeneity delineated by direct radioligand binding. Biochem Biophys Res Commun. 1984;121:346–54.
4. Hedberg A, Minneman KP, Molinoff PB. Differential distribution of beta-1 and beta-2 adrenergic receptors in cat and guinea-pig heart. J Pharmacol Exp Ther. 1980;212:503–8.
5. Gauthier C, Tavernier G, Charpentier F, Langin D, Le Marec H. Functional beta 3-adrenoceptor in the human heart. J Clin Invest. 1996;98:556–62.
6. Ping P, Hammond HK. Diverse G protein and beta-adrenergic receptor mRNA expression in normal and failing porcine hearts. Am J Physiol. 1994;267:H2079–85.
7. Bristow MR, Ginsburg R, Umans V, Fowler M, Minobe W, Rasmussen R, et al. Beta 1- and beta 2-adrenergic-receptor subpopulations in nonfailing and failing human ventricular myocardium: coupling of both receptor subtypes to muscle contraction and selective beta 1-receptor down-regulation in heart failure. Circ Res. 1986;59:297–309.
8. Motomura S, Reinhard-Zerkowski H, Daul A, Brodde OE. On the physiologic role of beta-2 adrenoceptors in the human heart: in vitro and in vivo studies. Am Heart J. 1990;119:608–19.
9. Bean BP. Two kinds of calcium channels in canine atrial cells: differences in kinetics, selectivity, and pharmacology. J Gen Physiol. 1985;86:1–30.
10. Hagiwara N, Irisawa H, Kameyama M. Contribution of two types of calcium currents to the pacemaker potentials of rabbit sine-atrial node cells. J Physiol. 1988;395:233–53.
11. Hartzell HC. Regulation of cardiac ion channels by catecholamines, acetylcholine and second messenger systems. Prog Biophys Mol Biol. 1988;52:165–247.
12. Shimoni Y. Hormonal control of cardiac ion channels and transporters. Prog Biophys Mol Biol. 1999;72:67–108.
13. De Jongh KS, Murphy BJ, Colvin AA, Hell JW, Takahashi M, Catterall WA. Specific phosphorylation of a site in the full-length form of the alpha 1 subunit of the cardiac L-type calcium channel by adenosine 3',5'-cyclic monophosphate dependent protein kinase. Biochemistry. 1996;35:10392–402.
14. McDonald TF, Pelzer S, Trautwein W, Pelzer DJ. Regulation and modulation of calcium channels in cardiac, skeletal, and smooth muscle cells. Physiol Rev. 1994;74:365–507.
15. Gerhardstein BL, Puri TS, Chien AJ, Hosey MM. Identification of the sites phosphorylated by cyclic AMP-dependent protein kinase on the $\beta_2$ subunit of L-type voltage-dependent calcium channels. Biochemistry. 1999;38:10361–70.
16. Yatani A, Codina J, Imoto Y, Reeves JP, Birnbaumer L, Brown AM. A G protein directly regulates mammalian cardiac calcium channels. Science. 1987;238:1288–92.
17. Luo W, Chu G, Sato Y, Zhou Z, Kadambi VJ, Kranias EG. Transgenic approaches to define the functional role of dual site phospholamban phosphorylation. J Biol Chem. 1998;273:4734–9.

18. Takimoto E, Soergel DG, Janssen PM, Stull LB, Kass DA, Murphy AM. Frequency- and afterload-dependent cardiac modulation in vivo by troponin I with constitutively active protein kinase A phosphorylation sites. Circ Res. 2004;94:496–504.
19. Yasuda S, Coutu P, Sadayappan S, Robbins J, Metzger JM. Cardiac transgenic and gene transfer strategies converge to support an important role for troponin I in regulating relaxation in cardiac myocytes. Circ Res. 2007;101:377–86.
20. Solaro RJ, Rosevear P, Kobayashi T. The unique functions of cardiac troponin I in the control of cardiac muscle contraction and relaxation. Biochem Biophys Res Commun. 2008;369:82–7.
21. Christ T, Galindo-Tovar A, Thoms M, Ravens U, Kaumann AJ. Inotropy and L-type $Ca^{2+}$ current, activated by $\beta_1$- and $\beta_2$-adrenoceptors, are differently controlled by phosphodiesterases 3 and 4 in rat heart. Br J Pharmacol. 2009;156:62–83.
22. Pitcher JA, Touhara K, Payne ES, Lefkowitz RJ. Pleckstrin homology domain-mediated membrane association and activation of the beta-adrenergic receptor kinase requires coordinate interaction with G beta gamma subunits and lipid. J Biol Chem. 1995;270:11707–10.
23. Nantel F, Bonin H, Emorine LJ, Zilberfarb V, Strosberg AD, Bouvier M, et al. The human beta 3-adrenergic receptor is resistant to short term agonist-promoted desensitization. Mol Pharmacol. 1993;43:548–55.
24. Goodman Jr OB, Krupnick JG, Santini F, Gurevich VV, Penn RB, Gagnon AW, et al. Beta-arrestin acts as a clathrin adaptor in endocytosis of the beta2-adrenergic receptor. Nature. 1996;383:447–50.
25. Zhang J, Barak LS, Winkler KE, Caron MG, Ferguson SS. A central role for beta-arrestins and clathrin-coated vesicle-mediated endocytosis in beta2-adrenergic receptor resensitization. Differential regulation of receptor resensitization in two distinct cell types. J Biol Chem. 1997;272:27005–14.
26. Gauthier C, Langin D, Balligand JL. $\beta_3$-Adrenoceptors in the cardiovascular system. Trends Pharmacol Sci. 2000;21:426–31.
27. Endoh M. Cardiac alpha(1)-adrenoceptors that regulate contractile function: subtypes and subcellular signal transduction mechanisms. Neurochem Res. 1996;21:217–29.
28. Ruffolo RR, Hollinger MA, editors. G-Protein coupled transmembrane signaling mechanisms. Boca Raton: CRC Press; 1995. p. 1–34.
29. Perrella MA, Maki T, Prasad S, Pimental D, Singh K, Takahashi N, et al. Regulation of heparin-binding epidermal growth factor-like growth factor mRNA levels by hypertrophic stimuli in neonatal and adult rat cardiac myocytes. J Biol Chem. 1994;269:27045–50.
30. Rump LC, Riera-Knorrenschild G, Schwertfeger E, Bohmann C, Spillner G, Schollmeyer P. Dopaminergic and α-adrenergic control of neurotransmission in human right atrium. J Cardiovasc Pharmacol. 1995;26:462–70.
31. Wang Z, Shi H, Wang H. Functional $M_3$ muscarinic acetylcholine receptors in mammalian hearts. Br J Pharmacol. 2004;142:395–408.
32. Brodde OE, Broede A, Daul A, Kunde K, Michel MC. Receptor systems in the non-failing human heart. Basic Res Cardiol. 1992;87 Suppl 1:1–14.
33. Reuter H. Calcium channel modulation by neurotransmitters, enzymes and drugs. Nature. 1983;301:569–74.
34. Han X, Kubota I, Feron O, Opel DJ, Arstall MA, Zhao Y-Y, et al. Muscarinic cholinergic regulation of cardiac myocyte $I_{Ca-L}$ is absent in mice with targeted disruption of endothelial nitric oxide synthase (eNOS). Proc Natl Acad Sci USA. 1998;95:6510–5.
35. Feron O, Zhao Y-Y, Kelly RA. The ins and outs of caveolar signaling. m2 muscarinic cholinergic receptors and eNOS activation versus neuregulin and ErbB4 signaling in cardiac myocytes. Ann N Y Acad Sci. 1999;874:11–9.
36. Hazeki O, Ui M. Modification by islet-activating protein of receptor-mediated regulation of cyclic AMP accumulation in isolated rat heart cells. J Biol Chem. 1981;256:2856–62.
37. Hescheler J, Kameyama M, Trautwein W. On the mechanism of muscarinic inhibition of the cardiac Ca current. Pflugers Arch. 1986;407:182–9.
38. Warrier S, Belevych AE, Ruse M, Eckert RL, Zaccolo M, Pozzan T, et al. β-Adrenergic- and muscarinic receptor-induced changes in cAMP activity in adult cardiac myocytes detected with FRET-based biosensor. Am J Physiol Cell Physiol. 2005;289:C455–61.
39. Feron O, Smith TW, Michel T, Kelly RA. Dynamic targeting of the agonist-stimulated m2 muscarinic acetylcholine receptor to caveolae in cardiac myocytes. J Biol Chem. 1997;272:17744–8.
40. Herzig S, Meier A, Pfeiffer M, Neumann J. Stimulation of protein phosphatases as a mechanism of the muscarinic-receptor-mediated inhibition of cardiac L-type $Ca^{2+}$ channels. Pflugers Arch. 1995;429:531–8.
41. Shen JB, Pappano AJ. On the role of phosphatase in regulation of cardiac L-type calcium current by cyclic GMP. J Pharmacol Exp Ther. 2002;301:501–6.
42. Shi W, Wymore R, Yu H, Wu J, Wymore RT, Pan Z, et al. Distribution and prevalence of hyperpolarization activated cation channel (HCN) mRNA expression in cardiac tissues. Circ Res. 1999;85:e1–6.
43. Ludwig A, Budde T, Stieber J, Moosmang S, Wahl C, Holthoff K, et al. Absence epilepsy and sinus dysrhythmia in mice lacking the pacemaker channel HCN2. EMBO J. 2003;22:216–24.
44. DiFrancesco D, Tortora P. Direct activation of cardiac pacemaker channels by intracellular cyclic AMP. Nature. 1991;351:145–7.
45. DiFrancesco D, Tromba C. Muscarinic control of the hyperpolarization-activated current ($i_f$) in rabbit sino-atrial node myocytes. J Physiol. 1988;405:493–510.
46. Hutter OF, Trautwein W. Vagal and sympathetic effects on the pacemaker fibers in the sinus venosus of the heart. J Gen Physiol. 1955;39:715–33.
47. Yamada M, Inanobe A, Kurachi Y. G Protein regulation of potassium ion channels. Pharmacol Rev. 1998;50:723–57.
48. Krapivinsky G, Gordon EA, Wickman K, Velimirovic B, Krapivinsky L, Clapham DE. The G-protein-gated atrial $K^+$ channel $I_{KACh}$ is a heteromultimer of two inwardly rectifying $K^+$ channel protein. Nature. 1995;374:135–41.
49. Tucker SJ, Pessia M, Adelman JP. Muscarine-gated $K^+$ channel: subunit stoichiometry and structural domains essential for G protein stimulation. Am J Physiol. 1996;271:H379–85.
50. Reuveny E, Slesinger PA, Inglese J, Molales JM, Iniguez-Lluhi JA, Lefkowitz RJ, et al. Activation of the cloned muscarinic potassium channel by G protein beta gamma subunits. Nature. 1994;370:143–6.
51. Slesinger PA, Reuveny E, Jan YN, Jan LY. Identification of structural elements involved in G protein gating of the GIRK1 potassium channel. Neuron. 1995;15:1145–56.
52. Dascal N, Doupnik CA, Ivanina T, Bausch S, Wang W, Lin C, et al. Inhibition of function in *Xenopus* oocytes of the inwardly rectifying G-protein-activated atrial K channel (GIRK1) by overexpression of a membrane-attached form of the C-terminal tail. Proc Natl Acad Sci USA. 1995;92:6758–62.
53. Huang C-L, Feng S, Hilgemann DW. Direct activation of inward rectifier potassium channels by $PIP_2$ and its stabilization by $G_{\beta\gamma}$. Nature. 1998;391:803–6.
54. Wickman KD, Iniguez-Lluhi JA, Davenport PA, Taussig R, Krapivinsky GB, Linder ME, et al. Recombinant $G_{\beta\gamma}$ activates the muscarinic-gated atrial potassium channel $I_{KACh}$. Nature. 1994;368:255–7.
55. Doupnik CA, Davidson N, Lester HA, Kofuji P. RGS proteins reconstitute the rapid gating kinetics of gbetagamma-activated inwardly rectifying K1 channels. Proc Natl Acad Sci USA. 1997;94:10461–6.
56. Verma SC, McNeill JH. Cardiac histamine receptors: differences between left and right atria and right ventricle. J Pharmacol Exp Ther. 1977;200:352–62.

57. Arrang JM, Garbarg M, Schwartz JC. Auto-inhibition of brain histamine release mediated by a novel class ($H_3$) of histamine receptor. Nature. 1983;302:832–7.
58. Olsson RA, Pearson JD. Cardiovascular purinoceptors. Physiol Rev. 1990;70:761–845.
59. Villalon CM, Centurion D. Cardiovascular responses produced by 5-hydroxytriptamine: a pharmacological update on the receptors/mechanisms involved and therapeutic implications. Naunyn Schmiedebergs Arch Pharmacol. 2007;376:45–63.
60. Yusuf S, Al-Saady N, Camm AJ. 5-Hydroxytryptamine and atrial fibrillation: how significant is this piece in the puzzle? J Cardiovasc Electrophysiol. 2003;14:209–14.
61. Ramage AG. Central cardiovascular regulation and 5-hydroxytryptamine receptors. Brain Res Bull. 2001;56:425–39.
62. Brattelid T, Qvigstad E, Lynham JA, Molenaar P, Aass H, Geiran O, et al. Functional serotonin 5-HT4 receptors in porcine and human ventricular myocardium with increased 5-HT4 mRNA in heart failure. Naunyn Schmiedebergs Arch Pharmacol. 2004;370:157–66.
63. Qvigstad E, Brattelid T, Sjaastad I, Andressen KW, Krobert KA, Birkeland JA, et al. Appearance of a ventricular 5-HT4 receptor-mediated inotropic response to serotonin in heart failure. Cardiovasc Res. 2005;65:869–78.
64. Saxena PR, Villalon CM. Cardiovascular effects of serotonin agonists and antagonists. J Cardiovasc Pharmacol. 1990;15 Suppl 7:S17–34.
65. Weihe E, Reinecke M. Peptidergic innervation of the mammalian sinus nodes: vasoactive intestinal polypeptide, neurotensin, substance P. Neurosci Lett. 1981;26:283–8.
66. Hokfelt T, Elfvin LG, Elde R, Schultzberg M, Goldstein M, Luft R. Occurrence of somatostatin-like immunoreactivity in some peripheral sympathetic noradrenergic neurons. Proc Natl Acad Sci USA. 1977;74:3587–91.
67. Jacques D, Sader S, Perreault C, Fournier A, Pelletier G, Beck-Sickinger AG, et al. Presence of neuropeptide Y and the Y1 receptor in the plasma membrane and nuclear envelope of human endocardial endothelial cells: modulation of intracellular calcium. Can J Physiol Pharmacol. 2003;81:288–300.
68. Jacques D, Abdel-Samad D. Neuropeptide Y (NPY) and NPY receptors in the cardiovascular system: implication in the regulation of intracellular calcium. Can J Physiol Pharmacol. 2007;85:43–53.
69. Balasubramaniam A, Grupp I, Matlib MA, Benza R, Jackson RL, Fischer JE, et al. Comparison of the effects of neuropeptide Y (NPY) and 4-norleucine-NPY on isolated perfused rat hearts; effects of NPY on atrial and ventricular strips of rat heart and on rabbit heart mitochondria. Regul Pept. 1988;21:289–99.
70. Wei Y, Mojsov S. Tissue specific expression of different human receptor types for pituitary adenylate cyclase activating polypeptide and vasoactive intestinal polypeptide: implications for their role in human physiology. J Neuroendocrinol. 1996;8:811–7.
71. Chatelain P, Robberecht P, Waelbroeck M, De Neef P, Camus JC, Huu AN, et al. Topographical distribution of the secretin- and VIP-stimulated adenylate cyclase system in the heart of five animal species. Pflugers Arch. 1983;397:100–5.
72. Bell D, McDermott BJ. Secretin and vasoactive intestinal peptide are potent stimulants of cellular contraction and accumulation of cyclic AMP in rat ventricular cardiomyocytes. J Cardiovasc Pharmacol. 1994;23:959–69.
73. Accili EA, Redaelli G, DiFrancesco D. Activation of the hyperpolarization-activated current ($i_f$) in sino-atrial node myocytes of the rabbit by vasoactive intestinal peptide. Pflugers Arch. 1996;431:803–5.
74. Pelaprat D. Interactions between neurotensin receptors and G proteins. Peptides. 2006;27:2476–87.
75. Osadchii O, Norton G, Deftereos D, Muller D, Woodiwiss A. Impact of chronic β-adrenoceptor activation on neurotensin-induced myocardial effects in rats. Eur J Pharmacol. 2006;553:246–53.
76. Siehler S, Hoyer D. Characterization of human recombinant somatostatin receptors: modulation of adenylate cyclase activity. Naunyn Schmiedebergs Arch Pharmacol. 1999;360:510–21.
77. Tallent M, Liapakis G, O'Carroll AM, Lolait SJ, Dichter M, Reisine T. Somatostatin receptor subtypes SSTR2 and SSTR5 couple negatively to an L type calcium channel current in the pituitary cell line AtT-20. Neuroscience. 1996;71:1073–81.
78. Kreienkamp HJ, Honck HH, Richter D. Coupling of rat somatostatin receptor subtypes to a G-protein gated inwardly rectifying potassium channel. FEBS Lett. 1997;419:92–4.
79. Akbar M, Okajima F, Tomura H, Majid MA, Yamada Y, Seino S, et al. Phospholipase C activation and calcium mobilization by cloned human somatostatin receptor subtypes 1–5 in transfected cos-7 cells. FEBS Lett. 1994;348:192–6.
80. Bell D, Zhao Y, McMaster B, McHenry EM, Wang X, Kelso EJ, et al. SRIF receptor subtype expression and involvement in positive and negative contractile effects of somatostatin-14 (SRIF-14) in ventricular cardiomyocytes. Cell Physiol Biochem. 2008;22:653–64.
81. Schwabe W, Brennan MB, Hochgeschwender U. Isolation and characterization of the mouse (*Mus musculus*) somatostatin receptor type 4-encoding gene (mSSTR4). Gene. 1996;168:233–5.
82. Smith WHT, Nair RU, Adamson D, Kearney MT, Ball SG, Balmforth AJ. Somatostatin receptor subtype expression in the human heart: differential expression by myocytes and fibroblasts. J Endocrinol. 2005;187:379–86.
83. Murray F, Bell D, Kelso EJ, Millar BC, McDermott BJ. Positive and negative contractile effects of somatostatin-14 on rat ventricular cardiomyocytes. J Cardiovasc Pharmacol. 2001;37:324–32.
84. Boehm S, Huck S. Receptors controlling transmitter release from sympathetic neurons in vitro. Prog Neurobiol. 1997;51:225–42.
85. Endoh M, Murayama M, Taira N. Modification by islet-activating protein of direct and indirect inhibitory actions of adenosine on rat atrial contraction in relation to cyclic nucleotide metabolism. J Cardiovasc Pharmacol. 1983;5:131–42.
86. Leung E, Johnston CI, Woodcock EA. A comparison between the adenosine receptors mediating adenylate cyclase inhibition and cardiac depression in the guinea pig heart. J Cardiovasc Pharmacol. 1986;8:1003–8.
87. Chiba S, Himori N. Different inotropic responses to adenosine on the atrial and ventricular muscle of the dog heart. Jpn J Pharmacol. 1975;25:489–91.
88. Kubalak SW, Newman W, Webb JG. Differential effect of pertussis toxin on adenosine and muscarinic inhibition of cyclic AMP accumulation in canine ventricular myocytes. J Mol Cell Cardiol. 1991;23:199–205.
89. Trautwein W, Cavalie A, Flockerzei V, Hofmann F, Pelzer D. Modulation of calcium channel function by phosphorylation in guinea pig ventricular cells and phospholipids bilayer membranes. Circ Res. 1987;61(Suppl I):I-17–23.
90. Walsh KB, Kass RS. Regulation of a heart potassium channel by protein kinase A and C. Science. 1988;242:67–9.
91. Martynyuk AG, Kane KA, Cobbe SM, Rankin AC. Nitric oxide mediates the anti-adrenergic effect of adenosine on calcium current in isolated rabbit atrioventricular nodal cells. Pflugers Arch. 1996;431:452–7.
92. Hove-Madsen L, Prat-Vidal C, Llach A, Ciruela F, Casado V, Lluis C, et al. Adenosine $A_{2A}$ receptors are expressed in human atrial myocytes and modulate spontaneous sarcoplasmic reticulum calcium release. Cardiovasc Res. 2006;72:292–302.
93. Xu H, Stein B, Liang B. Characterization of a stimulatory adenosine A2a receptor in adult rat ventricular myocyte. Am J Physiol. 1996;270:H16550–61.
94. Richardt G, Waas W, Kranzhofer R, Mayer E, Schomig A. Adenosine inhibits exocytotic release of endogenous noradrenaline in rat heart: a protective mechanism in early myocardial ischemia. Circ Res. 1987;61:117–23.

95. Thomas WG, Thekkumkara TJ, Baker KM. Cardiac effects of AIIAT1A receptor signaling, desensitization, and internalization. Adv Exp Med Biol. 1996;396:59–69.
96. Nio Y, Matsubara H, Murasawa S, Kanasaki M, Inada M. Regulation of gene transcription of angiotensin II receptor subtypes in myocardial infarction. J Clin Invest. 1995;95:46–54.
97. Ono K, Eto K, Sakamoto A, Masaki T, Shibata K, Sada T, et al. Negative chronotropic effect of endothelin 1 mediated through $ET_A$ receptors in guinea pig atria. Circ Res. 1995;76:284–92.
98. Humbert M, Simonneau G. Drug insight: endothelin-receptor antagonists for pulmonary arterial hypertension in systemic rheumatic diseases. Nat Clin Pract Rheumatol. 2005;1:93–101.
99. McGoon MD, Frost AE, Oudiz RJ, Badesch DB, Galie N, Olschewski H, et al. Ambrisentan therapy in patients with pulmonary arterial hypertension who discontinued bosentan or sitaxsentan due to liver function test abnormalities. Chest. 2009;135:122–9.
100. Ng LL, Loke IW, O'Brien RJ, Squire IB, Davies JE. Plasma urocortin in human systolic heart failure. Clin Sci (Lond). 2004;106:383–8.
101. Dibb KM, Graham HK, Venetucci LA, Eisner DA, Trafford AW. Analysis of cellular calcium fluxes in cardiac muscle to understand calcium homeostasis in the heart. Cell Calcium. 2007;42:503–12.
102. Quednau BD, Nicoll DA, Philipson KD. Tissue specificity and alternative splicing of the $Na^+/Ca^{2+}$ exchanger isoforms NCX1, NCX2, and NCX3 in rat. Am J Physiol. 1997;272:C1250–61.
103. Sasaki N, Mitsuiye T, Noma A. Effects of mechanical stretch on membrane currents of single ventricular myocytes of guinea-pig heart. Jpn J Physiol. 1992;42:957–70.
104. Coulombe A, Lefever IA, Baro I, Coraboeuf E. Barium- and calcium-permeable channels open at negative membrane potentials in rat ventricular myocytes. J Membr Biol. 1989;111:57–67.
105. Uehara A, Yasukochi M, Imanaga I, Nishi M, Takeshima H. Store-operated $Ca^{2+}$ entry uncoupled with ryanodine receptor and junctional membrane complex in heart muscle cells. Cell Calcium. 2002;31:89–96.
106. Satoh S, Tanaka H, Ueda Y, Oyama J, Sugano M, Sumimoto H, et al. Transient receptor potential (TRP) protein 7 acts as a G protein-activated $Ca^{2+}$ channel mediating angiotensin II-induced myocardial apoptosis. Mol Cell Biochem. 2007;294:205–15.
107. Asahi M, Nakayama H, Tada M, Otsu K. Regulation of sarco(endo) plasmic reticulum $Ca^{2+}$ adenosine triphosphatase by phospholamban and sarcolipin: implication for cardiac hypertrophy and failure. Trends Cardiovasc Med. 2003;13:152–7.
108. Cartwright EJ, Schuh K, Neyses L. Calcium transport in cardiovascular health and disease – the sarcolemmal calcium pump enters the stage. J Mol Cell Cardiol. 2005;39:403–6.
109. Kockskämper J, Zima AV, Roderick HL, Pieske B, Blatter LA, Bootman MD. Emerging roles of inositol 1,4,5-trisphosphate signaling in cardiac myocytes. J Mol Cell Cardiol. 2008;45:128–47.
110. Sasse P, Zhang J, Cleemann L, Morad M, Hescheler J, Fleischmann BK. Intracellular $Ca^{2+}$ oscillations, a potential pacemaking mechanism in early embryonic heart cells. J Gen Physiol. 2007;130:133–44.
111. Zima AV, Bare DJ, Mignery GA, Blatter LA. $IP_3$-dependent nuclear $Ca^{2+}$ signalling in the mammalian heart. J Physiol. 2007;584:601–11.
112. Wu X, Zhang T, Bossuyt J, Li X, McKinsey TA, Dedman JR, et al. Local $InsP_3$-dependent perinuclear $Ca^{2+}$ signaling in cardiac myocyte excitation-transcription coupling. J Clin Invest. 2006;116: 675–82.
113. Carafoli E. Historical review: mitochondria and calcium: ups and downs of an unusual relationship. Trends Biochem Sci. 2003;28: 175–81.
114. Bell CJ, Bright NA, Rutter GA, Griffiths EJ. ATP regulation in adult rat cardiomyocytes: time-resolved decoding of rapid mitochondrial calcium spiking imaged with targeted photoproteins. J Biol Chem. 2006;281:28058–67.
115. Beutner G, Sharma VK, Lin L, Ryu SY, Dirksen RT, Sheu SS. Type 1 ryanodine receptor in cardiac mitochondria: transducer of excitation–metabolism coupling. Biochim Biophys Acta. 2005;1717:1–10.
116. Jo H, Noma A, Matsuoka S. Calcium-mediated coupling between mitochondrial substrate dehydrogenation and cardiac workload in single guinea-pig ventricular myocytes. J Mol Cell Cardiol. 2006;40:394–404.
117. Balaban RS. The role of $Ca^{2+}$ signaling in the coordination of mitochondrial ATP production with cardiac work. Biochim Biophys Acta. 2009;1787:1334–41.
118. Denton RM. Regulation of mitochondrial dehydrogenases by calcium ions. Biochim Biophys Acta. 2009;1787:1309–16.
119. Hopper RK, Carroll S, Aponte AM, Johnson DT, French S, Shen RF, et al. Mitochondrial matrix phosphoproteome: effect of extra mitochondrial calcium. Biochemistry. 2006;45:2524.
120. Yamada EW, Huzel NJ. The calcium binding ATPase inhibitor protein from bovine heart mitochondria. Purification and properties. J Biol Chem. 1988;263:11498.
121. Boerries M, Most P, Gledhill JR, Walker JE, Katus HA, Koch WJ, et al. $Ca^{2+}$-dependent interaction of S100A1 with F1-ATPase leads to an increased ATP content in cardiomyocytes. Mol Cell Biol. 2007;27:4365.
122. Garcia-Perez C, Hajnoczky G, Csordas G. Physical coupling supports the local $Ca^{2+}$ transfer between SR subdomains and the mitochondria in heart muscle. J Biol Chem. 2008;283: 32771–80.
123. Sanchez JA, Garcia MC, Sharma VK, Young KC, Matlib MA, Sheu S-S. Mitochondria regulate inactivation of L-type $Ca^{2+}$ channels in rat heart. J Physiol. 2001;536:387–96.
124. Metzger JM, Westfall MV. Covalent and noncovalent modification of thin filament action. The essential role of troponin in cardiac muscle regulation. Circ Res. 2004;94:146–58.
125. Kabaeva ZT, Perrot A, Wolter B, Dietz R, Cardim N, Correia JM, et al. Systematic analysis of the regulatory and essential myosin light chain genes: genetic variants and mutations in hypertrophic cardiomyopathy. Eur J Hum Genet. 2002;10:741–8.
126. Santamaria-Kisiel L, Rintala-Dempsey AC, Shaw GS. Calcium-dependent and -independent interactions of the S100 protein family. Biochem J. 2006;396:201–14.
127. Most P, Remppis A, Pleger ST, Katus HA, Koch WJ. S100A1: a novel inotropic regulator of cardiac performance. Transition from molecular physiology to pathophysiological relevance. Am J Physiol Regul Integr Comp Physiol. 2007;293:R568–77.

# Chapter 5
# Growth Factors Signaling

**Abstract** Cardiac myocytes rapidly proliferate in the embryo but exit the cell cycle shortly after birth, with growth shifting from hyperplastic to hypertrophic. Intensive research efforts have focused on identifying mitogenic stimuli and signaling pathways that mediate these distinct growth processes in isolated cells and with *in vivo* hearts. Although the molecular mechanisms underlying the proliferative growth of embryonic myocardium *in vivo* and adult cardiac myocyte hypertrophy *in vivo* remain largely undetermined, considerable progress has been made using post-genomic analysis methodologies, including manipulation of the murine genome in concert with mutational analysis of these signaling and growth control pathways *in vivo* and in cardiomyocytes grown *in vitro*, including the use of gene transfer/knockout.

Numerous growth factors exert a variety of actions in the cardiovascular system. For example, growth factors, such as fibroblast growth factor 2, significantly promote neonatal cardiac myocyte proliferation. Also, a number of growth factors protect the cardiomyocyte from the detrimental effects of acute ischemia–reperfusion injury, through the activation of a variety of cell-surface receptors and recruitment of intracellular signal transduction pathways, including components of the reperfusion injury salvage kinase pathway.

In this chapter, we review growth factors that actively participate in cardiogenesis and coronary vasculogenesis as well as factor that play a role in cardioprotection and apoptosis.

**Keywords** Growth factors • Kinase receptors • Insulin • RISK pathway

## Introduction

A variety of growth factors (GFs) have been identified to participate in the signaling pathways underlying normal cardiac growth and development. For example, overexpression of the fibroblast growth factor 2 receptor (FGF-R1) in neonatal rats results in marked cardiomyocytes proliferation [1]. Also, several growth factors are released by cardiomyocytes during myocardial ischemia, suggesting a potential role in myocardial repair and myocardial angiogenesis. Furthermore, exogenous administration of several of these growth factors has been reported to recruit specific intracellular signal transduction pathways linked to cardioprotection (CP), for example, the signaling components of the reperfusion injury salvage kinase (RISK) pathway (a group of prosurvival kinases, phosphatidylinositol 3-kinase (PI3K)-Akt, mitogen-activated protein kinase kinase (MEK)1/2, and extracellular signal-regulated kinase(Erk)1/2. Activation of these signaling cascades may protect the myocardium from the detrimental effects of acute ischemia–reperfusion injury.

A number of growth factors-signaling pathways involved in cardiomyocytes growth and proliferation are discussed in this chapter.

## Protein Tyrosine Kinase Receptors

In this section, growth factors which bind to protein tyrosine kinase (PTK) receptors are reviewed, including fibroblast growth factor (FGF), vascular endothelial growth factor (VEGF), insulin, insulin-like growth factor (IGF), epidermal growth factor (EGF), platelet-derived growth factor (PDGF), and neurotrophin (tropomyosin kinase receptor). The activation of the PTK receptor by its specific growth factor results in the autophosphorylation of tyrosine residues, leading to the recruitment of PI3K to the membrane, which is activated by direct binding to phosphotyrosine residues of the growth factor receptor. PI3K then generates the lipid product phosphatidylinositol-3,4,5-trisphosphate, which in turn recruits signaling proteins with pleckstrin homology domains to the membrane, including the protein serine–threonine kinase, Akt. In addition, growth factor binding to the receptor tyrosine kinase (RTK) results in the activation of Ras, leading to the recruitment of Raf to the membrane, and the subsequent activation of the MEK1/2-Erk1/2 kinase cascade.

J. Marín-García, *Signaling in the Heart*,
DOI 10.1007/978-1-4419-9461-5_5, © Springer Science+Business Media, LLC 2011

### Fibroblast Growth Factor Family

FGF family includes 23 members which regulate a variety of effects, including embryogenesis, angiogenesis, growth, and cell survival. Numerous experimental studies have been focused on the actions of FGF-1 and FGF-2. FGF-1 and FGF-2 exert their regulatory effects by binding to specific plasma membrane tyrosine kinase FGF receptor, which result in the recruitment of a number of different signal transduction pathways.

Both FGF-1 and FGF-2 are secreted by cardiomyocytes in response to myocardial ischemia and they bind to FGF-1 receptor, which is known to be present on cardiomyocytes [2, 3].

FGFs constitute at least one component of the embryonic epicardial and myocardial signal that induces myocyte proliferation and formation of different components of the coronary vasculature in the developing heart (see section "GFs and Development" below). Another important function of FGF-1 and FGF-2 is the promotion of endothelial cell proliferation and the physical organization of endothelial cells into tube-like structures. Thus, they promote angiogenesis, the growth of new blood vessels from the preexisting vasculature (see section on "Angiogenesis" below).

A number of experimental studies have demonstrated an acute cardioprotective effect with FGF which is independent of its angiogenic/arteriogenic actions (see section on "Cardioprotection" below).

### Vascular Endothelial Growth Factor

VEGF, a 45 kDa polypeptide, controls blood vessel development during embryogenesis (see section "GFs and Development" below) and is a major regulator of angiogenesis in the heart (see section "Angiogenesis" below). VEGF is generated in response to myocardial ischemia and binds to two high-affinity tyrosine kinase receptors, the VEGFR1 (Flt-1) and the VEGFR2 (Flk-1/KDR), which are preferentially distributed on vascular endothelial cells but are also known to be present in cardiomyocytes [4, 5]. The activation of the VEGF receptor leads to the activation of RISK-related signal transduction pathways: MEK1/2-Erk1/2-p90$^{rsk}$ (in the cardiomyocytes) and PI3K-Akt-endothelial nitric oxide synthase (eNOS) (in endothelial cells) [6, 7] (Fig. 5.1).

Diverse pleiotropic effects of VEGF include coronary vasculogenesis in the developing heart, an acute cardioprotective effect and participation in revascularization after myocardial ischemia (see below).

To add complexity to VEGF-dependent signaling, VEGF receptors can interact with coreceptors, neuropilins (NRPs)-1 and -2. NRPs are transmembrane glycoproteins, which do not transduce signals themselves, but mediate functional responses as a result of complex formation with VEGF receptors. Mouse studies show that NRP1 plays a role in angiogenesis and neurogenesis and that it is essential for neuronal and cardiovascular development [8–10]. NRPs comprise a large extracellular region, a single transmembrane domain and a small cytoplasmic domain. The cytoplasmic domain of neuropilins contains PDZ domain-binding motif for association with PDZ domain neuropilin interacting protein-1 (NIP1) [11]. Two extracellular domains (ECDs), b1 and b2, are essential for optimal binding of several splice isoforms of VEGF. According to current model, optimal binding of VEGF to NRP1 promotes its dimerization with VEGFR2 which is essential for VEGF-induced migration of endothelial cells [12]. The enhanced function of VEGFR2 in the presence of NRP1 is related to enhanced VEGFR2 signaling when it complexes with NRP1, rather than to an increase in the affinity of VEGF for its receptors: inhibition of complex NRP1-VEGFR2 is associated with reduced VEGFR2 phosphorylation, intracellular signaling, mitogenesis, cell migration, and angiogenesis [13–17]. Nonetheless, NRP1 is not essential for the full spectrum of VEGF-regulated signaling pathways and biological responses, but is rather required for optimal VEGF-induced VEGFR2 signaling through some pathways [18]. Several mechanisms could account for the NRP1-dependent enhancement of VEGFR2 signaling. First, when in complex with NRP1, VEGFR2 may be stabilized at the cell surface, it is less prone to endocytosis and therefore be more effective in terms of receptor activation and signaling. Second, NRP1 may itself trigger intracellular signals, which enhance signaling via VEGFR2. Thus, the expression of NRP1 missing SEA motif and NIP1 knockdown disrupt vessel formation in zebrafish and human endothelial cell migration, which supports a functional role for the NRP1 in association with NIP1 in angiogenesis. Binding of NIP1 to the C-terminus of NRP1 raises the possibility that NRP1 takes part in independent cytoplasmic-signaling pathways [19] (Fig. 5.1).

### Platelet-Derived Growth Factor

PDGFs are products of four different genes that produce peptides, designated PDGF-A, -B, -C, and -D. Homo- and heterodimerization of PDGF-peptides produces five forms of active PDGF ligand: AA, AB, BB, CC, and DD. Two different highly sensitive transmembrane tyrosine kinase receptors bind active PDGF ligands: PDGF receptor (PDGFR)α (binds all the PDGF dimers except for PDGF-DD) and PDGFRβ (binds only to -BB and -DD homodimers) [20]. PDGFs produced by embryonic myocardium, may play a role in epicardial cells undergoing epithelial to mesenchymal transformation (EMT), and are important for the development of

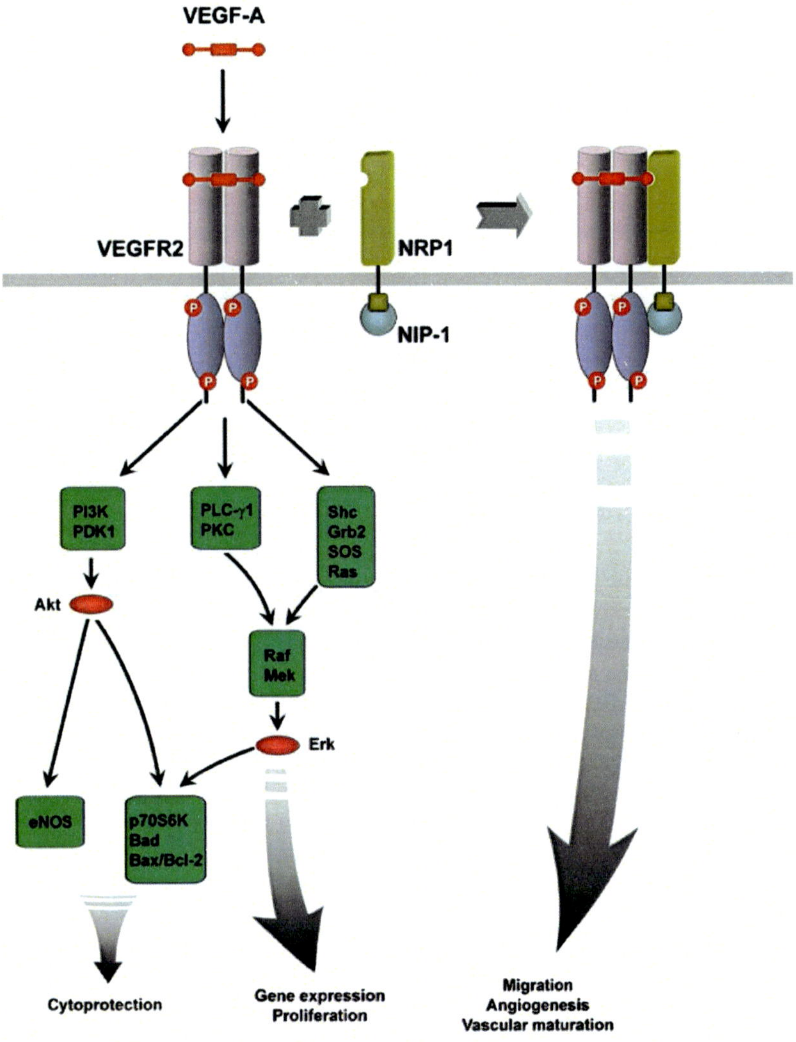


**Fig. 5.1** Vascular endothelial growth factor signaling. Vascular endothelial growth factor (VEGF) A binding to VEGF receptor (VEGFR) 2 activates intracellular signaling kinase cascades, including Ras-Raf-Mek-Erk and phosphoinositide 3-kinase (PI3K)-Akt. VEGFR2 forms complex with neuropilin (NRP) 1 which is important for optimal VEGFR2 functioning in endothelial cells. The NRP1 associates with the PDZ protein *n*europilin-*i*nteracting *p*rotein-*1* (NIP-1) and this is important for the role of NRP1 in VEGF-A. *Bad* Bcl-2-*a*ssociated *d*eath promoter protein, *Bax* Bcl-2-*a*ssociated *X* protein, *Bcl-2 B*-cell *l*ymphoma *2* protein, *eNOS e*ndothelial *n*itric *o*xide *s*ynthase, *Erk* extracellular signal-regulated kinase, *Grb2 g*rowth factor *r*eceptor-*b*ound protein *2*, *PDK1* 3-*p*hosphoinositide-*d*ependent protein *k*inase *1*, *PKC* protein kinase C, *PLC-γ1* phospholipase Cγ 1, *Ras* small GTPase "*Ra*t *s*arcoma," *SOS* guanine nucleotide exchange factor "*S*on *o*f *S*evenless," *p70S6K* ribosomal protein *S6 k*inase

coronary vascular smooth muscle (see section "GFs and Development" below). Recent studies suggested that PDGF plays an essential role in the activation of myofibroblasts and contributes to cardiac fibrosis [see section on "GFs and Myocardium Pathophysiology: Cardiac Fibrosis"].

## Epidermal Growth Factor Family

EGF is a low-molecular weight (6 kDa) polypeptide first purified from the mouse submandibular gland, but since then it has been found in many human tissues. EGF is the founding member of the EGF-family of proteins which have similar structural and functional characteristics. All EGF-family members contain one or more conserved amino acid sequence with six cysteines "$CX_7CX_{4-5}CX_{10-13}CXCX_8GXRC$" ("X" stands for any amino acid) essential for high-affinity binding of the EGF-family GFs to their receptors. Besides EGF itself, EGF-family includes heparin-binding EGF-like growth factor (HB-EGF), transforming growth factor-α (TGF-α), amphiregulin (AR; AREG), epiregulin (EPR), epigen, betacellulin (BTC) and neuregulins 1–4 (NRG1–NRG4). Growth factors from the EGF-family are important in the development of the embryonic heart, and play an important role in the regulation of cell proliferation, differentiation, and survival.

EGF-family GFs bind with high affinity to cell surface EGF receptors and stimulate intrinsic PTK activity of the receptor (RTK). Once activated, RTK initiates a signal transduction cascade(s) leading to a variety of functional changes within the cell. Receptors to EGF-related ligands is an ErbB family of transmembrane proteins which includes four members: EGF receptor itself (EGFR or ErbB1), ErbB2, ErbB3, and ErbB4. ErbB receptors are large proteins (molecular masses of ~180 kDa) with domain structure and glycosylated on their ECDs. ErbBs consist of the N-terminal ligand-binding ECD, a single transmembrane domain, and an intracellular domain (ICD). The ICD in turn consists of three subdomains: juxtamembrane subdomain, PTK subdomain, and C-terminal regulatory region. There are several Tyr-residues in the C-terminal region which become phosphorylated in activated receptor, as an important step in downstream effector binding. GF binds to a monomeric receptor and promotes receptor homo- or heterodimerization which is necessary for the activation of the receptor's PTK activity. Active PTKs in the stimulated ErbB dimer transphosphorylate Tyr-residues in the C-terminal region of receptor molecules, thus building docking sites for effectors (Fig. 5.2).

EGF, HB-EGF, TGF-α, AR, EPR, BTC, and epigen bind to the EGFR. Whereas EGF, TGF-α, AR, and epigen bind exclusively to EGFR, the other three ligands (HB-EGF, EPR, and BTC) also activate the ErbB4 receptor. Neuregulins are a subfamily of the EGF-family, and they are the products of four different genes. The actual number of ligands that belong to this subfamily is much higher than four. Thus, about 15 subtypes of NRG1 are known, and they are the products of different promoters and alternative splicing of the *nrg1* gene. NRG2 also can be alternatively spliced [21]. The current view is that the NRG1- and NRG2-related GFs activate ErbB3 and ErbB4, whereas NRG3 and NRG4 bind only to ErbB4. Interestingly, ErbB2 cannot bind ligands, but contains catalytically active PTK and is able to function as a coreceptor: it is the preferred heterodimerization partner of all other ErbB receptors and increases their ligand-binding affinity. Moreover, ErbB2-containing heterodimers are stable and do not degrade once internalized.

Ligand-dependent homo- and heterodimerization of ErbB receptors with subsequent transautophosphorylation create phosphotyrosine binding (PTB) sites on the C-terminal region of the receptors for numerous adaptor/signaling proteins carrying PTB or Src homology 2 (SH2) motifs. Moreover, PTK receptors can phosphorylate proteins bound to them, thus changing their functional activity and creating additional SH2/PTB binding sites. High-throughput screening [22, 23] demonstrated that many signaling proteins interacting with ErbB isoforms have different affinities and specificity to individual sites [24]. Moreover, the PTK-domain in ErbB3 is catalytically inactive, so it is unable to transphosphorylate C-terminal tyrosines on the partner molecule, in the active dimeric receptor.

Ligands from the EGF-family via ErbB receptors activate the Erk1/2 signaling pathway. This EGF signal-transducing cascade starts from the binding of two adaptor proteins, Grb2 and Shc, to specific phosphotyrosine sites in the activated ErbB receptor, via their SH2- and PTB-domains. ErbB can Tyr-phosphorylate Shc, and this creates additional binding sites for Grb2. Grb2 exists in the cell as a tight complex with guanine nucleotide exchange factor Sos. Thus, Grb2 plays in important role in the EGF signaling, as it delivers Sos to the plasma membrane, where it activates the membrane-bound small G protein Ras by enhancing GTP/GDP exchange. Activated Ras stimulates plasma membrane-localized MAPK kinase kinases (MKKKs) Rafs, which subsequently results in phosphorylation/activation of MAPK kinases (MEK1, MEK2) and MAPKs (Erk1/2). Erks phosphorylate many target proteins, involved in cell growth and survival. EGF-dependent activation of the Erk1/2 pathway plays a role in adaptive myocardial hypertrophic growth and survival (see below).

Another important signaling molecule recruited to and activated by ligand-bound ErbB is the lipid kinase PI3K. PI3K is an oligomeric protein consisting of dimeric catalytic subunit (p110α, p110β, or p110δ) and regulatory subunit (p85α, p85δ, or p55γ). The regulatory subunit contains SH2-domain and is the cause of the association of PI3K with Tyr-phosphorylated ErbB receptor, and the starting point of another signaling cascade activated by GFs from the EGF-family, PI3K-Akt. PI3K activated by ErbB phosphorylates phosphoinositides in the plasma membrane leading to the synthesis of a very important signaling phospholipid, phosphatidylinositol (3,4,5)-trisphosphate ($PIP_3$). Increased concentration of $PIP_3$ is the cause for translocation of Akt to plasma membrane because it contains $PIP_3$-binding pleckstrin homology (PH) domain. Translocation of Akt is a very important step in the activation of PI3K-Akt-signaling pathway: the plasma membrane is the compartment, where Akt undergoes phosphorylation/activation by membrane-localized 3-phosphoinositide-dependent protein kinase 1 (PDK1). Activated Akt phosphorylates a number of substrate proteins, including glycogen synthase kinase 3 (GSK3) and FoxO (Forkhead) transcription factors. Activation of PI3K-Akt pathway promotes growth and survival of cardiac cells (see below) (Fig. 5.2).

Interestingly, ErbB4 receptor contains a unique fragment located next to the N-terminus of the transmembrane domain that makes it sensitive to proteolytic cleavage by membrane metalloproteases (MMPs). The first cleavage (release of ErbB4 ECD) is followed by a second cleavage which releases ICD. Some observations indicated that the 80 kDa-ICD may translocate to the nucleus and regulate transcription [25, 26].

## *Insulin-Like Growth Factor*

Insulin-like growth factor I is a 72 kDa protein originally identified as somatomedin C. It is produced in the liver in response to growth hormone, and circulates in the complex with IGF-binding factors. IGF-I and another IGF, IGF-II, bind to tyrosine kinase receptors IGF-IR and IGF-IIR.

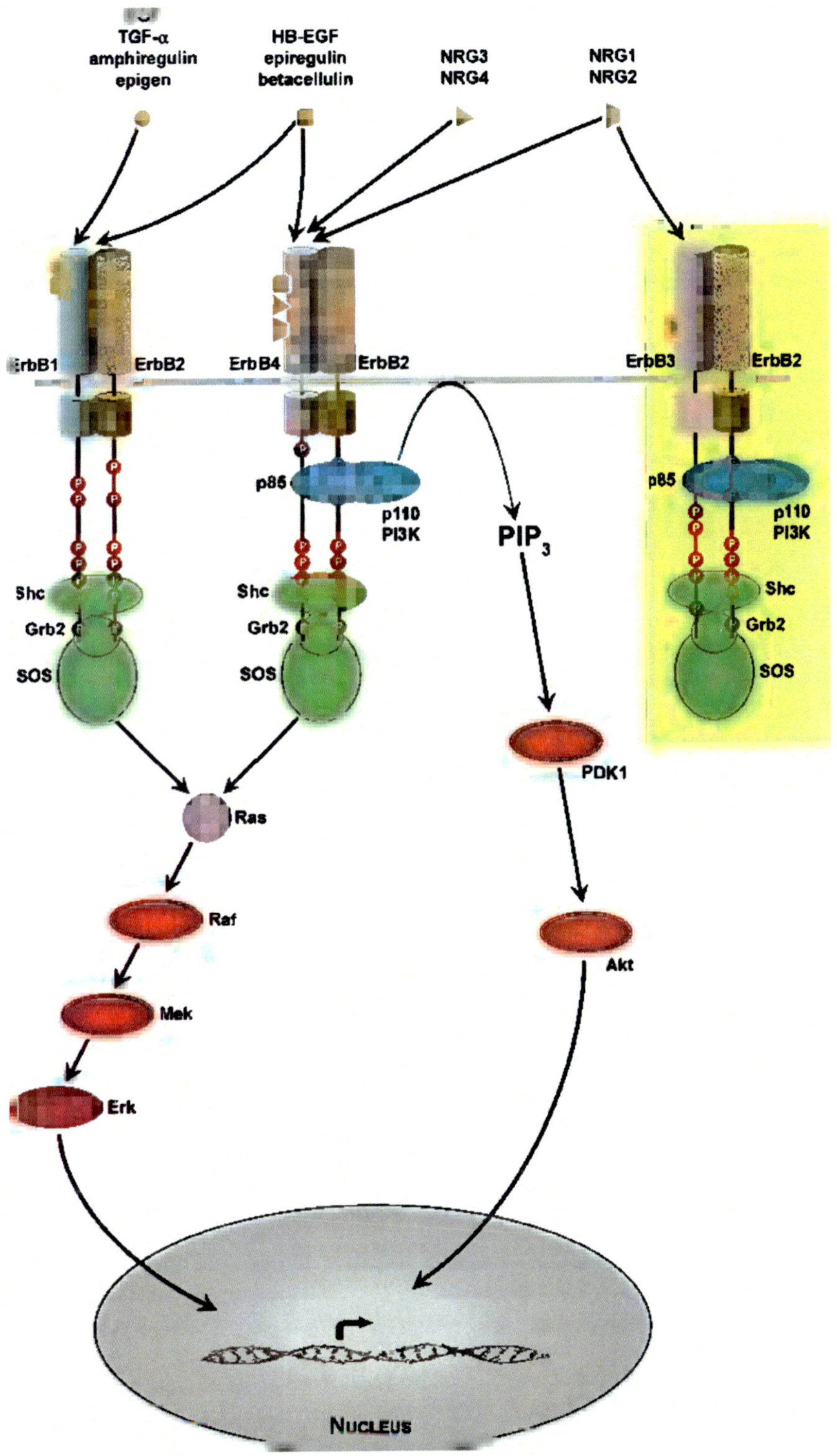


**Fig. 5.2** Epidermal growth factor family signal transduction pathways. Binding of ligands to ErbB1, ErbB3, and ErbB4 receptors stimulates their homo- and heterodimerization (only heterodimerization of each of them with ErbB2 is shown). Within a dimer, partners induce mutual transphosphorylation of Tyr-residues (P). Phosphotyrosine residues serve as a docking sites for signaling molecules containing Src homology 2 (SH2) and/or phosphotyrosine-binding (PTB) motifs. This leads to the activation of downstream-signaling cascades. In adult cardiomyocytes, ErbB3 is not expressed (*yellow background*), so only ErbB4 connects to the PI3K-Akt pathway. *EGF* epidermal growth factor, *ErbB* epidermal growth factor receptor, *Erk* extracellular signal-regulated kinase, *Grb2* growth factor receptor-bound protein 2, *HB-EGF* heparin-binding epidermal growth factor-like growth factor, *NRG* neuregulin, *PDK1* 3-*p*hosphoinositide-*d*ependent protein *k*inase *1*, *PI3K* phosphoinositide 3-kinase, *PIP₃* phosphatidylinositol (3,4,5)-trisphosphate, *Ras* small GTPase "*Ra*t *s*arcoma," *SOS* guanine nucleotide exchange factor "Son of Sevenless," *TGF-α* transforming growth factor α

Binding of IGF-I to IGF-IR activates the intrinsic RTK with subsequent autophosphorylation of tyrosine residues and phosphorylation of serine residues. This in turn leads to IRS-1 and IRS-2 phosphorylation followed by the activation of downstream PI3K-Akt and MEK1/2-Erk1/2-signaling pathways. In particular, the activation of Akt involves the interaction of IRS-1 with and the activation of PI3K. IRS-I-dependent activation of Grb2/SOS is necessary for stimulation of Ras-Raf-MEK1/2-Erk1/2 kinase cascade. As we already mentioned earlier, PI3K-Akt pathway recruits downstream antiapoptotic events, whereas MEK1/2-Erk1/2 pathway recruits prosurvival pathways. Cardioprotective function of IGF is described below.

### *Insulin*

Similar to many GFs described above, binding of insulin to its specific PTK receptor (IR) results in the autophosphorylation of the receptor's intracellular tyrosine residues, recruitment/phosphorylation of IR substrate (IRS) and generation of SH2 domain-binding sites followed by the recruitment of PI3K and subsequent activation of Akt. Also, translocation of Grb2/SOS to the IR/IRS-complex initiates Ras-Raf-MEK1/2-Erk1/2 signaling cascade. The intricate involvement of insulin in the growth and metabolism of the heart is described in more detail in Chap. 16.

## Protein Serine/Threonine Kinase Receptors

### *Transforming Growth Factor-β Superfamily*

Transforming growth factor β (TGF-β) refers to a superfamily of extracellular growth factors comprising nearly 40 members which have been implicated in a variety of cellular processes, including cardiac development, vascular fibrosis, apoptosis, and inflammation [27, 28]. TGF-β family ligands form homo- or heterodimers which bind to and activate two types of transmembrane receptors with intrinsic serine/threonine kinase activity. Receptors then stimulate downstream regulatory Smad proteins to translocate them from the cytosol to the nucleus where they regulate transcription.

The TGF-β superfamily includes TGF-βs, bone morphogenetic proteins (BMPs), growth differentiation factors (GDFs, including myostatin), activins and inhibins, Müllerian inhibiting substance (MIS), nodal, and leftys. GFs from the TGF-β superfamily are translated as nonactive precursors. During secretion endopeptidase cleaves the so-called latency associated peptide (LAP) from mature domain of GF, but LAP remains associated with the mature domain and therefore masks the activity of GF. In the case of TGF-β subfamily GFs (and probably other members of the superfamily), they become active after dissociation or proteolysis of LAP. Most members of the TGF-β superfamily contain characteristic 6–9 conserved cysteine residues in the mature domain. These cysteines form important intra/intermolecular disulfide bonds and stabilize the dimeric structure of active GF.

TGF-β-related GFs signal via single-pass transmembrane receptors with intrinsic serine/threonine kinase activity. There are two subfamilies of the TGF-β receptors, which differ structurally and functionally: type I and type II receptors. Type I and type II receptors interact upon GF binding. Type I receptors contain unique 30-amino acid GS domain upstream of kinase domain which becomes phosphorylated by the GF-activated type II receptor. This phosphorylation is required for transduction of signal from receptor to downstream components of the signaling cascade. Activin receptor linked- or activin receptor-like kinases (ALKs) 1–7 belong to the type I receptors. Some of them transduce signal from activins, such as ALK2 (ActR1A/ACVR1), ALK4 (ActR1B), and ALK7; some – from BMPs, such as ALK2, ALK3 (BMPR1A), and ALK6 (BMPR1B); whereas ALK5 (TβR1) is the receptor for TGF-βs. Subfamily of type II receptors include those which are involved in the signaling from activins (ACVR2/ActRIIA and ACVR2B/ActRIIB), BMPs (BMPR2/BMPRII), MIS (AMHR2/MISRII), and TGF-βs (TGFBR2 TβR2).

Activation of an intracellular-signaling cascade by TGF-βs and activins starts from their binding to corresponding type II receptor (it is important to emphasize that TGF-βs and activins do not interact with type I receptors). GF-bound/activated type II receptor binds to the type I receptor. This yields to the formation of a tetrameric receptor complex, formed by two type I and two type II receptors. In contrast, both type I and type II BMP receptors can bind their ligands leading to the active GF/type I/type II receptor complex. Once an active oligomeric receptor complex is formed, type II receptor phosphorylates/activates type I receptor, which subsequently phosphorylates downstream Smad proteins that transmit the signal to the nucleus.

Interestingly, endothelial cells, smooth muscle cells (SMCs), fibroblasts, and activated macrophages express a type III receptor called endoglin. This high-molecular weight homodimer (180 kDa) can bind TGF-βs, activin-A, some BMPs, and interact with TGF-β type I and type II receptors [29, 30]. This receptor has been shown to play a role in the development of the cardiovascular system and in vascular remodeling [31].

Receptor-associated Smad proteins (R-Smads: Smad1, Smad2, Smad3, Smad5, and Smad8) which are the targets for phosphorylation by activated type I receptor, belong to the family of Smad proteins playing different roles in TGF-β superfamily signaling. Different Smads are coupled to different receptors. Thus, Smad2 and Smad3 are C-terminally phosphorylated and translocated to the nucleus upon stimulation by activins and TGF-βs [32]. Smad1, Smad5, and Smad8 are phosphorylated and translocated to the nucleus upon stimulation by BMPs [33, 34]. Another member of the Smad family, common Smad (Smad4), differs from R-Smads by the absence of the C-terminal site of phosphorylation. Smad4 forms a complex with phosphorylated R-Smad, and oligomeric complex "R-Smad/Smad4" translocates into the nucleus to activate transcription of target genes. Another two members of Smad family, Smad6 and Smad7, comprise a group of inhibitory Smads. Smad6 inhibits GF-triggered signaling by competing with Smad4 for binding to activated R-Smads and forms an inactive complex "R-Smad/Smad6." Smad7 occupies the type I receptors and thus prevents receptor-dependent phosphorylation/activation of R-Smads (Fig. 5.3).

Smads can bind to DNA directly, but for high-affinity binding they require interactions with a number of DNA-binding protein coactivators, such as cAMP response element-binding protein (CREB)-binding protein (CBP), p300, FoxH1 [35, 36]. There is also a group of protein repressors that inhibit Smad-regulated pathways. SMAD nuclear interacting protein 1 (SNIP1), transforming growth 3′ interacting factors (TGIF, TGIF2), SnoN- and Ski-proteins prevent the association of Smad4 with coactivators [37, 38]. In addition, SnoN, Ski, TGIF, and TGIF2 can recruit histone deacetylase [39]. Ubiquitin ligases SMURF1 and SMURF2 inhibit BMP-signaling pathway via targeting Smad1 for ubiquitination/degradation [40, 41].

It is important to mention that TGF-β can activate a variety of noncanonical (non-Smad)-signaling pathways, including Ras-MEK1/2-Erk-1/2; p38; c-Jun N-terminal kinase (JNK)-focal adhesion kinase (FAK)/TGF-β activated kinase (TAK); PI3K-Akt; and PP2A.

Similar to other signaling pathways the TGF-β signaling exhibits cross talk with several second messenger-regulating systems. For example, EGF receptor and hepatocyte growth factor receptor phosphorylate Smad2, and induce its nuclear translocation; Erk and the $Ca^{2+}$/calmodulin-dependent protein kinase II can phosphorylate/activate R-Smads.

## G Protein-Coupled Receptors

### Urocortin

Urocortin is a 40 amino acid peptide member of the corticotrophin-releasing hormone (CRH) family. In the vascular endothelial cells and the heart, it binds to the G protein-coupled receptor, CRH-receptor 2 (CRH-R2). The binding of urocortin to its receptor leads to the recruitment to the membrane of PI3K-γ (class IB), followed by the membrane recruitment and activation of Akt. Also PI3K-γ is able to activate MEK1/2-Erk1/2 cascade. Some studies have demonstrated an acute cardioprotective effect with urocortin (see section on "Cardioprotection" below).

### Adrenomedullin

Adrenomedullin (AM) was initially identified as a vasodilator. Other effects of AM include increasing the tolerance of cells to oxidative stress and hypoxic injury, and angiogenesis. AM is seen as a positive influence in diseases, such as chronic obstructive pulmonary disease, hypertension, myocardial infarction, and other cardiovascular diseases, whereas it can be seen as a negative factor in potentiating cancerous cells to extend their blood supply and cause cell proliferation.

Calcitonin receptor-like (CALCRL) (also known as the calcitonin receptor-like receptor, CRLR) is GPCR. When complexes with single transmembrane domain receptor activity-modifying protein 2 (RAMP2), it functions as an AM receptor ($AM_1$). When CALCRL complexes with RAMP3, it functions as an AM receptor ($AM_2$) and also as calcitonin gene-related peptide (CGRP) receptor.

AM receptors are linked to the $G_s$ protein, which activates adenylate cyclase. The outcome of AM stimulation of its receptor is the cellular production of both cyclic AMP (cAMP) and nitric oxide (NO).

## GFs and Development

The establishment of the coronary circulation is critical for the development of the embryonic heart. Coronary development depends on a complex communication between the epicardium, the subepicardial mesenchyme, and the myocardium mediated in part by secreted growth factors.

An important class of signaling molecules produced by embryonic heart epicardium and myocardium is FGFs. A number of FGF family members have been reported to be expressed in the epicardium, including FGF-1, -2, -4, -9, -16, and -20 [42, 43]. Embryonic cardiomyocytes express two FGF receptor subtypes, FGFR1 and FGFR2c [42]. Cardiomyocyte-specific ablation of them results in severe hypoproliferative embryonic myocardium, suggesting that FGF signaling is required for cardiomyocyte proliferation. According to Lavine et al. [44], FGF-9 is an important regulator of myocardial proliferation in embryo: FGF-9-deficient mice die at birth because of decreased embryonic myocyte proliferation.

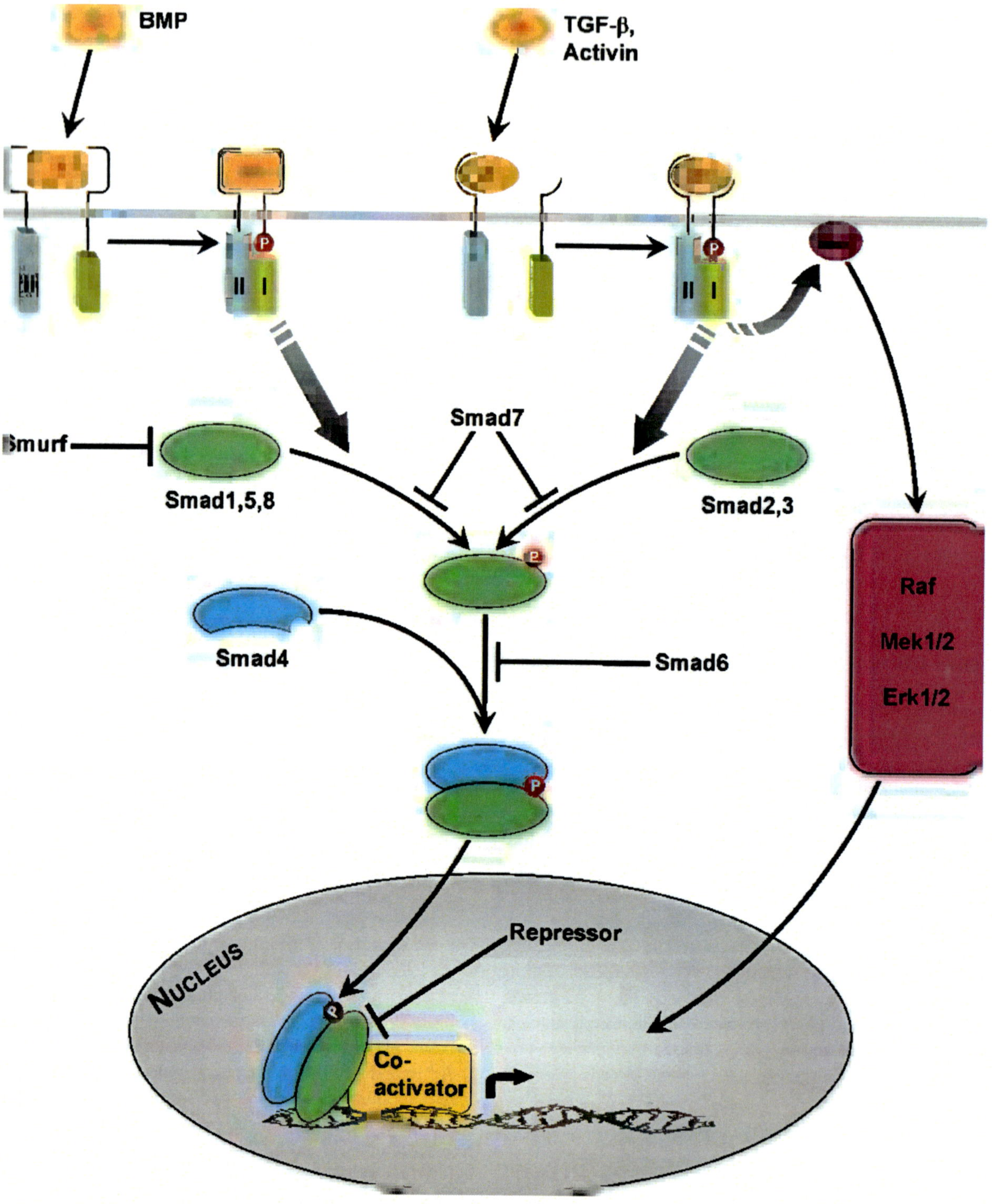


**Fig. 5.3** TGF-β superfamily signal transduction pathway. Binding of BMPs to either type I or type II receptor stimulates the formation of a ternary complex. The type II receptor phosphorylates and activates the type I receptor. The type I receptor then phosphorylates BMP-related R-Smads: Smad1, Smad5, or Smad8. Activins and TGF-βs first bind type II receptors and then form a ternary complex. Their phosphorylated/activated type I receptor phosphorylates TGF-β/activin-related R-Smads: Smad2, or Smad3. Phosphorylated R-Smads interact with Smad4, migrate to the nucleus, interact with nuclear coactivators and stimulate specific target genes. Smad6 and Smad7 are negative regulatory Smads that interfere R-Smad interaction with Smad4 and type I receptor, respectively. Ubiquitin ligases Smurfs cause degradation of Smad1. Nuclear repressors interfere with R-Smad-Smad4 complex binding to coactivator. Smad independent signaling pathways contribute to diversify responses to TGF-β1 (MAPK-signaling pathway is shown)

Interestingly, there is also evidence suggesting that the myocardium may signal back to the epicardium via FGFs. FGFR1 is expressed in embryonic epicardium and is upregulated in response to myocardial FGF [45]. FGF signals from myocardium induce the differentiation of the subepicardial mesenchyme into the different components of the coronary vasculature [45, 46].

VEGF have been shown to control blood vessel development during embryogenesis [47, 48]. One result of VEGF-mediated signaling may be essential for coronary vasculogenesis in the developing heart: when added to an *in vitro* heart culture system, VEGF-A has been shown to promote transformation of epicardial cells into precursor cells that become coronary vascular smooth muscle and perivascular fibroblasts [process which is called epithelial–mesenchymal transformation (see Chap. 9)] [49]. In addition, VEGF induces coronary endothelial cell proliferation and migration and tube formation in conjunction with FGF-2 [46, 50]. Blocking VEGF function significantly reduces tubulogenesis of coronary vascular bed [50].

Receptor cofactor of VEGFR, NRP, plays an essential role in cardiovascular development. Endothelial-specific NRP1 knockout mice exhibit embryonic lethality, a poorly developed vasculature and multiple defects in the major arteries [10]. In contrast with NRP1 mutant mice, NRP2-null mice have no obvious cardiovascular abnormalities [51]. For further discussion on VEGF during cardiovascular development, see Chap. 2.

Another growth factor which promotes epicardial EMT during heart embryogenesis, is PDGF. Epicardial cells express two subtypes of PDGFR and are able to elongate and migrate in response to PDGF ligands, when cultured *in vitro* [52, 53]. Experimental data suggest that PDGFRβ is the major receptor subtype regulating EMT: PDGF-BB homodimer induces EMT activation much more potently compare to either PDGF-AB or PDGF-AA [53]. Moreover, in the subepicardial space and myocardium of transgenic PDGFRβ-lacking mice embryos, there are much less epicardial-derived cells indicating a defect in EMT [54].

In addition to its role in regulating epicardial EMT, PDGF is also a required signal for the differentiation and recruitment of SMCs and pericytes to the coronary vascular bed. PDGF-induced SMC differentiation is mediated through PDGFRβ: only PDGF-BB (neither PDGF-AA nor -AB) induces SMS differentiation in explanted proepicardium [53]; *Pdgfβ*$^{-/-}$ mice show massive reduction in vascular SMCs in the developing myocardium [54, 55].

ErbB receptors and several of their ligands are other important regulators of development of the embryonic heart [56, 57]. Gene targeting strategies in mice have shown the significance of the ErbB4 receptor in cardiac development: mice embryos deficient in ErbB4 die because of abnormal trabeculation [58], but this lethal ErbB4 "knock-out" phenotype can be rescued by cardiomyocyte-specific expression of ErbB4 during embryogenesis [59]. ErbB3, not detectable in postnatal cardiomyocytes, is essential for myocardium development: ErbB3-null mice embryos die because of defective cardiac cushion formation [60]. One of the GFs regulating trabeculation during development through the ErbB4 receptor and the ErbB2 coreceptor is NRG1, which is released from the endocardium [56, 61]. Also HB-EGF is essential for heart embryogenesis, and HB-EGF-deficient mice display mesenchymal hyperplasia in cushion tissue [62].

TGF-β superfamily signaling is essential for heart development (see Chap. 9). Genetic studies have shown that BMP-2 is required for the initial formation of the cardiac primordium because *Bmp2* knockout mice develop a very retarded and malformed heart [63]. TGF-β1–3 are expressed in embryonic epicardium and myocardium, and stimulate EMT of proepicardial and epicardial cells, as it was demonstrated by *in vitro* collagen gel invasion assays [64, 65]. Surprisingly, mice lacking individual TGF-β isoforms develop hearts without obvious morphological or physiological defects [66, 67]. One possible explanation is the functional redundancy of TGF-β isoforms: expressed isoforms can compensate the absence of one particular isoform.

GFs from TGF-β super family also regulate coronary SMC recruitment and differentiation in the developing heart. Thus, a deletion of the ALK5 which signals through the Smad2/Smad3 pathway, results in mice with coronary vessels that contain less SMCs [68]. Interestingly, for proper SMC differentiation during coronary vessel remodeling, activation of TGF-β-ALK5-Smad2-Smad3-signaling pathway should be accompanied by Smad6-dependent downregulation of signaling triggered by another member of TGF-β super family, BMP [69]. For further discussion on other growth factors during cardiovascular development, see Chap. 9.

## GFs and Myocardium Pathophysiology: Cardioprotection

Since signaling pathways in CP are discussed in detail in Chap. 20, here it is suffice to note that among the several mechanisms participating in CP, activation of components of the RISK pathway, Akt and Erk1/2 attenuates myocardial reperfusion injury and limits myocardial infarct size [70]. Moreover, as a consequence of this activation other potential mechanistic effects include: inhibition of the mitochondrial permeability transition pore (MTP); reduction of calcium-induced MTP opening via potentiation of calcium uptake by sarco(endo)plasmic reticulum $Ca^{2+}$-ATPase (SERCA) into the sarcoplasmic reticulum; phosphorylation/inhibition of proapoptotic factors BAD and BAX; and the inhibition of cytochrome *c* release [71, 72].

GFs activate PI3K-Akt and MEK1/2-Erk1/2 components [73]. For instance, pretreatment with FGF has infarct-limiting effects, which can be abolished by antagonists of GFs and tyrosine kinase, suggesting that the cardioprotective effect of FGF may be receptor-mediated [74]. Importantly, the cardioprotective effect of FGF-2 is independent of its mitogenic effects, as a nonmitogenic mutant form of FGF-2 can reduce myocardial infarct size in the perfused rat heart [75]. Direct protective effect of FGF-2 on cardiomyocyte is mediated through the activation of PKCε, PKCδ, and Erk1/2 [76, 77]. Particularly, the Erk pathway in FGF-2-mediated CP acts as an upstream activator of PKC [78]. At the same time, PI3K-Akt component of the RISK pathway does not seem to be linked to the FGF's CP effect, although it is known to be activated by FGF in the cardiomyocyte.

Analogous to FGF, acute administration of VEGF in isolated rat hearts results in improved functional recovery following a period of ischemia–reperfusion injury [79]. VEGF can activate cytoprotective MEK1/2-Erk1/2-p90$^{rsk}$-signaling pathway directly in cardiomyocyte [6] and/or be cardioprotective indirectly, via vasculatory effect.

IGFs which activate PI3K-Akt and MEK1/2-Erk1/2-signaling pathways via RTKs IGF-Rs are also implicated in CP. For instance, transgenic overexpression of IGF-I is cardioprotective in mice myocardial reperfusion model, and this effect of IGF-I requires Akt phosphorylation [80]. Several antiapoptotic actions of IGF-I target mitochondria: prevention of mitochondrial cytochrome *c* release, the inhibition of MTP opening, and the inhibition of reactive oxygen species (ROS) production [81, 82]. Moreover, pretreatment with IGF-I improves the function of isolated rat heart and survival of adult rat cardiomyocytes via inhibition of proapoptotic proteins, such as $Bcl_2$, Bax, and Caspase 3 [81, 83].

TGF-β1 protects isolated cardiomyocytes directly (in the absence of endothelium and neutrophils); and the cardioprotective effect of this GF is realized via MEK1/2-Erk1/2 component of the RISK signal transduction pathway [84, 85].

Okosi et al. [86] have reported an acute cardioprotective effect of urocortin. They found that pretreatment with urocortin reduce cell death in neonatal rat cardiomyocytes subjected to 6 h hypoxia. Cardioprotective effect of urocortin was abolished by CRH receptor antagonists [87]. Also, the cardioprotective effect of urocortin depends on the activation of prosurvival kinases of the RISK pathway, the PI3K-Akt and the MEK1/2-Erk1/2 [88, 89]. However, the cardioprotective mechanisms downstream of these kinase-controlled signaling pathways are unclear, although it appears to involve a number of signaling molecules in the cardiac mitochondria [90–94]. Peptide analogues of urocortin that specifically bind to CRH-R2 have been reported to have potential cardioprotective effect similar to urocortin [95, 96], making them very useful clinically because they do not influence brain-specific receptors, CRH-R1.

Finally, the cardioprotective function of GFs requires transient short-lived activation of cardioprotective cascades as opposed to their chronic activation, which can lead to unwanted cardiovascular side-effects, such as cardiac hypertrophy.

## GFs and Myocardial Pathophysiology: Cardiac Hypertrophy

The role that FGF-2 plays in the heart hypertrophic response has been well established, and FGF-2 has been found responsible for the hypertrophic remodeling occurring after myocardial infarct (MI) [97]. Also a mouse model lacking the *Fgf2* gene, subjected to pressure overload, had a reduced hypertrophic response [98]. Additionally, angiotensin II fails to induce compensatory hypertrophy in mice lacking FGF-2 [99].

In relation to the myocardium and GFs, there is evidence from studies both *in vivo* and *ex vivo* that both, survival following a cardiotoxic insult and adaptive cardiac hypertrophic growth can be promoted by the activation of the Erk1/2 cascade [100, 101], and also by the activation of PI3K and Akt and the subsequent downstream signaling [100, 102–105].

GFs from the EGF-family which activate Erk1/2- and PI3K-Akt cascades, can promote survival of cardiomyocytes. Thus, NRG1 was found to protect cardiomyocytes from β-AR-induced apoptosis [106–108]. Treatment with the PI3K inhibitor or overexpression of a dominant negative Akt isoform both negate the effect of NRG1 on cardiomyocyte survival, which indicates that the effect of NRG1 is mediated by the PI3K-Akt pathway [106, 108–110]. NRG1 can improve cardiac performance, attenuate pathologic changes, and prolong survival in rodent model of heart failure (secondary to ischemic, dilated, and viral cardiomyopathies) and also in the canine model of pacing-induced heart failure [111]. NRG1-dependent restoration of cardiac function after cardiac injury may include several other mechanisms in addition to the improvement of cardiomyocyte survival, such as ameliorating the sarcomeric structure or cell–cell adhesion or balancing $Ca^{2+}$ homeostasis [112].

Other aspects of GFs' functioning related to cardiac hypertrophy are addressed in Chap. 15.

## GFs and Myocardium Pathophysiology: Atherosclerosis

Since a detailed discussion on signaling pathways and atherosclerosis is presented in Chap. 18, here it suffices to say that the development of atherosclerosis occurs in areas of arteries where blood flow is disturbed, and an initial event for this development is endothelial cell dysfunction/cell death: the mechanisms of impaired endothelial cell death/proliferation in these altered blood flow areas involve different signal pathways.

The endothelial survival-signaling pathway involves VEGF receptor-PI3K-Akt activation, which phosphorylates endothelial nitric oxide synthase (eNOS) leading to NO production [113]. NO production can be induced through both the transcriptional upregulation of eNOS gene expression and the posttranslational modification of eNOS protein [114]. Laminar flow can directly activate VEGF receptors in endothelial cells followed by PI3K-Akt activation. This activation is mediated by c-Src dependent activation of VEGF receptor [115]. Rapid activation of eNOS involves phosphorylation at serine 635 and serine 1179 sites by PI3K-Akt dependent activation [116]. However, this signal pathway could be interrupted by disturbed flow, thus, leading to imbalanced signaling to survival and apoptosis, and development of atherosclerosis.

Neuregulin-1, which belongs to EGF-family, has an anti-atherosclerotic effect. In primary culture of human monocyte-derived macrophages, NRG1 reduces acetylated LDL-induced cholesterol ester accumulation by increasing the expression of ATP-binding cassette transporter, subfamily A, member 1 (ABCA1) and by reducing scavenger receptor SR-A and acyl-coenzyme A: cholesterol acyltransferase 1 (ACAT1) expression. These data obtained from *in vitro* experiments are in agreement with animal model data: chronic infusion of NRG1β suppressed the development of atherosclerotic lesions in $ApoE^{-/-}$ mice [117]. Overexpression of NRG1 in lesions of human coronary arteries, and during the development of carotid stenosis may reflect a protective antiatherosclerotic response of the organism [118, 119].

## GFs and Myocardial Pathophysiology: Cardiac Fibrosis

Cardiac fibrosis is an excessive deposition of scar tissue that significantly reduces heart function in patients with hypertension and heart failure. Numerous studies have suggested that fibrotic diseases develop as a result of abnormal persistent activation of tissue repair system. The latter one involves specialized type of fibroblasts (myofibroblasts) and pericytes. These cells are activated during connective tissue repair and migrate into the wound area, where they synthesize and remodel newly created extracellular matrix (ECM). In response to normal tissue injury, these cells disappear, thereafter, probably by apoptosis. In fibrotic disease, however, myofibroblasts persist, resulting in the excessive production and remodeling of ECM.

Several GFs, such as TGF-β, connective tissue growth factor (CCN2/CTGF), and PDGF contribute into myofibroblast differentiation and its persistence leading to fibrosis. Myofibroblasts express α-smooth muscle actin (α-SMA) and produce increased amounts of ECM. New evidence suggest that TGF-β and CCN2/CTGF cause activation/differentiation of myofibroblasts from mesenchymal fibroblasts [120, 121], whereas PDGF stimulates differentiation/recruitment of myofibroblasts from pericytes [122].

TGF-β plays a central role in fibroblast activation. In isolated cardiac fibroblasts TGF-β directly induces ECM gene expression and promotes ECM deposition [123], and this process involves Smad3 and Erk [123–126]. In addition, TGF-β induces fibroblasts to differentiate into α-SMA-expressing myofibroblasts via signaling pathway involving FAK, JNK, and TAK1 [121, 127–129]. It has been suggested that ALK5 is the predominant receptor mediating TGF-β signaling into the activation of fibroblast, since ALK5 inhibitor significantly reduced TGF-β activity and attenuates left ventricular remodeling [130].

An inhibitor of PDGFRβ, imatinib mesylate, potently inhibits fibroblast proliferation and migration via blocking pericyte recruitment [122]. In addition, PDGF stimulates fibroblasts to differentiate into myofibroblasts *in vitro* and may promote fibrosis by elevating levels of profibrotic TGF-β1 [131–133]. Interestingly, neutralization of another subtype of PDGFR (i.e., PDGFRα) with a specific antibody attenuated the atrial fibrosis underlying atrial fibrillation [134].

## GFs and Endothelium: Angiogenesis

Large MI in the left ventricle (LV) leads to expansion of the necrotic infarct zone and to the compensation of remodeling throughout the remainder of the LV. A new concept with significant clinical potential is the use of therapeutic angiogenesis in MI. Animal studies have confirmed stimulation of ischemic cardiac tissue in response to direct delivery of GFs (or of genes that encode for synthesis of GFs). Several GFs, such as FGFs, VEGFs, monocyte chemotactic protein-1 (MCP-1), and granulocyte-macrophage colony-stimulating factor (GM-CSF), have been shown to induce coronary collateral vessels formation under ischemic conditions by sprouting of new blood vessels from the existing one ("angiogenesis") and by the maturation of capillary blood vessels into mature arteriolar vessels ("arteriogenesis") [135].

FGF-1 and FGF-2 which promote endothelial cell proliferation and the physical organization of endothelial cells into tube-like structures are very potent angiogenic factors (more potent than VEGF or PDGF) [136]. Yanagisawa-Miwa et al. [137] infused FGF-2 into the coronary arteries of dogs and improved cardiac function after experimental MI. Similarly, in the pig model, injection of DNA-encoding FGF-1 or adenovirus-encoding FGF-5 increased collateral blood flow [138]. FGF-2 activates the Erk-1/2 pathway via the Grb2–SOS–Ras pathway, and activation of this pathway is required for FGF's angiogenic activity [139].

Another main growth factor involved in the process of angiogenesis is VEGF. Experimentally induced ischemia results in a dramatic increase in VEGF levels in the dog and rat myocardium, suggesting the possibility that VEGF may participate in revascularization after myocardial ischemia [140, 141]. VEGF was successfully applied to relieve ischemia in atherosclerotic peripheral vascular disease by initiation of angiogenesis after intra-arterial injection of VEGF naked cDNA [142]. Similarly, infusion of VEGF into the iliac artery promotes the development of collateral vessels in the ischemic rabbit hind limb [48]. VEGF appears to act by local upregulation of NO production: it induces the production of NO from rabbit, pig, bovine, and human vascular endothelial cells [143–146]; inhibition of NO production by eNOS inhibitors significantly inhibit the mitogenic and angiogenic effects of VEGF [147]. The precise signaling pathway(s) that mediate endothelial NO-synthesizing machinery responses to VEGF have not yet been elucidated; however, key enzymes activated by VEGF have been identified as phospholipase D and NO synthase, and their activation depends on PKC and its downstream target, MAPK [148, 149].

The evidence points to a role for receptor cofactor NRP1 in VEGFR-regulated endothelial-cell migration and adhesion. For instance, blocking of VEGF binding specifically to NRP1 leads to inhibition of the migratory response to VEGF, endothelial cell sprouting (*in vitro*), and neovascularization (*in vivo*) [18].

A number of alterations observed in microvessels of the ischemic heart are related to changes in vascular endothelial and SMC function, and induction of vessel wall remodeling. Recent observations showed that the development of pathological remodeling in mesenteric resistance arteries and coronary arterioles is highly dependent on EGFR Tyr-phosphorylation/activation in SMCs and endothelial cells [150, 151]. Moreover, structural wall remodeling was improved with EGFR tyrosine kinase inhibition, suggesting that EGFR is a key element in microvascular pathophysiology and could be used as a therapeutic target [151, 152].

## Conclusions

A number of GFs are involved in signaling in the heart, both in normal development/function and in pharmacological cardiotoxicity. They activate different receptors, including PTK receptors, G protein-coupled receptors, and serine–threonine kinase receptors present on the cellular plasma membrane, which then recruit a number of intracellular signal transduction pathways.

Multiple GF-controlled signaling pathways are involved in cardiogenesis and coronary vasculogenesis. In each step of heart morphogenesis, numerous signaling pathways are used in multiple roles. For example, FGFs released by the embryonic epicardium, act on the underlying myocardium, induce cardiomyocyte proliferation and likely participate in the elaboration of other growth factors that signal back to the epicardium. VEGF-pathway is important for the generation of vascular endothelial cells, whereas PDGF and TGF-β are important for the differentiation of coronary vascular smooth muscle.

In the adult heart, GFs exert a diverse array of cardiovascular effects. Interestingly, many of GFs are released by the cardiomyocyte during myocardial ischemia, suggesting that they may play a role in endogenous cardioprotection. In this respect, genetic ablation studies suggest that some of these GFs (FGF-2, VEGF, and urocortin) may actually confer endogenous protection against acute ischemia–reperfusion injury. Cardioprotective GFs activate cardioprotective intracellular signal transduction pathways, including the RISK pathway, many of which terminate in the mitochondria.

Many GFs have been safely used in different clinical settings, such as heart failure. Also, there are several potential clinical applications for both agonists and antagonists of different GFs. In this context, drugs affecting TGF-β and PDGF are being considered as antifibrotic therapies: there are a wide range of possible antifibrotic treatments that target the TGF-β, PDGF network. However, more clinical data are needed to properly evaluate the efficacy of them. In clinical cases of loss of functional myocardium, FGF agonists seem to restore myocardial mass and improve myocardial function. Similarly, neuregulin-1 activation in the heart may offer some degree of cardioprotection or enhance recovery from injury.

On the other hand, there are some limitations for GFs-based therapy. For example, the anticancer agent trastuzumab (herceptin) cardiotoxic side-effects are well-documented. Trastuzumab is an anti-ErbB2 monoclonal antibody which is effective in the treatment of breast cancer patients. It is possible that in the heart trastuzumab reduces ErbB signaling and thus interferes with ErbB-controlled cardiac cell-survival pathways. Given that the mammalian cardiomyocyte is a terminally differentiated nonproliferating cell, cardiotoxic side effects of anticancer agent result in the loss of cardiomyocytes. Similarly, FGF antagonists who are currently being evaluated for the treatment of different kinds of cancers may affect myocardial function and its vascular supply.

## Summary

- FGF-1 and FGF-2 are secreted by cardiomyocytes in response to myocardial ischemia. Important functions of FGFs include the formation of coronary vasculature in the developing heart and promotion of angiogenesis.
- VEGF controls blood vessel development, angiogenesis, and cardioprotection. The activation of the VEGF receptor

leads to the activation of RISK-related signal transduction pathways. VEGF receptors can interact with coreceptors, neuropilins (NRPs)-1 and -2. Several mechanisms could account for the NRP1-dependent enhancement of VEGFR2 signaling.

- PDGFs are products of four different genes that bind to two different receptors, regulate epicardial EMT and the development of coronary vascular SMCs, and contribute to cardiac fibrosis.
- Growth factors from the EGF-family are important during cardiac embryogenesis and play an important role in the regulation of cell proliferation, differentiation, and survival. ErbB family of transmembrane proteins are receptors for EGF-family GFs.
- Ligand-dependent autophosphorylation of ErbB receptors initiates interaction with many signaling proteins and leads to theactivation of Ras-Raf-MEK-Erk- and PI3K-Akt-signaling pathways. Proteolytic 80 kDa-fragment of ErbB4 translocates to the nucleus and regulates transcription.
- Binding of IGFs and insulin to their receptors results in the activation of Ras-Raf-MEK-Erk- and PI3K-Akt-signaling pathways.
- TGF-β family ligands form homo- or heterodimers which bind to and activate two types of transmembrane receptors with intrinsic serine/threonine kinase activity. There are two subfamilies of the TGF-β receptors, type I and type II. Type I receptor becomes phosphorylated/activated by the GF-activated type II receptor.
- The family of Smad proteins plays different roles in TGF-β superfamily signaling. R-Smads are targets for phosphorylation by activated type I receptor. Smad4 complexes with phosphorylated R-Smad and translocates into the nucleus to activate transcription of target genes. Smad6 and Smad7 are inhibitory Smads.
- Smads require interactions with a number of DNA binding protein coactivators for high-affinity binding to DNA. There is also a group of protein repressors that inhibit Smad-regulated pathways.
- TGF-β can activate a variety of noncanonical (non-Smad) signaling pathways and cross talk with several second messenger-regulating systems.
- Urocortin and adrenomedullin represent GFs that signal via G protein-coupled receptors. Some studies have demonstrated that urocortin has an acute cardioprotective effect. Adrenomedullin has a positive influence in several cardiovascular diseases.
- FGF signaling is required for embryonic cardiomyocyte proliferation. FGF signals from myocardium induce differentiation of the subepicardial mesenchyme into the different components of the coronary vasculature. Another GFs, which promote epicardial EMT during heart embryogenesis, are PDGF and TGF-β1–3.
- PDGF and GFs from TGF-β super family are also a required for the differentiation of SMCs. Coronary vasculogenesis in the developing heart is mediated by VEGF and cofactor of VEGFR, NRP.
- Gene targeting strategies in mice have shown the significance of the ErbB3 and ErbB4 receptors in cardiac development. For initial formation of the cardiac primordium, BMP-2 is required.
- Cardioprotection involves the activation of components of the RISK pathway, Akt and Erk1/2. A number of GFs that activate PI3K-Akt and MEK1/2-Erk1/2 components are cardioprotective: FGF-2, VEGF, IGF-I, TGF-β1, urocortin.
- FGF-2 and GFs from the EGF-family are involved in the compensatory hypertrophic remodeling of the heart after MI.
- VEGF plays a protective role during atherosclerosis because it activates the prosurvival signal pathway "PI3K-Akt-eNOS" in the endothelial cells. Antiatherosclerotic effect of neuregulin-1 is related to the reduction of cholesterol ester accumulation in macrophages.
- TGF-β, CTGF, and PDGF contribute to the development of cardiac fibrosis because they cause differentiation of myofibroblasts and promote ECM deposition. An inhibitor of PDGFRβ, imatinib mesylate, potently inhibits fibroblast proliferation and migration via blocking pericyte recruitment.
- Several growth factors, such as FGFs, VEGFs, MCP-1, and GM-CSF, have been shown to induce coronary collateral vessels formation under ischemic conditions by sprouting of new blood vessels from the existing one and by the maturation of capillary blood vessels into mature arteriolar vessels. Recent data suggest that EGFR is also a key element in microvascular pathophysiology and could be used as a therapeutic target.

## References


1. Sheikh F, Jin Y, Pasumarthi KB, Kardami E, Cattini PA. Expression of fibroblast growth factor receptor-1 in rat heart H9c2 myoblasts increases cell proliferation. Mol Cell Biochem. 1997;176:89–97.
2. Detillieux KA, Sheikh F, Kardami E, Cattini PA. Biological activities of fibroblast growth factor-2 in the adult myocardium. Cardiovasc Res. 2003;57:8–19.
3. Kardami E, Detillieux K, Ma X, et al. Fibroblast growth factor-2 and cardioprotection. Heart Fail Rev. 2007;12:267–77.
4. Ahn A, Frishman WH, Gutwein A, Passeri J, Nelson M. Therapeutic angiogenesis: a new treatment approach for ischemic heart disease – part I. Cardiol Rev. 2008;16:163–71.
5. Takahashi N, Seko Y, Noiri E, et al. Vascular endothelial growth factor induces activation and subcellular translocation of focal adhesion kinase (p125FAK) in cultured rat cardiac myocytes. Circ Res. 1999;84:1194–202.
6. Seko Y, Takahashi N, Tobe K, Ueki K, Kadowaki T, Yazaki Y. Vascular endothelial growth factor (VEGF) activates Raf-1, mitogen-activated protein (MAP) kinases, and S6 kinase (p90rsk) in cultured rat cardiac myocytes. J Cell Physiol. 1998;175:239–46.
7. Gerber HP, McMurtrey A, Kowalski J, et al. Vascular endothelial growth factor regulates endothelial cell survival through the

phosphatidylinositol 3'-kinase/Akt signal transduction pathway. Requirement for Flk-1/KDR activation. J Biol Chem. 1998;273:30336–43.

8. Kawasaki T, Kitsukawa T, Bekku Y, et al. A requirement for neuropilin-1 in embryonic vessel formation. Development. 1999;126:4895–902.
9. Lee P, Goishi K, Davidson AJ, Mannix R, Zon L, Klagsbrun M. Neuropilin-1 is required for vascular development and is a mediator of VEGF-dependent angiogenesis in zebrafish. Proc Natl Acad Sci USA. 2002;99:10470–5.
10. Gu C, Rodriguez ER, Reimert DV, et al. Neuropilin-1 conveys semaphorin and VEGF signaling during neural and cardiovascular development. Dev Cell. 2003;5:45–57.
11. Cai H, Reed RR. Cloning and characterization of neuropilin-1-interacting protein: a PSD-95/Dlg/ZO-1 domain-containing protein that interacts with the cytoplasmic domain of neuropilin-1. J Neurosci. 1999;19:6519–27.
12. Pan Q, Chathery Y, Wu Y, et al. Neuropilin-1 binds to VEGF121 and regulates endothelial cell migration and sprouting. J Biol Chem. 2007;282:24049–56.
13. Whitaker GB, Limberg BJ, Rosenbaum JS. Vascular endothelial growth factor receptor-2 and neuropilin-1 form a receptor complex that is responsible for the differential signaling potency of VEGF(165) and VEGF(121). J Biol Chem. 2001;276:25520–31.
14. Soker S, Miao HQ, Nomi M, Takashima S, Klagsbrun M. VEGF165 mediates formation of complexes containing VEGFR-2 and neuropilin-1 that enhance VEGF165-receptor binding. J Cell Biochem. 2002;85:357–68.
15. Oh H, Takagi H, Otani A, et al. Selective induction of neuropilin-1 by vascular endothelial growth factor (VEGF): a mechanism contributing to VEGF-induced angiogenesis. Proc Natl Acad Sci USA. 2002;99:383–8.
16. Jia H, Bagherzadeh A, Hartzoulakis B, et al. Characterization of a bicyclic peptide neuropilin-1 (NP-1) antagonist (EG3287) reveals importance of vascular endothelial growth factor exon 8 for NP-1 binding and role of NP-1 in KDR signaling. J Biol Chem. 2006;281:13493–502.
17. Wang L, Zeng H, Wang P, Soker S, Mukhopadhyay D. Neuropilin-1-mediated vascular permeability factor/vascular endothelial growth factor-dependent endothelial cell migration. J Biol Chem. 2003;278:48848–60.
18. Pan Q, Chanthery Y, Liang WC, et al. Blocking neuropilin-1 function has an additive effect with anti-VEGF to inhibit tumor growth. Cancer Cell. 2007;11:53–67.
19. Wang L, Mukhopadhyay D, Xu X. C terminus of RGS-GAIP-interacting protein conveys neuropilin-1-mediated signaling during angiogenesis. FASEB J. 2006;20:1513–5.
20. Fredriksson L, Li H, Eriksson U. The PDGF family: four gene products form five dimeric isoforms. Cytokine Growth Factor Rev. 2004;15:197–204.
21. Busfield SJ, Michnick DA, Chickering TW, et al. Characterization of a neuregulin-related gene, Don-1, that is highly expressed in restricted regions of the cerebellum and hippocampus. Mol Cell Biol. 1997;17:4007–14.
22. Blagoev B, Kratchmarova I, Ong SE, Nielsen M, Foster LJ, Mann M. A proteomics strategy to elucidate functional protein-protein interactions applied to EGF signaling. Nat Biotechnol. 2003;21:315–8.
23. Dengjel J, Akimov V, Olsen JV, et al. Quantitative proteomic assessment of very early cellular signaling events. Nat Biotechnol. 2007;25:566–8.
24. Fuller SJ, Sivarajah K, Sugden PH. ErbB receptors, their ligands, and the consequences of their activation and inhibition in the myocardium. J Mol Cell Cardiol. 2008;44:831–51.
25. Carpenter G. Nuclear localization and possible functions of receptor tyrosine kinases. Curr Opin Cell Biol. 2003;15:143–8.
26. Schlessinger J, Lemmon MA. Nuclear signaling by receptor tyrosine kinases: the first robin of spring. Cell. 2006;127:45–8.
27. Hermonat PL, Li D, Yang B, Mehta JL. Mechanism of action and delivery possibilities for TGFbeta1 in the treatment of myocardial ischemia. Cardiovasc Res. 2007;74:235–43.
28. Ruiz-Ortega M, Rodriguez-Vita J, Sanchez-Lopez E, Carvajal G, Egido J. TGF-beta signaling in vascular fibrosis. Cardiovasc Res. 2007;74:196–206.
29. Gougos A, Letarte M. Primary structure of endoglin, an RGD-containing glycoprotein of human endothelial cells. J Biol Chem. 1990;265:8361–4.
30. Lastres P, Letamendia A, Zhang H, et al. Endoglin modulates cellular responses to TGF-beta 1. J Cell Biol. 1996;133:1109–21.
31. Lopez-Novoa JM, Bernabeu C. The physiological role of endoglin in the cardiovascular system. Am J Physiol Heart Circ Physiol. 2010;299:H959–74.
32. Nakao A, Imamura T, Souchelnytskyi S, et al. TGF-beta receptor-mediated signalling through Smad2, Smad3 and Smad4. EMBO J. 1997;16:5353–62.
33. Kretzschmar M, Liu F, Hata A, Doody J, Massague J. The TGF-beta family mediator Smad1 is phosphorylated directly and activated functionally by the BMP receptor kinase. Genes Dev. 1997;11:984–95.
34. Watanabe TK, Suzuki M, Omori Y, et al. Cloning and characterization of a novel member of the human Mad gene family (MADH6). Genomics. 1997;42:446–51.
35. Janknecht R, Wells NJ, Hunter T. TGF-beta-stimulated cooperation of smad proteins with the coactivators CBP/p300. Genes Dev. 1998;12:2114–9.
36. Hoodless PA, Pye M, Chazaud C, et al. FoxH1 (Fast) functions to specify the anterior primitive streak in the mouse. Genes Dev. 2001;15:1257–71.
37. Kim RH, Wang D, Tsang M, et al. A novel smad nuclear interacting protein, SNIP1, suppresses p300-dependent TGF-beta signal transduction. Genes Dev. 2000;14:1605–16.
38. Sun Y, Liu X, Eaton EN, Lane WS, Lodish HF, Weinberg RA. Interaction of the Ski oncoprotein with Smad3 regulates TGF-beta signaling. Mol Cell. 1999;4:499–509.
39. Melhuish TA, Gallo CM, Wotton D. TGIF2 interacts with histone deacetylase 1 and represses transcription. J Biol Chem. 2001;276:32109–14.
40. Zhu H, Kavsak P, Abdollah S, Wrana JL, Thomsen GH. A SMAD ubiquitin ligase targets the BMP pathway and affects embryonic pattern formation. Nature. 1999;400:687–93.
41. Zhang Y, Chang C, Gehling DJ, Hemmati-Brivanlou A, Derynck R. Regulation of Smad degradation and activity by Smurf2, an E3 ubiquitin ligase. Proc Natl Acad Sci USA. 2001;98:974–9.
42. Lavine KJ, Yu K, White AC, et al. Endocardial and epicardial derived FGF signals regulate myocardial proliferation and differentiation in vivo. Dev Cell. 2005;8:85–95.
43. Pennisi DJ, Mikawa T. Normal patterning of the coronary capillary plexus is dependent on the correct transmural gradient of FGF expression in the myocardium. Dev Biol. 2005;279:378–90.
44. Lavine KJ, White AC, Park C, et al. Fibroblast growth factor signals regulate a wave of Hedgehog activation that is essential for coronary vascular development. Genes Dev. 2006;20:1651–66.
45. Pennisi DJ, Mikawa T. FGFR-1 is required by epicardium-derived cells for myocardial invasion and correct coronary vascular lineage differentiation. Dev Biol. 2009;328:148–59.
46. Tomanek RJ, Zheng W, Peters KG, Lin P, Holifield JS, Suvarna PR. Multiple growth factors regulate coronary embryonic vasculogenesis. Dev Dyn. 2001;221:265–73.
47. Fong GH, Rossant J, Gertsenstein M, Breitman ML. Role of the Flt-1 receptor tyrosine kinase in regulating the assembly of vascular endothelium. Nature. 1995;376:66–70.

48. Takeshita S, Zheng LP, Brogi E, et al. Therapeutic angiogenesis. A single intraarterial bolus of vascular endothelial growth factor augments revascularization in a rabbit ischemic hind limb model. J Clin Invest. 1994;93:662–70.
49. Nesbitt TL, Roberts A, Tan H, et al. Coronary endothelial proliferation and morphogenesis are regulated by a VEGF-mediated pathway. Dev Dyn. 2009;238:423–30.
50. Tomanek RJ, Ishii Y, Holifield JS, Sjogren CL, Hansen HK, Mikawa T. VEGF family members regulate myocardial tubulogenesis and coronary artery formation in the embryo. Circ Res. 2006;98:947–53.
51. Yuan L, Moyon D, Pardanaud L, et al. Abnormal lymphatic vessel development in neuropilin 2 mutant mice. Development. 2002;129: 4797–806.
52. Shinbrot E, Peters KG, Williams LT. Expression of the platelet-derived growth factor beta receptor during organogenesis and tissue differentiation in the mouse embryo. Dev Dyn. 1994;199:169–75.
53. Lu J, Landerholm TE, Wei JS, et al. Coronary smooth muscle differentiation from proepicardial cells requires rhoA-mediated actin reorganization and p160 rho-kinase activity. Dev Biol. 2001; 240:404–18.
54. Mellgren AM, Smith CL, Olsen GS, et al. Platelet-derived growth factor receptor beta signaling is required for efficient epicardial cell migration and development of two distinct coronary vascular smooth muscle cell populations. Circ Res. 2008;103: 1393–401.
55. Hellstrom M, Kalen M, Lindahl P, Abramsson A, Betsholtz C. Role of PDGF-B and PDGFR-beta in recruitment of vascular smooth muscle cells and pericytes during embryonic blood vessel formation in the mouse. Development. 1999;126:3047–55.
56. Garratt AN. "To erb-B or not to erb-B…" Neuregulin-1/ErbB signaling in heart development and function. J Mol Cell Cardiol. 2006;41:215–8.
57. Iwamoto R, Mekada E. ErbB and HB-EGF signaling in heart development and function. Cell Struct Funct. 2006;31:1–14.
58. Gassmann M, Casagranda F, Orioli D, et al. Aberrant neural and cardiac development in mice lacking the ErbB4 neuregulin receptor. Nature. 1995;378:390–4.
59. Tidcombe H, Jackson-Fisher A, Mathers K, Stern DF, Gassmann M, Golding JP. Neural and mammary gland defects in ErbB4 knockout mice genetically rescued from embryonic lethality. Proc Natl Acad Sci USA. 2003;100:8281–6.
60. Erickson SL, O'Shea KS, Ghaboosi N, et al. ErbB3 is required for normal cerebellar and cardiac development: a comparison with ErbB2-and heregulin-deficient mice. Development. 1997;124: 4999–5011.
61. Smith TK, Bader DM. Signals from both sides: control of cardiac development by the endocardium and epicardium. Semin Cell Dev Biol. 2007;18:84–9.
62. Iwamoto R, Yamazaki S, Asakura M, et al. Heparin-binding EGF-like growth factor and ErbB signaling is essential for heart function. Proc Natl Acad Sci USA. 2003;100:3221–6.
63. Zhang H, Bradley A. Mice deficient for BMP2 are nonviable and have defects in amnion/chorion and cardiac development. Development. 1996;122:2977–86.
64. Olivey HE, Mundell NA, Austin AF, Barnett JV. Transforming growth factor-beta stimulates epithelial-mesenchymal transformation in the proepicardium. Dev Dyn. 2006;235:50–9.
65. Austin AF, Compton LA, Love JD, Brown CB, Barnett JV. Primary and immortalized mouse epicardial cells undergo differentiation in response to TGFbeta. Dev Dyn. 2008;237:366–76.
66. Shull MM, Ormsby I, Kier AB, Pawlowski S, Diebold RJ, Yin M, et al. Targeted disruption of the mouse transforming growth factor-beta 1 gene results in multifocal inflammatory disease. Nature. 1992;359:693–9.
67. Kaartinen V, Voncken JW, Shuler C, et al. Abnormal lung development and cleft palate in mice lacking TGF-beta 3 indicates defects of epithelial-mesenchymal interaction. Nat Genet. 1995;11:415–21.
68. Sridurongrit S, Larsson J, Schwartz R, Ruiz-Lozano P, Kaartinen V. Signaling via the Tgf-beta type I receptor Alk5 in heart development. Dev Biol. 2008;322:208–18.
69. Galvin KM, Donovan MJ, Lynch CA, et al. A role for smad6 in development and homeostasis of the cardiovascular system. Nat Genet. 2000;24:171–4.
70. Hausenloy DJ, Yellon DM. Survival kinases in ischemic preconditioning and postconditioning. Cardiovasc Res. 2006;70:240–53.
71. Davidson SM, Hausenloy D, Duchen MR, Yellon DM. Signalling via the reperfusion injury signalling kinase (RISK) pathway links closure of the mitochondrial permeability transition pore to cardioprotection. Int J Biochem Cell Biol. 2006;38:414–9.
72. Hausenloy DJ, Yellon DM. Reperfusion injury salvage kinase signalling: taking a RISK for cardioprotection. Heart Fail Rev. 2007;12:217–34.
73. Yellon DM, Baxter GF. Reperfusion injury revisited: is there a role for growth factor signaling in limiting lethal reperfusion injury? Trends Cardiovasc Med. 1999;9:245–9.
74. Htun P, Ito WD, Hoefer IE, Schaper J, Schaper W. Intramyocardial infusion of FGF-1 mimics ischemic preconditioning in pig myocardium. J Mol Cell Cardiol. 1998;30:867–77.
75. Jiang ZS, Srisakuldee W, Soulet F, Bouche G, Kardami E. Non-angiogenic FGF-2 protects the ischemic heart from injury, in the presence or absence of reperfusion. Cardiovasc Res. 2004;62: 154–66.
76. Cuevas P, Carceller F, Martinez-Coso V, et al. Cardioprotection from ischemia by fibroblast growth factor: role of inducible nitric oxide synthase. Eur J Med Res. 1999;4:517–24.
77. Padua RR, Merle PL, Doble BW, et al. FGF-2-induced negative inotropism and cardioprotection are inhibited by chelerythrine: involvement of sarcolemmal calcium-independent protein kinase C. J Mol Cell Cardiol. 1998;30:2695–709.
78. Sheikh F, Sontag DP, Fandrich RR, Kardami E, Cattini PA. Overexpression of FGF-2 increases cardiac myocyte viability after injury in isolated mouse hearts. Am J Physiol Heart Circ Physiol. 2001;280:H1039–50.
79. Luo Z, Diaco M, Murohara T, Ferrara N, Isner JM, Symes JF. Vascular endothelial growth factor attenuates myocardial ischemia-reperfusion injury. Ann Thorac Surg. 1997;64:993–8.
80. Yamashita K, Kajstura J, Discher DJ, et al. Reperfusion-activated Akt kinase prevents apoptosis in transgenic mouse hearts overexpressing insulin-like growth factor-1. Circ Res. 2001;88:609–14.
81. Yamamura T, Otani H, Nakao Y, Hattori R, Osako M, Imamura H. IGF-I differentially regulates Bcl-xL and Bax and confers myocardial protection in the rat heart. Am J Physiol Heart Circ Physiol. 2001;280:H1191–200.
82. Pi Y, Goldenthal MJ, Marin-Garcia J. Mitochondrial involvement in IGF-1 induced protection of cardiomyocytes against hypoxia/reoxygenation injury. Mol Cell Biochem. 2007;301:181–9.
83. Wang L, Ma W, Markovich R, Chen JW, Wang PH. Regulation of cardiomyocyte apoptotic signaling by insulin-like growth factor I. Circ Res. 1998;83:516–22.
84. Yang BC, Zander DS, Mehta JL. Hypoxia-reoxygenation-induced apoptosis in cultured adult rat myocytes and the protective effect of platelets and transforming growth factor-beta(1). J Pharmacol Exp Ther. 1999;291:733–8.
85. Baxter GF, Mocanu MM, Brar BK, Latchman DS, Yellon DM. Cardioprotective effects of transforming growth factor-beta1 during early reoxygenation or reperfusion are mediated by p42/p44 MAPK. J Cardiovasc Pharmacol. 2001;38:930–9.
86. Okosi A, Brar BK, Chan M, et al. Expression and protective effects of urocortin in cardiac myocytes. Neuropeptides. 1998;32:167–71.

87. Brar BK, Stephanou A, Okosi A, et al. CRH-like peptides protect cardiac myocytes from lethal ischaemic injury. Mol Cell Endocrinol. 1999;158:55–63.
88. Brar BK, Stephanou A, Knight R, Latchman DS. Activation of protein kinase B/Akt by urocortin is essential for its ability to protect cardiac cells against hypoxia/reoxygenation-induced cell death. J Mol Cell Cardiol. 2002;34:483–92.
89. Brar BK, Jonassen AK, Stephanou A, et al. Urocortin protects against ischemic and reperfusion injury via a MAPK-dependent pathway. J Biol Chem. 2000;275:8508–14.
90. Lawrence KM, Chanalaris A, Scarabelli T, et al. K(ATP) channel gene expression is induced by urocortin and mediates its cardioprotective effect. Circulation. 2002;106:1556–62.
91. Gordon JM, Dusting GJ, Woodman OL, Ritchie RH. Cardioprotective action of CRF peptide urocortin against simulated ischemia in adult rat cardiomyocytes. Am J Physiol Heart Circ Physiol. 2003;284:H330–6.
92. Townsend PA, Davidson SM, Clarke SJ, et al. Urocortin prevents mitochondrial permeability transition in response to reperfusion injury indirectly by reducing oxidative stress. Am J Physiol Heart Circ Physiol. 2007;293:H928–38.
93. Lawrence KM, Scarabelli TM, Turtle L, et al. Urocortin protects cardiac myocytes from ischemia/reperfusion injury by attenuating calcium-insensitive phospholipase A2 gene expression. FASEB J. 2003;17:2313–5.
94. Lawrence KM, Townsend PA, Davidson SM, et al. The cardioprotective effect of urocortin during ischaemia/reperfusion involves the prevention of mitochondrial damage. Biochem Biophys Res Commun. 2004;321:479–86.
95. Chanalaris A, Lawrence KM, Stephanou A, et al. Protective effects of the urocortin homologues stresscopin (SCP) and stresscopin-related peptide (SRP) against hypoxia/reoxygenation injury in rat neonatal cardiomyocytes. J Mol Cell Cardiol. 2003;35:1295–305.
96. Brar BK, Jonassen AK, Egorina EM, et al. Urocortin-II and urocortin-III are cardioprotective against ischemia reperfusion injury: an essential endogenous cardioprotective role for corticotropin releasing factor receptor type 2 in the murine heart. Endocrinology. 2004;145:24–35. discussion 21–3.
97. Jiang ZS, Jeyaraman M, Wen GB, et al. High- but not low-molecular weight FGF-2 causes cardiac hypertrophy in vivo; possible involvement of cardiotrophin-1. J Mol Cell Cardiol. 2007;42:222–33.
98. Schultz JE, Witt SA, Nieman ML, et al. Fibroblast growth factor-2 mediates pressure-induced hypertrophic response. J Clin Invest. 1999;104:709–19.
99. Pellieux C, Foletti A, Peduto G, et al. Dilated cardiomyopathy and impaired cardiac hypertrophic response to angiotensin II in mice lacking FGF-2. J Clin Invest. 2001;108:1843–51.
100. Heineke J, Molkentin JD. Regulation of cardiac hypertrophy by intracellular signalling pathways. Nat Rev Mol Cell Biol. 2006;7: 589–600.
101. Wang Y. Mitogen-activated protein kinases in heart development and diseases. Circulation. 2007;116:1413–23.
102. Matsui T, Rosenzweig A. Convergent signal transduction pathways controlling cardiomyocyte survival and function: the role of PI 3-kinase and Akt. J Mol Cell Cardiol. 2005;38:63–71.
103. Shiojima I, Walsh K. Regulation of cardiac growth and coronary angiogenesis by the Akt/PKB signaling pathway. Genes Dev. 2006;20:3347–65.
104. Kerkela R, Woulfe K, Force T. Glycogen synthase kinase-3beta – actively inhibiting hypertrophy. Trends Cardiovasc Med. 2007;17: 91–6.
105. Sugden PH, Fuller SJ, Weiss SC, Clerk A. Glycogen synthase kinase 3 (GSK3) in the heart: a point of integration in hypertrophic signalling and a therapeutic target? A critical analysis. Br J Pharmacol. 2008;153 Suppl 1:S137–53.
106. Kuramochi Y, Lim CC, Guo X, Colucci WS, Liao R, Sawyer DB. Myocyte contractile activity modulates norepinephrine cytotoxicity and survival effects of neuregulin-1beta. Am J Physiol Cell Physiol. 2004;286:C222–9.
107. Okoshi K, Nakayama M, Yan X, et al. Neuregulins regulate cardiac parasympathetic activity: muscarinic modulation of beta-adrenergic activity in myocytes from mice with neuregulin-1 gene deletion. Circulation. 2004;110:713–7.
108. Fukazawa R, Miller TA, Kuramochi Y, et al. Neuregulin-1 protects ventricular myocytes from anthracycline-induced apoptosis via erbB4-dependent activation of PI3-kinase/Akt. J Mol Cell Cardiol. 2003;35:1473–9.
109. Miao W, Luo Z, Kitsis RN, Walsh K. Intracoronary, adenovirus-mediated Akt gene transfer in heart limits infarct size following ischemia-reperfusion injury in vivo. J Mol Cell Cardiol. 2000; 32:2397–402.
110. Fujio Y, Nguyen T, Wencker D, Kitsis RN, Walsh K. Akt promotes survival of cardiomyocytes in vitro and protects against ischemia-reperfusion injury in mouse heart. Circulation. 2000;101:660–7.
111. Liu X, Gu X, Li Z, et al. Neuregulin-1/erbB-activation improves cardiac function and survival in models of ischemic, dilated, and viral cardiomyopathy. J Am Coll Cardiol. 2006;48:1438–47.
112. Jiang Z, Zhou M. Neuregulin signaling and heart failure. Curr Heart Fail Rep. 2010;7:42–7.
113. Hemmings BA. Akt signaling: linking membrane events to life and death decisions. Science. 1997;275:628–30.
114. Jin ZG, Ueba H, Tanimoto T, Lungu AO, Frame MD, Berk BC. Ligand-independent activation of vascular endothelial growth factor receptor 2 by fluid shear stress regulates activation of endothelial nitric oxide synthase. Circ Res. 2003;93:354–63.
115. Chen KD, Li YS, Kim M, et al. Mechanotransduction in response to shear stress. Roles of receptor tyrosine kinases, integrins, and Shc. J Biol Chem. 1999;274:18393–400.
116. Davis ME, Cai H, Drummond GR, Harrison DG. Shear stress regulates endothelial nitric oxide synthase expression through c-Src by divergent signaling pathways. Circ Res. 2001;89:1073–80.
117. Xu G, Watanabe T, Iso Y, et al. Preventive effects of heregulin-beta1 on macrophage foam cell formation and atherosclerosis. Circ Res. 2009;105:500–10.
118. Sigala F, Georgopoulos S, Papalambros E, et al. Heregulin, cysteine rich-61 and matrix metalloproteinase 9 expression in human carotid atherosclerotic plaques: relationship with clinical data. Eur J Vasc Endovasc Surg. 2006;32:238–45.
119. Panutsopulos D, Arvanitis DL, Tsatsanis C, Papalambros E, Sigala F, Spandidos DA. Expression of heregulin in human coronary atherosclerotic lesions. J Vasc Res. 2005;42:463–74.
120. Hinz B, Phan SH, Thannickal VJ, Galli A, Bochaton-Piallat ML, Gabbiani G. The myofibroblast: one function, multiple origins. Am J Pathol. 2007;170:1807–16.
121. Leask A. Targeting the TGFbeta, endothelin-1 and CCN2 axis to combat fibrosis in scleroderma. Cell Signal. 2008;20:1409–14.
122. Rajkumar VS, Shiwen X, Bostrom M, et al. Platelet-derived growth factor-beta receptor activation is essential for fibroblast and pericyte recruitment during cutaneous wound healing. Am J Pathol. 2006;169:2254–65.
123. Bujak M, Ren G, Kweon HJ, et al. Essential role of Smad3 in infarct healing and in the pathogenesis of cardiac remodeling. Circulation. 2007;116:2127–38.
124. Verrecchia F, Chu ML, Mauviel A. Identification of novel TGF-beta/Smad gene targets in dermal fibroblasts using a combined cDNA microarray/promoter transactivation approach. J Biol Chem. 2001;276:17058–62.
125. Yang YC, Piek E, Zavadil J, et al. Hierarchical model of gene regulation by transforming growth factor beta. Proc Natl Acad Sci USA. 2003;100:10269–74.
126. Liu X, Sun SQ, Hassid A, Ostrom RS. cAMP inhibits transforming growth factor-beta-stimulated collagen synthesis via inhibition of extracellular signal-regulated kinase 1/2 and Smad signaling in cardiac fibroblasts. Mol Pharmacol. 2006;70:1992–2003.

127. Liu S, Xu SW, Kennedy L, et al. FAK is required for TGFbeta-induced JNK phosphorylation in fibroblasts: implications for acquisition of a matrix-remodeling phenotype. Mol Biol Cell. 2007;18:2169–78.
128. Thannickal VJ, Lee DY, White ES, et al. Myofibroblast differentiation by transforming growth factor-beta1 is dependent on cell adhesion and integrin signaling via focal adhesion kinase. J Biol Chem. 2003;278:12384–9.
129. Shi-wen X, Parapuram SK, Pala D, et al. Requirement of transforming growth factor beta-activated kinase 1 for transforming growth factor beta-induced alpha-smooth muscle actin expression and extracellular matrix contraction in fibroblasts. Arthritis Rheum. 2009;60:234–41.
130. Tan SM, Zhang Y, Connelly KA, Gilbert RE, Kelly DJ. Targeted inhibition of activin receptor-like kinase 5 signaling attenuates cardiac dysfunction following myocardial infarction. Am J Physiol Heart Circ Physiol. 2010;298:H1415–25.
131. Rhee S, Grinnell F. P21-activated kinase 1: convergence point in PDGF- and LPA-stimulated collagen matrix contraction by human fibroblasts. J Cell Biol. 2006;172:423–32.
132. Jinnin M, Ihn H, Mimura Y, Asano Y, Yamane K, Tamaki K. Regulation of fibrogenic/fibrolytic genes by platelet-derived growth factor C, a novel growth factor, in human dermal fibroblasts. J Cell Physiol. 2005;202:510–7.
133. Tuuminen R, Nykanen AI, Krebs R, et al. PDGF-A, -C, and -D but not PDGF-B increase TGF-beta1 and chronic rejection in rat cardiac allografts. Arterioscler Thromb Vasc Biol. 2009;29:691–8.
134. Liao CH, Akazawa H, Tamagawa M, et al. Cardiac mast cells cause atrial fibrillation through PDGF-A-mediated fibrosis in pressure-overloaded mouse hearts. J Clin Invest. 2010;120:242–53.
135. Yla-Herttuala S, Alitalo K. Gene transfer as a tool to induce therapeutic vascular growth. Nat Med. 2003;9:694–701.
136. Joseph-Silverstein J, Rifkin DB. Endothelial cell growth factors and the vessel wall. Semin Thromb Hemost. 1987;13:504–13.
137. Yanagisawa-Miwa A, Uchida Y, Nakamura F, et al. Salvage of infarcted myocardium by angiogenic action of basic fibroblast growth factor. Science. 1992;257:1401–3.
138. Tio RA, Tkebuchava T, Scheuermann TH, et al. Intramyocardial gene therapy with naked DNA encoding vascular endothelial growth factor improves collateral flow to ischemic myocardium. Hum Gene Ther. 1999;10:2953–60.
139. Simons M. Integrative signaling in angiogenesis. Mol Cell Biochem. 2004;264:99–102.
140. Ferrara N, Davis-Smyth T. The biology of vascular endothelial growth factor. Endocr Rev. 1997;18:4–25.
141. Neufeld G, Cohen T, Gengrinovitch S, Poltorak Z. Vascular endothelial growth factor (VEGF) and its receptors. FASEB J. 1999;13:9–22.
142. Isner JM, Pieczek A, Schainfeld R, et al. Clinical evidence of angiogenesis after arterial gene transfer of phVEGF165 in patient with ischaemic limb. Lancet. 1996;348:370–4.
143. Murohara T, Asahara T, Silver M, et al. Nitric oxide synthase modulates angiogenesis in response to tissue ischemia. J Clin Invest. 1998;101:2567–78.
144. Morbidelli L, Chang CH, Douglas JG, Granger HJ, Ledda F, Ziche M. Nitric oxide mediates mitogenic effect of VEGF on coronary venular endothelium. Am J Physiol. 1996;270:H411–5.
145. Parenti A, Morbidelli L, Cui XL, et al. Nitric oxide is an upstream signal of vascular endothelial growth factor-induced extracellular signal-regulated kinase1/2 activation in postcapillary endothelium. J Biol Chem. 1998;273:4220–6.
146. van der Zee R, Murohara T, Luo Z, et al. Vascular endothelial growth factor/vascular permeability factor augments nitric oxide release from quiescent rabbit and human vascular endothelium. Circulation. 1997;95:1030–7.
147. Papapetropoulos A, Garcia-Cardena G, Madri JA, Sessa WC. Nitric oxide production contributes to the angiogenic properties of vascular endothelial growth factor in human endothelial cells. J Clin Invest. 1997;100:3131–9.
148. Seymour LW, Shoaibi MA, Martin A, et al. Vascular endothelial growth factor stimulates protein kinase C-dependent phospholipase D activity in endothelial cells. Lab Invest. 1996;75:427–37.
149. Wood KW, Sarnecki C, Roberts TM, Blenis J. ras mediates nerve growth factor receptor modulation of three signal-transducing protein kinases: MAP kinase, Raf-1, and RSK. Cell. 1992;68:1041–50.
150. Zhai P, Galeotti J, Liu J, et al. An angiotensin II type 1 receptor mutant lacking epidermal growth factor receptor transactivation does not induce angiotensin II-mediated cardiac hypertrophy. Circ Res. 2006;99:528–36.
151. Palen DI, Matrougui K. Role of elevated EGFR phosphorylation in the induction of structural remodelling and altered mechanical properties of resistance artery from type 2 diabetic mice. Diabetes Metab Res Rev. 2008;24:651–6.
152. Benter IF, Benboubetra M, Hollins AJ, Yousif MH, Canatan H, Akhtar S. Early inhibition of EGFR signaling prevents diabetes-induced up-regulation of multiple gene pathways in the mesenteric vasculature. Vascul Pharmacol. 2009;51:236–45.

# Chapter 6
# Ion Signaling and Electrophysiological Function

**Abstract** The normal sequence and synchronous contraction of the heart is governed by the cardiac action potential generated by the highly regulated activity of the ion channels. Ion channels represent multiprotein complexes composed of homologous subunits circularly arranged around a water-filled pore through the plane of the membrane lipid bilayer. Based on their gating (the mechanism of opening and closing of a channel) properties, the ion channels can be divided into subclasses: the voltage-gated, ligand-gated, and mechanosensitive ion channels. In this chapter, the action potential properties and the function of major ion channels are presented.

**Keywords** Ion channels • Ion signaling • Action potential • Cyclic nucleotide

## Introduction

Ion channels function as macromolecular complexes assembled at specific sites within cellular membrane. These transmembrane assemblies control the flux of ions across the membrane of cardiomyocytes generating ionic currents responsible for fast depolarization and maintenance of impulse conduction in the heart. During the past 15 years, the genes encoding the major ion channels have been cloned and sequenced. Various mutations in ion channels and interacting proteins have been linked to different forms of arrhythmias. In this chapter, we discuss major properties of the action potential, structure and functional characteristics of cardiac $Na^+$, $K^+$, hyperpolarization-activated cyclic nucleotide-gated and $Cl^-$ channels. Cardiac $Ca^{2+}$ channels as well as abnormalities of expression or function of cardiac ion channels associated with cardiovascular diseases will be discussed in the corresponding separate chapters.

## Cardiac Action Potential

The classical model used to understand the cardiac action potential is the action potential of the ventricular myocytes. The propagating cardiac action potential enables rapid changes in heart rate and responds to the autonomic tone changes. In Fig. 6.1, the five phases of the cardiac action potential are schematically shown.

*Phase 4* is the resting membrane potential stable at around −85 mV. This phase of the action potential is associated with diastole of the chamber of the heart.

*Phase 0* is the rapid depolarization phase. The membrane potential switches rapidly into positive voltage range. This phase is due to the opening of the fast $Na^+$ channels causing a rapid influx of $Na^+$ ions into the cell, an $Na^+$ current ($I_{Na}$). This is known as cardiac muscle automaticity.

*Phase 1*, a rapid repolarization phase, sets the membrane potential for the next phase of the action potential. This phase is due to the inactivation of the fast $Na^+$ channels and outward flux of $K^+$ and $Cl^-$ ions ($I_{to1}$ and $I_{to2}$ currents, respectively).

*Phase 2* is a "plateau" phase, the longest phase of the action potential. This phase is unique among excitable cells and is sustained by a balance between influx of $Ca^{2+}$ through L-type $Ca^{2+}$ channels ($I_{Ca}$) and efflux of $K^+$ through the slow-delayed rectifier $K^+$ channels ($I_{Ks}$). The $Na^+$–$Ca^{2+}$ exchanger current ($I_{Na,Ca}$) and the $Na^+$/$K^+$ pump current ($I_{Na,K}$) also play minor roles during phase 2.

*Phase 3*, the phase of rapid repolarization, restores the resting membrane potential. During this phase, the L-type $Ca^{2+}$ channels close, while the slow-delayed rectifier ($I_{Ks}$) $K^+$ channels are still opened. This ensures a net outward current, corresponding to negative change in membrane potential, thus allowing more types of $K^+$ channels to open. These are primarily the rapid delayed rectifier $K^+$ channels ($I_{Kr}$) and the inwardly rectifying $K^+$ current, $I_{K1}$. The delayed rectifier $K^+$ channels close when the membrane potential is

J. Marín-García, *Signaling in the Heart*,
DOI 10.1007/978-1-4419-9461-5_6, © Springer Science+Business Media, LLC 2011

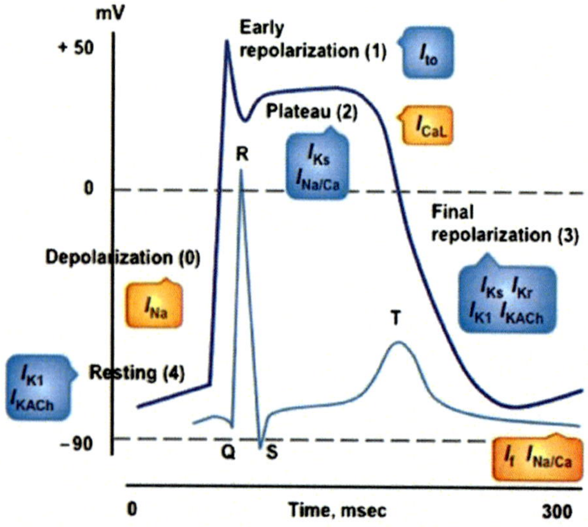


**Fig. 6.1** The normal action potential (AP) mediated by ion currents in cardiomyocytes [1]. The five phases of the AP (*dark blue line*) are schematically shown. Inward currents, $I_{Na}$, $I_{CaL}$, and $I_f$, are depicted in *orange boxes*. Outward currents, $I_{K1}$, $I_{KACh}$, $I_{to}$, $I_{Ks}$, and $I_{Kr}$ are depicted in *blue boxes*. The $Na^+$–$Ca^{2+}$ exchanger can generate inward or outward current, $I_{Na,Ca}$. In the resting state (phase 4, diastole), ion channels maintain a higher intracellular concentration of $K^+$ and a higher extracellular concentration of $Na^+$ and $Ca^{2+}$. Large inward $Na^+$ current ($I_{Na}$ ) through $Na^+$ channels causes rapid depolarization (upstroke, phase 0). $Ca^{2+}$ influx through the L-type $Ca^{2+}$ channel ($I_{CaL}$) contributes to the depolarization maintenance (plateau, phase 2). During the phase 2, inward $Ca^{2+}$ and outward $K^+$ currents are relatively balanced. Series of outward $K^+$ currents ($I_{K1}$, $I_{KACh}$, $I_{to}$, $I_{Ks}$, and $I_{Kr}$) are responsible for the repolarization (phase 3) of the AP and bring the cell back to its resting state. Typical ECG profile (*blue line*) denoting QRS depolarization and the ST-T repolarization is also shown

restored to about −80 to −85 mV, while $I_{K1}$ remains conducting throughout phase 4, contributing to set the resting membrane potential.

## General Properties of Ion Channels

The generation of the membrane action potential results from the selective permeability of ion channels specifically located within the cell membrane of cardiomyocytes. During the action potential, permeability of an ion channel changes allowing ion movement across cell membrane passively down the electrochemical gradients. Depending on the electrochemical gradient, an ion moves into the cell or out of the cell, depolarizing or repolarizing current for cations, respectively.

The ion channels represent multiprotein complexes composed of homologous subunits circularly arranged around a water-filled pore through the plane of the membrane lipid bilayer (Fig. 6.2). The pore-forming, primary subunits are

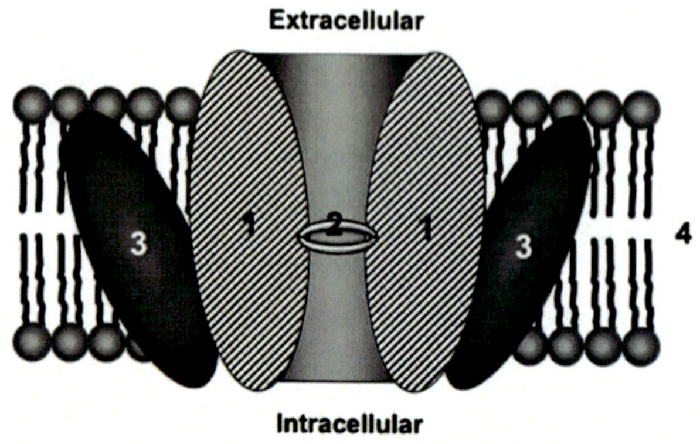


**Fig. 6.2** Scheme of a typical ion channel. Cross section of primary α-subunits (1) is shown. Four α-subunits spanning the membrane lipid bilayer (4) are circularly arranged around the pore region containing the selectivity filter (2). The pore-forming α-subunits are associated with auxiliary β-subunits (3)

called the α-subunits, while the auxiliary, secondary subunits are denoted β, γ, and so on.

Ion channels possess two fundamental characteristics: the selective permeability and gating. The selective permeability of ion channels to specific ions depends on their size, valency, and hydration energy.

Gating is the mechanism of opening and closing of cardiac ion channels. Voltage-gated ion channels open or close in response to changes in membrane potential, while ligand-gated ion channels open or close depending on binding of ligands to the channel. Voltage-gated ion channels are the most common among ion channels. A majority of ion channels open upon membrane depolarization, while the pacemaker channels open in response to hyperpolarization.

The second main gating mechanism of ion channels is ligand-dependent gating. This class of ion channels opens in response to specific ligand molecules binding to the extracellular domain of the channel. Ligand binding results in a conformational change in the structure of the channel protein leading to the opening of the channel and subsequent ion flux across the cell membrane. The most studied example of such channels is the acetylcholine (Ach)-activated $K^+$ channel. Binding of acetylcholine to the M2 muscarinic receptor activates G protein signaling pathway leading to dissociation of $G_{i\alpha}$ and $G_{\beta\gamma}$ subunits. The $G_{\beta\gamma}$ subunit upregulates an inward-rectifying $K^+$ channel ($I_{KAch}$), which abbreviates the membrane action potential in pacemaker cells.

The least characterized class of ion channels is the mechanosensitive channels. They can modulate their activity in response to mechanical stress and mechanical stimuli such as tissue stretch.

In the past 15 years, it has been appreciated that ion channels are not randomly distributed within the cell membrane of cardiac myocytes. The cell membrane represents a highly

**Table 6.1** Main cardiac ion channels

| | α-Subunit | | β-Subunit/accessory protein | | | |
|---|---|---|---|---|---|---|
| Channel | Gene | Protein | Gene | Protein | Current | Gating |
| $Na^+$ | *SCN5A* | $Na_V1.5$ | *SCN1B* | β1 | $I_{Na}$ | Voltage depolarization |
| $K^+$, voltage-dependent | *KCND3* | $K_V4.3$ | *KCNE2* | MiRP1 | $I_{to,f}$ (transient outward, fast) | Voltage depolarization |
| | | | *KCNE3* | MiRP2 | | |
| | | | *DPP6* | DPP6 | | |
| | | | multiple genes | KChIPs | | |
| | *KCNA4* | $K_V1.4$ | *KCNB1* | $K_V\beta1$ | $I_{to,s}$ (transient outward, slow) | Voltage depolarization |
| | | | *KCNB2* | $K_V\beta2$ | | |
| | | | *KCNB3* | $K_V\beta3$ | | |
| | | | *KCNB4* | $K_V\beta4$ | | |
| | *KCNA5* | $K_V1.5$ | *KCNAB1* | $K_V\beta1$ | $I_{Kur}$ (delayed rectifier, ultrarapid) | Voltage depolarization |
| | | | *KCNAB2* | $K_V\beta2$ | | |
| | *KCNH2* | $K_V11.1$ (HERG) | *KCNE2* | MiRP1 | $I_{Kr}$ (delayed rectifier, fast) | Voltage depolarization |
| | *KCNQ1* | $K_V7.1$/($K_V$LQT1) | *KCNE1* | minK | $I_{Ks}$ (delayed rectifier, slow) | Voltage depolarization |
| | *KCNJ2/12* | $K_{ir}2.1/2.2$ | | | $I_{K1}$ (inward rectifier) | Voltage depolarization |
| $K^+$, voltage-independent | *KCNJ11* | $K_{ir}6.2$ | *ABCC9* | SUR2 | $I_{KATP}$ (ADP-activated) | Increased ADP/ATP ratio |
| | *KCNJ3/5* | $K_{ir}3.1/3.4$ | | | $I_{KAch}$ (acetylcholine-activated) | Acetylcholine |
| | *KCNK1/6* | TWIK-1/2 | | | $I_{KP}$ (background) | Metabolism, stretch |
| | *KCNK3* | TASK-1 | | | | |
| | *KCNK4* | TRAAK | | | | |
| $Ca^{2+}$, L-type | *CACNA1C* | $Ca_V1.2$ | *CACNB2* | $Ca_V\beta2$ | $I_{Ca,L}$ | Voltage depolarization |
| | | | *CACNA2D1* | $Ca_V\beta2\delta1$ | | |
| $Ca^{2+}$, T-type | *CACNA1G/H* | $Ca_V3.1/3.2$ | | | $I_{Ca,T}$ | Voltage depolarization |
| Hyperpolarization-activated cyclic nucleotides | *HCN1–4* | HCN1–4 | | | $I_f$ (pacemaker) | Voltage hyperpolarization |
| $Cl^-$ | *CFTR* (*ABCC7*) | CFTR | | | $I_{Cl,PKA}$ $I_{Cl,PKC}$ $I_{Cl,ATP}$ | PKA, PKC, and extracellular ATP |

organized heterogeneous mixture of various proteins, cholesterol, and phospho-, glycero-, and sphingolipids. Cholesterol associates laterally with sphingolipids forming membrane microdomains, known as lipid rafts. The cardiac ion channels have been localized to the lipids rafts. This specific localization of ion channels provides their efficient integration into macromolecular signaling complexes allowing precise highly localized regulation of the channels [2, 3].

The main ion channels governing the cardiac action potential have been cloned and sequenced (Table 6.1).

## Na⁺ Channels

The human cardiac Na⁺ channel $hNa_v1.5$ (encoded by *SCN5A*) is a member the family of voltage-gated Na⁺ channels ($hNa_v1$ to 9). It is responsible for fast depolarization in atrial and ventricular myocytes (phase 0, Fig. 6.1), and contributes to maintenance of impulse conduction in the heart. The channel consists of a primary pore-forming α-subunit and multiple ancillary modulatory β-subunits. The α-subunit encoded by the *SCN5A* gene consists of four homologous domains (DI–DIV), each made of six membrane-spanning segments (S1–S6) (Fig. 6.3) [4, 5]. The loops between S5 and S6 of each domains (P loops) form the channel pore and govern ion selectivity and permeation. Each S4 segment is positively charged and acts as voltage sensor.

Na⁺ channel inactivation demonstrates two distinct kinetic modes: fast inactivation occurs for milliseconds, and slow inactivation occurs when the membrane is depolarized for hundreds of milliseconds and is sustained for seconds [6, 7].

In humans, four β-subunits (β1–β4) are expressed in the heart. They share a common structure: N-terminal domain, single membrane-spanning segment, and intracellular C-terminal domain. Although β-subunits do not form the channel pore, they play essential roles in the modulation of the channel function and expression within the cell membrane [8, 9].

The cardiac Na⁺ channels are phosphorylated by various protein kinases including protein kinase C (PKC), protein kinase A (PKA), and $Ca^{2+}$-calmodulin kinase. Channel phosphorylation by PKC leads to a downregulation of $I_{Na}$, whereas data on the effects of PKA on the $I_{Na}$ are so far controversial [10–12].

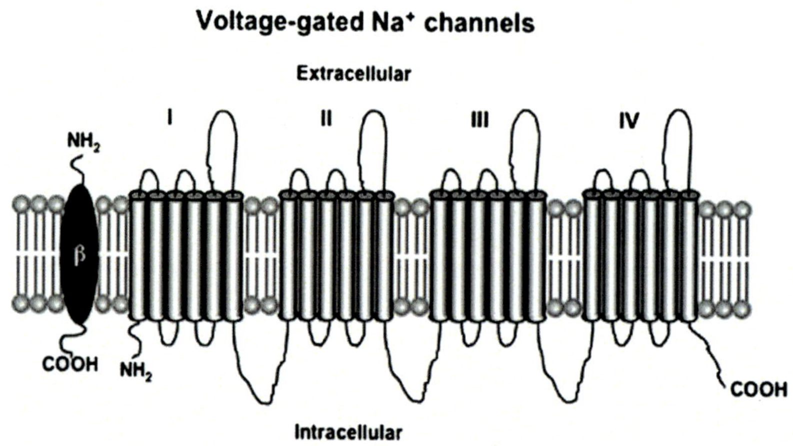


**Fig. 6.3** Transmembrane organization of voltage-gated $Na^+$ channel. A pore-forming α-subunit associated with auxiliary β-subunit is shown. The α-subunit is composed of four homologous domains (I–IV), which contain six transmembrane segments (S1–S6) and a pore loop. The amino and carboxyl termini are intracellular. S4 segment in each domain is positively charged and serves as a voltage sensor for channel activation. S5 and S6 segments and the intervening pore loop form the channel pore, whereas the intracellular loop between III and IV domains forms the inactivation gate. The β-subunit is a single membrane-spanning glycoprotein that can modulate channel function

It has been demonstrated recently that the cardiac $Na^+$ channels form a macromolecular complex containing in addition to the α- and β-subunits several other proteins such as caveolin, ankyrin, calmodulin, syntrophin, and Nedd4-like ubiquitin-protein ligases [13].

Most mutations which affect the cardiac $Na^+$ channel are localized on *SCN5A* gene encoding the α-subunit. They have been linked to long-QT syndrome (LQTS), Brugada syndrome, primary cardiac conduction system disease (PCCP), and dilated cardiomyopathy [14].

## $K^+$ Channels

Various types of $K^+$ channels are expressed in cardiomyocytes with distinct subunit composition and electrophysiological characteristics. Cardiac $K^+$ channels fall into three broad categories based on their composition and gating properties (Fig. 6.4):

1. Voltage-gated channels ($K_V$) underlying $I_{to}$, $I_{Kur}$, $I_{Kr}$, and $I_{Ks}$ currents.
2. Inward rectifier channels ($K_{ir}$) underlying $I_{K1}$, $I_{KAch}$, and $I_{KATP}$ currents.
3. Background $K^+$ currents (TASK-1 and TWIK-1/2). The significant variation in the expression of these channels is responsible for the unique shape of the action potential in the atria, ventricles, and transmurally across the myocardial wall.

Generally, $K^+$ channels contribute to maintaining the resting membrane potential and to repolarization during the action potential.

Voltage-gated $K^+$ channels are composed of primary α-subunits, multiple β-subunits, and the complementary $K_V$-channel interacting proteins, KChIP and KChAP. Most of α- and β-subunits have been cloned and sequenced. The main subtypes of α-subunits include $K_VN$.x ($N$=1–4), hERG (*KCNH2* gene), and $K_V$LQT1 (*KCNQ1* gene). They are responsible for generating outward current in the heart. The α-subunits that form different types of $K^+$ channels and their role in generating the cardiac action potential are summarized in Table 6.1. The complementary proteins KChIP (*KCHIP* gene) and KChAP (*KCHAP* gene) increase channel activity and modulate channel kinetics.

The transient outward current consists of a $K^+$ current $I_{to1}$, and a $Ca^{2+}$-activated chloride current, $I_{to2}$. $K^+$ current $I_{to2}$ possesses fast ($I_{to,f}$) and slow ($I_{to,s}$) components. $I_{to1}$ represents the main subtype expressed in human atrium, while $I_{to,f}$ and $I_{to,s}$ are expressed in the ventricle. Myocardial regions with relatively short action potentials such as the epicardium, right ventricle, and the septum express higher levels of $I_{to}$. Activation of $I_{to}$ is fast (<10 ms) compared with other voltage-gated $K^+$ channels. α-Adrenergic stimulation inhibits $I_{to}$ in PKA-dependent fashion and also reduces channel expression.

The delayed rectifier $K^+$ currents $I_{Kur}$, $I_{Kr}$, and $I_{Ks}$ represent slowly activating outward currents. They are essential for the control of repolarization during phase 3 (Fig. 6.1). High levels of $I_{Kur}$ in atrial myocytes are responsible for the significantly shorter action potential duration (APD) in the atrium. $I_{Kr}$ is highly expressed in the left atrium and ventricular endocardium, whereas $I_{Ks}$ is expressed in all cell types. The α-subunits responsible for the delayed rectifier $K^+$ currents are listed in Table 6.1. The β-subunits modulate gating and response to sympathetic stimulation and drugs.

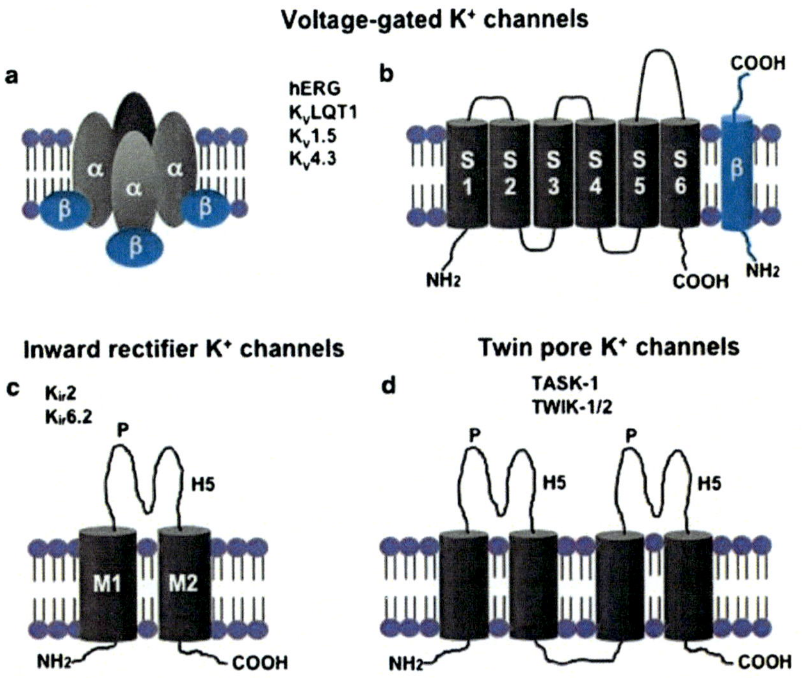


**Fig. 6.4** Schematic representation of K⁺ channels. (**a**) Functional voltage-gated K⁺ channels (hERG, K$_V$LQT1, K$_V$1.5, and K$_V$4.3) are formed by four primary α-subunits clustered around a channel pore. α-Subunits are physically associated with auxiliary β-subunits. (**b**) Each pore-forming α-subunit consists of six membrane-spanning domains (S1–6). (**c**) Inward rectifier K⁺ channels (K$_{ir}$2 and K$_{ir}$ 6.2) consist of two membrane-spanning domains (M1 and M2) connected by a loop with the conserved pore (P) and H5 segments. Four channel subunits assemble to form the functional channel. (**d**) Two pore K⁺ channels (TASK-1 and TWIK-1/2) contain four transmembrane domains and are responsible for "background" K⁺ currents and are activated by pH, mechanical stretch, heat, and coupling to G proteins

β-Adrenergic stimulation inhibits $I_{Kr}$ via PKA activation and activates it through elevation of cAMP levels. α-Adrenergic stimulation is inhibitory. For $I_{Ks}$, β-adrenergic stimulation induces the current via PKA-dependent phosphorylation. β-Adrenergic blockers are promising therapeutic treatments for the action potential prolongation observed in LQT1.

The inward rectifier channel current $I_{K1}$ is responsible for the resting membrane potential in the atria and ventricles. The channel is highly expressed in ventricular cells protecting them from pacemaker activity. $I_{K1}$ limits the outward current throughout phases 0, 1, and 2 and contributes significantly to phase 3 repolarization.

The acetylcholine-activated K⁺ channel belongs to the G protein-coupled inward rectifying K⁺ channels. Channel expression is much higher in the sinoatrial (SA) and atrioventricular (AV) nodes and atria than in ventricle. Activation of $I_{KAch}$ results in membrane hyperpolarization and abbreviation of the action potential (Fig. 6.1). Stimulation of the M2 muscarinic receptor with acetylcholine activates the $G_i$ protein causing its dissociation into $G_{i\alpha}$ and $G_{\beta\gamma}$ subunits. The $G_{\beta\gamma}$ subunit binds to and activates the channel. Stimulation of the P1 receptor with adenosine results in similar effect.

In the adult human heart, the ATP-sensitive K⁺ channel, $K_{ATP}$ is composed of $K_{ir}$6.2 and the sulfonylurea receptor 2 (SUR2) subunits. $I_{KATP}$ is essential for shortening of the action potential during ischemia [15]. Moreover, the channel contributes to the protective response of the myocardium in ischemic preconditioning [16].

## Hyperpolarization-Activated Cyclic Nucleotide-Gated Channels

The hyperpolarization-activated cyclic nucleotide-gated (HCN) ion channels generate the pacemaker current $I_f$ playing essential role in the regulation of rhythmic activity of cardiac pacemaker cells. $I_f$ possesses unique features: it is an inward current activated, unlike vast majority of cellular conductances, upon membrane hyperpolarization and carried by both Na⁺ and K⁺ ions. Moreover, $I_f$ is activated by cAMP in a direct, PKA-independent manner. Due to its unusual biophysical characteristics, the current was called "funny" current, $I_f$ [17].

The ion channels underlying $I_f$ current are encoded by four distinct genes HCN1–4 that have been cloned and sequenced in the past decade. They represent the subfamily of cyclic nucleotide-regulated cation channels within a large superfamily of the pore-loop cation channels [18]. In mammals, HCN1–4 share ~60% sequence identity with each other and are expressed mainly in the heart and brain.

Similar to other members of the pore-loop channel family, HCN channels are formed by four α-subunits arranged around the centrally located pore. HCN α-subunits are structurally analogous to those of the voltage-gated $K^+$ channels. Each HCN channel monomer is composed of three major structural domains: the transmembrane core and the intracellular $NH_2$-terminal and COOH-terminal regions (Fig. 6.5).

The transmembrane core is composed of six α-helical regions (S1–S6) including the positively charged voltage sensor (S4) and an ion-conducting pore loop between S5 and S6 (Fig. 6.5). The extracellular loop between S5 and the pore region is glycosylated which is essential for normal cell surface localization.

Under physiological conditions, HCN channels carry an inward $Na^+$ current and slightly select $K^+$ over $Na^+$. Surprisingly, however, the HCN pore loop is closely related to that of highly selective $K^+$ channels.

The proximal segment of the intracellular COOH terminus of HCN channels harbors a cyclic nucleotide-binding domain (CNBD) and the C-linker connecting the CNBD with the S6 region. The CNBD is evolutionary conserved domain found in variety of proteins ranging from bacterial transcription factors to cAMP- and cGMP-dependent protein kinases, cAMP-dependent guanine nucleotide exchange factors and cyclic nucleotide-gated [19–24].

In contrast to CNG channels that require cAMP binding to open, HCN channels are operated by membrane hyperpolarization, while cAMP and cGMP only facilitate their activation. The stimulatory effect of cAMP is mediated by its direct binding to the CNBD of HNC channels rather than by protein phosphorylation [25]. According to a recent model, binding of cAMP or cGMP promotes conformational change in the CNBD and removes the autoinhibition of the COOH terminus facilitating the opening of the channel [26, 27].

Four mammalian HCN channel, subtypes differ with respect to their corresponding activation time constants, their steady-state voltage dependence, and their modulation by cAMP or cGMP.

HCN1 is the subtype demonstrating the fastest activation kinetics in strongly voltage-dependent fashion with $\tau_{act}$ ranging from 30 to 300 ms at −140 to −90 mV, and the most positive $V_{0.5}$ value (−70 to −90 mV) [28, 29]. However, HCN1 shows a very modest facilitation of activation by cAMP, compared with HCN2 and HCN4 [26–28].

In contrast to HCN1, HCN4 is characterized by the slowest activation kinetics with $\tau_{act}$ ranging between a few milliseconds at highly hyperpolarized voltages (−140 mV) up to several seconds at resting potential (−70 mV) [30–32]. HCN2 demonstrates in this respect intermediate characteristics with $\tau_{act}$ in a range of 150 ms to 1 s. Both HCN2 and HCN4 possess $V_{0.5}$ between −70 and −100 mV, they are also very sensitive to cAMP modulation with shift of $V_{0.5}$ of 10–25 mV [26, 27].

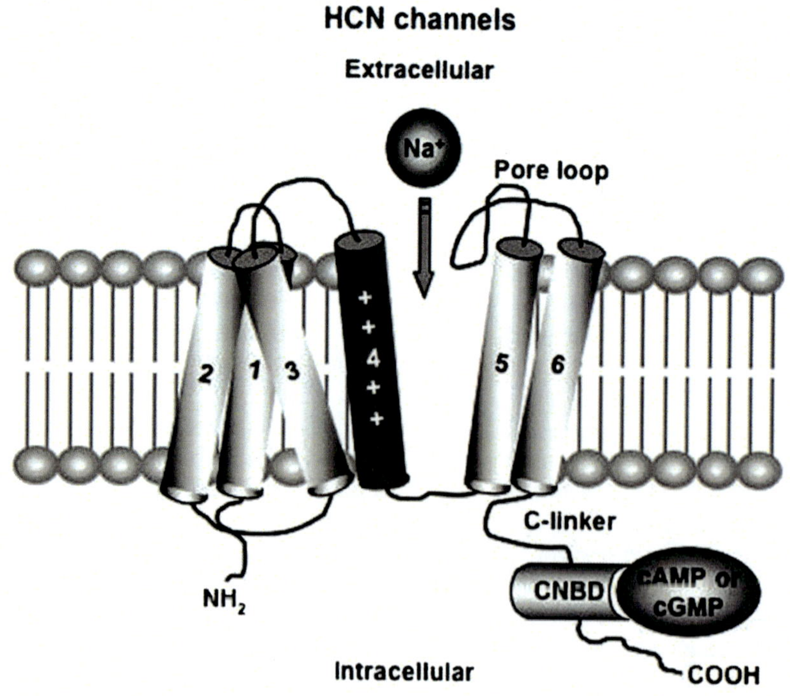


**Fig. 6.5** Structure of HCN channels. HCN channel subunit consists of six transmembrane domains (1–6) and a pore loop. The S4 domain (4) is positively charged and serves as the voltage sensor. The cytosolic carboxyl terminus contains the cyclic nucleotide-binding domain (CNBD) connected to the membrane-spanning channel core through the C-linker region

HCN3 appears to occupy an intermediate position between HCN2 and HCN4 with $\tau_{act}$ of 250–400 ms at −140 mV, and $V_{0.5}$ −80 and −90 mV. Unlike other HCN channels, activation of HCN3 is not modulated by cyclic nucleotides [28]. The structural basis for the lack of cAMP or cGMP sensitivity is unclear.

Expression of both major cardiac HCN channels, HCN2 and HCN4 is regional and developmentally regulated. HCN2 channels are primarily expressed in neonatal ventricular cardiomyocytes. Their expression decreases in adulthood. HCN4 is predominantly expressed in the sinus node, AV node, and ventricular conducting system.

The central role of cardiac HCN channels in pacemaker activity places them as promising targets for the development of pharmacological agents for the treatment of cardiac dysrhythmias and ischemia.

## $Cl^-$ Channels

Since the first discovery of a cAMP-activated $Cl^-$ current in the heart several types of $Cl^-$ channels have been identified in different regions of the heart (Fig. 6.6). All cardiac $Cl^-$ channels can be divided into the following subfamilies:

1. The cystic fibrosis conductance regulator (CFTR) is a member of the adenosine triphosphate-binding cassette (ABC) transporter superfamily responsible for the $Cl^-$ currents activated by PKA ($I_{Cl, PKA}$), PKC ($I_{Cl, PKC}$), and extracellular ATP ($I_{Cl, ATP}$) [33–38]. Under basal conditions, CFTR channels are closed and are activated upon increase of PKA and PKC activity. Main physiologic function of activation of CFTR channels is to shorten the APD and to protect the heart against the development of early afterdepolarization (EAD).
2. ClC-2 is a member of the ClC voltage-gated $Cl^-$ channel superfamily responsible for the hyperpolarization and cell swelling-activated inwardly rectifying $Cl^-$ current ($I_{Cl, ir}$) [39–41]. Activation of ClC-2 channels by hyperpolarization, cell swelling, and acidosis results in $Cl^-$ efflux at negative membrane potentials causing depolarization of the resting membrane potential of cardiomyocytes. It has been suggested that the $I_{Cl, ir}$ current controlled by ClC-2 channels plays more significant role in the SA and AV nodal regions of the heart. Moreover, ClC-2 sensitivity to

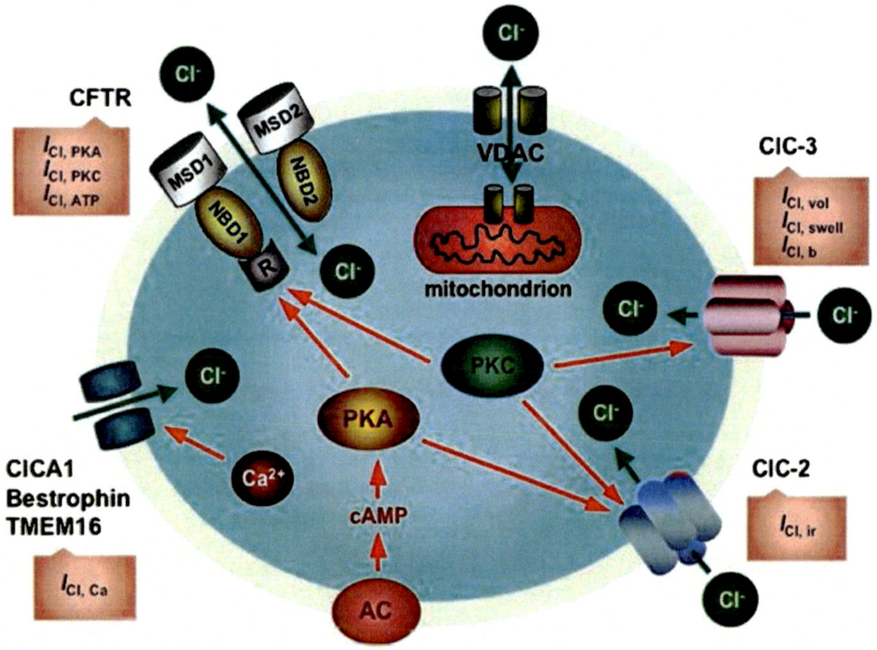


**Fig. 6.6** $Cl^-$ channels in cardiomyocytes. Cystic fibrosis transmembrane conductance regulator (CFTR) is $Cl^-$ channel activated by cAMP-protein kinase (PKA), protein kinase C (PKC), or extracellular ATP. It consists of two membrane-spanning domains (MSD1 and MSD2), two nucleotide binding domains (NBD1 and NBD2), and a regulatory subunit (R). ClC-2 and ClC-3 are members of voltage-gated $Cl^-$ channel family. ClC-2 is responsible for a volume-regulated and hyperpolarization-activated inward rectifying $Cl^-$ current ($I_{Cl, ir}$). ClC-3 is volume and cell swelling sensitive ($I_{Cl, ir}$ and $I_{Cl, swell}$). $I_{Cl, b}$ represents a basally activated ClC-3 $Cl^-$ current. Candidates for a $Cl^-$ current activated by increased intracellular $Ca^{2+}$ concentration are transmembrane protein 16 (TMEM16), bestrophin, and ClCA1. The voltage-dependent anion channel 1 (VDAC) is mainly expressed in the outer membrane of mitochondria and in the sarcolemmal membrane and may contribute to $Cl^-$ currents. *AC* adenylyl cyclase, an enzyme responsible for cellular cAMP production

cell volume and acidosis suggests importance of the $I_{Cl, ir}$ current during some pathological conditions.

3. ClC-3 is also a member of the ClC voltage-gated $Cl^-$ channel superfamily responsible for the volume-regulated outwardly rectifying $Cl^-$ current ($I_{Cl, vol}$) including the basally activated ($I_{Cl, b}$) and swelling-activated ($I_{Cl, swell}$) components [42–44]. Recent reports suggest that $I_{Cl, swell}$ and ClC-3 channels may play an essential role in the adaptive heart remodeling against pressure overload, apoptosis [45], and inflammation [46].
4. CLCA-1 is thought responsible for the $Ca^{2+}$-activated $Cl^-$ current ($I_{Cl, Ca}$) [47, 48].
5. Bestrophin is also a candidate for $I_{Cl, Ca}$ [49].
6. TMEM16 is a novel candidate for $I_{Cl, Ca}$ [50, 51].

Even though $Ca^{2+}$-activated $Cl^-$ channels are widely expressed in the heart and may contribute significantly to the regulation of cardiac action potential the molecular identity of these channels remains to be determined.

Additionally, it has been demonstrated recently that the voltage-dependent anion channel 1 (VDAC1) mainly expressed in the outer membrane of mitochondria and in the sarcolemmal membrane may contributes to $Cl^-$ currents [52].

Under normal physiological conditions, an extracellular $Cl^-$ concentration is about 145 mmol/l and intracellular $Cl^-$ concentration is 10–20 mmol/l. Thus, the $Cl^-$ equilibrium potential is more positive than the resting membrane potential and can be either negative or positive to the actual membrane potential throughout the cardiac cycle. Hence, unlike cation channels, cardiac $Cl^-$ channels possess unique ability to govern both inward and outward currents causing both depolarization and repolarization during the action potential. Therefore, their activation can modulate significantly cardiac action potential as well as pacemaker activity (Fig. 6.7).

The activation of cardiac ClC-3 and CFTR channels under normal physiologic conditions results in outward rectifying currents accelerating repolarization and leading to a shortening of Q-T interval (Fig. 6.7). These effects are more significant at positive potentials compared with smaller depolarization at negative potentials near the resting membrane potential. $K^+$ current, which controls normally the resting potential, opposes depolarization of cardiomyocytes caused by $Cl^-$ channels activation. Activation of cardiac $Cl^-$ channels may induce EAD and contribute to dysrhythmogenesis under pathological conditions.

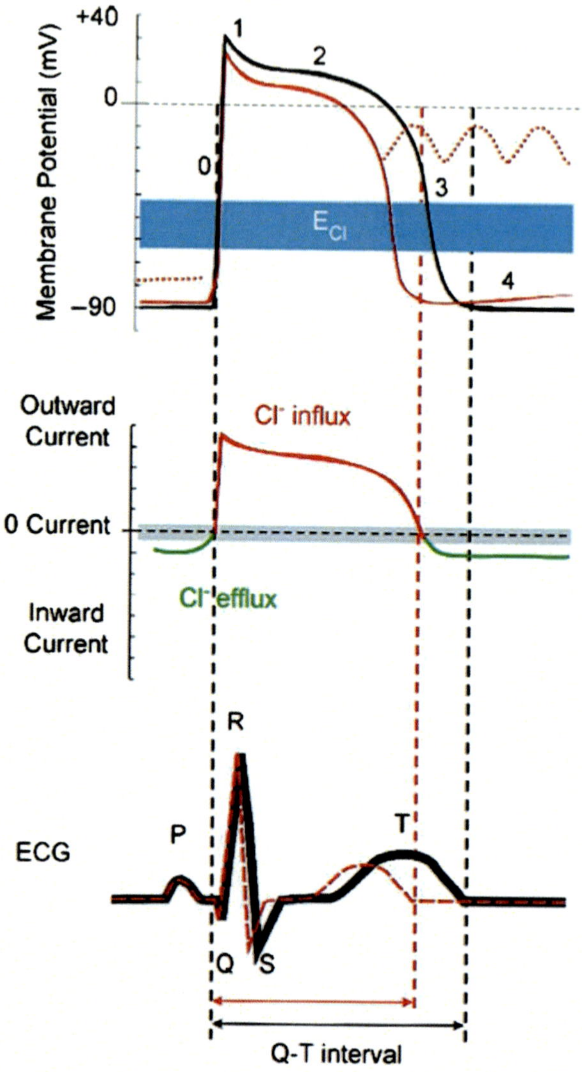


**Fig. 6.7** Modulation of the action potential by cardiac $Cl^-$ channels. *Top panel*, cardiac action potential under normal conditions (*black line*) and after activation of $Cl^-$ currents (*red line*) are shown. Numbers indicate phases of action potential. Range of $Cl^-$ equilibrium potential ($E_{Cl}$) under normal physiological conditions is depicted in *blue*. *Middle panel*, cardiac $Cl^-$ channels (CFTR or ClC-3) can generate both inward (*green line*) and outward (*red line*) current causing both depolarization and repolarization. Thus, activation of $Cl^-$ currents causes larger membrane depolarization and induction of early afterdepolarizations under reduced resting $K^+$ conductance. Range of zero-current corresponding to $E_{Cl}$ is indicated in *gray*. *Bottom panel*, ECG under normal physiological conditions (*black line*) and after activation of $Cl^-$ currents (*red line*) is depicted. Activation of $Cl^-$ currents result in a shortening of Q-T interval corresponding to the shortening of cardiac action potential

Recent evidence suggests that cardiac $Cl^-$ channels function as multiprotein complexes associated with various accessory proteins for modulating their activity and orchestrating their biological role under different conditions [54]. They contribute to plethora of cellular functions ranging from cell excitability, cell volume homeostasis, intracellular acidification to cell migration, proliferation, and differentiation [53]. However, our understanding of $Cl^-$ channel functional role in heart physiology has been hampered due to the complex expression pattern when various types of $Cl^-$ channels are expressed in the same cell and due to the lack of specific $Cl^-$ channel blockers.

## Conclusions and Future Perspectives

During the past two decades, intensive efforts have demonstrated a pivotal role of ion signaling and various cardiac ion channels in the regulation of systolic and diastolic function of the heart. The currents conducted by cardiac ion channels are critical for the generation, propagation, and maintenance of impulse conduction in the atria and ventricles. The genes encoding the main cardiac ion channels have been cloned and sequenced, and molecular structure of some ion channels has been described. On the other hand, the correlation of the structure with function has been done with a limited number of channels. Due to the landmark studies of MacKinnon and coworkers [55], the selectivity and gating of voltage-gated $K^+$ channels can be explained at the atomic level. The structures of the $Na^+$ and $Ca^{2+}$ pores, however, are still not available and our understanding of properties of these important channels require more detailed crystal structure studies. Less is also known about the functional roles of the intracellular domains of ion channels, their contribution to the channel surface expression, trafficking, and degradation.

Recent advances have made it possible to combine electrophysiology with functional genomics and gene-targeting approaches. In the past 10 years, invaluable information on the function of cardiac ion channels has been gained using various transgenic mouse models. An important finding in these studies has been complex compensatory changes in transgenic animals in response to gene targeting, which complicate understanding the phenotypes of the animals. Additional complication results from the heteromultimeric nature of the functional ion channels when various accessory subunits are associated with a main pore forming subunit. Therefore, a great caution should be taken to address phenotypes of gene-targeted animals. Some of these issues may be resolved in the future by using tissue-specific conditional and inducible transgenic models. Moreover, understanding the roles of epigenetic and environmental factors in the modulation of cardiac ion channels' function represents a challenging task.

In the first decade of the twenty-first century, we are also witnessing the results of several advanced techniques that have recently been applied to ion channel research such as nuclear magnetic resonance and single molecule fluorescence. The real knowledge of a cardiac ion channel, i.e., when its electrophysiological characteristics can be predicted from the molecular structure, will be undoubtedly achieved by further merging structural, functional, and computational analyses with the targeted genetic manipulation.

## Summary

- The normal sequence and synchronous contraction of the heart is governed by the cardiac action potential generated by the highly regulated activity of the ion channels. These transmembrane assemblies control the flux of ions across the membrane of cardiomyocytes generating ionic currents responsible for fast depolarization and maintenance of impulse conduction in the heart.
- The propagating cardiac action potential enables rapid changes in heart rate and responds to the autonomic tone changes. Resting (phase 0), depolarization or upstroke (phase 4), early repolarization (phase 1), plateau (phase 2), and final repolarization (phase 3) are the five phases of the action potential.
- The ion channels represent multiprotein complexes composed of homologous subunits circularly arranged around a water-filled pore through the plane of the membrane lipid bilayer. The pore-forming, primary subunits are called the α-subunits, while the auxiliary, secondary subunits are denoted β, γ. Based on their gating (the mechanism of opening and closing of a channel) properties, the ion channels can be divided into voltage-gated, ligand-gated, and mechanosensitive ion channels.
- The human cardiac $Na^+$ channel $hNa_v1.5$ encoded by the *SCN5A* gene is a member of the family of voltage-gated $Na^+$ channels ($hNa_v1$ to 9). The channel consists of a primary pore-forming α-subunit and multiple ancillary modulatory β-subunits. It is responsible for fast depolarization in atrial and ventricular myocytes (phase 0) and contributes to maintenance of impulse conduction in the heart. Mutations which affect the human cardiac $Na^+$ channel have been linked to long-QT syndrome, Brugada syndrome, primary cardiac conduction system disease and dilated cardiomyopathy.
- Cardiac $K^+$ channels fall into three broad categories based on their composition and gating properties: voltage-gated channels ($K_V$) underlying $I_{to}$, $I_{Kur}$, $I_{Kr}$, and $I_{Ks}$ currents; inward rectifier channels ($K_{ir}$) underlying $I_{K1}$, $I_{KAch}$, and $I_{KATP}$ currents; background $K^+$ currents (TASK-1 and TWIK-1/2). Generally, $K^+$ channels contribute to maintaining the resting membrane potential and to repolarization during the action potential. The significant variation in the expression of these channels is responsible for the unique shape of the action potential in the atria, ventricles and transmurally across the myocardial wall.
- The hyperpolarization-activated cyclic nucleotide-gated (HCN) ion channels generate the pacemaker ("funny") current $I_f$ playing essential role in the regulation of rhythmic activity of cardiac pacemaker cells. $I_f$ is an inward current activated, unlike vast majority of cellular conductances, upon membrane hyperpolarization and carried by

both $Na^+$ and $K^+$ ions. Moreover, $I_f$ is activated by cAMP in a direct, PKA-independent manner. The ion channels underlying $I_f$ current are encoded by four distinct genes HCN1–4 representing the subfamily of cyclic nucleotide-regulated cation channels. HCN2 and HCN4 are main cardiac HCN channels. HCN2 channels are primarily expressed in ventricular cardiomyocytes, whereas HCN4 is predominantly expressed in the sinus node, AV node, and ventricular conducting system.

- Several types of $Cl^-$ channels are expressed in different regions of the heart. Unlike cation channels, cardiac $Cl^-$ channels possess unique ability to govern both inward and outward currents causing both depolarization and repolarization during the action potential. Cardiac $Cl^-$ channels can modulate significantly cardiac action potential as well as pacemaker activity contributing to various cellular functions ranging from cell excitability, cell volume homeostasis, intracellular acidification to cell migration, proliferation, and differentiation.

## References


1. Nattel S, Carlsson L. Innovative approaches to anti-arrhythmic drug therapy. Nat Rev Drug Discov. 2006;5:1034–49.
2. Marx SO, Kurokawa J, Reiken S, et al. Requirement of a macromolecular signaling complex for beta adrenergic receptor modulation of the KCNQ1-KCNE1 potassium channel. Science. 2002;295:496–9.
3. Delmar M, Duffy HS, Sorgen PL, Taffet SM, Spray DC. Molecular organization and regulation of the cardiac gap junction channel connexin43. In: Zipes DP, Jalife J, editors. Cardiac electrophysiology: from cell to bedside. Philadelphia: Elsevier Inc.; 2004.
4. Li RA, Tomaselli F, Marbin E. Sodium channels. In: Zipes DP, Jalife J, editors. Cardiac electrophysiology: from cell to bedside. Philadelphia: Elsevier Inc.; 2004.
5. Noda M, Shimizu S, Tanabe T, et al. Primary structure of Electrophorus electricus sodium channel deduced from cDNA sequence. Nature. 1984;312:121–7.
6. Ruan Y, Liu N, Priori SG. Sodium channel mutations and arrhythmias. Nat Rev Cardiol. 2009;6:337–48.
7. Balser JR. The cardiac sodium channel: gating function and molecular pharmacology. J Mol Cell Cardiol. 2001;33:599–613.
8. Medeiros-Domingo A, Kaku T, Tester DJ, et al. SCN4B-encoded sodium channel beta4 subunit in congenital long-QT syndrome. Circulation. 2007;116:134–42.
9. Watanabe H, Koopmann TT, Le Scouarnec S, et al. Sodium channel beta1 subunit mutations associated with Brugada syndrome and cardiac conduction disease in humans. J Clin Invest. 2008;118:2260–8.
10. Ono K, Kiyosue T, Arita M. Isoproterenol, DBcAMP, and forskolin inhibit cardiac sodium current. Am J Physiol. 1989;256:C1131–7.
11. Kirstein M, Eickhorn R, Kochsiek K, Langenfeld H. Dose-dependent alteration of rat cardiac sodium current by isoproterenol: results from direct measurements on multicellular preparations. Pflugers Arch. 1996;431:395–401.
12. Frohnwieser B, Chen LQ, Schreibmayer W, Kallen RG. Modulation of the human cardiac sodium channel alpha-subunit by cAMP-dependent protein kinase and the responsible sequence domain. J Physiol. 1997;498(Pt 2):309–18.
13. Abriel H, Kass RS. Regulation of the voltage-gated cardiac sodium channel Nav1.5 by interacting proteins. Trends Cardiovasc Med. 2005;15:35–40.
14. Clancy CE, Kass RS. Inherited and acquired vulnerability to ventricular arrhythmias: cardiac $Na^+$ and $K^+$ channels. Physiol Rev. 2005;85:33–47.
15. Zingman LV, Alekseev AE, Hodgson-Zingman DM, Terzic A. ATP-sensitive potassium channels: metabolic sensing and cardioprotection. J Appl Physiol. 2007;103:1888–93.
16. Patel HH, Gross ER, Peart JN, Hsu AK, Gross GJ. Sarcolemmal KATP channel triggers delayed ischemic preconditioning in rats. Am J Physiol Heart Circ Physiol. 2005;288:H445–7.
17. Brown HF, DiFrancesco D, Noble SJ. How does adrenaline accelerate the heart? Nature. 1979;280:235–6.
18. Yu FH, Yarov-Yarovoy V, Gutman GA, Catterall WA. Overview of molecular relationships in the voltage-gated ion channel superfamily. Pharmacol Rev. 2005;57:387–95.
19. Biel M, Wahl-Schott C, Michalakis S, Zong X. Hyperpolarization-activated cation channels: from genes to function. Physiol Rev. 2009;89:847–85.
20. Pfeifer A, Ruth P, Dostmann W, Sausbier M, Klatt P, Hofmann F. Structure and function of cGMP-dependent protein kinases. Rev Physiol Biochem Pharmacol. 1999;135:105–49.
21. Kaupp UB, Seifert R. Cyclic nucleotide-gated ion channels. Physiol Rev. 2002;82:769–824.
22. Rehmann H, Wittinghofer A, Bos JL. Capturing cyclic nucleotides in action: snapshots from crystallographic studies. Nat Rev Mol Cell Biol. 2007;8:63–73.
23. Taylor SS, Kim C, Vigil D, et al. Dynamics of signaling by PKA. Biochim Biophys Acta. 2005;1754:25–37.
24. Weber IT, Gilliland GL, Harman JG, Peterkofsky A. Crystal structure of a cyclic AMP-independent mutant of catabolite gene activator protein. J Biol Chem. 1987;262:5630–6.
25. DiFrancesco D, Tortora P. Direct activation of cardiac pacemaker channels by intracellular cyclic AMP. Nature. 1991;351:145–7.
26. Wainger BJ, DeGennaro M, Santoro B, Siegelbaum SA, Tibbs GR. Molecular mechanism of cAMP modulation of HCN pacemaker channels. Nature. 2001;411:805–10.
27. Wang J, Chen S, Siegelbaum SA. Regulation of hyperpolarization-activated HCN channel gating and cAMP modulation due to interactions of COOH terminus and core transmembrane regions. J Gen Physiol. 2001;118:237–50.
28. Stieber J, Stockl G, Herrmann S, Hassfurth B, Hofmann F. Functional expression of the human HCN3 channel. J Biol Chem. 2005;280:34635–43.
29. Ishii TM, Takano M, Ohmori H. Determinants of activation kinetics in mammalian hyperpolarization-activated cation channels. J Physiol. 2001;537:93–100.
30. Seifert R, Scholten A, Gauss R, Mincheva A, Lichter P, Kaupp UB. Molecular characterization of a slowly gating human hyperpolarization-activated channel predominantly expressed in thalamus, heart, and testis. Proc Natl Acad Sci USA. 1999;96:9391–6.
31. Ishii TM, Takano M, Xie LH, Noma A, Ohmori H. Molecular characterization of the hyperpolarization-activated cation channel in rabbit heart sinoatrial node. J Biol Chem. 1999;274:12835–9.
32. Ludwig A, Zong X, Stieber J, Hullin R, Hofmann F, Biel M. Two pacemaker channels from human heart with profoundly different activation kinetics. EMBO J. 1999;18:2323–9.
33. Bahinski A, Nairn AC, Greengard P, Gadsby DC. Chloride conductance regulated by cyclic AMP-dependent protein kinase in cardiac myocytes. Nature. 1989;340:718–21.
34. Nagel G, Hwang TC, Nastiuk KL, Nairn AC, Gadsby DC. The protein kinase A-regulated cardiac Cl- channel resembles the cystic fibrosis transmembrane conductance regulator. Nature. 1992;360:81–4.
35. Collier ML, Hume JR. Unitary chloride channels activated by protein kinase C in guinea pig ventricular myocytes. Circ Res. 1995;76:317–24.

36. Walsh KB, Long KJ. Properties of a protein kinase C-activated chloride current in guinea pig ventricular myocytes. Circ Res. 1994;74:121–9.
37. Levesque PC, Hume JR. ATPo but not cAMPi activates a chloride conductance in mouse ventricular myocytes. Cardiovasc Res. 1995;29:336–43.
38. Yamamoto-Mizuma S, Wang GX, Hume JR. P2Y purinergic receptor regulation of CFTR chloride channels in mouse cardiac myocytes. J Physiol. 2004;556:727–37.
39. Duan D, Ye L, Britton F, Horowitz B, Hume JR. A novel anionic inward rectifier in native cardiac myocytes. Circ Res. 2000;86:E63–71.
40. Komukai K, Brette F, Orchard CH. Electrophysiological response of rat atrial myocytes to acidosis. Am J Physiol Heart Circ Physiol. 2002;283:H715–24.
41. Komukai K, Brette F, Pascarel C, Orchard CH. Electrophysiological response of rat ventricular myocytes to acidosis. Am J Physiol Heart Circ Physiol. 2002;283:H412–22.
42. Duan DY, Fermini B, Nattel S. Sustained outward current observed after I(to1) inactivation in rabbit atrial myocytes is a novel Cl- current. Am J Physiol. 1992;263:H1967–71.
43. Wang GX, Hatton WJ, Wang GL, et al. Functional effects of novel anti-ClC-3 antibodies on native volume-sensitive osmolyte and anion channels in cardiac and smooth muscle cells. Am J Physiol Heart Circ Physiol. 2003;285:H1453–63.
44. Yamamoto-Mizuma S, Wang GX, Liu LL, et al. Altered properties of volume-sensitive osmolyte and anion channels (VSOACs) and membrane protein expression in cardiac and smooth muscle myocytes from Clcn3–/– mice. J Physiol. 2004;557:439–56.
45. Guan YY, Wang GL, Zhou JG. The ClC-3 Cl- channel in cell volume regulation, proliferation and apoptosis in vascular smooth muscle cells. Trends Pharmacol Sci. 2006;27:290–6.
46. Volk AP, Heise CK, Hougen JL, et al. ClC-3 and IClswell are required for normal neutrophil chemotaxis and shape change. J Biol Chem. 2008;283:34315–26.
47. Xu Y, Dong PH, Zhang Z, Ahmmed GU, Chiamvimonvat N. Presence of a calcium-activated chloride current in mouse ventricular myocytes. Am J Physiol Heart Circ Physiol. 2002;283: H302–14.
48. Collier ML, Levesque PC, Kenyon JL, Hume JR. Unitary Cl- channels activated by cytoplasmic Ca2+ in canine ventricular myocytes. Circ Res. 1996;78:936–44.
49. Hartzell C, Putzier I, Arreola J. Calcium-activated chloride channels. Annu Rev Physiol. 2005;67:719–58.
50. Caputo A, Caci E, Ferrera L, et al. TMEM16A, a membrane protein associated with calcium-dependent chloride channel activity. Science. 2008;322:590–4.
51. Schroeder BC, Cheng T, Jan YN, Jan LY. Expression cloning of TMEM16A as a calcium-activated chloride channel subunit. Cell. 2008;134:1019–29.
52. Baker MA, Lane DJ, Ly JD, De Pinto V, Lawen A. VDAC1 is a transplasma membrane NADH-ferricyanide reductase. J Biol Chem. 2004;279:4811–9.
53. Duan D. Phenomics of cardiac chloride channels: the systematic study of chloride channel function in the heart. J Physiol. 2009;587:2163–77.
54. Jentsch TJ, Stein V, Weinreich F, Zdebik AA. Molecular structure and physiological function of chloride channels. Physiol Rev. 2002;82:503–68.
55. Long SB, Tao X, Campbell EB, MacKinnon R. Atomic structure of a voltage-dependent $K^+$ channel in a lipid membrane-like environment. Nature. 2007;450:376–82.

# Chapter 7
# Lipid Signaling Pathways in the Heart

**Abstract** Despite significant advances, multiple questions remain to be answered regarding the molecular mechanisms involved in cardiac lipid signaling and its regulation. A complex network of signal transduction pathways is a critical regulator of cellular function permitting adaptation to a wide range of physiological conditions. One of the major drawbacks in the field of lipid signaling is the lack of selective enzyme inhibitors or activators and receptor subtype-specific agonists and antagonists. Therefore, design of highly selective compounds will greatly facilitate our understanding of roles of lipid signaling in cardiac physiology. Furthermore, detailed knowledge of the precise changes in lipid metabolism induced by different cardiac pathological conditions is also still lacking. Thus, further mechanistic insights are necessary to better assess whether and how modulation of lipid signaling could become an efficient novel therapeutic approach in the treatment of coronary artery disease.

**Keywords** Lipid signaling • Cardiac phosphoinositide • Eicosanoid signaling • Cardiac cyclooxygenases

## Introduction

Intricate network of signal transduction pathways is a key regulator of cellular function permitting adaptation to a wide range of physiological conditions. In the previous chapters, we have discussed a crucial role, in which numerous ion channels and ion transporters with associated proteins play in the precise regulation of myocardial function. Cellular membranes integrate these membrane-associated proteins and modulate highly tuned interplay of their activities. Additionally, myocardial membranes serve as a surface for signaling scaffolds as well as an endogenous depot for the precursors of lipid-related second messengers.

Lipids are well known as constituents forming the core of cellular membranes. However, they are not inert structural components of cell membranes but substrates for enzymes generating mediators involved in complex cardiovascular signaling pathways.

In this chapter, we discuss recent advances in our understanding of the molecular mechanisms underlying action of major lipid-related mediators (e.g., phosphoinositides, sphingolipids, eicosanoids signaling) and their roles in cardiomyocytes under physiological and pathological conditions.

## Phosphoinositide Signaling in the Heart

Inositol phospholipids also known as phosphoinositides, short-lived phosphorylated derivatives of phosphatidylinositol (PI), are the most prominent among lipids acting as second messengers. In the past decade, accumulating evidence has indicated that in addition to their essential roles as mediators of signal transduction regulating the ion channel functions, $Ca^{2+}$ homeostasis, and protein kinase activities, they are also involved in the spatiotemporal control of membrane trafficking creating an interface between cell signaling and membrane dynamics [1, 2].

PI represents a minor phospholipid component in all eukaryotic cell membrane. In the sarcolemma, PI is phosphorylated on the 4- then on the 5-position by PI4-kinase and PI4P5-kinase, respectively (Fig. 7.1). Resultant product, phosphatidylinositol(4,5)bisphosphate ($PIP_2$) is the central intermediate in inositide signaling pathway regulating numerous cellular responses ranging from vascular tone and hormone secretion to cell growth and differentiation [2–4].

Stimulation of appropriate surface receptors results in the activation of PI-specific phospholipases C (PLC), which hydrolyze $PIP_2$ and produce the hydrophilic inositol(1,4,5) trisphosphate [Ins(1,4,5)$P_3$] and the neutral *sn*-1,2-diacylglycerol (DAG) [3] (Fig. 7.1). In cardiomyocytes, this response is initiated by the engagement of various G protein coupled receptors (GPCRs), coupled to the $G_q$ proteins, with $\alpha_1$-adrenergic agonists, purine nucleotides, angiotensin or endothelin, and activates PLCβ subtypes [5]. In response to growth factors, receptor tyrosine kinases (RTKs) can activate PLCγ isozymes [6]. Ins(1,4,5)$P_3$ can activate its own receptors, $IP_3$-R, intracellular $Ca^{2+}$ channels located in the sarcoplasmic reticulum (SR) and nuclear membrane, and trigger thereby

J. Marín-García, *Signaling in the Heart*,
DOI 10.1007/978-1-4419-9461-5_7, 

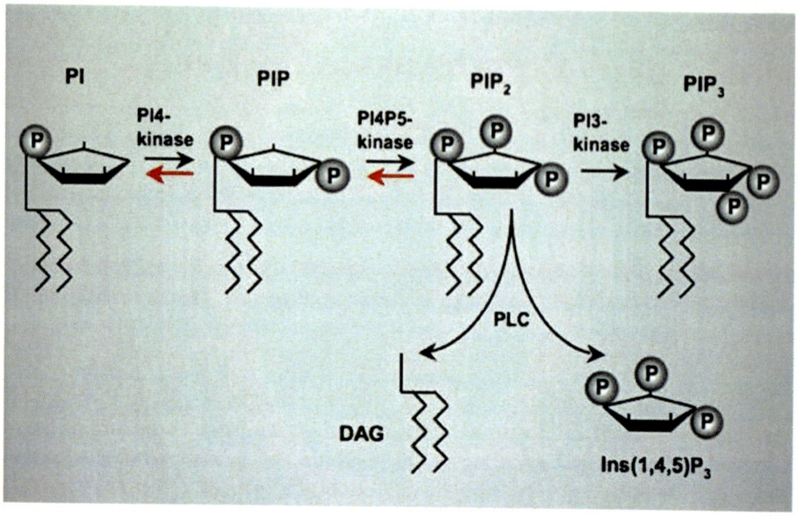


**Fig. 7.1** Inositol phospholipid metabolism. Phosphatidylinositol (PI) is phosphorylated by PI4- and PI4P5-kinases to generate phosphatidylinositol(4,5)bisphosphate ($PIP_2$). $PIP_2$ is hydrolyzed by phospholipase C (PLC) to produce inositol(1,4,5)trisphosphate ($Ins(1,4,5)P_3$) and *sn*-1,2-diacylglycerol (DAG). $PIP_2$ can be further phosphorylated to phosphatidylinositol(3,4,5) trisphosphate ($PIP_3$) by PI 3-kinases (PI3Ks). Actions of phosphatases are indicated by *red arrows*

intracellular $Ca^{2+}$ release from cellular stores resulting in the increase of cytosolic $Ca^{2+}$ [7]. DAG is a well-established co-activator of conventional and novel subtypes of protein kinase C [8]. Moreover, DAG is able to activate the canonical transient receptor potential channels (TRPC) independently of PKC activation [9].

In addition to PLC-catalyzed hydrolysis and PI 3-kinase-dependent 3′-phosphorylation, $PIP_2$ can be cleaved by phospholipase D to produce phospatidic acid, an activator of various signaling intermediates [10, 11]. Moreover, $PIP_2$ regulates various cardiac ion channels and ion exchangers as well as contributes to the membrane localization of numerous critical signaling proteins [12].

Most if not all of these metabolites have the potential to affect significantly heart's electrical activity and therefore to contribute to the development of dysrhythmias. However, only $PIP_2$ and $Ins(1,4,5)P_3$ have so far been associated with dysrhythmogenesis, therefore, these signal mediators will be discussed here.

## $PIP_2$ Signaling Associated with Dysrhythmias

$PIP_2$ appears to be tightly regulated and similar concentrations of $PIP_2$ have been detected in mouse, rat, and human atria and ventricle [13, 14]. Overexpression of PI4P5 kinase has not resulted in the elevated $PIP_2$ levels due to probably changes in other $PIP_2$-metabolizing enzymes [15]. On the other hand, it has recently been suggested that localized depletion of functionally different $PIP_2$ pools may be sufficient to affect adjacent ion channels [16–18]. List of currently known cardiac ion channels and ion exchangers regulated by $PIP_2$ is present in Table 7.1 [19].

All major cardiac inwardly rectifying $K^+$ channels, $K_{ir}2$, $K_{ir}3$, and $K_{ir}6$, are critically regulated by $PIP_2$ (Fig. 7.2). $K_{ir}2$ channels conduct the $I_{K1}$ current responsible for the resting action potential (AP) in atrial and ventricular myocytes. $K_{ir}3$ channels in atrial and pacemaker myocytes are responsible for the acetylcholine-activated $I_{KAch}$ current. $K_{ir}6$ channels underlay $I_{ATP}$ current and are regulated by intracellular ATP (see Chap. 6).

Inwardly rectifying $K^+$ channels require $PIP_2$ binding to the cytoplasmic domains for their opening, while $PIP_2$ depletion leads to closure of the channel [20]. Accordingly, mutant $K_{ir}2.1$ channel characterized by decreased affinity to $PIP_2$ binding has recently been identified in patients with Andersen–Tawil syndrome associated with ventricular dysrhythmia [21].

$K_{ir}3$ channels are regulated via activation of the heterotrimeric $G_i$ protein dissociating into $G_\alpha$ and $G_{\beta\gamma}$ subunits. Interruption of $PIP_2$ binding to the channel abrogates stimulatory effect of $G_{\beta\gamma}$ or $Na^+$ on channel function. Importantly, decreased activation of mutant channel characterized by low affinity to $PIP_2$ can be rescued by adding $G_{\beta\gamma}$ [22].

Unlike most $K_{ir}$ channels that are specific to $PI(4,5)P_2$, $K_{ir}6$ channels are relatively nonspecific activating not only by 4,5-$PIP_2$ but also by 3,4- and 3,5-isomers as well as by $PIP_3$. Furthermore, $K_{ir}6$ channels can be activated by other lipids particularly by long chain fatty acyl CoA derivatives acting similar to $PIP_2$ [23–25]. This activation may contribute to cardiac complications observed in type 2 diabetes where levels of long chain fatty acyl CoA derivatives are increased [25]. Therefore, these findings suggest the possibility of selective dietary manipulation of channel function since lipid intake can change composition of fatty acid derivatives [26].

The voltage-gated $K^+$ channels $K_V7.1$ and $K_V11.1$ (hERG) are responsible for the slow and rapid repolarization, respectively. Mutations in the genes encoding these channels are

**Table 7.1** Cardiac ion channels and exchangers regulated by $PIP_2$

| Current | $PIP_2$ effect | Channel protein | Gene | Ion | Function | Rel. to dysrhythmia | Associated disease |
|---|---|---|---|---|---|---|---|
| $I_{K1}$ (inward rectifier K current) | Channel opening | $K_{IR}2.1$ | KCNJ2 | $K^+$ | Maintain resting membrane potential | VF | Andersen–Tawil syndrome |
| $I_{KAch}$ (the acetylcholine-activated K current) | Modify the channel interaction with G protein βγ subunits | $K_{IR}3.1/4$ (GIRK1/4) | KCNJ3/5 | $K^+$ | In response to parasympathetic stimulation | Perpetuation AF and involved in AF remodeling | |
| $I_{KATP}$ | Decreases the channel affinity for ATP | $K_{IR}$ 6.2 and SUR2A | KCNJ11 | $K^+$ | The channel close by intracellular ATP. Metabolic and mechano-sensitive repolarization reserve | AF, VF | Sudden cardiac death (vein of Marshall adrenergic AF) |
| $I_{KR}$ (the rapidly activating delayed rectifier $K^+$ current) | Hyperpolarizing shifts in the voltage dependence of activation and slows deactivation. Also influences responses to cAMP and PKA | HERG (Kv 11.1) and mink/MiRP | KCNH2 KCNE1/2 | $K^+$ | Promote repolarization and maintaining pacemaker automaticity | Polymorphic ventricular tachycardia (*torsades de pointes*; TdP) ventricular fibrillation VF | Type 2 congenital long QT syndrome (LQT-2) |
| $I_{KS}$ (the slowly activating delayed rectifier $K^+$ current) | Prevents endogenous channel inhibition and suppresses the channel rundown | Kv7.1 and minK | KCNQ1, KCNE1 | $K^+$ | Repolarization and determine heart rate-dependent shortening of APD and contribute to the slow diastolic depolarization of pacemaker AP | Cardiac arrest AV block | LQT-1, LQT5 Jervell and Lange–Nielsen syndrome |
| $I_f$ (hyperpolarization-activated cation current) | Shifts the voltage dependence towards depolarized potentials | HCN | HCN4/2 | $Na^+$, $K^+$ | Spontaneously diastolic depolarization | Yes | AF, sick sinus syndrome |
| $I_{Na/Ca}$ ($Na^+/Ca^{2+}$ exchanger current) | Prevent auto-inhibition by binding to exchanger-inhibitory peptide | NCX | NCX1 | $Na^+$, $Ca^{2+}$ | $Ca^{2+}$ extrusion $Ca^{2+}$ uptake spontaneously diastolic depolarization | Yes | I/R |
| SOC/ROC (store or receptor-operated $Ca^{2+}$ channels) | Inhibits? | TRPC | TRP (C1–7) | $Ca^{2+}$, $Na^+$, $K^+$, or others | Pacemaker current? Stretch activated current? | Not known | ? |
| $NSC_{Ca}$ ($Ca^{2+}$ activated nonselective cation current) | Depletion of $PIP_2$ desensitizes the channel | TRPM4 | TRPM | $Na^+$, $K^+$ | Depolarization | Delayed after-depolarization | Cardiac hypertrophy |

*AF* atrial fibrillation; *HCN* hyperpolarization-activated cyclic nucleotide-gated channel; *VF* ventricular fibrillation; *I/R* ischemia/reperfusion

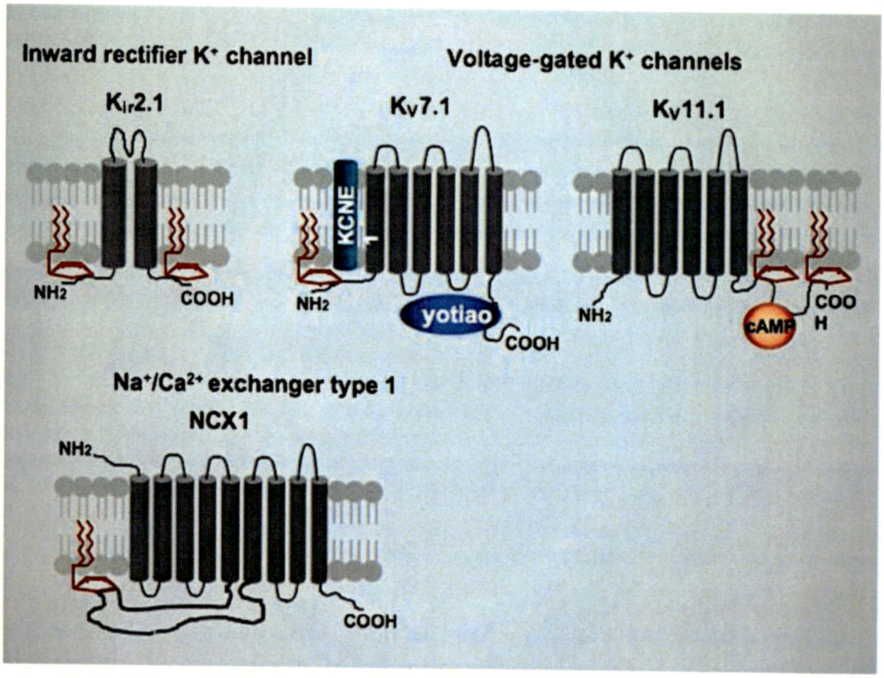


**Fig. 7.2** Interaction of $PIP_2$ with cardiac inwardly rectifying $K^+$ and voltage-gated $K^+$ channels, and the $Na^+/Ca^{2+}$ exchanger. Main channel protein subunits and accessory proteins are schematically shown. $PIP_2$ is depicted in *red*

associated with inherited dysrhythmias LQTS and SQTS (see Chap. 14). Both $K_V7.1$ and $K_V11.1$ interact with and are regulated by $PIP_2$, although these interactions are less well characterized than in case of inwardly rectifying $K^+$ channels [27, 28]. Some mutations in $K_V7.1$ associated with LQTS are located in the protein region involved in the interaction with $PIP_2$ [29].

$K_V11.1$ is regulated by direct cAMP binding and by PKA-dependent phosphorylation. Both cAMP-binding and PKA-dependent phosphorylation sites are located in the intracellular COOH terminus of $K_V11.1$. Importantly, the same protein region contains the putative $PIP_2$ interaction sites suggesting that $PIP_2$ can affect the $K_V11.1$ response to cAMP and PKA. Since $PIP_2$ induces delay of $K_V11.1$ deactivation, down regulation of $PIP_2$ after activation of PLC-coupled receptors could be responsible for the channel deactivation [30, 31].

The functional $K_V7.1$ channel requires the $PIP_2$ binding to its N-terminal intracellular region. The $PIP_2$ binding region is located within an auto-inhibitory domain of the $K_V7.1$ α-subunit and the $PIP_2$ binding prevents this inhibition [32]. Importantly, mutations in this region causing decreased $PIP_2$ binding have been associated with LQTS. The addition of excess of $PIP_2$ in model experiments has restored the activity of the mutant channel confirming the functional importance of the $K_V7.1$–$PIP_2$ interaction [29].

Finally, $PIP_2$ is also involved in the regulation of the hyperpolarization-activated cyclic nucleotide gated channels, HCNs, governing the pacemaker current $I_f$. $PIP_2$ shifts the voltage dependence of the HCN channels toward depolarized AP increasing the spontaneous firing rate [33]. Although the HCN channels have been reported to contribute to the development of various types of atrial fibrillation (AF) [34], currently there is no direct evidence that this involves $PIP_2$.

## $Ins(1,4,5)P_3$ Signaling in the Heart

$Ins(1,4,5)P_3$ engages specific receptors, $IP_3$-R, located in intracellular $Ca^{2+}$ stores stimulating $Ca^{2+}$-induced $Ca^{2+}$ release (CICR) and activating protein kinase pathways to further enhance $Ca^{2+}$ responses. $Ca^{2+}$ release mediated by Ins(1,4,5) $P_3$ via $IP_3$-R is independent of the CIRC controlled by the ryanodine receptors (RyR). In atrial myocytes, it has been reported that stimulation of the $IP_3$-R type 2 [$IP_3$-R (2)] can contribute to excitation–contraction coupling and therefore predispose to dysrhythmia by interfering with the RyR signaling [35]. Moreover, $Ca^{2+}$ signals generated by $IP_3$-R (2) localized close to the sarcolemma could affect sarcolemmal $Ca^{2+}$ channels and $Na^+/Ca^{2+}$ exchanger [36].

On the other hand, expression of $IP_3$-R in cardiomyocytes is very low compared to the RyR, and they located predominantly in the nuclear envelope therefore distant from both the

RyR and the generation sites of Ins(1,4,5)$P_3$ [37–39]. It has been reported higher $IP_3$-R levels in atrial than in ventricular myocytes suggesting that perturbations in Ins(1,4,5)$P_3$ signaling can contribute to the development of atrial rather than ventricular dysrhythmias [40]. Based on the different localization and subtypes of $IP_3$-R found in the working and conducting myocytes, it has also been suggested that Ins(1,4,5)$P_3$ signaling in the latter predominantly predispose to ventricular dysrhythmias [41, 42].

During ischemia and postischemic perfusion, elevated Ins(1,4,5)$P_3$ generation has been detected [43–45]. Accordingly, inhibitors of PLC, which inhibit Ins(1,4,5)$P_3$ generation, reduce ventricular dysrhythmias [46, 47]. Given low specificity of PLC inhibitors and their effects not only on Ins(1,4,5)$P_3$ generation but also on ion channels activities, it is presently not clear whether changes in Ins(1,4,5)$P_3$ or $PIP_2$ generation are associated with dysrhythmia during ischemia.

It has been suggested that both Ins(1,4,5)$P_3$ and $PIP_2$ can influence the $Na^+/Ca^{2+}$ exchanger (NCX1) activity. Ins(1,4,5)$P_3$ can affect sarcolemmal NCX1 by changing localized $Ca^{2+}$, whereas $PIP_2$ stimulates the exchanger by direct binding to its internal auto-inhibitory domain (Fig. 7.2) [48, 49]. Both mechanisms may contribute to NCX1-associated dysrhythmogenesis. Moreover, elevated expression of NCX1 and $IP_3$-R has been found in heart failure suggesting the possibility of a greater contribution of Ins(1,4,5)$P_3$ and $PIP_2$ under this condition [50–52].

$PIP_2$ also regulates TRPC channels, low conductance, relatively nonselective cation channels activated by receptors coupled to PLC [53, 54]. TRPC3, 6, and 7 are activated directly by DAG, while TRPC4 and 5 are regulated by PLC in DAG-independent manner [55, 56]. $PIP_2$ binding can inhibit TRPC4, but activate TRPC6 and 7 via enhancing $Ca^{2+}$ entry, and PLC-mediated depletion of $PIP_2$ can abrogate these effects [57, 58]. In cardiomyocytes, TRPC3 is associated with NCX1 and $Ca^{2+}$ entry mediated by this channel can interfere with the voltage-regulated $Ca^{2+}$ channels and/or NCX1 contributing to dysrhythmias. Moreover, it has recently been suggested that TRPC3 and closely related TRPC4 expressed in the plasma membrane of mouse pacemaker cells could contribute to pacemaker current modulated by $PIP_2$ and Ins(1,4,5)$P_3$ signaling [59, 60].

## Cardiac Phosphoinositide 3-Kinases and Phosphatase and Tensin Homolog

Phosphatidylinositol plays a central role in signal transduction, thus the enzymes which regulate its levels are critical determinants of cellular responses. Phosphoinositide 3-kinases (PI3Ks) are responsible for the generation of $PIP_3$, whereas PTEN (phosphatase and tensin homolog deleted on chromosome 10) is the major phosphatase responsible for hydrolysis of this lipid (Fig. 7.1).

### *PI3K Family*

The PI3K family consists of eight distinct enzymes characterized by dual protein and lipid kinase activity. They all share a common catalytic activity: they phosphorylate the D3 hydroxyl group of PI upon GPCR, RTK, or Ras activation. Members of PI3K family are divided into three functional classes based on their protein structure, substrate specificity, associated regulatory subunits, and activation mechanisms [61–63]. Class I PI3Ks are heterodimeric enzymes composed of a catalytic (p110α, β, δ, and γ) and a regulatory adaptor subunit (p85, p55, or p101 family). The class II enzymes are monomeric proteins, ubiquitously expressed PI3K-C2α and PI3K-C2β, and the liver specific PI3K-C2γ. Ubiquitous vacuolar protein sorting 34 (Vps34) is the sole class III enzyme.

*In vitro*, all class I PI3Ks are capable of phosphorylating PI, PI(4)P, and PI(4,5)$P_2$ to produce PI(3)P, PI(3,4)$P_2$ ($PIP_2$), and PI(3,4,5)$P_3$ ($PIP_3$), respectively. However, *in vivo*, phosphorylation of PI(4,5)$P_2$ to $PIP_3$ is the predominant pathway. Class II PI3Ks phosphorylate PI and PI(4)P *in vitro*, but not PI(4,5)$P_2$. In contrast to class I and II PI3Ks, class III PI3K, Vsp34, can phosphorylate only PI [63].

All catalytic subunits of class I PI3Ks are 110-kDa proteins and share common structural features: they contain four homology domains (HDs) including (from C- to N-terminus) the catalytic domain (HR1), PIK domain (HR2), C2 domain (HR3), and Ras-binding domain (RBD or HR4). The PIK and C2 domains are involved in protein–protein interactions and in phospholipid binding, respectively, while the RBD (HR4) domain binds the monomeric G protein Ras.

Adaptor subunits of class I enzymes p85α, p85β, p55γ, and p101γ bind corresponding catalytic subunits via p110-binding domains. p85α is predominant subunit expressed in the heart. p85α, p85β, and p55γ contain two SH2 domains, which bind phosphotyrosine on activated RTKs, mediating the catalytic subunit translocation to the cell membrane. p101 adapter contains a $G_{\beta\gamma}$ binding site involved in the activation of the p110γ catalytic subunit. However, this activation via p101 is not absolutely required as p110γ contains its own $G_{\beta\gamma}$ binding site and binds $G_{\beta\gamma}$ following GPCR stimulation.

The class I family of PI3Ks resides predominantly in the cytosol until activated and recruited to active signaling complexes while the class II enzymes are mainly constitutively associated with cell membranes including plasma membrane, intracellular membranes, and nuclear envelope. Currently, there is no clear mechanism of activation of the class II PI3Ks or clearly defined role for these kinases.

Class I PI3Ks are the best studied enzymes, whereas less is currently known about the two other PI3K classes. Three class I PI3K isoforms (p110α, β, and γ) are expressed in the heart, among them PI3Kα and PI3Kγ are the best characterized.

## PTEN

Mammalian PTEN, also known as MMAC1 (mutated in multiple advanced cancers) and TEP1 (TGFβ-regulated and epithelial cell-enriched phosphatase), is a 40–50-kDa protein with 3′-lipid phosphatase activity [64, 65]. It contains an N-terminal phospholipid-binding region, an N-terminal phosphatase domain, a C2 domain, and a C-terminal PSD-95/Dlg/ZO-1 (PDZ) homology domain. Although PTEN may possess some protein phosphatase activity, *in vivo* it functions mainly as a 3′-lipid phosphatase hydrolyzing its main physiologic substrate, membrane-bound $PIP_3$. Upon activation, PTEN is recruited to the plasma membrane due to electrostatic interaction of the polybasic N-terminal tail, catalytic and C2 domains of PTEN with $PIP_3$. Dephosphorylation of the phosphorylated C-terminal residues of PTEN (inactive form) contributes to its targeting to the plasma membrane while PDZ-binding domain of PTEN is involved in protein–protein interactions essential for the formation of an active multiprotein complex [66, 67].

Antagonistic action of PTEN on PI3K signaling is highly conserved and occurs in various mammalian cells, including cardiomyocytes, vascular smooth muscle cells, and endothelial cells [68, 69]. Molecular targets of the PI3K and PTEN signaling in the heart include phosphoinositide-dependent kinase-1 (PDK1), Akt (protein kinase B, PKB), glycogen synthase kinase-3 (GSK3), mammalian target of rapamycin (mTOR, also known as FRAP, FKBP-12 rapamycin-associated protein), and p70 ribosomal protein S6 kinase (p70S6K).

## PI3Kα Signaling and Myocardial Hypertrophy

PI3K activity regulates hypertrophy of the heart in physiological (e.g., exercise training) and pathological (e.g., pressure overload) conditions [70, 71]. Mice overexpressing a cardiac-specific constitutively active PI3Kα (CA-PI3Kα) have enlarged hearts as a result of cardiomyocyte hypertrophy, while mice expressing of dominant-negative PI3Kα (DN-PI3Kα) have smaller hearts with decreased cardiomyocyte size. The number of cardiomyocytes has been unchanged in both animal models suggesting that PI3Kα controls cell size but not the number [68].

PI3Kα regulates these changes in heart size via the alteration in the function of its downstream effectors (Fig. 7.3). Overexpression of CA-PI3Kα increases Akt/PKB, p70S6K, and GSK3β phosphorylation and activities, whereas overexpression of DN-PI3Kα abolishes this activation [70, 72]. Consistent with a critical role of Akt/PKB in this pathway, Akt1/PKBα-deficient mice have smaller hearts, whereas

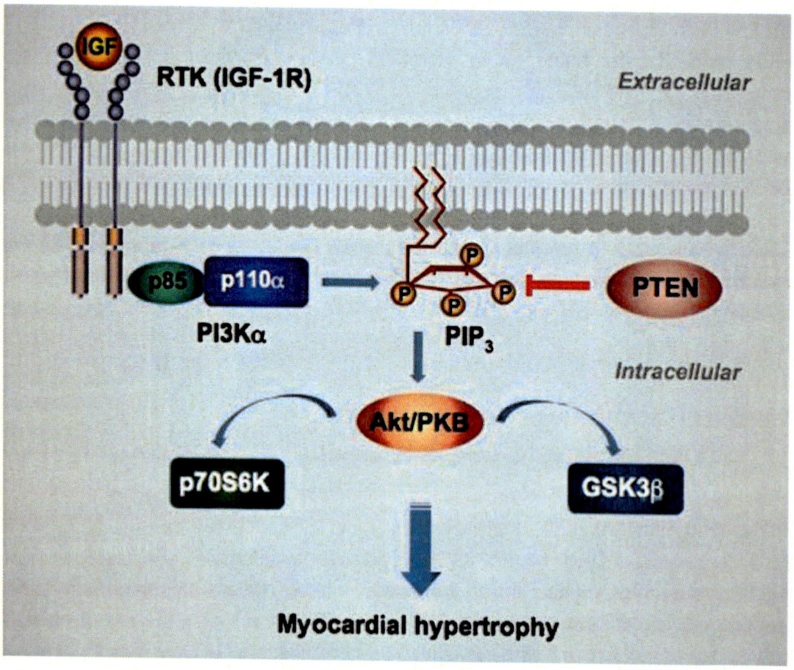


**Fig. 7.3** Cardiac PI3Kα signaling pathway. Engagement of receptor tyrosine kinases (RTK) (e.g., IGF-1R) activates PI3Kα via interaction of RTKs with its adaptor subunit (p85) to generate the second messenger $PIP_3$. Activation of PI3Kα induces Akt/PKB, p70S6K, and GSK3β phosphorylation and activities. In cardiomyocytes, the PI3Kα/Akt signaling pathway induces hypertrophy in physiologic conditions, without affecting contractility. PTEN degrades $PIP_3$ attenuating the PI3Kα signaling

mice overexpressing CA-Akt/PKB have large hearts [73–75]. Moreover, increased Akt/PKB activity has resulted in elevated both GSK3β phosphorylation and p70S6K activity [75]. Overexpression of Akt/PKB has resulted not only in myocardial hypertrophy but also in resistance to apoptosis.

PI3Kα signaling is largely attenuated by PTEN: inactivation of PTEN activates Akt/PKB and increases cardiomyocytes size, while PTEN overexpression abolishes these responses [68, 76].

Cardiac PI3Kα signaling is induced by various physiological stimuli including exercise and insulin-like growth factor-1 receptor (IGF-1R) (Fig. 7.3). Accordingly, cardiac-specific overexpression of IGF-1R results in myocardial hypertrophy and increased level of IGF-1 after exercise training is a major cause of elevated PI3Kα activity [70]. Moreover, some PI3Kα targets including GSK3β and mTOR are also affected in pathological cardiac hypertrophy [77, 78].

## *PI3Kγ Signaling: Myocardial Contractility and β-Adrenergic Signaling*

Levels of a second messenger, cAMP, is regulated by a highly tuned interplay between adenylyl cyclase (AC)-induced synthesis and phosphodiesterase (PDE)-mediated degradation. The main effector of cAMP is protein kinase A (PKA), a serine threonine kinase, which phosphorylates various target proteins. In cardiomyocytes, this response is mediated via proteins involved in excitation–contraction coupling including phospholamban (PLB), which contributes to the $Ca^{2+}$ reuptake by SR [79].

PI3Kγ$^{-/-}$ mice exhibit hyper-contractility associated with elevated levels of cAMP without changes in AC expression, while mice expressing a catalytically inactive PI3Kγ (PI3Kγ KD mice) have unchanged myocardial contractility and normal levels of cAMP [68, 71]. These findings suggest that PI3Kγ regulates cellular cAMP levels independently of its catalytic activity. PI3Kγ functions as a scaffold protein associated with two cAMP-degrading enzymes, PDE3B and PDE4. Current model suggests that PI3Kγ pathway regulates cAMP signaling in subcellular compartments within cardiomyocytes via complex formation with specific PDEs (Fig. 7.4) [80, 81].

PI3Kγ is also involved in the regulation of β-adrenergic receptor (β-AR) signaling essential for the fine-tuning of cardiac function. β-ARs belong to the subfamily of GPCRs stimulated by epinephrine and norepinephrine and comprise of three distinct isoforms, $\beta_1$, $\beta_2$, and $\beta_3$ [82, 83]. $\beta_1$-AR couples to $G_s$, whereas $\beta_2$-AR and $\beta_3$-AR can couple to $G_s$ or $G_{i/o}$. In cardiomyocytes, $\beta_1$- and $\beta_2$-ARs are predominantly expressed mediating sympathetic stimulation of the heart [84]. In response to agonist stimulation, GPCR kinase 2 (GRK2, also known as βARK1) phosphorylates β-AR triggering interaction between the receptor and PI-binding endocytic proteins β-arrestins and adaptor protein 2 (AP-2), which blocks coupling to G protein (desensitization) and ultimately leads to β-AR internalization [84].

In cardiomyocytes, β-AR signaling, predominantly via $\beta_2$-AR, activates PI3Kγ through $G_{\beta\gamma}$ subunits resulting in

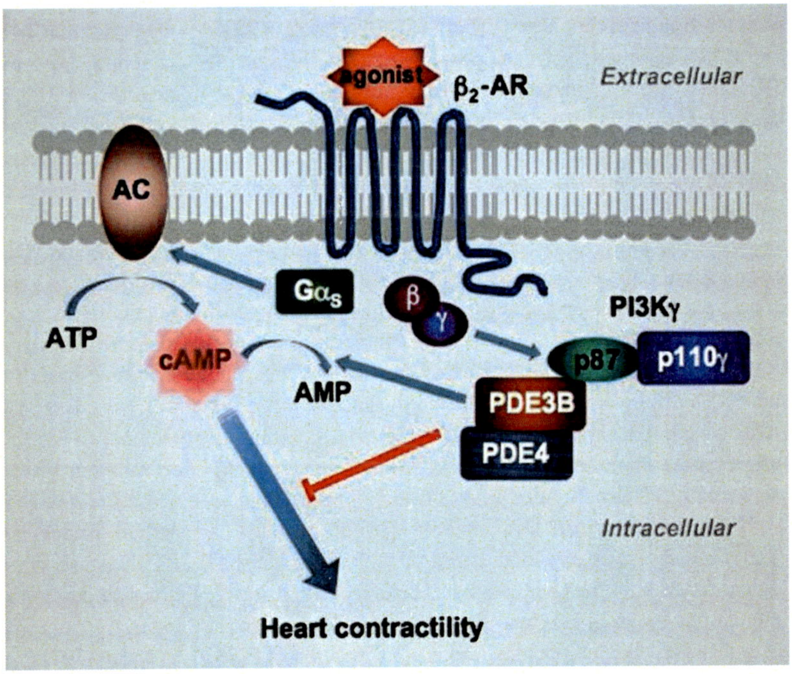


**Fig. 7.4** The scaffolding role of PI3Kγ in regulation of myocardial contractility. β-Adrenergic stimulation induces dissociation of the heterotrimeric G protein into $G\alpha_s$ and heterodimeric β, γ subunits. In cardiomyocytes, $G\alpha_s$ activates adenylyl cyclase (AC) to increase cAMP production in restricted microdomains, while β, γ subunits induce association of PI3Kγ with PDE3B and PDE4. This complex converts cAMP into AMP thereby regulating negatively cardiomyocyte contractility. In that way, PI3Kγ also contributes to regulation of cAMP concentration in distinct subcellular domains

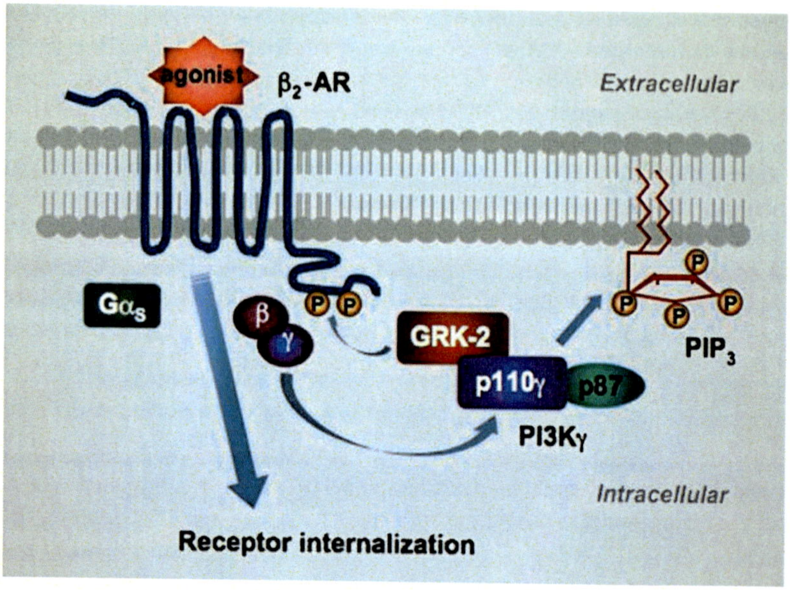


**Fig. 7.5** PI3Kγ turns off $\beta_2$-AR signaling. In cardiomyocytes, $\beta_2$-AR signaling activates PI3Kγ via $G_{\beta\gamma}$ subunits resulting in PI3Kγ binding to GRK2 and translocation of this complex to the plasma membrane. PI3Kγ/GRK2 complex induces $PIP_3$ production and also phosphorylates $\beta_2$-AR reducing its ability to relay signals in the presence of high concentration of the agonist (desensitization). Eventually, phosphorylated $\beta_2$-AR is targeted to internalization reducing the number of the receptors at the plasma membrane and thereby switching off the signal

PI3Kγ binding to GRK2 and translocation of this complex to plasma membrane (Fig. 7.5) [85, 86]. Relocation of PI3Kγ–GRK2 complex to membrane also contributes to PI3Kγ activation. Loss of PI3Kγ activity does not prevent downregulation of β-AR but rather inhibits β-AR internalization in response to chronic receptor stimulation. Elevated PTEN levels caused by β-AR stimulation may represent negative feedback on β-AR-mediated PI3Kγ activation [69, 87]. Furthermore, loss of PI3Kγ activity results in the downregulation of phosphodiesterase 4 (PDE4), increased phosphorylation of PLB, and sarcoplasmic reticulum $Ca^{2+}$ ATPase (SERCA2a) activation. These biochemical changes appear to occur in the specific SR compartment containing SERCA2a but not L-type $Ca^{2+}$ channels or RyR2 and lead to $Ca^{2+}$ leakage from the SR without change in $I_{Ca,L}$ current [88, 89].

PI3Kγ signaling also downregulates $\beta_2$-AR stimulated cAMP production and PKA activation [68, 90]. Accordingly, loss of PI3Kγ elevates intracellular cAMP levels and inhibits isoproterenol-induced upregulation of Akt/PTB activity and Erk1/2 phosphorylation leading to restoration of heart function [87]. In contrast, loss of PTEN results in significant reduction of intracellular cAMP concentration and is linked to reduced cardiomyocyte contractility [68]. Since reduction of β-AR signaling has been reported to be associated with heart failure [84], inhibition of PI3Kγ could be a promising therapeutic approach to normalize β-AR signaling in heart failure.

PI3Ks signaling can also regulate various cardiac ion channels and exchangers modulating thereby electrophysiological function of the heart. As it has been discussed above, interactions between $PIP_2$ and $K_{ir}$ channels enhance subunit assembly and channel opening, therefore activation of PI3K and/or PLC activity can deplete $PIP_2$ nearby the channel downregulating its activity [69]. This mechanism may underlie the LQT7 phenotype caused by the loss of function mutations altering $K_{ir}2.1$–$PIP_2$ interactions in cardiomyocytes [91]. Activity of the voltage-gated $K_V11.1$ channels can be also modulated by the PI3K-Akt signaling pathway [92]. Moreover, PI3Kγ appears to contribute to the M2 receptor-dependent activation of cardiac $K_{ir}3.x$ channels responsible for $I_{KAch}$ current [69].

Gene expression analysis of ventricular myocytes has revealed that several pathways regulated by PI3K-Akt signaling are activated in response to myocardial ischemia–reperfusion [93]. Enhanced PI3K and/or Akt/PKB signaling in animal models has been associated with increased cell survival and smaller infarct size in ischemia–reperfusion [69, 74, 94]. Accordingly, inhibition of PI3K blocks reductions in infarct size and associated enhanced phosphorylation of Akt/PKB and PDK1 [95, 96]. Importantly, mice with cardiac-specific PDK1 deletion display heart failure and higher sensitivity to hypoxia [97]. Although protective mechanism of Akt/PKB activation remains to be determined, it has been suggested that this may involve modulation of GSK3β activity as well as activation of mitochondrial $K_{ATP}$ channels [69, 98, 99]. Defect of PI3Kγ, but not PI3Kα, leads to decreased phosphorylation of GSK3β and Akt/PKB, elevated cell death, and compromised functional recovery in response to myocardial ischemia–reperfusion [100]. Haploinsufficiency of PTEN is associated with increased Akt/PKB phosphorylation and cardioprotection in ischemia–reperfusion confirming the role of PTEN as a negative regulator of PI3K signaling pathway [101].

Distinct PI3K isoforms contribute to different physiological and pathological responses in the heart. In this regard, one of the unanswered questions to be addressed in the future is the role of cardiac PI3K in the heart physiology. The refinement of the molecular mechanisms of PI3K signaling and the development of isoform-selective inhibitors will provide invaluable insight into this important pathway.

## Sphingolipid Signaling in Cardiomyocytes

Sphingolipids are widely distributed in all types of mammalian cells. Their role as structural components of cell membranes has been appreciated for many years. However, sphingolipid pathways are also involved in the generation of signal mediators serving both as intracellular second messengers and as ligands for specialized receptors. During past two decades, sphingolipids was demonstrated to mediate a wide variety of biological responses, such as cell proliferation, survival, migration, cytoskeleton rearrangements, and intracellular mobilization [102–104].

*De novo* synthesis of sphingolipids is initiated with the condensation of serine and palmitoyl-CoA catalyzed by serine palmitoyltransferase. This reaction occurs at the cytoplasmic surface of the endoplasmic reticulum (ER) and results in the generation of ceramide (Fig. 7.6). Ceramide is mainly produced as the result of hydrolysis of the membrane phospholipid sphingomyelin (SM) by sphingomyelinase. Ceramide can be converted to sphingosine by ceramidase. Sphingosine is phosphorylated by sphingosine kinases (SKs) to produce sphingosine-1-phosphate (S1P), one of the most prominent intracellular second messengers of this cascade. S1P can be dephosphorylated back to sphingosine by S1P phosphatase or be irreversibly hydrolyzed into phosphoethanolamine and hexadecanal by S1P lyase [105, 106].

S1P concentrations in plasma are typically in the range between 0.1 and 0.6 μM, while its concentrations in tissues are low [107]. S1P in plasma is mainly associated with high-density lipoproteins and albumin [108]. Erythrocytes and endothelial cells (e.g., blood platelets) contribute to plasma S1P. Although S1P levels within most other cells are low, its intracellular concentrations may be differentially modulated in different subcellular compartments.

### *Sphingomyelinases and Their Role in the Heart*

Sphingomyelinases (SMases) hydrolyze sphingomyelin into phosphocholine and ceramide, which is not only a bioactive sphingolipid but also a substrate for the generation of a cascade of other lipid mediators (Fig. 7.6). Ceramide can also be generated *de novo* via condensation of serine with palmitoyl-CoA by serine palmitoyltransferase. Ceramide can be further converted by ceramidase into sphingosine that, in turn, can be phosphorylated into S1P by sphingosine kinases. Importantly, these sphingolipids play opposite biological roles: ceramide

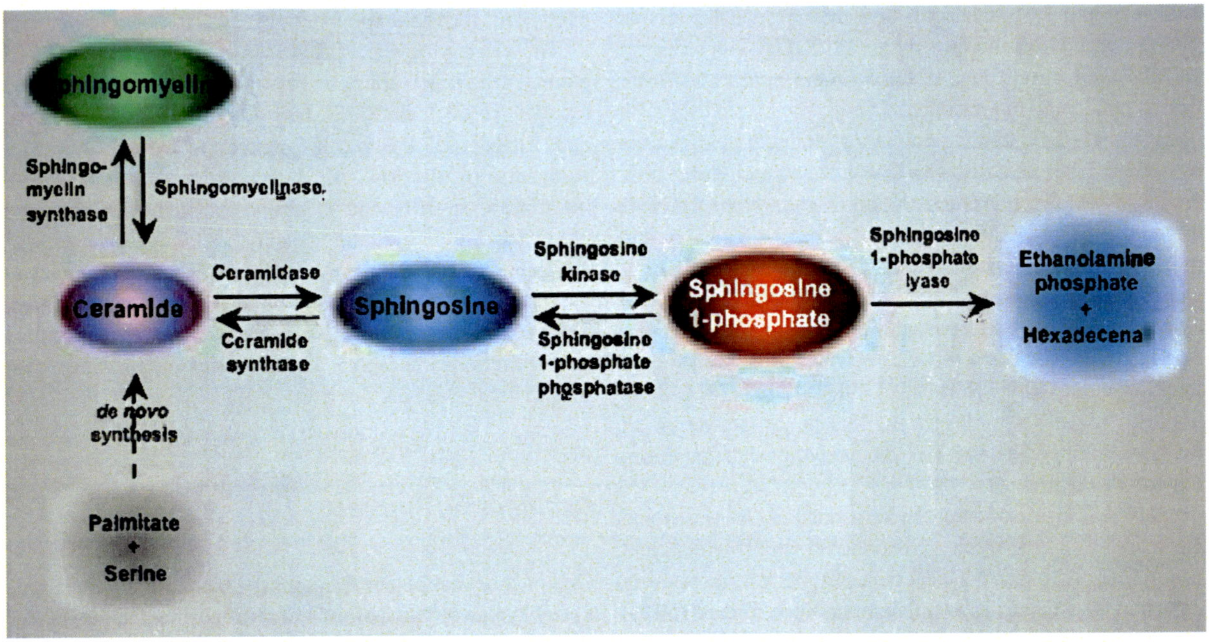


**Fig. 7.6** Scheme of sphingolipid metabolism. Major intermediates and enzymes catalyzing corresponding steps are shown. See text for the details

and sphingosine are mainly antiproliferative and pro-apoptotic mediators, while S1P mediates cell proliferation and prevents apoptosis. Thus, a cell's fate depends on the ratio between ceramide with sphingosine and S1P (the ceramide/S1P rheostat) rather than on their individual concentrations [109].

Based on their optimum pH, SMases can be divided into three groups: alkaline, neutral, and acid SMases. In humans, alkaline SMases have been found in the intestinal mucosa, bile, and liver, whereas neutral (NSMases) and acid (ASMases) play a crucial role in cardiovascular physiology [110]. Like many other enzymes involved in sphingolipid metabolism, SMases have specific subcellular localization: NSMases localize mainly in the endoplasmic reticulum and the Golgi apparatus, whereas ASMases are associated with the endosomes/lysosomes but can relocate to the outer leaflet of the plasma membrane [110].

Three distinct genes encoding NSMases have been cloned [111–113]. Role for NSMase1 (*Smpd2*) *in vivo* remains currently unclear [114, 115]. Ubiquitously expressed NSMase2 (*Smpd3*) is essential for skeletal development [116]. The most recently discovered NSMase3 (*Smpd4*) is a member of the C-tail-anchored membrane protein family. NSMase3 is an integral part of TNFα receptor type 1 (TNFR1) and adaptor protein factor associated with NSMase activation (FAN) signaling [113]. NSMases are responsible for the hydrolysis of SM within the inner leaflet of the plasma membrane. Activation of NSMases in various cardiovascular cells in response to variety of external signals plays a central regulatory role in ceramide-dependent apoptosis and cell proliferation [115]. TNFα-induced apoptosis in cardiomyocytes is mediated by NSMases [117, 118]. Activity of NSMase1 and 3 is inhibited by glutathione and their responsiveness to TNFα depends on this tripeptide [119, 120]. Accordingly, NSMase activation found in heart failure appears to result from a deficiency in glutathione [121].

In heart failure, cardiac deficiency in glutathione, caused by oxidative stress, along with TNFα elevation induces activation of NSMases. Administration of glutathione prevents the activation of NSMase-mediated apoptotic pathway and improves cardiac function [121]. Cardiac glutathione deficiency in ischemia–reperfusion is also a likely cause of rapid NSMase activation under this condition [122–124]. Moreover, NSMase mediates the negative effect on cardiomyocyte contractility induced by TNFα or IL1-β [118, 125]. Thus, these findings suggest the deleterious effects of activation of NSMase/sphingosine pathway on cardiomyocyte functioning and survival.

Both ASMase isoforms, the lysosomal (L-ASMase) and the secreted (S-ASMase), are encoded by the single *Smpd1* gene [126]. ASMase deficiency causes the rare recessively inherited lysosomal disorder, Niemann–Pick disease (NPD) characterized by multiorgan abnormalities resulting from lysosomal SM accumulation [127]. In addition to their essential housekeeping role in the lysosomes, ASMases contribute to cardiac responses in ischemia–reperfusion and upon stimulation of the TNF receptors.

Prolonged myocardial ischemia eventually induces cell death and infarct size depends on the duration of ischemia. Although reoxygenation due to reperfusion reduces damages, it leads to additional cell death [128]. Ischemic preconditioning (IPC), brief transient periods of mild ischemia before sustained ischemic event, can protect the heart from ischemic damage [99]. Postconditioning, transient brief ischemic episodes after sustained ischemia, has been suggested as a novel more relevant therapeutic approach [129].

Similar signaling pathways likely underlie both pre- and postconditioning cardioprotective effects. The dual role of ceramide has been suggested: it promotes cardioprotection in IPC, while induces apoptosis after ischemia–reperfusion [130]. ASMase-catalyzed accumulation of ceramide in the ischemic heart leads to apoptosis and administration of ASMase inhibitors before ischemia reproduces preconditioning protection. However, a limited accumulation of ceramide and S1P during preconditioning results in cardioprotective effect. Importantly, loss of sphingosine kinase-1 (SK1) abolishes both pre- and postconditioning cardioprotection [131, 132].

The *Smpd1* gene generates the single 75-kDa protein precursor, from which two functionally different ASMases can be produced: L-ASMase containing mannose-6-phosphate residues, and S-ASMase containing complex N-linked oligosaccharides [110]. Both isoforms require $Zn^{2+}$ for their optimum activation, however, L-ASMase is tightly associated with $Zn^{2+}$, whereas S-ASMase requires the addition of exogenous $Zn^{2+}$ for its maximal activity.

The elevated levels of S-ASMase activity in plasma of patients with heart failure, an important predictor of impaired survival, result from S-ASMase secretion from endothelial cells triggered by pro-inflammatory cytokines [133]. Activation of plasma S-ASMase positively correlates with the disease severity and is also associated with impaired blood flow and vasodilator capacity.

In conclusion, stress-induced SMase activation is involved in the development of various cardiovascular diseases. SMases represent potential targets for drug therapy against heart failure, postischemic injury, atherosclerosis, and age-related cardiovascular diseases.

## Sphingosine Kinases

One of the most potent sphingolipid mediators, S1P, is produced by one of two isoforms of sphingosine kinases SK1 or SK2, catalyzing phosphorylation of sphingosine. The complexity in the SK/S1P pathway interconnections enables cells to orchestrate cellular responses to the variety of external stimuli. This pathway is critical for cell proliferation and survival, cytoskeletal organization and vasculogenesis, lymphoid trafficking and immune response. However, until recently there was only limited published information regarding the role of SK/S1P signaling in cardiac physiology and pathophysiology. Essential role of SK/S1P pathway in the function of vascular endothelium has been discussed in recent reviews [134, 135].

SK1 and SK2 have been cloned and corresponding genes are located on different chromosomes [136–139]. Murine, rat, and human SK1 display significant homology. SK2 shows a high degree of homology to SK1 but contains approximately 240 additional amino acid residues in its center and at the N terminus accounting for its larger molecular size [138]. Both SK1 and SK2 have several splice variants [140]. Mice deficient in either SK1 or SK2 viable and develop normally, whereas loss of both isoforms leads to embryonic lethality caused by severe defects in neural and vascular development [141–143].

Both SK1 and SK2 are activated by phosphorylation mediated by Erk1/2. Human SK1 is phosphorylated at serine 225 by Erk1/2 and can be dephosphorylated by the phosphatase PP2A (Fig. 7.7) [144, 145]. SK2 is phosphorylated at two sites, serine 351 and threonine 578, by Erk1/2 (Fig. 7.8) [146]. The phosphorylation not only stimulates intrinsic activities of both SKs but it is also required for agonist-induced translocation of SK1 to the plasma membrane [144]. Both SK isoforms bind calmodulin and disruption of this interaction abrogates agonist-induced SK1 translocation to the plasma membrane but has no effect on its activity [147]. Another similarity is that SK1 and SK2 can also be activated by acidic phospholipids such as phosphatidylserine [138, 148]. Engagement of various GPCRs and RTKs with specific agonists, immunoglobulin receptor cross-linking, interleukins, estrogen, and activation of PKCε affect SK activity [139, 149].

Despite all these biochemical similarities, two SK isoforms have opposing actions: SK1 plays antiapoptotic role, while SK2 inhibits cell proliferation and induces apoptosis. Presently, there is no clear answer why SK2 producing S1P is pro-apoptotic. It has been suggested that SK2 ability to enhance apoptosis is linked to its regulation of ceramide levels. Consistent with pro-apoptotic role of SK2, its expression, but not SK1, increases pro-apoptotic ceramide levels. Recent studies have suggested a mechanism for how S1P catalyzed by SK2, but not by SK1, could be converted to ceramide (Fig. 7.8). An unusual recycling pathway in mammalian cells for the reconversion of sphingosine into pro-apoptotic ceramide requires SK2, but not SK1, functioning in concert with S1P phosphatase 1 [150]. Importantly, similar

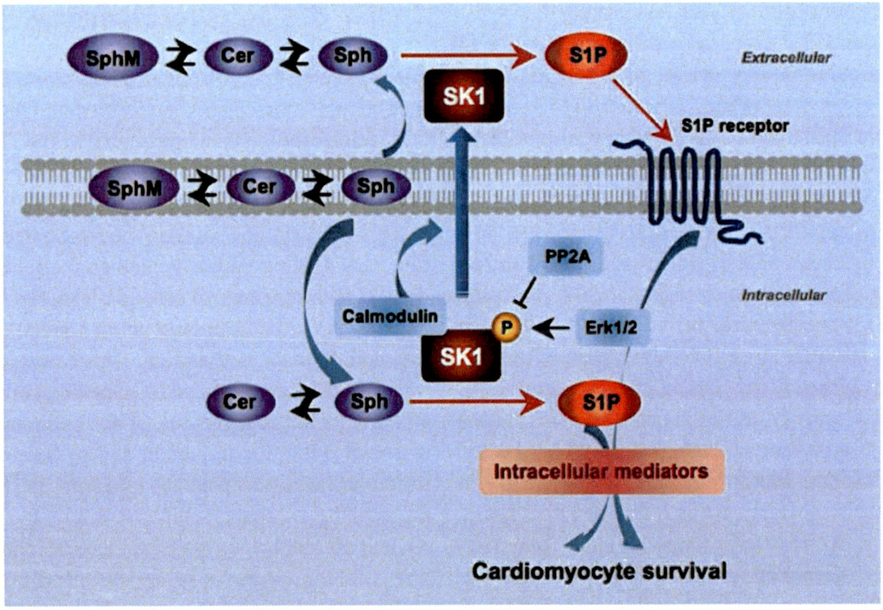


**Fig. 7.7** Role of sphingosine kinase-1 (SK1) in the heart. The ubiquitous membrane lipid sphingomyelin (SphM) is metabolized within the cell membrane and extracellularly by sphingomyelinases to ceramide (Cer). Ceramide is deacylated by neutral ceramidase to sphingosine (Sph), which is then converted into sphingosine-1-phosphate (S1P) by the action of SK1. SK1 is activated by Erk1/2-mediated phosphorylation at serine 225 while the phosphatase PP2A can dephosphorylate it abbreviating the activation. Association of SK1 with calmodulin is required for its translocation to the plasma membrane. The generated S1P can act as an intracellular second messenger as well as the high affinity ligand for S1P receptors to initiate the cascade resulting in cell survival and cardioprotection

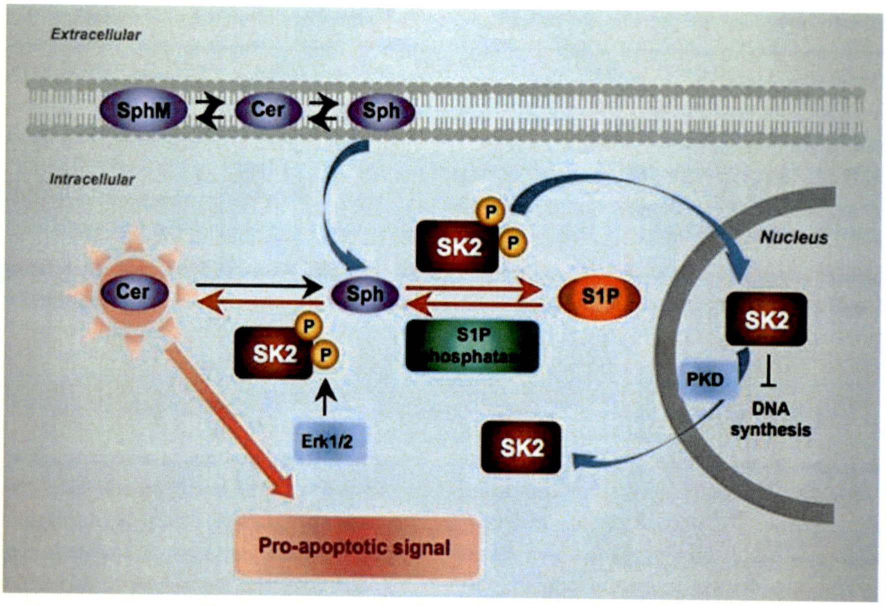


**Fig. 7.8** Pro-apoptotic role of cardiac SK2. A second SK isoform, SK2, also catalyzes S1P formation. Similar to SK1, SK2 is activated by phosphorylation at two sites, serine 351 and threonine 578, by Erk 1/2. SK2 can reside in the nucleus and its export is dependent on the protein kinase D (PKD) activity. The recycling pathway for the conversion of sphingosine (Sph) into pro-apoptotic ceramide (Cer) is dependent on the combined action of SK2 and S1P phosphatase 1. Nuclear localization of SK2 involved in inhibition of DNA synthesis may also contribute to its pro-apoptotic function

pathway has been identified in yeast suggesting its evolutionary conservation [151]. Reported nuclear localization of SK2 involved in inhibition of DNA synthesis may also contribute to its pro-apoptotic function [152, 153]. Additionally, it has been hypothesized that opposing roles of SK1 and SK2 might be related to distinct subcellular compartment of S1P production [154].

In isolated rodent hearts, exogenously applied S1P and the monoganglioside GM-1, which activates SKs via PKC, induce significant resistance to ischemia–reperfusion injury [155, 156]. It has been further suggested that PKCε is an upstream modulator of SK activity essential for cardioprotection induced by IPC [157]. Indeed, IPC stimulates both SK activity and its translocation to plasma membrane in wild-type hearts, whereas it only triggers SK membrane translocation without increase in SK activity in PKCε deficient hearts. Importantly, in PKCε deficient hearts IPC also does not decrease infarction size after ischemia–reperfusion [158]. Emerging evidence suggests that, like IPC, ischemic postconditioning cardioprotection is mediated through SK1 activation [132]. Based on discussed above findings that S1P, low-dose sphingosine, ischemic pre- and postconditioning mediated via SK1 are cardioprotective, combined therapy involving several elements of the SK/S1P pathway may be effectively used to prevent severe ischemic–reperfusion injury.

## Cardiac S1P Receptor Signaling

Although a central bioactive sphingolipid, S1P, can act as an intracellular messenger, there is no doubt that its most important actions are mediated through specific GPCRs. Family of GPCRs for which S1P is the high affinity ligand, currently consists of five receptors, $S1P_1$, $S1P_2$, $S1P_3$, $S1P_4$, and $S1P_5$ [159–162]. Binding of S1P to these receptors regulates diverse signaling pathways due to the receptor coupling to distinct heterotrimeric G proteins. The $S1P_1$ receptor couples exclusively to the $G_i$ protein, while the $S1P_2$ and $S1P_3$ receptors couple to $G_i$, $G_q$, and $G_{12/13}$ proteins, and the $S1P_4$ and $S1P_5$ receptors couple to both $G_i$ and $G_{12/13}$ proteins [163–167]. Upon receptor engagement, of the heterotrimeric G protein dissociates into the α subunit and βγ heterodimer both able to interact with downstream effectors. In the case of the $G_i$ protein, the α subunit inhibits adenylyl cyclase whereas the βγ heterodimer can regulate ion channels and various kinases. The effector for the α subunit of $G_q$ is PLC whereas that for the α subunit of $G_{12/13}$ is RhoA exchange factor. This complex signaling network by coupling to particular G proteins and downstream effector determines the nature of cellular response.

The $S1P_1$, $S1P_2$, and $S1P_3$ receptors are ubiquitously expressed including cardiomyocytes, while the $S1P_4$ and $S1P_5$ receptor expression is restricted to nervous and immune system [164, 167, 168]. Although the $S1P_1$, $S1P_2$, and $S1P_3$ receptors are found in the cardiovascular system, their relative levels differ significantly in various cardiac cell types (Table 7.2) [169].

In cardiomyocytes, the $S1P_1$ receptor is the predominant S1P receptor subtype, whereas the $S1P_2$ and $S1P_3$ receptors are expressed at significantly lower levels. Signaling through cardiac S1P receptors has been demonstrated to contribute to the regulation of myocardium contractility, induction of hypertrophy, intracellular $Ca^{2+}$ homeostasis, and cardioprotection from ischemia–reperfusion (Fig. 7.9).

Activation of the $S1P_1$ receptor antagonizes isoproterenol-induced upregulation of cAMP production in ventricular myocytes. It appears to be the predominant receptor that mediates cardiac negative inotropic effects of S1P, the $S1P_3$ receptor expressed at much lower levels may also play a minor role [170, 171]. Negative inotropy induced by the $S1P_1$ receptor activation may be mediated via both the α subunit of the $G_i$ protein to inhibit adenylyl cyclase cAMP production and the βγ heterodimer to affect $I_{KAch}$ current (Fig. 7.9). Thereby downregulated cAMP production can also contribute to decreased activation of L-type $Ca^{2+}$ channels mediated by PKA [170]. It has been suggested that the S1P effects on the cardiac inwardly rectifying $K^+$ channel, responsible for $I_{KAch}$ current, are mediated by the $S1P_3$ receptor since these effects could be inhibited by a putative $S1P_3$ receptor antagonist, suramin [172, 173]. However, one has to take into account that in addition to its action as a potential $S1P_3$ receptor antagonist, suramin induces a variety of nonspecific responses [174].

**Table 7.2** Relative expression of S1P receptor subtypes in various cell types [169]

| Cell type | Relative S1P receptor expression |
|---|---|
| Cardiac myocytes | $S1P_1 \gg S1P_3 > S1P_2$ |
| Cardiac fibroblasts | $S1P_3 \gg S1P_1 > S1P_2$ |
| Aortic smooth muscle cells | $S1P_2 > S1P_3 \gg S1P_1$ |
| Vascular endothelial cells | $S1P_1 > S1P_3 \gg S1P_2$ |

Using an antibody, which acts as an agonist specific to the $S1P_1$ receptor, it has been demonstrated that the $S1P_1$ receptor activation protects cardiomyocytes from hypoxia similar to exogenously administered S1P. The cardioprotection appears to be mediated via PI3K-Akt signaling pathway involving inhibition of GSK3β [175]. Moreover, the $S1P_3$ receptor may also contribute to this cardioprotection.

As we have previously discussed, S1P induces PKCε-mediated cardioprotection against ischemia–reperfusion. However, the protection by exogenously applied S1P observed in PKCε knockout mice suggests that intracellular generated S1P can be exported from the cell and activate the $S1P_1$ and $S1P_3$ receptors on the surface of cardiomyocytes

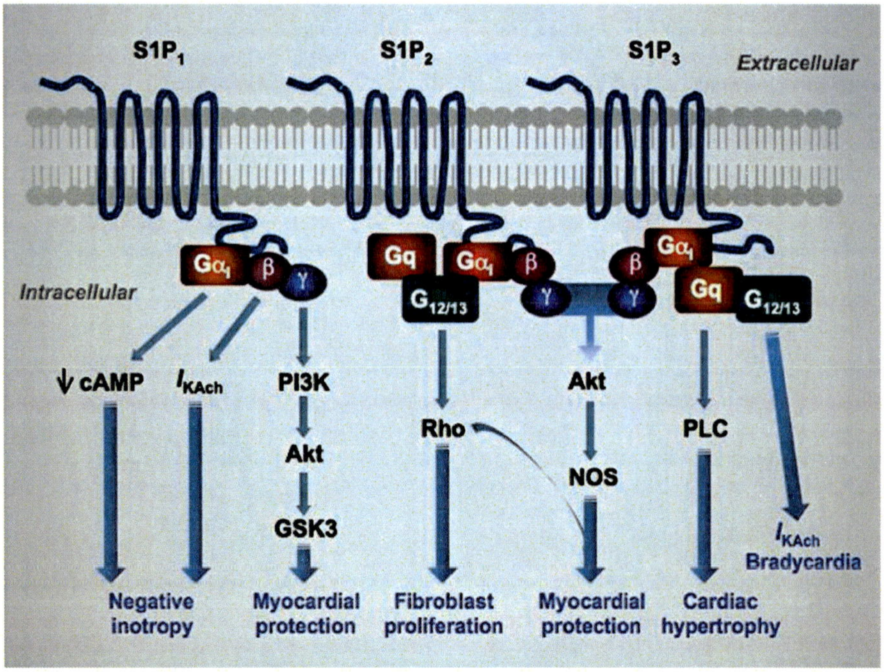


**Fig. 7.9** Cardiac S1P receptor signaling. Engagement of the $S1P_1$ receptor induces $G_{\alpha i}$-mediated downregulation of cAMP production and $G_{\beta\gamma}$-mediated inhibition of $I_{KAch}$ current leading to negative inotropy. $G_{\beta\gamma}$-mediated regulation of PI3K-Akt pathway may also contribute to cardioprotection. The $S1P_2$ receptor induces Rho-mediated fibroblast proliferation and myocardial protection. The $S1P_3$ receptor is coupled to PLC pathway resulting in cardiac hypertrophy. Activation of this receptor also may lead to bradycardia. Moreover, the $S1P_2$ and $S1P_3$ receptors appear to contribute to cardioprotection from ischemia reperfusion

[131, 155, 176]. Furthermore, the S1P-induced protection has been abolished upon inhibition of nitric oxide synthase (NOS) suggesting its involvement in this signaling pathway [177]. Finally, studies using the $S1P_{2/3}$ receptor double knockout mice have implicated also the $S1P_2$ and $S1P_3$ receptors in protection mediated via Akt and NOS from myocardial ischemia–reperfusion (Fig. 7.9) [178].

Although data regarding the *in vivo* role of S1P in cardiac hypertrophy are conflicting, it has been demonstrated that S1P-induced hypertrophy in neonatal rat cardiomyocytes appears to be mediated by the $S1P_1$ receptor via $G_i$-coupled signaling pathways and activation of MAP kinases, Akt, p70S6 kinase, and Rho [179]. Moreover, recent studies conducted with $S1P_2$ and $S1P_3$ receptor deficient myocytes suggest that the $S1P_3$ receptor/$G_q$ pathway rather than $S1P_2$ receptor, may contribute to the cardiac hypertrophy [169].

Considerable recent progress in uncovering the molecular mechanism of S1P signaling mediated through the S1P receptors has led to appreciation of essential role for this complex pathway in cardiovascular physiology and pathophysiology. However, further detailed analysis of subtype-specific S1P receptor functions in different cardiac cell types is necessary for design of efficient therapeutic strategy for cardiovascular disease.

## Eicosanoid Signaling in Cardiomyocytes

Eicosanoids, including prostaglandins and leukotrienes, are biologically active lipids that play essential roles in the regulation of myocardial bioenergetics, contractility, and various signaling pathways. The rate-determining step in the eicosanoid production in the myocardium is the release of arachidonic acid (AA), a polyunsaturated fatty acid, from the cellular membranes due to hydrolysis of glycophospholipids by cytoplasmic phospholipase $A_2$ ($PLA_2$) (Fig. 7.10) [180]. The released AA can be metabolized through three main pathways: the cyclooxygenase (COX), the lipoxygenase (LOX), and the cytochrome P450 monooxygenase (CYP) pathways generating a variety of lipid second messengers. COXs catalyze the conversion of AA into an intermediate

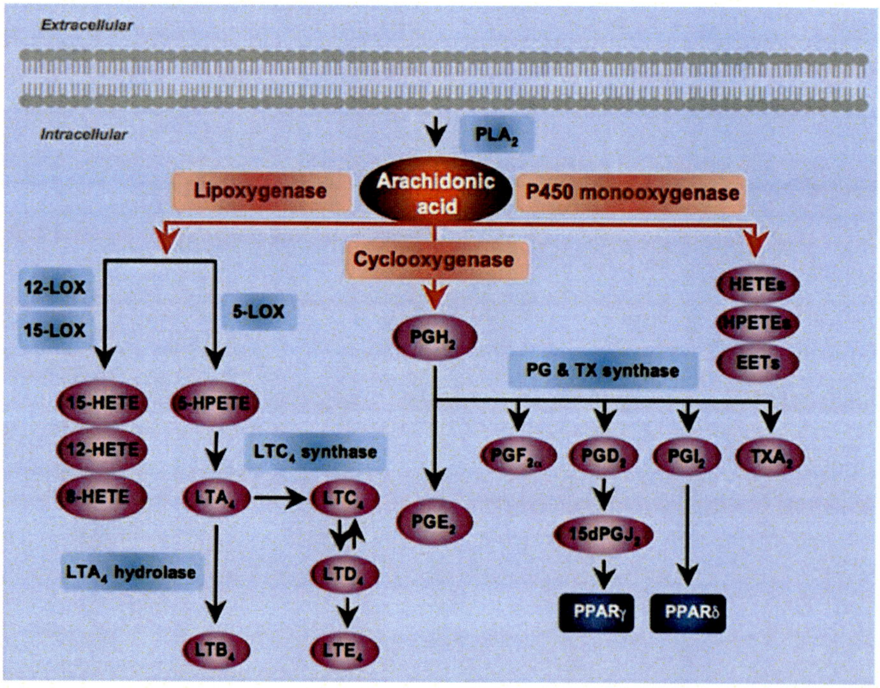


**Fig. 7.10** Myocardial eicosanoid metabolism. Arachidonic acid (AA) is liberated from the cellular membranes by cytoplasmic phospholipase $A_2$ ($PLA_2$). Free AA is metabolized to eicosanoids through three major pathways: the lipoxygenase (LOX), the cyclooxygenase (COX), and the cytochrome P450 monooxygenase (CYP) pathways. LOXs convert AA into biologically active leukotrienes and hydroxyeicosatetraenoic acids (HETEs). COXs convert AA into an intermediate $PGH_2$ which is sequentially metabolized to prostaglandins (PGs) and thromboxanes (TXs) by specific prostaglandin and thromboxane synthases. Finally, CYPs metabolize AA into epoxyeicosatrienoic acids (EETs), HETEs and hydroperoxyeicosatetraenoic acids (HPETEs). $PGI_2$ can transactivate the nuclear peroxisome proliferator-activated receptor-δ (PPARδ), and a $PGD_2$ dehydration product, $15dPGJ_2$, is a ligand for PPARγ

$PGH_2$, which is converted into prostaglandins (PGs) and thromboxanes (TXs) by corresponding specific prostaglandin and thromboxane synthases. LOXs metabolize AA into biologically active leukotrienes and hydroxyeicosatetraenoic acids (HETEs), while CYP metabolizes AA into epoxyeicosatrienoic acids (EETs), HETEs, and hydroperoxyeicosatetraenoic acids (HPETEs). In the LOX pathway, 5-LOX generates 5-HPETE intermediate that is further converted into the unstable leukotriene (LT) A4 (LTA4) and eventually into 5-HETE, LTB4, LTC4, LTD4, and LTE4. The prostaglandins and leukotrienes can serve as specific ligands for corresponding GPCRs to initiate various cellular responses. Moreover, $PGI_2$ can transactivate the nuclear peroxisome proliferator-activated receptor-δ (PPARδ), whereas $PGD_2$ dehydration generates $15dPGJ_2$, a natural ligand for PPARγ.

## Phospholipases

Phospholipase $A_2$ ($PLA_2$) enzymes are critical for eicosanoid generation as they hydrolyze membrane glycophospholipids to release AA from the cell membrane (Fig. 7.10). Intracellular $PLA_2$ can be divided into cytosolic $PLA_2$ ($cPLA_2$) and $Ca^{2+}$-independent $PLA_2$ ($iPLA_2$). $cPLA_2$s, with the exception of $cPLA_2\gamma$, require micromolar concentration of $Ca^{2+}$ for membrane association to catalyze phospholipolysis, while $iPLA_2$s do not require $Ca^{2+}$ for membrane association and catalytic activity [181–184].

Most cardiac $PLA_2$ enzymes belong to the $iPLA_2$ subfamily [185]. The major $iPLA_2$ present in canine and human myocardium was identified in the mid-1980s and was subsequently designated $iPLA_2\beta$ [181, 186]. Studies of normal and transgenic mice overexpressing $iPLA_2\beta$ in cardiomyocyte-specific manner have demonstrated that ischemia induces the rapid activation of $iPLA_2\beta$ leading to the liberation of fatty acids and the formation of lysolipids [187, 188]. This activation during myocardial ischemia not only initiates eicosanoid signaling pathways but also modulates cardiomyocyte membrane dynamics affecting myocardial metabolism and bioenergetics. Given $iPLA_2\beta$ ability to hydrolyze fatty acyl-CoAs and the rapid accumulation of acyl-CoA during ischemia, it has been suggested that acyl-CoA can act as both a substrate and activator of $iPLA_2\beta$ in myocardial ischemia [189].

A second major cardiac $iPLA_2$, $iPLA_2\gamma$, contains N-terminal mitochondrial localization and C-terminal peroxisomal localization sequences [190, 191]. Moreover, multiple splicing and proteolytically processed $iPLA_2\gamma$ variants have been identified. Immunochemical analysis has confirmed mitochondrial and peroxisomal localization of $iPLA_2\gamma$. Analysis of recently generated transgenic mice overexpressing $iPLA_2\gamma$ in cardiomyocyte-specific manner and $iPLA_2\gamma$ knockout mice has confirmed an essential role of this enzyme in the generation of 2-arachidonoyl lysophosphatidylcholine and its downstream bioactive molecules in myocardium [192, 193].

Cytosolic $PLA_2\alpha$ ($cPLA_2\alpha$) is present in the myocardium at lower levels compared to the $iPLA_2$s. Elevation in intracellular $Ca^{2+}$ leads to membrane translocation and activation of $cPLA_2\alpha$ [194, 195]. The enzyme can be also activated by phosphorylation mediated by various protein kinases, including MAPKs and Calmodulin-dependent protein kinase II (CaMKII) [196, 197]. Loss of $cPLA_2\alpha$ leads to significant myocardial hypertrophy increasing upon the stress pressure overload [198]. $cPLA_2\alpha$ deficiency appears to interfere with PDK-1 recruitment and activation of PKCζ, a negative regulatory pathway of IGF-1-mediated hypertrophy. Finally, it has been suggested that $cPLA_2\alpha$ may be involved in the inhibition of cardiac $\beta_2$-adrenergic receptor signaling [199].

## Cardiac Cyclooxygenases

Metabolism of AA by COX-1 and COX-2 is a central step in the generation of prostanoids, such as $PGI_2$, $PGD_2$, $PGE_2$, $PGF_{2\alpha}$, and thromboxane $A_2$ ($TXA_2$). Human cardiomyocytes express both COX-1 and COX-2 [200, 201]. COX-1 is constitutively expressed and appears to maintain cardiac homeostasis under normal physiological conditions, while expression of COX-2 is upregulated under various pathological states and can result in cardiac fibrosis [201, 202]. However, emerging evidence suggests that COX-2 can also contribute to cardioprotection. Indeed, COX-2 is induced followed by upregulation of $PGE_2$ and 6-keto-$PGF_{1\alpha}$ production during IPC and inhibition of COX-2 abbreviates this cardioprotective effect [203, 204].

Downstream prostanoids, such as $PGF_{2\alpha}$ and $PGI_2$, produced by both myocardial COX-1 and COX-2 modulate multiple signaling pathways in the heart. $PGF_{2\alpha}$ stimulates the $PGF_{2\alpha}$ receptor-mediated response, JNK1 and c-Jun signaling pathways involved in myocardial hypertrophy [205, 206]. $PGI_2$ receptor knockout mice exhibit severe cardiac fibrosis and elevated myocardial hypertrophy upon pressure overload [207, 208].

## Cardiac Lipoxygenases

LOXs catalyze oxidation of AA generating HPETEs converted into their hydroxyeicosatrienoic derivatives, HETEs. The cardiac LOX activity appears to be represented mainly by 12-LOX generating 12-HETE with significantly lower amounts of 15-LOX producing 15-HETE [209]. LOX-mediated pathway appears to contribute to maintaining cardiac physiology. Ischemia-induced LOX activation resulting in significantly elevated 12-HETE and 5-HETE levels has been demonstrated in rabbit myocardium and in cultured canine cardiomyocytes [210, 211]. 12-LOX overexpression in cardiac fibroblasts appears to contribute to hypertrophy [212, 213]. Finally, in ventricular myocytes, downregulation of LOXs inhibits insulin-stimulated glucose transport suggesting a role for this pathway in insulin response [214, 215].

### *Cytochrome P450 Monooxygenases*

Various CYPs, including CYP1B, CYP2A, CYP2B, CYP2C, CYP2E, CYP2J, CYP4A and CYP11 are expressed in the heart. CYP2C and CYP2J are predominant cardiac isoforms generating EETs [mainly 8(S),9(R)-EET and 14(R),15(S)-EET] and HETEs (mainly 16-HETE, with minor amounts of 20-HETE) [216, 217]. Cardiac CYP-mediated pathway is upregulated in heart failure and hypertrophy while its inhibition appears to result in decrease in ischemia–reperfusion injury [218–220]. Consistently, CYP2J2 overexpression improves postischemic functional recovery while CYP2J2 inhibition reverses the effect [221]. Increased levels of EETs due to CYP2J2 overexpression also stimulate mitochondrial $K_{ATP}$ channel activity and MAPK signaling. Unlike maybe EETs, HETEs, 20-HETE in particular, generated by CYP4 appear to increase myocardial ischemic damage and are detrimental to cardiac activity [222, 223]. Finally, AA and its EET metabolites may affect the activity of various cardiac ion channels, including delayed rectifier $K^+$, ATP-sensitive $K^+$, and L-type $Ca^{2+}$ channels [224–226].

## Conclusions

Recent advances in molecular genetics and pharmacological approaches reveal a highly complex and pronounced role of lipid signaling in myocardial physiology and pathophysiology. It is now well appreciated that lipid metabolism generates multiple signaling mediators acting as intracellular second messengers as well as ligands for cognate GPCRs that induce myriad cellular responses. However, one should consider that various cell types forming cardiovascular system respond to bioactive lipids in different manner. Additionally, spatial–temporal dynamics of this intricate signaling network adds additional level of complexity. Studies of some lipid-metabolizing enzymes have been further complicated by the presence of several isozymes catalyzing the same reaction but potentially localized in different cellular compartments.

In the last years, many new analytical and molecular tools have become available to study the dynamics and topology of lipid metabolism and signaling. Various enzymes involved in these pathways have been cloned and several specific antibodies have been produced. Several transgenic cellular and animal models allowing ablation and/or cardiac specific overexpression of various enzymes and receptors involved in lipid metabolism, such as SK1 and SK2, subtype-specific S1P receptors, $PLA_2$s, PI3Ks, PTEN, Akt/PKB, and GSK3s, have been developed. The use of modern highly sensitive mass-spectrometry techniques allow precise quantitative analysis of endogenous lipid intermediates generated in cardiomyocytes in response to various stimuli.

Despite the significant recent advances, multiple problems remain to be solved regarding molecular mechanisms of lipid signaling and its regulation in the heart. One of the major drawbacks in the field is the lack of selective enzyme inhibitors or activators and receptor subtype-specific agonists and antagonists. Therefore, design of highly selective compounds will greatly facilitate our understanding of roles of lipid signaling in cardiac physiology. Not much information is currently available on species differences regarding cardiac lipid metabolism, although it is well appreciated that the cardiac physiology of rodents (e.g., $Ca^{2+}$ handling) differs significantly from that of humans. Detailed knowledge on precise changes in lipid metabolism induced by different cardiac pathological conditions is also still lacking. Clearly, further mechanistic insights are necessary to better assess whether and how modulation of lipid signaling could become an efficient novel therapeutic approach for cardiovascular disease.

## Summary

- Inositol phospholipids also known as phosphoinositides, short-lived phosphorylated derivatives of phosphatidylinositol (PI), are the most prominent among lipids acting as second messengers. Accumulating evidence has indicated that they play essential roles as mediators of signal transduction regulating the ion channel functions, $Ca^{2+}$ homeostasis, protein kinase activities, and membrane trafficking creating an interface between cell signaling and membrane dynamics.
- PI represents a minor phospholipid component in all eukaryotic cell membrane. In the sarcolemma, PI is phosphorylated on the 4-, then on the 5-position by PI4-kinase and PI4P5-kinase, respectively. Resultant product, phosphatidylinositol(4,5)bisphosphate ($PIP_2$) is the central intermediate in inositide signaling pathway regulating numerous cellular responses ranging from vascular tone and hormone secretion to cell growth and differentiation.

- All major cardiac inwardly rectifying $K^+$ channels, $K_{ir}2$, $K_{ir}3$, and $K_{ir}6$, are critically regulated by $PIP_2$. $PIP_2$ is also involved in the regulation of both the hyperpolarization-activated cyclic nucleotide gated channels, HCN, governing the pacemaker current $I_f$ and TRPC channels, low conductance, relatively nonselective cation channels activated by receptors coupled to PLC.
- Ins(1,4,5)$P_3$ engages specific receptors, $IP_3$-Rs, located in intracellular $Ca^{2+}$ stores, stimulating $Ca^{2+}$-induced $Ca^{2+}$ release (CICR) and activating protein kinase pathways to further enhance $Ca^{2+}$ responses. In atrial myocytes, stimulation of the $IP_3$-R type 2 [$IP_3$-R (2)] can contribute to excitation–contraction coupling and therefore predispose to dysrhythmia by interfering with the RyR signaling. Moreover, $Ca^{2+}$ signals generated by $IP_3$-R (2) localized close to the sarcolemma could affect sarcolemmal $Ca^{2+}$ channels and $Na^+/Ca^{2+}$ exchanger.
- Elevated levels of Ins(1,4,5)$P_3$ and $IP_3$-R detected in ischemia and postischemic perfusion and heart failure, respectively, suggest that this intermediate may contribute to these conditions.
- PI3Ks are responsible for the generation of $PIP_3$, whereas PTEN (phosphatase and tensin homolog deleted on chromosome 10) is the major phosphatase responsible for hydrolysis of this lipid. The PI3K family consists of eight distinct enzymes characterized by dual protein and lipid kinase activity. They all share a common catalytic activity: they phosphorylate the D3 hydroxyl group of PI upon GPCR, RTK, or Ras activation. Members of PI3K family are divided into three functional classes based on their protein structure, substrate specificity, associated regulatory subunits, and activation mechanisms.
- Mammalian PTEN is a 40–50-kDa protein with 3′-lipid phosphatase activity hydrolyzing its main physiologic substrate membrane-bound $PIP_3$. It contains an N-terminal phospholipid-binding region, an N-terminal phosphatase domain, a C2 domain, and a C-terminal PSD-95/Dlg/ZO-1 (PDZ) homology domain. Upon activation, PTEN is recruited to the plasma membrane due to electrostatic interaction of the polybasic N-terminal tail, catalytic and C2 domains of PTEN with $PIP_3$.
- Antagonistic action of PTEN on PI3K signaling is highly conserved and occurs in various mammalian cells, including cardiomyocytes, vascular smooth muscle cells, and endothelial cells. Molecular targets of the PI3K and PTEN signaling in the heart include phosphoinositide-dependent kinase-1 (PDK1), Akt (protein kinase B, PKB), glycogen synthase kinase-3 (GSK3), mammalian target of rapamycin (mTOR), and p70 S6 kinase (p70S6K).
- Distinct PI3K isoforms contribute to different physiological and pathological responses in the heart. PI3K activity regulates hypertrophy of the heart in physiological (e.g., exercise training) and pathological (e.g., pressure overload) conditions. PI3Kα regulates these changes via the alteration in the function of its downstream effectors, Akt/PKB, p70S6K, and GSK3β. PI3Kα signaling is largely attenuated by PTEN: inactivation of PTEN activates Akt/PKB and increases cardiomyocytes size, while PTEN overexpression abolishes these responses.
- PI3Kγ regulates cellular cAMP levels independently of its catalytic activity via complex formation with two cAMP-degrading enzymes, PDE3B and PDE4. PI3Kγ is also involved in the regulation of β-AR signaling essential for the fine-tuning of cardiac function. In cardiomyocytes, β-AR signaling, predominantly via $\beta_2$-AR, activates PI3Kγ through $G_{\beta\gamma}$ subunits resulting in PI3Kγ binding to GRK2 and translocation of this complex to plasma membrane. Relocation of PI3Kγ–GRK2 complex to membrane also contributes to PI3Kγ activation. Elevated PTEN levels caused by β-AR stimulation may represent negative feedback on β-AR-mediated PI3Kγ activation.
- Sphingolipid metabolism is involved in the generation of signal mediators serving both as intracellular second messengers and as ligands for specialized receptors. During past two decades, sphingolipids was demonstrated to mediate a wide variety of biological responses, such as cell proliferation, survival, migration, cytoskeleton rearrangements, and intracellular mobilization.
- SMases hydrolyze sphingomyelin into phosphocholine and ceramide, which is not only a bioactive sphingolipid but also a substrate for the generation of a cascade of other lipid mediators. Based on their optimum pH, SMases can be divided into three groups: alkaline, neutral, and acid SMases. In humans, alkaline SMases have been found in the intestinal mucosa, bile, and liver, whereas neutral (NSMases) and acid (ASMases) play a crucial role in cardiovascular physiology.
- Activation of NSMases in various cardiovascular cells in response to variety of external signals plays a central regulatory role in ceramide-dependent apoptosis and cell proliferation. In heart failure, cardiac deficiency in glutathione, caused by oxidative stress, along with TNFα elevation induces activation of NSMases.
- ASMase deficiency causes the rare recessively inherited lysosomal disorder, Niemann–Pick disease (NPD) characterized by the multiorgan abnormalities resulting from lysosomal SM accumulation. In addition to their essential housekeeping role in the lysosomes, ASMases contribute to cardiac responses in ischemia–reperfusion and upon stimulation of the TNF receptors. ASMase-catalyzed accumulation of ceramide in the ischemic heart leads to apoptosis and administration of ASMase inhibitors before ischemia reproduces preconditioning protection.
- One of the most potent sphingolipid mediators, S1P, is produced by one of two isoforms of sphingosine kinases (SKs), SK1 and SK2, catalyzing phosphorylation of

sphingosine. The complexity in the SK/S1P pathway interconnections enables cells to orchestrate cellular responses to the variety of external stimuli. Two SK isoforms have opposing actions: SK1 plays antiapoptotic role, while SK2 inhibits cell proliferation and induces apoptosis. Presently, there is no clear answer why SK2 producing S1P is pro-apoptotic.

- Although a central bioactive sphingolipid, S1P, can act as an intracellular messenger, its most important actions are mediated through specific GPCRs. Family of GPCRs for which S1P is the high affinity ligand, currently consists of five receptors, $S1P_1$, $S1P_2$, $S1P_3$, $S1P_4$, and $S1P_5$. Binding of S1P to these receptors regulates diverse signaling pathways due to the receptor coupling to distinct heterotrimeric G proteins.
- Signaling through cardiac S1P receptors contributes to the regulation of myocardium contractility, induction of hypertrophy, intracellular $Ca^{2+}$ homeostasis, and cardioprotection from ischemia–reperfusion. The $S1P_1$ receptor activation protects cardiomyocytes from hypoxia similar to exogenously administered S1P. The cardioprotection appears to be mediated via PI3K-Akt signaling pathway involving inhibition of GSK3β. The $S1P_2$ and $S1P_3$ receptors are also implicated in Akt- and NOS-mediated protection from myocardial ischemia–reperfusion. Moreover, the $S1P_3$ receptor/$G_q$ pathway rather than $S1P_2$ receptor, may contribute to the cardiac hypertrophy.
- Eicosanoids are biologically active lipids that play essential roles in the regulation of myocardial bioenergetics, contractility, and various signaling pathways. The rate-determining step in the eicosanoid production in the myocardium is the release of arachidonic acid (AA) from the cellular membranes due to hydrolysis of glycophospholipids by cytoplasmic phospholipase $A_2$ ($PLA_2$). The released AA can be metabolized through three main pathways: the cyclooxygenase (COX), the lipoxygenase (LOX), and the cytochrome P450 monooxygenase (CYP) pathways generating a variety of lipid second messengers.
- Intracellular $PLA_2$ can be divided into cytosolic $PLA_2$ ($cPLA_2$) and $Ca^{2+}$-independent $PLA_2$ ($iPLA_2$). $cPLA_2$s, with the exception of $cPLA_2\gamma$, require micromolar concentration of $Ca^{2+}$ for membrane association to catalyze phospholipolysis, while $iPLA_2$s do not require $Ca^{2+}$ for membrane association and catalytic activity.
- Most cardiac $PLA_2$ enzymes belong to the $iPLA_2$ subfamily. Ischemia induces the rapid activation of $iPLA_2\beta$ leading to the liberation of fatty acids and the formation of lysolipids. $iPLA_2\gamma$ plays an essential role in the generation of 2-arachidonoyl lysophosphatidylcholine and its downstream bioactive molecules in the myocardium. Loss of $cPLA_2\alpha$ leads to significant myocardial hypertrophy increasing upon the stress pressure overload. $cPLA_2\alpha$ may also be involved in the inhibition of cardiac $\beta_2$-AR signaling.
- Metabolism of AA by COX-1 and COX-2 is a central step in the generation of prostanoids, such as $PGI_2$, $PGD_2$, $PGE_2$, $PGF_{2\alpha}$, and thromboxane $A_2$. COX-1 is constitutively expressed and appears to maintain cardiac homeostasis under normal physiological conditions, while expression of COX-2 is upregulated under various pathological states and can result in cardiac fibrosis. However, emerging evidence suggests that COX-2 can also contribute to cardioprotection.
- LOXs catalyze oxidation of AA generating HPETEs converted into their hydroxyeicosatrienoic derivatives (HETEs). LOX-mediated pathway appears to contribute to maintaining cardiac physiology. Ischemia-induced LOX activation results in significantly elevated 12-HETE and 5-HETE levels in rabbit and canine cardiomyocytes. 12-LOX overexpression in cardiac fibroblasts appears to contribute to hypertrophy.
- CYP2C and CYP2J are predominant cardiac CYP isoforms generating EETs and HETEs. Cardiac CYP-mediated pathway is upregulated in heart failure and hypertrophy while its inhibition appears to result in decrease in ischemia–reperfusion injury. AA and its EET metabolites may affect the activity of various cardiac ion channels, including delayed rectifier $K^+$, ATP-sensitive $K^+$, L-type $Ca^{2+}$, and mitochondrial $K_{ATP}$ channels.
- Recent advances in molecular genetics and pharmacological approaches reveal a highly complex and pronounced role of lipid signaling in myocardial physiology and pathophysiology. It is now well appreciated that lipid metabolism generates multiple signaling mediators acting as intracellular second messengers as well as ligands for cognate GPCRs that induce myriad cellular responses. However, our understanding of the mechanisms of lipid signaling and its regulation remains rather preliminary – detailed mechanistic insights are necessary to better assess whether and how modulation of these complex network could become an efficient novel therapeutic approach for cardiovascular disease.

## References


1. Martin TF. Phosphoinositide lipids as signaling molecules: common themes for signal transduction, cytoskeletal regulation, and membrane trafficking. Annu Rev Cell Dev Biol. 1998;14:231–64.
2. Krauss M, Haucke V. Phosphoinositide-metabolizing enzymes at the interface between membrane traffic and cell signalling. EMBO Rep. 2007;8:241–6.
3. Berridge MJ. Inositol trisphosphate and diacylglycerol: two interacting second messengers. Annu Rev Biochem. 1987;56:159–93.
4. Blero D, Payrastre B, Schurmans S, Erneux C. Phosphoinositide phosphatases in a network of signalling reactions. Pflugers Arch. 2007;455:31–44.
5. Woodcock EA, Matkovich SJ. Ins(1,4,5)P3 receptors and inositol phosphates in the heart-evolutionary artefacts or active signal transducers? Pharmacol Ther. 2005;107:240–51.

6. Guse AH, Kiess W, Funk B, Kessler U, Berg I, Gercken G. Identification and characterization of insulin-like growth factor receptors on adult rat cardiac myocytes: linkage to inositol 1,4,5-trisphosphate formation. Endocrinology. 1992;130:145–51.
7. Streb H, Irvine RF, Berridge MJ, Schulz I. Release of $Ca^{2+}$ from a nonmitochondrial intracellular store in pancreatic acinar cells by inositol-1,4,5-trisphosphate. Nature. 1983;306:67–9.
8. Nishizuka Y. Intracellular signaling by hydrolysis of phospholipids and activation of protein kinase C. Science. 1992;258:607–14.
9. Onohara N, Nishida M, Inoue R, et al. TRPC3 and TRPC6 are essential for angiotensin II-induced cardiac hypertrophy. EMBO J. 2006;25:5305–16.
10. van Dijk MC, Postma F, Hilkmann H, Jalink K, van Blitterswijk WJ, Moolenaar WH. Exogenous phospholipase D generates lysophosphatidic acid and activates Ras, Rho and $Ca^{2+}$ signaling pathways. Curr Biol. 1998;8:386–92.
11. Ross EM, Mateu D, Gomes AV, Arana C, Tran T, Litosch I. Structural determinants for phosphatidic acid regulation of phospholipase C-beta1. J Biol Chem. 2006;281:33087–94.
12. Raucher D, Stauffer T, Chen W, et al. Phosphatidylinositol 4,5-bisphosphate functions as a second messenger that regulates cytoskeleton-plasma membrane adhesion. Cell. 2000;100:221–8.
13. Hilgemann DW. Local PIP(2) signals: when, where, and how? Pflugers Arch. 2007;455:55–67.
14. Amirahmadi F, Turnbull L, Du XJ, Graham RM, Woodcock EA. Heightened alpha1A-adrenergic receptor activity suppresses ischaemia/reperfusion-induced Ins(1,4,5)P3 generation in the mouse heart: a comparison with ischaemic preconditioning. Clin Sci (Lond). 2008;114:157–64.
15. Padron D, Wang YJ, Yamamoto M, Yin H, Roth MG. Phosphatidylinositol phosphate 5-kinase Ibeta recruits AP-2 to the plasma membrane and regulates rates of constitutive endocytosis. J Cell Biol. 2003;162:693–701.
16. Cho H, Kim YA, Yoon JY, et al. Low mobility of phosphatidylinositol 4,5-bisphosphate underlies receptor specificity of Gq-mediated ion channel regulation in atrial myocytes. Proc Natl Acad Sci USA. 2005;102:15241–6.
17. Cho H, Lee D, Lee SH, Ho WK. Receptor-induced depletion of phosphatidylinositol 4,5-bisphosphate inhibits inwardly rectifying $K^+$ channels in a receptor-specific manner. Proc Natl Acad Sci USA. 2005;102:4643–8.
18. Mao YS, Yin HL. Regulation of the actin cytoskeleton by phosphatidylinositol 4-phosphate 5 kinases. Pflugers Arch. 2007; 455:5–18.
19. Woodcock EA, Kistler PM, Ju YK. Phosphoinositide signalling and cardiac arrhythmias. Cardiovasc Res. 2009;82:286–95.
20. Huang CL, Feng S, Hilgemann DW. Direct activation of inward rectifier potassium channels by PIP2 and its stabilization by Gbetagamma. Nature. 1998;391:803–6.
21. Ma D, Tang XD, Rogers TB, Welling PA. An Andersen-Tawil syndrome mutation in Kir2.1 (V302M) alters the G-loop cytoplasmic $K^+$ conduction pathway. J Biol Chem. 2007;282:5781–9.
22. Sui JL, Petit-Jacques J, Logothetis DE. Activation of the atrial KACh channel by the betagamma subunits of G proteins or intracellular $Na^+$ ions depends on the presence of phosphatidylinositol phosphates. Proc Natl Acad Sci USA. 1998;95:1307–12.
23. Rohacs T, Chen J, Prestwich GD, Logothetis DE. Distinct specificities of inwardly rectifying K(+) channels for phosphoinositides. J Biol Chem. 1999;274:36065–72.
24. Rohacs T, Lopes CM, Jin T, Ramdya PP, Molnar Z, Logothetis DE. Specificity of activation by phosphoinositides determines lipid regulation of Kir channels. Proc Natl Acad Sci USA. 2003;100:745–50.
25. Rapedius M, Soom M, Shumilina E, et al. Long chain CoA esters as competitive antagonists of phosphatidylinositol 4,5-bisphosphate activation in Kir channels. J Biol Chem. 2005;280:30760–7.
26. Anderson KE, Du XJ, Sinclair AJ, Woodcock EA, Dart AM. Dietary fish oil prevents reperfusion Ins(1,4,5)P3 release in rat heart: possible antiarrhythmic mechanism. Am J Physiol. 1996; 271:H1483–90.
27. Li Y, Gamper N, Hilgemann DW, Shapiro MS. Regulation of Kv7 (KCNQ) $K^+$ channel open probability by phosphatidylinositol 4,5-bisphosphate. J Neurosci. 2005;25:9825–35.
28. Bian JS, McDonald TV. Phosphatidylinositol 4,5-bisphosphate interactions with the HERG K(+) channel. Pflugers Arch. 2007;455:105–13.
29. Park KH, Piron J, Dahimene S, et al. Impaired KCNQ1-KCNE1 and phosphatidylinositol-4,5-bisphosphate interaction underlies the long QT syndrome. Circ Res. 2005;96:730–9.
30. Bian J, Cui J, McDonald TV. HERG K(+) channel activity is regulated by changes in phosphatidyl inositol 4,5-bisphosphate. Circ Res. 2001;89:1168–76.
31. Thomas D, Wu K, Wimmer AB, et al. Activation of cardiac human ether-a-go-go related gene potassium currents is regulated by alpha(1A)-adrenoceptors. J Mol Med. 2004;82:826–37.
32. Oliver D, Lien CC, Soom M, Baukrowitz T, Jonas P, Fakler B. Functional conversion between A-type and delayed rectifier $K^+$ channels by membrane lipids. Science. 2004;304:265–70.
33. Zolles G, Klocker N, Wenzel D, et al. Pacemaking by HCN channels requires interaction with phosphoinositides. Neuron. 2006;52:1027–36.
34. Zorn-Pauly K, Schaffer P, Pelzmann B, et al. If in left human atrium: a potential contributor to atrial ectopy. Cardiovasc Res. 2004;64:250–9.
35. Li X, Zima AV, Sheikh F, Blatter LA, Chen J. Endothelin-1-induced arrhythmogenic $Ca^{2+}$ signaling is abolished in atrial myocytes of inositol-1,4,5-trisphosphate(IP3)-receptor type 2-deficient mice. Circ Res. 2005;96:1274–81.
36. Roderick HL, Bootman MD. Pacemaking, arrhythmias, inotropy and hypertrophy: the many possible facets of IP3 signalling in cardiac myocytes. J Physiol. 2007;581:883–4.
37. Tovey SC, Dyer JL, Godfrey RE, et al. Subtype identification and functional properties of inositol 1,4,5-trisphosphate receptors in heart and aorta. Pharmacol Res. 2000;42:581–90.
38. Lipp P, Laine M, Tovey SC, et al. Functional InsP3 receptors that may modulate excitation–contraction coupling in the heart. Curr Biol. 2000;10:939–42.
39. Bare DJ, Kettlun CS, Liang M, Bers DM, Mignery GA. Cardiac type 2 inositol 1,4,5-trisphosphate receptor: interaction and modulation by calcium/calmodulin-dependent protein kinase II. J Biol Chem. 2005;280:15912–20.
40. Mackenzie L, Bootman MD, Laine M, et al. The role of inositol 1,4,5-trisphosphate receptors in Ca(2+) signalling and the generation of arrhythmias in rat atrial myocytes. J Physiol. 2002;541:395–409.
41. Kleber AG, Rudy Y. Basic mechanisms of cardiac impulse propagation and associated arrhythmias. Physiol Rev. 2004;84:431–88.
42. Lopera G, Stevenson WG, Soejima K, et al. Identification and ablation of three types of ventricular tachycardia involving the his-purkinje system in patients with heart disease. J Cardiovasc Electrophysiol. 2004;15:52–8.
43. Mouton R, Huisamen B, Lochner A. The effect of ischaemia and reperfusion on sarcolemmal inositol phospholipid and cytosolic inositol phosphate metabolism in the isolated perfused rat heart. Mol Cell Biochem. 1991;105:127–35.
44. Anderson KE, Dart AM, Woodcock EA. Inositol phosphate release and metabolism during myocardial ischemia and reperfusion in rat heart. Circ Res. 1995;76:261–8.
45. Jacobsen AN, Du XJ, Dart AM, Woodcock EA. Ins(1,4,5)$P_3$ and arhythmogenic responses during myocardial reperfusion: evidence for receptor specificity. Am J Physiol. 1997;42:H1119–25.
46. Du XJ, Anderson KE, Jacobsen A, Woodcock EA, Dart AM. Suppression of ventricular arrhythmias during ischemia–reperfusion

by agents inhibiting Ins(1,4,5)P3 release. Circulation. 1995;91:2712–6.
47. Jacobsen AN, Du XJ, Lambert KA, Dart AM, Woodcock EA. Arrhythmogenic action of thrombin during myocardial reperfusion via release of inositol 1,4,5-triphosphate. Circulation. 1996;93:23–6.
48. Hilgemann DW, Ball R. Regulation of cardiac $Na^+$, $Ca^{2+}$ exchange and KATP potassium channels by PIP2. Science. 1996;273:956–9.
49. He Z, Feng S, Tong Q, Hilgemann DW, Philipson KD. Interaction of PIP(2) with the XIP region of the cardiac Na/Ca exchanger. Am J Physiol Cell Physiol. 2000;278:C661–6.
50. Go LO, Moschella MC, Watras J, Handa KK, Fyfe BS, Marks AR. Differential regulation of two types of intracellular calcium release channels during end-stage heart failure. J Clin Invest. 1995;95:888–94.
51. Houser SR, Piacentino III V, Weisser J. Abnormalities of calcium cycling in the hypertrophied and failing heart. J Mol Cell Cardiol. 2000;32:1595–607.
52. Bootman MD, Higazi DR, Coombes S, Roderick HL. Calcium signalling during excitation–contraction coupling in mammalian atrial myocytes. J Cell Sci. 2006;119:3915–25.
53. Putney Jr JW. The enigmatic TRPCs: multifunctional cation channels. Trends Cell Biol. 2004;14:282–6.
54. Rohacs T. Regulation of TRP channels by PIP(2). Pflugers Arch. 2007;453:753–62.
55. Dietrich A, Mederos y Schnitzler M, Kalwa H, Storch U, Gudermann T. Functional characterization and physiological relevance of the TRPC3/6/7 subfamily of cation channels. Naunyn Schmiedebergs Arch Pharmacol. 2005;371:257–65.
56. Plant TD, Schaefer M. Receptor-operated cation channels formed by TRPC4 and TRPC5. Naunyn Schmiedebergs Arch Pharmacol. 2005;371:266–76.
57. Runnels LW, Yue L, Clapham DE. The TRPM7 channel is inactivated by PIP(2) hydrolysis. Nat Cell Biol. 2002;4:329–36.
58. Jardin I, Redondo PC, Salido GM, Rosado JA. Phosphatidylinositol 4,5-bisphosphate enhances store-operated calcium entry through hTRPC6 channel in human platelets. Biochim Biophys Acta. 2008;1783:84–97.
59. Ju YK, Allen DG. Store-operated $Ca^{2+}$ entry and TRPC expression; possible roles in cardiac pacemaker tissue. Heart Lung Circ. 2007;16:349–55.
60. Demion M, Bois P, Launay P, Guinamard R. TRPM4, a $Ca^{2+}$-activated nonselective cation channel in mouse sino-atrial node cells. Cardiovasc Res. 2007;73:531–8.
61. Stephens LR, Jackson TR, Hawkins PT. Agonist-stimulated synthesis of phosphatidylinositol(3,4,5)-trisphosphate: a new intracellular signalling system? Biochim Biophys Acta. 1993;1179:27–75.
62. Katso R, Okkenhaug K, Ahmadi K, White S, Timms J, Waterfield MD. Cellular function of phosphoinositide 3-kinases: implications for development, homeostasis, and cancer. Annu Rev Cell Dev Biol. 2001;17:615–75.
63. Foster FM, Traer CJ, Abraham SM, Fry MJ. The phosphoinositide (PI) 3-kinase family. J Cell Sci. 2003;116:3037–40.
64. Maehama T, Taylor GS, Dixon JE. PTEN and myotubularin: novel phosphoinositide phosphatases. Annu Rev Biochem. 2001;70:247–79.
65. Leslie NR, Downes CP. PTEN: the down side of PI 3-kinase signalling. Cell Signal. 2002;14:285–95.
66. Das S, Dixon JE, Cho W. Membrane-binding and activation mechanism of PTEN. Proc Natl Acad Sci USA. 2003;100:7491–6.
67. Campbell RB, Liu F, Ross AH. Allosteric activation of PTEN phosphatase by phosphatidylinositol 4,5-bisphosphate. J Biol Chem. 2003;278:33617–20.
68. Crackower MA, Oudit GY, Kozieradzki I, et al. Regulation of myocardial contractility and cell size by distinct PI3K-PTEN signaling pathways. Cell. 2002;110:737–49.
69. Oudit GY, Sun H, Kerfant BG, Crackower MA, Penninger JM, Backx PH. The role of phosphoinositide-3 kinase and PTEN in cardiovascular physiology and disease. J Mol Cell Cardiol. 2004;37:449–71.
70. McMullen JR, Shioi T, Zhang L, et al. Phosphoinositide 3-kinase(p110alpha) plays a critical role for the induction of physiological, but not pathological, cardiac hypertrophy. Proc Natl Acad Sci USA. 2003;100:12355–60.
71. Patrucco E, Notte A, Barberis L, et al. PI3Kgamma modulates the cardiac response to chronic pressure overload by distinct kinase-dependent and -independent effects. Cell. 2004;118:375–87.
72. DeBosch B, Treskov I, Lupu TS, et al. Akt1 is required for physiological cardiac growth. Circulation. 2006;113:2097–104.
73. Condorelli G, Drusco A, Stassi G, et al. Akt induces enhanced myocardial contractility and cell size in vivo in transgenic mice. Proc Natl Acad Sci USA. 2002;99:12333–8.
74. Matsui T, Li L, Wu JC, et al. Phenotypic spectrum caused by transgenic overexpression of activated Akt in the heart. J Biol Chem. 2002;277:22896–901.
75. Shioi T, McMullen JR, Kang PM, et al. Akt/protein kinase B promotes organ growth in transgenic mice. Mol Cell Biol. 2002;22:2799–809.
76. Schwartzbauer G, Robbins J. The tumor suppressor gene PTEN can regulate cardiac hypertrophy and survival. J Biol Chem. 2001;276:35786–93.
77. Antos CL, McKinsey TA, Frey N, et al. Activated glycogen synthase-3 beta suppresses cardiac hypertrophy in vivo. Proc Natl Acad Sci USA. 2002;99:907–12.
78. Shioi T, McMullen JR, Tarnavski O, et al. Rapamycin attenuates load-induced cardiac hypertrophy in mice. Circulation. 2003;107:1664–70.
79. MacLennan DH, Kranias EG. Phospholamban: a crucial regulator of cardiac contractility. Nat Rev Mol Cell Biol. 2003;4:566–77.
80. Warrier S, Ramamurthy G, Eckert RL, Nikolaev VO, Lohse MJ, Harvey RD. cAMP microdomains and L-type $Ca^{2+}$ channel regulation in guinea-pig ventricular myocytes. J Physiol. 2007;580:765–76.
81. Damilano F, Perino A, Hirsch E. PI3K kinase and scaffold functions in heart. Ann N Y Acad Sci. 2010;1188:39–45.
82. Brodde OE, Bruck H, Leineweber K. Cardiac adrenoceptors: physiological and pathophysiological relevance. J Pharmacol Sci. 2006;100:323–37.
83. Weis WI, Kobilka BK. Structural insights into G-protein-coupled receptor activation. Curr Opin Struct Biol. 2008;18:734–40.
84. Rockman HA, Koch WJ, Lefkowitz RJ. Seven-transmembrane-spanning receptors and heart function. Nature. 2002;415:206–12.
85. Nienaber JJ, Tachibana H, Naga Prasad SV, et al. Inhibition of receptor-localized PI3K preserves cardiac beta-adrenergic receptor function and ameliorates pressure overload heart failure. J Clin Invest. 2003;112:1067–79.
86. Naga Prasad SV, Jayatilleke A, Madamanchi A, Rockman HA. Protein kinase activity of phosphoinositide 3-kinase regulates beta-adrenergic receptor endocytosis. Nat Cell Biol. 2005;7:785–96.
87. Oudit GY, Crackower MA, Eriksson U, et al. Phosphoinositide 3-kinase gamma-deficient mice are protected from isoproterenol-induced heart failure. Circulation. 2003;108:2147–52.
88. Kerfant BG, Zhao D, Lorenzen-Schmidt I, et al. PI3Kgamma is required for PDE4, not PDE3, activity in subcellular microdomains containing the sarcoplasmic reticular calcium ATPase in cardiomyocytes. Circ Res. 2007;101:400–8.
89. Oudit GY, Kassiri Z. Role of PI3 kinase gamma in excitation–contraction coupling and heart disease. Cardiovasc Hematol Disord Drug Targets. 2007;7:295–304.
90. Jo SH, Leblais V, Wang PH, Crow MT, Xiao RP. Phosphatidylinositol 3-kinase functionally compartmentalizes the concurrent G(s) signaling during beta2-adrenergic stimulation. Circ Res. 2002;91:46–53.

91. Lopes CM, Zhang H, Rohacs T, Jin T, Yang J, Logothetis DE. Alterations in conserved Kir channel-PIP2 interactions underlie channelopathies. Neuron. 2002;34:933–44.
92. Zhang Y, Wang H, Wang J, Han H, Nattel S, Wang Z. Normal function of HERG $K^+$ channels expressed in HEK293 cells requires basal protein kinase B activity. FEBS Lett. 2003;534:125–32.
93. Arab S, Konstantinov IE, Boscarino C, et al. Early gene expression profiles during intraoperative myocardial ischemia–reperfusion in cardiac surgery. J Thorac Cardiovasc Surg. 2007;134(74–81):e1–2.
94. Yamashita K, Kajstura J, Discher DJ, et al. Reperfusion-activated Akt kinase prevents apoptosis in transgenic mouse hearts overexpressing insulin-like growth factor-1. Circ Res. 2001;88:609–14.
95. Murphy E, Tong H, Steenbergen C. Preconditioning: is the Akt-ion in the PI3K pathway? J Mol Cell Cardiol. 2003;35:1021–5.
96. Kis A, Yellon DM, Baxter GF. Second window of protection following myocardial preconditioning: an essential role for PI3 kinase and p70S6 kinase. J Mol Cell Cardiol. 2003;35:1063–71.
97. Mora A, Davies AM, Bertrand L, et al. Deficiency of PDK1 in cardiac muscle results in heart failure and increased sensitivity to hypoxia. EMBO J. 2003;22:4666–76.
98. Tong H, Imahashi K, Steenbergen C, Murphy E. Phosphorylation of glycogen synthase kinase-3beta during preconditioning through a phosphatidylinositol-3-kinase-dependent pathway is cardioprotective. Circ Res. 2002;90:377–9.
99. Yellon DM, Downey JM. Preconditioning the myocardium: from cellular physiology to clinical cardiology. Physiol Rev. 2003;83:1113–51.
100. Ban K, Cooper AJ, Samuel S, et al. Phosphatidylinositol 3-kinase gamma is a critical mediator of myocardial ischemic and adenosine-mediated preconditioning. Circ Res. 2008;103:643–53.
101. Siddall HK, Warrell CE, Yellon DM, Mocanu MM. Ischemia–reperfusion injury and cardioprotection: investigating PTEN, the phosphatase that negatively regulates PI3K, using a congenital model of PTEN haploinsufficiency. Basic Res Cardiol. 2008;103:560–8.
102. Ghosh TK, Bian J, Gill DL. Intracellular calcium release mediated by sphingosine derivatives generated in cells. Science. 1990;248:1653–6.
103. Spiegel S, Milstien S. Sphingosine-1-phosphate: an enigmatic signalling lipid. Nat Rev Mol Cell Biol. 2003;4:397–407.
104. Alvarez SE, Milstien S, Spiegel S. Autocrine and paracrine roles of sphingosine-1-phosphate. Trends Endocrinol Metab. 2007;18:300–7.
105. Futerman AH, Hannun YA. The complex life of simple sphingolipids. EMBO Rep. 2004;5:777–82.
106. Le Stunff H, Milstien S, Spiegel S. Generation and metabolism of bioactive sphingosine-1-phosphate. J Cell Biochem. 2004;92:882–99.
107. Yatomi Y. Plasma sphingosine 1-phosphate metabolism and analysis. Biochim Biophys Acta. 2008;1780:606–11.
108. Okajima F. Plasma lipoproteins behave as carriers of extracellular sphingosine 1-phosphate: is this an atherogenic mediator or an anti-atherogenic mediator? Biochim Biophys Acta. 2002;1582:132–7.
109. Alewijnse AE, Peters SL, Michel MC. Cardiovascular effects of sphingosine-1-phosphate and other sphingomyelin metabolites. Br J Pharmacol. 2004;143:666–84.
110. Goni FM, Alonso A. Sphingomyelinases: enzymology and membrane activity. FEBS Lett. 2002;531:38–46.
111. Tomiuk S, Hofmann K, Nix M, Zumbansen M, Stoffel W. Cloned mammalian neutral sphingomyelinase: functions in sphingolipid signaling? Proc Natl Acad Sci USA. 1998;95:3638–43.
112. Hofmann K, Tomiuk S, Wolff G, Stoffel W. Cloning and characterization of the mammalian brain-specific, $Mg^{2+}$-dependent neutral sphingomyelinase. Proc Natl Acad Sci USA. 2000;97:5895–900.
113. Krut O, Wiegmann K, Kashkar H, Yazdanpanah B, Kronke M. Novel tumor necrosis factor-responsive mammalian neutral sphingomyelinase-3 is a C-tail-anchored protein. J Biol Chem. 2006;281:13784–93.
114. Zumbansen M, Stoffel W. Neutral sphingomyelinase 1 deficiency in the mouse causes no lipid storage disease. Mol Cell Biol. 2002;22:3633–8.
115. Clarke CJ, Snook CF, Tani M, Matmati N, Marchesini N, Hannun YA. The extended family of neutral sphingomyelinases. Biochemistry. 2006;45:11247–56.
116. Stoffel W, Jenke B, Holz B, et al. Neutral sphingomyelinase (SMPD3) deficiency causes a novel form of chondrodysplasia and dwarfism that is rescued by Col2A1-driven smpd3 transgene expression. Am J Pathol. 2007;171:153–61.
117. Krown KA, Page MT, Nguyen C, et al. Tumor necrosis factor alpha-induced apoptosis in cardiac myocytes. Involvement of the sphingolipid signaling cascade in cardiac cell death. J Clin Invest. 1996;98:2854–65.
118. Oral H, Dorn II GW, Mann DL. Sphingosine mediates the immediate negative inotropic effects of tumor necrosis factor-alpha in the adult mammalian cardiac myocyte. J Biol Chem. 1997;272:4836–42.
119. Liu B, Andrieu-Abadie N, Levade T, Zhang P, Obeid LM, Hannun YA. Glutathione regulation of neutral sphingomyelinase in tumor necrosis factor-alpha-induced cell death. J Biol Chem. 1998;273:11313–20.
120. Defer N, Azroyan A, Pecker F, Pavoine C. TNFR1 and TNFR2 signaling interplay in cardiac myocytes. J Biol Chem. 2007;282:35564–73.
121. Adamy C, Mulder P, Khouzami L, et al. Neutral sphingomyelinase inhibition participates to the benefits of N-acetylcysteine treatment in post-myocardial infarction failing heart rats. J Mol Cell Cardiol. 2007;43:344–53.
122. Blaustein A, Deneke SM, Stolz RI, Baxter D, Healey N, Fanburg BL. Myocardial glutathione depletion impairs recovery after short periods of ischemia. Circulation. 1989;80:1449–57.
123. Hernandez OM, Discher DJ, Bishopric NH, Webster KA. Rapid activation of neutral sphingomyelinase by hypoxia-reoxygenation of cardiac myocytes. Circ Res. 2000;86:198–204.
124. Ozer MK, Parlakpinar H, Cigremis Y, Ucar M, Vardi N, Acet A. Ischemia–reperfusion leads to depletion of glutathione content and augmentation of malondialdehyde production in the rat heart from overproduction of oxidants: can caffeic acid phenethyl ester (CAPE) protect the heart? Mol Cell Biochem. 2005;273:169–75.
125. Radin MJ, Holycross BJ, Dumitrescu C, Kelley R, Altschuld RA. Leptin modulates the negative inotropic effect of interleukin-1beta in cardiac myocytes. Mol Cell Biochem. 2008;315:179–84.
126. Schuchman EH, Levran O, Pereira LV, Desnick RJ. Structural organization and complete nucleotide sequence of the gene encoding human acid sphingomyelinase (SMPD1). Genomics. 1992;12:197–205.
127. Brady RO, Kanfer JN, Mock MB, Fredrickson DS. The metabolism of sphingomyelin. II. Evidence of an enzymatic deficiency in Niemann-Pick disease. Proc Natl Acad Sci USA. 1966;55:366–9.
128. Garcia-Dorado D, Vinten-Johansen J, Piper HM. Bringing preconditioning and postconditioning into focus. Cardiovasc Res. 2006;70:167–9.
129. Thibault H, Piot C, Staat P, et al. Long-term benefit of postconditioning. Circulation. 2008;117:1037–44.
130. Cui J, Engelman RM, Maulik N, Das DK. Role of ceramide in ischemic preconditioning. J Am Coll Surg. 2004;198:770–7.
131. Jin ZQ, Zhang J, Huang Y, Hoover HE, Vessey DA, Karliner JS. A sphingosine kinase 1 mutation sensitizes the myocardium to ischemia/reperfusion injury. Cardiovasc Res. 2007;76:41–50.
132. Jin ZQ, Karliner JS, Vessey DA. Ischaemic postconditioning protects isolated mouse hearts against ischaemia/reperfusion injury via sphingosine kinase isoform-1 activation. Cardiovasc Res. 2008;79:134–40.
133. Doehner W, Bunck AC, Rauchhaus M, et al. Secretory sphingomyelinase is upregulated in chronic heart failure: a second messenger

system of immune activation relates to body composition, muscular functional capacity, and peripheral blood flow. Eur Heart J. 2007; 28:821–8.
134. Igarashi J, Michel T. Sphingosine-1-phosphate and modulation of vascular tone. Cardiovasc Res. 2009;82:212–20.
135. Sattler K, Levkau B. Sphingosine-1-phosphate as a mediator of high-density lipoprotein effects in cardiovascular protection. Cardiovasc Res. 2009;82:201–11.
136. Kohama T, Olivera A, Edsall L, Nagiec MM, Dickson R, Spiegel S. Molecular cloning and functional characterization of murine sphingosine kinase. J Biol Chem. 1998;273:23722–8.
137. Melendez AJ, Carlos-Dias E, Gosink M, Allen JM, Takacs L. Human sphingosine kinase: molecular cloning, functional characterization and tissue distribution. Gene. 2000;251:19–26.
138. Liu H, Sugiura M, Nava VE, et al. Molecular cloning and functional characterization of a novel mammalian sphingosine kinase type 2 isoform. J Biol Chem. 2000;275:19513–20.
139. Alemany R, van Koppen CJ, Danneberg K, Ter Braak M, Meyer Zu Heringdorf D. Regulation and functional roles of sphingosine kinases. Naunyn Schmiedebergs Arch Pharmacol. 2007;374:413–28.
140. Imamura T, Ohgane J, Ito S, et al. CpG island of rat sphingosine kinase-1 gene: tissue-dependent DNA methylation status and multiple alternative first exons. Genomics. 2001;76:117–25.
141. Allende ML, Sasaki T, Kawai H, et al. Mice deficient in sphingosine kinase 1 are rendered lymphopenic by FTY720. J Biol Chem. 2004;279:52487–92.
142. Michaud J, Kohno M, Proia RL, Hla T. Normal acute and chronic inflammatory responses in sphingosine kinase 1 knockout mice. FEBS Lett. 2006;580:4607–12.
143. Mizugishi K, Yamashita T, Olivera A, Miller GF, Spiegel S, Proia RL. Essential role for sphingosine kinases in neural and vascular development. Mol Cell Biol. 2005;25:11113–21.
144. Pitson SM, Moretti PA, Zebol JR, et al. Activation of sphingosine kinase 1 by ERK1/2-mediated phosphorylation. EMBO J. 2003;22: 5491–500.
145. Barr RK, Lynn HE, Moretti PA, Khew-Goodall Y, Pitson SM. Deactivation of sphingosine kinase 1 by protein phosphatase 2A. J Biol Chem. 2008;283:34994–5002.
146. Hait NC, Bellamy A, Milstien S, Kordula T, Spiegel S. Sphingosine kinase type 2 activation by ERK-mediated phosphorylation. J Biol Chem. 2007;282:12058–65.
147. Sutherland CM, Moretti PA, Hewitt NM, Bagley CJ, Vadas MA, Pitson SM. The calmodulin-binding site of sphingosine kinase and its role in agonist-dependent translocation of sphingosine kinase 1 to the plasma membrane. J Biol Chem. 2006;281:11693–701.
148. Olivera A, Rosenthal J, Spiegel S. Effect of acidic phospholipids on sphingosine kinase. J Cell Biochem. 1996;60:529–37.
149. Cavallini L, Venerando R, Miotto G, Alexandre A. Ganglioside GM1 protection from apoptosis of rat heart fibroblasts. Arch Biochem Biophys. 1999;370:156–62.
150. Le Stunff H, Giussani P, Maceyka M, Lepine S, Milstien S, Spiegel S. Recycling of sphingosine is regulated by the concerted actions of sphingosine-1-phosphate phosphohydrolase 1 and sphingosine kinase 2. J Biol Chem. 2007;282:34372–80.
151. Funato K, Lombardi R, Vallee B, Riezman H. Lcb4p is a key regulator of ceramide synthesis from exogenous long chain sphingoid base in Saccharomyces cerevisiae. J Biol Chem. 2003;278:7325–34.
152. Igarashi N, Okada T, Hayashi S, Fujita T, Jahangeer S, Nakamura S. Sphingosine kinase 2 is a nuclear protein and inhibits DNA synthesis. J Biol Chem. 2003;278:46832–9.
153. Okada T, Ding G, Sonoda H, et al. Involvement of N-terminal-extended form of sphingosine kinase 2 in serum-dependent regulation of cell proliferation and apoptosis. J Biol Chem. 2005;280:36318–25.
154. Maceyka M, Sankala H, Hait NC, et al. SphK1 and SphK2, sphingosine kinase isoenzymes with opposing functions in sphingolipid metabolism. J Biol Chem. 2005;280:37118–29.
155. Jin ZQ, Zhou HZ, Zhu P, et al. Cardioprotection mediated by sphingosine-1-phosphate and ganglioside GM-1 in wild-type and PKC epsilon knockout mouse hearts. Am J Physiol Heart Circ Physiol. 2002;282:H1970–7.
156. Lecour S, Smith RM, Woodward B, Opie LH, Rochette L, Sack MN. Identification of a novel role for sphingolipid signaling in TNF alpha and ischemic preconditioning mediated cardioprotection. J Mol Cell Cardiol. 2002;34:509–18.
157. Gray MO, Zhou HZ, Schafhalter-Zoppoth I, Zhu P, Mochly-Rosen D, Messing RO. Preservation of base-line hemodynamic function and loss of inducible cardioprotection in adult mice lacking protein kinase C epsilon. J Biol Chem. 2004;279:3596–604.
158. Jin ZQ, Goetzl EJ, Karliner JS. Sphingosine kinase activation mediates ischemic preconditioning in murine heart. Circulation. 2004;110:1980–9.
159. Hla T, Maciag T. An abundant transcript induced in differentiating human endothelial cells encodes a polypeptide with structural similarities to G-protein-coupled receptors. J Biol Chem. 1990; 265:9308–13.
160. Lee MJ, Van Brocklyn JR, Thangada S, et al. Sphingosine-1-phosphate as a ligand for the G protein-coupled receptor EDG-1. Science. 1998;279:1552–5.
161. Ancellin N, Hla T. Switching intracellular signaling pathways to study sphingosine 1-phosphate receptors. Ann N Y Acad Sci. 2000; 905:260–2.
162. Hla T, Lee MJ, Ancellin N, Paik JH, Kluk MJ. Lysophospholipids–receptor revelations. Science. 2001;294:1875–8.
163. Windh RT, Lee MJ, Hla T, An S, Barr AJ, Manning DR. Differential coupling of the sphingosine 1-phosphate receptors Edg-1, Edg-3, and H218/Edg-5 to the G(i), G(q), and G(12) families of heterotrimeric G proteins. J Biol Chem. 1999;274:27351–8.
164. Im DS, Heise CE, Ancellin N, et al. Characterization of a novel sphingosine 1-phosphate receptor, Edg-8. J Biol Chem. 2000; 275:14281–6.
165. Van Brocklyn JR, Graler MH, Bernhardt G, Hobson JP, Lipp M, Spiegel S. Sphingosine-1-phosphate is a ligand for the G protein-coupled receptor EDG-6. Blood. 2000;95:2624–9.
166. Ishii I, Friedman B, Ye X, et al. Selective loss of sphingosine 1-phosphate signaling with no obvious phenotypic abnormality in mice lacking its G protein-coupled receptor, LP(B3)/EDG-3. J Biol Chem. 2001;276:33697–704.
167. Graler MH, Grosse R, Kusch A, Kremmer E, Gudermann T, Lipp M. The sphingosine 1-phosphate receptor S1P4 regulates cell shape and motility via coupling to Gi and G12/13. J Cell Biochem. 2003;89:507–19.
168. Ishii I, Fukushima N, Ye X, Chun J. Lysophospholipid receptors: signaling and biology. Annu Rev Biochem. 2004;73:321–54.
169. Means CK, Brown JH. Sphingosine-1-phosphate receptor signalling in the heart. Cardiovasc Res. 2009;82:193–200.
170. Landeen LK, Dederko DA, Kondo CS, et al. Mechanisms of the negative inotropic effects of sphingosine-1-phosphate on adult mouse ventricular myocytes. Am J Physiol Heart Circ Physiol. 2008;294:H736–49.
171. Means CK, Miyamoto S, Chun J, Brown JH. S1P1 receptor localization confers selectivity for Gi-mediated cAMP and contractile responses. J Biol Chem. 2008;283:11954–63.
172. Ancellin N, Hla T. Differential pharmacological properties and signal transduction of the sphingosine 1-phosphate receptors EDG-1, EDG-3, and EDG-5. J Biol Chem. 1999;274:18997–9002.
173. Himmel HM, Meyer Zu Heringdorf D, Graf E, et al. Evidence for Edg-3 receptor-mediated activation of I(K.ACh) by sphingosine-1-phosphate in human atrial cardiomyocytes. Mol Pharmacol. 2000;58:449–54.
174. Voogd TE, Vansterkenburg EL, Wilting J, Janssen LH. Recent research on the biological activity of suramin. Pharmacol Rev. 1993;45:177–203.

175. Zhang J, Honbo N, Goetzl EJ, Chatterjee K, Karliner JS, Gray MO. Signals from type 1 sphingosine 1-phosphate receptors enhance adult mouse cardiac myocyte survival during hypoxia. Am J Physiol Heart Circ Physiol. 2007;293:H3150–8.
176. Tao R, Zhang J, Vessey DA, Honbo N, Karliner JS. Deletion of the sphingosine kinase-1 gene influences cell fate during hypoxia and glucose deprivation in adult mouse cardiomyocytes. Cardiovasc Res. 2007;74:56–63.
177. Theilmeier G, Schmidt C, Herrmann J, et al. High-density lipoproteins and their constituent, sphingosine-1-phosphate, directly protect the heart against ischemia/reperfusion injury in vivo via the S1P3 lysophospholipid receptor. Circulation. 2006;114:1403–9.
178. Means CK, Xiao CY, Li Z, et al. Sphingosine 1-phosphate S1P2 and S1P3 receptor-mediated Akt activation protects against in vivo myocardial ischemia–reperfusion injury. Am J Physiol Heart Circ Physiol. 2007;292:H2944–51.
179. Robert P, Tsui P, Laville MP, et al. EDG1 receptor stimulation leads to cardiac hypertrophy in rat neonatal myocytes. J Mol Cell Cardiol. 2001;33:1589–606.
180. Wang D, Dubois RN. Eicosanoids and cancer. Nat Rev Cancer. 2010;10:181–93.
181. Wolf RA, Gross RW. Identification of neutral active phospholipase C which hydrolyzes choline glycerophospholipids and plasmalogen selective phospholipase A2 in canine myocardium. J Biol Chem. 1985;260:7295–303.
182. Clark JD, Lin LL, Kriz RW, et al. A novel arachidonic acid-selective cytosolic PLA2 contains a Ca(2+)-dependent translocation domain with homology to PKC and GAP. Cell. 1991;65:1043–51.
183. Ghosh M, Tucker DE, Burchett SA, Leslie CC. Properties of the Group IV phospholipase A2 family. Prog Lipid Res. 2006;45: 487–510.
184. Cedars A, Jenkins CM, Mancuso DJ, Gross RW. Calcium-independent phospholipases in the heart: mediators of cellular signaling, bioenergetics, and ischemia-induced electrophysiologic dysfunction. J Cardiovasc Pharmacol. 2009;53:277–89.
185. McHowat J, Creer MH. Catalytic features, regulation and function of myocardial phospholipase A2. Curr Med Chem Cardiovasc Hematol Agents. 2004;2:209–18.
186. Hazen SL, Gross RW. Human myocardial cytosolic Ca(2+)-independent phospholipase A2 is modulated by ATP. Concordant ATP-induced alterations in enzyme kinetics and mechanism-based inhibition. Biochem J. 1991;280(Pt 3):581–7.
187. Ford DA, Hazen SL, Saffitz JE, Gross RW. The rapid and reversible activation of a calcium-independent plasmalogen-selective phospholipase A2 during myocardial ischemia. J Clin Invest. 1991;88:331–5.
188. Mancuso DJ, Abendschein DR, Jenkins CM, et al. Cardiac ischemia activates calcium-independent phospholipase A2beta, precipitating ventricular tachyarrhythmias in transgenic mice: rescue of the lethal electrophysiologic phenotype by mechanism-based inhibition. J Biol Chem. 2003;278:22231–6.
189. Jenkins CM, Yan W, Mancuso DJ, Gross RW. Highly selective hydrolysis of fatty acyl-CoAs by calcium-independent phospholipase A2beta. Enzyme autoacylation and acyl-CoA-mediated reversal of calmodulin inhibition of phospholipase A2 activity. J Biol Chem. 2006;281:15615–24.
190. Mancuso DJ, Jenkins CM, Gross RW. The genomic organization, complete mRNA sequence, cloning, and expression of a novel human intracellular membrane-associated calcium-independent phospholipase A(2). J Biol Chem. 2000;275:9937–45.
191. Mancuso DJ, Jenkins CM, Sims HF, Cohen JM, Yang J, Gross RW. Complex transcriptional and translational regulation of iPLAgamma resulting in multiple gene products containing dual competing sites for mitochondrial or peroxisomal localization. Eur J Biochem. 2004;271:4709–24.
192. Mancuso DJ, Han X, Jenkins CM, et al. Dramatic accumulation of triglycerides and precipitation of cardiac hemodynamic dysfunction during brief caloric restriction in transgenic myocardium expressing human calcium-independent phospholipase A2gamma. J Biol Chem. 2007;282:9216–27.
193. Mancuso DJ, Sims HF, Han X, et al. Genetic ablation of calcium-independent phospholipase A2gamma leads to alterations in mitochondrial lipid metabolism and function resulting in a deficient mitochondrial bioenergetic phenotype. J Biol Chem. 2007;282:34611–22.
194. Channon JY, Leslie CC. A calcium-dependent mechanism for associating a soluble arachidonoyl-hydrolyzing phospholipase A2 with membrane in the macrophage cell line RAW 264.7. J Biol Chem. 1990;265:5409–13.
195. Nalefski EA, Sultzman LA, Martin DM, et al. Delineation of two functionally distinct domains of cytosolic phospholipase A2, a regulatory Ca(2+)-dependent lipid-binding domain and a Ca(2+)-independent catalytic domain. J Biol Chem. 1994;269:18239–49.
196. Shimizu M, Nakamura H, Hirabayashi T, Suganami A, Tamura Y, Murayama T. Ser515 phosphorylation-independent regulation of cytosolic phospholipase A2alpha (cPLA2alpha) by calmodulin-dependent protein kinase: possible interaction with catalytic domain A of cPLA2alpha. Cell Signal. 2008;20:815–24.
197. Pavicevic Z, Leslie CC, Malik KU. cPLA2 phosphorylation at serine-515 and serine-505 is required for arachidonic acid release in vascular smooth muscle cells. J Lipid Res. 2008;49:724–37.
198. Haq S, Kilter H, Michael A, et al. Deletion of cytosolic phospholipase A2 promotes striated muscle growth. Nat Med. 2003;9:944–51.
199. Ait-Mamar B, Cailleret M, Rucker-Martin C, et al. The cytosolic phospholipase A2 pathway, a safeguard of beta2-adrenergic cardiac effects in rat. J Biol Chem. 2005;280:18881–90.
200. Testa M, Rocca B, Spath L, et al. Expression and activity of cyclooxygenase isoforms in skeletal muscles and myocardium of humans and rodents. J Appl Physiol. 2007;103:1412–8.
201. Zidar N, Dolenc-Strazar Z, Jeruc J, et al. Expression of cyclooxygenase-1 and cyclooxygenase-2 in the normal human heart and in myocardial infarction. Cardiovasc Pathol. 2007;16:300–4.
202. Zhang Z, Vezza R, Plappert T, et al. COX-2-dependent cardiac failure in Gh/tTG transgenic mice. Circ Res. 2003;92:1153–61.
203. Shinmura K, Tang XL, Wang Y, et al. Cyclooxygenase-2 mediates the cardioprotective effects of the late phase of ischemic preconditioning in conscious rabbits. Proc Natl Acad Sci USA. 2000;97:10197–202.
204. Xuan YT, Guo Y, Zhu Y, et al. Mechanism of cyclooxygenase-2 upregulation in late preconditioning. J Mol Cell Cardiol. 2003;35:525–37.
205. Kunapuli P, Lawson JA, Rokach JA, Meinkoth JL, FitzGerald GA. Prostaglandin F2alpha (PGF2alpha) and the isoprostane, 8, 12-iso-isoprostane F2alpha-III, induce cardiomyocyte hypertrophy. Differential activation of downstream signaling pathways. J Biol Chem. 1998;273:22442–52.
206. Adams JW, Sah VP, Henderson SA, Brown JH. Tyrosine kinase and c-Jun NH2-terminal kinase mediate hypertrophic responses to prostaglandin F2alpha in cultured neonatal rat ventricular myocytes. Circ Res. 1998;83:167–78.
207. Francois H, Athirakul K, Howell D, et al. Prostacyclin protects against elevated blood pressure and cardiac fibrosis. Cell Metab. 2005;2:201–7.
208. Hara A, Yuhki K, Fujino T, et al. Augmented cardiac hypertrophy in response to pressure overload in mice lacking the prostaglandin I2 receptor. Circulation. 2005;112:84–92.
209. Breitbart E, Sofer Y, Shainberg A, Grossman S. Lipoxygenase activity in heart cells. FEBS Lett. 1996;395:148–52.
210. Hughes H, Gentry DL, McGuire GM, Taylor AA. Gas chromatographic-mass spectrometric analysis of lipoxygenase products in

post-ischemic rabbit myocardium. Prostaglandins Leukot Essent Fatty Acids. 1991;42:225–31.
211. Kuzuya T, Hoshida S, Kim Y, et al. Free radical generation coupled with arachidonate lipoxygenase reaction relates to reoxygenation induced myocardial cell injury. Cardiovasc Res. 1993;27:1056–60.
212. Wen Y, Gu J, Liu Y, Wang PH, Sun Y, Nadler JL. Overexpression of 12-lipoxygenase causes cardiac fibroblast cell growth. Circ Res. 2001;88:70–6.
213. Wen Y, Gu J, Peng X, Zhang G, Nadler J. Overexpression of 12-lipoxygenase and cardiac fibroblast hypertrophy. Trends Cardiovasc Med. 2003;13:129–36.
214. Dransfeld O, Rakatzi I, Sasson S, et al. Eicosanoids participate in the regulation of cardiac glucose transport by contribution to a rearrangement of actin cytoskeletal elements. Biochem J. 2001;359:47–54.
215. Dransfeld O, Rakatzi I, Sasson S, Eckel J. Eicosanoids and the regulation of cardiac glucose transport. Ann N Y Acad Sci. 2002;967:208–16.
216. Tsao CC, Coulter SJ, Chien A, et al. Identification and localization of five CYP2Cs in murine extrahepatic tissues and their metabolism of arachidonic acid to regio- and stereoselective products. J Pharmacol Exp Ther. 2001;299:39–47.
217. Wang H, Zhao Y, Bradbury JA, et al. Cloning, expression, and characterization of three new mouse cytochrome p450 enzymes and partial characterization of their fatty acid oxidation activities. Mol Pharmacol. 2004;65:1148–58.
218. Delozier TC, Kissling GE, Coulter SJ, et al. Detection of human CYP2C8, CYP2C9, and CYP2J2 in cardiovascular tissues. Drug Metab Dispos. 2007;35:682–8.
219. Zordoky BN, El-Kadi AO. Modulation of cardiac and hepatic cytochrome P450 enzymes during heart failure. Curr Drug Metab. 2008;9:122–8.
220. Granville DJ, Tashakkor B, Takeuchi C, et al. Reduction of ischemia and reperfusion-induced myocardial damage by cytochrome P450 inhibitors. Proc Natl Acad Sci USA. 2004;101:1321–6.
221. Seubert J, Yang B, Bradbury JA, et al. Enhanced postischemic functional recovery in CYP2J2 transgenic hearts involves mitochondrial ATP-sensitive $K^+$ channels and p42/p44 MAPK pathway. Circ Res. 2004;95:506–14.
222. Nithipatikom K, Gross ER, Endsley MP, et al. Inhibition of cytochrome P450omega-hydroxylase: a novel endogenous cardioprotective pathway. Circ Res. 2004;95:e65–71.
223. Seubert JM, Zeldin DC, Nithipatikom K, Gross GJ. Role of epoxyeicosatrienoic acids in protecting the myocardium following ischemia/reperfusion injury. Prostaglandins Other Lipid Mediat. 2007;82:50–9.
224. Gubitosi-Klug RA, Yu SP, Choi DW, Gross RW. Concomitant acceleration of the activation and inactivation kinetics of the human delayed rectifier $K^+$ channel (Kv1.1) by Ca(2+)-independent phospholipase A2. J Biol Chem. 1995;270:2885–8.
225. Lu T, Hoshi T, Weintraub NL, Spector AA, Lee HC. Activation of ATP-sensitive K(+) channels by epoxyeicosatrienoic acids in rat cardiac ventricular myocytes. J Physiol. 2001;537:811–27.
226. Chen J, Capdevila JH, Zeldin DC, Rosenberg RL. Inhibition of cardiac L-type calcium channels by epoxyeicosatrienoic acids. Mol Pharmacol. 1999;55:288–95.

# Part III
# Mitochondria

# Chapter 8
# Heart Mitochondria: A Receiver and Integrator of Signals

**Abstract** Besides their essential bioenergetic role in supplying ATP, heart mitochondria play a central role in the regulatory and signaling events that occur in response to physiological stresses, including but not limited to heart failure (HF), myocardial ischemia and reperfusion (I/R), hypoxia, oxidative stress (OS), and hormonal and cytokine stimuli. Research on both intact cardiac and skeletal muscle tissue and cultured cardiomyocytes has just begun to probe the nature and the extent of mitochondrial involvement in interorganelle communication, hypertrophic growth, and cell death. In this chapter, we examine heart mitochondria under the perspective of a receiver/integrator and transmitter of signals, dissecting the multiple and interrelated signaling pathways playing at both the molecular and biochemical levels with particular focus on nuclear and cytoplasmic factors involved in the shaping of the organelles' responses, and gauging the effect that mitochondria have (as a receiver, integrator, and transmitter of signals) on cardiac phenotype.

**Keywords** Mitochondria • Bioenergetic • ROS • Apoptosis • Cell death • Calcium signaling

## Introduction

Mitochondrial signaling is a process by which the organelle communicates with its environment as a transmitter and receiver of signals. Evidence has shown that mitochondria act as a dynamic receiver and integrator of numerous translocated signaling proteins (including protein kinases and transcription factors), regulatory $Ca^{2+}$ fluxes and membrane phospholipids, as well as transmission of mitochondrial-generated reactive oxygen species (ROS) and energy-related signaling. The contribution that mitochondria make to cardiac function extends well beyond their critical role as a bioenergetic supplier of ATP. The organelle plays an integral part in the regulatory and signaling events that occur in response to physiological stresses including but not limited to HF (see Chap. 14), myocardial I/R, hypoxia, OS, and hormonal and cytokine stimuli. Research on intact cardiac muscle tissue and cultured cardiomyocytes has just begun to probe the nature and the extent of mitochondrial involvement in interorganelle communication, hypertrophic growth, and cell death. Together with their central role in cardiac and skeletal muscle apoptosis, mitochondria are essential players in the proliferative signaling pathways, nutrient sensing, interorganellar communication and in the responses of cells to metabolic changes and physiological stresses. The participation of mitochondria in numerous and interrelated signaling pathways is discussed in this chapter, gauging the effect that mitochondria have as a receiver, integrator, and transmitter of signals on the cardiac phenotype and on their potential impact on future treatments for cardiac diseases.

## Mitochondria Signaling

Without receptor molecules, mitochondria, the powerhouse of the cell, would be isolated and unaware of its environment. Signaling pathways allow these organelles to respond to heart energy demands as well as to cardiomyocytes growth, death, and a variety of physiological stimuli and stresses. In order to understand mitochondrial signaling in the context of the diverse and numerous intracellular events, a brief background is presented, highlighting mitochondrial bioenergetics and biogenesis and its involvement in ROS generation and apoptosis followed by an analysis of the multiple roles that mitochondria play as a receiver, integrator, and transmitter of signals within this complex assemblage, including interacting signaling from numerous translocated proteins such as protein kinases, receptors and transcription factors, regulatory $Ca^{2+}$ fluxes and membrane lipids, mitochondrial-generated ROS, and energy-related signaling to other cellular compartments. Novel experimental approaches used to probe mitochondrial signaling using metabolic inhibitors and genetic stresses [e.g., mitochondrial DNA (mtDNA) depletion] are also presented.

J. Marín-García, *Signaling in the Heart*,
DOI 10.1007/978-1-4419-9461-5_8, © Springer Science+Business Media, LLC 2011

## *Mitochondrial Bioenergetics*

Mitochondrial energy production depends on both nuclear and mtDNA-encoded genetic factors which modulate normal mitochondrial function, including enzyme activity and cofactor availability, and on environmental factors such as substrate availability (e.g., sugars, fats, and proteins) and oxygen. Several interacting bioenergetic pathways contribute to mitochondrial energy metabolism including pyruvate oxidation, the tricarboxylic acid (TCA) cycle, fatty acid β-oxidation (FAO), and the common final pathway of oxidative phosphorylation (OXPHOS), which generates most of the cellular ATP. OXPHOS is performed by complexes of proteins located at the mitochondrial inner membrane [1] including the electron transport chain (ETC)/respiratory complexes I–IV, ATP synthase (complex V), and the adenine nucleotide translocator (ANT). In order to be effectively utilized for bioenergetic production via mitochondrial FAO, fatty acids need to be transported into the cell and subsequently into the mitochondria, a process requiring several transport proteins including the carnitine shuttle (composed of carnitine acyltransferase and two carnitine palmitoyl transferases as well as carnitine). FAO and carbohydrate oxidation, via the TCA cycle, generate the majority of intramitochondrial NADH and $FADH_2$, the direct source of electrons for the ETC.

## *Mitochondrial Biogenesis*

Animal mitochondria have their own genome (mtDNA), a double-strand DNA circular molecule, present in each cell in multiple copies (10–1,000 per cell depending on cell type), and encoding 13 proteins which constitute a portion of the multienzyme complexes involved in ETC and OXPHOS [2]. The protein-encoding mtDNA genes are transcribed into specific mRNAs which are translated on a mitochondrial-specific ribosome/protein synthesis apparatus (Fig. 8.1). The mtDNA also encodes part of the mitochondrial protein synthesis machinery including 2 ribosomal RNAs (rRNAs) and 22 transfer RNAs (tRNAs) [3]. Mitochondrial biogenesis is increased in specific cell types (e.g., cardiac and skeletal muscle) during cell hypertrophy, treatment with a variety of agents, e.g., thyroid hormone (TH), exercise, glucocorticoids, xenobiotics, and electrical stimulation [4–10]. While it has long been recognized that mtDNA copy number is highly regulated within a given cell or tissue type [11], the mechanisms that regulate specific mtDNA levels and overall mitochondrial number have not yet been well delineated. Conversely, the nuclear genome encodes the entire complement of proteins involved in mtDNA replication and transcription, protein components of mitochondrial ribosomes, multiple structural and transport proteins of the mitochondrial membranes, the remaining peptide subunits of the respiratory complexes (other than the 13 mtDNA-encoded peptide subunits), and the mitochondrial enzymes involved in mitochondrial lipid metabolism and the TCA cycle [12]. These nuclear-encoded proteins are synthesized on cytosolic ribosomes, targeted to mitochondria and imported by a complex but well-characterized process involving signal peptide recognition, membrane receptors, proteases, and an array of molecular chaperones [3]. Specific regulation of a number of the nuclear-encoded OXPHOS proteins is mediated by variable gene expression sensitive to a variety of physiological (e.g., hypoxia) and developmental stimuli [13]. Tissue-specific isoforms for specific peptides (e.g., cardiac/skeletal muscle specific isoforms exist for genes encoding cytochrome *c* oxidase subunits VIa, VIIa, and VIII) show entirely different patterns of gene expression in adult compared to fetal stages of development [13, 14].

# Signaling at the Mitochondria

## *ROS Generation and Signaling*

A critical by-product of mitochondrial bioenergetic activity is the generation of ROS including superoxide, hydroxyl radicals, and hydrogen peroxide ($H_2O_2$) (Fig. 8.2). Side reactions of mitochondrial ETC enzymes with oxygen directly generate the superoxide anion radical. The primary sites for mitochondrial ROS generation, as a by-product of normal metabolism, are at complex I, II, and III of the respiratory chain; either excessive or diminished electron flux at these sites can stimulate the auto-oxidation of flavins and quinones (including coenzyme Q) producing superoxide radicals [15]. The superoxide radicals can react with nitric oxide (NO) to form peroxynitrite, which is a highly reactive and deleterious free radical species or can be converted by superoxide dismutase (SOD) to $H_2O_2$ that can further react to form hydroxyl radicals. Generation of the hydroxyl radical (the most reactive and deleterious form of ROS) is primarily responsible for the damage to cellular macromolecules such as proteins, DNA, and lipids. The highly reactive hydroxyl radical is generated from reactions involving other ROS species (e.g., the Fenton reaction) in which ubiquitous metal ions, such as Fe (II) or Cu (I), react with $H_2O_2$. The high reactivity of the hydroxyl radical and its extremely short physiological half-life of $10^{-9}$ s restrict its damage to a small radius from its origin, since it is too short-lived to diffuse a considerable distance [16]. In contrast, the less reactive superoxide radicals produced in mitochondria can be delivered to the cytosol through anion channels [e.g., voltage-dependent

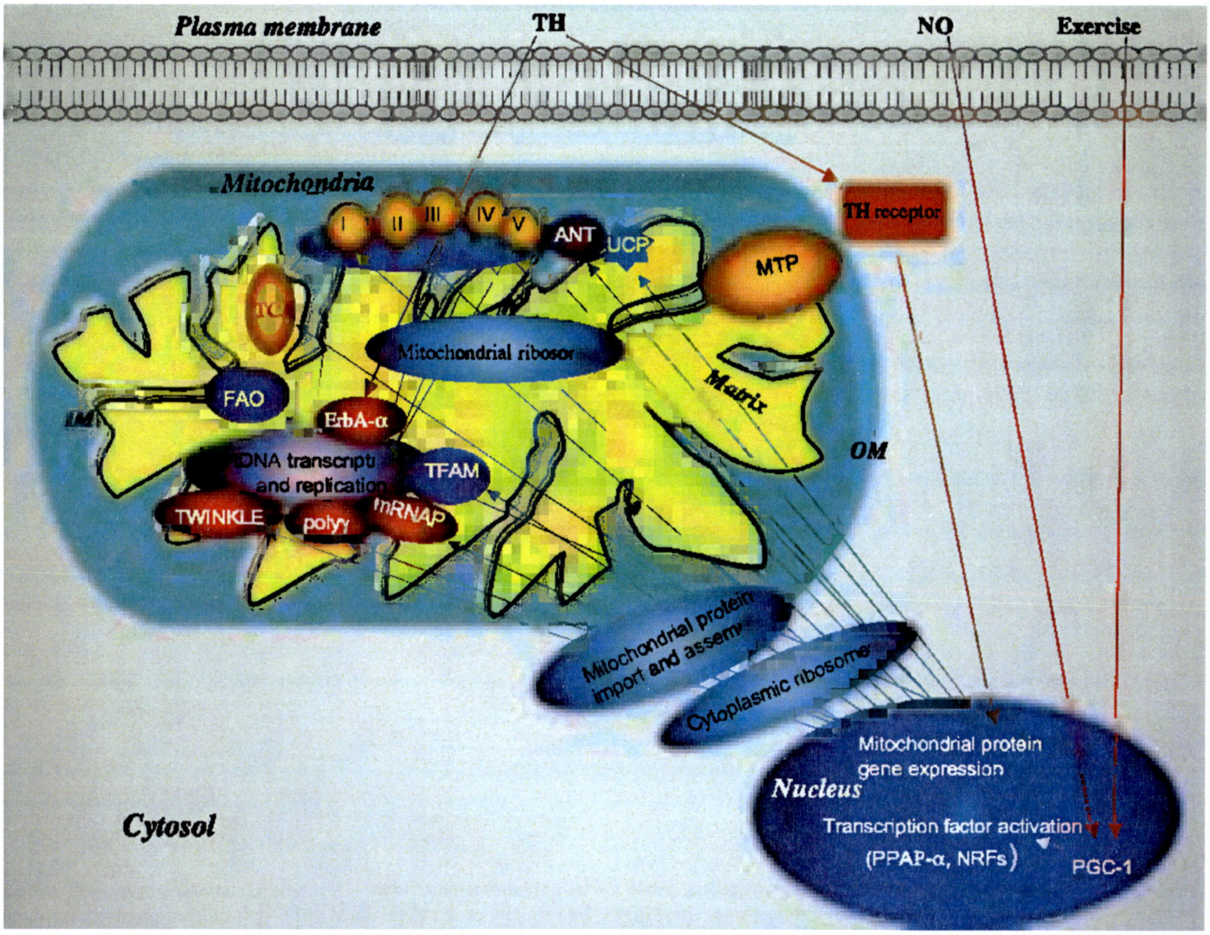


**Fig. 8.1** Biogenesis and bioenergetic pathways of myocardial mitochondria are signal driven. The major mitochondrial bioenergetic pathways including the matrix-localized TCA cycle, the inner membrane (IM), associated fatty acid oxidation (FAO) pathway and the respiratory complexes (I–V) with closely associated adenine nucleotide translocator (ANT), involved in ATP/ADP mitochondria transport, and uncoupling proteins (UCP) affecting proton transport are shown. Thyroid hormone (TH), exercise and nitric oxide (NO)-dependent activation of transcription factors in the nucleus (e.g., PGC1, PPAR, and NRF1), and a subset of factors which activate mtDNA transcription by interaction with D-loop sequences promoting mitochondrial biogenesis are also shown. All but 13 mitochondrial proteins are encoded in the nucleus, translated on cytoplasmic ribosomes, and incorporated through the complex outer membrane (OM) by a mitochondrial import apparatus

anion channel (VDAC)], and thereby may impact sites far from its generation, including activation of transcription factors such as nuclear factor B (NF-κB) among other effects [17]. Similarly, the relatively stable $H_2O_2$ produced by mitochondria can freely diffuse to the cytosol to act as a signaling entity impacting on cytosolic events.

Under normal physiological conditions, the primary source of ROS is the ETC located in the mitochondrial inner membrane, where oxygen can be activated to form superoxide radicals by a nonenzymatic process. Mitochondrial ROS generation can be amplified in cells with abnormal respiratory chain function as well as under physiological and pathological conditions where oxygen consumption is increased.

## Negative Effects of ROS

Increased ROS generation resulting from myocardial I/R, inflammation, impaired antioxidant defenses, and aging may cause profound effects on cells including elevated lipid peroxidation targeting membrane phospholipids and proteins. Protein modifications, such as carbonylation, nitration, and the formation of lipid peroxidation adducts, e.g., 4-hydroxynonenal (HNE), are products of oxidative damage secondary to ROS [18]. ROS-mediated nitration, carbonylation, and HNE adduct formation reduces the enzymatic activity of myocardial respiratory complexes I –V as shown with *in vitro* studies [19]. Superoxide is also particularly damaging to the Fe-S centers of enzymes such as complex I, aconitase,

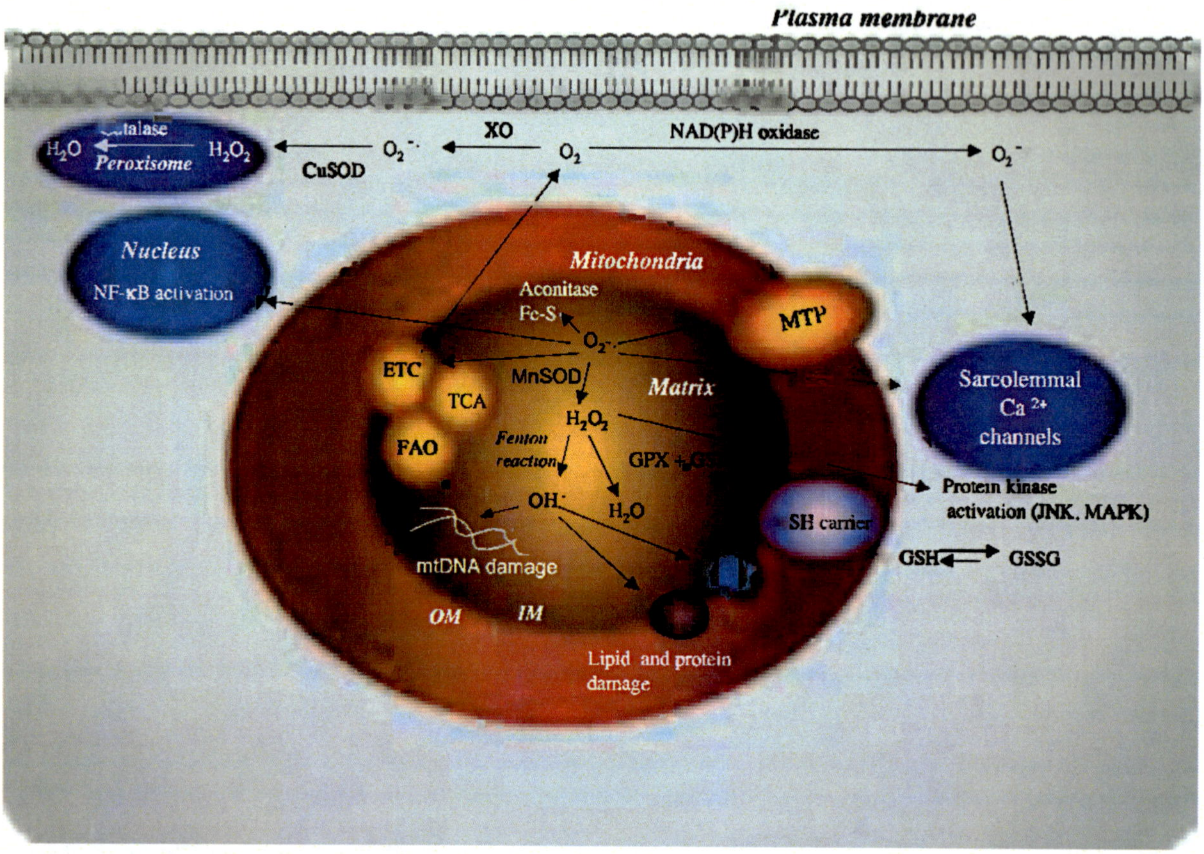


**Fig. 8.2** Mitochondria generate reactive oxygen species (ROS) including superoxide, hydroxyl radicals, and hydrogen peroxide ($H_2O_2$). Superoxide $O_2^{\cdot-}$radical [via complexes I, II, and III of the electron transport chain (ETC)] and cytosolic $O_2^{\cdot-}$ radical generation (by NAD(P)H oxidase or xanthine oxidase (OX)) are shown. Also Mn-SOD (in mitochondria) and Cu-SOD (in cytosol) to form $H_2O_2$ are displayed. The $H_2O_2$ is then either further neutralized in the mitochondria by glutathione peroxidase (GPx) and glutathione, in the peroxisome by catalase, or in the presence of $Fe^{2+}$ via the Fenton reaction, which forms the highly reactive $OH^-$ radical, which can cause severe lipid peroxidation and extensive oxidative damage to proteins and mtDNA. Superoxide radicals produced in mitochondria may impact sites far from its generation, including activation of transcription factor NF-κB

and Succinate dehydrogenase (SDH) causing inhibition of mitochondrial bioenergetic function. Moreover, the inactivation of mitochondrial aconitase by superoxide, which generates Fe (II) and $H_2O_2$, also increases hydroxyl radical formation through the Fenton reaction [20], thereby amplifying the deleterious effects of ROS production. Lipids and in particular the mitochondrial-specific phospholipid cardiolipin serve as a focal target for ROS damage. A large accumulation of superoxide radicals produced *in vitro*, with submitochondrial particles from heart, resulted in extensive cardiolipin peroxidation with a parallel loss of cytochrome *c* oxidase activity [21, 22]. Oxidative damage also targets nucleic acids, and in particular, mtDNA by inducing single- and double-strand breaks, base damage, and modification (including 8-oxoguanosine formation), resulting in the generation of point mutations and deletions in mtDNA. Inhibition of mitochondrial respiration by NO can result in further increases in mitochondrial ROS production; interaction with NO enhances the potency of superoxide as an inhibitor of respiration [23]. In addition, the highly reactive peroxynitrite irreversibly impairs mitochondrial respiration [24], since it inhibits complex I activity, largely by tyrosine nitration of several targeted subunits [25, 26], modifies cytochrome *c* structure and function [27], affects cytochrome *c* oxidase (COX) activity, inhibits mitochondrial aconitase [28], and causes induction of the mitochondrial permeability transition pore (MTP) [29]. A number of peroxynitrite effects on its mitochondrial targets (e.g., the MTP) are potentiated by increased calcium levels [30] and can be clearly distinguished from the effects of NO which often are reversible [24].

Not surprisingly, mitochondria (a major site of intracellular ROS generation) are also a primary *locus* of its damaging effects. ROS-induced damage to mtDNA induces abnormalities in the mtDNA-encoded polypeptides of the respiratory complexes located in the inner membrane, with consequent decrease of electron transfer and further production of ROS, thus establishing a vicious cycle of OS, mitochondrial function, and bioenergetic decline.

It is worth noting that ROS produced from other cellular sources, besides mitochondria, can have substantial effects on cardiovascular function. Superoxide radicals are generated from reactions of oxygen with microsomal cytochrome p450, which has an endogenous NAD(P)H oxidase activity, usually in the presence of metal ions. Phagocytic cells (present at sites of active inflammation), vascular endothelial cells, and smooth muscle cells (SMCs) have a NAD(P)H oxidase activity that can be induced by certain stimuli such as angiotensin II [31], tumor necrosis factor (TNF)-α [32], and thrombin [33] to generate ROS. NAD(P)H oxidase also produces ROS in response to endothelin-1 in vascular SMCs and cardiac muscle cells. As a result, NAD(P)H oxidases may be a key source of ROS that participate in vascular oxidant-related signaling mechanisms under physiological and pathophysiological conditions. In addition, xanthine oxidase (XO), a primarily cytosolic enzyme involved in purine metabolism, is also a source of the superoxide radical. Notably, XO activity and its superoxide generation are markedly increased in the heart after I/R damage. Its location within the human myocardium is primarily in the endothelial cells of capillaries and smaller vessels [34]. Ischemia and hypoxia promote the accumulation of XO substrates, hypoxanthine, and xanthine. Numerous studies have shown that the XO inhibitor allopurinol can provide protection against the cardiac damage resulting from anoxia. Recently, a provocative link was proposed between XO activity and abnormal cardiac energy metabolism in patients with idiopathic dilated cardiomyopathy (DCM), since inhibition of XO with allopurinol significantly improved myocardial function [35]. These toxic metabolic by-products, which are potent cell-damaging oxidants, are normally neutralized by antioxidant enzymes, some of which are mitochondrially located (e.g., Mn-SOD and glutathione peroxidase), while others are cytosolic (e.g., Cu-SOD and catalase).

## Role of ROS in Cell Signaling

In addition to their cell-damaging effects, ROS generation and OS play a critical role in cell regulation and signaling. Oxidative species such as $H_2O_2$ and the superoxide anion can be deployed as potent signals sent from mitochondria to other cellular sites rapidly and reversibly triggering an array of intracellular cascades leading to diverse physiological end points for the cardiomyocyte, some negative (e.g., apoptosis and necrosis) and others positive (e.g., cardioprotection and cell proliferation). Mitochondrial-produced $H_2O_2$ exported to the cytosol is involved in several signal transduction pathways, including the activation of JNK1 and MAPK activities [36, 37], and can impact the regulation of redox-sensitive $K^+$ channels affecting arteriole constriction [38]. The release of $H_2O_2$ from mitochondria and its subsequent cellular effects are increased in cardiomyocytes treated with antimycin and high $Ca^{2+}$, and further enhanced by treatment with CoQ. CoQ plays a dual role in the mitochondrial generation of intracellular redox signaling, by acting both as a prooxidant involved in ROS generation and as an antioxidant [39]. Increased mitochondrial $H_2O_2$ generation and signaling also occur with NO modulation of the ETC [40] as well as with the induction of myocardial mitochondrial NO production, resulting from treatment with enalapril [41]. Furthermore, ROS plays a fundamental role in the cardioprotective signaling pathways of ischemic preconditioning (IPC), in oxygen sensing, and in the induction of stress responses that promote cell survival.

## Mitochondrial $K_{ATP}$ Channel

ATP-sensitive potassium channels of the inner mitochondrial membrane (mito$K_{ATP}$) are blocked by ATP and have been implicated as potential mediators of cardioprotective mechanisms such as IPC [42]. This cardioprotective effect is partially mediated by attenuating $Ca^{2+}$ overloading in the mitochondrial matrix and by increased ROS generation during preconditioning, further leading to protein kinase activation and decreased ROS levels generated during reperfusion [43]. The mito$K_{ATP}$ channel is also regulated by a variety of ligands (e.g., adenosine, opioids, bradykinin, and acetylcholine), which bind sarcolemmal G protein-coupled receptors, with subsequent activation of calcium flux, tyrosine protein kinases, and the phosphatidylinositol 3-kinase (PI3K)-Akt (protein kinase B) pathway [44, 45]. In addition, marked changes in mitochondrial matrix volume associated with mito$K_{ATP}$ channel opening may play a contributory role in the cytoprotection process [45], although this has been challenged [46]. Drugs such as diazoxide and nicorandil specifically activate the mito$K_{ATP}$ channel opening and can also inhibit $H_2O_2$-induced apoptotic progression in cardiomyocytes, suggesting that mito$K_{ATP}$ channels may also play a significant role in mediating OS signals in the mitochondrial apoptotic pathway [47, 48]. Another ion channel (i.e., the calcium-activated $K^+$ channel) has been identified on the mitochondrial inner membrane and has shown to have a cardioprotective function [49]. Nevertheless, the precise temporal order of events in the mitochondrial cardioprotection (CP) cascade and the exact molecular nature of the mito$K_{ATP}$ channel remain to be defined [50]. Further discussion on the relationship of the mito$K_{ATP}$ channel to CP is presented in Chap. 19.

## Mitochondrial Permeability Transition Pore

The opening of another mitochondrial membrane mega-channel, the MTP, located at contact sites between the inner and outer membranes, has been suggested to cause a number of important changes in mitochondrial structure and metabolism, including increased mitochondrial matrix volume (leading to mitochondrial swelling), release of matrix $Ca^{2+}$, altered cristae, cessation of ATP production, primarily due to uncoupling of the ETC, and dissipation of the mitochondrial membrane potential. At the onset of reperfusion following an episode of myocardial ischemia, opening of this nonspecific pore is a critical determinant of myocyte death. Besides its role in the mitochondrial pathway of apoptosis, the opening of the MTP, if unrestrained, leads to the loss of ionic homeostasis and ultimately to necrotic cell death [51]. The MTP appears to be composed of several mitochondrial membrane proteins including the VDAC/porin, peripheral benzodiazepine receptor (PBR), the ANT, cytosolic proteins (e.g., hexokinase II and glycerol kinase), matrix proteins [e.g., cyclophilin D (CypD)], and from proteins of the intermembrane space such as creatine kinase. Fatty acids, high-matrix $Ca^{2+}$ levels, prooxidants, metabolic uncouplers, NO, and excessive mitochondrial ROS production (primarily from respiratory complex I and III) promote the opening of the MTP.

The MTP may be also an important target of CP. New observations have showed that suppressing MTP opening at the onset of reoxygenation can protect human myocardium against lethal hypoxia–reoxygenation injury [52]; the inhibition of MTP opening can be mediated either directly by Cyclosporin A (CsA) and Sanglifehrin A (SfA) or indirectly by decreasing calcium loading and ROS levels.

## Mitochondrial Kinases

Mitochondria contain multiple phosphoprotein substrates for protein kinases, and a number of protein kinases translocate into heart mitochondria. This suggests that protein phosphorylation within the mitochondria is a critical component of the mitochondrial signaling pathways [53]. Protein kinases identified in heart mitochondria include pyruvate dehydrogenase (PDH) kinase, protein kinase A (PKA), protein kinase C (PKC) δ and ε isoforms, and JNK kinase (Table 8.1). Characterization of these proteins has provided new insights into the fundamental mechanisms regulating the mitochondrial response to diverse physiological stimuli and stresses.

In cardiomyocytes, isoforms of PKC (PKCs δ and ε) translocate from the cytoplasm to mitochondria for subsequent signal transduction [57]. PKCε after translocation forms a "signaling module" by complexing with specific MAP kinases (e.g., Erk, p38, and JNK) resulting in phosphorylation of the proapoptotic protein Bad. Also, PKCε forms physical interaction with components of the cardiac MTP, in particular, VDAC and ANT. This interaction may inhibit pathological opening of the pore, including $Ca^{2+}$-induced opening, and subsequent mitochondrial swelling, contributing to PKCε-induced CP [63]. Activation of PKC in CP likely precedes mito$K_{ATP}$ channel opening; nevertheless, a direct interaction of these kinases with the mito$K_{ATP}$ channels has not yet been proved. Following diazoxide treatment, PKCδ is translocated to cardiac mitochondria which triggers mito$K_{ATP}$ channel opening leading to CP [56]; however, other studies have shown that PKCδ does not play a contributory role in the CP provided by IPC, although the role of PKCε translocation has been confirmed.

**Table 8.1** Myocardial mitochondrial-located protein kinases

| Protein kinase | References |
|---|---|
| PDH kinase | [54] |
| PKA | [55] |
| PKCδ | [56] |
| PKCε | [57] |
| PKD | [58] |
| PKG | [59] |
| JNK | [60, 61] |
| p38 MAPK | [57] |
| Stress-activated protein kinase-3 | [62] |
| Erk 1/2 | [57] |
| MAP kinase | [61] |

Although the serine/threonine protein kinase D (PKD) family members compared to other protein kinases have been of limited interest to cardiovascular researchers, the mitochondrial ROS (mROS)-activated PKD regulates a radical-sensing signaling pathway, which relays mROS production to the induction of nuclear genes that mediate cellular detoxification and survival. According to Storz [58], this PKD-regulated signaling pathway is mitochondrial located and mitochondrially regulated antioxidant system that protects the organelle and cell from OS-mediated damage or cell death. The release of mROS activates a signal relay pathway in which PKD activates the NF-κB transcription factor, leading to induction of SOD2. On the other hand, the FOXO3 transcription factor appears to be dispensable for mROS-induced SOD2 induction. PKD-mediated $Mn^{2+}$-superoxide dismutase (Mn-SOD) expression promotes increased survival of cells upon release of mROS, suggesting that mitochondria-to-nucleus signaling is necessary for detoxification mechanisms and cellular viability [64]. Recently, Cowell et al. [65] reported that the formation of mitochondrial diacylglycerol (DAG) and its binding to PKD1 is the means by which PKD1 is localized to the mitochondria in response to ROS. Interestingly, DAG to which PKD1 is recruited in this pathway is formed downstream of phospholipase D1 (PLD1), and a lipase-inactive PLD1 or inhibition of PLD1 by pharmacological inhibitors blocked PKD1 activation under OS. To date it has been viewed that monosaturated and saturated DAG formed via PLD1 have no signaling function. However, their findings suggest a role for PLD1-induced DAG as a competent second messenger at the mitochondria that relays ROS to PKD1-mediated mitochondria-to-nucleus signaling.

Also, a PKD1-dependent mechanism that links OS to decreased CREB protein abundance have been predicted to contribute to the pathogenesis of HF by influencing cardiac growth and apoptosis responses [66]. Identification of this and other intracellular protective signaling pathways may allow the manipulation of mROS and might be critical in targeting HF, aging, and the age-related diseases with mitochondrial dysfunctions.

A mitochondrial cAMP-dependent protein kinase A (mtPKA) as well as its protein substrates has been localized to the matrix side of the inner mitochondrial membrane [55]. In cardiomyocytes, mtPKA phosphorylates the 18 kDa subunit of complex I (NDUFS4), and increased levels of cAMP promote NDUFS4 phosphorylation enhancing both complex I activity and NAD-linked mitochondrial respiration [67]. These posttranslational changes can be reversed by dephosphorylation mediated by a mitochondrial-localized phosphatase. In addition, PKA-dependent phosphorylation of several subunits of COX (COXI, III, and Vb) at serine residues modulates the activity of this important respiratory enzyme [68] and is considered to be a critical element of respiratory control. This cAMP-dependent phosphorylation occurs with high ATP/ADP ratios, resulting in the allosteric inhibition of COX activity. In the resting state, this regulatory control results in reduced membrane potential and more efficient energy transduction. Conversely, increases in mitochondrial phosphatase ($Ca^{2+}$-induced) reverse the allosteric COX inhibition/respiratory control, resulting in increased membrane potential and ROS formation. Similarly, various stress stimuli leading to increased $Ca^{2+}$ flux (activating the phosphatase) result in increased membrane potential and ROS formation.

New techniques of proteomic analysis have led to the identification of mitochondrial phosphoprotein targets for these kinases. Interestingly, a group of proteins constituting a mitochondrial phosphoprotein proteome has been identified using a proteomic approach in bovine heart and is characterized as protein targets of kinase-mediated phosphorylation [69]. The majority of the identified phosphoproteins were involved in mitochondrial bioenergetic pathways, including the TCA cycle (e.g., aconitase, isocitrate, and PDH), and mitochondrial respiratory complexes, including NDUFA 10 (complex I), the flavoprotein subunit of SDH (complex II), core I and III subunits (complex III), α and β subunits of complex V while others are essential elements for the homeostasis of mitochondrial bioenergetics (e.g., creatine kinase and ANT).

## Mitochondrial-related Translocations

An important subject concerning cell signaling and activation includes the stimuli-dependent translocation to and incorporation of specific cytosolic proteins into the cardiomyocyte mitochondria. A growing list of such translocated molecules include several of the proapoptotic proteins (e.g., Bax and Bid), as well as some of the aforementioned protein kinases. Many of these proteins target or interact with specific proteins on the outer mitochondria membrane, and others are imported as preproteins by virtue of recognizing a small set of specific receptors (*t*ranslocases) on the mitochondrial *o*uter *m*embrane (TOM). Mitochondrial protein import is often mediated by heat shock proteins (HSPs) (e.g., HSP60 and HSP70), which specifically interact with a complex mitochondrial protein import apparatus, including matrix proteases. In addition, physiological stimuli and stresses, including temperature changes and hormone treatment (e.g., thyroid hormone), affect the regulation of the heart mitochondrial import apparatus [70, 71].

Nuclear transcription factors have been described and characterized in other tissues/cell types while translocating to the mitochondria, including p53, NF-κB, peroxisome proliferator activated receptor (PPAR)-α, retinoid X receptor (RXR), and TR3, although they have not yet been detected in heart mitochondria. Furthermore, no specific mitochondrial receptors have yet been found in cardiomyocytes or myocardium that bind TNF-α or the various cytokines known to effect cardiac mitochondrial function.

## Mitochondrial Retrograde Signaling

Mitochondrial retrograde signaling is a pathway of communication from mitochondria to the nucleus that influences many cellular activities under both normal and pathophysiological conditions. In both yeast and animal cells, retrograde signaling is linked to mammalian target of rapamycin (mTOR) signaling, but the precise connections in cardiomyocytes have not yet been determined. In mammalian cells, mitochondrial dysfunction sets off signaling cascades through altered $Ca^{2+}$ dynamics including calcineurin activation, which activate several protein kinase pathways (e.g., PKC and MAPK) and transcription factors such as NF-κB, calcineurin-dependent NFAT, CREB, and ATF leading to stress protein expression (e.g., chaperone proteins) and activities [72]. These can result in alterations in both cell morphology and phenotype including proliferative growth, apoptotic signaling, and glucose metabolism.

## Endoplasmic Reticulum

The endoplasmic reticulum (ER) is a multifunctional signaling organelle that contributes to the regulation of cellular processes such as the entry and release of $Ca^{2+}$, sterol biosynthesis, apoptosis, and the release of arachidonic acid [73]. One of its primary functions is as a source of the $Ca^{2+}$ signals that are released through either IP3 or ryanodine receptors (RyRs), which are themselves $Ca^{2+}$-sensitive. Another significant function of the cardiomyocyte ER is to regulate apoptosis by operating in tandem with mitochondria. Antiapoptotic regulators of apoptosis such as Bcl-2 may act by reducing the ebb and flow of $Ca^{2+}$ through ER/mitochondrial cross-talk.

The capability of ER in spreading signals throughout the cell is mediated by a process of $Ca^{2+}$-induced $Ca^{2+}$ release and is particularly important in the control of cardiomyocyte function. The role of ER as an internal reservoir of $Ca^{2+}$ is coordinated with its role in protein synthesis since a constant luminal level of $Ca^{2+}$ is essential for protein folding. In order to achieve this regulation, the ER also contains several stress signaling pathways that can activate transcriptional cascades to regulate the luminal content of the $Ca^{2+}$-dependent chaperones responsible for the folding and packaging of secretory proteins.

## Mitochondria and Apoptosis Pathways

In both animal models and human clinical studies, apoptosis may be causally linked to the myocardial dysfunction stemming from I/R, MI, and HF. Furthermore, cultured cardiomyocytes undergo apoptosis in response to a variety of stimuli including hypoxia (particularly when followed by reoxygenation), acidosis, increased serum deprivation, glucose deprivation and metabolic inhibition, β1-adrenergic agonists, TNF-α, Fas ligand, and doxorubicin. The apoptotic death process is mediated by two central pathways, an extrinsic pathway featuring cell-surface receptors and an intrinsic pathway involving mitochondria and the ER [74]. The two pathways share a number of components and appear to be substantially intertwined; in both pathways, signaling leads to the activation of a family of cysteine proteases called caspases that trigger cell breakdown and death. At this time, we will briefly summarize the pivotal signaling events in both pathways.

Extrinsic signaling is initiated by the binding of a death ligand (usually present as a trimer) to its cognate cell-surface receptor (Fig. 8.3). The death ligand may be an integral membrane protein on the surface of a neighboring cell (e.g., Fas ligand) or a soluble extracellular protein (e.g., TNF-α). Death receptors are single transmembrane spanning proteins with domains containing cysteine-rich repeats and cytoplasmic regions including a death domain sequence of approximately 80 amino acids. Ligand binding to the death receptor initiates the formation of a multiprotein complex termed the death-inducing signaling complex (DISC). Upon binding of the death ligand, conformational change and trimerization of the death receptors result, as does the recruitment of an adaptor protein, e.g., FADD (Fas-associated death domain) through interactions involving death domains in each of the proteins. The formation of this complex signals the recruitment of procaspase-8 into the DISC resulting in procaspase-8 dimerization and activation. Once activated, caspase-8 cleaves and activates downstream procaspase-3 and Bid (a proapoptotic Bcl-2-related protein), which links the extrinsic and intrinsic pathways.

A variety of extracellular and intracellular signals can contribute to the initiation of the intrinsic pathway. Extracellular stimuli include deficiencies in survival/trophic factors/nutrients, radiation, and chemical (e.g., doxorubicin) and physical stresses, whereas intracellular stimuli include OS or ROS, DNA damage, protein misfolding, and changes in intracellular $Ca^{2+}$ which can be directed in part by the ER. This myriad of signals converges on the mitochondria leading to pronounced changes in the membrane organization and dysfunction of this organelle, the release of apoptogenic proteins, and the subsequent activation of caspases.

While in many cases, the precise interaction of the heterogenous and diverse apoptotic signaling with the mitochondria remains not well defined, a common element that has emerged involves the Bcl-2 protein family which has both proapoptotic and antiapoptotic elements. While these are present in both the ER and in the cytoplasm, the translocation and presence in the mitochondrial outer membrane is a key element governing apoptotic progression.

A change in mitochondrial membrane integrity is regulated by the complex and dynamic interactions of different members of the Bcl-2 family, including Bax, Bid, Bcl-2, and Bcl-$X_L$. Proteins of the Bcl-2 family share one or several Bcl-2 homology (BH) regions and behave as either pro- or antiapoptotic proteins. The highly conserved BH domains (BH1–4) are essential for homo- and heterocomplex formation as well as to induce cell death. Proapoptotic homologs can be subdivided into two major subtypes, the multidomain Bax subfamily (e.g., Bax and Bak) which possesses BH1–3 domains and the BH3-only subfamily (e.g., Bad and Bid). Both proapoptotic subtypes promote cell-death signaling by targeting mitochondrial membranes, albeit by different

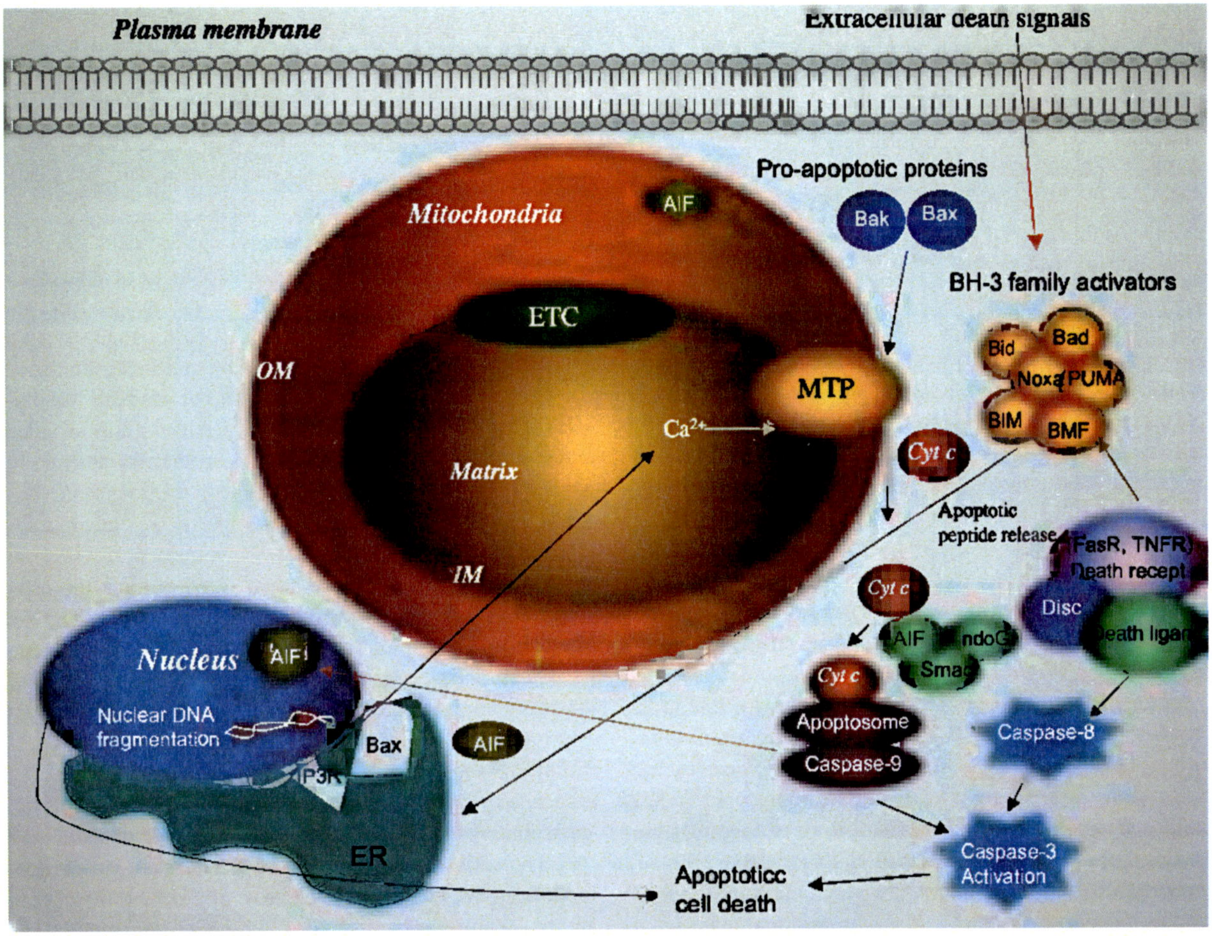


**Fig. 8.3** Cell signaling and apoptosis. An array of extracellular and intracellular signals triggers the intrinsic apoptotic pathway, which is regulated by proapoptotic proteins (e.g., Bax, Bid, and Bak) binding to the outer mitochondrial membrane leading to outer membrane permeabilization and MTP opening. Elevated levels of mitochondrial $Ca^{2+}$ as well as ETC-generated ROS also promote MTP opening. This is followed by the release of cytochrome *c* (Cyt *c*), Smac, EndoG, and AIF from the mitochondria intermembrane space to the cytosol and apoptosome formation (with Cyt *c*) leading to caspase-9 activation, DNA fragmentation (with nuclear translocation of AIF and EndoG), and inhibition of IAP (by Smac), further stimulating activation of caspases-9 and -3. Bax- and Bid-mediated mitochondrial membrane permeabilization and apoptogen release are prevented by antiapoptogenic proteins (e.g., Bcl-2). Intracellular stimuli trigger ER release of $Ca^{2+}$ through both Bax and BH3–protein interactions. Endogenous myocardial factors including apoptosis repressor with caspase recruitment domain (ARC) can target discrete *loci* impacting mitochondrial-based apoptotic progression. Also, depicted are some components (e.g., DISC) of the extrinsic pathway

mechanisms [75]. Proapoptotic membrane-binding proteins (e.g., Bax) upon translocation from the cytosol to mitochondria potentiate cytochrome *c* release, presumably by forming channels in the outer membrane. This is supported by data showing that Bax can form channels and release cytochrome *c* from artificial membranes or liposomes [76]. The BH3-only proteins (e.g., Bid) act by activating the multidomain proapoptotic proteins or by binding and antagonizing the antiapoptotic proteins. Activation of proapoptotic proteins such as Bax to oligomerize, translocate, and bind to the mitochondria represents a critical control point for apoptosis. This process requires extensive conformational changes, in response to a multitude of death signals involving the binding of several factors (e.g., BH3-only proteins) and phosphorylation by several kinases including p38 MAP kinase. Cytoplasmic p53 can directly activate Bax and trigger apoptosis by functioning similarly to the BH3-only proteins [77].

The antiapoptotic proteins Bcl-2 and Bcl-$X_L$ display conservation in all four BH1–4 domains and act to preserve mitochondrial outer membrane integrity by binding and sequestering proapoptotic activating factors (e.g., Bad or Bid), antagonizing the events of channel formation and cytochrome *c* release. Bcl-2 prevents the functional association of Bax with the mitochondria and interferes with the release of apoptogenic peptides (e.g., cytochrome *c* and AIF) from the mitochondria (Fig. 8.3).

An early event in the mitochondrial apoptotic pathway is the release of a group of proteins from the intermembrane space into the cytosol (e.g., cytochrome *c*, Smac/diablo, AIF, and endonuclease G) [78]. These mitochondrial proteins are involved in triggering the subsequent activation of downstream caspases, initiating cell self-digestion (e.g., cytochrome *c* and Smac/diablo), and nuclear DNA fragmentation by endonucleases (e.g., endonuclease G and AIF) leading to apoptotic cell death [79]. Caspases, which are normally inactive enzymes, require specific proteolytic cleavage for their activation. This is achieved by the formation of large cytosolic complexes termed apoptosomes in the cytosol, which incorporate released cytochrome *c*, Apaf-1, and recruited caspases (e.g., caspase-9). The assembly and function of the apoptosome are regulated by Smac/diablo, intracellular $K^+$ levels, and a class of proteins termed IAPs. The release of the mitochondrial intermembrane peptides to the cytosol occurs primarily as a result of the disruption of the mitochondrial outer membrane.

Protein release from the mitochondrial cristae, where the majority of cytochrome *c* is located, may also be associated with the opening of the MTP [80]. However, the role of the transient opening of the MTP in the release of cytochrome *c* is not yet fully understood, since cytochrome *c* release occurs prior to any discernable mitochondrial swelling. Nevertheless, MTP opening is an early requisite feature of apoptosis preceding the activation of caspases. Both Bax and Bcl-2 directly interact with VDAC/porin, both as a component of the MTP and as a major contributor to the mitochondrial outer membrane permeability [81]. The efflux of cytochrome *c* is therefore coordinated with the activation of a mitochondrial remodeling pathway characterized by changes in inner mitochondrial membrane morphology and organization, ensuring the complete release of cytochrome *c* and the onset of mitochondrial dysfunction.

Parenthetically, besides MTP, proapoptotic protein-releasing pores can also be formed by an alternative mechanism, independent of the MTP, and these are referred to as mitochondrial apoptosis-inducing channels (MACs). Unlike the MTP, MACs are composed exclusively of Bax and Bak. Following apoptotic stimulation, Bax and Bak translocate from the sarcoplasm and oligomerize on the outer membrane (OM) of the mitochondria to form MACs independent of the MTP components [82]. In comparison to the MTP, MACs are smaller in size, and only able to mediate the release of small proteins, such as cytochrome *c* [82]. The precise signal transduction pathway that controls either Bax- and Bak-dependent formation of the MACs or rather interaction with the MTP components has not been established. Although, OM pore formation by Bax and Bak is facilitated by proapoptotic BH3 domain-only factors, including cardiac-expressed BNip3 and its homolog Nix. Recently, Dorn [83] reported that Nix not only stimulates cardiomyocyte apoptosis, but also induces mitochondrial autophagy (mitophagy) and indirectly activates the MTP, causing cell necrosis. It appears that Nix and BNip3 have a critical function in the cardiomyocyte, "mitochondrial pruning," that controls mitochondrial proliferation and without which an age-dependent mitochondrial cardiomyopathy develops.

Similar to apoptosis occurring in cardiac cells, there is evidence that apoptosis plays a critical role in skeletal muscle degeneration albeit apoptosis in skeletal muscle has specific and distinctive characteristics. According to Hood et al. [84, 85], since skeletal muscle is multinucleated, the decay of one myonucleus by apoptosis will not produce "wholesale" muscle cell death, but it does result in a loss of gene expression within the local myonuclear domain, potentially leading to cellular atrophy. Nevertheless, the precise mechanisms involved in apoptosis, particularly skeletal muscle-related apoptosis and the actual involvement in skeletal muscle nuclei loss, are not known. To determine the relevance of apoptosis to skeletal muscle homeostasis and the possible role of inhibitor of differentiation-2 (Id2), a basic helix-loop-helix protein that acts as a negative regulator of the myogenic regulatory transcription factor family of Id2, Always et al. [86] assessed Id2 during skeletal muscle hypertrophy and subsequent atrophy, and whether this protein was associated with any alterations in skeletal muscle apoptosis. Their findings in young adult quails suggested that Id2 may play a potential role in apoptosis-induced loss of muscle during unloading. Increases in Id2 were of a similar magnitude and time course as the increases in caspase and PARP apoptotic markers. Although these results do not prove a causative role for Id2 in apoptosis in skeletal muscle following unweighting, they show to play at least a partial role in muscle damage and disease.

Skeletal muscle, as well as cardiac muscle, contains two morphologically and biochemically distinct subfractions of mitochondria, subsarcolemmal (SS) and intermyofibrillar (IMF), that exist in different regions of the fiber and this could produce regional differences in the sensitivity to apoptotic stimuli within the cell [87]. In addition, skeletal muscle being a malleable tissue is capable of changing its mitochondrial content and/or composition in response to chronic alterations in muscle use or disuse. This variability in mitochondrial content and/or composition can undoubtedly influence the degree of organelle-directed apoptotic signaling in skeletal and heart muscles. Interestingly, chronic contractile activity seems to induce predominantly antiapoptotic adaptations in both mitochondrial subfractions suggesting that chronic contractile activity can exert a protective effect on mitochondrially mediated apoptosis in skeletal muscle likely by attenuation of both cytochrome *c* and AIF release, despite the presence of higher levels of these proteins within the mitochondrial subfractions [88].

## Mitochondrial Signaling Defects and Cardiomyopathies

A number of clinical studies have shown that pathogenic point mutations and large-scale deletions in cardiac mtDNA have severe consequences for the heart. Specific mtDNA mutations with associated mitochondrial respiratory dysfunction have been reported in isolated cases of cardiomyopathies as well as in systemic encephalomyopathies with cardiac involvement including Leigh disease, MELAS (mitochondrial myopathy, encephalopathy, lactic acidosis, and stroke like episodes), and MERRF (myoclonic epilepsy and ragged-red fibers) [89]. In addition, depletion of cardiac mtDNA levels concomitant with myocardial mitochondrial respiratory dysfunction has been found in patients with both dilated and hypertropic cardiomyopathy (HCM) and in patients and animal models treated with zidovudine [90, 91]. While the dependence of cardiac homeostasis on functional mitochondria is primarily attributed to needed ATP derived from OXPHOS for maintaining myocardial contractility, the role of cardiac mitochondria in responding to a variety of intracellular and extracellular signals, metabolic substrates, and physiological stresses has recently become of increasing interest. Examination of cardiomyocytes containing specific pathogenic mtDNA point mutations and deletions or depleted mtDNA levels for their effect(s) on mitochondrial signaling may provide further information on the role of mitochondrial cytopathy in cardiac disease pathogenesis.

Mutations at several different nuclear gene *loci* have also been reported in conjunction with mitochondrial OXPHOS deficiencies in association with cardiomyopathies associated with Leigh syndrome, cytochrome *c* oxidase deficiency, and Friedreich ataxia [92–94]. In addition, mutations in nuclear genes (involved in mitochondrial biogenesis) that contribute to the observed cardiac disease-associated mitochondrial enzyme and mtDNA defects (including large-scale mtDNA deletions and mtDNA depletion) have been reported [95]. The involvement of both genomes in mitochondrial biogenesis and mitochondria-based pathogenesis serves as an important rationale for examining the cross-talk and regulatory signaling between both genomes, as well as to expand the search for mutations involved in mitochondrial biogenesis and in the regulation of cardiac mitochondrial function. Mutation and physiological insults targeting various mitochondrial pathways (other than mitochondrial OXPHOS function) are contributory to cardiac disease. Defects in mitochondrial carnitine, fatty acid transport, and FAO have a crucial role in cardiac sudden death, bioenergetic dysfunction, cardiac dysrhythmias, and cardiomyopathy [89]. In transgenic mice, the disruption of specific nuclear genes encoding mitochondrial proteins (engaged in a broad array of functions) leads to cardiomyopathy and HF. Ablation of genes (i.e., knockout mutations) encoding the ANT, Mn-SOD, the frataxin gene associated with Friedreich ataxia, and the mitochondrial transcription factor A (TFAM) leads to phenotypic cardiomyopathy and cardiac failure [96, 97]. These findings underscore the importance of mitochondria in the overall maintenance of a functional cardiac phenotype. In particular, the mitochondrial role in myocardial apoptosis (i.e., ANT is an essential component of the MTP, which mediates early apoptotic progression) and mitochondrial antioxidant response (e.g., Mn-SOD) to OS, a critical element of myocardial ischemia and mitochondrial cytoprotection, need to be emphasized.

## Mitochondrial Signaling in Myocardial Ischemia and Cardioprotection

When the supply of oxygen is disrupted such as occurs with myocardial ischemia, mitochondrial ETC flux, and OXPHOS decline, the pool of high-energy phosphates is rapidly depleted, pyruvate oxidation decreases, and ATP production is impaired. The hydrolysis of glycolytically derived ATP and resulting accumulation of lactate and pyruvate lead to intracellular acidosis (which has a direct effect on cardiac contractile function) and to the accumulation of myocardial sodium and calcium. Moreover, the energy deficit occurring as a result of ATP depletion is further compounded by the deployment of ATP to reestablish the disturbed myocardial ionic balance rather than to fuel contraction.

AMP and other intermediates also accumulate with subsequent mitochondrial swelling and degeneration. In addition, activity levels of respiratory complexes IV and V decrease in myocardial ischemia and lead to increased levels of mtDNA deletions [98, 99]. While ultimately, sustained myocardial ischemia leads to ATP depletion and necrotic cell death, there is ample evidence that both ischemia and hypoxia can activate cardiomyocyte mitochondrial death pathway with opening of the MTP concomitant to mitochondrial membrane depolarization, eventual disruption of the mitochondrial membranes, and the release of cytochrome *c* [78].

Increased mitochondria function can also exacerbate ischemic damage, especially at the onset of reperfusion where fatty acid influx increases and unbalanced FAO occurs [100]. Excess acetyl-coenzyme A (CoA) is produced saturating the TCA cycle at the expense of glucose and pyruvate oxidation. Increased OXPHOS elevates mitochondrial ROS production and myocardial lipid peroxidation resulting in cardiolipin depletion with severe effect on both complex I and IV activities [101]. These enzyme activities can be restored to normal levels by adding exogenous cardiolipin or by induction of antioxidants, Mn-SOD, and catalase. IPC has been demonstrated when brief periods of myocardial ischemia are applied prior to a more prolonged ischemic insult. Cardioprotection is partially mediated by $Ca^{2+}$ overloading in the mitochondrial matrix, and increased mitochondrial ROS generation leading to further protein kinase activation. It can be initiated by the binding of variety of ligands (e.g., adenosine, opioids, bradykinin, and acetylcholine) to sarcolemmal G protein-coupled receptors with subsequent activation of calcium flux, tyrosine protein kinases, and the PI3K-Akt pathway (Fig. 8.4). While considerable attention has focused on the opening of the cardiomyocyte mito$K_{ATP}$ channel [42, 46, 102, 103] as the primary regulatory event in mitochondrial cardioprotective signaling, this view has recently been challenged [46]. In addition, marked changes in mitochondrial matrix volume associated with mito$K_{ATP}$ channel opening may also play a contributory role in cytoprotection [104, 105]. Drugs such as diazoxide and nicorandil specifically activate mito$K_{ATP}$ channel opening and can also inhibit $H_2O_2$-induced apoptotic progression in cardiomyocytes, suggesting that mito$K_{ATP}$ channels also play a significant mediative role in OS signals in mitochondrial apoptosis [47]. The precise temporal order of events in the mitochondrial CP cascade and the exact molecular nature of the mito$K_{ATP}$ channel remain to be defined [102, 103].

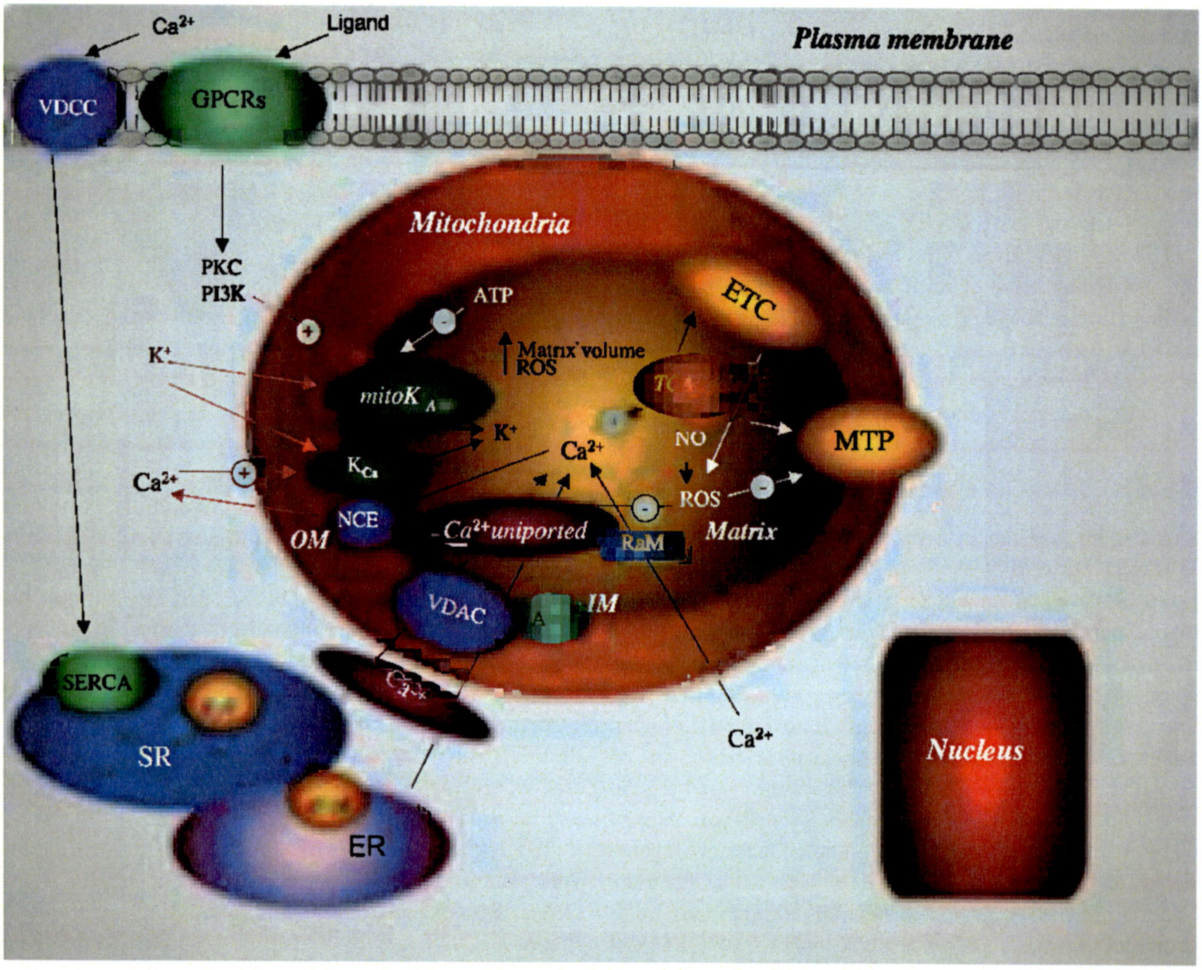


**Fig. 8.4** Myocardial mitochondria ion channels are critical signal transducers. The cardioprotection pathway involving the mitochondrial $K^+$ channels (mito$K_{ATP}$) is located in the inner membrane (IM) and is mediated by ROS, ligand binding to membrane GPCRs, and protein kinases (PKC, PI3K). Calcium enters the mitochondria in response to stimuli via a complex of membrane proteins, including the outer membrane (OM), voltage-dependent anion carrier (VDAC), the IM proteins, RAM, and the $Ca^{2+}$ uniporter; high $Ca^{2+}$ levels can be supplied from ryanodine receptors (RyR) and inositol 1,4,5-trisphosphate (IP3R) receptors located in the sarcoplasmic (SR) and endoplasmic (ER) reticulum, respectively, and is coupled to the entry of calcium into the cardiomyocyte through the voltage-dependent $Ca^{2+}$ channel (VDCC). Entry of high calcium levels into the mitochondria can increase the activities of several enzymes of TCA cycle, electron transport chain (ETC), and complex V. Calcium also modulates the opening of the mitochondrial permeability transition pore (MTP). $Ca^{2+}$ efflux is primarily managed by the $Na^+/Ca^{2+}$ exchanger (NCE)

Cardioprotection associated with mitochondrial signaling has also been demonstrated with brief periods of hypothermia prior to a prolonged ischemic insult [106]. The preservation of myocardial function and ATP levels are accompanied by increased expression of stress proteins (e.g., HSP70) and constitutive mitochondrial proteins (e.g., ANT and ATP synthase β-subunit).

## Mitochondrial Signaling and Myocardial Hypertrophy

The concept that cardiac cells have the ability to grow in number is controversial. The consensus view remains that cardiomyocytes are postmitotic with limited proliferative capacity and that their regulation may be disturbed in hearts undergoing severe remodeling (such as in late-stage HF) [107]. An increase in cardiac cell size and not in cell number is more commonly accepted [108]. Stimuli that provoke myocardial hypertrophy include increases in mechanical and hemodynamic loads (volume/pressure overload/mechanical stretch), inflammatory cytokines, peptide growth factors, neuroendocrine factors (e.g., norepinephrine and angiotensin), and OS.

Molecular features of the hypertrophic response include: changes in the gene expression program (fetal gene transcription) with a large array of nuclear transcription factors and activators identified, multiple signaling cascade pathways featuring an array of protein kinases [109] with resulting effects on cellular protein synthesis (including ribosomes), activated membrane ATPase pumps and calcium handlers (e.g., sarco/endoplasmic reticulum $Ca^{2+}$ ATPase), and induced protein synthesis of the sarcomeric contractile apparatus (e.g., specific myosin isoforms). Cardiac hypertrophy resulting from primarily physiological stimuli generally does not lead to HF, in contrast to cardiac hypertrophy resulting from pathological stimuli (decompensation) which often does [110]. The former tends to involve concentric hypertrophy manifested largely by cardiac myocyte thickening, while eccentric hypertrophy is characterized by cell elongation. Recent evidence has demonstrated that these growth responses are mediated by different signal transduction pathways. In addition, cardiac cells undergoing hypertrophy from pathological stimuli display both an increased sensitivity to apoptotic stimuli and an expression pattern favoring proapoptotic regulation of Fas, Bcl-2 protein family, and caspases [111].

In cardiac hypertrophy, the effects on heart mitochondria are manifold. There is a downregulation of mitochondrial pathways involving FAO/fatty acid transport system occurring as part of a shift in cardiac bioenergetic substrate utilization from fatty acid to glucose (glycolytic pathways) [112]. For instance, levels of medium-chain acyl-CoA dehydrogenase, a key enzyme in FAO, have been shown to be both transcriptionally and translationally downregulated in the rat pressure-overload model and during hypertrophic HF, in part mediated by modulation of levels of the global transcription factor PPAR [113].

Cardiac hypertrophy (both due to physiological and pathological stimuli) is accompanied by increased mitochondrial number resulting from increased mitochondrial biogenesis and protein synthesis [114]. Stimuli ranging from electrical stimulation and exercise to thyroid hormone treatment elicit cardiac hypertrophy with increased mitochondrial biogenesis and function. In addition, increased mitochondrial number has been reported in both experimental animal and transgenic models of HCM and in clinical cases of mitochondria-based cardiac disease [89, 115, 116]. Aberrant mitochondrial accumulation, abnormal mitochondrial function, and myocardial hypertrophy have been widely recognized in patients with an HCM phenotype presenting either as isolated cardiomyopathy or in systemic neuropathologies such as MELAS, MERFF, and Leigh disease. The increase in mitochondria is considered to be a compensatory response to mitochondrial bioenergetic dysfunction, observable in animal models and patients, and is often discernable by increased ragged-red fiber staining in cardiomyocytes [115]. In addition, the precise elements involved in signaling the events leading to mitochondrial biogenesis during cardiac hypertrophy and HCM remain to be identified. Cardiac hypertrophy is also associated with shifts in mitochondrial metabolism elicited by signaling proteins (e.g., PI3K-Akt-mTOR pathway), which coordinate hypertropic growth responses to a variety of cellular physiological stimuli (e.g., glucose and serum deprivation). The critical role played by bioenergetic substrates/products (e.g., fatty acids, ATP, pyruvate, and phosphocreatine) in myocardial hypertrophy, and the commonality of many of the signaling elements in the hypertrophic and apoptotic pathways, further support a pivotal mitochondrial role in committing the myocardial cell to growth/hypertrophy or to cell death (by apoptosis or necrosis) [109, 112]. While structural gene mutations in sarcomeric contractile proteins such as myosin, myosin binding protein, cardiac troponin T and I, and tropomyosin have been found in several cases of familial HCM, cardiac energy depletion rather than depressed sarcomeric contraction may be the underlying cause [117]. This is supported by the fact that mutations in different sarcomeric proteins can lead to inefficient ATP utilization. Pronounced mitochondrial respiratory enzymatic dysfunction has been demonstrated in patients containing mutations in myosin structural genes known to cause HCM, in transgenic mice with missense cardiac troponin T alleles, and more recently in a transgenic mouse model of HCM with mutant myosin heavy chain alleles [116, 118, 119] Similarly, transgenic mice containing cardiac-specific overexpression of calcineurin exhibited severe cardiac hypertrophy (that progresses to HF), marked mitochondrial respiratory dysfunction, and superoxide generation [120]. In addition, mutations in the regulatory subunit of AMP-activated protein kinase (AMPK), a key sensor and mediator in cellular energy metabolism, have been found in a subset of cases of HCM; these AMPK mutations aberrantly signal cardiac energy depletion. In addition, a number of mtDNA point mutations have been identified that lead to an HCM phenotype [116]. Assuming that mitochondria energy deficiency may act as an initiating signal for myocardial hypertrophy is useful as a unifying framework and would allow understanding of a number of clinical features of HCM such as heterogeneity and variable onset.

## Key Players in Mitochondrial Signaling

### *Nuclear Gene Activation*

Nuclear transcriptional modulators have been identified that govern the expression of a wide array of mitochondrial proteins in response to diverse cellular stimuli and signals. For instance, nuclear transcription factors such as nuclear respiratory factors (NRFs) 1 and 2 are implicated in activation of mitochondrial biogenesis [10, 121, 122]. These factors exert a direct effect on the synthesis of specific nuclear-encoded subunits of the mitochondrial respiratory enzymes as well as indirectly by upregulating levels of TFAM, involved in both mtDNA replication and transcription (see Fig. 8.1). In addition, a master transcription coactivator (PGC-1) activates expression of transcription factors NRF1 and NRF2 [123].

The nucleus also contains global regulatory transcription factors such as PPARs and their transcriptional coactivators (i.e., PPAR, RXR, which also play a pivotal regulatory role in the expression of mitochondrial FAO pathways, integral to bioenergetic metabolism) [124–126]. Transcriptional control by these activators is affected by hypoxia, ischemia, and HF [127, 128]. Interestingly, PGC-1 expression and mitochondrial biogenesis are modulated by the activation of a calcium/calmodulin-dependent protein kinase, indicating that the calcium-regulated signaling pathway plays a significant role in transcriptional activation of genes governing mitochondrial biogenesis [129]. Moreover, these nuclear activators are modulated during cardiac development [130, 131]; activation of the nuclear factor activated T cell gene has been shown to be crucial in early cardiac development and is required to maintain myocardial mitochondrial oxidative function. Targeted cardiac gene disruption of nuclear factor activated T cell genes results in cardiomyocyte mitochondria ETC dysfunction, reduced ventricular size, and aberrant cardiomyocyte structure in mice embryo [131]. These nuclear transcriptional factors modulate the cardiac phenotype in transgenic animals bearing either mutations in specific transcriptional factors or overexpressed genes [123–131]. Also, it has been found that specific nuclear transcription factors are essential for normal cardiac phenotype and mitochondrial function. For instance, cardiac-specific PGC-1 overexpression in transgenic mice results in uncontrolled mitochondrial proliferation and extensive loss of sarcomeric structure leading to DCM [123]. Myocardial overexpression of PPAR may lead to severe cardiomyopathy with both increased myocardial fatty acid uptake and mitochondrial FAO [132]. Similarly, mutations targeting TFAM produce inactivation of myocardial mitochondrial gene expression and ETC dysfunction resulting in DCM and atrioventricular conduction defects [97].

### *Protein Kinases*

Evidence that mitochondria contain multiple phosphoprotein substrates for protein kinases and that a number of protein kinases are translocated into heart mitochondria strongly suggests that protein phosphorylation within the mitochondria is a critical component of mitochondrial signaling pathways [53]. However, it is also important to note that detection of a protein in a phosphorylated state does not mean that such phosphorylation plays a regulatory role. Many proteins can be phosphorylated *in vitro* by protein kinases yet show no changes in activity. Thus far, protein kinases identified in heart mitochondria include PDH kinase, branched-chain keto acid dehydrogenase kinase, PKA, PKC isoforms, and c-Jun N-terminal kinase [55–57, 60, 133, 134] Their characterization has offered new insights into the fundamental mechanisms regulating mitochondrial responses to diverse physiological stimuli and stresses.

In cardiomyocytes, PKC after translocation to the mitochondria forms a signaling module by complexing with specific mitogen-activated protein kinases (e.g., extracellular signal regulated kinase, p38, and c-Jun N-terminal kinase) resulting in phosphorylation of the proapoptotic protein Bad [57]. Also, PKC forms physical interaction with components of the cardiac MTP (in particular, VDAC and ANT) [63]. This interaction may inhibit the pathological opening of the pore (including $Ca^{2+}$-induced opening and subsequent mitochondrial swelling) contributing to PKC-induced CP. While PKC activation in CP likely precedes mito$K_{ATP}$ channel opening, its direct interaction with the mito$K_{ATP}$ channels has not been demonstrated. Following treatment with diazoxide, PKC is also translocated to cardiac mitochondria [56]; however, several studies have shown that PKC does not play a contributory role in CP provided by IPC [135, 136].

A mtPKA as well as its protein substrates has been localized to the matrix side of the inner mitochondrial membrane [55]. In cardiomyocytes, the NDUFS4 is phosphorylated by mtPKA, and increased levels of cAMP promote NDUFS4 phosphorylation, enhancing both complex I activity and NAD-linked mitochondrial respiration [55, 67, 137]. These posttranslational changes can be reversed by dephosphorylation via a mitochondria-localized phosphatase. Phosphorylation of several subunits of COX, including COXI, III, and Vb, occurs at serine residues by mtPKA, modulates COX activity [68] and has been considered to be a critical element of respiratory control. This cAMP-dependent phosphorylation occurs with high ATP/ADP ratios and results in allosteric inhibition of COX activity; at the same time this regulatory control can result in reduced membrane potential and more efficient energy transduction in the resting state. Conversely, increase in mitochondrial phosphatase ($Ca^{2+}$-induced) reverses allosteric COX inhibition/respiratory control, resulting in increased membrane potential and ROS formation; interestingly, various stress stimuli leading to increased $Ca^{2+}$ flux result in increased potential and mitochondrial ROS formation. With the development of new technologies, including kinase inhibitor assays and proteomic analysis, a variety of mitochondrial phosphoprotein targets for these kinases have been reported. A set of proteins has been identified in the mitochondrial phosphoprotein proteome of bovine heart as protein targets of kinase-mediated phosphorylation [69]. The majority of identified phosphoproteins were involved in mitochondrial bioenergetic function either in the TCA cycle (e.g., aconitase, isocitrate, and PDH), as respiratory complex subunits (e.g., NDUFA 10 of complex I, succinate dehydrogenase flavoprotein subunit of complex II, core I and core III subunits of complex III, and subunits of complex V) or as essential players in the homeostasis of mitochondrial bioenergetics (e.g., creatine kinase, ANT). In addition, during myocardial ischemia, phosphorylation of the elongation factor Tu, a key regulatory protein of the cardiac mitochondrial protein translation apparatus, is modulated [138].

## Calcium Signaling

Since a comprehensive discussion on calcium signaling is presented in Chap. 5, here it is suffice to say that the import of $Ca^{2+}$ from cytosol into cardiac mitochondria is an important regulatory event in cell signaling. The organelle couples cellular metabolic state with $Ca^{2+}$ transport processes; therefore, it controls not only its own intraorganelle $Ca^{2+}$, but it also influences the entire cellular network of cellular $Ca^{2+}$ signaling, including the endoplasmic reticulum, the plasma membrane, and the nucleus [139]. These organelle interactions are critical in $Ca^{2+}$ homeostasis and signaling of the cardiomyocyte and thus the numerous cellular functions regulated by the cation [140, 141]. Bidirectional signaling mediated by $Ca^{2+}$ provides a framework by which mitochondrial biogenesis, structure, and metabolic activity will dictate a number of adaptive mechanisms during cell proliferation and cellular stress [142].

Mitochondrial calcium flux, particularly in cultured cardiomyocytes, has become detectable using advanced cell imaging techniques with fluorescent dyes, confocal microscopy, and recombinantly derived $Ca^{2+}$-sensitive photoprobes [143, 144]. Mitochondrial calcium influx is primarily provided by a $Ca^{2+}$ pump uniporter (see Fig. 8.4) located in the inner membrane and driven by the mitochondrial membrane potential as well as by low-matrix $Ca^{2+}$ levels, and can be blocked by ruthenium red [145].

Mitochondrial $Ca^{2+}$ uptake is significantly and rapidly elevated in cardiomyocytes during physiological $Ca^{2+}$ signaling and is often accompanied by a highly localized transient mitochondrial depolarization [144]. Efflux of $Ca^{2+}$ from cardiomyocyte mitochondria is mediated by an $Na^{+}/Ca^{2+}$ exchanger (NCE) linked to ETC proton pumping, although calcium efflux also occurs with MTP opening. Activation of the MTP and mitochondrial $Ca^{2+}$ flux also occur in early myocardial apoptosis and I/R and are involved in the generation of a calcium wave delivering system between adjacent mitochondria [146]. A major consequence of increased mitochondrial $Ca^{2+}$ uptake is the upregulation of energy metabolism and stimulation of mitochondrial OXPHOS. Elevated mitochondrial $Ca^{2+}$ levels allosterically stimulate the activity of three TCA cycle enzymes including pyruvate, isocitrate, and 2-oxoglutarate dehydrogenases [147, 148]. Activation of these enzymes by $Ca^{2+}$ results in increased $NADH/NAD^{+}$ ratios and ultimately leads to increased mitochondrial ATP synthesis. A thermokinetic model of cardiac bioenergetics showed calcium-dependent activation of the dehydrogenases as the rate-limiting determinant of respiratory flux regulating myocardial oxygen consumption, proton efflux, and NADH and ATP synthesis [149]. In cardiomyocytes, mitochondrial ATP synthase activity can be directly modulated by increased mitochondrial $Ca^{2+}$ levels [150, 151].

Intracompartment $Ca^{2+}$ signaling is recognized as a key mode of signal transduction and amplification in mitochondria [143, 144]. Using IP3 as second messenger, a variety of cell-surface hormones and neurotransmitters signal the release of $Ca^{2+}$ from endoplasmic reticulum (ER) and Golgi apparatus into the cytosol. The proximity of mitochondria to ER membranes appears to be a significant factor for ER $Ca^{2+}$ release and mitochondrial $Ca^{2+}$ uptake [152]. This dramatic increase in mitochondrial $Ca^{2+}$ is rapidly mobilized from the ER-IP3 receptor (Fig. 8.4) when in close contact to mitochondria, albeit the precise molecular mechanism of this transfer has not been fully established. Similarly, the sarcoplasmic reticulum ryanodine receptors are also located near the cardiomyocyte mitochondria undergoing calcium release [153]. Proposed mechanisms for the rapid calcium mitochondrial import include the involvement of diffusible cytosolic factors that stimulate the $Ca^{2+}$ uniporter, activation of rapid mode of uptake (RaM), and enhanced uptake by mitochondrial analogues of ryanodine receptors residing in the inner membrane [154–157]. A voltage-dependent anion carrier has also been identified as a component in $Ca^{2+}$ transport from ER through the outer mitochondrial membrane [158].

## Mitochondrial Receptors

Few well-characterized heart mitochondrial receptors have been detected despite the large number of receptors that have been identified in other tissues/cell types. The TH receptor ErbA-α, which was identified as an orphan receptor [159, 160] involved in interaction with mtDNA during targeted hormonal stimulation, has not yet been documented in cardiac tissue, despite the known marked effect of TH in heart mitochondria. Also, a large number of nuclear transcription factors characterized in many other tissues/cell types as translocating to mitochondria, including p53, NF-κB, PPAR, RXR, and TH receptor 3, have also not yet been documented in heart mitochondria. No specific mitochondrial receptors have yet been found that bind TNF or various cytokines known to effect cardiac mitochondrial function despite several recent studies demonstrating that TNF impacts cardiomyocyte mitochondria. Moreover, in contrast to other tissues, there has been no characterization in heart mitochondria of anchoring proteins, which bind and concentrate protein kinases.

A common theme concerning signaling and activation includes the stimuli-generated translocation of specific cytosolic proteins into the mitochondria; the growing list of such translocated entities includes many of the proapoptotic proteins (e.g., Bax and Bid) as well as protein kinases. Many of these appear to target specific proteins on the outer mitochondria membrane; others are imported as preproteins by recognizing a small set of specific receptors, TOMs. The import of proteins into mitochondria is often mediated by HSPs (e.g., HSP60 and HSP70) which specifically interact with a complex mitochondrial protein import apparatus (including matrix proteases). Interestingly, a number of physiological stimuli and stresses, including temperature changes and hormone treatment (including TH), can result in regulation of the heart mitochondrial import apparatus [71, 161].

## Signals of Survival and Stress Impact Heart Mitochondria

The list of extracellular influences and intracellularly generated signals which impact the mitochondrial organelle is growing, as reflected in Table 8.2. In addition to hormonal and cytokines stimuli (e.g., TH, TNF-α, interleukins), there are also pro/antiapoptotic modulators, nutrients, serum, growth and mitotic factors, as well as stress and metabolic stimuli which we describe in more detail in this section.

**Table 8.2** Stimuli signaling myocardial mitochondrial function

| Stimuli | Signaling pathway | Cardiac myocyte phenotype | Mitochondrial effect | References |
|---|---|---|---|---|
| IL-1β | NO production | Cardiac dysfunction | Decreased respiration | [162] |
| TNF-α | Ceramide pathway | Cell death | Reduced activity levels of PDH, complexes I and II | [163] |
| Heat stress | Increased levels of HSP 32,60,72 | Improved cardiac function after I/R | Increased complex I–V activities | [164] |
| Low glucose | Myocardial apoptosis | Cell death | Cyt *c* release | [165] |
| Low serum | Myocardial apoptosis | Cell death | Cyt *c* release | [165] |
| Palmitate | Myocardial apoptosis; ceramide increase | Cell death | Reduced complex III and membrane potential; increased cyt *c* release, UCP and swollen mitochondria | [166] |
| Ceramide | Ceramide pathway | Cell death | Decreased complex III activity | [167] |
| Electrical stimulation | NRF-1 activation | Hypertrophy | Mitochondrial proliferation | [47] |
| Nitric oxide | Peroxynitrite formation | Myocardial $O_2$ uptake decline; increased $H_2O_2$ | Complexes I and IV decrease; increased cyt *c* release | [26, 168] |
| Thyroid hormone (T3/T4) | Receptor-mediated nuclear and mtDNA gene activation | Hypertrophy | Mitochondrial proliferation; increased UCP and uncoupled OXPHOS | [169, 170] |

Cyt *c* cytochrome *c*, *HSP* heat shock protein, *I/R* ischemia/reperfusion, *IL* interleukin, *NO* nitric oxide, *NRF-1* nuclear respiratory factor-1, *OXPHOS* oxidative phosphorylation, *PDH* pyruvate dyhydrogenase, *T3* triiodothyronine, *T4* thyroxine, *TNF-α* tumor necrosis factor-α, *UCP* uncoupling protein

## Survival Signals/Apoptosis

The PI3K-Akt pathway promotes cell survival primarily by intervening in the mitochondrial apoptosis cascade at events before cytochrome *c* release and caspase activation occur (Fig. 8.5). Akt activation inhibits changes in the inner mitochondrial membrane potential that occur in apoptosis (suppressing apoptotic progression and cytochrome *c* release induced by several proapoptotic proteins). While Akt also contributes to the phosphorylation and inactivation of the proapoptogenic protein Bad, it remains unclear whether Bad phosphorylation is the mechanism by which Akt ensures cell survival and mitochondrial integrity since other mitochondrial targets of Akt remain to be identified [171]. With regard to the heart, PI3K-Akt signaling promotes glucose uptake and growth and survival of cardiomyocytes and has been directly implicated in heart growth [172]. Growth factors known to effect cardiomyocyte growth [e.g., type 1 insulin-like growth factor (IGF-1) receptor] signal through the PI3K-Akt pathway [173]. Recently, microarray analysis of cardiomyocytes demonstrated that treatment with IGF-1 results in the differential expression of genes involved in cellular signaling and mitochondrial function and confirmed that this IGF-1-mediated gene regulation requires extracellular signal regulated kinase and PI3K activation [174]. In transgenic mice with cardiac-specific expression of activated Akt, IGF-binding protein is upregulated (consistent with its growth signaling/antiapoptotic role) and both PGC-1 and PPAR-α (activating mitochondrial FAO and mitochondrial biogenesis) are downregulated, presumably shifting cardiomyocytes toward glycolytic metabolism [175].

Deprivation of nutrients (e.g., amino acids), glucose, and growth factor, which can lead to cardiomyocyte apoptosis [165], have been found to signal via the mitochondrial-associated mTOR protein (Fig. 8.5) [176]. Moreover, both the Akt pathway and the downstream mTOR protein impact cardiomyocyte survival and cell size largely through increased cytoplasmic protein synthesis by mediating activation of translational initiation factors and ribosomal proteins. In addition, serotonin binding to the serotonin 2B receptor protects cardiomyocytes against serum deprivation-induced apoptosis via the PI3K pathway impact ANT and Bax expression. Transgenic mice harboring serotonin 2B receptor null mutations manifest pronounced myocardial mitochondrial defects in addition to altered mitochondrial ETC activities (complexes II and IV), ANT-1, and Bax expression [177].

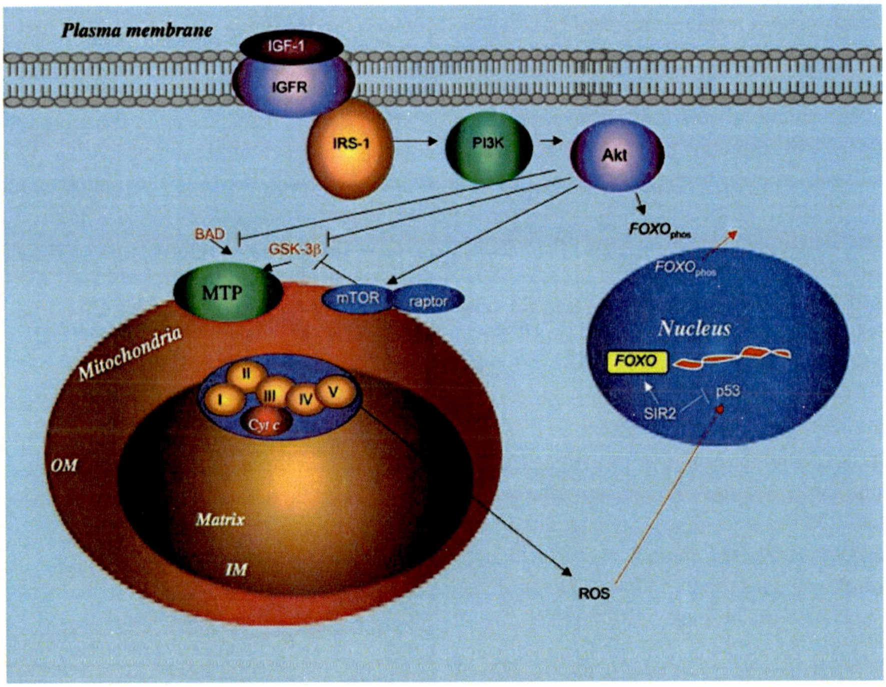


**Fig. 8.5** The myocardial phosphatidylinositol 3-kinase (PI3K)-Akt signaling pathway has a mitochondrial component. Interacting signaling pathways, including insulin-like growth factor (IGF), mTOR, and SIR2 are shown. Downstream of the receptor, the signal is transmitted to kinases (first PI3K and later Akt). Activation of Akt results in the negative regulation of proapoptotic protein activation (e.g., Bad), apoptotic suppression, and upregulation of glucose uptake. It also attenuates glycogen synthase kinase-3β (GSK-3β) activity reducing MTP. Furthermore, Akt activates mTOR signaling, part of the mitochondrial retrograde pathway. Akt phosphorylates FOXO, inactivating it, and increasing its translocation from the nucleus to the cytosol. On the other hand, the sirtuin (SIR2) activates FOXO transcriptional activity by reversing its acetylation. Similarly, SIR2 inactivates p53 by deacetylation and attenuates its apoptotic program

Finally, Akt signaling also provides CP against ischemic injury in response to diverse treatments including cardiotrophin-1, acetylcholine, adenosine, and bradykinin-mediated preconditioning [178–180] although the precise target of Akt action in mitochondria-based cardioprotection remains undetermined since Akt does not associate directly with mito$K_{ATP}$ channels.

## Stress Signals

Stresses in cardiac hypertrophy (e.g., mechanical) and ischemia/hypoxia (e.g., oxidative) elicit a variety of adaptive responses at the tissue, cellular, and molecular levels. A current model displaying the cardiac physiological response to hypoxia suggests that mitochondria function as $O_2$ sensors both by increasing ROS generation during hypoxia and via their abundant heme-containing proteins (e.g., COX) which reversibly bind oxygen [181]. Oxidant signals such as ROS act as second messengers initiating signaling cascades and are prominent features in both adaptive responses to hypoxia and mechanically stressed heart. Downregulation of COX activity contributes to the increased ROS generation and signaling observed in cardiomyocytes during hypoxia [182]. Also, hypoxia stimulates NO synthesis in cardiomyocytes [183] and NO downregulates COX activity with subsequent mitochondrial $H_2O_2$ production. This event has been proposed to provide a mitochondria-generated signal for further regulating redox-sensitive signaling pathways, including apoptosis, and can proceed even in the absence of marked changes in ATP levels [168]. Interestingly, nitric oxide synthase has been identified in heart mitochondria although its role in regulating OXPHOS is not clear [184, 185]. Mitochondrial ROS has also been shown to activate p38 kinase in hypoxic cardiomyocytes [186]. Longer-term responses to hypoxia have been shown to involve increased gene expression of hypoxia-induced factors and the activation of transcription factors such as NF-κB which has also been implicated in the complex regulation of cardiac hypertrophy and inflammatory cytokines (TNF-α and interleukin 1). While increased ROS has been shown to be an important element in NF-κB gene activation, there is recent evidence showing that cardiomyocyte hypoxia-induced factor gene activation can also occur in the absence of ROS [187].

## Metabolic Signals and UCPs

Mitochondria respond to changes in cellular levels of key metabolites such as adenosine, ATP, ADP, oxygen, and NADH as well as numerous substrates and coenzymes. After birth, cardiac FAO becomes critical as a bioenergetic substrate and source of electrons/NADH for the TCA cycle and ETC function [188]. Fatty acids also physically interact with mitochondria, impacting on membrane structure and function such as transport and excitability. In addition, amphiphilic long-chain fatty acids have detergent-like properties and have a variety of toxic effects on electrophysiological properties of cardiac cell membranes including disturbed ion transport and impaired gap junction activity [116]. Increased accumulation of intermediary metabolites of fatty acids, which occurs with defective mitochondrial FAO and transport, is considered responsible for dysrhythmias and also contribute to cardiac failure and sudden death [189]. Long-chain fatty acids (e.g., palmitate) can modulate inner membrane proton conductance (increased uncoupling) and affect MTP opening, determining apoptogenic protein release into the cytosol [166]. Another major mitochondrial target of hormone signaling (TH) as well as of long-chain fatty acids (e.g., palmitate) is the family of uncoupling proteins (UCP1–UCP5 and KMCP1) (Table 8.3) that appear to be involved in TH-modulation of cardiac function.

Mitochondrial UCPs, members of a family of mitochondrial anion carrier proteins (MACP), are nuclear-encoded transmembrane transporter proteins located in the mitochondrial inner membrane [190]. UCP1, mainly expressed in brown adipose tissue (BAT), was the first to be discovered and is responsible for thermogenesis in animals; UCP2, originally thought to play a role in nonshivering thermogenesis, obesity, and diabetes, its main function appears to be in the control of mitochondria-derived ROS. Another uncoupling protein homolog, the UCP3, is mainly expressed in skeletal muscle and BAT, and its gene is transcribed from tissue-specific promoters in humans but not in rodents. All the members of this protein family possess a common feature of shunting protons across the mitochondrial inner membrane and reduce ATP synthesis; however, this common mechanism of action is used to carry out different functions by the different UCPs. The distribution and abundance of UCPs are tissue specific (Table 8.3), which is also reflected into the processes that these proteins are thought to be participating. UCPs other than UCP1 are involved in several biological

**Table 8.3** Uncoupling proteins homologs

| Protein | Presence and abundance |
|---|---|
| UCP1 | Brown adipose tissue |
| UCP2 | Macrophages, pancreatic β-cells, skeletal muscle |
| UCP3 | Heart, skeletal muscle |
| UCP4 | Nervous system |
| UCP5 (BMCP1) | Brain |
| KMCP1 | Kidney |

*UCP* uncoupling proteins, *KMCP1* kidney mitochondrial carrier protein 1, *BMCP1* brain mitochondrial carrier protein 1

processes such as fatty acid metabolism, insulin secretion, OS, heart pathophysiology, and macrophage activation.

New discoveries are advancing our understanding of UCP's roles in cardiovascular physiology. These inner membrane-localized carrier proteins function to dissipate the proton gradient across the membrane. Expression of uncoupling proteins is upregulated transcriptionally with either palmitate or TH treatment [191–193]. Interestingly, cardiac expression of one of the uncoupling protein genes (UCP3) has also been reported to be PPAR dependent [193]. In addition, increased expression of uncoupling proteins in cardiac muscle results in increased uncoupling of OXPHOS from respiration, decreased myocardial efficiency, and mitochondrial membrane potential [192]. Potential biochemical and physiological processes where UCPs are involved or affected are presented in Fig. 8.6. Upregulation of UCPs 2 and 3 mRNA expressions in human skeletal muscle mitochondria by TH occurs without coordinated induction of respiratory chain genes. Using a whole animal/whole organ-heart model, Barbe et al. [194] administered TH to Wistar rats for 7 days. Within 24 h after the last dose, heart mitochondria were isolated, and UCPs levels were determined. UCP2 and UCP3 increased by about 40%, and mitochondrial uncoupling, as measured by oligomycin insensitive respiration rate, increased twofold in the presence of palmitate. In the isolated working heart, the presence of palmitate significantly reduced cardiac output and efficiency by about 36% in the TH-treated rats [192]. Thus, increased UCPs in hyperthyroid rats are associated with increased uncoupling and decreased myocardial efficiency in the presence of palmitate.

In muscle cells, UCP3 augments FAO and decreases ROS [195]. In L6 muscle cells, when compared with the effect of an uncoupler agent like dinitrophenol, UCP3 seems to be preferentially promoting FAO rather than glucose utilization. Moreover, UCP3 reduces ROS production without significant increase in oxygen consumption. This function of UCP3 may be useful particularly in managing type-2 diabetes where impaired fatty acid metabolism and ROS handling set the stage for muscular insulin resistance.

Interestingly, UCP1 confers resistance against hypoxia–reoxygenation in a specific heart cell line [196]. H9c2 cell line transfected with UCP1 showed that overexpression of UCP1 was not compromising cell viability. In hypoxia–reoxygenation experiments, compared to control cells, the UCP1 expressing cells show a moderate decrease in OXPHOS capacity, but significantly higher survival with largely preserved subcellular ultra-structure. It is worth noting that the surge in ROS production was significantly reduced.

UCP2 from the mitochondrial matrix side is activated by superoxide. Also reactive aldehydes, such as 4-hydroxy-2-noneal produced from peroxidation of membrane phospholipids, as a result of oxidative damage induce mitochondrial uncoupling through UCPs [197–199].

In cardiomyocytes under OS induced by 100 μM/L $H_2O_2$, UCP2 overexpression inhibits the mitochondrial death pathway. Furthermore, early apoptotic events (i.e., decrease in DeltaPsi or membrane potential), increased ROS generation,

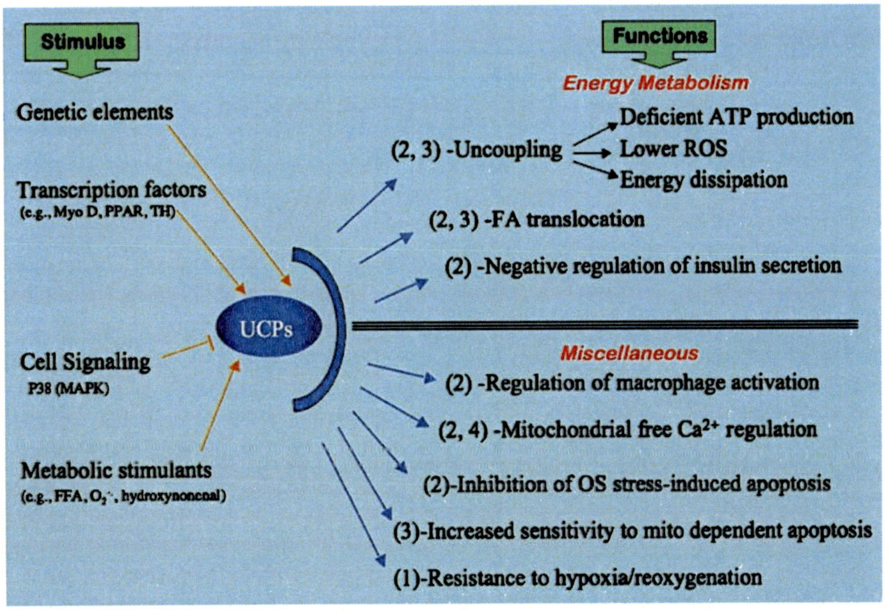


**Fig. 8.6** Flow chart showing an overview of UCPs functions. The chart enlists the effectors as "Stimulus" for UCP activity and the resulting activities as "Functions" of UCPs. Number in parenthesis indicates the specific UCP 1, 2, 3, or 4 related to that activity

$Ca^{2+}$ overload and late phase apoptotic events as detected by terminal deoxynucleotidyl transferase dUTP nick end labeling (TUNEL), and caspase-3 activity have been assessed in cells transfected with an adenoviral vector containing human UCP2. UCP2 overexpression significantly suppressed the apoptotic events caused by OS secondary to $H_2O_2$ exposure. These data suggest that UCP2 may mitigate cell death occurring in ischemia/reperfusion injury by preventing apoptotic events, through preserving membrane potential and lowering ROS production [200].

Hypoxia decreased UCP3 expression in rat cardiomyocytes without changing mitochondrial respiratory coupling, while UCP2 expression was unchanged. A role for UCP3 in the regulation of heart FAO has been proposed instead of a role in uncoupling mitochondria. Hypoxia-induced regulation indicates a distinct mitochondrial regulatory function in the heart in response to metabolic stress [201]. Hypoxia as well as exercise increases the UCP3 levels of rat skeletal muscle by four- to sixfold compared to controls. Furthermore, AMPK, known to be stimulated during exercise and hypoxia, induces UCP3 expression [202]. These observations present further evidence supporting UCPs tissue-specific behavior. Finally, the role that UCP2 and UCP3 play in cardiac pathophysiology appears to be dependent on the surrounding conditions (Fig. 8.6). For example, in ischemia preconditioning UCPs support the survival of heart tissue by playing the crucial role of preventing oxidative damage. On the other hand, in the presence of elevated levels of circulating fatty acid, disruption in energy metabolism is actually worsened by uncoupling with the depletion of ATP, as it may occur in the diabetic heart. In these circumstances, the question of "is UCPs a Friend or a Foe?" seems appropriate; however, to answer this question further research is warranted.

## Future Prospects: Therapeutic Targets and Directions

Exploiting the recognition that mitochondria play an essential role in cytoprotective signaling and CP may stimulate further collaboration among cardiologists and other researchers in the fields of drug discovery to carry out successfully the pharmacological manipulation of mitochondrial metabolism and signaling in cardiovascular diseases. Research in mitochondria-based CP may reveal potential target molecules (e.g., mitochondrial receptor, signaling kinase or channel) for highly specific pharmacological intervention, although, as a caveat, a greater understanding of the network of complex interacting pathways will be needed. For example, despite the recent achievements in identifying genetic and signaling defects causing cardiac dysrhythmias, the development of effective drugs (e.g., specific ion channel blockers) to substantially reduce the mortality associated with severe dysrhythmias has thus far shown little success, underscoring the complex circuitry involved in evolving cardiac disease phenotypes [203]. Moreover, pharmacological agents which are cardioprotective in animal studies can have variable effects in clinical settings; diazoxide has shown negative effects while nicorandil has proved more successful in limited clinical trials [204]. Nevertheless, knowledge of the specific molecular and biochemical nature of mitochondrial bioenergetic defects has provided a rationale for treatment with metabolic intermediates (e.g., succinate), coenzymes and vitamins serving as electron donors, transporters and cofactors (e.g., vitamin K, thiamine, and ascorbate) in order to bypass specific defects in OXPHOS and increase ATP production [205]. For instance, coenzyme Q10 and its analogue idebenone have shown beneficial effects in the treatment of cardiomyopathy associated with Friedreich ataxia [206]. Shifting myocardial oxidative substrates from fatty acid to glucose can be used to prevent the accumulation of long-chain acylcarnitines and in improving myocardial energy efficiency in HF [207]. In addition, dichloroacetate has shown promise in stemming the lactic acidosis and declining PDH activity noted in myocardial ischemia and reperfusion injury [205]. Dietary therapies including replacement of normal dietary fat by medium-chain triglycerides and increased carbohydrates and carnitine supplementation are effective in cardiomyopathies due to mitochondrial long-chain FAO disorders and mitochondrial carnitine transport deficiencies and in lowering acyl-CoA accumulation [208]. Increased free polyunsaturated fatty acids can provide significant cardioprotective effect against both ischemia-related ventricular fibrillation and a dysrhythmias [209].

Interestingly, the participation of UCPs in a number of biological processes, including a role in insulin secretion and fatty acid metabolism might be considered. Potential target, for example, is the targeting of UCP2 to treat diabetes.

Although some studies have been reported the changes in UCPs expression, UCPs mRNA expression does not always correlate to its protein levels; therefore, other unknown mechanisms may be present in the regulation of UCPs during translation. Other than in UCP1, there are a number of unanswered questions about the function of the new UCPs. For example, the quantity of the new UCPs (nUCPs) in normal cells is about 100-fold lower compared to UCP1 levels. At such low concentration, it is questionable if they can really alter ATP synthesis in the cell. Furthermore, we do not know if these nUCPs do uncoupling passively or in the presence of an activator, together with an increase in protein molecules. On the other hand, nUCPs may never function as uncouplers but rather as special type of transporters that carry ions like $Ca^{2+}$ or fatty acids across the mitochondrial inner membrane. Finally, it is evident that UCPs are truly versatile proteins that play a plurality of roles, including their uncoupling activity and partial control of ATP synthesis that in turn regulate insulin secretion and ROS generation; participation in the relief of OS, with regulation of apoptosis pathway and helping the cell immune defense against microorganisms; activity as a transporter that facilitate the traffic of specific ions; and potential involvement in processes such as metabolic switching from glycolytic to oxidative energy pathway.

Future therapeutic modalities to treat mitochondrial cardiac disorders may also exploit technologies such as gene therapy and stem cell transplantation, which while showing exciting potential in animal models ameliorating myocardial function in the damaged hearts [210, 211], it remains to be proven both effective and safe in human.

## Conclusions

A myocardial mitochondria's perspective as a receiver, integrator, and transmitter of signals has been presented in this chapter, as well as an outline of their multiple and interrelated signaling transduction pathways at the molecular and biochemical levels, and a number of nuclear and cytoplasmic factors that are involved in shaping the organelle response. Increasing evidence supports the concept that mitochondria act as a dynamic receiver and integrator of numerous translocated signaling proteins that include protein kinases, uncoupling proteins, nuclear transcription factors, regulatory $Ca^{2+}$ fluxes, membrane phospholipids as well as transmission of mitochondria-generated OS and energy-related signaling contributing to the overall cardiomyocyte response to myocardial ischemia and hypertrophy. Besides their critical role on myocardial apoptosis and in cardiac remodeling, myocardial mitochondria signaling also plays critical roles in the cardiomyocytes proliferative pathways, nutrient and $O_2$ sensing, bioenergetic metabolite/substrate selection, interorganellar cross-talk, and in cardiomyocytes response to metabolic transition and physiological stresses; roles that only recently are coming to light and merit further investigation. Understanding of mitochondrial signaling pathways might lead to identification of the molecular mechanisms that underlie the pathogenesis of cardiovascular diseases and improve our knowledge of the ever-widening spectrum of abnormal human cardiac phenotypes with mitochondrial dysfunction.

## Summary

- Mitochondria play a critical role in the regulatory and signaling events that occur in response to physiological stresses, including heart failure, myocardial ischemia and reperfusion, hypoxia, oxidative stress, and hormonal and cytokine stimuli.
- Heart mitochondria are receiver/integrator and transmitter of signals, with multiple and interrelated signaling pathways at both the molecular and biochemical levels.
- Signaling pathways allow these organelles to respond to heart energy demands as well as to cardiomyocytes growth, death, and a variety of physiological stimuli and stresses.
- Increased ROS generation resulting from myocardial I/R, inflammation, impaired antioxidant defenses, and aging may cause profound alterations on cells including elevated lipid peroxidation targeting membrane phospholipids and proteins.
- Inhibition of mitochondrial respiration by NO can result in further increases in mitochondrial ROS production; interaction with NO enhances the potency of superoxide as an inhibitor of respiration.
- ROS produced from other cellular sources, besides mitochondria, can have substantial effects on cardiovascular function.
- Besides their cell-damaging effects, ROS generation and oxidative stress play a critical role in cell regulation and signaling.
- Oxidative species such as $H_2O_2$ and the superoxide anion can be deployed as signals sent from mitochondria to other cellular sites rapidly, and reversibly triggering an array of intracellular signaling cascades leading to diverse physiological end points in the cardiomyocyte, some negative (e.g., apoptosis and necrosis) and others positive (e.g., cardioprotection and cell proliferation).
- Drugs such as diazoxide and nicorandil specifically activate the mitoK$_{ATP}$ channel opening and can also inhibit $H_2O_2$-induced apoptotic progression in cardiomyocytes, suggesting that mitoK$_{ATP}$ channels may also play a significant role in mediating oxidative-stress signals in the mitochondrial apoptotic pathway.
- Mitochondria contain multiple phosphoprotein substrates for protein kinases, and a number of protein kinases translocate into heart mitochondria. This suggests that protein phosphorylation within the mitochondria is a critical component of the mitochondrial signaling pathways.
- PKD-regulated signaling pathway is a mitochondrially located and mitochondrially regulated antioxidant system that protects the organelle and cells from oxidative stress-mediated damage and cell death.
- Stimuli-dependent translocation to and incorporation of specific cytosolic proteins into cardiomyocyte mitochondria are important factors in cell signaling.
- Mitochondrial retrograde signaling is a pathway of communication from mitochondria to the nucleus that influences many cellular activities under both normal and pathophysiological conditions.

- In mammalian cells, mitochondrial dysfunction sets off signaling cascades through abnormal $Ca^{2+}$ dynamics including calcineurin activation, which activate several protein kinase pathways (e.g., PKC and MAPK) and transcription factors such as NF-κB, calcineurin-dependent NFAT, CREB, and ATF leading to stress protein expression (e.g., chaperone proteins) and activities.
- One of the ER primary functions is as a source of the $Ca^{2+}$ signals which are released through either IP3 or ryanodine receptors, themselves $Ca^{2+}$-sensitive.
- The apoptotic death process is mediated by two central pathways, an extrinsic pathway featuring cell-surface receptors and an intrinsic pathway involving mitochondria and the ER.
- A myriad of signals converges on the mitochondria leading to significant changes in membrane organization and dysfunction of the organelle, the release of apoptogenic proteins, and subsequent activation of caspases.
- Mutation in mitochondrial DNA and physiological insults targeting various mitochondrial pathways (other than mitochondrial OXPHOS function) are contributory to cardiac disease.
- In cardiac hypertrophy, downregulation of the mitochondrial pathways involves the FAO/fatty acid transport system, occurring as part of a shift in cardiac bioenergetic substrate utilization from fatty acid to glucose (glycolytic pathways).
- Nuclear transcriptional modulators govern the expression of a wide array of mitochondrial proteins in response to diverse cellular stimuli and signals.
- PGC-1 expression and mitochondrial biogenesis are modulated by the activation of a calcium/calmodulin-dependent protein kinase, indicating that the calcium-regulated signaling pathway plays a significant role in transcriptional activation of genes governing mitochondrial biogenesis.
- The import of $Ca^{2+}$ from cytosol into cardiac mitochondria is an important regulatory event in cell signaling.
- A common theme concerning signaling and activation includes the stimuli-generated translocation of specific cytosolic proteins into the mitochondria; the growing list of such translocated entities includes many of the proapoptotic proteins (e.g., Bax and Bid) as well as protein kinases.
- The PI3K-Akt pathway promotes cell survival primarily by intervening in the mitochondrial apoptosis cascade at events before the release of cytochrome *c* and the occurrence of caspase activation.
- A major mitochondrial target of hormone signaling (TH) and long-chain fatty acids (e.g., palmitate) is a family of uncoupling proteins (UCP1 UCP5 and KMCP1) that appear to be involved in TH-modulation of cardiac function.
- Upregulation of UCPs 2 and 3 mRNA expressions in human skeletal muscle mitochondria by TH occurs without coordinated induction of respiratory chain genes.
- Understanding of mitochondrial signaling pathways might lead to identification of the molecular mechanisms that underlie the pathogenesis of cardiovascular diseases.

## References


1. Hatefi Y. The mitochondrial electron transport and oxidative phosphorylation system. Annu Rev Biochem. 1985;54:1015–69.
2. Anderson S, Bankier AT, Barrell BG, et al. Sequence and organization of the human mitochondrial genome. Nature. 1981;290:457–65.
3. Attardi G, Schatz G. Biogenesis of mitochondria. Annu Rev Cell Biol. 1988;4:289–333.
4. Hood DA. Invited review: contractile activity-induced mitochondrial biogenesis in skeletal muscle. J Appl Physiol. 2001;90: 1137–57.
5. Totland GK, Madsen L, Klementsen B, et al. Proliferation of mitochondria and gene expression of carnitine palmitoyltransferase and fatty acyl-CoA oxidase in rat skeletal muscle, heart and liver by hypolipidemic fatty acids. Biol Cell. 2000;92:317–29.
6. Lundgren B, Meijer J, DePierre JW. Induction of cytosolic and microsomal epoxide hydrolases and proliferation of peroxisomes and mitochondria in mouse liver after dietary exposure to p-chlorophenoxyacetic acid, 2,4-dichlorophenoxyacetic acid and 2,4,5-trichlorophenoxyacetic acid. Biochem Pharmacol. 1987;36: 815–21.
7. Weber K, Bruck P, Mikes Z, Kupper JH, Klingenspor M, Wiesner RJ. Glucocorticoid hormone stimulates mitochondrial biogenesis specifically in skeletal muscle. Endocrinology. 2002;143: 177–84.
8. Williams RS, Garcia-Moll M, Mellor J, Salmons S, Harlan W. Adaptation of skeletal muscle to increased contractile activity. Expression nuclear genes encoding mitochondrial proteins. J Biol Chem. 1987;262:2764–7.
9. Nelson BD. Thyroid hormone regulation of mitochondrial function. Comments on the mechanism of signal transduction. Biochim Biophys Acta. 1018;1990:275–7.
10. Xia Y, Buja LM, Scarpulla RC, McMillin JB. Electrical stimulation of neonatal cardiomyocytes results in the sequential activation of nuclear genes governing mitochondrial proliferation and differentiation. Proc Natl Acad Sci USA. 1997;94:11399–404.
11. Bogenhagen D, Clayton DA. The number of mitochondrial deoxyribonucleic acid genomes in mouse L and human HeLa cells. Quantitative isolation of mitochondrial deoxyribonucleic acid. J Biol Chem. 1974;249:7991–5.
12. Shadel GS, Clayton DA. Mitochondrial DNA maintenance in vertebrates. Annu Rev Biochem. 1997;66:409–35.
13. Kadenbach B, Stroh A, Becker A, Eckerskorn C, Lottspeich F. Tissue- and species-specific expression of cytochrome c oxidase isozymes in vertebrates. Biochim Biophys Acta. 1990;1015: 368–72.
14. Lenka N, Vijayasarathy C, Mullick J, Avadhani NG. Structural organization and transcription regulation of nuclear genes encoding the mammalian cytochrome c oxidase complex. Prog Nucleic Acid Res Mol Biol. 1998;61:309–44.
15. McLennan HR, Degli Esposti M. The contribution of mitochondrial respiratory complexes to the production of reactive oxygen species. J Bioenerg Biomembr. 2000;32:153–62.

16. Pryor WA. Oxy-radicals and related species: their formation, lifetimes, and reactions. Annu Rev Physiol. 1986;48:657–67.
17. Han D, Antunes F, Canali R, Rettori D, Cadenas E. Voltage-dependent anion channels control the release of the superoxide anion from mitochondria to cytosol. J Biol Chem. 2003;278:5557–63.
18. Stadtman ER, Berlett BS. Reactive oxygen-mediated protein oxidation in aging and disease. Drug Metab Rev. 1998;30:225–43.
19. Choksi KB, Boylston WH, Rabek JP, Widger WR, Papaconstantinou J. Oxidatively damaged proteins of heart mitochondrial electron transport complexes. Biochim Biophys Acta. 2004;1688:95–101.
20. Vasquez-Vivar J, Kalyanaraman B, Kennedy MC. Mitochondrial aconitase is a source of hydroxyl radical. An electron spin resonance investigation. J Biol Chem. 2000;275:14064–9.
21. Paradies G, Petrosillo G, Pistolese M, Ruggiero FM. Reactive oxygen species affect mitochondrial electron transport complex I activity through oxidative cardiolipin damage. Gene. 2002;286:135–41.
22. Petrosillo G, Ruggiero FM, Pistolese M, Paradies G. Reactive oxygen species generated from the mitochondrial electron transport chain induce cytochrome c dissociation from beef-heart submitochondrial particles via cardiolipin peroxidation. Possible role in the apoptosis. FEBS Lett. 2001;509:435–8.
23. Wolin MS, Ahmad M, Gupte SA. Oxidant and redox signaling in vascular oxygen sensing mechanisms: basic concepts, current controversies, and potential importance of cytosolic NADPH. Am J Physiol Lung Cell Mol Physiol. 2005;289:L159–73.
24. Brown GC. Nitric oxide and mitochondrial respiration. Biochim Biophys Acta. 1999;1411:351–69.
25. Murray J, Taylor SW, Zhang B, Ghosh SS, Capaldi RA. Oxidative damage to mitochondrial complex I due to peroxynitrite: identification of reactive tyrosines by mass spectrometry. J Biol Chem. 2003;278:37223–30.
26. Riobo NA, Clementi E, Melani M, et al. Nitric oxide inhibits mitochondrial NADH:ubiquinone reductase activity through peroxynitrite formation. Biochem J. 2001;359:139–45.
27. Cassina AM, Hodara R, Souza JM, et al. Cytochrome c nitration by peroxynitrite. J Biol Chem. 2000;275:21409–15.
28. Castro L, Rodriguez M, Radi R. Aconitase is readily inactivated by peroxynitrite, but not by its precursor, nitric oxide. J Biol Chem. 1994;269:29409–15.
29. Packer MA, Scarlett JL, Martin SW, Murphy MP. Induction of the mitochondrial permeability transition by peroxynitrite. Biochem Soc Trans. 1997;25:909–14.
30. Brookes PS, Darley-Usmar VM. Role of calcium and superoxide dismutase in sensitizing mitochondria to peroxynitrite-induced permeability transition. Am J Physiol Heart Circ Physiol. 2004;286:H39–46.
31. Griendling KK, Minieri CA, Ollerenshaw JD, Alexander RW. Angiotensin II stimulates NADH and NADPH oxidase activity in cultured vascular smooth muscle cells. Circ Res. 1994;74:1141–8.
32. De Keulenaer GW, Alexander RW, Ushio-Fukai M, Ishizaka N, Griendling KK. Tumour necrosis factor alpha activates a p22phox-based NADH oxidase in vascular smooth muscle. Biochem J. 1998;329(Pt 3):653–7.
33. Patterson C, Ruef J, Madamanchi NR, et al. Stimulation of a vascular smooth muscle cell NAD(P)H oxidase by thrombin. Evidence that p47(phox) may participate in forming this oxidase in vitro and in vivo. J Biol Chem. 1999;274:19814–22.
34. Hellsten-Westing Y. Immunohistochemical localization of xanthine oxidase in human cardiac and skeletal muscle. Histochemistry. 1993;100:215–22.
35. Cappola TP, Kass DA, Nelson GS, et al. Allopurinol improves myocardial efficiency in patients with idiopathic dilated cardiomyopathy. Circulation. 2001;104:2407–11.
36. Nemoto S, Takeda K, Yu ZX, Ferrans VJ, Finkel T. Role for mitochondrial oxidants as regulators of cellular metabolism. Mol Cell Biol. 2000;20:7311–8.
37. Bogoyevitch MA, Ng DC, Court NW, Draper KA, Dhillon A, Abas L. Intact mitochondrial electron transport function is essential for signalling by hydrogen peroxide in cardiac myocytes. J Mol Cell Cardiol. 2000;32:1469–80.
38. Archer SL, Wu XC, Thebaud B, Moudgil R, Hashimoto K, Michelakis ED. O2 sensing in the human ductus arteriosus: redox-sensitive $K^+$ channels are regulated by mitochondria-derived hydrogen peroxide. Biol Chem. 2004;385:205–16.
39. Yamamura T, Otani H, Nakao Y, et al. Dual involvement of coenzyme Q10 in redox signaling and inhibition of death signaling in the rat heart mitochondria. Antioxid Redox Signal. 2001;3:103–12.
40. Brookes PS, Levonen AL, Shiva S, Sarti P, Darley-Usmar VM. Mitochondria: regulators of signal transduction by reactive oxygen and nitrogen species. Free Radic Biol Med. 2002;33:755–64.
41. Boveris A, D'Amico G, Lores-Arnaiz S, Costa LE. Enalapril increases mitochondrial nitric oxide synthase activity in heart and liver. Antioxid Redox Signal. 2003;5:691–7.
42. O'Rourke B. Myocardial K(ATP) channels in preconditioning. Circ Res. 2000;87:845–55.
43. Lebuffe G, Schumacker PT, Shao ZH, Anderson T, Iwase H, Vanden Hoek TL. ROS and NO trigger early preconditioning: relationship to mitochondrial KATP channel. Am J Physiol Heart Circ Physiol. 2003;284:H299–308.
44. Ardehali H, O'Rourke B. Mitochondrial K(ATP) channels in cell survival and death. J Mol Cell Cardiol. 2005;39:7–16.
45. Garlid KD, Dos Santos P, Xie ZJ, Costa AD, Paucek P. Mitochondrial potassium transport: the role of the mitochondrial ATP-sensitive K(+) channel in cardiac function and cardioprotection. Biochim Biophys Acta. 2003;1606:1–21.
46. Das M, Parker JE, Halestrap AP. Matrix volume measurements challenge the existence of diazoxide/glibencamide-sensitive KATP channels in rat mitochondria. J Physiol. 2003;547:893–902.
47. Akao M, Teshima Y, Marban E. Antiapoptotic effect of nicorandil mediated by mitochondrial atp-sensitive potassium channels in cultured cardiac myocytes. J Am Coll Cardiol. 2002;40:803–10.
48. Nagata K, Obata K, Odashima M, et al. Nicorandil inhibits oxidative stress-induced apoptosis in cardiac myocytes through activation of mitochondrial ATP-sensitive potassium channels and a nitrate-like effect. J Mol Cell Cardiol. 2003;35:1505–12.
49. Xu W, Liu Y, Wang S, et al. Cytoprotective role of $Ca^{2+}$-activated $K^+$ channels in the cardiac inner mitochondrial membrane. Science. 2002;298:1029–33.
50. Hanley PJ, Daut J. K(ATP) channels and preconditioning: a re-examination of the role of mitochondrial K(ATP) channels and an overview of alternative mechanisms. J Mol Cell Cardiol. 2005;39:17–50.
51. Halestrap AP, Clarke SJ, Javadov SA. Mitochondrial permeability transition pore opening during myocardial reperfusion–a target for cardioprotection. Cardiovasc Res. 2004;61:372–85.
52. Shanmuganathan S, Hausenloy DJ, Duchen MR, Yellon DM. Mitochondrial permeability transition pore as a target for cardioprotection in the human heart. Am J Physiol Heart Circ Physiol. 2005;289:H237–42.
53. Thomson M. Evidence of undiscovered cell regulatory mechanisms: phosphoproteins and protein kinases in mitochondria. Cell Mol Life Sci. 2002;59:213–9.
54. Sugden MC, Orfali KA, Fryer LG, Holness MJ, Priestman DA. Molecular mechanisms underlying the long-term impact of dietary fat to increase cardiac pyruvate dehydrogenase kinase: regulation by insulin, cyclic AMP and pyruvate. J Mol Cell Cardiol. 1997;29:1867–75.

55. Technikova-Dobrova Z, Sardanelli AM, Stanca MR, Papa S. cAMP-dependent protein phosphorylation in mitochondria of bovine heart. FEBS Lett. 1994;350:187–91.
56. Wang Y, Hirai K, Ashraf M. Activation of mitochondrial ATP-sensitive K(+) channel for cardiac protection against ischemic injury is dependent on protein kinase C activity. Circ Res. 1999;85:731–41.
57. Baines CP, Zhang J, Wang GW, et al. Mitochondrial PKCepsilon and MAPK form signaling modules in the murine heart: enhanced mitochondrial PKCepsilon-MAPK interactions and differential MAPK activation in PKCepsilon-induced cardioprotection. Circ Res. 2002;90:390–7.
58. Storz P. Mitochondrial ROS–radical detoxification, mediated by protein kinase D. Trends Cell Biol. 2007;17:13–8.
59. Garlid KD, Costa AD, Cohen MV, Downey JM, Critz SD. Cyclic GMP and PKG activate mito $K_{ATP}$ channels in isolated mitochondrial. Cardiovasc J S Afr. 2004;15 Suppl 1:S5.
60. He H, Li HL, Lin A, Gottlieb RA. Activation of the JNK pathway is important for cardiomyocyte death in response to simulated ischemia. Cell Death Differ. 1999;6:987–91.
61. Aoki H, Kang PM, Hampe J, et al. Direct activation of mitochondrial apoptosis machinery by c-Jun N-terminal kinase in adult cardiac myocytes. J Biol Chem. 2002;277:10244–50.
62. Court NW, Kuo I, Quigley O, Bogoyevitch MA. Phosphorylation of the mitochondrial protein Sab by stress-activated protein kinase 3. Biochem Biophys Res Commun. 2004;319:130–7.
63. Baines CP, Song CX, Zheng YT, et al. Protein kinase Cepsilon interacts with and inhibits the permeability transition pore in cardiac mitochondria. Circ Res. 2003;92:873–80.
64. Storz P, Doppler H, Toker A. Protein kinase D mediates mitochondrion-to-nucleus signaling and detoxification from mitochondrial reactive oxygen species. Mol Cell Biol. 2005;25:8520–30.
65. Cowell CF, Doppler H, Yan IK, Hausser A, Umezawa Y, Storz P. Mitochondrial diacylglycerol initiates protein-kinase D1-mediated ROS signaling. J Cell Sci. 2009;122:919–28.
66. Ozgen N, Guo J, Gertsberg Z, Danilo Jr P, Rosen MR, Steinberg SF. Reactive oxygen species decrease cAMP response element binding protein expression in cardiomyocytes via a protein kinase D1-dependent mechanism that does not require Ser133 phosphorylation. Mol Pharmacol. 2009;76:896–902.
67. Papa S. The NDUFS4 nuclear gene of complex I of mitochondria and the cAMP cascade. Biochim Biophys Acta. 2002;1555:147–53.
68. Lee I, Bender E, Kadenbach B. Control of mitochondrial membrane potential and ROS formation by reversible phosphorylation of cytochrome c oxidase. Mol Cell Biochem. 2002;234–235: 63–70.
69. Schulenberg B, Aggeler R, Beechem JM, Capaldi RA, Patton WF. Analysis of steady-state protein phosphorylation in mitochondria using a novel fluorescent phosphosensor dye. J Biol Chem. 2003;278:27251–5.
70. Hood DA, Joseph AM. Mitochondrial assembly: protein import. Proc Nutr Soc. 2004;63:293–300.
71. Colavecchia M, Christie LN, Kanwar YS, Hood DA. Functional consequences of thyroid hormone-induced changes in the mitochondrial protein import pathway. Am J Physiol Endocrinol Metab. 2003;284:E29–35.
72. Biswas G, Guha M, Avadhani NG. Mitochondria-to-nucleus stress signaling in mammalian cells: nature of nuclear gene targets, transcription regulation, and induced resistance to apoptosis. Gene. 2005;354:132–9.
73. Berridge MJ. The endoplasmic reticulum: a multifunctional signaling organelle. Cell Calcium. 2002;32:235–49.
74. Crow MT, Mani K, Nam YJ, Kitsis RN. The mitochondrial death pathway and cardiac myocyte apoptosis. Circ Res. 2004;95:957–70.
75. Danial NN, Korsmeyer SJ. Cell death: critical control points. Cell. 2004;116:205–19.
76. Epand RF, Martinou JC, Montessuit S, Epand RM, Yip CM. Direct evidence for membrane pore formation by the apoptotic protein Bax. Biochem Biophys Res Commun. 2002;298:744–9.
77. Chipuk JE, Kuwana T, Bouchier-Hayes L, et al. Direct activation of Bax by p53 mediates mitochondrial membrane permeabilization and apoptosis. Science. 2004;303:1010–4.
78. Regula KM, Ens K, Kirshenbaum LA. Mitochondria-assisted cell suicide: a license to kill. J Mol Cell Cardiol. 2003;35:559–67.
79. Kroemer G. Mitochondrial control of apoptosis: an introduction. Biochem Biophys Res Commun. 2003;304:433–5.
80. Scorrano L, Ashiya M, Buttle K, et al. A distinct pathway remodels mitochondrial cristae and mobilizes cytochrome c during apoptosis. Dev Cell. 2002;2:55–67.
81. Belzacq AS, Vieira HL, Verrier F, et al. Bcl-2 and Bax modulate adenine nucleotide translocase activity. Cancer Res. 2003;63: 541–6.
82. Dejean LM, Martinez-Caballero S, Kinnally KW. Is MAC the knife that cuts cytochrome c from mitochondria during apoptosis? Cell Death Differ. 2006;13:1387–95.
83. Dorn II GW. Mitochondrial pruning by Nix and BNip3: an essential function for cardiac-expressed death factors. J Cardiovasc Transl Res. 2010;3:374–83.
84. Primeau AJ, Adhihetty PJ, Hood DA. Apoptosis in heart and skeletal muscle. Can J Appl Physiol. 2002;27:349–95.
85. Adhihetty PJ, O'Leary MF, Hood DA. Mitochondria in skeletal muscle: adaptable rheostats of apoptotic susceptibility. Exerc Sport Sci Rev. 2008;36:116–21.
86. Alway SE, Martyn JK, Ouyang J, Chaudhrai A, Murlasits ZS. Id2 expression during apoptosis and satellite cell activation in unloaded and loaded quail skeletal muscles. Am J Physiol Regul Integr Comp Physiol. 2003;284:R540–9.
87. Marín-García J, Goldenthal MJ, Damle S, Pi Y, Moe GW. Regional distribution of mitochondrial dysfunction and apoptotic remodeling in pacing-induced heart failure. J Card Fail. 2009;15:700–8.
88. Adhihetty PJ, Ljubicic V, Hood DA. Effect of chronic contractile activity on SS and IMF mitochondrial apoptotic susceptibility in skeletal muscle. Am J Physiol Endocrinol Metab. 2007;292:E748–55.
89. Marin-Garcia J, Goldenthal MJ. Understanding the impact of mitochondrial defects in cardiovascular disease: a review. J Card Fail. 2002;8:347–61.
90. Marin-Garcia J, Ananthakrishnan R, Goldenthal MJ, Pierpont ME. Biochemical and molecular basis for mitochondrial cardiomyopathy in neonates and children. J Inherit Metab Dis. 2000;23: 625–33.
91. Lewis W, Dalakas MC. Mitochondrial toxicity of antiviral drugs. Nat Med. 1995;1:417–22.
92. Benit P, Slama A, Cartault F, et al. Mutant NDUFS3 subunit of mitochondrial complex I causes Leigh syndrome. J Med Genet. 2004;41:14–7.
93. Papadopoulou LC, Sue CM, Davidson MM, et al. Fatal infantile cardioencephalomyopathy with COX deficiency and mutations in SCO2, a COX assembly gene. Nat Genet. 1999;23:333–7.
94. Lodi R, Cooper JM, Bradley JL, et al. Deficit of in vivo mitochondrial ATP production in patients with Friedreich ataxia. Proc Natl Acad Sci USA. 1999;96:11492–5.
95. Zeviani M, Spinazzola A, Carelli V. Nuclear genes in mitochondrial disorders. Curr Opin Genet Dev. 2003;13:262–70.
96. Graham BH, Waymire KG, Cottrell B, Trounce IA, MacGregor GR, Wallace DC. A mouse model for mitochondrial myopathy and cardiomyopathy resulting from a deficiency in the heart/muscle isoform of the adenine nucleotide translocator. Nat Genet. 1997;16:226–34.
97. Wang J, Wilhelmsson H, Graff C, et al. Dilated cardiomyopathy and atrioventricular conduction blocks induced by heart-specific inactivation of mitochondrial DNA gene expression. Nat Genet. 1999;21:133–7.

98. Corbucci GG. Adaptive changes in response to acute hypoxia, ischemia and reperfusion in human cardiac cell. Minerva Anestesiol. 2000;66:523–30.
99. Corral-Debrinski M, Stepien G, Shoffner JM, Lott MT, Kanter K, Wallace DC. Hypoxemia is associated with mitochondrial DNA damage and gene induction. Implications for cardiac disease. JAMA. 1991;266:1812–6.
100. Gottlieb RA, Burleson KO, Kloner RA, Babior BM, Engler RL. Reperfusion injury induces apoptosis in rabbit cardiomyocytes. J Clin Invest. 1994;94:1621–8.
101. Paradies G, Petrosillo G, Pistolese M, Di Venosa N, Serena D, Ruggiero FM. Lipid peroxidation and alterations to oxidative metabolism in mitochondria isolated from rat heart subjected to ischemia and reperfusion. Free Radic Biol Med. 1999;27:42–50.
102. Schulz R, Cohen MV, Behrends M, Downey JM, Heusch G. Signal transduction of ischemic preconditioning. Cardiovasc Res. 2001;52:181–98.
103. Marin-Garcia J, Goldenthal MJ. Mitochondria play a critical role in cardioprotection. J Card Fail. 2004;10:55–66.
104. Halestrap AP. Regulation of mitochondrial metabolism through changes in matrix volume. Biochem Soc Trans. 1994;22:522–9.
105. Garlid KD, Paucek P, Yarov-Yarovoy V, et al. Cardioprotective effect of diazoxide and its interaction with mitochondrial ATP-sensitive K+channels. Possible mechanism of cardioprotection. Circ Res. 1997;81:1072–82.
106. Ning XH, Xu CS, Song YC, et al. Hypothermia preserves function and signaling for mitochondrial biogenesis during subsequent ischemia. Am J Physiol. 1998;274:H786–93.
107. Nadal-Ginard B, Kajstura J, Leri A, Anversa P. Myocyte death, growth, and regeneration in cardiac hypertrophy and failure. Circ Res. 2003;92:139–50.
108. Colucci WS. Molecular and cellular mechanisms of myocardial failure. Am J Cardiol. 1997;80:15L–25.
109. Hunter JJ, Chien KR. Signaling pathways for cardiac hypertrophy and failure. N Engl J Med. 1999;341:1276–83.
110. Katz AM. Maladaptive growth in the failing heart: the cardiomyopathy of overload. Cardiovasc Drugs Ther. 2002;16:245–9.
111. Kang PM, Yue P, Liu Z, Tarnavski O, Bodyak N, Izumo S. Alterations in apoptosis regulatory factors during hypertrophy and heart failure. Am J Physiol Heart Circ Physiol. 2004;287: H72–80.
112. Sack MN, Kelly DP. The energy substrate switch during development of heart failure: gene regulatory mechanisms (Review). Int J Mol Med. 1998;1:17–24.
113. Lehman JJ, Kelly DP. Gene regulatory mechanisms governing energy metabolism during cardiac hypertrophic growth. Heart Fail Rev. 2002;7:175–85.
114. Zak R, Rabinowitz M, Rajamanickam C, Merten S, Kwiatkowska-Patzer B. Mitochondrial proliferation in cardiac hypertrophy. Basic Res Cardiol. 1980;75:171–8.
115. Lucas DT, Aryal P, Szweda LI, Koch WJ, Leinwand LA. Alterations in mitochondrial function in a mouse model of hypertrophic cardiomyopathy. Am J Physiol Heart Circ Physiol. 2003;284: H575–83.
116. Marin-Garcia J, Goldenthal MJ, Moe GW. Mitochondrial pathology in cardiac failure. Cardiovasc Res. 2001;49:17–26.
117. Blair E, Redwood C, Ashrafian H, et al. Mutations in the gamma(2) subunit of AMP-activated protein kinase cause familial hypertrophic cardiomyopathy: evidence for the central role of energy compromise in disease pathogenesis. Hum Mol Genet. 2001;10: 1215–20.
118. Tardiff JC, Hewett TE, Palmer BM, et al. Cardiac troponin T mutations result in allele-specific phenotypes in a mouse model for hypertrophic cardiomyopathy. J Clin Invest. 1999;104:469–81.
119. Fananapazir L, Dalakas MC, Cyran F, Cohn G, Epstein ND. Missense mutations in the beta-myosin heavy-chain gene cause central core disease in hypertrophic cardiomyopathy. Proc Natl Acad Sci USA. 1993;90:3993–7.
120. Sayen MR, Gustafsson AB, Sussman MA, Molkentin JD, Gottlieb RA. Calcineurin transgenic mice have mitochondrial dysfunction and elevated superoxide production. Am J Physiol Cell Physiol. 2003;284:C562–70.
121. Scarpulla RC. Nuclear activators and coactivators in mammalian mitochondrial biogenesis. Biochim Biophys Acta. 2002;1576:1–14.
122. Goffart S, Wiesner RJ. Regulation and co-ordination of nuclear gene expression during mitochondrial biogenesis. Exp Physiol. 2003;88:33–40.
123. Lehman JJ, Barger PM, Kovacs A, Saffitz JE, Medeiros DM, Kelly DP. Peroxisome proliferator-activated receptor gamma coactivator-1 promotes cardiac mitochondrial biogenesis. J Clin Invest. 2000;106:847–56.
124. Gilde AJ, van der Lee KA, Willemsen PH, et al. Peroxisome proliferator-activated receptor (PPAR) alpha and PPARbeta/delta, but not PPARgamma, modulate the expression of genes involved in cardiac lipid metabolism. Circ Res. 2003;92:518–24.
125. Barger PM, Kelly DP. PPAR signaling in the control of cardiac energy metabolism. Trends Cardiovasc Med. 2000;10:238–45.
126. Djouadi F, Brandt JM, Weinheimer CJ, Leone TC, Gonzalez FJ, Kelly DP. The role of the peroxisome proliferator-activated receptor alpha (PPAR alpha) in the control of cardiac lipid metabolism. Prostaglandins Leukot Essent Fatty Acids. 1999;60:339–43.
127. Garnier A, Fortin D, Delomenie C, Momken I, Veksler V, Ventura-Clapier R. Depressed mitochondrial transcription factors and oxidative capacity in rat failing cardiac and skeletal muscles. J Physiol. 2003;551:491–501.
128. Huss JM, Levy FH, Kelly DP. Hypoxia inhibits the peroxisome proliferator-activated receptor alpha/retinoid X receptor gene regulatory pathway in cardiac myocytes: a mechanism for O2-dependent modulation of mitochondrial fatty acid oxidation. J Biol Chem. 2001;276:27605–12.
129. Wu H, Kanatous SB, Thurmond FA, et al. Regulation of mitochondrial biogenesis in skeletal muscle by CaMK. Science. 2002;296:349–52.
130. Sack MN, Harrington LS, Jonassen AK, Mjos OD, Yellon DM. Coordinate regulation of metabolic enzyme encoding genes during cardiac development and following carvedilol therapy in spontaneously hypertensive rats. Cardiovasc Drugs Ther. 2000;14:31–9.
131. Bushdid PB, Osinska H, Waclaw RR, Molkentin JD, Yutzey KE. NFATc3 and NFATc4 are required for cardiac development and mitochondrial function. Circ Res. 2003;92:1305–13.
132. Finck BN, Lehman JJ, Leone TC, et al. The cardiac phenotype induced by PPARalpha overexpression mimics that caused by diabetes mellitus. J Clin Invest. 2002;109:121–30.
133. Orfali KA, Fryer LG, Holness MJ, Sugden MC. Long-term regulation of pyruvate dehydrogenase kinase by high-fat feeding. Experiments in vivo and in cultured cardiomyocytes. FEBS Lett. 1993;336:501–5.
134. Doering CB, Danner DJ. Amino acid deprivation induces translation of branched-chain alpha-ketoacid dehydrogenase kinase. Am J Physiol Cell Physiol. 2000;279:C1587–94.
135. Fryer RM, Schultz JE, Hsu AK, Gross GJ. Importance of PKC and tyrosine kinase in single or multiple cycles of preconditioning in rat hearts. Am J Physiol Heart Circ Physiol. 1999;276:H1229–35.
136. Chen L, Hahn H, Wu G, et al. Opposing cardioprotective actions and parallel hypertrophic effects of delta PKC and epsilon PKC. Proc Natl Acad Sci USA. 2001;98:11114–9.
137. Sardanelli AM, Technikova-Dobrova Z, Scacco SC, Speranza F, Papa S. Characterization of proteins phosphorylated by the cAMP-dependent protein kinase of bovine heart mitochondria. FEBS Lett. 1995;377:470–4.
138. He H, Chen M, Scheffler NK, Gibson BW, Spremulli LL, Gottlieb RA. Phosphorylation of mitochondrial elongation factor Tu in

ischemic myocardium: basis for chloramphenicol-mediated cardioprotection. Circ Res. 2001;89:461–7.
139. Lebiedzinska M, Szabadkai G, Jones AW, Duszynski J, Wieckowski MR. Interactions between the endoplasmic reticulum, mitochondria, plasma membrane and other subcellular organelles. Int J Biochem Cell Biol. 2009;41:1805–16.
140. Vandecasteele G, Szabadkai G, Rizzuto R. Mitochondrial calcium homeostasis: mechanisms and molecules. IUBMB Life. 2001;52:213–9.
141. Szabadkai G, Duchen MR. Mitochondria: the hub of cellular $Ca^{2+}$ signaling. Physiology (Bethesda). 2008;23:84–94.
142. Bianchi K, Vandecasteele G, Carli C, Romagnoli A, Szabadkai G, Rizzuto R. Regulation of $Ca^{2+}$ signalling and $Ca^{2+}$-mediated cell death by the transcriptional coactivator PGC-1alpha. Cell Death Differ. 2006;13:586–96.
143. Rutter GA, Rizzuto R. Regulation of mitochondrial metabolism by ER Ca++ release: an intimate connection. Trends Biochem Sci. 2000;25:215–22.
144. Duchen MR. Contributions of mitochondria to animal physiology: from homeostatic sensor to calcium signalling and cell death. J Physiol. 1999;516:1–17.
145. Griffiths EJ. Use of ruthenium red as an inhibitor of mitochondrial Ca(2+) uptake in single rat cardiomyocytes. FEBS Lett. 2000;486:257–60.
146. Pacher P, Hajnoczky G. Propagation of the apoptotic signal by mitochondrial waves. EMBO J. 2001;20:4107–21.
147. McCormack JG, Halestrap AP, Denton RM. Role of calcium ions in regulation of mammalian intramitochondrial metabolism. Physiol Rev. 1990;70:391–425.
148. Robb-Gaspers LD, Burnett P, Rutter GA, Denton RM, Rizzuto R, Thomas AP. Integrating cytosolic calcium signals into mitochondrial metabolic responses. EMBO J. 1998;17:4987–5000.
149. Cortassa S, Aon MA, Marban E, Winslow RL, O'Rourke B. An integrated model of cardiac mitochondrial energy metabolism and calcium dynamics. Biophys J. 2003;84:2734–55.
150. Das AM, Harris DA. Control of mitochondrial ATP synthase in rat cardiomyocytes: effects of thyroid hormone. Biochim Biophys Acta. 1991;1096:284–90.
151. Territo PR, Mootha VK, French SA, Balaban RS. Ca(2+) activation of heart mitochondrial oxidative phosphorylation: role of the F(0)/F(1)-ATPase. Am J Physiol Cell Physiol. 2000;278: C423–35.
152. Rizzuto R, Pinton P, Carrington W, et al. Close contacts with the endoplasmic reticulum as determinants of mitochondrial $Ca^{2+}$ responses. Science. 1998;280:1763–6.
153. Csordas G, Thomas AP, Hajnoczky G. Calcium signal transmission between ryanodine receptors and mitochondria in cardiac muscle. Trends Cardiovasc Med. 2001;11:269–75.
154. Gunter TE, Gunter KK. Uptake of calcium by mitochondria: transport and possible function. IUBMB Life. 2001;52:197–204.
155. Buntinas L, Gunter KK, Sparagna GC, Gunter TE. The rapid mode of calcium uptake into heart mitochondria (RaM): comparison to RaM in liver mitochondria. Biochim Biophys Acta. 2001;1504: 248–61.
156. Crompton M, Costi A, Hayat L. Evidence for the presence of a reversible $Ca^{2+}$-dependent pore activated by oxidative stress in heart mitochondria. Biochem J. 1987;245:915–8.
157. Hajnoczky G, Csordas G, Yi M. Old players in a new role: mitochondria-associated membranes, VDAC, and ryanodine receptors as contributors to calcium signal propagation from endoplasmic reticulum to the mitochondria. Cell Calcium. 2002;32:363–77.
158. Rapizzi E, Pinton P, Szabadkai G, et al. Recombinant expression of the voltage-dependent anion channel enhances the transfer of $Ca^{2+}$ microdomains to mitochondria. J Cell Biol. 2002;159:613–24.
159. Casas F, Rochard P, Rodier A, et al. A variant form of the nuclear triiodothyronine receptor c-ErbAalphal plays a direct role in regulation of mitochondrial RNA synthesis. Mol Cell Biol. 1999;19:7913–24.
160. Scheller K, Seibel P, Sekeris CE. Glucocorticoid and thyroid hormone receptors in mitochondria of animal cells. Int Rev Cytol. 2003;222:1–61.
161. Schneider JJ, Hood DA. Effect of thyroid hormone on mtHsp70 expression, mitochondrial import and processing in cardiac muscle. J Endocrinol. 2000;165:9–17.
162. Oddis CV, Finkel MS. Cytokine-stimulated nitric oxide production inhibits mitochondrial activity in cardiac myocytes. Biochem Biophys Res Commun. 1995;213:1002–9.
163. Zell R, Geck P, Werdan K, Boekstegers P. TNF-alpha and IL-1 alpha inhibit both pyruvate dehydrogenase activity and mitochondrial function in cardiomyocytes: evidence for primary impairment of mitochondrial function. Mol Cell Biochem. 1997;177:61–7.
164. Sammut IA, Harrison JC. Cardiac mitochondrial complex activity is enhanced by heat shock proteins. Clin Exp Pharmacol Physiol. 2003;30:110–5.
165. Bialik S, Cryns VL, Drincic A, et al. The mitochondrial apoptotic pathway is activated by serum and glucose deprivation in cardiac myocytes. Circ Res. 1999;85:403–14.
166. Sparagna GC, Hickson-Bick DL, Buja LM, McMillin JB. Fatty acid-induced apoptosis in neonatal cardiomyocytes: redox signaling. Antioxid Redox Signal. 2001;3:71–9.
167. Gudz TI, Tserng KY, Hoppel CL. Direct inhibition of mitochondrial respiratory chain complex III by cell-permeable ceramide. J Biol Chem. 1997;272:24154–8.
168. Poderoso JJ, Peralta JG, Lisdero CL, et al. Nitric oxide regulates oxygen uptake and hydrogen peroxide release by the isolated beating rat heart. Am J Physiol. 1998;274:C112–9.
169. Wiesner RJ, Hornung TV, Garman JD, Clayton DA, O'Gorman E, Wallimann T. Stimulation of mitochondrial gene expression and proliferation of mitochondria following impairment of cellular energy transfer by inhibition of the phosphocreatine circuit in rat hearts. J Bioenerg Biomembr. 1999;31:559–67.
170. Tanaka T, Morita H, Koide H, Kawamura K, Takatsu T. Biochemical and morphological study of cardiac hypertrophy. Effects of thyroxine on enzyme activities in the rat myocardium. Basic Res Cardiol. 1985;80:165–74.
171. Kennedy SG, Kandel ES, Cross TK, Hay N. Akt/Protein kinase B inhibits cell death by preventing the release of cytochrome c from mitochondria. Mol Cell Biol. 1999;19:5800–10.
172. Shioi T, McMullen JR, Kang PM, et al. Akt/protein kinase B promotes organ growth in transgenic mice. Mol Cell Biol. 2002;22:2799–809.
173. Condorelli G, Drusco A, Stassi G, et al. Akt induces enhanced myocardial contractility and cell size in vivo in transgenic mice. Proc Natl Acad Sci USA. 2002;17:12333–8.
174. Liu T, Lai H, Wu W, Chinn S, Wang PH. Developing a strategy to define the effects of insulin-like growth factor-1 on gene expression profile in cardiomyocytes. Circ Res. 2001;88:1231–8.
175. Cook SA, Matsui T, Li L, Rosenzweig A. Transcriptional effects of chronic Akt activation in the heart. J Biol Chem. 2002;277:22528–33.
176. Edinger AL, Thompson CB. Akt maintains cell size and survival by increasing mTOR-dependent nutrient uptake. Mol Biol Cell. 2002;13:2276–88.
177. Nebigil CG, Etienne N, Messaddeq N, Maroteaux L. Serotonin is a novel survival factor of cardiomyocytes: mitochondria as a target of 5-HT2B receptor signaling. FASEB J. 2003;17:1373–5.
178. Matsui T, Tao J, del Monte F, et al. Akt activation preserves cardiac function and prevents injury after transient cardiac ischemia in vivo. Circulation. 2001;104:330–5.
179. Krieg T, Qin Q, McIntosh EC, Cohen MV, Downey JM. ACh and adenosine activate PI3-kinase in rabbit hearts through transactivation

of receptor tyrosine kinases. Am J Physiol Heart Circ Physiol. 2002;283:H2322–30.

180. Li Y, Sato T. Dual signaling via protein kinase C and phosphatidylinositol 3′-kinase/Akt contributes to bradykinin B2 receptor-induced cardioprotection in guinea pig hearts. J Mol Cell Cardiol. 2001;33:2047–53.
181. Chandel NS, Schumacker PT. Cellular oxygen sensing by mitochondria: old questions, new insight. J Appl Physiol. 2000;88:1880–9.
182. Duranteau J, Chandel NS, Kulisz A, Shao Z, Schumacker PT. Intracellular signaling by reactive oxygen species during hypoxia in cardiomyocytes. J Biol Chem. 1998;273:11619–24.
183. Kacimi R, Long CS, Karliner JS. Chronic hypoxia modulates the interleukin-1beta-stimulated inducible nitric oxide synthase pathway in cardiac myocytes. Circulation. 1997;96:1937–43.
184. French S, Giulivi C, Balaban RS. Nitric oxide synthase in porcine heart mitochondria: evidence for low physiological activity. Am J Physiol Heart Circ Physiol. 2001;280:H2863–7.
185. Kanai AJ, Pearce LL, Clemens PR, et al. Identification of a neuronal nitric oxide synthase in isolated cardiac mitochondria using electrochemical detection. Proc Natl Acad Sci USA. 2001;98:14126–31.
186. Kulisz A, Chen N, Chandel NS, Shao Z, Schumacker PT. Mitochondrial ROS initiate phosphorylation of p38 MAP kinase during hypoxia in cardiomyocytes. Am J Physiol Lung Cell Mol Physiol. 2002;282:L1324–9.
187. Enomoto N, Koshikawa N, Gassmann M, Hayashi J, Takenaga K. Hypoxic induction of hypoxia-inducible factor-1alpha and oxygen-regulated gene expression in mitochondrial DNA-depleted HeLa cells. Biochem Biophys Res Commun. 2002;297:346–52.
188. Lopaschuk GD, Collins-Nakai RL, Itoi T. Developmental changes in energy substrate use by the heart. Cardiovasc Res. 1992;26:1172–80.
189. Bonnet D, Martin D, De Pascale L, et al. Arrhythmias and conduction defects as presenting symptoms of fatty acid oxidation disorders in children. Circulation. 1999;100:2248–53.
190. Damle S, Marín-García J. Mitochondrial uncoupler proteins. Curr Enz Inhib. 2010;6:1–10.
191. Lanni A, De Felice M, Lombardi A, et al. Induction of UCP2 mRNA by thyroid hormones in rat heart. FEBS Lett. 1997; 418:171–4.
192. Boehm EA, Jones BE, Radda GK, Veech RL, Clarke K. Increased uncoupling proteins and decreased efficiency in palmitate-perfused hyperthyroid rat heart. Am J Physiol Heart Circ Physiol. 2001;280:H977–83.
193. Young ME, Patil S, Ying J, et al. Uncoupling protein 3 transcription is regulated by peroxisome proliferator-activated receptor (alpha) in the adult rodent heart. FASEB J. 2001;15:833–45.
194. Barbe P, Larrouy D, Boulanger C, et al. Triiodothyronine-mediated up-regulation of UCP2 and UCP3 mRNA expression in human skeletal muscle without coordinated induction of mitochondrial respiratory chain genes. FASEB J. 2001;15:13–5.
195. MacLellan JD, Gerrits MF, Gowing A, Smith PJ, Wheeler MB, Harper ME. Physiological increases in uncoupling protein 3 augment fatty acid oxidation and decrease reactive oxygen species production without uncoupling respiration in muscle cells. Diabetes. 2005;54:2343–50.
196. Bienengraeber M, Ozcan C, Terzic A. Stable transfection of UCP1 confers resistance to hypoxia/reoxygenation in a heart-derived cell line. J Mol Cell Cardiol. 2003;35:861–5.
197. Echtay KS, Roussel D, St-Pierre J, et al. Superoxide activates mitochondrial uncoupling proteins. Nature. 2002;415:96–9.
198. Echtay KS, Murphy MP, Smith RA, Talbot DA, Brand MD. Superoxide activates mitochondrial uncoupling protein 2 from the matrix side. Studies using targeted antioxidants. J Biol Chem. 2002;277:47129–35.
199. Echtay KS, Esteves TC, Pakay JL, et al. A signalling role for 4-hydroxy-2-nonenal in regulation of mitochondrial uncoupling. EMBO J. 2003;22:4103–10.
200. Teshima Y, Akao M, Jones SP, Marban E. Uncoupling protein-2 overexpression inhibits mitochondrial death pathway in cardiomyocytes. Circ Res. 2003;93:192–200.
201. Essop MF, Razeghi P, McLeod C, Young ME, Taegtmeyer H, Sack MN. Hypoxia-induced decrease of UCP3 gene expression in rat heart parallels metabolic gene switching but fails to affect mitochondrial respiratory coupling. Biochem Biophys Res Commun. 2004;314:561–4.
202. Zhou M, Lin BZ, Coughlin S, Vallega G, Pilch PF. UCP-3 expression in skeletal muscle: effects of exercise, hypoxia, and AMP-activated protein kinase. Am J Physiol Endocrinol Metab. 2000;279:E622–9.
203. Sanguinetti MC, Bennett PB. Antiarrhythmic drug target choices and screening. Circ Res. 2003;93:491–9.
204. Ito H, Taniyama Y, Iwakura K, et al. Intravenous nicorandil can preserve microvascular integrity and myocardial viability in patients with reperfused anterior wall myocardial infarction. J Am Coll Cardiol. 1999;33:654–60.
205. Shoffner JM, Wallace DC. Oxidative phosphorylation diseases and mitochondrial DNA mutations: diagnosis and treatment. Annu Rev Nutr. 1994;14:535–68.
206. Rustin P, von Kleist-Retzow JC, Chantrel-Groussard K, Sidi D, Munnich A, Rotig A. Effect of idebenone on cardiomyopathy in Friedreich's ataxia: a preliminary study. Lancet. 1999;354:477–9.
207. Wallhaus TR, Taylor M, DeGrado TR, et al. Myocardial free fatty acid and glucose use after carvedilol treatment in patients with congestive heart failure. Circulation. 2001;103:2441–6.
208. Pollitt RJ. Disorders of mitochondrial long-chain fatty acid oxidation. J Inherit Metab Dis. 1995;18:473–90.
209. Pepe S, Tsuchiya N, Lakatta EG, Hansford RG. PUFA and aging modulate cardiac mitochondrial membrane lipid composition and $Ca^{2+}$ activation of PDH. Am J Physiol. 1999;276:H149–58.
210. Ennis IL, Li RA, Murphy AM, Marban E, Nuss HB. Dual gene therapy with SERCA1 and Kir2.1 abbreviates excitation without suppressing contractility. J Clin Invest. 2002;109:393–400.
211. Beltrami AP, Barlucchi L, Torella D, et al. Adult cardiac stem cells are multipotent and support myocardial regeneration. Cell. 2003;114:763–76.

# Part IV
# Pediatric Cardiology

# Chapter 9
# Signaling Pathways in Cardiovascular Development

**Abstract** Cardiogenesis is an active and complex process with many and coordinated phases of pattern formation, morphogenesis, and regulation of gene expression. Although the field of signaling in general and in particular signaling in cardiovascular development is still in its infancy, it is clear that signaling pathways are participatory from a number of surrounding cell types, including the endoderm and epicardium and from the anterior field constituting a complex milieu of inductive signaling. The information available on signaling pathways has come to light mainly as a by-product of research effort to understand the mechanisms of cardiac hypertrophy, cell growth and proliferation, cell death, and myocardial remodeling. While our understanding of cardiac pathology has grown rapidly in recent years, the basic underlying mechanisms of many specific cardiovascular diseases, including the signaling pathways, remain largely unknown. In this chapter, we discuss recent progress that is making possible to understand the molecular circuitry that governs complex network of signaling pathways, which are responsive to both physiological stimuli and genetic programming, and also the critical role that these signaling pathways play in morphogenesis and regulation of gene expression during cardiovascular development.

**Keywords** Cardiac development • Gene expression • Transcription factors • Protein kinases • Marker genes • Cardiac precursors

## Introduction

A functional heart is the first organ to form during progressive embryonic development. The synthesis and integration of precisely patterned modular elements are all necessary during the development of the heart chambers and valves. Knowledge of the pathways involved in both the genesis and integration of these individual modular elements is fundamental to our understanding of congenital heart defects (CHDs), which mainly result from defects in specific structural components of the developing heart.

Normal formation and integration of the heart modular elements requires a complex interplay of many genes and transcriptional factors whose cell type-specific expression is highly organized and precisely regulated at the spatial and temporal levels. These regulatory transcription factors can operate differently in various cell types and respond to a variety of intracellular and extracellular signals that modulate the precise integration of gene expression and morphological development. Their importance is underscored by the fact that the great majority of identified mutations, which lead to specific CHD, reside in genes encoding transcription factors. Furthermore, the heart through its signaling pathways functions both as a transmitter and a dynamic receiver of multiple intracellular and intercellular stimuli and also as an integrator of numerous interacting transducers that include protein kinases and effectors, the G proteins, and small G protein activators which are profoundly influenced by their location in the cell. Indeed, the targeting and localization of signaling factors and enzymes to discrete subcellular compartments or substrates become important regulatory mechanisms that secure specificity of signaling events in response to local stimuli. These systems merit examination both from a subcellular/organellar and from a functional standpoint under both physiological and pathophysiological conditions. The focus of this chapter is mainly a discussion on cell signaling pathways during cardiac growth and development.

## Cardiac Development and Gene Expression

During embryonic development, the genesis of somatic, visceral, and heart muscle is all derived from mesoderm progenitor cells, and it requires the coordinated participation of multiple genes and signaling pathways. Gene disruption in mice and large-scale mutagenesis screens in zebrafish have proved valuable in delineating fundamental genetic pathways governing early heart patterning and differentiation. A number of genes have been identified that play important and selective roles in cardiac valve development, in endothelial and cardiac cell proliferation, and in the early development of the

J. Marín-García, *Signaling in the Heart*,
DOI 10.1007/978-1-4419-9461-5_9, © Springer Science+Business Media, LLC 2011

**Table 9.1** Participating genes in heart development

| Cardiac crescent | Linear heart tube | Looping heart | Chamber formation | Chamber maturation/septation. Valve formation |
|---|---|---|---|---|
| *Nkx2-5* | *GATA4* | *Nkx2-5* | *Nkx2-5* | *Nkx2-5* |
| *Myocardin* | *Mespl* | *MEF2C* | *Irx4* | *RXR-a* |
| *SRF* | *Mesp2* | *Tbx5* | *Tbx5* | *Tbx5* |
| *Mespl* | | *dHAND* | *dHAND* | *NFATc3* |
| *Mesp2* | | *eHAND* | *eHAND* | *GATA4* |
| | | *Irx4* | *Pitx2* | *Fog-2* |
| | | *RA* | *Smad6* | *ErbB* |
| | | | | *Sox4* |
| | | | | *HF-1B* |
| | | | | *PitX2* |
| | | | | *CITED2* |
| | | | | *PAX3* |

cardiac pacemaking and conduction system (Table 9.1). In addition, several distinctive transcriptional programs, as seen in the patterning of expression of a number of genes, direct the compartmentalization of the forming heart.

## Signaling During Cardiac Development

Studies of the developing heart have unveiled a complex molecular circuitry in which gene regulation governs a network of signaling pathways responsive to both physiological stimuli and genetic programming [1]. Signaling stimuli such as bone morphogenetic protein (BMP) acting on receptors (BMPRs) and specific transcription factors (e.g., Nkx2-5 and GATA) are critical elements in early cardiac development.

The GATA family of transcription factors was initially identified as an essential regulator of the two natriuretic peptide genes, ANP and BNP, which are important markers of cardiomyocyte differentiation, with distinct spatial, developmental, and hormonal regulation [2]. The analysis of the ANP and BNP promoters in association with GATA binding has led to the delineation of combinatorial interactions of multiple factors required for proper regulation of the cardiac genes as a key regulator of myocardial development [2, 3].

Interestingly, the role played by the GATA family of transcription factors (Fig. 9.1) emerged from using a variety of methodological approaches. The use of standard molecular biology techniques, gene transfections, and transgenic animals proved to be complementary in these studies [4–6]. GATA4 is also an essential component of the retinoid-mediated cardiogenic pathway involved in the morphogenesis of the posterior heart tube and the development of the cardiac inflow tract [7]. The importance of GATA as a developmental factor also was established from evidence [8] of its early expression in a variety of developmental models and was further underscored by clinical studies demonstrating that specific mutations in GATA were associated with severe CHD [9] and atrial septal defects [10]. Other studies employing antisense constructs to block GATA expression further supported an important role that GATA has in cardiac development. More recently, GATA4 has emerged as the nuclear effector of several signaling pathways involved in cardiac growth, development, and differentiation whose transcriptional activity function can be modulated by posttranslational modifications [e.g., p38 mitogen-activated protein kinase (MAPK)-mediated phosphorylation and acetyltransferase p300-mediated acetylation] and protein–protein interactions [e.g., interaction with transcriptional repressor Jumonji (JMJ)] [11–15]. BMP, an indispensable factor for cardiomyocyte differentiation, acts in part by inducing the expression of two cardiac transcription factors, Csx/Nkx2-5 and GATA4 [16]. Moreover, dynorphin B, a product of the prodynorphin gene, has been found to promote cardiogenesis in embryonic cells by inducing the expression of GATA4 and Nkx2-5 (Fig. 9.1) [17].

Recent studies in vertebrates have demonstrated the vital nature of calcium/calcineurin/nuclear factor of activated T-cells (NFAT) signaling in cardiac muscle development. Inhibition, mutation, or forced expression of calcineurin pathway genes result in defects or alterations in cardiomyocyte maturation, heart valve formation, vascular development, skeletal muscle differentiation, fiber-type switching, and cardiac hypertrophy. Interestingly, a family of proteins previously considered under the domain of interest of neuroscientists has been shown to possess essential cardiovascular functions. The neurotrophin family and their receptors are essential factors in the formation of the heart and in the regulation of vascular development. It appears that the survival of endothelial cells, vascular smooth muscle cells (SMCs), and cardiomyocytes are under neurotrophin control, regulating angiogenesis and vasculogenesis by autocrine and paracrine mechanisms. Moreover, via their tropomyosin-kinase receptors, neurotropins seem to promote therapeutic neovascularization in animal models of hindlimb ischemia. On the other hand,

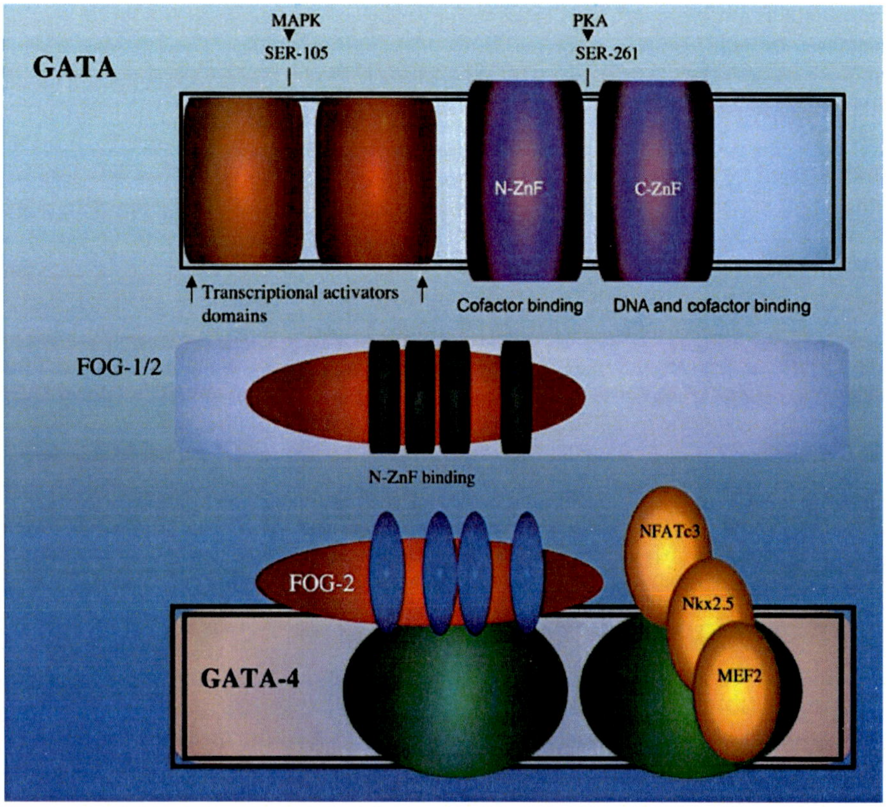


**Fig. 9.1** GATA family of transcription factors

the neurotrophin low-affinity p75$^{NTR}$ receptor induces apoptosis of endothelial cells and vascular SMCs and impairs angiogenesis. Nerve growth factors may hold promise in the treatment of the microvascular complications of diabetes or reducing cardiomyocyte apoptosis in the infarcted heart [18].

The serendipitous discovery of the convergence of several signaling pathways that regulate endothelial proliferation and differentiation in developing and postnatal heart valves has provided insight into the roles played by highly conserved signaling elements including vascular endothelial growth factor (VEGF), NFATc1, Notch, Wnt/β-catenin, BMP/transforming growth factor beta (TGF-β), epidermal growth factor (EGF) receptor ErbB, and neurofibromin 1 (NF-1)/Ras.

## Signaling the Cardiomyocyte During Physiological Growth

Several signal transduction pathways regulate cardiomyocyte growth and/or proliferation. These are redundant mechanisms converging on one or several serine/threonine kinases. Several G protein-coupled receptors (GPCRs) such as β-adrenergic receptors (β-ARs), and receptors for angiotensin II, and endothelin-1 (ET-1) are able to activate these signaling cascades and induce changes in cell growth and proliferation. A general scheme involves the following: external signals received at the plasma membrane by GPCRs are transmitted via G proteins/second messengers to a wide-spectrum of protein kinases and phosphatases, which are in turn activated. These activated protein modifiers may lead to the activation and/or deactivation of specific transcription factors, which modulate specific gene expression affecting a broad spectrum of cellular events, or they can target directly proteins involved in metabolic pathways, transport, regulation, and handling of various ions, which affect contractility and excitability, as well as the pathways of cardiomyocyte apoptosis and/or cell survival.

Cardiomyocytes rapidly proliferate in the embryo but exit the cell-cycle irreversibly shortly after birth, with the predominant form of growth shifting from hyperplastic to hypertrophic. Extensive research has focused on identifying the mitogenic stimuli and signaling pathways that mediate these distinct growth processes in isolated cells and *in vivo* hearts. The molecular mechanisms underlying the proliferative growth of embryonic myocardium *in vivo* and adult cardiac myocyte hypertrophy remain largely unknown, although considerable progress has been made using postgenomic analysis including studies involving manipulation of the murine genome in concert with mutational analysis of these signaling and growth control pathways *in vivo* and in cardiomyocytes grown *in vitro*. For instance, cell-cycle control can be mediated by p38 MAPK which regulates the expression of genes required for mitosis in cardiomyocytes including cyclin A and cyclin B. Cardiac-specific p38 MAPK knockout mice show a 92% increase in neonatal cardiomyocyte mitosis. Furthermore, inhibition of p38 MAPK promotes cytokinesis in adult cardiomyocytes [19]. Collective observations have indicated that cyclin D1, a cell-cycle regulator involved in promoting the $G_1$-to-S phase progression via phosphorylation of the retinoblastoma (Rb) protein, is localized in the nucleus of fetal cardiomyocytes but is primarily cytoplasmic in neonatal and adult cardiomyocytes (concomitant with Rb underphosphorylation). Ectopic expression of a variant of cyclin D1 equipped with nuclear localization signals dramatically promoted neonatal cardiomyocyte proliferation and Rb phosphorylation [20]. Growth factors such as fibroblast growth factor (FGF)-2 significantly promote neonatal cardiac myocyte proliferation [21]. Similarly, expression of FGF2 receptor (FGFR1) in rat heart H9c2 myoblasts increased cell proliferation [22]. Cardiotrophin (CT-1), an interleukin 6-related cytokine, has been shown to promote both the survival and proliferation of cultured neonatal cardiac myocytes [23]. This is likely mediated by the phosphatidylinositol 3-kinase (PI3K)-Akt pathway since CT-1 phosphorylates and activates Akt [24]. These diverse approaches have confirmed the importance of suspected pathways as well as implicated unexpected pathways leading to new paradigms for the control of cardiac growth.

## Cell Differentiation and Mesoderm Development

The formation of somatic, visceral and heart muscle from mesoderm progenitor cells requires a coordinated involvement of numerous genes and signals. Subsequently, several distinctive transcriptional programs, as seen in the patterning of expression of a number of genes, direct the compartmentalization of the forming heart.

While the initiation of cardiac differentiation has been a topic of extensive investigation, no single transcription factor has yet been identified that is solely responsible for the differentiation of lateral plate mesoderm into cardiac cells. Rather, several factors appear to play a central role.

In vertebrates, the homologue of the *tinman* gene (originally discovered in the fruit fly *Drosophila melanogaster*), the cardiac gene *Nkx2-5*, has been shown to be crucial for cardiac differentiation processes including the establishment or maintenance of a ventricular gene expression program. Very early pattern of *Nkx2-5* expression in the developing myocardium made it initially a primary candidate for a regulator of cardiogenesis; however, it is not essential for the specification of the heart cell lineage. Gene targeting of *Nkx2-5* showed that this gene is not required for heart tube formation (it is normal in mutant mice), but is required for the completion of the looping morphogenesis of the heart [25].

Factors that regulate the expression of Nkx2-5 (as well as other transcription factors) are important in early cardiac differentiation. Serum-response factor (SRF) has been shown to be an obligatory transcription factor, required for the formation of vertebrate mesoderm leading to the origin of the cardiovascular system. SRF is required for the serum induction of immediate early genes such as c-fos and for the downstream expression of many muscle- and cardiac-specific genes. The difficulty of studying in transgenic mice factors such as SRF, which are required for mesoderm formation (knockout resulting in embryonic lethality) has been overcome by generating a conditional mutant of SRF using a Cre-LoxP strategy [26]. Mice containing a heart-specific deletion of SRF displayed lethal cardiac defects between embryonic day E10.5 and E13.5, as evidenced by abnormally thin myocardium, dilated cardiac chambers, poor trabeculation, and a disorganized interventricular septum preceded at E9.5 (prior to overt maldevelopment) by a marked reduction in the expression of transcription regulators including Nkx2-5, GATA4, and myocardin. In addition, SRF null embryonic stem (ES) cells used as a model system to investigate the specification of multiple embryonic lineages, including cardiac myocytes, exhibited the absence of myogenic α-actin and myocardin expression and the failure to form beating cardiac myocytes [27].

Myocardin belongs to the SAP (SAF-A/B, Acinus, and PIAS) domain family of nuclear proteins and activates cardiac muscle promoters by associating with SRF. Experiments in *Xenopus* embryos with the use of a dominant-negative myocardin molecule indicate that it may be necessary for early stages of cardiac differentiation, including high-level gene expression of *Nkx2-5* [28]. Myocardin-related family of transcription factors plays a critical role in the transcriptional program regulating SMC differentiation. The function of myocardin-related transcription factor (MRTF)-B has been examined in mice harboring a conditional insertional mutation [29]. These studies found that MRTF-B plays a critical role in regulating differentiation of cardiac neural crest cells (NCCs) into SMC and showed that NCC-derived SMC differentiation is specifically required for normal cardiovascular morphogenesis. Further support for myocardin participation in cardiac differentiation comes from recent studies in which inhibition of myocardin function in the teratocarcinoma cell line P19CL6 prevents differentiation into cardiac myocytes [30]. On the other hand, forced expression of myocardin was not found to be sufficient for induction of SMC differentiation in multipotential embryonic cells since overexpression of myocardin induced only a subset of SMC marker genes [31].

It is important to point out that cardiac development intrinsically depends not only on cardiac lineages but also on extracardiac factors. Davidson and Levine [32] in their analysis of cardiac lineage in the ascidian *Ciona intestinalis* noted that cardiac lineage shares a common origin with the germ line, and zygotic transcription is first detected in the heart progenitors only after its separation from the germ line at the 64-cell stage. Germ-line determinants may influence the specification of the cardiac mesoderm, both by inhibiting inductive signals required for the development of noncardiac mesoderm lineages and by providing a localized source of Wnt5 and other signals required for heart development. Likely, the germ line influences the specification of the vertebrate heart. Moreover, on their migration, NCCs pass the second field and the interaction with a variety of cell population results in a complex differentiation schedule that may lead to cardiac defects. NCCs are important in the differentiation of the pharyngeal arch arteries, outflow tract (OFT), valves formation, and conduction system [33]. On the other hand, besides NCC other extracardiac cell populations contribute to cardiovascular development such as the epicardium-derived cells (EPDCs) that will be further discussed later.

As with many of the cardiac transcription factors, interactions with other cofactors can modulate their function. A novel SRF cofactor, called p49/STRAP, for SRF-dependent transcription regulation-associated protein, was recently identified [34]. This protein interacts mainly with the transcription activation domain of the SRF protein and binds to SRF or to the complex of SRF and another cofactor, such as myocardin or Nkx2-5. The expression of p49/STRAP differentially affected the promoter activity of SRF target genes, activating myosin light chain 2V (MLC2V) and cardiac actin promoters upon co-transfection with SRF, while repressing atrial natriuretic factor (ANF) expression which was strongly induced by myocardin. While strongly expressed in fetal heart, this cofactor role in early cardiac differentiation has not yet been determined.

The GATA family of zinc finger-containing transcription factors comprise another significant group of transcription factors that appears to contribute to the activation of the cardiac-specific gene program involved in cardiac cell differentiation. Three GATA family genes have been identified as being expressed in the developing heart: *gata4*, *gata5*, and *gata6*. A characteristic feature shared by the GATA factors (Fig. 9.1) is a central domain composed of two adjacent zinc fingers; the N-terminal zinc finger is involved in binding of some protein cofactors (e.g., friend of GATA [FOG] transcription factors), whereas the C-terminal zinc finger contains the DNA sequence recognition domain (binding to specific promoters) and the binding site for the majority of cofactors (e.g., dHAND, GATA6, myocyte enhancer factor [MEF]-2, Nkx2-5, p300, and SRF) [35].

GATA family members have been implicated as key regulators of cardiogenesis in several model systems [36]. Transgenic mice with inactivation of the *GATA4* gene die during embryonic development due to failure of ventral morphogenesis and heart tube formation [37]. Embryos of *GATA4*$^{-/-}$ mice developed splanchnic mesoderm differentiated into primitive cardiac myocytes that expressed contractile protein but failed to form a linear heart tube indicating that *GATA4* is not essential for the specification of the cardiac cell lineages while suggesting a critical role for *GATA4* in early cardiac morphogenesis [38]. These studies also noted that upregulation of endogenous *GATA6* mRNA could potentially compensate for the lack of *GATA4*. Using the cardiac differentiation model of pluripotent P19 embryonal carcinoma cells, which can be differentiated into beating cardiac muscle cells, inhibition of *GATA4* with antisense oligonucleotides blocked differentiation to beating cardiac muscle cells and interfered with the expression of cardiac muscle markers [39]. In the absence of *GATA4*, differentiation is blocked at the precardiac (cardioblast) stage and cells are lost through extensive apoptosis; however, these studies also demonstrated that the mesoderm commitment was not affected in GATA4-deficient cells as gauged by specific marker levels (brachyury and goosecoid). Overexpression of GATA4 increases differentiation of P19 cells to beating cardiomyocytes; however, this effect requires cell aggregation suggesting that GATA4 may relay on cell-contact generated signals required for cardiomyocyte differentiation and survival [40]. There is also evidence that GATA function is shared by other members of the redundant GATA family suggesting that GATA5 and GATA6 can at least partially compensate for a lack of GATA4 [41, 42]. The mechanisms determining GATA factor specificity are not fully understood and may involve interaction among GATA factors, or interactions between GATA factors and other cofactors differentially controlled at various stages of cardiogenesis. A number of highly conserved cofactors which interact with GATA have been identified, many via the yeast 2-hybrid technology. These include the *gata4* homologue *pannier* required for normal proliferation of cardiogenic precursors and the multitype zinc-finger proteins *U-shaped* (*USH*) and *FOGs-1* and *-2*, the latter are also required for cardiac development but more likely in the outlet tract and atrioventricular valves of the heart as we shall shortly examine [42].

In addition, recent evidence indicates that a posttranslational modification of the transcription factors can modulate cardiac differentiation programming, a potential form of epigenetic control of differentiation. The histone acetyltransferase p300 is required for the acetylation and DNA binding of GATA4 and for its full transcriptional activity as well as for promotion of a transcriptionally active chromatin configuration.

In an embryonic stem cell model of developing embryoid bodies, an acetylated forms of GATA4, histones 3 and 4 are increased along with the expression of p300 during the differentiation of ES cells into cardiac myocytes [13]. Treatment of embryonic cells with trichostatin A, a specific histone deacetylase (HDAC) inhibitor, augmented the increase in an acetylated form of GATA4 and its DNA binding during the ES cell differentiation, and increased the expression of green fluorescence protein under the control of the cardiac-specific *Nkx2-5* promoter and of endogenous cardiac β-myosin heavy chain suggesting that acetylation of GATA4 is involved in cardiac cell differentiation.

## *Cardiac Precursors Differentiation*

Formation of the embryonic heart tube requires the determination or specification of heart precursor cells within defined areas of the mesoderm, their intracellular organization (often in an epithelial monolayer), medial migration, and the merger of bilateral precursor populations. Increasing evidence supports the notion that mechanisms and molecular players involved in these highly complex processes are highly conserved [43, 44]. With the advent of exciting new molecular and cellular technologies for the study of events of early cardiac development, many of the initially well-established facts have had to be substantially revised. For instance, the concept of the symmetrical nature of the bilateral cardiac precursors had to be modified. Moreover, anterior and posterior polarity appears to be established at the early stages of cardiac differentiation. Anterior heart progenitors differ from the posterior heart progenitors in their myosin isoform gene expression [45].

The heart is derived from the anterior splanchnic mesoderm. It forms from two crescent-like cardiogenic plates that already express cardiac-specific genes like *Nkx2-5* and *GATA4*. It is well established that heart induction in both vertebrates and nonvertebrates such as *Drosophila* is primarily dependent on the interaction of the heart primordia with the Spemann organizer. Signals from the Spemann organizer during gastrulation are a primary determinant of the dorsal mesoderm specification. Several genes encoding transcription factors are expressed specifically in the Spemann organizer region of the gastrula. Expression of one of these genes, the homeobox gene *goosecoid*, has been shown to be sufficient to elicit the formation of a dorsal axis in the embryo [46]. The genes encoding the BMP antagonists noggin, a small polypeptide that can induce dorsal development in *Xenopus* embryos, and chordin are primary dorsalizing signals from the organizer [47]. In *Xenopus* as well as other organisms, Nodal-related TGF-β family signals including XNr-1, a *Xenopus* homologue of the mouse Nodal protein, play a key role in organizer formation and function [48–50]. Evidence from studies in several organisms suggests that the gastrula organizer is populated by a succession of cell populations with different fates as opposed to population with a single lineage [51]. Activation of the BMP pathway prior to the onset of gastrulation can promote the suppression of the Spemann organizer formation in *Xenopus* embryos [52].

In addition to signals from the organizer, inductive signals derived from surrounding tissues (e.g., endoderm and ectoderm) during early gastrulation contribute to mesoderm differentiation. For instance, the presence of deep endoderm can dramatically enhance heart formation in explants of heart primordia, both in the presence and absence of organizer [53] while ablation of the entire endoderm can decrease the frequency of heart formation in embryos that retain organizer activity [54].

An early and important event in the regional subdivision of the mesoderm is the restriction of *Nkx2-5* expression to dorsal mesodermal cells. An inductive signal from dorsal ectodermal cells is required for the activation of *Nkx2-5*, in the underlying dorsal mesoderm in *Drosophila*; decapentaplegic (Dpp), a member of the TGF-β family, serves as a pivotal signaling molecule in this process [55]. A second secreted signaling molecule, wingless (Wg), has also been shown to be involved in cardiac mesoderm specification [56]. These studies have also been extended to demonstrate that various combinations of inductive signals and mesoderm-intrinsic transcription factors can cooperate to induce the progenitors of heart precursors at precisely defined positions within the mesoderm layer. Positive Dpp signals and antagonistic Wg inputs are integrated by virtue of combinatorial binding sites on a mesoderm-specific NK homeobox gene that functions as an early regulator of mesoderm development. Binding sites for Dpp-activated Smad proteins have been identified as having adjacent binding sites for the FoxG transcription factors which are direct targets of the Wg signaling cascade. Binding to the second site blocks the activity of activated Smads [57]. Recent studies have shown that the GATA transcription factor, Pannier, mediates as well as maintains the cardiogenic Dpp signal in both germ layers [58].

In chick embryos, endoderm-secreted activin A and FGF2 have been shown to regulate early cardiac differentiation [59]. Combined treatment with BMP2 and fibroblast FGF4 can induce cardiogenic events culminating in full cardiac differentiation of nonprecardiac mesoderm explanted from stage 6 avian embryos [60]. BMP2, like its *Drosophila* homologue Dpp, is an important signaling molecule for specification of cardiogenic mesoderm in vertebrates. The induction of cardiac lineage markers in central mesendoderm (*Nkx2-5*, *GATA4*, *eHAND*, *MEF2A*, and *vMHC*) exhibited a distinct time course with respect to BMP2-dependent induction; *Nkx2-5*, *GATA4*, and *MEF2A* were induced within 6 h of BMP2 treatment, *eHAND* and *dHAND* required 12 h while structural markers *vMHC* and *titin* were induced at significant levels only after 48 h of BMP2 addition [61]. BMPs apparently function, in part, by affecting the levels of these transcription factors and work in parallel to FGF signaling from the cardiac mesoderm.

FGF8 signaling also contributes to the heart-inducing properties of the endoderm [62]. *Fgf8* is highly expressed in endoderm adjacent to the precardiac mesoderm. The rapid downregulation of cardiac markers, including Nkx2-5 and MEF2C which results from the removal of endoderm can be reversed by supplying exogenous FGF8. Expression of cardiac markers is increased only in regions where BMP signaling is also present, suggesting that cardiogenesis occurs in regions exposed to both FGF and BMP signaling. These studies also showed that *Fgf8* expression is regulated by BMP signaling with low levels of BMP2 resulting in ectopic expression of *Fgf8*, while with higher levels of BMP2 resulting in repression of *Fgf8* expression. The Heartless (Htl) FGF receptor is required for the differentiation of a variety of mesodermal tissues in the *Drosophila* embryo [63]. Null *htl* mutant embryos display irregular migration and spreading of the mesoderm over the ectoderm. A common role for Htl in both directional mesoderm cell migration and pattern formation underlies the pleiotropic defects of the *htl* mutation. Studies have indicated that the GATA-related *USH* cofactor interacts directly with Htl receptor in determining mesoderm migration [64].

Two novel *Fgf* genes, *thisbe* (*ths*) and *pyramus* (*pyr*), have been recently identified which may encode the elusive ligands for the Htl [65]. The two genes exhibit dynamic patterns of expression in epithelial tissues adjacent to Htl-expressing mesoderm. Embryos that lack *ths* and *pyr* display defects related to those seen in *htl* mutants, including delayed mesodermal migration during gastrulation and a loss of cardiac tissues.

Wnt signals are transduced via the canonical pathway (Fig. 9.2a) for cell fate determination and via the noncanonical pathway (Fig. 9.2b) to control cell movement and tissue polarity. The canonical Wnt pathway has been shown to play a significant role in mesoderm specification, and it appears to be primarily active in the OFT of the heart, whereas noncanonical Wnt signaling is present in the ventricular and atrial regions.

Thus, noncanonical Wnt11 seems to exhibit a prominent bi-phasic expression pattern influencing key events associated with cardiac development and suggesting that the heart is composed of many Wnt signaling pathway components and that Wnt11 may promote cardiac differentiation through suppression of the canonical Wnt pathway [66]. In an interesting review on transcriptional pathways in second heart field (SHF) development, Black [67] noted that SHF gives rise to the right ventricle (RV) and OFT (Fig. 9.3) in birds and mammals [68–71].

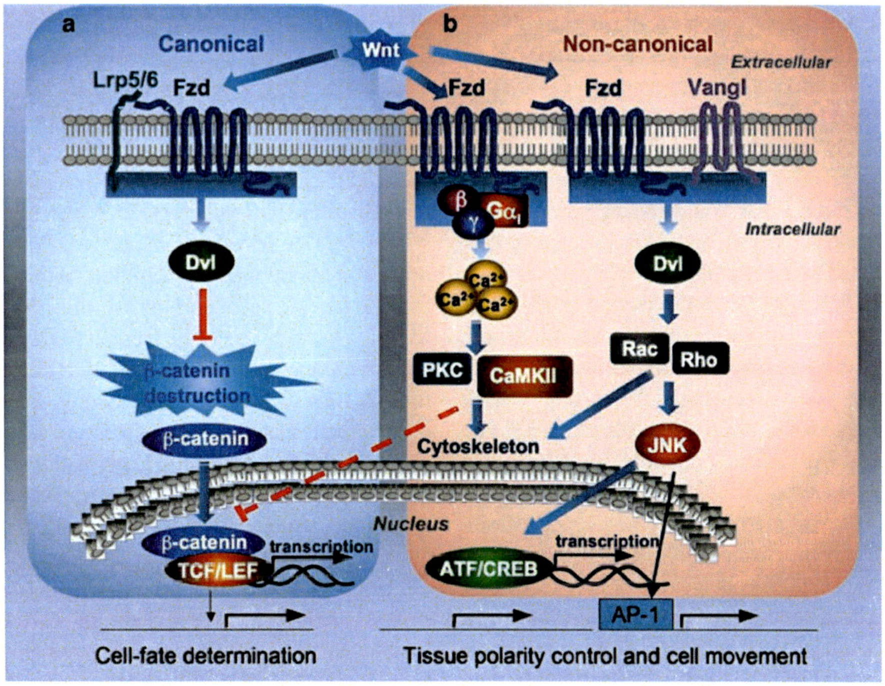


**Fig. 9.2** Wnt signals are transduced via the canonical pathway (**a**) for cell fate determination and via the noncanonical pathway (**b**) to control cell movement and tissue polarity

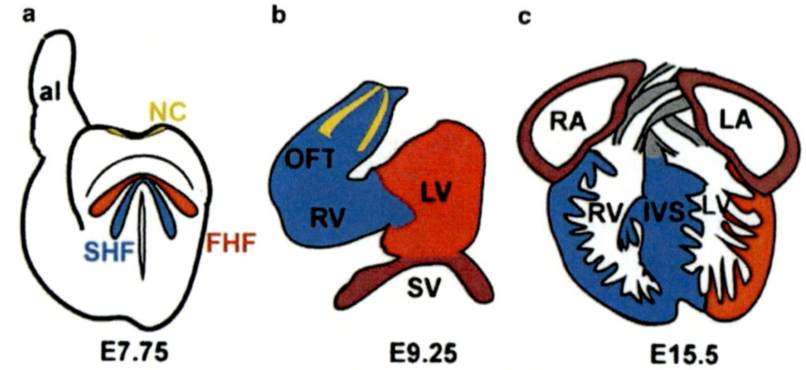


**Fig. 9.3** Second heart field (SHF) development (see text for further details). Reprinted from Black [67]. With kind permission from Elsevier

This field referred to as the anterior heart field (AHF), the SHF, provided a potential explanation for the observations that many genes and transgenes are expressed in the RV and OFT, but not in the left ventricle (LV) or atria, and that there are several mutations in mice and diseases in humans that selectively impact the RV [72, 73]. This has led to numerous studies addressing the molecular, genetic, and embryological aspects of this novel field. The SHF progenitor cells that are initially sequestered outside the heart, following migration into the heart, give rise to endocardium, myocardium, and smooth muscle. Because of its distinct development, the SHF is likely subjected to a variety of signals from the first heart field (FHF). Cells within the FHF specifically contribute to the left ventricular myocardium, whereas the SHF contributes to OFT and RV myocardium, endocardium, and smooth muscle of the great vessels [68, 74–77]. It has been shown that the canonical Wnt signaling negatively regulates FHF specification. After inactivation of the obligate canonical Wnt effector β-catenin using a β-catenin conditional null allele and the MEF2C AHF-Cre driver that directs cre activity specifically in SHF, Ai et al. [78] have expressed a stabilized form of β-catenin to model continuous Wnt signaling in SHF. Findings indicated that Wnt signaling plays a major positive role in promoting growth and diversification of SHF precursors into RV and interventricular myocardium.

The presence of Wnt antagonists Dickkopf-1 (Dkk-1) and Crescent (a Frizzled-related protein that inhibits Wnt8) can induce heart formation in explants of ventral marginal zone mesoderm. Wnt3A and Wnt8, but not Wnt5A or Wnt11, inhibited endogenous heart induction [79]. Others have shown that the inhibition of Wnt signaling promotes heart formation in the anterior lateral mesoderm, whereas active Wnt signaling in the posterior lateral mesoderm promotes blood development [80]. Presently little is known about the downstream effectors of these secreted Wnt antagonists or the mechanism by which they activate heart formation. A screen for downstream mediators has revealed that Dkk-1 and other inhibitors of the Wnt pathway induce the homeodomain transcription factor Hex, normally expressed in endoderm underlying the presumptive cardiac mesoderm in amphibian, bird, and mammalian embryos. Loss of Hex function blocks both endogenous heart development and heart induction by ectopic Dkk-1. As with the Wnt pathway antagonists, ectopic Hex induces expression of cardiac markers. Thus, in order to initiate cardiogenesis, Wnt antagonists act on endoderm to upregulate Hex, which, in turn, mediates production of a diffusible heart-inducing factor. This novel function for Hex suggests an etiology for the cardiac malformations in Hex mutant mice and enables the isolation of factors that induce heart directly in the mesoderm [81]. These findings suggest that the specification of heart precursors is a result of multiple tissue and cell–cell interactions that involve both temporal and spatial integrated programs of inductive signaling events.

## Migration of the Cardiac Precursors

The basic helix-loop-helix (bHLH) transcription factors Mesp1 and Mesp2 are required for the migration of cardiac precursors. Mesp1 is expressed in the early mesoderm that is destined to become the cranial-cardiac mesoderm, and is one of the earliest molecular markers expressed in cardiac precursor cells as revealed by lineage study [54]. Unlike the other aforementioned transcription factors involved in heart morphogenesis such as GATA4 and Nkx2-5, Mesp1 and Mesp2 are expressed in the nascent heart precursor cells and not during later heart morphogenesis. They therefore have a limited role in heart tube formation, but are involved in mesodermal specification. Disruption of the *Mesp1* gene resulted in a morphogenetic abnormality of the heart, cardia bifida [82]. In the absence of Mesp1, mesodermal cells fated to become cardiac myocytes fail to migrate normally out of the primitive streak during gastrulation and, consequently, fall behind the morphogenetic movements of the rest of the embryo, resulting in complete or partial cardia bifida. Somitogenesis is not disrupted in these embryos because of normal expression of the related *Mesp2* gene. Mice lacking both *Mesp1* and *Mesp2* exhibit a complete block in migration of the mesoderm from the primitive streak, resulting in a complete lack of cardiac and other mesodermal derivatives and die around E9.5 days. A major defect in this *Mespl/Mesp2* double-knockout embryo was the absence of any mesodermal layer between the endoderm and ectoderm [83].

The transcriptional pathway involving GATA factors may also be involved in the movements of the paired progenitor pools that coalesce to form the linear heart tube. During early mouse development, while GATA4 is expressed in cardiogenic splanchnic mesoderm and associated endoderm, endodermal GATA4 expression is required for ventral morphogenesis [84]. Both GATA4 and GATA5 have been implicated in the regulation of normal formation of the endoderm underlying the myocardial precursors [85]. With antisense-mediated reduction of GATA4/5/6 function in chick embryos [86], endodermal cells do not normally differentiate, their ventral migration is inhibited, preventing the concomitant movement of myocardial cells. This severe block in GATA expression results in a high percentage of the embryos developing abnormal hearts including the development of cardia bifida. A mix-like homeodomain transcription factor (bon) has also been reported in conjunction with the impaired endodermal differentiation observed in the zebrafish cardia bifida mutants [87].

Cardiac precursor migration also involves formation of a coherent epithelial layer. Specific targeting GATA4 directly in cardiac mesodermal cells with siRNA led to the development of cardia bifida in chick embryo and selective suppression of the cell adhesion protein N-cadherin expression [88]. Observations in zebrafish cardiogenesis implicate cell adhesion molecules as playing an important role in maintaining cardiac epithelial integrity [89]. As precardiac mesoderm cells epithelialize, they become stably committed by the activation of several cell–matrix and intracellular signal transduction pathways. Two different families of cell adhesion molecules are primarily involved; the calcium-dependent cadherins, specifically N-cadherin, and the extracellular matrix glycoprotein, fibronectin [90]. N-cadherin acts by binding to the intracellular protein β-catenin while fibronectin acts by binding to integrins at focal adhesion sites. Both are involved in the regulation of gene expression by their association with the cytoskeleton and through signal transduction pathways. Cross-talk between the adhesion signaling pathways initiates a number of characteristic phenotypic changes associated with cardiomyocyte differentiation, electrical activity, and myofibrillar organization.

## Coordination of Signaling Pathways and Progenitor Cells Functionality in Cardiogenesis

Functional coordination of a number of cell types including cardiac myocytes, SMCs, endothelial cells, and connective tissue elements is required for normal cardiogenesis, and these different cell types appear to arise from a pool of common progenitor cardiac cells. This have been documented by the previously unrecognized contribution to the developing heart of a population of progenitor cells in the pharyngeal mesoderm that gives rise to myocardium at the arterial pole. Lineage-tracing experiments have defined the extent to which pharyngeal progenitor cells colonize the heart, revealing a contribution to venous, as well as arterial, pole myocardium [91]. In the pharyngeal myocardial progenitor cells, a role for transcription regulators including Forkhead, GATA, LIM homeodomain, MEF2, Smad, and T-box (Tbx) transcription factors has been suggested [92]. In addition, Isl1, a LIM homeodomain transcription factor, is an important marker for a distinct population of undifferentiated cardiac progenitor cells, which proliferates prior to differentiation and gives rise to the cardiac segments [77].

Furthermore, the discovery of the convergency of several signaling pathways that regulate endothelial proliferation and differentiation in developing and postnatal heart valves has provided important insights into the roles played by highly conserved signaling elements including VEGF, NFATc1, Notch, Wnt-β-catenin, BMP-TGF-β, ErbB, NF-1/Ras, and newcomers including retinoic acid (RA). BMPs are members of the TGF-β cytokine superfamily of which BMP2, BMP4, TGF-β2, and TGF-β3 have been implicated in cardiac development. Some BMPs are expressed before cardioblast formation and throughout the late stages of heart development. BMPs play crucial roles in a broad range of biological events, including cellular proliferation, differentiation, migration, and apoptosis during organ development. At least six *BMP* genes (*BMP2*, *4*, *5*, *6*, *7*, and *10*) are expressed in the heart, where they have both distinct and redundant functions [93]. Observations in genetically modified mice showed the significant role that BMP signaling plays after the mid-gestation stage in heart morphogenesis. In fact, BMPs play a dual role in heart development and precise regulation of BMP inhibition, and BMP stimulation is required for normal heart development during the early stage of cardiomyocyte differentiation. Importantly, knowledge from embryo-genesis is often used in studies of stem cell biology. For example, through the application of BMP signaling regulation in cardiac development, new systems are being developed for the differentiation of ES cells into cardiomyocytes. These results reveal the crucial role of temporal and spatial regulation of BMPs in heart development. This was further confirmed by McCulley et al. [94] showing that BMP4 is required in the SHF by inactivating *BMP4* in the *MEF2C*-AHF-Cre expression domain. *MEF2C*-AHF-Cre is expressed exclusively in the SHF and its derivatives within the myocardium and endocardium of the RV and OFT [95]. Inactivation of *BMP4* in the SHF resulted in lethality at birth due to defects in endocardial cushion development. In addition, it was noted that *BMP4* SHF conditional knockout mice have significant defects in OFT septation and in the formation of the membranous portion of the interventricular septum. Moreover, BMP4 is required for normal expansion and remodeling of the endocardial cushions within the OFT and for normal semilunar valve formation and remodeling. Taken together, these findings establish that SHF-derived myocardium is an essential source of BMP4 that is required for OFT septation, cushion remodeling, and semilunar valve maturation. Moreover, other key intercellular signaling pathways that control the various steps in cardiovascular development have been recently identified. Ryckebusch et al. [96] have reported on the role that RA, the active derivative of vitamin A, plays during cardiovascular development. The retinaldehyde dehydrogenase 2 (Raldh2) enzyme catalyzes the second oxidative step in RA biosynthesis and its loss of function creates a severe embryonic RA deficiency. *Raldh2*$^{-/-}$ knockout embryos fail to undergo heart looping and have impaired atrial and sinus venosus development. To understand the mechanism(s) producing these changes, they examined the contribution of the SHF to pharyngeal mesoderm, atria, and OFT in *Raldh2*$^{-/-}$ embryos. RA deficiency alters SHF gene expression in two ways. First, *Raldh2*$^{-/-}$ embryos exhibited a posterior expansion of anterior markers of the SHF, including *Tbx1*, *Fgf8*, and the *Mlc1v-nlacZ-24/Fgf10* reporter transgene as well as of *Isl1*. This occurred at early somite stages, when cardiac defects became irreversible in an avian vitamin A-deficiency model, indicating that endogenous RA is required to restrict the SHF posteriorly. Explant studies showed that this expanded progenitor population cannot differentiate properly. Furthermore, RA upregulated cardiac BMP expression levels at the looping stage. Thus, contribution of the SHF to both inflow and outflow poles was perturbed under RA deficiency, creating a disorganization of the heart tube. Interestingly, they also investigated genetic cross-talk between Nkx2-5 and RA signaling by generating double mutant mice and found that Nkx2-5 deficiency was able to rescue molecular defects in the posterior region of the *Raldh2*$^{-/-}$ mutant heart, in a gene dosage-dependent manner.

Hoover et al. [97] have also reviewed the significance of retinoid signaling during cardiac development. Experiments performed over 50 years ago revealed that too much or too little maternal intake of vitamin A proved detrimental for embryos, resulting in predictable cardiac developmental defects. Germ line and conditional knockout mice showed which molecular players in the vitamin A signaling cascade are potentially responsible for regulating specific developmental events, and many of these molecules have been temporally and spatially characterized. It is evident that intact and controlled retinoid signaling is necessary for each stage of cardiac development to proceed normally, including cardiac lineage determination, heart tube formation, looping, epicardium formation, ventricular maturation, chamber and OFT septation, and coronary arteriogenesis.

## Proliferation of Progenitor Cardiac Cells in Cardiac Development

The proliferation and deployment of SHF/AHF progenitor cells and the signaling pathways controlling its development have been recently reviewed by Kelly et al. [98]. Factors such as Wnt, FGF, BMP, Hedgehog, and RA have helped to identify the ligand sources and responding cell types that control SHF development. These observations underlined the importance of signals from the pharyngeal mesoderm itself, as well as the critical input from adjacent pharyngeal epithelium and NCCs. Proliferation of progenitor cardiac cells is a critical factor in the regulation of SHF development, and characterization of the signaling pathways that maintain, expand, and regulate differentiation of these cardiac progenitor cells is paramount to understand both the etiology of CHD and the application of stem cells to repair the damaged heart. In this vein, using two independent transgenic and gene-targeting approaches in human embryonic stem cell lines, Bu et al. [99] have shown that purified Isl1(+) primordial progenitors are capable of self-renewal and expansion before differentiation into the three major cell types in the heart laying the foundation for the generation of human model systems for cardiovascular disease as well as novel approaches for human regenerative cardiovascular medicine.

### *Tube Looping and Segmentation*

The next step of cardiovascular development involves the looping of the heart with tightening of the inner curvature and the completion of an arterial and a venous pole, together with a complicated set of septa separating veins, atria, ventricles, and the great arteries. Differentiation of the cardiac chambers and myocardium of the transitional or intersegmental zones takes place during the rightward looping process. The latter process is regulated by a cascade of genes that are essential for left–right programming. Disturbances in this programming will lead to abnormal looping that can vary from random, anterior to leftward looping. The consequences for human development are found in the abnormal atrial situs (situs inversus or isomerism), dextrocardia, and heterotaxy.

As shown in Fig. 9.4, ventricular precursors from the primary and secondary heart fields migrate medially and coalesce to form the linear heart tube [100]. The anteroposterior (AP) organization of these precursors is maintained in the heart tube, with RV precursors anterior to LV precursors and the future atria. Looping leads to the appropriate juxtaposition of these heart regions and the formation of an inner

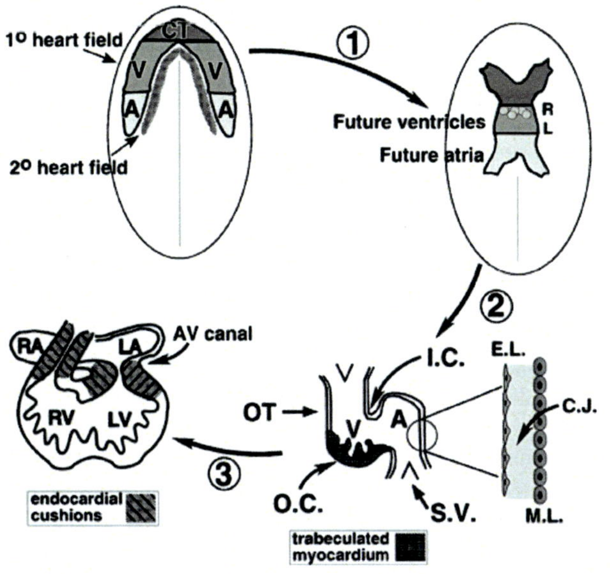


**Fig. 9.4** Early heart development and ventricle formation [103]. *A* atria, *C.J.* cardiac jelly, *CT* conotruncus, *E.L.* endocardial cell layer, *LA* left atrium, *LV* left ventricle, *M.L.* myocardial cell layer, *RA* right atrium, *RV* right ventricle, *SV* sinus venosus, *V* ventricle. With permission of the Royal Society of Medicine Press

curvature (IC) and a trabeculated outer curvature (OC), the future ventricular myocardium [101]. Progressive changes in the myocardium and septation into right and left atria and ventricles precede the synchronized pumping of blood between the atria and ventricles and the formation of one-way valves to control blood flow [102]. These valve precursors are the endocardial cushions that form in the atrioventricular (AV) canal (AVC).

In the looped heart tube, several transitional zones and intervening primitive cardiac chambers can be discerned. The transitional zones will become part of the septa, valves, conduction system, and fibrous heart skeleton, and will be partly incorporated in the cardiac chambers during formation of the definitive right and left atrium and their ventricular counterparts.

Several of the transitional zones, such as the AVC and OFT, develop endocardial cushions, whereas others do not. The process of looping aligns the primordial chambers such that they face the outer curvature, whereas the transitional zones are brought together in the inner curvature of the heart tube. Myocardium of the inner curvature, as well as that of inflow tract, AVC, and OFT, retains the molecular signature originally found in linear heart tube myocardium. Specific gene defects affecting the morphologic characteristics can result in deficient remodeling of the inner curvature leading to a leftward movement of the OFT over the AVC, resulting in a spectrum of OFT abnormalities. Examples include the formation of a double-outlet RV with an obligatory ventricular septal defect or, in less extreme cases, just a ventricular septal defect, malformations more common in transgenic mice.

The formation of the cardiac chambers from the primary heart tube has recently received a great deal of attention. The initial linear heart, composed of primary myocardium, shows a polarity both in phenotype and gene expression along its anteroposterior and dorsoventral axes. Specialized ventricular chamber myocardium is specified at the ventral surface of the linear heart tube, while distinct left and right atrial myocardium forms more caudally on laterodorsal surfaces. Each chamber has specific biochemical and physiological properties that are important for heart function which are established and maintained by chamber-specific gene expression. The spatial pattern of gene expression bears a strong relationship to morphogenesis. This involves various promoter elements of genes such as *ANF*, sarcoplasmic reticulum calcium ATPase (*SERCA2a*), myosin light chain 2v (*Mlc2v*), and β-myosin heavy chain (*β-MHC*), which restrict the expression of genes to the atrial or ventricular compartments as well as the transcription factors responsible for the compartment restricted gene expression, a number of which have been identified.

Several factors have been identified which contribute to patterning and chamber formation in the developing heart by regulating gene programs at specific sites within the heart tube and integrating positional information with respect to AP patterning. These include RA, Iroquois homeobox protein 4 (Irx4), Tbx T-box genes (including Tbx5, Tbx2, and Tbx20), and GATA factors.

RA is a powerful effector of AP patterning, chamber specification, and morphogenesis. Transient exposure of zebrafish to RA either during or shortly after gastrulation results in heart tube truncation, initiating with the OFT and ventricle at low doses, and further along the AP axis with higher doses [104]. In several species including chicken and mouse, excess of RA promotes increased expression in the anterior region of genes normally expressed only in the posterior region. Conversely, RA deficiency brought about by using an anti-RA monoclonal antibody causes underdevelopment of the posterior structures of the heart, most notably the sinus venosus and the atria [105].

To further assess the role of RA in early cardiac development, the expression of the key enzyme involved in endogenous RA synthesis, Raldh2, has been studied [106]. In avian embryos, Raldh2 expression was observed exclusively in posterior mesoderm and in posterior heart precursors. In mouse, Raldh2 expression was initiated in the posterior mesoderm shortly after gastrulation. Mice whose synthesis of endogenous RA was blocked as a result of disrupted *Raldh2* alleles displayed severe heart abnormalities and died *in utero*. Analysis of the embryos revealed that although the embryonic heart tube formed normally, they exhibited failure of upward right looping and in the development of the posterior chambers (atria, sinus venosus). In addition, these embryos displayed an anterior-based defect and the proper differentiation of ventricular cardiomyocytes, which tended to be prematurely differentiated.

Gathered observations from molecular and functional studies revealed that RA regulates the expression of a variety of cardiogenic transcription factors including GATA4 as well as several heart asymmetry genes [7, 107]. During the crucial RA requiring developmental window, RA transduces its signals to genes for heart morphogenesis via the RA receptors RARα2, RARγ, and RXRα. RARs and RXRs are ligand-activated transcription factors that have been implicated in many aspects of heart development, including ventricular maturation and cardiac septation [108–110]. In avian embryos, blocking the expression of RARγ, RARα2, and RXRα using antisense oligonucleotides recapitulates the complete RA-deficient phenotype [111]. Similar to mice with synthesis of RA blocked, mice lacking either the RAR coreceptor (RXR) or RAR have defective ventricular maturation, related to accelerated cardiomyocyte differentiation [112]. Therefore, RA is required for delay in differentiation.

The *Irx4* gene is a member of the *Iroquois* family of homeodomain-containing transcription factor genes, which have been implicated in chamber-specific gene expression. When the tubular heart is formed, Irx4 expression is confined to a restricted segment which excludes the inflow and OFT, and the AVC and ventricular myocardium including the inner curvature after looping, resembling the pattern of MLC2V [113]. In later stages, Irx4 expression is largely confined to ventricular myocardium in all species examined [114, 115].

Irx4 expression is reduced in mice lacking *Nkx2-5* or *dHAND*, in which ventricular differentiation is compromised [116]. Irx4 modulates specific ventricular gene expression as shown by Irx4-dependent activation of the expression of the ventricle myosin heavy chain-1 (VMHC1) and suppression of the expression of the atrial myosin heavy chain-1 (AMHC1) in the ventricles in chick embryos [114]. It has not yet been determined how Irx4 functions as a positive activator of VMHC1 since most studies have indicated that transcription factors belonging to the Iroquios family have been characterized as acting primarily as repressors [117]. Evidence that Irx4 plays a role in chamber-specific gene expression in the ventricles has also come from the analysis of the promoter of AMHC1 homologue, atrium-specific *slow myosin heavy chain 3* (*SMyHC3*), which is repressed by Irx4 in chicken ventricular myocardium and upregulated in *Irx4*$^{-/-}$ embryonic ventricles [118]. The regulatory elements of the *SMyHC3* gene are functional in quail and mouse [119, 120], and in both species, the transcriptional elements controlling the chamber specificity of this promoter are under the control of Irx4; interestingly, this promoter also contains both a functional GATA binding element and a vitamin D response element (VDRE). These findings coupled with the fact that Irx4 does not bind directly to the *SMyHC3* promoter elements required for ventricular repression, strongly suggest the involvement of additional factors in this compartment-specific transcriptional control. In fact, interaction of Irx4 with a RAR/vitamin D receptor complex has been reported [121]. However, evidence does not support Irx4 as a global regulator of ventricle-specific gene expression, since mice with a targeted disruption of *Irx4* have only a partial disturbance of ventricle-specific gene expression [118]. The identification of five additional *Irx* genes present in the developing heart has indicated the possibility of some genetic redundancy [122]. However, their spatial and temporal patterns of expression in development are markedly different from *Irx4* suggesting that they can only partially compensate for *Irx4*. *Irx4*-deficient mice have impaired cardiac function and develop cardiomyopathy, underscoring the important role of chamber-specific gene expression for proper cardiac function [118].

Present in all metazoans, the T-box family of transcription factors are involved in early embryonic cell fate decisions, regulation of the development of extraembryonic structures, embryonic patterning, and many aspects of organogenesis. T-box genes include *Tbx1*, *Tbx2*, *Tbx3*, *Tbx5*, *Tbx6*, *Tbx18*, and *Tbx20*, all of which exhibit complex patterns of temporal and spatial regulation in developing cardiac structures. T-box transcription factors function in many different signaling pathways, notably in BMP and FGF pathways. The downstream target genes that have been identified thus far indicate a wide range of downstream effectors. Moreover, mutations in the T-box genes are responsible for developmental dysmorphic syndromes with striking cardiac abnormalities in humans (as we shall discuss in the following chapter), and have also been implicated in regulation of cell proliferation in cancer.

*Tbx5* is expressed initially throughout the cardiac mesoderm in its earliest stages. In the linear heart tube, its expression pattern displays an AP gradient [123, 124]. At mid-gestation, it is restricted to the atria and LV. *Tbx5* mRNA levels decrease in the LV during subsequent stages of development, such that by late gestation and adulthood, low levels of *Tbx5* transcripts can be detected equivalently in both LV and RV in mice and humans [125]. *Tbx5* similar to *GATA4* is upregulated in RA-treated chicken embryos; in the aforementioned *Raldh2*-deficient embryos; moreover, both *Tbx5* and *GATA4* are downregulated in the posterior section of the tubular heart [106]. In zebrafish, *Tbx5* deficiency results in the *heartstrings* mutation, characterized by a failure in heart tube looping with extensive deterioration of both the atrium and ventricle [126]. Lack of Tbx5 in mice results in severe hypoplasia of posterior structures such as atria, with RV and OFT growth remaining intact [127]. This suggests that Tbx5 is required for formation of the posterior heart. Tbx5 deficiency also results in a marked downregulation of *Nkx2-5* and *GATA4* as well as anteriorly expressed genes including *Irx4*, *Mlcv*, and *Hey2*. *Tbx5* haploinsufficiency also markedly decreased *atrial natriuretic factor (ANF)* and *connexin 40 (cx40)*.

Overall ventricular differentiation is impaired in *Tbx5*-deficient embryos, including decreased expression of the ventricle-specific genes *Mlc2v*, *Irx4*, and *Hey2*. This is perhaps due to early pleiotropic effects of the absence of *Tbx5* on cardiac differentiation, including decreased *GATA4* and *Nkx2-5* expression. Consistent with these observations, *Tbx5* has been shown to accelerate cardiac differentiation of P19Cl6 cell line, including increased *Nkx2-5* expression [128], and inhibition of *Tbx5* in *Xenopus* embryos also leads to hypoplasia of cardiac tissues and decreased *Nkx2-5* mRNA levels [129].

In embryos in which *Tbx5* is misexpressed, the ventricular septum was not formed, resulting in a single ventricle [130]. In such heart, LV-specific *ANF* gene was induced. Transgenic overexpression of *Tbx5* in tubular hearts of chicken or mice results in thinned and hypoproliferative ventricular myocardium, with retardation of ventricular chamber morphogenesis and loss of anterior gene expression [124, 131]. These findings are consistent with the role of *Tbx5* in AP patterning, imposing a posterior identity on the heart tube. *Tbx5*-mediated inhibition of myocardial growth is suggestive that *Tbx5* is involved in the downregulation of cell proliferation. It should also be noted that *Tbx5* haploinsufficiency in mice also causes distinct morphological and functional defects in the atrioventricular and bundle-branch conduction systems [132]. Its critical role in the development and maturation of the cardiac conduction system (CCS) will be addressed in a later section of this chapter.

## Other T-Box Factors

Tbx2 functions as a primary determinant in the local repression of chamber-specific gene expression and chamber differentiation [133]. The pattern of *Tbx2* mRNA and protein expression displayed a temporal and spatial profile parallel to that found with chamber myocardium-specific genes *Nppa*, *Cx40*, *Cx43*, and *Chisel* in both mouse and human hearts and human. *In vitro*, Tbx2 repressed the activity of regulatory fragments of *Cx40*, *Cx43*, and *Nppa*. Hearts of transgenic embryos that expressed Tbx2 in the prechamber myocardium completely failed to form chambers and to express the chamber myocardium-specific genes *Nppa*, *Cx40*, and *Chisel*, whereas other cardiac genes were normally expressed. Tbx2 has been proposed to inhibit chamber-specific gene expression via competition with the positive factor Tbx5.

The murine T-box transcription factor Tbx20 also plays a central role in the genetic hierarchy guiding lineage and chamber specification. Disruption of *Tbx20* in mice results in a severely altered cardiac transcriptional program, reduced cardiac progenitors, and blocked chamber differentiation leading to embryonic death at mid-gestation with grossly abnormal heart morphogenesis [134]. Moreover, *Tbx20*-null embryos display activation of the repressive *Tbx2* across the whole heart myogenic field, which is largely responsible for the cardiac phenotype in *Tbx20*-null mice, placing *Tbx20* upstream of *Tbx2*. In addition, Tbx2 directly binds to the promoter and represses the cell proliferation gene, *Nmyc1* in developing heart, which likely contributes to the observed cardiac hypoplasia in *Tbx20* mutants [135]. *Nmyc1* is required for growth and development of multiple organs, including the heart. This suggests a model by which T-box proteins regulate regional differences in Nmyc1 expression to affect organ morphogenesis; in normal chamber myocardium, Tbx20 represses Tbx2, preventing repression of Nmyc1 and resulting in relatively high level of proliferation.

Observations on two mutant alleles of *Tbx6* have shown that this factor is involved in both the specification and patterning of the somites along the entire length of the embryo and the null allele, *Tbx6(tm1Pa)*, causes abnormal patterning of the cervical somites and improper specification of more posterior paraxial mesoderm [136].

## MEF2C and HAND Proteins

eHAND (also termed Hand 1) and dHAND (Hand 2) are basic helix-loop-helix transcription factors that play critical roles in cardiac development (Fig. 9.5). The *HAND* genes have a complementary left–right cardiac asymmetry of expression with *Hand2/dHAND* predominantly on the right side and *Hand1/eHAND* on the left side of the looped heart tube; their expression is downregulated in the adult mouse [137].

*Hand2/dHAND* is expressed in cardiac precursors throughout the cardiac crescent and the linear heart tube, before becoming restricted to the right ventricular chamber at the onset of looping morphogenesis. *Hand2/dHAND* is a direct transcriptional target of GATA and Nkx2-5 during RV development [138, 139]. Since GATA is not chamber-restricted, these findings suggest the existence of positive and/or negative coregulators that cooperate with GATA to control RV-specific gene expression in the developing heart. Mice that lack *Hand2/dHAND* die at E10.5 from right ventricular hypoplasia and vascular defects. The RV region of the heart and OFT are presently thought to arise from the secondary (or anterior) heart field in contrast to the atria and LV chambers which are believed to arise from the primary heart field. The absence of *Hand2/dHAND* resulting in the

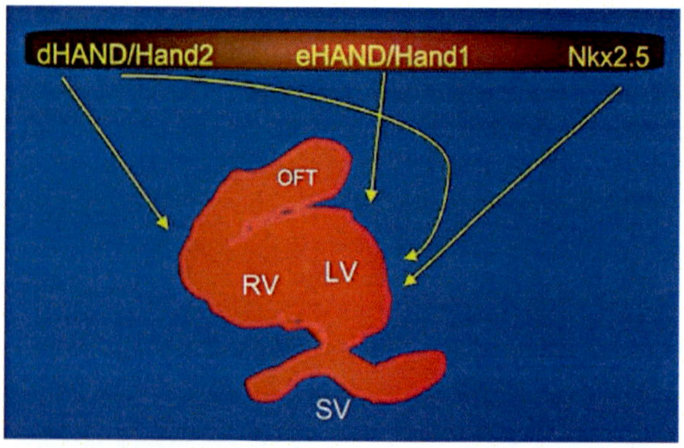


**Fig. 9.5** Asymmetric pattern of *HAND* expression and their role in developing myocardium. Knockout mice lacking dHAND/Hand2 showed the essential role of this gene in the formation of the RV. Knockout mice lacking Nkx2.5 revealed a loss of the LV as well as downregulation of eHAND/Hand1. Knockout mice lacking eHAND/Hand1 showed minor abnormalities in the LV, implying that loss of eHAND/Hand1 alone is not enough to account for the more severe LV defects in Nkx2.5 mutant embryos. Knockout mice lacking both eHAND/Hand1 and dHAND/Hand2 demonstrated severe ventricular hypoplasia, suggesting redundant functions of the HAND genes in the developing LV

deletion of the RV regions of the heart suggests that it is an essential component of the pathway for development of the SHF [140].

Although *Handl/eHAND* is asymmetrically expressed along the AP and dorsal-ventral embryonic axes, it is symmetrically expressed along the left–right axis at early stages of embryonic and cardiac development. After cardiac looping, expression of *Handl/eHAND* is restricted largely to the LV (systemic) and the atria, a pattern identical to that of the *Nppa* gene, which encodes ANF, the major secretory product of the heart. Mice that lack *Handl/eHAND* die at E8.5 from placental and extraembryonic abnormalities that preclude analysis of its potential role in later stages of heart development [137]. To examine the role of *Handl/eHAND* on chamber specification and morphogenesis, *Handl/eHAND* knock-in mice were generated, in which the *Handl/eHAND* cDNA was placed under the control of the *Mlc2v* promoter, which is fully expressed in ventricular myocardium throughout development [141]. Embryos with knock-in *Handl/eHAND* had a morphologically single ventricle, but exhibited distinctive LV and RV expression at the molecular level. Forced expression of *Handl/eHAND* resulted in the marked expansion of outer curvature of the LV and RV with limited interventricular groove or septum formation between the two ventricles. Furthermore, these mice displayed altered expression patterns of molecular markers of the working myocardium (e.g., Chisel and ANF) and *Hand2/dHAND* in the RV but did not affect *Tbx5* expression. These findings indicate that *Handl/eHAND* involved in the ventricular wall expansion is unlikely to act as a master regulatory gene required to specify LV myocyte lineage and that *Handl/eHAND* expression may be critical in the proper formation of the interventricular groove and septum. More recently, mice were generated that contained a conditional *Handl/eHAND*-null allele flanked by Cre recombinase loxP recognition sites and specifically deleted the allele in the developing heart by cardiac-specific expression of Cre recombinase [142]. Embryos homozygous for the cardiac *Handl/eHAND* gene deletion displayed defects in the LV and endocardial cushions and exhibited dysregulated ventricular gene expression, but nevertheless survived until the perinatal period when they succumb to a spectrum of CHDs. Combination of the conditional *Handl/eHAND* mutation with a *Hand2/dHAND* loss-of-function mutation revealed dose-sensitive effects in the control of cardiac morphogenesis and ventricular gene expression. The cardiac phenotype resulting from cardiac deletion of *Handl/eHAND* was much less severe than that of *Hand2/dHAND* mutant embryos, in which the entire RV region of the heart was absent. *Handl/eHAND* is expressed specifically in the outer curvatures of the embryonic LV, RV, and OFT in contrast to *Hand2/dHAND* expression, which occurs throughout the LV and RV chambers, although highest in the latter [140, 143, 144]. In the absence of *Handl/eHAND*, residual *Hand2/dHAND* expression in the LV and OFT is likely to compensate partially for the loss of *Handl/eHAND*, in contrast to the absence of *Hand2/dHAND* which results in a complete lack of Hand factors in the presumptive RV [140].

Also the less severe phenotypes of either the *Handl/eHAND* or *Handl/eHAND/Hand2/dHAND* double knockout mice contrast with mice lacking *Nkx2-5*, in which the LV chamber fails to expand following cardiac looping, and expression of several markers of cardiac differentiation is reduced throughout the remaining myocardium [25]. *Handl/eHAND 1* expression is abolished in the hearts of *Nkx2-5* mutant embryos [143] and the loss of *Handl/eHAND* contributes but is unlikely to be solely responsible for the abnormal cardiac morphogenesis of *Nkx2-5* mutant hearts. These findings demonstrate that Hand factors play both pivotal and partially redundant roles in cardiac morphogenesis, cardiomyocyte differentiation, and cardiac-specific transcription.

In lower vertebrates, such as frogs and fish, a single cardiac *Hand* gene is found (more closely related to *Hand2/dHAND*) responsible for the morphogenesis of the single ventricle found in these species [145, 146]. Similarly, *Tbx5* is expressed throughout the single ventricle of frogs and fish and is involved in the morphogenesis of the entire heart in these species [145]. This suggests that with the acquisition of pulmonary circulation and an RV chamber, gene duplication and specialization of *Hand* and *Tbx5* function evolved along with their restricted chamber-specific expression.

In the human adult heart, *Hand2/dHAND* expression was observed in all four chambers but was diminished in the right atrium in contrast to *Hand1/eHAND*, which was expressed in the LV and RV, but downregulated in both atrial chambers. Expression of *Hand1/eHAND* and not *Hand2/dHAND* has been reported to be significantly downregulated in hearts of ischemic and dilated cardiomyopathy (DCM) suggesting a correlation between *Hand1/eHAND* dysregulation and the evolution of a subset of cardiomyopathies [147].

As previously noted, the RV chamber and OFT are derived primarily from a population of progenitors known as the SHF. These regions of the heart are severely hypoplastic in mutant mice lacking either MEF2C or BOP transcription factors, suggesting that these cardiogenic regulatory factors may act in a common pathway for development of the SHF and its derivatives [148, 149]. *Bop* expression in the developing heart depends on the direct binding of MEF2C to a MEF2-response element in the *Bop* promoter that is necessary and sufficient to recapitulate endogenous *Bop* expression in the SHF and its cardiac derivatives during mouse development [149]. *Bop* has been identified as an essential downstream effector gene of *MEF2C* in the developing heart. MEF2C also interacts with the Hand and GATA proteins in order to function [150].

Members of the MEF2 family of transcription factors are known to be important for cardiac muscle formation. In mice, there are four *MEF2* genes (*MEF2A*, *B*, *C*, and *D*), each of which being expressed during some stage of cardiac development [151]. Several downstream myogenic genes including *Mlc2*, *α-myosin heavy chain*, and *ANF* have promoters or enhancers that bind these proteins. In mice lacking *MEF2C*, these proteins are not expressed, and while the linear heart tube forms, cardiac looping is defective and anterior heart structures and not posterior structures are malformed and hypoplastic. The latter feature is similar to the cardiac phenotypes obtained by targeted deletion of *Nkx2-5*, *Hand2/dHAND*, and *Bop* genes, in which embryos are affected more at the anterior than at the posterior side. The resulting defect in ventricular chamber morphogenesis and expansion is likely due to a failure in downstream cardiac gene expression including the failure to activate the *HAND* genes with which MEF2 proteins interact. In mice with null mutations of *MEF2C*, the expression of *Hand2/dHAND* is absent, while *Hand1/eHAND* is present in both the LV and RV [144, 152]. Its ventricular-specific requirement therefore indicates that MEF2C is a necessary cofactor for ventricle-specific factors, and a cooperative interaction of Hand2/dHAND with MEF2C has been suggested as a pivotal event in formation of the anterior region of the heart [153]. Recent studies have further established a relationship between MEF2 and Hand factors: Hand1/eHAND is recruited to the cardiac ANF promoter via physical interaction with MEF2 proteins [154]. This interaction results in the synergistic activation of MEF2-dependent promoters, and MEF2 binding sites are sufficient to mediate this synergy. In addition to a variety of cofactors and co-activators that appear to play a role in MEF2C transactivation, posttranscriptional modification also may play a substantial role in the MEF2 regulatory pathway. In skeletal and cardiac muscle, negative regulation of MEF2 function by HDACs has been revealed as an important mechanism modulating MEF2 activity [155]. The E1A-binding protein p300, which functions as an acetylase, is involved in the regulation of MEF2C cardiac transcription during development [156]. MEF2C activity has been shown to be regulated by phosphorylation [157] and activity of a *MEF2C*-dependent transgene in the heart is stimulated by $Ca^{2+}$/calmodulin protein kinase activation [158].

## Generation of Left–Right Identity

The acquisition of left versus right (LR) identity in the developing heart is presently an area of intensive investigation. The LR identity of the heart-forming regions and the direction of looping of the atrial and ventricular regions of the tubular heart are highly conserved in evolution, and both are dependent on LR signaling in the embryo [159]. Defects in the determination of laterality can result in serious clinical cardiac malformations – a topic covered in depth in the following chapter.

The importance of LR signaling is particularly evident in the atria, which are initially positioned in a LR arrangement, unlike the ventricles, for which the left and right derive from an initial anteroposterior arrangement. The analysis of transcription profiles using reporter transgenes further supported the notion that the molecular specification of left and right atrial chambers but not ventricular chambers is dependent on left–right signaling cues [160]. Atrial identity is essential for the proper alignment of septal and valve structures and for the normal connections of pulmonary veins. Abnormalities in these processes lead to severe defects, such as common AVC or total anomalous pulmonary venous return.

A cascade of signaling molecules that regulate the establishment of LR identity of the embryo culminates in the expression of the paired-domain homeodomain transcription factor Pitx2 on the left side of the visceral organs, including the heart. The mouse nodal gene and its homologues in chick and *Xenopus* are among the first genes known to be asymmetrically expressed along the LR axis. A key event in this pathway is the restriction of *Nodal* expression to the left side of the lateral plate mesoderm and its repression in the right lateral plate mesoderm [161]. In addition to *Nodal* involvement, several TGF-β family signaling proteins including secreted extracellular factors (e.g., lefty-1, lefty-2, and BMP4), membrane receptors (e.g., activin receptor type IIB), membrane-associated proteins encoded by *EFC-CFC* genes (e.g., cryptic, Notch receptor, and BMP type I receptor ACVRI), intracellular mediators (e.g., sonic hedgehog (SHH) Smads), and an array of transcription factors (e.g., Zic3, SnR, Hand1, Nkx2-5, CITED, β-catenin) have been implicated in LR axis determination [144, 162–175].

Asymmetric *Pitx2* expression seems to be sufficient for establishing LR identity of the heart and represents a major determinant in establishing atrial identity. Mice lacking *Pitx2* have a single large atrium with right atrial morphology, including abnormal connection of venae cavae and pulmonary veins [107, 176]. The resulting defects resemble complete atrioventricular septal defects, with a single atrioventricular valve, ventricular and atrial septal defects, and double-outlet RV. Decreased dosage of *Pitx2* leads to relatively normal chamber formation, but septal and valve defects occur, which are perhaps due to the misalignment of structures during development [176]. In addition to previously discussed RA involvement in early mesoderm differentiation and AP patterning, it is also required for the LR asymmetry pathway, which is responsive to either RA excess or deficiency [108, 177]. RA controls both the level and location of expression of components of the LR signaling pathway (lefty, nodal, and Pitx2).

## Proepicardium

The proepicardium (PE) is a transient structure that forms at the venous pole of the embryonic vertebrate heart and gives rise to several cell types of the mature heart. Schlueter et al. [178] have analyzed the expression pattern of the marker genes *Tbx18*, *Wt1*, and *Cfc* in the development of PE in the chick embryos. During PE induction, these three marker genes displayed LR asymmetric expression patterns with the higher expression on the right side than on the left side. The observed LR asymmetric gene expression was in accord with the asymmetric formation of the PE in the chick embryo. While initially the marker genes were expressed in the primitive sinus horn, subsequently, expression became confined to the PE mesothelium. To determine the signaling factors involved in PE development, *Bmp2* and *Bmp4* expression were assessed. While *Bmp2* was bilaterally expressed in the sinus venosus, *Bmp4* initially was expressed unilaterally in the right sinus horn and later in the PE. BMP signaling function was experimentally modulated by supplying exogenous BMP2 and by inhibiting endogenous BMP signaling through the addition of Noggin. Interestingly, both supplying BMP and blocking BMP signaling resulted in a loss of PE marker gene expression and both experimental situations lead to cardiomyocyte formation in the PE cultures. Titration experiments with exogenously added BMP2 or Noggin showed that PE-specific marker gene expression depends on a low level of BMP signaling. Implantation of BMP2-secreting cells or beads filled with Noggin protein into the right sinus horn of HH stage 11 embryos resulted in downregulation of Tbx18 expression, corresponding to the results of the explant assay. Taken together, it appears that a distinct level of BMP signaling is required for PE formation in the chick embryo.

More recently, Schlueter and Brand [179] have studied the control of asymmetrical development of the PE in the chick embryo and observed that this cardiac progenitor cell population gave rise to the epicardium, coronary vasculature, and fibroblasts. In the chicken embryo, the PE displays LR asymmetry and develops only on the right side, while a vestigial PE is only formed on the left side and subsequently it gets lost by apoptosis. Experimental manipulation of left-side determinants such as *Shh*, *Nodal*, and *Cfc* as well as forced expression of *Pitx2* had no effect on the sidedness of PE development. In contrast, inhibition of early acting regulators of LR axis formation such as $H^+/K^+$-ATPase or primitive streak apoptosis affected the sidedness of PE development. Experimental interference with the right-side determinants FGF8 or Snail prevented PE formation, whereas ectopic left-sided expression of FGF8 or Snail resulted in bilateral PE development. Thus, this study provided novel insights into the molecular control of asymmetric morphogenesis and suggests that also the right side harbors a signaling pathway that is involved in the control of PE development. This pathway might be of significance for setting up LR asymmetries at the venous pole of the heart.

Multiple cell biological processes coordinate the LR determination in mammals; for example, fluid dynamics has been applied to developmental biology and the effect of fluid dynamics and dynamical systems seem to have a role in the left–right symmetry breaking in vertebrates [180]. Hirokawa et al. [181] have reported that the leftward movement of fluid at the ventral node, called nodal flow, is the central process in symmetry breaking on the LR axis. Nodal flow is autonomously generated by the rotation of posteriorly tilted cilia that are built by transport via KIF3 motor on cells of the ventral node. It is possible that the leftward movement of sheathed lipidic particles, known as nodal vesicular parcels (NVPs), may result in the activation of the noncanonical hedgehog signaling pathway, an asymmetric elevation in intracellular $Ca^{2+}$ and changes in gene expression. These authors noted that although the human body is apparently bilaterally symmetrical on the surface, the visceral organs are arranged asymmetrically in a stereotyped manner. The heart, spleen, and pancreas reside on the left side of the body, whereas the gall bladder and most of the liver are on the right side. Because the human body is formed from a spherically symmetrical egg (oocyte), symmetry breakdown is one of the fundamental processes of development.

## Chamber Growth and Maturation

Maturation of the heart into fully functional trabeculated chambers and septation of the atria and ventricles from one another and between their left and right sides are important processes that require precise integration of growth and differentiation signals.

There has been considerable effort to identify the components of the pathways involved in cardiomyocyte growth during embryogenesis especially since these pathways are at least partly recapitulated during both physiological and pathological cardiac hypertrophy in the adult heart as well as for their involvement in the proliferation of potential regenerative cells. Growth of the heart from a thin-walled structure with the atrial and ventricular chambers molecularly specified involves proliferation of myocytes along the walls of the heart tube and within the developing interventricular septum. The most highly proliferative cardiomyocytes are situated along the outer surface of the heart, in a highly mitotic area designated as the compact zone. As the wall thickens, cardiomyocytes along the inner wall become organized into fingerlike projections, or trabeculae, which enhances oxygen and nutrient exchange and force generation. The thin epicardial layer of cells, surrounding the heart, provides a source of mitogenic signals (e.g., RA) that are necessary and sufficient to stimulate proliferation of cardiomyocytes. Mice lacking the RXR die during embryogenesis from a failure in proliferative expansion of ventricular cardiomyocytes resulting in a thin-walled ventricle [108, 112]. Interestingly, this effect is observed only with epicardium-specific RXR deletion but not with cardiomyocyte-specific one indicating that the effects of RA on cardiac growth are primarily nonmyocyte autonomous [182, 183]. The epicardium also expresses high levels of the RA synthesizing enzyme Raldh2 [184]. Another epicardial-derived trophic signal affecting cardiomyocyte proliferation is erythropoietin (epo); blockade of either RA or epo signaling from the epicardium inhibits cardiac myocyte proliferation and survival [185]. A number of studies have suggested that these epicardial-derived signals do not act directly on the myocardium but rather regulate the production of an unidentified soluble epicardial-derived mitogen [182, 183].

Downstream signaling pathways that are elicited in response to this epicardial-derived factor include the activation of PI3K and extracellular-related kinase (Erk) pathways which are required for a proliferative response [186].

The mitogenic signals emanating from the epicardium are essential for the maintenance of the correct amount of myocyte proliferation in the compact myocardium. By using microsurgical inhibition of epicardium formation in the embryonic chick, it was shown that levels of expression of FGF2 and its receptor FGFR1 in myocyte are dependent on the presence of epicardial-derived signals [187]. Recently, FGFs expressed in the epicardium have been identified including FGF9, FGF16, and FGF20, which are RA inducible and contribute to the regulation of cardiomyocyte proliferation during mid-gestation [188]. These findings have led to the proposal that FGFs constitute all or part of the epicardial signal regulating myocardial growth and differentiation.

Growth signals originating from the endocardium, the specialized endothelial lining of the heart, are also critical. The neuregulin family of peptide growth factors and their tyrosine kinase receptors, ErbBs, have been shown to promote growth of embryonic cardiomyocytes *in vivo* [189]. Knockout mice lacking ErbB2, Erb4, or neuregulin-1 (NRG-1) die from cardiac growth defects characterized by the absence of trabeculae [190, 191]. This abnormality can be ascribed to the lack of signaling between the endocardium and myocardium. Cardiac-specific deletion of ErbB2 results in DCM with ventricular wall thinning in adult animals [192]. Interestingly, in response to RA treatment, cells in culture activate PI3K and Erk pathways and are required for a proliferative response [186].

### Nuclear Regulators of Chamber Growth and Maturation

A number of transcription factors and nuclear regulatory factors have been identified in the control of cardiac growth and chamber maturation during embryogenesis. The forkhead transcription factor Foxp1 has been shown to play a role in myocyte proliferation and maturation. *Foxp1*-deleted embryos display a thin ventricular myocardial compact zone caused by defects in myocyte maturation and proliferation [193].

The role of the GATA4 regulation in ventricular maturation has been demonstrated using Cre-LoxP technology to conditionally delete *GATA4* in the myocardium of mice at an early embryonic stage. The *GATA4* deletion resulted in hearts with striking myocardial thinning and with reduced cardiomyocyte proliferation more so in RV as compared to LV, leading to selective hypoplasia of the RV [12]. Another previously discussed transcription factor, Tbx5, has a contributory role in cardiomyocyte proliferation during embryonic development. Tbx5 overexpression in embryonic chick hearts *in vivo* inhibits myocardial growth and trabeculation largely as a result of suppressing embryonic cardiomyocyte proliferation [131]. Mice with targeted disruption of both *nfatc3* and *nfatc4* genes demonstrated early embryonic lethality (after E10.5) and exhibited thin ventricles and a reduction in ventricular myocyte proliferation [194]. A role for mitochondrial energy metabolism in early cardiomyocyte proliferation was suggested by a pronounced defect in mitochondrial structure (e.g., abnormal cristae) and function (e.g., OXPHOS complex II and IV activities) in mice containing these deleted genes. The cardiac-specific expression of constitutively active NFATc4 in nfatc3$^{-/-}$ and nfatc4$^{-/-}$ embryos prolonged embryonic viability to E12 and preserved ventricular myocyte proliferation, compact zone density, and trabecular formation, with enhanced cardiac mitochondrial ultrastructure and complex II enzyme activity in the rescued embryos.

Homeodomain-only protein (Hop) is a small divergent protein that lacks certain conserved residues required for DNA binding, initiates gene expression early in cardiogenesis, and is involved in the control of cardiac growth during embryogenesis and early prenatal development [195, 196]. During mid-embryogenesis, *Hop* is expressed predominantly in the trabecular region of the myocardium (where cardiomyocyte proliferation is diminished). *Hop* modulates cardiac growth and proliferation by inhibiting the transcriptional activity of SRF in cardiomyocytes by recruiting HDAC activity, forming an HDAC2-containing complex and affecting chromatin remodeling [197].

Mice deficient in the proto-oncogene transcription factor Nmyc also have defective trabeculation and thinned ventricular myocardium [198]. The negative regulation of Nmyc (and cardiomyocyte proliferation) by Tbx2 discussed earlier is relevant in this context [137]. Repression of Nmyc1 by aberrantly regulated Tbx2 accounts in part for the observed cardiac hypoplasia in Tbx20 mutants.

## Chamber Septation

Septation occurs at three levels: the atrium, the ventricle, and the arterial pole, and requires correct looping for normal septation. Cell populations extrinsic to the developing heart, including the neural crest, influence the process of ventricular septation and OFT formation through inductive interactions with neighboring tissues. With proper septation of the various levels, the previously described transitional zones are incorporated into the chambers leading to the formation of the definitive cardiac atria and ventricles. The molecular mechanisms responsible for various stages of ventricular septation, in particular, the processes that mediate morphogenetic movements and fusion between opposing structures remain largely unknown.

The primary heart tube consists of a myocardial outer mantle with an endocardial inner lining. Between these two concentric epithelial cell layers, a cellular matrix is found which is generally referred to as the cardiac jelly. During cardiac looping, the cardiac jelly basically disappears from the chamber-forming regions of the cardiac tube but accumulates in the junction between the atria and ventricles, the atrioventricular junction (AVJ), as well as in the developing OFT. This results in the formation of the endocardial cushion tissues in the AVJ and OFT. The subsequent maturation of these cushions is achieved when a subpopulation of endothelial cells overlying the cushions, triggered by growth factor signaling, undergoes a transformation from epithelial to mesenchymal cells followed by migration into the extracellular matrix of the cushions. NCCs also contribute to the mesenchyme of the OFT, while epicardial-derived cells contribute to the AV cushions. These endocardial cushion tissues constitute the major building blocks of the septal structures in the heart. The cushion tissues in the AVJ contribute to the formation of AV septal structures and AV valves, and the cushion tissues in the OFT take part in its septation and in the formation of the semilunar valves of aorta and pulmonary artery. Defects in the formation of these important developmental entities play a critical role in the etiology of variety of clinical congenital heart abnormalities. Mouse models of endocardial cushion defects have been generated by mutations in neurofibromin-1 [199], hyaluronan synthase-2 (Has2) [200], and RXR-knockout mice [110], and are generally associated with fetal lethality. While defects effecting endocardial cushion formation in mice generally affect both AV and OFT cushions, some defects (e.g., NFATc and Sox4) affect only OFT cushions [201].

## Atrioventricular Junction and Formation of the Atrioventricular Cushions

The endocardial jelly in the AVJ forms the base material for the AV cushions. Initially two prominent endocardial cushions develop opposing sides of the common AV canal. These comprise the inferior (or dorsal) AV cushion and the superior (or ventral) AV cushion. As development proceeds, the leading edges of the cushions fuse thereby separating the common AV canal into a left and right AV orifice. Subsequently, smaller endocardial cushions develop in the lateral AVJ. These lateral cushions contribute to valvuloseptal morphogenesis and are involved in the formation of anterosuperior leaflet of the tricuspid valve and the mural leaflet of the mitral valve.

After fusion, the major AV cushion-derived tissue basically forms a large mesenchymal "bridge" contiguous with the mesenchymal cap (also endocardial cushion material) that covers the leading edge of the forming primary atrial septum. As development progresses, the mesenchymal cap on the primary septum and the fused cushions merge completely, thereby closing this primary foramen. The mesenchymal remodeling at this point basically consists of the fused major AV cushions and the mesenchymal cap eventually leads to the formation of the membranous AV septum, the septal leaflet of the tricuspid valve, and the aortic leaflet of the mitral valve. Although the endocardial cushion tissues are important in the formation of the valves, AV valve morphogenesis also involves a number of myocardial remodeling events [202].

### Clinical Studies

Defects in the genesis, interaction, and fate of the endocardial cushions account for the majority of congenital heart malformations in humans, including atrial and ventricular septal defects (ASDs and VSDs, respectively), tetralogy of Fallot (TOF), common AVC, and double-outlet RV. While much information has been provided from animal models, the genetic analysis of cardiac septation defects has proved to be a major stimulus in demonstrating that specific transcription factors (e.g., Nkx2-5 and Tbx5) are required at both specific times and dosage for defining septal morphogenesis.

Dominant mutations in *Nkx2-5* have been found in patients with ASDs, VSDs, TOF, and Ebstein's anomaly of the tricuspid valve with associated conduction defects [203–205]. *Nkx2-5* haploinsufficiency appears to underlie these defects [206], although mice lacking one copy of *Nkx2-5* do not exhibit as severe cardiac defects as found with human mutations [207].

The identification of *Tbx5* mutations in Holt-Oram syndrome (HOS), a rare inherited disease characterized mainly by upper limb and CHD, has also provided insight into mechanisms of cardiac septation [208, 209]. The cardiac malformations in HOS are similar to those caused by *Nkx2-5* mutations: ASDs and VSDs, with occasional reports of TOF often combined with conduction system disease. As noted with *Nkx2-5* mutations, *Tbx5* haploinsufficiency is responsible for HOS [208, 209]. Accordingly a deletion of one copy of *Tbx5* in the mouse recapitulates a HOS phenotype [127]. Patients with missense *Tbx5* mutations have variable phenotypes, some mutations have associated with severe cardiac defects, others associated with milder defects [210, 211]. Functional studies of *Tbx5* missense mutations indicate that some result in a nonfunctional protein, whereas others have altered Tbx5 function (e.g., binding to specific DNA sites) [212].

## *Formation of the AV Valves*

The formation of the cardiac cushions is a complex event characterized by endothelial-to-mesenchymal transdifferentiation (EMT) of a subset of endothelial cells specified in the cushion-forming regions to invade the cardiac jelly, where they subsequently proliferate and complete their differentiation into mesenchymal cells. The cushions protrude from the underlying myocardium, and by a complex and relatively unknown mechanism form thin, tapered leaflets with a single endothelial cell layer and a central matrix composed of collagen, elastin, and glycosaminoglycans [213]. These delamination and remodeling events depend on further cell differentiation, apoptosis, and ECM remodeling. The final AVC valves (mitral and tricuspid) are derived entirely from endocardial cushion tissue [214]. Recent lineage analysis have documented that the leaflets and tendinous cords of the mitral and tricuspid valves, as well as the atrioventricular fibrous continuity, and the leaflets of the OFT valves are exclusively generated from mesenchyme derived from the endocardium, with no substantial contribution from cells of the myocardial and neural crest lineages.

A number of signaling molecules originating from both the AV myocardium and the endothelium participate in the formation of the AV valves and in EMT, a critical step of cardiac cushion tissue formation. A variety of techniques have implicated their involvement in AV valve formation including gene disruption in mice (over a dozen genes cause AV valve defective phenotype), spatiotemporal expression profiles, and the use of specific inhibitors. These signaling factors (listed in Table 9.2) include numerous ligands, membrane receptors,

**Table 9.2** Factors implicated in cardiac AV valve formation

| Factor | Phenotype |
|---|---|
| NFATc1 | Disruption of *NFATc1* gene leads to selective absence of the aortic and pulmonary valves |
| VEGF | Myocardial overexpression of VEGF results in failure of AVC and OFT cardiac cushion formation |
| Cx45 | *Cx45* knockout leads to decreased and delayed EC formation |
| Notch1 | Disruptions of *Notch1* leads to hypoplastic EC |
| NF-1 | Disruption of *Nf1* leads to markedly enlarged EC |
| ErbB3 | *ErbB3*$^{-/-}$ embryos exhibit EC abnormalities and defective valve formation |
| BMP ligand | *Bmp6*, *Bmp7* double mutants result in hypoplastic EC and delayed in OFT formation |
| BMPR (Alk3) | *Alk3* disruption leads to hypoplastic EC |
| Smad6 | *Smad6* disruption results in thickened and gelatinous AV and semilunar valves |
| Wnt-β-catenin-APC | APC disruption leads to thickened valves |
| HB-EGF | *HB-EGF*$^{-/-}$ mice have enlarged AVC and OFT valves |
| Has-2 | *Has2*$^{-/-}$ mice are unable to form EC |
| Hesr2 | Disruption of the Notch signaling target *Hesr2* results in dysplastic AV valves |
| TGF-β | *TGF-β1* inhibits valve myofibroblast proliferation; *TGF-β2* knockout results in AV and semilunar valve thickening |

*EC* endocardial cushion, *OFT* outflow tract, *AV* atrioventricular, *APC* adenomatous polyposis coli (gene), *AVC* atrioventricular canal, *HB-EGF* heparin-binding epidermal growth factor, *Hesr2* hairy and enhancer of split-related transcriptional repressor 2

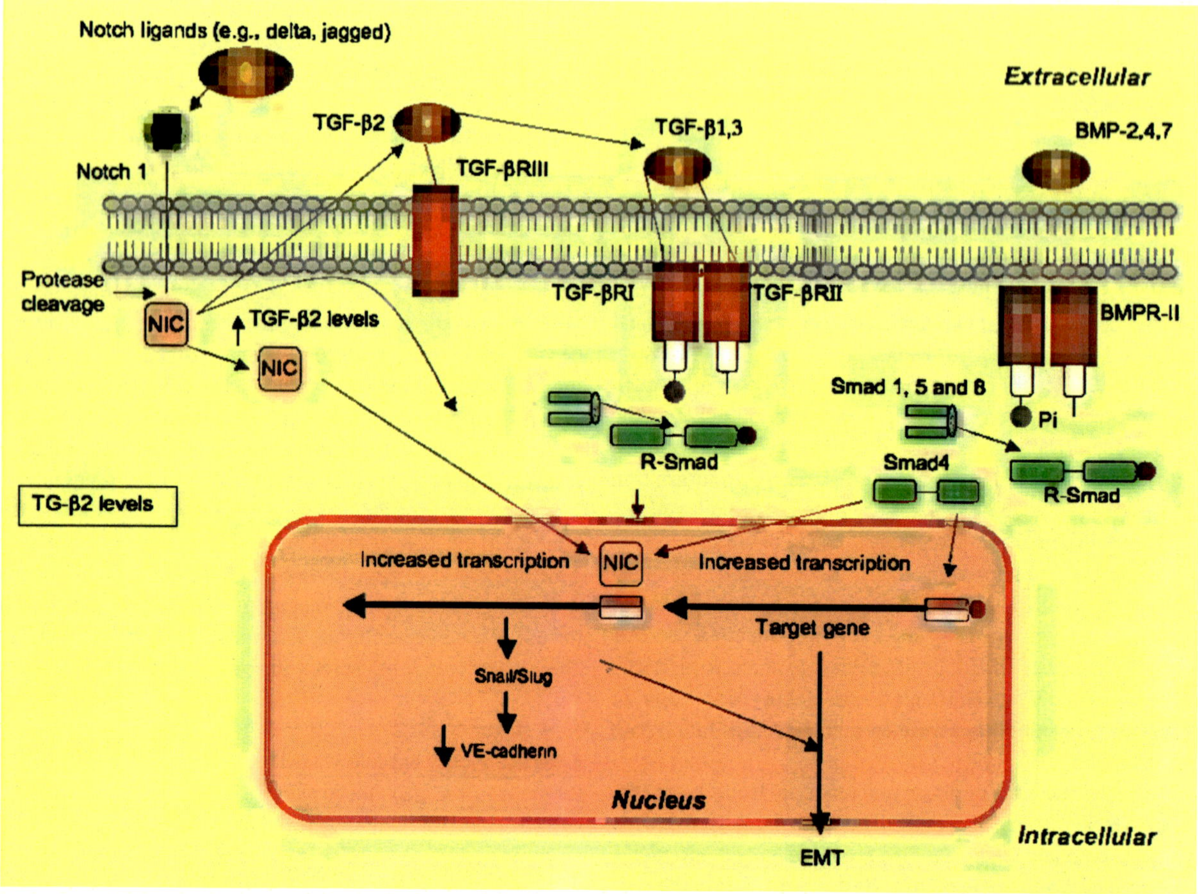


**Fig. 9.6** Pathway of TGF-β/BMP/Notch signaling in the initiation of EMT. Type III TGF-β receptor (TGF-β-RIII) presents TGF-β2 to TGF-β-RII in the developing heart valve. TGF-β signaling is mediated through snail/slug transcription factors and can result in reduced expression of VE-cadherin. BMPs (e.g., BMP2, 4, and 7) in the developing valve signal via the BMPR-II receptors, which are crucial for valve development and intracellular Smad mediators. Smad6 antagonizes the interaction of Smad1 with Smad4, thereby decreasing BMP signaling. Synergy between TGF-β and BMP signaling in cardiac cushion explants has been shown to facilitate EMT. Also shown is the Notch signaling pathway in which the transmembrane receptor Notch, upon binding specific ligands, is cleaved to form NIC, which subsequently translocates to the nucleus activating specific gene transcription leading to EMT

and transcription regulators such as BMP/TGF-β, Notch, VEGF, NFATc1, Wnt/β-catenin, ErbB, and NF-1-Ras.

As we have noted elsewhere, the BMPs are members of the TGF-β cytokine superfamily, of which BMP2, BMP4, TGF-β2, and TGF-β3 have been implicated in heart development. All TGF-β family members are homodimeric proteins that interact with transmembrane TGF-β receptors. Ligand binding activates type II receptors to transphosphorylate type I receptors within the ligand–receptor complexes. The phosphorylated type I receptor then acts as a serine/threonine kinase to phosphorylate and activate cytosolic Smad proteins, which are the major intracellular mediators of TGF-β signaling. TGF-β and BMP are the most extensively studied signaling partners in endocardial cushion formation and their pathways are shown in Fig. 9.6.

Expression levels of BMP2 in myocardial cells parallel the segmental pattern of cushion formation [215]. BMP2 protein is localized in AV myocardium in mice before the onset of AV mesenchymal cell formation, but absent from ventricular myocardium throughout these stages. After the subsequent cellularization of the AV cushion, BMP2 protein expression is reduced in AV myocardium, while initiated and maintained in cushion mesenchymal cells even during later stages of development. During valuvulogenesis, there is intense BMP2 expression in the valve tissue, maintained in adult mice. *In vitro* studies using cultured AV endocardial endothelium showed that BMP2 protein addition promoted the formation of cushion mesenchymal cells in the absence of AV myocardium, with enhanced expression of the mesenchymal marker, smooth muscle (SM) α-actin, loss of the endothelial marker, platelet-endothelial cell adhesion molecule (PECAM-1), and elevated levels of TGF-β2. Upon treatment with noggin, a specific antagonist of BMPs, applied together with BMP2 to the culture medium, AV endothelial cells remained as an epithelial monolayer with reduced expression of SM α-actin and TGF-β2, and normal expression of PECAM-1. These data indicate that BMP signaling is necessary and sufficient for the myocardial segmental regulation of AV endocardial cushion mesenchymal cell formation in mice.

Further evidence that BMP signaling plays an important role in the AV myocardium during the maturation of AV valves from the cushions has been obtained with the Cre-LoxP technique to target the deletion of a *Alk3* allele, gene encoding the type IA receptor for BMPs in cardiac myocytes of the AVC [216]. Cardiac myocytes of the AVC was shown by lineage analysis to contribute to the formation of the tricuspid mural and posterior leaflets, the mitral septal leaflet, and the atrial border of the annulus fibrosus. With *Alk3* deletion in these cells, defects were seen in the tricuspid mural leaflet and mitral septal leaflet, the tricuspid posterior leaflet was displaced, and the annulus fibrosus was disrupted resulting in ventricular preexcitation. In addition these findings provided support for the potential role of Alk3 in human CHD, such as Ebstein's anomaly.

The foregoing results concluded that the BMP ligand and/or receptor disruption results in specific valve phenotypes that decrease valve formation. The Smad proteins are intracellular mediators of signaling initiated by TGF-β superfamily ligands, which either can positively trigger further downstream transcriptional responses (Smad 1, 2, 3, 5, and 8) or function as inhibitory transcriptional regulators (Smad 6 and 7). Smad6, an inhibitor of Smad signaling downstream of Alk2, inhibited EMT in AV cushion endocardial cells [217]. Mice with disrupted *Madh6*$^{-/-}$, which encodes Smad6, display marked hyperplasia of the cardiac valves, and also have hypertension and decreased endothelial cell-mediated vasodilation [218]. Thus, inhibitory Smad6 signaling may play a physiologic feedback role in the heart development by limiting the number of endocardial cells that undergo EMT.

Besides the Smad transducers, the downstream effectors of TGF-β and BMP signaling in the developing cardiac cushions are still not entirely defined but may include the Snail/Slug family members, a group of zinc-finger transcription factors that primarily act as transcriptional repressors [219]. Inactivation of Slug with antisense oligonucleotides impairs epithelial-to-mesenchymal transdifferentiation [220]. The expression patterns of *Slug/Snail* in the developing heart suggest that these transcription factors could be involved in the regulation of EMT in cardiac cushion formation. In chick hearts, Slug proteins are highly expressed within the mesenchyme of developing cushions and in a subset of endocardial cells overlying the cushions [221], where it is thought to be a target of TGF-β2 during EMT [222]. Similarly in the embryonic mouse, *Snail* is expressed in the mesenchyme of many regions that undergo epithelial-to-mesenchymal transdifferentiation and in the endocardium and mesenchyme of developing heart valves [223]. VE-cadherin expression appears to be reciprocal to Snail, as revealed in studies of Notch signaling mutants.

Data derived from zebrafish studies suggest that the evolutionary conserved Notch signaling pathway may have a significant role in endocardial EMT, as expression of the receptor N*otch1b* is localized within the endocardium of the presumptive AV valve region both prior to and during the stages when EMT occurs [224]. Embryos deficient in Notch signaling exhibit severely attenuated cardiac *Snail* expression, abnormal maintenance of intercellular endocardial adhesion complexes, and abortive EMT. Transient ectopic expression of activated Notch1 in zebrafish embryos led to hypercellular cardiac valves, whereas inhibition of Notch prevented valve development [225].

Disruption of *HESR2*, a downstream target of Notch signaling, was found to result in mice with tricuspid and mitral valve regurgitation, dysplastic AV valves, a perimembranous ventricular septal defect, and a secundum atrial septal defect with a majority of mice dying from resultant congestive heart failure [226]. These observations support the view that the Notch signaling pathway, including the targeted *HESR2*, plays an important role in the formation and function of the AV valves. Notch1 mutants retain strong VE-cadherin, while fail to express Snail, and fail to undergo EMT [225]. Notch promotes EMT during normal cardiac development partially via the transcriptional induction of the Snail repressor, a potent and evolutionarily conserved mediator of EMT in many tissues types. In the embryonic heart, Notch functions via lateral induction of a selective TGF-β-mediated EMT that leads to cellularization of developing cardiac valvular primordia. Embryos that entirely lack Notch signaling exhibit severely attenuated cardiac Snail expression, abnormal maintenance of intercellular endocardial adhesion complexes, and abortive endocardial EMT *in vivo* and *in vitro*. A possible model to explain these findings would be that TGF-β induced by Notch signaling in the developing cushion activates Snail, which decreases the expression of cell adhesion molecules and thereby downregulates endocardial cell–cell adhesion promoting endocardial cells to initiate invasion into the cardiac jelly.

The biological effects of VEGF are mediated by two receptor tyrosine kinases (RTKs), VEGFR-1 and VEGFR-2, which differ considerably in signaling properties. Cell proliferation, vascular permeability, chemotaxis, and survival in endothelial cells in the developing embryo are regulated by VEGF [227]. The downstream components of VEGFR signaling cascade are not entirely known, but may include inositol 1,4,5-trisphosphate/diacylglycerol, and Erk/MAPK pathways and in valve endothelial cells, NFATc1 [228, 229].

While broadly expressed in early endocardial cells, VEGF expression becomes restricted to a subset of endocardial cells lining the AVC, a first indication that VEGF has a role in endocardial cushion formation [230].

It remains unclear whether the VEGF-expressing endothelial cells in the cushion-forming region represent a unique subpopulation of endothelial cells predetermined to undergo EMT or whether the VEGF-producing cells induce proliferation or increased permeability of adjacent endothelial cells in the developing cardiac cushions to undergo EMT [213].

Selective myocardial overexpression of VEGF in the early embryo (between E3.5 and E9.5) resulted in failure of cardiac cushion formation at the AVC and OFT [231]. These embryos also had multilayered endocardium, suggesting a dysregulation of the differentiation process and overexpression of an endothelial phenotype. These findings were confirmed *ex vivo* using the collagen explant system, in which myocardium and endocardium of the developing cushion are explanted onto a type I collagen gel to recapitulate the EMT events [232]. Addition of exogenous VEGF inhibited EMT in the forming AVC cushions. Together, these data confirm that VEGF levels are tightly regulated during normal heart development and that even moderate increases in VEGF expression can have profound developmental consequences, possibly by inhibiting endothelial cell differentiation and thereby negatively regulating EMT.

Growing evidence demonstrates that both hypoxia and hyperglycemia can regulate VEGF expression in developing heart valves. Studies employing tissue explants have shown that hypoxia decreases cardiac cushion EMT, reversible by the addition of soluble VEGFR1 [233]. Moreover, under hypoxic conditions, VEGF expression was increased nearly tenfold in cardiac cushions. These results suggest that fetal hypoxia may increase VEGF expression in the cushion-forming areas inhibiting EMT and may contribute to CHD in the cardiac valves and interatrial septum.

In the developing mouse, hyperglycemia reduces VEGF expression [234]. Employing the tissue explant system, it has been shown that elevated glucose levels inhibit AVC cushion ability to undergo EMT [235]. Adding back VEGF165 abrogated the effect of hyperglycemia by allowing normal cushion EMT. These results suggest that decreased VEGF expression during development inhibits cushion formation, potentially by inhibiting endothelial migration into the cardiac jelly, and underscores the importance of highly controlled VEGF levels during cardiac cushion formation, as either over- or under-expression of VEGF causes hypoplastic cardiac cushions. It is therefore of interest that neonates born to diabetic mothers have an approximately 3-fold increase in CHD, with roughly a 10- to 20-fold increased risk of rare abnormalities such as double-outlet RV and truncus arteriosus [236]. Dramatic reduction in the occurrence of congenital defects in children born to diabetic mothers with strict glycemic control during pregnancy suggests that hyperglycemia has a direct teratogenic effect [237].

Members of the NFAT family, which mediate transcriptional responses of the $Ca^{2+}$/calmodulin-dependent protein phosphatase calcineurin, have been implicated in cardiovascular development almost exclusively in vertebrates [238]. This pathway is shown schematically in Fig. 9.7.

Genetic inactivation of NFATc1 in the mouse demonstrated that NFATc1 is required for cardiac valve formation [239, 240], with greater effect on the pulmonary and aortic valves [239]. Consistent with a function in formation of these endocardial-derived structures, NFATc1 has been found to be exclusively expressed in the endocardium at the initiation of endocardial differentiation in the primary heart-forming field. Within the endocardium, specific inductive events appear to activate NFATc: it is localized to the nucleus only in endocardial cells that are adjacent to the interface with the cardiac jelly and myocardium, which are thought to give the inductive stimulus to the valve primordia. Treatment of wild-type embryos with FK506, a specific calcineurin inhibitor, prevents nuclear localization of NFATc [239]. Recent evidence has indicated that NFATc1 mediates VEGF signaling of proliferation of human pulmonary valve endothelial cells [229]. The calcineurin-specific peptide inhibitor reduced both VEGF-induced human pulmonary valve endothelial cell proliferation and abrogated VEGF-induced NFATc1 nuclear translocation suggesting a functional role for NFATc1 in endothelial growth. Others have shown that the initiation of heart valve morphogenesis in mice requires calcineurin-NFAT to repress VEGF expression in the myocardium underlying the site of prospective valve formation. This repression of VEGF at E9 is essential for endocardial cells to transform into mesenchymal cells [241]. Interestingly, an enhancer element located within the first intron of the mouse *NFATc1* gene has been recently identified to be responsible for the high-level *NFATc1* gene expression found in provalve endocardial cells of both the AVC and OFT during valvulogenesis [242].

It has also been reported that Down syndrome critical region 1 [DSCR1 alias MCIP1 (modulatory calcineurin interacting protein 1)], located in region 21q22.1–q22.2, may play a contributory role in the cardiac defects (mainly involving the endocardial cushion tissue) present in Down syndrome. Overexpression of DSCR1 in cardiomyocytes inhibits $Ca^{2+}$-dependent nuclear translocation of NFAT [243]. In addition, DSCR1 functions as an endogenous calcineurin inhibitor [244]. DSCR1 may play an important role in the development of endocardial cushion defects as suggested by its increased expression in regions that correlate with areas of defective endocardial cushion development [245].

Gap junction channels are necessary during early cardiogenesis. Connexins (Cxs) are a group of transmembrane proteins

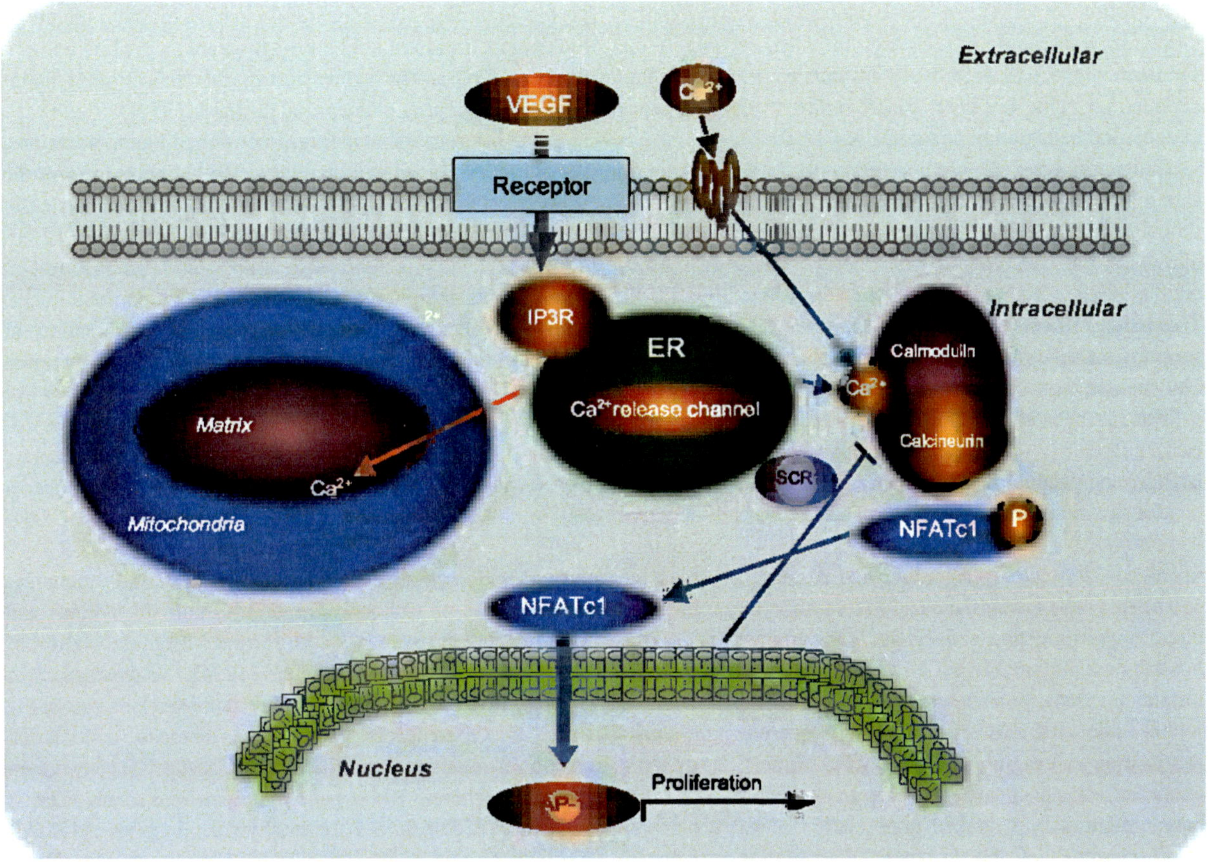


**Fig. 9.7** Model for NFATc1 as a transcriptional regulator of endothelial cell fate. VEGF signaling through NFATc1 increases the proliferation of pulmonary valve endothelial cells. In the developing cardiac cushion, $Ca^{2+}$ may enter the endothelial cell through connexin-45 gap junctions and activate calcineurin. Calcineurin, in turn, dephosphorylates NFAT isoforms, including NFATc1. NFATc1 is then transported into the nucleus, where it interacts with transcriptional regulators, including AP-1, to affect gene transcription. The endogenous calcineurin inhibitor DSCR1, a target of NFATc1, may establish a negative feedback loop by inhibiting calcineurin

that form gap junctions between cells. Products of several connexin genes have been identified in the mammalian heart (e.g., Cx45, Cx43, Cx40, and Cx37), and their expression has been shown to be regulated during the development of the myocardium. Cx45 is the first connexin expressed in the developing heart. In the mouse E9.5 developmental stage, Cx45 is markedly upregulated in cells in the AVC and OFT suggesting a potential role in the development of the endocardial tissue [246]. Whereas disruption of cardiac *Cx40* and *Cx43* genes cause heart defects such as conduction defects and dysrhythmias, but not embryonic death, disruption of *Cx45* gene does [247, 248]. Cx45-deficient mice died of heart failure at around E10 and harbored defective endocardial cushions. These defects are caused by impairment of the EMT of the cardiac endothelium [249]. Activation of this endothelium depends on the presence of the Cx45 gap junctions since signaling through $Ca^{2+}$/calcineurin and NFATc1 (originally named NFATc) was disrupted in the mutant mice hearts. These findings indicate the need for gap junction channels during early cardiogenesis and implicate Cx45 in the development of a number of CHD. However, caution is necessary in this mice model since it often features extensive atrioventricular conduction defects in addition to endocardial cushion defects. It has been noted that $Cx45^{-/-}$ mice die at E10 3 days earlier than mice with congestive heart failure secondary to lack or failure of valve development suggesting that the formation of the endocardial cushion in this model may in fact be delayed rather than disrupted, and that the cause of death may be secondary to a hypocontractile heart.

As previously discussed, the involvement of Wnt signaling in early cardiac development was demonstrated by studies showing the requirement for Wnt inhibition in early cardiac specification and revealing the role of Wnt signaling in the posterior mesoderm in repressing cardiogenesis and patterning in the chick embryo [80]. These studies employed the specific inhibition of canonical Wnt/β-catenin signaling by Dickkopf-1 (Dkk-1) or Crescent which resulted in the initiation of cardiogenesis in vertebrate embryos. Evidence suggests that Wnt signaling may also closely regulate valve development.

Induction of EMT in the endocardial cells requires β-catenin transcriptional activity. In mice, in which the *β-catenin* gene was selectively inactivated in endothelial/endocardial cells, the heart cushions fail to develop since endocardial cells do not undergo EMT [250].

Endocardial cushion-specific expression of Wnt-β-catenin signaling was also recently identified in the developing mouse heart. At E11.5, Wnt signaling is restricted to a subset of cells in the AVC and OFT, with only a subset of mesenchymal cells staining for Wnt-β-catenin signaling [251]. This pattern of expression, which is conserved in mammals, supports the role of the Wnt-β-catenin pathway in endocardial cushion development.

This has been also demonstrated in zebrafish, in which an early missense mutation in the *adenomatous polyposis coli* (*APC*) gene product resulted in embryonic lethality by 96 h postfertilization, timing which corresponds to cardiac valve formation. Loss-of-function mutations in the *APC* gene lead to increased β-catenin signaling and resulted in excessive and misplaced endocardial cushion formation throughout the heart, suggesting that endocardial cells lining most of the heart had undergone EMT [224]. Moreover, in contrast to the intranuclear location of β-catenin found in wild-type embryos within endocardial valve-forming cells overlying the cushion-forming areas, homozygous *APC* truncation mutants exhibited nuclear-localized β-catenin throughout the heart concomitant with a marked upregulation of valve markers reinforcing the view that an expanded population of endocardial cells was capable of proliferating and competent to undergo EMT. Overexpression of APC or inhibition of Wnt by Dickkopf 1 reversed this phenotype. These findings support the idea of a prominent role of the Wnt-catenin pathway in determining endocardial cell fate and EMT progression.

Other components of the Wnt signaling pathway have been shown to mediate mesenchymal cell proliferation leading to proper AV canal cushion outgrowth and remodeling in the developing heart. Within the chick heart, Wnt9a expression is restricted to AV endocardial cushions, primarily in the AV canal endocardial cells, while Wnt antagonist, the Frizzled-related protein gene (*Frzb*) expression is detected in both endocardial and transformed mesenchymal cells of the developing AV cardiac cushions [252]. Wnt9a stimulates β-catenin-responsive transcription and promotes cell proliferation in the AV canal cells; overexpression of Wnt9a results in enlarged endocardial cushions and AV inlet obstruction. Functional studies also revealed that Frzb inhibits Wnt9a-mediated cell proliferation in endocardial cushions indicating that a dynamic balance between Wnt9a and Frzb is involved in the regulation of mesenchymal cell proliferation in endocardial cushion formation.

Studies directed at both cataloguing a set of molecular markers encompassing all stages of cardiac valve development and providing an analysis of their gene expression profile identified 13 endocardial cushion markers including the demonstration of active Wnt-β-catenin signaling components as well as the expression of the transcription factor Fog1 in developing endocardial cushions [251].

In addition to its signaling functions, β-catenin acts as a structural link between β-actin and VE-cadherin to form the adherens junction, a molecular scaffold that mediates cell–cell adhesion and cell polarity in endothelial cells [253]. Phosphorylated β-catenin also associates with PECAM-1/CD31, a transmembrane protein of the immunoglobulin (Ig) family involved in cell–cell contact. The association of β-catenin with cadherins and PECAM-1/CD31 in dynamic junctional complexes may provide a cellular reservoir for phosphorylated β-catenin to regulate the level of free, cytosolic β-catenin and modulate its availability for nuclear translocation [254].

The interactions between β-catenin, PECAM-1/CD31, and VEGF may be a critical juncture in the control of EMT during endocardial cushion formation. VEGF signaling increases the phosphorylation of β-catenin in a time- and dose-dependent manner, leading to increased association of β-catenin with PECAM-1/CD31 [254]. Sequestration of β-catenin by VEGF signaling may represent one means by which increased levels of VEGF can decrease endocardial cushion formation. During EMT, PECAM-1/CD31 is downregulated, whereas SM α-actin is upregulated as endocardial cells differentiate into a mesenchymal phenotype [255]. If EMT is disrupted, PECAM-1/CD31 levels persist [235]. A hypothesis has been proposed that as PECAM-1/CD31 is downregulated, the cytosolic levels of β-catenin increase and activate proliferation of cells undergoing EMT; β-catenin could thus serve as the link between activation of the mesenchymal program and population of the cardiac jelly with mesenchymal cells; moreover, reduced PECAM-1/CD31 levels promote cell motility enabling their ability to migrate and penetrate the extracellular matrix. Consistent with this hypothesis, in *PECAM-1/CD31*$^{-/-}$ mice, the cushion-forming areas remain competent to undergo EMT even in the presence of hyperglycemia, which reduces the VEGF levels [235].

## Extracardiac Contribution to Normal and Abnormal Cardiac Development

### *Epicardium-Derived Cells*

During heart development, cells of the primary and SHF give rise to the myocardial component of the heart. The neural crest and epicardium provide the heart with a considerable amount of nonmyocardial cells that are indispensable for correct heart development. During the past two decades, the importance of EPDCs in heart formation became increasingly clear. The epicardium is embryologically formed by the outgrowth of proepicardial cells over the naked heart tube. Following epithelial–mesenchymal transformation, EPDCs form the subepicardial mesenchyme and subsequently migrate into the myocardium, and differentiate into SMCs and fibroblasts. They contribute to the media of the coronary arteries, the atrioventricular valves, and the fibrous heart skeleton. Furthermore, they are important for the myocardial architecture of the ventricular walls and for the induction of Purkinje fiber formation. Although the signaling cascades that participate in EPDC migration and function are not yet known, recent discoveries have shown several factors that are involved in EPDC migration and specialization as well as in the cross-talk between EPDCs and other cardiovascular cells during development. Among these factors are the Ets transcription factors Ets-1 and Ets-2. Data obtained with lentiviral antisense constructs targeting Ets-1 and Ets-2 specifically in the epicardium have shown that these factors are independently involved in the migratory behavior of EPDCs. Ets-2 seems to be especially important for the migration of EPDCs into the myocardial wall and to subendocardial positions in the atrioventricular cushions and the trabeculae. Furthermore, correct EPDC development appears to correlate with normal coronary arteriogenesis and also plays a role in cardiac looping, myocardial architecture, development of valves, and CHD [256].

## ErbB: Integration of Extracellular Matrix Signals

The ErbB family of receptors plays an important role in the regulation of cushion remodeling and valve formation (Fig. 9.8) [257]. The ErbB family of proteins is RTKs and includes ErbB1/EGFR/HER1, ErbB2/Neu/HER2, ErbB3/HER3, and ErbB4/HER4. The four ErbB proteins bind a wide range of ligands with varying affinities that EGF members of the heregulin/neuregulin (HRG) family, heparin-binding epidermal growth factor (HB-EGF), TGF, amphiregulins, β-cellulin, and epiregulin.

The requirement for ErbB/EGFR signaling in cushion development has been demonstrated by observations on cardiac-specific effects of ligand knockouts. HB-EGF is a widely expressed growth factor of the EGF family that can bind erbB1 and erbB4 [258]. In the developing mouse heart, HB-EGF is strongly expressed in the endocardium overlying the cushion-forming area [259]. *HB-EGF*$^{-/-}$ mice display markedly enlarged AVC and OFT valves and die shortly after birth. Consistent with this finding, mice deficient in *t*umor necrosis factor-α-converting *e*nzyme (TACE/ADAM-17, a processing enzyme that cleaves HB-EGF precursor to generate its active form) also possess enlarged AVC and OFT valves. *HB-EGF*$^{-/-}$ and *TACE*$^{-/-}$ mice display normal cushion development through E13.5, but have thickened valves by E14.5. These findings suggest that *HB-EGF*$^{-/-}$ and *TACE*$^{-/-}$ mice exhibit normal EMT initiation, but uncontrolled mesenchymal proliferation during remodeling resulted in defective cardiac valvulogenesis. *HB-EGF*$^{-/-}$ mice also displayed dramatic increases in activated Smad1, 5, and 8 suggesting that ErbB/EGFR signaling is required to regulate the BMP pathway. Moreover, developing valves in *HB-EGF*$^{-/-}$ and *TACE*$^{-/-}$ mice show increased staining for bromodeoxyuridine (BrdU, a marker of cell proliferation), but little evidence in cellular apoptosis. It seems likely that HB-EGF, after activation by TACE, signals through EGFR to limit mesenchymal cell proliferation, because *ErbB4* knockout mice display only moderately decreased endocardial cushion size, but have no defects in fully developed valves [191].

The crucial importance of the cardiac jelly in providing a signal that initiates endocardial differentiation is well recognized. While initial studies focused on soluble factors present in the jelly [260], subsequent investigations have centered on the involvement of the ECM in regulating growth factor activity. An area of recent focus in this regard has been the role of hyaluronic acid (HA) in mediating ErbB signaling [261]. HA is a glycosaminoglycan composed of alternating glucuronic acid and *N*-acetylglucosamine (NAG) residues and is present as a hydrated gel in the ECM, functioning to expand the extracellular space and regulate ligand availability.

Three *Has* genes, termed *Has1*, *Has2*, and *Has3*, are present in mammals [262]. *Has2* encodes the major enzyme responsible for HA synthesis during development. *Has2*$^{-/-}$ mice exhibit severe cardiac and vascular abnormalities including pericardial edema, disordered vessel growth, and complete absence of cardiac jelly and die by mid-gestation (between E9.5 and E10) [200]. In the absence of cardiac jelly, no EMT occurs and the endocardial cushions are unable to form. In addition to its structural role, HA can modulate cell signaling events [263]. Endocardial cells overlying the cushion-forming region in *Has2*$^{-/-}$ mice display reduced EMT and migration, an effect that can be blocked by gene rescue (adding back exogenous HA), or by transfection with constitutively active Ras [200]. Transfection with a dominant-negative Ras blocked the ability of HA to promote EMT. Further studies revealed that heregulin (a ligand for ErbB3) rescued the *Has2*$^{-/-}$ phenotype in *ex vivo* cushion explant models [264]. Furthermore, *Has2*$^{-/-}$ mice possess decreased ErbB2/ErbB3 phosphorylation in endocardial cushion as compared to wild-type embryos. Addition of HA to *Has2*$^{-/-}$ tissue explants restored ErbB3 phosphorylation. *ErbB3*$^{-/-}$ mice die by E13.5 and have hypoplastic cardiac cushions, with decreased mesenchyme content [265]. Notably, ErbB3 is expressed by endocardial cushion cells and mesenchymal

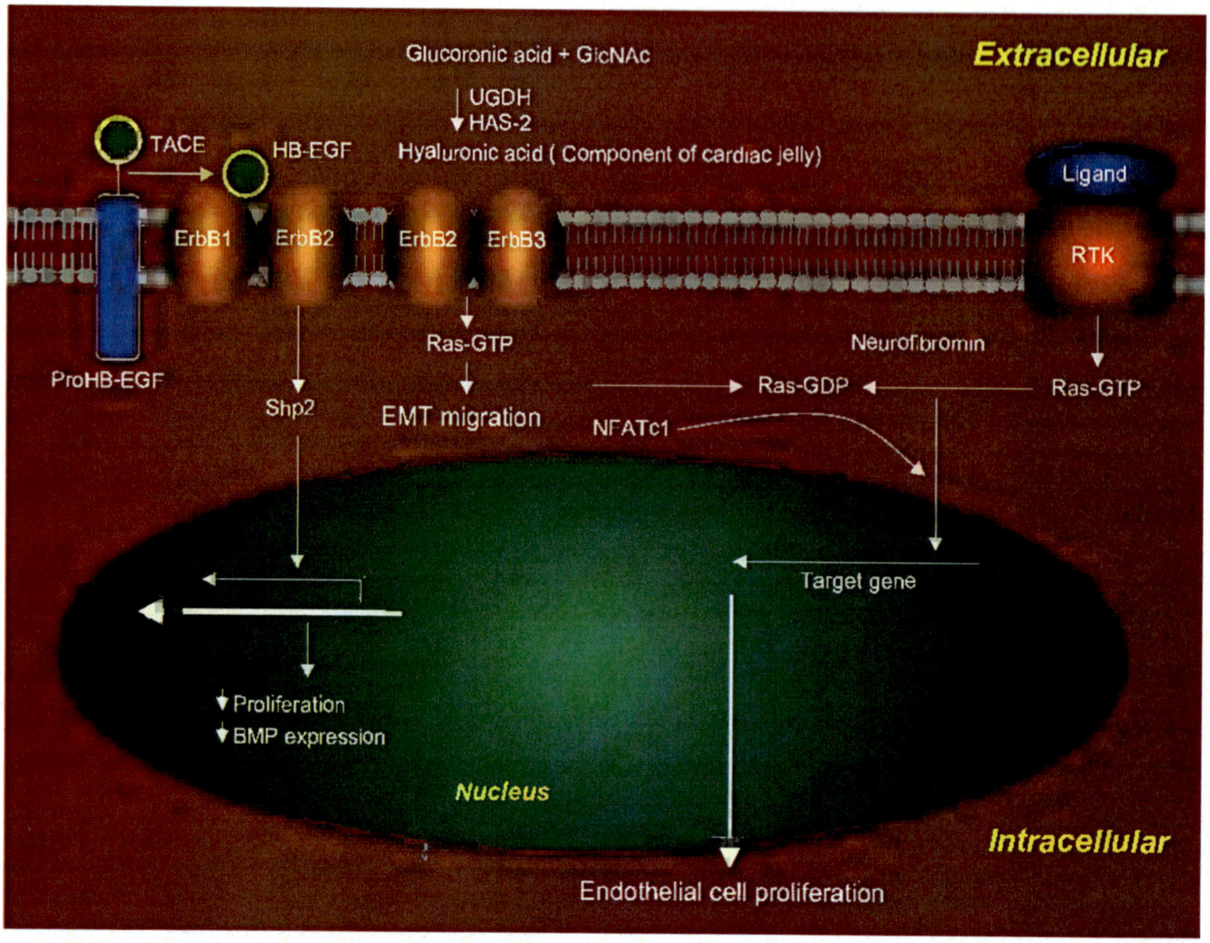


**Fig. 9.8** ErbB signaling is involved in the integration of extracellular matrix signals. Transmembrane precursor pro-HB-EGF is cleaved by TACE to HB-EGF, a ligand for membrane-localized ErbB1 and ErbB4. Binding of HB-EGF to ErbB1 and possibly formation of a heterodimer with ErbB2 appear to reduce downstream EMT, proliferation, and BMP expression. Activation of ErbB2/3 heterodimers by the extracellular matrix polysaccharide hyaluronic acid (HA) increases EMT and migration, an effect that is likely to be mediated by Ras signaling. Synthesis of HA from glucuronic acid and *N*-acetylglucosamine (GlcNAc) is dependent on the enzymes UDP-glucose dehydrogenase (UGDH) and hyaluronic acid synthase-2 (HAS-2). Also, depicted is the involvement of neurofibromin in EMT regulation. Neurofibromin is a Ras-specific GTPase activating protein (GAP) that cycles Ras from an active GTP-bound state to an inactive GDP-bound state. Ras signaling is activated by receptor tyrosine kinases (RTKs), which bind a wide range of ligands and transduce the activation of downstream targets to increase mesenchymal proliferation. Downstream signaling targets of Ras interact with NFATc1 to alter gene transcription. Neurofibromin may therefore decrease endothelial and/or mesenchymal cell proliferation by modulating Ras signaling

cells undergoing EMT. In contrast, EGF, ErbB2, and ErbB4 expression is limited largely to cardiomyocytes during the critical cushion-forming window (E9.5–E10.5). It remains to be assessed how HA signaling interacts with other ErbB ligands, such as neuregulin and HB-EGF.

In zebrafish, the *jekyll* mutant results in failure of cardiac cushion formation and deficient in the initiation of heart valve formation [266, 267]. The *jekyll* mutant contains a point mutation in the *ugdh* gene (encoding UDP-glucose dehydrogenase, UGDH) residing in the enzyme active site. UGDH is required for heparan sulfate, chondroitin sulfate, and hyaluronic acid production [268]. The close phenotypic correlation between the *jekyll* mutant and *Has2*$^{-/-}$ mice suggests that these two mutations interrupt the same pathway affecting cushion formation and subsequent valvulogenesis.

The expression of the ECM proteoglycan versican is associated with valvulogenesis in the developing mouse heart [269]. *Versican* (or *PG-M*) gene encodes a chondroitin sulfate proteoglycan expressed in the pathways of NCC migration and in prechondrogenic areas of the developing chick and mouse [270, 271]. This protein is nonpermissive for cell migration and appears in association with slow cell proliferation and differentiation. Versican may be a key participant in cardiogenesis, responding to the many diffusible signals that mediate interactions between the developing endocardium and myocardium.

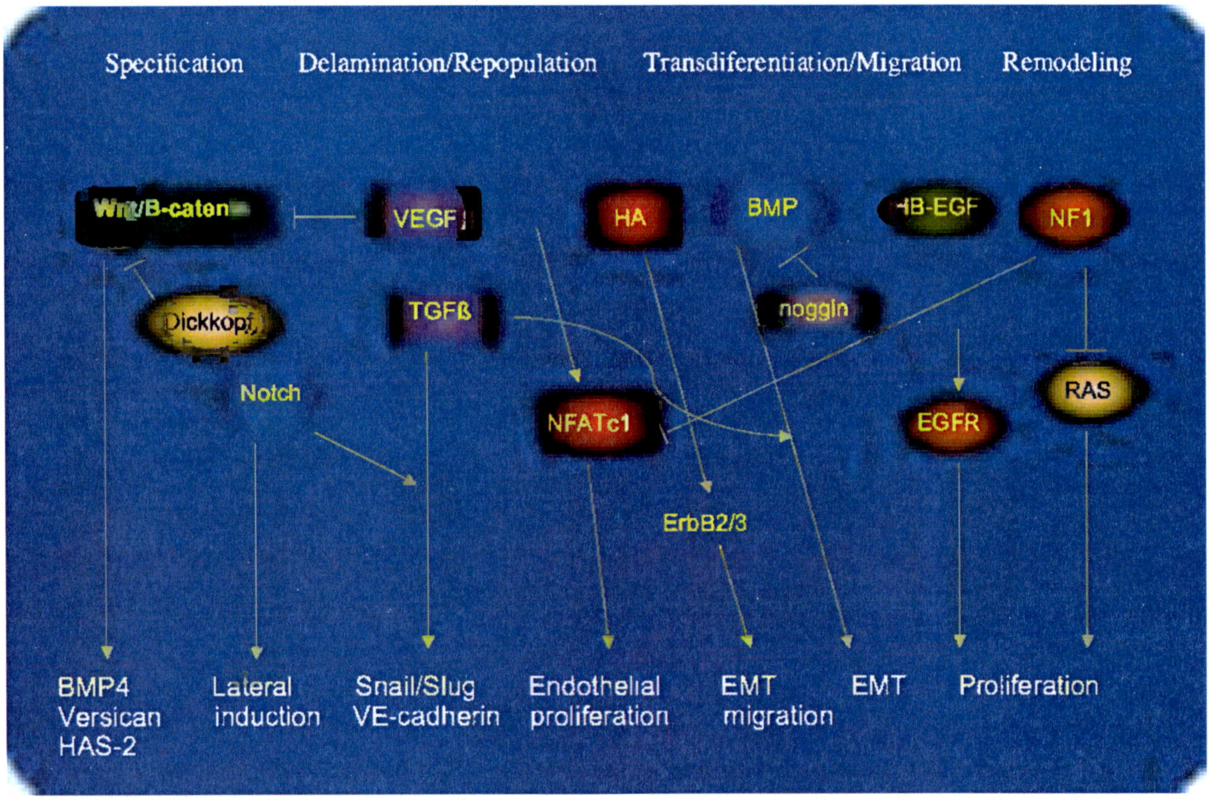


**Fig. 9.9** Interaction of signaling factors in valve development and remodeling. Many signaling pathways and transcriptional regulators coordinately regulate the process of heart valve formation. *Yellow arrows* denote positive/synergistic interactions between pathways. *Blunt arrows* denote inhibitory effects between pathways

In this section, we have described a number of apparently diverse regulatory pathways that contribute to early mesenchyme differentiation, cardiac jelly, migration and formation of endocardial cushions, EMT as well as to the final remodeling of the mesenchymal cushions into atrioventricular valves. Gathered observations indicate that a considerable interaction and integration exist between these signaling pathways (Fig. 9.9).

Other regulatory factors will undoubtedly be identified which serve to integrate these pathways and will further illuminate their functions. For instance, gene expression of the regulator Sox9 is activated when endocardial endothelial cells undergo mesenchymal transformation and migrate into an ECM, called cardiac jelly, to form endocardial cushions. This expression pattern suggested that Sox9 might play a contributory role in the pathway that controls the formation of cardiac valves and septa. In *Sox9*-null mutants, endocardial cushions are markedly hypoplastic [272]. In these mutants, NFATc1 is ectopically expressed and no longer restricted to endothelial cells, and *Sox9*-deficient endocardial mesenchymal cells fail to express ErbB3, which is required for endocardial cushion cell differentiation and proliferation. Another class of regulators recently identified in conjunction with atrial valve defects is the FOG-1 and FOG-2 zinc-finger proteins which are cofactors that interact with GATA4. These cofactors can act as either enhancers or repressors of GATA transcriptional activity, depending on the cell and promoter context [273]. Mice containing disrupted *FOG-2* alleles display tricuspid atresia, suggesting a genetic basis for the clinical entity [274].

Other cardiovascular defects occur in these mice including ventricular and atrioseptal defects reminiscent of TOF, accompanied by a general failure of coronary vessel formation [275]. Transgenic reexpression of FOG-2 in cardiomyocytes rescues the $FOG\text{-}2^{-/-}$ vascular phenotype, demonstrating that FOG-2 function in myocardium is required and sufficient for coronary vessel development. $FOG\text{-}1^{-/-}$ mice die at E14.5 with cardiac defects that include double-outlet RV and a common atrioventricular valve [45]. Conditional inactivation of *FOG-1* established that endocardial-derived rather than neural crest-derived tissues are the site of expression for this regulatory factor. These findings revealed a succession of molecular steps in the pathway of endocardial cushion development and suggest novel epistatic relationships within these interfacing pathways.

### Formation of Aortic and Pulmonic Valves

While the number of animal models and the information available regarding the embryonic development of the AV valves is rather impressive, the source of development for the semilunar valves has been more slowly forthcoming. Although the mechanisms that regulate the semilunar and AV valves were previously thought to be analogous, specific defects found in the semilunar valves suggest that their development may be differently regulated.

The OFT in the looped heart tube develops as a transitional zone lined on the inside by two endocardial OFT cushions, containing proximal (bulbar) and distal (truncal) regions which are often difficult to distinguish in humans and mammals. The distal cushions take part in semilunar valve formation, whereas the proximal cushions eventually become the muscular OFT septum [276]. The subsequent fusion of the cushions from distal to proximal has been shown to involve the participation of extracardiac NCCs, which contribute to the central (or condensed) mesenchyme incorporated within the OFT cushion mass. At the proximal OFT cushions, the NCCs are in close contact with the myocardial cells and in part invade the myocardium. The majority of these cells undergo apoptosis as a feature of the remodeling needed for valvulogenesis and are replaced during "myocardialization" [277].

The development of the OFT semilunar valves (aortic and pulmonic) is mediated by a subpopulation of ectodermally derived NCCs derived from the branchial arches migrating to the distal OFT, which is required for aortopulmonary septation [278]. However, recent cell lineage studies have definitively demonstrated the endothelial origin of resident cells throughout the leaflets of aortic and pulmonic valves, as well as the AVC valves [214, 279]. While precise role of the NCC in the development of the semilunar valves and OFT septation remains to be determined, critical elements of neural crest signaling are necessary for normal OFT and valve development indicating that these cells play an important inductive role. Either ablation of the NCC or neural crest-targeted mutations in specific resident signaling genes can lead to abnormal OFT and valve development as well as to CHD. Defects in the secreted factor, FGF15, and neural crest localized BMPR cause abnormal cardiac NCC migration and OFT malformation [280–282]. Moreover, neural crest produced neurotrophin-3 (NT-3) is present within the OFT throughout cardiac development even when NCCs are not detectable suggesting that NT-3 interacts with cells in the OFT that are not of neural crest origin [283]. This has elicited the hypothesis that neural crest signaling (e.g., NT-3) may function by its interaction with cells in the OFT such as those arising from the SHF, which has been shown to provide both essential myocardium and SM to the developing OFT [71, 73, 76]. This hypothesis has been strongly supported by the demonstration that in cardiac neural-crest-ablated embryos, the SHF (derived from splanchnic mesoderm) fails to add myocardial cells to the OFT, and elongation of the tube is deficient [284]. Defects in other signaling factors (not produced in NCC) including SHH and Tbx1 lead to defects in NCCs migration and positioning, which at least partly may mediate their overall effects on OFT and valve development [285, 286].

In addition to their role in OFT development, cardiac NCCs also contribute extensively to VSM differentiation. Recent studies have revealed that the MRTF-B targeted in NCC plays a critical role in regulating differentiation of cardiac NCC into SM and is required for normal cardiovascular morphogenesis [29].

Other significant signaling influences on OFT and semilunar valve development have been identified using genetic modification of the expression of genes *in vivo*. The Sox family is a family of transcription factors expressed at the neural plate border in response to neural crest-inducing signals, and Sox proteins [e.g., Sox8, Sox9, Sox10, long form of Sox5 (L-Sox5), Sox4, and Sox11] appear to function in the regulation of multiple aspects of neural crest development in many tissues [287] albeit little information elucidating this interaction is available in cardiac valve development. Expression of Sox8, Sox9, and Sox10 in the developing heart correlates with heart septation and with the differentiation of the connective tissue of the valve leaflets [288]. As noted in the previous section, Sox9 function is involved in the mesenchymal transformation from endocardial endothelial cells required for cushion and valve formation. In *Sox9*-null mutants, endocardial cushions are markedly hypoplastic. Moreover, in embryos in which the inactivation of Sox9 was specifically targeted in cardiac NCC using *Cre* technology, the resultant abnormality of the endocardial cushions appears only in the distal part of the OFT, whereas the proximal portion of the OFT and the AV cushions are normal. Thus, Sox9 is required for the development of endocardial cushions derived from both NCCs and endothelial cells of the heart. Sox4 has been found to be abundantly present in the developing heart of chick, mouse, and human. Abundant expression was also detected in tissues of neural crest origin including the pharyngeal arch [289]. Using targeted gene disruption, *Sox4*$^{-/-}$ embryos were found to succumb to circulatory failure at E14 [290]. This resulted from impaired development of the endocardial ridges (a specific site of Sox4 expression) into the semilunar valves and the outlet portion of the muscular ventricular septum. Subsequently, these findings have suggested the existence of a *Sox4-deficiency syndrome* defined as defective function of the endocardial tissue of the OFT, leading to a lack of development and/or fusion of the endocardial ridges and the semilunar valves, and an arrangement of the ventriculo-arterial connection corresponding with transposition of the great arteries (TGA) [201]. The Sox4-deficiency syndrome is due to a defective function of the endocardially derived tissue and structures whose fate could be followed, thanks to the presence of the highly characteristic rods of condensed mesenchyme. The restriction of the malformations to the arterial pole, even though *Sox4* is equally expressed in the endocardially derived tissue of the AVC, suggests that interaction between the endocardially derived tissue of the OFT and the neural crest-derived myofibroblasts could determine proper development of the arterial pole. The abundant expression of Sox4 in neural crest tissues supports a potential role in the cardiac pathology detected in Sox4 mutant mice.

Four genes are known to encode proteins belonging to the NFAT complex (*NFATc*, *NFATp*, *NFAT3*, and *NFAT4*). These genes are expressed in several tissues including the immune system, in which an enhanced response in mice lacking the transcription factor NFATp has been reported without definitive abnormalities in cardiac development [291]. In two independent studies, it has been shown that *NFATc3*-mutant mice embryos played a particular role in the development of the semilunar valves and septa [239, 240]. The specificity of *NFATc* expression in the endothelial lining is rather unique in heart development, as *NFATc* is the only endocardial-specific transcription factor thus far described. In both studies, by E13.5, the embryos displayed abnormalities of valve structure leading to death in utero from congestive heart failure. Although septation into RV (pulmonary) and LV (aortic), OFT was normal, both semilunar valves were underdeveloped. Interestingly, blocking calcineurin activity results in inhibition of *NFATc* in these structures, suggesting that *NFATc* may be responsive to calcium signaling. Moreover, the involvement of NFAT transcription factors as downstream effectors of calcineurin signaling has also been reported in cardiac hypertrophy [292].

A phenotype of hyperplastic semilunar valves present in the *EGF* receptor mutants has been reported in mice containing disrupted alleles of phospholipase Cε (PLCε), an isoform of phosphoinositide-specific PLC, and a downstream effector of RAS signaling [293]. Mice homozygous for the targeted *PLC* allele exhibit defective aortic and pulmonary valve. This malformation involves marked thickening of the valve leaflets, likely caused by a defect in valve remodeling at the late stages of semilunar valvulogenesis, and is accompanied by inhibition of Smad activation and BMP signaling.

Null mutation in the gene for NF-1 in mice causes markedly enlarged endocardial cushions, double-outlet RV, and other noncardiac defects, including sympathetic ganglia as well as liver and kidney abnormalities and die by E14.5 [294]. Using tissue-specific gene inactivation, it has been demonstrated that endothelial-specific *Nf1* inactivation recapitulates key aspects of the complete null phenotype, including multiple cardiovascular abnormalities involving the endocardial cushions and myocardium [295]. This phenotype is associated with an elevated level of Ras signaling in $Nf1^{-/-}$ endothelial cells and greater nuclear localization of the transcription factor NFATc (placing NF-1 upstream of NFAT). In contrast, *Nf1* inactivation targeted to NCC did not promote cardiac defects but resulted in tumors of neural-crest origin resembling those seen in humans with NF-1. Therefore, NF-1 plays an essential role in endothelial cells and confirms the requirement for neurofibromin in the neural crest. Although the physiological role of neurofibromin in cardiac cushion development is not yet fully established, it may be to limit the extent of NFAT signaling and thereby attenuate expression of *NFAT* gene targets and availability of transcriptional co-activators [213].

## Cardiac Conduction System

### *Formation of the Cardiac Conduction System*

While substantial progress has been made in the understanding of the signaling pathways regulating various aspects of cardiogenesis, the molecular mechanisms that regulate the formation of the CCS are not well understood. Notwithstanding, in the last decade, as the number of mouse mutants with defects in CCS has increased, the potential to decipher the molecular pathways controlling the formation and function of this cellular network has increased. Several genes involved in the function of the mature conduction system have been identified, although their association with development of specific subcomponents of the CCS remains questionable. A number of transcription factors, including homeodomain proteins and T-box proteins, are necessary for CCS development as well as for the activation or repression of key regulatory genes, and the loss of transcriptional regulation during cardiac development has detrimental effects on cardiogenesis eventually leading to dysrhythmias [296].

CCS is formed from the differentiation of cardiac cells into specialized conduction cells [297]. The earliest nonrhythmic contractions of the embryonic chick heart occur at the 10-somite stage [298] while rhythmic base-to-apex electric activation that persists just before ventricular septation is evident prior to Hamburger–Hamilton stage 31 of chick cardiogenesis [299]. In the embryonic chick, a group of specialized cardiomyocytes located in the sinoatrial region of the primitive heart tube initiate the heartbeat at stage 31 and during the final stages of outflow septation [299]. During mouse cardiogenesis, before ventricular septation is completed, the activation sequence is reversed to the apex-to-base pattern of a mature heart [300]. Albeit the processes that make conduction cells different from the surrounding myocardium are not completely understood, it appears that a number of transcription factors play a critical role in their differentiation.

The atrioventricular node defects, observed in humans and mice with dominant mutations in these transcription factors, have suggested that Nkx2-5 and Tbx5 play a role in the formation of a functional conduction system [127, 203–208]. Compared to the myocardium, increased expression of Nkx2-5 in specialized conduction fibers has been found and this may be the basis for the high sensitivity of these cells to decreased Nkx2-5 dosage [301]. Furthermore, a direct downstream target of Tbx5 and Nkx2-5 is Cx40, which shows decreased expression levels in Tbx5-deficient mice and in mice expressing a mutant Nkx2-5 protein. A ring of atrioventricular conduction tissue that develops at 40 h postfertilization has been found in the zebrafish heart, and analysis of the cloche (clo) mutant reveals the need for endocardial signals in the formation of this tissue [302]. Interestingly, differentiation of these specialized cells, unlike that of adjacent endocardial cushions and valves, is not dependent on blood flow or cardiac contraction. In addition, both neuregulin and Notch1b appear to be necessary for the development of atrioventricular conduction tissue.

To further understand the molecular determinants of the vertebrate CCS, Chi et al. [303] have employed a cardiac-specific fluorescent calcium indicator zebrafish transgenic line, *Tg(cmlc2:gCaMP)*$^{s878}$, that allows *in vivo* optical mapping analysis in intact animals. Four distinct stages of cardiac conduction development that correspond to cellular and anatomical changes of the developing heart were identified. In addition, it was observed that epigenetic factors, such as hemodynamic flow and contraction, regulate the fast conduction network of this specialized electrical system. To identify novel regulators of the CCS, a new, physiology-based, forward genetic screen was designed, which identified, probably for the first time, 17 conduction-specific mutations. Positional cloning of *hobgoblin*$^{s634}$ revealed that *tcf2*, a homeobox transcription factor gene involved in mature onset diabetes of the young and familial glomerulocystic kidney disease, also regulates conduction between the atrium and the ventricle. This combination of the *Tg(cmlc2:gCaMP)*$^{s878}$ line/*in vivo* optical mapping technique and characterization of cardiac conduction mutants may further advance our understanding of the molecular determinants of the vertebrate CCS.

## Connexins and CCS

Electrical coupling of myocytes is mediated by gap junctions, which establish via membrane channels, a connection between the cytoplasm of adjacent cells permitting intercellular current flow and the transfer of depolarizing action potentials, a prerequisite for CCS. These gap junction channels are formed by alignment of two hexameric hemichannels, each composed of Cx subunits. At least five different cardiac Cx types are expressed in the mammalian heart, Cx37, Cx40, Cx43, Cx45, and Cx46. Gap junctions and their Cx components are scarce in the developing sinoatrial and atrioventricular nodes, corresponding with the slow conduction in these regions. Cx37 is primarily found in endocardium [304], Cx40 is mainly found in the atrium and conduction system [305–309], Cx43 is located in the atrial and ventricular myocardium and in the distal parts of the CCS [307–309], and Cx45 is present throughout the heart in small amounts with some evidence of increased expression in the conduction system [310, 311]. The specific association of ventricular Cx45 with Cx40-expressing myocytes reported in mouse and rat may indicate that Cx45 contributes to the modulation of electrophysiological properties in the ventricular conduction system and needs to be further investigated in other species [312]. Cx45 appears to be the isoform most continuously expressed by conduction tissues and constitutes a defining feature of the heterogenous nature of the tissues comprising the CCS of the rodent heart [313].

In mice, a continuity between the common bundle and the septum is present, and Cx40 deficiency results in right bundle-branch block and impaired left bundle-branch [314]. Moreover, mice with null *Cx40* alleles exhibit reduced atrial but not ventricular conduction velocity [315]. *Cx40* knockout mice have prolonged P waves on ECG, suggesting reduced atrial conduction velocity. Also, the PR interval is prolonged, suggesting AV nodal dysfunction, and the QRS complexes are prolonged [248, 316, 317]. Impaired function of the specific conduction system could explain the prolonged QRS duration because Cx40 is normally not expressed in working ventricular myocardium. Cx40 and Cx45 have been found to be the major protein subunits of gap junction channels in the conduction system of mammals [314].

## Other Signaling Pathways and Transcriptional Regulators

In addition to the regulatory control of the Cx channels and the conduction system exerted by Tbx5 and Nkx2-5, several signaling pathways including endothelin, neuregulin, Wnts, and BMPs [318–325], several other transcriptional regulators including Tbx3 and Tbx2, the vertebrate muscle segment-related homeobox factor Msx-2, the SP1-related factor HF-1b, and Hop are operative in the development and functioning of the CCS.

The zinc-finger SP1-related transcription factor HF-1b has been shown to be critical in establishing conduction system identity. Mice lacking HF-1b exhibit a wide variety of conduction system defects, including spontaneous ventricular tachycardia and a high incidence of AV block [326]. In the absence of HF-1b, the myocardium surrounding specialized conduction fibers is characterized by a heterogenous expression profile of genes resulting in the decreased levels and mislocalization of Cx, as well as a marked increase in action potential heterogeneity. HF-1b appears, therefore, to be required for establishing a molecular or physical identity between conduction and nonconduction cells in the developing myocardium, although the direct targets and precise mechanism of HF-1b remains to be determined. Interestingly, HF-1b, Cx40, and Cx43 have been found misexpressed and/or mislocalized in mice homozygous for the *LMNA* allele N195K/N195K resulting in death at an early age due to cardiac dysrhythmias [327]. In addition, clinical laminopathies are associated with missense mutations in the *LMNA* gene, encoding lamin A/C nuclear intermediate filament protein serving as a major component of the nuclear envelope, and which result in DCM with prominent conduction defects [328].

The T-box transcription factor Tbx3 has been found to be an accurate marker for the murine CCS, and the regulatory function of Tbx3 in Nppa and Cx40 promoter activity has been studied *in vitro* [329]. In the formed heart, Tbx3 is expressed in the sinoatrial (SN) and AV nodes, bundle and proximal bundle branches, as well as the internodal regions and the AV region. Throughout cardiac development, Tbx3 is expressed in an uninterrupted myocardial region that extends from the SN node to the AV region and this expression domain is present in the looping heart tube from E8.5 onwards. Tbx3 is able to repress Nppa and Cx40 promoter activity and abolish the synergistic activation of the Nppa promoter by Tbx5 and Nkx2-5. Therefore, Tbx3 has a significant role repressing a chamber-specific program of gene expression in regions, from which the diverse components of the CCS are subsequently formed.

In addition to its critical role in morphogenesis of the heart, the transcription factor Hop is a molecular component of the CCS regulation. Hop, the unusual homeodomain-only regulatory protein, which does not bind DNA, apparently functions downstream of Nkx2-5 and has been implicated in the regulation of myocyte growth and proliferation via its antagonism of SRF activity in a process involving the recruitment of HDAC activity. In the mouse heart, the Hop transcript is strongly expressed throughout the developing myocardium prior to 11.5 days postcoitum and subsequently becomes restricted to the trabecular zone [330]. Using a knock-in strategy to place a lacZ reporter gene under the transcriptional control of the Hop locus, Ismat et al. [331] reported that in adult heart, expression of Hop was restricted to the AV node, His bundle and bundle branches, as well as more broadly within the atria. Hop inactivation in adult mice displays conduction defects below the AV node as determined by invasive electrophysiological testing including increased P-wave duration, minor prolongation of the atrial-His bundle interval, prolonged atrial refractoriness, widening of the QRS complex, and prolongation of the His bundle-ventricular interval, and is associated with the decreased Cx40 expression. Unlike the case with Nkx2-5-deficiency, the AV node and His bundle do not appear to be atrophied in surviving homozygous Hop mutant mice, suggesting Hop plays a primary role in the maintenance of CCS function, rather than in CCS specification.

## Epigenetic Factors and CCS

Besides the Hop factor, whose role in HDAC regulation suggests to be an epigenetic factor, the environment also impacts on cardiac embryonic gene expression signaling during development of the CCS. Biophysical factors that play a significant role in the differentiation of specialized cardiac CS tissues include the effects of physical conditioning [332, 333] and of hemodynamic-induced molecular signaling cascades [297, 334]. For instance, the timing of apex-first activation in chick embryo shows a striking dependence on hemodynamic load; conversion from immature to a mature pattern of ventricular activation was accelerated by increased loading and delayed by decreased load reinforcing the importance of biophysical forces in the differentiation of the His-Purkinje system (HPS) *in vivo* [335].

## Endothelin-1/Neuregulin and CCS

Endothelin-1, a shear stress-sensitive cytokine prominently expressed by endothelial tissues, induces embryonic chick myocytes to express specific markers of Purkinje fiber differentiation [318, 319]. The ET-1 secretion by endothelial cells is particularly increased by a combination of pulsatile shear stress and increased blood pressure [336, 337], hemodynamic changes particularly pronounced in conotruncal banding hearts [335].

Additional aspects of ET-1 signaling including alterations in ET-1 secretion and in other ET-1 signaling pathway components such as endothelin-converting enzyme-1 (ECE-1) in response to altered hemodynamic load remain to be determined. Evidence has also been presented that a second factor secreted by endothelial cells, NRG-1, also plays key roles in both trabeculae formation and the differentiation of CCS cells [323, 338, 339]. However, at this time, there is no definitive evidence that NRG-1 expression or secretion is directly modulated by mechanical factors such as strain or fluid shear stress and pressure. Interestingly, it has been reported that ET-1 treatment increases NRG-1 expression in cultured endothelial cells [189], indicating cross-talk between these two pathways and suggesting an indirect mechanism for upregulation of NRG-1 by physical force. However, the molecular mechanism controlling the functional maturation of HPS remains to be elucidated.

## Markers of CCS Development

What are the best markers for studying the early embryonic development of the CCS? Much of the conduction system including both the SN and AV node in early embryonic hearts is composed of small myocytes with poorly developed actin and myosin filaments and sarcoplasmic reticulum, difficult to distinguish from the surrounding myocardium [340].

Current techniques that facilitate delineation of CCS development include the use of specific antibodies and transgenic mice specifically expressing reporter genes. Several studies have employed the monoclonal antibody against human natural killer-1 protein (HNK1) which recognizes a carbohydrate epitope in cell surface glycoproteins and glycolipids. Expression of HNK1 epitopes during development has been reported in various species to be present in the cardiomyocytes generally thought to be an origin for CCS; however, the presence of HNK1 epitopes in migrating NCCs has raised the question of its involvement. Analysis of the spatiotemporal expression pattern of HNK1 in early chick cardiogenesis revealed that HNK1 expressed in premyocardium and that precardiac mesoderm generated HNK1-positive cardiomyocytes with morphological features similar to those of conduction cardiomyocytes [341]. Antibodies to GlN2 have also been used to delineate the developing conduction system in a number of species; in humans and rats, HNK1 and GlN2 share almost the same spatiotemporal distribution in the heart [342, 343]. The GlN2 epitope reacted with an antibody raised against an extract from the chicken nodose ganglion and was originally used to identify migrating NCCs [344]. Antibodies directed against the cell adhesion protein, NCAM, have also been found to be useful markers of CCS components; in the developing human heart, NCAM is highly expressed in the nodal areas [345]. It is important to note that while HNK1 and GlN2 have proved valuable as identifying markers of conduction regions during cardiac development, there is presently little information concerning their functional significance in mediating cell–cell interaction, adhesion, or differentiation in contrast to the well-defined involvement of NCAM [340, 346].

LacZ reporter constructs have been used as informative developmental markers monitoring the expression of the lacZ transgene under the direction of various cardiac promoter constructs (e.g., GATA6, mink, engrailed, and cardiac troponin I) [347, 348]. In a pioneering study, the analysis of lacZ expression during sequential stages of cardiogenesis provided a detailed view of the maturation of the conductive network and demonstrated that CCS patterning occurred very early in embryogenesis beginning in transgenic mice at 8.25 days postconception [300]. The same study also employed optical mapping of cardiac electrical activity using a voltage-sensitive dye and confirmed that cells identified by the lacZ reporter gene were CCS components and that a murine HPS was functioning well before septation completed. Using the same CCS-lacZ strain, the developing conduction system has been identified in regions previously not formerly appreciated including Bachmann';s bundle, the pulmonary veins, and sinus venosus derived internodal structures, regions associated with the occurrence of cardiac dysrhythmias in adult patients [349]. Similar constructs have been employed in order to examine the relationship of NCC with the developing conduction system. Wnt1-Cre/R26R conditional reporter mice were used that express β-galactosidase from ROSA26 upon Cre-mediated recombination [350]. This study identified two subpopulations of NCCs found in the myocardium of the early embryonic heart, one adjacent to the bundle branches at the arterial pole, while the other positioned contiguous to the nodes at the venous pole. The expression of a minK-lacZ construct has also been applied to the detection of conduction components in the early embryonic heart as well as in the postnatal and adult heart [351].

## Generation of the CCS

There is presently little evidence to support the idea of primary extracardiac contributions to the CCS; the primary myocardium can generate and conduct adult-like ECGs well before NCC enters into embryonic heart. The consensus is that both nodes arise from existing myocardium, and the ventricular conduction system is generated from a distinct region/transcriptional domain, the trabecular component of the ventricle, distinguished from compact myocardium [340]. The latter conclusion has been drawn from molecular, functional, and morphological analysis [340], and has been underscored by demonstration that loss-of-function mutations in genes encoding the peptide growth factor neuregulin and its receptors ErbB2 and ErbB4 effectively shut down the initiation of trabecular development but do not affect the development of the atrial myocardium and the compact myocardium zone [190, 191, 338]. In addition, the trabecular ventricular component has a substantially different gene expression profile than the compact ventricular component with higher levels of atrial isoforms (e.g., ANF and α-MHC) and displays different electrophysiological function. As embryonic development proceeds, the area of ventricle involving the trabecular component markedly decreases presumably by a gradual remodeling into compact myocardium.

Finally, using replication-defective retroviruses, encoding recombinant β-galactosidase, cell lineage analysis demonstrated that cells of the peripheral Purkinje conduction system have a myogenic origin [352].

## Conclusions

Signaling pathways have a built-in specificity, reversibility, and a redundancy of its components, which while making their analysis a very complicated provides the cardiac cells with great plasticity to respond to insult as well as to growth stimuli. The complex process of cardiogenesis involves multiple and precisely coordinated phases of pattern formation and morphogenesis. Taking advantage of new technology and the availability of a number of animal models, signaling pathways, and key regulators of the cardiac patterning and morphogenesis are progressively discovered. The targeting of signaling factors to discrete subcellular compartments or substrates is important regulatory mechanism, making sure the specificity of signaling events in response to local stimuli. Since most of cardiac defects occur early in embryogenesis, the sooner a defect is diagnosed the better outcomes of therapeutic treatment can be expected. Understanding of the signaling cascades that are active during cardiac development will advance progress in the identification and manipulation of embryonic and adult cardiac stem cells for the treatment of adult and children with cardiac diseases. Furthermore, understanding of fetal programs may be very useful to develop novel options in the treatment of heart failure and cardiac hypertrophy.

## Summary

- The complex process of cardiogenesis requires the sophisticated interplay of multiple genes whose cell type-specific expression is highly organized, and precisely regulated at the spatial and temporal levels by numerous transcription factors.
- The heart through its signaling pathways functions both as a transmitter and a dynamic receiver of various intracellular and intercellular stimuli as well as an integrator of numerous interacting transducers.
- These critical regulatory factors can operate differently in various cell types and respond to a variety of intracellular and extracellular signals to modulate the precise integration of gene expression and morphological development. The great majority of the mutations identified, which lead to specific CHDs, reside in genes encoding transcription factors.
- Factors that play a central role in the specification and differentiation of lateral plate mesoderm into cardiac cells include tinman (i.e., Nkx2-5), SRF, myocardin, and GATA. These factors interact with each other and with a variety of cofactors to affect the programming of cardiac differentiation via transcriptional regulatory changes.
- Signaling pathways originate from a number of surrounding cell types including the endoderm and epicardium, and from the anterior field constituting a complex milieu of inductive signaling. Elements of the TGF-β BMP and Wnt families play crucial roles as does FGF signaling.
- A number of transcription factors (e.g., Mesp1, Mesp2, and GATA) are involved in the migration of the cardiac precursors, with both intracellular signaling and cell adhesion pathways activated in this process. Substantial cross-talk between the components is involved in their regulation.
- Myocardial development involves the looping of the heart, definition of a L–R identity, and formation of the cardiac chambers.
- An early defined anterior-posterior polarity in the expression of specific genes contributes to the development of chamber-specific expression. Key factors in defining this early polarity of expression include RA, *Irx4*, *T-box* genes including *Tbx5*, *Tbx2*, *and Tbx20*, and GATA factors.
- Downstream markers of chamber specificity include ANF and MLC. Further definition of chamber-specific maturation as well as establishing a dorsal-ventral polarity involves regulation by dHAND and eHAND bHLH transcription factors with considerable interaction with Nkx2-5 and MEF2.
- L–R identity is achieved by a cascade of signaling molecules culminating in the expression of the paired-domain homeodomain transcription factor Pitx2 on the left side of the visceral organs, including the heart. In addition to Nodal involvement, this pathway involves several TGF-β family signaling proteins including secreted extracellular factors (i.e., lefty-1, lefty-2, and BMP4), membrane receptors (e.g., activin receptor type IIB), members of the membrane-associated proteins encoded by *EFC-CFC* genes [e.g., *cryptic*], Notch receptor, BMP type I receptor (ACVRI), intracellular mediators [e.g., sonic hedgehog (SHH), Smads), and an array of transcription factors (e.g., Zic3, SnR, Hand1, Nkx2-5, CITED, and β-catenin]. In addition to its profound effect on anterior-posterior polarity, RA controls both levels of expression and location of LR signaling pathway components. Both RA and FGF are critical signalers of the proliferative events with both epicardium and endocardium serving as important growth inducers.
- The neuregulin family of peptide growth factors and their tyrosine kinase receptors (ErbBs), the forkhead transcription factor Foxp1, GATA4, Tbx5, the Hop regulator, and

Nmyc play critical roles in myocyte proliferation and maturation. The formation of chamber septae and the valves is linked, involving multiple common molecular players. The formation of EMT and migration is a pivotal event in the formation of the endocardial cushions which precedes the formation of valves and septae. This is followed by a series of delamination and remodeling events requiring further cell differentiation, apoptosis, and remodeling of the extracellular matrix. The signaling factors involved include BMP-TGF-β, Notch, VEGF, NFATc1, Wnt-β-catenin, ErbB, Sox, Fog1-GATA, and NF-1-Ras. The formation of the semilunar valves and OFT involves several pathways including Sox and NFAT, as well as NCC.

- Development of the conduction system depends on *Nkx2-5*, *Hop*, *Tbx5*, and the *Cx* genes and ErbB. Sp-1 like factor HF-1b, which can be modulated by VEGF, hypoxia, and hyperglycemia are also involved.
- Defining the molecular markers of early stages of developing conduction system is a work of great significance.

## References

1. Temsah R, Nemer M. GATA factors and transcriptional regulation of cardiac natriuretic peptide genes. Regul Pept. 2005;128:177–85.
2. Nemer G, Nemer M. Regulation of heart development and function through combinatorial interactions of transcription factors. Ann Med. 2001;33:604–10.
3. Kim TG, Chen J, Sadoshima J, Lee Y. Jumonji represses atrial natriuretic factor gene expression by inhibiting transcriptional activities of cardiac transcription factors. Mol Cell Biol. 2004;24:10151–60.
4. Small EM, Krieg PA. Transgenic analysis of the atrialnatriuretic factor (ANF) promoter: Nkx2-5 and GATA-4 binding sites are required for atrial specific expression of ANF. Dev Biol. 2003;261:116–31.
5. Pu WT, Ishiwata T, Juraszek AL, Ma Q, Izumo S. GATA4 is a dosage-sensitive regulator of cardiac morphogenesis. Dev Biol. 2004;275:235–44.
6. Heicklen-Klein A, McReynolds LJ, Evans T. Using the zebrafish model to study GATA transcription factors. Semin Cell Dev Biol. 2005;16:95–106.
7. Kostetskii I, Jiang Y, Kostetskaia E, Yuan S, Evans T, Zile M. Retinoid signaling required for normal heart development regulates GATA-4 in a pathway distinct from cardiomyocyte differentiation. Dev Biol. 1999;206:206–18.
8. Morin S, Paradis P, Aries A, Nemer M. Serum response factor-GATA ternary complex required for nuclear signaling by a G-protein-coupled receptor. Mol Cell Biol. 2001;21:1036–44.
9. Pehlivan T, Pober BR, Brueckner M, et al. GATA4 haploinsufficiency in patients with interstitial deletion of chromosome region 8p23.1 and congenital heart disease. Am J Med Genet. 1999;83:201–6.
10. Hirayama-Yamada K, Kamisago M, Akimoto K, et al. Phenotypes with GATA4 or NKX2.5 mutations in familial atrial septal defect. Am J Med Genet A. 2005;135:47–52.
11. Kelley C, Blumberg H, Zon LI, Evans T. GATA-4 is a novel transcription factor expressed in endocardium of the developing heart. Development. 1993;118:817–27.
12. Zeisberg EM, Ma Q, Juraszek AL, et al. Morphogenesis of the right ventricle requires myocardial expression of Gata4. J Clin Invest. 2005;115:1522–31.
13. Kawamura T, Ono K, Morimoto T, et al. Acetylation of GATA-4 is involved in the differentiation of embryonic stem cells into cardiac myocytes. J Biol Chem. 2005;280:19682–8.
14. Dai YS, Markham BE. p300 Functions as a coactivator of transcription factor GATA-4. J Biol Chem. 2001;276:37178–85.
15. Crispino JD, Lodish MB, Thurberg BL, et al. Proper coronary vascular development and heart morphogenesis depend on interaction of GATA-4 with FOG cofactors. Genes Dev. 2001;15:839–44.
16. Monzen K, Shiojima I, Hiroi Y, et al. Bone morphogenetic proteins induce cardiomyocyte differentiation through the mitogen-activated protein kinase kinase kinase TAK1 and cardiac transcription factors Csx/Nkx-2.5 and GATA-4. Mol Cell Biol. 1999;19:7096–105.
17. Ventura C, Zinellu E, Maninchedda E, Maioli M. Dynorphin B is an agonist of nuclear opioid receptors coupling nuclear protein kinase C activation to the transcription of cardiogenic genes in GTR1 embryonic stem cells. Circ Res. 2003;92:623–9.
18. Caporali A, Emanueli C. Cardiovascular actions of neurotrophins. Physiol Rev. 2009;89:279–308.
19. Engel FB, Schebesta M, Duong MT, et al. p38 MAP kinase inhibition enables proliferation of adult mammalian cardiomyocytes. Genes Dev. 2005;19:1175–87.
20. Tamamori-Adachi M, Ito H, Sumrejkanchanakij P, et al. Critical role of cyclin D1 nuclear import in cardiomyocyte proliferation. Circ Res. 2003;92:e12–9.
21. Pasumarthi KB, Kardami E, Cattini PA. High and low molecular weight fibroblast growth factor-2 increase proliferation of neonatal rat cardiac myocytes but have differential effects on binucleation and nuclear morphology. Evidence for both paracrine and intracrine actions of fibroblast growth factor-2. Circ Res. 1996;78:126–36.
22. Sheikh F, Jin Y, Pasumarthi KB, Kardami E, Cattini PA. Expression of fibroblast growth factor receptor-1 in rat heart H9c2 myoblasts increases cell proliferation. Mol Cell Biochem. 1997;176:89–97.
23. Sheng Z, Pennica D, Wood WI, Chien KR. Cardiotrophin-1 displays early expression in the murine heart tube and promotes cardiac myocyte survival. Development. 1996;122:419–28.
24. Kuwahara K, Saito Y, Kishimoto I, et al. Cardiotrophin-1 phosphorylates akt and BAD, and prolongs cell survival via a PI3K-dependent pathway in cardiac myocytes. J Mol Cell Cardiol. 2000;32:1385–94.
25. Lyons I, Parsons LM, Hartley L, et al. Myogenic and morphogenetic defects in the heart tubes of murine embryos lacking the homeo box gene Nkx2-5. Genes Dev. 1995;9:1654–66.
26. Parlakian A, Tuil D, Hamard G, et al. Targeted inactivation of serum response factor in the developing heart results in myocardial defects and embryonic lethality. Mol Cell Biol. 2004;24:5281–9.
27. Niu Z, Yu W, Zhang SX, et al. Conditional mutagenesis of the murine serum response factor gene blocks cardiogenesis and the transcription of downstream gene targets. J Biol Chem. 2005;280:32531–8.
28. Wang D, Chang PS, Wang Z, et al. Activation of cardiac gene expression by myocardin, a transcriptional cofactor for serum response factor. Cell. 2001;105:851–62.
29. Li J, Zhu X, Chen M, et al. Myocardin-related transcription factor B is required in cardiac neural crest for smooth muscle differentiation and cardiovascular development. Proc Natl Acad Sci USA. 2005;102:8916–21.
30. Ueyama T, Kasahara H, Ishiwata T, Nie Q, Izumo S. Myocardin expression is regulated by Nkx2.5, and its function is required for cardiomyogenesis. Mol Cell Biol. 2003;23:9222–32.
31. Yoshida T, Kawai-Kowase K, Owens GK. Forced expression of myocardin is not sufficient for induction of smooth muscle differentiation in multipotential embryonic cells. Arterioscler Thromb Vasc Biol. 2004;24:1596–601.

32. Davidson B, Levine M. Evolutionary origins of the vertebrate heart: specification of the cardiac lineage in *Ciona intestinalis*. Proc Natl Acad Sci USA. 2003;100:11469–73.
33. Poelmann RE, Gittenberger-de Groot AC. Cardiac development. ScientificWorldJournal. 2008;8:855–8.
34. Zhang X, Azhar G, Zhong Y, Wei JY. Identification of a novel serum response factor cofactor in cardiac gene regulation. J Biol Chem. 2004;279:55626–32.
35. Pikkarainen S, Tokola H, Kerkela R, Ruskoaho H. GATA transcription factors in the developing and adult heart. Cardiovasc Res. 2004;63:196–207.
36. Charron F, Nemer M. GATA transcription factors and cardiac development. Semin Cell Dev Biol. 1999;10:85–91.
37. Molkentin JD, Lin Q, Duncan SA, Olson EN. Requirement of the transcription factor GATA4 for heart tube formation and ventral morphogenesis. Genes Dev. 1997;11:1061–72.
38. Kuo CT, Morrisey EE, Anandappa R, et al. GATA4 transcription factor is required for ventral morphogenesis and heart tube formation. Genes Dev. 1997;11:1048–60.
39. Grepin C, Robitaille L, Antakly T, Nemer M. Inhibition of transcription factor GATA-4 expression blocks in vitro cardiac muscle differentiation. Mol Cell Biol. 1995;15:4095–102.
40. Grepin C, Nemer G, Nemer M. Enhanced cardiogenesis in embryonic stem cells overexpressing the GATA-4 transcription factor. Development. 1997;124:2387–95.
41. Narita N, Bielinska M, Wilson DB. Cardiomyocyte differentiation by GATA-4-deficient embryonic stem cells. Development. 1997;124:3755–64.
42. Charron F, Paradis P, Bronchain O, Nemer G, Nemer M. Cooperative interaction between GATA-4 and GATA-6 regulates myocardial gene expression. Mol Cell Biol. 1999;19:4355–65.
43. Zaffran S, Astier M, Gratecos D, Guillen A, Semeriva M. Cellular interactions during heart morphogenesis in the Drosophila embryo. Biol Cell. 1995;84:13–24.
44. Olson EN, Srivastava D. Molecular pathways controlling heart development. Science. 1996;272:671–6.
45. Katz SG, Williams A, Yang J, et al. Endothelial lineage-mediated loss of the GATA cofactor Friend of GATA 1 impairs cardiac development. Proc Natl Acad Sci USA. 2003;100:14030–5.
46. Dawid IB. Mesoderm induction and axis determination in Xenopus laevis. Bioessays. 1992;14:687–91.
47. Smith WC, Knecht AK, Wu M, Harland RM. Secreted noggin protein mimics the Spemann organizer in dorsalizing Xenopus mesoderm. Nature. 1993;361:547–9.
48. Schier AF, Talbot WS. Nodal signaling and the zebrafish organizer. Int J Dev Biol. 2001;45:289–97.
49. Smith WC, McKendry R, Ribisi Jr S, Harland RM. A nodal-related gene defines a physical and functional domain within the Spemann organizer. Cell. 1995;82:37–46.
50. Vincent SD, Dunn NR, Hayashi S, Norris DP, Robertson EJ. Cell fate decisions within the mouse organizer are governed by graded Nodal signals. Genes Dev. 2003;17:1646–62.
51. Kinder SJ, Tsang T, Wakamiya M, Sasaki H, Behringer RR, Nagy A, et al. The organizer of the mouse gastrula is composed of a dynamic population of progenitor cells for the axial mesoderm. Development. 2001;128:3623–34.
52. Marom K, Levy V, Pillemer G, Fainsod A. Temporal analysis of the early BMP functions identifies distinct anti-organizer and mesoderm patterning phases. Dev Biol. 2005;282:442–54.
53. Schultheiss TM, Xydas S, Lassar AB. Induction of avian cardiac myogenesis by anterior endoderm. Development. 1995;121:4203–14.
54. Nascone N, Mercola M. An inductive role for the endoderm in Xenopus cardiogenesis. Development. 1995;121:515–23.
55. Frasch M. Induction of visceral and cardiac mesoderm by ectodermal Dpp in the early Drosophila embryo. Nature. 1995;374:464–7.
56. Lockwood WK, Bodmer R. The patterns of wingless, decapentaplegic, and tinman position the Drosophila heart. Mech Dev. 2002;114:13–26.
57. Lee HH, Frasch M. Nuclear integration of positive Dpp signals, antagonistic Wg inputs and mesodermal competence factors during Drosophila visceral mesoderm induction. Development. 2005;132:1429–42.
58. Klinedinst SL, Bodmer R. Gata factor Pannier is required to establish competence for heart progenitor formation. Development. 2003;130:3027–38.
59. Ladd AN, Yatskievych TA, Antin PB. Regulation of avian cardiac myogenesis by activin/TGFbeta and bone morphogenetic proteins. Dev Biol. 1998;204:407–19.
60. Kruithof BP, van Wijk B, Somi S, et al. BMP and FGF regulate the differentiation of multipotential pericardial mesoderm into the myocardial or epicardial lineage. Dev Biol. 2006;295:507–22.
61. Schlange T, Andree B, Arnold HH, Brand T. BMP2 is required for early heart development during a distinct time period. Mech Dev. 2000;91:259–70.
62. Alsan BH, Schultheiss TM. Regulation of avian cardiogenesis by Fgf8 signaling. Development. 2002;129:1935–43.
63. Beiman M, Shilo BZ, Volk T. Heartless, a Drosophila FGF receptor homolog, is essential for cell migration and establishment of several mesodermal lineages. Genes Dev. 1996;10:2993–3002.
64. Fossett N, Zhang Q, Gajewski K, Choi CY, Kim Y, Schulz RA. The multitype zinc-finger protein U-shaped functions in heart cell specification in the Drosophila embryo. Proc Natl Acad Sci USA. 2000;97:7348–53.
65. Stathopoulos A, Tam B, Ronshaugen M, Frasch M, Levine M. Pyramus and thisbe: FGF genes that pattern the mesoderm of Drosophila embryos. Genes Dev. 2004;19:687–99.
66. Abdul-Ghani MA, Megeney L. Wnt signaling in the mouse heart: effect of Wnt-11 on cardiac differentiation. Circulation. 2006;114:II_233.
67. Black BL. Transcriptional pathways in second heart field development. Semin Cell Dev Biol. 2007;18:67–76.
68. Buckingham M, Meilhac S, Zaffran S. Building the mammalian heart from two sources of myocardial cells. Nat Rev Genet. 2005;6:826–35.
69. Kelly RG, Brown NA, Buckingham ME. The arterial pole of the mouse heart forms from Fgf10-expressing cells in pharyngeal mesoderm. Dev Cell. 2001;1:435–40.
70. Mjaatvedt CH, Nakaoka T, Moreno-Rodriguez R, et al. The outflow tract of the heart is recruited from a novel heart-forming field. Dev Biol. 2001;238:97–109.
71. Waldo KL, Kumiski DH, Wallis KT, et al. Conotruncal myocardium arises from a secondary heart field. Development. 2001;128:3179–88.
72. Srivastava D. Making or breaking the heart: from lineage determination to morphogenesis. Cell. 2006;126:1037–48.
73. Kelly RG, Buckingham ME. The anterior heart-forming field: voyage to the arterial pole of the heart. Trends Genet. 2002;18:210–6.
74. Meilhac SM, Esner M, Kerszberg M, Moss JE, Buckingham ME. Oriented clonal cell growth in the developing mouse myocardium underlies cardiac morphogenesis. J Cell Biol. 2004;164:97–109.
75. Meilhac SM, Esner M, Kelly RG, Nicolas JF, Buckingham ME. The clonal origin of myocardial cells in different regions of the embryonic mouse heart. Dev Cell. 2004;6:685–98.
76. Waldo KL, Hutson MR, Ward CC, et al. Secondary heart field contributes myocardium and smooth muscle to the arterial pole of the developing heart. Dev Biol. 2005;281:78–90.
77. Cai CL, Liang X, Shi Y, et al. Isl1 identifies a cardiac progenitor population that proliferates prior to differentiation and contributes a majority of cells to the heart. Dev Cell. 2003;5:877–89.
78. Ai D, Fu X, Wang J, et al. Canonical Wnt signaling functions in second heart field to promote right ventricular growth. Proc Natl Acad Sci USA. 2007;104:9319–24.

79. Schneider VA, Mercola M. Wnt antagonism initiates cardiogenesis in Xenopus laevis. Genes Dev. 2001;15:304–15.
80. Marvin MJ, Di Rocco G, Gardiner A, Bush SM, Lassar AB. Inhibition of Wnt activity induces heart formation from posterior mesoderm. Genes Dev. 2001;15:316–27.
81. Foley AC, Mercola M. Heart induction by Wnt antagonists depends on the homeodomain transcription factor Hex. Genes Dev. 2005;19:387–96.
82. Saga Y, Miyagawa-Tomita S, Takagi A, Kitajima S, Miyazaki J, Inoue T. MesP1 is expressed in the heart precursor cells and required for the formation of a single heart tube. Development. 1999;126:3437–47.
83. Kitajima S, Takagi A, Inoue T, Saga Y. MesP1 and MesP2 are essential for the development of cardiac mesoderm. Development. 2000;127:3215–26.
84. Narita N, Bielinska M, Wilson DB. Wild-type endoderm abrogates the ventral developmental defects associated with GATA-4 deficiency in the mouse. Dev Biol. 1997;189:270–4.
85. Reiter JF, Alexander J, Rodaway A, et al. Gata5 is required for the development of the heart and endoderm in zebrafish. Genes Dev. 1999;13:2983–95.
86. Jiang Y, Tarzami S, Burch JB, Evans T. Common role for each of the cGATA-4/5/6 genes in the regulation of cardiac morphogenesis. Dev Genet. 1998;22:263–77.
87. Kikuchi Y, Trinh LA, Reiter JF, Alexander J, Yelon D, Stainier DY. The zebrafish bonnie and clyde gene encodes a Mix family homeodomain protein that regulates the generation of endodermal precursors. Genes Dev. 2000;14:1279–89.
88. Zhang H, Toyofuku T, Kamei J, Hori M. GATA-4 regulates cardiac morphogenesis through transactivation of the N-cadherin gene. Biochem Biophys Res Commun. 2003;312:1033–8.
89. Glickman NS, Yelon D. Coordinating morphogenesis: epithelial integrity during heart tube assembly. Dev Cell. 2004;6:311–2.
90. Linask KK, Manisastry S, Han M. Cross talk between cell–cell and cell–matrix adhesion signaling pathways during heart organogenesis: implications for cardiac birth defects. Microsc Microanal. 2005;11:200–8.
91. Kelly RG. Molecular inroads into the anterior heart field. Trends Cardiovasc Med. 2005;15:51–6.
92. Kathiriya IS, Srivastava D. Left-right asymmetry and cardiac looping: implications for cardiac development and congenital heart disease. Am J Med Genet. 2000;97:271–9.
93. Yuasa SFK. Multiple roles for BMP signaling in cardiac. Drug Discov Today Ther Strateg. 2008;5:209–14.
94. McCulley DJ, Kang JO, Martin JF, Black BL. BMP4 is required in the anterior heart field and its derivatives for endocardial cushion remodeling, outflow tract septation, and semilunar valve development. Dev Dyn. 2008;237:3200–9.
95. Verzi MP, McCulley DJ, De Val S, Dodou E, Black BL. The right ventricle, outflow tract, and ventricular septum comprise a restricted expression domain within the secondary/anterior heart field. Dev Biol. 2005;287:134–45.
96. Ryckebusch L, Wang Z, Bertrand N, et al. Retinoic acid deficiency alters second heart field formation. Proc Natl Acad Sci USA. 2008;105:2913–8.
97. Hoover LL, Burton EG, Brooks BA, Kubalak SW. The expanding role for retinoid signaling in heart development. ScientificWorldJournal. 2008;8:194–211.
98. Rochais F, Mesbah K, Kelly RG. Signaling pathways controlling second heart field development. Circ Res. 2009;104:933–42.
99. Bu L, Jiang X, Martin-Puig S, et al. Human ISL1 heart progenitors generate diverse multipotent cardiovascular cell lineages. Nature. 2009;460:113–7.
100. Wagner M, Siddiqui M. Signal transduction in early heart development (I): cardiogenic induction and heart tube formation. Exp Biol Med. 2007;232:852–65.
101. Brand T. Heart development: molecular insights into cardiac specification and early morphogenesis. Dev Biol. 2003;258:1–19.
102. Zaffran S, Kelly RG, Meilhac SM, Buckingham ME, Brown NA. Right ventricular myocardium derives from the anterior heart field. Circ Res. 2004;95:261–8.
103. Wagner MSM. Signal transduction in early heart development (II): ventricular chamber specification, trabeculation, and heart valve formation. Exp Biol Med (Maywood). 2007;232:866–80.
104. Stainier DY, Fishman MC. Patterning the zebrafish heart tube: acquisition of anteroposterior polarity. Dev Biol. 1992;153:91–101.
105. Twal W, Roze L, Zile MH. Anti-retinoic acid monoclonal antibody localizes all-trans-retinoic acid in target cells and blocks normal development in early quail embryo. Dev Biol. 1995;168:225–34.
106. Niederreither K, Vermot J, Messaddeq N, Schuhbaur B, Chambon P, Dolle P. Embryonic retinoic acid synthesis is essential for heart morphogenesis in the mouse. Development. 2001;128:1019–31.
107. Zile MH, Kostetskii I, Yuan S, et al. Retinoid signaling is required to complete the vertebrate cardiac left/right asymmetry pathway. Dev Biol. 2000;223:323–38.
108. Sucov HM, Dyson E, Gumeringer CL, Price J, Chien KR, Evans RM. RXR alpha mutant mice establish a genetic basis for vitamin A signaling in heart morphogenesis. Genes Dev. 1994;8:1007–18.
109. Kastner P, Grondona JM, Mark M, et al. Genetic analysis of RXR alpha developmental function: convergence of RXR and RAR signaling pathways in heart and eye morphogenesis. Cell. 1994;78:987–1003.
110. Gruber PJ, Kubalak SW, Pexieder T, Sucov HM, Evans RM, Chien KR. RXR alpha deficiency confers genetic susceptibility for aortic sac, conotruncal, atrioventricular cushion, and ventricular muscle defects in mice. J Clin Invest. 1996;98:1332–43.
111. Romeih M, Cui J, Michaille JJ, Jiang W, Zile MH. Function of RARgamma and RARalpha2 at the initiation of retinoid signaling is essential for avian embryo survival and for distinct events in cardiac morphogenesis. Dev Dyn. 2003;228:697–708.
112. Kastner P, Messaddeq N, Mark M, et al. Vitamin A deficiency and mutations of RXRalpha, RXRbeta and RARalpha lead to early differentiation of embryonic ventricular cardiomyocytes. Development. 1997;124:4749–58.
113. Christoffels VM, Habets PE, Franco D, et al. Chamber formation and morphogenesis in the developing mammalian heart. Dev Biol. 2000;223:266–78.
114. Bao ZZ, Bruneau BG, Seidman JG, Seidman CE, Cepko CL. Regulation of chamber-specific gene expression in the developing heart by Irx4. Science. 1999;283:1161–4.
115. Garriock RJ, Vokes SA, Small EM, Larson R, Krieg PA. Developmental expression of the Xenopus Iroquois-family homeobox genes, Irx4 and Irx5. Dev Genes Evol. 2001;211:257–60.
116. Bruneau BG, Bao ZZ, Tanaka M, et al. Cardiac expression of the ventricle-specific homeobox gene Irx4 is modulated by Nkx2-5 and dHand. Dev Biol. 2000;217:266–77.
117. Cavodeassi F, Modolell J, Gomez-Skarmeta JL. The Iroquois family of genes: from body building to neural patterning. Development. 2001;128:2847–55.
118. Bruneau BG, Bao ZZ, Fatkin D, et al. Cardiomyopathy in Irx4-deficient mice is preceded by abnormal ventricular gene expression. Mol Cell Biol. 2001;21:1730–6.
119. Xavier-Neto J, Neville CM, Shapiro MD, et al. A retinoic acid-inducible transgenic marker of sino-atrial development in the mouse heart. Development. 1999;126:2677–87.
120. Wang GF, Nikovits W, Schleinitz M, Stockdale FE. Atrial chamber-specific expression of the slow myosin heavy chain 3 gene in the embryonic heart. J Biol Chem. 1996;271:19836–45.
121. Wang GF, Nikovits Jr W, Bao ZZ, Stockdale FE. Irx4 forms an inhibitory complex with the vitamin D and retinoic X receptors to

regulate cardiac chamber-specific slow MyHC3 expression. J Biol Chem. 2001;276:28835–41.
122. Christoffels VM, Keijser AG, Houweling AC, Clout DE, Moorman AF. Patterning the embryonic heart: identification of five mouse Iroquois homeobox genes in the developing heart. Dev Biol. 2000;224:263–74.
123. Bruneau BG, Logan M, Davis N, et al. Chamber-specific cardiac expression of Tbx5 and heart defects in Holt-Oram syndrome. Dev Biol. 1999;211:100–8.
124. Liberatore CM, Searcy-Schrick RD, Yutzey KE. Ventricular expression of tbx5 inhibits normal heart chamber development. Dev Biol. 2000;223:169–80.
125. Hatcher CJ, Goldstein MM, Mah CS, Delia CS, Basson CT. Identification and localization of TBX5 transcription factor during human cardiac morphogenesis. Dev Dyn. 2000;219:90–5.
126. Garrity DM, Childs S, Fishman MC. The heartstrings mutation in zebrafish causes heart/fin Tbx5 deficiency syndrome. Development. 2002;129:4635–45.
127. Bruneau BG, Nemer G, Schmitt JP, et al. A murine model of Holt-Oram syndrome defines roles of the T-box transcription factor Tbx5 in cardiogenesis and disease. Cell. 2001;106:709–21.
128. Hiroi Y, Kudoh S, Monzen K, et al. Tbx5 associates with Nkx2-5 and synergistically promotes cardiomyocyte differentiation. Nat Genet. 2001;28:276–80.
129. Horb ME, Thomsen GH. Tbx5 is essential for heart development. Development. 1999;126:1739–51.
130. Takeuchi JK, Ohgi M, Koshiba-Takeuchi K, et al. Tbx5 specifies the left/right ventricles and ventricular septum position during cardiogenesis. Development. 2003;130:5953–64.
131. Hatcher CJ, Kim MS, Mah CS, et al. TBX5 transcription factor regulates cell proliferation during cardiogenesis. Dev Biol. 2001;230:177–88.
132. Moskowitz IP, Pizard A, Patel VV, et al. The T-Box transcription factor Tbx5 is required for the patterning and maturation of the murine cardiac conduction system. Development. 2004;131:4107–16.
133. Christoffels VM, Hoogaars WM, Tessari A, Clout DE, Moorman AF, Campione M. T-box transcription factor Tbx2 represses differentiation and formation of the cardiac chambers. Dev Dyn. 2004;229:763–70.
134. Stennard FA, Costa MW, Lai D, et al. Murine T-box transcription factor Tbx20 acts as a repressor during heart development, and is essential for adult heart integrity, function and adaptation. Development. 2005;132:2451–62.
135. Cai CL, Zhou W, Yang L, et al. T-box genes coordinate regional rates of proliferation and regional specification during cardiogenesis. Development. 2005;132:2475–87.
136. Chapman DL, Cooper-Morgan A, Harrelson Z, Papaioannou VE. Critical role for Tbx6 in mesoderm specification in the mouse embryo. Mech Dev. 2003;120:837–47.
137. Srivastava D. HAND proteins: molecular mediators of cardiac development and congenital heart disease. Trends Cardiovasc Med. 1999;9:11–8.
138. McFadden DG, Charite J, Richardson JA, Srivastava D, Firulli AB, Olson EN. A GATA-dependent right ventricular enhancer controls dHAND transcription in the developing heart. Development. 2000;127:5331–41.
139. Yamagishi H, Yamagishi C, Nakagawa O, Harvey RP, Olson EN, Srivastava D. The combinatorial activities of Nkx2.5 and dHAND are essential for cardiac ventricle formation. Dev Biol. 2001;239:190–203.
140. Srivastava D, Thomas T, Lin Q, Kirby ML, Brown D, Olson EN. Regulation of cardiac mesodermal and neural crest development by the bHLH transcription factor, dHAND. Nat Genet. 1997;16:154–60.
141. Togi K, Kawamoto T, Yamauchi R, Yoshida Y, Kita T, Tanaka M. Role of Hand1/eHAND in the dorso-ventral patterning and interventricular septum formation in the embryonic heart. Mol Cell Biol. 2004;24:4627–35.
142. McFadden DG, Barbosa AC, Richardson JA, Schneider MD, Srivastava D, Olson EN. The Hand1 and Hand2 transcription factors regulate expansion of the embryonic cardiac ventricles in a gene dosage-dependent manner. Development. 2005;132:189–201.
143. Biben C, Harvey RP. Homeodomain factor Nkx2-5 controls left/right asymmetric expression of bHLH gene eHand during murine heart development. Genes Dev. 1997;11:1357–69.
144. Thomas T, Yamagishi H, Overbeek PA, Olson EN, Srivastava D. The bHLH factors, dHAND and eHAND, specify pulmonary and systemic cardiac ventricles independent of left-right sidedness. Dev Biol. 1998;196:228–36.
145. Yelon D, Ticho B, Halpern ME, et al. The bHLH transcription factor hand2 plays parallel roles in zebrafish heart and pectoral fin development. Development. 2000;127:2573–82.
146. Angelo S, Lohr J, Lee KH, et al. Conservation of sequence and expression of Xenopus and zebrafish dHAND during cardiac, branchial arch and lateral mesoderm development. Mech Dev. 2000;95:231–7.
147. Natarajan A, Yamagishi H, Ahmad F, et al. Human eHAND, but not dHAND, is down-regulated in cardiomyopathies. J Mol Cell Cardiol. 2001;33:1607–14.
148. Lin Q, Schwarz J, Bucana C, Olson EN. Control of mouse cardiac morphogenesis and myogenesis by transcription factor MEF2C. Science. 1997;276:1404–7.
149. Phan D, Rasmussen TL, Nakagawa O, et al. BOP, a regulator of right ventricular heart development, is a direct transcriptional target of MEF2C in the developing heart. Development. 2005;132:2669–78.
150. Morin S, Charron F, Robitaille L, Nemer M. GATA-dependent recruitment of MEF2 proteins to target promoters. EMBO J. 2000;19:2046–55.
151. Naya FJ, Wu C, Richardson JA, Overbeek P, Olson EN. Transcriptional activity of MEF2 during mouse embryogenesis monitored with a MEF2-dependent transgene. Development. 1999;126:2045–52.
152. Overbeek PA. Right and left go dHAND and eHAND. Nat Genet. 1997;16:119–21.
153. Olson EN, Black B. Control of cardiac development by the MEF2 family of transcription factors (Chap. 8). In: Harvey RP, Rosenthal N, editors. Heart development. San Diego, CA: Academic; 1999. p. 131–54.
154. Morin S, Pozzulo G, Robitaille L, Cross J, Nemer M. MEF2-dependent recruitment of the HAND1 transcription factor results in synergistic activation of target promoters. J Biol Chem. 2005;280:32272–8.
155. McKinsey TA, Zhang CL, Olson EN. Control of muscle development by dueling HATs and HDACs. Curr Opin Genet Dev. 2001;11:497–504.
156. Slepak TI, Webster KA, Zang J, et al. Control of cardiac-specific transcription by p300 through myocyte enhancer factor-2D. J Biol Chem. 2001;276:7575–85.
157. Molkentin JD, Li L, Olson EN. Phosphorylation of the MADS-Box transcription factor MEF2C enhances its DNA binding activity. J Biol Chem. 1996;271:17199–204.
158. Passier R, Zeng H, Frey N, et al. CaM kinase signaling induces cardiac hypertrophy and activates the MEF2 transcription factor in vivo. J Clin Invest. 2000;105:1395–406.
159. Capdevila J, Vogan KJ, Tabin CJ, Izpisua Belmonte JC. Mechanisms of left-right determination in vertebrates. Cell. 2000;101:9–21.
160. Franco D, Kelly R, Moorman AF, Lamers WH, Buckingham M, Brown NA. MLC3F transgene expression in iv mutant mice reveals the importance of left-right signalling pathways for the acquisition

of left and right atrial but not ventricular compartment identity. Dev Dyn. 2001;221:206–15.
161. Lohr JL, Danos MC, Yost HJ. Left-right asymmetry of a nodal-related gene is regulated by dorsoanterior midline structures during Xenopus development. Development. 1997;124:1465–72.
162. Bamforth SD, Braganca J, Farthing CR, et al. Cited2 controls left-right patterning and heart development through a Nodal-Pitx2c pathway. Nat Genet. 2004;36:1189–96.
163. Lowe LA, Yamada S, Kuehn MR. Genetic dissection of nodal function in patterning the mouse embryo. Development. 2001;128:1831–43.
164. Tsukui T, Capdevila J, Tamura K, et al. Multiple left-right asymmetry defects in Shh(–/–) mutant mice unveil a convergence of the shh and retinoic acid pathways in the control of Lefty-1. Proc Natl Acad Sci USA. 1999;96:11376–81.
165. Kishigami S, Yoshikawa S, Castranio T, Okazaki K, Furuta Y, Mishina Y. BMP signaling through ACVRI is required for left-right patterning in the early mouse embryo. Dev Biol. 2004;276:185–93.
166. Raya A, Kawakami Y, Rodriguez-Esteban C, et al. Notch activity induces Nodal expression and mediates the establishment of left-right asymmetry in vertebrate embryos. Genes Dev. 2003;17:1213–8.
167. Gaio U, Schweickert A, Fischer A, et al. A role of the cryptic gene in the correct establishment of the left-right axis. Curr Biol. 1999;9:1339–42.
168. Chang H, Zwijsen A, Vogel H, Huylebroeck D, Matzuk MM. Smad5 is essential for left-right asymmetry in mice. Dev Biol. 2000;219:71–8.
169. Grinberg I, Millen KJ. The ZIC gene family in development and disease. Clin Genet. 2005;67:290–6.
170. Schilling TF, Concordet JP, Ingham PW. Regulation of left-right asymmetries in the zebrafish by Shh and BMP4. Dev Biol. 1999;210:277–87.
171. Chen Y, Mironova E, Whitaker LL, Edwards L, Yost HJ, Ramsdell AF. ALK4 functions as a receptor for multiple TGF beta-related ligands to regulate left-right axis determination and mesoderm induction in Xenopus. Dev Biol. 2004;268:280–94.
172. Rodriguez-Esteban C, Capdevila J, Kawakami Y, Izpisua Belmonte JC. Wnt signaling and PKA control Nodal expression and left-right determination in the chick embryo. Development. 2001;128:3189–95.
173. Isaac A, Sargent MG, Cooke J. Control of vertebrate left-right asymmetry by a snail-related zinc finger gene. Science. 1997;275:1301–4.
174. Meno C, Shimono A, Saijoh Y, et al. lefty-1 is required for left-right determination as a regulator of lefty-2 and nodal. Cell. 1998;94:287–97.
175. Gage PJ, Suh H, Camper SA. Dosage requirement of Pitx2 for development of multiple organs. Development. 1999;126:4643–51.
176. Kitamura K, Miura H, Miyagawa-Tomita S, et al. Mouse Pitx2 deficiency leads to anomalies of the ventral body wall, heart, extra- and periocular mesoderm and right pulmonary isomerism. Development. 1999;126:5749–58.
177. Chazaud C, Chambon P, Dolle P. Retinoic acid is required in the mouse embryo for left-right asymmetry determination and heart morphogenesis. Development. 1999;126:2589–96.
178. Schlueter J, Manner J, Brand T. BMP is an important regulator of proepicardial identity in the chick embryo. Dev Biol. 2006;295:546–58.
179. Schlueter J, Brand T. A right-sided pathway involving FGF8/Snail controls asymmetric development of the proepicardium in the chick embryo. Proc Natl Acad Sci USA. 2009;106:7485–90.
180. Cartwright JH, Piro N, Piro O, Tuval I. Fluid dynamics of nodal flow and left-right patterning in development. Dev Dyn. 2008;237:3477–90.
181. Hirokawa N, Tanaka Y, Okada Y, Takeda S. Nodal flow and the generation of left-right asymmetry. Cell. 2006;125:33–45.
182. Chen J, Kubalak SW, Chien KR. Ventricular muscle-restricted targeting of the RXRalpha gene reveals a non-cell-autonomous requirement in cardiac chamber morphogenesis. Development. 1998;125:1943–9.
183. Chen TH, Chang TC, Kang JO, et al. Epicardial induction of fetal cardiomyocyte proliferation via a retinoic acid-inducible trophic factor. Dev Biol. 2002;250:198–207.
184. Moss JB, Xavier-Neto J, Shapiro MD, et al. Dynamic patterns of retinoic acid synthesis and response in the developing mammalian heart. Dev Biol. 1998;199:55–71.
185. Stuckmann I, Evans S, Lassar AB. Erythropoietin and retinoic acid, secreted from the epicardium, are required for cardiac myocyte proliferation. Dev Biol. 2003;255:334–49.
186. Kang JO, Sucov HM. Convergent proliferative response and divergent morphogenic pathways induced by epicardial and endocardial signaling in fetal heart development. Mech Dev. 2005;122:57–65.
187. Pennisi DJ, Ballard VL, Mikawa T. Epicardium is required for the full rate of myocyte proliferation and levels of expression of myocyte mitogenic factors FGF2 and its receptor, FGFR-1, but not for transmural myocardial patterning in the embryonic chick heart. Dev Dyn. 2003;228:161–72.
188. Lavine KJ, Yu K, White AC, et al. Endocardial and epicardial derived FGF signals regulate myocardial proliferation and differentiation in vivo. Dev Cell. 2005;8:85–95.
189. Zhao YY, Sawyer DR, Baliga RR, et al. Neuregulins promote survival and growth of cardiac myocytes. Persistence of ErbB2 and ErbB4 expression in neonatal and adult ventricular myocytes. J Biol Chem. 1998;273:10261–9.
190. Lee KF, Simon H, Chen H, Bates B, Hung MC, Hauser C. Requirement for neuregulin receptor erbB2 in neural and cardiac development. Nature. 1995;378:394–8.
191. Gassmann M, Casagranda F, Orioli D, et al. Aberrant neural and cardiac development in mice lacking the ErbB4 neuregulin receptor. Nature. 1995;378:390–4.
192. Crone SA, Zhao YY, Fan L, et al. ErbB2 is essential in the prevention of dilated cardiomyopathy. Nat Med. 2002;8:459–65.
193. Wang B, Weidenfeld J, Lu MM, et al. Foxp1 regulates cardiac outflow tract, endocardial cushion morphogenesis and myocyte proliferation and maturation. Development. 2004;131:4477–87.
194. Bushdid PB, Osinska H, Waclaw RR, Molkentin JD, Yutzey KE. NFATc3 and NFATc4 are required for cardiac development and mitochondrial function. Circ Res. 2003;92:1305–13.
195. Shin CH, Liu ZP, Passier R, et al. Modulation of cardiac growth and development by HOP, an unusual homeodomain protein. Cell. 2002;110:725–35.
196. Chen F, Kook H, Milewski R, et al. Hop is an unusual homeobox gene that modulates cardiac development. Cell. 2002;110:713–23.
197. Kook H, Lepore JJ, Gitler AD, et al. Cardiac hypertrophy and histone deacetylase-dependent transcriptional repression mediated by the atypical homeodomain protein Hop. J Clin Invest. 2003;112:863–71.
198. Tran CM, Sucov HM. The RXRalpha gene functions in a non-cell-autonomous manner during mouse cardiac morphogenesis. Development. 1998;125:1951–6.
199. Lakkis MM, Epstein JA. Neurofibromin modulation of ras activity is required for normal endocardial-mesenchymal transformation in the developing heart. Development. 1998;125:4359–67.
200. Camenisch TD, Spicer AP, Brehm-Gibson T, et al. Disruption of hyaluronan synthase-2 abrogates normal cardiac morphogenesis and hyaluronan-mediated transformation of epithelium to mesenchyme. J Clin Invest. 2000;106:349–60.

201. Ya J, Schilham MW, de Boer PA, Moorman AF, Clevers H, Lamers WH. Sox4-deficiency syndrome in mice is an animal model for common trunk. Circ Res. 1998;83:986–94.
202. Wessels A, Sedmera D. Developmental anatomy of the heart: a tale of mice and man. Physiol Genomics. 2003;15:165–76.
203. Schott JJ, Benson DW, Basson CT, et al. Congenital heart disease caused by mutations in the transcription factor NKX2-5. Science. 1998;281:108–11.
204. Benson DW, Silberbach GM, Kavanaugh-McHugh A, et al. Mutations in the cardiac transcription factor NKX2.5 affect diverse cardiac developmental pathways. J Clin Invest. 1999;104:1567–73.
205. Goldmuntz E, Geiger E, Benson DW. NKX2.5 mutations in patients with tetralogy of fallot. Circulation. 2001;104:2565–8.
206. Kasahara H, Lee B, Schott JJ, et al. Loss of function and inhibitory effects of human CSX/NKX2.5 homeoprotein mutations associated with congenital heart disease. J Clin Invest. 2000;106:299–308.
207. Biben C, Weber R, Kesteven S, et al. Cardiac septal and valvular dysmorphogenesis in mice heterozygous for mutations in the homeobox gene Nkx2-5. Circ Res. 2000;87:888–95.
208. Basson CT, Bachinsky DR, Lin RC, et al. Mutations in human TBX5 [corrected] cause limb and cardiac malformation in Holt-Oram syndrome. Nat Genet. 1997;15:30–5.
209. Li QY, Newbury-Ecob RA, Terrett JA, et al. Holt-Oram syndrome is caused by mutations in TBX5, a member of the Brachyury (T) gene family. Nat Genet. 1997;15:21–9.
210. Basson CT, Huang T, Lin RC, et al. Different TBX5 interactions in heart and limb defined by Holt-Oram syndrome mutations. Proc Natl Acad Sci USA. 1999;96:2919–24.
211. Cross SJ, Ching YH, Li QY, et al. The mutation spectrum in Holt-Oram syndrome. J Med Genet. 2000;37:785–7.
212. Ghosh TK, Packham EA, Bonser AJ, Robinson TE, Cross SJ, Brook JD. Characterization of the TBX5 binding site and analysis of mutations that cause Holt-Oram syndrome. Hum Mol Genet. 2001;10:1983–94.
213. Armstrong EJ, Bischoff J. Heart valve development: endothelial cell signaling and differentiation. Circ Res. 2004;95:459–70.
214. de Lange FJ, Moorman AF, Anderson RH, et al. Lineage and morphogenetic analysis of the cardiac valves. Circ Res. 2004;95: 645–54.
215. Sugi Y, Yamamura H, Okagawa H, Markwald RR. Bone morphogenetic protein-2 can mediate myocardial regulation of atrioventricular cushion mesenchymal cell formation in mice. Dev Biol. 2004;269:505–18.
216. Gaussin V, Morley GE, Cox L, et al. Alk3/Bmpr1a receptor is required for development of the atrioventricular canal into valves and annulus fibrosus. Circ Res. 2005;97:219–26.
217. Desgrosellier JS, Mundell NA, McDonnell MA, Moses HL, Barnett JV. Activin receptor-like kinase 2 and Smad6 regulate epithelial-mesenchymal transformation during cardiac valve formation. Dev Biol. 2005;280:201–10.
218. Galvin KM, Donovan MJ, Lynch CA, et al. A role for smad6 in development and homeostasis of the cardiovascular system. Nat Genet. 2000;24:171–4.
219. Nieto MA. The snail superfamily of zinc-finger transcription factors. Nat Rev Mol Cell Biol. 2002;3:155–66.
220. Nieto MA, Sargent MG, Wilkinson DG, Cooke J. Control of cell behavior during vertebrate development by Slug, a zinc finger gene. Science. 1994;264:835–9.
221. Carmona R, Gonzalez-Iriarte M, Macias D, Perez-Pomares JM, Garcia-Garrido L, Munoz-Chapuli R. Immunolocalization of the transcription factor Slug in the developing avian heart. Anat Embryol (Berl). 2000;201:103–9.
222. Romano LA, Runyan RB. Slug is an essential target of TGFbeta2 signaling in the developing chicken heart. Dev Biol. 2000;223:91–102.
223. Carver EA, Jiang R, Lan Y, Oram KF, Gridley T. The mouse snail gene encodes a key regulator of the epithelial-mesenchymal transition. Mol Cell Biol. 2001;21:8184–8.
224. Hurlstone AF, Haramis AP, Wienholds E, et al. The Wnt/beta-catenin pathway regulates cardiac valve formation. Nature. 2003;425:633–7.
225. Timmerman LA, Grego-Bessa J, Raya A, et al. Notch promotes epithelial-mesenchymal transition during cardiac development and oncogenic transformation. Genes Dev. 2004;18:99–115.
226. Kokubo H, Miyagawa-Tomita S, Tomimatsu H, et al. Targeted disruption of hesr2 results in atrioventricular valve anomalies that lead to heart dysfunction. Circ Res. 2004;95:540–7.
227. Ferrara N, Gerber HP, LeCouter J. The biology of VEGF and its receptors. Nat Med. 2003;9:669–76.
228. Zachary I, Gliki G. Signaling transduction mechanisms mediating biological actions of the vascular endothelial growth factor family. Cardiovasc Res. 2001;49:568–81.
229. Johnson EN, Lee YM, Sander TL, et al. NFATc1 mediates vascular endothelial growth factor-induced proliferation of human pulmonary valve endothelial cells. J Biol Chem. 2003;278:1686–92.
230. Miquerol L, Gertsenstein M, Harpal K, Rossant J, Nagy A. Multiple developmental roles of VEGF suggested by a LacZ-tagged allele. Dev Biol. 1999;212:307–22.
231. Dor Y, Camenisch TD, Itin A, et al. A novel role for VEGF in endocardial cushion formation and its potential contribution to congenital heart defects. Development. 2001;128:1531–8.
232. Bernanke DH, Markwald RR. Migratory behavior of cardiac cushion tissue cells in a collagen-lattice culture system. Dev Biol. 1982;91:235–45.
233. Dor Y, Klewer SE, McDonald JA, Keshet E, Camenisch TD. VEGF modulates early heart valve formation. Anat Rec A Discov Mol Cell Evol Biol. 2003;271:202–8.
234. Pinter E, Haigh J, Nagy A, Madri JA. Hyperglycemia-induced vasculopathy in the murine conceptus is mediated via reductions of VEGF-A expression and VEGF receptor activation. Am J Pathol. 2001;158:1199–206.
235. Enciso JM, Gratzinger D, Camenisch TD, Canosa S, Pinter E, Madri JA. Elevated glucose inhibits VEGF-A-mediated endocardial cushion formation: modulation by PECAM-1 and MMP-2. J Cell Biol. 2003;160:605–15.
236. Ferencz C, Rubin JD, McCarter RJ, Clark EB. Maternal diabetes and cardiovascular malformations: predominance of double outlet right ventricle and truncus arteriosus. Teratology. 1990;41:319–26.
237. Kitzmiller JL, Gavin LA, Gin GD, Jovanovic-Peterson L, Main EK, Zigrang WD. Preconception care of diabetes. Glycemic control prevents congenital anomalies. JAMA. 1991;265:731–6.
238. Crabtree GR, Olson EN. NFAT signaling: choreographing the social lives of cells. Cell. 2002;109(Suppl):S67–79.
239. Ranger AM, Grusby MJ, Hodge MR, et al. The transcription factor NF-ATc is essential for cardiac valve formation. Nature. 1998;392:186–90.
240. de la Pompa JL, Timmerman LA, Takimoto H, et al. Role of the NF-ATc transcription factor in morphogenesis of cardiac valves and septum. Nature. 1998;392:182–6.
241. Chang CP, Neilson JR, Bayle JH, et al. A field of myocardial-endocardial NFAT signaling underlies heart valve morphogenesis. Cell. 2004;118:649–63.
242. Zhou B, Wu B, Tompkins KL, Boyer KL, Grindley JC, Baldwin HS. Characterization of Nfatc1 regulation identifies an enhancer required for gene expression that is specific to pro-valve endocardial cells in the developing heart. Development. 2005;132:1137–46.
243. Rothermel B, Vega RB, Yang J, Wu H, Bassel-Duby R, Williams RS. A protein encoded within the Down syndrome critical region is enriched in striated muscles and inhibits calcineurin signaling. J Biol Chem. 2000;275:8719–25.

244. Yang J, Rothermel B, Vega RB, et al. Independent signals control expression of the calcineurin inhibitory proteins MCIP1 and MCIP2 in striated muscles. Circ Res. 2000;87:E61–8.
245. Lange AW, Molkentin JD, Yutzey KE. DSCR1 gene expression is dependent on NFATc1 during cardiac valve formation and colocalizes with anomalous organ development in trisomy 16 mice. Dev Biol. 2004;266:346–60.
246. Delorme B, Dahl E, Jarry-Guichard T, et al. Expression pattern of connexin gene products at the early developmental stages of the mouse cardiovascular system. Circ Res. 1997;81:423–37.
247. Reaume AG, de Sousa PA, Kulkarni S, et al. Cardiac malformation in neonatal mice lacking connexin43. Science. 1995;267:1831–4.
248. Kirchhoff S, Nelles E, Hagendorff A, Kruger O, Traub O, Willecke K. Reduced cardiac conduction velocity and predisposition to arrhythmias in connexin40-deficient mice. Curr Biol. 1998;8:299–302.
249. Kumai M, Nishii K, Nakamura K, Takeda N, Suzuki M, Shibata Y. Loss of connexin45 causes a cushion defect in early cardiogenesis. Development. 2000;127:3501–12.
250. Liebner S, Cattelino A, Gallini R, et al. Beta-catenin is required for endothelial-mesenchymal transformation during heart cushion development in the mouse. J Cell Biol. 2004;166:359–67.
251. Gitler AD, Lu MM, Jiang YQ, Epstein JA, Gruber PJ. Molecular markers of cardiac endocardial cushion development. Dev Dyn. 2003;228:643–50.
252. Person AD, Garriock RJ, Krieg PA, Runyan RB, Klewer SE. Frzb modulates Wnt-9a-mediated beta-catenin signaling during avian atrioventricular cardiac cushion development. Dev Biol. 2005;278:35–48.
253. Perez-Moreno M, Jamora C, Fuchs E. Sticky business: orchestrating cellular signals at adherens junctions. Cell. 2003;112:535–48.
254. Ilan N, Mahooti S, Rimm DL, Madri JA. PECAM-1 (CD31) functions as a reservoir for and a modulator of tyrosine-phosphorylated beta-catenin. J Cell Sci. 1999;112(Pt 18):3005–14.
255. Nakajima Y, Mironov V, Yamagishi T, Nakamura H, Markwald RR. Expression of smooth muscle alpha-actin in mesenchymal cells during formation of avian endocardial cushion tissue: a role for transforming growth factor beta3. Dev Dyn. 1997;209:296–309.
256. Lie-Venema H, van den Akker NM, Bax NA, et al. Origin, fate, and function of epicardium-derived cells (EPDCs) in normal and abnormal cardiac development. ScientificWorldJournal. 2007;7:1777–98.
257. Schroeder JA, Jackson LF, Lee DC, Camenisch TD. Form and function of developing heart valves: coordination by extracellular matrix and growth factor signaling. J Mol Med. 2003;81:392–403.
258. Raab G, Klagsbrun M. Heparin-binding EGF-like growth factor. Biochim Biophys Acta. 1997;1333:F179–99.
259. Jackson LF, Qiu TH, Sunnarborg SW, et al. Defective valvulogenesis in HB-EGF and TACE-null mice is associated with aberrant BMP signaling. EMBO J. 2003;22:2704–16.
260. Krug EL, Mjaatvedt CH, Markwald RR. Extracellular matrix from embryonic myocardium elicits an early morphogenetic event in cardiac endothelial differentiation. Dev Biol. 1987;120:348–55.
261. McDonald JA, Camenisch TD. Hyaluronan: genetic insights into the complex biology of a simple polysaccharide. Glycoconj J. 2002;19:331–9.
262. Weigel PH, Hascall VC, Tammi M. Hyaluronan synthases. J Biol Chem. 1997;272:13997–4000.
263. Turley EA, Noble PW, Bourguignon LY. Signaling properties of hyaluronan receptors. J Biol Chem. 2002;277:4589–92.
264. Camenisch TD, Schroeder JA, Bradley J, Klewer SE, McDonald JA. Heart-valve mesenchyme formation is dependent on hyaluronan-augmented activation of ErbB2-ErbB3 receptors. Nat Med. 2002;8:850–5.
265. Erickson SL, O';Shea KS, Ghaboosi N, et al. ErbB3 is required for normal cerebellar and cardiac development: a comparison with ErbB2-and heregulin-deficient mice. Development. 1997;124:4999–5011.
266. Stainier DY, Fouquet B, Chen JN, et al. Mutations affecting the formation and function of the cardiovascular system in the zebrafish embryo. Development. 1996;123:285–92.
267. Walsh EC, Stainier DY. UDP-glucose dehydrogenase required for cardiac valve formation in zebrafish. Science. 2001;293:1670–3.
268. Lander AD, Selleck SB. The elusive functions of proteoglycans: in vivo veritas. J Cell Biol. 2000;148:227–32.
269. Henderson DJ, Copp AJ. Versican expression is associated with chamber specification, septation, and valvulogenesis in the developing mouse heart. Circ Res. 1998;83:523–32.
270. Landolt RM, Vaughan L, Winterhalter KH, Zimmermann DR. Versican is selectively expressed in embryonic tissues that act as barriers to neural crest cell migration and axon outgrowth. Development. 1995;121:2303–12.
271. Henderson DJ, Ybot-Gonzalez P, Copp AJ. Over-expression of the chondroitin sulphate proteoglycan versican is associated with defective neural crest migration in the Pax3 mutant mouse (splotch). Mech Dev. 1997;69:39–51.
272. Akiyama H, Chaboissier MC, Behringer RR, et al. Essential role of Sox9 in the pathway that controls formation of cardiac valves and septa. Proc Natl Acad Sci USA. 2004;101:6502–7.
273. Robert NM, Tremblay JJ, Viger RS. Friend of GATA (FOG)-1 and FOG-2 differentially repress the GATA-dependent activity of multiple gonadal promoters. Endocrinology. 2002;143:3963–73.
274. Svensson EC, Huggins GS, Lin H, et al. A syndrome of tricuspid atresia in mice with a targeted mutation of the gene encoding Fog-2. Nat Genet. 2000;25:353–6.
275. Tevosian SG, Deconinck AE, Tanaka M, et al. FOG-2, a cofactor for GATA transcription factors, is essential for heart morphogenesis and development of coronary vessels from epicardium. Cell. 2000;101:729–39.
276. Gittenberger-de Groot AC, Bartelings MM, Deruiter MC, Poelmann RE. Basics of cardiac development for the understanding of congenital heart malformations. Pediatr Res. 2005;57:169–76.
277. Poelmann RE, Gittenberger-de Groot AC. A subpopulation of apoptosis-prone cardiac neural crest cells targets to the venous pole: multiple functions in heart development? Dev Biol. 1999;207:271–86.
278. Kirby ML, Gale TF, Stewart DE. Neural crest cells contribute to normal aorticopulmonary septation. Science. 1983;220:1059–61.
279. Lincoln J, Alfieri CM, Yutzey KE. Development of heart valve leaflets and supporting apparatus in chicken and mouse embryos. Dev Dyn. 2004;230:239–50.
280. Vincentz JW, McWhirter JR, Murre C, Baldini A, Furuta Y. Fgf15 is required for proper morphogenesis of the mouse cardiac outflow tract. Genesis. 2005;41:192–201.
281. Kaartinen V, Dudas M, Nagy A, Sridurongrit S, Lu MM, Epstein JA. Cardiac outflow tract defects in mice lacking ALK2 in neural crest cells. Development. 2004;131:3481–90.
282. Stottmann RW, Choi M, Mishina Y, Meyers EN, Klingensmith J. BMP receptor IA is required in mammalian neural crest cells for development of the cardiac outflow tract and ventricular myocardium. Development. 2004;131:2205–18.
283. Bernd P, Miles K, Rozenberg I, Borghjid S, Kirby ML. Neurotrophin-3 and TrkC are expressed in the outflow tract of the developing chicken heart. Dev Dyn. 2004;230:767–72.
284. Waldo KL, Hutson MR, Stadt HA, Zdanowicz M, Zdanowicz J, Kirby ML. Cardiac neural crest is necessary for normal addition of the myocardium to the arterial pole from the secondary heart field. Dev Biol. 2005;281:66–77.

285. Washington Smoak I, Byrd NA, Abu-Issa R, et al. Sonic hedgehog is required for cardiac outflow tract and neural crest cell development. Dev Biol. 2005;283:357–72.
286. Moraes F, Novoa A, Jerome-Majewska LA, Papaioannou VE, Mallo M. Tbx1 is required for proper neural crest migration and to stabilize spatial patterns during middle and inner ear development. Mech Dev. 2005;122:199–212.
287. Hong CS, Saint-Jeannet JP. Sox proteins and neural crest development. Semin Cell Dev Biol. 2005;16:694–703.
288. Montero JA, Giron B, Arrechedera H, et al. Expression of Sox8, Sox9 and Sox10 in the developing valves and autonomic nerves of the embryonic heart. Mech Dev. 2002;118:199–202.
289. Maschhoff KL, Anziano PQ, Ward P, Baldwin HS. Conservation of Sox4 gene structure and expression during chicken embryogenesis. Gene. 2003;320:23–30.
290. Schilham MW, Oosterwegel MA, Moerer P, et al. Defects in cardiac outflow tract formation and pro-B-lymphocyte expansion in mice lacking Sox-4. Nature. 1996;380:711–4.
291. Xanthoudakis S, Viola JP, Shaw KT, et al. An enhanced immune response in mice lacking the transcription factor NFAT1. Science. 1996;272:892–5.
292. Wilkins BJ, De Windt LJ, Bueno OF, et al. Targeted disruption of NFATc3, but not NFATc4, reveals an intrinsic defect in calcineurin-mediated cardiac hypertrophic growth. Mol Cell Biol. 2002;22:7603–13.
293. Tadano M, Edamatsu H, Minamisawa S, et al. Congenital semilunar valvulogenesis defect in mice deficient in phospholipase C epsilon. Mol Cell Biol. 2005;25:2191–9.
294. Brannan CI, Perkins AS, Vogel KS, et al. Targeted disruption of the neurofibromatosis type-1 gene leads to developmental abnormalities in heart and various neural crest-derived tissues. Genes Dev. 1994;8:1019–29.
295. Gitler AD, Zhu Y, Ismat FA, et al. Nf1 has an essential role in endothelial cells. Nat Genet. 2003;33:75–9.
296. Hatcher CJ, Basson CT. Specification of the cardiac conduction system by transcription factors. Circ Res. 2009;105:620–30.
297. Gourdie RG, Kubalak S, Mikawa T. Conducting the embryonic heart: orchestrating development of specialized cardiac tissues. Trends Cardiovasc Med. 1999;9:18–26.
298. Patten BM, Kramer T. The initiation of contraction in the embryonic chick heart. Am J Anat. 1933;53:349–75.
299. Chuck ET, Freeman DM, Watanabe M, Rosenbaum DS. Changing activation sequence in the embryonic chick heart. Implications for the development of the His-Purkinje system. Circ Res. 1997;81:470–6.
300. Rentschler S, Vaidya DM, Tamaddon H, et al. Visualization and functional characterization of the developing murine cardiac conduction system. Development. 2001;128:1785–92.
301. Thomas PS, Kasahara H, Edmonson AM, et al. Elevated expression of Nkx-2.5 in developing myocardial conduction cells. Anat Rec. 2001;263:307–13.
302. Milan DJ, Giokas AC, Serluca FC, Peterson RT, MacRae CA. Notch1b and neuregulin are required for specification of central cardiac conduction tissue. Development. 2006;133:1125–32.
303. Chi NC, Shaw RM, Jungblut B, et al. Genetic and physiologic dissection of the vertebrate cardiac conduction system. PLoS Biol. 2008;6:e109.
304. Verheule S, van Kempen MJ, te Welscher PH, Kwak BR, Jongsma HJ. Characterization of gap junction channels in adult rabbit atrial and ventricular myocardium. Circ Res. 1997;80:673–81.
305. Bastide B, Neyses L, Ganten D, Paul M, Willecke K, Traub O. Gap junction protein connexin40 is preferentially expressed in vascular endothelium and conductive bundles of rat myocardium and is increased under hypertensive conditions. Circ Res. 1993;73:1138–49.
306. Gros D, Jarry-Guichard T, Ten Velde I, et al. Restricted distribution of connexin40, a gap junctional protein, in mammalian heart. Circ Res. 1994;74:839–51.
307. Gourdie RG, Severs NJ, Green CR, Rothery S, Germroth P, Thompson RP. The spatial distribution and relative abundance of gap-junctional connexin40 and connexin43 correlate to functional properties of components of the cardiac atrioventricular conduction system. J Cell Sci. 1993;105(Pt 4):985–91.
308. Davis LM, Rodefeld ME, Green K, Beyer EC, Saffitz JE. Gap junction protein phenotypes of the human heart and conduction system. J Cardiovasc Electrophysiol. 1995;6:813–22.
309. Van Kempen MJ, Vermeulen JL, Moorman AF, Gros D, Paul DL, Lamers WH. Developmental changes of connexin40 and connexin43 mRNA distribution patterns in the rat heart. Cardiovasc Res. 1996;32:886–900.
310. Kanter HL, Saffitz JE, Beyer EC. Cardiac myocytes express multiple gap junction proteins. Circ Res. 1992;70:438–44.
311. Darrow BJ, Laing JG, Lampe PD, Saffitz JE, Beyer EC. Expression of multiple connexins in cultured neonatal rat ventricular myocytes. Circ Res. 1995;76:381–7.
312. Coppen SR, Dupont E, Rothery S, Severs NJ. Connexin45 expression is preferentially associated with the ventricular conduction system in mouse and rat heart. Circ Res. 1998;82:232–43.
313. Coppen SR, Severs NJ, Gourdie RG. Connexin45 (alpha 6) expression delineates an extended conduction system in the embryonic and mature rodent heart. Dev Genet. 1999;24:82–90.
314. van Rijen HV, van Veen TA, van Kempen MJ, et al. Impaired conduction in the bundle branches of mouse hearts lacking the gap junction protein connexin40. Circulation. 2001;103:1591–8.
315. Verheule S, van Batenburg CA, Coenjaerts FE, Kirchhoff S, Willecke K, Jongsma HJ. Cardiac conduction abnormalities in mice lacking the gap junction protein connexin40. J Cardiovasc Electrophysiol. 1999;10:1380–9.
316. Hagendorff A, Schumacher B, Kirchhoff S, Luderitz B, Willecke K. Conduction disturbances and increased atrial vulnerability in Connexin40-deficient mice analyzed by transesophageal stimulation. Circulation. 1999;99:1508–15.
317. Simon AM, Goodenough DA, Paul DL. Mice lacking connexin40 have cardiac conduction abnormalities characteristic of atrioventricular block and bundle branch block. Curr Biol. 1998;8:295–8.
318. Gourdie RG, Wei Y, Kim D, Klatt SC, Mikawa T. Endothelin-induced conversion of embryonic heart muscle cells into impulse-conducting Purkinje fibers. Proc Natl Acad Sci USA. 1998;95:6815–8.
319. Takebayashi-Suzuki K, Yanagisawa M, Gourdie RG, Kanzawa N, Mikawa T. In vivo induction of cardiac Purkinje fiber differentiation by coexpression of preproendothelin-1 and endothelin converting enzyme-1. Development. 2000;127:3523–32.
320. Hyer J, Johansen M, Prasad A, et al. Induction of Purkinje fiber differentiation by coronary arterialization. Proc Natl Acad Sci USA. 1999;96:13214–8.
321. Hall CE, Hurtado R, Hewett KW, et al. Hemodynamic-dependent patterning of endothelin converting enzyme 1 expression and differentiation of impulse-conducting Purkinje fibers in the embryonic heart. Development. 2004;131:581–92.
322. Kanzawa N, Poma CP, Takebayashi-Suzuki K, Diaz KG, Layliev J, Mikawa T. Competency of embryonic cardiomyocytes to undergo Purkinje fiber differentiation is regulated by endothelin receptor expression. Development. 2002;129:3185–94.
323. Rentschler S, Zander J, Meyers K, et al. Neuregulin-1 promotes formation of the murine cardiac conduction system. Proc Natl Acad Sci USA. 2002;99:10464–9.
324. Patel R, Kos L. Endothelin-1 and Neuregulin-1 convert embryonic cardiomyocytes into cells of the conduction system in the mouse. Dev Dyn. 2005;233:20–8.

325. Bond J, Sedmera D, Jourdan J, et al. Wnt11 and Wnt7a are up-regulated in association with differentiation of cardiac conduction cells in vitro and in vivo. Dev Dyn. 2003;227:536–43.
326. Nguyen-Tran VT, Kubalak SW, Minamisawa S, et al. A novel genetic pathway for sudden cardiac death via defects in the transition between ventricular and conduction system cell lineages. Cell. 2000;102:671–82.
327. Mounkes LC, Kozlov SV, Rottman JN, Stewart CL. Expression of an LMNA-N195K variant of A-type lamins results in cardiac conduction defects and death in mice. Hum Mol Genet. 2005;14:2167–80.
328. Hershberger RE, Hanson EL, Jakobs PM, et al. A novel lamin A/C mutation in a family with dilated cardiomyopathy, prominent conduction system disease, and need for permanent pacemaker implantation. Am Heart J. 2002;144:1081–6.
329. Hoogaars WM, Tessari A, Moorman AF, et al. The transcriptional repressor Tbx3 delineates the developing central conduction system of the heart. Cardiovasc Res. 2004;62:489–99.
330. Fishman GI. Understanding conduction system development: a hop, skip and jump away? Circ Res. 2005;96:809–11.
331. Ismat FA, Zhang M, Kook H, et al. Homeobox protein Hop functions in the adult cardiac conduction system. Circ Res. 2005;96:898–903.
332. Thompson RP, Lindroth J, Wong YM. Regional differences in DNA-synthetic activity in the preseptation myocardium of the chick. In: Clark EB, Takao A, editors. Developmental cardiology. Mount Kisco, NY: Futura; 1990. p. 219–34.
333. Thompson RP, Reckova M, DeAlmeida A, Bigelow M, Stanley CP, Spruill JB, et al. The oldest, toughest cells in the heart. In: Chadwick DJ, Goode J, editors. Development of the cardiac conduction system. Chichester, UK: Wiley; 2003. p. 157–76.
334. Gourdie RG, Harris BS, Bond J, et al. Development of the cardiac pacemaking and conduction system. Birth Defects Res C Embryo Today. 2003;69:46–57.
335. Reckova M, Rosengarten C, deAlmeida A, et al. Hemodynamics is a key epigenetic factor in development of the cardiac conduction system. Circ Res. 2003;93:77–85.
336. Ziegler T, Bouzourene K, Harrison VJ, Brunner HR, Hayoz D. Influence of oscillatory and unidirectional flow environments on the expression of endothelin and nitric oxide synthase in cultured endothelial cells. Arterioscler Thromb Vasc Biol. 1998;18:686–92.
337. Markos F, Hennessy BA, Fitzpatrick M, O';Sullivan J, Snow HM. The effect of tezosentan, a non-selective endothelin receptor antagonist, on shear stress-induced changes in arterial diameter of the anaesthetized dog. J Physiol. 2002;544:913–8.
338. Meyer D, Birchmeier C. Multiple essential functions of neuregulin in development. Nature. 1995;378:386–90.
339. Hertig CM, Kubalak SW, Wang Y, Chien KR. Synergistic roles of neuregulin-1 and insulin-like growth factor-I in activation of the phosphatidylinositol 3-kinase pathway and cardiac chamber morphogenesis. J Biol Chem. 1999;274:37362–9.
340. Moorman AF, de Jong F, Denyn MM, Lamers WH. Development of the cardiac conduction system. Circ Res. 1998;82:629–44.
341. Nakajima Y, Yoshimura K, Nomura M, Nakamura H. Expression of HNK1 epitope by the cardiomyocytes of the early embryonic chick: in situ and in vitro studies. Anat Rec. 2001;263: 326–33.
342. Wessels A, Vermeulen JL, Verbeek FJ, et al. Spatial distribution of "tissue-specific" antigens in the developing human heart and skeletal muscle. III. An immunohistochemical analysis of the distribution of the neural tissue antigen G1N2 in the embryonic heart; implications for the development of the atrioventricular conduction system. Anat Rec. 1992;232:97–111.
343. Ito H, Iwasaki K, Ikeda T, Sakai H, Shimokawa I, Matsuo T. HNK-1 expression pattern in normal and bis-diamine induced malformed developing rat heart: three dimensional reconstruction analysis using computer graphics. Anat Embryol (Berl). 1992;186:327–34.
344. Barbu M, Ziller C, Rong PM, Le Douarin NM. Heterogeneity in migrating neural crest cells revealed by a monoclonal antibody. J Neurosci. 1986;6:2215–25.
345. Gordon L, Wharton J, Moore SE, et al. Myocardial localization and isoforms of neural cell adhesion molecule (N-CAM) in the developing and transplanted human heart. J Clin Invest. 1990;86:1293–300.
346. Hoffman S, Crossin KL, Prediger EA, Cunningham BA, Edelman GM. Expression and function of cell adhesion molecules during the early development of the heart. Ann N Y Acad Sci. 1990;588:73–86.
347. Davis DL, Edwards AV, Juraszek AL, Phelps A, Wessels A, Burch JB. A GATA-6 gene heart-region-specific enhancer provides a novel means to mark and probe a discrete component of the mouse cardiac conduction system. Mech Dev. 2001;108:105–19.
348. Wessels A, Phelps A, Trusk TC, et al. Mouse models for cardiac conduction system development. Novartis Found Symp. 2003;250:44–59. discussion 59–67, 276–279.
349. Jongbloed MR, Schalij MJ, Poelmann RE, et al. Embryonic conduction tissue: a spatial correlation with adult arrhythmogenic areas. J Cardiovasc Electrophysiol. 2004;15:349–55.
350. Poelmann RE, Jongbloed MR, Molin DG, et al. The neural crest is contiguous with the cardiac conduction system in the mouse embryo: a role in induction? Anat Embryol (Berl). 2004;208:389–93.
351. Kondo RP, Anderson RH, Kupershmidt S, Roden DM, Evans SM. Development of the cardiac conduction system as delineated by minK-lacZ. J Cardiovasc Electrophysiol. 2003;14:383–91.
352. Gourdie RG, Mima T, Thompson RP, Mikawa T. Terminal diversification of the myocyte lineage generates Purkinje fibers of the cardiac conduction system. Development. 1995;121:1423–31.

# Chapter 10
# Signaling in Congenital Heart Disease

**Abstract** While congenital heart disease (CHD), cardiomyopathy, dysrhythmias, and acquired cardiac diseases are common causes of mortality and morbidity in infants and children, the basic underlying mechanisms of many specific pediatric cardiovascular diseases still remain undetermined. Breakthroughs in molecular biology and genetic technology have just begun to be applied in pediatric cardiology stemming from the use of chromosomal mapping and the identification of genes involved in both the primary etiology and as significant risk factors in the development of cardiac and vascular abnormalities. This chapter focuses on information obtained thus far by molecular biology, in particular on numerous signaling pathways and genetic analysis to diagnosis, treatment and overall understanding of pediatric cardiovascular disease pathogenesis, mainly CHD.

**Keywords** Congenital heart defects • Molecular biology • MicroRNA • Gene expression

## Introduction

Congenital heart disease (CHD) refers to defects in the cardiac structure and/or function, which arise in the prenatal period. CHD occurs often and represents the most common type of birth defect, affecting 4–10 of every 1,000 live births [1]. In the USA, CHD affects almost 36,000 infants every year [2]. Despite recent advances in diagnostics and treatment, CHD is the leading noninfectious cause of mortality and morbidity in infants and children in the first year of life.

Abnormalities causing CHD can affect almost all parts of the heart (Fig. 10.1) [3]. They can be broadly divided into three types: cyanotic heart disease, left-sided defects, and septation defects.

Cyanotic heart disease results from the mixing of oxygenated and deoxygenated blood. This condition can be secondary to transposition of the great arteries (TGA), tetralogy of Fallot (TOF), tricuspid atresia (TA), pulmonary atresia (PA), Ebstein's anomaly of the tricuspid valve, double outlet right ventricle (DORV), persistent truncus arteriosus (PTA), and total anomalous pulmonary venous return (TAPVR). Left-sided obstructive defects include hypoplastic left heart syndrome (HLHS), mitral stenosis (MS), aortic stenosis, aortic coarctation, and interrupted aortic arch (IAA). Septation abnormalities include atrial septal defects (ASDs), ventricular septal defects (VSDs), and atrioventricular septal defects (AVSDs). Two additional CHD forms, which do not fit into the major types, are bicuspid aortic valve (BAV) and patent ductus arteriosus (PDA). BAV represents the most common CHD, whereas septation defects are the next most common [1].

## Etiology of CHD

Many CHD with monogenic inheritance are associated with multiple noncardiac birth defects constituting syndromic forms of CHD. Well-established examples of these types of CHD are Alagille syndrome, Costello syndrome, DiGeorge syndrome, Holt–Oram syndrome, Noonan syndrome, and many others [1, 4, 5]. However, most CHD occur sporadically representing nonsyndromic CHD [6, 7]. Although familial cases for both syndromic and nonsyndromic CHD have been identified, families with monogenic mode of inheritance of the latter are infrequent. This suggests that a complex interplay of numerous genetic and environmental risk factors results in the development of CHD. The multifactorial nature of CHD is schematically summarized in Fig. 10.2.

Great advances in developmental biology and molecular genetics have demonstrated a leading role of genetic factors in the development of CHD [1, 5, 6, 8]. Recent breakthroughs in DNA mutation and polymorphism analysis, in cytogenetic and fluorescence in situ hybridization techniques, in the sequencing of the human genome, and also in the generation of transgenic animal models have highlighted that defects in specific genes interfere with embryonic heart development

J. Marín-García, *Signaling in the Heart*,
DOI 10.1007/978-1-4419-9461-5_10, © Springer Science+Business Media, LLC 2011

leading to different CHD phenotypes. In addition to gene mutations, chromosomal structural abnormalities (e.g., deletions and duplications), single-nucleotide polymorphisms (SNPs), abnormal RNAs, and epigenetics also contribute to the development of CHD.

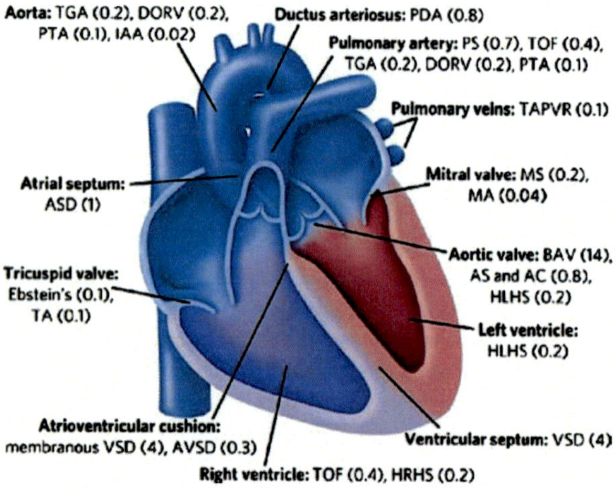


**Fig. 10.1** Scheme of congenital heart defects. Estimated incidence of each defect per 1,000 live births are shown in parenthesis. *AC* aortic coarctation, *AS* aortic stenosis, *ASD* atrial septal defect, *AVSD* atrioventricular septal defect, *BAV* bicuspid aortic valve, *DORV* double outlet right ventricle, *Ebstein's* Ebstein's anomaly of the tricuspid valve, *HLHS* hypoplastic left heart syndrome, *HRHS* hypoplastic right heart syndrome; *IAA* interrupted aortic arch, *MA* mitral atresia, *MS* mitral stenosis, *PDA* patent ductus arteriosus, *PS* pulmonary artery stenosis, *PTA* persistent truncus arteriosus, *TA* tricuspid atresia, *TAPVR* total anomalous pulmonary venous return, *TGA* transposition of the great arteries, *TOF* tetralogy of Fallot, *VSD* ventricular septal defect (reproduced from Bruneau [3]. With kind permission from Nature Publishing Group)

Numerous environmental factors, such as viral infections (e.g., *rubella*), exposure to various chemical teratogens [e.g., dilantin, halogenated hydrocarbon, lithium, retinoic acid (RA), and angiotensin-converting-enzyme inhibitors], maternal diseases (e.g., diabetes and lupus erythematosus), and abnormal hemodynamics, may also contribute during embryonic development to increased risk of CHD [9, 10] (Fig. 10.2)

## Molecular Mechanisms of CHD

With the impressive progress in molecular genetics, developmental biology and molecular cardiology over 40 different genes that are responsible for inherited and sporadic CHD have been identified [1, 7] (Table 10.1). The majority of these genes encode transcription and signal transduction factors, which control specific events in cardiogenesis. In addition to the devastating effects that such abnormalities can present in the neonate, it is becoming clear that genetic mutations that cause developmental malformations may result in cardiac dysfunction later in life [4, 11]. The molecular mechanisms underlying the well-characterized forms of CHD will be discussed in this chapter.

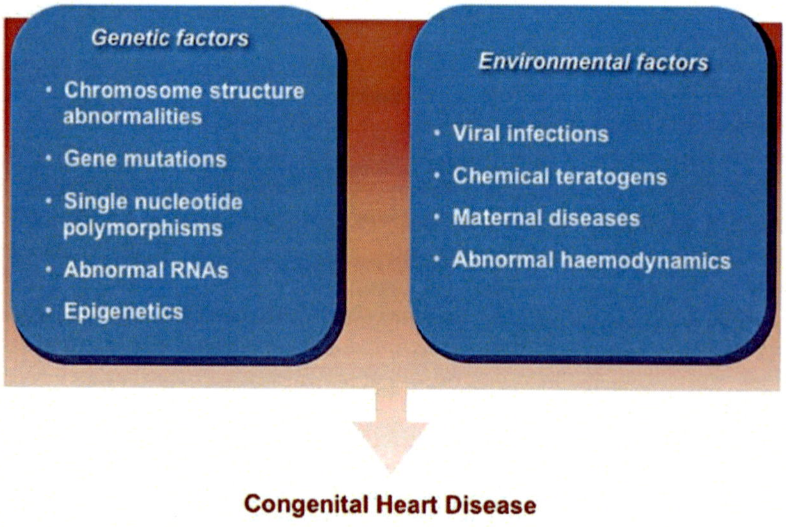


**Fig. 10.2** Complex interplay of genetic and environmental factors leads to CHD. See text for the details

**Table 10.1** Genes linked to CHD

| Gene | Locus | Encoded protein | Disorder/cardiac phenotype |
|---|---|---|---|
| *Syndromic CHD* | | | |
| *PTPN11* | 12q24.1 | SHP-2, protein tyrosine phosphatase | Noonan syndrome/AVSD, PS |
| *KRAS* | 12p1.21 | KRAS, small GTPase | |
| *SOS1* | 2p22.1 | SOS1, RasGEF | |
| *RAF1* | 3p25.1 | RAF1/CRAF, Ser/Thr protein kinase | |
| *PTPN11* | 12q24.1 | SHP-2, protein tyrosine phosphatase | LEOPARD syndrome/AVSD, PS |
| *RAF1* | 3p25.1 | RAF1/CRAF, Ser/Thr protein kinase | |
| *HRAS* | 11p15.5 | HRAS, small GTPase | Costello syndrome |
| *BRAF* | 7q34 | BRAF, Ser/Thr protein kinase | Cardio-facio-cutaneous syndrome |
| *KRAS* | 12p12.1 | KRAS, small GTPase | |
| *MEK1/MAP2K1* | 15q22.1–q22.33 | MEK1/MAP2K1, Ser/Thr protein kinase | |
| *MEK2/MAP2K2* | 19p13.3 | MEK2/MAP2K2, Ser/Thr protein kinase | |
| *JAG1* | 20p12.1–p11.23 | JAG1, NOTCH1 ligand | Alagille syndrome/PS, TOF |
| *NOTCH2* | 1p13–p11 | NOTCH2 receptor | |
| *TBX1* | 22q11.2 | TBX1, T-box transcription factor | DiGeorge syndrome/TOF, PTA, VSD, IAA |
| *ELN* | 7q11.23 | Elastin | Williams syndrome/SVAS, PS |
| *CDH7* | 8q12.2 | Chromodomain helicase DNA-binding protein 7 | CHARGE syndrome/PDA, septal defects |
| *TBX5* | 12q24.1 | TBX5, T-box transcription factor | Holt–Oram syndrome/AVSD |
| *SALL1* | 16q12.1 | SALL1, Zn-finger transcription factor | Townes–Brocks syndrome/ASD, VSD, TOF |
| *SALL4* | 20q13.2 | SALL4, Zn-finger transcription factor | Okihiro syndrome/ASD, VSD, TOF |
| *TFAP2B* | 6p12 | Transcription factor, activating enhancer-binding protein-2beta | Char syndrome/PDA |
| *JAM3/JAMC* | 11q25 | Junctional adhesion protein | Jacobsen syndrome,/HLSHS, DORV, TGA |
| *ETS-1* | 11q23.3 | ETS1, helix-turn-helix transcription factor | |
| *EVC* | 4p16 | Leucine zipper transmembrane protein | Ellis–van Cleveld syndrome/ASD |
| *EVC2* | 4p16.2-p16.1 | Protein involved in skeletal development | |
| *FBN1* | 15q21.1 | Fibrillin 1, extracellular matrix glycoprotein | Marfan syndrome/ASD |
| *TGFBR2* | 3p22 | TGF-β receptor type 2B, Ser/Thr protein kinase | Marfan-like syndrome/ASD |
| *Nonsyndromic CHD* | | | |
| *NKX2.5* | 5q34 | NKX2.5, NK2 family homeobox transcription factor | Familial ASD with progressive AV block |
| *NKX2.6* | 8p21.2 | NKX2.6, NK2 family transcription factor | PTA |
| *GATA4* | 8p23.1–p22 | GATA4, Zn-finger transcription factor | Familial ASD without progressive AV block, TOF, VSD |
| *GATA6* | 18q11.1–q11.2 | GATA6, Zn-finger transcription factor | PTA, PS |
| *NODAL* | 10q22.1 | NODAL, TGF-β family regulator | Heterotaxy |
| *ZIC3* | Xq26 | ZIC3, Zn-finger transcription factor | |
| *CFC1* | 2q21.1 | CRYPTIC, NODAL co-receptor | |
| *ACVR2B* | 3p22 | Activin A receptor type IIB, TGF-β family protein | |
| *LEFTY2* | 1q42.1 | L–R determination factor 2 | |
| *ACVR1/ALK2* | 2q23-q24 | Activin A receptor type I, TGF-β family protein | ASD, TGA, AVSD |
| *CFC1* | 2q21.1 | CRYPTIC, NODAL co-receptor | Heterotaxy, TGA, TOF, AVSD, DORV |
| *TDGF1* | 3p21.31 | CRIPTO, NODAL co-receptor | TOF |
| *FOXH1* | 8q24.3 | FOXH1, forkhead transcription factor | |
| *NOTCH1* | 9q34.3 | NOTCH1 receptor | BAV, VSD, TOF |
| *MED13L/THRAP2* | 12q24.21 | Mediator complex subunit 13-like protein | TGA |
| *FOG2/ZFPM2* | 8q23 | FOG2, Zn-finger transcription factor | TOF |
| *NKX2.5* | 5q34–q35 | NKX2.5, NK2 transcription factor-related protein | |
| *CRELD1* | 3p25.3 | CRELD1, EGF-like domain protein 1 | AVSD |
| *TBX20* | 7p14–p15 | TBX20, T-box transcription factor | ASD, VSD, valve defects |
| *ANKRD1* | 10q23.31 | Ankyrin repeat containing transcription factor | TAPVR |
| *ACTC1* | 15q11–q14 | α-Cardiac actin | ASD, VSD |
| *MYBPC3* | 11p11.2 | Myosin-binding protein C | |
| *MYH6* | 14q12 | α-Myosin heavy chain | ASD |
| *MYH7* | | β-Myosin heavy chain | |
| *MYH11* | 16p13.11 | Myosin heavy chain 11 | PDA, aorta aneurysm |

*ASD* atrial septal defect, *AV* atrioventricular, *AVSD* atrioventricular septal defect, *BAV* bicuspid aortic valve, *DORV* double outlet right ventricle, *IAA* interrupted aortic arch, *PDA* patent ductus arteriosus, *PS* pulmonary valve stenosis, *PTA* persistent truncus arteriosus, *SVAS* supravalvular aortic stenosis, *TAPVR* total anomalous pulmonary venous return, *TGA* transposition of the great arteries, *TOF* tetralogy of Fallot, *VSD* ventricular septal defect

## Alterations of Signaling Pathways Associated with Valve Abnormalities

Cardiac valve formation is a complex highly regulated process involving complex interactions between a variety of signaling cascades in the myocardium as well as a cross-talk between the myocardium and the overlying endocardium (see Chap. 9). Mutations in genes encoding various signaling proteins and downstream effectors can result in CHD of varying severity.

The etiology of one subgroup of CHD that include Noonan (NS), LEOPARD, Costello, and cardio-facio-cutaneous (CFC) syndromes has been shown to be caused by dysregulation of the RAS/mitogen-activated protein kinase (MAPK) pathway. For this reason, these syndromes are also called "RASopathies" [12]. The RAS/MAPK signaling pathway transduces growth factor signals into the intracellular environment and plays a key role in the control of cell differentiation, proliferation, and senescence (Fig. 10.3). Impairment of this pathway leads to significant defects in heart development. Affected individuals have a high incidence of heart disease (e.g., 80–90% in NS) including pulmonic stenosis, hypertrophic cardiomyopathy (HCM), AVSDs, and coarctation of the aorta.

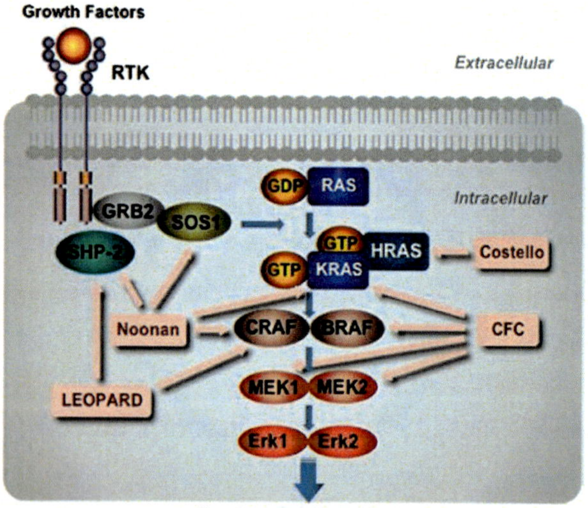


**Fig. 10.3** RAS/MAPK signaling pathway. Activation of receptor tyrosine kinases (RTK) by the binding of growth factors leads to the receptor autophosphorylation and interaction with the adaptor protein GRB2. GRB2 binds to SOS1, a guanine nucleotide exchange factor, which promotes conversion of the inactive GDP-bound RAS into the active GTP-bound form. Activated GTP-bound RAS stimulates RAF kinases (BRAF, CRAF) that in turn activate MAPK and PI3K-Akt (not shown) signaling cascades. Erk1/2 are the ultimate effectors of the RAS/MAPK signaling, which phosphorylate a wide range of downstream nuclear and cytosolic targets including transcription factors, protein kinases, and membrane proteins. Defects in components of the RAS/MAPK signaling pathway, associated with Noonan, LEOPARD, Costello, and cardio-facio-cutaneous (CFC) syndromes, are indicated

### *Noonan Syndrome*

Noncardiac features of NS, the most common of these conditions, include short statue, typical facial dysmorphism, webbed neck, chest deformity, cryptorchidism, bleeding diathesis, and developmental delay [1, 13, 14].

Currently, mutations in four distinct genes have been identified to be responsible for NS: *PTPN11*, *KRAS*, *SOS1*, and *RAF1* [15–21]. *PTPN11* encodes a non-receptor protein tyrosine phosphatase SHP-2, an important signaling protein involved in various biological processes including cardiac semilunar valvulogenesis [22, 23]. Mutations in *PTPN11* are found in almost 50% of patients with NS and are predominant among familial cases and among NS patients displaying pulmonary valve stenosis [24]. SHP-2 consists of $NH_2$- and COOH-terminal Src homology 2 (SH2) domains and a catalytic protein tyrosine phosphatase (PTP) domain. Autoinhibitory interaction between the $NH_2$-SH2 and PTP blocks the SHP-2 phosphatase activity [25]. Most of the *PTPN11* mutations associated with NS affect residues involved in this interaction. These mutations impair the enzyme transition from the catalytically active to inactive form leading to the activation of the RAS/MAPK signaling [26–28].

Mutations in the *KRAS* gene encoding small GTPase KRAS have been found in a small fraction (<2%) of patients with NS [17]. The described *KRAS* mutations result either in decreased GTPase activity or in altered guanine nucleotide binding [17, 29]. Both mechanisms lead to an increase in the active GTP-associated form of KRAS, which contributes to the activation of the RAS/MAPK signaling pathway (Fig. 10.3).

The second most common cause of NS (approximately 13% of cases) are missense mutations in *SOS1*, encoding guanine nucleotide exchange factor (GEF) that acts on the RAS, "son of sevenless" 1 (SOS1) [18, 19]. SOS1 is involved in the conversion of the inactive GDP-bound RAS into the active GTP-bound RAS. The majority of NS-causing mutations are gain-of-function ones leading to an elevation of the active GTP-RAS levels and activation of RAS/MAPK signaling.

Finally, *RAF1* mutations have been reported to also cause NS [20, 21]. The protein product of this gene is a serine/threonine protein kinase RAF1, also known as CRAF, the direct downstream effector of RAS. The majority of NS-associated *RAF1* mutations result in a gain-of-function of RAF1.

### *LEOPARD Syndrome*

LEOPARD syndrome, which shares many phenotypic features with NS, is an acronym describing its features: numerous *L*entigines, *E*KG abnormalities, *O*cular hypertelorism, *P*ulmonary stenosis, *A*bnormal genitalia, *R*etardation of growth and sensorineural *D*eafness. LEOPARD syndrome is caused by specific *PTPN11* mutations, which unlike NS are loss-of-function mutations and have a dominant negative effect [28, 30]. The most common LEOPARD syndrome-causing mutations cluster in residues within the catalytic PTP domain and lead to reduced SHP-2 activity [31, 32]. Mutations in *RAF1* and *BRAF* (the latter encoding serine/threonine-protein kinase BRAF) have also been identified in some patients with this disorder [20, 33, 34].

## Costello and Cardio-Facio-Cutaneous Syndromes

The phenotypically related Costello and CFC syndromes are the most severe of this subgroup of CHD. Affected individuals have severe feeding problem, failure to thrive, developmental delay, facial dysmorphisms, ectodermal abnormalities and cardiac malformations. While no *PTPN11* mutations have been detected in patients with these conditions, they are also characterized by dysregulated RAS/MAPK signaling pathway [35].

Over 83% of Costello syndrome cases have gain-of-function missense mutations in the *HRAS* proto-oncogene encoding small GTPase HRAS [36, 37]. Most of *HRAS* mutations have been identified in codons affecting amino acid residues 12 or 13 located at the GTP binding site and resulting in the constitutively active protein [37, 38]. Importantly, such mutations overlap with those identified in cancers explaining a correlation between disease genotype and risk of developing both benign and malignant neoplasms [39, 40]. While *HRAS* mutations in codon 12, 13, or 61 have been found in approximately 20% of all tumors, no mutations in codon 61 have been identified so far in patients with Costello syndrome [12, 41].

CFC syndrome, like Costello syndrome, is a rare developmental disease. CFC patients typically display more severe developmental delay and worse long-term neurological outcomes than Costello syndrome-affected individuals. Mutations in four genes encoding proteins in the RAS/MAPK signaling pathway are responsible for CFC syndrome: *BRAF*, *MEK1* (*MAP2K1*), *MEK2* (*MAP2K2*), and *KRAS* [42, 43]. Heterozygous *BRAF* mutations are the most common (approximately 75% of cases) in CFC syndrome; however, up to 40% of CFC patients have no molecular diagnosis currently [44]. *BRAF* encodes a serine/threonine protein kinase, which is one of the direct downstream effectors of RAS. *BRAF* mutations affect residues in the cysteine-rich domain in exon 6 and in the protein kinase domain. Most BRAF mutant proteins have increased kinase activity resulting in the activation of the MAPK signaling [42, 43]. A few BRAF mutants display impaired protein kinase activity which leads to dysregulation of the RAS/MAPK signaling pathway, possibly through the BRAF–CRAF interaction [45, 46].

Heterozygous missense *MEK1* and *MEK2* mutations have been identified in approximately 25% of patients with CFC in which a gene mutation has been found [42]. MEK1 (MAP2K1) and MEK2 (MAP2K2) are serine/threonine protein kinases, which phosphorylate and activate Erk1/2 in response to various stimuli (Fig. 10.3). Functional analysis of the mutant MEK1 and MEK2 associated with CFC syndrome has revealed that they are activated forms of kinases [42, 46].

## NOTCH Signaling Pathway in CHD

The highly conserved NOTCH signaling represents another notable example of signaling pathway, which impairment can lead to valve and outflow tract defects and vascular abnormalities. Components of the NOTCH pathway are robustly expressed in the endocardium and large vessels of the heart and play a critical role in multiple stages of heart development [47–50].

Four NOTCH receptors (NOTCH1-4) and five ligands, subdivided into two groups [delta-like 1, 3, 4 and JAGGED (*JAG*) 1, 2], are expressed in mammals. Both NOTCH receptors and ligands are single-pass transmembrane proteins, which enable the NOTCH signaling pathway to transduce signals between neighbor cells. Ligand binding to the NOTCH receptor induces a series of proteolytic cleavage events of the receptor, releasing the NOTCH intracellular domain (NICD). The NICD translocates into the nucleus where it interacts with various transcription regulators to control transcription of targeted genes. Several direct transcription targets of NOTCH, including *c-Myc*, the *Hairy enchancer of split* (*Hes*) and *Hes-related* (*Hrt*) family of genes, and other target genes, have been described [51, 52].

The critical role for NOTCH signaling in cardiac valve development has been highlighted by the identification of mutations in the *NOTCH1* gene associated with hereditary valve disease. Patients with *NOTCH1* haploinsufficiency demonstrate predominantly structural defects of the aortic valves, such as BAV, the most common form of CHD. In addition, they also have aortic stenosis, VSDs, TOF, and calcification of aortic valves [48, 53]. Involvement of *NOTCH1* in the regulation of bone-related pathways may explain the latter defect. Using transgenic mouse model, High et al. [54] have recently demonstrated that the inhibition of *NOTCH1* or deletion of its ligand *JAG1* leads to the downregulation of fibroblast growth factor 8 (*Fgf8*) in the second heart field (SHF) [54]. Although the precise mechanism of *NOTCH1–JAG1–FGF8* axis remains to be determined, these findings suggest that this pathway plays a role during outflow tract development.

Alagille syndrome is also caused by defective NOTCH signaling pathway. It is an autosomal dominant disorder characterized by highly variable liver disease in conjunction with cardiovascular, skeletal or ocular anomalies, or typical facial features. More than 90% of affected patients have cardiovascular abnormalities, including peripheral pulmonary hypoplasia, TOF, and pulmonary valve stenosis, whereas left-sided lesions and septal defects are less common [55]. These cardiovascular anomalies are responsible for a high morbidity and mortality in individuals with Alagille syndrome [56].

Alagille syndrome is caused by mutations in the *JAG1* gene, which encodes the NOTCH ligand jagged 1 (JAG1) [57, 58]. *JAG1* mutations have also been found in nonsyndromic cases of CHD [59, 60]. In addition, *NOTCH2* mutations have been identified in two families with Alagille syndrome [61]. Transgenic mice with defects in genes encoding components of the NOTCH signaling cascade confirm the importance of this pathway in cardiac valve and outflow tract development.

## Conotruncal and Outflow Tract Defects

Conotruncal and outflow tract defects accounts for 20–30% of CHD [1]. These abnormalities (e.g., TOF) are frequently accompanied by "accessory" cardiac disorders, such as persistent right-sided aortic arch. It has been suggested that altered hemodynamics, resulted from these anomalies, is an essential epigenetic factor linking defective outflow tract morphogenesis and branchial-arch artery remodeling [62, 63].

### *DiGeorge Syndrome*

The most common genetic cause of these anomalies is the 22q11.2 microdeletion encompassing DiGeorge, velo-cardio-facial, and conotruncal-face syndromes [64]. Clinical characteristics of the 22q11 deletion syndromes are highly variable between affected individuals [65]. The most common features are cardiovascular and palate anomalies, feeding disorders, speech and learning disabilities, renal anomalies, and behavioral disorders. The most common cardiovascular abnormalities include TOF, PTA, conoventricular VSDs, and IAA [66].

The *TBX1* gene, encoding a T-box family transcription factor (Tbx1), accounts for much of the human phenotype. The identification of *TBX1* missense mutations in patients with features of the 22q11 deletion syndromes, but without a deletion, has further supported this conclusion [67]. Mouse models have confirmed a role for the *TBX1* gene in the 22q11 microdeletion syndromes [68]. In addition to microdeletion, the 22q11.2 microduplication occurs with a frequency half of that of microdeletion [69, 70]. Although the phenotype of patients with the 22q11.2 microduplication is highly variable, it overlaps with the features of the 22q11 deletion syndromes [71].

In addition to *TBX1*, other genes within the 22q11 region may contribute to the 22q11 deletion syndromes. One of such candidate genes is *Crkl*. *Crkl*-deficient mice have similar defects to those of Tbx1-deficient mice [72, 73].

### *Williams Syndrome*

Williams syndrome (Williams–Beuren syndrome) represents an autosomal dominant disorder characterized by cardiovascular, skeletal and renal abnormalities, infantile hypercalcemia, cognitive deficits, and elfin faces. The most common cardiovascular defects include supravalvar aortic stenosis and peripheral pulmonary stenosis [74, 75].

Approximately 90% of patients with Williams syndrome have the 7q11.23 microdeletion [76, 77]. Molecular analysis has revealed that the deletion of one copy of the *ELN* gene, encoding elastin, a component of elastic fibers, is responsible for vascular anomalies of Williams syndrome [76]. Larger deletions, visible cytogenetically, are typically associated with more severe clinical manifestations of this disorder.

### *CHARGE Syndrome*

CHARGE syndrome is an acronym for a wide range of developmental abnormalities including *c*olobomata, *h*eart defects, choanal *a*tresia, *r*etardation of growth and development, *g*enital and *e*ar abnormalities. Cardiovascular defects are variable and include conotruncal and left-sided obstructive defects, although PDA and septal defects are observed in some patients. Mutations in the *CHD7* gene, found in approximately 70% of affected individuals, cause this disorder [78–81]. Protein product of *CHD7* belongs to the chromodomain helicase DNA-binding protein family. A positive correlation between *CHD7* mutations and the presence of cardiovascular defects, colobomata and facial asymmetry has been reported [81]. The precise molecular mechanism underlying CHARGE syndrome is currently unclear.

### *Jacobsen Syndrome*

Approximately 56% of patients with Jacobsen syndrome, associated with the 11q terminal deletion, display conotruncal defects including HLSHS, DORV, and TGA [82, 83]. Several candidate genes, mutations in which may be responsible for heart anomalies associated with this disorder, have been suggested. *JAM3* (*JamC* in mice), encoding a junction adhesion protein, has previously been suggested as causative gene of Jacobsen syndrome [82]. More recently, however, it has been reported that *JamC*-deficient mice do not have cardiac defects [84].

Another candidate gene, located in the cardiac critical region of 11q, is *ETS-1*, which encodes ETS1, a helix-turn-helix transcription factor, implicated in vascular development and angiogenesis [85]. Importantly, loss of its mouse analogue *Ets-1* has resulted in VSD and abnormal ventricular morphology [86]. The precise role of *ETS-1* in CHD remains to be determined.

## Cardiac Septation Abnormalities

The most common form of CHD, accounting for almost 50% of all CHD cases, is septation defects including ASDs, VSDs, and AVSDs [1].

### *Holt–Oram Syndrome*

Holt–Oram syndrome is an autosomal dominant syndrome characterized by secundum ASDs and VSDs in patient with upper limb deformities [87]. This syndrome is caused by mutations, mainly nonsense or frameshift, in the *TBX5* gene leading to haploinsufficiency [88, 89]. Mouse models have confirmed the importance of gene dosage demonstrating that gene dosage correlates with the severity of cardiac defects [90, 91]. *TBX5* encodes the T-box transcription factor that is a critical regulator of gene expression during embryonic development and its deficiency impairs heart and limb development. Importantly, TBX5 interacts physically with two other transcription factors, the NK2 family transcription factor NKX2.5 and Zn-finger transcription factor GATA4, to activate its downstream targets [90, 92, 93].

NKX2.5 belongs to homeobox transcription factors and is involved in the regulation of various pathways during heart development via interactions with other transcriptional regulators of cardiogenesis. GATA4 is a member of the GATA transcription factor family (GATA1-6), which recognizes the consensus sequence (T/A)GATA(A/G) in target genes and plays a key role in the control of developmental processes, including cardiovascular development.

In mice, loss of *Nkx2.5* results in the lack of the primordium of the AV node, while its ventricular-specific deficiency leads to progressive cardiomyopathy and complete heart block [94, 95]. *Gata4*$^{-/-}$ mice die in utero, whereas mice harboring heterozygous *Gata4* mutations have septal defects and endocardial cushion anomalies [96–98]. The importance of TBX5–NKX2.5–GATA4 interactions has been highlighted by identifying mutations in *NKX2.5* and *GATA4* in patients with ASDs and VSDs as well as with other CHD [93, 99–101].

Recent studies have identified mutations in other members of the TBX–NKX2–GATA axis. Mutations in *TBX20* have been found in patients with familial ASDs [102, 103]. Importantly, a gain-of-function *TBX20* mutation, associated with familial ASDs spanning three generations of kindred, has resulted in increased transcriptional activity of associated genes, *NKX2.5* and *GATA4/5* [103]. A single mutation in *NKX2.6* has been identified in a family with PTA [104]. Intriguingly, loss of *Nkx2.6* in mice has not led to cardiac abnormality [105]. Mutations in *GATA6*, a gene homologous to *GATA4*, have been found in patients with CHD including PTA and PS [106].

In addition to interactions with TBX and NKX factors, the GATA transcription factors interact with several other nuclear proteins, such as the FOG and NFAT family members and p300 and CREB-binding protein (CBP) transcriptional coactivators [107, 108]. Therefore, a complex set of tightly regulated interactions between TBX, NKX2, GATA and other transcription regulators is essential for cardiac morphogenesis.

### *Okihiro and Townes–Brocks Syndromes*

Phenotypically related Okihiro and Townes–Brocks syndromes are characterized by VSDs, ASDs, and TOF; in the latter, pulmonary atresia and truncus arteriosus can also be found. Mutations in the *SALL* family genes, *SALL1* and *SALL4*, are responsible for the phenotypes of Townes–Brocks syndrome and Okihiro syndrome, respectively [109–112]. Interestingly, patients with Okihiro syndrome have in addition to CHD limb defects almost identical to those in Holt–Oram syndrome. The *SALL* genes encode Zn-finger transcription factors. In mice, SALL4 interacts physically with TBX5 and this interaction is essential for the heart and limb development [113].

Missense mutations in the *MYH6* gene, which encodes α-myosin heavy chain, have been found in a large family with inherited ASDs [114]. Expression of *MYH6* is significantly activated in the developing atria. Moreover, its transcription is induced by normal TBX5 but markedly reduced by mutant TBX5 found in Holt–Oram syndrome. In addition, *GATA4* mutations responsible for familial ASDs are associated with reduced MYH6 transactivation [93].

It has been suggested that defects in the *CITED2* gene are associated with cardiac septal abnormalities [115]. *Cite2*-deficient mice die in utero due to heart malformations, such as ASDs, VSDs, and conotruncal defects. Importantly, these mice display also deficiency in *Pitx2c*, a target gene of the Nodal signaling, confirming that they are components of this signaling pathway. CITE2 (CBP/p300-interacting transactivator with E/D-rich C-terminal domain, type 2) is a transcription factor, which upon interaction with p300 and CBP, coactivates the neural crest-related TFAP2 transcription factor. Mutations in genes encoding CITE2-interacting proteins are responsible for syndromes associated with CHDs. Mutations in genes encoding p300 and CBP cause Rubinstein–Taybi syndrome characterized by PDA and septation defects, while mutant *TFAP2B* causes Char syndrome characterized mainly by PDA [116–118].

Taken together, these findings highlight the importance of the intricate network of interactions of multiple transcription factors for highly tuned regulation of heart development. However, only a few downstream cellular targets responsible for CHD have been identified and their identification represents a major challenge.

## ErbB Signaling in CHD

The ErbB family of receptor tyrosine kinases (RTKs) represents a family of highly conserved and structurally homologous type 1 tyrosine kinase transmembrane glycoproteins that transduce extracellular stimuli to the cell nucleus to promote differentiation, proliferation, and migration of cells essential for heart development [119, 120].

This family consists of four members including the ErbB1, also known as the epidermal growth factor (EGF) receptor (EGFR) or HER1, the ErbB2 (HER2), the ErbB3 (HER3), and the ErbB4 (HER4). The ErbB ligands include EGF-like ligands (EGF, transforming growth factor (TGF)-α, heparin-binding EGF-like growth factor (HB-EGF), βcellulin, epigen, and epiregulin) and neuregulins 1–4 (NRG 1–4). EGF-like ligands bind and activate ErbB1, while NRG 1–4 bind to ErbB3 and ErbB4. In addition, ErbB3 can be activated by EGF and TGF-α, whereas ErbB4 can bind HB-EGF, βcellulin, and epiregulin [121–123]. No ligands for ErbB2 has been identified, but it is able to form heterodimers with other ErbBs to transduce extracellular signals.

Ligand binding to extracellular domain (ECD) of an ErbB induces homo- or heterodimerization of the ErbB family members leading to the activation of their intracellular tyrosine kinase domain and transphosphorylation of their intracellular tyrosines. Transphosphorylation in turn recruits a variety of signaling proteins involved in various signaling cascades, such as RAS/MAPK, phosphatidylinositol 3-kinase (PI3K)/protein kinase Akt, Src/focal adhesion kinase (FAK), and NKX2.5 pathway, which play critical roles in cardiogenesis (Fig. 10.4). The ErbB3 lacks a functional tyrosine kinase domain; however, similar to ErbB2, it acts as a co-receptor and heterodimerizes with other ErbBs.

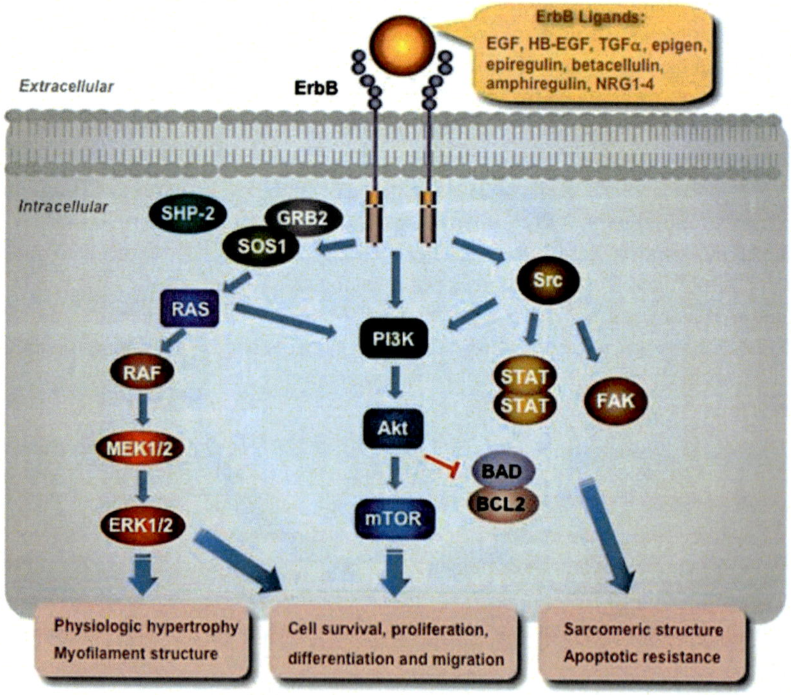


**Fig. 10.4** ErbB signaling in cardiac cells. Engagement of ErbB with specific ligands (shown in *yellow box*) leads to receptor homo- or heterodimerization and subsequent activation of various signaling cascades such as RAS/MAPK, PI3K/Akt, Src/FAK, and others. This complex signaling network plays critical roles in the heart development and function

The ErbB receptor signaling is mediated by various effector proteins containing SH2 domains. They directly interact with phosphorylated tyrosine residues in the intracellular region of activated ErbB receptors. The presence of multiple tyrosines, which can be phosphorylated, as well as the involvement of adaptor SH2-containing proteins in this pathway explains a wide range of effector proteins activating by the ErbB signaling [124–126].

A notable example of defects in SH2-mediated interactions, mutant protein tyrosine phosphatase SHP-2, associated with Noonan syndrome, has been discussed above. Another signal transduction pathway induced by the engaged ErbB receptor is PI3K/Akt signaling. SH2 domain of PI3K mediates its interaction with the ErbB receptor and its recruitment to the plasma membrane, where it generates $PIP_3$. $PIP_3$ in turn recruits Akt to the membrane to be phosphorylated and activated by phosphoinositide-dependent kinase-1 (PDK1) [127]. NRG 1 engagement of ErbB4 promotes PI3K-mediated activation of Akt pathway essential for cardiomyocyte survival. Interestingly, although PI3K can also be activated by ErbB2, this route is not required for Akt activation [128]. Moreover, emerging evidence suggests that Akt signaling play a central role in endocardial–mesenchymal transition, a crucial process for valve development [129].

Loss of ErbB2, ErbB3, ErbB4, or NRG 1 in mice results in early embryonic lethality due to severe heart defects including the lack of ventricular trabeculation and disruption of endocardial cushion mesenchyme formation [23, 130–135]. In addition to a key role in heart development, ErbB2 is also essential for the development of neuromuscular junctions [136]. Mice deficient in ErbB ligand, HB-EGF, have myxomatous enlarged valves, and loss of a membrane-bound metalloprotease ADAM17/TACE, which cleaves pro-HB-EGF to generate mature HB-EGF and results in similar phenotypes as HB-EGF knockout [135]. These mouse models support the essential roles of the ErbB signaling in cardiogenesis, especially in cardiac valve development.

## NODAL Signaling Pathway

The NODAL signaling pathway plays a key role in the left–right (L–R) axis specification [137]. NODAL is a member of TGF-β family of developmental regulators. NODAL, LEFTY1/LEFTY2, and GDF1 act as ligands for a membrane multisubunit receptor composed by ACVR1B/ACVR2A/ACVR2B, CFC1 and TDGF1, which controls through transcription factors, such as FOXH1 and ZIC3, cardiac-specific effector genes. The NODAL signal transduction pathway represents a paradigm of multifactorial inheritance of CHD. Mutations in the *NODAL, CFC1*, *ACVR2B*, *LEFTY2*, *GDF1*, *FOXH1*, and *TDGF1* genes, encoding different components of the NODAL pathway, have been identified in patients with heterotaxy and looping defects, such as TGA and DORV [138–145].

A NODAL mutation (G260R) has been identified in approximately 5% of patients with heterotaxy or heterotaxy-related CHD (TGA and DORV) [144]. Interestingly, this mutation has also been found in a control and in one unaffected parent and therefore, has been considered as a rare variant with low penetrance. *Nodal*$^{-/-}$ mice die before the establishment of the L–R axis, have defects in the mesoderm formation, and lack the primitive streak [146, 147].

The *EGF-CFC* gene family encodes a group of structurally related extracellular proteins that play critical roles in intercellular signaling during embryogenesis in *Xenopus*, zebrafish, mice, and humans. They are essential for the formation of mesoderm and endoderm, the establishment of the anterior–posterior, and the L–R axis [148, 149]. EGF-CFC proteins (CRYPTIC, CRIPTO, and Frl1) function as membrane-bound NODAL co-receptors [137, 150].

Transgenic mice harboring homozygous mutations in the *EGF-CFC* gene, *Cfc1*, develop visceral laterality defects and complex cardiac malformations analogous to human heterotaxy. In particular, homozygous mutant mice frequently have malposition of the great arteries, such as TGA and DORV [151]. Loss-of-function mutations in human *CFC1*, encoding the CRYPTIC protein, have been identified in patients with heterotaxic phenotypes characterized by randomized organ positioning [140]. Analysis of the *CFC1* gene in patients with DORV and TGA has identified three *CFC1* mutations in 86 patients [141]. One of these mutations, a single base pair deletion (G174 del1), had previously been reported in two patients with heterotaxy and has been predicted to result in a functionally different protein [140]. Immunohistochemical analysis has confirmed that prediction: while the wild-type protein is retained on the cell surface, the G174del1 mutant protein is not localized on the cell surface, but accumulates intracellularly. Finally, a missense alteration (R78W) in *CFC1* has been recently reported in patients with laterality defects and AVSDs [152, 153]. Given that the altered amino acid is not conserved among different species its biological significance presently remains unclear.

Other genes of the NODAL pathway with defects, associated with CHD, include *ACVR2B* and homologous *ACVR1*, encoding activin A receptors, members of the TGF-β family [139, 154]; *LEFTY1* and *2*, encoding L–R determination factor 1 and 2, TGF-β-related proteins, which function as ligands in the NODAL signaling pathway [138, 155]; *GDF1*, encoding a growth differentiation factor, belonging to the TGF-β family (ACVR2 ligand) [142]; *FOXH1*, encoding a forkhead transcription factor [143]; *TDGF1*, encoding an ECF-CFC family protein, CRIPTO, that functions as a

NODAL co-receptor [143]; and *ZIC3*, encoding a Zn-finger transcription factor, which functions as an enhancer of the NODAL signaling pathway [156]. However, the precise roles of these components of the NODAL signal transduction pathway in the development of CHD remain to be determined.

### *Marfan and Marfan-Like Syndromes*

Marfan syndrome is an autosomal dominant hereditary disorder characterized by implication of three main systems: skeletal, cardiovascular, and ocular [157–159]. In untreated patients with Marfan syndrome, the leading cause of premature death is acute aortic dissection, followed by a period of progressive dilatation of the ascending aorta.

Mutations in the *FBN1* gene encoding fibrillin-1 are responsible for this disease [160, 161]. *FBN1* is a large (230 kb) gene, which contains 65 coding exons [162]. To date, more than 600 mutations distributed in all exons have been reported (http://www.umd.be). Subsequently, mutations in the gene encoding fibrillin-2 (*FBN2*) have been identified in individuals with a phenotypically related disease, congenital contractural arachnodactyly [163].

The fibrillins are multidomain proteins with both structural and nonstructural functions in the extracellular matrix. They form extracellular microfibrils and also contribute to the activation of cytokines, such as bone morphogenetic protein 7 (BMP7) and TGF-β [164–167]. Thus, the fibrillins link the extracellular matrix interactions to TGF-β signaling pathway.

The molecular mechanism underlying Marfan syndrome has initially been attributed to structural defects in the extracellular fibrillin-rich microfibrils. However, growing evidence suggests that abnormalities in TGF-β signaling contribute to the pathogenesis of this disorder. Subsequent discovery of mutations in the genes encoding the TGF-β receptor type I and II (*TGFBR1* and *TGFBR2*) in the related Loeys–Dietz aortic aneurysm and Marfan-like syndromes, respectively, has confirmed a direct link between the extracellular matrix and TGF-β signaling abnormalities [168, 169]. Intriguingly, in contrast to *in vitro* experiments showing loss-of-function of the mutant TGF-β receptors, mouse models as well as Marfan syndrome and LDAS patients demonstrate paradoxical upregulation of TGF-β signaling, rather than repression [168–172]. While no strict genotype–phenotype correlations for *TGFBR* mutations have been reported, recent data suggest the possibility of distinct biochemical effects associated with different *TGFBR2* mutations [173, 174]. Further studies are required to identify more germline mutations in *TGFBR1* and *TGFBR2* and to elucidate the role of TGF-β signaling in the cardiac extracellular matrix.

## CHD-Causing Genes with Elusive Molecular Mechanism(s)

Defects in a number of genes have been reported to contribute to various forms of CHD; however, corresponding molecular mechanisms underlying these cardiovascular abnormalities are currently less characterized.

### *CRELD1 Gene Mutations*

*CRELD1* encodes a cell surface protein containing $Ca^{2+}$-binding (CB) EGF-like domains that likely functions as a cell adhesion molecule. Three missense *CRELD1* mutations have been detected in 50 unrelated individuals with AVSD [175]. One *CRELD1* mutation has also been associated with heterotaxy with dextrocardia, pulmonary atresia, and a right aortic arch. Interestingly, two of these mutations are localized to EGF-like domains, a type of domain present in various extracellular proteins. Heterozygous mutations affecting CB-EGF domains have been associated with a number of cardiovascular genetic disorders, including Marfan syndrome, hemophilia B, familial hypercholesterolemia, and heterotaxy. A C4201T nucleotide change results in a substitution of cysteine for a highly conserved arginine (R329C) in the second CB-EGF domain. Of note, the same mutation has recently been found among patients with Down syndrome and complete AVSD, supporting further the role for *CRELD1* as an AVSD susceptibility gene that likely acts in concert with additional modifier genes and/or environmental factors [176]. Whether these changes affect cell–cell signaling or calcium signaling in the developing heart remains presently unclear.

### *FLNA Gene Mutations*

Emerging evidence suggests that filamins, large cytoplasmic actin-binding proteins, forming delicate three-dimensional filament structures and integrating architectural and signaling functions, are implicated in organogenesis, including cardiovascular development. Filamins interact with a broad range (>70) of cellular proteins including, in addition to F-actin, transmembrane receptors, signaling molecules, and transcription factors, providing a highly dynamic scaffold, and integrating diverse cellular functions [177, 178]. Three members of the mammalian filamin family, filamin A, B, and C (human FLNA, FLNB, and FLNC), are encoded by three highly conserved genes [179]. FLNA appears to be the major filamin implicated in cardiovascular development.

Mutations in the *FLNA* gene in patients with periventricular heterotopias, a neuronal migration disorder, have also been associated with CHD, such as PDA and aortic aneurysms [180, 181]. Moreover, specific *FLNA* mutations have recently been linked to familial cardiac valvular dystrophy, characterized by mitral valve prolapsed and mitral and aortic regurgitation [182].

Loss of *Flna* in mice results in embryonic lethality due to severe cardiac structural anomalies involving ventricles, atria and outflow tracts as well as widespread aberrant vascular patterning [183]. Mice deficient in *Flnb* or *Flnc* appear to have no cardiac phenotypes but display severe defects in skeletal or muscle development, respectively [184–186]. Thus, murine models support a critical role for FLNA in cardiovascular development. However, the mechanism, by which mutations in filamins result in cardiovascular abnormalities, remains to be determined.

### Mutations in Other Genes

Mutations in *ANKRD1*, which encodes a transcription regulator belonging to the muscle ankyrin repeat protein family, have been identified in two sporadic cases displaying total anomalous pulmonary venous return [187].

In addition to *MYH6*, mutations in other genes that encode cardiac sarcomeric proteins have been reported to be associated with CHD. They include the β-myosin heavy chain gene *MYH7* and myosin heavy chain 11 gene *MYH11* [188, 189]; the myosin-binding protein C gene *MYBPC3* [190, 191]; and the cardiac actin gene *ACTC1* [192–194]. Although currently it is unclear how defects in cardiac contractile proteins can lead to CHD, changes in hemodynamics may play an important role in this process.

## CHD Associated with MicroRNA Dysregulation

In the past decade, extensive research has focused on small (21-nucleotide) noncoding RNAs referred to microRNAs (miRNAs) [195, 196]. Mature miRNAs are generated from 70-nucleotide precursor miRNAs. miRNAs control gene expression by binding to target messenger RNAs, leading to either suppression of their translation or to their degradation [197]. Although the specific functions of miRNAs are poorly understood, they appear to play important roles in mammalian development including heart development during embryogenesis [198, 199].

miR-1, encoded by two distinct genes *miR-1-1* and *miR-1-2*, has been reported to be essential in embryonic cardiogenesis [200, 201]. Both genes are expressed in the developing heart and are controlled by serum response factor. miR-1 targets the cardiac transcription factor HAND2 that contributes to the control of cardiac growth during embryogenesis. Importantly, *miR-1-2*-deficient mice have heart abnormalities including septal defects [201]. Thus, dysfunction of miRNAs, essential for heart development, might represent a novel molecular mechanism underlying CHD and will likely be the subject of future studies.

### Gene Expression Profiling in CHD

Gene expression profiling using DNA microarrays is a novel powerful tool to study the highly complex nature of CHD. Recently, myocardial gene profiling has been carried out in cardiomyocytes derived from individuals with HLHS, which accounts for nearly 25% of cardiac deaths in the first year of life. Transcript analysis, conducted by quantitative PCR, has been combined with proteomic analysis, performed by two-dimensional gel electrophoresis followed by mass spectroscopy [202]. This study has demonstrated that both the left and right ventricles display the fetal or "heart failure" expression pattern of some cardiac-specific proteins. Importantly, HLHS cardiomyocytes inappropriately express the platelet-endothelial cell adhesion molecule-1 (PECAM-1), a member of the cell adhesion molecule family that has a primary role in the regulation of tissue morphogenesis. Given that cardiomyocytes have never been demonstrated to express PECAM-1 at any stage of development or in any cardiac disease, these findings suggest that PECAM-1 or a related gene under the same regulatory control may be responsible for HLHS.

In another study, distinct gene expression profiles have been identified in a variety of CHD, including TOF, VSDs, and right ventricular hypertrophy (RVH) [203]. The cases with TOF have displayed differential expression of genes involved in cell cycle and cardiogenesis (upregulation of *SNIP*, *A2BP1*, and *KIAA1437* and downregulation of *STK33*, *BRDG1*, and *TEKT2*), and exhibited upregulation of ribosomal genes (*S6*, *L37a*, *S3A*, *S14*, and *L13A*). The RVH group displayed expression profile of genes primarily involved in stress response, cell proliferation, and metabolism and included the upregulation of *ADD2*. VSD cases exhibited a specific signature consisting of marked downregulation (primarily in the right atrium) of genes encoding ribosomal proteins *S11*, *L18A*, *L36*, *LP0*, *L31*, and *MRPS7*, genes involved in cell proliferation, differentiation, and apoptosis (e.g., *AMD1*, *RIPK3*, *EGLN1*, *SIAHBP1*, and *ARVCF*). Several ion channel genes, including *SLC26A8*, *SLC16A5*, *SLC4A7*, *KCNS2*, and *KCNN3*, have also been found to be differentially expressed in patients with VSD.

A large-scale transcriptional analysis of the left ventricle (LV) and right ventricle (RV) in patients with CHD has demonstrated significant differences in the expression of 70 genes [204]. Expression of genes encoding angiotensin, adrenergic receptors, G proteins, cytoskeletal and contractile proteins has been lower, and expression of maladaptive factors, such as FGF, TGF-β, caspases, ubiquitin, has been higher in the RV than in the LV. The authors have suggested that downregulation in the transcription of adaptive remodeling factors detected in the RV compared with the LV may be responsible for the lower ability of the RV to adapt to hemodynamic load in CHD.

Recently, microarray analysis containing 1,421 miRNAs have been performed to examine miRNAs in patients with aortic stenosis (AS) and aortic insufficiency (AI) [205]. Differences in seven miRNAs have been reported between the AS and AI patients. They include significant reduction in the miR-26a, miR-30b, and miR-195 levels in patient with AS. Interestingly, these miRNAs can modulate calcification-related genes and may be implicated in BAV, the most common form of CHD.

Two recent gene expression analyses have been carried out using mouse models of CHD. A subset of genes has been shown to be differentially regulated by Nkx2-5, a transcription factor implicated in human CHD, during outflow tract morphogenesis [206]. One of these genes, the transcriptional regulator *Jarid2*, appears to be a direct downstream target of Nkx2-5 contributing to outflow tract morphogenesis. In another study, gene chip analysis has revealed a Zn-finger transcription factor Zac1 as an additional player in the Nkx2-5 pathway [207]. Zac1 is strongly expressed in the developing heart. Zac1 binds Nkx2-5 through Zn-fingers 5 and 6 in Zac1 and the homeodomain in Nkx2-5 and is directly associated with the ANF promoter. Importantly, mice harboring an interruption in the *Zac1* locus display ASD and VSD as well as thin ventricular walls supporting a role for Zac1 in the development of CHD.

Taken together, gene expression profiling, allowing to examine the expression of thousands of genes simultaneously, holds promise for identifying new candidate genes implicated in various forms of CHD; however, the significance of these findings have to be further carefully evaluated.

## Conclusions and Future Progress

In the past two decades, tremendous advances have been made in fetal echocardiography and angiography, pediatric cardiac surgery techniques and more complete knowledge of newborn physiology and pathophysiology. The Congenital Surgery Database has recently been created to allow the computerized outcomes analysis for over 40,000 CHD cases [208, 209]. Development of new surgical techniques and miniaturized electronic devises allows their successful use in neonates for total correction of many CHD. All these advances have resulted in an estimated million adults living with CHD, which required surgical treatment in the neonatal period [210].

Despite significant improvements in the diagnosis and surgical treatment of infants born with CHD, insights into the molecular mechanisms of these developmental disorders just have begun to be gained. The clinical heterogeneity of various forms of CHD results from genetic defects in a diverse set of molecular targets. Affected factors have been localized in multiple subcellular compartments including the cytoplasm, nucleus, plasma membrane, endoplasmic reticulum, mitochondria, and lysosomes. In addition, these molecules (transcription factors and signaling proteins, receptors and ion channels) frequently play multiple roles in the intricate network of cellular pathways contributed to the response of the developing heart to external stimuli. The extensive communications between numerous players of this network have made the unraveling of its nature more difficult than it had previously been anticipated.

Embryonic heart development is driven by a highly regulated cascade of ligand–receptor interactions, extensively cross-talking downstream signal transduction regulators and numerous transcription factors [7] (Fig. 10.5). This intricate network tightly controls the expression of cardiac-specific genes to ensure that the correct cell lineages proliferate and differentiate in the correct location and at the correct time. Defects in components of this network are associated with various forms of CHD.

A wide range of extracellular ligands, such as hormones, growth factors, and cytokines, activates membrane receptors located on the cell surface of the developing cardiac cells. These ligand–receptor interactions comprise EGF-RTKs, JAGGED-NOTCH, TGF-β/BMP-TGFBR, NODAL/LEFTYs/GDF1-ACVRs/CFC1/TDGF1, and VEGF-FLK1/FLT1. Activated receptors trigger various signaling pathways, including RAS/MAPK, PI3K/Akt, Src/FAK, and cJUN pathways, which converge on a complex network of transcription factors and regulators (Figs. 10.4 and 10.5). They include the GATA family transcription factors (GATA4 and FOG2), homeobox transcription factors (NKX2-5 and NKX26), T-box transcription factors (TBX1, TBX5, and TBX20), nuclear factor of activated T cells (NFAT), basic helix-loop-helix (bHLH) transcription factors (HAND1 and HAND2), and SMAD transcription factors. These transcription factors govern the expression of a variety of cardiac genes. Only a few of them, such as the cardiac actin gene *ACTC1*, α-myosin heavy chain gene *MYH6*, atrial natriuretic factor gene *ANF*, and β-type natriuretic peptide gene *BNP*, are currently known. One of the major challenges is to identify specific cardiac effectors downstream of transcription factors to decipher the molecular mechanisms underlying various forms of CHD.

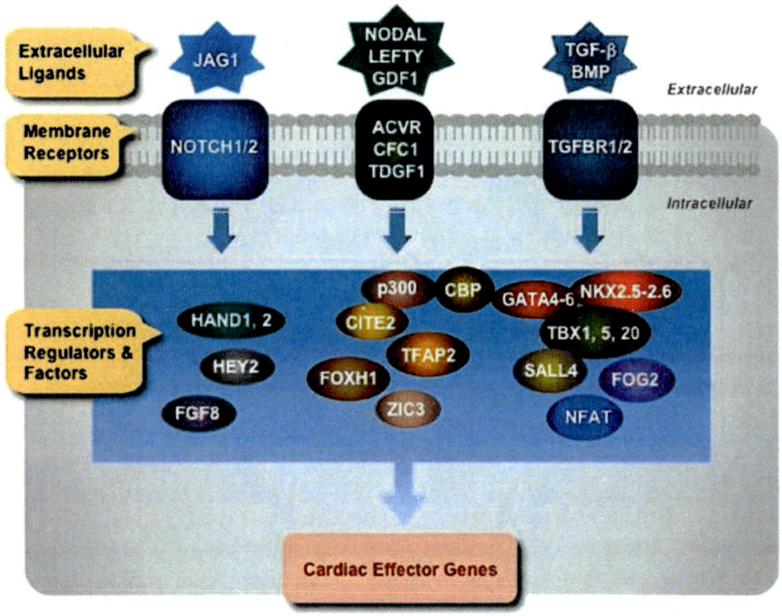


**Fig. 10.5** Scheme of the intricate network of extracellular ligands, membrane receptors, and interacting transcription factors involved in the control of heart development. Major players implicated in various forms of CHD are shown. Extensive interplay between the different signaling cascades exists. See text for the details

While the critical roles of transcriptional regulators in CHD are well appreciated, the role of factors, implicated in the modification of chromatin structure, is less understood. For example, deficiency in BAF60C, a subunit of Swi/Snf-like chromatin-remodeling complex, leads to severe impairment of heart development [211]. Other chromatin-remodeling proteins, which modify histones, also contribute to cardiac morphogenesis. It has recently been reported that cardiac histone deacetylases 1 and 2 regulate cardiac morphogenesis, growth, and contractility [212]. Elucidation of the precise role of various chromatin-remodeling factors in the development of CHD deserves intense investigation.

Recent breakthroughs in molecular genetics, developmental biology and cardiology, applied to studies of CHD, have resulted in the identification of multiple genes as the primary causes as well as risk factors in the development of these anomalies. Significant improvements in genotyping and gene expression profiling techniques as well as better understanding of genetic variations provide the basis for genome-wide studies to identify susceptibility genes for CHD. Identification of novel genes involved in cardiovascular development as well as interactions between known genes will serve as an important foundation for our understanding how specific gene defects translate into cardiac malformations.

Functional analysis of recombinant proteins modified at the sites mutated in CHD complements genetic studies and will provide further insights into the molecular basis of these multifaceted disorders. Generation of recombinant proteins with deletion or modification of specific functional domains will also be a very helpful tool to gain insights into the precise roles of multifunctional proteins. Such approach is particularly useful to study proteins performing their functions through numerous interactions with diverse cellular factors (e.g., NKX, TBX, and filamins).

Mouse models harboring mutations in genes linked to CHD have often been proved to accurately mimic the cardiac defects found in patients with CHD. Genetically modified mouse models with mutations in specific functional domains of the candidate genes will be particularly informative in elucidating their precise roles in the development of CHD. However, differences in mouse and human heart physiology have to be carefully considered in these studies. For example, significant differences in sensitivity to gene dosage of *NKX2-5*, *TBX5*, and *TBX20* between humans and mice, which lead to differing manifestations of CHD, have been reported [90, 213–215]. Nevertheless, animal models for cardiovascular disease will continue to provide invaluable information and will undoubtedly lead to identifying novel genes and mechanisms underlying various forms of CHD.

In the postgenomic era, bioinformatics is becoming a very powerful tool in exploring CHD. Various computerized methods are employed to search the growing list of databases [216]. This approach combined with the power of reverse genetics and subsequent cloning of novel genes/cDNAs of interest, followed by the characterization of spatial–temporal patterns of specific gene expression in the developing embryo will continue to be a research priority.

The related fields of pharmacogenomics and pharmacogenetics hold the promise of improved drug development, and the tailoring of therapies based on the individual's factors specifying the transport, metabolism and targeting of the drugs. For example, a subset of single-nucleotide polymorphisms (SNPs) identified in human genes, encoding β-adrenoceptor (β-AR) and angiotensin-converting enzyme, have been associated with significant changes in the clinical response to medications used in the treatment of CVD [217]. Individualized therapy may be particularly critical in the calculation of drug dosages and efficacies in children with CHD, a population for which pharmacokinetics has proven to be poorly defined and often unpredictable. Further research in this area will reveal potential targets for highly specific pharmacological intervention.

Insight into the cardiovascular consequences of abnormal gene expression and function should ultimately impact on the development of targeted therapeutic strategies and disease management for children with inherited and acquired heart disorders, and may replace less effective treatment modalities directed solely at rectifying structural cardiac defects and temporal improvement of function. As our knowledge of the genes that are responsible for both early and late cardiac abnormalities increases, we may be able to modify the effects of mutated genes by modifying the environment in which genes act. Evaluation of CHD at the molecular level will allow a more effective stratification of patients, leading to the optimization of patient-specific therapy. Hopefully, new drugs will be generated to alter or prevent the adverse effect of mutant genes, and heart transplantation will be replaced by cell-based therapy. Bioengineered valves will be developed to replace mechanical valves for the treatment of cardiac malformations in children. Valves generated from the patient's own stem or progenitor cells will grow as the patient grows, and, therefore, will not need to be replaced. Thus, new more efficient and successful therapeutic strategies to treat and/or prevent CHD are on the horizon.

## Summary

- CHD refers to defects in the cardiac structure and/or function, which arise in the prenatal period. CHD occurs often and represents the most common type of birth defect, affecting 4–10 of every 1,000 live births. Abnormalities causing CHD can affect almost all parts of the heart and can be broadly divided into three types: cyanotic heart disease, left-sided defects, and septation defects.
- Many CHD with monogenic inheritance are associated with multiple noncardiac birth defects constituting syndromic forms of CHD. However, most CHD occur sporadically representing nonsyndromic CHD. A complex interplay of numerous genetic and environmental risk factors results in the development of CHD.
- With the impressive progress in molecular genetics, developmental biology, and molecular cardiology, over 40 different genes responsible for inherited and sporadic CHD have been identified. The majority of these genes encodes transcription and signal transduction factors, which tightly control specific events in cardiogenesis.
- The etiology of one subgroup of CHD that include Noonan (NS), LEOPARD, Costello, and cardio-facio-cutaneous (CFC) syndromes has been shown to be caused by dysregulation of the RAS/mitogen-activated protein kinase (MAPK) pathway. Affected individuals have a high incidence of heart disease (e.g., 80–90% in NS) including pulmonic stenosis, hypertrophic cardiomyopathy, AVSDs, and coarctation of the aorta.
- Currently, mutations in four distinct genes have been identified to be responsible for NS, *PTPN11*, *KRAS*, *SOS1*, and *RAF1*. Mutations in *PTPN11*, encoding a nonreceptor protein tyrosine phosphatase SHP-2, are found in almost 50% of patients with NS and are predominant among familial cases and among NS patients displaying pulmonary valve stenosis. These mutations impair the enzyme transition from the catalytically active to inactive form leading to the activation of the RAS/MAPK signaling.
- LEOPARD syndrome, which shares many phenotypic features with NS, is caused by specific *PTPN11* mutations, which unlike NS are loss-of-function mutations and have a dominant negative effect. Mutations in *RAF1* and *BRAF* have also been identified in some patients with this disorder.
- The phenotypically related Costello and CFC syndromes are the most severe of this subgroup of CHD. While no *PTPN11* mutations have been detected in patients with these conditions, they are also characterized by dysregulated RAS/MAPK signaling pathway. Over 83% of Costello syndrome cases have gain-of-function missense mutations in the *HRAS* proto-oncogene. Mutations in four genes encoding proteins in the RAS/MAPK signaling pathway are responsible for CFC syndrome: *BRAF*, *MEK1 (MAP2K1)*, *MEK2* (*MAP2K2*), and *KRAS*.
- The highly conserved NOTCH signaling represents another notable example of signaling pathway, which impairment can lead to valve and outflow tract defects and vascular abnormalities. Patients with *NOTCH1* haploinsufficiency demonstrate predominantly structural defects of the aortic valves, such as BAV, the most common form of CHD. In addition, they also have aortic stenosis, VSDs, TOF, and calcification of aortic valves.
- Alagille syndrome is also caused by defective NOTCH signaling pathway. More than 90% of affected patients have cardiovascular abnormalities, including peripheral pulmonary hypoplasia, TOF, and pulmonary valve stenosis. Alagille syndrome is caused by mutations in the *JAG1* gene, which encodes the NOTCH ligand jagged 1 (JAG1).

*JAG1* mutations have also been found in nonsyndromic cases of CHD. In addition, *NOTCH2* mutations have been identified in two families with Alagille syndrome.

- The 22q11.2 microdeletion causes DiGeorge, velo-cardio-facial, and conotruncal-face syndromes. The most common cardiovascular abnormalities include TOF, PTA, conoventricular VSDs, and IAA. The *TBX1* gene, encoding a T-box family transcription factor, accounts for much of the human phenotype. In addition to microdeletion, the 22q11.2 microduplication occurs with a frequency half of that of microdeletion.
- The most common cardiovascular defects associated with Williams syndrome include supravalvar aortic stenosis and peripheral pulmonary stenosis. Molecular analysis has revealed that the deletion of one copy of the *ELN* gene, encoding elastin, a component of elastic fibers, is responsible for this disease.
- Cardiovascular defects in patients with CHARGE syndrome are variable and include conotruncal and left-sided obstructive anomalies, although PDA and septal defects are observed in some patients. Mutations in the *CHD7* gene, encoding the chromodomain helicase DNA-binding protein 7, found in approximately 70% of affected individuals, cause this disorder.
- Approximately 56% of patients with Jacobsen syndrome, associated with the 11q terminal deletion, display conotruncal defects including HLHS, DORV, and TGA. Several candidate genes, mutations in which may be responsible for heart anomalies associated with this disorder, have been suggested. They include *JAM3*, encoding a junction adhesion protein, and *ETS-1*, which encodes ETS1, a helix-turn-helix transcription factor, implicated in vascular development and angiogenesis.
- Holt–Oram syndrome is an autosomal dominant syndrome characterized by secundum ASDs and VSDs in patient with upper limb deformities. This syndrome is caused by mutations in the *TBX5* gene leading to haploinsufficiency. *TBX5* encodes the T-box transcription factor that is a critical regulator of gene expression during embryonic development.
- TBX5 interacts physically with two other transcription factors, the NK2 family transcription factor NKX2-5 and Zn-finger transcription factor GATA4, to activate its downstream targets. The importance of TBX5–NKX2.5–GATA4 interactions has been highlighted by identifying mutations in *NKX2-5* and *GATA4* in patients with ASD and VSD as well as with other CHD. Recent studies have identified mutations in other members of the TBX–NKX2–GATA axis, including *TBX20*, *NKX2.6*, and *GATA6*, associated with CHD.
- Phenotypically related Okihiro and Townes–Brocks syndromes are characterized by VSD, ASD, and TOF. Mutations in the *SALL* family genes, *SALL1* and *SALL4*, are responsible for the phenotypes of Townes–Brocks syndrome and Okihiro syndrome, respectively. The *SALL* genes encode Zn-finger transcription factors. In mice, SALL4 interacts physically with TBX5, and this interaction is essential for the heart and limb development.
- Defects in the *CITED2* gene are associated with cardiac septal abnormalities. CITE2 is a transcription factor, which upon interaction with p300 and CBP, coactivates the neural crest-related TFAP2 transcription factor. Mutations in genes encoding p300 and CBP cause Rubinstein–Taybi syndrome characterized by PDA and septation defects, while mutant *TFAP2B* causes Char syndrome characterized mainly by PDA.
- The ErbB family of receptor tyrosine kinases is a highly conserved family of transmembrane glycoproteins that transduce extracellular stimuli to the cell nucleus to promote differentiation, proliferation, and migration of cells essential for heart development. Activation of ErbB recruits a variety of signaling proteins involved in various signaling cascades, such as RAS/MAPK, PI3K/AKT, Src/FAK, and NKX2-5 pathway, which play critical roles in cardiogenesis. Loss of ErbB members in mice results in early embryonic lethality due to severe heart defects including the lack of ventricular trabeculation and disruption of endocardial cushion mesenchyme formation.
- The NODAL signaling pathway plays a key role in the L–R axis specification. NODAL, LEFTY1/LEFTY2, and GDF1 act as ligands for a membrane multisubunit receptor composed by ACVR1B/ACVR2A/ACVR2B, CFC1 and TDGF1, which controls through transcription factors, such as FOXH1 and ZIC3, cardiac-specific effector genes. The NODAL signal transduction pathway represents a paradigm of multifactorial inheritance of CHD. Mutations in the *NODAL*, *CFC1*, *ACVR2B*, *LEFTY2*, *GDF1*, *FOXH1*, and *TDGF1* genes have been identified in patients with heterotaxy and looping defects such as TGA and DORV.
- Defects in a number of less characterized genes have been reported to be associated with various forms of CHD. They include *CRELD1* encoding a cell surface protein containing $Ca^{2+}$-binding EGF-like domains that likely functions as a cell adhesion molecule; *FLNA* encoding a large cytoplasmic actin-binding protein filamin A; *ANKRD1*, which encodes a transcription regulator belonging to the muscle ankyrin repeat protein family; several genes that encode cardiac sarcomeric proteins, *ACTC1*, *MYH6*, *MYH7*, and *MYH11*.
- miRNAs are essential for heart development, and their dysfunction might represent a novel molecular mechanism underlying CHD. Differences in seven miRNAs have been reported between AS and AI patients. They include significant reduction in the miR-26a, miR-30b, and miR-195 levels in patient with AS. These miRNAs

can modulate calcification-related genes and may be implicated in BAV, the most common form of CHD.

- Despite significant improvements in the diagnosis and surgical treatment of infants born with CHD, insights into the molecular mechanisms of these developmental disorders just have begun to be gained. The clinical heterogeneity of various forms of CHD results from genetic defects in a diverse set of molecular targets localized in multiple subcellular compartments. The numerous extensive communications between different players (transcription factors and signaling proteins, receptors and ion channels) of this network have made the unraveling of its nature more difficult than it had previously been anticipated.

## References


1. Pierpont ME, Basson CT, Benson Jr DW, et al. Genetic basis for congenital heart defects: current knowledge: a scientific statement from the American Heart Association Congenital Cardiac Defects Committee, Council on Cardiovascular Disease in the Young: endorsed by the American Academy of Pediatrics. Circulation. 2007;115:3015–38.
2. Association AH. Heart disease and stroke statistics – 2005 update. Dallas, TX: American Heart Association; 2005.
3. Bruneau BG. The developmental genetics of congenital heart disease. Nature. 2008;451:943–948.
4. Weismann CG, Gelb BD. The genetics of congenital heart disease: a review of recent developments. Curr Opin Cardiol. 2007;22: 200–6.
5. Wolf M, Basson CT. The molecular genetics of congenital heart disease: a review of recent development. Curr Opin Cardiol. 2010; 25:192–7.
6. Garg V. Insights into the genetic basis of congenital heart disease. Cell Mol Life Sci. 2006;63:1141–8.
7. Wessels MW, Willems PJ. Genetic factors in non-syndromic congenital heart malformations. Clin Genet. 2010;78:103–23.
8. Bruneau BG. The developmental genetics of congenital heart disease. Nature. 2008;451:943–8.
9. Cooper WO, Hernandez-Diaz S, Arbogast PG, et al. Major congenital malformations after first-trimester exposure to ACE inhibitors. N Engl J Med. 2006;354:2443–51.
10. Jenkins KJ, Correa A, Feinstein JA, et al. Noninherited risk factors and congenital cardiovascular defects: current knowledge: a scientific statement from the American Heart Association Council on Cardiovascular Disease in the Young: endorsed by the American Academy of Pediatrics. Circulation. 2007;115:2995–3014.
11. Srivastava D. Genetic regulation of cardiogenesis and congenital heart disease. Annu Rev Pathol. 2006;1:199–213.
12. Tidyman WE, Rauen KA. The RASopathies: developmental syndromes of Ras/MAPK pathway dysregulation. Curr Opin Genet Dev. 2009;19:230–6.
13. Noonan JA. Noonan syndrome. An update and review for the primary pediatrician. Clin Pediatr (Phila). 1994;33:548–55.
14. Marino B, Digilio MC, Toscano A, Giannotti A, Dallapiccola B. Congenital heart diseases in children with Noonan syndrome: an expanded cardiac spectrum with high prevalence of atrioventricular canal. J Pediatr. 1999;135:703–6.
15. Jamieson CR, van der Burgt I, Brady AF, et al. Mapping a gene for Noonan syndrome to the long arm of chromosome 12. Nat Genet. 1994;8:357–60.
16. Tartaglia M, Mehler EL, Goldberg R, et al. Mutations in PTPN11, encoding the protein tyrosine phosphatase SHP-2, cause Noonan syndrome. Nat Genet. 2001;29:465–8.
17. Schubbert S, Zenker M, Rowe SL, et al. Germline KRAS mutations cause Noonan syndrome. Nat Genet. 2006;38:331–6.
18. Roberts AE, Araki T, Swanson KD, et al. Germline gain-of-function mutations in SOS1 cause Noonan syndrome. Nat Genet. 2007;39:70–4.
19. Tartaglia M, Pennacchio LA, Zhao C, et al. Gain-of-function SOS1 mutations cause a distinctive form of Noonan syndrome. Nat Genet. 2007;39:75–9.
20. Pandit B, Sarkozy A, Pennacchio LA, et al. Gain-of-function RAF1 mutations cause Noonan and LEOPARD syndromes with hypertrophic cardiomyopathy. Nat Genet. 2007;39:1007–12.
21. Razzaque MA, Nishizawa T, Komoike Y, et al. Germline gain-of-function mutations in RAF1 cause Noonan syndrome. Nat Genet. 2007;39:1013–7.
22. Feng GS. Shp-2 tyrosine phosphatase: signaling one cell or many. Exp Cell Res. 1999;253:47–54.
23. Chen B, Bronson RT, Klaman LD, et al. Mice mutant for Egfr and Shp2 have defective cardiac semilunar valvulogenesis. Nat Genet. 2000;24:296–9.
24. Tartaglia M, Kalidas K, Shaw A, et al. PTPN11 mutations in Noonan syndrome: molecular spectrum, genotype-phenotype correlation, and phenotypic heterogeneity. Am J Hum Genet. 2002;70:1555–63.
25. Hof P, Pluskey S, Dhe-Paganon S, Eck MJ, Shoelson SE. Crystal structure of the tyrosine phosphatase SHP-2. Cell. 1998;92: 441–50.
26. Keilhack H, David FS, McGregor M, Cantley LC, Neel BG. Diverse biochemical properties of Shp2 mutants. Implications for disease phenotypes. J Biol Chem. 2005;280:30984–93.
27. Niihori T, Aoki Y, Ohashi H, et al. Functional analysis of PTPN11/SHP-2 mutants identified in Noonan syndrome and childhood leukemia. J Hum Genet. 2005;50:192–202.
28. Tartaglia M, Martinelli S, Stella L, et al. Diversity and functional consequences of germline and somatic PTPN11 mutations in human disease. Am J Hum Genet. 2006;78:279–90.
29. Schubbert S, Bollag G, Lyubynska N, et al. Biochemical and functional characterization of germ line KRAS mutations. Mol Cell Biol. 2007;27:7765–70.
30. Kontaridis MI, Swanson KD, David FS, Barford D, Neel BG. PTPN11 (Shp2) mutations in LEOPARD syndrome have dominant negative, not activating, effects. J Biol Chem. 2006;281: 6785–92.
31. Digilio MC, Conti E, Sarkozy A, et al. Grouping of multiple-lentigines/LEOPARD and Noonan syndromes on the PTPN11 gene. Am J Hum Genet. 2002;71:389–94.
32. Legius E, Schrander-Stumpel C, Schollen E, Pulles-Heintzberger C, Gewillig M, Fryns JP. PTPN11 mutations in LEOPARD syndrome. J Med Genet. 2002;39:571–4.
33. Sarkozy A, Carta C, Moretti S, et al. Germline BRAF mutations in Noonan, LEOPARD, and cardiofaciocutaneous syndromes: molecular diversity and associated phenotypic spectrum. Hum Mutat. 2009;30:695–702.
34. Sarkozy A, Digilio MC, Dallapiccola B. Leopard syndrome. Orphanet J Rare Dis. 2008;3:13.
35. Wright EM, Kerr B. RAS-MAPK pathway disorders: important causes of congenital heart disease, feeding difficulties, developmental delay and short stature. Arch Dis Child. 2010;95:724–30.
36. Aoki Y, Niihori T, Kawame H, et al. Germline mutations in HRAS proto-oncogene cause Costello syndrome. Nat Genet. 2005;37: 1038–40.
37. Gripp KW, Lin AE, Stabley DL, et al. HRAS mutation analysis in Costello syndrome: genotype and phenotype correlation. Am J Med Genet A. 2006;140:1–7.

38. Estep AL, Tidyman WE, Teitell MA, Cotter PD, Rauen KA. HRAS mutations in Costello syndrome: detection of constitutional activating mutations in codon 12 and 13 and loss of wild-type allele in malignancy. Am J Med Genet A. 2006;140:8–16.
39. Kerr B, Delrue MA, Sigaudy S, et al. Genotype-phenotype correlation in Costello syndrome: HRAS mutation analysis in 43 cases. J Med Genet. 2006;43:401–5.
40. Tidyman WE, Rauen KA. Noonan, Costello and cardio-facio-cutaneous syndromes: dysregulation of the Ras-MAPK pathway. Expert Rev Mol Med. 2008;10:e37.
41. Bos JL. ras oncogenes in human cancer: a review. Cancer Res. 1989;49:4682–9.
42. Rodriguez-Viciana P, Tetsu O, Tidyman WE, et al. Germline mutations in genes within the MAPK pathway cause cardio-facio-cutaneous syndrome. Science. 2006;311:1287–90.
43. Niihori T, Aoki Y, Narumi Y, et al. Germline KRAS and BRAF mutations in cardio-facio-cutaneous syndrome. Nat Genet. 2006; 38:294–6.
44. Armour CM, Allanson JE. Further delineation of cardio-facio-cutaneous syndrome: clinical features of 38 individuals with proven mutations. J Med Genet. 2008;45:249–54.
45. Wan PT, Garnett MJ, Roe SM, et al. Mechanism of activation of the RAF-ERK signaling pathway by oncogenic mutations of B-RAF. Cell. 2004;116:855–67.
46. Anastasaki C, Estep AL, Marais R, Rauen KA, Patton EE. Kinase-activating and kinase-impaired cardio-facio-cutaneous syndrome alleles have activity during zebrafish development and are sensitive to small molecule inhibitors. Hum Mol Genet. 2009;18:2543–54.
47. Timmerman LA, Grego-Bessa J, Raya A, et al. Notch promotes epithelial-mesenchymal transition during cardiac development and oncogenic transformation. Genes Dev. 2004;18:99–115.
48. Garg V, Muth AN, Ransom JF, et al. Mutations in NOTCH1 cause aortic valve disease. Nature. 2005;437:270–4.
49. Jain R, Rentschler S, Epstein JA. Notch and cardiac outflow tract development. Ann N Y Acad Sci. 2010;1188:184–90.
50. Marin-Garcia J. Advances in molecular genetics of congenital heart disease. Rev Esp Cardiol. 2009;62:242–5.
51. Weng AP, Millholland JM, Yashiro-Ohtani Y, et al. c-Myc is an important direct target of Notch1 in T-cell acute lymphoblastic leukemia/lymphoma. Genes Dev. 2006;20:2096–109.
52. High FA, Epstein JA. The multifaceted role of Notch in cardiac development and disease. Nat Rev Genet. 2008;9:49–61.
53. Mohamed SA, Aherrahrou Z, Liptau H, et al. Novel missense mutations (p.T596M and p.P1797H) in NOTCH1 in patients with bicuspid aortic valve. Biochem Biophys Res Commun. 2006;345: 1460–5.
54. High FA, Jain R, Stoller JZ, et al. Murine Jagged1/Notch signaling in the second heart field orchestrates Fgf8 expression and tissue-tissue interactions during outflow tract development. J Clin Invest. 2009;119:1986–96.
55. McElhinney DB, Krantz ID, Bason L, et al. Analysis of cardiovascular phenotype and genotype-phenotype correlation in individuals with a JAG1 mutation and/or Alagille syndrome. Circulation. 2002;106:2567–74.
56. Kamath BM, Spinner NB, Emerick KM, et al. Vascular anomalies in Alagille syndrome: a significant cause of morbidity and mortality. Circulation. 2004;109:1354–8.
57. Li L, Krantz ID, Deng Y, et al. Alagille syndrome is caused by mutations in human Jagged1, which encodes a ligand for Notch1. Nat Genet. 1997;16:243–51.
58. Oda T, Elkahloun AG, Pike BL, et al. Mutations in the human Jagged1 gene are responsible for Alagille syndrome. Nat Genet. 1997;16:235–42.
59. Krantz ID, Smith R, Colliton RP, et al. Jagged1 mutations in patients ascertained with isolated congenital heart defects. Am J Med Genet. 1999;84:56–60.
60. Eldadah ZA, Hamosh A, Biery NJ, et al. Familial Tetralogy of Fallot caused by mutation in the jagged1 gene. Hum Mol Genet. 2001;10:163–9.
61. McDaniell R, Warthen DM, Sanchez-Lara PA, et al. NOTCH2 mutations cause Alagille syndrome, a heterogeneous disorder of the notch signaling pathway. Am J Hum Genet. 2006;79:169–73.
62. Hove JR, Koster RW, Forouhar AS, Acevedo-Bolton G, Fraser SE, Gharib M. Intracardiac fluid forces are an essential epigenetic factor for embryonic cardiogenesis. Nature. 2003;421:172–7.
63. Yashiro K, Shiratori H, Hamada H. Haemodynamics determined by a genetic programme govern asymmetric development of the aortic arch. Nature. 2007;450:285–8.
64. Momma K, Kondo C, Matsuoka R, Takao A. Cardiac anomalies associated with a chromosome 22q11 deletion in patients with conotruncal anomaly face syndrome. Am J Cardiol. 1996;78: 591–4.
65. Digilio MC, Angioni A, De Santis M, et al. Spectrum of clinical variability in familial deletion 22q11.2: from full manifestation to extremely mild clinical anomalies. Clin Genet. 2003;63:308–13.
66. Marino B, Digilio MC, Toscano A, et al. Anatomic patterns of conotruncal defects associated with deletion 22q11. Genet Med. 2001;3:45–8.
67. Yagi H, Furutani Y, Hamada H, et al. Role of TBX1 in human del22q11.2 syndrome. Lancet. 2003;362:1366–73.
68. Lindsay EA, Vitelli F, Su H, et al. Tbx1 haploinsufficieny in the DiGeorge syndrome region causes aortic arch defects in mice. Nature. 2001;410:97–101.
69. Edelmann L, Pandita RK, Spiteri E, et al. A common molecular basis for rearrangement disorders on chromosome 22q11. Hum Mol Genet. 1999;8:1157–67.
70. Ou Z, Berg JS, Yonath H, et al. Microduplications of 22q11.2 are frequently inherited and are associated with variable phenotypes. Genet Med. 2008;10:267–77.
71. Ensenauer RE, Adeyinka A, Flynn HC, et al. Microduplication 22q11.2, an emerging syndrome: clinical, cytogenetic, and molecular analysis of thirteen patients. Am J Hum Genet. 2003;73: 1027–40.
72. Guris DL, Duester G, Papaioannou VE, Imamoto A. Dose-dependent interaction of Tbx1 and Crkl and locally aberrant RA signaling in a model of del22q11 syndrome. Dev Cell. 2006; 10:81–92.
73. Moon AM, Guris DL, Seo JH, et al. Crkl deficiency disrupts Fgf8 signaling in a mouse model of 22q11 deletion syndromes. Dev Cell. 2006;10:71–80.
74. Wessel A, Pankau R, Kececioglu D, Ruschewski W, Bursch JH. Three decades of follow-up of aortic and pulmonary vascular lesions in the Williams-Beuren syndrome. Am J Med Genet. 1994;52:297–301.
75. Eronen M, Peippo M, Hiippala A, et al. Cardiovascular manifestations in 75 patients with Williams syndrome. J Med Genet. 2002;39:554–8.
76. Ewart AK, Morris CA, Atkinson D, et al. Hemizygosity at the elastin locus in a developmental disorder, Williams syndrome. Nat Genet. 1993;5:11–6.
77. Wu YQ, Nickerson E, Shaffer LG, Keppler-Noreuil K, Muilenburg A. A case of Williams syndrome with a large, visible cytogenetic deletion. J Med Genet. 1999;36:928–32.
78. Vissers LE, van Ravenswaaij CM, Admiraal R, et al. Mutations in a new member of the chromodomain gene family cause CHARGE syndrome. Nat Genet. 2004;36:955–7.
79. Aramaki M, Udaka T, Kosaki R, et al. Phenotypic spectrum of CHARGE syndrome with CHD7 mutations. J Pediatr. 2006;148: 410–4.
80. Jongmans MC, Admiraal RJ, van der Donk KP, et al. CHARGE syndrome: the phenotypic spectrum of mutations in the CHD7 gene. J Med Genet. 2006;43:306–14.

81. Lalani SR, Safiullah AM, Fernbach SD, et al. Spectrum of CHD7 mutations in 110 individuals with CHARGE syndrome and genotype-phenotype correlation. Am J Hum Genet. 2006;78:303–14.
82. Phillips HM, Renforth GL, Spalluto C, et al. Narrowing the critical region within 11q24-qter for hypoplastic left heart and identification of a candidate gene, JAM3, expressed during cardiogenesis. Genomics. 2002;79:475–8.
83. Grossfeld PD, Mattina T, Lai Z, et al. The 11q terminal deletion disorder: a prospective study of 110 cases. Am J Med Genet A. 2004;129A:51–61.
84. Ye M, Hamzeh R, Geddis A, Varki N, Perryman MB, Grossfeld P. Deletion of JAM-C, a candidate gene for heart defects in Jacobsen syndrome, results in a normal cardiac phenotype in mice. Am J Med Genet A. 2009;149A:1438–43.
85. Sato Y. Role of ETS family transcription factors in vascular development and angiogenesis. Cell Struct Funct. 2001;26:19–24.
86. Ye M, Coldren C, Liang X, et al. Deletion of ETS-1, a gene in the Jacobsen syndrome critical region, causes ventricular septal defects and abnormal ventricular morphology in mice. Hum Mol Genet. 2010;19:648–56.
87. Holt M, Oram S. Familial heart disease with skeletal malformations. Br Heart J. 1960;22:236–42.
88. Basson CT, Bachinsky DR, Lin RC, et al. Mutations in human TBX5 [corrected] cause limb and cardiac malformation in Holt-Oram syndrome. Nat Genet. 1997;15:30–5.
89. Li QY, Newbury-Ecob RA, Terrett JA, et al. Holt-Oram syndrome is caused by mutations in TBX5, a member of the Brachyury (T) gene family. Nat Genet. 1997;15:21–9.
90. Bruneau BG, Nemer G, Schmitt JP, et al. A murine model of Holt-Oram syndrome defines roles of the T-box transcription factor Tbx5 in cardiogenesis and disease. Cell. 2001;106:709–21.
91. Mori AD, Zhu Y, Vahora I, et al. Tbx5-dependent rheostatic control of cardiac gene expression and morphogenesis. Dev Biol. 2006;297:566–86.
92. Hiroi Y, Kudoh S, Monzen K, et al. Tbx5 associates with Nkx2-5 and synergistically promotes cardiomyocyte differentiation. Nat Genet. 2001;28:276–80.
93. Garg V, Kathiriya IS, Barnes R, et al. GATA4 mutations cause human congenital heart defects and reveal an interaction with TBX5. Nature. 2003;424:443–7.
94. Jay PY, Harris BS, Maguire CT, et al. Nkx2-5 mutation causes anatomic hypoplasia of the cardiac conduction system. J Clin Invest. 2004;113:1130–7.
95. Pashmforoush M, Lu JT, Chen H, et al. Nkx2-5 pathways and congenital heart disease; loss of ventricular myocyte lineage specification leads to progressive cardiomyopathy and complete heart block. Cell. 2004;117:373–86.
96. Kuo CT, Morrisey EE, Anandappa R, et al. GATA4 transcription factor is required for ventral morphogenesis and heart tube formation. Genes Dev. 1997;11:1048–60.
97. Molkentin JD, Lin Q, Duncan SA, Olson EN. Requirement of the transcription factor GATA4 for heart tube formation and ventral morphogenesis. Genes Dev. 1997;11:1061–72.
98. Rajagopal SK, Ma Q, Obler D, et al. Spectrum of heart disease associated with murine and human GATA4 mutation. J Mol Cell Cardiol. 2007;43:677–85.
99. Schott JJ, Benson DW, Basson CT, et al. Congenital heart disease caused by mutations in the transcription factor NKX2-5. Science. 1998;281:108–11.
100. Benson DW, Silberbach GM, Kavanaugh-McHugh A, et al. Mutations in the cardiac transcription factor NKX2.5 affect diverse cardiac developmental pathways. J Clin Invest. 1999;104:1567–73.
101. Goldmuntz E, Geiger E, Benson DW. NKX2.5 mutations in patients with tetralogy of fallot. Circulation. 2001;104:2565–8.
102. Hirayama-Yamada K, Kamisago M, Akimoto K, et al. Phenotypes with GATA4 or NKX2.5 mutations in familial atrial septal defect. Am J Med Genet A. 2005;135:47–52.
103. Posch MG, Gramlich M, Sunde M, et al. A gain-of-function TBX20 mutation causes congenital atrial septal defects, patent foramen ovale and cardiac valve defects. J Med Genet. 2010;47:230–5.
104. Heathcote K, Braybrook C, Abushaban L, et al. Common arterial trunk associated with a homeodomain mutation of NKX2.6. Hum Mol Genet. 2005;14:585–93.
105. Tanaka M, Yamasaki N, Izumo S. Phenotypic characterization of the murine Nkx2.6 homeobox gene by gene targeting. Mol Cell Biol. 2000;20:2874–9.
106. Kodo K, Nishizawa T, Furutani M, et al. GATA6 mutations cause human cardiac outflow tract defects by disrupting semaphorin-plexin signaling. Proc Natl Acad Sci USA. 2009;106:13933–8.
107. Durocher D, Nemer M. Combinatorial interactions regulating cardiac transcription. Dev Genet. 1998;22:250–62.
108. Mackay JP, Crossley M. Zinc fingers are sticking together. Trends Biochem Sci. 1998;23:1–4.
109. Al-Baradie R, Yamada K, St Hilaire C, et al. Duane radial ray syndrome (Okihiro syndrome) maps to 20q13 and results from mutations in SALL4, a new member of the SAL family. Am J Hum Genet. 2002;71:1195–9.
110. Kohlhase J, Heinrich M, Schubert L, et al. Okihiro syndrome is caused by SALL4 mutations. Hum Mol Genet. 2002;11:2979–87.
111. Bohm J, Munk-Schulenburg S, Felscher S, Kohlhase J. SALL1 mutations in sporadic Townes-Brocks syndrome are of predominantly paternal origin without obvious paternal age effect. Am J Med Genet A. 2006;140:1904–8.
112. Borozdin W, Steinmann K, Albrecht B, et al. Detection of heterozygous SALL1 deletions by quantitative real time PCR proves the contribution of a SALL1 dosage effect in the pathogenesis of Townes-Brocks syndrome. Hum Mutat. 2006;27:211–2.
113. Koshiba-Takeuchi K, Takeuchi JK, Arruda EP, et al. Cooperative and antagonistic interactions between Sall4 and Tbx5 pattern the mouse limb and heart. Nat Genet. 2006;38:175–83.
114. Ching YH, Ghosh TK, Cross SJ, et al. Mutation in myosin heavy chain 6 causes atrial septal defect. Nat Genet. 2005;37:423–8.
115. Sperling S, Grimm CH, Dunkel I, et al. Identification and functional analysis of CITED2 mutations in patients with congenital heart defects. Hum Mutat. 2005;26:575–82.
116. Petrij F, Giles RH, Dauwerse HG, et al. Rubinstein-Taybi syndrome caused by mutations in the transcriptional co-activator CBP. Nature. 1995;376:348–51.
117. Roelfsema JH, White SJ, Ariyurek Y, et al. Genetic heterogeneity in Rubinstein-Taybi syndrome: mutations in both the CBP and EP300 genes cause disease. Am J Hum Genet. 2005;76:572–80.
118. Satoda M, Zhao F, Diaz GA, et al. Mutations in TFAP2B cause Char syndrome, a familial form of patent ductus arteriosus. Nat Genet. 2000;25:42–6.
119. Schroeder JA, Jackson LF, Lee DC, Camenisch TD. Form and function of developing heart valves: coordination by extracellular matrix and growth factor signaling. J Mol Med. 2003;81:392–403.
120. Sanchez-Soria P, Camenisch TD. ErbB signaling in cardiac development and disease. Semin Cell Dev Biol. 2010;21(9):929–35.
121. Pinkas-Kramarski R, Soussan L, Waterman H, et al. Diversification of Neu differentiation factor and epidermal growth factor signaling by combinatorial receptor interactions. EMBO J. 1996;15:2452–67.
122. Elenius K, Paul S, Allison G, Sun J, Klagsbrun M. Activation of HER4 by heparin-binding EGF-like growth factor stimulates chemotaxis but not proliferation. EMBO J. 1997;16:1268–78.

123. Riese 2nd DJ, Komurasaki T, Plowman GD, Stern DF. Activation of ErbB4 by the bifunctional epidermal growth factor family hormone epiregulin is regulated by ErbB2. J Biol Chem. 1998;273: 11288–94.
124. Schulze WX, Deng L, Mann M. Phosphotyrosine interactome of the ErbB-receptor kinase family. Mol Syst Biol. 2005;1:2005.0008.
125. Dengjel J, Akimov V, Blagoev B, Andersen JS. Signal transduction by growth factor receptors: signaling in an instant. Cell Cycle. 2007;6:2913–6.
126. Fuller SJ, Sivarajah K, Sugden PH. ErbB receptors, their ligands, and the consequences of their activation and inhibition in the myocardium. J Mol Cell Cardiol. 2008;44:831–54.
127. Morgensztern D, McLeod HL. PI3K/Akt/mTOR pathway as a target for cancer therapy. Anticancer Drugs. 2005;16:797–803.
128. Fukazawa R, Miller TA, Kuramochi Y, et al. Neuregulin-1 protects ventricular myocytes from anthracycline-induced apoptosis via erbB4-dependent activation of PI3-kinase/Akt. J Mol Cell Cardiol. 2003;35:1473–9.
129. Meadows KN, Iyer S, Stevens MV, et al. Akt promotes endocardial-mesenchyme transition. J Angiogenes Res. 2009;1:2.
130. Lee KF, Simon H, Chen H, Bates B, Hung MC, Hauser C. Requirement for neuregulin receptor erbB2 in neural and cardiac development. Nature. 1995;378:394–8.
131. Sibilia M, Wagner EF. Strain-dependent epithelial defects in mice lacking the EGF receptor. Science. 1995;269:234–8.
132. Erickson SL, O'Shea KS, Ghaboosi N, et al. ErbB3 is required for normal cerebellar and cardiac development: a comparison with ErbB2-and heregulin-deficient mice. Development. 1997;124: 4999–5011.
133. Camenisch TD, Schroeder JA, Bradley J, Klewer SE, McDonald JA. Heart-valve mesenchyme formation is dependent on hyaluronan-augmented activation of ErbB2-ErbB3 receptors. Nat Med. 2002;8:850–5.
134. Chan R, Hardy WR, Laing MA, Hardy SE, Muller WJ. The catalytic activity of the ErbB-2 receptor tyrosine kinase is essential for embryonic development. Mol Cell Biol. 2002;22:1073–8.
135. Jackson LF, Qiu TH, Sunnarborg SW, et al. Defective valvulogenesis in HB-EGF and TACE-null mice is associated with aberrant BMP signaling. EMBO J. 2003;22:2704–16.
136. Lin AE, Birch PH, Korf BR, et al. Cardiovascular malformations and other cardiovascular abnormalities in neurofibromatosis 1. Am J Med Genet. 2000;95:108–17.
137. Schier AF, Shen MM. Nodal signalling in vertebrate development. Nature. 2000;403:385–9.
138. Kosaki K, Bassi MT, Kosaki R, et al. Characterization and mutation analysis of human LEFTY A and LEFTY B, homologues of murine genes implicated in left-right axis development. Am J Hum Genet. 1999;64:712–21.
139. Kosaki R, Gebbia M, Kosaki K, et al. Left-right axis malformations associated with mutations in ACVR2B, the gene for human activin receptor type IIB. Am J Med Genet. 1999;82:70–6.
140. Bamford RN, Roessler E, Burdine RD, et al. Loss-of-function mutations in the EGF-CFC gene CFC1 are associated with human left-right laterality defects. Nat Genet. 2000;26:365–9.
141. Goldmuntz E, Bamford R, Karkera JD, dela Cruz J, Roessler E, Muenke M. CFC1 mutations in patients with transposition of the great arteries and double-outlet right ventricle. Am J Hum Genet. 2002;70:776–80.
142. Karkera JD, Lee JS, Roessler E, et al. Loss-of-function mutations in growth differentiation factor-1 (GDF1) are associated with congenital heart defects in humans. Am J Hum Genet. 2007;81:987–94.
143. Roessler E, Ouspenskaia MV, Karkera JD, et al. Reduced NODAL signaling strength via mutation of several pathway members including FOXH1 is linked to human heart defects and holoprosencephaly. Am J Hum Genet. 2008;83:18–29.
144. Mohapatra B, Casey B, Li H, et al. Identification and functional characterization of NODAL rare variants in heterotaxy and isolated cardiovascular malformations. Hum Mol Genet. 2009;18:861–71.
145. De Luca A, Sarkozy A, Consoli F, et al. Familial transposition of the great arteries caused by multiple mutations in laterality genes. Heart. 2010;96:673–7.
146. Conlon FL, Lyons KM, Takaesu N, et al. A primary requirement for nodal in the formation and maintenance of the primitive streak in the mouse. Development. 1994;120:1919–28.
147. Lowe LA, Supp DM, Sampath K, et al. Conserved left-right asymmetry of nodal expression and alterations in murine situs inversus. Nature. 1996;381:158–61.
148. Saloman DS, Bianco C, Ebert AD, et al. The EGF-CFC family: novel epidermal growth factor-related proteins in development and cancer. Endocr Relat Cancer. 2000;7:199–226.
149. Shen MM, Schier AF. The EGF-CFC gene family in vertebrate development. Trends Genet. 2000;16:303–9.
150. Gritsman K, Zhang J, Cheng S, Heckscher E, Talbot WS, Schier AF. The EGF-CFC protein one-eyed pinhead is essential for nodal signaling. Cell. 1999;97:121–32.
151. Gaio U, Schweickert A, Fischer A, et al. A role of the cryptic gene in the correct establishment of the left-right axis. Curr Biol. 1999; 9:1339–42.
152. Ozcelik C, Bit-Avragim N, Panek A, et al. Mutations in the EGF-CFC gene cryptic are an infrequent cause of congenital heart disease. Pediatr Cardiol. 2006;27:695–8.
153. Selamet Tierney ES, Marans Z, Rutkin MB, Chung WK. Variants of the CFC1 gene in patients with laterality defects associated with congenital cardiac disease. Cardiol Young. 2007;17:268–74.
154. Harrison CA, Gray PC, Fischer WH, Donaldson C, Choe S, Vale W. An activin mutant with disrupted ALK4 binding blocks signaling via type II receptors. J Biol Chem. 2004;279:28036–44.
155. Meno C, Takeuchi J, Sakuma R, et al. Diffusion of nodal signaling activity in the absence of the feedback inhibitor Lefty2. Dev Cell. 2001;1:127–38.
156. Gebbia M, Ferrero GB, Pilia G, et al. X-linked situs abnormalities result from mutations in ZIC3. Nat Genet. 1997;17:305–8.
157. Pyeritz RE. The Marfan syndrome. Annu Rev Med. 2000;51: 481–510.
158. Boileau C, Jondeau G, Mizuguchi T, Matsumoto N. Molecular genetics of Marfan syndrome. Curr Opin Cardiol. 2005;20: 194–200.
159. Robinson PN, Arteaga-Solis E, Baldock C, et al. The molecular genetics of Marfan syndrome and related disorders. J Med Genet. 2006;43:769–87.
160. Lee B, Godfrey M, Vitale E, et al. Linkage of Marfan syndrome and a phenotypically related disorder to two different fibrillin genes. Nature. 1991;352:330–4.
161. Dietz HC, Cutting GR, Pyeritz RE, et al. Marfan syndrome caused by a recurrent de novo missense mutation in the fibrillin gene. Nature. 1991;352:337–9.
162. Corson GM, Chalberg SC, Dietz HC, Charbonneau NL, Sakai LY. Fibrillin binds calcium and is coded by cDNAs that reveal a multidomain structure and alternatively spliced exons at the 5′ end. Genomics. 1993;17:476–84.
163. Putnam EA, Zhang H, Ramirez F, Milewicz DM. Fibrillin-2 (FBN2) mutations result in the Marfan-like disorder, congenital contractural arachnodactyly. Nat Genet. 1995;11:456–8.
164. Sakai LY, Keene DR, Glanville RW, Bachinger HP. Purification and partial characterization of fibrillin, a cysteine-rich structural component of connective tissue microfibrils. J Biol Chem. 1991; 266:14763–70.
165. Carta L, Pereira L, Arteaga-Solis E, et al. Fibrillins 1 and 2 perform partially overlapping functions during aortic development. J Biol Chem. 2006;281:8016–23.

166. Gregory KE, Ono RN, Charbonneau NL, et al. The prodomain of BMP-7 targets the BMP-7 complex to the extracellular matrix. J Biol Chem. 2005;280:27970–80.
167. Chaudhry SS, Cain SA, Morgan A, Dallas SL, Shuttleworth CA, Kielty CM. Fibrillin-1 regulates the bioavailability of TGFbeta1. J Cell Biol. 2007;176:355–67.
168. Loeys BL, Chen J, Neptune ER, et al. A syndrome of altered cardiovascular, craniofacial, neurocognitive and skeletal development caused by mutations in TGFBR1 or TGFBR2. Nat Genet. 2005; 37:275–81.
169. Mizuguchi T, Collod-Beroud G, Akiyama T, et al. Heterozygous TGFBR2 mutations in Marfan syndrome. Nat Genet. 2004;36: 855–60.
170. Grady WM, Myeroff LL, Swinler SE, et al. Mutational inactivation of transforming growth factor beta receptor type II in microsatellite stable colon cancers. Cancer Res. 1999;59:320–4.
171. Neptune ER, Frischmeyer PA, Arking DE, et al. Dysregulation of TGF-beta activation contributes to pathogenesis in Marfan syndrome. Nat Genet. 2003;33:407–11.
172. Ng CM, Cheng A, Myers LA, et al. TGF-beta-dependent pathogenesis of mitral valve prolapse in a mouse model of Marfan syndrome. J Clin Invest. 2004;114:1586–92.
173. Loeys BL, Schwarze U, Holm T, et al. Aneurysm syndromes caused by mutations in the TGF-beta receptor. N Engl J Med. 2006;355:788–98.
174. Horbelt D, Guo G, Robinson PN, Knaus P. Quantitative analysis of TGFBR2 mutations in Marfan-syndrome-related disorders suggests a correlation between phenotypic severity and Smad signaling activity. J Cell Sci. 2010;123(Pt 24):4340–50.
175. Robinson SW, Morris CD, Goldmuntz E, et al. Missense mutations in CRELD1 are associated with cardiac atrioventricular septal defects. Am J Hum Genet. 2003;72:1047–52.
176. Maslen CL, Babcock D, Robinson SW, et al. CRELD1 mutations contribute to the occurrence of cardiac atrioventricular septal defects in Down syndrome. Am J Med Genet A. 2006;140:2501–5.
177. Stossel TP, Condeelis J, Cooley L, et al. Filamins as integrators of cell mechanics and signalling. Nat Rev Mol Cell Biol. 2001;2: 138–45.
178. Zhou AX, Hartwig JH, Akyurek LM. Filamins in cell signaling, transcription and organ development. Trends Cell Biol. 2010;20: 113–23.
179. van der Flier A, Sonnenberg A. Structural and functional aspects of filamins. Biochim Biophys Acta. 2001;1538:99–117.
180. Fox JW, Lamperti ED, Eksioglu YZ, et al. Mutations in filamin 1 prevent migration of cerebral cortical neurons in human periventricular heterotopia. Neuron. 1998;21:1315–25.
181. Parrini E, Ramazzotti A, Dobyns WB, et al. Periventricular heterotopia: phenotypic heterogeneity and correlation with Filamin A mutations. Brain. 2006;129:1892–906.
182. Kyndt F, Gueffet JP, Probst V, et al. Mutations in the gene encoding filamin A as a cause for familial cardiac valvular dystrophy. Circulation. 2007;115:40–9.
183. Feng Y, Chen MH, Moskowitz IP, et al. Filamin A (FLNA) is required for cell-cell contact in vascular development and cardiac morphogenesis. Proc Natl Acad Sci USA. 2006;103:19836–41.
184. Lu J, Lian G, Lenkinski R, et al. Filamin B mutations cause chondrocyte defects in skeletal development. Hum Mol Genet. 2007;16: 1661–75.
185. Zhou X, Tian F, Sandzen J, et al. Filamin B deficiency in mice results in skeletal malformations and impaired microvascular development. Proc Natl Acad Sci USA. 2007;104:3919–24.
186. Dalkilic I, Schienda J, Thompson TG, Kunkel LM. Loss of FilaminC (FLNc) results in severe defects in myogenesis and myotube structure. Mol Cell Biol. 2006;26:6522–34.
187. Cinquetti R, Badi I, Campione M, et al. Transcriptional deregulation and a missense mutation define ANKRD1 as a candidate gene for total anomalous pulmonary venous return. Hum Mutat. 2008; 29:468–74.
188. Budde BS, Binner P, Waldmuller S, et al. Noncompaction of the ventricular myocardium is associated with a de novo mutation in the beta-myosin heavy chain gene. PLoS One. 2007; 2:e1362.
189. Zhu L, Vranckx R, Khau Van Kien P, et al. Mutations in myosin heavy chain 11 cause a syndrome associating thoracic aortic aneurysm/aortic dissection and patent ductus arteriosus. Nat Genet. 2006;38:343–9.
190. Lekanne Deprez RH, Muurling-Vlietman JJ, Hruda J, et al. Two cases of severe neonatal hypertrophic cardiomyopathy caused by compound heterozygous mutations in the MYBPC3 gene. J Med Genet. 2006;43:829–32.
191. Zahka K, Kalidas K, Simpson MA, et al. Homozygous mutation of MYBPC3 associated with severe infantile hypertrophic cardiomyopathy at high frequency among the Amish. Heart. 2008;94: 1326–30.
192. Olson TM, Doan TP, Kishimoto NY, Whitby FG, Ackerman MJ, Fananapazir L. Inherited and de novo mutations in the cardiac actin gene cause hypertrophic cardiomyopathy. J Mol Cell Cardiol. 2000;32:1687–94.
193. Monserrat L, Hermida-Prieto M, Fernandez X, et al. Mutation in the alpha-cardiac actin gene associated with apical hypertrophic cardiomyopathy, left ventricular non-compaction, and septal defects. Eur Heart J. 2007;28:1953–61.
194. Matsson H, Eason J, Bookwalter CS, et al. Alpha-cardiac actin mutations produce atrial septal defects. Hum Mol Genet. 2008;17: 256–65.
195. Couzin J. Breakthrough of the year. Small RNAs make big splash. Science. 2002;298:2296–7.
196. Zhang B, Wang Q, Pan X. MicroRNAs and their regulatory roles in animals and plants. J Cell Physiol. 2007;210:279–89.
197. Ambros V. The functions of animal microRNAs. Nature. 2004;431: 350–5.
198. Lee CT, Risom T, Strauss WM. MicroRNAs in mammalian development. Birth Defects Res C Embryo Today. 2006;78: 129–39.
199. van Rooij E, Olson EN. MicroRNAs: powerful new regulators of heart disease and provocative therapeutic targets. J Clin Invest. 2007;117:2369–76.
200. Zhao Y, Samal E, Srivastava D. Serum response factor regulates a muscle-specific microRNA that targets Hand2 during cardiogenesis. Nature. 2005;436:214–20.
201. Zhao Y, Ransom JF, Li A, et al. Dysregulation of cardiogenesis, cardiac conduction, and cell cycle in mice lacking miRNA-1-2. Cell. 2007;129:303–17.
202. Bohlmeyer TJ, Helmke S, Ge S, et al. Hypoplastic left heart syndrome myocytes are differentiated but possess a unique phenotype. Cardiovasc Pathol. 2003;12:23–31.
203. Kaynak B, von Heydebreck A, Mebus S, et al. Genome-wide array analysis of normal and malformed human hearts. Circulation. 2003;107:2467–74.
204. Kaufman BD, Desai M, Reddy S, et al. Genomic profiling of left and right ventricular hypertrophy in congenital heart disease. J Card Fail. 2008;14:760–7.
205. Nigam V, Sievers HH, Jensen BC, et al. Altered microRNAs in bicuspid aortic valve: a comparison between stenotic and insufficient valves. J Heart Valve Dis. 2010;19:459–65.
206. Barth JL, Clark CD, Fresco VM, et al. Jarid2 is among a set of genes differentially regulated by Nkx2.5 during outflow tract morphogenesis. Dev Dyn. 2010;239:2024–33.

207. Yuasa S, Onizuka T, Shimoji K, et al. Zac1 is an essential transcription factor for cardiac morphogenesis. Circ Res. 2010;106:1083–91.
208. Jacobs JP, Maruszewski B. Computerized outcomes analysis for congenital heart disease. Curr Opin Pediatr. 2005;17:586–91.
209. Zannini L, Borini I. State of the art of cardiac surgery in patients with congenital heart disease. J Cardiovasc Med (Hagerstown). 2007;8:3–6.
210. Dorfman AT, Marino BS, Wernovsky G, et al. Critical heart disease in the neonate: presentation and outcome at a tertiary care center. Pediatr Crit Care Med. 2008;9:193–202.
211. Lickert H, Takeuchi JK, Von Both I, et al. Baf60c is essential for function of BAF chromatin remodelling complexes in heart development. Nature. 2004;432:107–12.
212. Montgomery RL, Davis CA, Potthoff MJ, et al. Histone deacetylases 1 and 2 redundantly regulate cardiac morphogenesis, growth, and contractility. Genes Dev. 2007;21:1790–802.
213. Biben C, Weber R, Kesteven S, et al. Cardiac septal and valvular dysmorphogenesis in mice heterozygous for mutations in the homeobox gene Nkx2-5. Circ Res. 2000;87:888–95.
214. Takeuchi JK, Mileikovskaia M, Koshiba-Takeuchi K, et al. Tbx20 dose-dependently regulates transcription factor networks required for mouse heart and motoneuron development. Development. 2005;132:2463–74.
215. Stennard FA, Costa MW, Lai D, et al. Murine T-box transcription factor Tbx20 acts as a repressor during heart development, and is essential for adult heart integrity, function and adaptation. Development. 2005;132:2451–62.
216. Todd AK. Bioinformatics approaches to quadruplex sequence location. Methods. 2007;43:246–51.
217. Daley GQ, Cargill M. The heart SNPs a beat: polymorphisms in candidate genes for cardiovascular disease. Trends Cardiovasc Med. 2001;11:60–6.

# Part V
# Aging

# Chapter 11
# Signaling in the Aging Heart

**Abstract** Cardiac aging is a complex multifactorial process not dependent on a unique, singular, pathway, or determined gene(s) or gene products. Rather, a number of specific and nonspecific pathways and genes play a role in the general regulation/modulation of life span, and in particular cardiac aging. Multiple molecular mechanisms interact in cardiac aging either in parallel or in series, including the involvement of somatic mutations, telomere loss, defects in protein turnover, protein functional decline with accumulation of defective proteins (i.e., impaired induction of heat shock proteins and decline in chaperone function), and mitochondrial defects. Most of these mechanisms produce significant damage to cardiac macromolecules.

Mechanisms that appear to play a critical role in aging are the molecular stresses that defective mitochondrial bioenergetics and biogenesis may bring to cardiomyocytes, as well as defects in hormonal and inflammatory signaling and telomere shortening. However, it may not be possible to identify the totality of the mechanisms/pathways that contribute to cardiac aging; intensive research in this regard is currently in process.

**Keywords** Aging heart • Rodent models • Telomeres • Cellular damage • Neuroendocrine signaling

## Introduction

The processes of biological cardiovascular (CV) aging are mechanistically connected to reduced physiological reserve, abnormal drug handling, and pharmacodynamic responses. During the normal aging process, changes occur in the autonomic control of the cardiovascular system, favoring increased cardiac sympathetic tone with parasympathetic withdrawal and blunted cardiovagal baroreflex sensitivity. Furthermore, changes in cardiac phenotype with diastolic dysfunction, reduced contractility, left ventricular (LV) hypertrophy and heart failure (HF), all increase in incidence with age. The aging human heart displays alterations in the histology of the vasculature and hemodynamics, including the development of large resistance vessels with intima-media-thickening and increasing deposition of matrix substance, which ultimately leads to reduced compliance and increased vessel stiffness and endothelial dysfunction [1, 2]. Furthermore, with aging, there is increased left ventricular mass relative to chamber volume, decreased diastolic function [3, 4] and decreased β-adrenergic sympathetic responsiveness [5]. While interactions between advanced age, disease and physical inactivity need to be considered when interpreting age-associated changes in cardiovascular function, the "aging process" in itself occurs independently of changes on cardiac structure and performance, such as cardiac hypertrophy and prolonged myocardial contractility. Cellular mechanisms thought to be involved in cardiac aging, include prolonged action potential (AP) duration, altered myosin heavy chain (MHC) isoform expression and sarcoplasmic reticulum (SR) function, all of which may lead to changes in cardiac excitation–contraction (E–C) coupling. Cardiac E–C coupling cycle or cardiac cycle has been shown to be prolonged with increased age, probably due to cytosolic $Ca^{2+}$ overload-induced dysregulation [5]. The cytosolic $Ca^{2+}$ load is dependent on multiple factors, including membrane structure and permeability, regulatory proteins within the membrane, and reactive oxygene species (ROS) levels, which affect both membrane structure and function. However, the link between advanced age and altered cardiac E–C coupling is not yet fully understood.

### *Animal Studies*

Most studies demonstrating diminished cardiac function and altered signaling have been performed with rodent models [6–9]. Aged animals had elevated left ventricular end-diastolic pressure and decreased dP/dt suggesting that a significant impairment in ventricular function occurred with senescence. Further observations have shown that hypertrophic, fibrotic, and cell-death pathways underlying extensive remodeling appear to be triggered in the aging rodent heart [10, 11]. Nevertheless, given their profound difference in

J. Marín-García, *Signaling in the Heart*,
DOI 10.1007/978-1-4419-9461-5_11, © Springer Science+Business Media, LLC 2011

size and life-span compared to human, the relevance of rodents to humans remains uncertain.

Observations in a primate model, *Macaca fascicularis*, revealed that in old male there was a fourfold increase in the frequency of myocyte apoptosis, a nonuniform increase (51%) in mean myocyte cross-sectional area and no increase in proliferation-capable myocytes resulting in a significant decline in LV weight/body weight [12]. In contrast, myocyte apoptosis was unchanged in old female monkey hearts and no significant changes were noted in either mean myocyte cross-sectional area or in the LV mass. The decreased in heart mass in aging male monkeys differed from the findings of cardiac enlargement and associated hyperplasia found in aging rodents [9]. Data from human studies support similar apoptotic and hypertrophic remodeling occurring in the aging heart. Olivetti et al. [13] found evidence supporting the proliferation of myocyte nuclei (with potential hyperplasia) and connective tissue accumulation as contributory factors of ventricular remodeling in the hypertrophic senescent heart. In the aging human heart, a distinct subpopulation of myocytes that undergoes hypertrophy was found, with another subpopulation that undergoes apoptosis and necrosis and yet another subpopulation containing cycling myocytes capable of DNA replication and mitosis [14]. These distinct cell populations were noted to differ in their ability to react to growth stimuli. Whether myocytes response to increased load is replication, hypertrophy, or cell death has been suggested to be largely influenced by its size that in turn, reflects the age of the cell. Large myocytes tended to be old, do not react to growth stimuli and have lost the potential to proliferate, express more inhibitors of the cell cycle and senescence-associated proteins, such as p16$^{INK4a}$ [15], and are more prone to activate the cell death signaling pathway. In contrast, smaller cells are younger, are less susceptible to cell death and possess the ability to divide and enlarge, a challenge to the accepted paradigm of the adult heart as a postmitotic organ containing terminally differentiated myocytes that cannot reenter the cell cycle. This apparent contradiction has been largely resolved by recent identification of small population(s) of resident regenerative cells within the heart, termed cardiac stem cells (CSCs).

## Cardiac Stem and Progenitor Cells in Aging

Several research groups have confirmed the existence of CSCs although presently there are striking phenotypic differences between these different isolates. These cells are activated in response to pathological or physiological stimuli, whereby they enter the cell cycle and differentiate into new myocytes (and coronary vessels) that can significantly contribute to changes in myocardial mass [16]. Hence, the small cells that were initially thought to be cardiomyocytes that had reentered the cell cycle are most likely the products of CSC differentiation that can undergo several rounds of replication before becoming terminally differentiated [17]. Unexpectedly, aging appears to affect the growth and differentiation potential of CSCs interfering not only with their ability to sustain physiological cell turnover, but also with their capacity to adapt to increases in pressure and volume loads [18]. In the aging heart, CSCs have been found to undergo replicative senescence with loss of self-renewing capacity and commitment and differentiation. This is in part underlied by CSC telomere dysfunction with alterations in telomeric-binding proteins in combination with the expression of p53 and p16$^{INK4a}$, activating the death program, and further decreasing the size of the CSC pool in the heart [19]. Nevertheless, delineation of the signaling factors enhancing the activation of the CSC pool, their mobilization, and translocation may reveal novel approaches by which the detrimental effects of aging on the heart might be abrogated or reversed. For instance, Rota et al. [20] have shown that ablation of the adaptor protein p66shc had remarkable beneficial consequences on the viability and function of CSCs, positively interfering with death stimuli and the aging-associated inhibition of CSC growth and differentiation. Similarly, using myocardial-specific expression of nuclear-targeted Akt (Akt/nuc), Gude et al. [21] showed that myocardial stem and progenitor cell populations proliferation is enhanced, suggesting that the implementation of Akt activity as a molecular interventional approach may be effective in the treatment of cardiomyopathic damage resulting from the debilitating changes of aging.

Age-associated decline in cardiac and vascular regenerative capacity may contribute to the progressive deterioration of cardiovascular health. Therefore, understanding the mechanisms which underlie the dysregulation of cardiac stem and progenitor cells may lead to the identification of novel targets and approaches to reverse this decline [22]. Importantly, recognition that the adult heart possesses a pool of resident cardiac progenitor cells (CPCs), which are self-renewing, clonogenic, and multipotent [23–25], has encouraged the development of a new area of research that may allow to harvest cells, which are primed to acquire a cardiac phenotype and, therefore, might be optimally suited for repair of the damaged heart.

In an interesting review on "Aging and Disease" as modifiers of the efficacy of cell therapy, Dimmeler and Leri [26] pointed out that adult stem and progenitor cells from various sources have experimentally shown to improve functional recovery after ischemia, and several clinical trials confirmed that autologous cell therapy using bone marrow-derived or circulating blood-derived progenitor cells was safe and effective, although aging and risk factors for coronary artery disease (CAD) affect the functionality of the endogenous stem/progenitor cell pools, and partially limited the therapeutic potential of the applied cells. In addition, age and disease affect the tissue environment, in which the cells are infused or injected.

Observational evidence has shown that cardiovascular risk factors interfere with circulating progenitor and proangiogenic cells. Whether the impairment of circulating progenitor cells and proangiogenic cells in patients with high-risk factors load is caused by a depletion of stem cell reserves and stem cell exhaustion in the bone marrow, or might be related to signaling defects and increased apoptosis of circulating cells is not clear. It is worth noting that old rats undergone telomeric shortening in CPCs, which by necessity generate a differentiated progeny, rapidly acquires the senescent phenotype [27]. The daughter cells inherit the shortened telomeres of the maternal CPCs and after a few divisions they express the senescence-associated protein $p16^{INK4a}$. The pool of old cardiomyocytes progressively decreases and ventricular function is impaired. However, telomerase competent CPCs with long telomeres are present in the regions of storage in the atria and apex, and these cells after activation by growth factors migrate to areas of damage where they create a population of young myocytes, reversing to some extent the aging myopathy structurally and functionally. On the other hand, in the rat model the senescent heart phenotype is partially corrected and improvement in cardiac hemodynamics results in prolongation of maximum lifespan [27].

The loss of CPC function with aging is mediated partly by an imbalance between factors promoting growth, migration, and survival, and factors enhancing oxidative stress (OS), telomere attrition, and death. Three growth-factor receptor systems appear to play a major role in the development of CPC senescence and myocardial aging: Insulin-like growth factor 1(IGF-1)-IGF-1 receptor (IGF-1R). Hepatocyte growth factor (HGF)/c-Met, and the renin angiotensin system (RAS). In the heart, the IGF-1–IGF-1R induces CPC division, upregulates telomerase activity, hinders replicative senescence, and preserves the pool of functionally competent CPCs [28–30]. The expression of IGF-1R and the synthesis of IGF-1 are attenuated in aging CPCs, and these negative variables diminish the ability of IGF-1 to activate cell growth and interfere with oxidative damage and telomeric shortening [31]. Additionally, the expression and secretion of HGF in CPCs decreases as a function of age and this modification has a major impact on the migration of CPCs [32–35] and on the ability of these cells to translocate spontaneously to areas of damage and promote cardiac repair. Defects in these two autocrine–paracrine effector pathways of CPCs may have profound physiological consequences and may account for the chronological increase in myocyte death, myocardial scarring, and depressed performance of the aging heart.

Documentation that various components of RAS are present in CPCs and the formation of angiotensin (Ang) II is enhanced in old cells provides evidence in support of the role of this octapeptide in CPC senescence and death. Ang II may be a significant contributor of the age-dependent accumulation of oxidative damage in the heart [28, 36]. Inhibition of Ang II function positively interferes with heart failure and prolongs life in humans [37]. Ang II generates ROS and sustained oxidative stress (OS) triggers telomeric shortening and uncapping [38]. Conversely, IGF-1 interferes with the generation of ROS [28], decreases oxidative damage in the myocardium with age [29], and can repair oxidative DNA damage by homologous recombination [39]. Collectively, these findings suggest that cardiac aging is associated with a dysfunction of the endogenous stem cell pool that is dictated partly by the imbalance between RAS and IGF-1/HGF.

It is likely that during aging and in chronic cardiac diseases, stem cells tend to become quiescent. While quiescence in the young, active progenitor cells are modulated by $p21^{Cip1}$[40] in old/diseased stem cells, and irreversible growth arrest is regulated by $p16^{INK4a}$ [41]. Loss of telomerase, critical telomere shortening, and increased nuclear expression of p53 and $p16^{INK4a}$ may all occur resulting in the loss of growth reserve [42]. Early stem cell depletion may induce premature aging while replenishment of stem cells in depleted organs may reverse aging and disease, promoting positive remodeling and recovery of function. Notwithstanding translation of results from simple postmitotic organisms to large mammals and particularly to humans should be considered with caution because the life and death of most somatic organs in mammals is regulated by a stem cell compartment, which plays a critical role in aging and in the response of organs to disease. Understanding the complex relationship between stem cells and their environment is critical for the success of cell therapy. As such the impact of disease and age on endogenous stem and progenitor cells and on the environment may limit the benefit of cell therapy in the chronically ill patients; nevertheless, this also opens a new horizon for therapeutic strategies to counteract the dysregulated cell intrinsic and extrinsic signaling pathways.

## Signaling the Endothelium in Aging

Tumor necrosis factor-α (TNF-α), is a proinflammatory cytokine expressed in ischemic tissue and is known to modulate angiogenesis. Goukassian et al. [43] have reported that endothelial cell (EC)/endothelial progenitor cell (EPC) survival, vascular endothelial growth factor (VEGF) expression, EPC mobilization from bone marrow, EPC differentiation, and ultimately ischemia-induced collateral vessel development are dependent on signaling through Tumor Necrosis Factor-Receptor p75 (TNF-R2/p75), and since TNF-R2/p75 becomes an age-related limiting factor in postischemic recovery, it may be a potential gene target for treatment. These investigators evaluated neovascularization in the hindlimb ischemia model in young and old TNFR2/p75 knockout (p75KO) and wild-type age-matched controls. Poor blood flow recovery in p75KO mice was associated with increased endothelial cell apoptosis, decreased capillary density, and significant reduction in the expression of VEGF and basic fibroblast growth factor-2 mRNA transcripts in ischemic tissue and in circulating endothelial progenitor cells.

Transplantation of wild-type bone marrow mononuclear cells into irradiated old p75KO mice 1 month before hindlimb surgery prevented limb loss.

Aging impairs EPC trafficking to sites of ischemia because of the failure of aged tissues to normally activate the hypoxia-inducible factor (HIF) 1α-mediated hypoxia response. Because EPCs participate in the impair neovascularization occurring in aging, Chang et al. [44] have examined the effects of aging on EPC recruitment and vascular incorporation. Murine neovascularization was analyzed using an ischemic flap model that in aged mice (19–24 months) showed decreased EPC mobilization that resulted in impaired gross tissue survival compared with young mice (2–6 months). This decrease correlated with diminished tissue perfusion ($P<0.005$) and decreased CD31+ vascular density ($P<0.005$). Gender-mismatched bone marrow transplantation revealed significantly fewer chimeric vessels in aged mice ($P<0.05$) confirming a deficit in bone marrow-mediated vasculogenesis. Age had no effect on total EPC number in mice or humans. Reciprocal bone marrow transplantations showed that impaired neovascularization results from defects in the response of aged tissue to hypoxia and not from intrinsic defects in EPC function. Aging decreased HIF-1α stabilization in ischemic tissues because of increased prolyl hydroxylase-mediated hydroxylation [*oxidation by a process that introduces one or more hydroxyl groups (–OH) into a compound (or radical)*] ($P<0.05$) and proteasomal degradation. This resulted in a diminished hypoxia response, including decreased stromal cell-derived factor 1 ($P<0.005$) and VEGF ($P<0.0004$). This effect can be reversed with the iron chelator deferoxamine, which results in HIF-1α stabilization and increased tissue survival.

The ability of cardiac myocytes to respond to inotropic agents is reduced with aging. Recently, in our institution we studied if HIF-1α would improve the functional and $Ca^{2+}$ transient responses of aging myocytes to inotropic agents, and if this would act, at least in part, through altered mitochondrial activity [45]. Young (3–4 months) and older (18–20 months) Fischer 344 rats were used. HIF-1α was upregulated with ciclopirox olamine (CPX, 50 mg/kg/2 days). HIF-1α upregulation was detected by Western blot. Cardiomyocyte contraction and $Ca^{2+}$ transients were measured at baseline and after forskolin and ouabain. We also measured mitochondrial complex activities and ROS production. In the young group, forskolin (31%) and ouabain (31%) significantly increased percentage shortening. Similar changes were observed in the young + CPX group. $Ca^{2+}$ transients also responded similarly. However, in the older group, forskolin (12%) and ouabain (6%) did not significantly increase myocyte contractility or $Ca^{2+}$ transients. In the older + CPX group, the effects of forskolin (34%) and ouabain (29%) were restored. In the young + CPX group, there was an increase in ROS production and mitochondrial complex I and III activity compared to the young control group (Fig. 11.1).

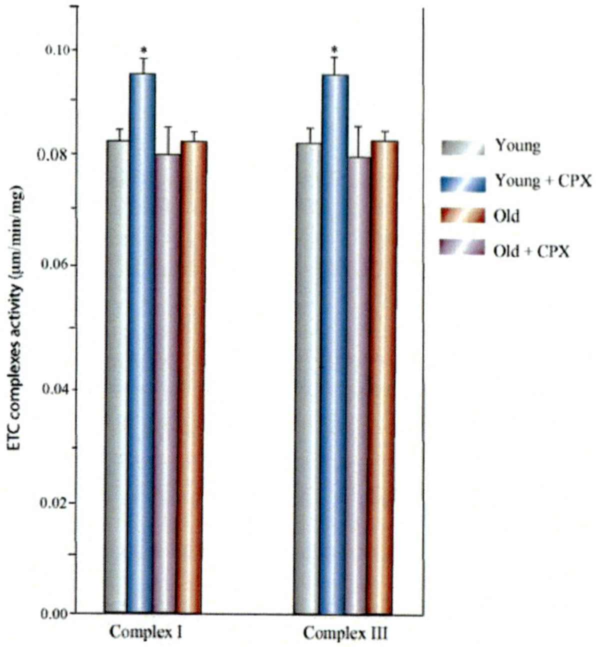


**Fig. 11.1** Mitochondrial electron transport chain (ETC) complex I and III activities in control conditions and in response to HIF-1 upregulation by ciclopirox olamine (CPX) injection among four groups of rat myocytes. Complex I and III activities were significantly (*$P<0.05$) elevated in CPX-treated young rats, but not in older rats

These differences were not observed in older groups. Our findings demonstrated impaired functional and $Ca^{2+}$ effect of positive inotropic agents in older myocytes. Upregulation of HIF-1α restored this blunted response, but this was not related to changes in mitochondrial activity induced by HIF-1α. Thus, HIF-1α improves inotropy in older myocytes without requiring mitochondrial activity changes. Furthermore, in our study HIF-1α induced increases in malonyl dialdehyde (MDA) levels in young rats but not in old rats (Fig. 11.2) concur with recent observations of an age-dependent decline in brain HIF-1α accumulation and activation of HIF-1α target genes in response to hypoxia [46]. The inducible response was significantly attenuated in the cerebral cortex of 18-month-old Fischer 344 rat but virtually absent in the cerebral cortex of 24-month-old Fischer 344 rat, and may be related to increased expression of prolyl-4-hydroxylase that continuously destroys HIF-1α and makes it unavailable for gene induction. Mitochondrial complex-III activity generates ROS that can stabilize HIF-1α through specific signaling pathways [47]. However, the reverse, i.e., HIF-1α mediated increase in ROS has not been reported yet. On the other hand, it has been shown that HIF-1α activation can take place independent of complex-III activity [48]. Thus, HIF-1α mediated (CPX treated) increases in ROS are observed in young rats (through mechanisms that remain

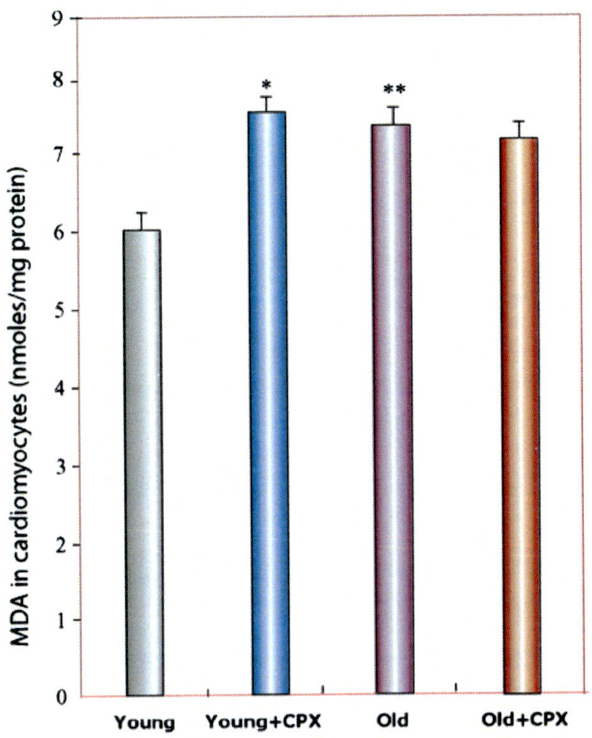


**Fig. 11.2** Measurements of the lipid peroxidation product malonyl dialdehyde (MDA) in four groups of rat cardiomyocytes [young and older rats, with and without ciclopirox olamine (CPX)]. This was used as an index of ROS production. Note the significant increase in MDA levels in the young rats treated with CPX. The CPX had no effect in the older rats. *$P<0.05$ Young+CPX versus Young; **$P<0.05$ Old versus Young

unknown) while in old rats; ROS levels are elevated and remain unchanged.

## Telomeres and CV Aging

Most of the current theories on CV aging stand by the principle that accumulation of nuclear and mtDNA damage, of variable severity, results in senescence of cardiomyocytes since DNA can be easily oxidized and damaged by a number of insults, including diet, toxins, pollution, environment, as well as other epigenetic influences and lifestyle. Despite a variety of DNA repair systems, DNA damage including base modification, large-scale rearrangements, single-strand and double-strand breaks produced over a lifetime accumulate in aging. Evidence of this aging-associated DNA damage is most evident in mtDNA, presumably because of its proximity to the formation of ROS as well as to the less extensive protection provided by mitochondrial-based DNA repair systems.

Telomeres are particular aging-sensitive areas of nuclear chromosomes. Located at the end of chromosomes, telomeres are DNA–protein structures which have been shown to be indispensable in maintaining not only the stability and integrity of the genome but are integral to the regulation of lifespan of both cultured cardiomyocytes and the entire organism [49]. The telomeric complex is composed of noncoding double-stranded repeats of G-rich tandem DNA sequences (TTAGGG in humans) that are extended several kilobase pairs, the enzyme telomerase [consisting of a catalytic telomerase reverse transcriptase (TERT) and a telomerase RNA component (Terc) serving as a template for the synthesis of new telomere DNA repeats] and numerous associated proteins with structural and regulatory roles that participate in the control of telomere length and capping. In most somatic cells, including vascular ECs, smooth muscle cells (SMCs), and cardiomyocytes, telomerase activity is insufficient, and telomere shortening occurs with increasing cell division, resulting in irreversible cell growth arrest and cellular senescence. For instance, ECs from human abdominal aorta display age-dependent telomere shortening and increased frequency of aneuploidy [50]. Moreover, a greater rate of age-associated telomere loss has been found in ECs from intimal regions of human iliac arteries [51, 52]. Age-dependent telomere shortening may trigger defective cellular function, vascular dysfunction and compromised viability of the aged organism (as illustrated in Fig. 11.3). Moreover, there is increasing evidence suggesting an association between telomere length and cardiovascular disease (CVD). For instance, several studies have documented telomere attrition and evidence of cellular senescence in vascular cells in CAD and human atherosclerosis [53–55]. Vascular SMCs from atherosclerotic fibrous caps expressed markers of senescence (e.g., senescence-associated β-galactosidase) not seen in normal vessels, and markedly shorter telomeres than normal vessels compared with normal medial vascular SMCs and telomere shortening was closely associated with increasing severity of atherosclerosis [53]. Average telomere length in leukocytes of ten patients with severe coronary artery disease (CAD) was significantly shorter than in 20 controls with normal coronary angiograms after adjustment for age and sex [54]. Telomere length was shorter in white blood cells from hypertensive men with carotid artery plaques compared to hypertensive men without plaques, and multivariate analysis indicated that telomere length was a significant predictor of the presence of carotid artery plaques [55].

Interestingly, in the Newcastle 85+ study telomere length was found to be a predictor of LV ejection fraction (EF) in the oldest old subjects (85+ years of age) [56], in which longer telomeres were associated with a significant increase in EF as measured by echocardiography. Nevertheless, whether telomere shortening is a direct cause of the cardiovascular pathology of aging or a consequence is not known.

Since telomere dysfunction and telomere shortening have been observed with aging in SMCs, endothelial, and white blood cells, they may be the primary factors in predisposing

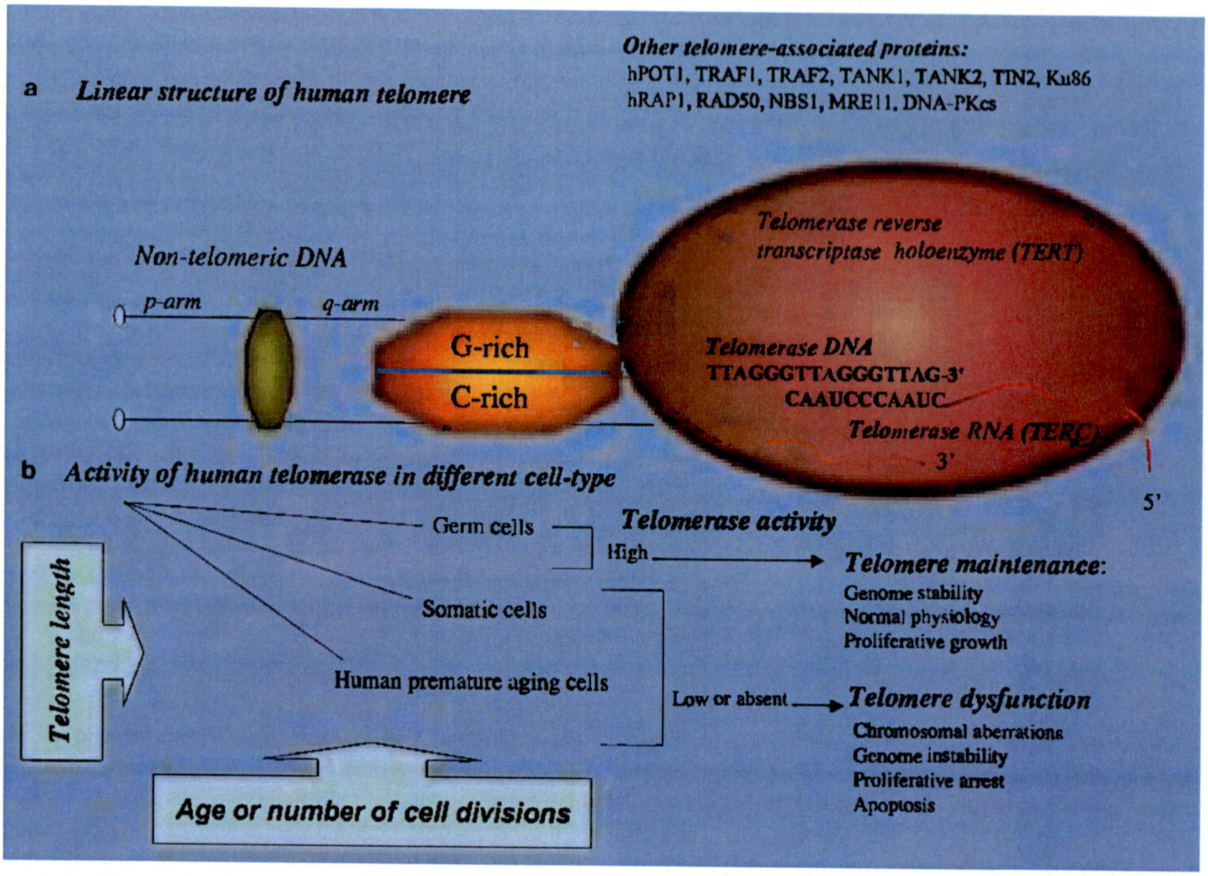


**Fig. 11.3** Telomeres, telomerase, and cell proliferation. (**a**) Linear structure of telomeres present at the ends of the chromosomes is composed of a tandem repeated sequence in telomeric DNA, a telomerase ribonucleoprotein complex, including a catalytic telomerase reverse transcriptase holoenzyme (TERT), an RNA component and additional telomeric proteins (*listed on top*). (**b**) Telomere length decreases with aging in somatic cells but not in germ cells. Progressive telomere erosion occurs with each mitotic division during normal aging. Accelerated telomere attrition is associated with human premature aging. Phenotypic changes associated with telomere length are shown

vascular tissues to atherosclerosis, and to a decreasing capacity for neovascularization [49]. Interestingly, attrition of telomere length is most prominent under conditions of high OS, which prevalence is likely linked in hypertensive and diabetic individuals, and in individuals with CAD. Under conditions of mild chronic OS, the loss of telomere integrity is a major trigger for the onset of premature senescence [57]. Kurz et al. recently demonstrated that glutathione-dependent redox homeostasis plays a key role in the preservation of telomere function in endothelial cells [58]. Increasing intracellular OS in human umbilical vein ECs (HUVECs) by treating cells with inhibitors of glutathione synthesis caused premature EC senescence with diminished telomere integrity. Incubation of ECs with $H_2O_2$ induced the nuclear export of TERT into the cytosol and loss of nuclear telomerase activity, and the onset of EC replicative senescence [59]. Similarly, cultivation of ECs resulted in an increased endogenous ROS formation starting after 29 population doublings concomitant with nuclear TERT export and senescence onset. Incubation of ECs with the antioxidant *N*-acetylcysteine, or with the statin atorvastatin, reduced the intracellular ROS formation and delayed nuclear export of TERT protein, loss in the overall TERT activity, and the onset of replicative senescence.

Recent observations have demonstrated that telomere biology also plays a significant role in the functional augmentation of by statins [60]. The *ex vivo* culturing of EPCs leads to premature replicative senescence brought about by uncapping of telomeres associated with the loss of telomere repeat-binding factor (TRF2) and telomeric dysfunction rather than telomere shortening. Cotreatment of the cultured EPCs with statins delayed their premature senescence, in part by induction of TRF2 expression at the posttranslational level. While the ability of EPCs to sustain ischemic tissue repair may be limited in the aging heart, estrogens which have been shown to accelerate recovery of the endothelium after vascular injury, were able to inhibit the onset of EPC senescence and significantly increase EPC telomerase activity [61]. Incubation of *ex vivo* cultivated EPCs with 17β-estradiol in a dose-dependent fashion significantly increased TERT transcript levels as gauged by RT-PCR. The finding that this antisenescent effect of estrogen was significantly inhibited by pharmacological blockers of the phosphoinositide 3-kinase (PI3K) activity confirmed the role of PI3K-Akt pathway in mediating TERT induction by estrogen also suggested by elevated Akt phosphorylation in estrogen-treated EPCs. In addition, EPCs treated with 17β-estradiol exhibited significantly enhanced mitogenic potential and the release of VEGF protein; moreover, EPCs treated with both 17β-estradiol and VEGF were more likely to integrate into the network formation than those treated with VEGF alone. Others have shown that estrogen also promotes cardiovascular protection via nonnuclear effects of estrogen receptor signaling leading to endothelial nitric oxide synthase (eNOS) activation also utilizing the PI3K-Akt pathway [62]. The production of nitric oxide (NO) by cultured vascular ECs which could be induced by repeated exposure to the NO donor *S*-nitroso-penicillamine reduced EC senescence and delayed age-dependent inhibition of telomerase activity [63].

Telomerase inactivation during aging may also be related to the oxidized low-density lipoprotein (LDL)-accelerated onset of EPC senescence, which leads to the impairment of proliferative capacity and network formation [64]. Oxidized LDL-induced EPC senescence was accompanied by a 50% reduction in telomerase activity. The cardiovascular protective effects of estrogens promoted via indirect actions on lipoprotein metabolism, and through direct effects on vascular ECs and SMCs, likely contribute to the lower incidence of CVD observed in premenopausal women compared with men. In this regard, it is noteworthy that women have a decelerated rate of age-dependent telomere attrition over men [65].

Several observations primarily in studies with human fibroblasts support the concept that the average telomere length is better maintained in conditions of low OS generated by either the addition of antioxidants, low oxygen, or overexpression of antioxidant enzymes [66–68]. Selective targeting of antioxidants directly to the mitochondria can counteract telomere shortening and increase lifespan in fibroblasts under mild OS [69]. These studies have led to the proposal that mitochondrial dysfunction may be a key mechanism underlying the loss of cellular proliferative capacity characterizing replicative senescence through enhanced telomere shortening (Fig. 11.4), and the generation of ROS may signal the nucleus to limit cell proliferation through telomere shortening and telomeres as sensors to damaged mitochondria [70].

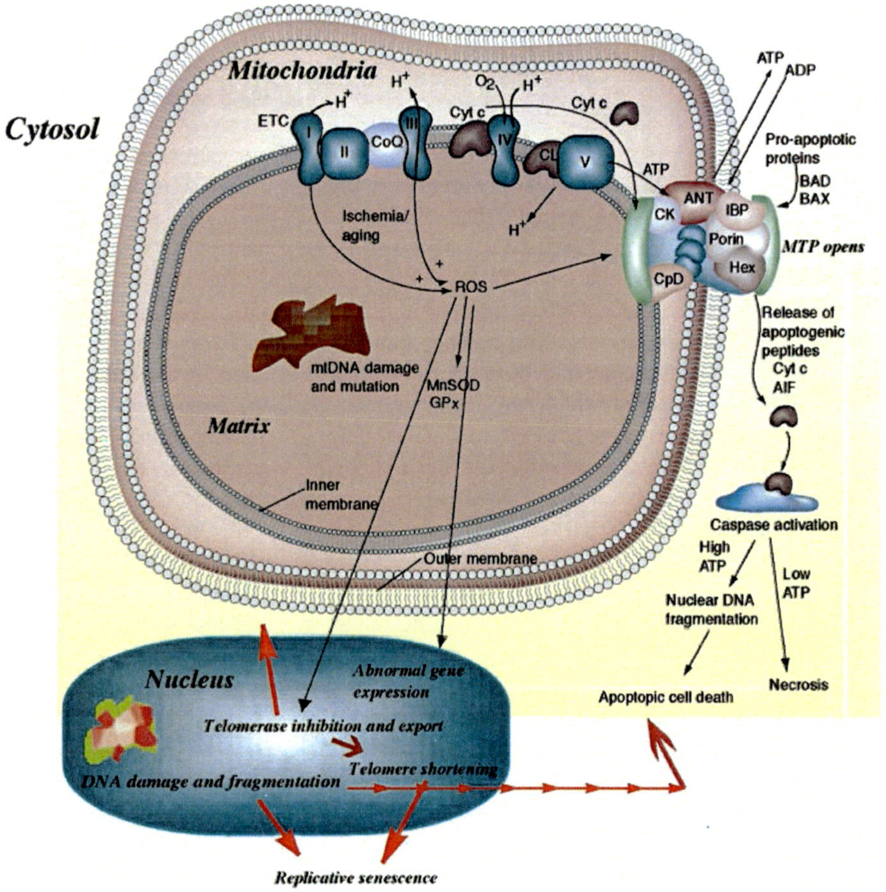


**Fig. 11.4** Mitochondrial ROS production contributes to telomere-dependent replicative senscence. Production of ROS by aberrant mitochondrial respiratory complexes thought to occur in aging leads to mtDNA and protein damage and lipid peroxidation, MTP opening and apoptotic progression (via cytochrome c release, membrane permeabilization, apoptosome formation, and nuclear DNA fragmentation). ROS also affects the nucleus by reducing telomerase activity and increasing its export from the nucleus, causing direct DNA damage and activating specific gene expression – all of which contribute to the signaling of replicative senescence

**Fig. 11.5** Telomeres and telomerase in human with cardiovascular disease (CVD). Telomere and telomerase phenotype in different tissues from patients with indicated CVD

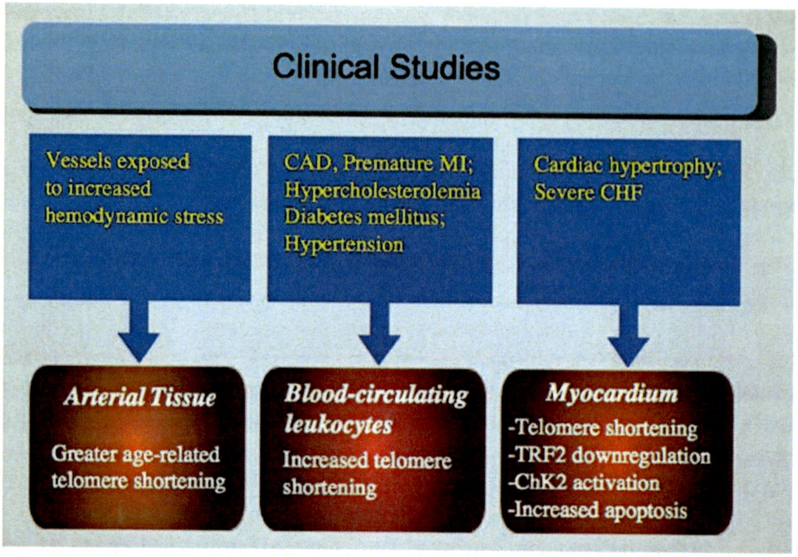


Furthermore, there is increasing evidence that telomere dysfunction is an important factor in the pathogenesis of human and experimental CVDs associated with aging, including hypertension, atherosclerosis, and HF (Fig. 11.5), and is increasingly viewed as an important potential therapeutic target in their treatment.

## Cellular Damage/Cell Loss, Mitochondria, and CV Aging

Cell damage occurs at random in any organ or tissue, including the heart, although the number of damaged cells required to affect cardiac function is unknown. When assessing cell damage in the heart (or any other similar organ), one should be mindful of the significant difference between using *in vitro* compared to *in vivo* approaches, with cells in culture reaching a limit in their potential for cell division/differentiation, which may not occur in *in vivo* [71]. Therefore, caution should be exercised in the interpretation of data collected from different methodologies. Previous observations have shown that under physiological and pathological conditions, ROS are abundantly generated in mitochondria, mainly during aging and in ischemia/reperfusion (I/R) of the heart. Moreover, mitochondrial dysfunction and ROS generation may play a significant part in the loss of postmitotic cells, such as cardiomyocytes. Confirming this, *in vitro* studies of $H_2O_2$-treated cardiomyocytes in our laboratory and others have shown that increased mitochondrial OS and declining mitochondrial energy production lead to the activation of apoptotic pathways [72, 73]. However, whether this also occurs in the aging heart *in vivo* is not known.

Although the role of apoptosis in normal myocardial aging is presently under considerable debate, cardiomyocyte apoptosis has been confirmed by data demonstrating that the aging rat heart had significantly elevated levels of cytochrome *c* release from mitochondria, as well as dec-reased levels of the antiapoptotic protein Bcl-2, whereas levels of the proapoptotic protein Bax were unchanged [74]. Moreover, preliminary evidence from both animal models and human clinical studies suggests that apoptosis also plays a contributory role in aging-associated cardiovascular diseases, including cardiomyopathy and HF, increased susceptibility to ischemia and myocardial infarction (MI), and atherosclerosis [75–77].

The mitochondrial-mediated intrinsic apoptotic pathway, which has been documented in the aging and failing heart, may play a significant role in the development of cardiac dysfunction and pathogenesis. This pathway also features an extensive dialogue between the mitochondria, the nucleus and other subcellular organelles (described in more detail in Chap. 8). Central to the early triggering events in the intrinsic apoptotic pathway, the release to the cytosol of a number of mitochondrial-specific proteins from the mitochondrial intermembrane space, including cytochrome *c*, endonuclease G (EndoG), apoptosis inducing factor (AIF), and Smac/Diablo, subsequently leads to downstream caspase activation, nuclear DNA fragmentation, and cell death [78]. The release of EndoG and AIF, and their subsequent translocation to the nucleus, specifically affect degradation of nuclear DNA, even in the absence of caspase activation [79, 80]. In addition, AIF loss-of-function studies in transgenic mice have recently revealed that this mitochondrial flavoprotein also plays an essential function in mitochondrial respiration and aerobic energy and loss of this mitochondrial-based protein leads to impaired cardiac contractility and oxidative phosphorylation (OXPHOS) deficiency [81, 82].

Studies employing knockdown of EndoG using lentiviral-delivered RNAi have demonstrated that EndoG function is essential for ischemia-mediated nuclear DNA degradation in postnatal cardiomyocytes and was associated with an increase to caspase-independent pathways occurring during early cardiac differentiation [83].

Both Smac/Diablo and cytochrome *c* are released from the mitochondria and become subsequently involved in cytosolic caspase activation. Smac binds and inhibits cytosolic anti-apoptotic signaling complexes [e.g., inhibitor of apoptosis proteins (IAPs)] that modulate apoptosis, whereas cytosolic cytochrome *c* binds apoptotic protease activating factor 1 (Apaf-1) along with dATP and promotes the recruitment of procaspase-9 into the apoptosome, a multiprotein complex resulting in cytosolic caspase activation [83, 84]. The release of these mitochondrial peptides primarily involves an outer membrane permeabilization mediated by proapoptotic cytosolic factors Bax, Bak, and tBID. In addition, a group of BH3-only proteins termed executioner proteins have been identified, including Bnip3, PUMA, Bid, Bad, HGTD-P, and Noxa, which in response to I/R in heart and/or brain also serve this function [85]. In response to both external pro-death signals largely provided by the extrinsic apoptosis pathway (e.g., ischemia/hypoxia, cytotoxic cytokine TNF-$\alpha$, and Fas ligand) and nuclear signals (e.g., p53), these proapoptotic factors translocate to mitochondria where they bind outer membrane proteins [e.g., voltage-dependent anion channel (VDAC)] leading to channel and pore formation in the outer membrane [78]. Activation of mitochondrial apoptosis-induced channels (MAC) in the outer membrane, which provide specific pores for the passage of intermembrane proteins, in particular cytochrome *c*, to the cytosol, is also regulated by Bcl-2 family proteins [86]. On the other hand, and concurring with Martinez-Caballero et al. [87] although Bcl-2 family proteins control intrinsic apoptosis, the mechanisms underlying this regulation are incompletely understood. They reported that patch clamp studies of mitochondria isolated from cells deficient in one or both of the proapoptotic proteins Bax and Bak show that at least one of the proteins must be present for the formation of the cytochrome *c*-translocating channel, MAC, and that the single channel behaviors of MACs containing exclusively Bax or Bak are similar. Truncated Bid catalyzes MAC formation in isolated mitochondria containing Bax and/or Bak with a time course of minutes and does not require VDAC1 or VDAC3. Mathematical analysis of the stepwise changes in conductance associated with MAC formation is consistent with pore assembly by a barrel-stave model. Assuming the staves are two transmembrane $\alpha$-helices in Bax and Bak, mature MAC pores would typically contain approximately nine monomers and have diameters of 5.5–6 nm. The mitochondrial permeability data are inconsistent with the formation of lipidic pores capable of transporting megadalton-sized macromolecules as observed with recombinant Bax in liposomes.

Protein release from the intermembrane space and cristae, where the majority of cytochrome *c* is located, also appears to be closely associated with the opening of the voltage-sensitive mitochondrial permeability transition pore (MTP) located at the contact sites between inner and outer membranes, which is responsive to membrane potential changes, mitochondrial ROS and $Ca^{2+}$ overload, prooxidant accumulation and NO [88]. Recent studies in the postischemic heart have noted that MTP opening in reperfused hearts increased along with reperfusion time and concurs with cytochrome *c* release from mitochondria [89]. However, unlike MAC, MTP is nonspecific to apoptotic cell death and is a feature of necrotic cell death.

The composition of the MTP, while still controversial, is thought to involve several key components of mitochondrial bioenergetic metabolism, including the adenine nucleotide translocator (ANT), mitochondrial creatine kinase, the outer membrane porin molecule and inner membrane cyclophilin D. Opening of the MTP promotes significant changes in mitochondrial structure and metabolism, including increased mitochondrial matrix volume leading to mitochondrial swelling, release of matrix calcium, altered cristae, and cessation of ATP production secondary to electron transport chain (ETC) uncoupling, and dissipation of the mitochondrial membrane potential [88, 90]. Scorrano et al. have proposed that cytochrome *c* efflux in apoptosis is coordinated with the activation of a mitochondrial remodeling pathway characterized by changes in inner mitochondrial membrane morphology and organization, ensuring the complete release of cytochrome *c* as well as the onset of mitochondrial dysfunction, which might further contribute to the aging and/or failing heart phenotype [91].

Studies with the multifaceted apoptosis repressor with a caspase recruitment domain (ARC), which acts on targets in both the intrinsic and extrinsic pathway, have demonstrated that ARC is cardioprotective stemming the release of cytochrome *c* and hypoxia-induced injury in the heart, although its role in aging has not yet been assessed [92].

The endoplasmic reticulum (ER) has been recently recognized as an important organelle in the intrinsic pathway mediating cell death elicited by a subset of stimuli, such as OS [93]. Similar to their roles in transducing upstream signals to the mitochondria, proapoptotic proteins appear to relay upstream death signals to the ER triggering the release of $Ca^{2+}$, which in turn can rapidly accumulate in mitochondria promoting MTP opening [94, 95].

Prosurvival factors from the growth factor signaling pathways (e.g., IGF-1) can inhibit the progression of the apoptotic pathway.

Experimental models have suggested that mitochondria-related apoptosis is a contributory mechanism of aging. Compared to myocytes from younger animals, myocytes derived from the hearts of old mice displayed increased levels of markers of cell death and senescence [29]. While it is possible that apoptosis may be a protective mechanism in the aging heart to get rid of those damaged, potentially dangerous cells in a mechanistic effort to incline the balance toward healthy cells, the marked accumulation of mtDNA mutations in strains of transgenic mice overexpressing proofreading-deficient polymerase γ was unexpectedly not associated with increased markers of OS or defective cellular proliferation, but rather correlated with the induction of apoptotic markers, particularly in tissues characterized by rapid cellular turnover. The levels of apoptotic markers which are increased during aging in normal mice, increased further in these transgenic mouse strains and correlated with the accumulation of mtDNA mutations in the heart [96]. Analysis of senescence in cultured human fetal cardiomyocytes has shown growth arrest and morphological changes in association with senescence associated β-galactosidase activity [97]. Studies with glioma cells stained for senescence-associated β-galactosidase activity, a biomarker specific for senescent cells, showed that the enlarged cells gave a distinctive positive staining reaction [98]. This senescence phenotype appears to be dependent on the continuous expression of $p16^{INK4a}$. It has been proposed that the induced expression of $p16^{INK4a}$ in these cells reverted their immortal phenotype and caused immediate cellular senescence. Interestingly, increased expression of $p16^{INK4a}$ has also been detected in aging cardiomyocytes [29]. Interestingly, mice containing an IGF-1 transgene had attenuated levels of senescence-associated gene products (e.g., p27Kip1, p53, $p16^{INK4a}$, and p19ARF) and Akt phosphorylation in myocytes, and compared to wild-type mice exhibited decreased levels of myocyte DNA damage and cell death. Unfortunately, neither myocyte mitochondrial structure nor function was evaluated [29].

## Reactive Oxidative Species Generation and CV Aging

ROS, by-products of normal metabolic processes are highly reactive, small oxygen-containing molecules that play an essential role on OS levels and signaling of the aging heart. In the mitochondria, ROS are generated from electrons produced (or leaked) from the ETC at complexes I and III although nonmitochondrial sources of ROS generation are both active and physiologically relevant in the heart. The inefficiency of electron transfer through the mitochondrial ETC and altered level of antioxidant defenses underlie the accumulation of ROS and OS in the aging heart [99]. A causative role for ROS in the aging process, known as the free radical theory of aging (FRTA), presupposes that in biological systems ROS attack molecules and cause a decline in the function of organ systems, eventually leading to organ failure and death. In addition, the bioenergetic dysfunction that occurs with aging further increases the accumulation of ROS. Mitochondrial ROS generation is increased in cells with abnormal ETC function as well as under physiological and pathological conditions, where oxygen consumption is increased with respiratory complex I in particular playing a major role in the formation of superoxide radicals [100]. A decline in complex I activity (in concert with increased state 4 respiration) elevates ROS production during aging. This promotes the generation of prooxidant compounds, leading to the modulation of MTP opening, abnormal mitochondrial membrane potential, and induction of cell death.

Among its many targets in the cardiomyocyte, ROS causes extensive damage in particular to mitochondrial macromolecules (e.g., carbohydrates, DNA, lipids, and proteins) proximal to its source as well as reduced fluidity in the mitochondrial inner-membrane. Age-associated mitochondrial membrane changes include increased membrane rigidity, elevated levels of cholesterol, phosphatidylcholine, omega-6 polyunsaturated fatty acids (PUFA), and 4-hydroxy-2-nonenal with decreased levels and oxidative modifications in omega-3 PUFA and cardiolipin, unique inner membrane phospholipids [101]. These changes are potentially responsible for the increased susceptibility of the aging heart to the damaging effects of I/R, including mitochondrial $Ca^{2+}$ overload, opening of the MTP and cell death. Under the appropriate conditions, the aging myocardium may exhibit increased permeability of the inner mitochondrial membrane to solutes causing mitochondrial swelling, "proton leak," reduced ETC efficiency and uncoupling of OXPHOS from respiration which would limit net ATP production [102, 103]. Studies from our laboratory have confirmed increased MTP sensitivity to $Ca^{2+}$ in senescent rat myocardial mitochondria, and these findings are consistent with previous observations of enhanced $Ca^{2+}$ vulnerability and $Ca^{2+}$-induced damage in mitochondria from senescent animals' hearts [104].

Oxidative modification of proteins, such as carbonylation, nitration, and the formation of lipid peroxidation adducts, such as 4-hydroxynonenal (HNE) and malondialdehyde (MDA), are by-products of oxidative damage secondary to ROS [105, 106]. Several studies support across-the-board or global increases in levels of some protein oxidative modifications [107]. For instance, protein carbonyls have been reported to increase exponentially with age, particularly in the last third of lifespan reaching a level such that on average one out of every three protein molecules carries the modification and the modified proteins are likely dysfunctional either as enzymes or structural proteins [108]. Other studies have focused on specific proteins (and specific residues within those proteins) that are modified with age. While modifications in membrane-localized ANT and in the protein subunits of respiratory complexes I–V secondary to ROS-mediated nitration, carbonylation, HNE and MDA adduct formation and an associated decline in enzymatic activity *in vitro* have been reported [109, 110]. Moreover, studies carried out on bovine heart submitochondrial particles found that proteins sustaining oxidative damage generated from *in vivo* basal level of ROS were primarily localized in the mitochondrial matrix [111]. Superoxide is also especially damaging to the Fe-S centers of metabolic enzymes (e.g., complex I, aconitase, and succinate dehydrogenase). Inactivation of the Krebs cycle enzyme, mitochondrial aconitase by superoxide, which generates Fe (II) and $H_2O_2$, increases formation of the highly reactive hydroxyl radical [109, 112], and in aging mouse heart, this enzyme was also a prominent target of MDA-adduct formation resulting in significant age-related decline in its activity [109]. Also decline in mitochondrial aconitase activity and attenuation of hydroxyl radical generation have been reported in cardiac H9c2 cells after heat shock and this might play an important role in reducing the myocardial ischemic injury, observed in heat shock-treated animals [113].

Another type of reactive free radical, peroxynitrite formed from NO reacting with superoxide radicals is a strong oxidant associated with age-related modification of specific proteins generally detected as tyrosine nitration of targeted subunits. Viner et al. [114] have demonstrated significant age-dependent loss in the sarcoplasmic reticulum $Ca^{2+}$-ATPase (SERCA) activity in parallel with the accumulation of significant amount of nitrotyrosine in skeletal muscle. Significant increases in nitrotyrosine in SERCA were also found in aging rat heart [115] and in atherosclerotic plaques and vessels from rabbits and from human [116]. While it has been demonstrated that peroxynitrite reacts in vitro with mitochondrial membranes from bovine heart to significantly inhibit the activities of respiratory complexes I and V as well as resulting in 3-nitrotyrosine accumulation in selective complex I and V subunits [117], this has not been demonstrated in aging. However, a recent analysis of the aging cardiac proteome in rats with regards to 3-nitrotyrosine has identified substantial modification in over 48 proteins, including many involved in bioenergetics, such as cytosolic proteins of glycolysis (i.e., α-enolase, α-aldolase, GAPDH), mitochondrial proteins involved in electron transport, TCA cycle, fatty acid β-oxidation and OXPHOS [i.e., 3-ketoacyl-CoA thiolase, acetyl-CoA acetyltransferase, malate dehydrogenase, creatine kinase, electron-transfer flavoprotein, F1-ATPase (ATP synthase), and VDAC/porin], as well as proteins involved in cardiomyocyte structural integrity (i.e., desmin) [118].

Another central premise of the mitochondrial theory of aging suggests that somatic mutations in mtDNA, induced by oxygen free radicals, are a primary cause of energy decline. Evidence has shown that oxidative damage affects nucleic acids, and in particular mtDNA, by the induction of single- and double-strand breaks, base damage, and modification (including 8-hydroxyguanosine formation) resulting in the generation of point mutations and deletions [119, 120].

The role of point mutations and deletions in mtDNA in aging has been tested experimentally by several independent groups using different transgenic mouse strains and constructs of the proofreading-deficient version of Polγ, the nuclear-encoded catalytic subunit of mtDNA polymerase [96, 121]. These mutant mice developed between a three- and eightfold increase in the levels of mtDNA point mutations, as well as increased amounts of deleted mtDNA associated with a reduced life span and premature onset of aging-related phenotypes, including cardiomegaly, providing a causative link between mtDNA mutations and aging. However, no significant accumulation of ROS or of ROS damage, changes of antioxidant enzymes nor markers of OS were found in tissues of the mice [122]. The finding of respiratory deficiency and increased apoptosis in these transgenic strains rather than ROS suggest their potential involvement as primary inducers of this aging phenotype. Thus, the overall relevance of these transgenic models to normal aging processes including in human has been questioned [123].

## Inflammation Signaling Pathways and CV Aging

Major chronic aging-related diseases, such as atherosclerosis, arthritis, dementia, osteoporosis, and CVDs, are inflammation-related, and inflammation is likely the underlying basis for the molecular alterations that link aging and age-related diseases.

OS and redox dysfunction, which appear to be inextricably linked with aging, are critical risk factors for age-related inflammation [124]. Much of this is mediated by the activation of redox-sensitive transcription factors and dysregulated gene expression in the nucleus by age-related OS. Besides ROS and NO other molecules and signal transduction pathways, such as *inflammatory signaling*, are actively involved in aging. Key players involved in the inflammatory process are the age-related upregulation of NF-κB, IL-1β, IL-6, TNF-α, cyclooxygenase-2, adhesion molecules, and inducible NO synthase. Through this increased expression of inflammatory components, a series of profound subcellular changes occur in aging outside the nucleus of the cardiovascular cells and in the membranes and membrane-bound organelles (e.g., mitochondria), which dictate the electrophysiological and metabolic functioning of cardiovascular cells, as well as relating to the further control and generation of OS.

Inflammatory markers have been identified as significant independent risk indicators for cardiovascular events. While adults over the age of 65 have experienced a high proportion of such events, the available epidemiological data comes mainly from middle-aged subjects. Kritchevsky et al. [125] have examined the role that inflammatory markers play as predictors of the incidence of CVD specifically in older adults. Interestingly, IL-6, TNF-α, and IL-10 levels appear to predict cardiovascular outcomes in adults <65 years. Data on C-reactive protein (CRP) was rather inconsistent and appeared to be less reliable in old age than in middle age. Also, fibrinogen levels have some value in predicting mortality but in a nonspecific manner. Thus, the elderly inflammatory markers are nonspecific measures of health and may predict both disability and mortality, even in the absence of clinical CVD, and methodology designed to prevent CVD through the modulation of inflammation may be helpful in reducing disability and mortality. The role of increased inflammatory markers, such as IL-6 and IL-1β as risk factors in aging, and in the development of MI has also been reported [126]. Analysis of polymorphisms in *IL-6* gene promoter (−174 G→C) has indicated that elderly patients with acute coronary syndrome (ACS) carrying *IL-6* −174 GG genotypes exhibited a marked increase in 1-year follow-up mortality rate, suggesting that *IL-6* −174 G→C polymorphisms can be added to the other clinical markers, such as CRP serum levels and a history of CAD, useful in identifying elderly male patients at higher risk of death after ACS [127].

## Neuroendocrine Signaling in CV Aging

The neuroendocrine theory of aging elaborates mainly on the wear and tear occurring in the aging neuroendocrine system. This system is a complicated network of signals governed by the release of hormones largely regulated by the hypothalamus, which controls a large assortment of chain reactions in numerous target organs and which in turn regulates other glands to release their hormones. It also responds to the body hormone levels as a guide to modulate overall hormonal activity. For example, if cortisol damages the hypothalamus, then over time it becomes a vicious cycle of continued hypothalamic damage, leading to an ever-increasing degree of cortisol production and thus more hypothalamic damage. This damage could then lead to hormonal imbalance as the hypothalamus loses its ability to control the system. The cause–effect relationship of specific elements of neuroendocrine regulation, including cardiac β-adrenergic receptor (β-AR) and G protein-coupled receptors, thyroid hormone (TH), growth hormone (GH) and IGF-1, and their role in the dysfunction of the aging heart is discussed in this subsection.

### *Adrenergic and Muscarinic Receptors in the Aging Heart*

Cardiac β-AR responsiveness in model systems and in humans *in vivo* decreases with aging, and the mechanisms by which this may occur include the downregulation and decreased agonist binding of β1-receptors, uncoupling of β2-receptors, and abnormal G protein-mediated signal transduction [128]. While age-dependent changes in human adrenergic receptors are well established, little is known about possible age-dependent alterations in human cholinergic receptors [129]. In the human heart, there are muscarinic acetylcholine receptors (mAChRs) that are predominantly of the $M_2$ subtype and that couple to the inhibitory G protein, $G_i$ [130, 131]. Stimulation of these receptors causes inhibition of adenylyl cyclase activity and a decrease in heart rate as well as in β-adrenoceptor-mediated increases in ventricular contractility [132, 133].

In healthy humans, ganglionic blockade unmasks a clear age-related decrease in cardiac responses to isoproterenol but not to epinephrine. It has been hypothesized by Turner et al. [134] that an age-related decrease in neuronal uptake (which affects epinephrine but not isoproterenol) may offset a parallel decrease in β-receptor-mediated responses. Healthy human aging was associated with decreased cardiac responsiveness to the β-agonist epinephrine, and this decrease is balanced by concomitant decreases in buffering of these responses by neuronal uptake and the arterial baroreflex. With aging, a decline in cardiac function, in part due to decreased α- and β-AR-mediated contractility occurs. While defects in β-AR signaling are known to occur in the aging heart, which components of the α1-AR signaling cascade are responsible for the aging-associated deficit in α1-AR contractile function, have just begun to be identified. These include protein kinase C (PKC) and associated anchoring proteins, including receptors for activated C kinase (RACKs).

Age can significantly influence the cardiovascular responses to α-adrenergic stimulation, and phenylephrine, by acutely increasing afterload, has been shown to be effective in revealing the left ventricular systolic dysfunction occurring with aging. Moreover, it appears that the increase in systolic blood pressure in response to an α-adrenergic challenge is significantly influenced not only by age, but also by gender [134]. Hees et al. [135] have analyzed the effects of β-adrenergic stimulation on LV filling, and its major determinant, relaxation, in an aging population. They found that aging was accompanied by a blunted inotropic but preserved chronotropic response to steady-state dobutamine infusion and although LV filling reserve declines with age, relaxation reserve does not.

Korzick et al. [136] using a heart Langendorff-perfused model, have measured cardiac contractility (dP/dt) in 5-month adult and 24-month-old aging Wistar rats following maximal α1-AR stimulation with phenylephrine. Evaluation of the subcellular distribution of PKCα and PKCε, and their respective anchoring proteins RACK1 and RACK2 by Western immunoblot analysis revealed that the subcellular translocation of PKCα and PKCε, in response to α1-AR stimulation, was disrupted in the aging myocardium. Moreover, age-related reductions in RACK1 and RACK2 levels were also observed, suggesting that alterations in PKC-anchoring proteins may contribute to impaired PKC translocation and defective α1-AR contraction in the aged rat heart. In addition, these investigators subsequently sought to determine whether age-related defects in α1-AR contraction could be reversed by chronic exercise training (treadmill) in 4-month adult and 24-month-old rats and found that the age-related decrease in α1-AR contractility in the rat heart can be partially reversed by exercise, suggesting that alterations in PKC levels underlie, at least in part, exercise training-induced improvements in α1-AR contraction [137]. Moreover, in the Fischer 344 rat both aging and gender mediate substantial alterations in various PKC isoform interactions with other signaling factors, such as Erk1/2 to form signaling modules (SMS) [138]. Senescence was associated with increased cytosolic and mitochondrial PKCα levels in both males and females, whereas increases in cytosolic PKCα-Erk1/2 SMS were only observed in aged females. Mitochondrial PKCδ and PKCδ-Erk1/2 SMS increased in both males and females with age; however, increases in cytosolic PKCδ were only observed in aged males. Nuclear and mitochondrial PKCδ-Erk1/2 SMS were 3.5- and 4.8-fold greater in males compared to females, respectively, and increases in mitochondrial-PKCε-Erk1/2 SMS were also specific to aged males. It is thought that these substantial age and gender-associated differences in the magnitude and distribution of cardiac PKC-Erk1/2 SMS may in part underlie the age-related reductions in ischemic stress reserves, particularly apparent in aged women as well as differences in cardioprotection with aging.

The cardiac effects of α1-adrenergic stimulation, both in cardiomyocyte $Ca^{2+}$-transient and cardiac PKC activity have been assessed by Montagne et al. [139] in 3- and 24-month-old Wistar rats. Their findings suggested that the negative effect of α1-adrenergic stimulation on cardiomyocyte $Ca^{2+}$ transient observed in old rats could be related to the absence of α1-adrenergic-induced PKC translocation.

The effect of aging on the human sympathetic nervous system is at the present time a subject of great interest mainly because diverse cardiac pathologies, including essential hypertension, CAD, HF, and dysrhythmias increase with age, and the involvement of the sympathetic nervous system may be an important pathophysiological component [140].

## *Cardiac G Protein-Coupled Receptors*

Cardiac G protein-coupled receptors (GPCRs) that function through stimulatory G protein $G\alpha_s$, such as β1- and β2-ARs, play a key role in cardiac contractility. $G\alpha_{i2}$ levels increases with age in both human atria resulting in diminished levels of both basal and receptor-mediated adenylyl cyclase activity [141]. Similarly, significant increases were found in $G\alpha_{i2}$ levels by immunoblot analysis in left ventricles of 24-month-old Fischer 344 rats with reduced levels of both basal and receptor-mediated adenylyl cyclase activity [142]. These elevated levels of $G_i$ may subsequently increase the receptor-mediated activation of $G_i$ through multiple GPCRs. Increased $G_i$ activity is likely to have an adverse effect on heart function since $G_i$-coupled signaling pathways in the heart reduce both the rate and force of contraction, which is in part mediated through cAMP-mediated phosphorylation of phospholamban [143]. Investigation of the effects of age on GPCR signaling in human atrial tissue also showed that the density of atrial mAChR increases with age but reaches statistical significance only in patients with diabetes [144]. Interestingly, in elderly subjects of similar ages, those with diabetes have 1.7-fold higher levels of $G\alpha_{i2}$ and 2-fold higher levels of $G\beta_1$. In contrast, earlier studies have reported that right atrial mAChR density significantly decreased in advanced age [129]. The differences between these studies could be explained by differences in age between patients groups; one study examined only adults with an age range from 41 to 85 years [1] while the other study group's age ranged from 5 days to 76 years. In this regard, it is interesting to note that Oberhause et al. [145] found that acetylcholine release in human atria which is controlled by mAChR $M_2$ is significantly reduced in atria of patients >70 years of age and patients with late diabetic complications suggesting that locally impaired parasympathetic activity may be a contributory risk factor in sudden cardiac death in the elderly and in diabetic patients.

Although structurally closely related to insulin, the relaxin family peptides (RXFPs) act on a group of four GPCRs now known as RXFP receptors [146]. Interestingly, relaxin and its receptor are often involved in pathologies that are considered to be age-related, such as fibrosis, wound healing and in response to myocardial infarction. Relaxin affects collagen metabolism, inhibiting collagen synthesis and enhancing its breakdown by increasing matrix metalloproteinases. It also enhances angiogenesis and is a potent renal vasodilator. Activation of two of the leucine-rich receptors, RXFP1 or RXFP2, causes increased cAMP accumulation and the initial response for both receptors is the result of $G_s$-mediated activation and $G_{oB}$-mediated inhibition of adenylate cyclase; RXFP1 has a higher affinity for relaxin while RXFP2 primarily binds to insulin-like peptide 3. Since drugs acting at RXFP1 may have potential clinical applications in diseases involving tissue fibrosis, such as cardiac and renal failure, the relaxin systems may represent an important pharmacological target in clinical management of aging, and in particular age-related cardiac pathologies [147].

## Thyroid Hormone/SERCA in the Aging Heart

The sarcoplasmic reticulum $Ca^{2+}$-ATPase (SERCA) is a transmembrane protein that pumps cytoplasmic $Ca^{2+}$ into the SR and in this context plays an important role in regulating the concentration of calcium around the contractile elements in cardiomyocytes. In the aging/senescent heart, the expression of the SERCA gene is downregulated, contributing to abnormal calcium homeostasis and impairment of cardiac function. However, the molecular mechanisms that regulate the SERCA gene expression in the heart during aging are still unclear. In our laboratory, we have found evidence that implicates a decrease in TH responsiveness in the aging heart (unpublished data). This decrease in large part involves the binding of the TH receptor (TR) and retinoid X receptor (RXR) heterodimer to TH-responsive elements (TREs) located in the *SERCA* and cardiac *MHC* gene promoters.

Age-associated changes in the TR and RXRs could explain the age-associated changes in SERCA and MHC expression. Long et al. [148] have reported that although no significant changes in *RXRα* or *RXRβ* mRNA levels occur in the aging rat heart, both α1 and *α2 TR* mRNA levels decreased significantly between 2 and 6 months of age. During this time period, the mRNA levels of *α-MHC* declined by more than half, whereas *β-MHC* mRNA levels remained low and unchanged. On the other hand, between 6 and 24 months, when mRNA levels of *β-MHC* increased and *α-MHC* continued to decrease, there was a significant decline in *TRβ1* and *RXRγ* mRNA levels accompanied by a reduction in the TRβ1 and RXRγ protein levels. These findings suggest that the decline in *α-MHC* gene expression may be biphasic and in part due to a decline in α1 (and possibly α2) TR levels between 2 and 6 months of age, and a decline in TRβ1 and RXRγ levels at later age.

The aging-mediated downregulation of MHC and SERCA mediated by myocardial TH/TR signaling-mediated transcriptional control can be reversed with exercise [149]. While the expression of myocardial TRα1 and TRβ1 proteins is significantly lower in sedentary aged rats than in sedentary young rats, their expression is significantly higher in exercise-trained aged rats than in sedentary aged rats. Furthermore, the activity of TR binding to the TRE transcriptional regulatory elements in the *α-MHC* and *SERCA* promoters and the myocardial expression of α-MHC and SERCA (both mRNA and protein) were upregulated with exercise training in the aging heart, in association with changes in the myocardial TR protein levels. In addition, plasma 3,5,3'-triiodothyronine (T3) and thyroxine (T4) levels, which decrease in aging [150, 151], are increased with exercise training. The reversal of aging-induced downregulation of myocardial TR signaling-mediated transcription of MHC and SERCA genes by exercise training appears to be related to the cardiac functional improvement observed in trained aged hearts.

The identification of the specific mechanisms contributing to decreased TH signaling in the aging heart may provide significant insights into possible therapeutic keeping in mind that in the aging heart decreased TH activity may be a physiological adaptation. It is possible that therapies which increase SERCA activity might improve cardiac function in the senescent heart. On the other hand, it has been shown that a decrease in SERCA activity contributes to the functional abnormalities observed in senescent hearts, and that $Ca^{2+}$ cycling proteins can be targeted to improve cardiac function in senescence [152]. The well-established decline in myocardial SERCA content with age may also contribute to the increased development of impaired function after I/R in aging subjects. Furthermore, the ratio of SERCA to either phospholamban or calsequestrin decreased in the senescent human myocardium [153]. Decreased rates of $Ca^{2+}$ transport mediated by the SERCA isoform are responsible for the slower sequestration of cytosolic $Ca^{2+}$ and consequent prolonged muscle relaxation times in the aging heart.

SERCA is a prominent target of oxidative/nitrative damage in aging. Knyushko et al. [115] have found that senescent Fischer 344 rat heart showed a 60% decrease in SERCA activity relative to that of young adult hearts, and this functional reduction in activity could be attributed, in part, to both a lower abundance of SERCA protein, and increased 3-nitrotyrosine modifications of multiple tyrosines within the cardiac SERCA protein. Nitration in the senescent heart was found to increase by more than two nitrotyrosines per $Ca^{2+}$-ATPase molecule, coinciding with the appearance of partial nitrated Tyr(294), Tyr(295), and Tyr(753) residues. In contrast, skeletal muscle SERCA exhibited a homogeneous pattern of nitration, with full site nitration of Tyr(753) in the young, with additional nitration of Tyr(294) and Tyr(295) in the senescent muscle. The nitration of these latter sites correlates with diminished transport function in both types of muscle, suggesting that these sites have a potential role in the downregulation of ATP utilization by the $Ca^{2+}$-ATPase under conditions of nitrative stress.

## Insulin, Growth Hormone and Other Interdependent Signaling Molecules

Reduced signaling of insulin and highly conserved insulin-like peptides can profoundly affect organismal life span. Mutations in genes involved in the insulin/IGF-I signal response pathway have been reported to significantly extend life span in diverse species, including yeast, nematodes, fruit flies, and rodents. Intriguingly, the long-lived mutants, share important phenotypic characteristics, including reduced insulin signaling, enhanced sensitivity to insulin, and reduced IGF-I plasma levels. In the nematode and the fly, secondary hormones downstream of insulin-like signaling also appear to regulate aging. However, the relative order and significance in which the hormones act in mammals has been difficult to resolve since there is a complex network of interacting and interdependent signaling molecules, including insulin, IGF-1, GH, and TH, affecting multiple interacting cellular pathways [154].

Several mutant mouse strains have been instrumental in providing models of lifespan and aging modulation. These include GH-deficient/resistant animals which have a prolonged lifespan compared with their normal siblings. The Ames and Snell dwarf mice strain and GH receptor/GH binding protein knockout (GHR-KO) mice do not experience aging at the same rate as their normal siblings but are subject to delayed aging. The Snell and Ames Dwarf mice are homozygous for recessive mutations at the pituitary-1 (pit1) *Pit-1*, or *Prop-1 locus*, respectively, which encode transcription factors controlling pituitary development [155]. Both Snell and Ames Dwarf mice demonstrate increased longevity (50% in males and 64% in females) compared to their wild-type controls, which has been generally attributed to GH/IGF-I deficiency [156]. Mice homozygous for such a mutation are deficient in serum GH, thyroid-stimulating hormone (TSH), and prolactin as well as IGF-I.

While the mechanism of increased lifespan has not been fully delineated, there is increased support for the centrality of insulin signaling in the control of mammalian aging and for the involvement of this pathway in extending the life span of IGF-I-deficient mice. In the Snell dwarf mouse, GH deficiency leads to reduced insulin release and alterations in insulin signaling, including a decreased IRS-2 pool level, a reduction in PI3K activity and its association with IRS-2, decreased docking of p85 to IRS-2, and preferential docking of IRS-2 to p85–p110 leading to reduced insulin levels, enhanced insulin sensitivity, alterations in carbohydrate and lipid metabolism, reduced generation of ROS, enhanced resistance to stress, reduced oxidative damage, and delayed onset of age-related disease [155, 157, 158]. These alterations would establish a physiological homeostasis that favors longevity. Mouse longevity is also increased by fat-specific disruption of the insulin receptor gene *FIRKO* [159].

While a lower level of circulating growth hormone and an enhanced life span was found in transgenic mice expressing bovine growth hormone (bGH) [159], mouse mutant models containing high plasma GH but a 90% lower IGF-I also live longer than wild-type mice. This suggests that reduction in plasma IGF-I levels may be primarily responsible for a major portion of the life span increase in dwarf, GH-deficient, and GHRBP-null mice [158]. Further evidence for the direct role of IGF-I signaling in the control of mammalian aging has also been provided by mouse strains in which the loss of a single copy of the *igf1r* gene (encoding the IGF receptor) results in a 26% increase in mouse life span [160].

Moreover, IGF-I/GH/IGF-1 receptor system not only plays an important role in determining organism development and lifespan, but also it is in itself affected by age. IGF-I decreased linearly with age in both sexes, with significantly higher levels in men than women [161]. The decrease in GH-induced IGF-I secretion in the elderly suggests that resistance to the action of GH may be a secondary contributing factor in the low plasma IGF-1 concentrations [162].

It has also been argued that decreased IGF-I level with age may contribute to the increase in cardiac disease found in the elderly, including HF [163]. Findings from the Framingham Heart Study in a prospective, community-based investigation indicated that serum IGF-I level was inversely related to the risk for HF in elderly people without a previous MI, suggesting that the maintenance of an optimal IGF-I levels in the elderly may reduce the risk for HF [164]. In addition, this study revealed that greater levels or production of the catabolic cytokines TNF-α and interleukin 6 were associated with increased mortality in community-dwelling elderly adults, whereas IGF-I levels had the opposite effect [165].

In aged animals and humans, the secretion of GH and the response of GH to the administration of GH-releasing hormone (GHRH) are lower than in young adults [166]. In rodents, a twofold increase in GH receptors has been observed with age but this increase fails to compensate for the reduction in GH secretion [167, 168]. Further investigation revealed that the apparent size of the GH receptor was not altered with age, whereas the capacity of GH to induce *IGF-1* gene expression and secretion was 40–50% less in old than in young animals [163].

Amelioration of the aging-like phenotypes in Klotho-deficient mice was observed by perturbing insulin and IGF-I signaling, suggesting that Klotho-mediated inhibition of insulin and IGF-I signaling contributes to its antiaging properties [169]. On the other hand, because Klotho induces IGF-1 and insulin resistance, it has been suggested that the above findings seem to contradict previous evidence for increased life span of dwarf mice with reduced IGF-1 and insulin levels and enhanced insulin sensitivity. Nevertheless, since activation of signaling molecules downstream from IGF-1 and insulin receptors is reduced in both Klotho and dwarf mice, a common mechanism of delayed aging is suggested [170]. Furthermore, it has been reported that the Klotho protein increases resistance to OS at the cellular and organismal level in mammals through activation of the FoxO forkhead transcription factors that are negatively regulated by insulin/IGF-1 signaling, thereby inducing expression of manganese-superoxide dismutase

(Mn-SOD). This facilitates the removal of ROS and confers OS resistance, likely contributing to the antiaging properties of Klotho [171].

## Pro-death and Prosurvival Signaling Pathways in Aging

In cardiac diseases associated with aging, such as ischemic heart disease and cardiomyopathy, intralysosomal degradation of cells plays an essential role in the renewal of cardiac myocytes. The interaction of mitochondria and lysosomes in cellular homeostasis is of great significance, since both organelles suffer significant age-related alterations in postmitotic cells [172]. Many mitochondria undergo enlargement and structural disorganization, and since lysosomes responsible for mitochondrial turnover experience a loss of function the rate of total mitochondrial protein turnover declines with age [173]. Coupled mitochondrial and lysosomal defects contribute to irreversible functional impairment and cell death.

Under pathophysiological conditions, autophagy may have a protective role or may contribute to cell damage [174]. Nutrient depletion classically induces autophagy in order to provide amino acids for the synthesis of essential proteins, thus prolonging cell survival [175], and may up to a point neutralize the apoptotic or programmed cell death stimuli (PCD type I) [176, 177]. However, other observations suggest that autophagy can act as an alternative form of PCD, termed PCD type II [178]. Apoptosis and autophagy being closely regulated biological processes, they play a central role in tissue homeostasis, development, and disease. Pattingre et al. [179] have shown that the antiapoptotic protein, Bcl-2, interacts with the evolutionarily conserved autophagy protein, Beclin 1, and that the wild-type Bcl-2 antiapoptotic proteins, but not Beclin 1 binding defective mutants of Bcl-2, inhibit Beclin 1-dependent autophagy in yeast and mammalian cells. Moreover, cardiac Bcl-2 transgenic expression inhibited autophagy in mouse heart muscle. In addition, Beclin 1 mutants that cannot bind to Bcl-2 induce more autophagy than wild-type Beclin 1, and differently from the wild-type Beclin 1, promote cell death [180]. Besides its function as an antiapoptotic protein, Bcl-2 also operates as an antiautophagy protein through its inhibitory interaction with Beclin 1. This latter function of Bcl-2 may be helpful to keep autophagy in check, at levels that are compatible with cell survival, rather than cell death.

Furthermore, mitochondrial interaction with other functional compartments of the cardiac cell (e.g., the ER for $Ca^{2+}$ metabolism, peroxisomes for the interchange of antioxidant enzymes essential in the production and decomposition of $H_2O_2$) must be kept in check since defects in communication between these organelles may accelerate autophagy and the aging process.

## Gene Induction in Cardiac Aging

Genes, that otherwise may remain in a quiescent state, are stimulated with aging under adequate environmental conditions, to express transcription factors/proteins that may facilitate the development of cardiac pathologies, e.g., upon physiological stress, families of stress-response genes are activated as natural defense mechanisms. Induction of specific inflammatory genes is significantly deregulated and altered in the heart of aged versus young mice when challenged with the bacterial endotoxin lipopolysaccharide (LPS), suggesting that endotoxin-mediated induction of specific inflammatory genes in cardiovascular tissues is abnormal with aging, and this may be causally related to the increased susceptibility of aged animals to endotoxic stress [181].

The *Klotho*-deficient mice [KL(−/−)] develop a syndrome resembling accelerated human aging, with significant and accelerated atherosclerosis [182]. Moreover, a number of advanced aging-like KL(−/−) phenotypes could be restored to normal whenever Klotho expression was induced. On the other hand, decreasing Klotho expression in these rescued KL(−/−) mice induced several aging-like KL(−/−) phenotypes [183]. Therefore, Klotho may be effective in the prevention and treatment of age-related disorders. The *Klotho* gene encodes a single-pass transmembrane protein that functions in signaling pathways that suppress aging and which has α-glucuronidase activity. In humans, a functional variant of *Klotho* termed *KL-VS* has been found, common in the general population (frequency 0.157) and individuals homozygous for *KL-VS* manifest reduced human longevity [184]. The *KL-VS* variant harbors three mutations in the coding region, of which one is silent, and two codes for missense mutations F352V and C370S, which substantially alter Klotho metabolism. The *KL-VS* allele influences the trafficking and catalytic activity of Klotho, and the variations in Klotho function contributed to heterogeneity in the onset and severity of human age-related phenotypes [184], and early onset of occult CAD [185]. Furthermore, recent cross-sectional and prospective studies have confirmed a genetic model in which the *KL-VS* allele confers a heterozygous advantage in conjunction with a marked homozygous disadvantage for high-density lipoprotein cholesterol (HDL-C) levels, systolic blood pressure, stroke, and longevity [186].

Wessells et al. [187] have reported that characteristic age-related changes in *Drosophila* decreased or are absent in long-lived flies when systemic levels of insulin-like peptides are reduced by mutations of the receptor, InR, or its substrate, chico [188]. Furthermore, the age-related decline in cardiac performance was prevented by interfering with InR signaling exclusively in the heart, by overexpressing the phosphatase dPTEN or the forkhead transcription factor dFOXO. Taken together, this suggests that in addition to its systemic effect on lifespan, insulin-IGF signaling influences age-dependent organ physiology and senescence directly and autonomously.

Ocorr et al. [189] have evaluated heart function in *Drosophila* and found that the fly's cardiac performance, as in human, deteriorates with age. The aging fruit flies exhibit a progressive increase in electrical pacing-induced HF as well as in dysrhythmias.

While it is clear that in *Drosophila* the insulin receptor and associated pathways have a dramatic and heart-autonomous influence on age-related cardiac performance, altered *KCNQ* and $K_{ATP}$ ion channel functions (besides their conserved role in protecting against dysrrhythmias and hypoxia/ischemia, respectively), also seem to contribute to the decline in heart performance in the aging flies. It is possible that both mechanisms may be operative in the regulation of cardiac aging in vertebrates.

## Epigenetics and Environmental Factors in Cardiac Aging

A number of epigenetic factors, including increased caloric intake, inadequate diet, alcohol intake, smoking, obesity, and lack of adequate aerobic exercise may contribute to the development of diverse cardiac pathologies (e.g., CAD, hypertension, etc.) in aging. Intrinsically, in the normal human aging heart there is a significant decrease in the chronotropic and inotropic responses to catecholamine stimulation, compromising cardiac function. An age-associated reduction in cardiovascular β-adrenergic responsiveness has been noted in Fischer 344 rats, corresponding with alterations in postreceptor adrenergic signaling rather than with a decrease in LV β-AR number [190]. Interestingly, chronic dynamic exercise partially attenuated these reductions through alterations in postreceptor elements of cardiac signal transduction. Moreover, exercise training improves the aging-induced downregulation of myocardial peroxisome proliferator activated receptor (PPAR) α-mediated metabolic pathways, and contributes to an amelioration in fatty acid metabolic enzyme activity in rats [190].

Moreover, endothelial function deteriorates with aging in human, and exercise training appears to improve the function of vascular endothelial cells. Regular aerobic-endurance exercise has been found to reduce plasma endothelin (ET) 1 concentration and to increase NO production in previously sedentary older women, with probable beneficial effects on the cardiovascular system, i.e., prevention of progression of hypertension and/or atherosclerosis by endogenous ET-1 and the potent vasodilatory effects of NO [191, 192]. Also, regular aerobic exercise may prevent the age-associated loss in endothelium-dependent vasodilation and restore the levels in previously sedentary middle aged and older healthy men. This may be an important mechanism by which aerobic exercise lowers CVD risk in this population [193]. Furthermore, endothelial release of tissue-type plasminogen activator a primary regulator of fibrinolysis and part of the endogenous defense mechanism against thrombosis, decreases with age in sedentary men, and regular aerobic exercise may not only prevent it, but could also reverse the age-related loss in endothelial fibrinolytic function [194].

## Conclusions

The molecular biology of CV aging encompasses changes in signaling pathways, cumulative cellular and molecular damage mediated through a variety of insults, OS, nonenzymatic glycation, and inflammation and changes in cardiovascular gene expression. The critical involvement of mitochondrial pathways in myocardial bioenergetic regulation, the balance of oxidants and antioxidants, and the progression of apoptosis are being increasingly considered as contributory to the cardiac dysfunction and remodeling found both in the aging and in the failing heart. These pathways involve considerable cross talk between both nuclear and mitochondrial components and they represent potential targets not only for the treatment of HF, but also for reversing the cardiac dysfunction occurring with aging. Furthermore, the development of novel strategies, by either targeting these factors directly (e.g., apoptotic factors), promoting or redirecting bioenergetic resources, or activating mitochondrial responses against apoptosis and/or OS, holds a great promise for providing cardioprotection to the aging heart.

## Summary

- Increased left ventricular mass relative to chamber volume, decreased diastolic function, and decreased β-adrenergic sympathetic responsiveness occur with aging.
- Prolonged AP duration, altered MHC isoform expression, and SR dysfunction may lead to changes in cardiac excitation–contraction (E–C) coupling which is prolonged with increased age, likely as a result of cytosolic $Ca^{2+}$ overload-induced dysregulation.
- Accumulation of nuclear and mtDNA damage results in senescent cardiomyocytes in the aging heart since DNA is

easily oxidized and damaged by insults from diet, toxins, pollution, environment, as well as other epigenetic influences.

- Despite extensive DNA repair systems, DNA damage, including base modification, large-scale rearrangements single-strand and double-strand breaks, and mutations produced over a lifetime accumulate in aging. This aging-associated DNA damage is most evident in mtDNA probably because of its proximity to ROS, and decreased protection of the mitochondrial-based DNA repair systems.
- Significant changes in telomere structure, overall length, DNA, and function have been found in cardiovascular cells, including endothelial cells, vascular SMCs, EPCs, and cardiomyocytes, associated with replicative senescence both *in vivo* and *in vitro*. Changes in telomeric length and function have also been reported in association with cardiovascular diseases associated with aging, including atherosclerosis and CAD.
- OS and ROS can mediate both telomeric dysfunction and length (in part by regulating telomerase presence in the nucleus) leading to replicative senescence as well as can trigger senescence by telomere-independent pathways.
- Cell loss and remodeling of the heart occur in aging in part due to increased apoptosis. Apoptosis also plays a contributory role in aging-associated CVDs, including cardiomyopathy and heart failure. In addition, susceptibility to ischemia and atherosclerosis increases with aging.
- Apoptosis involves an extrinsic signaling pathway triggered by multiple stimuli targeting death receptors and stimulating a cascade of caspases activation.
- An intrinsic signaling pathway mediated largely by mitochondrial permeabilization via MAC formation, MTP activation, and the release of mitochondrial peptide factors (e.g., AIF, endoG, smac/Diablo, and cytochrome *c*) to the cytosol leading to caspase activation and nuclear DNA degradation as well as mitochondrial bioenergetic dysfunction. The intrinsic pathway is regulated by prooxidants and ROS, intracellular and mitochondrial calcium, proapoptotic factors of the BH3-family, and antiapoptotic factors of the Bcl-2 family.
- Both apoptotic pathways exhibit extensive cross talk, involve multiple organelles (nucleus, mitochondria, ER) and are highly regulated by both proapoptotic factors, such as BH3-proteins and specific endogenous apoptotic inhibitors (e.g., XIAP, ARC). Factors, such as p53 and p66Shc, also can impact on apoptotic progression, and survival-promoting pathways promoted by stimuli, such as IGF-1, can attenuate apoptotic progression.
- Increased level of mtDNA mutations (such as generated in transgenic mice containing proofreading deficient DNA polymerase $\gamma$) result in both accelerated aging and increased apoptosis.
- Accumulation of ROS, ROS-mediated damage to proteins, DNA, and membrane lipids are significantly increased in aging. Age-mediated oxidative damage, including nitration, has been described in specific residues of cardiac proteins (SERCA, ANT) as well as more globally affecting proteins (e.g., protein carbonyl modification). Both point mutations and deletions in mtDNA have been reported to increase with age and may impact on further mitochondrial energetic dysfunction.
- Inflammatory mediators and neuroendocrine regulators, including adrenergic, cholinergic and thyroid hormone receptors, G protein, and signaling modulators (kinases and associated anchoring proteins), are altered in aging underlying significant changes in cardiac responsiveness to multiple hormonal and physiological stimuli and downstream changes in components of calcium signaling pathways (e.g., SERCA).
- Aging-mediated changes IGF/GH signaling not only affects longevity, but also is associated with changes in cardiovascular function.
- The removal of oxidative damaged macromolecules and organelles by autophagy, lysosomal, and proteosomal degradation is less efficient in aging. Moreover, autophagy in aging can lead to elevated cell death.
- Epigenetic and environmental factors, including diet and exercise, not only play a critical role in cardiovascular aging and longevity phenotypes, but also can reverse age-mediated damage and cardiovascular dysfunction.

## References


1. Roman MJ, Ganau A, Saba PS, Pini R, Pickering TG, Devereux RB. Impact of arterial stiffening on left ventricular structure. Hypertension. 2000;36:489–94.
2. Taddei S, Virdis A, Mattei P, et al. Lack of correlation between microalbuminuria and endothelial function in essential hypertensive patients. J Hypertens. 1995;13:1003–8.
3. Lakatta EG, Gerstenblith G, Angell CS, Shock NW, Weisfeldt ML. Prolonged contraction duration in aged myocardium. J Clin Invest. 1975;55:61–8.
4. Merillon JP, Motte G, Masquet C, et al. Changes in the physical properties of the arterial system and left ventricular performance with age and in permanent arterial hypertension: their interrelation. Arch Mal Coeur Vaiss. 1982;75 Spec No:127–32.
5. Lakatta EG. Cardiovascular regulatory mechanisms in advanced age. Physiol Rev. 1993;73:413–67.
6. Lakatta EG. Arterial and cardiac aging: major shareholders in cardiovascular disease enterprises. Circulation. 2003;107:490–7.
7. Lakatta EG, Levy D. Arterial and cardiac aging: major shareholders in cardiovascular disease enterprises. Part II: the aging heart in health: links to heart disease. Circulation. 2003;107:346.
8. Lakatta EG, Levy D. Arterial and cardiac aging: major shareholders in cardiovascular disease enterprises. Part I: aging arteries: a "Set Up" for vascular disease. Circulation. 2003;107:139.
9. Anversa P, Palackal T, Sonnenblick EH, Olivetti G, Meggs LG, Capasso JM. Myocyte cell loss and myocyte cellular hyperplasia in the hypertrophied aging rat heart. Circ Res. 1990;67:871–85.

10. Swynghedauw B, Besse S, Assayag P, et al. Molecular and cellular biology of the senescent hypertrophied and failing heart. Am J Cardiol. 1995;76:2D–7.
11. Anversa P, Olivetti G. Cellular basis of physiological and pathological myocardial growth. New York, NY: Oxford University Press; 2002.
12. Zhang XP, Vatner SF, Shen YT, et al. Increased apoptosis and myocyte enlargement with decreased cardiac mass; distinctive features of the aging male, but not female, monkey heart. J Mol Cell Cardiol. 2007;43:487–91.
13. Olivetti G, Melissari M, Balbi T, Quaini F, Sonnenblick EH, Anversa P. Myocyte nuclear and possible cellular hyperplasia contribute to ventricular remodeling in the hypertrophic senescent heart in humans. J Am Coll Cardiol. 1994;24:140–9.
14. Nadal-Ginard B, Kajstura J, Leri A, Anversa P. Myocyte death, growth, and regeneration in cardiac hypertrophy and failure. Circ Res. 2003;92:139–50.
15. Kajstura J, Pertoldi B, Leri A, et al. Telomere shortening is an in vivo marker of myocyte replication and aging. Am J Pathol. 2000;156:813–9.
16. Elisson GM, Galuppo V. Cardiac stem and progenitor cell identification: different markers for the same cell? Front Biosci. 2010;2:641–52.
17. Torella D, Ellison GM, Nadal-Ginard B, Indolfi C. Cardiac stem and progenitor cell biology for regenerative medicine. Trends Cardiovasc Med. 2005;15:229–36.
18. Anversa P, Rota M. Myocardial aging – a stem cell problem. Basic Res Cardiol. 2005;100:482–93.
19. Chimenti C, Kajstura J, Torella D, et al. Senescence and death of primitive cells and myocytes lead to premature cardiac aging and heart failure. Circ Res. 2003;93:604–13.
20. Rota M, LeCapitaine N, et al. Diabetes promotes cardiac stem cell aging and heart failure, which are prevented by deletion of the p66shc gene. Circ Res. 2006;99:42–52.
21. Gude N, Muraski J. Akt promotes increased cardiomyocyte cycling and expansion of the cardiac progenitor cell population. Circ Res. 2006;99:381–8.
22. Kawamoto A, Asahara T. Role of progenitor endothelial cells in cardiovascular disease and upcoming therapies. Catheter Cardiovasc Interv. 2007;70(4):477–84.
23. Baltrami AP, Barlucchi L. Adult cardiac stem cells are multipotent and support myocardial regeneration. Cell. 2003;114: 763–76.
24. Oh H, Bradfute SB, Gallardo TD, et al. Cardiac progenitor cells from adult myocardium: homing, differentiation, and fusion after infarction. Proc Natl Acad Sci USA. 2003;100:12313–8.
25. Pfister O, Mouquet F, Jain M, et al. CD31- but not CD31+ cardiac side population cells exhibit functional cardiomyogenic differentiation. Circ Res. 2005;97:52–61.
26. Dimmeler S, Leri A. Aging and disease as modifiers of efficacy of cell therapy. Circ Res. 2008;102:1319–30.
27. Gonzalez A, Rota M, Nurzynska D, et al. Activation of cardiac progenitor cells reverses the failing heart senescent phenotype and prolongs lifespan. Circ Res. 2008;102:597–606.
28. Kajstura J, Fiordaliso F, Andreoli AM, et al. IGF-1 overexpression inhibits the development of diabetic cardiomyopathy and angiotensin II-mediated oxidative stress. Diabetes. 2001;50:1414–24.
29. Torella D, Rota M, Nurzynska D, et al. Cardiac stem cell and myocyte aging, heart failure, and insulin-like growth factor-1 overexpression. Circ Res. 2004;94:514–24.
30. Davis ME, Hsieh PC, Takahashi T, et al. Local myocardial insulin-like growth factor 1 (IGF-1) delivery with biotinylated peptide nanofibers improves cell therapy for myocardial infarction. Proc Natl Acad Sci USA. 2006;103:8155–60.
31. Kenyon C. The plasticity of aging: insights from long-lived mutants. Cell. 2005;120:449–60.
32. Urbanek K, Torella D, Sheikh F, et al. Myocardial regeneration by activation of multipotent cardiac stem cells in ischemic heart failure. Proc Natl Acad Sci USA. 2005;102:8692–7.
33. Kollet O, Shivtiel S, Chen YQ, et al. HGF, SDF-1, and MMP-9 are involved in stress-induced human CD34+ stem cell recruitment to the liver. J Clin Invest. 2003;112:160–9.
34. Linke A, Muller P, Nurzynska D, et al. Stem cells in the dog heart are self-renewing, clonogenic, and multipotent and regenerate infarcted myocardium, improving cardiac function. Proc Natl Acad Sci USA. 2005;102:8966–71.
35. Rosu-Myles M, Stewart E, Trowbridge J, Ito CY, Zandstra P, Bhatia M. A unique population of bone marrow cells migrates to skeletal muscle via hepatocyte growth factor/c-met axis. J Cell Sci. 2005;118:4343–52.
36. Fiordaliso F, Leri A, Cesselli D, et al. Hyperglycemia activates p53 and p53-regulated genes leading to myocyte cell death. Diabetes. 2001;50:2363–75.
37. Pfeffer MA, McMurray JJ, Velazquez EJ, et al. Valsartan, captopril, or both in myocardial infarction complicated by heart failure, left ventricular dysfunction, or both. N Engl J Med. 2003;349: 1893–906.
38. Kawanishi S, Oikawa S. Mechanism of telomere shortening by oxidative stress. Ann N Y Acad Sci. 2004;1019:278–84.
39. Yang S, Yang S, Chintapalli J, et al. Activated IGF-1R inhibits hyperglycemia-induced DNA damage and promotes DNA repair by homologous recombination. Am J Physiol Renal Physiol. 2005;289:F1144–52.
40. Cheng T, Rodrigues N, Shen H, et al. Hematopoietic stem cell quiescence maintained by p21cip1/waf1. Science. 2000;287:1804–8.
41. Janzen V, Forkert R. Stem-cell ageing modified by the cyclin-dependent kinase inhibitor p16INK4a. Nature. 2006;443:421–6.
42. Sharpless NE, DePinho RA. How stem cells age and why this makes us grow old. Nat Rev Mol Cell Biol. 2007;8:703–13.
43. Goukassian DD, Qin G, Dolan C, et al. Tumor necrosis factor-alpha receptor p75 is required in ischemia-induced neovascularization. Circulation. 2007;115:752–62.
44. Chang EI, Loh SA, Ceradini DJ, et al. Age decreases endothelial progenitor cell recruitment through decreases in hypoxia-inducible factor 1 stabilization during ischemia. Circulation. 2007;116: 2818–29.
45. Tan T, Marin-Garcia J, Damle S, Weiss, H. HIF and mitochondria. Exp Physiol. 2010;95:712–22.
46. Ndubuizu OI, Chavez CJ, LaManna JC. Increased prolyl 4-hydroxylase expression and differential regulation of hypoxia-inducible factors in the aged rat brain. Am J Physiol Regul Integr Comp Physiol. 2009;297:R158–65.
47. Yoshida K, Kirito K, Yongzhen H, Ozawa K, Kaushansky K, Komatsu N. Thrombopoietin (TPO) regulates HIF-1 alpha levels through generation of mitochondrial reactive oxygen species. Int J Hematol. 2008;88:43–51.
48. Vaux EC, Metzen E, Yeates KM, Ratcliffe PJ. Regulation of hypoxia-inducible factor is preserved in the absence of a functioning mitochondrial respiratory chain. Blood. 2001;98: 296–302.
49. Edo MD, Andres V. Aging telomeres and atherosclerosis. Cardiovasc Res. 2005;66:213–21.
50. Aviv H, Khan MY, Skurnick J, et al. Age dependent aneuploidy and telomere length of the human vascular endothelium. Atherosclerosis. 2001;159:281–7.
51. Chang E, Harley CB. Telomere length as a biomarker in human vascular tissues. Proc Natl Acad Sci USA. 1995;92:11190–4.
52. Okuda K, Khan MY, Skurnick J, Kimura M, Aviv H, Aviv A. Telomere attrition of the human abdominal aorta: relationships with age and atherosclerosis. Atherosclerosis. 2000;152:391–8.
53. Matthews C, Gorenne I, Scott S, et al. Vascular smooth muscle cells undergo telomere-based senescence in human atherosclerosis.

Effects of telomerase and oxidative stress. Circ Res. 2006; 2006(99):156–64.
54. Samani NJ, Boultby R, Butler R, Thompson JR, Goodall AH. Telomere shortening in atherosclerosis. Lancet. 2001;358:472–3.
55. Benetos A, Gardner JP, Zureik M, et al. Short telomeres are associated with increased carotid atherosclerosis in hypertensive subjects. Hypertension. 2004;43:182–5.
56. Collerton J, Martin-Ruiz C, Kenny A, et al. Telomere length is associated with left ventricular function in the oldest old: the Newcastle 85+ study. Eur Heart J. 2007;28:172–6.
57. von Zglinicki T. Oxidative stress shortens telomeres. Trends Biochem Sci. 2002;27:339–44.
58. Kurz DJ, Decary S, Hong Y, Trivier E, Akhmedov A, Erusalimsky JD. Chronic oxidative stress compromises telomere integrity and accelerates the onset of senescence in human endothelial cells. J Cell Sci. 2004;117:2417–26.
59. Haendeler J, Hoffmann J, Diehl JF, et al. Antioxidants inhibit nuclear export of telomerase reverse transcriptase and delay replicative senescence of endothelial cells. Circ Res. 2004;94:768–75.
60. Spyridopoulos I, Haendeler J, Urbich C, et al. Statins enhance migratory capacity by upregulation of the telomere repeat-binding factor TRF2 in endothelial progenitor cells. Circulation. 2004;110: 3136–42.
61. Imanishi T, Hano T, Nishio I. Estrogen reduces endothelial progenitor cell senescenes through augmentation of telomerase activity. J Hypertens. 2005;23:1699–706.
62. Simoncini T, Hafezi-Moghadam A, Brazil DP, Ley K, Chin WW, Liao JK. Interaction of oestrogen receptor with the regulatory subunit of phosphatidylinositol-3-OH kinase. Nature. 2000;407:538–41.
63. Vasa M, Breitschopf K, Zeiher AM, Dimmeler S. Nitric oxide activates telomerase and delays endothelial cell senescence. Circ Res. 2000;87:540–2.
64. Imanishi T, Hano T, Sawamura T, Nishio I. Oxidized low-density lipoprotein induces endothelial progenitor cell senescence, leading to cellular dysfunction. Clin Exp Pharmacol Physiol. 2004; 31:407–13.
65. Serrano AL, Andrés V. Telomeres and cardiovascular disease: does size matter? Circ Res. 2004;94:575–84.
66. von Zglinicki T, Pilger R, Sitte N. Accumulation of single-strand breaks is the major cause of telomere shortening in human fibroblasts. Free Radic Biol Med. 2000;28:64–74.
67. Forsyth NR, Evans AP, Shay JW, Wright WE. Developmental differences in the immortalization of lung fibroblasts by telomerase. Aging Cell. 2003;2:235–43.
68. Serra V, von Zglinicki T, Lorenz M, Saretzki G. Extracellular superoxide dismutase is a major antioxidant in human fibroblasts and slows telomere shortening. J Biol Chem. 2003;278:6824–30.
69. Saretzki G, Murphy MP, von Zglinicki T. MitoQ counteracts telomere shortening and elongates lifespan of fibroblasts under mild oxidative stress. Aging Cell. 2003;2:141–3.
70. Passos JF, von Zglinicki T. Mitochondria, telomeres and cell senescence. Exp Gerontol. 2005;40:466–72.
71. Kirkwood TB. Understanding the odd science of aging. Cell. 2005;120:437–47.
72. Cook SA, Sugden PH, Clerk A. Regulation of bcl-2 family proteins during development and in response to oxidative stress in cardiac myocytes: association with changes in mitochondrial membrane potential. Circ Res. 1999;85:940–9.
73. Long X, Goldenthal MJ, Wu GM, Marin-Garcia J. Mitochondrial $Ca^{2+}$ flux and respiratory enzyme activity decline are early events in cardiomyocyte response to $H_2O_2$. J Mol Cell Cardiol. 2004; 37:63–70.
74. Pollack M, Phaneuf S, Dirks A, Leeuwenburgh C. The role of apoptosis in the normal aging brain, skeletal muscle, and heart. Ann N Y Acad Sci. 2002;959:93–107.
75. Marin-Garcia J, Pi Y, Goldenthal MJ. Mitochondrial-nuclear cross-talk in the aging and failing heart. Cardiovasc Drugs Ther. 2006;20:477–91.
76. Narula J, Haider N, Arbustini E, Chandrashekhar Y. Mechanisms of disease: apoptosis in heart failure–seeing hope in death. Nat Clin Pract Cardiovasc Med. 2006;3:681–8.
77. Madamanchi NR, Runge MS. Mitochondrial dysfunction in atherosclerosis. Circ Res. 2007;100:460–73.
78. Danial NN, Korsmeyer SJ. Cell death: critical control points. Cell. 2004;116:205–19.
79. Li LY, Luo X, Wang X. Endonuclease G is an apoptotic DNase when released from mitochondria. Nature. 2001;412:95–9.
80. Susin SA, Lorenzo HK, Zamzami N, et al. Molecular characterization of mitochondrial apoptosis-inducing factor. Nature. 1999;397:441–6.
81. Joza N, Oudit GY, Brown D, et al. Muscle-specific loss of apoptosis-inducing factor leads to mitochondrial dysfunction, skeletal muscle atrophy, and dilated cardiomyopathy. Mol Cell Biol. 2005;25:10261–72.
82. Vahsen N, Candé C, Briere JJ, et al. AIF deficiency compromises oxidative phosphorylation. EMBO J. 2004;23:4679–89.
83. Bahi N, Zhang J, Llovera M, Ballester M, Comella JX, Sanchis D. Switch from caspase-dependent to caspase-independent death during heart development: essential role of endonuclease G in ischemia-induced DNA processing of differentiated cardiomyocytes. J Biol Chem. 2006;281:22943–52.
84. Acehan D, Jiang X, Morgan DG, Heuser JE, Wang X, Akey CW. Three-dimensional structure of the apoptosome: implications for assembly, procaspase-9 binding, and activation. Mol Cell. 2002;9: 423–32.
85. Webster KA, Graham RM, Thompson JW, et al. Redox stress and the contributions of BH3-only proteins to infarction. Antioxid Redox Signal. 2006;8:1667–76.
86. Kinnally KW, Antonsson B. A tale of two mitochondrial channels, MAC and PTP, in apoptosis. Apoptosis. 2007;12:857–68.
87. Martinez-Caballero S, Dejean LM, Kinnally MS, Oh KJ, Mannella CA, Kinnally KW. Assembly of the mitochondrial apoptosis-induced channel, MAC. J Biol Chem. 2009;284: 12235–45.
88. Kroemer G. Mitochondrial control of apoptosis: an introduction. Biochem Biophys Res Commun. 2003;304:433–5.
89. Correa F, Soto V, Zazueta C. Mitochondrial permeability transition relevance for apoptotic triggering in the post-ischemic heart. Int J Biochem Cell Biol. 2007;39:787–98.
90. Marzo I, Brenner C, Zamzami N, et al. The permeability transition pore complex: a target for apoptosis regulation by caspases and Bcl-2 related proteins. J Exp Med. 1998;187:1261–7.
91. Scorrano L, Ashiya M, Buttle K, et al. A distinct pathway remodels mitochondrial cristae and mobilizes cytochrome c during apoptosis. Dev Cell. 2002;2:55–67.
92. Ekhterae D, Lin Z, Lundberg MS, Crow MT, Brosius III FC, Nunez G. ARC inhibits cytochrome c release from mitochondria and protects against hypoxia-induced apoptosis in heart-derived H9c2 cells. Circ Res. 1999;85:e70–7.
93. Scorrano L, Oakes SA, Opferman JT, et al. BAX and BAK regulation of endoplasmic reticulum $Ca^{2+}$: a control point for apoptosis. Science. 2003;300:135–9.
94. Hajnoczky G, Csordas G, Das S, et al. Mitochondrial calcium signalling and cell death: approaches for assessing the role of mitochondrial $Ca^{2+}$ uptake in apoptosis. Cell Calcium. 2006;40: 553–60.
95. Jacobson J, Duchen MR. Mitochondrial oxidative stress and cell death in astrocytes – requirement for stored $Ca^{2+}$ and sustained opening of the permeability transition pore. J Cell Sci. 2002;115: 1175–88.

96. Kujoth GC, Hiona A, Pugh TD, et al. Mitochondrial DNA mutations, oxidative stress, and apoptosis in mammalian aging. Science. 2005;309:481–4.
97. Ball AJ, Levine F. Telomere-independent cellular senescence in human fetal cardiomyocytes. Aging Cell. 2005;4:21–30.
98. Uhrbom L, Nister M, Westermark B. Induction of senescence in human malignant glioma cells by p16INK4A. Oncogene. 1997;15:505–14.
99. Melov S. Mitochondrial oxidative stress. Physiologic consequences and potential for a role in aging. Ann N Y Acad Sci. 2000;908:219–25.
100. Lenaz G, D'Aurelio M, Merlo Pich M, et al. Mitochondrial bioenergetics in aging. Biochim Biophys Acta. 2000;1459:397–404.
101. Pepe S. Effect of dietary polyunsaturated fatty acids on age-related changes in cardiac mitochondrial membranes. Exp Gerontol. 2005;40:751–8.
102. Hansford RG, Tsuchiya N, Pepe S. Mitochondria in heart ischaemia and aging. Biochem Soc Symp. 1999;66:141–7.
103. Harper ME, Bevilacqua L, Hagopian K, Weindruch R, Ramsey JJ. Ageing, oxidative stress, and mitochondrial uncoupling. Acta Physiol Scand. 2004;182:321–31.
104. DiLisa F, Bernardi P. Mitochondrial function and myocardial aging. A critical analysis of the role of permeability transition. Cardiovasc Res. 2005;66:222–32.
105. Russell LK, Finck BN, Kelly DP. Mouse models of mitochondrial dysfunction and heart failure. J Mol Cell Cardiol. 2005;38:81–91.
106. Chakravarti B, Chakravarti DN. Oxidative modification of proteins: age-related changes. Gerontology. 2007;53:128–39.
107. Levine RL, Stadtman ER. Oxidative modification of proteins during aging. Exp Gerontol. 2001;36:1495–502.
108. Stadtman ER, Levine RL. Protein oxidation. Ann N Y Acad Sci. 2000;899:191–208.
109. Yarian CS, Rebrin I, Sohal RS. Aconitase and ATP synthase are targets of malondialdehyde modification and undergo an age-related decrease in activity in mouse heart mitochondria. Biochem Biophys Res Commun. 2005;330:151–6.
110. Yan LJ, Sohal RS. Mitochondrial adenine nucleotide translocase is modified oxidatively during aging. Proc Natl Acad Sci USA. 1998;95:12896–901.
111. Choksi KB, Boylston WH, Rabek JP, Widger WR, Papaconstantinou J. Oxidatively damaged proteins of heart mitochondrial electron transport complexes. Biochim Biophys Acta. 2004;1688:95–101.
112. Vasquez-Vivar J, Kalyanaraman B, Kennedy MC. Mitochondrial aconitase is a source of hydroxyl radical. An electron spin resonance investigation. J Biol Chem. 2000;275:14064–9.
113. Ilangovan G, Venkatakrishnan CD, Bratasz A, et al. Heat shock-induced attenuation of hydroxyl radical generation and mitochondrial aconitase activity in cardiac H9c2 cells. Am J Physiol Cell Physiol. 2006;290:C313–24.
114. Viner RI, Ferrington DA, Williams TD, Bigelow DJ, Schoneich C. Protein modification during biological aging: selective tyrosine nitration of the SERCA2a isoform of the sarcoplasmic reticulum $Ca^{2+}$-ATPase in skeletal muscle. Biochem J. 1999;340(Pt 3):657–69.
115. Knyushko TV, Sharov VS, Williams TD, Schoneich C, Bigelow DJ. 3-Nitrotyrosine modification of SERCA2a in the aging heart: a distinct signature of the cellular redox environment. Biochemistry. 2005;44:13071–81.
116. Xu S, Ying J, Jiang B, et al. Detection of sequence-specific tyrosine nitration of manganese SOD and SERCA in cardiovascular disease and aging. Am J Physiol Heart Circ Physiol. 2006;290: H2220–7.
117. Murray J, Taylor SW, Zhang B, Ghosh SS, Capaldi RA. Oxidative damage to mitochondrial complex I due to peroxynitrite: identification of reactive tyrosines by mass spectrometry. J Biol Chem. 2003;278:37223–30.
118. Kanski J, Behring A, Pelling J, Schoneich C. Proteomic identification of 3-nitrotyrosine-containing rat cardiac proteins: effects of biological aging. Am J Physiol Heart Circ Physiol. 2005;288:H371–81.
119. LeDoux SP, Wilson GL. Base excision repair of mitochondrial DNA damage in mammalian cells. Prog Nucleic Acid Res Mol Biol. 2001;68:273–84.
120. Yakes FM, Van Houten B. Mitochondrial DNA damage is more extensive and persists longer than nuclear DNA damage in human cells following oxidative stress. Proc Natl Acad Sci USA. 1997; 94:514–9.
121. Trifunovic A, Wredenberg A, Falkenberg M, et al. Premature ageing in mice expressing defective mitochondrial DNA polymerase. Nature. 2004;429:417–23.
122. Trifunovic A, Hansson A, Wredenberg A, et al. Somatic mtDNA mutations cause aging phenotypes without affecting reactive oxygen species production. Proc Natl Acad Sci USA. 2005;102: 17993–8.
123. Loeb LA, Wallace DC, Martin GM. The mitochondrial theory of aging and its relationship to reactive oxygen species damage and somatic mtDNA mutations. Proc Natl Acad Sci USA. 2005;102: 18769–70.
124. Chung HY, Sung B, Jung KJ, Zou Y, Yu BP. The molecular inflammatory process in aging. Antioxid Redox Signal. 2006;8:572–81.
125. Kritchevsky SB, Cesari M, Pahor M. Inflammatory markers and cardiovascular health in older adults. Cardiovasc Res. 2005;66: 265–75.
126. Deten A, Marx G, Briest W, Christian Volz H, Zimmer HG. Heart function and molecular biological parameters are comparable in young adult and aged rats after chronic myocardial infarction. Cardiovasc Res. 2005;66:364–73.
127. Antonicelli R, Olivieri F, Bonafe M, et al. The interleukin-6–174 G>C promoter polymorphism is associated with a higher risk of death after an acute coronary syndrome in male elderly patients. Int J Cardiol. 2005;103:266–71.
128. White M, Roden R, Minobe W, et al. Age-related changes in beta-adrenergic neuroeffector systems in the human heart. Circulation. 1994;90:1225–38.
129. Brodde OE, Konschak U, Becker K, et al. Cardiac muscarinic receptors decrease with age. In vitro and in vivo studies. J Clin Invest. 1998;101:471–8.
130. Giraldo E, Martos F, Gomez A, et al. Characterization of muscarinic receptor subtypes in human tissues. Life Sci. 1988;43: 1507–15.
131. Deighton NM, Motomura S, Borquez D, Zerkowski HR, Doetsch N, Brodde OE. Muscarinic cholinoceptors in the human heart: demonstration, subclassification, and distribution. Naunyn Schmiedebergs Arch Pharmacol. 1990;341:14–21.
132. Bohm M, Gierschik P, Jakobs KH, et al. Increase of Gi alpha in human hearts with dilated but not ischemic cardiomyopathy. Circulation. 1990;82:1249–65.
133. Landzberg JS, Parker JD, Gauthier DF, Colucci WS. Effects of intracoronary acetylcholine and atropine on basal and dobutamine-stimulated left ventricular contractility. Circulation. 1994;89: 164–8.
134. Turner MJ, Mier CM, Spina RJ, Ehsani AA. Effects of age and gender on cardiovascular responses to phenylephrine. J Gerontol A Biol Sci Med Sci. 1999;54:M17–24.
135. Hees PS, Fleg JL, Mirza ZA, Ahmed S, Siu CO, Shapiro EP. Effects of normal aging on left ventricular lusitropic, inotropic, and chronotropic responses to dobutamine. J Am Coll Cardiol. 2006;47:1440–7.
136. Korzick DH, Holiman DA, Boluyt MO, et al. Diminished $alpha_1$-adrenergic-mediated contraction and translocation of PKC in senescent rat heart. Am J Physiol Heart Circ Physiol. 2001; 281:H581–9.

137. Korzick DH, Hunter JC, McDowell MK, Delp MD, Tickerhoof MM, Carson LD. Chronic exercise improves myocardial inotropic reserve capacity through α1-adrenergic and protein kinase C-dependent effects in Senescent rats. J Gerontol A Biol Sci Med Sci. 2004;59:1089–98.
138. Hunter JC, Korzick DH. Age- and sex-dependent alterations in protein kinase C (PKC) and extracellular regulated kinase 1/2 (ERK1/2) in rat myocardium. Mech Ageing Dev. 2005;126: 535–50.
139. Montagne O, Le Corvoisier P, Guenoun T, Laplace M, Crozatier B. Impaired alpha1-adrenergic responses in aged rat hearts. Fundam Clin Pharmacol. 2005;19:331–9.
140. Esler M, Kaye D. Sympathetic nervous system activation in essential hypertension, cardiac failure and psychosomatic heart disease. J Cardiovasc Pharmacol. 2000;35:S1–7.
141. Kilts JD, Akazawa T, El-Moalem HE, Mathew JP, Newman MF, Kwatra MM. Age increases expression and receptor-mediated activation of G alpha i in human atria. J Cardiovasc Pharmacol. 2003;42:662–70.
142. Kilts JD, Akazawa T, Richardson MD, Kwatra MM. Age increases cardiac Galpha(i2) expression, resulting in enhanced coupling to G protein-coupled receptors. J Biol Chem. 2002;277:31257–62.
143. Brodde OE, Michel MC. Adrenergic and muscarinic receptors in the human heart. Pharmacol Rev. 1999;51:651–90.
144. Richardson MD, Kilts JD, Kwatra MM. Increased expression of Gi-coupled muscarinic acetylcholine receptor and Gi in atrium of elderly diabetic subjects. Diabetes. 2004;53:2392–6.
145. Oberhauser V, Schwertfeger E, Rutz T, Beyersdorf F, Rump LC. Acetylcholine release in human heart atrium: influence of muscarinic autoreceptors, diabetes, and age. Circulation. 2001;103:1638–43.
146. Halls ML, van der Westhuizen ET, Bathgate RA, Summers RJ. Relaxin family peptide receptors – former orphans reunite with their parent ligands to activate multiple signalling pathways. Br J Pharmacol. 2007;150:677–91.
147. Bathgate RA, Ivell R, Sanborn BM, Sherwood OD, Summers RJ. International Union of Pharmacology LVII: recommendations for the nomenclature of receptors for relaxin family peptides. Pharmacol Rev. 2006;58:7–31.
148. Long X, Boluyt MO, O'Neill L, et al. Myocardial retinoid X receptor, thyroid hormone receptor, and myosin heavy chain gene expression in the rat during adult aging. J Gerontol A Biol Sci Med Sci. 1999;54:B23–7.
149. Iemitsu M, Miyauchi T, Maeda S, et al. Exercise training improves cardiac function-related gene levels through thyroid hormone receptor signaling in aged rats. Am J Physiol Heart Circ Physiol. 2004;286:H1696–705.
150. Tang F. Effect of sex and age on serum aldosterone and thyroid hormones in the laboratory rat. Horm Metab Res. 1985;17:507–9.
151. Buttrick P, Malhotra A, Factor S, Greenen D, Leinwand L, Scheuer J. Effect of aging and hypertension on myosin biochemistry and gene expression in the rat heart. Circ Res. 1991;68:645–52.
152. Schmidt U, del Monte F, Miyamoto MI, et al. Restoration of diastolic function in senescent rat hearts through adenoviral gene transfer of sarcoplasmic reticulum Ca(2+)-ATPase. Circulation. 2000;101:790–6.
153. Cain BS, Meldrum DR, Joo KS, et al. Human SERCA2a levels correlate inversely with age in senescent human myocardium. J Am Coll Cardiol. 1998;32:458–67.
154. Tatar M, Bartke A, Antebi A. The endocrine regulation of aging by insulin-like signals. Science. 2003;299:1346–51.
155. Bartke A. Minireview: role of the growth hormone/insulin-like growth factor system in mammalian aging. Endocrinology. 2005;146:3718–23.
156. Brown-Borg HM, Borg KE, Meliska CJ, Bartke A. Dwarf mice and the ageing process. Nature. 1996;384:33.
157. Hsieh CC, DeFord JH, Flurkey K, Harrison DE, Papaconstantinou J. Effects of the Pit1 mutation on the insulin signaling pathway: implications on the longevity of the long-lived Snell dwarf mouse. Mech Ageing Dev. 2002;123:1245–55.
158. Barbieri M, Bonafe M, Franceschi C, Paolisso G. Insulin/IGF-I-signaling pathway: an evolutionarily conserved mechanism of longevity from yeast to humans. Am J Physiol Endocrinol Metab. 2003;285:E1064–71.
159. Bluher M, Kahn BB, Kahn CR. Extended longevity in mice lacking the insulin receptor in adipose tissue. Science. 2003;299: 572–4.
160. Holzenberger M, Dupont J, Ducos B, et al. IGF-1 receptor regulates lifespan and resistance to oxidative stress in mice. Nature. 2003;421:182–7.
161. Goodman-Gruen D, Barrett-Connor E. Epidemiology of insulin-like growth factor-I in elderly men and women. The Rancho Bernardo Study. Am J Epidemiol. 1997;145:970–6.
162. Lieberman SA, Mitchell AM, Marcus R, Hintz RL, Hoffman AR. The insulin-like growth factor I generation test: resistance to growth hormone with aging and estrogen replacement therapy. Horm Metab Res. 1994;26:229–33.
163. Khan AS, Sane DC, Wannenburg T, Sonntag WE. Growth hormone, insulin-like growth factor-1 and the aging cardiovascular system. Cardiovasc Res. 2002;54:25–35.
164. Vasan RS, Sullivan LM, D'Agostino RB, et al. Serum insulin-like growth factor I and risk for heart failure in elderly individuals without a previous myocardial infarction: the Framingham Heart Study. Ann Intern Med. 2003;139:642–8.
165. Roubenoff R, Parise H, Payette HA, et al. Cytokines, insulin-like growth factor 1, sarcopenia, and mortality in very old community-dwelling men and women: the Framingham Heart Study. Am J Med. 2003;115:429–35.
166. Ghigo E, Arvat E, Gianotti L, et al. Human aging and the GH-IGF-I axis. J Pediatr Endocrinol Metab. 1996;9 Suppl 3:271–8.
167. Takahashi S, Meites J. GH binding to liver in young and old female rats: relation to somatomedin-C secretion. Proc Soc Exp Biol Med. 1987;186:229–33.
168. Xu X, Bennett SA, Ingram RL, Sonntag WE. Decreases in growth hormone receptor signal transduction contribute to the decline in insulin-like growth factor I gene expression with age. Endocrinology. 1995;136:4551–7.
169. Kurosu H, Yamamoto M, Clark JD, et al. Suppression of aging in mice by the hormone Klotho. Science. 2005;309:1829–33.
170. Bartke A. Long-lived Klotho mice: new insights into the roles of IGF-1 and insulin in aging. Trends Endocrinol Metab. 2006;17: 33–5.
171. Yamamoto M, Clark JD, Pastor JV, et al. Regulation of oxidative stress by the anti-aging hormone klotho. J Biol Chem. 2005;280: 38029–34.
172. Brunk UT, Terman A. The mitochondrial-lysosomal axis theory of aging: accumulation of damaged mitochondria as a result of imperfect autophagocytosis. Eur J Biochem. 2002;269: 1996–2002.
173. Rooyackers OE, Adey DB, Ades PA, Nair KS. Effect of age on in vivo rates of mitochondrial protein synthesis in human skeletal muscle. Proc Natl Acad Sci USA. 1996;93:15364–9.
174. Hamacher-Brady A, Brady NR, Gottlieb RA. The interplay between pro-death and pro-survival signaling pathways in myocardial ischemia/reperfusion injury: apoptosis meets autophagy. Cardiovasc Drugs Ther. 2006;20:445–62.
175. Klionsky DJ, Emr SD. Autophagy as a regulated pathway of cellular degradation. Science. 2000;290:1717–21.
176. Breckenridge DG, Germain M, Mathai JP, Nguyen M, Shore GC. Regulation of apoptosis by endoplasmic reticulum pathways. Oncogene. 2003;22:8608–18.

177. Ravikumar B, Berger Z, Vacher C, O'Kane CJ, Rubinsztein DC. Rapamycin pre-treatment protects against apoptosis. Hum Mol Genet. 2006;15:1209–16.
178. Canu N, Tufi R, Serafino AL, Amadoro G, Ciotti MT, Calissano P. Role of the autophagic-lysosomal system on low potassium-induced apoptosis in cultured cerebellar granule cells. J Neurochem. 2005;92:1228–42.
179. Pattingre S, Tassa A, Qu X, et al. Bcl-2 antiapoptotic proteins inhibit Beclin 1-dependent autophagy. Cell. 2005;122:927–39.
180. Shimizu S, Kanaseki T, Mizushima N, et al. Role of Bcl-2 family proteins in a non-apoptotic programmed cell death dependent on autophagy genes. Nat Cell Biol. 2004;6:1221–8.
181. Saito H, Papaconstantinou J. Age-associated differences in cardiovascular inflammatory gene induction during endotoxic stress. J Biol Chem. 2001;276:29307–12.
182. Kuro-o M, Matsumura Y, Aizawa H, et al. Mutation of the mouse klotho gene leads to a syndrome resembling ageing. Nature. 1997;390:45–51.
183. Masuda H, Chikuda H, Suga T, Kawaguchi H, Kuro-o M. Regulation of multiple ageing-like phenotypes by inducible klotho gene expression in klotho mutant mice. Mech Ageing Dev. 2005;126:1274–83.
184. Arking DE, Krebsova A, Macek Sr M, et al. Association of human aging with a functional variant of klotho. Proc Natl Acad Sci USA. 2002;99:856–61.
185. Arking DE, Becker DM, Yanek LR, et al. KLOTHO allele status and the risk of early-onset occult coronary artery disease. Am J Hum Genet. 2003;72:1154–61.
186. Arking DE, Atzmon G, Arking A, Barzilai N, Dietz HC. Association between a functional variant of the KLOTHO gene and high-density lipoprotein cholesterol, blood pressure, stroke, and longevity. Circ Res. 2005;96:412–8.
187. Wessells RJ, Fitzgerald E, Cypser JR, Tatar M, Bodmer R. Insulin regulation of heart function in aging fruit flies. Nat Genet. 2004;36:1275–81.
188. Paternostro G, Vignola C, Bartsch DU, Omens JH, McCulloch AD, Reed JC. Age-associated cardiac dysfunction in Drosophila melanogaster. Circ Res. 2001;88:1053–8.
189. Ocorr K, Akasaka T, Bodmer R. Age-related cardiac disease model of Drosophila. Mech Ageing Dev. 2007;128:112–6.
190. Roth DA, White CD, Podolin DA, Mazzeo RS. Alterations in myocardial signal transduction due to aging and chronic dynamic exercise. J Appl Physiol. 1998;84:177–84.
191. Maeda S, Tanabe T, Miyauchi T, et al. Aerobic exercise training reduces plasma endothelin-1 concentration in older women. J Appl Physiol. 2003;95:336–41.
192. Maeda S, Tanabe T, Otsuki T, et al. Moderate regular exercise increases basal production of nitric oxide in elderly women. Hypertens Res. 2004;27:947–53.
193. DeSouza CA, Shapiro LF, Clevenger CM, et al. Regular aerobic exercise prevents and restores age-related declines in endothelium-dependent vasodilation in healthy men. Circulation. 2000; 102:1351–7.
194. Smith DT, Hoetzer GL, Greiner JJ, Stauffer BL, DeSouza CA. Effects of ageing and regular aerobic exercise on endothelial fibrinolytic capacity in humans. J Physiol. 2003;546:289–98.

# Part VI
# Signaling in Cardiovascular Disease

# Chapter 12
# Signaling in Endomyocarditis

**Abstract** Myocarditis is an inflammatory disorder of the myocardium that is associated with cardiac dysfunction. Many viruses have been implicated as causes of myocarditis, most commonly adenoviruses and enteroviruses, such as the coxsackieviruses. Viral myocarditis has been recognized as a cause of congestive heart failure (HF).

Endocarditis is an infection of the endocardium that may involve cardiac valves and adjacent structures and may be caused by wide spectrum of bacteria and fungi. In this chapter, we focus on the host cell-signaling systems which are involved in the different phases of viral and/or bacterial infection, including viral entry into the cell and the development of innate immunity and in a number of signaling molecules (both host and pathogen originated) that are involved in different aspects of pathogenesis of myocarditis/infective endocarditis. Also signaling pathways participating in the activation of adaptive immunity toward the viral/bacteria pathogen is presented.

**Keywords** Endomyocardial signaling • Myocarditis • Endocarditis • Innate immunity • Nonviral myocarditis

## Introduction

Myocarditis is a complex disease in which distinct immunopathogenic mechanisms cause myocardial inflammation and tissue injury. Clinical presentation range from nonspecific systemic symptoms (fever, myalgias, palpitations, or exertional dyspnea) to fulminant hemodynamic collapse and sudden death, and it has been identified as a significant cause of dilated cardiomyopathy (DCM).

Although a broad array of etiologies have been implicated as causes of myocarditis, viral myocarditis remains the primary cause except in countries in which Chagas disease or diphtheria is common. Enteroviruses, specifically Coxsackie group B serotypes, are the dominant viral causes although with the advent of improved molecular diagnostic techniques, such as polymerase chain reaction (PCR) amplification and *in situ* hybridization, other viral entities, including parvovirus, hepatitis C, and adenovirus, have been recognized as relevant pathogens in myocarditis etiology. In addition, a high incidence of myocarditis and left ventricular dysfunction has been reported in patients with asymptomatic and advanced human immunodeficiency virus (HIV) albeit the role of the HIV itself, the effects of antiretroviral medications used in its treatment, the presence of coinfectioning agents in the myocarditis and the observed left ventricular systolic dysfunction remain undefined.

In a subset of cases, autoimmunity has been shown to be a subsequent major pathogenic factor and constitute a second phase of the disease. Both cellular and humoral autoimmunity in myocarditis is underlied by cross-reactivity between viral and myosin epitopes suggesting that the genetics of the host as well as the virus are determinants of disease pathogenicity at this stage. In the third phase of the disease, DCM develops which appears to be largely a result of viral and autoimmune injury but it may progress after cessation of injury.

Infective endocarditis has been classified as acute, subacute, and chronic on the basis of the timing severity of the clinical presentation and disease progression. Vegetations are the characteristic lesions in endocarditis, and they are composed of platelets, fibrin, microorganisms, and inflammatory cells. It may involve not only the cardiac valves, but other structures like the interventricular septum, the chordae tendineae, intracardiac devices, and the mural endocardium.

In this chapter, we discuss the molecular correlates and signaling pathways that participate in the inflammatory and immune response in both myocarditis, endocarditis, and other disorders involving viral/bacterial infections of the heart.

## Viral Entry into the Cardiac Myocyte

During the period of active viremia, cardiotropic RNA viruses, such as Coxsackievirus B (CVB) 3 or encephalomyocarditis virus (EMCV) are taken into myocytes by receptor-mediated endocytosis and are directly translated intracellularly to produce viral protein. A common mammalian viral receptor, the coxsackie-adenoviral receptor (CAR), has

J. Marín-García, *Signaling in the Heart*,
DOI 10.1007/978-1-4419-9461-5_12, 

been identified for both coxsackieviruses and adenoviruses which allow internalization of the coxsackieviral genome after attachment, a critical step for viral infection. Similarly, CAR protein is a facilitatory receptor for adenoviruses two and five fiber protein [1, 2]. The multifunctional internalizing receptor CAR belongs to the immunoglobulin superfamily, plays a role in cell–cell adhesion, functions as a transmembrane component of specialized intercellular junctions, including the myocardial intercalated disc, and utilizes coreceptors to enhance binding efficiency in host cell targeting. For instance, CVB utilizes the complement deflecting protein decay accelerating factor (DAF, CD55), as its coreceptor, whereas adenovirus uses integrins $\alpha_v\beta_3$ and $\alpha_v\beta_5$ as its coreceptors [3]. Functioning as a coreceptor, DAF increases the binding efficiency of coxsackievirus and facilitates internalization of the virus by CAR [4]. Recent studies have suggested that CAR provides positive survival signals to cardiomyocytes that are essential for normal heart development [5]. Moreover, CAR-deficient mice causes embryonic death with degeneration of the myocardial wall and thoracic hemorrhaging [5] and hyperplasia of the left ventricular myocardium, distention of the cardinal veins, and abnormalities of sinoatrial valves; [6] analysis of cardiomyocyte ultrastructure revealed a marked reduction of myofibril density [7]. CAR expression appears to be highly induced after myocardial infarction in the rat [8], during the active phase of experimental autoimmune myocarditis in adult rats [9], in human DCM [10] and HF [11]. An age-dependent decline in cardiomyocyte CAR expression has been found to explain the higher efficiency of adenovirus-assisted exogenous $Ca^{2+}$ ATPase (SERCA) and reporter (EGFP) gene expression in primary cultures of myocytes from neonatal rat hearts, as compared to primary cultures of myocytes from adult rat hearts [12]. The high level of CAR expression in the young may explain the higher susceptibility to myocarditis in children.

## Innate Immunity

The role of innate immunity in stimulating a nonspecific immune response in myocarditis has been the subject of intensive investigation [13]. In the first 4–5 days after cardiomyopathic virus infection, innate immunity plays a central role in the heart to minimize virus replication and propagation. Cytokines, important mediators of the innate immune response and main players of innate immunity in cardiomyocytes, are several proteins which belong to Toll-like receptor (TLR) family of pattern recognition receptors that respond to pathogen-associated molecular models (Fig. 12.1).

Cardiomyocytes express a number of TLRs (TLR2, TLR3, TLR4, TLR5, TLR7, TLR9). Some of them are involved in the initiation of early myocardial inflammatory and functional responses to danger signals arising from ischemia-reperfusion and inflammatory stimuli. For example, using transfection of a nuclear factor κB (NF-κB)-luciferase reporter plasmid, Boyd et al. [14] demonstrated that TLR2, TLR4, and TLR5 in murine cardiac cells signal through the NF-κB pathway to express physiologically relevant proinflammatory cytokines (interleukin (IL)-6), chemokines (KC and MIP-2) and cell surface adhesion molecules, intercellular adhesion molecule 1 (ICAM-1) also known as CD54 (Cluster of Differentiation 54). In this chapter, our focus is on TLR3 and TLR7/8, which can be activated by double-stranded RNA (dsRNA) and single-stranded RNA, respectively, because CVB and EMCV carry positive, single-stranded RNA. Moreover, after entry into the cell, the viral genome replicates using the positive-stranded RNA as its template. This results in the formation of dsRNA intermediates. Therefore, both single-stranded RNA and dsRNA are present in virally infected cardiac cells.

TLR3 knockout mice infected with EMCV demonstrate significantly earlier mortality in association with increased viral replication and myocardial injury in the heart compared with wild-type mice [15]. This indicates that the activation of TLR3-mediated antiviral mechanisms is an important innate immune mechanism to suppress virus replication in the heart after EMCV infection. Similarly, deficiency in another isoform of receptor, TLR4, significantly increased titer of another myocarditis causing virus, CVB3, in the mouse heart 2 days after infection [16].

Cytokines, such as interferon (IFN)-α/β, IFN-γ, and IL-6, exert their effect by binding to specific receptors in the cell membrane. IL-6-related cytokines share glycoprotein 130 (gp130) as the signal-transducing protein. Downstream of gp130, two signal-transducing pathways have been identified in cardiac myocytes, the Janus kinase (JAK)-signal transducer and activator of transcription (STAT) pathway, and the Ras-mitogen-activated protein kinase (MAPK) pathway. STAT proteins (acting without the interplay of second messengers) directly transduce signals from the plasma membrane to the nucleus where they modulate target gene transcription. Within a plasma membrane located receptor complex, ligand binding to the extracellular domain of cytokine receptors leads to receptor dimerization, activation of JAK with subsequent active JAK-catalyzed phosphorylation of a single tyrosine residue within the C-termini of STAT protein. Phosphorylated STATs form dimers and translocate to the nucleus where they bind specific *cis* DNA sequences in the target gene promoters and modulate gene transcription. Evidence suggests that decreased JAK and STAT3 phosphorylation are found in both myocarditis and DCM [17], whereas tyrosine phosphorylated STAT1 was detected in the nuclei of cardiomyocytes shortly after intraperitoneal injections of CVB3 [18]. In addition, recent findings of Yajima et al. [19] showed that increased cardiomyocyte survival and innate defense mechanisms against CVB3 infection

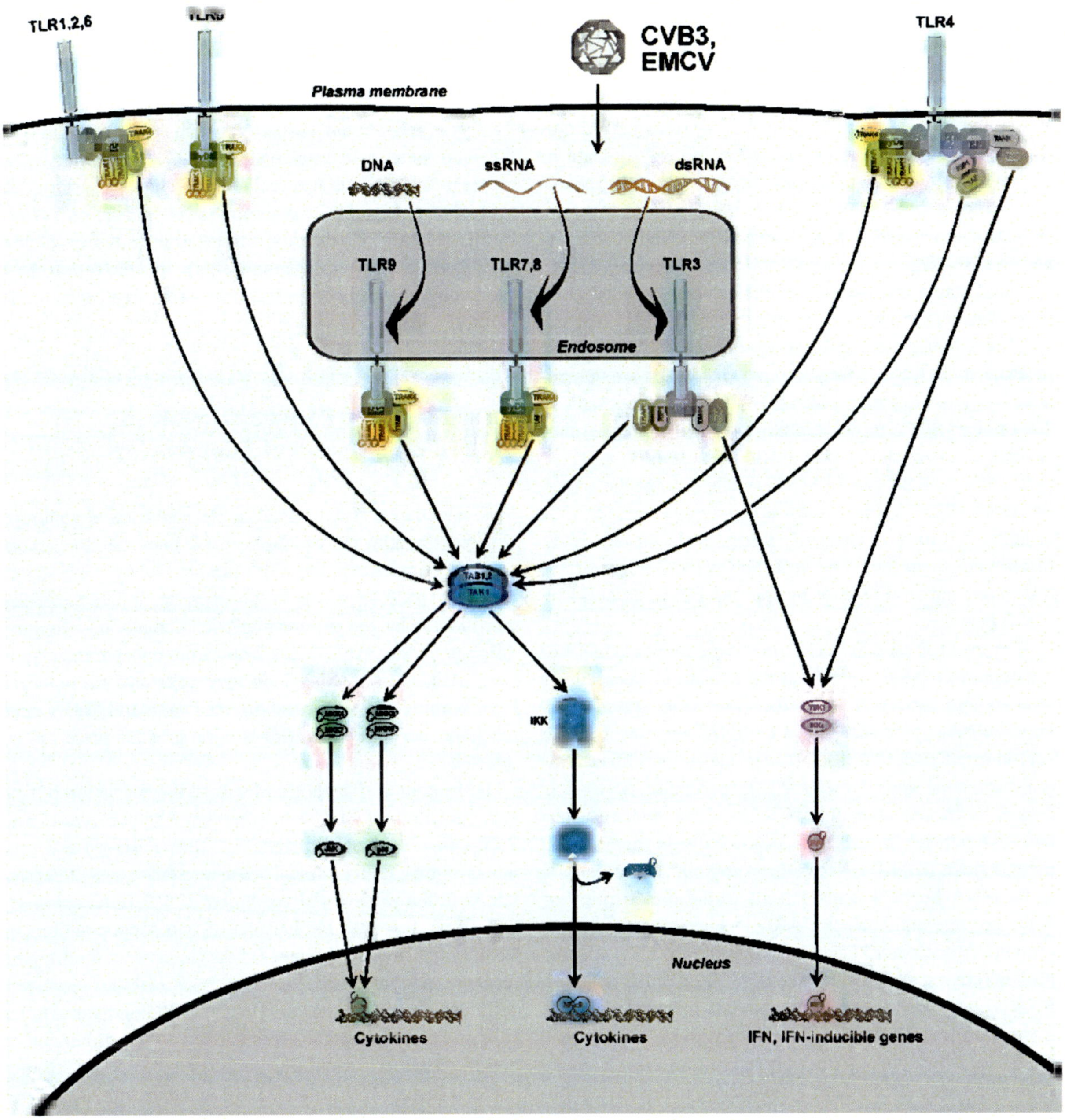


**Fig. 12.1** Virus-activated Toll-like receptor signaling pathways. After entering myocarditis-causing virus, both single-stranded RNAs and double-stranded RNAs are present in the virally infected cell. They activate endosomal TLRs. Once activated, TLRs recruit specific repertoire adapter proteins MyD88, or TRIF, recruit and activate IRAKs and TRAFs. This leads to the activation of IKK complex and the subsequent phosphorylation/degradation of IκBα, the inhibitory subunit of NF-κB. Free NF-κB translocates from the cytosol to the nucleus and activates NF-κB-dependent cytokine genes. TAK1-mediated activation of MAPKs, JNK and p38, also leads to the induction of inflammatory cytokine genes. TLR3 via TRIF/TRAF3 activates the noncanonical IKKs TBK1 and IKKε, resulting in the activation of IRF3 and the transcription of IFN and IFN-inducible genes. Abbreviations: *AP1* transcription factor activator protein 1, *CVB3* coxsackievirus B3, *dsRNA* double-stranded RNA, EMCV encephalomyocarditis virus, *IFN* interferon, *IκB* inhibitor of κB protein, *IKK* IκB kinase, *IRAK* IL-1R-associated kinase, *IRF* interferon regulatory factor, *JNK* c-jun N-terminal kinase, *MAPK* mitogen-activated protein kinase, *MKK* MAPK kinase, *MyD88* myeloid differentiation primary response gene 88, *NF-κB* nuclear factor κB, *p38* p38 MAPK, *RIP* receptor-interacting protein, *ssRNA* single-stranded RNA, *TAB* TAK1-binding protein, *TAK1* transforming growth factor β-activated kinase 1, *TANK* TNFR-associated factor family member-associated NF-κB activator, *TBK1* TANK-binding kinase 1, *TIRAP* TIR-domain containing adaptor protein, *TLR* Toll-like receptor, *TRAF* TNFR-associated factor, *TRAM* TRIF-related adaptor molecule, *TRIF* Toll-IL-1 receptor domain-containing adaptor inducing interferon-β

requires gp130 signal transducer and STAT3. These results suggest that the IL-6/gp130 receptor system and its main downstream mediator STAT3 act to promote cardiomyocyte survival, induce hypertrophy, modulate cardiac extracellular matrix and cardiac function, and play a key role in cardioprotection [20]. Cytokines can also activate death-domain or ceramide-mediated signaling pathways as part of the remodeling process [21]. In later stages of immune activation, cytokines play a leading role in adverse remodeling and progressive HF [22].

The duration and intensity of the cytokine-induced JAK-STAT signaling in cardiac cell is tightly regulated via suppressor of cytokine signaling (SOCS)-dependent negative-feedback mechanism. SOCS is a family of proteins which are upregulated by cytokine-induced JAK-STAT signaling mechanism. The induced SOCS proteins inhibit JAK-mediated phosphorylation of the cytokine receptor (in the case of SOCS3) or JAK activity (in the case of SOCS1), which stops the activation of STAT [23]. Cardiac-specific overexpression of SOCSs in transgenic mice dramatically increases the susceptibility of cardiomyocyte to CVB3 infection, which demonstrates that innate antiviral defense mechanism can be affected by SOCS [19].

Besides TLR-based mechanism of viral nucleic acids recognition/inactivation, additional mechanism of intracellular viral dsRNA recognition was recently found. Intracellular viral dsRNA can be recognized by RNA helicases, retinoic acid-induced protein I (RIG-I), and melanoma differentiation-associated gene 5 (MDA-5). In particular, MDA-5 is critical for the detection of EMCV [24]. RNA helicases contain RNA helicase domain and caspase activation and recruitment domain (CARD). Recognition and binding of dsRNA to the RNA helicase domain leads to the dimerization of RNA helicase and enables CARD to interact with adaptor proteins. Mitochondrial antiviral signaling (MAVS) adaptor protein (other names include IPS-1, VISA, Cardif) connects RNA helicase to downstream antiviral mechanisms. The interaction of dsRNA/MDA-5 complex with MAVS is essential for the activation of transcriptional factors (NF-κB, interferon regulatory factors 3 and 7) which lead to innate immune reactions [25].

## Virus-Mediated Myocardial Injury

Viruses that were not destroyed by the innate immune system start replicating. Some of viral proteins can injure host cardiac cell through the activation of apoptosis pathway. Thus, CVB3 produces proteases 2A and 3C which are shown to induce apoptosis by the activation of caspase-8-mediated pathway and mitochondria-mediated intrinsic apoptosis pathway. More specifically, overexpression of proteases 2A and 3C in a model cell system induces cleavage of procaspase-8 and further cleavage of target protein Bid to yield a truncated form of Bid. Truncated Bid causes the release of cytochrome *c* from the mitochondria, the activation of caspase-9, and finally the activation of caspase-3 which is a central player of the programed cell death pathway [26]. In addition, viral protease 2A disrupts host cell sarcolemma via dystrophin proteolysis and inhibits cardiomyocyte translation system by degradation of eukaryotic translation initiation factor-4G [26–28]. Viral protease-inhibiting agents seem to be a promising approach to prevent and treat myocarditis and cardiomyopathy.

## Virus-Mediated Pathways Involved in the Development of Adaptive Immunity

Viral infection is accompanied by the activation of a number of signaling pathways, which play a role in the development of postinfectious myocarditis. For example, pathogen-associated molecular patterns present in the virus, upregulate molecules essential for antigen presentation, such as major histocompatibility complex (MHC) class I on cardiomyocyte membrane [29], MHC class II, B7-1/2 on mast cells and macrophages [13]. Within 7 days of infection, the cytotoxic $CD8^+$ T cells recognize degraded fragments of viral proteins presented by infected cells.

Infiltration of T cells also leads to the formation of two subpopulations of $CD4^+$ T helper cells, T-helpers types 1 and 2 ($T_h1$ and $T_h2$, respectively). $T_h1$ (interferon-$\gamma^+$) cell promotes cardiac injury while disease resistance tends to correlate with the preferential activation of $T_h2$ (interleukin-$4^+$) cell. Thus, the dynamic balance of two subpopulations of $CD4^+$ lymphocytes contributes to the character of the immune response in myocarditis.

Infiltration of inflammatory cells, including natural killer cells and macrophages is accompanied by the expression of proinflammatory cytokines, such as IFN-γ, IL-4, tumor necrosis factor (TNF)-α and IL-1β which in turn has several regulatory consequences. For example, TFN-α activates endothelial cells, recruits inflammatory cells and enhances the production of inflammatory cytokines; IFN-γ induces upregulation of the MHC antigens on the surface of the myocytes. In addition, TNF-α and INF-γ induce upregulation of intercellular adhesion molecules on the surface of the infected myocyte, which allows effective lysis of virus-infected cardiomyocytes by cytotoxic T lymphocytes (Fig. 12.2).

The balance between subpopulations of T cells appears to be particularly important at the initial stages of immunization. The cell-mediated immune system is designed to attack and eliminate viral-infected cells, but disorder in the elimination of activated lymphocytes' clones, the degree of which

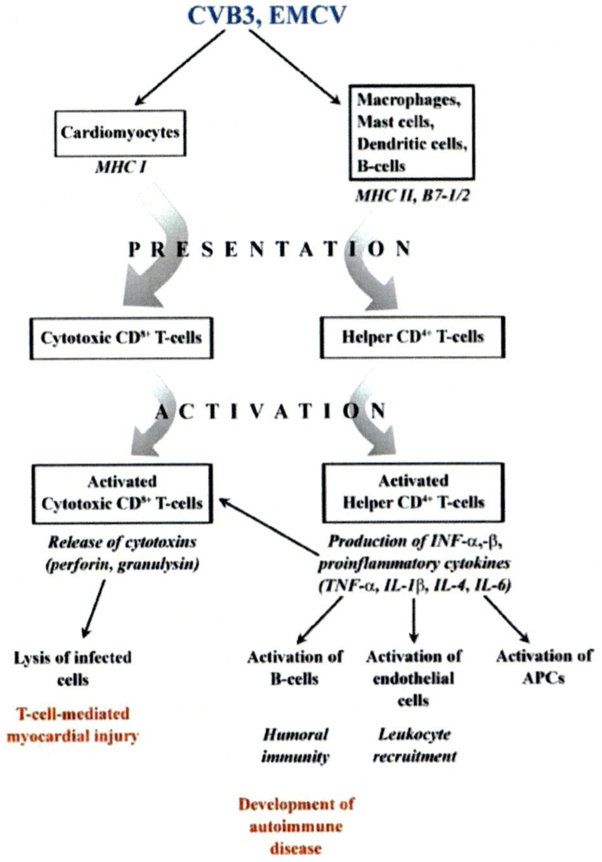


**Fig. 12.2** Virus-activated immunity mechanisms. Abbreviations: *APCs* antigen-presenting cells, *B7-1/2* proteins CD80 and CD86, *CVB3* coxsackievirus B3, *EMCV* encephalomyocarditis virus, *IL* interleukin, *INF* interferon, *MHC* major histocompatibility complex, *TNF* tumor necrosis factor. Events associated with disease development are shown in *red*

correlates with the degree of cardiac insufficiency, may be one of the mechanisms of myocarditis progression. Therefore, the immune reaction can be viewed as a two-edged sword, with appropriate activation of the immune system capable of viral clearance, but excessive immune activation can lead to a chronic inflammatory process that triggers the remodeling of the heart and consequent clinical HF. Massive infiltration of inflammatory cells during the chronic phase of myocarditis leads to the damage of terminally differentiated nondividing cardiomyocytes and results in large regions of fibrosis, and necrosis in the myocardium. $CD4^+$ T cell/$CD8^+$ T cell double-knockout mice have less mortality with less myocarditis after CVB3 infection [30].

It has been hypothesized that normal immunoresponsiveness facilitates viral clearing and allows healing to occur, whereas abnormal immunologic activity is the basis of autoimmune myocarditis which can develop long after viral infection. The activation of potentially autoreactive T lymphocytes under conditions when viral infection triggers imbalance and defects in the regulatory cytokine network system have been described in both murine and human myocarditis [31, 32]. For example, in patients with DCM, there were found circulating autoantibodies directed against contractile (e.g., myosin heavy chain) [33], sarcolemmal (e.g., membrane muscarinic receptor) [34], structural [35], and mitochondrial proteins, such as the ADP/ATP translocator [36, 37].

These autoantibodies may exert direct cytopathic effects on energy metabolism, calcium homeostasis, and signal transduction. Thus, autoantibodies against $\beta_1$-adrenergic receptor function as agonists and cause abnormalities in β-adrenergic receptor coupling *in vitro* [38]. Similarly, in animal model of autoimmune myocarditis (experimental autoimmune myocarditis, EAM), soluble factors released by cells immunized with heart antigens, decreased heart contractility similarly to a muscarinic agonist [39]. Interestingly, in the latter case the contractile effects of immune cell-originated factors were mediated by IFN-γ since they can be blocked by an anti-IFN-γ monoclonal antibody [39]. Therefore, IFN-γ released by autoreactive T cells has an additional contribution to the impaired cardiac function in autoimmune myocarditis. Autoantibodies also can induce complement activation, leading to lysis of antibody-coated cells. The removal of circulating autoantibodies by immunoadsorption has been shown to improve cardiac function and decrease myocardial inflammation.

The autoimmune phase of myocarditis is also characterized by inflammatory cellular infiltration, including natural killer cells and macrophages, with the subsequent expression of proinflammatory cytokines, particularly interleukin-1, interleukin-2, TNF-α, and interferon-γ [40]. TNF-α activates endothelial cells, recruits additional inflammatory cells, further enhances local cytokine production, and has direct negative inotropic effects. Cytokines also activate inducible nitric oxide (NO) synthase in cardiac myocytes. NO can inhibit viral replication by targeting specific viral proteases, and peroxynitrate formation has potent antiviral effects. Mice deficient in NO synthase (NOS) have greater viral titers, a higher viral mRNA load, and more widespread myocyte necrosis. Conversely, in experimental myosin-induced autoimmune myocarditis, NOS expression in myocytes and macrophages is associated with more intense inflammation, whereas NOS inhibitors have been shown to reduce myocarditis severity.

According to Marsland et al. [41], serine/threonine kinase PKCθ may be required for the development of T-cell autoimmune responses in EAM animal model. They showed that PKCθ-deficient mice do not develop EAM. However, systemic administration of the TLR ligand CpG restores EAM in PKCθ-deficient mice indicating that TLR-mediated activation of T cells can directly overcome the requirement for PKCθ signaling and can promote the development of autoimmunity.

Recent studies suggest that persistent T-cell activation can also be induced by endogenous myocardial antigens that cross-react with myocarditis-causing viral peptides. This process is known as molecular mimicry. According to this concept, CVB contains epitopes that are immunologically similar to cardiac myosin [42, 43]. Under conditions, when the cell-mediated immune system is overactivated, it may attack cardiac myosin after the virus is gone. Thus, molecular mimicry may be involved in the pathogenesis of autoimmune myocarditis.

In addition to their roles in the elimination of virus/viral-infected cells and the development of abnormal immunologic activity, T and B cells macrophages provide a reservoir for viral RNA during acute and persistent infections. CVB3 uses two surface receptors expressed on lymphocytes for attachment, internalization, and productive infection: CAR and DAF. DAF and Src family tyrosine kinase $p56^{lck}$ are essential factors controlling the pathogenicity and replication of CVB3 in T cells *in vivo* [44].

## Other Viruses

With the increased use of molecular techniques, besides those described above, a number of other viruses have been found associated with myocarditis. For example, a link between mumps infection and endocardial fibroelastosis (EFE) has been suggested on the basis of the dramatic decline in the incidence of EFE in infants since the mumps vaccine became available [45], although direct evidence of a viral etiology for EFE is lacking. Furthermore, some observations associate myocarditis with respiratory tract viruses, such as Epstein-Barr and influenza viruses [46, 47]. It is worth to note here that the detection of a virus (or viral genome) cannot serve as a proof of its pathogenicity since such an agent may be present as "innocent bystander," or as a result of secondary infection. Several lines of evidence support the assignment of parvovirus B19 (PVB19) as a possible cause of myocarditis and HF. Gathered observations have suggested that the primary targets for PVB19 infection are endothelial cells in small cardiac vessels. Virus plays an important role in the induction of endothelial dysfunction leading to the impairment of myocardial microcirculation with secondary myocyte necrosis [48–50]. Furthermore, PVB19 DNA has been found in fetal myocardial cells, although there is little evidence that cardiomyocytes are the direct target for this virus [49, 51].

HIV disease has been also recognized as a cause of myocarditis and DCM although the direct role and mechanism of HIV in promoting myocarditis and DCM remain unclear [52, 53]. According to one current model of HIV pathogenesis, HIV infects and replicates in inflammatory cells, leading to the induction of cardiomyocyte apoptosis through the activation of apoptotic ligands (multifunctional cytokines, i.e., TNF-α), inducible NOS and by gp120-proapoptotic signaling [54, 55].

## Nonviral Infective Myocarditis

Myocarditis caused by nonviral infectious agents is less common, although more prevalent in immunosuppressed or compromised individuals. The etiology can be either bacterial or parasitic.

An example of bacteria-mediated myocarditis is the one induced by the diphtheria pathogen, *Corynebacterium diphtheriae*, streptococcal, and meningococcal infections [56, 57].

Also, it has been well documented, by both clinical observations and mouse models, that spirochete *Borrelia burgdorferi,* the causative agent of Lyme disease, can result in myocarditis [58, 59]. Myocarditis that develops in a number of patients with Rocky Mountain spotted fever has been found linked to tick-borne pathogenic organism *Ricketssia* [60, 61]. In addition, isolated cases of myocarditis have been related to infections by *Salmonella*, *campylobacter*, *yersinia*, *mycoplasma*, and *Chlamydia* [56, 62–65].

Similar to virus-caused diseases, bacterial infective myocarditis can arise due to autoimmunity. Thus, streptococcal M protein induces immune response resulting in antibodies and anticardiac myosin immunocompetent cells (streptococcal M protein-reactive T lymphocytes) that attack the heart [66]. Molecular mimicry between OspA protein of bacterium *B. burgdorferi* and host myosin also contributes to $CD4^+$-T cell-mediated autoimmunity and to the development of myocarditis [67, 68].

Frequently, patients with Chagas disease may develop myocarditis. The disease is endemic in several areas of South America and is caused by infection with the parasite *Trypanosoma cruzi*. It has been suggested that the neural cell adhesion molecule (NCAM, CD56) functions as a receptor for tissue targeting and cardiomyocyte invasion by *T. cruzi*. Enhanced parasite multiplication in myocytes causes disruption and liberation of several inflammatory mediators, the CC chemokines [69]. CC chemokines, especially CCL5/RANTES and CCL3/macrophage inflammatory protein-1 (MIP-1), function as chemoattractants to recruit $CD4^+$ (helper) and $CD8^+$ (cytotoxic-suppressor) T cells in endomyocardium [70]. Antagonists which can selectively block CC chemokine receptors CCR1 and CCR5 were shown to significantly reduce the numbers of $CD4^+$ and $CD8^+$ T cells, $CCR5^+$, and interleukin-$4^+$ cells invading the heart, resulting in increased survival of *T. cruzi*-infected mice [71]. These observations indicate that CC chemokine receptors may be an attractive therapeutic target in Chagas disease.

## Endocarditis

Cardiac valves and other endocardial surfaces are the targets for bacteria or fungi circulating in the bloodstream. Pathogenic infection leads to infective endocarditis – inflammation of the endocardium. *Streptococcus viridans*, commonly found in the mouth, is responsible for about 50% of all bacterial endocarditis cases. Other common culprits include *Staphylococcus aureus* and enterococcus. *S. aureus* can infect normal heart valves, and is the most common cause of infectious endocarditis in intravenous drug users and in patients with implanted devices. Infective endocarditis due to enterococci infection has been increasingly reported, and is estimated to account for about 10% of the cases. The HACEK group of fastidious gram-negative aerobic bacteria (*Haemopihilus* species, *Actinobacillus actinomycetemcomitans, Cardiobacterium hominis, Eikenella corrodens,* and *Kingella* species) also can cause endocarditis.

Several surface structures of bacteria were identified as factors of virulence. For instance, the fibronectin-binding proteins (FnBPs) encoded by the *S. aureus* gene have been shown to be involved in increased adherence to damaged heart valves. They enhance the attachment of bacteria to damaged tissue providing a coagulum platform for pathogen (the latter consists of fibrin, fibronectin, plasma proteins, and platelet proteins) [72]. Recently, two more cell surface proteins, the DltABCD and MprF proteins, have been identified in *S. aureus*. Mutation in the DltABCD protein significantly reduced the ability of bacteria to bind to endothelial cells *in vitro* [73].

*Streptococcus* also produces surface molecules that contribute to its pathogenicity: glucans (gtf and ftf), ECM adhesins (e.g., fibronectin-binding proteins, FimA), platelet aggregating factors (phase I and phase II antigens, pblA, pblB, and pblT) [74]. One of *S. sanguis* platelet-aggregating factors, the platelet aggregation-associated protein (PAAP), directly contributes to the development of experimental endocarditis. PAAP is a 115 kDa cell wall glycoprotein containing a collagen-like platelet-interactive domain. It interacts with a signal-transducing receptor complex on platelets. Thus, on injured heart valves PAAP enhances platelet accumulation into a fibrin-enmeshed thrombus (vegetation), where *S. sanguis* colonizes [75].

A number of signaling molecules (both host and pathogen originated) are involved in different aspects of infective endocarditis pathogenesis. For instance, the coagulum platform which forms in the damaged valve tissue induces the production of cytokines and procoagulant factors, which contribute to enlargement of the pathogen vegetation; [76] biological cues in serum modulate the enterococcal virulence at sites of infection [77]. On the other hand, *S. aureus*-encoded protein FnBP can trigger host cell apoptosis [78]. Finally, *Enterococcus faecalis*-derived sex pheromones, which are chemotactic for polymorphonuclear leukocytes, contribute to the host inflammatory response often associated with *enterococcus* infection [79].

## Conclusions

Viral infection of the myocardium has a profound and significant effect on the initiation and progression of viral myocarditis. Previously, the treatment of myocarditis primarily included the management of HF and dysrhythmias with angiotensin receptor blockers, angiotensin I-converting enzyme inhibitors, β-blockers, amiodarone, and diuretics; also, heart transplantation in severe cases. Recent observations from animal studies have led to delineation of the major cellular processes that are responsible for myocardial dysfunction in individual patients: viral replication, humoral immunity, or cellular immunity. Such efforts provide new opportunities for the development of specific therapeutic algorithms. For instance, in the initial phase of viral myocarditis potentially effective therapies include boosters of the intrinsic immune defenses (immunoglobulin, interferon) and blockers of viral entry through CAR (WIN class of drugs that inhibit viral attachment). The contribution of some cytokines (TNFs, ILs), adhesion molecules with potential therapeutic implications, needs to be further investigated.

Finally, a variety of immunomodulatory therapies targeting various signaling molecules have been proposed for the autoimmune phase of myocarditis, including immunosuppression, manipulation of cytokines, anti-T-cell-receptor vaccines. Tyrosine kinase $p56^{lck}$ which is responsible for intracellular signal amplification during T-cell activation could be a target for future antiinflammatory drug development.

Infective endocarditis, another pathogen-caused cardiac disease, has not significantly decreased over the past decade despite improvements in health care because new risk factors for infective endocarditis have emerged (haemodialysis, intravenous drug use, use of prosthetic valves, sclerotic valve disease in elderly patients, etc.). A large number of infectious agents have been implicated in infective endocarditis; however, to develop new, effective therapies better understanding of its pathogenesis is required. This understanding may be greatly improved by further investigation on both the infectious agents, and the host response at the molecular level. At present, the penicillins, often in combination with gentamicin, remain the cornerstones in the treatment of endocarditis; for the HACEK group of bacteria, the preferred treatment is ceftriaxone alone, or a combination of ampicillin with gentamicin.

## Summary


- A common mammalian viral receptor (CAR) has been identified for both coxsackieviruses and adenoviruses which allows internalization of the coxsackieviral genome after attachment. CVB utilizes the complement deflecting protein DAF as its coreceptor, whereas adenovirus uses integrin $\alpha_v\beta_3$ and $\alpha_v\beta_5$ as its coreceptors.
- The high level of CAR expression in the young heart may explain the higher susceptibility to myocarditis in children.
- Main players of innate immunity in cardiomyocytes are several proteins which belong to the TLR family of pattern recognition receptors. TLR3 and TLR7/8 can be activated by dsRNA and single-stranded RNA, respectively.
- Cytokines, such as IFN-α/β, IFN-γ, and IL-6, exert their effect by binding to specific receptors in the cell membrane. Downstream of cytokine receptor, the JAK-STAT pathway has been identified in cardiac myocytes. IL-6/gp130 receptor system and its main downstream mediator STAT3 play a key role in cardioprotection after CVB3 infection.
- Besides TLR-based mechanism, viral dsRNA can be recognized by RNA helicases. MAVS adaptor protein connects RNA helicase to downstream antiviral mechanisms.
- Viral proteases 2A and 3C can injure cardiomyocytes through the activation of apoptosis pathways.
- Pathogen-associated molecular patterns present in virus upregulate molecules essential for antigen presentation. The cytotoxic T cells recognize degraded viral proteins shown by infected cells. The balance of subpopulations of T-helper cells contributes to the character of the immune response in myocarditis.
- Massive infiltration of inflammatory cells during the chronic phase of myocarditis leads to the damage of non-dividing cardiomyocytes and results in large regions of necrosis in the myocardium.
- Abnormal immunologic activity is the basis of autoimmune myocarditis which can develop long after viral infection. Molecular mimicry may be involved in the pathogenesis of autoimmune myocarditis.
- Thanks to improved molecular diagnostic techniques, a number of viral agents have been recognized as relevant pathogens in myocarditis, such as Epstein-Barr, influenza viruses, and PVB19.
- HIV disease is often recognized as a cause of myocarditis, although the direct role of HIV in myocarditis remains unclear.
- In some cases, myocarditis can be caused by nonviral infectious agents. Similar to virus-caused diseases, bacterial infective myocarditis can arise due to autoimmunity. Myocarditis may accompany Chagas disease which is caused by the parasite *T. cruzi*. Gathered observations indicate that CC chemokine receptors may be a potential therapeutic target in Chagas disease.
- A large number of infectious agents have been implicated in the pathogenesis of infective endocarditis (inflammatory disease of the endocardium). Several surface structures of bacteria were identified as factors of virulence: fibronectin-binding proteins, DltABCD and MprF proteins. A number of signaling molecules (both host and pathogen originated) are involved in the pathogenesis of infective endocarditis.


## References


1. Bergelson JM, Cunningham JA, Droguett G, et al. Isolation of a common receptor for Coxsackie B viruses and adenoviruses 2 and 5. Science. 1997;275:1320–3.
2. Martino T, Petric M, Weingartl H, et al. The coxsackie-adenovirus receptor (CAR) is used by reference strains and clinical isolates representing all 6 serotypes of coxsackievirus group B, and by swine vesicular disease virus. Virology. 2000;271:99–108.
3. Wickham TJ, Mathias P, Cheresh DA, Nemerow GR. Integrins $\alpha_{v\beta3}$ and $\alpha_{v\beta5}$ promote adenovirus internalization but not virus attachment. Cell. 1993;73:309–19.
4. Martino TA, Petric M, Brown M, et al. Cardiovirulent coxsackieviruses and the decay-accelerating factor (CD55) receptor. Virology. 1998;244:302–14.
5. Asher DR, Cerny AM, Weiler SR, et al. Coxsackievirus and adenovirus receptor is essential for cardiomyocyte development. Genesis. 2005;42:77–85.
6. Chen JW, Zhou B, Yu QC, et al. Cardiomyocyte-specific deletion of the coxsackievirus and adenovirus receptor results in hyperplasia of the embryonic left ventricle and abnormalities of sinuatrial valves. Circ Res. 2006;98:923–30.
7. Dorner AA, Wegmann F, Butz S, et al. Coxsackievirus-adenovirus receptor (CAR) is essential for early embryonic cardiac development. J Cell Sci. 2005;118:3509–21.
8. Fechner H, Noutsias M, Tschoepe C, et al. Induction of coxsackievirus-adenovirus-receptor expression during myocardial tissue formation and remodeling: identification of a cell-to-cell contact-dependent regulatory mechanism. Circulation. 2003;107:876–82.
9. Ito M, Kodama M, Masuko M, et al. Expression of coxsackievirus and adenovirus receptor in hearts of rats with experimental autoimmune myocarditis. Circ Res. 2000;86:275–80.
10. Noutsias M, Fechner H, de Jonge H, Wang X, Dekkers D, Houtsmuller AB, et al. Human coxsackie-adenovirus receptor is colocalized with integrins $\alpha_v\beta_3$ and $\alpha_v\beta_5$ on the cardiomyocyte sarcolemma and upregulated in dilated cardiomyopathy: implications for cardiotropic viral infections. Circulation. 2001;104:275–80.
11. Sasse A, Wallich M, Ding Z, Goedecke A, Schrader J. Coxsackie-and-adenovirus receptor mRNA expression in human heart failure. J Gene Med. 2003;5:876–82.
12. Sumbilla C, Ma H, Seth M, Inesi G. Dependence of exogenous SERCA gene expression on coxsackie adenovirus receptor levels in neonatal and adult cardiac myocytes. Arch Biochem Biophys. 2003; 415:178–83.
13. Fairweather D, Frisancho-Kiss S, Rose NR. Viruses as adjuvants for autoimmunity: evidence from Coxsackievirus-induced myocarditis. Rev Med Virol. 2005;15:17–27.

14. Boyd JH, Mathur S, Wang Y, Bateman RM, Walley KR. Toll-like receptor stimulation in cardiomyoctes decreases contractility and initiates an NF-κB dependent inflammatory response. Cardiovasc Res. 2006;72:384–93.
15. Hardarson HS, Baker JS, Yang Z, et al. Toll-like receptor 3 is an essential component of the innate stress response in virus-induced cardiac injury. Am J Physiol Heart Circ Physiol. 2007;292: H251–8.
16. Fairweather D, Yusung S, Frisancho S, et al. IL-12 receptor β1 and Toll-like receptor 4 increase IL-1β- and IL-18-associated myocarditis and coxsackievirus replication. J Immunol. 2003;170:4731–7.
17. Podewski EK, Hilfiker-Kleiner D, Hilfiker A, et al. Alterations in Janus kinase (JAK)-signal transducers and activators of transcription (STAT) signaling in patients with end-stage dilated cardiomyopathy. Circulation. 2003;107:798–802.
18. Ruppert V, Meyer T, Pankuweit S, Jonsdottir T, Maisch B. Activation of STAT1 transcription factor precedes up-regulation of coxsackievirus-adenovirus receptor during viral myocarditis. Cardiovasc Pathol. 2008;17:81–92.
19. Yajima T, Yasukawa H, Jeon ES, et al. Innate defense mechanism against virus infection within the cardiac myocyte requiring gp130-STAT3 signaling. Circulation. 2006;114:2364–73.
20. Fischer P, Hilfiker-Kleiner D. Survival pathways in hypertrophy and heart failure: the gp130-STAT3 axis. Basic Res Cardiol. 2007; 102:393–411.
21. Liu P, Sole MJ. What is the relevance of apoptosis to the myocardium? Can J Cardiol. 1999;15:8B–10.
22. Nakamura H, Yamamoto T, Yamamura T, et al. Repetitive coxsackievirus infection induces cardiac dilatation in post-myocarditic mice. Jpn Circ J. 1999;63:794–802.
23. Darnell Jr JE. STATs and gene regulation. Science. 1997;277: 1630–5.
24. Gitlin L, Barchet W, Gilfillan S, et al. Essential role of mda-5 in type I IFN responses to polyriboinosinic:polyribocytidylic acid and encephalomyocarditis picornavirus. Proc Natl Acad Sci USA. 2006; 103:8459–64.
25. Hiscott J, Lin R, Nakhaei P, Paz S. MasterCARD: a priceless link to innate immunity. Trends Mol Med. 2006;12:53–6.
26. Chau DH, Yuan J, Zhang H, et al. Coxsackievirus B3 proteases 2A and 3C induce apoptotic cell death through mitochondrial injury and cleavage of eIF4GI but not DAP5/p97/NAT1. Apoptosis. 2007;12:513–24.
27. Badorff C, Lee GH, Lamphear BJ, et al. Enteroviral protease 2A cleaves dystrophin: evidence of cytoskeletal disruption in an acquired cardiomyopathy. Nat Med. 1999;5:320–6.
28. Lamphear BJ, Yan R, Yang F, et al. Mapping the cleavage site in protein synthesis initiation factor eIF-4 gamma of the 2A proteases from human Coxsackievirus and rhinovirus. J Biol Chem. 1993; 268:19200–3.
29. Herskowitz A, Ahmed-Ansari A, Neumann DA, et al. Induction of major histocompatibility complex antigens within the myocardium of patients with active myocarditis: a nonhistologic marker of myocarditis. J Am Coll Cardiol. 1990;15:624–32.
30. Opavsky MA, Penninger J, Aitken K, et al. Susceptibility to myocarditis is dependent on the response of αβ T lymphocytes to coxsackieviral infection. Circ Res. 1999;85:551–8.
31. Schulze K, Becker B, Schultheiss HP. Antibodies to the ADP/ATP carrier, an autoantigen in myocarditis and dilated cardiomyopathy, penetrate into myocardial cells and disturb energy metabolism in vivo. Circ Res. 1989;64:179–92.
32. Schwimmbeck PL, Badorff C, Schultheiss HP, et al. Transfer of human myocarditis into severe combined immunodeficiency mice. Circ Res. 1994;75:156–64.
33. Rose NR, Beisel KW, Herskowitz A, et al. Cardiac myosin and autoimmune myocarditis. Ciba Found Symp. 1987;129:3–24.
34. Fu LX, Magnusson Y, Bergh CH, et al. Localization of a functional autoimmune epitope on the muscarinic acetylcholine receptor-2 in patients with idiopathic dilated cardiomyopathy. J Clin Invest. 1993;91:1964–8.
35. Maisch B, Wedeking U, Kochsiek K. Quantitative assessment of antilaminin antibodies in myocarditis and perimyocarditis. Eur Heart J. 1987;8(Suppl J):233–5.
36. Klein R, Maisch B, Kochsiek K, Berg PA. Demonstration of organ specific antibodies against heart mitochondria (anti-M7) in sera from patients with some forms of heart diseases. Clin Exp Immunol. 1984;58:283–92.
37. Schultheiss HP, Bolte HD. Immunological analysis of auto-antibodies against the adenine nucleotide translocator in dilated cardiomyopathy. J Mol Cell Cardiol. 1985;17:603–17.
38. Magnusson Y, Wallukat G, Waagstein F, Hjalmarson A, Hoebeke J. Autoimmunity in idiopathic dilated cardiomyopathy: characterization of antibodies against the $\beta_1$-adrenoceptor with positive chronotropic effect. Circulation. 1994;89:2760–7.
39. Perez Leirós C, Goren N, Sterin-Borda L, Borda ES. Myocardial dysfunction in an experimental model of autoimmune myocarditis: role of IFN-γ. Neuroimmunomodulation. 1997;4:91–7.
40. Matsumori A, Yamada T, Suzuki H, Matoba Y, Sasayama S. Increased circulating cytokines in patients with myocarditis and cardiomyopathy. Br Heart J. 1994;72:561–6.
41. Marsland BJ, Nembrini C, Grün K, et al. TLR ligands act directly upon T cells to restore proliferation in the absence of protein kinase C-q signaling and promote autoimmune myocarditis. J Immunol. 2007;178:3466–73.
42. Davies JM. Molecular mimicry: can epitope mimicry induce autoimmune disease? Immunol Cell Biol. 1997;75:113–26.
43. Fairweather D, Lawson CM, Chapman AJ, et al. Wild isolates of murine cytomegalovirus induce myocarditis and antibodies that cross-react with virus and cardiac myosin. Immunology. 1998;94: 263–70.
44. Liu P, Aitken K, Kong YY, et al. The tyrosine kinase $p56^{lck}$ is essential in coxsackievirus B3-mediated heart disease. Nat Med. 2000;6: 429–34.
45. Ni J, Bowles NE, Kim YH, et al. Viral infection of the myocardium in endocardial fibroelastosis. Molecular evidence for the role of mumps virus as an etiologic agent. Circulation. 1997;95:133–9.
46. Kuhl U, Pauschinger M, Seeberg B, Lassner D, Noutsias M, Poller W, et al. Viral persistence in the myocardium is associated with progressive cardiac dysfunction. Circulation. 2005;112:1965–70.
47. Bowles NE, Ni J, Kearney DL, et al. Detection of viruses in myocardial tissues by polymerase chain reaction. Evidence of adenovirus as a common cause of myocarditis in children and adults. J Am Coll Cardiol. 2003;42:466–72.
48. Kuhl U, Pauschinger M, Bock T, et al. Parvovirus B19 infection mimicking acute myocardial infarction. Circulation. 2003;108: 945–50.
49. Bultmann BD, Klingel K, Sotlar K, et al. Fatal parvovirus B19-associated myocarditis clinically mimicking ischemic heart disease: an endothelial cell-mediated disease. Hum Pathol. 2003;34:92–5.
50. Klingel K, Sauter M, Bock CT, Szalay G, Schnorr JJ, Kandolf R. Molecular pathology of inflammatory cardiomyopathy. Med Microbiol Immunol (Berl). 2004;193:101–7.
51. O'Malley A, Barry-Kinsella C, Hughes C, et al. Parvovirus infects cardiac myocytes in hydrops fetalis. Pediatr Dev Pathol. 2003;6: 414–20.
52. Zareba KM, Miller TL, Lipshultz SE. Cardiovascular disease and toxicities related to HIV infection and its therapies. Expert Opin Drug Saf. 2005;4:1017–25.
53. Barbaro G. Reviewing the cardiovascular complications of HIV infection after the introduction of highly active antiretroviral therapy. Curr Drug Targets Cardiovasc Haematol Disord. 2005;5:337–43.

54. Kan H, Xie Z, Finkel MS. HIV gp120 enhances NO production by cardiac myocytes through p38 MAP kinase-mediated NF-κB activation. Am J Physiol Heart Circ Physiol. 2000;279:H3138–43.
55. Fiala M, Popik W, Qiao JH, et al. HIV-1 induces cardiomyopathy by cardiomyocyte invasion and gp120, Tat, and cytokine apoptotic signaling. Cardiovasc Toxicol. 2004;4:97–107.
56. Calabrese F, Thiene G. Myocarditis and inflammatory cardiomyopathy: microbiological and molecular biological aspects. Cardiovasc Res. 2003;60:11–25.
57. Hardman JM, Earle KM. Myocarditis in 200 fatal meningococcal infections. Arch Pathol. 1969;87:318–25.
58. Nagi KS, Joshi R, Thakur RK. Cardiac manifestations of Lyme disease: a review. Can J Cardiol. 1996;12:503–6.
59. Defosse DL, Duray PH, Johnson RC. The NIH-3 immunodeficient mouse is a model for Lyme borreliosis myositis and carditis. Am J Pathol. 1992;141:3–10.
60. Marin-Garcia J, Mirvis DM. Myocardial disease in Rocky Mountain spotted fever: clinical, functional, and pathologic findings. Pediatr Cardiol. 1984;5:149–54.
61. Marin-Garcia J, Gooch 3rd WM, Coury DL. Cardiac manifestations of Rocky Mountain spotted fever. Pediatrics. 1981;67:358–61.
62. Burian J, Buser P, Eriksson U. Myocarditis: the immunologist's view on pathogenesis and treatment. Swiss Med Wkly. 2005;135: 359–64.
63. Uzoigwe C. Campylobacter infections of the pericardium and myocardium. Clin Microbiol Infect. 2005;11:253–5.
64. Saikku P. Chlamydia pneumoniae and cardiovascular diseases. Clin Microbiol Infect. 1996;1 Suppl 1:S19–22.
65. Paz A, Potasman I. Mycoplasma-associated carditis. Case reports and review. Cardiology. 2002;97:83–8.
66. Galvin JE, Hemric ME, Kosanke SD, Factor SM, Quinn A, Cunningham MW. Induction of myocarditis and valvulitis in Lewis rats by different epitopes of cardiac myosin and its implications in rheumatic carditis. Am J Pathol. 2002;160:297–306.
67. Raveche ES, Steven E, Schutzer SE, Fernandes H. Evidence of *Borrelia* autoimmunity-induced component of lyme carditis and arthritis. J Clin Microbiol. 2005;43:850–6.
68. McKisic MD, Redmond WL, Barthold SW. Cutting edge: T cell-mediated pathology in murine Lyme borreliosis. J Immunol. 2000; 164:6096–9.
69. Pinto AY, Valente SA, Valente Vda C. Emerging acute Chagas disease in Amazonian Brazil: case reports with serious cardiac involvement. Braz J Infect Dis. 2004;8:454–60.
70. Fuenmayor C, Higuchi ML, Carrasco H, et al. Acute Chagas' disease: immunohistochemical characteristics of T cell infiltrate and its relationship with T. cruzi parasitic antigens. Acta Cardiol. 2005;60:33–7.
71. Marino AP, da Silva A, dos Santos P, et al. Regulated on activation, normal T cell expressed and secreted (RANTES) antagonist (Met-RANTES) controls the early phase of Trypanosoma cruzi-elicited myocarditis. Circulation. 2004;110:1443–9.
72. Minhas T, Ludlam HA, Wilks M, Tabaqchali S. Detection by PCR and analysis of the distribution of a fibronectin-binding protein gene (fbn) among staphylococcal isolates. J Med Microbiol. 1995;42:96–101.
73. Weidenmaier C, Peschel A, Kempf VA, et al. DltABCD- and MprF-Mediated cell envelope modifications of *Staphylococcus aureus* confer resistance to platelet microbicidal proteins and contribute to virulence in a rabbit endocarditis model. Infect Immun. 2005;73:8033–8.
74. Moreillon P, Que YA, Bayer AS. Pathogenesis of streptococcal and staphylococcal endocarditis. Infect Dis Clin North Am. 2002; 16:297–318.
75. Erickson PR, Herzberg MC. Altered expression of the platelet aggregation-associated protein from *Streptococcus sanguis* after growth in the presence of collagen. Infect Immun. 1995;63:1084–8.
76. Tak T, Shukla SK. Molecular diagnosis of infective endocarditis: a helpful addition to the Duke criteria. Clin Med Res. 2004;2:206–8.
77. Jett BD, Huycke MM, Gilmore MS. Virulence of enterococci. Clin Microbiol Rev. 1994;7:462–78.
78. Dziewanowska K, Carson AR, Patti JM, Deobald CF, Bayles KW, Bohach GA. Staphylococcal fibronectin binding protein interacts with heat shock protein 60 and integrins: role in internalization by epithelial cells. Infect Immun. 2000;68:6321–8.
79. Johnson AP. The pathogenicity of enterococci. J Antimicrob Chemother. 1994;33:1083–9.

# Chapter 13
# Signaling in Hypertension

**Abstract** It is well established that signaling through G protein-coupled receptors, including those that respond to angiotensin II ($AT_1$-receptors), endothelin (ET1B receptors) and to epinephrine and norepinephrine (β-adrenergic receptors) are intimately implicated in the regulation of normal cardiovascular function, including mediation in peripheral artery resistance, vasodilation, contraction, and vascular tone. Given the key role of angiotensin II, endothelin-1, and adrenergic agonists and their signaling pathways in stimulating vascular smooth muscle cell proliferation and modifying endothelial cell function, it is not surprising that alterations in these signaling pathways are increasingly implicated as contributory factors in vascular pathologies, including hypertension and atherosclerosis.

**Keywords** Hypertension signaling • Endothelin receptors • Renin–angiotensin system • Sympathetic activity

## Introduction

Hypertension is a leading cause of morbidity and mortality. In the USA, it is estimated that more than 65 million people have hypertension. Continual hypertension is one of the risk factors for strokes, myocardial infarction (MI), heart failure (HF), and arterial aneurysm, and is a leading cause of chronic kidney failure resulting in significant mortality. In about 90–95% of cases, hypertension is a multifactorial disorder which develops because of the interaction of multiple genes with environmental and/or behavioral factors. Medical cause of high blood pressure (BP), in this situation of "essential hypertension" can hardly be found. Mechanisms associated with essential hypertension are not well understood, but it is well documented by scientific and clinical literature that several signaling systems play an essential role in the pathophysiology of this illness.

For instance, an overactive renin–angiotensin–aldosterone system (RAAS) leads to vasoconstriction and retention of sodium and water. The resulting increase in blood volume causes hypertension. The major bioactive peptide of the RAAS is angiotensin II (Ang II). Mechanistic insights by which Ang II contributes to the pathophysiology of hypertension include the activation of signaling pathways leading to vascular remodeling and endothelial dysfunction. In addition, recent findings suggest the existence of a cross talk between Ang II and another component of RAAS, aldosterone, that plays a role in cardiovascular disease. Pharmacologic blockade of the RAAS has been available for almost 25 years, and there is extensive documentation of its effectiveness in the treatment of hypertension.

A dominant factor in blood pressure control is the sympathetic nervous system (SNS). Increased SNS activity is implicated in the early events of hypertension development, including increased vascular resistance, endothelial dysfunction, and vascular remodeling. Recently, new agents which reduce SNS activity in peripheral vasculature through antagonism of specific β- and $\alpha_1$-adrenergic receptor subtypes were generated. This leads to vasodilation of the peripheral vasculature and results in blood pressure reduction (decrease in peripheral vascular resistance).

Natriuretic peptides (NPs) are a family of cardiac- and vascular-derived hormones that play an important role in cardiovascular homeostasis through the regulation of electrolytes and water balance (diuretic and natriuretic effects in kidney) as well as vasodilatory activity. Natriuretic peptides exert their functions by binding to the guanylyl cyclase-linked receptors. All biological effects of natriuretic peptides involve stimulation of cGMP as a second intracellular messenger. Natriuretic peptides might be the base for the development of novel therapeutic strategies with potential great benefit in patients with cardiovascular disease.

In this chapter, we review the different signaling pathways and molecules, including atrial natriuretic peptide (ANP) and b-type natriuretic peptide (BNP), that are implicated in hypertension pathogenesis, in particular in essential/idiopathic hypertension. Information from these findings may result not only in enhanced effectiveness of known drugs, but also in the development of new therapeutic modalities, including gene therapy.

J. Marín-García, *Signaling in the Heart*,
DOI 10.1007/978-1-4419-9461-5_13, © Springer Science+Business Media, LLC 2011

# Renin–Angiotensin–Aldosteron System

## *Angiotensin*

Ang II is the major peptide hormone of the RAAS. It plays a critical role in the control of cardiovascular homoeostasis, including mediation of peripheral artery resistance, vasodilation, contraction, and vascular tone. Given the key role of the Ang II-dependent signaling pathway in the stimulation of vascular smooth muscle cell (SMC) proliferation and modification of endothelial cell function, it is not surprising that alterations in this signaling pathway are increasingly implicated as contributory factors in vascular pathologies, such as hypertension and atherosclerosis.

Two G protein-coupled receptors (GPCRs) mediate Ang II function, the $AT_1$ and $AT_2$ receptors. The $AT_1$ receptor has been shown to mediate most of the pathophysiological effects of Ang II. Via $AT_1$ receptor, Ang II activates several cytoplasmic signaling pathways, which contribute to vascular remodeling, inducing hypertrophy, hyperplasia, and migration of vascular SMCs, and endothelial dysfunction. The major acute function of Ang II is vasoconstriction. Ang II-bound $AT_1$ receptors via $G_{\alpha q/11}$, $G_{\alpha 12/13}$, and $G_{\beta\gamma}$ complexes activate downstream effectors – phospholipases (phospholipase C [PLC], phospholipase D [PLD], and phospholipase A2 [PLA2]). The PLC/PLD pathways which cause muscle contraction and cell growth are augmented in hypertensive rats compared to controls, suggesting that alterations in the G protein-activated second messengers may play a role in the pathogenesis of hypertension [1].

An important role of heterotrimeric G proteins in hypertension was shown in studies of polymorphism in the gene *GNB3*, which encodes the $\beta_3$ subunit of heterotrimeric G proteins. Single nucleotide polymorphic variant of human *GNB3* (C825T) encodes the protein with increased biological activity that enhances G protein signaling [2]. Several population-based and case-control studies of different ethnicities suggested that the 825T allele carriers have an increased risk for hypertension combined with features of the metabolic syndrome, such as dyslipidemia, hypercholesterolemia, insulin resistance, and obesity [3, 4].

Very important downstream components of Ang II/$AT_1$ receptor signaling pathway are MAP kinases (MAPK), including extracellular signal-regulated kinases (Erk) 1/2, p38MAPK, and c-Jun $NH_2$-terminal kinase (JNK). They are implicated in vascular SMC growth and hypertrophy. Activation of Erk1/2 is a complex process involving active PLC/protein kinase C (PKC) as well as transactivation of epidermal growth factor receptor (EGFR) (see below) with subsequent activation of Raf, Ras, and MAPK kinase (Mek) [5]. Several mechanisms of Ang II-mediated activation of p38MAPK and JNK have been reported and they provide links between oxidative stress (OS), hypertension, and vascular remodeling [6]. Ang II-mediated MAPK activation is followed by an increase in *c-fos* (activated by Erk1/2) and *c-jun* (activated by JNK) gene expression, as well as increased AP-1 activity. AP-1 is a transcription factor complex (formed from dimerization of c-Jun and c-Fos) that controls a number of cellular processes, including cell differentiation, proliferation, and apoptosis [7]. In addition, Erk1/2 participates in Ang II-induced cellular growth and protein synthesis via regulation of inhibitor of eukaryotic initiation factor-4E (PHAS-1) [8]. Ang II-induced enhanced activation of vascular MAP kinases, such as Erk1/2, has been also implicated in hypertension [9].

Ang II through the $AT_1$ receptor activates small GTP-binding proteins. Small GTP-binding proteins from the Rho family have been identified as major regulators of numerous vascular cell functions, including, migration, proliferation, differentiation, and gene transcription. Small GTP-binding proteins appear to play important roles in mediating cardiovascular remodeling induced by Ang II. Thus, in vascular SMCs, Ang II activates RhoA and induces its translocation to the particulate fraction [10, 11], and this plays a critical role in Ang II-induced vascular remodeling associated with hypertension [12]. It is very likely that the activation of Rho is achieved through guanine nucleotide-exchange factors (GEFs) of Rho, which are sensitive to $AT_1$ receptor-$G_{12/13}$-$G_q$ protein-signaling cascade [13, 14]. Ang II-activated Rho triggers a cascade of downstream molecules, including Rho kinase (ROCK) and JNK, leading to vascular migration and hypertrophy [15]. There is evidence for a negative regulatory effect of RhoA and its downstream effector, ROCK, on endothelial nitric oxide synthase (eNOS) gene expression (by destabilizing eNOS mRNA), negative effect on vasodilation and a positive regulatory effect on preproET-1 transcription favoring vascular occlusion [16]. Another important role of small GTP-binding proteins, mainly RhoA-ROCK, in both vascular SMCs and endothelial cells is the regulation of endothelial barrier dysfunction and enhanced SMC contraction (the latter via enhanced myosin light chain phosphorylation) [13]. Thus, recent observations have demonstrated that the activation of Rho proteins appears to be a common component for the pathogenesis of hypertension and vascular proliferative disorders. These findings open new opportunities for hypertension management by targeting Rho-ROCK-pathway in vascular SMCs.

Recently, a novel mechanism of Ang II-promoted pathophysiological hypertrophy of vascular SMCs has been reported: transactivation of EGFR. Ang II through $AT_1$ receptor rapidly *trans*-activates the EGFR leading to the Ras-Raf-Erk, phosphatidylinositol 3-kinase (PI3K)-3-phosphoinositide dependent protein kinase-1 (PDK1)-Akt, p70 S6 kinase, and p38 mitogen-activated protein kinase activation, induction of c-Fos, and subsequent growth of vascular SMCs [17, 18]. EGFR activation is required for Ang II-mediated hypertension. In rats, Ang II-mediated hypertension is attenuated when animals are treated with an antisense oligodeoxynucleotide to EGFR [19]. Mechanism for EGFR transactivation by $AT_1$ receptor involves a metalloprotease-dependent EGFR ligand production from its membrane-bound pro-form [20]. Pharmacological inhibition or genetic elimination of metalloprotease blocks Ang II-stimulated hypertrophy of SMCs, which suggests that metalloprotease-mediated EGFR transactivation plays a critical role in the cardiovascular remodeling induced by Ang II [20, 21]. EGFR ligand (heparin-binding EGF) induces conformational changes in EGFRs, allowing them to dimerize and autophosphorylate on tyrosine. Once activated, EGFRs serve as a docking site for Grb2/Shc/Sos complexes, inducing two major transduction pathways: the PI3K-PDK1-Akt cascade and the Ras-Raf-Erk pathway, which leads to cellular growth, remodeling, hypertrophy, and inflammation.

Besides EGFR, Ang II regulates several nonreceptor tyrosine kinases. Recently, Src has emerged as an important player in Ang II-mediated cellular effects. Src can be activated by $G_{\beta\gamma}$ in a reactive oxygen species (ROS)-dependent manner, and participates in the integration of several downstream pathways, including serine/treonine kinases (Ras-, PLC-related pathways) and tyrosine kinases [focal adhesion kinase (FAK)-proline-rich tyrosine kinase 2 (Pyk2)-related pathways]. Ang II-Src-dependent activation of FAK and Pyk2 initiates in vascular SMCs focal adhesion complex formation and cell growth [22, 23]. C-terminus of activated $AT_1$ receptor contains a docking site for another nonreceptor tyrosine kinase, janus kinase 2 (JAK2). Interaction with activated $AT_1$ receptor stimulates JAK2 autophosphorylation/activation, followed by JAK2-dependent dimerization of signal transducer and activator of transcription (STAT) proteins, their translocation to the nucleus and induction of early growth response genes, *c-fos* and *c-myc* [24, 25]. Thus, JAK-STAT-signaling cascade is another system, involved in Ang II-mediated growth, migration, and remodeling of vascular SMC.

Ang II participates in the development of endothelial dysfunction, which is a hallmark of the hypertensive patient. So far, the main cause of hypertension-related endothelial dysfunction in humans has been identified with an increase in nitric oxide (NO) breakdown, as a consequence of increased production of ROS (see section on "Redox" below).

One important abnormality associated with hypertension is the proliferation of cardiac fibroblasts and increased deposition of the extracellular matrix (ECM) protein, collagen. This results in hypertensive heart disease that is characterized by extensive myocardial fibrosis and myocardial stiffness with cardiac dysfunction [26]. Ang II, aldosterone, and endothelins play a central role in the remodeling of the ECM in hypertension. Binding of Ang II to $AT_1$-receptors activates multiple-signaling pathways in fibroblasts, including phospholipase Cβ, with the subsequent release of $Ca^{2+}$ from intracellular stores and the activation of PKC, MAP kinases, tyrosine kinases, PLD, phosphatidic acid formation, and the STAT family of transcription factors. As a result, fibroblasts respond to Ang II with hyperplastic/hypertrophic growth, and increased expression of collagen, fibronectin, and integrins [27]. Ang II also promotes the association of scaffolding proteins, such as paxillin, talin, and p130Cas, leading to focal adhesion and ECM formation. The change in ECM modifies the signals that cardiac myocytes receive from their scaffolding environment, leading to changes in gene expression associated with hypertrophy and contractile dysfunction. Also the activation of RAAS and increased secretion and activation of transforming growth factor $\beta_1$ (TGF-$\beta_1$) recruit SMCs, monocytes, and fibroblasts and stimulate a genetic program of wound repair and ECM deposition, leading to perivascular fibrosis and amplification of the profibrotic state [28].

It should be noted that some of the $AT_1$ functions on the cardiovascular system are mediated through other organs, such as the kidney and brain, and not by a direct action on the vasculature [29, 30].

## Renin

The mechanisms controlling the formation and degradation of Ang II are important in determining its final physiological effect. Ang II is formed from enzymatic cleavage of angiotensinogen to angiotensin I (Ang I) by the aspartyl protease renin, with subsequent conversion of Ang I to Ang II by angiotensin converting enzyme (ACE). Proteolytic conversion of angiotensinogen into Ang I is the rate-limiting step in the synthesis of Ang II. Renin, in turn, is produced through the activation of its enzymatically inactive precursor, pro-renin. Pro-renin may be rendered enzymatically active in two ways, proteolytic and nonproteolytic activation. Proteolytic activation occurs via the actual removal of the pro-peptide chain. Most proteolytic activation of pro-renin occurs in kidney leading to the production of active renin. Nonproteolytic activation involves unfolding of the pro-peptide chain away from the catalytic domain followed by an additional conformational change.

Importantly, in 2002 Nguyen et al. [31] discovered a receptor for renin and pro-renin, (pro)renin receptor. (Pro) renin receptor serves several functions. First, interaction with (pro)renin receptor increases the enzymatic activity of renin, and the production of Ang I on the cell surface accelerates. Second, binding to (pro)renin receptor induces nonproteolytic activation of pro-renin, and contributes to Ang I production. Third, renin and pro-renin via binding to (pro) renin receptor exert physiological effects not related to Ang II. They involve the activation of signaling pathways (Erk1/2, p38 MAP kinase) and induce DNA synthesis and stimulation TGF-β [32].

On the basis of molecular modeling of renin, Wood et al. [33] developed aliskiren, a potent low-molecular-weight, hydrophilic nonpeptide, that functions as a direct renin inhibitor. In patients with hypertension, aliskiren suppresses plasma renin activity and reduces blood pressure [34]. When taken with an angiotensin receptor blocker, aliskiren attains significant additional blood pressure reduction, indicating a more complimentary pharmacology and more complete renin–angiotensin system blockade [35].

### *Angiotensin-Converting Enzyme*

The ACE (or kininase II) is a bivalent dipeptidyl carboxyl metallopeptidase, which is a membrane enzyme in endothelial, epithelial, and neuroepithelial cells and in the brain; it is also present in a soluble form in blood and numerous body fluids. ACE cleaves the C-terminal dipeptide from Ang I and bradykinin, thus interacting with RAAS and the kallikrein–kinin system simultaneously. An increase in ACE activity disturbs this delicate balance and promotes vasoconstrictive and salt-retention Ang II.

The ACE inhibitors (ACEIs) restore this balance by decreasing the formation of Ang II and the degradation of bradykinin. In addition, ACEIs reduce the effects of Ang II on the kidney, resulting in natriuresis and diuresis. This decreases the circulating blood volume and cardiac output. According to the Joint National Committee (JNC) VII, ACEIs are some of the first-line drugs for hypertension [36]. One of the characteristics of ACEIs is that they lower peripheral vascular resistance (PVR) without causing a compensatory increase in heart rate [37].

### *Aldosterone*

Adrenal corticosteroid aldosterone has emerged as an important component and mediator of the effects of RAAS. Obesity is associated with increased production of aldosterone, and increased mineralocorticoid activity is one of the mechanisms by which obesity induces hypertension [38, 39]. There are two mechanisms of aldosterone actions on target cells. First, aldosterone binds to the mineralocorticoid receptor (MR). MR is a member of the steroid hormone receptor superfamily; thus, aldosterone-bound MR functions as a ligand-dependent transcription factor and activates protein synthesis. This mechanism underlies, for example, aldosterone-dependent modulation of vascular tone: it occurs at least in part through the upregulation of Ang II receptors under conditions where availability of endothelial NO is reduced. Second, aldosterone excess promotes collagen deposition in blood vessels, enhancing vascular remodeling at the expense of compliance. Furthermore, peripheral blood monocytes and vascular SMCs are both influenced by aldosterone to produce ROS, which further leads to the activation of the transcription factor NF-κB, and to enhanced expression of its target genes. Chronic aldosterone administration has also been shown to stimulate perivascular and interstitial cardiac fibrosis, and cardiac hypertrophy. Interestingly, there is evidence from several clinical trials that aldosterone blockade (e.g., eplerenone) offers significant benefit in postmyocardial infarction and HF patients.

There is evidence that aldosterone-mediated rapid effects lead to the activation of ion channels and other signaling pathways. Recent observations by Ishizawa et al. [40], demonstrated that aldosterone/MR-induced vascular injury involves the activation of a new MAP kinase, big mitogen-activated protein kinase 1 (BMK1) [40]. Interestingly, activated MR can also stimulate two other MAPK in vascular SMCs – Erk1/2 and p38, and cross-activation of EGFR- and Src-dependent signaling pathways are required for these events [41, 42]. Furthermore, exposure of vascular SMC to aldosterone amplifies signaling of Ang II, which is critically involved in processes associated with hypertension (see above). Gathered observations suggest that interaction between aldosterone/MR- and Ang II-induced signaling pathways leads to modulation of ROS production, transactivation of EGFR, stimulation of Erk1/2, and affects proliferation of vascular SMC [43–45]. Taken together, above data suggest that blockade or inhibition of both Ang II and aldosterone can result in greater therapeutical benefit than either one of them alone. Moreover, recent experimental data validate this concept: combination of aldosterone antagonists with captopril (an angiotensin-converting enzyme inhibitor) or $AT_1$ receptor blocker results in efficient blockade of the RAAS and better cardioprotection in a rat model of hypertension [45–47].

## Sympathetic Overactivity

The SNS is a dominant factor in arterial blood pressure control. Sympathetic nerves are continuously active so innervated blood vessels remain under some degree of continuous constriction control. By rapid regulation via SNS, the degree of vasoconstriction in the blood vessels can be altered. This in turn changes blood flow affecting the function of the organ, peripheral resistance, and arterial pressure.

In essential hypertension, there is a disproportionate increase in sympathetic activity to the heart and kidneys, with approximately half of the increase in sympathetic neuron neurotransmitter, norepinephrine, being accounted for by increased sympathetic nerve activity (SNA) to these organs [48, 49].

The role of increased SNA in the development of the vascular changes associated with hypertension includes increased vascular resistance, endothelial dysfunction, and vascular remodeling [50]. Reduction of SNS activity not only lowers blood pressure, but may also decrease the risk of developing the vascular changes associated with hypertension.

In the kidney, overactivated SNA affects systems that maintain body fluid balance and therefore blood pressure. Normally, the kidneys respond to changes in arterial pressure by altering the renal excretion of salt and water. The "pressure – natriuresis" relationship is critical to keep a proper balance of body fluid and BP, and changes in this relationship is important in the development of hypertension. One mechanism by which increased renal SNA could increase blood pressure is by alteration in the "pressure – natriuresis" relationship: chronic infusion of norepinephrine directly into the renal artery causes the retention of sodium and water and produce sustained increases in arterial blood pressure [51]. In support of this mechanism, Grisk et al. [52] have observed the resetting of "pressure – natriuresis" curve to a lower pressure after renal denervation.

Traditional β-adrenoceptor blockers (e.g., atenolol, metoprolol, propranolol) are sympatholytic drugs which inhibit postganglionic functioning of the SNS. They affect only β-adrenergic receptors, and reduce blood pressure in hypertension mainly through decreasing cardiac output (CO), with little effect on the peripheral vasculature [53]. Newer vasodilatory β-adrenoceptor blockers (carvedilol, labetalol, nebivolol) are mixed β-blockers with β- and α-adrenoceptor blocking properties. They reduce blood pressure predominantly by decreasing PVR, with little effect on CO [53]. Moreover, vasodilatory β-adrenoceptor antagonists have been shown to improve endothelial function. In addition, ACEIs and angiotensin receptor antagonists may exert at least some of their action through a reduction in SNA to different organs [54].

## Natriuretic Peptides

Natriuretic peptides control vascular SMC relaxation and kidney function, and therefore play an important role in the regulation of blood pressure. ANP rapidly lowers BP and promotes sodium and water excretion through a cGMP-signaling pathway. ANP is mainly produced in the atria and stored in granules as a proANP form. ProANP secretion is stimulated by atrial stretching after volume overload, and active ANP is produced under the action of corin, a transmembrane serine protease acting as a converting enzyme. Another member of natriuretic peptide family, BNP is actively produced by the ventricles upon atrial wall stretching (regulated by the transcription factor GATA-4).

Receptors for these peptides, natriuretic peptide receptors (NPRs), are composed of two members: NPR-A and NPR-B. They possess guanylyl cyclase (GC) activity required for signal transduction. In the basal state, NPRs are inactive, and conversion of GTP to cGMP by their GC domain is low. Activation of these receptors by the binding of NPs stimulates their GC activity which leads to cGMP production. cGMP interacts with several target proteins, including protein kinase G (PKG), ion-gated channels, and phosphodiesterases (PDEs).

Major NPR/GC-signaling effects which influence BP include the relaxation of vascular SMCs and antagonism of renal RAAS. In vascular SMCs, ANP via NPR-GC-PKG-I leads to phosphorylation of several ion channels and pumps with subsequent changings in $Ca^{2+}$ homeostasis and SMCs relaxation [55–58]. In adrenal gland, ANP decreases the secretion of aldosterone probably through activation of cyclic GMP-stimulated PDE2 and degradation of cAMP necessary for aldosterone synthesis [59]. RAAS is also antagonized by natriuretic peptides at the kidney level: ANP inhibits secretion of renin and Ang II-dependent reabsorption of sodium and water [60, 61]. Thus, NPs participate in the maintenance of vessel volume by increasing fluid and electrolyte excretion.

Pathophysiologically, hypertension is often associated with low levels of NPs, and several natural variations in human ANP and BNP genes (single nucleotide polymorphism) are strongly associated with circulating levels of NPs and the risk of developing hypertension [62]. Several studies have been conducted to clinically restore normal levels of NPs. In clinical studies as well as in animal models, ANP injection decreases BP with parallel increase in natriuresis and diuresis [63, 64]. Similar results were obtained in hypertensive individuals with a synthetic form of ANP, anaritide [65].

With the development of molecular genetics, the use of gene transfer and gene therapy strategies for long-term control of hypertension have become possible. A single intravenous injection of human ANP DNA construct produced reduction in blood pressure in young, spontaneously hypertensive rats (SHRs) [64]. Similarly, sustained ANP expression after administration of adenoviral vector was associated with a decrease in systolic blood pressure [66]. These findings provide the basis for ANP gene delivery as a strategy for treating essential hypertension. Furthermore, studies on mice models have confirmed the role of NPs in BP regulation at the level of NPR. Thus, the deletion of

NPR-A causes salt-resistant hypertension [67], whereas NPR-A overexpression decreases BP and protects against salt-sensitive hypertension [68]. Interestingly, among the pharmacotherapeutical approaches with promising results is the lowering of natriuretic peptide degradation. For example, omapatrilat, an inhibitor of neutral endopeptidases, was shown to elevate circulating NPs and have hypotensive effect [69].

## RedOx Signaling

An important characteristic of hypertension is increased peripheral resistance. This resistance increases due to the reduction of vessels' diameter. Structural and functional changes in vessels in hypertension which contribute to a decrease in their diameter include impaired endothelial function, vascular smooth muscle growth, increased contractile activity, and increased extracellular matrix. All of these processes are regulated by ROS, which are elevated in hypertension.

Major sources of ROS in vascular disease and hypertension are xanthine oxidase, uncoupled endothelial NO synthase and NAD(P)H oxidase. For example, elevated levels of xanthine oxidase activity in association with increased arteriolar tone were demonstrated in SHRs [70]. Also, in mice with experimentally induced hypertension, uncoupling of endothelial NOS has been shown to be associated with increased production of $O_2^{\bullet -}$. When uncoupling of eNOS was reversed by tetrahydrobiopterin, blood pressure improved [71]. Moreover, vasoactive agents, such as Ang II, and mechanical forces, such as shear stress activate NAD(P)H oxidase, which is the major source of ROS in vascular cells (Fig. 13.1) [72, 73].

The ability of ROS to affect the vasculature is based on targeting several signaling cascades that control cell proliferation, differentiation, and cell death. Specifically, one of the targets of ROS is mitogen-activated protein kinase p38. Accumulating evidence suggests that ROS-dependent activation

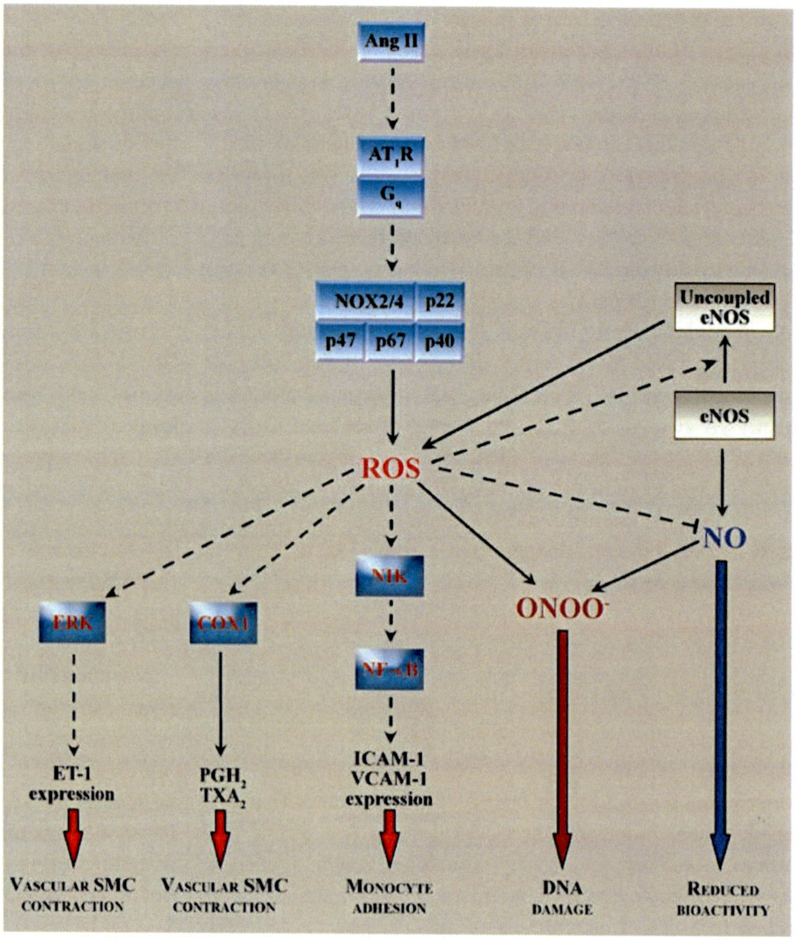


**Fig. 13.1** Angiotensin II-related signaling pathways involved in endothelial dysfunction. Enzymatic and nonenzymatic synthesis of molecules (and eNOS uncoupling) are indicated by *solid black arrows*; activatory/inhibitory events are indicated by *dashed lines*. Abbreviations: *Ang II* angiotensin II, *$AT_1R$* Ang II type 1 receptor, *COX1* cyclooxygenase 1, *eNOS* endothelial nitric oxide synthase, *ERK* extracellular signal-regulated kinase, *ET-1* endothelin-1, *$G_q$* heterotrimeric $G_q$ protein, *ICAM-1* intercellular adhesion molecule-1, *NF-κB* nuclear factor κB, *NIK* NF-κB inducing kinase, *NO* nitric oxide, *NOX2/4* NAD(P)H oxidase isoforms 2 and 4, *$ONOO^-$* peroxynitrite, *p22* p22 phagocytic oxidase (phox), *p40* p40 phox, *p47* p47 phox, *p67* p67 phox, *$PGH_2$* prostaglandin $H_2$, *ROS* reactive oxygen species, *SMC* smooth muscle cell, *$TXA_2$* thromboxane $A_2$, *VCAM-1* vascular cell adhesion molecule-1

of p38 mediates the growth of vascular SMC and collagen synthesis in SHR [74, 75]. The constrictor effect of Ang II seen in hypertension can also be explained (at least in part) by the activation of p38 via NAD(P)H oxidase-$H_2O_2$-pathway [75–78]. Stimulation of a number of redox-sensitive signaling cascades in hypertension may involve ROS-mediated activation of nonreceptor tyrosine kinase Src, activation of Erk1/2, and also Erk1/2-dependent expression of the early response gene *c-fos* in rat small arteries [79], or Ang II-induced activation of NAD(P)H oxidase. In the latter case, Src stimulates phosphorylation/translocation of cytosolic subunit of NAD(P)H oxidase, p47phox [80]. Moreover, there is now a body of evidence indicating that in vascular SMCs, ROS can mediate Ang II-dependent transactivation of receptor tyrosine kinase (EGFR) and this process involves Src [81, 82]. Thus, the activation of redox-sensitive tyrosine kinases may mediate some of the vascular changes that occur in hypertension.

Direct molecular targets of ROS are protein tyrosine phosphatases (PTPs). The level of tyrosine phosphorylation in cells is controlled by a tightly regulated balance between protein tyrosine kinases and PTPs. PTPs are susceptible to oxidation/inactivation by ROS. Exposure to oxidants induces tyrosine phosphorylation due to PTP inactivation. Inactivation of PTPs (receptor PTP-D2, CD45, SHP-1, HePTP) is involved in OS-induced activation of several tyrosine kinases, such as EGFR, insulin receptor, Lck, Fyn, and Ang II-mediated EGFR transactivation [83]. Thus, the activation of vascular MAPK by Ang II may be partially mediated through ROS-dependent inactivation of PTPs.

Redox-regulated pathways are also involved in the remodeling of arteries during hypertension. In particular, ROS-activated transcription factors induce the expression of proinflammatory molecules leading to recruitment of inflammatory cells and vascular inflammation. Thus, in vascular SMCs, Ang II, platelet-derived growth factor induce the expression of interleukin-6 and other proinflammatory genes, and this requires NAD(P)H oxidase/ROS-dependent activation of transcription factor NF-κB and JAK-STAT pathway [84, 85]. In addition, studies *in vivo* (hypertensive rats) and *in vitro* (cultured cells) demonstrated that under pro-hypertensive conditions (cyclic strain, increased Ang II) endothelial cells and fibroblasts express elevated amounts of molecules adhesive for inflammatory cells – such as intercellular adhesion molecule-1 and vascular cell adhesion molecule-1; and this expression is mediated by ROS/NF-κB (Fig. 13.1) [86–88].

Under hypertension conditions, ROS are key players in the development of endothelial dysfunction, as follows: (1) Elevated Ang II via $AT_1$-receptor activates NAD(P)H oxidase. (2) Ang II-mediated activation of NAD(P)H oxidase involves the upstream mediators Src-EGFR-PI3K-Rac-1, as well as PLD-PKC-dependent phosphorylation of 47 kDa-cytosolic subunit of NAD(P)H oxidase (p47phox), and leads to the generation of vascular superoxide anion [73, 89]. (3) Ang II can uncouple endothelial NOS and activate xanthine oxidase, both processes also leading to enhanced superoxide production [90]. (4) Increased oxidant levels inactivate NO and generate peroxynitrite; the latter product also participates in eNOS uncoupling. In summary, during the hypertensive phase, elevated ROS reduce endothelial NO bioactivity leading to endothelial dysfunction (see below). Interestingly, in an animal model, the endothelial responses can be restored with superoxide dismutase therapy or the $AT_1$ receptor antagonist – losartan [91, 92]. Inhibition of $AT_1$ receptor by losartan also reverses endothelial dysfunction in patients with atherosclerosis by improving NO availability (Fig. 13.1) [93].

## Mitochondrial Dysfunction

Increased blood pressure has been increasingly associated with mitochondrial dysfunction. According to Postnov [94], the pathogenesis of arterial hypertension correlates with mitochondrial energy deficiency and calcium overload. In agreement with several clinical observations, decreased mitochondrial energy metabolism and alterations in calcium metabolism has been identified in the myocardium of SHRs [95, 96]. In mice, increased OS, hypertension, and dietary atherosclerosis are consequences of mild respiratory uncoupling in arterial SMCs [97].

Both in experimental and human hypertension, it has been reported the association of hypertension and mitochondrial uncoupling proteins (UCPs). UCPs are mitochondrial anion transporters located in the inner mitochondrial membrane. The activation of these anion transporters allows protons to leak back into the mitochondrial matrix, thus decreasing mitochondrial membrane potential and ROS generation. In particular, mice overexpressing UCP1 in the arterial walls develop hypertension and dietary atherosclerosis [97]. Also, a common polymorphism of the UCP2 gene has been associated with hypertension in a Japanese population, and with hypertension in Caucasians [98].

The electron transport chain (ETC) is a main site of mitochondrial ROS formation. However, there are several other mitochondrial factors that contribute to ROS generation. For instance, monoamine oxidase (MAO) activity in mitochondria results in production of significant amounts of $H_2O_2$. MAO is located within the outer mitochondrial membrane and is responsible for the degradation of endogenous monoamine neurotransmitters and dietary amines, which may cause hypertensive crises if not properly catabolized [99]. The Role of MAO isoforms, MAO-A and MAO-B, in terminating the action of neurotransmitters in the nervous system and in the oxidation of dietary amines in nonneuronal tissues, is well known. However, less attention has been paid to the fact that one of the products of MAO activity is $H_2O_2$, one of the three main ROS (i.e., $H_2O_2$, superoxide and hydroxyl radicals). In the last 5 years, the MAO isoforms, especially MAO-A, have been shown to be an important source of ROS in the receptor-independent apoptotic effects of serotonin in isolated cardiomyocytes and in postischemic myocardial injury [100, 101], and play an important role in myocardial injury caused by postischemic reperfusion [100]. Atherogenic effects of MAO-A are based on the promotion of apoptosis through ROS-dependent sphingosine kinase inhibition (which results in the accumulation of proapoptotic ceramide) [102], as well as vascular wall remodeling through ROS-dependent activation of the metalloproteinase MMP-2 which induces mitogenic signaling in SMCs [103].

Another mitochondrial protein involved in mitochondrial ROS production (overproduction) is p66Shc protein, which also favors cytochrome *c* release, the dissipation of mitochondrial transmembrane potential and apoptosis [104]. The relationship of p66Shc to ROS formation was elucidated showing that it partially localizes within the mitochondria, where it binds to cytochrome *c* and acts as oxidoreductase catalyzing electron transfer from cytochrome *c* to oxygen, thus generating ROS [104]. A recent study by Pinton et al. [105] suggests that initial OS triggers PKCβ-dependent phosphorylation of Serine-36, located in amino-terminal proline-rich region of p66Shc. Ser-36-phosphorylated p66Shc translocates to mitochondria, where it increases mitochondrial ROS formation leading to organelle dysfunction and cell death. p66Shc-deficient mice subjected to experimental hypercholesterolemia displayed reduced levels of oxidized low-density lipoprotein (LDL), a decreased number of foam cells, and a reduction of both apoptosis and early atherogenic lesions, which demonstrates the p66Shc implication in the signaling pathways that lead to apoptosis, vascular wall remodeling and atherosclerosis [106].

In structurally and functionally intact mitochondria, ROS generation is balanced by the organelle antioxidant defense capacity, and the net ROS availability is limited. Mitochondrial antioxidant defense system contains both nonenzymatic and enzymatic components. Nonenzymatic components include hydrophilic and lipophilic radical scavengers (cytochrome *c*, α-tocopherol, ascorbic acid, reduced coenzyme $Q_{10}$, glutathione). Enzymatic components include manganese superoxide dismutase (SOD2), catalase, glutathione peroxidase, phospholipid hydroperoxide, glutathione peroxidase, and several reductases. Decreased antioxidant defense capacity is a prerequisite for increased ROS generation and OS. Thus, in SOD2-deficient mice arterial blood pressure increases with aging or high-salt diet [107]. On the other hand, neurogenic hypertension that develops as a result of oxidative impairment of mitochondrial ETC, can be attenuated by administration of coenzyme $Q_{10}$ which restores ETC [108].

Recently, a few pharmacological tools to counter mitochondrial OS have been described. Some of them are designed to target antioxidants to mitochondria. For example, antioxidants α-tocopherol or ubiquinone can be effectively delivered into mitochondria by conjugation of them to positively charged lipophilic substances which can selectively accumulate in mitochondria [109]. Also mitochondrial metabolites (acetyl-L-carnitine, lipoic acid) and antioxidants (α-phenyl-*N*-*t*-butyl nitrone, *N*-*t*-butyl hydroxylamine) have been shown to improve mitochondrial ultrastructure and function [110]. Furthermore, considering MAO an important source of $H_2O_2$, MAO inhibitors can be viewed as an important tool in the treatment of vascular pathologies that depend on OS.

## Signaling in Dysfunctioning Endothelium

Under physiologic conditions, endothelial cells are able to synthesize and secrete NO which diffuses to surrounding tissues and cells and exerts its cardiovascular protective roles, including relaxation of vascular SMCs and a number of antiatherosclerotic actions. In the presence of pathological conditions, such as hypertension, endothelium undergoes functional and structural alterations, and loses its normal function. This process, as previously discussed, is referred to as "endothelial dysfunction" (both, Ang II and ROS play an important role in the development of pathological changes of endothelium). Dysfunctional endothelium is characterized by significantly reduced NO availability (reduced the production of NO by eNOS and increased breakdown of NO by ROS). As a result, the vasodilatory capacity of dysfunctional endothelium is reduced. Also, impaired endothelial cells, in response to a number of agents and physical stimuli, become a source of endothelium-derived contracting factors (EDCFs), including endothelin-1 (ET-1), Ang II, cyclooxygenase (COX)-derived prostanoids (thromboxane $A_2$, prostaglandin $H_2$) and ROS (Fig. 13.1).

In humans, hypertension-related endothelial dysfunction has been identified with an increased in NO breakdown. Moreover, various sources of ROS are activated not only in endothelial cells, but also in SMCs and inflammatory cells within the arterial wall. These sources include NAD(P)H oxidase, xanthine oxidase, COX and uncoupled eNOS. ROS, mainly superoxide anions, are highly reactive and destroy NO, thus reducing its bioavailability and producing peroxynitrites. Administration of the antioxidant vitamin C to patients with essential hypertension restores NO bioavailability and normal endothelium-dependent vasodilation [111].

EDCFs, which play an important role in the pathophysiology of endothelium-dependent contraction are COX-derived products. In hypertensive rats, EDCFs can be released following endothelial stimulation with acetylcholine or in response to shear stress [112, 113]. Two main COX-derived EDCFs have been identified as thromboxane $A_2$ and prostaglandin $H_2$ [114]. Once produced, EDCFs diffuse to the vascular SMCs and induce contraction through the activation of thromboxane receptors (TPs) [115]. Accordingly, in hypertension, the COX inhibitor (indomethacin) and TP-receptor antagonists increase the vasodilatation response to acetylcholine and inhibit COX-mediated endothelium-dependent contraction of SMCs [116, 117]. It is worth noting that COX-dependent production of EDCFs occurs in the process of natural vascular aging, and that vascular aging is accelerated in essential hypertension [118].

Another endothelium-derived factor which contributes to the development of vascular disease in hypertension is ET-1. The role of ET-1 was first shown in several models of experimental hypertension (high salt and Ang II-caused animal models); and also that ET-receptor antagonists have potent antihypertensive effects. Specifically, chronic treatment with $ET_A$-receptor antagonist improves endothelium-dependent vasodilation both in animal model and clinical studies [119, 120]. These observations prove that increased ET-1 activity contributes to the vasomotor dysfunction associated with hypertension. At the vascular level, ET-1 binds to ET-receptor subtypes $ET_A$ and $ET_B$. The $ET_A$-receptors are mainly localized in the SMC and stimulate vascular contraction; on the other hand, $ET_B$-receptors are abundant on endothelial cells and mediate NO release thus inhibiting vasoconstriction. In the dysfunctioning endothelium, the activation of endothelial $ET_B$-receptors fails to increase NO-mediated vasodilation, and the overall "net"-effect of ET-1 results in enhancement of SMC contraction via stimulation of $ET_A$-receptors (Fig. 13.1) [121].

## Conclusions

Components of the Ang II signaling could play a specific role in mediating vascular remodeling or endothelial dysfunction, leading to hypertension. Inhibition of $AT_1$-receptor activation may be a primary treatment goal in patients with cardiovascular risk factors. However, it is still unclear whether Ang II-regulated cascades can be applied to explain human disease since most of the collected data are from experiments at the cellular level. Nonetheless, an increasing number of observations have underlined the importance of cross talk between Ang II and aldosterone to activate specific signaling pathways, and this knowledge seems to justify the combined use of Ang II and mineralocorticoid receptor blockade in the treatment of hypertensive cardiovascular disease. However, further research on Ang II signal transduction helps to better understand the mechanism of hypertension.

The sympathetic nervous system-related signaling cascades represent another promising target for therapeutic intervention in hypertension. The new generation of vasodilatory β-blockers, such as labetalol, carvedilol, and nebivolol, decreases PVR with little side effect on cardiac output, control blood pressure effectively and improve endothelial dysfunction.

The natriuretic peptide system through cGMP generation has vasodilatory, natriuretic, and diuretic actions. In hypertension, several defects in signaling pathways lead to low levels of natriuretic peptides and decreased sensitivity of the natriuretic peptide system. Nevertheless, further studies are needed to identify the mechanisms of impaired NPR ratio, lowered NPR-A/B content and increased natriuretic peptide clearance in order to target the natriuretic peptide system pharmacologically and genetically, which may benefit hypertension patients.

In hypertension, increased formation of ROS takes place as a result of activation of prooxidant enzymes, such as NAD(P)H oxidase, NOS, and xanthine oxidase. ROS function as second messengers activating tyrosine phosphatases, tyrosine kinases, MAPK, and many other signaling molecules. Major stimuli that activate pro-oxidant systems involve Ang II, and mechanical forces (shear stress). Ang II/oscillatory shear stress-induced redox signaling (OS) contributes to the development of a hypertensive vascular phenotype: endothelial dysfunction, increased vascular tone, cell growth, ECM protein deposition, activation of matrix metalloproteinases and inflammation. Pharmacological strategies to reduce ROS bioavailability may prevent the vascular injury and hypertension.

Interestingly, mitochondrial ETC under physiological conditions is a major source of superoxide and other ROS. ROS generation in mitochondria is influenced by multiple factors, including the efficiency of the ETC, oxygen concentration, availability of electron donors, such as NADH and FADH2, the activity of UCPs and cytokines, and the activity of antioxidant defenses. This mechanism may be triggered by risk factors, with subsequent endothelial dysfunction, vascular wall remodeling, and atherogenesis.

One of the OS-induced changes is a decline in the vasodilatory function of endothelium, in the coronary and peripheral arteries, associated with a progressive decrease in NO bioavailability and increase in the production of EDCFs. Among EDCFs, COX-derived prostanoids and superoxide anions are the most important signaling molecules for the impaired agonist-stimulated vasodilation. Also, ET-1 contributes to endothelial dysfunction by its vasoconstrictor activity during the development of hypertension. Thus, ET-1-regulated signaling pathway system appears to be an important therapeutic target to improve endothelial function and thereby to reduce cardiovascular disease morbidity and mortality.

## Summary


- Ang II is the major peptide hormone of the RAAS. The $AT_1$ receptor has been shown to mediate most of the pathophysiological effects of Ang II. The major acute function of Ang II is vasoconstriction.
- Important downstream components of Ang II-$AT_1$ receptor-signaling pathway are the MAPK. They are implicated in vascular SMC growth and hypertrophy.
- Ang II through the $AT_1$ receptor activates small GTP-binding proteins. It has been shown that the activation of Rho proteins is involved in the pathogenesis of hypertension and vascular proliferative disorders.
- The mechanism for EGFR transactivation by $AT_1$ receptor involves a metalloprotease-dependent EGFR ligand production from its membrane-bound pro-form. EGFR transactivation plays a critical role in cardiovascular remodeling induced by Ang II.
- Ang II regulates several nonreceptor tyrosine kinases: Src, FAK, Pyk2, JAK2.
- Ang II, aldosterone, and endothelins play a central role in ECM remodeling in hypertension.
- Proteolytic conversion of angiotensinogen into Ang I, catalyzed by renin, is the rate-limiting step in the synthesis of Ang II. (Pro)renin receptor serves several functions.
- Nonpeptide direct renin inhibitor, aliskiren, produces significant blood pressure reduction.
- Increased ACE activity promotes vasoconstrictive and salt-retentive Ang II. The ACEIs lower PVR without causing a compensatory increase in heart rate.
- Adrenal corticosteroid aldosterone as an important component of RAAS. There are two aspects of aldosterone actions on target cells. First, aldosterone/mineralocorticoid receptor function as a ligand-dependent transcription factor upregulating Ang II receptor and enhancing vascular remodeling. In addition, there is evidence of aldosterone-mediated rapid effects.
- Sympathetic nerves are continuously active so innervated blood vessels remain under some degree of continuous constriction. In essential hypertension, increased SNA increased vascular resistance, endothelial dysfunction, and vascular remodeling.
- New vasodilatory β-adrenoceptor blockers decrease peripheral vascular resistances, with little effect on cardiac output.
- ANP rapidly lowers blood pressure and promotes sodium and water excretion through cGMP-signaling pathway.
- Hypertension is often associated with low levels of natriuretic peptides. Several findings have provided the basis for ANP gene delivery, as a strategy for treating essential hypertension patients.
- ROS are elevated in the hypertension. Ang II and mechanical forces (shear stress) activate NAD(P)H oxidase which is the major source of ROS in vascular cells.
- The ability of ROS to affect the vasculature is based on the targeting of several signaling cascades, such as MAP kinases, Src, and NAD(P)H oxidase.
- Direct molecular targets of ROS are PTPs.
- Under pro-hypertensive conditions, redox-regulated pathways induce the expression of proinflammatory molecules leading to recruitment of inflammatory cells and vascular inflammation.
- Mitochondrial dysfunction has been implicated in increased arterial blood pressure and the development of hypertension. Both in experimental and human hypertension, it has been reported an the association of hypertension with mitochondrial UCPs.
- ETC is the main site of mitochondrial ROS formation. In addition, MAO activity in mitochondria results in the production of significant amounts of $H_2O_2$. Atherogenic effects of MAO-A are based on the promotion of apoptosis and vascular wall remodeling.
- p66Shc protein partially localizes within mitochondria, where it acts as oxidoreductase catalyzing generation of ROS by cytochrome *c*.
- Experimental data demonstrate p66Shc implication in signaling pathways that lead to apoptosis, vascular wall remodeling, and atherosclerosis.
- In intact mitochondria, ROS generation is balanced by the activity of the antioxidant defense system. Decreased antioxidant defense capacity is a prerequisite for increased ROS generation and OS.
- Several recent pharmacological tools to counter mitochondrial OS include antioxidants targeted to mitochondria, mitochondrial metabolites that improve mitochondrial ultrastructure and function, and MAO inhibitors.
- Under hypertension conditions, ROS are key players in the development of endothelial dysfunction. In dysfunctioning endothelium, NO is inactivated by superoxide and endothelial NO synthase is uncoupled. As a result, the vasodilatory activity of the dysfunctional endothelium is reduced.
- Dysfunctioning endothelial cells are also a source of EDCFs. Two main EDCFs are COX-derived products, thromboxane $A_2$ and prostaglandin $H_2$.
- ET-1 is another endothelium-derived factor which contributes to the development of vascular disease in hypertension. In dysfunctioning endothelium, the activation of endothelial $ET_B$-receptors fails to increase NO-mediated vasodilation and the overall net effect of ET-1 results in the enhancement of SMC contraction via stimulation of $ET_A$-receptors.

## References


1. Touyz RM, Schiffrin EL. Signal transduction mechanisms mediating the physiological and pathophysiological actions of angiotensin II in vascular smooth muscle cells. Pharmacol Rev. 2000;52:639–72.
2. Rosskopf D, Koch K, Habich C, et al. Interaction of Gβ3s, a splice variant of the G-protein Gβ3, with Gγ- and Gα-proteins. Cell Signal. 2003;15:479–88.
3. Hengstenberg C, Schunkert H, Mayer B, et al. Association between a polymorphism in the G protein β3 subunit gene (GNB3) with arterial hypertension but not with myocardial infarction. Cardiovasc Res. 2001;49:820–7.
4. Brand E, Wang JG, Herrmann SM, et al. An epidemiological study of blood pressure and metabolic phenotypes in relation to the Gβ3 C825T polymorphism. J Hypertens. 2003;21:729–37.
5. Sugden PH, Clerk A. Regulation of the ERK subgroup of MAP kinase cascades through G protein-coupled receptors. Cell Signal. 1997;9:337–51.
6. Mehta PK, Griendling KK. Angiotensin II cell signaling: physiological and pathological effects in the cardiovascular system. Am J Physiol Cell Physiol. 2007;292:C82–97.
7. Touyz RM. Reactive oxygen species and angiotensin II signaling in vascular cells – implications in cardiovascular disease. Braz J Med Biol Res. 2004;37:1263–73.
8. Rocic P, Jo H, Lucchesi PA. A role for PYK2 in ANG II-dependent regulation of the PHAS-1-eIF4E complex by multiple signaling cascades in vascular smooth muscle. Am J Physiol Cell Physiol. 2003;285:C1437–44.
9. Ishida M, Ishida T, Thomas SM, Berk BC. Activation of extracellular signal-regulated kinases (ERK1/2) by angiotensin II is dependent on c-Src in vascular smooth muscle cells. Circ Res. 1998;82:7–12.
10. Seko T, Ito M, Kureishi Y, et al. Activation of RhoA and inhibition of myosin phosphatase as important components in hypertension in vascular smooth muscle. Circ Res. 2003;92:411–8.
11. Yamakawa T, Tanaka S, Numaguchi K, et al. Involvement of Rho-kinase in angiotensin II-induced hypertrophy of rat vascular smooth muscle cells. Hypertension. 2000;35:313–8.
12. Ohtsu H, Suzuki H, Nakashima H, et al. Angiotensin II signal transduction through small GTP-binding proteins: mechanism and significance in vascular smooth muscle cells. Hypertension. 2006;48:534–40.
13. Lee DL, Webb RC, Jin L. Hypertension and RhoA/Rho-kinase signaling in the vasculature: highlights from the recent literature. Hypertension. 2004;44:796–9.
14. Lutz S, Freichel-Blomquist A, Yang Y, et al. The guanine nucleotide exchange factor p63RhoGEF, a specific link between $G_{q/11}$-coupled receptor signaling and RhoA. J Biol Chem. 2005;280:11134–9.
15. Ohtsu H, Mifune M, Frank GD, et al. Signal-crosstalk between Rho/ROCK and c-Jun $NH_2$-terminal kinase mediates migration of vascular smooth muscle cells stimulated by angiotensin II. Arterioscler Thromb Vasc Biol. 2005;25:1831–6.
16. Barandier C, Ming XF, Yang Z. Small G proteins as novel therapeutic targets in cardiovascular medicine. News Physiol Sci. 2003;18:18–22.
17. Eguchi S, Inagami T. Signal transduction of angiotensin II type 1 receptor through receptor tyrosine kinase. Regul Pept. 2000;91:13–20.
18. Eguchi S, Dempsey PJ, Frank GD, Motley ED, Inagami T. Activation of MAPKs by angiotensin II in vascular smooth muscle cells. Metalloprotease-dependent EGF receptor activation is required for activation of ERK and p38 MAPK but not for JNK. J Biol Chem. 2001;276:7957–62.
19. Kagiyama S, Eguchi S, Frank GD, Inagami T, Zhang YC, Phillips MI. Angiotensin II-induced cardiac hypertrophy and hypertension are attenuated by epidermal growth factor receptor antisense. Circulation. 2002;106:909–12.
20. Ohtsu H, Dempsey PJ, Frank GD, et al. ADAM17 mediates epidermal growth factor receptor transactivation and vascular smooth muscle cell hypertrophy induced by angiotensin II. Arterioscler Thromb Vasc Biol. 2006;26:e133–7.
21. Saito S, Frank GD, Motley ED, et al. Metalloprotease inhibitor blocks angiotensin II-induced migration through inhibition of epidermal growth factor receptor transactivation. Biochem Biophys Res Communs. 2002;294:1023–9.
22. Berk BC, Corson MA. Angiotensin II signal transduction in vascular smooth muscle: role of tyrosine kinases. Circ Res. 1997;80:607–16.
23. Ishida T, Ishida M, Suero J, Takahashi M, Berk BC. Agonist-stimulated cytoskeletal reorganization and signal transduction at focal adhesions in vascular smooth muscle cells require c-Src. J Clin Invest. 1999;103:789–97.
24. Marrero MB, Schieffer B, Paxton WG, et al. Direct stimulation of Jak/STAT pathway by the angiotensin II AT1 receptor. Nature. 1995;375:247–50.
25. Madamanchi NR, Li S, Patterson C, Runge MS. Reactive oxygen species regulate heat-shock protein 70 via the JAK/STAT pathway. Arterioscler Thromb Vasc Biol. 2001;21:321–6.
26. Booz GW, Baker KM. Molecular signalling mechanisms controlling growth and function of cardiac fibroblasts. Cardiovasc Res. 1995;30:537–43.
27. Berk BC, Fujiwara K, Lehoux S. ECM remodeling in hypertensive heart disease. J Clin Invest. 2007;117:568–75.
28. Iaccarino G, Ciccarelli M, Sorriento D, et al. AKT participates in endothelial dysfunction in hypertension. Circulation. 2004;109: 2587–93.
29. Crowley SD, Gurley SB, Herrera MJ, et al. Angiotensin II causes hypertension and cardiac hypertrophy through its receptors in the kidney. Proc Natl Acad Sci USA. 2006;103:17985–90.
30. Veerasingham SJ, Raizada MK. Brain renin-angiotensin system dysfunction: recent advances and perspectives. Br J Pharmacol. 2003;139:191–202.
31. Nguyen G, Delarue F, Burcklé C, Bouzhir L, Giller T, Sraer J-D. Pivotal role of the renin/pro-renin receptor in angiotensin II production and cellular responses to renin. J Clin Invest. 2002;109: 1417–27.
32. Nguyen G. Renin/pro-renin receptors. Kidney Int. 2006;69: 1503–6.
33. Wood JM, Maibaum J, Rahuel J, et al. Structure-based design of aliskiren, a novel orally effective renin inhibitor. Biochem Biophys Res Commun. 2003;308:698–705.
34. Oh BH, Mitchell J, Herron JR, et al. Aliskiren, an oral renin inhibitor, provides dose-dependent efficacy and sustained 24-h blood pressure control in patients with hypertension. J Am Coll Cardiol. 2007;49:1157–63.
35. Oparil S, Yarows SA, Patel S, et al. Efficacy and safety of combined use of aliskiren and valsartan in patients with hypertension: a randomized double-blind trial. Lancet. 2007;370:221–9.
36. Chobian AV, Barkis GL, Black HR, et al. The seventh report of the joint national committee on prevention, detection, evaluation and treatment of high blood pressure. JAMA. 2003;289:2560–71.
37. Dunn GF, Oigman W, Ventura HO, Messerli FH, Kobrin I, Frolich ED. Enalapril improves systemic and renal hemodynamics and allows regression of left ventricular mass in essential hypertension. Am J Cardiol. 1984;53:105–8.
38. Hiramatsu K, Yamada T, et al. Changes in endocrine activities relative to obesity in patients with essential hypertension. J Am Geriatr Soc. 1981;29:25–30.

39. Dustan HP. Mechanisms of hypertension associated with obesity. Ann Intern Med. 1983;98:860–4.
40. Ishizawa K, Izawa Y, Ito H, et al. Aldosterone stimulates vascular smooth muscle cell proliferation via big mitogen-activated protein kinase 1 activation. Hypertension. 2005;46:1046–52.
41. Grossmann C, Benesic A, Krug AW, et al. Human mineralocorticoid receptor expression renders cells responsive for nongenotropic aldosterone actions. Mol Endocrinol. 2005;19:1697–710.
42. Callera GE, Touyz RM, Tostes RC, et al. Aldosterone activates vascular p38MAP kinase and NADPH oxidase via c-Src. Hypertension. 2005;45:773–9.
43. Xiao F, Puddefoot JR, Vinson GP. Aldosterone mediates angiotensin II-stimulated rat vascular smooth muscle cell proliferation. J Endocrinol. 2000;165:533–6.
44. Mazak I, Fiebeler A, Muller DN, et al. Aldosterone potentiates angiotensin II-induced signaling in vascular smooth muscle cells. Circulation. 2004;109:2792–800.
45. Lemarié CA, Simeone SMC, Nikonova A, et al. Aldosterone-induced activation of signaling pathways requires activity of angiotensin type 1a receptors. Circ Res. 2009;105:852–9.
46. Kambara A, Holycross BJ, Wung P, et al. Combined effects of low-dose oral spironolactone and captopril therapy in a rat model of spontaneous hypertension and heart failure. J Cardiovasc Pharmacol. 2003;41:830–7.
47. Tanabe A, Naruse M, Hara Y, et al. Aldosterone antagonist facilitates the cardioprotective effects of angiotensin receptor blockers in hypertensive rats. J Hypertens. 2004;22:1017–23.
48. Anderson EA, Sinkey CA, Lawton WJ, Mark AL. Elevated sympathetic nerve activity in borderline hypertensive humans: evidence from direct intraneural recordings. Hypertension. 1989;14:177–83.
49. Esler M, Jennings G, Biviano B, Lambert G, Hasking G. Mechanism of elevated plasma noradrenaline in the course of essential hypertension. J Cardiovasc Pharmacol. 1986;8 Suppl 5:S39–43.
50. King AJ, Osborn JW, Fink GD. Splanchnic circulation is a critical neural target in angiotensin II salt hypertension in rats. Hypertension. 2007;50:547–56.
51. Cowley AW, Lohmeier TE. Changes in renal vascular sensitivity and arterial pressure associated with sodium intake during long-term intrarenal norepinephrine infusion in dogs. Hypertension. 1979;1:549–58.
52. Grisk O, Rose HJ, Lorenz G, Rettig R. Sympathetic-renal interaction in chronic arterial pressure control. Am J Physiol Regul Integr Comp Physiol. 2002;283:R441–50.
53. Pedersen ME, Cockcroft JR. The vasodilatory beta-blockers. Curr Hypertens Rep. 2007;9:269–77.
54. Neumann J, Ligtenberg G, Klein IH, et al. Sympathetic hyperactivity in hypertensive chronic kidney disease patients is reduced during standard treatment. Hypertension. 2007;49:506–10.
55. Swayze RD, Braun AP. A catalytically inactive mutant of type I cGMP-dependent protein kinase prevents enhancement of large conductance, calcium-sensitive $K^+$ channels by sodium nitroprusside and cGMP. J Biol Chem. 2001;276:19729–37.
56. Jiang LH, Gawler DJ, Hodson N, et al. Regulation of cloned cardiac L-type calcium channels by cGMP-dependent protein kinase. J Biol Chem. 2000;275:6135–43.
57. Cornwell TL, Pryzwansky KB, Wyatt TA, Lincoln TM. Regulation of sarcoplasmic reticulum protein phosphorylation by localized cyclic GMP-dependent protein kinase in vascular smooth muscle cells. Mol Pharmacol. 1991;40:923–31.
58. Komalavilas P, Lincoln TM. Phosphorylation of the inositol 1,4,5-trisphosphate receptor. Cyclic GMP-dependent protein kinase mediates cAMP and cGMP dependent phosphorylation in the intact rat aorta. J Biol Chem. 1996;271:21933–8.
59. MacFarland RT, Zelus BD, Beavo JA. High concentrations of a cGMP-stimulated phosphodiesterase mediate ANP induced decreases in cAMP and steroidogenesis in adrenal glomerulosa cells. J Biol Chem. 1991;266:136–42.
60. Gambaryan S, Wagner C, Smolenski A, et al. Endogenous or overexpressed cGMP-dependent protein kinases inhibit cAMP dependent renin release from rat isolated perfused kidney, microdissected glomeruli, and isolated juxtaglomerular cells. Proc Natl Acad Sci USA. 1998;95:9003–8.
61. Harris PJ, Thomas D, Morgan TO. Atrial natriuretic peptide inhibits angiotensin-stimulated proximal tubular sodium and water reabsorption. Nature. 1987;326:697–8.
62. Newton-Cheh C, Larson MG, Vasan RS, et al. Association of common variants in NPPA and NPPB with circulating natriuretic peptides and blood pressure. Nat Genet. 2009;41:348–53.
63. Cusson JR, Thibault G, Cantin M, Larochelle P. Prolonged low dose infusion of atrial natriuretic factor in essential hypertension. Clin Exp Hypertens A. 1990;12:111–35.
64. Lin KF, Chao J, Chao L. Human atrial natriuretic peptide gene delivery reduces blood pressure in hypertensive rats. Hypertension. 1995;26:847–53.
65. Weder AB, Sekkarie MA, Takiyyuddin M, Schork NJ, Julius S. Antihypertensive and hypotensive effects of atrial natriuretic factor in men. Hypertension. 1987;10:582–9.
66. Schillinger KJ, Tsai SY, Taffet GE, et al. Regulatable atrial natriuretic peptide gene therapy for hypertension. Proc Natl Acad Sci USA. 2005;102:13789–94.
67. Oliver PM, Fox JE, Kim R, et al. Hypertension, cardiac hypertrophy, and sudden death in mice lacking natriuretic peptide receptor A. Proc Natl Acad Sci USA. 1997;94:14730–5.
68. Oliver PM, John SW, Purdy KE, et al. Natriuretic peptide receptor 1 expression influences blood pressures of mice in a dose-dependent manner. Proc Natl Acad Sci USA. 1998;95:2547–51.
69. Kostis JB, Packer M, Black HR, Schmieder R, Henry D, Levy E. Omapatrilat and enalapril in patients with hypertension: the Omapatrilat Cardiovascular Treatment vs. Enalapril (OCTAVE) trial. Am J Hypertens. 2004;17:103–11.
70. Suzuki H, DeLano FA, Parks DA, et al. Xanthine oxidase activity associated with arterial blood pressure in spontaneously hypertensive rats. Proc Natl Acad Sci USA. 1998;95:4754–9.
71. Landmesser U, Dikalov S, Price SR, et al. Oxidation of tetrahydrobiopterin leads to uncoupling of endothelial cell nitric oxide synthase in hypertension. J Clin Invest. 2003;111:1201–9.
72. Lassegue B, Clempus RE. Vascular NAD(P)H oxidases: specific features, expression, and regulation. Am J Physiol Regul Integr Comp Physiol. 2003;285:R277–97.
73. Seshiah PN, Weber DS, Rocic P, Valppu L, Taniyama Y, Griendling KK. Angiotensin II stimulation of NAD(P)H oxidase activity: upstream mediators. Circ Res. 2002;91:406–13.
74. Touyz RM, He G, El Mabrouk M, Schiffrin EL. p38 Map kinase regulates vascular smooth muscle cell collagen synthesis by angiotensin II in SHR but not in WKY. Hypertension. 2001;37:574–80.
75. Ushio-Fukai M, Alexander RW, Akers M, Griendling KK. p38 mitogen-activated protein kinase is a critical component of the redox-sensitive signaling pathways activated by angiotensin II. Role in vascular smooth muscle cell hypertrophy. J Biol Chem. 1998;273:15022–9.
76. Viedt C, Soto U, Krieger-Brauer HI, et al. Differential activation of mitogen-activated protein kinases in smooth muscle cells by angiotensin II: involvement of p22phox and reactive oxygen species. Arterioscler Thromb Vasc Biol. 2000;20:940–8.
77. Meloche S, Landry J, Huot J, Houle F, Marceau F, Giasson E. p38 MAP kinase pathway regulates angiotensin II-induced contraction of rat vascular smooth muscle. Am J Physiol Heart Circ Physiol. 2000;279:H741–51.

78. Torrecillas G, Boyano-Adanez MC, Medina J, et al. The role of hydrogen peroxide in the contractile response to angiotensin II. Mol Pharmacol. 2001;59:104–12.
79. Wesselman JP, Dobrian AD, Schriver SD, Prewitt RL. Src tyrosine kinases and extracellular signal-regulated kinase 1/2 mitogen activated protein kinases mediate pressure-induced c-fos expression in cannulated rat mesenteric small arteries. Hypertension. 2001;37: 955–60.
80. Touyz RM, Yao G, Schiffrin EL. c-Src induces phosphorylation and translocation of p47phox: role in superoxide generation by angiotensin II in human vascular smooth muscle cells. Arterioscler Thromb Vasc Biol. 2003;23:981–7.
81. Ushio-Fukai M, Griendling KK, Becker PL, Hilenski L, Halleran S, Alexander RW. Epidermal growth factor receptor transactivation by angiotensin II requires reactive oxygen species in vascular smooth muscle cells. Arterioscler Thromb Vasc Biol. 2001;21: 489–95.
82. Frank GD, Mifune M, Inagami T, et al. Distinct mechanisms of receptor and nonreceptor tyrosine kinase activation by reactive oxygen species in vascular smooth muscle cells: role of metalloprotease and protein kinase C-δ. Mol Cell Biol. 2003;23:1581–9.
83. Stoker AW. Protein tyrosine phosphatases and signalling. J Endocrinol. 2005;185:19–33.
84. Schieffer B, Luchtefeld M, Braun S, Hilfiker A, Hilfiker-Kleiner D, Drexler H. Role of NAD(P)H oxidase in angiotensin II-induced JAK/STAT signaling and cytokine induction. Circ Res. 2000;87: 1195–201.
85. Simon AR, Rai U, Fanburg BL, Cochran BH. Activation of the JAK-STAT pathway by reactive oxygen species. Am J Physiol. 1998;275:C1640–52.
86. Tummala PE, Chen XL, Sundell CL, et al. Angiotensin II induces vascular cell adhesion molecule-1 expression in rat vasculature: a potential link between the renin–angiotensin system and atherosclerosis. Circulation. 1999;100:1223–9.
87. Cheng JJ, Wung BS, Chao YJ, Wang DL. Cyclic strain-induced reactive oxygen species involved in ICAM-1 gene induction in endothelial cells. Hypertension. 1998;31:125–30.
88. Pueyo ME, Gonzalez W, Nicoletti A, Savoie F, Arnal JF, Michel JB. Angiotensin II stimulates endothelial via nuclear factor-κB activation induced by intracellular oxidative stress. Arterioscler Thromb Vasc Biol. 2000;20:645–51.
89. Touyz RM, Yao G, Quinn MT, Pagano PJ, Schiffrin EL. p47phox associates with the cytoskeleton through cortactin in human vascular smooth muscle cells: role in NAD(P)H oxidase regulation by angiotensin II. Arterioscler Thromb Vasc Biol. 2005;25:512–8.
90. Wassmann S, Wassmann K, Nickenig G. Modulation of oxidant and antioxidant enzyme expression and function in vascular cells. Hypertension. 2004;44:381–6.
91. Laursen JB, Rajagopalan S, Galis Z, Tarpey M, Freeman BA, Harrison DG. Role of superoxide in angiotensin II–induced but not catecholamine-induced hypertension. Circulation. 1997;95: 588–93.
92. Rajagopalan S, Kurz S, Munzel T, et al. Angiotensin II-mediated hypertension in the rat increases vascular superoxide production via membrane NADH/NADPH oxidase activation. J Clin Invest. 1996;97:1916–23.
93. Prasad A, Tupas-Habib T, Schenke WH, et al. Acute and chronic angiotensin-1 receptor antagonism reverses endothelial dysfunction in atherosclerosis. Circulation. 2000;101:2349–54.
94. Postnov IuV. The role of mitochondrial calcium overload and energy deficiency in pathogenesis of arterial hypertension. Arkh Patol. 2001;63:3–10.
95. Chen L, Tian X, Song L. Biochemical and biophysical characteristics of mitochondria in the hypertrophic hearts from hypertensive rats. Chin Med J Engl. 1995;108:361–6.
96. Atlante A, Seccia TM, Pierro P, et al. ATP synthesis and export in heart left ventricle mitochondria from spontaneously hypertensive rat. Int J Mol Med. 1998;1:709–16.
97. Bernal-Mizrachi C, Gates AC, Weng S, et al. Vascular respiratory uncoupling increases blood pressure and atherosclerosis. Nature. 2005;435:502–6.
98. Ji Q, Ikegami H, Fujisawa T, et al. A common polymorphism of uncoupling protein 2 gene is associated with hypertension. J Hypertens. 2004;22:97–102.
99. Youdim MB, Edmondson D, Tipton KF. The therapeutic potential of monoamine oxidase inhibitors. Nat Rev Neurosci. 2006;7: 295–309.
100. Bianchi P, Kunduzova O, Masini E, et al. Oxidative stress by monoamine oxidase mediates receptor-independent cardiomyocyte apoptosis by serotonin and postischemic myocardial injury. Circulation. 2005;112:3297–305.
101. Bianchi P, Pimentel DR, Murphy MP, Colucci WS, Parini A. A new hypertrophic mechanism of serotonin in cardiac myocytes: receptor-independent ROS generation. FASEB J. 2005;19:641–3.
102. Pchejetski D, Kunduzova O, Dayon A, et al. Oxidative stress-dependent sphingosine kinase-1 inhibition mediates monoamine oxidase A-associated cardiac cell apoptosis. Circ Res. 2007; 100:41–9.
103. Coatrieux C, Sanson M, Negre-Salvayre A, et al. MAO-A-induced mitogenic signaling is mediated by reactive oxygen species, MMP-2, and the sphingolipid pathway. Free Radic Biol Med. 2007; 43:80–9.
104. Giorgio M, Migliaccio E, Orsini F, et al. Electron transfer between cytochrome c and p66Shc generates reactive oxygen species that trigger mitochondrial apoptosis. Cell. 2005;122:221–33.
105. Pinton P, Rimessi A, Marchi S, et al. Protein kinase C β and prolyl isomerase 1 regulate mitochondrial effects of the life-span determinant p66Shc. Science. 2007;315:659–63.
106. Napoli C, Martin-Padura I, de Nigris F, et al. Deletion of the p66Shc longevity gene reduces systemic and tissue oxidative stress, vascular cell apoptosis, and early atherogenesis in mice fed a high-fat diet. Proc Natl Acad Sci USA. 2003;100:2112–6.
107. Rodriguez-Iturbe B, Sepassi L, Quiroz Y, Ni Z, Vaziri ND. Association of mitochondrial sod deficiency with salt-sensitive hypertension and accelerated renal senescence. J Appl Physiol. 2007;102:255–60.
108. Chan SH, Wu KL, Chang AY, Tai MH, Chan JY. Oxidative impairment of mitochondrial electron transport chain complexes in rostral ventrolateral medulla contributes to neurogenic hypertension. Hypertension. 2009;53:217–27.
109. Smith RA, Kelso GF, Blaikie FH, et al. Using mitochondria-targeted molecules to study mitochondrial radical production and its consequences. Biochem Soc Trans. 2003;31:1295–9.
110. Liu J, Atamna H, Kuratsune H, Ames BN. Delaying brain mitochondrial decay and aging with mitochondrial antioxidants and metabolites. Ann N Y Acad Sci. 2002;959:133–66.
111. Taddei S, Virdis A, Ghiadoni L, Magagna A, Salvetti A. Vitamin C improves endothelium-dependent vasodilation by restoring nitric oxide activity in essential hypertension. Circulation. 1998; 97:2222–9.
112. Luscher TF, Vanhoutte PM. Endothelium-dependent contractions to acetylcholine in the aorta of the spontaneously hypertensive rat. Hypertension. 1986;8:344–8.
113. Huang A, Sun D, Koller A. Shear stress-induced release of prostaglandin $H_2$ in arterioles of hypertensive rats. Hypertension. 2000;35:925–30.
114. Luscher TF, Boulanger CM, Dohi Y, Yang ZH. Endothelium derived contracting factors. Hypertension. 1992;19:117–30.
115. Vanhoutte PM, Feletou M, Taddei S. Endothelium-dependent contractions in hypertension. Br J Pharmacol. 2005;144:449–58.

116. Taddei S, Virdis A, Mattei P, Salvetti A. Vasodilation to acetylcholine in primary and secondary forms of human hypertension. Hypertension. 1993;21:929–33.
117. Virdis A, Colucci R, Fornai M, et al. Cyclooxygenase-1 is involved in endothelial dysfunction of mesenteric small arteries from angiotensin II-infused mice. Hypertension. 2007;49:679–86.
118. Taddei S, Virdis A, Mattei P, et al. Hypertension causes premature aging of endothelial function in humans. Hypertension. 1997;29: 736–43.
119. Barton M. Endothelial dysfunction and atherosclerosis: endothelin receptor antagonists as novel therapeutics. Curr Hypertens Rep. 2000;2:84–91.
120. Verhaar MC, Strachan FE, Newby DE, et al. Endothelin-A receptor antagonist-mediated vasodilatation is attenuated by inhibition of nitric oxide synthesis and by endothelin-B receptor blockade. Circulation. 1998;97:752–6.
121. Penna C, Rastaldo R, Mancardi D, et al. Effect of endothelins on the cardiovascular system. J Cardiovasc Med. 2006;7:645–52.

# Chapter 14
# Gene Expression and Signaling Pathways in Myocardial Ischemia

**Abstract** Myocardial ischemia and myocardial infarction (MI) are endemic health problems all over the world with very high mortality and morbidity. When assessing the severity of myocardial ischemia/MI, factors like the extension of collateral circulation, vasoconstriction, and spontaneous thrombosis must be considered. Risk factors such as diabetes mellitus, arterial hypertension, and hypercholesterolemia contribute to the development of the disease and although each risk factor by itself is partly under genetic control, a positive family history is an independent predictor, which suggests that there are additional susceptibility genes. A number of myocardial ischemia/MI-associated genes and signaling pathways have been identified in both human and animal studies suggesting that gene profiling and signal transduction analysis of animal MI and ischemia models may have relevance for human. In addition, the availability of the human genome and technologies, such as microarray and pharmacogenomics, are already changing our understanding of global and specific patterns of gene expression and protein defects, which will greatly impact the diagnosis and treatment of these conditions. Identification of individuals at risk for the disease may permit its prevention by a systematic genetic and signaling pathways screening of the proband and family members. Furthermore, identification and modification of certain environmental factors should be a high priority.

**Keywords** Myocardial ischemia • Stress signaling • Metabolic signaling • Myocardial signaling

## Introduction

Coronary artery disease (CAD) including its most important manifestation, MI, is a multifactorial polygenic disorder caused by multiple genetic and environmental factors and complex gene–environment interactions modulated by modifiers such as diabetes, dyslipidemias, and hypertension. Postgenomic studies of myocardial ischemia and infarction are being carried out to identify which genes contribute to the polygenic nature of CAD and MI, their expression, and their involvement in the pathophysiological mechanism and phenotype of both myocardial ischemia/MI and events that occur after MI, and their interaction(s) with a number of genetic and environmental risk factors [1, 2].

Myocardial ischemia has a large number of effects on cardiac physiology [1], and its lethality most likely stems from its deleterious effect on metabolism, ultimately depriving the cardiomyocyte of the bioenergy necessary to provide pumping energy and electrical signaling. Early ischemic damage shares with other physiological stresses (e.g., heat shock) severe alterations of the mitochondrial phenotype, such as increased mitochondrial swelling, and uncoupling of respiration and oxidative phosphorylation (OXPHOS). Furthermore, myocardial ischemia results in a selective depletion of the mitochondrial inner-membrane phospholipid, cardiolipin (CL), which is involved in cytochrome *c* insertion, retention, and electron transport chain (ETC) function [2]. Moreover, in sustained ischemia, ATP depletion occurs with subsequent de-energization of the cell resulting in both necrotic cell death and signaling apoptosis or programmed cell death [3, 4]. While the focus will be mainly on stress, metabolic, mitochondria, and inflammatory signaling pathways in myocardial ischemia, its genetics and cell death pathways are also discussed in this chapter.

## Genetics of Myocardial Ischemia

A number of genes has been identified in both human and animal myocardial ischemia studies suggesting that gene profiling of animal models of MI and ischemia may be relevant in human. The availability of the human genome is already changing our understanding of global and specific patterns of gene expression that will impact the diagnosis and treatment of these conditions. Identification of individuals at risk for the disease may allow its prevention by genetic screening of family members of the proband. And modification of a number of environmental factors should be of high significance.

J. Marín-García, *Signaling in the Heart*,
DOI 10.1007/978-1-4419-9461-5_14, © Springer Science+Business Media, LLC 2011

Using genetic case–control studies, a number of genes involved in increasing the risk of myocardial ischemia/MI has been identified [5]. Genes associated with MI include those involved in thrombosis such as thrombospondin (*TSP*) genes *TSP-1*, *TSP-2*, *TSP-3*, *TSP-4*, plasminogen activator inhibitor (*PAI-1*), platelet glycoprotein IIIa, fibrinogen, factors V and VII encoding inflammatory mediators (including several cytokines and the *LTA* gene encoding lymphotoxin-α), other atherosclerotic factors like stromelysin-1, apolipoprotein E (ApoE), cholesterol ester transfer protein (CETP), and other factors, including angiotensin-converting enzyme (ACE), angiotensinogen, and endothelial nitric oxide synthase (eNOS). While this genetic information has been primarily obtained from human clinical studies, the availability of animal models of myocardial ischemia/MI allowed the identification of a number of novel genes, the temporal studies of gene expression, the testing of the effects of null mutations in the genetic *loci*, and also preclinical therapeutic trials. Expression of these and other genes in myocardial ischemia and MI have been assessed in limited clinical studies and in animal models using both global gene profiling and specific gene expression approaches. The gene profiling has resulted in the identification of genes that predispose to MI and ischemia susceptibility (Fig. 14.1), and also in identifying useful biomarkers for diagnosis. The impact of gene variants on pharmaceutical treatment and the use of pharmacogenomic information may allow a personalized and potentially more effective type of therapies that can be tailored to the individual's genomic profile [6].

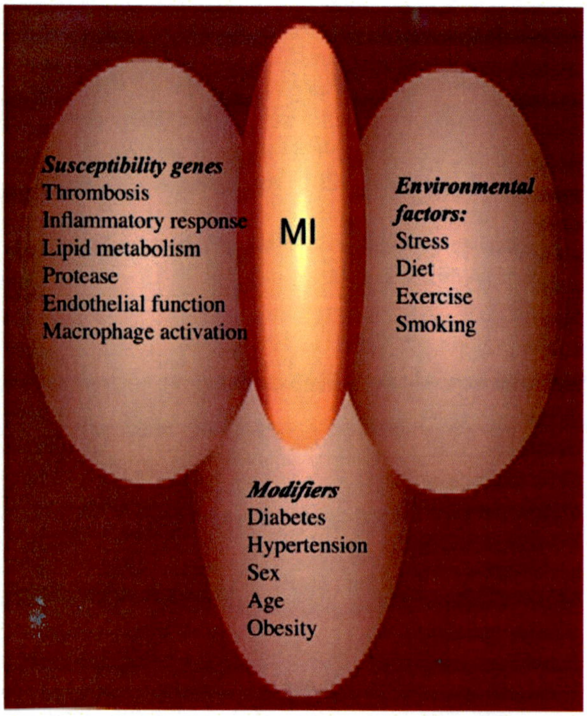


**Fig. 14.1** Intersection of susceptibility genes, environmental factors, and modifiers in the genesis of myocardial ischemia/infarction

Importantly, the genetic predisposition for MI can be modulated by the interactions of several components of the inflammatory signaling pathway (see below). Variants in several genes involved in cell adhesion function in the atherosclerotic process have been associated with increased MI risk. Specific haplotypes of the P-selectin gene defined by specific polymorphisms located in the coding region of the P-selectin gene (SELP) have been associated with increased risk for MI; association with MI risk was significantly modulated by different polymorphic variants within the P-selectin coding region [7]. In Japanese patients, a Ser128Arg polymorphism has been found in the E-selectin gene (*SELE*) in association with MI; this variant appears to modulate kinase activities in endothelial signaling pathways and may functionally alter leukocyte–endothelial interactions [8]. Platelet endothelial cell adhesion molecule-1 (PECAM1/CD31) plays a pivotal role in the migration of circulating leukocytes during vascular inflammation. Polymorphisms in the PECAM1, Asn563Ser and Gly670Arg, were associated with acute MI (AMI) and with a stronger association in young male Japanese patients [9]. Others investigators have reported a significantly higher frequency of Gly670Arg PECAM1 variant in young adult male Italian patients with AMI [10].

Gene variants involved in cardiovascular physiology outside of the atherosclerotic pathways have been implicated in MI. A common Arg389Gly polymorphism in the $\beta_1$-adrenoreceptor ($\beta_1$-AR) gene has been associated with AMI; the prevalence of the Arg389 homozygote genotype was significantly higher in patients with AMI than in control subjects [11]. The site of the polymorphism lies within a region important for receptor-$G_s$ protein coupling and subsequent agonist-stimulated adenylyl cyclase activation. A suggested mechanism underlying the $\beta_1$-AR-mediated AMI is that augmented sympathetic activity might trigger AMI via enhanced hemodynamic or mechanical forces through β-AR activation. The Arg 389 polymorphism when present with an $\alpha_{2c}$-adrenoceptor ($\alpha_{2c}$-AR) variant is a risk factor for human heart failure (HF) [12]. Individuals with a combination of the Arg389 allele and the $\alpha_{2c}$-AR variant (α2CDel322-325) show a tenfold risk for HF development. Moreover, variants at this position have also been associated with increased diastolic blood pressure and heart rate (in individuals homozygous for the Arg389 allele) [13], and differential therapeutic responses to $\beta_1$-AR antagonists (greater in patients with Arg389 compared to those carrying Gly389 alleles) [14]. Interestingly, functional studies of transgenic mice containing either the Arg389 or Gly389 variants demonstrated that the Arg389 variant provided both enhanced ventricular function and blunted β-AR signaling, and elicited improved myocardial recovery after ischemic injury, which in the mouse model was age-dependent [15].

## Stress Signaling

Stresses in myocardial ischemia/hypoxia (e.g., oxidative stress) elicit a variety of adaptive responses at the tissue, cellular, and molecular levels. A model displaying the cardiac physiological response to hypoxia suggests the existence of a mitochondrial $O_2$ sensor coupled to a signal transduction system, which in turn activates a functional response [16]. By increasing the generation of reactive oxygen species (ROS) and heme proteins [e.g., cytochrome *c* oxidase (COX)], which reversibly bind oxygen, myocardial mitochondria may function as $O_2$ sensors with ROS playing a central role in the mitochondrial signaling during hypoxic response. Acting as second messenger, ROS initiates a signaling cascades in both adaptive responses to hypoxia and in mechanically stressed heart. Downregulation of COX activity contributes to ROS generation and signaling that are observed in cardiomyocytes during hypoxia [17]. Also, hypoxia stimulates nitric oxide (NO) synthesis in cardiomyocytes [18], and NO downregulates COX activity with subsequent mitochondrial $H_2O_2$ production (see further NO discussion later in this chapter). This event may provide a mitochondrial-generated signal for further regulation of redox-sensitive signaling pathways, including apoptosis and may proceed even in the absence of significant changes in ATP levels [18]. Furthermore, mitochondrial ROS has been shown to activate p38 MAPK in hypoxic cardiomyocytes [19]. Longer-term responses to hypoxia include increasing gene expression of hypoxia-induced factors (HIFs) and activation of transcription factors such, as nuclear transcription factor κB (NF-κB) which, have also been implicated in the complex regulation of myocardial hypertrophy (see Chap. 15) and inflammatory cytokines [e.g., tumor necrosis factor-α (TNF-α) and interleukin (IL)-1]. While increased ROS has been shown to be an important element in NF-κB gene activation, there is evidence that cardiomyocyte HIF gene activation can also occur in the absence of ROS [20].

## Metabolic Signaling

Myocardial function is dependent on a constant supply of oxygen from the coronary circulation and also responds to alterations in cardiomyocyte levels of essential metabolites, such as ATP, ADP, NADH, as well as numerous substrates and coenzymes. A reduction in oxygen supply due to coronary stenosis results in myocardial ischemia, which may lead to cardiac dysfunction. While reperfusion of the ischemic myocardium is essential for tissue survival, the restoration of blood flow can also lead to reperfusion injury, resulting in myocyte death (see Chap. 20). Thus, an imbalance between oxygen supply and metabolic demand leads to functional, metabolic, morphologic, and electrophysiologic alterations in the myocardium, and depending on the duration of injury as well as the quantity of oxygen delivery during reperfusion this can lead to extensive cell death [21]. Collective evidence has shown that myocardial ischemia–reperfusion (I/R) injury is a primary contributor to the morbidity and mortality associated with CAD. Cellular events leading to I/R-induced cellular injury include I/R-induced ROS production, disturbances in calcium homeostasis, neutrophil activation and infiltration, the activation of cellular proteases, and injury to mitochondria, which are the final arbitrators of I/R-induced cell death and whether cardiomyocyte will die from necrosis or apoptosis.

While most of these injuries lead to irreversible cellular damage, reversible myocyte injury largely involves reduced ATP synthesis, with an attendant decline in ATP phosphorylation potential, intracellular acidosis (resulting from ATP hydrolysis), and phosphate accumulation [22]. Myocardial complex V, which is both the mitochondrial $F_1F_0$-ATPase and the ATP synthase, catalyzes greater ATP hydrolysis during myocardial ischemia, switching back from ATP hydrolysis to ATP synthesis after reperfusion [23], and contributing further to both increasing acidosis and reduction in cardiac energy reserve during ischemia. Intracellular acidosis occurring mainly during myocardial ischemia is related to increased rate of fatty acid oxidation (FAO), which leads to uncoupling of glycolysis from glucose oxidation as well as to inhibition of glucose oxidation [24].

Along with a decrease in the rate of oxygen consumption and ATP production, metabolism in the ischemic myocardium is characterized by an increase in cardiac uptake of glucose and by glycolytic activation. However, in contrast to conditions with normal aerobic metabolism, in the ischemic heart glucose is not readily oxidized to pyruvate in the mitochondria, but is rather converted to lactate resulting in lactate accumulation and further decrease in cytosolic pH that worsens the ischemic damage partially by decreasing cardiac efficiency. Elevated rate of myocardial FAO results in the inhibition of glucose oxidation (the Randle reaction), primarily targeting the mitochondrial pyruvate dehydrogenase (PDH) activity. Also, in myocardial ischemia, rapid activation of the stress-signaling protein AMP-activated protein kinase (AMPK) results in stimulation of glucose uptake, glycolysis, and FAO which have both beneficial adaptive function (e.g., increased glycolytic ATP production to compensate for the loss of mitochondrial ATP) and deleterious effects (i.e., increased uncoupling of glycolysis from glucose oxidation and enhanced protons and lactate production) on the myocardium. The AMPK-mediated increase in FAO (mediated in part by AMPK inhibitory effect on the production and promoting the degradation of malonyl-CoA levels, a critical regulatory block of mitochondrial fatty acid uptake) is further augmented during I/R with increased levels in circulating fatty acids. Albeit the precise mechanism of ischemia-mediated AMPK activation has not been established, it appears to involve allosteric stimulation (and direct conformational changes) by increased levels of AMP and phosphorylation at specific sites by the AMPK kinases, including liver kinase B1(LKB1) and calmodulin-dependent protein kinase kinase (CAMKK) [25]. In addition, while AMPK can mediate translocation of pro-apoptotic protein Bax to mitochondria in neonatal cardiomyocytes [26], most studies have demonstrated an anti-apoptotic effects of AMPK [27–29]. Moreover, AMPK activation inhibits cardiac protein synthesis and myocardial Akt-induced hypertrophic growth signaling pathways including the mTOR, p70S6 kinase, ribosomal protein S6, and the eEF2 kinase/eEF2 pathways [30–32], and often is also associated with adiponectin signaling [33]. Although modulation of the ischemic-induced AMPK activation might be found more or less beneficial [34], prevention of the downstream decrease in malonyl-CoA levels has been proposed as a potential therapeutic target for the management of ischemic heart disease [35].

The alteration of glucose metabolism, either by inhibition of an excessive rate of glycolysis or by the stimulation of glucose oxidation, has been increasingly considered as a potential drug target to reduce $H^+$ production and to limit $Na^+$ and $Ca^{2+}$ accumulation and thereby preventing postischemic left ventricle (LV) dysfunction [36]. For instance, ranolazine, an antianginal medication, has been found to reduce ischemia by blocking sodium-induced calcium overload in myocardial cells. Several studies have found that ranolazine stimulates glucose oxidation (increasing PDH activation) and partially decreases FAO while maintaining the coupling of glycolysis to glucose oxidation during ischemia, thus reducing tissue acidosis [37, 38]. On the other hand, recent observations have shown that ranolazine behaves as a cell membrane inhibitor of the late sodium current, thereby attenuating $Ca^{2+}$ overload during ischemia [39].

Alteration of the activity of sarcolemmal and sarcoplasmic reticulum-localized ion channels and pumps (in part redox-mediated) is directly involved in I/R injury. Alterations in the activity of the major components of $[Ca^{2+}]_i$ regulation, such as ryanodine receptor (RyR), $Na^+$–$Ca^{2+}$ exchanger (NCX), and $Ca^{2+}$ ATPase, appear to play an important role in ischemia-related $Ca^{2+}$ overload [40]. During ischemia and reperfusion, increased intracellular $Na^+$ resulting from decreased $Na^+/K^+$ pumping during ischemia and increased $Na^+$–$H^+$ exchange following reperfusion (triggered in part by the proton accumulation affecting the pH-regulated $Na^+$–$H^+$ exchanger [NHE-1]) lead to a decrease in $Ca^{2+}$ efflux and enhanced $Ca^{2+}$ influx via the cardiac NCX [36]. The resultant $Ca^{2+}$ overload represents one of the major pathophysiological mechanisms for I/R injury. Supraphysiological $Ca^{2+}$ levels elicit hypercontracture and cellular damage. In addition, increased mitochondrial membrane permeability transition (MTP) pore opening occurs during exposure to supraphysiological $Ca^{2+}$ concentrations or mitochondrial calcium overload.

## Myocardial Ischemia and Mitochondria Signaling

In addition to the profound effects on mitochondrial fatty acid uptake and oxidation described above, ischemia also markedly impacts mitochondrial OXPHOS. Primarily from animal models, there is considerable evidence implicating progressive damage to the ETC and phosphorylation apparatus during ischemia. Early ischemic damage appears to profoundly reduce complex I, complex V, and adenine nucleotide translocator (ANT) activities [41, 42]. With more prolonged ischemic damage, myocardial complex III and IV (cytochrome *c* oxidase) activities are also decreased [1, 43].

Ischemic damage to complex I has been associated with a marked decrease in the levels of flavin mononucleotide (FMN) cofactor [41]. Moreover, there is evidence of myocardial complex I activity modulation by posttranslational modifications, such as *S*-nitrosylation, tryptophan nitration, and phosphorylation, albeit their role in I/R has not been established [44–46]. Importantly, the resulting damage to the NADH dehydrogenase component of complex I can increase electron leak and stimulate the production of ROS, a potential trigger of many of the downstream events of ischemia. Ischemia damages complex III by functional inactivation of the redox-active iron-sulfur cluster within the 22 kDa iron-sulfur protein (ISP) subunit [1]. Regarding complex IV (cytochrome *c* oxidase), ischemia appears to be associated with a selective decrease in the content of the inner-membrane phospholipid, cardiolipin, required for optimal activity of complex IV [1]. Studies with ischemic and ischemic–reperfused rat heart have shown a parallel loss of complex I and III activities with the loss of cardiolipin content (reduced by 25 and 50% in mitochondria isolated from the ischemic and the reperfused rat heart, respectively) and also have shown that these activities could be completely restored by exogenous addition of cardiolipin in the reperfused heart [47, 48]. While early studies found no evidence of subunit changes in complex IV with myocardial I/R, more recent observations in myocardial ischemia have documented that protein kinase A modulates complex IV activity and have identified phosphorylation at specific residues within COX subunits I (at Ser115 and Ser116), IV (at Thr52), and Vb (at Ser40) [49, 50].

It is noteworthy to point out that subsarcolemmal mitochondria (SSM), located near the plasma membrane, appear to be more selectively damaged in ischemia than mitochondria located near the myofibrils (IFM), with increasing susceptibility to calcium overload, increased cytochrome *c* release, more rapid ischemic damage, and cardiolipin loss [51, 52].

ROS generation is increased in myocardial I/R, and this is supported by findings in transgenic mice with genetic impairment of the antioxidant systems and showing increased susceptibility to myocardial I/R injury [53]; antioxidants (e.g., alpha-lipoic acid, dihydrolipoic acid, and synthetic cell-permeable peptide antioxidants) administered during I/R confer protection against subsequent cell death [54, 55]; and antioxidant enzyme overexpression [e.g., copper–zinc superoxide dismutase(Cu–ZnSOD), nitric oxide synthase (NOS), and heme oxygenase 1 (HO-1)] confers protection against I/R injury [56–58]. Several studies have shown that a burst of ROS during the first moments of reperfusion is associated with changes in mitochondrial permeability transition (MTP) pore opening and myocardial injury [59, 60]. However, the source of this ROS production has thus far remained uncertain [61]. Since the ROS burst occurring upon reperfusion was not attenuated by treatment with a number of mitochondrial ETC inhibitors, the likelihood of a mitochondrial source of reperfusion-mediated ROS generation is unlikely.

Besides mitochondria, other cellular sources for the generation of superoxide radicals include the reactions of oxygen with microsomal cytochrome p450 (CYP) and with reduced flavins (e.g., NADPH), usually in the presence of metal ions. A potential role of CYP has been suggested by the substantial reduction in ischemia and reperfusion-induced myocardial damage and ROS production resulting after treatment with CYP inhibitors [62]. Furthermore, given the association between I/R injury, early vascular dysfunction, and cardiac allograft vasculopathy (CAV), it has been hypothesized that CYP 2C may contribute to the onset of CAV. Data from Lewis-to-Fisher rat heterotopic heart transplants demonstrated that CYP 2C contributes to smooth muscle cell proliferation and CAV without affecting general immune infiltration [63]. Moreover, xanthine oxidase (XO), a primarily cytosolic enzyme involved in purine metabolism, is also a source of superoxide radical. Zweier et al. [64] identified endothelial cells as a major source of ROS at reperfusion that could be inhibited by XO blockers and react with iron to form the very reactive hydroxyl radical. Later studies have confirmed that XO is a significant source of oxidative injury, which occurs upon reperfusion of the ischemic rat heart [65]. Also, there is evidence that the XO inhibitor allopurinol provides protection in chronic intermittent hypoxia-associated OS, myocardial dysfunction, and apoptosis in rats [66].

ROS generation has been implicated in ischemia without reperfusion. Although somewhat counterituitive given the decreased levels of oxygen available to the myocardium, it has been shown that there is enough molecular oxygen in the ischemic heart to generate ROS. Ischemia-generated ROS appears to play a critical signaling role, which contributes to cellular oxidant damage and appears to be the same ROS that triggers preconditioning [67, 68]. Using cardiomyocytes exposed to simulated ischemia, Becker et al. [69] detected significant superoxide generation during myocardial ischemia prior to reperfusion. The sensitivity of this oxidants generation to treatment with specific mitochondrial ETC inhibitors, including amytal, rotenone (complex I inhibitors), and myxothiazol (complex III inhibitors), suggested that the site(s) of ROS generation were upstream from the ubisemiquinone site of the mitochondrial ETC and also suggested the involvement of both complex I and complex III in ischemic ROS generation. It has subsequently been confirmed that ischemia increases significantly ROS levels in isolated guinea pig hearts [70]. In this animal model, intracellular superoxide levels increased by 35% within 1 min of ischemia, and further increased to 95% after 20 min of ischemia, although the source of ROS was not investigated in this study.

## Ischemia and Cell Death

Periods of ischemia longer than 20 min inevitably result in irreversible cardiac damage with subsequent cell loss. Following myocardial I/R, cell death occurs by both necrosis and apoptosis, two distinct modes of death (see Chap. 20) [71]. Necrosis is a process in which cellular integrity is lost and the release of cytosolic contents (such as creatine kinase MB, troponins, and other proteins) provokes an inflammatory response characterized by leukocyte infiltration and phagocytosis, and subsequent damage or death to neighboring cells. While the necrosis-associated inflammatory response can contribute to tissue repair and scar formation, there is ample evidence from both experimental models and clinical studies that it can worsen the myocardial injury, including weakening the connective tissue scaffolding of the heart [72]. Characteristic features of the necrotic cell include swelling of the cell and its organelles, extensive disruption of the mitochondria, plasma membrane blebbing and rupture, and ultimately cell lysis. Plasma membrane disruption is the basis of several assays of necrosis detection involving the use of exclusion dyes or large antibodies. Unlike apoptosis which is an energy-dependent process, necrosis is in fact typically the consequence of acute metabolic perturbations resulting in excessive ATP depletion such as those occur in myocardial I/R and in acute drug-induced toxicity. Ischemia tends to induce necrosis in a subset of myocardial cells with necrotic cells localized primarily in the central zone of the infarcted area and the extent of necrotic cell loss increases with the duration of the ischemic insult. While necrosis was formerly considered to be an unregulated catastrophic event, some investigators have challenged this view [73], and have shown more overlap between apoptosis and necrosis than previously seen. Diverse stimuli have been identified (e.g., cytokines, ischemia, heat, irradiation, and DNA damage) that can cause both apoptosis and necrosis in the same cell population, and a number of signaling pathways including death receptors, kinase cascades, and mitochondria appears to participate in both processes. Modulation of these pathways may provide a switch between apoptosis and necrosis. Moreover, there is evidence that anti-apoptotic factors [e.g., Bcl-2/Bcl-x proteins and heat shock proteins (HSPs)] and prosurvival factors (e.g., Akt, MAP kinase Erk) can be equally effective in protecting against apoptosis and necrosis [74, 75]. For example, Akt overexpression reduced the necrotic zone formation in ischemic myocardium [76], and I/R-induced Erk activation has shown to be protective against necrosis in cardiomyocytes *in vitro* [77] and *in vivo* [78]. In addition, programmed cell death induced by simulated ischemia [i.e., hypoxia–reoxygenation (H/R)] induces the expression of Bnip3, a pro-apoptotic member of the Bcl-2 family of death-regulating proteins involved in the intrinsic cell death signaling pathway (Fig. 14.2), and has features of both apoptosis and necrosis [79, 80]. Interestingly, Bnip3 is expressed in mitochondria and is loosely associated with the mitochondrial membrane in normal tissue, but fully integrates into the mitochondrial outer membrane, with the N terminus in the cytoplasm and the C terminus in the membrane, during induction of cell death.

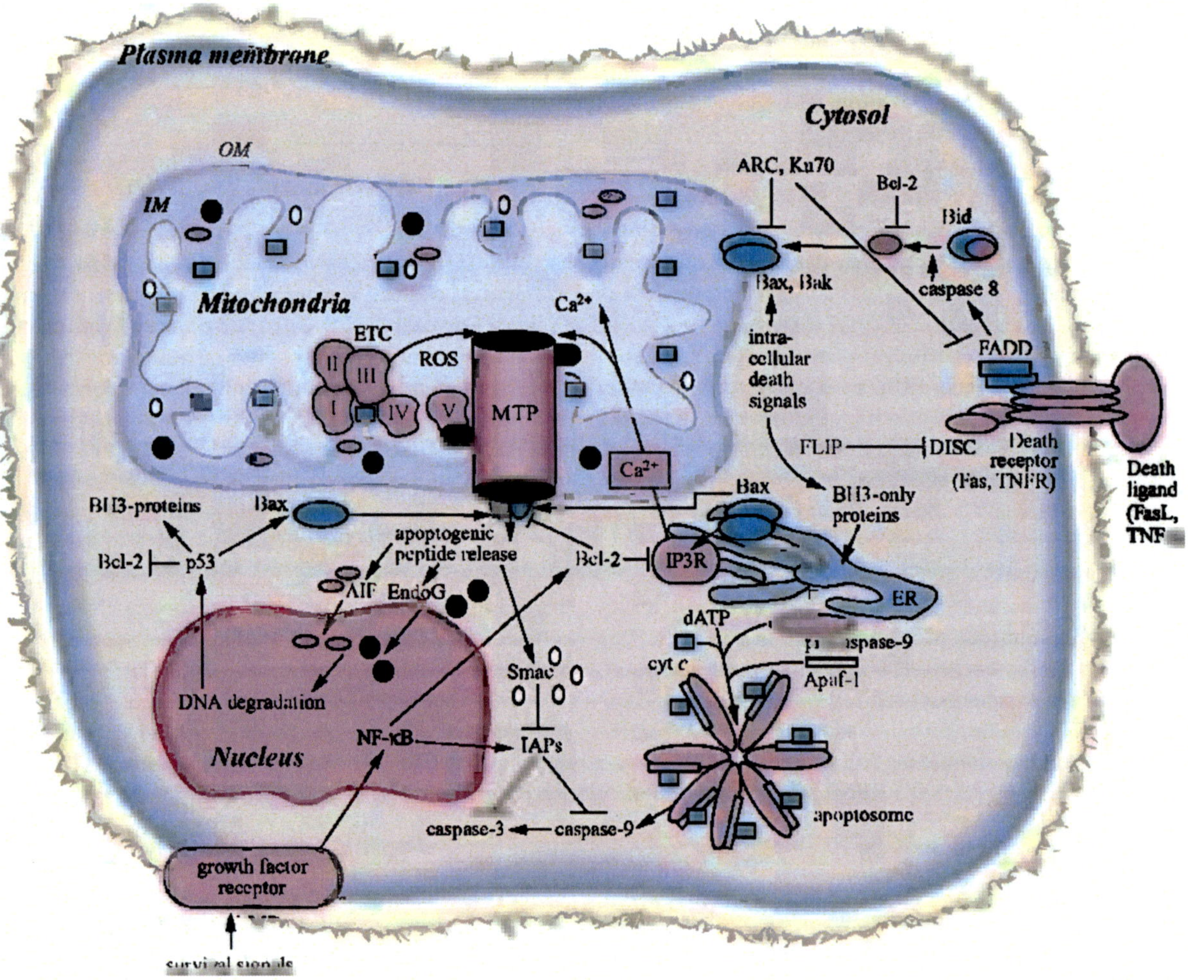


**Fig. 14.2** The extrinsic apoptotic and survival pathways. The extrinsic apoptotic pathway is initiated by ligand binding to death receptors leading to recruitment of FADD and DISC formation which stimulates the activation of caspase-8 resulting in caspase-3 activation and Bid cleavage (a C-terminal fragment of Bid targets mitochondria). FLIP, ARC, and Ku70 can stem this pathway's progression at specific points. Intracellular stimuli trigger ER release of $Ca^{2+}$ through both Bax and BH3-protein interactions. Also depicted is the survival pathway triggered by survival stimuli, mediated by growth factor receptors, transcription factor activation (e.g., NF-κB), and enhanced expression of IAPs and Bcl-2. Also shown is the intrinsic apoptotic pathway. See text for details

Using Erk1 nullizygous gene-targeted mice, Erk2 heterozygous gene-targeted mice, and transgenic mice with activated MEK1–ERK1/2 signaling in the heart, Lips et al. [81] searched for a causal relationship between ERK1/2 signaling and cardioprotection. While MEK1 transgenic mice were mostly resistant to ischemia–reperfusion injury, Erk2+/– gene-targeted mice showed increased infarction areas, DNA laddering, and terminal deoxynucleotidyl transferase-mediated dUTP biotin nick-end labeling (TUNEL) compared with littermate controls. On the other hand, enhanced MEK1–ERK1/2 signaling protected the hearts from DNA laddering and TUNEL, and they preserved their hemodynamic function as assessed by pressure–volume loop recordings after ischemia–reperfusion injury. Taken together these findings are the first to demonstrate that ERK2 signaling is required to protect the myocardium from ischemia–reperfusion injury *in vivo*.

The role of mitochondria in myocardial ischemia-related necrosis has received increasing attention. The MTP pore opening which leads to increased inner-membrane permeability, matrix swelling, and outer membrane disruption with attendant mitochondrial membrane depolarization and dysfunction, is considered to be one of the key early events in apoptosis and necrosis. During ischemia, MTP pore opening is suppressed by acidosis, elevated $Mg^{2+}$ and depressed electron transport, and minimal ROS production. In contrast, during reperfusion, with reoxygenation and resumption of electron transport triggering a burst of ROS at the same time that intracellular $Ca^{2+}$ is elevated, adenine nucleotides are depleted, acidosis is reversed, and MTP is promoted [82]. The MTP inhibitors cyclosporin A (CsA) and sanglifehrin A both protect against I/R injury. Mice lacking cyclophilin-D (CypD), a mitochondrial matrix protein considered to be a component of the MTP pore, exhibited resistance to necrotic (but not apoptotic) cell death induced by ROS and $Ca^{2+}$ overload and showed a high level of resistance to I/R-induced cardiac injury [83, 84].

It has been proposed that in response to stresses such as I/R, the MTP pore becomes a part of the process by which the myocyte will commit to the cell-death pathway. If the extent of MTP is minimal, the cell may recover; if moderate, the cell may undergo apoptosis; if severe, the cell may die from necrosis due to inadequate energy production [85]. Recent evidence obtained from *in vitro* studies of cardiomyocytes exposed to high OS (e.g., $H_2O_2$) suggest that the paradigm of MTP pore and mitochondrial ATP being primary factors in the progression to stress-induced cardiomyocyte necrosis is not invariable. Cascy et al. [86] showed that cardiomyocyte necrotic death in response to elevated OS involved neither MTP pore, reduced mitochondrial membrane potential, or an extensive decline in ATP, but was associated with elevated lipid peroxidation. Moreover, inhibition of lipid peroxidation in $H_2O_2$-treated cardiomyocytes prevented necrotic cell death.

In contrast to necrosis, apoptosis requires energy in the form of ATP for its successful completion, and apoptosis does not provoke an inflammatory response. Apoptotic cell death is characterized by cell shrinkage and membrane blebbing (without the loss of membrane integrity) and nuclear chromatin condensation and degradation (evident as nucleosomal DNA fragmentation). It also appears to be, at least in part, genetically programmed and highly conserved with respect to evolution. While both highly regulated intrinsic pathways (involving both mitochondrial and ER events) and extrinsic pathways with receptors to death ligand (to be discussed more extensively later) have been increasingly found in myocardial apoptosis, under most circumstances both pathways converge by providing signals for execution to a family of ubiquitously expressed cysteine proteases termed caspases. Caspases are generally constitutively expressed as inactive zymogens or procaspases, and after activation by proteolysis at aspartic residues, they target/degrade numerous cytosolic proteins including structural cytoskeletal and contractile proteins, regulatory signaling proteins (e.g., protein kinases), and DNA repair proteins contributing to cell death progression. The apoptotic cell is eventually broken into small membrane-enclosed pieces (apoptotic bodies), which are removed *in vivo* by macrophages or taken up by neighboring cells preventing the release of cellular compounds and thereby ensuring that an inflammatory response is not provoked.

In I/R injury, apoptotic cell loss manifests itself during the reperfusion period due to the slowly orchestrated execution of the apoptotic cell death program and is more apparent at the marginal zone of the infarct area. Gottlieb et al. [87] have reported myocyte and endothelial cell apoptosis in ischemic–reperfused rabbit hearts exposed to 30 min of ischemia followed by 4 h of reperfusion, but not in normal or continuously ischemic myocardium. Freude et al. [88] demonstrated the cleavage and activation of caspase-3 in myocytes, which are otherwise TUNEL negative, after ischemia alone, and they suggested that ischemia per se is sufficient to initiate apoptosis, although reperfusion is required to complete the process to the stage of DNA fragmentation.

The extrinsic pathway (Fig. 14.2) is mediated by death receptors at the cell surface (including TNF-α and Fas/CD95), transducing signaling from external stimuli.

Ligand binding to these receptors induces their trimerization and recruitment of death domain adaptor molecules, e.g., TNF-receptor 1-associated death domain (TRADD) protein and Fas-associated death domain (FADD) protein to form a death-inducing signaling complex (DISC) [89]. Formation of this complex in turn recruits the initiator caspase, caspase-8 in its zymogen form. Subsequent autoproteolysis activates caspase-8, which then cleaves and activates downstream targets such as the more broadly acting executioner caspase, caspase-3, initiating a caspase cascade [90]. Elevated levels of TNF-α have been reported during I/R, although both TNF-α-activated pro- and anti-apoptotic pathways has been shown with I/R [91]. *In vivo* studies support a role of Fas apoptotic signaling in I/R. Fas receptor deficient mice displayed marked reduction in apoptotic cell death with smaller infarcts compared to wild-type animals, and the levels of soluble Fas ligand expression during I/R in rat hearts correlated with infarct size and cell death [92]. However, *in vitro* studies examining the susceptibility of cardiomyocytes to Fas-mediated apoptosis have noted limited Fas-induced cell death, suggesting that hypoxia or ischemia may sensitize cardiomyocytes to the effects of Fas ligand, a notion confirmed in neonatal rat cardiomyocytes exposed to hypoxia [93].

Endogenous inhibitors of the death receptor pathway including the caspase-8 inhibitor c-FLIP and the apoptosis repressor with CARD domain (ARC) may also contribute to apoptotic susceptibility in I/R. Following I/R injury *in vivo*, significantly lower levels of c-FLIP were found in myocardial infarcts where TUNEL-positive myocytes and active caspase-3 expression were prominent; in contrast to abundant c-FLIP expression (and a caspase-3 deficit) present in surrounding unaffected cardiac tissues that suggest reciprocal regulation of these pro- and anti-apoptotic molecules *in vivo* [94]. Similarly, in a murine model of cardiac ischemia, 50% reduction in ARC protein levels was detected in early ischemia relative to levels in sham-operated left ventricles [95]. Nam et al. [96] have showed that endogenous ARC protein levels significantly decrease in response to death stimuli in a variety of cell contexts, as well as in a murine model of myocardial I/R; this ARC decline was mediated by increased ARC protein degradation affected via the ubiquitin-proteasomal pathway.

As previously discussed in Chap. 8, the intrinsic apoptotic pathway is primarily mediated through the mitochondria (hence is termed the mitochondrial pathway), although the endoplasmic reticulum have been implicated as well as contributory to this response.

In this section, it will suffice to mention that in myocytes the intrinsic pathway is activated by diverse cellular stimuli (including hypoxia, ROS, OS, and calcium) with the involvement of diverse cytosolic signaling elements (e.g., kinases) and stress proteins (e.g., HSPs). These pro-apoptotic signals induce the previously mentioned MTP pore opening, characterized by increased permeability of the outer and inner mitochondrial membranes. The components of this nonspecific pore spanning both membranes include the voltage-dependent anion channel (VDAC) in the outer membrane, the ANT in the inner membrane, and cyclophilin-D in the matrix. Permeabilization of the mitochondrion results in mitochondrial swelling contributing to disruption of the outer membrane and subsequent release of inter-membrane proteins most notably cytochrome *c*, Smac/DIABLO, endonuclease G (Endo G), and apoptosis-inducing factor (AIF). Once released, cytochrome *c* binds to the cytosolic protein apaf1 facilitating formation of the apoptosome complex, which results in caspase-9 activation that in turn facilitates caspase-3 activation. Cytosolic Smac/DIABLO indirectly activates caspases by sequestering caspase-inhibitory proteins. The release of Endo G and AIF (caspase-independent factors) from mitochondria results in their translocation to the nucleus where they facilitate DNA fragmentation.

Apoptosis in patients with AMI have been assessed for the rate of cardiomyocyte apoptosis relative to indices of structural LV remodeling. Within the infarct area and in areas remote from the site of injury, the rate of myocardial apoptosis increased and was strongly associated with maladaptive LV remodeling, in addition to adverse clinical outcomes [97]. Activation of leptin signaling in the heart has been found to reduce cardiac morbidity and mortality after MI. Recently, it has been reported that after MI intact leptin signaling acting through STAT-3 increases anti-apoptotic Bcl-2 and survivin genes expression and reduces caspase-3 activity, which is consistent with a cardioprotective role of leptin in the setting of chronic ischemic injury [98].

## Inflammatory Signaling Pathway

Inflammatory signaling pathway is actively involved in myocardial ischemia/MI. For example, a cytokines cascade and chemokine upregulation (see later on this chapter) may be triggered by complement activation and free radical generation. Also, IL-8 and fragment of complement component 5a (C5a) are released in the ischemic myocardium, and may have a crucial role in neutrophil recruitment [99]. Extravasated neutrophils may induce potent cytotoxic effects through the release of proteolytic enzymes and adhesion with Intercellular Adhesion Molecule (ICAM)-1 expressing cardiomyocytes; although, despite these potentially injurious effects, the postreperfusion inflammatory response may significantly enhance healing. Also, reperfused myocardial infarcts show enhanced inflammatory reaction that promotes improved cardiac repair and patient survival [99].

Inflammatory markers have been identified as prominent independent risk indicators for cardiovascular events. While adults over the age of 65 have experienced a high proportion of such events, the available epidemiological data come mainly from middle-aged subjects. Kritchevsky et al. [100] have examined the role that inflammatory markers play in predicting the incidence of CVD, specifically in older adults. Interestingly, IL-6, TNF-α, and IL-10 levels appear to predict cardiovascular outcomes in adults <65 years. Data on C-reactive protein (CRP) was rather inconsistent and appeared to be less reliable in old age than in middle age. In addition, fibrinogen levels have some value in predicting mortality but in a nonspecific manner. They concluded that in older adults, inflammatory markers are nonspecific measures of health and may predict both disability and mortality, even in the absence of clinical CVD. Interventions designed to prevent CVD through the modulation of inflammation may be helpful in reducing disability and mortality. The role of increased inflammatory markers such as IL-6 and IL-1β as a risk factor in the development of MI has also been reported [32]. Polymorphisms in IL-6 gene promoter (−174 G→C) present mainly in older patients with acute coronary syndrome (ACS) carrying IL-6 −174 GG genotypes, showed a marked increase at 1 year follow-up mortality rate, suggesting that IL-6 −174 G→C polymorphisms can be added to the other clinical markers such as CRP serum levels and a history of CAD, useful mainly in male patients at higher risk of death after ACS [101].

After myocardial injury by myocardial infarction, hemodynamic overload, or inflammation, various secondary mediators such as cytokines through self-amplification pathways, and neurohormones act on the myocardium and stimulate myocardial remodeling through myocyte hypertrophy, apoptosis, and altered gene expression in cardiac myocytes. Acutely, the elaboration of TNF-α, IL-1, and IL-6, transforming growth factor (TGF) families of cytokines, contribute to survival or death of myocytes, modulation of cardiac contractility, alterations of vascular endothelium, and recruitment of additional circulating cells of inflammation to the injured myocardium. This leads to further local OS and remodeling but also initiates the processes of wound healing.

Several hypotheses have been proposed to address the source of proinflammatory cytokines in myocardial ischemia as well as in myocarditis (see Chap. 12) and HF. One hypothesis is that activation of the immune system is responsible for cytokine elaboration, which happens in response to some forms of tissue injury (e.g., myocardial ischemia) or possibly by some unknown stimulus to the immune system. A second hypothesis is that myocardial ischemia/MI/HF may be the source of TNF-α production and that elevated levels of TNF-α represent spillover of cytokines that were produced locally within the myocardium, leading to secondary activation of the immune system [102] that is capable of amplifying the cytokine signal in the periphery. The third hypothesis is that underperfusion of systemic tissues leads to the elaboration of TNF-α. A further extension of this hypothesis is that gut wall edema allows translocation of endotoxin, which activates cytokine production [103].

Specific receptors on the cardiomyocyte initiate the production of cytokines. In the case of TNF-α, ligand–receptor signaling is initiated by binding of TNF-α to a lower-affinity, 55-kDa receptor (TNFR1), or to a higher affinity, 75-kDa receptor (TNFR2). In HF, TNF receptors are capable of receptor shedding; that is, they are cleaved from the cell membrane and subsequently exist in the circulation as circulating soluble receptors, referred to as sTNFR1 and sTNFR2. In cell cultures, these soluble receptors have been shown to retain their ability to bind to TNF-α and inhibit its cytotoxic activity. The role of circulating soluble binding proteins (such as sTNF-binding proteins) *in vivo* is to serve as biologic buffers capable of neutralizing the highly cytotoxic activities of cytokines [102]. Moreover, chronically, sustained presence of cytokines leads to myocyte phenotype transition and activation of matrix metalloproteinases that modify interstitial matrix, augmenting further the remodeling process. This in turn alters the local collagen composition and also the integrins that constitute the interface between myocytes and the matrix. These processes may promote angiogenesis and cellular regeneration.

It is possible that modulation of cytokine signaling pathways with new therapies may improve the healing and the cardiac remodeling of myocardial ischemia/MI [104]. The transient receptor potential vanilloid (TRPV1) channels expressed in sensory afferent fibers innervating the heart may be activated by protons or endovanilloids released during myocardial ischemia. Recently, Huang et al. [105] using TRPV1-null mutant and wild-type (WT) mice subjected to left anterior descending coronary ligation or sham operation tested the hypothesis that TRPV1 modulates inflammatory and early remodeling processes to prevent cardiac functional deterioration after MI. These investigators found that the infarct size was greater in TRPV1$^{-/-}$ than in WT mice ($P<0.001$) 3 days after MI, and the mortality rate was higher in TRPV1(−/−) than in WT mice ($P<0.05$) 7 days after MI. Plasma levels of cardiac troponin I, cytokines, including TNF-α, interleukin-1β, and interleukin-6, chemokines, including monocyte chemoattractant protein-1 (MCP-1) and macrophage inflammatory protein-2, and infiltration of inflammatory cells, including neutrophils, macrophages, and myofibroblasts as well as collagen content were greater in the infarct area in TRPV1(−/−) than in WT mice ($P<0.05$) on days 3 and 7 after MI. In addition, pathological LV remodeling occurred in TRPV1(−/−) compared with WT mice. Taken together these findings suggest that TRPV1 gene deletion results in excessive inflammation, disproportional left ventricular remodeling, and deteriorated cardiac function after MI, and indicated that by inhibiting inflammation and abnormal tissue remodeling, TRPV1 may prevent increase in infarct size and cardiac injury.

## Other Participants' Molecules in the Inflammatory Signaling Pathways

### *Nuclear Transcription Factor Kappa B*

The NF-κB controls the expression of hundreds of genes involved in the regulation of the inflammatory response, cardiovascular development, and apoptosis. Since NF-κB signaling pathways has been largely discussed within the context of cardiovascular development (Chap. 9) and stem cells (Chap. 19), here we will only mention that besides controlling a variety of cellular responses, NF-κB signaling has been linked to multiple physiological and pathological processes in the myocardium, including hypertrophy and myocardial ischemia/MI. Notwithstanding, the intracellular mechanisms that govern NF-κB activity in the myocardial cells remain unclear. Gathered observations on the regulation of NF-κB signaling in nonmyocyte systems suggest that the activity of the NF-κB pathway is tightly regulated by a diversity of stress-activated signaling intermediates through direct posttranslational modification of various components of the NF-κB pathway [106].

It has been found that ROS enhance the cytoplasmic signaling pathways leading to NF-κB nuclear translocation, but reduction/oxidation (redox) also controls a number of important steps in the nuclear phase of the NF-κB signaling, including chromatin remodeling, recruitment of co-activators, and DNA binding [107]. Moreover, the survival pathway triggered by survival stimuli is also mediated by NF-κB, growth factors receptors, and enhanced expression of initiator of IAPs and Bcl-2.

## *Toll-Like Receptor Signaling Pathways in Myocardial Ischemia*

Since the role of Toll-like receptor (TLR)-mediated signaling pathway has been amply discussed in Chap. 8, here it is suffice to briefly present recently available information regarding its participation in the pathophysiology of the ischemic heart.

Recent evidence has shown that TLRs play a critical role as a modulator of both cell survival and myocardial ischemia. According to Chao [108], in animal models of I/R injury or in hypoxic cardiomyocytes *in vitro*, the administration of a sublethal dose of lipopolysaccharide, which signals through TLR4, reduces the size of subsequent MI, improves cardiac functions, and attenuates cardiomyocyte apoptosis. On the other hand, a systemic deficiency of TLR2, TLR4, or myeloid differentiation primary-response gene 88, an adaptor critical for all TLR signaling, except TLR3, leads to an attenuated myocardial inflammation, a smaller infarction size, and a better preserved ventricular function with reduced ventricular remodeling after ischemia. Although loss-of-function studies suggest that both TLRs contribute to myocardial inflammation and ischemia, the exact contribution of cardiac (vs. circulatory cell) TLRs is not clear.

Also, Gao et al. [109] have studied the expression of TLR2 and the modulatory role of Glycogen synthase kinase (GSK)-3β inhibitor on TLR2/NF-κB signaling in rats following myocardial I/R. Administration of the GSK-3β inhibitor 4-benzyl-2-methyl-1,2,4-thiadiazolidine-3,5-dione (TDZD-8) 5 min prior to reperfusion and following 1 h reperfusion showed downregulation of mRNA levels of TLR2 and downstream proinflammatory cytokines, decreased activity of NF-κB and the size of MI. Taken together these findings demonstrated that TLR2 and its signaling components are activated by I/R and that the administration of TDZD-8 attenuates TLR2/NF-κB signaling, further suggesting a possible mechanism for GSK-3β, through inhibition of inflammation and apoptosis, to improve the outcome of I/R.

Compared with WT mice, TLR4-deficient mice sustain significantly smaller infarctions in response to myocardial injury. In a study by Takeishi and Kubota [110], TLR4-deficient mice compared to control mice have increased serum levels of IL-1β, IL-6, and TNF-α after I/R and also TLR2 signaling appeared to contribute to cardiac dysfunction following I/R. The protective effects of TLR signaling inhibition in I/R injury was attributed to MyD88, a key adaptor protein for TLR signaling. Furthermore, TLR4 gene polymorphism (Asp299Gly) decreases innate immune responsiveness, reducing the risk for CAD, and appears to increase longevity. These findings seem to infer that the immune system (as in myocarditis) is involved in the pathogenesis of I/R injury and its target may be a new therapeutic modality.

## *Chemokines*

Chemokines are a large family of small chemotactic cytokines secreted by the cell. These proteins are classified according to their structural characteristics such as small size and the presence of four cysteine residues in conserved locations, which are critical to forming their three-dimensional shape. Most chemokines belong to one of two major subfamilies, the CXC and CC subfamilies, based on the arrangement of the first two of the four cysteine residues. It has been reported that levels of CC chemokine ligand (CCL)-5 and CCL-18 are raised during episodes of unstable angina pectoris (UAP) and that increased levels of both chemokines suggest refractory ischemic symptoms. Levels of both chemokines and cognate receptor in circulating peripheral blood mononuclear cells were found elevated during myocardial ischemia, suggesting their involvement in the pathophysiology of UAP and/or post-UAP responses [111].

Also, it was reported that defective CC chemokine MCP-1 signaling inhibits the development of ischemic fibrotic cardiomyopathy in mice, and this may be related to the profibrotic actions of MCP-1 associated with decreased macrophage recruitment, and may not involve direct effects on cardiac fibroblasts [112]. Hence, it is clear that understanding the role of chemokines in myocardial ischemia is critical to developing new approaches to treat these patients.

## Other Signaling Pathways in Myocardial Ischemia

### Connexin and Transforming Growth Factors Signaling

Connexin represents an essential protein forming myocardial channels and gap junctions in the myocardium and has been implicated in tissue repair after injury, dysrhythmia susceptibility, and cardiac conduction defects. Connexin-formed channels are also involved in I/R injury, in ischemic hypercontracture, infarct development, post-MI remodeling, and in ischemic preconditioning. The risk of MI in a large case–control study of Japanese population was found to be associated with C1019T polymorphism in the *GJA4* gene encoding connexin 37, most significantly in men [113]. Interestingly, the MI risk-associated TT genotype was also more prevalent in African-American compared to European Americans [114].

After MI, connexin 43 (Cx43), the major gap junction protein expressed in the heart, shows altered expression and distribution, although the effects of Cx43 on wound healing and ventricular dysfunction are incompletely understood. Recently, to test the hypothesis that reduced expression of Cx43 after MI influences wound healing, fibrosis, and ventricular remodeling, Zhang et al. [115] quantified the progression of infarct healing in Cx43-deficient and wild-type mice subjected to proximal ligation of the anterior descending coronary artery. They measured neutrophil expression, collagen content, and myofibroblast expression. TGF-β, a profibrotic cytokine, was dramatically upregulated in MI hearts, but its phosphorylated comediator (pSmad) was significantly downregulated in the nuclei of Cx43-deficient post-MI hearts, suggesting that downstream signaling of TGF-β is significantly diminished in Cx43-deficient hearts. This decrease in profibrotic TGF-β signaling resulted in attenuation of pathological structural remodeling as assessed by echocardiography. These findings suggest that until the role of Cx43 in infarct healing is better understood, enhancement of Cx43 expression to maintain intercellular coupling or reduce susceptibility to dysrhythmias should be approached with caution.

As noted in the subsection on Inflammatory Signaling, TRPV1 channels expressed in the heart may be activated by protons or endovanilloids (putative endogenous ligands and activators of transient receptor potential vanilloid type 1 channels) released during myocardial ischemia/MI. Recently, Hwang et al. [116] evaluated whether TRPV1 channels can modulate post-MI fibrosis and matrix formation via the TGF-β-Smad signaling pathway, to preserve cardiac function and geometrical regeneration. RPV1-null mutant (TRPV1) and wild-type mice underwent ligation of the left anterior descending coronary artery or sham operation. Compared to wild-type mice, TRPV1 mice showed increased infarct size and mortality rate at 1 week post-MI. TGF-β1, vascular endothelial growth factor, and matrix metalloproteinase-2 expression were more upregulated in TRPV1$^{-/-}$ than in wild-type mice after MI. Also Smad2 expression was enhanced in TRPV1 compared with post-MI wild-type mice. These findings suggest that ablation of TRPV1 markedly enhances post-MI fibrosis and impairs myocardial contractile performance, leading to increased incidence of HF and mortality possibly via stimulation of the TGF-β-Smad2 signaling pathway. Also, these data suggest that TRPV1 plays a protective role in MI healing and regeneration.

### NO Signaling

Reoxygenation of the hypoxic cardiac tissue H/R is one of the prime mechanisms underlying cell and tissue damage in pathological conditions including ischemic heart disease. Moreover, recent evidence has implicated mitochondria as a primary source of oxidative species and subsequent myocardial oxidative injury. During H/R, mitochondrial NOS-derived NO and peroxynitrite formation cause oxidative modification of mitochondrial macromolecules [117] and stimulate the release of the mitochondrial pro-apoptotic protein, cytochrome *c* [118]. Furthermore, H/R-induced cardiac injury is decreased when the production of peroxynitrite is inhibited [119]. Using an *in vitro* model of isolated rat heart mitochondria subjected to H/R treatment, Zenebe et al. [120] demonstrated that H/R but not hypoxia treatment alone markedly increased mitochondrial $Ca^{2+}$ levels, which stimulated mtNOS activity and NO production, and showed evidence of peroxynitrite oxidative damage (i.e., diminished levels of aconitase and mitochondrial creatine kinase (mtCK) activity, lipid peroxidation, and protein carbonyls) and cytochrome *c* release. Taken together these observations suggest that heart mitochondrial calcium homeostasis and mitochondrial NOS have a crucial role in OS induced by HR. Therefore, the identification of a critical mitochondrial signaling pathway involved in the cardiac injury occurring with oxygenation following ischemic insult may ultimately lead to new therapeutic strategies (e.g., selective inhibition of mtNOS) to attenuate the damage and cell death associated with restoration of blood flow to the heart.

Other signaling pathways that participate in myocardial ischemia such as phosphatidylinositol 3-kinase (PI3K) and MAPK are discussed elsewhere in this volume.

## Conclusions

Genetic and profiling studies of myocardial ischemia and MI have shown to be useful in the identification of genes that contribute to the susceptibility to pathogenic processes of myocardial ischemia/MI. Postgenomic approaches presently available have implicated an increasing number of genes and their products as being critical to cardiomyocyte signaling. These data should be useful in the elucidation of pathways and in defining the temporal sequence of events, involved in myocardial ischemia/MI. Clearly, the reconstruction of such events occurring at a variety of time points and spatial *loci* require more than a reductionist approach of the molecular biologist or biochemist, rather it will need significant contributions from an integrative system biology approach, using a combined approach of proteomics, genomics, physiological, and statistical modeling. This will also allow the development of a new arrays of biomarkers that can be applied to screening for disease susceptibility (e.g., which vascular plaques might be most vulnerable for rupture) as well as to assess both progressive ischemic damage, the viability of the cellular response to myocardial ischemic injury, and the ability to repair the damage. Improved understanding of the precise order of intracellular events, their downstream consequences, the overall inter-relatedness, and regulation of these pathways will be necessary for the discovery of therapeutic modalities, e.g., the development and characterization of reagents with high specificity to heart signaling pathways and the development of new technologies for the inhibition of stress signaling (e.g., specific kinase inhibitors) in cardiac and vascular cells.

## Summary


- Myocardial infarction (MI) is a multifactorial polygenic disorder caused by multiple genetic and environmental factors and complex gene–environment interactions modulated by modifiers such as diabetes, dyslipidemias, and hypertension.
- A diverse variety of environmental factors such as diet and smoking can modulate the effects of specific genetic loci on MI risk.
- Linkage studies have provided evidence of genetic susceptibility loci involved in MI as well as the identification of specific genes that appear to be involved in myocardial ischemia/MI.
- The use of single nucleotide polymorphisms and haplotype analysis have allowed for the identification of gene variants that contribute to MI.
- To identify genes involved in increased risk of CAD and susceptibility to MI, genetic case–control studies have proved increasingly powerful.
- The availability of animal models of myocardial ischemia/MI has allowed both the identification of novel genes by using gene profiling techniques, the temporal studies of gene expression, the testing of the effects of null mutations in the genetic *loci*, and finally preclinical therapeutic trials.
- Animal models circumvent many of the problems associated with human studies including genetic heterogeneity, difficult to control variables such as diet, medicines, and other diseases.
- Some events in the phenotype and progression of MI may be different in animals and humans. Nevertheless, some specific gene expression patterns have been found to be similar in both animal models of MI and patient profiles.
- Stresses in myocardial ischemia/hypoxia (e.g., oxidative stress) elicit a variety of adaptive responses at the tissue, cellular, and molecular levels.
- During hypoxia, myocardial mitochondria may function as $O_2$ sensors by increasing the generation of ROS and heme proteins (e.g., cytochrome *c* oxidase [COX]) which reversibly bind oxygen. ROS play a central role in the mitochondrial signaling during hypoxic response.
- In myocardial ischemia, rapid activation of the stress-signaling protein AMP-activated protein kinase (AMPK) results in stimulation of glucose uptake, glycolysis, and FAO which have both beneficial adaptive function (e.g., increased glycolytic ATP production to compensate for the loss of mitochondrial ATP) and deleterious effects (i.e., increased uncoupling of glycolysis from glucose oxidation and enhanced protons and lactate production) on the myocardium.
- Although modulation of the ischemia-induced AMPK activation might prove more or less beneficial, preventing the downstream decrease in malonyl-CoA levels has been proposed as a potential therapeutic target in the management of ischemic heart disease.
- Alterations in the activity of the major components of $[Ca^{2+}]_i$ regulation, such as RyR, $Na^+$–$Ca^{2+}$ exchanger, and $Ca^{2+}$ ATPase, appear to play an important role in ischemia-related $Ca^{2+}$ overload.
- In addition to the profound effects on mitochondrial fatty acid uptake and oxidation, ischemia also markedly impacts mitochondrial OXPHOS.
- Ischemic damage to complex I has been associated with a marked decrease in the levels of FMN cofactor.
- Ischemia-generated ROS appears to play a critical signaling role, which contributes to cellular oxidant damage and appears to be the same ROS that triggers preconditioning.
- Diverse stimuli have been identified (e.g., cytokines, ischemia, heat, irradiation, and DNA damage) that can cause both apoptosis and necrosis in the same cell population and a number of signaling pathways including death receptors, kinase cascades, and mitochondria appear to participate in both processes.
- Akt overexpression reduced the necrotic zone formation in ischemic myocardium, and I/R-induced Erk activation has shown to be protective against necrosis in cardiomyocytes *in vitro* and *in vivo*.

- It has been proposed that in response to stresses such as I/R the MTP pore becomes part of the process by which the myocyte will commit to the cell-death pathway.
- If the extent of MTP pore is minimal, the cell may recover; if moderate, the cell may undergo apoptosis; if severe, the cell may die from necrosis due to inadequate energy production.
- Activation of leptin signaling in the heart has been found to reduce cardiac morbidity and mortality after MI.
- After IM, intact leptin signaling acts through STAT-3 to increase anti-apoptotic Bcl-2 and survivin gene expression, and reduces caspase-3 activity that is consistent with a cardioprotective role of leptin in the setting of chronic ischemic injury.
- Inflammatory signaling pathways are actively involved in myocardial ischemia/MI.
- A cytokines cascade and chemokine upregulation may be triggered by complement activation and free radical generation. Also, IL-8 and C5a are released in the ischemic myocardium and may have a crucial role in neutrophil recruitment.
- Acutely, the elaboration of TNF-α, IL-1, and IL-6, TGF families of cytokines, contributes to survival or deaths of myocytes, modulation of cardiac contractility, alterations of vascular endothelium, and recruitment of additional circulating cells of inflammation to the injured myocardium.
- Levels of CC chemokine ligand (CCL)-5 and CCL-18 have been found elevated during episodes of UAP and increased levels of both chemokines suggest refractory ischemic symptoms.
- Connexin-formed channels are also involved in I/R injury, in ischemic hypercontracture, infarct development, post-MI remodeling, and in ischemic preconditioning.
- TGF-β, a profibrotic cytokine, was dramatically upregulated in MI hearts, but its phosphorylated comediator (pSmad) was significantly downregulated in the nuclei of Cx43-deficient hearts post-MI, suggesting that downstream signaling of TGF-β is significantly diminished in Cx43-deficient hearts.
- During H/R, mtNOS-derived NO and peroxynitrite formation cause oxidative modification of mitochondrial macromolecules and stimulate the release of the mitochondrial pro-apoptotic protein, cytochrome *c*. H/R-induced cardiac injury is diminished when production of peroxynitrite is inhibited.

## References


1. Lesnefsky EJ, Moghaddas S, Tandler B, Kerner J, Hoppel CL. Mitochondrial dysfunction in cardiac disease: ischemia–reperfusion, aging, and heart failure. J Mol Cell Cardiol. 2001; 33:1065–89.
2. Paradies G, Petrosillo G, Pistolese M, Di Venosa N, Serena D, Ruggiero FM. Lipid peroxidation and alterations to oxidative metabolism in mitochondria isolated from rat heart subjected to ischemia and reperfusion. Free Radic Biol Med. 1999;27:42–50.
3. Eefting F, Rensing B, Wigman J, et al. Role of apoptosis in reperfusion injury. Cardiovasc Res. 2004;61:414–26.
4. Honda HM, Korge P, Weiss JN. Mitochondria and ischemia/reperfusion injury. Ann N Y Acad Sci. 2005;1047:248–58.
5. Jefferson BK, Topol EJ. Molecular mechanisms of myocardial infarction. Curr Probl Cardiol. 2005;30:333–74.
6. Winkelmann BR, Hager J. Genetic variation in coronary heart disease and myocardial infarction: methodological overview and clinical evidence. Pharmacogenomics. 2000;1:73–94.
7. Tregouet DA, Barbaux S, Escolano S, et al. Specific haplotypes of the P-selectin gene are associated with myocardial infarction. Hum Mol Genet. 2002;11:2015–23.
8. Yoshida M, Takano Y, Sasaoka T, Izumi T, Kimura A. E-Selectin polymorphism associated with myocardial infarction causes enhanced leukocyte–endothelial interactions under flow conditions. Arterioscler Thromb Vasc Biol. 2003;23:783–8.
9. Sasaoka T, Kimura A, Hohta SA, Fukuda N, Kurosawa T, Izumi T. Polymorphisms in the platelet-endothelial cell adhesion molecule-1 (PECAM-1) gene, Asn563Ser and Gly670Arg, associated with myocardial infarction in the Japanese. Ann N Y Acad Sci. 2001;947:259–69. discussion 269–70.
10. Listi F, Candore G, Lio D, et al. Association between platelet endothelial cellular adhesion molecule 1 (PECAM-1/CD31) polymorphisms and acute myocardial infarction: a study in patients from Sicily. Eur J Immunogenet. 2004;31:175–8.
11. Iwai C, Akita H, Kanazawa K, et al. Arg389Gly polymorphism of the human beta1-adrenergic receptor in patients with nonfatal acute myocardial infarction. Am Heart J. 2003;146:106–9.
12. Small KM, Wagoner LE, Levin AM, Kardia SL, Liggett SB. Synergistic polymorphisms of beta1- and alpha2C-adrenergic receptors and the risk of congestive heart failure. N Engl J Med. 2002;347:1135–42.
13. Bengtsson K, Melander O, Orho-Melander M, et al. Polymorphism in the beta(1)-adrenergic receptor gene and hypertension. Circulation. 2001;104:187–90.
14. Liu J, Liu ZQ, Tan ZR, et al. Gly389Arg polymorphism of beta1-adrenergic receptor is associated with the cardiovascular response to metoprolol. Clin Pharmacol Ther. 2003;74:372–9.
15. Akhter SA, D'Souza KM, Petrashevskaya NN, Mialet-Perez J, Liggett SB. Myocardial beta1-adrenergic receptor polymorphisms affect functional recovery after ischemic injury. Am J Physiol Heart Circ Physiol. 2006;290:H1427–32.
16. Chandel NS, Schumacker PT. Cellular oxygen sensing by mitochondria: old questions, new insight. J Appl Physiol. 2000;88: 1880–9.
17. Duranteau J, Chandel NS, Kulisz A, Shao Z, Schumacker PT. Intracellular signaling by reactive oxygen species during hypoxia in cardiomyocytes. J Biol Chem. 1998;273:11619–24.
18. Kacimi R, Long CS, Karliner JS. Chronic hypoxia modulates the interleukin-1beta-stimulated inducible nitric oxide synthase pathway in cardiac myocytes. Circulation. 1997;96:1937–43.
19. Kulisz A, Chen N, Chandel NS, Shao Z, Schumacker PT. Mitochondrial ROS initiate phosphorylation of p38 MAP kinase during hypoxia in cardiomyocytes. Am J Physiol Lung Cell Mol Physiol. 2002;282:L1324–9.
20. Enomoto N, Koshikawa N, Gassmann M, Hayashi J, Takenaga K. Hypoxic induction of hypoxia-inducible factor-1alpha and oxygen-regulated gene expression in mitochondrial DNA-depleted HeLa cells. Biochem Biophys Res Commun. 2002;297: 346–52.
21. Kutala VK, Khan M, Angelos MG, Kuppusamy P. Role of oxygen in postischemic myocardial injury. Antioxid Redox Signal. 2007;9:1193–206.

22. Zucchi R, Ghelardoni S, Evangelista S. Biochemical basis of ischemic heart injury and of cardioprotective interventions. Curr Med Chem. 2007;14:1619–37.
23. Grover GJ, Atwal KS, Sleph PG, et al. Excessive ATP hydrolysis in ischemic myocardium by mitochondrial F1F0-ATPase: effect of selective pharmacological inhibition of mitochondrial ATPase hydrolase activity. Am J Physiol Heart Circ Physiol. 2004;287:H1747–55.
24. Dyck JR, Lopaschuk GD. Glucose metabolism, $H^+$ production and $Na^+/H^+$-exchanger mRNA levels in ischemic hearts from diabetic rats. Mol Cell Biochem. 1998;180:85–93.
25. Dolinsky VW, Dyck JR. Role of AMP-activated protein kinase in healthy and diseased hearts. Am J Physiol Heart Circ Physiol. 2006;291:H2557–69.
26. Capano M, Crompton M. Bax translocates to mitochondria of heart cells during simulated ischaemia: involvement of AMP-activated and p38 mitogen-activated protein kinases. Biochem J. 2006;395:57–64.
27. Hickson-Bick DL, Buja LM, McMillin JB. Palmitate-mediated alterations in the fatty acid metabolism of rat neonatal cardiac myocytes. J Mol Cell Cardiol. 2000;32:511–9.
28. Russell III RR, Li J, Coven DL, et al. AMP-activated protein kinase mediates ischemic glucose uptake and prevents postischemic cardiac dysfunction, apoptosis, and injury. J Clin Invest. 2004;114:495–503.
29. Shibata R, Sato K, Pimentel DR, et al. Adiponectin protects against myocardial ischemia-reperfusion injury through AMPK- and COX-2-dependent mechanisms. Nat Med. 2005;11:1096–103.
30. Chan AY, Soltys CL, Young ME, Proud CG, Dyck JR. Activation of AMP-activated protein kinase inhibits protein synthesis associated with hypertrophy in the cardiac myocyte. J Biol Chem. 2004;279:32771–9.
31. Browne GJ, Finn SG, Proud CG. Stimulation of the AMP-activated protein kinase leads to activation of eukaryotic elongation factor 2 kinase and to its phosphorylation at a novel site, serine 398. J Biol Chem. 2004;279:12220–31.
32. Browne GJ, Proud CG. A novel mTOR-regulated phosphorylation site in elongation factor 2 kinase modulates the activity of the kinase and its binding to calmodulin. Mol Cell Biol. 2004;24:2986–97.
33. Shibata R, Ouchi N, Ito M, et al. Adiponectin-mediated modulation of hypertrophic signals in the heart. Nat Med. 2004;10:1384–9.
34. Arad M, Seidman CE, Seidman JG. AMP-activated protein kinase in the heart: role during health and disease. Circ Res. 2007;100:474–88.
35. Lopaschuk GD, Stanley WC. Malonyl-CoA decarboxylase inhibition as a novel approach to treat ischemic heart disease. Cardiovasc Drugs Ther. 2006;20:433–9.
36. Clanachan AS. Contribution of protons to post-ischemic Na(+) and Ca(2+) overload and left ventricular mechanical dysfunction. J Cardiovasc Electrophysiol. 2006;17 Suppl 1:S141–8.
37. Opie LH, Sack MN. Metabolic plasticity and the promotion of cardiac protection in ischemia and ischemic preconditioning. J Mol Cell Cardiol. 2002;34:1077–89.
38. Wolff AA, Rotmensch HH, Stanley WC, Ferrari R. Metabolic approaches to the treatment of ischemic heart disease: the clinicians' perspective. Heart Fail Rev. 2002;7:187–203.
39. Hasenfuss G, Maier LS. Mechanism of action of the new anti-ischemia drug ranolazine. Clin Res Cardiol. 2008;97:222–6.
40. Zima AV, Blatter LA. Redox regulation of cardiac calcium channels and transporters. Cardiovasc Res. 2006;71:310–21.
41. Rouslin W. Mitochondrial complexes I, II, III, IV, and V in myocardial ischemia and autolysis. Am J Physiol Heart Circ Physiol. 1983;244:H743–8.
42. Asimakis G, Conti VR. Myocardial ischemia: correlation of mitochondrial adenine nucleotide and respiratory function. J Mol Cell Cardiol. 1984;16:439–47.
43. Lesnefsky EJTB, Ye J, Slabe TJ, Turkaly J, Hoppel CL. Myocardial ischemia decreases oxidative phosphorylation through cytochrome oxidase in subsarcolemmal mitochondria. Am J Physiol Heart Circ Physiol. 1997;273:H1544–54.
44. Burwell LS, Nadtochiy SM, Tompkins AJ, Young S, Brookes PS. Direct evidence for S-nitrosation of mitochondrial complex I. Biochem J. 2006;394:627–34.
45. Murray J, Taylor SW, Zhang B, Ghosh SS, Capaldi RA. Oxidative damage to mitochondrial complex I due to peroxynitrite: identification of reactive tyrosines by mass spectrometry. J Biol Chem. 2003;278:37223–30.
46. Chen R, Fearnley IM, Peak-Chew SY, Walker JE. The phosphorylation of subunits of complex I from bovine heart mitochondria. J Biol Chem. 2004;279:26036–45.
47. Paradies G, Petrosillo G, Pistolese M, Di Venosa N, Federici A, Ruggiero FM. Decrease in mitochondrial complex I activity in ischemic/reperfused rat heart: involvement of reactive oxygen species and cardiolipin. Circ Res. 2004;94:53–9.
48. Petrosillo G, Ruggiero FM, Di Venosa N, Paradies G. Decreased complex III activity in mitochondria isolated from rat heart subjected to ischemia and reperfusion: role of reactive oxygen species and cardiolipin. FASEB J. 2003;17:714–6.
49. Fang JK, Prabu SK, Sepuri NB, et al. Site specific phosphorylation of cytochrome c oxidase subunits I, IVi1 and Vb in rabbit hearts subjected to ischemia/reperfusion. FEBS Lett. 2007;581: 1302–10.
50. Prabu SK, Anandatheerthavarada HK, Raza H, Srinivasan S, Spear JF, Avadhani NG. Protein kinase A-mediated phosphorylation modulates cytochrome c oxidase function and augments hypoxia and myocardial ischemia-related injury. J Biol Chem. 2006; 281:2061–70.
51. Palmer JW, Tandler B, Hoppel CL. Heterogeneous response of subsarcolemmal heart mitochondria to calcium. Am J Physiol. 1986;250:H741–8.
52. Lesnefsky EJCQ, Slabe TJ, Stoll MS, et al. Ischemia, rather than reperfusion, inhibits respiration through cytochrome oxidase in the isolated, perfused rabbit heart: role of cardiolipin. Am J Physiol Heart Circ Physiol. 2004;287:H258–67.
53. Jain MCL, Brenner DA, Wang B, et al. Increased myocardial dysfunction after ischemia-reperfusion in mice lacking glucose-6-phosphate dehydrogenase. Circulation. 2004;109:898–903.
54. Schonheit K, Gille L, Nohl H. Effect of alpha-lipoic acid and dihydrolipoic acid on ischemia/reperfusion injury of the heart and heart mitochondria. Biochim Biophys Acta. 1995;1271:335–42.
55. Zhao K, Zhao GM, Wu D, et al. Cell-permeable peptide antioxidants targeted to inner mitochondrial membrane inhibit mitochondrial swelling, oxidative cell death, and reperfusion injury. J Biol Chem. 2004;279:34682–90.
56. Chen Z, Oberley TD, Ho Y, et al. Overexpression of CuZnSOD in coronary vascular cells attenuates myocardial ischemia/reperfusion injury. Free Radic Biol Med. 2000;29:589–96.
57. Janssens S, Pokreisz P, Schoonjans L, et al. Cardiomyocyte-specific overexpression of nitric oxide synthase 3 improves left ventricular performance and reduces compensatory hypertrophy after myocardial infarction. Circ Res. 2004;94:1256–62.
58. Pachori AS, Melo LG, Zhang L, Solomon SD, Dzau VJ. Chronic recurrent myocardial ischemic injury is significantly attenuated by pre-emptive adeno-associated virus heme oxygenase-1 gene delivery. J Am Coll Cardiol. 2006;47:635–43.
59. Hess ML, Manson NH. Molecular oxygen: friend and foe. The role of the oxygen free radical system in the calcium paradox, the oxygen paradox and ischemia/reperfusion injury. J Mol Cell Cardiol. 1984;16:969–85.
60. Bolli R, Jeroudi MO, Patel BS, et al. Marked reduction of free radical generation and contractile dysfunction by antioxidant therapy begun at the time of reperfusion. Evidence that myocardial

"stunning" is a manifestation of reperfusion injury. Circ Res. 1989;65:607–22.
61. Becker LB. New concepts in reactive oxygen species and cardiovascular reperfusion physiology. Cardiovasc Res. 2004;61: 461–70.
62. Granville DJ, Tashakkor B, Takeuchi C, et al. Reduction of ischemia and reperfusion-induced myocardial damage by cytochrome P450 inhibitors. Proc Natl Acad Sci USA. 2004;101:1321–6.
63. Hunter AL, Kerjner A, Mueller KJ, McManus BM, Granville DJ. Cytochrome p450 2C enzymes contribute to peritransplant ischemic injury and cardiac allograft vasculopathy. Am J Transplant. 2008;8:1631–8.
64. Zweier JL, Kuppusamy P, Lutty GA. Measurement of endothelial cell free radical generation: evidence for a central mechanism of free radical injury in postischemic tissues. Proc Natl Acad Sci USA. 1988;85:4046–50.
65. Thompson-Gorman SL, Zweier JL. Evaluation of the role of xanthine oxidase in myocardial reperfusion injury. J Biol Chem. 1990;265:6656–63.
66. Williams AL, Chen L, Scharf SM. Effects of allopurinol on cardiac function and oxidant stress in chronic intermittent hypoxia. Sleep Breath. 2010;14:51–7.
67. Vanden Hoek TL, Becker LB, Shao Z, Li C, Schumacker PT. Reactive oxygen species released from mitochondria during brief hypoxia induce preconditioning in cardiomyocytes. J Biol Chem. 1998;273:18092–8.
68. Vanden Hoek T, Becker LB, Shao ZH, Li CQ, Schumacker PT. Preconditioning in cardiomyocytes protects by attenuating oxidant stress at reperfusion. Circ Res. 2000;86:541–8.
69. Becker LB, vandenHoek TL, Shao ZH, Li CQ, Schumacker PT. Generation of superoxide in cardiomyocytes during ischemia before reperfusion. Am J Physiol. 1999;277:H2240–6.
70. Kevin LG, Camara AK, Riess ML, Novalija E, Stowe DF. Ischemic preconditioning alters real-time measure of $O_2$ radicals in intact hearts with ischemia and reperfusion. Am J Physiol Heart Circ Physiol. 2003;284:H566–74.
71. Kajstura J, Cheng W, Reiss K, et al. Apoptotic and necrotic myocyte cell deaths are independent contributing variables of infarct size in rats. Lab Invest. 1996;74:86–107.
72. Scarabelli TM, Gottlieb RA. Functional and clinical repercussions of myocyte apoptosis in the multifaceted damage by ischemia/reperfusion injury: old and new concepts after 10 years of contributions. Cell Death Differ. 2004;11 Suppl 2:S144–52.
73. Syntichaki P, Tavernarakis N. Death by necrosis. Uncontrollable catastrophe, or is there order behind the chaos? EMBO Rep. 2002;3:604–9.
74. Brocheriou V, Hagege AA, Oubenaissa A, et al. Cardiac functional improvement by a human Bcl-2 transgene in a mouse model of ischemia/reperfusion injury. J Gene Med. 2000;2:326–33.
75. Jaattela M. Heat shock proteins as cellular lifeguards. Ann Med. 1999;31:261–71.
76. Matsui T, Tao J, del Monte F, et al. Akt activation preserves cardiac function and prevents injury after transient cardiac ischemia in vivo. Circulation. 2001;104:330–5.
77. Punn A, Mockridge JW, Farooqui S, Marber MS, Heads RJ. Sustained activation of p42/p44 mitogen-activated protein kinase during recovery from simulated ischaemia mediates adaptive cytoprotection in cardiomyocytes. Biochem J. 2000;350(Pt 3):891–9.
78. Yue TL, Wang C, Gu JL, et al. Inhibition of extracellular signal-regulated kinase enhances ischemia/reoxygenation-induced apoptosis in cultured cardiac myocytes and exaggerates reperfusion injury in isolated perfused heart. Circ Res. 2000;86:692–9.
79. Graham RM, Thompson JW, Wei J, Bishopric NH, Webster KA. Regulation of Bnip3 death pathways by calcium, phosphorylation, and hypoxia-reoxygenation. Antioxid Redox Signal. 2007;9: 1309–15.
80. Vande Velde C, Cizeau J, Dubik D, et al. BNIP3 and genetic control of necrosis-like cell death through the mitochondrial permeability transition pore. Mol Cell Biol. 2000;20:5454–68.
81. Lips DJ, Bueno OF, Wilkins BJ, et al. MEK1-ERK2 signaling pathway protects myocardium from ischemic injury in vivo. Circulation. 2004;109:1938–41.
82. Weiss JN, Korge P, Honda HM, Ping P. Role of the mitochondrial permeability transition in myocardial disease. Circ Res. 2003;93: 292–301.
83. Nakagawa T, Shimizu S, Watanabe T, et al. Cyclophilin D-dependent mitochondrial permeability transition regulates some necrotic but not apoptotic cell death. Nature. 2005;434:652–8.
84. Baines CP, Kaiser RA, Purcell NH, et al. Loss of cyclophilin D reveals a critical role for mitochondrial permeability transition in cell death. Nature. 2005;434:658–62.
85. Honda HM, Ping P. Mitochondrial permeability transition in cardiac cell injury and death. Cardiovasc Drugs Ther. 2006;20:425–32.
86. Casey TM, Arthur PG, Bogoyevitch MA. Necrotic death without mitochondrial dysfunction-delayed death of cardiac myocytes following oxidative stress. Biochim Biophys Acta. 2007;1773: 342–51.
87. Gottlieb RA, Burleson KO, Kloner RA, Babior BM, Engler RL. Reperfusion injury induces apoptosis in rabbit cardiomyocytes. J Clin Invest. 1994;94:1621–8.
88. Freude B, Masters TN, Robicsek F, et al. Apoptosis is initiated by myocardial ischemia and executed during reperfusion. J Mol Cell Cardiol. 2000;32:197–208.
89. Peter ME, Krammer PH. The CD95(APO-1/Fas) DISC and beyond. Cell Death Differ. 2003;10:26–35.
90. Logue SEMS. Caspase activation cascades in apoptosis. Biochem Soc Trans. 2008;36:1–9.
91. Logue SE, Gustafsson AB, Samali A, Gottlieb RA. Ischemia/reperfusion injury at the intersection with cell death. J Mol Cell Cardiol. 2005;38:21–33.
92. Jeremias I, Kupatt C, Martin-Villalba A, et al. Involvement of CD95/Apo1/Fas in cell death after myocardial ischemia. Circulation. 2000;102:915–20.
93. Yaniv G, Shilkrut M, Larisch S, Binah O. Hydrogen peroxide predisposes neonatal rat ventricular myocytes to Fas-mediated apoptosis. Biochem Biophys Res Commun. 2005;336:740–6.
94. Rasper DM, Vaillancourt JP, Hadano S, et al. Cell death attenuation by 'Usurpin', a mammalian DED-caspase homologue that precludes caspase-8 recruitment and activation by the CD-95 (Fas, APO-1) receptor complex. Cell Death Differ. 1998;5:271–88.
95. Lyn D, Bao S, Bennett NA, Liu X, Emmett NL. Ischemia elicits a coordinated expression of pro-survival proteins in mouse myocardium. ScientificWorldJournal. 2002;2:997–1003.
96. Nam YJ, Mani K, Wu L, et al. The apoptosis inhibitor ARC undergoes ubiquitin-proteasomal-mediated degradation in response to death stimuli: identification of a degradation-resistant mutant. J Biol Chem. 2007;282:5522–8.
97. Abbate A, Biondi-Zoccai GG, Bussani R, et al. Increased myocardial apoptosis in patients with unfavorable left ventricular remodeling and early symptomatic post-infarction heart failure. J Am Coll Cardiol. 2003;41:753–60.
98. McGaffin KR, Zou B, McTiernan CF, O'Donnell CP. Leptin attenuates cardiac apoptosis after chronic ischaemic injury. Cardiovasc Res. 2009;83:313–24.
99. Ren G, Dewald O, Frangogiannis NG. Inflammatory mechanisms in myocardial infarction. Curr Drug Targets Inflamm Allergy. 2003;2:242–56.
100. Kritchevsky SB, Cesari M, Pahor M. Inflammatory markers and cardiovascular health in older adults. Cardiovasc Res. 2005;66: 265–75.
101. Antonicelli R, Olivieri F, Bonafe M, et al. The interleukin-6 –174 G>C promoter polymorphism is associated with a higher risk of

death after an acute coronary syndrome in male elderly patients. Int J Cardiol. 2005;103:266–71.
102. Kapadia S, Dibbs Z, Kurrelmeyer K, et al. The role of cytokines in the failing human heart. Cardiol Clin. 1998;16:645–56. viii.
103. Rauchhaus M, Coats AJ, Anker SD. The endotoxin-lipoprotein hypothesis. Lancet. 2000;356:930–3.
104. Arnold JM, Liu P, Demers C, et al. Canadian Cardiovascular Society consensus conference recommendations on heart failure 2006: diagnosis and management. Can J Cardiol. 2006;22:23–45.
105. Huang W, Rubinstein J, Prieto AR, Thang LV, Wang DH. Transient receptor potential vanilloid gene deletion exacerbates inflammation and atypical cardiac remodeling after myocardial infarction. Hypertension. 2009;53:243–50.
106. Hall G, Hasday JD, Rogers TB. Regulating the regulator: NF-kappaB signaling in heart. J Mol Cell Cardiol. 2006;41:580–91.
107. Gloire G, Piette J. Redox regulation of nuclear post-translational modifications during NF-kappaB activation. Antioxid Redox Signal. 2009;11:2209–22.
108. Chao W. Toll-like receptor signaling: a critical modulator of cell survival and ischemic injury in the heart. Am J Physiol Heart Circ Physiol. 2009;296:H1–12.
109. Gao HK, Yin Z, Zhang RQ, Zhang J, Gao F, Wang HC. GSK-3beta inhibitor modulates TLR2/NF-kappaB signaling following myocardial ischemia-reperfusion. Inflamm Res. 2009;58:377–83.
110. Takeishi Y, Kubota I. Role of Toll-like receptor mediated signaling pathway in ischemic heart. Front Biosci. 2009;14:2553–8.
111. Kraaijeveld AO, de Jager SC, de Jager WJ, et al. CC chemokine ligand-5 (CCL5/RANTES) and CC chemokine ligand-18 (CCL18/PARC) are specific markers of refractory unstable angina pectoris and are transiently raised during severe ischemic symptoms. Circulation. 2007;116:1931–41.
112. Frangogiannis NG, Dewald O, Xia Y, et al. Critical role of monocyte chemoattractant protein-1/CC chemokine ligand 2 in the pathogenesis of ischemic cardiomyopathy. Circulation. 2007;115:584–92.
113. Yamada Y, Izawa H, Ichihara S, et al. Prediction of the risk of myocardial infarction from polymorphisms in candidate genes. N Engl J Med. 2002;347:1916–23.
114. Lanfear DE, Marsh S, Cresci S, Shannon WD, Spertus JA, McLeod HL. Genotypes associated with myocardial infarction risk are more common in African Americans than in European Americans. J Am Coll Cardiol. 2004;44:165–7.
115. Zhang Y, Wang H, Kovacs A, Kanter EM, Yamada KA. Reduced expression of Cx43 attenuates ventricular remodeling after myocardial infarction via impaired TGF-beta signaling. Am J Physiol Heart Circ Physiol. 2010;298:H477–87.
116. Huang W, Rubinstein J, Prieto AR, Wang DH. Enhanced postmyocardial infarction fibrosis via stimulation of the transforming growth factor-beta-Smad2 signaling pathway: role of transient receptor potential vanilloid type 1 channels. J Hypertens. 2010;28:367–76.
117. Ronson RS, Nakamura M, Vinten-Johansen J. The cardiovascular effects and implications of peroxynitrite. Cardiovasc Res. 1999;44:47–59.
118. Ghafourifar P, Schenk U, Klein SD, Richter C. Mitochondrial nitric-oxide synthase stimulation causes cytochrome c release from isolated mitochondria. Evidence for intramitochondrial peroxynitrite formation. J Biol Chem. 1999;274:31185–8.
119. Xie YW, Wolin MS. Role of nitric oxide and its interaction with superoxide in the suppression of cardiac muscle mitochondrial respiration. Involvement in response to hypoxia/reoxygenation. Circulation. 1996;94:2580–6.
120. Zenebe WJ, Nazarewicz RR, Parihar MS, Ghafourifar P. Hypoxia/reoxygenation of isolated rat heart mitochondria causes cytochrome c release and oxidative stress; evidence for involvement of mitochondrial nitric oxide synthase. J Mol Cell Cardiol. 2007;43:411–9.

# Chapter 15
# Signaling in Hypertrophy and Heart Failure

**Abstract** In human, clinical cardiac hypertrophy is characterized by increased myocardial mass secondary to cardiomyocyte proliferation. At the cellular level, on the other hand, cardiac hypertrophy is characterized by increased cardiomyocyte size. In the mammalian embryo, cardiac myocytes rapidly proliferate but shortly after birth irreversibly exit the cell cycle, shifting the predominant form of growth from hyperplasia (proliferation) to hypertrophy (increase in size). Significant research efforts have been directed to identify the mitogenic stimuli, and the signaling pathways that mediate these distinct growth processes not only in isolated cells, but also in *in vivo* hearts; however, at this point if cardiac hypertrophy is a "good" mechanism to stimulate or a "bad" process to prevent remains a matter of discussion. A broad analysis of the cardiovascular signaling pathways that play a central role in hypertrophy, either functioning singly or in synchronous mode, is presented in this chapter.

**Keywords** Myocardial hypertrophy • Heart failure • Second messengers signaling • Kinases • Phosphatases

## Introduction

Cardiac hypertrophy is a well-recognized remodeling response to increased hemodynamic load [1], as well as a mechanism designed to improve pump function by expanding the number of contractile units, while simultaneously reducing wall stress by increasing wall thickness. Postnatal growth, exercise, and pregnancy promote physiologic growth of the heart, whereas neurohumoral activation, hypertension, valvular disease, and myocardial injury can cause pathologic hypertrophic growth. Hypertrophy can directly result in heart failure (HF), as evidenced by patients with hypertrophic cardiomyopathy (HCM) [2, 3]. Furthermore, in the failing heart, excessive left ventricle (LV) hypertrophy (LVH) is associated with reduced myocardial compliance, myocardial fibrosis, and often lethal dysrhythmias [4].

While LVH is considered by some investigators as an adaptive response to increased workload and is accompanied by alterations in metabolism [5], for others if LVH is adaptive or maladaptive is rather a controversial concept, as suggested by the different patterns of signaling pathways in transgenic models and clinical findings in aortic stenosis [6]. On the other hand, the time frame of transition from adaptive to maladaptive hypertrophy is not known.

The heart might respond to environmental stimuli not only by growth with increasing myocardial mass, but also by shrinkage (atrophy) with a dynamic range that has been estimated to be of at least 100% [7]. Hypertrophy is often present in cardiac remodeling prior to chamber dilatation and progressive impairment in function [8]. Indeed, progressive cardiac remodeling results in the onset of LV dilatation with relatively proportional increases in LV mass, a process sometimes named eccentric hypertrophy [9]. These events are characteristic of patients with compensated LV dysfunction with the development of clinically overt HF, mainly triggered by marked ventricular dilatation, that occurs once the resources from the myocardial hypertrophic response are exhausted [10]. With end-stage disease and clinically overt HF, severe LV dilatation occurs without comparable levels of hypertrophy, resulting in relative wall thinning from the disproportional increases in LV volume. While in concentric cardiac hypertrophy cardiomyocytes thickened, in eccentric hypertrophy cardiomyocytes will mainly elongate. This suggests that cardiomyocytes growth responses are mediated by different molecular processes, including different signaling pathways. Interestingly, these intracellular signaling pathways involved in the development of cardiac hypertrophy may operate together in a coordinated mode triggered by different mechanisms whose delineation and understanding are critical in the prevention and treatment of pathological or maladaptive hypertrophy. Current experimental evidence indicates that targeted increases in wall thickness in the absence of concomitant increases in LV volume can have beneficial effects on cardiac performance and can halt the progression to HF [11]. In this chapter, the contribution of signaling pathways to cardiac hypertrophy and potential therapies are discussed.

J. Marín-García, *Signaling in the Heart*,
DOI 10.1007/978-1-4419-9461-5_15, 

## Triggers of Cardiac Hypertrophy

Cardiomyocyte hypertrophy occurs in response to many extrinsic and intrinsic stimuli that force augmented biomechanical stress. Although hemodynamic overload is usually the first stimulus to induce cardiomyocytes hypertrophy, neurohormones and cytokines also play an important role in its maintenance [12, 13]. Transduction of mechanical stress and other environmental signals is believed to occur through integrin proteins, transmembrane receptors that couple extracellular matrix components directly to the intracellular cytoskeleton and nucleus [14, 15] and are major players in the transmission of mechanical force through the plasma membrane and sensing the workload in cardiomyocytes (see below). In general, the signals for hypertrophy are mediated by a complex cascade of integrated signaling systems within the cardiomyocyte (Fig. 15.1), resulting in gene reprogramming [16]. Diverse neurohumoral signals acting through many interrelated signaling transduction pathways lead to pathological or maladaptive cardiac hypertrophy and HF [16]. In addition, activated hypertrophy-related genes induce synthesis of new contractile proteins which are organized into new sarcomeres. Then, extensive remodeling of the complete intracellular contractile apparatus occurs in hypertrophy and in the failing heart [17]; although, in specific instances, hypertrophy may functionally help the heart to adapt to increased workload.

## Promoters of the Hypertrophic Response

The hypertrophic process may be initiated by factors both extrinsic and intrinsic to the cardiac myocyte. Extrinsic stimuli include vasoactive peptides (e.g., angiotensin II and endothelin-1), adrenergic agonists (e.g., norepinephrine, epinephrine, and phenylephrine), activators of protein kinase C (PKC) (e.g., tumor-producing phorbol esters), peptide growth factors [e.g., insulin-like growth factor (IGF) and fibroblast growth factor (FGF)], cytokines (e.g., cardiotrophin-1), arachidonate metabolites (e.g., prostaglandin F2), mechanical stretch, and cell contact. Intrinsic stimuli include elevated $[Ca^{2+}]_i$, the heterotrimeric G protein $G_q$, as well as activated small G proteins, kinases, phosphatases, and numerous transcriptional factors [18, 19]. These extrinsic and intrinsic factors initiate a complex cascade of intracellular signaling pathways, termed the "hypertrophic response," which results in increased myocardial mass, altered spatial relationships between myocytes and other cellular and extracellular components of the myocardium, reprogramming of myocardial gene expression, and apoptosis. When overexpressed, an increasing number of genes have been found capable to elicit cardiac hypertrophy in transgenic mice; determination of which of these genes are clinically relevant in human hypertrophy is important.

In some cases of pathological (or maladaptive) hypertrophy, as in HCM, the stimuli for hypertrophy have not been

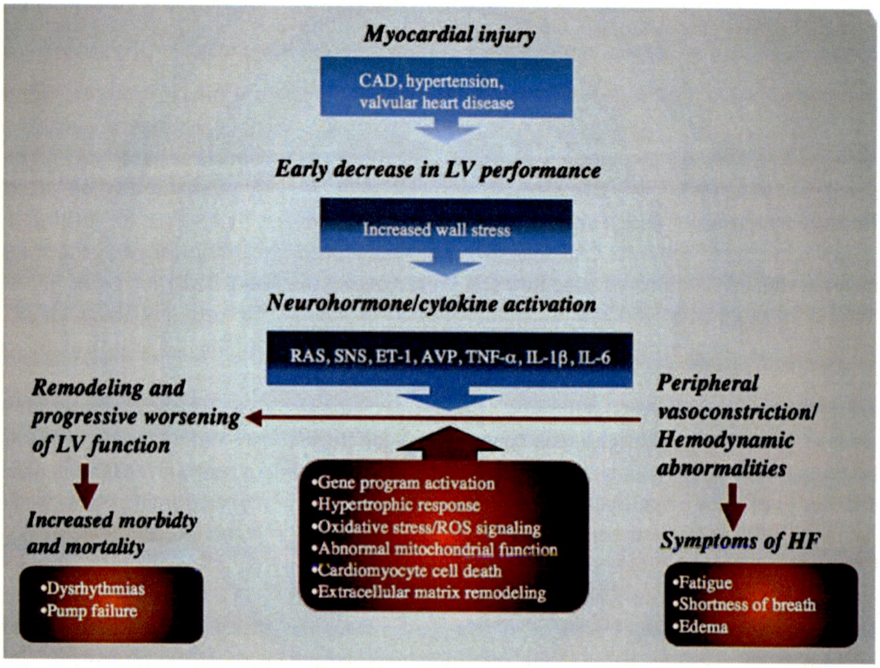


**Fig. 15.1** Chart model showing the participation of neurohormonal and cytokine activation in hypertrophy. *CAD* coronary artery disease, *LV* left ventricle, *RAAS* renin-angiotensin-aldosterone system, *SNS* sympathetic nervous system, *ET-1* endothelin-1, *TNF-α* tumor necrosis factor-α, *IL-1β* interleukin-1β

clearly identified. While some studies have shown that sarcomere protein gene mutations perturb sarcomere structure and/or function, the precise consequences of sarcomere protein gene mutations may differ according to the type of model studied, with both reduced and augmented motor function as described for individual HCM disease genes. However, a shared feature among these models is mechanical dysfunction of the sarcomere that also is a potential stimulus for hypertrophy.

While the molecular mechanisms underlying the proliferative growth of embryonic myocardium and adult cardiac myocyte hypertrophy *in vivo* remain mostly undetermined, considerable progress has been made using postgenomic analysis. This includes studies on the manipulation of murine genome together with mutational analysis of signaling and growth control pathways *in vivo*, and in cardiomyocytes grown *in vitro*, including the use of gene transfer/knock-out.

## Common Signaling Pathways in Physiological and Pathological Cardiomyocyte Growth

Physiological and pathological cardiomyocyte growth and/or proliferation are regulated by several signal transduction pathways. While in physiological hypertrophy signaling pathways behave in a relative state of equilibrium (homeostatic), in pathological hypertrophy cardiomyocyte signaling is rather disorganized, not functioning in a normal orderly way. In general, these pathways are redundant mechanisms that converge on one or several serine/threonine kinases. Several G protein-coupled receptors (GPCRs) such as β-adrenoceptors (β-ARs), angiotensin II receptors, and endothelin-1 receptors can activate these signaling cascades and induce changes in cell growth and proliferation. A general framework involves the following: signals received at the plasma membrane receptors are transmitted via GPCRs-G proteins-second messengers to a wide-spectrum of protein kinases and phosphatases, which become in turn activated (Fig. 15.2). These activated protein modifiers may lead to the activation and/or deactivation of specific transcription factors, which modulate specific gene expression affecting a broad spectrum of cellular events or they can directly target proteins involved in metabolic pathways, ion transport, $Ca^{2+}$ regulation, and handling, which affect contractility and excitability, as well as the pathways of cardiomyocyte apoptosis and/or cell survival.

## Second Messengers Signaling Pathways

A number of second messenger-producing intracellular signaling pathways behave as important transducers of the hypertrophic response, including α and β adrenergic receptor proteins, G protein isoforms, low-molecular-weight GTPases (Ras, RhoA, and Rac), kinases, phosphatases, calcineurin-calmodulin, and many others as we will see below.

### *Adrenergic Signaling*

Although neurohormonal signaling is comprehensively dealt with in Chap. 5; here it is suffice to say that the adrenergic receptors (ARs), members of the GPCR superfamily, interface between the sympathetic nervous system and the cardiovascular

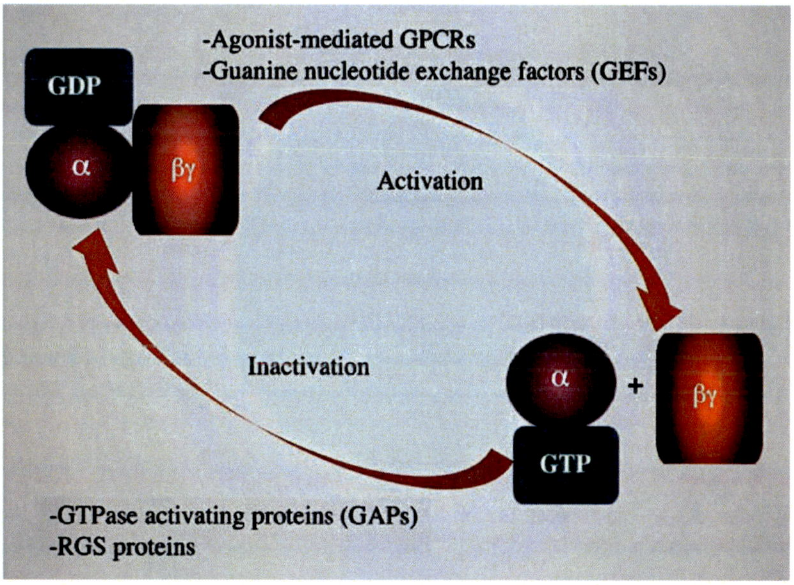


**Fig. 15.2** G protein-coupled receptors (GPCRs). Upon activation with the appropriate ligands, GPCRs are converted into the active conformation and are able to complex with and activate heterotrimeric G proteins

system, with integral roles in the rapid regulation of myocardial function. Chronic catecholamine stimulation of ARs has been associated with cardiac remodeling, cardiomyocyte apoptosis, and hypertrophy. Stimulation of β1-ARs (the predominant subtype found in both neonatal and adult ventricular myocytes) results in the activation of $G_s$ protein and adenylyl cyclase (AC) with increased intracellular cyclic AMP (cAMP), and activation of protein kinase A (PKA). This leads to the downstream phosphorylation of key target proteins, including L-type calcium channels, phospholamban, troponin I, and ryanodine receptors (RyRs) resulting in the modulation of cardiomyocyte growth and cardiac contractility in both neonatal and adult heart (Fig. 15.3).

In contrast, β2-ARs (which play a critical role in the inotropic support of the failing and aging heart) are linked to different signaling pathways in neonatal and adult ventricular myocytes. For instance, the pathway linking β2-ARs to cAMP-dependent changes in contractile function is only expressed in the neonate [20].

Among the protein targets for PKA are the β-ARs. Phosphorylation of the β-ARs promotes uncoupling and desensitization of the receptors. Phosphorylation via the β-adrenergic receptor kinase (βARK) also caused uncoupling and reduced the β-adrenergic responsiveness [21]. The uncoupling of the receptor is the prerequisite for receptor internalization, in which the receptor is translocated from the sarcolemmal membrane into cytosolic compartments. Chronic β-AR stimulation causes downregulation of the receptors. During this process of desensitization, the expression of the β-AR at the mRNA and protein level is reduced.

Members of the α1-adrenoceptor (α1-AR) family contain seven transmembrane spanning domains and are linked to G proteins. Upon stimulation with agonists such as noradrenaline and adrenaline, α1-ARs activate $G_q$ proteins and subsequently activate phospholipase Cβ (PLC-β) resulting in increased levels of second messengers, inositol trisphosphate (IP3) and diacylglycerol (DAG) which promote an increase in intracellular $Ca^{2+}$ levels, and PKC activation which modulate myocardial contractility [22]. However, activation of the α-ARs and the $G_q$-PLC-β-PKC pathway has limited acute effect on instantaneous myocardial contractility secondary to β-AR regulation, but rather serves as a potent stimulus for cardiac hypertrophy. Also, PKC-dependent activation of extracellular signal-regulated kinase (Erk) has also been implicated in the α1-AR-dependent stimulation of cardiomyocyte hypertrophy.

The normal heart contains a relatively small number of α1-ARs with the β/α-AR ratio of 10:1. On the other hand, in the hypertrophic and failing heart, with diminished β-AR levels, there is an increase of the α1-AR-to-β-AR ratio, suggesting that α1-AR may assume a greater functional role in the failing heart by providing a secondary inotropic system.

The α2-adrenoceptors (α2-ARs) are receptors to endogenous catecholamine agonists (e.g., norepinephrine and epinephrine) that mediate a number of physiological and pharmacological responses such as changes in blood pressure and heart rate. Three distinct subtypes, denoted as α2A-, α2B-, and α2C-AR, have been characterized and cloned [23]. Screening of human populations from various ethnic backgrounds has shown that α2-AR genes are polymorphic.

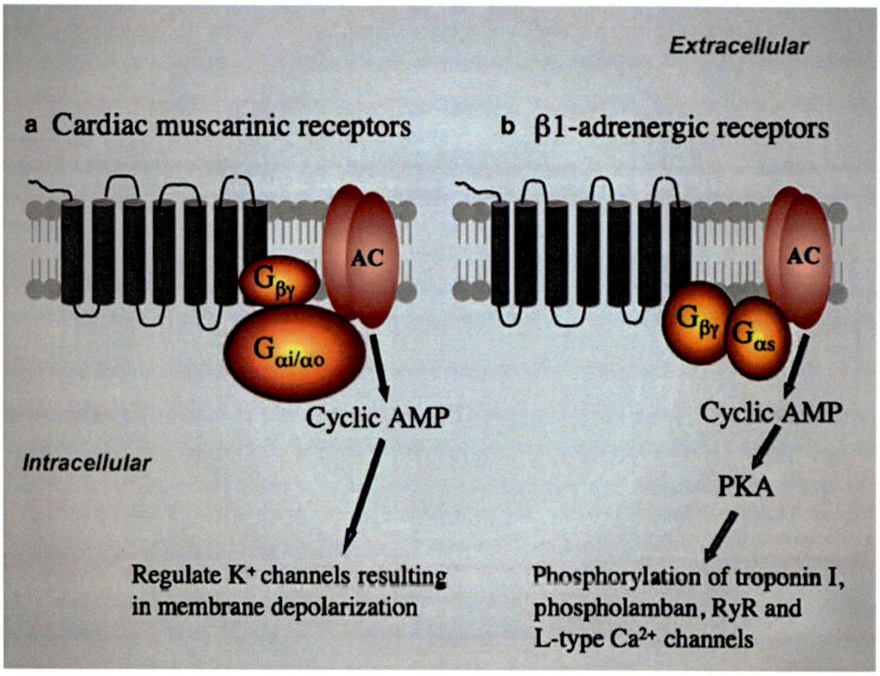


**Fig. 15.3** Stimulation of cardiac muscarinic and β1-adrenergic receptors. (**a**) The activation of muscarinic acetylcholine receptors which are coupled to $G_{i/o}$ proteins leads to a variety of cellular responses, including inhibition of adenylyl cyclase (AC) and modulation of $K^+$ channels. (**b**) β1-adrenoceptor-dependent activation of $G_s$, and AC with increased intracellular cAMP, and activation of protein kinase A are depicted. This pathway leads to the phosphorylation of key target proteins, including L-type calcium channels, phospholamban, troponin I, and ryanodine receptors (RyR)

Functional changes in G protein coupling, in agonist-promoted receptor phosphorylation and desensitization, have been found in heterologous systems such as CHO and COS-7 cells which express these genetic polymorphisms in comparison to wild-type receptors [24]. Moreover, in the failing heart, several changes occur in cardiac AR-regulated signal transduction pathways, the most striking of which include downregulation of β1-AR number and function, uncoupling of β-ARs, and increased activity of the inhibitory G protein, $G_i$. Most of these changes appear to be related to increased activity of the adrenergic nervous system, i.e., increased exposure to norepinephrine. In addition, β-AR desensitization and uncoupling are further induced by HF-induced increases in the level of G protein-coupled receptor kinases (GRKs) [25]. Increased myocardial levels of the most abundant GRK, GRK2 (also known as βARK-1), precede HF development in several animal models and are present in HF patients, and may be a key factor in the transition from compensatory cardiac hypertrophy to overt HF [26]. Moreover, it has been suggested that components of the AC-PKA pathway are sensitized by chronically increased βARK-1 activity that may increase mouse cardiomyocyte contractile function in the absence of exogenous agonist. Changes in contractile function in the failing heart may reflect, in part, the effects of chronic upregulation of βARK-1 on the cAMP-PKA pathway [27].

## Muscarinic Receptors

Muscarinic acetylcholine receptors (mAChRs) are G protein-coupled acetylcholine receptors that mediate a variety of cellular responses, including inhibition of AC (Fig. 15.3), modulation of $K^+$ channels, and increased phosphoinositide breakdown [22]. These diverse effects of mAChR activation elicit both negative and positive inotropic and chronotropic effects in the heart. Positive inotropic effects of cholinergic agonists are present only at high agonist concentration (>10 μmol/L) and tend to be pertussis toxin (PTX)-insensitive in contrast to the negative inotropic effects observed at lower agonist concentrations, which are sensitive to inactivation by PTX. These dual effects of mAChR activation in the heart may be a result of the presence of multiple subtypes of mAChRs [23]. Thus far, five mAChR subtypes (M1–M5) have been identified, and each subtype is encoded by a different gene. The mAChR proteins contain seven transmembrane spanning domains and are coupled to G proteins of the $G_i$ and $G_q$ families to inhibit AC and activate PLC, respectively. During compensatory cardiac hypertrophy in the rat, hemodynamic overload induced a parallel decrease in the densities of both β1-ARs and M2-mAChRs in the left ventricle, although the total number of receptors remained unchanged [28]. M2-mAChRs have been considered to be the only functional mAChRs in the myocardium; however, some observations revealed that M3-mAChRs are also present in the hearts of various species [29]. Furthermore, stimulation of mAChR results in the activation of an inward rectifier $K^+$ current termed $I_{KACh}$ in cardiac myocytes, primarily mediated by the M2 subtype of mAChR. However, a novel delayed rectifier-like $K^+$ current designated $I_{KM3}$ has been recently identified which is distinct from $I_{KACh}$ and other known $K^+$ currents and which is mediated by the activation of the cardiac M3-mAChRs [30]. While $I_{KACh}$ is known to be a $G_i$ protein-gated $K^+$ channel, $I_{KM3}$ represents the first $G_q$ protein-coupled $K^+$ channel described in cardiomyocytes.

## Cyclic GMP

In cardiac hypertrophy, myocardial $O_2$ balance is essentially normal [31]. However, during increased work, this balance can be more adversely affected than in normal hearts [32]. One of the key responses to increased cardiac needs is an enhanced sympathetic activity and increases in myocardial cyclic AMP. However, with most forms of pressure-overload cardiac hypertrophy, usually, there is a loss of β-ARs, a reduction in muscarinic receptor levels [33, 34], and also increases in cyclic GMP (cGMP) [35, 36]. This increase in cyclic GMP may be a protective response to increases in cyclic AMP. Furthermore, the interaction between cyclic AMP and cyclic GMP appears to be altered in pressure-overload hypertrophy [37], and the ability of the heart to respond to stresses is reduced under these circumstances.

The significance and actions of natriuretic peptides (NPs) and cyclic GMP in cardiac hypertrophy and failure have been of great research interest, and circulating levels of cyclic GMP and NP can be used as an index of the degree in human HF [35, 36, 38]. Also, increased levels of cyclic GMP have been reported in experimental models of hypertrophy and HF [39–41]. These changes may be a protective response secondary to the increased levels of cyclic AMP and sympathetic activity in HF and hypertrophy. In addition, after cardiac hypertrophy, there are changes in the ability of cyclic GMP that affect myocardial function [42].

Understanding the effect of cyclic GMP is important in the management of hypertrophy and HF [43–46]. In experimental models of cardiac hypertrophy, increases in the level of nitric oxide (NO) or NPs may prove to be beneficial [44, 47–49]. While blockade of cyclic GMP production can cause cardiac hypertrophy [50], increasing the level of atrial natriuretic peptide (ANP) or cyclic GMP has been reported to lead to regression of pressure-overload cardiac hypertrophy in mice [51, 52]. While changes in the ability of NO and NPs to affect function in the failing cardiac myocytes occur, it is possible that control of the NO-NP-cyclic GMP signal transduction system will be of significant benefit in the treatment of decompensated hypertrophy.

Loss of the ability of cyclic GMP-dependent protein kinase (PKG) to affect myocardial function and phosphorylate proteins may occur during decompensated cardiac hypertrophy [53, 54]. Therefore, in spite of high levels of ANP and cyclic GMP, this signaling pathway system's effects are significantly depressed in decompensated cardiac hypertrophy [53, 55]. This would affect the system's ability to increase phosphorylation of L-type calcium channels, RyRs, and phospholamban, and may cause significant loss of the downstream signaling and function of the system. Also the functions of the cyclic GMP-dependent cyclic AMP phosphodiesterases are decreased in decompensated cardiac hypertrophy [56], and the role of cyclic GMP in the control of the level of cyclic AMP is reduced. Losses of downstream effects in this system may partly explain the higher levels of this second messenger in HF. Understanding the role that cyclic GMP signal transduction system plays in the development of HF will be of great significance to develop new therapies. Potential therapeutic approaches to improve cardiac myocyte function in the failing heart may include changes in the guanylyl cyclase (GC), cyclic GMP-dependent protein kinase, or the cyclic GMP-dependent cyclic AMP phosphodiesterases.

## Endothelin

A comprehensive discussion on Endothelin-1 (ET-1) is presented in Chap. 4. Here, it suffices to say that three ET signaling peptides have been identified: ET-1, ET-2, and ET-3 with well-established effects on cardiomyocyte, including modulation of contractile function and growth stimulation [57]. ET-1 binds to the ET (A) receptor on the cell surface. ET (A) receptor is coupled to the $G_q$ class of GTP binding proteins and stimulates hydrolysis of phosphatidylinositol 4,5-bisphosphate to DAG and IP3. DAG remains in the plane of the membrane causing translocation and activation of PKCδ- and ε-isoforms. This is followed by the activation of small G protein Ras and an Erk cascade. Over a longer time course, two protein kinase cascades related to the Erk1/2 cascade, the c-Jun N-terminal kinase (JNK) and p38 MAP kinase cascades, also become activated. Downstream activation of nuclear transcription factors (e.g., GATA-4 and c-Jun), protein kinases (e.g., 90-kDa ribosomal protein S6 kinase and MAPK-activated protein kinase 2), and ion exchangers/channels (e.g., the $Na^+/H^+$ exchanger 1) follows JNK and p38 activation. These changes are responsible for the overall biological effects of ET isopeptides on the cardiomyocyte [58].

## Angiotensin

Angiotensin-converting enzyme (ACE), a central element of the renin-angiotensin system, converts the decapeptide angiotensin I to the potent pressor octapeptide angiotensin II (Ang II), mediating peripheral vascular tone as well as glomerular filtration in the kidney (see Chaps. 4 and 13). In addition to its direct effect on blood flow, Ang II directly causes changes in cell phenotype, cell growth and apoptosis, and regulates gene expression of a broad range of bioactive molecules (e.g., vasoactive hormones, growth factors, extracellular matrix components, and cytokines). In addition, Ang II activates multiple intracellular signaling cascades, involving numerous transduction components such as MAP kinases, tyrosine kinases and various transcription factors in cardiomyocytes, fibroblasts, vascular endothelial cells, smooth muscle cells (SMCs), and renal cells [59]. Ang II also promotes increasing cardiomyocyte size and protein synthesis, as well as hypertrophy-associated alterations in the cardiac gene expression program through specific cellular receptor subtypes, AT1 and AT2. AT1 is the more abundant in the adult heart and has been linked to control both hypertrophy and apoptosis in the cardiomyocyte. While several studies have found AT2 receptor to act in opposition to AT1, the myocardial AT2 receptor functioning is less well understood. Through the $G_q$ proteins, the AT1 receptor is coupled to a variety of intracellular signals, including the generation of oxygen free radicals, the activation of Ras, and the Erk-MAPK protein kinase family [60]. Ang II activates NF-κB-dependent transcription in other cell types, likely through its effects on the cellular redox state [61]. These actions contribute to the pathophysiology of cardiac hypertrophy and remodeling, HF, vascular thickening, and atherosclerosis [62].

## Growth Factors

The Erk1/2 are activated in cardiomyocytes by $G_q$ protein-coupled receptors and are associated with the induction of hypertrophy. In primary cardiomyocyte cultures, platelet-derived growth factor (PDGF), epidermal growth factor (EGF), and FGF promoted receptor-coupled Erk1 activation and significantly increased cardiomyocyte size in contrast to insulin, insulin-like growth factor-1 (IGF-1), and nerve growth factor (NGF), which had little effect. Peptide growth factors activate phospholipase Cγ1 (PLC-γ1) and PKC. In cardiomyocytes, only PDGF stimulated tyrosine phosphorylation of PLC-γ1 and PKCδ. Furthermore, activation of Erk1/2 by PDGF, but not EGF, required PKC activity. In contrast, EGF substantially increased Ras-GTP with rapid activation of c-Raf, whereas stimulation of Ras-GTP loading by PDGF was minimal, and activation of c-Raf was delayed suggesting differential coupling of PDGF and EGF receptors to the Erk1/2 cascade [63].

## Protease-Activated Receptors

A novel class of protease-activated receptors (PAR-1, PAR-2, PAR-3, and PAR-4), containing seven transmembrane spanning domains and G protein-coupled, has been identified. These receptors are activated by cleavage with serine proteases such as thrombin and trypsin [64, 65]. Extracellular proteolytic activation of protease-activated receptors results in: (1) cleavage of specific sites in the extracellular domain; (2) formation of a new N-terminus (often containing the sequence SFLLRN) which functions as a tethered ligand and binds to an exposed site in the second transmembrane loop triggering G protein binding; (3) intracellular signaling.

In cardiomyocytes expressing PAR-1, a high-affinity receptor for thrombin, agonist binding, and activation of PAR leads to IP3 accumulation, stimulation of Erk, and modulated contractile function. Coexpression in cardiomyocytes of PAR-2, activated by trypsin/tryptase but not thrombin, with PAR-1 leads to a more extensive signaling response including IP3 accumulation, stimulation of MAP kinases (both Erk and p38-MAP kinase), elevated $Ca^{2+}$ levels and contractile function as well as the activation of JNK and Akt associated with growth and/or survival pathways, and induction of both cardiomyocyte hypertrophy and hyperplasia [65]. Furthermore, PAR-1, PAR-2, and PAR-4 have been implicated in vascular development, as well as in a variety of other biological processes, including apoptosis and remodeling [66].

## G Proteins

Upon activation with the appropriate ligands, GPCRs are converted into the active conformation and are able to complex with and activate heterotrimeric G proteins [67]. The heterotrimeric G proteins are composed of three subunits: the α subunit which carries the guanine-nucleotide binding site, and the β and γ subunits which form a tightly bound dimer. Inactive G proteins are heterotrimers composed of a GDP-bound α subunit associated with the $G_{\beta\gamma}$ dimer, which serves to anchor the heterotrimeric G protein to the membrane. The activated GPCRs function as GDP/GTP exchange factors and promote the release of GDP and the binding of GTP to the α subunits leading to dissociation of the α subunit and the $G_{\beta\gamma}$ dimer. Both GTP-$G_{\alpha}$ and $G_{\beta\gamma}$ can interact with a variety of effectors such as AC and PLC in order to modulate cellular signaling pathways. The deactivation of GPCR signaling occurs at several levels. Importantly, the $G_{\alpha}$ subunit has an innate GTPase activity which hydrolyses GTP to GDP and promotes reassociation with $G_{\beta\gamma}$ to form the inactive heterotrimer. In addition, ligand dissociation from the GPCRs converts the receptors back to their inactive state (see Fig. 15.2).

Heterotrimeric G proteins are classified into subclasses according to the α subunit, with each subfamily designated by its corresponding downstream signaling effect. The $G_{\alpha s}$ or more simply $G_s$ proteins are stimulatory regulators of AC linking ligand stimulation (e.g., β-AR) to the accumulation of the second messenger. The $G_{\alpha s}$ protein is a target of covalent modification by cholera toxin (CTX) which slows GTP hydrolysis, locking $G_{\alpha s}$ in an active GTP-bound form that constitutively stimulates AC. In contrast, the $G_{\alpha i/o}$ or $G_{i/o}$ proteins inhibit AC activity. These proteins are targets for ADP-ribosylation by the PTX which prevents their interaction with receptors and inhibits their downstream signaling. The other two subfamilies, $G_q$- and $G_{12/13}$ proteins, are insensitive to PTX and CTX. The GTP-bound $G_{\alpha q}$ protein activates PLC-β, leading to generation of IP3 and DAG, by mobilization of calcium and the activation of PKC, respectively. Dissociated $G_{\beta\gamma}$ can activate small GTP binding protein Ras and initiate a tyrosine kinase cascade leading to the activation of MAPK. Furthermore, $G_q$ may activate MAPK independently of $G_{\beta\gamma}$ via a mechanism that is PKC dependent [68].

Hormones such as angiotensin II, endothelin 1, and norepinephrine which bind and activate cardiomyocyte membrane receptors coupled to the $G_q$ proteins have been implicated in the development and finally decompensated cardiac hypertrophy [69].

Regulators of G protein signaling (RGS) proteins are a family of proteins that accelerate intrinsic GTP hydrolysis on α subunits of heterotrimeric G proteins [70]. They play crucial roles in the physiological regulation of G protein-mediated cell signaling. In addition, the small G proteins are a superfamily of guanine-nucleotide binding proteins with a size ranging from 20 to 25 kDa including several subfamilies such as Ras, Rho, Rab, Ran, and ADP-ribosylation factor(s). These small G proteins behave as molecular switches regulating numerous cellular responses including cardiac myocyte hypertrophy and cell survival associated with cell growth and division, multiple changes in the cytoskeleton, vesicular transport, and myofibrillar apparatus. They share some features with the heterotrimeric G proteins including activation by the exchange of GDP to GTP and inactivation by their return to a GDP-bound state which is enhanced by GTPase activating proteins. Not surprisingly, there are regions of homology shared between these proteins and the $G_{\alpha}$ subunit. Modification of these proteins by isoprenylation (addition of hydrophobic molecules) promotes their attachment to the membrane. However, the activation of the small G proteins differs from that of the heterotrimeric G proteins in one critical respect. With the heterotrimeric G proteins, ligand binding to a GPCR is the primary stimulus that promotes GDP release from, and GTP binding to the α subunit, whereas an association of agonist-occupied receptors with small G proteins (e.g., Ras and Rho) is not found. Instead, activation by the release of GDP from the small G proteins is primarily mediated by the activation of guanine-nucleotide exchange factors (GEFs).

Hearts from transgenic mice expressing activated Ras develop features consistent with myocardial hypertrophy, whereas mice overexpressing RhoA develop lethal HF. In isolated neonatal rat cardiac myocytes, transfection or infection with activated Ras, RhoA, or Rac1 induces features of myocardial hypertrophy.

The G proteins and second messenger pathways function differently in cardiac fibroblasts than those in cardiac myocytes. Cardiac fibroblasts are an important cellular component of the myocardial responses to injury and to hypertrophic stimuli. In cardiac fibroblasts, agonists such as bradykinin stimulate IP3 production and increased intracellular $Ca^{2+}$ levels, while endothelin-1 and norepinephrine do not, in contrast to their action in cardiac myocytes. Cardiac fibroblasts express functional G protein-linked receptors that couple to $G_q$ and $G_s$, with little or no coupling to $G_i$. The expression of receptors and their coupling to $G_q$ but not to $G_i$-linked responses distinguishes the signaling in cardiac fibroblasts from that in myocytes. Furthermore, agonists that activate $G_q$ in fibroblasts also potentiate the stimulation of $G_s$, an example of signaling cross talk not previously observed in adult cardiomyocytes [71].

## Cyclin Signaling

A discussion on cyclin signaling pathways has already been presented in Chap. 2. In this chapter, it suffices to mention that cell-cycle control can be mediated by p38 MAP kinase activity, which regulates the expression of genes required for mitosis in cardiomyocytes including cyclin A and cyclin B. Cardiac-specific p38 MAPK knock-out mice show a 92% increase in neonatal cardiomyocyte mitosis. Furthermore, inhibition of p38 MAPK promotes cytokinesis in adult cardiomyocytes [72].

Moreover, cyclin D1, a cell-cycle regulator involved in promoting the $G_1$-to-S phase progression via phosphorylation of the retinoblastoma (Rb) protein, is localized in the nucleus of fetal cardiomyocytes but is primarily cytoplasmic in neonatal and adult cardiomyocytes (concomitant with Rb underphosphorylation). Ectopic expression of a variant of cyclin D1 equipped with nuclear localization signals dramatically promoted neonatal cardiomyocyte proliferation and Rb phosphorylation [73].

## Adenylyl Cyclase

The amount of AC sets a limit on cardiac β-adrenergic signaling *in vivo*, and increased AC, independent of β-AR number and G protein content, provides a means to regulate cardiac responsiveness to β-AR stimulation [74]. Overexpress an effector, such as AC does not alter transmembrane signaling except when receptors are activated, in contrast to receptor/G protein overexpression, which promotes continuous activation with detrimental consequences [75, 76]. These data suggest AC overexpression as a novel target for safely increasing cardiac responsiveness to β-AR stimulation [74].

Expression of type AC isoform is restricted to the heart and brain and generates the major AC isoform found in the adult heart. Type AC is potently activated through PKC-mediated phosphorylation. The degree of this activation is greater than that achieved by forskolin, the most potent AC agonist. Furthermore, the two PKC isoenzymes are additive in their capacity to activate AC. In contrast, PKA-mediated phosphorylation inhibits type AC. Thus, type AC is subject to dual regulation by phosphorylation: activation by PKC and inhibition by PKA, mediated via phosphorylation at unique residues within the type AC5 molecule [77].

PKA-mediated inactivation of AC creates a feedback system within the cAMP-signaling pathway, analogous to PKC-mediated inhibition of the phospholipase C pathway. Catecholamine stimulation in the heart activates both the phospholipase C-PKC pathway via α-adrenoreceptors and the AC-PKA pathway via β-adrenergic receptors. Dual regulation of AC by PKC and PKA may play a major role in integrating these two principal signal transduction pathways, modulating neuronal and hormonal input to the heart.

## Phospholipase C

Diverse and distinct hormonal stimuli engage specific surface receptors of the cardiomyocyte to initiate the hydrolysis of inositol phospholipids, mediated by the effector PLC, while changes in intracellular levels of IP3 and inositol 1,3,4,5-tetrakisphosphate, DAG, and $Ca^{2+}$ result in the specific phosphorylation of cellular proteins by various protein kinases such as the PKC family, $Ca^{2+}$-calmodulin-dependent kinase, and MAPK. Four classes of PLC isozymes are considered to underlie these signaling responses [78].

A myriad of seven transmembrane spanning receptors activate isozymes of the PLC-β class through the release of α-subunits of the $G_q$ family of heterotrimeric G proteins. A subset of PLC-β isozymes can also be activated by $G_{\beta\gamma}$. PLC-γ isozymes are activated by protein phosphorylation following the activation of receptor tyrosine kinases (RTKs). Another class of PLC isozymes, PLC-ε, has been found to be involved in signaling [79] and exhibits a novel pattern of regulation mediated by the Ras oncoprotein and $G_{\alpha 12}$-subunit of heterotrimeric G protein.

### Nitric Oxide

Among endothelium-released factors, NO exerts multifactorial effects on various cell types in the heart and may play a role in growth of the vasculature and myocardial hypertrophy [80]. Changes in NO and its myocardial effects are linked to hypertrophy and HF [81, 82], where significant endothelial dysfunction and reduced NO levels may be present [83]. Both NO and NPs increase myocardial cyclic GMP levels. NO and NPs have emerged as potent endogenous inhibitors of maladaptive hypertrophy signaling. Overall, it appears that cardiac hypertrophy is controlled by an interplay of pro- and antihypertrophic signaling networks. This delicate balance can tip toward adaptation or heart failure. In the future, patients living with cardiac disease may benefit from therapeutic strategies targeting maladaptive hypertrophy signaling pathways [84].

## Kinases and Phosphatases

### Protein Kinase A

This enzyme is structurally organized as a heterotetramer composed of two regulatory (R) subunits that upon binding the two catalytic (C) subunits maintain the overall complex in a dormant state. The binding of two cyclic AMP molecules to tandem sites on each R subunit results in the release of the C subunits and the activation of their enzymatic activity. The dissociated C subunits phosphorylate serine or threonine residues on target proteins in the nucleus and cytoplasm leading to changes in cardiomyocyte metabolism, ion channel function, growth, and gene expression. The catalytic subunits are encoded by three different genes (Cα, Cβ, and Cγ), while the regulatory subunits are encoded by four genes (RIα, RIβ, RIIα, and RIIβ). The regulatory subunit contains an N-terminal dimerization domain, an autophosphorylation site that also comprises the primary site for catalytic subunit binding, and two tandem cAMP binding sites.

Compartmentalization of these enzymes can be achieved through association with anchoring or adaptor proteins that target them to subcellular organelles or tether them directly to target substrates via protein–protein interactions. Specific PKA anchoring proteins (AKAPs) serve as important regulators of PKA function and signaling by directing the subcellular localization of PKA, by binding to its R subunits, in effect concentrating PKA at specific intracellular locations. Using a variety of experimental approaches, including yeast two-hybrid screening, proteomic analysis and interaction cloning, two major anchoring proteins for PKA, MAP2, and AKAP75, and over 13 different AKAPs have been found in the heart [84]. Targeting of AKAPs to specific sites within the cell is governed by sequences in the AKAP. Interestingly, despite their diverse structure, the AKAPs all contain an amphipathic helical region of 14–18 amino acids that bind to the N-terminus of the RII subunit, underlying their interaction with PKA. Besides PKA, AKAPs also interact with other signaling components, including phosphodiesterase inhibitors, phosphatases, and PKA substrates [85]. Myocyte AKAPs have been identified in association with specific plasma membrane ion channels (e.g., hKCNQ1 and the L-type $Ca^{2+}$ channel), the β-AR complex, and the sodium-calcium exchanger (NCX1), in association with RyR at both the sarcoplasmic reticulum (SR) and T-tubule junction. In addition, AKAPs have been found on the nuclear membrane and in association with the mitochondrial outer membrane. The precise functional role of a number of the identified AKAPs in the cardiac myocyte has not yet been established.

Also a novel role for leucine zipper motifs in targeting kinases and phosphatases via anchoring proteins has been identified. Several cardiac ion channels contain a domain to anchor phosphorylation modulatory proteins to the channel, essentially allowing the formation of a scaffolding structure for regulatory proteins. Ion channels such as the RyR, the dihydropyridine receptor (L-type $Ca^{2+}$ channel), and the delayed rectifier $K^+$ channel subunit KCNQ1, all contain a modified leucine zipper termed as LIZ (leucine/isoleucine zipper) which promotes protein–protein interaction and protein oligomerization [86].

### Protein Kinase B (PKB/Akt) and Phosphoinositide 3-Kinase

Key signaling pathways in cardiomyocyte proliferative growth and cell survival involve phosphoinositide 3-kinase (PI3K) and protein kinase B (PKB), also known as Akt. In the heart, Akt contributes to both pathological and physiological cardiac growth, myocyte survival, and contractile function [87]. Activation of Akt in transgenic models induced cardiac hypertrophy primarily by increasing the size of cardiomyocytes [88]. In addition, Akt expression confers protection from ischemia-induced cell death and cardiac dysfunction.

The class I PI3Ks phosphorylate PI $(4,5)P_2$ on the 3′ position of the inositol ring to form PI $(3,4,5)P_3$ otherwise termed as PIP3 which can be mediated through RTK or GPCR signaling. Class IA PI3Ks are composed of a p110 catalytic subunit (either α, β, or γ) in association with a p85 regulatory protein [89]. Activation of PI3K is promoted by the direct interaction of p85 with specific phosphorylated tyrosine residues present on growth factor receptors or by Ras activation. Another PI3K type, class IB PI3K, contains a p110γ catalytic subunit in association with a p101 adaptor molecule, and its activation proceeds via interaction of the γ isoform of the p110 catalytic subunit with heterotrimeric G protein βγ subunits, resulting from GPCR stimulation.

As a highly potent signaling lipid involved in a multiplicity of cellular processes, the levels of PIP3 are tightly regulated not only at the level of synthesis by PI3K activity, but also by rapid dephosphorylation by lipid phosphatases, PTEN, and SHP enzymes (for further discussion on lipid signaling, see Chap. 7). Downstream of the PIP3 production at the plasma membrane, phosphoinositide-dependent kinase 1 (PDK-1) becomes activated in part by its translocation to the plasma membrane in proximity to its substrates, which include Akt (PKB) (Fig. 15.4).

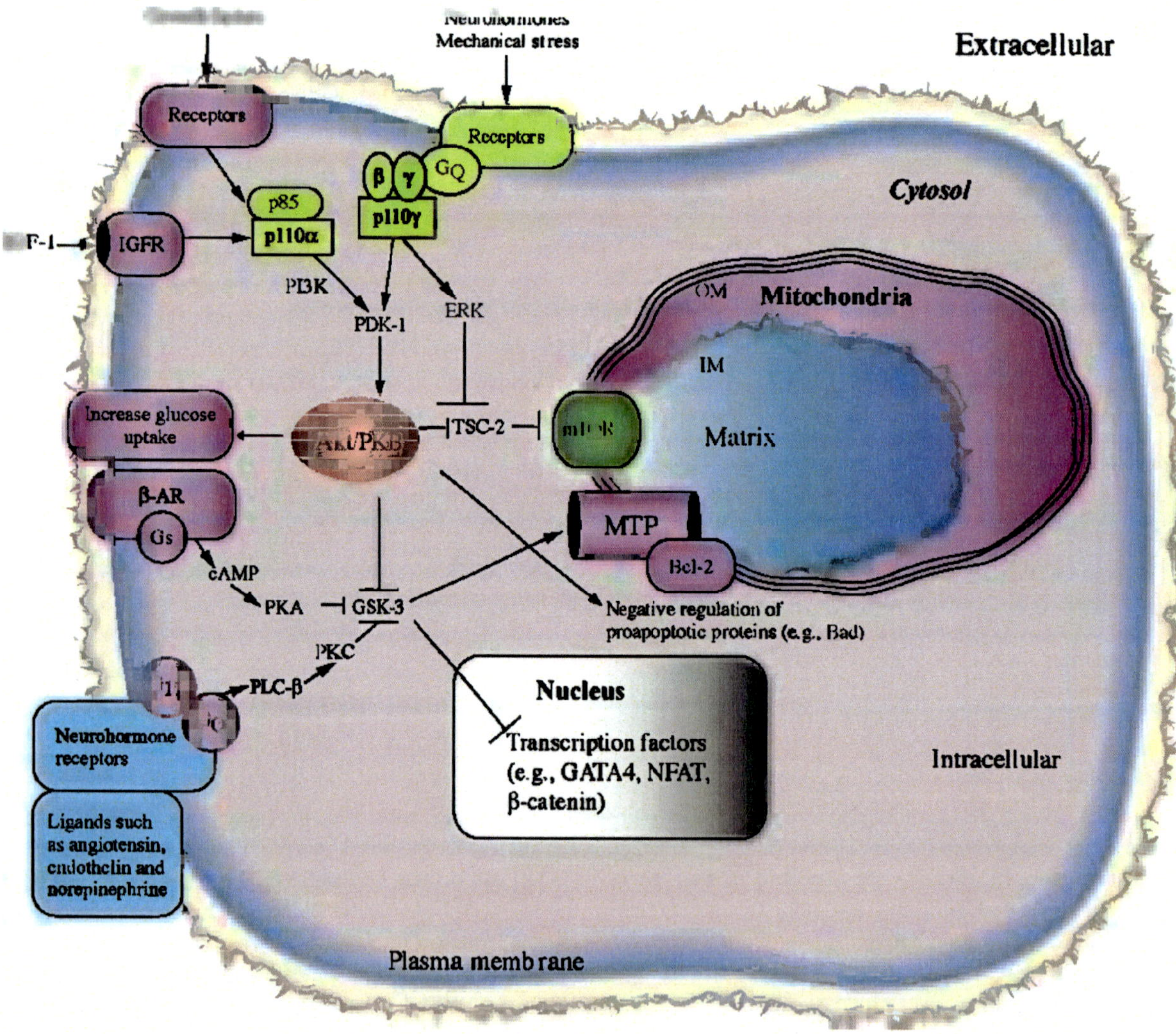


**Fig. 15.4** Akt signaling. Akt is positioned at a signaling cascade branch point. One branch leads to mammalian target of rapamycin (mTOR) and the activation of the protein synthetic machinery, which is essential for all forms of hypertrophy. Downstream effects of Akt kinase activity include changes in myocardial bioenergetic substrates, affected by increasing glucose uptake and by downregulating FAO metabolism via direct effect on transcription regulators (e.g., PPAR-α and RXR-α). It may also impact mitochondrial OXPHOS activities, which decline in parallel with increased Akt activity. Both mTOR and Akt modulate cytoplasmic protein synthesis by activation of translation initiation factor (IF) and of ribosomal proteins (not shown). Also depicted are peptide growth factors (IGF) and other downstream targets of mTOR. See text for details

Akt represents a family of serine and threonine kinases which includes Akt1, Akt2, and Akt3 encoded by three distinct genetic *loci* with extensive homology (approximately 80%) [90]. There are considerable differences in both the expression and function of the Akt isoforms with only Akt1 and Akt2 being highly expressed in the heart. All three Akt isoforms contain a kinase domain (with structural homology to PKA and PKC), which contains the primary site ($Thr^{308}$) of phosphorylation by PDK-1. Although $Thr^{308}$ phosphorylation partially activates Akt, subsequent phosphorylation of Akt at a C-terminal site, $Ser^{473}$ is required for its full activation; under some conditions, this phosphorylation may be produced by PDK-1, another kinase (PDK-2), or by autophosphorylation.

The stimuli that result in Akt activation in the heart are shown in Table 15.1. Upon activation, myocardial Akt can phosphorylate a number of downstream targets, including cardioprotective factors involved in glucose and mitochondrial metabolism, apoptosis, and regulators of protein synthesis, as discussed below. The regulation of Akt is also achieved by its dephosphorylation brought about by the protein phosphatase PP2A. Inhibitors of PP2A activity, including okadaic acid, increase Akt activity, while ceramide, which is involved in enhanced apoptotic signaling, stimulates PP2A. In the heart, Akt signaling has a pronounced antiapoptotic effect, significantly increases cardiomyocyte growth, and enhanced function.

Apoptotic progression of hypoxic cardiomyocytes is abrogated with IGF-1 treatment, which activates PI3K and Akt [93]. Overexpression of constitutively active transgenes, either PI3K or Akt, in cultured hypoxic cardiomyocytes reduces apoptosis. Moreover, after transient ischemia *in vivo*, gene transfer of constitutively active Akt to the heart resulted in reduced apoptosis and infarct size [103]. Several mechanisms and/or effectors by which the PI3K-Akt pathway stems apoptosis have been identified, and it appears likely that these effectors might be enhanced when applied in combination. These include the phosphorylation and inactivation of the proapoptotic protein Bad, NF-κB activation, enhanced NO release, endothelial NO Synthase (eNOS) activation, changes in the mitochondrial membrane pores, and membrane potential suppressing apoptotic progression, as well as cytochrome *c* release induced by several proapoptotic proteins. Recent observations have also shown that Akt's effect on apoptosis and cell survival is mediated by its phosphorylation of the Forkhead transcription factor, FOXO3, that in turn reduces the transcription of specific proapoptotic molecules [104].

**Table 15.1** Stimuli activating myocardial Akt

| Stimuli | References |
|---|---|
| Insulin | [91, 92] |
| IGF-1 | [93, 94] |
| Cardiotrophin-1 | [95] |
| LIF | [96] |
| β-AR agonists | [88] |
| Angiotensin II | [97] |
| Endothelin-1 | [98] |
| Acetylcholine | [99] |
| Adrenomodulin | [100] |
| Pressure overload | [101] |
| Ischemia (hypoxia) | [102] |

Both PI3K and Akt can modulate cardiomyocyte hypertrophy with significant effects on both cell and organ size. Transgenic mice with cardiac-specific overexpression of either the constitutively active or dominant-negative alleles of PI3K exhibited an increase or decrease in cardiomyocyte size, respectively [105]. The constitutive activation of PI3K led to adaptive hypertrophy and did not change into a maladaptive hypertrophy, consistent with a critical role for the PI3K-PDK-1-Akt pathway in regulation of normal cardiac growth. In addition, Akt expression confers protection from ischemia-induced cell death and cardiac dysfunction. Furthermore, Akt1 knock-out mice weigh approximately 20% less than wild-type littermates and have a proportional reduction in the size of all somatic tissues, including the heart. In contrast, Akt2 knock-out mice have only a modest reduction in organ size. Thus, data from the available Akt knock-out models support a critical role specifically for Akt1 in the normal growth of the heart. On the other hand, Akt1/Akt2 double-knock-out mice suffer marked growth deficiency and striking defect in cell proliferation.

Interestingly, activation of Akt in these transgenic models induced cardiac hypertrophy primarily by increasing the size of cardiomyocytes [106]. It is noteworthy that none of the transgenic models in which PI3K or Akt activation were associated with increased heart size demonstrated increased cardiomyocyte proliferation, in contrast to findings in transgenic mice with cardiac-specific expression of IGF-1. This suggests that PI3K activation is not sufficient to induce cardiomyocyte proliferation, which likely involves the coordination of other signaling pathways downstream of IGF-1 and upstream of PI3K.

Akt is positioned at a signaling cascade branch point (Fig. 15.5) [89]. One branch leads to mammalian target of rapamycin (mTOR) and the activation of the protein synthetic machinery, which is essential for all forms of hypertrophy. It has been reported that insulin rapidly activates the 70-kDa ribosomal protein S6 kinase (p70S6k), and this effect is inhibited by both rapamycin and inhibitors of PI3K [107]. Peptide growth factors (e.g., growth hormone [GH] and IGF) are primary activators of mTOR in mammalian cells and activate mTOR primarily via Akt. Interestingly, downstream activation of p70S6k is mediated by a signaling pathway involving mTOR, a molecule that responds to the nutritional status and amino acid availability, and is centrally involved in cell growth and proliferation. One downstream target of

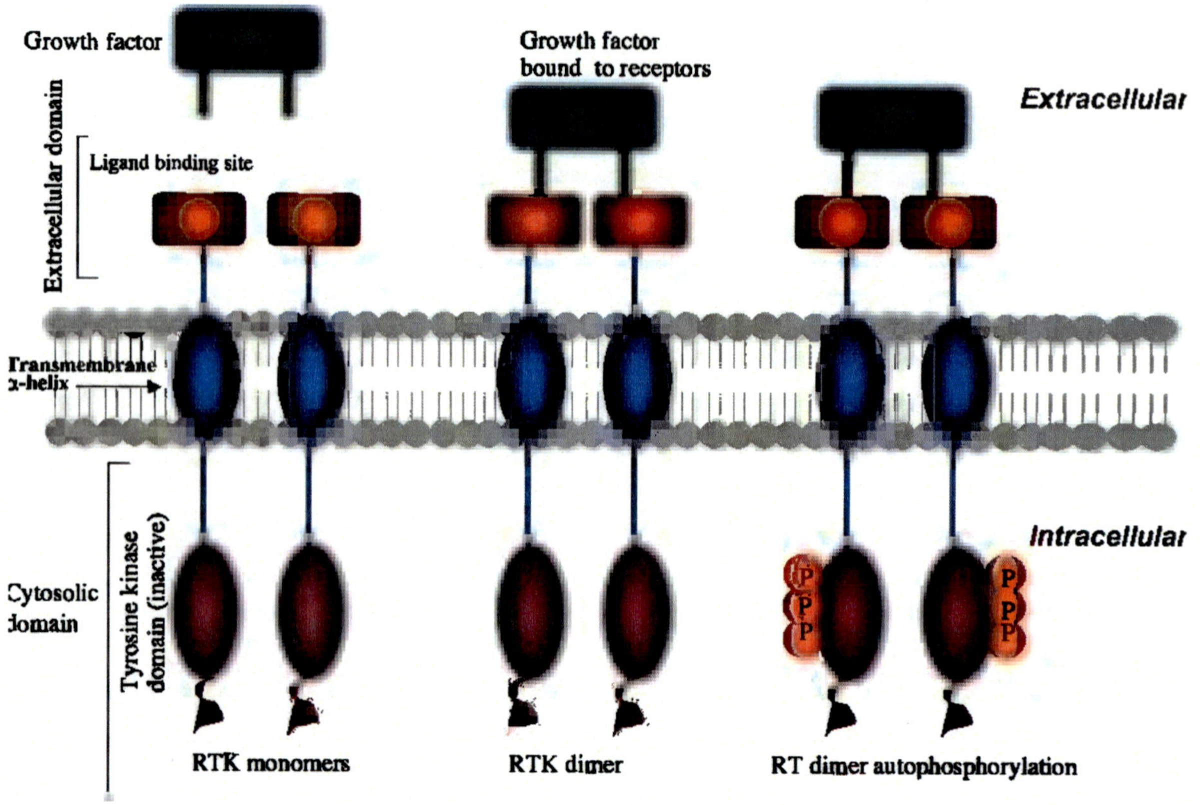


**Fig. 15.5** Receptor tyrosine kinases. Ligand binding to the extracellular portion of these receptors results in receptor dimerization, which facilitates the transautophosphorylation of specific tyrosine residues in the highly conserved cytoplasmic portion. RTKs are phosphorylated in response to stimulation by cytokines, cell adhesion, and stress stimuli

mTOR signaling is the 4E binding protein (4E-BP), a translational repressor that directly regulates the activity of the eIF4E translation initiation factor. In addition, activated mTOR is able to phosphorylate p70S6k, which can inactivate eEF2 kinase (regulating translation elongation) as well as phosphorylate the 40S ribosomal protein S6. In several cell types, activation of the TSC1–TSC2 complex (Tuberous sclerosis gene) negatively regulates p70S6k; it also inhibits mTOR signaling, reducing cell growth (and insulin signaling), and is inhibited by Akt-dependent phosphorylation [108, 109]. The TSC complex mediates its effect on mTOR signaling by targeting Rheb, which is a small Ras-homologous GTPase implicated with mTOR in the activation of p70S6k [110, 111]. However, the identification and the contribution of the TSC complex and Rheb to the signaling events occurring in cardiomyocyte growth, and cardiac hypertrophy remain to be fully determined.

A second branch of the signaling cascade (Fig. 15.4) leads to glycogen synthase kinase-3 (GSK-3), which also regulates the general protein translational machinery, as well as specific transcription factors implicated in both normal and pathologic cardiac growth. GSK-3β, which was among the first negative regulators of cardiac hypertrophy to be identified, blocks cardiomyocyte hypertrophy in response to ET-1, isoproteronol, and Fas signaling [112–114]. In addition, GSK-3β has been found to be a negative regulator of both normal and pathologic stress-induced growth (e.g., pressure overload) [115]. GSK-3β plays a key inhibitory role in both insulin signaling and the Wnt signaling pathway which has been implicated in early cardiomyocyte differentiation as well as in myocardial hypertrophic growth responses. In unstimulated cells, GSK-3β phosphorylates the N-terminal domain of β-catenin, thereby targeting it for ubiquitylation and proteasomal degradation. GSK-3β is constitutively active unlike most kinases; it is turned "off" by cell stimulation by growth factors and hypertrophic agonists. GSK-3β negatively regulates most of its substrates, including the protein translation initiation factor eIF2B [116] as well as transcription factors implicated in cardiac growth, including c-Myc, GATA-4, and β-catenin [117–119]. In addition, GSK-3β is a counter-regulator of calcineurin-nuclear factor of activated T-cells (NFAT) signaling pathway. Whereas calcineurin dephosphorylates/promotes nuclear translocation of NFAT; GSK-3β phosphorylates/maintains inative NFAT in cytosol [117]. Also, Akt phosphorylates GSK-3β at serine-9 thus inhibiting its activity.

Inhibition of GSK-3β releases a number of transcription factors from tonic inhibition and also releases eIF2β, allowing translation activation. Transgenic mice overexpressing GSK-3β in the heart exhibit significantly defective postnatal cardiomyocyte growth as well as markedly abnormal cardiac contractile function related to downregulation of ATP-dependent $Ca^{2+}$ pump ($Ca^{2+}$ ATPase or SERCA) expression (resulting in abnormal calcium handling) and severe diastolic dysfunction with progressive HF [120]. Moreover, it has been suggested that a family of dimeric phophoserine-binding molecules, the 14-3-3 proteins (which are implicated in cell-cycle control and the stress response) participate in the regulation of GSK-3β phosphorylation [121].

It is noteworthy that the activation of protein translation affected by both of these signaling branches can also be regulated by stress-activated mechanisms that are independent of Akt. For instance, AMP-activated protein kinase (AMPK), a key regulator of cellular energy homeostasis, is involved in modulating the activity of mTOR and can affect the translational response in cardiac hypertrophy [122]. Moreover, hypoxia can rapidly and reversibly trigger mTOR hypophosphorylation and mediate changes in its effectors such as 4E-BP1 and p70S6K, independent of Akt or AMPK signaling [123]. Also, the TSC complex can be phosphorylated and inactivated by stress-mediated stimuli, and the Erk pathway is independent of Akt signaling. The Akt-independent mechanism of activation of mTOR may be particularly relevant to pathological stress-induced growth.

Inactivation of GSK-3β by Ser-9 phosphorylation may also occur independent of the PI3K-Akt pathway, and this includes involvement of growth factors such as EGF and PDGF, that stimulate the GSK-3β-inactivating kinase p90 ribosomal protein S6 kinase (RSK) through MAP kinases, activators of cAMP-activated PKA, and PKC. Moreover, exposure of cells to Wnt protein ligands leads to inactivation of GSK-3β by an undefined mechanism [124].

### Protein Kinase C

The serine/threonine kinase PKC has been implicated as the intracellular mediator of a variety of factors acting through multiple signal transduction pathways. The PKC family of isozymes is increasingly recognized as playing a pivotal role in the cardiac phenotype expressed during postnatal growth and development and in response to pathologic stimuli and in the development of cardiac hypertrophy and HF. The expression of multiple PKC isoforms contributes to both a broad spectrum of adaptive and maladaptive cardiac responses with significantly different responses provided by each isoform [125]. While over 12 isoforms of PKC have been reported, in the heart the four most functionally significant members of the PKC family are PKCα and β (both calcium- and DAG-activated) and PKCδ and ε (DAG-activated with no requirement for calcium). These PKC isoforms are activated by membrane receptors coupled to PLC via $G_q/G_{11}$ heterotrimeric G proteins. Activation of PKC resulted in translocation of it from the cytosol to the site of action (e.g., to the plasma membrane and to the mitochondria). PKC has an N-terminal regulatory region and a C-terminal catalytic region; protein–lipid interactions are implicated in PKC (N-terminus is required for PKC interaction with the second messenger DAG and $Ca^{2+}$).

Anchoring proteins recruit PKC to specific sites, and receptors for activated C-kinase (RACKs) have been found to be operative during PKCε-dependent growth signaling in cardiomyocyte [126].

### Protein Kinase G

See later section on "Antihypertrophic Signaling Pathways."

## $Ca^{2+}$-Mediated Kinase Signaling

### Calcineurin/Calmodulin

$Ca^{2+}$ is a major intracellular messenger involved in the activation of $Ca^{2+}$-dependent signaling pathways where it regulates cardiac growth and function by the activation of kinases and phosphatases as well as playing a pivotal role in excitation–contraction coupling (ECC). The $Ca^{2+}$ signal inducing contraction in cardiac muscle originates from two sources. In response to depolarization of the sarcolemma, $Ca^{2+}$ enters the cell through voltage-dependent L-type $Ca^{2+}$ channels (VDCCs). This $Ca^{2+}$ binds to and activates $Ca^{2+}$ release channels (RyRs) of the SR through a $Ca^{2+}$-induced $Ca^{2+}$ release (CICR) process increasing intracellular $Ca^{2+}$ concentration by more than tenfold to induce contraction. Entry of $Ca^{2+}$ with each contraction requires an equal amount of $Ca^{2+}$ extrusion within a single heartbeat to maintain $Ca^{2+}$ homeostasis and to ensure relaxation. Removal of $Ca^{2+}$ from the cytosol is mainly affected by the sarcolemmal $Na^+/Ca^{2+}$ exchanger and by the SR ATP-dependent $Ca^{2+}$ pump, SERCA. These transport systems are important determinants of the intracellular $Ca^{2+}$ level and cardiac contractility. Altered intracellular $Ca^{2+}$ handling is one factor contributing to impaired contractility in heart failure.

$Ca^{2+}$-associated stress response pathways also control cardiac gene expression by modulating the activities of chromatin-remodeling enzymes, which have been shown to act as global regulators of the cardiac genome during pathological remodeling of the heart. Deacetylation of nucleosomal histones in chromatin by histone deacetylases (HDACs) results in transcriptional repression due to chromatin condensation. Several lines of evidence from transgenic mice have strongly implicated class II HDACs in preventing myocyte hypertrophy in response to diverse agonists [127]. Moreover, in stressed animals, class II HDACs are shuttled out of the nucleus, dependent on phosphorylation by GPCR-activated kinases including the calcium/calmodulin-dependent protein kinase (CaMK) and the protein kinase D (PKD). In addition to kinase-dependent regulation of cardiac gene expression controlling hypertrophy through negative effects on HDACs, protein phosphatases also play an equally important role in the regulation of chromatin structure and gene expression during cardiac remodeling. For example, the calcium and calmodulin-dependent protein phosphatase calcineurin is activated in response to cardiac stress signaling, and its activation has been shown to be sufficient for pathological cardiac hypertrophy. Calcineurin dephosphorylates members of the NFAT family of transcription factors, which enables them to translocate into the nucleus, where they activate transcription together with other transcription factors, including myocyte enhancer factor-2 (MEF2) and GATA4. NFAT factors activate myocardial gene expression, in part, by recruiting histone acetyltransferases (HATs) to gene regulatory elements containing NFAT and MEF2 binding sites [128, 129].

Many of the actions of $Ca^{2+}$ are mediated via its interaction with calmodulin (CaM), which is an intracellular $Ca^{2+}$ sensor and selectively activates downstream signaling pathways in response to local changes in $Ca^{2+}$ (see Chap. 4) [130].

Signaling pathways that are both directly activated by $Ca^{2+}$ and able to promote cardiac hypertrophy include $Ca^{2+}$/CaM-regulated kinase and phosphatase pathways, and PKC isozymes [131]. Major cardiac $Ca^{2+}$/CaM-dependent enzymes include CaMK, myosin light chain kinase (MLCK), and the phosphatase calcineurin [131]. In the heart, the main CaMKII isoform is CaMKIIδ, which is localized in the nucleus, whereas other isoforms are localized in the SR [132]. Increased intracellular [$Ca^{2+}$] results in autophosphorylation of CaMKII, which switches it to a $Ca^{2+}$ -independent state and prolongs its activation. CaMKI, which is ubiquitously expressed, and CaMKIV, mainly expressed in testis and brain, are activated by upstream $Ca^{2+}$/CaM-dependent protein kinases [133]. CaMKII isoforms associated with SR are capable of phosphorylating $Ca^{2+}$-cycling proteins, and hence altering $Ca^{2+}$ re-uptake and release. In cultured cardiac myocytes, pharmacological inhibitors of CaMKII have been found to attenuate both endothelin-1-induced hypertrophy [133] and mechanically stretch activated B-type natriuretic peptide (BNP) gene transcription [134]. Moreover, overexpression of CaMKIIδ in the heart of transgenic mice is sufficient to promote hypertrophic growth [135] as well as cardiac-specific overexpression of either CaMKI or CaMKIV [136]. Furthermore, MLCK is activated by $Ca^{2+}$/CaM leading to subsequent phosphorylation of its single main substrate, ventricular-specific isoform of myosin light chain-2 (MLC-2).

Besides cardiac kinases, $Ca^{2+}$/CaM activates phosphatases, including calcineurin [137]. Activation of calcineurin by the $Ca^{2+}$/CaM complex results in dephosphorylation of its substrates including NFAT or Ets-like gene-1 (Elk-1). Interaction of $Ca^{2+}$/CaM and calcineurin has been found to increase in the failing human heart, and increased levels of cardiac calcineurin activity have been reported in experimental animal models of cardiac hypertrophy [138]. In addition, calcineurin is sufficient to produce cardiac hypertrophy when overexpressed in the heart of transgenic mice [137]. In rodent models, pharmacological inhibition of calcineurin for the prevention of cardiac hypertrophy has yielded conflicting results [139, 140]. This may be due to the absence of well-tolerated calcineurin inhibitors and also that calcineurin is not a cardiac-restricted enzyme. In comparison to normal controls, the human hypertrophied heart exhibits higher calcineurin activity [141], and patients maintained with partially inhibited calcineurin activity still can develop cardiac hypertrophy. This is probably related to the presence of additional pathways making calcineurin signaling not indispensable for hypertrophy [128, 142].

In response to disease-causing stimuli, endocrine growth factors and/or cytokines induce ventricular remodeling, cardiomyocyte hypertrophy, and alterations in the viability of myocytes. Many of these endocrine factors (e.g., angiotensin II and endothelin-1) signal through GPCRs on cardiac myocytes to induce phospholipase C activation, which in turn generates IP3 and DAG. IP3 generation in turn leads to the release of $Ca^{2+}$ from the endoplasmic and sarcoplasmic reticulum through the IP3R channels and ryanodine receptor/intracellular $Ca^{2+}$ release channels. Moreover, neuroendocrine factors and cytokines have been implicated in the development of cardiac hypertrophy and failure, in some degree by promoting the activation of a number of intracellular signaling pathways in cardiac myocytes, including MAPK, PKC, calcineurin, CaMK, and IGF-1 pathway constituents. Interestingly, three of these signaling factors, calcineurin, CaMK, and PKC, require increases in $Ca^{2+}$ to become activated, and both calcineurin and CaMK are potent inducers of the myocardial hypertrophic response. Chronic hyperactivity of the β-adrenergic signaling pathway results in PKA-dependent hyperphosphorylation of the cardiac

RyR/intracellular $Ca^{2+}$ release channels. It has been proposed that the increase in $Ca^{2+}$ that drives the larger contractions may be responsible for switching on a second process of signalosome remodeling to downregulate the $Ca^{2+}$ signaling pathway and that the $Ca^{2+}$ transient convey information responsible for remodeling of the cardiac gene transcription program that leads first to hypertrophy and then to heart failure [143].

## G Protein Regulated Kinases

The regulation of myocardial adrenergic receptors, like that of most GPCRs involves a desensitization mechanism characterized by a rapid loss of receptor responsiveness despite the continued presence of agonist. The desensitization process has been particularly well characterized using the β2-AR system promoted by a phosphorylation event targeting only the agonist-occupied receptors by a serine/threonine kinase known as β-ARK. β-ARK1 and a highly homologous isozyme β-ARK2 are two of the most studied members of GRK family, which currently consists of six members. Desensitization of GPCRs requires not only GRK-mediated phosphorylation but also the binding of a second class of inhibitory proteins, the β-arrestins (β-arrestin-1 and β-arrestin-2), which bind to phosphorylated receptors and sterically restrict the further activation of G proteins, in part by their displacement from the receptors resulting in receptor/G protein uncoupling [144]. The GRKs shown to be expressed in the heart are β-ARK1 (the most abundant), β-ARK2, GRK5, and GRK6. Since β1-AR is the most critical receptor mediating acute changes in cardiac rate and contractility, it is important to realize that three of these GRKs (β-ARK1, β-ARK2, and GRK5) have been shown to phosphorylate and desensitize β1-ARs *in vitro* [145].

Like most GRKs, β-ARK1 is a cytosolic enzyme that has to be translocated to the plasma membrane in order to phosphorylate the activated receptor substrate. The mechanism for translocation of β-ARK1 and β-ARK2 involves the physical interaction between the kinase and the membrane-bound β subunits of G proteins ($G_\beta$). $G_\beta$, anchored to the membrane through a lipid modification on the C-terminus of the subunit (termed prenylation), is available to interact with β-ARK after G protein activation and dissociation. The region of β-ARK responsible for binding $G_\beta$ has been mapped to a 125-amino acid domain located within the C-terminus of the enzyme. Recently, peptides derived from the $G_\beta$-binding domain of β-ARK have been shown to act as *in vitro* β-ARK inhibitors by competing for $G_\beta$ and preventing translocation [146].

A pathophysiological role for GRKs can be inferred from studies on heart failure as well as from the observation that chronic treatment with various agonists or antagonists for GPCRs results in alterations of GRK expression [147, 148].

## MAP Kinases

The MAPK cascade consists of a series of successively acting protein kinases that include three well-characterized branches, the Erks, the JNKs, and the p38 MAPKs. Signaling through each of these MAPK branches is initiated by diverse stress and mitogenic stimuli localized to the cell membrane or within the cytoplasm. Activation of Erks, JNKs, and p38 MAPKs facilitates the phosphorylation of multiple transcription regulators such as MEF2, activating transcription factor-2 (ATF-2), p53, NFAT, c-Jun, and c-Myc. MAPK-mediated phosphorylation of these and other transcriptional regulators profoundly influences adaptive and inducible gene expression in many cell types. Differential activation of MAPKs has been found to correlate with the severity of cardiac hypertrophy (due to mild and to severe pressure overload), and this may contribute to differential expression of myocardial gene expression resulting in a distinctive cardiac phenotype associated with the severity of hemodynamic overload [149]. The role that p38 MAPK pathway plays in cardiac hypertrophy remains questionable. Recently, the contribution of MAPK-activated protein kinase-2 (MK2), a well-established p38 downstream kinase, to the pathological cardiac remodeling induced by p38 has been investigated by Streicher et al. [150]. They used a cardiomyocyte-specific and inducible transgenic approach to determine the functional and molecular impact of acute activation of the p38 pathway in the heart in either a MK2 wild-type or a MK2-null background. p38 activation in wild-type mice led to a rapid onset of lethal cardiomyopathy associated with cardiomyocyte hypertrophy, interstitial fibrosis, and contractile dysfunction. Inactivation of MK2 partially reduced cardiomyocyte hypertrophy, improved contractile performance, and prevented early lethality. Interestingly, MK2 inactivation had no effect on mRNA levels of hypertrophic marker genes nor on proinflammatory gene cyclooxygenase (COX)-2, although MK2 had a major role in COX-2 protein synthesis without affecting the mRNA level or protein stability. Taken together, it appears that p38 activity can contribute to pathological or maladaptive cardiomyocyte hypertrophy and remodeling in adult heart, and that MK2 is an important downstream molecule responsible for specific features of p38-induced cardiac pathology. Thus, members of the MAPK signaling cascade are important regulators of cardiomyocyte hypertrophy; although, further research is needed to identify the downstream transcriptional mechanisms that alter cardiac gene expression.

## Integrating Responses: Transcription Factors and Translational Control

### Role of Growth Factors

Since Chap. 5 presents a detailed account of growth factors (GFs) signaling, here it will suffice to note that GFs such as FGF-2 significantly promote neonatal cardiac myocyte proliferation [151], and overexpression of the FRF-2 receptor (FGF-R1) in neonatal rat cardiomyocytes results in marked proliferation [152]. Cardiotrophin (CT-1), an interleukin 6-related cytokine, has been shown to promote both the survival and proliferation of cultured neonatal cardiac myocytes [153]. This likely is mediated by the PI3K-Akt pathway since CT-1 phosphorylates and activates Akt [154]. These diverse approaches have confirmed the importance of suspected pathways and implicate as well unexpected pathways leading to new paradigms for the control of cardiac growth. Furthermore, in the regulation of cardiomyocyte growth and/or proliferation, there are several signal transductions acting as redundant mechanisms converging on one or several serine/threonine kinases. Several GPCRs such as α-AR, β-AR, angiotensin II receptor, and endothelin-1 receptor are able to activate these signaling cascades and induce changes in cell growth and proliferation. A general scheme involves the following: signals received at the plasma membrane receptors are transmitted via GPCR-G-proteins-second messengers pathways to a wide-spectrum of protein kinases and phosphatases, which are in turn activated. These activated protein modifiers may lead to activation and/or deactivation of specific transcription factors, which modulate specific gene expression affecting a broad spectrum of cellular events, or they can target directly proteins involved in metabolic pathways, ion transport, $Ca^{2+}$ regulation and handling, influencing cardiac contractility and excitability, as well as the pathways of cardiomyocyte apoptosis and/or cell survival. Also, transcriptional networks implicated in mitochondrial biogenesis and function [155] include PPAR-γ coactivator-1 (PGC-1), the ensemble of downstream nuclear receptor partners (e.g., PPARs and estrogen-related receptors), converging molecular signals such as $Ca^{2+}$ (see below in this chapter), NO, MAPKs, β-adrenergic mechanisms, cAMP, and signal transduction pathways (e.g., Erks).

At least some of the cytoplasmic signaling pathways thought to be responsible for pathological hypertrophy are mediated through increased levels of growth factors signaling through GPCRs [156]. In addition, ANP, through its guanylyl cyclase-A (GC-A) receptor, locally moderates cardiomyocyte growth. To characterize the antihypertrophic effects of ANP, the possible contribution of $Na^+/H^+$ exchanger (NHE-1) to cardiac remodeling was recently examined in a model of GC-A-deficient [GC-A (-/-)] mice [157]. Fluorometric measurements in the cardiomyocytes demonstrated that cardiac hypertrophy in GC-A (-/-) mice was associated with enhanced NHE-1 activity, alkalinization of intracellular pH, and increased $Ca^{2+}$ levels. Chronic treatment of GC-A (-/-) mice with the NHE-1 inhibitor cariporide normalized cardiomyocyte pH and $Ca^{2+}$ levels and regressed cardiac hypertrophy and fibrosis. Four prohypertrophic signaling pathways, MAPK, Akt, calcineurin-NFAT, and CaMKII, were activated in GC-A (-/-) mice, but only CaMKII and Akt activity regressed upon reversal of the hypertrophic phenotype following cariporide. By contrast, the MAPK and calcineurin-NFAT signaling pathways remained activated during regression of hypertrophy. These observations suggest that the ANP-GC-A system modulates the cardiac growth response to pressure overload during remodeling by preventing excessive activation of NHE-1 and subsequent increases in cardiomyocyte intracellular pH, $Ca^{2+}$, and CaMKII as well as Akt activity.

### Receptor Tyrosine Kinases

A large family containing over 20 RTK classes has been identified, all of which share a similar structure that includes a ligand binding extracellular domain, a single transmembrane domain, and an intracellular tyrosine kinase domain. This large protein family includes the receptors for many growth factors and for insulin. Most of the RTK subfamilies are defined by the extracellular region containing the ligand binding domains (LBDs), which exhibit variable length and subdomain composition with highly conserved structural motifs, including domains that are immunoglobulin-binding, cysteine-rich, ephrin-binding, and fibronectin repeats [158]. Except in the case of the insulin receptor which exists as a dimer in the absence of ligand, ligand binding to the extracellular portion of these receptors results in receptor dimerization, which facilitates the transautophosphorylation of specific tyrosine residues in the highly conserved cytoplasmic portion (Fig. 15.5). The phosphotyrosine residues enhance the receptor catalytic activity and can provide docking sites for downstream signaling proteins. The creation of docking sites allows the recruitment to the receptor kinase complex of a variety of proteins containing specific binding domains, such as Src homology 2 and 3 or phosphotyrosine binding domains, which can broaden the signaling capacity of the RTKs. For instance, GPCRs which lack an intrinsic kinase activity possessed by RTKs, such as PDGFR or EGFR have been shown to activate RTKs in response to stimulation by cytokines, cell adhesion, and stress stimuli. In this way, RTKs can function in integrating a large array of stimuli from diverse environmental and intracellular inputs.

Phosphorylation, although necessary, may not be sufficient to fully activate many RTKs. Oligomerization-induced conformational changes may be necessary to modulate the kinetic properties of RTKs, and render them fully functional. Because of the critical roles played by RTKs in cellular signaling processes, their catalytic activity is normally under tight control by a variety of intrinsic regulatory mechanisms as well as by protein tyrosine phosphatases.

A number of RTKs have been shown to play essential roles in early cardiac development as well as in the growth, repair, and survival of adult cardiomyocytes as part of a signaling network. For example, the erbB2 RTK is known to have a critical role in cardiac development. In addition, erbB2 participates in an important pathway that involves neuregulins, cell–cell signaling proteins that are ligands for RTKs of the ErbB family, and the neuregulin receptor erbB4. Two of the neuregulins (NRG1 and NRG2) and their receptors (erbB2 and erbB4) are essential for normal cardiac development and can mediate hypertrophic growth and enhance the survival of embryonic, postnatal, and adult ventricular cardiomyocytes. Targeting the neuregulin receptors to caveolae microdomains, within cardiac myocytes, has been shown to be a viable approach to regulating neuregulin signaling in the heart [159, 160].

RTKs also play a pivotal role in the growth responses of vascular cells. RTKs include the VEGF receptors, Eph receptors, Tie1, and Tie2, all of which are expressed on vascular endothelial cells, as well as the PDGF receptors which are expressed on vascular SMCs [160, 161]. While all of these RTKs activate many similar effector molecules, some of the initiated signals appear to be distinct. This could explain, at least in part, how different RTKs expressed in the developing vasculature can direct unique biological functions. Furthermore, RTK-mediated signaling requires the engagement and activation of the MAPK cascade composed of Raf, MEK, and Erk kinases. Progress has been made in the understanding of the complex network of regulatory mechanisms that control the Erk1/2-signaling cascade. Erk1/2 is a mediator of cardiac hypertrophy, which is a critical risk factor for the development of heart failure and sudden death, although this signaling cascade also has beneficial effects on myocardial cell death and ischemic injury [162]. Potentially, the targeting of these rather uncertain Erk functions may be a successful tool in the treatment of cardiac disease.

## NF-κB

NF-κB is a pleiotropic family of transcription factors implicated in the regulation of diverse biological phenomena, including apoptosis, cell survival, cell growth, division and differentiation, innate immunity, and responses to stress, hypoxia, stretch, and ischemia. In the heart, NF-κB is activated in atherosclerosis, myocarditis, during transplant rejection, after myocardial ischemia/reperfusion (I/R), in decompensated HF, dilated cardiomyopathy (DCM), after ischemic and pharmacological preconditioning, heat shock, as well as in the hypertrophy of isolated cardiomyocytes. In addition to being activated by cytokine-mediated pathways, NF-κB is modulated by many of the signal transduction cascades associated with the development of cardiac hypertrophy and in response to oxidative stress. Many of these signaling cascades activate NF-κB by activating the IκB kinase (IKK) complex. These signaling interactions primarily involve the MAPK/Erk kinase kinases (MEKKs) that, as previously noted, are components of MAPK signaling pathways. In addition, other signaling factors directly activate NF-κB via IκB or via direct phosphorylation of NF-κB subunits. Combinatorial interactions have been reported at the level of the promoter between NF-κB, its co-activators, and other transcription factors, several of which are activated by MAPK and cytokine signaling pathways. In addition to being a major mediator of cytokine effects in the heart, NF-κB represents a signaling integrator, functioning as a key regulator of cardiac gene expression programs, downstream of multiple signal transduction cascades in a variety of physiological, and pathophysiological states. Genetic blockade of NF-κB can reduce the size of infarcts resulting from I/R in the murine heart, consistent with its role as a major determinant of cell death after I/R, and suggests that NF-κB may constitute an important therapeutic target in specific cardiovascular diseases [163].

## Peroxisome Proliferator-Activated Receptors α and γ and Co-Factors (RXR and PGC)

Nuclear receptor transcription factors are important regulatory players governing the cardiac metabolic gene program (Table 15.2) [164]. This superfamily of receptors was originally described as ligand-dependent transcription factors, which is in fact the case for nearly 50% of those characterized so far [165]. Such ligand-activated receptors include the classical endocrine receptors that respond to steroid or thyroid hormones. Also, a number of receptors have been identified (without prior insight to their ligands) that respond to dietary-derived lipid intermediates, including long-chain fatty acids (LCFAs) and bile acids. These receptors generally participate in the regulation of pathways involved in the metabolism of the activating ligands. On the other hand, a group of "orphan" nuclear receptors (some of them implicated in cell-cycle regulation, and apoptosis) have no known endogenous ligands, although modulating ligands may have been identified for some of these receptors.

**Table 15.2** Myocardial nuclear receptor transcription factors and co-activators

| Receptor | Cardiac function | Target genes | References |
|---|---|---|---|
| Peroxisome proliferator-activated receptor-α (PPAR-α) | Lipid utilization | FAO-CPT1, MLC-2, PDK-4, UCP3 | [164, 166, 167] |
| Peroxisome proliferator-activated receptor-β (PPAR-β) | Lipid homeostasis | FAO | [164] |
| Peroxisome proliferator-activated receptor-γ (PPAR-γ) | Lipid storage<br>Increased glucose oxidation | iNOS | [168] |
| Peroxisome proliferator-activated receptor-δ (PPAR-δ) | Lipid homeostasis | Same as PPAR-α | [169] |
| Peroxisome proliferator-activated receptor-γ coactivator-1α (PGC-1) | Cardiac metabolism | FAO, ETC, UCP, NRF1, NRF2, mtTFA (or TFAM), Glut4 | [170, 171] |
| Retinoid X receptor (RXR-α) | Cardiac development | GATA-4, FAO, ANF, SERCA | [172] |
| Thyroid hormone receptor (TR) | Cardiac hypertrophy | BNP, β-AR, SERCA | [173, 174] |
| Vitamin D receptor (VDR) | Cardiac morphogenesis | ANP | [175] |
| Estrogen-related receptor (ERRα) | Orphan receptor | TRα, FA uptake,<br>FAO, ETC | [167, 176] |
| Estrogen-receptor α (ER-α) | I/R<br>Cardioprotection | Nd<br>ANP, eNOS/iNOS | [177–179] |
| Estrogen-receptor β (ER-β) | Mitochondrial localization | Nd | [180] |
| Retinoic acid receptor (RAR) | Cardiac morphogenesis | GATA-4 | [181] |
| Glucocorticoid receptor (GR) | Decreased growth | β-AR | [182] |
| Androgen receptor (AR) | Cardiac hypertrophy | Nd | [183] |
| Mineralocorticoid receptor (MR) | Cardioprotective<br>Electrical remodeling | Nd<br>Nd | [184, 185] |

*Nd* Not determined

Because the heart must adapt to continuously changing energy demands, but has limited capacity for storing fatty acids or glucose, myocardial energy substrate flux must be tightly matched with the demand. Therefore, ligand-activated nuclear receptors, as metabolite sensors, participate in a rapidly activated program of gene expression in response to fluctuating substrate levels. Of particular interest are the peroxisome proliferator-activated receptors (PPARs), fatty acid-activated nuclear receptors, increasingly recognized as key regulators of cardiac fatty acid metabolism. The PPAR-α isoform has been characterized as the central regulator of mitochondrial fatty acid catabolism including fatty acid oxidation (FAO), whereas PPAR-γ primarily regulates lipid storage [122]. In addition, a selected group of orphan nuclear receptors have been identified which serve new roles in the regulation of cardiac energy metabolism (Table 15.2).

The nuclear receptors have a conserved modular domain structure [164–166]. These proteins contain an N-terminal region containing a ligand-independent transcriptional activation function (AF-1) domain, a conserved DNA binding domain (DBD) containing two highly conserved zinc-finger sequences that target the receptor to specific regulatory regions termed hormone response elements (HREs), and a composite C-terminal region that includes the ligand-binding domain (LBD), a dimerization domain, and a conserved ligand-dependent activation function domain (AF-2). The nuclear receptors bind to regulatory DNA elements in target genes as homodimers, heterodimers, or in some cases as monomers. Unlike the classic steroid receptors that function as homodimers, a number of nuclear receptors have been involved in nutrient sensing and metabolic regulation [e.g., PPARs, thyroid hormone receptor (TR), retinoic acid receptor (RAR) heterodimerize with the retinoid X receptor (RXR)]. These receptors interact with regulatory DNA elements within the 5′ regulatory region of their target genes that are composed of variably spaced hexameric half-sites (AGGTCA) arranged as direct, indirect, or everted repeats. Once bound to their specific response element, the receptors recruit coactivator proteins often in concert with the displacement of corepressor proteins. For ligand-activated receptors, it is through ligand binding that the receptor adopts a permissive conformation for the recruitment of coactivator proteins to enable transcriptional activation. More than 100 coactivator proteins have been identified for nuclear receptors, including ATP-dependent chromatin-remodeling complexes, histone acetylases, histone methyltransferases, and RNA polymerase II recruiting complexes that open up the chromatin structure and facilitate the binding of basal transcription factors and RNA polymerase II. Histone-modifying proteins are often recruited into complexes by adapter proteins, which lack catalytic activity. One such adapter/coactivator, the PPAR-γ-coactivator-1 (PGC-1), serves as a key link between physiological cues and metabolic regulation in heart.

## Toll-Like Receptors

Transmembrane signaling proteins of the Toll-like receptor (TLR) family constitute key signaling elements in both macrophages and in atherosclerotic lesions (see Chap. 18). In this chapter, it suffices to mention that the Toll-like receptor 4 (TLR4) is highly expressed in the heart, and this expression is strongly upregulated in mice with cardiac ischemia (relative to controls) as well as in patients with idiopathic DCM [186], and is consistent with the activation of signaling pathways leading to the expression of proinflammatory cytokines, which have been implicated in the etiology of DCM. Moreover, myocardial TLR4 levels have been found positively correlated with levels of enteroviral replication in DCM [187]. The role of TLR4 in cardiac hypertrophy *in vivo* has been studied by Ha et al. [188]. Following aortic banding, the ratio of heart weight/body weight (HW/BW) and cell size significantly increased (by 33.9% and 68.4%, respectively) in wild-type (WT) mice (compared with sham controls), and only by 10.00% and by 11.8% in TLR4-deficient mice. Furthermore, NF-κB binding activity and phospho-p70S6K levels, compared to sham controls, were increased (by 182.6% and 115.2% respectively) in the aortic-banded WT mice and by 78.0 and 162.0% in aortic-banded TLR4-deficient mice. Interestingly, in rapamycin-treated aortic-banded mice, the ratio of HW/BW was increased by 18.0% in WT mice and by 3.5% in TLR4-deficient mice compared to sham controls. Thus, TLR4 contributes to the development of cardiac hypertrophy in an *in vivo* model, and both the TLR4-mediated pathway and PI3K-Akt-mTOR signaling participate in the development of cardiac hypertrophy. TLR4 also has a proinflammatory role in I/R injury as suggested by observations that TLR4-deficient mice sustain smaller infarctions and exhibit less inflammation after myocardial I/R injury [189]. Interestingly, systemic administration of lipopolysaccharide (LPS), a TLR4 agonist, confers a cardioprotective effect against ischemic injury and myocardial infarction (MI) [190].

## Thyroid Hormone

Thyroid hormone (TH) is a key regulator of metabolism in tissues such as heart and liver, and changes in thyroid status have been associated with profound alterations in their biochemical and physiological functioning. In the heart, hyperthyroidism is associated with increased metabolic rate, augmented cardiac muscle contractility, and structural hypertrophy; thus, cardiac hypertrophy, and therefore remodeling is the usual response to TH. Also, models of cardiac hypertrophy have shown TH-induced increases in total tissue RNA and ventricular weight, as well as in the levels of both cytosolic and mitochondrial ribosomes. TH can also modulate mitochondrial enzyme activities. One way that TH regulates mitochondria is by modulating the transcriptional activation of nuclear genes encoding mitochondrial proteins, including components of the respiratory pathway [191]. This regulatory effect occurs as a result of many of the nuclear-encoded mitochondrial genes having TH-sensitive promoter elements [192]. TH treatment *in vivo* may cause elevated levels of markers of mitochondrial biogenesis including myocardial mtDNA, specific mtDNA-encoded proteins, and transcripts, as well as nuclear-encoded regulators including mitochondrial transcription factor A (mtTFA or TFAM) and PPAR-α [193]. In general, the majority of the observed TH-induced changes in myocardial mitochondria will follow the onset of cardiac hypertrophy. However, little is presently known about the myocardial signaling events preceding TH-induced mitochondrial biogenesis. Of interest, myocardial mitochondria more distally located in relationship to the capillaries (i.e., with increased oxygen diffusion distance) have been identified as the organelles apparently undergoing the most biogenesis, based on morphometric and ultrastructural analysis [194]. This would suggest that relative oxygen deprivation or hypoxia might provide part of the initial stimulus for myocardial mitochondrial biogenesis and remodeling events. Moreover, reactive oxidative species (ROS) generation by mitochondria appears to be a critical signaling event in myocardial hypoxia [195]. The role of oxygen and free radicals in myocardial mitochondrial signaling with thyroxine (T4) induction has been proposed but not well defined. In our laboratory, although ROS levels was measured only indirectly, we detected significant T4-induced increase in myocardial antioxidant response provided by increased levels of mitochondrial manganese superoxide dismutase (MnSOD). The peroxisomal enzyme catalase (gauged both at the level of activity and content) showed no change in cardiac tissues. Levels of other antioxidants [e.g., glutathione, glutathione reductase and peroxidase, and SOD1 (copper/zinc superoxide dismutase 1) were not evaluated]. In our experience, the long-term effect of TH, requiring both nuclear expression and protein synthesis, was the development of cardiac hypertrophy (already significantly present in the 10-day T4-treated animals, albeit at lower levels than with the 15-day group). This could be dissociated from cardiac mitochondrial biogenesis, as gauged by unchanged myocardial enzyme activity levels in the 10-day TH-treated rats [193], and suggested that myocardial mitochondrial biogenesis and bioenergetic remodeling may occur as an adaptive response to cardiac hypertrophy. In the myocardium, TH also exerts a short-term influence (occurring within minutes of hormone treatment) in addition to its more delayed effects. Recently, it has been suggested that calcineurin, a participant in mitochondrial biogenesis which activates NFAT, is the target of calcineurin inhibitors such as cyclosporine-A (CsA), and TH pathways are interconnected. Bigard et al. [196] have studied the specific and combined effects of pharmacological calcineurin inhibition (using CsA administration) and TH deficiency in muscle phenotype. CsA effects were only observed if TH was available, while TH deficiency totally blunted the muscle responses to calcineurin inhibition. Interestingly, in the presence of TH deficiency, there was no response to the pharmacological inhibition of calcineurin that is known to induce a slow-to-fast IIA transition associated with an enhancement of mitochondrial biogenesis in normothyroid rats. Moreover, thyroid deficiency markedly decreased the expression of endogenous calcineurin inhibitors MCIP-1 and MCIP-2, mRNA, and protein.

## Insulin

### Hypertrophy/Heart Failure and Glucose Transporters

As in all other cells, the entry of glucose into cardiac myocytes is dependent on the transmembrane glucose gradient and is facilitated by members of the GLUT family of facilitative glucose transporters [197]. The GLUT-1 transporter, which is localized on the plasma membrane under basal conditions, is thought to be the primary mediator of basal glucose uptake in the heart [198]. Its myocardial expression is stably increased within hours of ischemia or induction of cardiac hypertrophy. The most abundant glucose transporter in the heart is the insulin-responsive GLUT-4 transporter. Insulin mediates the translocation of GLUT-4 to the plasma membrane from a pool of intracellular vesicles and represents a critical control point by which the net flux of glucose is regulated. A variety of physiological stimuli including ischemia and cardiac work overload can induce this translocation, thereby increasing glucose uptake and glycolytic metabolism [199]. Insulin-independent stimulation of GLUT-4 translocation, and enhance glucose uptake may occur in cardiac tissue and myocytes in response to catecholamines, calcium and exercise [200, 201].

Moreover, studies on mice containing muscle-specific deletions in the insulin receptor gene revealed that normal expression of muscle insulin receptors is not needed for the exercise-mediated increase in glucose uptake and glycogen synthase activity *in vivo* [202]. Moreover, GLUT-4 translocation to the plasma membrane can also be stimulated by the activation of AMPK which occurs during exercise stress [199]. Studies with transgenic mice containing an inactive form of AMPK have shown normal GLUT-4 levels and glucose uptake, but no increase in glucose uptake, glycolysis, or FAO during ischemia [199]. These findings have led to the conclusion that AMPK is involved in mediating glucose uptake and glycolysis during ischemia presumably as a protective adaptation. Gathered observations have also demonstrated that changes in myocardial AMPK activity in exercise-trained rats (which increased in proportion to exercise intensity) were associated with physiological AMPK effects (e.g., GLUT-4 translocation to the myocardial sarcolemma) and are consistent with AMPK as a key mediator in the cardiac response to exercise, as previously demonstrated with skeletal muscle [203].

Defects in the ability of insulin to regulate GLUT-4 translocation may be contributory (as one of multiple effects of insulin) to the development of insulin resistance and noninsulin-dependent type 2 diabetes. Mice heterozygous for a null *GLUT-4* allele display reduced muscle glucose uptake, insulin resistance, and diabetes [204]. In contrast, mice homozygous for a null *GLUT-4* were growth retarded and exhibited decreased longevity associated with cardiac hypertrophy but displayed little effect on muscle glucose uptake in either fasted or fed state; however, GLUT-4-deficient animals had postprandial hyperinsulinemia, indicating possible insulin resistance. A more recent study targeting muscle-specific GLUT-4 demonstrated a profound reduction in basal glucose transport and near-absence of stimulation by insulin or contraction demonstrating severe insulin resistance and glucose intolerance from an early age [205].

In a model of type 2 diabetes utilizing the Goto-Kakizaki (GK) rat, insulin-stimulated glucose uptake was 50% ($p<0.03$) lower in GK rat hearts compared with their Wistar controls with marked GLUT-4 protein depletion [206]. Moreover, these animals exhibited significant decreases in levels of the myocardial insulin receptor substrate-1 (IRS-1) as well as reduced IRS-1 association with PI3K, all key upstream events in the insulin signaling pathway. While this study also revealed decreased levels of the myocardial insulin receptor in GK rats, other studies have found conflicting information about its role in muscle insulin resistance [207]. Interestingly, while most transgenic studies addressing the complicity of the insulin signaling pathway on the development of insulin resistance have utilized the target of muscle-specific components such as GLUT-4 and the insulin receptor, there has been limited examination of the myocardial-specific inactivation of these critical genes. In transgenic mice utilizing the Cre/LoxP-generated construct of GLUT-4 directed to the heart, reduction of the GLUT-4 to a level as low as 15% of wild-type levels was sufficient to allow normal levels of insulin-stimulated glucose uptake, which was markedly reduced with further reduction of GLUT-4 levels. If GLUT-4 levels were reduced to 5% of wild-type, cardiac hypertrophy resulted [208].

HF patients showed marked downregulation of myocardial GLUT transporters that limited both glucose uptake and oxidation, and contributed to the heart's inability to generate much needed ATP [209]. Moreover, GLUT-4 transcripts have been found markedly downregulated in the human failing heart [210]. In diabetic cardiomyopathy, decreased glucose utilization results in an almost exclusive utilization of fatty acids as the myocardial energy source [211, 212]. The relatively recent finding of increased myocardial insulin resistance accompanying advanced DCM limiting both glucose uptake and oxidation may provide critical targets for therapeutic intervention [213, 214].

Insulin resistance characterized by lower myocardial insulin-sensitive glucose uptake and a reduced GLUT-4 protein level has been reported in patients with severe cardiac hypertrophy in the absence of hypertension, diabetes, and CAD [215, 216]. Interestingly, myocardial insulin-independent glucose uptake (and basal glucose uptake) as well as glycolytic metabolism are enhanced in patients with hypertension or in experimental animals with cardiac hypertrophy, the latter in striking contrast to insulin-stimulated glucose uptake which is depressed in these animals, and suggests that hearts subjected to pressure overload appear to be resistant to the metabolic effects of insulin [217, 218]. This may have important therapeutic consequences in the setting of individuals with hypertension/hypertrophy, in whom treatment with glucose–potassium–insulin aimed at altering myocardial glucose utilization may be less effective [219].

Regulation of myocardial glucose transporter levels is primarily exerted transcriptionally [220]. Observational data suggest that in the failing heart a fetal metabolic gene profile is established largely by downregulating adult gene transcripts of metabolic proteins (e.g., GLUT-4) rather than by upregulating fetal genes (e.g., GLUT-1) [221]. Similar findings of re-activation of a fetal metabolic program, which involves the downregulation of adult but not fetal isoforms as well as a re-expression of growth factors and protooncogenes, has been described in both the hypertrophied heart and with mechanical unloading of the heart [222–224]. Reactivation of these fetal genes includes a pivotal metabolic switch from fat to glucose oxidation, which though initially adaptive, ultimately results in a loss of insulin sensitivity and hence a loss of metabolic flexibility. Recent evidence suggests that this loss of flexibility then becomes an early feature of metabolic dysregulation in the failing heart, which also exhibits all the features of insulin resistance [221].

### Insulin Resistance

Although in experimental models insulin has shown cardioprotective effects, a potential link exists between insulin resistance (hyperinsulinemia) and HF. Shimizu et al. [225] have reported that excessive cardiac insulin signaling in rodents increases systolic dysfunction secondary to pressure overload. Chronic pressure overload induced hepatic insulin resistance and plasma insulin level elevation. On the other hand, cardiac insulin signaling is upregulated by chronic pressure overload because of mechanical stretch-induced activation of cardiomyocyte insulin receptors and upregulation of insulin receptor and IRS-1 expression. Moreover, chronic pressure overload increased the mismatch between cardiomyocyte size and vascularity, thereby inducing myocardial hypoxia and cardiomyocyte death. Significantly, inhibition of hyperinsulinemia ameliorated pressure overload-induced cardiac dysfunction, improving myocardial hypoxia and decreasing cardiomyocyte death. Moreover, cardiomyocyte-specific reduction of insulin receptor expression prevented cardiac ischemia and hypertrophy and also ameliorated systolic dysfunction due to pressure overload. In contrast, treatment of type 1 diabetic mice with insulin improved hyperglycemia during pressure overload, but increased myocardial ischemia and cardiomyocyte death, thereby inducing HF. Promoting angiogenesis restored the cardiac dysfunction induced by insulin treatment. Taken together, the use of insulin to control hyperglycemia could be harmful in the setting of pressure overload and modulation of insulin signaling appears to be crucial in the management of HF.

For further discussion on insulin signaling and insulin resistance, the reader should refer to Chap. 16.

## Translation Control

Activation of p70s6k has been identified as another key step in stimulation of protein synthesis under β-adrenoceptor stimulation and downstream of PI3K activation [226]. The p70s6k phosphorylates the S6 protein of the 40S ribosomal subunit, which may modulate overall translational activity. Inhibition of p70s6k attenuates the growth effect and increases β1- and α1-AR-stimulated protein synthesis in adult cardiomyocytes. Another factor that contributes to the increase in translational activity is the activation of the peptide chain initiation factor eIF-4E. The phosphorylation and activation of this translation initiation factor also depends on the activation of PKC [227] and on the presence of oxidative stress [228]. Therefore, activation of PKC, PI3K, and p70s6k represents key steps of the intracellular signaling that controls protein synthesis in adult cardiomyocytes.

In resting cardiomyocytes, activation of p70s6k can recruit a subset of the available ribosomes to participate in protein synthesis. In active cardiomyocytes, however, virtually all ribosomes appear to be functionally active and the *de novo* synthesis of ribosomal RNA is required for a significant acceleration of protein synthesis. Upon both β1- and α1-adrenoceptor stimulation, or direct stimulation of PKC, RNA polymerase I is activated, and increased levels of the ribosomal DNA transcription factor UBF are detected resulting in the increased synthesis of rRNA [229, 230]. The stimulation in RNA synthesis is also mediated through PI3K and p70s6k suggesting that p70s6k has other targets in addition to the S6 protein of ribosomes. These observations also support the premise that increased translational efficiency of existing ribosomes alone is insufficient to account for the hypertrophic growth of cardiomyocytes and that synthesis of new functional ribosomes must occur. Moreover, gene expression in vascular and endothelial cells can also be similarly regulated at the level of translation [231].

## Other Signaling Pathways in Hypertrophy and Heart Failure

### *Apoptosis Signaling*

A conceptual framework has been developed for apoptosis, a highly regulated cell suicide process that is hard-wired into all metazoan cells. Cardiac remodeling response and the transition to hypertrophy and overt HF have been associated with modestly increased apoptosis [232, 233]. However, the actual burden of chronic cell loss attributable to apoptosis is not clear.

The regulation of apoptosis is complex and not completely defined. An important signaling pathway for apoptosis in the heart during the transition from hypertrophy to HF is mediated by the balance of the proapoptotic protein Bax relative to the antiapoptotic protein Bcl-2 [80]. These proteins can stimulate or suppress the action of the caspases, which carry out the characteristic biochemical and morphological changes of apoptosis. Other key pathways include the expression of the death receptor, Fas, which is upregulated in failing cardiomyocytes and activates downstream caspases, resulting in apoptosis [81, 82]. On the other hand, while the proapoptotic multidomain proteins function as effectors, the proapoptotic Bcl-2-related proteins that possess only Bcl-2 homology domain 3 (BH3 domain-only proteins) appear to regulate the multidomain proteins by sensing stress signals and undergoing increased expression, activation, or mitochondrial translocation [234]. Nix, an proapoptotic, inducible cardiac-expressed BH3-only protein, is transcriptionally upregulated in hypertrophy due to forced increases in $G_q$ signaling, in experimental pressure overload hypertrophy, and in human hypertensive heart disease [235]. Nix induction demonstrated an important role for PKC that is also activated in reactive hypertrophy [89, 236]. Nix and other factors induced during the cardiomyocyte hypertrophic growth may sensitize cardiomyocytes to apoptosis, particularly when normal protective factors are diminished [237].

As previously discussed, apoptosis is mediated by two evolutionarily conserved central death pathways: the extrinsic pathway, which utilizes cell surface death receptors; and the intrinsic pathway, involving mitochondria and the ER (see Chap. 8) [238]. In the extrinsic pathway, death ligands (e.g., FasL) initiate apoptosis by binding their cognate receptors [239]. This stimulates the recruitment of the adaptor protein Fas-associated via death domain (FADD), which then recruits procaspase-8 into the death-inducing signaling complex (DISC) [240]. Procaspase-8 is activated by dimerization within this complex and subsequently cleaves and activates procaspase-3 and other downstream procaspases [241].

The intrinsic pathway transduces a wide variety of extracellular and intracellular stimuli including loss of survival/trophic factors, toxins, radiation, hypoxia, oxidative stress, myocardial I/R injury, and DNA damage. Although a number of peripheral pathways connect these signals with the central death machinery, each ultimately feeds into a variety of proapoptotic BH3-only proteins and the proapoptotic multidomain Bcl-2 proteins Bax and Bak [242]. These proteins undergo activation through diverse mechanisms to trigger the release of mitochondrial apoptogens, such as cytochrome *c*, Smac, EndoG, and AIF into the cytoplasm [243–246]. Once in the cytoplasm, cytochrome *c* binds Apaf-1 along with dATP. This stimulates Apaf-1 to homo-oligomerize and recruit procaspase-9 into the multiprotein complex called the apoptosome [247–250]. Within the apoptosome, procaspase-9 is activated by dimerization, after which it cleaves and activates downstream procaspases. Bid, a BH3-only protein, unites the extrinsic and intrinsic pathways: following cleavage by caspase-8, Bid's C-terminal portion translocates to the mitochondria, and triggers further apoptogen release [251, 252].

The extrinsic and intrinsic pathways are regulated by a variety of endogenous inhibitors of apoptosis (Fig. 15.5). FLICE-like (Fas-associated death domain protein-like-interleukin-1-converting enzyme-like) inhibitory protein (FLIP), whose expression is highly enriched in striated muscle, binds to and inhibits procaspase-8 in the DISC [253]. Antiapoptotic Bcl-2 proteins, such as Bcl-2 and Bcl-xL, inhibit mitochondrial apoptogen release through biochemical mechanisms that are still incompletely understood. Ku-70 and humanin bind Bax and block its conformational activation and translocation to the mitochondria [254]. X-linked inhibitor of apoptosis (XIAP) and related proteins that contain baculovirus inhibitor of apoptosis repeats bind to and inhibit already activated caspases-9, -3, and -7, as well as interfere with procaspase-9 dimerization and activation [255, 256]. Each of these inhibitors acts on circumscribed portions of either the extrinsic or intrinsic pathway. By contrast, the apoptosis repressor with a CARD (ARC), which is expressed preferentially in striated muscle and some neurons, antagonizes both the intrinsic and extrinsic apoptosis pathways [257]. The extrinsic pathway is inhibited by ARC's direct interactions with Fas, FADD, and procaspase-8, which prevent DISC assembly, while the intrinsic pathway is inhibited by ARC's direct binding and inhibition of Bax's interaction with the mitochondrial membrane [257, 258]. More recently, the endoplasmic reticulum has been recognized as an important organelle in the intrinsic pathway. In addition to its role in mediating cellular responses to traditional ER stresses, such as misfolded proteins, this organelle appears to be critical in mediating cell death elicited by a subset of stimuli originating outside of the ER, such as oxidative stress [259]. Similar to their roles in transducing upstream signals to the mitochondria, BH3-only proteins appear to relay upstream death signals to the ER [260].

In the failing human heart, the expression of protooncogenes that regulate programmed cell death is increased and is associated with increased cardiomyocyte apoptosis [261]. Myocyte death and regeneration have been considered homeostatic mechanisms intrinsic to both the normal and diseased heart [262, 263], and estimates of the rates of cell death suggest that an innate system of cardiomyocyte replacement might be present to account for the maintenance of cardiac muscle mass over a lifetime [264]. In contrast, some investigators believe that significant regeneration is not observed in conditions where myocytes are extensively lost; myocardial tumors are rare, and the evidence supporting cell renewal in the heart is rather controversial [265]. Findings of cyclin and cyclin-dependent kinase upregulation in both normal and pathological myocardium, as well as increased telomerase activity in cardiomyocytes, have been proposed to be evidence of myocyte renewal; although opposing viewpoints dismiss these observations as nonspecific biochemical events of hypertrophy and not as evidence of cardiomyocyte hyperplasia [265].

## Caveolae

Caveolae are 50–100-nm flask-shaped specialized subdomains of the plasma membrane. They are particularly abundant in cells of the cardiovascular system, including endothelial cells, SMCs, macrophages, cardiac myocytes, and fibroblasts. In these cell types, caveolae function both in protein trafficking and signal transduction. Caveolins (primarily caveolin-2 and caveolin-3 in cardiomyocytes) are structural proteins that are both necessary and sufficient for the formation of caveolae. In a number of ways, caveolins serve both to compartmentalize and to concentrate key signaling proteins, thereby regulating cardiomyocyte signaling. Multiple components of signaling cascades, including β-ARs, G proteins, AC, the Rho family of small GTPases, PKCα, and PKCε, and Erk have been localized to caveolae [266–268]. Co-localization of G protein pathway signaling molecules may be a contributory factor in both the spatial and temporal regulation of cardiomyocyte signal transduction.

Recent observations using caveolin-deficient mouse models have demonstrated that caveolae and caveolins can promote pathological phenotypes, including atherosclerosis, cardiac hypertrophy, and cardiomyopathy.

## Integrin Signaling

Several lines of evidence have shown that transduction of mechanical stress and other environmental signals are believed to occur through integrin proteins, transmembrane receptors that couple extracellular matrix components directly to the intracellular cytoskeleton and nucleus [14, 15]. In general, the signal for hypertrophy is mediated by a complex cascade of signaling systems within the cardiomyocyte, resulting in gene reprogramming [16]. These signaling mechanisms include diverse neurohumoral signals acting through many interrelated signal transduction pathways that lead to pathological cardiac hypertrophy and HF [16]. Moreover, activated hypertrophy-related genes induce the synthesis of new contractile proteins that are organized into new sarcomeres. Thus, extensive remodeling of the complete intracellular contractile apparatus is characteristic of the failing heart.

## Hypertrophic Cardiac Remodeling

The mechanisms of cardiac remodeling, particularly those underlying the transition from stable hypertrophy to cardiac dilatation and ultimately to overt HF, remain unclear. Many factors including neurohormonal and cytokine activation and impaired $Ca^{2+}$ handling are thought to play a contributory role, particularly following myocardial infarct (Fig. 15.1) [269, 270]. In addition, with increased work load, the adult heart develops pathological growth and activation of the fetal program of gene expression followed by LV dilatation, thinning of the wall, and pump failure. In a transgenic mouse model, it has been demonstrated that activation of the fetal gene program and pathological remodeling of the heart can be achieved with overexpression of MEF2D, a transcription factor that serve as target of the signaling pathway that drive pathological remodeling [271]. Also, it has been demonstrated that class II HDACs suppress stress-dependent remodeling of the heart through their association with MEF2 transcription factor and that PKD, a stress-responsive kinase that phosphorylates class II HDACs, dissociates MEF2 resulting in activation of MEF2 target genes. A mouse model with a conditional PKD1-null allele has been recently developed by Fielitz et al. [272]. Mice with cardiac-specific deletion of PKD1 exhibited improved cardiac function and diminished hypertrophy in response to pressure overload and angiotensin II signaling, confirming that PKD1 is a key transducer of stress stimuli involved in the abnormal remodeling of the heart. Previously, Harrison et al. [273] have observed that activation of PKD1 in cardiomyocytes occurs through PKC-dependent and -independent mechanisms, and *in vivo* cardiac PKD1 is activated in rodent models of pathological cardiac remodeling. Moreover, PKD1 activation correlates with phosphorylation-dependent nuclear export of HDAC5, and reduction of endogenous PKD1 expression with siRNA suppresses HDAC5 shuttling and associated cardiomyocyte growth. In contrast, ectopic overexpression of constitutively active PKD1 in mouse heart leads to DCM.

## Myocardial Metabolism and Neurohormonal Signaling in Cardiac Remodeling

### *Insights from Transgenic Models*

The creation of transgenic mice with altered expression of genes involved in carbohydrate and lipid metabolism has provided unique insights into the fine balance within the heart to maintain energy status and function, as well as to evaluate the cause–effect relationships between mitochondrial function and myocardial disease [274]. A list of transgenic models of metabolic modification in the heart that associated with cardiac dysfunction and/or HF phenotype is shown in Table 15.3.

Loss-of-function model systems, which disrupt mitochondrial metabolism, exhibit different cardiac phenotypes. One genetically engineered mouse model having a causal relationship between a mitochondrial energetic defect and cardiomyopathy is the *Ant1* null mouse. The adenine nucleotide translocators (ANTs) are a family of proteins that exchange mitochondrial ATP for cytosolic ADP, providing new ADP substrate to the mitochondria while delivering ATP to the cytoplasm for cellular work. Mice express two isoforms of this enzyme; with tissue-specific expression patterns [275]. Ant1 is expressed in skeletal muscle, heart, and the brain, while Ant2 is expressed in all tissues except for skeletal muscle. The Ant1 gene-deficient mice exhibit mitochondrial abnormalities including a partial deficit in ADP-stimulated respiration, consistent with impairment in the translocation of ADP into mitochondria in both skeletal muscle and heart. The skeletal muscle respiratory defect was profound, presumably because Ant2 can partially compensate for the loss of Ant1 in heart, but not in skeletal muscle. Ant1 (–/–) mice exhibit a progressive cardiac hypertrophic phenotype coincident with the proliferation of mitochondria [276]. The mitochondrial biogenic response has therefore been hypothesized to be a compensatory mechanism to correct the energy deficit, but could also contribute to cardiac remodeling.

**Table 15.3** Transgenic models of metabolopathies associated with cardiac remodeling and HF

| Gene | Manipulation | Cardiac phenotype |
|---|---|---|
| GLUT-4 | Global KO | Hypertrophy, ↑ expression of MCAD and LCAD, ↑ glycogen synthesis, interstitial fibrosis |
| | Heterozygous KO | Diabetic cardiomyopathy |
| | Cardiac-specific KO | Hypertrophy, ↑ BNP, glucose uptake |
| IR | Cardiac-specific KO | ↓ Myocyte size, ↓ basal glucose uptake, impaired cardiac function |
| IGF-R | Cardiac overexpression | Cardiac hypertrophy |
| IGF-1 | Overexpression | Cardiac hypertrophy, attenuation of cardiac dysfunction following myocardial infarct |
| PI3Kα | Constitutively active | Cardiac hypertrophy with no alterations in cardiac function or fibrosis |
| | Dominant negative | Smaller hearts with normal cardiac function |
| PTEN | Cardiac-specific KO | Cardiac hypertrophy and impaired contractility |
| PTEN | Constitutively active | Cardiac hypertrophy, concentric LV hypertrophy, ↓ infarct size following I/R, altered contractility |
| PDK-1 | Cre/LoxP KO | Sudden death due to HF |
| AMPKα2 | Inactive kinase | ↓ Heart weight and *in vivo* LV dP/dt, impaired glucose uptake, glycolysis and FAO, ↑ apoptosis and impaired LV recovery in response to ischemia |
| MEF2 | Overexpression | Pathological cardiac remodeling |
| PFK2 | Cardiac-specific kinase deficient | Multiple cardiac pathologies, fibrosis, ↓ contractility, impaired glycolysis |
| G6PDH | Global KO | Myocardial dysfunction |

*KO* Knock-out, *IR* insulin receptor, *IGF* insulin-like growth factor, *MCAD* medium chain acycl CoA dehydrogenase, *PDK* pyruvate dehydrogenase kinase, *I/R* ischemia/reperfusion, *LCAD* long-chain acyl CoA dehydrogenase, *MEF2* myocite enhancer factor-2, *PFK2* phosphofructokinase 2, *G6PDH* glucose 6 phosphate dehydrogenase, *AMPK* adenosine monophosphate kinase, *PTEN* phosphatase and tensin homolog deleted on chromosome 10, *GLUT* glucose transporter, *BNP* B-type or brain natriuretic peptide, *PI3K* phosphatidylinositol-3-kinase, *FAO* fatty acid oxidation

Unlike mice with PKCβ-2 overexpression, transgenic mice with cardiac-specific overexpression of a constitutively active mutant of the PKCε isoform driven by an α-myosin heavy chain promoter develops concentric cardiac hypertrophy with normal *in vivo* cardiac function, suggesting that PKC isoforms may play differential functional roles in maladaptive cardiac hypertrophy and failure [277]. That PKC isoforms which play differential roles was recently confirmed by Liu et al. [278]. They studied mice lacking PKCα, PKCβ, and PKCγ for effects on cardiac contractility and HF susceptibility. PKCα (–/–) mice, but not PKCβ/γ (–/–) mice, showed increased cardiac contractility, myocyte cellular contractility, $Ca^{2+}$ transients, and sarcoplasmic reticulum $Ca^{2+}$ load. PKCα (–/–) mice were less susceptible to HF following long-term pressure-overload stimulation or 4 weeks after myocardial infarction injury, whereas PKCβ/γ (–/–) mice showed more severe failure. Of interest, infusion of ruboxistaurin (LY333531), an orally available PKCα/β/γ inhibitor, increased cardiac contractility in wild-type and PKCβ/γ (–/–) mice, but not in PKCα (–/–) mice. Significantly, ruboxistaurin prevented death in wild-type mice throughout 10 weeks of pressure-overload stimulation, reduced ventricular dilation, enhanced ventricular performance, reduced fibrosis, and reduced pulmonary edema comparable to or better than metoprolol treatment. Ruboxistaurin was also administered to PKCβ/γ (–/–) mice subjected to pressure overload, resulting in decreasing incidence of death and HF, suggesting that PKCα as the primary target of this drug in alleviating heart disease. Taken together, PKCα isoform differs functionally from PKCβ and PKCγ in regulating cardiac contractility and HF, and broad-acting PKC inhibitors such as ruboxistaurin may be a novel approach in the treatment human HF.

### *Neurohormonal Changes and Cytokines*

Besides abnormal lipid metabolism (see Chap. 7), diverse neurohormonal signals acting through interwoven signal transduction pathways can lead to pathological cardiac hypertrophy and HF. Many of such agonists act through cell surface receptors coupled with G proteins to mobilize intracellular calcium, with consequent activation of downstream kinases and the calcium- and calmodulin-dependent phosphatase calcineurin. MAPK signaling pathways are also interconnected at multiple levels with calcium-dependent kinases and calcineurin [279]. β-Adrenergic agonists influence cardiac growth and function through the generation of cAMP, which activates PKA and other downstream effectors [280]. These signaling pathways target a variety of substrates in the cardiomyocyte, including components of the contractile apparatus, calcium channels, and their regulatory proteins.

## Cardiac Hypertrophy and Hypertension. Gender Differences

While signaling pathways in hypertension is mainly discussed in Chap. 13, here it is suffice to say that cardiac hypertrophy frequently accompanied by cardiac fibrosis, and myocardial dysfunction, is associated with gender-based differences, with higher mortality in men. Female patients with aortic stenosis exhibit better preservation of systolic function and increased LV hypertrophy than males [281–283]. Nevertheless, it is still unclear whether sex hormones, and by how much, are responsible for these differences in LV function and structural remodeling/hypertrophy [284]. For females, it may be particularly important to regulate the development of cardiac hypertrophy, since female patients with a similar degree of nonischemic hypertrophy as males exhibit higher mortality rates [285].

Gender has a significant effect on arterial blood pressure (ABP), with premenopausal women having a lower ABP than age-matched men. Also, compared with premenopausal women, postmenopausal women have higher ABP, implying a modulating role in ABP for ovarian hormones [286]. In spite of that, it is not known whether sex hormones are responsible for the gender-associated differences in ABP and whether ovarian hormones account for differences in blood pressure in premenopausal compared to postmenopausal women. On the other hand, sex hormone receptors have been identified in vascular endothelium and smooth muscle, and sex hormone interaction with cytosolic/nuclear receptors initiates long-term genomic effects that stimulate endothelial cell growth, but inhibit smooth muscle proliferation. Khalil et al. [287] noted that the activation of sex hormone receptors on the plasma membrane triggers nongenomic effects that stimulate endothelium-dependent vascular relaxation via NO/cGMP, prostacyclin-cAMP, and hyperpolarization pathways. In addition, sex hormones cause endothelium-independent inhibition of vascular smooth muscle contraction, $[Ca^{2+}]_i$, and PKC. These vasorelaxant/vasodilator effects suggested that sex hormone therapy may have a positive vascular effect in natural and surgically induced hypogonadism. Nevertheless, in some clinical trials, hormone replacement therapy (HRT) showed none or minimal benefits in postmenopausal women with hypertension; whether this was due to the type/dose of sex hormone, subject's age, or other factors remains unclear.

The potential relationship between estrogen-related gene polymorphisms and blood pressure had been further evaluated in a Framingham Heart Study offspring cohort [288]. Untreated cross-sectional and longitudinal blood pressure was correlated with polymorphisms in genes encoding the estrogene receptor (ER)α or (ER1), ERβ or (ER2), aromatase (CYP19A1), and nuclear receptor coactivator 1 (NCOA1).

In men, systolic blood pressure and pulse pressure were associated with two polymorphisms in ER1, while pulse pressure was also associated with variations in NCOA1 and CYP19A1. Polymorphisms in ER1, CYP19A1, and NCOA1 were associated with diastolic blood pressure in women. Although the underlying relations between genes involved in estrogen action and hypertension are incompletely understood, these findings suggest that there is a gender-specific contribution of estrogen-related genes to blood pressure variation.

## Antihypertrophic Signaling Pathways

Although the signal transduction pathways promoting cardiomyocyte hypertrophy have been increasingly well characterized, information concerning signaling pathways that oppose cardiomyocyte hypertrophy is more limited. Negative regulators or antihypertrophic molecules might be important targets in the treatment of cardiac remodeling/hypertrophy. Molecules such as ATP, expression of a cardiac-specific RNA helicase CHAMP (cardiac-specific helicase activated by MEF2), and retinoic acid (RA) inhibit hypertrophy when introduced exogenously or are overexpressed. Others are endogenous molecules whose essential activity is constantly active at baseline [289].

CHAMP is specifically expressed in the heart during development and adulthood [290]. At embryonic day 15, when ventricular cardiomyocytes form trabeculae, CHAMP appears to be expressed preferentially in the trabecular region where the proliferative rate is diminished relative to the adjacent compact zone, suggesting that CHAMP might play an active role in cell cycle arrest [290]. These findings indicate that intrauterine growth restriction induces global gene expression changes in the heart which act in concert to reduce cardiomyocyte proliferation [291].

### *Calcineurin Inhibitors*

In rodent models, pharmacological inhibition of calcineurin for the prevention of cardiac hypertrophy has yielded conflicting results [139, 140]. This may be due to the absence of well-tolerated calcineurin inhibitors, and also that calcineurin is not a cardiac-restricted enzyme. In comparison to normal control, the human hypertrophied heart exhibits higher calcineurin activity [141], and patients maintained with partially inhibited calcineurin activity still can develop cardiac hypertrophy. This is probably related to the presence of additional pathways making calcineurin signaling not indispensable for hypertrophy [128, 142].

### *Nitric Oxide/PKG I/Calcium*

Nitric oxide, through activation of soluble GC and cGMP formation, attenuates the hypertrophic response to growth factor stimulation in cardiomyocytes. In addition to its antihypertrophic effect, NO promotes apoptosis in cardiomyocytes in a dose-dependent manner. On the other hand, the role of cGMP in the proapoptotic effects of NO is rather controversial since cGMP analogs may or may not induce cardiomyocyte apoptosis. Moreover, activation of PKG I by NO/cGMP suppressed NFAT transcriptional activity, BNP induction, and cell enlargement. PKG I inhibits cardiomyocyte hypertrophy by targeting the calcineurin-NFAT signaling pathway and provides a framework for understanding how NO inhibits cardiomyocyte hypertrophy [292].

In general, cGMP effectors include cGMP-regulated phosphodiesterases, cGMP-regulated ion channels, and cGMP-dependent protein kinases [293]. Two PKG genes have been identified in mammalian cells, encoding PKG type I (including α- and β-splice variants) and PKG type II. In cardiomyocytes, PKG I has been suggested to mediate negative inotropic effects of NO/cGMP, possibly through regulation of the L-type $Ca^{2+}$ channel and troponin I, thereby reducing $Ca^{2+}$ influx and myofilament $Ca^{2+}$ sensitivity. However, a role for PKG I in controlling cardiomyocyte hypertrophy and/or apoptosis has not been reported.

Potential targets for PKG I in cardiomyocytes include $Ca^{2+}$-dependent signaling pathways, RhoA, and vasodilator-stimulated phosphoprotein (VASP) [294, 295]. Localization of VASP at intercalated discs in cardiomyocytes suggests that VASP may be involved in PKG I regulation of electrical coupling.

$Ca^{2+}$-dependent signaling pathways, such as calcineurin and $Ca^{2+}$/calmodulin-dependent kinases, are crucial regulators of the hypertrophic response in cardiomyocytes. PKG I regulates intracellular $Ca^{2+}$ at multiple levels, including the L-type $Ca^{2+}$ channel and the IP3 receptor. PKG I-dependent inhibition of the L-type $Ca^{2+}$ current may mediate negative inotropic effects of NO/cGMP in cardiomyocytes. Importantly, $Ca^{2+}$ influx through the L-type $Ca^{2+}$ channel has also been implicated in the regulation of cardiomyocyte hypertrophy. Therefore, antihypertrophic effects of PKG I may be mediated in part through inhibition of $Ca^{2+}$-dependent signaling pathways in cardiomyocytes. The low-molecular-weight GTPase RhoA, which is required for α1-AR signaling in cardiomyocytes, may represent an additional PKG I target. PKG I has recently been shown to phosphorylate RhoA and inhibits its biological activity in vascular SMCs, suggesting that inhibition of RhoA may also contribute to the antihypertrophic effects of PKG I. Notwithstanding the above observations, further research is needed to identify which of its many molecular targets PKG I uses to inhibit cardiomyocyte hypertrophy.

## *Diacylglycerol*

Diacylglycerol, a lipid second messenger, accumulates in cardiomyocytes when stimulated by $G_{\alpha q}$ protein-coupled receptor agonists such as angiotensin II and phenylephrine. GPCR signaling pathway, which includes DAG and PKC, plays a critical role in the development of cardiac hypertrophy and HF. DAG kinase (DGK) phosphorylates DAG and controls cellular DAG levels, thus acting as a regulator of GPCR signaling. Furthermore, DGK inhibits GPCR agonist-induced activation of the DAG-PKC signaling and subsequent cardiomyocyte hypertrophy. To determine whether DGK modifies the development of cardiac hypertrophy induced by pressure overload, Harada et al. [296] assessed the effect of thoracic aortic stenosis on the heart of transgenic mice with cardiac-specific overexpression of DGKξ (DGKξ-TG), and in wild-type mice. While the WT mice showed increases in interventricular septal thickness, LV dilatation, decreases in LV systolic function, cardiac fibrosis, and upregulation of profibrotic genes, such as transforming growth factor-β1, collagen type I, and collagen type III at 4 weeks postsurgery, the DGKξ-TG mice heart weight appears to decrease. Furthermore, cardiac fibrosis and gene induction of type I and type III collagens, but not transforming growth factor-β1, were blocked in the DGKξ-TG mice. Taken together, these data confirm that DGKξ suppresses cardiac hypertrophy and fibrosis and prevents LV systolic dysfunction caused by pressure overload. Interestingly, in neonatal rat cardiomyocytes, DGKξ inhibits PKC activation and subsequent hypertrophic programs in response to endothelin-1, DGKξ blocks cardiac hypertrophy induced by GPCR agonists and pressure overload *in vivo*. Moreover, DGKξ attenuates ventricular remodeling and improves survival after myocardial infarction [297].

Pressure and volume overload have been known to be accompanied by increased glucose uptake and glycolysis and decreased FAO [298]. Based on observations from animal studies, alterations in energy status and metabolism have been proposed to play a role in the transition from stable cardiac hypertrophy to overt HF [5, 299, 300].

## miRNA and Hypertrophy

Through gain- and loss-of-function studies in mice, it appears that microRNAs (miRNA) play specific and critical roles during cardiac hypertrophy, angiogenesis, apoptosis, and contractility. The powerful effects of modifying microRNA levels genetically have resulted in the rapid progression of oligo-based regulation of miRNAs as a new class of cardiovascular therapeutics [301]. Gathered experimental observations have shown that the introduction of functional miRNA (e.g., miRNA-21 or miRNA-18b) into cardiomyocytes represses myocyte hypertrophy. The role miRNAs play in cardiac remodeling and HF was recently reviewed by Divakaran and Mann [301]. Alterations in the expression or activity, or both, of myofilament regulatory proteins are potential mechanisms for the decrease in cardiac contractile function occurring in HF, including changes in myosin light chains and the troponin–tropomyosin complex. While some studies have implicated a role for miRNA-21 in the regulation of cardiomyocytes hypertrophic growth, others did not. Discrepancy between studies could be explained by findings that the modulation of hypertrophic growth by miRNA-21 in myocytes is achieved through an indirect mechanism, rather than a direct targeting effect of miRNA-21 on hypertrophy-related genes. So far, miRNA-21 has no confirmed gene targets that are related to cardiomyocyte hypertrophy. Therefore, its role in cardiomyocyte hypertrophy remains unknown, since exogenously administered premiRNA-21 (i.e., primary transcripts to short stem-loop structures that become a messenger RNA after processing) fails to be processed in myocytes, resulting in a lack of capacity to overexpress mature miRNA-21 in these cells.

## Conclusions

As the heart undergoes the transition from adaptive to maladaptive hypertrophy and to HF, significant changes in the signaling pathways systems occur, impacting on their ability to affect cardiac function. While significant progress has been made in the identification of the downstream targets of a number of genes and the pathways they constitute, the majority of targets and the complex interrelationships between the signaling pathways involved still remain largely undetermined. In particular, a large number of signaling interactions between cellular organelles remain to be identified. The identification of kinases, ligands, and second messengers, the entire panoply of docking proteins and scaffolding proteins which organize the aggregate structure and numerous modulators which affect the signaling processes will eventually allow a three-dimensional architecture of cardiomyocyte signaling. Manipulation of these components in the heart of transgenic animals and also in cardiomyocytes grown *in vitro* is currently possible using sophisticated techniques of gene transfer and RNAi to target specific gene expression.

Several important caveats pertain to the application of targeting specific signaling molecules within the clinical setting since numerous signaling molecules participate in multiple pathways. For instance, inhibiting ROS production might prove helpful in reducing the negative consequences of myocardial I/R, but could prove counterproductive in cardioprotection and in oxygen sensing. Moreover, the existence of redundant signaling pathways that will trigger maladaptive hypertrophy and later HF also poses significant challenges for therapeutic intervention. Understanding the signaling pathways and potential biochemical events that determine whether hypertrophy is adaptive or maladaptive, the precise order of intracellular events, their downstream consequences, the overall interrelatedness, and regulation of these pathways will be necessary to devise new therapies enabling the development and characterization of reagents with high specificity to heart signaling pathways and the arrival of new technologies for the inhibition of stress signaling (e.g., specific kinase inhibitors, oligo-based regulation of miRNAs) in cardiac and vascular cells.

## Summary

- Multiple signaling pathways regulate cardiomyocyte growth and proliferation.
- Several GPCRs including the α- and β-adrenergic receptors, angiotensin II, and endothelin-1 are able to activate these signaling cascades. Other receptors pivotal in cardiac and cardiovascular signaling include protease-activated, TLRs, RTKs, and muscarinic receptors.
- The modulation of gene expression can be fostered as well by effectors that modulate chromatin structure and histone proteins (e.g., acetylases and deacetylases).
- These effectors also can directly target proteins involved in metabolic pathways, ion transport, $Ca^{2+}$ regulation, and handling, which affect contractility and excitability, as well as cardiomyocyte apoptosis and cell survival.
- Regulatory control of the receptors is exerted by phosphorylation (by either PKA or other kinases), which can lead to receptor internalization and desensitization to the agonist. Chronic β-AR stimulation causes downregulation of the receptors.
- Upon stimulation with agonists such as noradrenaline and adrenaline, α1-ARs activate $G_q$ proteins and subsequently activate phospholipase Cβ resulting in increased levels of second messengers IP3 and DAG which promote an increase in intracellular $Ca^{2+}$ levels and PKC activation. They are involved in both modulating cardiac contractility and cardiomyocyte hypertrophy.
- Caveolae and specific proteins can organize the signaling complexes within microdomains at the plasma membrane. Other proteins serve to enhance and orchestrate the signaling complexes including anchoring, adaptor, and scaffolding proteins interacting with diverse stimuli.
- Stress, metabolic, and mitotic stimuli can act through signaling pathways and share a variety of signaling components. Survival pathways involving Akt and PI3K, PKA, and PKC can be utilized by these different stimuli.
- Besides abnormal lipid metabolism, diverse neurohormonal signals acting through interwoven signal transduction pathways can lead to pathological cardiac hypertrophy and HF.
- Female patients with aortic stenosis exhibit better preservation of systolic function and increased LV hypertrophy than males.
- Negative regulator or antihypertrophic molecules might be important targets in the treatment of cardiac remodeling/hypertrophy.
- Nitric oxide (NO), through activation of soluble GC and cGMP formation, attenuates the hypertrophic response to growth factor stimulation in cardiomyocytes.
- PKG I inhibits cardiomyocyte hypertrophy by targeting the calcineurin-NFAT signaling pathway and provides a framework for understanding how NO inhibits cardiomyocyte hypertrophy.
- DGKzeta inhibits GPCR agonist-induced activation of the DAG-PKC signaling and subsequent cardiomyocyte hypertrophy.
- Experimental observations have shown that the introduction of functional miRNA (e.g., miRNA-21 or miRNA-18b) into cardiomyocytes represses myocyte hypertrophy.
- Discrepancy between studies could be explained by findings that the modulation of hypertrophic growth by miRNA-21 in myocytes is achieved through an indirect mechanism, rather than a direct targeting effect of miR-21 on hypertrophy-related genes.

## References


1. Houser SR, Margulies KB. Is depressed myocyte contractility centrally involved in heart failure? Circ Res. 2003;92:350–8.
2. Davies CH, Davia K, Bennett JG, Pepper JR, Poole-Wilson PA, Harding SE. Reduced contraction and altered frequency response of isolated ventricular myocytes from patients with heart failure. Circulation. 1995;92:2540–9.
3. Maron BJ. Hypertrophic cardiomyopathy: a systematic review. JAMA. 2002;287:1308–20.
4. Swynghedauw B. Molecular mechanisms of myocardial remodeling. Physiol Rev. 1999;79:215–62.
5. Sambandam N, Lopaschuk GD, Brownsey RW, Allard MF. Energy metabolism in the hypertrophied heart. Heart Fail Rev. 2002;7:161–73.
6. Opie LH, Commerford PJ, Gersh BJ, Pfeffer MA. Controversies in ventricular remodelling. Lancet. 2006;367:356–67.
7. Hill JA, Olson EN. Cardiac plasticity. N Engl J Med. 2008;358:1370–80.
8. Cohn JN, Ferrari R, Sharpe N. Cardiac remodeling–concepts and clinical implications: a consensus paper from an international forum on cardiac remodeling. Behalf of an International Forum on Cardiac Remodeling. J Am Coll Cardiol. 2000;35:569–82.
9. Carabello BA. Concentric versus eccentric remodeling. J Card Fail. 2002;8:S258–63.

10. Brower GL, Janicki JS. Contribution of ventricular remodeling to pathogenesis of heart failure in rats. Am J Physiol Heart Circ Physiol. 2001;280:H674–83.
11. Litwin SE, Raya TE, Anderson PG, Litwin CM, Bressler R, Goldman S. Induction of myocardial hypertrophy after coronary ligation in rats decreases ventricular dilatation and improves systolic function. Circulation. 1991;84:1819–27.
12. Asakura M, Kitakaze M, Takashima S, et al. Cardiac hypertrophy is inhibited by antagonism of ADAM12 processing of HB-EGF: metalloproteinase inhibitors as a new therapy. Nat Med. 2002;8:35–40.
13. Liao JK. Shedding growth factors in cardiac hypertrophy. Nat Med. 2002;8:20–1.
14. Ross RS, Borg TK. Integrins and the myocardium. Circ Res. 2001;88:1112–9.
15. Terracio L, Rubin K, Gullberg D, et al. Expression of collagen binding integrins during cardiac development and hypertrophy. Circ Res. 1991;68:734–44.
16. Frey N, Olson EN. Cardiac hypertrophy: the good, the bad, and the ugly. Annu Rev Physiol. 2003;65:45–79.
17. Fedak PW, Verma S, Weisel RD, Li RK. Cardiac remodeling and failure: from molecules to man (Part I). Cardiovasc Pathol. 2005;14:1–11.
18. Dorn 2nd GW, Hahn HS. Genetic factors in cardiac hypertrophy. Ann NY Acad Sci. 2004;1015:225–37.
19. Frey N, Katus HA, Olson EN, Hill JA. Hypertrophy of the heart: a new therapeutic target? Circulation. 2004;109:1580–9.
20. Rybin VO, Pak E, Alcott S, Steinberg SF. Developmental changes in beta2-adrenergic receptor signaling in ventricular myocytes: the role of Gi proteins and caveolae microdomains. Mol Pharmacol. 2003;63:1338–48.
21. Rockman HA, Koch WJ, Milano CA, Lefkowitz RJ. Myocardial beta-adrenergic receptor signaling in vivo: insights from transgenic mice. J Mol Med. 1996;74:489–95.
22. Brodde OE, Michel MC. Adrenergic and muscarinic receptors in the human heart. Pharmacol Rev. 1999;51:651–90.
23. O'Rourke MF, Iversen LJ, Lomasney JW, Bylund DB. Species orthologs of the alpha-2A adrenergic receptor: the pharmacological properties of the bovine and rat receptors differ from the human and porcine receptors. J Pharmacol Exp Ther. 1994;271:735–40.
24. Flordellis C, Manolis A, Scheinin M, Paris H. Clinical and pharmacological significance of alpha2-adrenoceptor polymorphisms in cardiovascular diseases. Int J Cardiol. 2004;97:367–72.
25. Penela P, Murga C, Ribas C, Tutor AS, Peregrin S, Mayor Jr F. Mechanisms of regulation of G protein-coupled receptor kinases (GRKs) and cardiovascular disease. Cardiovasc Res. 2006; 69:46–56.
26. Hata JA, Koch WJ. Phosphorylation of G protein-coupled receptors: GPCR kinases in heart disease. Mol Interv. 2003;3:264–72.
27. Mueller EE, Grandy SA, Howlett SE. Protein kinase A-mediated phosphorylation contributes to enhanced contraction observed in mice that overexpress beta-adrenergic receptor kinase-1. J Pharmacol Exp Ther. 2006;319:1307–16.
28. Mondry A, Bourgeois F, Carre F, Swynghedauw B, Moalic JM. Decrease in beta 1-adrenergic and M2-muscarinic receptor mRNA levels and unchanged accumulation of mRNAs coding for G alpha i-2 and G alpha s proteins in rat cardiac hypertrophy. J Mol Cell Cardiol. 1995;27:2287–94.
29. Wang Z, Shi H, Wang H. Functional M3 muscarinic acetylcholine receptors in mammalian hearts. Br J Pharmacol. 2004;142: 395–408.
30. Shi H, Wang H, Yang B, Xu D, Wang Z. The M3 receptor-mediated K(+) current (IKM3), a G(q) protein-coupled K(+) channel. J Biol Chem. 2004;279:21774–8.
31. Cimini CM, Upsher ME, Weiss HR. Myocardial O2 supply and consumption in early cardiac hypertrophy of renal hypertensive rabbits. Basic Res Cardiol. 1989;84:13–21.
32. Cimini CM, Weiss HR. Isoproterenol and myocardial O2 supply/consumption in hypertension-induced myocardial hypertrophy. Am J Physiol. 1990;259:H346–51.
33. Brodde OE, Leineweber K. Autonomic receptor systems in the failing and aging human heart: similarities and differences. Eur J Pharmacol. 2004;500:167–76.
34. Scholz PM, Grover GJ, Mackenzie JW, Weiss HR. Regional oxygen supply and consumption balance in experimental left ventricular hypertrophy. Basic Res Cardiol. 1990;85:575–84.
35. Jakob G, Mair J, Pichler M, Puschendorf B. Ergometric exercise testing and sensitivity of cyclic guanosine 3′,5′-monophosphate (cGMP) in diagnosing asymptomatic left ventricular dysfunction. Br Heart J. 1995;73:145–50.
36. Lubien E, DeMaria A, Krishnaswamy P, et al. Utility of B-natriuretic peptide in detecting diastolic dysfunction: comparison with Doppler velocity recordings. Circulation. 2002;105:595–601.
37. Kotchi Kotchi E, Weisselberg T, Rohnert P, et al. Nitric oxide inhibits isoprenaline-induced positive inotropic effects in normal, but not in hypertrophied rat heart. Naunyn Schmiedebergs Arch Pharmacol. 1998;357:579–83.
38. Morrison LK, Harrison A, Krishnaswamy P, Kazanegra R, Clopton P, Maisel A. Utility of a rapid B-natriuretic peptide assay in differentiating congestive heart failure from lung disease in patients presenting with dyspnea. J Am Coll Cardiol. 2002;39:202–9.
39. Guo X, Kedem J, Weiss HR, Tse J, Roitstein A, Scholz PM. Effect of cyclic GMP reduction on regional myocardial mechanics and metabolism in experimental left ventricular hypertrophy. J Cardiovasc Pharmacol. 1996;27:392–400.
40. Roitstein A, Kedem J, Cheinberg B, Weiss HR, Tse J, Scholz PM. The effect of intracoronary nitroprusside on cyclic GMP and regional mechanics is altered in a canine model of left ventricular hypertrophy. J Surg Res. 1994;57:584–90.
41. Seymour AA, Burkett DE, Asaad MM, Lanoce VM, Clemons AF, Rogers WL. Hemodynamic, renal, and hormonal effects of rapid ventricular pacing in conscious dogs. Lab Anim Sci. 1994;44:443–52.
42. Tajima M, Bartunek J, Weinberg EO, Ito N, Lorell BH. Atrial natriuretic peptide has different effects on contractility and intracellular pH in normal and hypertrophied myocytes from pressure-overloaded hearts. Circulation. 1998;98:2760–4.
43. Paulus WJ, Frantz S, Kelly RA. Nitric oxide and cardiac contractility in human heart failure: time for reappraisal. Circulation. 2001;104:2260–2.
44. Rosenkranz AC, Hood SG, Woods RL, Dusting GJ, Ritchie RH. B-type natriuretic peptide prevents acute hypertrophic responses in the diabetic rat heart: importance of cyclic GMP. Diabetes. 2003;52:2389–95.
45. Silberbach M, Roberts Jr CT. Natriuretic peptide signalling: molecular and cellular pathways to growth regulation. Cell Signal. 2001;13:221–31.
46. Simko F, Simko J. The potential role of nitric oxide in the hypertrophic growth of the left ventricle. Physiol Res. 2000;49:37–46.
47. Fagan JM, Rex SE, Hayes-Licitra SA, Waxman L. L-arginine reduces right heart hypertrophy in hypoxia-induced pulmonary hypertension. Biochem Biophys Res Commun. 1999;254:100–3.
48. Padilla F, Garcia-Dorado D, Agullo L, et al. Intravenous administration of the natriuretic peptide urodilatin at low doses during coronary reperfusion limits infarct size in anesthetized pigs. Cardiovasc Res. 2001;51:592–600.
49. Wollert KC, Drexler H. Regulation of cardiac remodeling by nitric oxide: focus on cardiac myocyte hypertrophy and apoptosis. Heart Fail Rev. 2002;7:317–25.
50. Devlin AM, Brosnan MJ, Graham D, et al. Vascular smooth muscle cell polyploidy and cardiomyocyte hypertrophy due to chronic NOS inhibition in vivo. Am J Physiol. 1998;274:H52–9.
51. Bubikat A, De Windt LJ, Zetsche B, et al. Local atrial natriuretic peptide signaling prevents hypertensive cardiac hypertrophy in

endothelial nitric-oxide synthase-deficient mice. J Biol Chem. 2005;280:21594–9.
52. Takimoto E, Champion HC, Belardi D, et al. cGMP catabolism by phosphodiesterase 5A regulates cardiac adrenergic stimulation by NOS3-dependent mechanism. Circ Res. 2005;96:100–9.
53. Su J, Zhang Q, Moalem J, Tse J, Scholz PM, Weiss HR. Functional effects of C-type natriuretic peptide and nitric oxide are attenuated in hypertrophic myocytes from pressure-overloaded mouse hearts. Am J Physiol Heart Circ Physiol. 2005;288:H1367–73.
54. Yan L, Zhang Q, Scholz PM, Weiss HR. Cyclic GMP protein kinase activity is reduced in thyroxine-induced hypertrophic cardiac myocytes. Clin Exp Pharmacol Physiol. 2003;30:943–50.
55. Katz E, Zhang Q, Weiss HR, Scholz PM. T4-induced cardiac hypertrophy disrupts cyclic GMP mediated responses to brain natriuretic peptide in rabbit myocardium. Peptides. 2006;27:2276–83.
56. Zhang Q, Lazar M, Molino B, et al. Reduction in interaction between cGMP and cAMP in dog ventricular myocytes with hypertrophic failure. Am J Physiol Heart Circ Physiol. 2005; 289:H1251–7.
57. Giannessi D, Del Ry S, Vitale RL. The role of endothelins and their receptors in heart failure. Pharmacol Res. 2001;43:111–26.
58. Sugden PH. An overview of endothelin signaling in the cardiac myocyte. J Mol Cell Cardiol. 2003;35:871–86.
59. Dinh DT, Frauman AG, Johnston CI, Fabiani ME. Angiotensin receptors: distribution, signalling and function. Clin Sci (Lond). 2001;100:481–92.
60. Saito Y, Berk BC. Angiotensin II-mediated signal transduction pathways. Curr Hypertens Rep. 2002;4:167–71.
61. Brasier AR, Jamaluddin M, Han Y, Patterson C, Runge MS. Angiotensin II induces gene transcription through cell-type-dependent effects on the nuclear factor-kappaB (NF-kappaB) transcription factor. Mol Cell Biochem. 2000;212:155–69.
62. Chen Y, Arrigo AP, Currie RW. Heat shock treatment suppresses angiotensin II-induced activation of NF-kappaB pathway and heart inflammation: a role for IKK depletion by heat shock? Am J Physiol Heart Circ Physiol. 2004;287:H1104–14.
63. Clerk A, Aggeli IK, Stathopoulou K, Sugden PH. Peptide growth factors signal differentially through protein kinase C to extracellular signal-regulated kinases in neonatal cardiomyocytes. Cell Signal. 2006;18:225–35.
64. Coughlin SR, Camerer E. PARticipation in inflammation. J Clin Invest. 2003;111:25–7.
65. Sabri A, Muske G, Zhang H, et al. Signaling properties and functions of two distinct cardiomyocyte protease-activated receptors. Circ Res. 2000;86:1054–61.
66. Barnes JA, Singh S, Gomes AV. Protease activated receptors in cardiovascular function and disease. Mol Cell Biochem. 2004;263:227–39.
67. Hamm HE. The many faces of G protein signaling. J Biol Chem. 1998;273:669–72.
68. Adams JW, Brown JH. G-proteins in growth and apoptosis: lessons from the heart. Oncogene. 2001;20:1626–34.
69. Dorn 2nd GW, Brown JH. Gq signaling in cardiac adaptation and maladaptation. Trends Cardiovasc Med. 1999;9:26–34.
70. Clerk A, Sugden PH. Small guanine nucleotide-binding proteins and myocardial hypertrophy. Circ Res. 2000;86:1019–23.
71. Meszaros JG, Gonzalez AM, Endo-Mochizuki Y, Villegas S, Villarreal F, Brunton LL. Identification of G protein-coupled signaling pathways in cardiac fibroblasts: cross talk between G(q) and G(s). Am J Physiol Cell Physiol. 2000;278:C154–62.
72. Engel FB, Schebesta M, Duong MT, et al. p38 MAP kinase inhibition enables proliferation of adult mammalian cardiomyocytes. Genes Dev. 2005;19:1175–87.
73. Tamamori-Adachi M, Ito H, Sumrejkanchanakij P, et al. Critical role of cyclin D1 nuclear import in cardiomyocyte proliferation. Circ Res. 2003;92:e12–9.
74. Gao MH, Lai NC, Roth DM, et al. Adenylylcyclase increases responsiveness to catecholamine stimulation in transgenic mice. Circulation. 1999;99:1618–22.
75. Lai NC, Roth DM, Gao MH, et al. Intracoronary adenovirus encoding adenylyl cyclase VI increases left ventricular function in heart failure. Circulation. 2004;110:330–6.
76. Roth DM, Gao MH, Lai NC, et al. Cardiac-directed adenylyl cyclase expression improves heart function in murine cardiomyopathy. Circulation. 1999;99:3099–102.
77. Iwami G, Kawabe J, Ebina T, Cannon PJ, Homcy CJ, Ishikawa Y. Regulation of adenylyl cyclase by protein kinase A. J Biol Chem. 1995;270:12481–4.
78. Rhee SG. Regulation of phosphoinositide-specific phospholipase C. Annu Rev Biochem. 2001;70:281–312.
79. Wing MR, Bourdon DM, Harden TK. PLC-epsilon: a shared effector protein in Ras-, Rho-, and G alpha beta gamma-mediated signaling. Mol Interv. 2003;3:273–80.
80. Tirziu D, Simons M. Endothelium-driven myocardial growth or nitric oxide at the crossroads. Trends Cardiovasc Med. 2008;18:299–305.
81. Burger AJ, Horton DP, LeJemtel T, et al. Effect of nesiritide (B-type natriuretic peptide) and dobutamine on ventricular arrhythmias in the treatment of patients with acutely decompensated congestive heart failure: the PRECEDENT study. Am Heart J. 2002;144:1102–8.
82. Winter WE, Elin RJ. The role and assessment of ventricular peptides in heart failure. Clin Lab Med. 2004;24:235–74.
83. Parodi O, De Maria R, Roubina E. Redox state, oxidative stress and endothelial dysfunction in heart failure: the puzzle of nitrate-thiol interaction. J Cardiovasc Med (Hagerstown). 2007;8:765–74.
84. Kempf T, Wollert KC. Nitric oxide and the enigma of cardiac hypertrophy. Bioessays. 2004;26:608–15.
85. Dodge KL, Khouangsathiene S, Kapiloff MS, et al. mAKAP assembles a protein kinase A/PDE4 phosphodiesterase cAMP signaling module. EMBO J. 2001;20:1921–30.
86. Hulme JT, Scheuer T, Catterall WA. Regulation of cardiac ion channels by signaling complexes: role of modified leucine zipper motifs. J Mol Cell Cardiol. 2004;37:625–31.
87. Shiojima I, Walsh K. Regulation of cardiac growth and coronary angiogenesis by the Akt/PKB signaling pathway. Genes Dev. 2006;20:3347–65.
88. Chesley A, Lundberg MS, Asai T, et al. The beta(2)-adrenergic receptor delivers an antiapoptotic signal to cardiac myocytes through G(i)-dependent coupling to phosphatidylinositol 3′-kinase. Circ Res. 2000;87:1172–9.
89. Dorn 2nd GW, Force T. Protein kinase cascades in the regulation of cardiac hypertrophy. J Clin Invest. 2005;115:527–37.
90. Sugden PH. Ras, Akt, and mechanotransduction in the cardiac myocyte. Circ Res. 2003;93:1179–92.
91. Aikawa R, Nawano M, Gu Y, et al. Insulin prevents cardiomyocytes from oxidative stress-induced apoptosis through activation of PI3 kinase/Akt. Circulation. 2000;102:2873–9.
92. Gao F, Gao E, Yue TL, et al. Nitric oxide mediates the antiapoptotic effect of insulin in myocardial ischemia-reperfusion: the roles of PI3-kinase, Akt, and endothelial nitric oxide synthase phosphorylation. Circulation. 2002;105:1497–502.
93. Matsui T, Li L, del Monte F, et al. Adenoviral gene transfer of activated phosphatidylinositol 3′-kinase and Akt inhibits apoptosis of hypoxic cardiomyocytes in vitro. Circulation. 1999;100:2373–9.
94. Yamashita K, Kajstura J, Discher DJ, et al. Reperfusion-activated Akt kinase prevents apoptosis in transgenic mouse hearts overexpressing insulin-like growth factor-1. Circ Res. 2001;88:609–14.
95. Craig R, Wagner M, McCardle T, Craig AG, Glembotski CC. The cytoprotective effects of the glycoprotein 130 receptor-coupled cytokine, cardiotrophin-1, require activation of NF-kappa B. J Biol Chem. 2001;276:37621–9.

96. Negoro S, Oh H, Tone E, et al. Glycoprotein 130 regulates cardiac myocyte survival in doxorubicin-induced apoptosis through phosphatidylinositol 3-kinase/Akt phosphorylation and Bcl-xL/caspase-3 interaction. Circulation. 2001;103:555–61.
97. Tian B, Liu J, Bitterman P, Bache RJ. Angiotensin II modulates nitric oxide-induced cardiac fibroblast apoptosis by activation of AKT/PKB. Am J Physiol Heart Circ Physiol. 2003;285:H1105–12.
98. Clerk A, Sugden PH. Activation of protein kinase cascades in the heart by hypertrophic G protein-coupled receptor agonists. Am J Cardiol. 1999;83:64H–9.
99. Krieg T, Landsberger M, Alexeyev MF, Felix SB, Cohen MV, Downey JM. Activation of Akt is essential for acetylcholine to trigger generation of oxygen free radicals. Cardiovasc Res. 2003;58:196–202.
100. Yin H, Chao L, Chao J. Adrenomedullin protects against myocardial apoptosis after ischemia/reperfusion through activation of Akt-GSK signaling. Hypertension. 2004;43:109–16.
101. Naga Prasad SV, Esposito G, Mao L, Koch WJ, Rockman HA. Gbetagamma-dependent phosphoinositide 3-kinase activation in hearts with in vivo pressure overload hypertrophy. J Biol Chem. 2000;275:4693–8.
102. Mockridge JW, Marber MS, Heads RJ. Activation of Akt during simulated ischemia/reperfusion in cardiac myocytes. Biochem Biophys Res Commun. 2000;270:947–52.
103. Matsui T, Tao J, del Monte F, et al. Akt activation preserves cardiac function and prevents injury after transient cardiac ischemia in vivo. Circulation. 2001;104:330–5.
104. Brunet A, Bonni A, Zigmond MJ, et al. Akt promotes cell survival by phosphorylating and inhibiting a Forkhead transcription factor. Cell. 1999;96:857–68.
105. Shioi T, McMullen JR, Kang PM, et al. Akt/protein kinase B promotes organ growth in transgenic mice. Mol Cell Biol. 2002;22: 2799–809.
106. Matsui T, Li L, Wu JC, et al. Phenotypic spectrum caused by transgenic overexpression of activated Akt in the heart. J Biol Chem. 2002;277:22896–901.
107. Wang L, Wang X, Proud CG. Activation of mRNA translation in rat cardiac myocytes by insulin involves multiple rapamycin-sensitive steps. Am J Physiol Heart Circ Physiol. 2000;278: H1056–68.
108. Gao X, Zhang Y, Arrazola P, et al. Tsc tumour suppressor proteins antagonize amino-acid-TOR signalling. Nat Cell Biol. 2002;4: 699–704.
109. Marygold SJ, Leevers SJ. Growth signaling: TSC takes its place. Curr Biol. 2002;12:R785–7.
110. Zhang Y, Gao X, Saucedo LJ, Ru B, Edgar BA, Pan D. Rheb is a direct target of the tuberous sclerosis tumour suppressor proteins. Nat Cell Biol. 2003;5:578–81.
111. Tee AR, Manning BD, Roux PP, Cantley LC, Blenis J. Tuberous sclerosis complex gene products, Tuberin and Hamartin, control mTOR signaling by acting as a GTPase-activating protein complex toward Rheb. Curr Biol. 2003;13:1259–68.
112. Badorff C, Ruetten H, Mueller S, et al. Fas receptor signaling inhibits glycogen synthase kinase 3 beta and induces cardiac hypertrophy following pressure overload. J Clin Invest. 2002;109:373–81.
113. Haq S, Choukroun G, Kang ZB, et al. Glycogen synthase kinase-3beta is a negative regulator of cardiomyocyte hypertrophy. J Cell Biol. 2000;151:117–30.
114. Morisco C, Zebrowski D, Condorelli G, Tsichlis P, Vatner SF, Sadoshima J. The Akt-glycogen synthase kinase 3beta pathway regulates transcription of atrial natriuretic factor induced by beta-adrenergic receptor stimulation in cardiac myocytes. J Biol Chem. 2000;275:14466–75.
115. Antos CL, McKinsey TA, Frey N, et al. Activated glycogen synthase-3 beta suppresses cardiac hypertrophy in vivo. Proc Natl Acad Sci USA. 2002;99:907–12.
116. Proud CG. Ras, PI3-kinase and mTOR signaling in cardiac hypertrophy. Cardiovasc Res. 2004;63:403–13.
117. Haq S, Michael A, Andreucci M, et al. Stabilization of beta-catenin by a Wnt-independent mechanism regulates cardiomyocyte growth. Proc Natl Acad Sci USA. 2003;100:4610–5.
118. Morisco C, Seta K, Hardt SE, Lee Y, Vatner SF, Sadoshima J. Glycogen synthase kinase 3beta regulates GATA4 in cardiac myocytes. J Biol Chem. 2001;276:28586–97.
119. Xiao G, Mao S, Baumgarten G, et al. Inducible activation of c-Myc in adult myocardium in vivo provokes cardiac myocyte hypertrophy and reactivation of DNA synthesis. Circ Res. 2001;89: 1122–9.
120. Michael A, Haq S, Chen X, et al. Glycogen synthase kinase-3beta regulates growth, calcium homeostasis, and diastolic function in the heart. J Biol Chem. 2004;279:21383–93.
121. Liao W, Wang S, Han C, Zhang Y. 14-3-3 proteins regulate glycogen synthase 3beta phosphorylation and inhibit cardiomyocyte hypertrophy. FEBS J. 2005;272:1845–54.
122. Chan AY, Soltys CL, Young ME, Proud CG, Dyck JR. Activation of AMP-activated protein kinase inhibits protein synthesis associated with hypertrophy in the cardiac myocyte. J Biol Chem. 2004;279:32771–9.
123. Arsham AM, Howell JJ, Simon MC. A novel hypoxia-inducible factor-independent hypoxic response regulating mammalian target of rapamycin and its targets. J Biol Chem. 2003;278:29655–60.
124. van Noort M, Meeldijk J, van der Zee R, Destree O, Clevers H. Wnt signaling controls the phosphorylation status of beta-catenin. J Biol Chem. 2002;277:17901–5.
125. Sabri A, Steinberg SF. Protein kinase C isoform-selective signals that lead to cardiac hypertrophy and the progression of heart failure. Mol Cell Biochem. 2003;251:97–101.
126. Mochly-Rosen D, Wu G, Hahn H, et al. Cardiotrophic effects of protein kinase C epsilon: analysis by in vivo modulation of PKCepsilon translocation. Circ Res. 2000;86:1173–9.
127. Kook H, Lepore JJ, Gitler AD, et al. Cardiac hypertrophy and histone deacetylase-dependent transcriptional repression mediated by the atypical homeodomain protein Hop. J Clin Invest. 2003;112: 863–71.
128. McKinsey TA, Olson EN. Toward transcriptional therapies for the failing heart: chemical screens to modulate genes. J Clin Invest. 2005;115:538–46.
129. Im SH, Rao A. Activation and deactivation of gene expression by Ca2+/calcineurin-NFAT-mediated signaling. Mol Cells. 2004; 18:1–9.
130. Chin D, Means AR. Calmodulin: a prototypical calcium sensor. Trends Cell Biol. 2000;10:322–8.
131. Frey N, McKinsey TA, Olson EN. Decoding calcium signals involved in cardiac growth and function. Nat Med. 2000;6: 1221–7.
132. Zhang T, Johnson EN, Gu Y, et al. The cardiac-specific nuclear delta(B) isoform of Ca2+/calmodulin-dependent protein kinase II induces hypertrophy and dilated cardiomyopathy associated with increased protein phosphatase 2A activity. J Biol Chem. 2002; 277:1261–7.
133. Zhu W, Zou Y, Shiojima I, et al. Ca2+/calmodulin-dependent kinase II and calcineurin play critical roles in endothelin-1-induced cardiomyocyte hypertrophy. J Biol Chem. 2000;275:15239–45.
134. Liang F, Wu J, Garami M, Gardner DG. Mechanical strain increases expression of the brain natriuretic peptide gene in rat cardiac myocytes. J Biol Chem. 1997;272:28050–6.
135. Zhang T, Maier LS, Dalton ND, et al. The deltaC isoform of CaMKII is activated in cardiac hypertrophy and induces dilated cardiomyopathy and heart failure. Circ Res. 2003;92:912–9.
136. Passier R, Zeng H, Frey N, et al. CaM kinase signaling induces cardiac hypertrophy and activates the MEF2 transcription factor in vivo. J Clin Invest. 2000;105:1395–406.

137. Molkentin JD. Calcineurin-NFAT signaling regulates the cardiac hypertrophic response in coordination with the MAPKs. Cardiovasc Res. 2004;63:467–75.
138. Olson EN, Molkentin JD. Prevention of cardiac hypertrophy by calcineurin inhibition: hope or hype? Circ Res. 1999;84:623–32.
139. Bueno OF, Brandt EB, Rothenberg ME, Molkentin JD. Defective T cell development and function in calcineurin A beta -deficient mice. Proc Natl Acad Sci USA. 2002;99:9398–403.
140. McKinsey TA, Olson EN. Cardiac hypertrophy: sorting out the circuitry. Curr Opin Genet Dev. 1999;9:267–74.
141. Haq S, Choukroun G, Lim H, et al. Differential activation of signal transduction pathways in human hearts with hypertrophy versus advanced heart failure. Circulation. 2001;103:670–7.
142. Chen M, Li X, Dong Q, Li Y, Liang W. Neuropeptide Y induces cardiomyocyte hypertrophy via calcineurin signaling in rats. Regul Pept. 2005;125:9–15.
143. Berridge MJ. Remodelling Ca2+ signalling systems and cardiac hypertrophy. Biochem Soc Trans. 2006;34:228–31.
144. Pi M, Oakley RH, Gesty-Palmer D, et al. Beta-arrestin- and G protein receptor kinase-mediated calcium-sensing receptor desensitization. Mol Endocrinol. 2005;19:1078–87.
145. Hata JA, Williams ML, Koch WJ. Genetic manipulation of myocardial beta-adrenergic receptor activation and desensitization. J Mol Cell Cardiol. 2004;37:11–21.
146. Koch WJ, Rockman HA, Samama P, et al. Cardiac function in mice overexpressing the beta-adrenergic receptor kinase or a beta ARK inhibitor. Science. 1995;268:1350–3.
147. Metaye T, Gibelin H, Perdrisot R, Kraimps JL. Pathophysiological roles of G protein-coupled receptor kinases. Cell Signal. 2005; 17:917–28.
148. Vinge LE, Oie E, Andersson Y, Grogaard HK, Andersen G, Attramadal H. Myocardial distribution and regulation of GRK and beta-arrestin isoforms in congestive heart failure in rats. Am J Physiol Heart Circ Physiol. 2001;281:H2490–9.
149. Sopontammarak S, Aliharoob A, Ocampo C, Arcilla RA, Gupta MP, Gupta M. Mitogen-activated protein kinases (p38 and c-Jun NH2-terminal kinase) are differentially regulated during cardiac volume and pressure overload hypertrophy. Cell Biochem Biophys. 2005;43:61–76.
150. Streicher JM, Ren S, Herschman H, Wang Y. MAPK-activated protein kinase-2 in cardiac hypertrophy and cyclooxygenase-2 regulation in heart. Circ Res. 2010;106:1434–43.
151. Pasumarthi KB, Kardami E, Cattini PA. High and low molecular weight fibroblast growth factor-2 increase proliferation of neonatal rat cardiac myocytes but have differential effects on binucleation and nuclear morphology. Evidence for both paracrine and intracrine actions of fibroblast growth factor-2. Circ Res. 1996;78:126–36.
152. Sheikh F, Jin Y, Pasumarthi KB, Kardami E, Cattini PA. Expression of fibroblast growth factor receptor-1 in rat heart H9c2 myoblasts increases cell proliferation. Mol Cell Biochem. 1997;176:89–97.
153. Sheng Z, Pennica D, Wood WI, Chien KR. Cardiotrophin-1 displays early expression in the murine heart tube and promotes cardiac myocyte survival. Development. 1996;122:419–28.
154. Kuwahara K, Saito Y, Kishimoto I, et al. Cardiotrophin-1 phosphorylates akt and BAD, and prolongs cell survival via a PI3K-dependent pathway in cardiac myocytes. J Mol Cell Cardiol. 2000;32:1385–94.
155. Huss JM, Kelly DP. Mitochondrial energy metabolism in heart failure: a question of balance. J Clin Invest. 2005;115:547–55.
156. Molkentin JD, Dorn 2nd GW. Cytoplasmic signaling pathways that regulate cardiac hypertrophy. Annu Rev Physiol. 2001;63:391–426.
157. Kilic A, Velic A, De Windt LJ, et al. Enhanced activity of the myocardial Na+/H+exchanger NHE-1 contributes to cardiac remodeling in atrial natriuretic peptide receptor-deficient mice. Circulation. 2005;112:2307–17.
158. Grassot J, Mouchiroud G, Perriere G. RTKdb: database of Receptor Tyrosine Kinase. Nucleic Acids Res. 2003;31:353–8.
159. Lee HJ, Koh GY. Shear stress activates Tie2 receptor tyrosine kinase in human endothelial cells. Biochem Biophys Res Commun. 2003;304:399–404.
160. Zhao YY, Sawyer DR, Baliga RR, et al. Neuregulins promote survival and growth of cardiac myocytes. Persistence of ErbB2 and ErbB4 expression in neonatal and adult ventricular myocytes. J Biol Chem. 1998;273:10261–9.
161. Becker E, Huynh-Do U, Holland S, Pawson T, Daniel TO, Skolnik EY. Nck-interacting Ste20 kinase couples Eph receptors to c-Jun N-terminal kinase and integrin activation. Mol Cell Biol. 2000; 20:1537–45.
162. Lorenz K, Schmitt JP, Vidal M, Lohse MJ. Cardiac hypertrophy: targeting Raf/MEK/ERK1/2-signaling. Int J Biochem Cell Biol. 2009;41:2351–5.
163. Jones WK, Brown M, Ren X, He S, McGuinness M. NF-kappaB as an integrator of diverse signaling pathways: the heart of myocardial signaling? Cardiovasc Toxicol. 2003;3:229–54.
164. Huss JM, Kelly DP. Nuclear receptor signaling and cardiac energetics. Circ Res. 2004;95:568–78.
165. Chawla A, Repa JJ, Evans RM, Mangelsdorf DJ. Nuclear receptors and lipid physiology: opening the X-files. Science. 2001;294: 1866–70.
166. van Bilsen M, van der Vusse GJ, Gilde AJ, Lindhout M, van der Lee KA. Peroxisome proliferator-activated receptors: lipid binding proteins controlling gene expression. Mol Cell Biochem. 2002;239:131–8.
167. Brandt JM, Djouadi F, Kelly DP. Fatty acids activate transcription of the muscle carnitine palmitoyltransferase I gene in cardiac myocytes via the peroxisome proliferator-activated receptor alpha. J Biol Chem. 1998;273:23786–92.
168. Liu HR, Tao L, Gao E, et al. Anti-apoptotic effects of rosiglitazone in hypercholesterolemic rabbits subjected to myocardial ischemia and reperfusion. Cardiovasc Res. 2004;62:135–44.
169. Muoio DM, MacLean PS, Lang DB, et al. Fatty acid homeostasis and induction of lipid regulatory genes in skeletal muscles of peroxisome proliferator-activated receptor (PPAR) alpha knock-out mice. Evidence for compensatory regulation by PPAR delta. J Biol Chem. 2002;277:26089–97.
170. Lehman JJ, Barger PM, Kovacs A, Saffitz JE, Medeiros DM, Kelly DP. Peroxisome proliferator-activated receptor gamma coactivator-1 promotes cardiac mitochondrial biogenesis. J Clin Invest. 2000;106:847–56.
171. Puigserver P, Spiegelman BM. Peroxisome proliferator-activated receptor-gamma coactivator 1 alpha (PGC-1 alpha): transcriptional coactivator and metabolic regulator. Endocr Rev. 2003;24: 78–90.
172. Clabby ML, Robison TA, Quigley HF, Wilson DB, Kelly DP. Retinoid X receptor alpha represses GATA-4-mediated transcription via a retinoid-dependent interaction with the cardiac-enriched repressor FOG-2. J Biol Chem. 2003;278:5760–7.
173. Kahaly GJ, Dillmann WH. Thyroid hormone action in the heart. Endocr Rev. 2005;26:704–28.
174. Bahouth SW, Cui X, Beauchamp MJ, Park EA. Thyroid hormone induces beta1-adrenergic receptor gene transcription through a direct repeat separated by five nucleotides. J Mol Cell Cardiol. 1997;29:3223–37.
175. Chen S, Nakamura K, Gardner DG. 1,25-dihydroxyvitamin D inhibits human ANP gene promoter activity. Regul Pept. 2005;128:197–202.
176. Sladek R, Bader JA, Giguere V. The orphan nuclear receptor estrogen-related receptor alpha is a transcriptional regulator of the human medium-chain acyl coenzyme A dehydrogenase gene. Mol Cell Biol. 1997;17:5400–9.
177. Booth EA, Obeid NR, Lucchesi BR. Activation of estrogen receptor-alpha protects the in vivo rabbit heart from ischemia-reperfusion injury. Am J Physiol Heart Circ Physiol. 2005;289: H2039–47.

178. Jankowski M, Rachelska G, Donghao W, McCann SM, Gutkowska J. Estrogen receptors activate atrial natriuretic peptide in the rat heart. Proc Natl Acad Sci USA. 2001;98:11765–70.
179. Nuedling S, Karas RH, Mendelsohn ME, et al. Activation of estrogen receptor beta is a prerequisite for estrogen-dependent upregulation of nitric oxide synthases in neonatal rat cardiac myocytes. FEBS Lett. 2001;502:103–8.
180. Yang SH, Liu R, Perez EJ, et al. Mitochondrial localization of estrogen receptor beta. Proc Natl Acad Sci USA. 2004;101: 4130–5.
181. Romeih M, Cui J, Michaille JJ, Jiang W, Zile MH. Function of RARgamma and RARalpha2 at the initiation of retinoid signaling is essential for avian embryo survival and for distinct events in cardiac morphogenesis. Dev Dyn. 2003;228:697–708.
182. Sato A, Sheppard KE, Fullerton MJ, Funder JW. cAMP modulates glucocorticoid-induced protein accumulation and glucocorticoid receptor in cardiomyocytes. Am J Physiol. 1996;271: E827–33.
183. Marsh JD, Lehmann MH, Ritchie RH, Gwathmey JK, Green GE, Schiebinger RJ. Androgen receptors mediate hypertrophy in cardiac myocytes. Circulation. 1998;98:256–61.
184. Perrier E, Kerfant BG, Lalevee N, et al. Mineralocorticoid receptor antagonism prevents the electrical remodeling that precedes cellular hypertrophy after myocardial infarction. Circulation. 2004;110:776–83.
185. Le Menuet D, Viengchareun S, Muffat-Joly M, Zennaro MC, Lombes M. Expression and function of the human mineralocorticoid receptor: lessons from transgenic mouse models. Mol Cell Endocrinol. 2004;217:127–36.
186. Frantz S, Kobzik L, Kim YD, et al. Toll4 (TLR4) expression in cardiac myocytes in normal and failing myocardium. J Clin Invest. 1999;104:271–80.
187. Satoh M, Nakamura M, Akatsu T, Shimoda Y, Segawa I, Hiramori K. Toll-like receptor 4 is expressed with enteroviral replication in myocardium from patients with dilated cardiomyopathy. Lab Invest. 2004;84:173–81.
188. Ha T, Li Y, Hua F, et al. Reduced cardiac hypertrophy in toll-like receptor 4-deficient mice following pressure overload. Cardiovasc Res. 2005;68:224–34.
189. Oyama J, Blais Jr C, Liu X, et al. Reduced myocardial ischemia-reperfusion injury in toll-like receptor 4-deficient mice. Circulation. 2004;109:784–9.
190. Wang YP, Sato C, Mizoguchi K, Yamashita Y, Oe M, Maeta H. Lipopolysaccharide triggers late preconditioning against myocardial infarction via inducible nitric oxide synthase. Cardiovasc Res. 2002;56:33–42.
191. Tavener SA, Long EM, Robbins SM, McRae KM, Van Remmen H, Kubes P. Immune cell Toll-like receptor 4 is required for cardiac myocyte impairment during endotoxemia. Circ Res. 2004; 95:700–7.
192. Nelson BD, Luciakova K, Li R, Betina S. The role of thyroid hormone and promoter diversity in the regulation of nuclear encoded mitochondrial proteins. Biochim Biophys Acta. 1995;1271: 85–91.
193. Goldenthal MJ, Weiss HR, Marin-Garcia J. Bioenergetic remodeling of heart mitochondria by thyroid hormone. Mol Cell Biochem. 2004;265:97–106.
194. Seiden D, Navidad P, Weiss HR. Oxygen diffusion distance in thyroxine-induced hypertrophic rabbit myocardium. J Mol Cell Cardiol. 1988;20:917–30.
195. Lebuffe G, Schumacker PT, Shao ZH, Anderson T, Iwase H, Vanden Hoek TL. ROS and NO trigger early preconditioning: relationship to mitochondrial KATP channel. Am J Physiol Heart Circ Physiol. 2003;284:H299–308.
196. Bigard AX, Koulmann N, Bahi L, Sanchez H, Ventura-Clapier R. Thyroid hormones and muscle phenotype: involvement of new signaling pathways. J Soc Biol. 2008;202:93–100.
197. Abel ED. Glucose transport in the heart. Front Biosci. 2004;9: 201–15.
198. Flier JS, Mueckler MM, Usher P, Lodish HF. Elevated levels of glucose transport and transporter messenger RNA are induced by ras or src oncogenes. Science. 1987;235:1492–5.
199. Russell 3rd RR, Li J, Coven DL, et al. AMP-activated protein kinase mediates ischemic glucose uptake and prevents postischemic cardiac dysfunction, apoptosis, and injury. J Clin Invest. 2004;114:495–503.
200. Chou SW, Chiu LL, Cho YM, et al. Effect of systemic hypoxia on GLUT4 protein expression in exercised rat heart. Jpn J Physiol. 2004;54:357–63.
201. Rattigan S, Appleby GJ, Clark MG. Insulin-like action of catecholamines and Ca2+ to stimulate glucose transport and GLUT4 translocation in perfused rat heart. Biochim Biophys Acta. 1991; 1094:217–23.
202. Wojtaszewski JF, Higaki Y, Hirshman MF, et al. Exercise modulates postreceptor insulin signaling and glucose transport in muscle-specific insulin receptor knockout mice. J Clin Invest. 1999; 104:1257–64.
203. Coven DL, Hu X, Cong L, et al. Physiological role of AMP-activated protein kinase in the heart: graded activation during exercise. Am J Physiol Endocrinol Metab. 2003;285:E629–36.
204. Katz EB, Stenbit AE, Hatton K, DePinho R, Charron MJ. Cardiac and adipose tissue abnormalities but not diabetes in mice deficient in GLUT4. Nature. 1995;377:151–5.
205. Zisman A, Peroni OD, Abel ED, et al. Targeted disruption of the glucose transporter 4 selectively in muscle causes insulin resistance and glucose intolerance. Nat Med. 2000;6:924–8.
206. Desrois M, Sidell RJ, Gauguier D, King LM, Radda GK, Clarke K. Initial steps of insulin signaling and glucose transport are defective in the type 2 diabetic rat heart. Cardiovasc Res. 2004;61: 288–96.
207. Bruning JC, Michael MD, Winnay JN, et al. A muscle-specific insulin receptor knockout exhibits features of the metabolic syndrome of NIDDM without altering glucose tolerance. Mol Cell. 1998;2:559–69.
208. Kaczmarczyk SJ, Andrikopoulos S, Favaloro J, et al. Threshold effects of glucose transporter-4 (GLUT4) deficiency on cardiac glucose uptake and development of hypertrophy. J Mol Endocrinol. 2003;31:449–59.
209. Razeghi P, Young ME, Ying J, et al. Downregulation of metabolic gene expression in failing human heart before and after mechanical unloading. Cardiology. 2002;97:203–9.
210. Razeghi P, Young ME, Alcorn JL, Moravec CS, Frazier OH, Taegtmeyer H. Metabolic gene expression in fetal and failing human heart. Circulation. 2001;104:2923–31.
211. Razeghi P, Young ME, Cockrill TC, Frazier OH, Taegtmeyer H. Downregulation of myocardial myocyte enhancer factor 2 C and myocyte enhancer factor 2 C-regulated gene expression in diabetic patients with nonischemic heart failure. Circulation. 2002;106: 407–11.
212. Stanley WC, Lopaschuk GD, McCormack JG. Regulation of energy substrate metabolism in the diabetic heart. Cardiovasc Res. 1997;34:25–33.
213. Nikolaidis LA, Sturzu A, Stolarski C, Elahi D, Shen YT, Shannon RP. The development of myocardial insulin resistance in conscious dogs with advanced dilated cardiomyopathy. Cardiovasc Res. 2004;61:297–306.
214. Shah A, Shannon RP. Insulin resistance in dilated cardiomyopathy. Rev Cardiovasc Med. 2003;4 Suppl 6:S50–7.
215. Paternostro G, Pagano D, Gnecchi-Ruscone T, Bonser RS, Camici PG. Insulin resistance in patients with cardiac hypertrophy. Cardiovasc Res. 1999;42:246–53.
216. Nuutila P, Maki M, Laine H, et al. Insulin action on heart and skeletal muscle glucose uptake in essential hypertension. J Clin Invest. 1995;96:1003–9.

217. Paternostro G, Clarke K, Heath J, Seymour AM, Radda GK. Decreased GLUT-4 mRNA content and insulin-sensitive deoxyglucose uptake show insulin resistance in the hypertensive rat heart. Cardiovasc Res. 1995;30:205–11.
218. Nascimben L, Ingwall JS, Lorell BH, et al. Mechanisms for increased glycolysis in the hypertrophied rat heart. Hypertension. 2004;44:662–7.
219. Brownsey RW, Boone AN, Allard MF. Actions of insulin on the mammalian heart: metabolism, pathology and biochemical mechanisms. Cardiovasc Res. 1997;34:3–24.
220. Santalucia T, Boheler KR, Brand NJ, et al. Factors involved in GLUT-1 glucose transporter gene transcription in cardiac muscle. J Biol Chem. 1999;274:17626–34.
221. Taegtmeyer H, Golfman L, Sharma S, Razeghi P, van Arsdall M. Linking gene expression to function: metabolic flexibility in the normal and diseased heart. Ann NY Acad Sci. 2004;1015:202–13.
222. Young LH, Coven DL, Russell 3rd RR. Cellular and molecular regulation of cardiac glucose transport. J Nucl Cardiol. 2000; 7:267–76.
223. Depre C, Shipley GL, Chen W, et al. Unloaded heart in vivo replicates fetal gene expression of cardiac hypertrophy. Nat Med. 1998;4:1269–75.
224. Doenst T, Goodwin GW, Cedars AM, Wang M, Stepkowski S, Taegtmeyer H. Load-induced changes in vivo alter substrate fluxes and insulin responsiveness of rat heart in vitro. Metabolism. 2001;50:1083–90.
225. Shimizu I, Minamino T, Toko H, et al. Excessive cardiac insulin signaling exacerbates systolic dysfunction induced by pressure overload in rodents. J Clin Invest. 2010;120:1506–14.
226. Colombo F, Gosselin H, El-Helou V, Calderone A. Beta-adrenergic receptor-mediated DNA synthesis in neonatal rat cardiac fibroblasts proceeds via a phosphatidylinositol 3-kinase dependent pathway refractory to the antiproliferative action of cyclic AMP. J Cell Physiol. 2003;195:322–30.
227. Tuxworth Jr WJ, Saghir AN, Spruill LS, Menick DR, McDermott PJ. Regulation of protein synthesis by eIF4E phosphorylation in adult cardiocytes: the consequence of secondary structure in the 5′-untranslated region of mRNA. Biochem J. 2004;378:73–82.
228. Pham FH, Sugden PH, Clerk A. Regulation of protein kinase B and 4E-BP1 by oxidative stress in cardiac myocytes. Circ Res. 2000;86:1252–8.
229. Hannan RD, Luyken J, Rothblum LI. Regulation of rDNA transcription factors during cardiomyocyte hypertrophy induced by adrenergic agents. J Biol Chem. 1995;270:8290–7.
230. Hannan KM, Brandenburger Y, Jenkins A, et al. mTOR-dependent regulation of ribosomal gene transcription requires S6K1 and is mediated by phosphorylation of the carboxy-terminal activation domain of the nucleolar transcription factor UBF. Mol Cell Biol. 2003;23:8862–77.
231. Lindemann SW, Weyrich AS, Zimmerman GA. Signaling to translational control pathways: diversity in gene regulation in inflammatory and vascular cells. Trends Cardiovasc Med. 2005;15:9–17.
232. Mani K, Kitsis RN. Myocyte apoptosis: programming ventricular remodeling. J Am Coll Cardiol. 2003;41:761–4.
233. Moe GW, Naik G, Konig A, Lu X, Feng Q. Early and persistent activation of myocardial apoptosis, bax and caspases: insights into mechanisms of progression of heart failure. Pathophysiology. 2002;8:183–92.
234. Strasser A. The role of BH3-only proteins in the immune system. Nat Rev Immunol. 2005;5:189–200.
235. Yussman MG, Toyokawa T, Odley A, et al. Mitochondrial death protein Nix is induced in cardiac hypertrophy and triggers apoptotic cardiomyopathy. Nat Med. 2002;8:725–30.
236. Galvez AS, Brunskill EW, Marreez Y, et al. Distinct pathways regulate proapoptotic Nix and BNip3 in cardiac stress. J Biol Chem. 2006;281:1442–8.
237. Diwan A, Dorn 2nd GW. Decompensation of cardiac hypertrophy: cellular mechanisms and novel therapeutic targets. Physiology (Bethesda). 2007;22:56–64.
238. Danial NN, Korsmeyer SJ. Cell death: critical control points. Cell. 2004;116:205–19.
239. Ashkenazi A, Dixit VM. Death receptors: signaling and modulation. Science. 1998;281:1305–8.
240. Muzio M, Chinnaiyan AM, Kischkel FC, et al. FLICE, a novel FADD-homologous ICE/CED-3-like protease, is recruited to the CD95 (Fas/APO-1) death-inducing signaling complex. Cell. 1996;85:817–27.
241. Boatright KM, Renatus M, Scott FL, et al. A unified model for apical caspase activation. Mol Cell. 2003;11:529–41.
242. Crow MT, Mani K, Nam YJ, Kitsis RN. The mitochondrial death pathway and cardiac myocyte apoptosis. Circ Res. 2004;95:957–70.
243. Du C, Fang M, Li Y, Li L, Wang X. Smac, a mitochondrial protein that promotes cytochrome c-dependent caspase activation by eliminating IAP inhibition. Cell. 2000;102:33–42.
244. Li LY, Luo X, Wang X. Endonuclease G is an apoptotic DNase when released from mitochondria. Nature. 2001;412:95–9.
245. Liu X, Kim CN, Yang J, Jemmerson R, Wang X. Induction of apoptotic program in cell-free extracts: requirement for dATP and cytochrome c. Cell. 1996;86:147–57.
246. Susin SA, Lorenzo HK, Zamzami N, et al. Molecular characterization of mitochondrial apoptosis-inducing factor. Nature. 1999;397:441–6.
247. Acehan D, Jiang X, Morgan DG, Heuser JE, Wang X, Akey CW. Three-dimensional structure of the apoptosome: implications for assembly, procaspase-9 binding, and activation. Mol Cell. 2002;9:423–32.
248. Hu Y, Ding L, Spencer DM, Nunez G. WD-40 repeat region regulates Apaf-1 self-association and procaspase-9 activation. J Biol Chem. 1998;273:33489–94.
249. Qin H, Srinivasula SM, Wu G, Fernandes-Alnemri T, Alnemri ES, Shi Y. Structural basis of procaspase-9 recruitment by the apoptotic protease-activating factor 1. Nature. 1999;399:549–57.
250. Zou H, Henzel WJ, Liu X, Lutschg A, Wang X. Apaf-1, a human protein homologous to *C. elegans* CED-4, participates in cytochrome c-dependent activation of caspase-3. Cell. 1997;90:405–13.
251. Gross A, Yin XM, Wang K, et al. Caspase cleaved BID targets mitochondria and is required for cytochrome c release, while BCL-XL prevents this release but not tumor necrosis factor-R1/Fas death. J Biol Chem. 1999;274:1156–63.
252. Luo X, Budihardjo I, Zou H, Slaughter C, Wang X. Bid, a Bcl2 interacting protein, mediates cytochrome c release from mitochondria in response to activation of cell surface death receptors. Cell. 1998;94:481–90.
253. Peter ME. The flip side of FLIP. Biochem J. 2004;382:e1–3.
254. Guo B, Zhai D, Cabezas E, et al. Humanin peptide suppresses apoptosis by interfering with Bax activation. Nature. 2003;423: 456–61.
255. Shiozaki EN, Chai J, Rigotti DJ, et al. Mechanism of XIAP-mediated inhibition of caspase-9. Mol Cell. 2003;11:519–27.
256. Sun C, Cai M, Meadows RP, et al. NMR structure and mutagenesis of the third Bir domain of the inhibitor of apoptosis protein XIAP. J Biol Chem. 2000;275:33777–81.
257. Nam YJ, Mani K, Ashton AW, et al. Inhibition of both the extrinsic and intrinsic death pathways through nonhomotypic death-fold interactions. Mol Cell. 2004;15:901–12.
258. Gustafsson AB, Tsai JG, Logue SE, Crow MT, Gottlieb RA. Apoptosis repressor with caspase recruitment domain protects against cell death by interfering with Bax activation. J Biol Chem. 2004;279:21233–8.
259. Scorrano L, Oakes SA, Opferman JT, et al. BAX and BAK regulation of endoplasmic reticulum Ca2+: a control point for apoptosis. Science. 2003;300:135–9.

260. Morishima N, Nakanishi K, Tsuchiya K, Shibata T, Seiwa E. Translocation of Bim to the endoplasmic reticulum (ER) mediates ER stress signaling for activation of caspase-12 during ER stress-induced apoptosis. J Biol Chem. 2004;279:50375–81.
261. Olivetti G, Abbi R, Quaini F, et al. Apoptosis in the failing human heart. N Engl J Med. 1997;336:1131–41.
262. Nadal-Ginard B, Kajstura J, Leri A, Anversa P. Myocyte death, growth, and regeneration in cardiac hypertrophy and failure. Circ Res. 2003;92:139–50.
263. Nadal-Ginard B, Kajstura J, Anversa P, Leri A. A matter of life and death: cardiac myocyte apoptosis and regeneration. J Clin Invest. 2003;111:1457–9.
264. Anversa P, Nadal-Ginard B. Myocyte renewal and ventricular remodelling. Nature. 2002;415:240–3.
265. Soonpaa MH, Field LJ. Survey of studies examining mammalian cardiomyocyte DNA synthesis. Circ Res. 1998;83:15–26.
266. Rybin VO, Xu X, Steinberg SF. Activated protein kinase C isoforms target to cardiomyocyte caveolae: stimulation of local protein phosphorylation. Circ Res. 1999;84:980–8.
267. Head BP, Patel HH, Roth DM, et al. G protein-coupled receptor signaling components localize in both sarcolemmal and intracellular caveolin-3-associated microdomains in adult cardiac myocytes. J Biol Chem. 2005;280:31036–44.
268. Williams TM, Lisanti MP. The Caveolin genes: from cell biology to medicine. Ann Med. 2004;36:584–95.
269. del Monte F, Hajjar RJ. Targeting calcium cycling proteins in heart failure through gene transfer. J Physiol. 2003;546:49–61.
270. Fedak PW, Verma S, Weisel RD, Li RK. Cardiac remodeling and failure From molecules to man (Part II). Cardiovasc Pathol. 2005;14:49–60.
271. Kim Y, Phan D, van Rooij E, et al. The MEF2D transcription factor mediates stress-dependent cardiac remodeling in mice. J Clin Invest. 2008;118:124–32.
272. Fielitz J, Kim MS, Shelton JM, et al. Requirement of protein kinase D1 for pathological cardiac remodeling. Proc Natl Acad Sci USA. 2008;105:3059–63.
273. Harrison BC, Kim MS, van Rooij E, et al. Regulation of cardiac stress signaling by protein kinase d1. Mol Cell Biol. 2006;26:3875–88.
274. Hartil K, Charron MJ. Genetic modification of the heart: transgenic modification of cardiac lipid and carbohydrate utilization. J Mol Cell Cardiol. 2005;39:581–93.
275. Stepien G, Torroni A, Chung AB, Hodge JA, Wallace DC. Differential expression of adenine nucleotide translocator isoforms in mammalian tissues and during muscle cell differentiation. J Biol Chem. 1992;267:14592–7.
276. Graham BH, Waymire KG, Cottrell B, Trounce IA, MacGregor GR, Wallace DC. A mouse model for mitochondrial myopathy and cardiomyopathy resulting from a deficiency in the heart/muscle isoform of the adenine nucleotide translocator. Nat Genet. 1997;16:226–34.
277. Takeishi Y, Ping P, Bolli R, Kirkpatrick DL, Hoit BD, Walsh RA. Transgenic overexpression of constitutively active protein kinase C epsilon causes concentric cardiac hypertrophy. Circ Res. 2000;86:1218–23.
278. Liu Q, Chen X, Macdonnell SM, et al. Protein kinase C{alpha}, but not PKC{beta} or PKC{gamma}, regulates contractility and heart failure susceptibility: implications for ruboxistaurin as a novel therapeutic approach. Circ Res. 2009;105:194–200.
279. Sugden PH, Clerk A. "Stress-responsive" mitogen-activated protein kinases (c-Jun N-terminal kinases and p38 mitogen-activated protein kinases) in the myocardium. Circ Res. 1998;83:345–52.
280. Rockman HA, Koch WJ, Lefkowitz RJ. Seven-transmembrane-spanning receptors and heart function. Nature. 2002;415:206–12.
281. Rienstra M, Van Veldhuisen DJ, Hagens VE, et al. Gender-related differences in rhythm control treatment in persistent atrial fibrillation: data of the Rate Control Versus Electrical Cardioversion (RACE) study. J Am Coll Cardiol. 2005;46:1298–306.
282. Rogge C, Geibel A, Bode C, Zehender M. Cardiac arrhythmias and sudden cardiac death in women. Z Kardiol. 2004;93:427–38.
283. Aurigemma GP, Silver KH, McLaughlin M, Mauser J, Gaasch WH. Impact of chamber geometry and gender on left ventricular systolic function in patients > 60 years of age with aortic stenosis. Am J Cardiol. 1994;74:794–8.
284. Douglas PS, Otto CM, Mickel MC, Labovitz A, Reid CL, Davis KB. Gender differences in left ventricle geometry and function in patients undergoing balloon dilatation of the aortic valve for isolated aortic stenosis. NHLBI Balloon Valvuloplasty Registry. Br Heart J. 1995;73:548–54.
285. Carroll JD, Carroll EP, Feldman T, et al. Sex-associated differences in left ventricular function in aortic stenosis of the elderly. Circulation. 1992;86:1099–107.
286. Liao Y, Cooper RS, Mensah GA, McGee DL. Left ventricular hypertrophy has a greater impact on survival in women than in men. Circulation. 1995;92:805–10.
287. Khalil RA. Sex hormones as potential modulators of vascular function in hypertension. Hypertension. 2005;46:249–54.
288. Peter I, Shearman AM, Zucker DR, et al. Variation in estrogen-related genes and cross-sectional and longitudinal blood pressure in the Framingham Heart Study. J Hypertens. 2005;23:2193–200.
289. Hardt SE, Sadoshima J. Negative regulators of cardiac hypertrophy. Cardiovasc Res. 2004;63:500–9.
290. Liu ZP, Nakagawa O, Nakagawa M, et al. CHAMP, a novel cardiac-specific helicase regulated by MEF2C. Dev Biol. 2001; 234:497–509.
291. Porrello ER, Widdop RE, Delbridge LM. Early origins of cardiac hypertrophy: does cardiomyocyte attrition programme for pathological 'catch-up' growth of the heart? Clin Exp Pharmacol Physiol. 2008;35:1358–64.
292. Fiedler B, Lohmann SM, Smolenski A, et al. Inhibition of calcineurin-NFAT hypertrophy signaling by cGMP-dependent protein kinase type I in cardiac myocytes. Proc Natl Acad Sci USA. 2002;99:11363–8.
293. Layland J, Li JM, Shah AM. Role of cyclic GMP-dependent protein kinase in the contractile response to exogenous nitric oxide in rat cardiac myocytes. J Physiol. 2002;540:457–67.
294. Becker EM, Schmidt P, Schramm M, et al. The vasodilator-stimulated phosphoprotein (VASP): target of YC-1 and nitric oxide effects in human and rat platelets. J Cardiovasc Pharmacol. 2000;35:390–7.
295. Sporbert A, Mertsch K, Smolenski A, et al. Phosphorylation of vasodilator-stimulated phosphoprotein: a consequence of nitric oxide- and cGMP-mediated signal transduction in brain capillary endothelial cells and astrocytes. Brain Res Mol Brain Res. 1999;67:258–66.
296. Harada M, Takeishi Y, Arimoto T, et al. Diacylglycerol kinase zeta attenuates pressure overload-induced cardiac hypertrophy. Circ J. 2007;71:276–82.
297. Takeishi Y, Goto K, Kubota I. Role of diacylglycerol kinase in cellular regulatory processes: a new regulator for cardiomyocyte hypertrophy. Pharmacol Ther. 2007;115:352–9.
298. Leong HS, Brownsey RW, Kulpa JE, Allard MF. Glycolysis and pyruvate oxidation in cardiac hypertrophy–why so unbalanced? Comp Biochem Physiol A Mol Integr Physiol. 2003;135: 499–513.
299. Garnier A, Fortin D, Delomenie C, Momken I, Veksler V, Ventura-Clapier R. Depressed mitochondrial transcription factors and oxidative capacity in rat failing cardiac and skeletal muscles. J Physiol. 2003;551:491–501.
300. Ananthakrishnan R, Moe GW, Goldenthal MJ, Marin-Garcia J. Akt signaling pathway in pacing-induced heart failure. Mol Cell Biochem. 2005;268:103–10.
301. Divakaran V, Mann DL. The emerging role of microRNAs in cardiac remodeling and heart failure. Circ Res. 2008;103:1072–83.

# Chapter 16
# Signaling in Diabetes and Metabolic Syndrome

**Abstract** Diabetes mellitus is a condition in which an organism contains elevated blood sugar. This pathological state could be a result of two major abnormalities both related to the functioning of insulin, a hormone that regulates glucose metabolism. Type 1 diabetes (T1D) develops when β cells fail to produce insulin. The most common diabetes affecting 90–95% of the US diabetes population is Type 2 diabetes (T2D), which results from "insulin resistance": cells lose sensitivity and respond weakly (or stop responding) to the insulin that is produced.

This chapter focuses on the signaling pathways triggered by insulin under normal conditions and describes the changes leading to the development of resistance of target cells to insulin. Also are described consequences of insulin resistance for the development of obesity, a number of signaling systems affected under other metabolic disorders that increase the risk of developing cardiovascular disease (components of metabolic syndrome (MetSyn), such as dyslipidemia and hyperinsulinemia) and genes polymorphisms associated with diabetes, resistance to insulin and MetSyn.

**Keywords** Diabetes signaling • Metabolic syndrome • Insulin • Advance glycation end-products

## Introduction

Classical metabolic action of insulin is a promotion of glucose uptake in skeletal muscle and adipose tissue. This action is mediated by specific cell surface insulin receptors (IRs) which regulate translocation of the insulin-responsive glucose transporter GLUT 4. In addition, insulin plays an important role in the coupling of metabolic and cardiovascular physiology.

Under a number of pathological complications (Type 2 diabetes (T2D), metabolic syndrome (MetSyn)), an insulin resistance develops which impairs some insulin-dependent pathways and changes the balance between separate branches of insulin-regulated signaling cascades. These changes contribute to reciprocal relationships between insulin resistance and endothelial dysfunction in metabolic and cardiovascular tissues. The clinical consequences of insulin resistance include hyperglycemia-induced tissue damage, hypertension, dyslipidemia, MetSyn, and cardiovascular disease. Another aspect of insulin resistance includes the development of compensatory hyperinsulinemia. Compensatory hyperinsulinemia may stimulate or even overstimulate certain aspects of insulin action in various cells and tissues.

Some mediators synthesized from cells of the adipose tissue are critically involved in the regulation of insulin action. In obesity, an increased adipose tissue tumor necrosis factor-α (TNF-α) expression and reduced levels of adiponectin contribute to the impaired insulin sensitivity and elevated free fatty acids (FFAs).

Hyperglycemia and insulin resistance accompanying diabetes affect many biochemical pathways in various tissues. Some of them, including glucose oxidation, the formation of advanced glycation end-products (AGEs), the activation of the renin-angiotensin-aldosterone system (RAAS), and subsequent elevations in angiotensin II (Ang II) and aldosterone, are associated with the generation of reactive oxygen species (ROS) and lead to increased oxidative stress and cardiovascular disease.

An understanding of the cardiovascular actions of insulin in health and disease has important implications for the explanation of frequent associations between metabolic and cardiovascular diseases and for developing novel therapeutic strategies to improve metabolic and cardiovascular health simultaneously.

## Insulin

Cardiovascular diseases are the leading cause of morbidity and mortality in insulin-resistant individuals. Insulin-resistance is a cardinal feature of diabetes, obesity, and dyslipidemia, and is also a prominent component of hypertension, coronary heart disease, and atherosclerosis. Insulin resistance is typically defined as decreased sensitivity and/or responsiveness to metabolic actions of insulin. In insulin-resistant

J. Marín-García, *Signaling in the Heart*,
DOI 10.1007/978-1-4419-9461-5_16, © Springer Science+Business Media, LLC 2011

conditions, impairment of shared insulin-signaling pathways in metabolic and cardiovascular tissues contributes to reciprocal relationships between insulin resistance and endothelial dysfunction that explains frequent associations between metabolic and cardiovascular diseases exemplified by the MetSyn.

The biological actions of insulin are mediated by specific cell surface receptors. The IR is a ligand-activated tyrosine kinase which consists of two α-subunits and two β-subunits. The β-subunits pass through the cellular membrane and are linked by disulfide bonds. IR phosphorylates intracellular substrates, including IR substrate (IRS) proteins and Shc that serve as docking proteins for downstream-signaling molecules. There are two major signaling cascades triggered by insulin-activated IR (Fig. 16.1).

One of them involves IRS adaptor proteins. Tyrosine phosphorylation of IRS at multiple sites creates Src homology 2 (SH2) domain-binding motifs for SH2 domain-containing phosphatidylinositol 3-kinase (PI3K). PI3K is a heterodimer composed of a regulatory and a catalytic subunits. Several isoforms of catalytic subunit (p110α, p110β, p110δ) bind to the major regulatory isoforms (p85α, p55α, p50α). When SH2 domains of PI3K regulatory subunit bind to tyrosine-phosphorylated motifs on IRS, this allosterically activates the preassociated catalytic subunit of PI3K to generate the lipid product phosphatidylinositol 3,4,5-trisphosphate [PI(3,4,5)$P_3$] from the substrate phosphatidylinositol 4,5-bisphosphate ($PIP_2$). PI(3,4,5)$P_3$ in turn, is an activator of 3-phosphoinositide-dependent protein kinase-1 (PDK-1): binding of PI(3,4,5)$P_3$ to pleckstrin-homology domain of PDK-1 leads to its phosphorylation/activation and subsequent phosphorylation of downstream serine–threonine kinases, Akt and atypical protein kinase C (PKC) isoforms. PI3K-PDK-1-dependent branch of the insulin-signaling pathway results in numerous biological insulin actions, including vascular relaxation, glucose uptake, glycogen synthesis, and gluconeogenesis (Fig. 16.1).

Activated IR also phosphorylates adaptor protein Shc which serves as a starting point for another signaling pathway. SH2 domain-binding motif on Shc attracts growth factor receptor-bound protein-2 (Grb-2) and activates GTP exchange factor Sos preassociated with Grb-2. Sos facilitates the formation of active GTP-containing small GTP-binding protein, Ras, which then triggers a signaling cascade involving Raf kinase, MAP kinase kinase (MKK1), and the extracellular regulated kinases-1 and -2 (Erk1/2). This branch of insulin signaling regulates cell growth, mitogenesis, and differentiation (Fig. 16.1).

Recent observations have provided evidence that in addition to the "classical" actions (regulation of glucose uptake through insulin-responsive glucose transporter GLUT4), there are "non-classical" actions of insulin that play an important role in cardiovascular physiology. Very important cardiovascular action of insulin under healthy conditions is the stimulation of vasodilation and increase of blood flow. This function of insulin is based on the stimulation of production of the potent vasodilator nitric oxide (NO) from vascular endothelium. Insulin-dependent activation of the IR tyrosine kinase, followed by phosphorylation/activation of PI3K-PDK-1, leads to the activation of Akt. Akt directly phosphorylates and activates endothelial NO synthase (eNOS), leading to increased production of NO [1–3]. Similarly, insulin stimulates the release of NO in vascular smooth muscle cells (VSMCs) via the IR tyrosine kinase activation and subsequent activation of PI3K [4]. Thus, in addition to endothelial NO, VSMC-derived NO autocrinally regulates vasorelaxation in response to insulin. Some evidences also suggest that insulin can directly stimulate the production of another endothelial vasodilator, prostacyclin, but pathway(s)-regulating prostacyclin production are yet to be elucidated [5].

Mechanisms of insulin-dependent attenuation of VSMC contractility are realized via IR-PI3K-Akt-NO-cascade and target one of the major proteins of VSMC contractile machinery, myosin light chain (MLC). In particular, increased NO in response to insulin activates VSMC cGMP-dependent protein kinase 1 (cGKI) α. Once activated, cGKIα mediates phosphorylation of Ser-188 on a small GTP-binding protein, RhoA. This is an inhibitory modification: phosphorylated RhoA is not able to activate Rho kinase (ROK) α, and ROKα does not phosphorylate/inhibit MLC phosphatase (ROK-dependent phosphorylation of Thr-697 and Thr-855 attenuates MLC phosphatase activity) [6]. Moreover, insulin-activated cGKIα activates MLC phosphatase via activatory phosphorylation [7]. Thus, MLC phosphatase stays active under insulin treatment conditions, MLC phosphorylation is decreased and vascular tone is lowered because of impeded interaction of myosin with F-actin. Decreased vasoconstrictor tone is also a result of insulin-dependent decrease of intracellular $Ca^{2+}$ (again, via IR-PI3K-Akt-NO-pathway) [8].

In myocardium, insulin regulates metabolism, contractility, and cardiac growth. For instance, about 30% of energy is supplied to the heart in the form of glucose and lactate, and insulin regulates glucose uptake by glucose transporter GLUT4 via PI3K-Akt pathway. Insulin-activated Akt also inhibits the activity of glycogen synthase kinase 3 (GSK3) and AMP-activated protein kinase (AMPK) which results in accumulation of cardiac glycogen [9]. Besides its well-known role in glucose transport and metabolism, insulin also has a key role in regulating the transport of long-chain fatty acids (LCFAs). According to Chabowski et al. [10] insulin increases the expression of LCFA transporter (fatty acid translocase; FAT/CD36) and targets FAT/CD36 to the sarcolemma. The insulin-PI3K-Akt-signaling pathway must be activated to stimulate the expression of FAT/CD36. Interestingly, the same studies have also shown that insulin

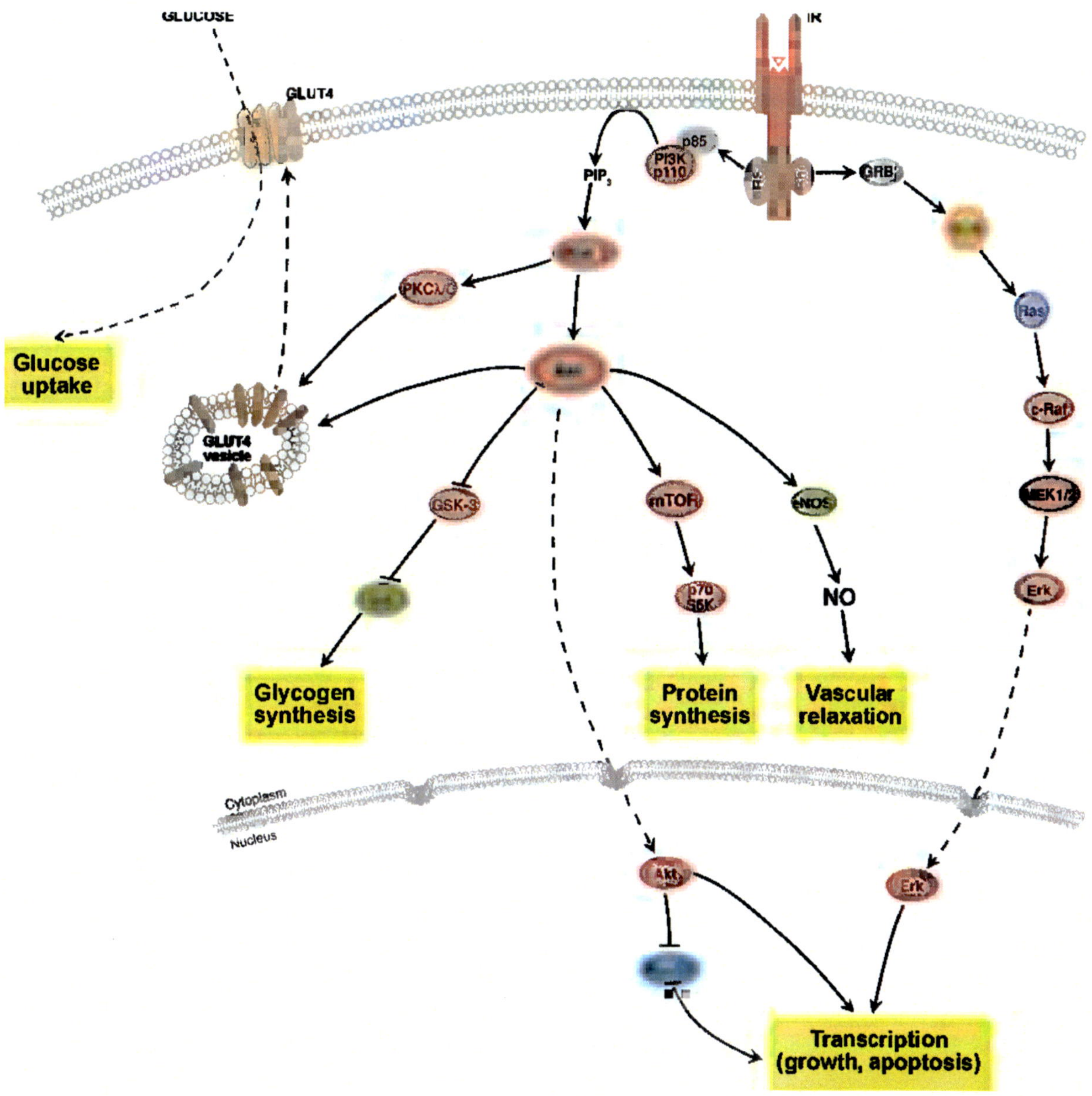


**Fig. 16.1** Insulin signal transduction pathways. PI3K branch of insulin signaling regulates glucose metabolism (in cardiac muscle, skeletal muscle, adipose tissue, and liver), vasodilation (via stimulation of NO production in vascular endothelium). Insulin also stimulates protein synthesis by Akt-mediated activation of mTOR and p70S6K, and stimulates glycogene synthesis by Akt-mediated inhibition of GSK-3. Insulin/Akt-dependent inhibition of FOXO leads to inhibition of protein breakdown. MAPK branch of insulin signaling regulates growth and mitogenesis. Abbreviations: *eNOS* endothelial NO synthase; *Erk* extracellular signal-regulated kinase; *FOXO* forkhead dependent transcription factor; *GLUT* glucose transporter; *GRB* growth factor receptor-bound protein; *GS* glycogen synthase; *GSK* glycogen synthase kinase; *IR* insulin receptor; *IRS* IR substrate; *MAPK* mitogen-activated protein kinase; *MEK* Erk kinase; *mTOR* mammalian target of rapamycin; *NO* nitric oxide; *PDK1* phosphoinositide-dependent protein kinase-1; *PI3K* phosphatidylinositol 3-kinase; *$PIP_3$* phosphatidylinositol 3,4,5-trisphosphate; *PKC* protein kinase C; *p70 S6K* ribosomal p70 S6 kinase; *SOS* Son of Sevenless

does not regulate the expression and translocation of another LCFA transporter, fatty acid-binding protein (FABPpm). The functional consequence of the insulin-induced increase in FAT/CD36 expression is an increased rate of LCFA transport across the plasma membrane.

Positive inotropic effect of insulin on myocardium was described to be associated with the enhancement of $Ca^{2+}$ influx through the activation of L-type $Ca^{2+}$ channels and $Na^{+}/Ca^{2+}$-exchanger [11]. This triggering $Ca^{2+}$ stimulates the release of $Ca^{2+}$ from the sarcoplasmic reticulum (SR) via ryanodine receptors, which leads to myofilament activation and contraction. Involvement of PI3K-Akt pathway in this regulation was confirmed by experiments, where PI3K inhibitors inhibited the inotropic actions of insulin, and where overexpression of Akt increased cytoplasmic $Ca^{2+}$ in cardiomyocytes through the activation of L-type $Ca^{2+}$ channels and $Ca^{2+}$ release from sarcoplasmic reticulum [12, 13].

The regulation of physiological growth of cardiomyocytes by insulin involves the activation of Akt as well. In this case, Akt activates mammalian target of rapamycin (mTOR)-dependent pathway (Fig. 16.1) and suppresses cell atrophy programs via inhibition of GSK3β and transcription factor Forkhead box O (FOXO) [14]. In animal model, chronic insulin infusion (mimicking hyperinsulinemic human disease state, such as T2D mellitus) causes cardiomyocyte hypertrophy and increases interstitial fibrosis without cardiomyocyte proliferation, and this leads to increased left ventricle mass, relative wall thickness, and reduced cardiac output [15]. Several rodent models have been generated with either global, cardiac-specific or muscle-specific deletions in individual components of insulin-triggered pathways, and they well characterize the importance of IRS-1, p85α/β-p110 PI3Kα, and Akt1 in insulin-mediated cardiac growth [16].

One of the major factors in the development of MetSyn (T2D, obesity, hypertension) is an insulin resistance. Insulin resistance is characterized by pathway-specific impairment in PI3K-dependent signaling. Metabolic insulin resistance is usually accompanied by compensatory hyperinsulinemia to maintain euglycemia. As a result, under the disease conditions imbalance between PI3K (affected) and MAPK (unaffected)-dependent functions of insulin takes place. Thus, in vascular endothelium under conditions of insulin resistance vasodilatory antihypertensive effects of insulin (stimulation of NO production) are reduced, whereas vasoconstrictory prohypertensive effects of insulin (secretion of endothelin-1, expression of adhesion molecules, such as E-selectin, intracellular adhesion molecule, vascular cellular adhesion molecule) are elevated – characteristics of endothelial dysfunction.

A distinct subset of the insulin-signaling pathways may be related to another prohypertensive event – increased proliferatory activity of VSMCs under the conditions of compensatory hyperinsulinemia. When the level of insulin is increased, it continues to activate prenyltransferases. These enzymes catalyze prenylation of small GTP-binding proteins, farnesylation of Ras (by farnesyltransferase, FTase), and geranylgeranylation and Rho (by geranylgeranyltransferase I, GGTase I) [17]. Prenylated G proteins translocate to the plasma membrane, where various growth factors activate them. For instance, hyperinsulinemia-induced prenylated Ras and Rho proteins enhance cellular response to angiotensin II (Ang II) which transactivates nuclear factor NF-κB in VSMC. Importantly, hyperinsulinemia-dependent activation of NF-κB can be blocked by an inhibitor of GGTase I [18]. Insulin-dependent signaling cascade is mediated via the Erk kinase, includes Shc, Grb-2/Sos, Ras, and Raf. Finally, phosphorylation of the prenyltransferase α-subunit increases the activity of the enzyme.

## Advanced Glycation End-Products

Advanced AGEs are the result of a series of chemical reactions which start from initial nonenzymatic glycosylation (glycation) of proteins or lipids. Typically, glycation starts from fructose or glucose bonding to a protein or lipid molecule with subsequent modification attached sugar moiety via Maillard reaction, formation of a Schiff base, and Amadori rearrangement and oxidation. Intracellular AGE formation can cause generalized cellular dysfunction, and AGE modification of proteins can lead to alterations in normal cell function by inducing cross-linking of extracellular matrices.

The formation and accumulation of AGEs have been known to progress at an accelerated rate in diabetes. A growing body of evidence has implicated AGEs and their receptor (RAGE) as a contributory factor in the pathogenesis of diabetic vascular complications. AGE formation in the diabetic myocardium appears to be further enhanced by increased glucose flux to the polyol pathway via its first and key enzyme, aldose reductase (AR). In this alternative route of glucose metabolism, AR catalyzes the reduction of the aldehyde form of glucose converting it to sorbitol, using NADPH as a cofactor. Sorbitol is in turn converted to fructose by sorbitol dehydrogenase (this enzyme uses $NAD^+$ as a cofactor). Enhancement of polyol pathway results in impaired glycolysis under normoxic and ischemic conditions in diabetic myocardium. Moreover, the fructose produced by the enhanced flux through the AR-mediated polyol pathway promotes glycation. The critical involvement of AR in both cardiovascular and ventricular damage in diabetes has been further supported by numerous studies showing that many of the deleterious effects of diabetes on vascular and cardiac function are reversed by the treatment with AR inhibitors.

The mechanism by which AR is activated during diabetes/hyperglycemia remains unclear. Using a murine model of streptozotocin-induced hyperglycemia, Iwata et al. [19] recently showed that AR activity is significantly elevated in the cardiac ventricles of diabetic mice without a concomitant increase in either the AR protein or transcript level. The authors suggested that posttranslational modification of the AR protein particularly at a redox-sensitive cysteine residue (Cys-298) previously shown to modulate AR enzyme activity [20] may underlie the AR activation in diabetes although this hypothesis has not yet been proven.

AGEs can mediate their effects via RAGE receptors, activating diverse signal transduction cascades and downstream pathways, including the generation of ROS and oxidative stress [21], resulting in cytokine production, subsequently inflammatory responses evoked in various types of cells, leading to the development and progression of diabetic micro- and macroangiopathy [22, 23]. This pathway has also been suggested to play a key role in the increased sensitivity of diabetic myocardium to ischemic episodes, the development of diabetic cardiomyopathy [24] and progression of heart failure [25, 26]. In addition to targeting collagen and extracellular matrix proteins with increased cross-linking (causing stiffness in arteries) [27, 28], diabetic-associated AGEs have also been implicated in targeting cardiac ryanodine receptors/$Ca^{2+}$-release channels [29] resulting in dysfunction of the type 2 ryanodine receptor $Ca^{2+}$-release channel (RyR2) and reduced cardiac contractility and may be linked to the predominant left ventricular diastolic dysfunction and cardiac relaxation abnormalities associated with T2D [30]. This cardiac dysfunction is also mediated in part by the development of cardiac fibrosis stemming from AGE-stimulated profibrotic growth factor signaling, promoting collagen deposition, and increasing inflammation [31, 32].

## Lipotoxicity

One characteristic metabolic disturbance evident in diabetic states is hyperlipidemia which is usually present in the form of increased triglycerides and nonesterified fatty acids (NEFAs). NEFAs play a central role in altering cellular insulin signaling through several mechanisms leading to insulin resistance and compensatory hyperinsulinemia. Elevated plasma NEFAs activate novel protein kinase Cθ (PKCθ), a serine/threonine kinase that phosphorylates and subsequently activates another serine/threonine kinase, IκB kinase. IκB kinase phosphorylates serine residues on insulin receptor substrate-1, thus inhibiting its ability to activate PI3K and therefore impairing insulin signal transduction [33]. In addition, increased levels of intracellular NEFAs can alter insulin signaling without affecting IRS-1/PI3K activation. As natural ligands for the nuclear receptor, peroxisome proliferator-activated receptor (PPAR), NEFAs can induce the upregulation of the tumor suppressor phosphatase PTEN, which dephosphorylates $PI(3,4,5)P_3$. Decreased levels of $PI(3,4,5)P_3$ affect signaling pathways which are downstream from IRS-1/PI3K. Specifically, low level of $PI(3,4,5)P_3$ prevents the activation of PDK-1 which is responsible for phosphorylation of Akt-1, thus preventing its activation [34].

In addition to disturbances in insulin-related signaling cascades, elevated NEFAs can alter several cardiomyocyte functions regulated independently to insulin. For example, increased NEFA flux into the myocardium directly alters myocardial contractility. As demonstrated by Liu et al. [35] increased long-chain fatty acyl coenzyme A esters may modulate the contractile state of isolated guinea pig ventricular myocytes by the opening of the sarcolemmal $K_{ATP}$ channel: the activation of the $K_{ATP}$ channel leads to shortening of the action potential and reduces transsarcolemmal calcium flux and subsequently myocardial contractility. Intracellular accumulation of NEFAs under circumstances in which they do not undergo β-oxidation are lipotoxic and can directly contribute to cell death by damaging mitochondria (see subsection "Mitochondrial Dysfunction" below).

It is well established that phospholipid-related signaling pathways (a comprehensive discussion is presented in the Chap. 7) are central to the pathology of cardiac hypertrophy and congestive heart failure (CHF), diabetic cardiomyopathy, as well as in ischemia/reperfusion (I/R) [36]. Cardiac sarcolemmal phospholipids serve as substrates for several phospholipases (e.g., $PLA_2$, PLC, and PLD) which produce important lipid-signaling molecules. Diverse and distinct metabolic and hormonal stimuli ranging from muscarinic, $\alpha_1$-adrenergic, Ang II, endothelin-1, thrombin, adenine nucleotide, and opioid peptides arriving at the cardiomyocyte engage specific surface receptors to initiate hydrolysis of inositol phospholipids (e.g., $PIP_2$) by PLC to produce diacylglycerol (DAG) and inositol 1,4,5-trisphosphate ($IP_3$). While $IP_3$ stimulates $Ca^{2+}$ release from SR, DAG activates PKC, $Ca^{2+}$-calmodulin-dependent kinase and mitogen activated kinases stimulating myriad-signaling pathways impacting a wide array of cellular functions, such as ion transport, myofibrillar $Ca^{2+}$ sensitivity, "cross-talk" between phospholipases C and D, gene expression, protein synthesis, and hypertrophic cell growth. Alterations in the sarcolemmal fatty acid composition, particularly the polyunsaturated fatty acids, modify the phosphoinositide response induced by hormones. For instance, cultured cardiomyocytes, incubated in media containing the fatty acids 18:2 ($n-6$) or 20:5 ($n-3$) [but neither 18:0 nor 18:1 ($n-9$)], show a decrease in the PLC responses mediated by $\alpha_1$-adrenoceptors [37].

Oxidants under pathophysiological conditions, including diabetic cardiomyopathy, ischemic heart disease, and CHF, in cardiomyocytes and vascular cells regulate activity of yet another phospholipase, PLD [38]. This enzyme hydrolyzes phosphatidylcholine (PC) to form phosphatidic acid (PA). PA acts as a stimulator of cardiac sarcolemmal membrane and SR $Ca^{2+}$ transport systems involved in increasing intracellular levels of free $Ca^{2+}$ in adult cardiomyocytes and augmenting cardiac contractile activity of the normal heart [39]. PA also appears to be involved in the signal transduction under cardiac hypertrophy conditions [40].

Finally, another two phospholipid derivatives, arachidonic acid (AA) and lysophosholipid, are implicated as downstream effectors in signal pathways, membrane remodeling and lipid metabolism. They are formed from PC hydrolyzed by phospholipases $A_2$. Three phospholipase $A_2$ isoforms have been characterized in the cardiac sarcolemma: extracellular or secretory ($sPLA_2$), cytosolic ($cPLA_2$) and a $Ca^{2+}$-independent ($iPLA_2$) isoforms. Full activation of $cPLA_2$ is mediated by both MAP kinase-dependent phosphorylation (e.g., p38 and Erks1/2) and $I_{Ca2+}$ increase, and leads to $cPLA_2$ translocation from the cytosol to intracellular membranes, such as endoplasmic/sarcoplasmic reticulum, Golgi apparatus, and nuclear envelope. Recent studies have demonstrated that $\beta_2$-adrenergic receptors ($\beta_2$-ARs) are biochemically coupled not only to the classical adenylyl cyclase(AC)-$G_s$-cyclic AMP pathway, but also to the $cPLA_2$ pathway associated with $G_i$ [41]. These studies also revealed that in human tissue, $\beta_2$-AR activates both AC and $cPLA_2$, with the $\beta_2$-AR-$cPLA_2$ pathway being favored when $\beta_2$-AR-AC-PKA signaling branch is altered. However, it remains largely unclear whether $\beta_2$-AR-mediated $cPLA_2$ signaling in the heart is protective or deleterious with regards to cardiac disease. Engelbrecht and Ellis [42] have recently shown that $cPLA_2$ plays a contributory role in apoptosis induction in simulated ischemia/reperfusion (SI/R)-induced injury in neonatal cardiomyocytes. Inhibition of $cPLA_2$ with the specific $cPLA_2$ inhibitor AACOCF3 significantly improved cell viability during SI/R and the increase in cell viability was associated with a significant decrease in caspase-3 and PARP-cleavage during SI/R.

The activation of $iPLA_2$ can also be linked with cardiac injury in diabetic cardiomyopathy and myocardial ischemia. Thus, a significant increase was reported in both $iPLA_2$ mRNA and $iPLA_2$ activity with diabetes in rat myocardium [43]. Similarly, $iPLA_2$ activity is also elevated in myocardial ischemia, an increase which is further amplified in genetically engineered mice overexpressing $iPLA_2$. In experiments of Mancuso et al. [44] coronary artery occlusion in Langendorff perfused hearts from transgenic mice resulted in a large increase in $iPLA_2$ activation (indicated by a 22-fold increase in fatty acid release and a fourfold increase in lysolipid accumulation) in intact myocardium, and in the induction of malignant ventricular tachyarrhythmias within minutes of ischemia. Interestingly, these arrhythmias were not present in either normally perfused transgenic or ischemic wild-type hearts. Pretreatment of perfused transgenic hearts with the $iPLA_2$-specific inhibitor, bromoenol lactone (BEL), prior to the induction of ischemia completely ablated fatty acid release and lysolipid accumulation and rescued transgenic hearts from malignant ventricular tachyarrhythmias. In addition, recent studies indicate that the $iPLA_2$ increase in both diabetes and myocardial ischemia can affect mitochondria (see subsection "Mitochondrial Dysfunction").

## Adipocytokines

Recent studies have shown that adipocyte-secreted factors called adipocytokines contribute to pathogenesis of obesity-associated diseases, including hyperlipidemia, diabetes, hypertension, MetSyn, atherosclerosis, and heart failure through their abilities to modulate inflammatory signaling and metabolic processes. Adipocytokines, such as adiponectin, are markedly decreased in obesity-associated metabolic and vascular disorders and have been shown to have significant antiinflammatory and antiatherogenic effects. On the other hand, levels of several adipocytokines, including leptin, tumor necrosis factor α, interleukins (e.g., IL-1β, IL-6, and IL-8), and plasminogen activator inhibitor-1, are upregulated in obesity and tend to be proinflammatory.

### *Cytokines*

FFAs are important determinants of TNF-α activity and expression in adipose tissue [45]. This proinflammatory cytokine contributes to insulin resistance via a number of potential biochemical mechanisms.

FFAs which are elevated in diabetes increase ROS in adipocytes with subsequent activation of NF-κB, positive regulator of TNF-α and IL-6 expression [46, 47]. In endothelial cells, TNF-α activates c-Jun N-terminal kinase (JNK), which leads to increased phosphorylation of IRS-1 at Ser-307 [48]. As we mentioned earlier, Ser-phosphorylation of IRS reduces the response of PI3K-Akt-eNOS signaling cascade to insulin. In addition to modulating eNOS activity, TNF-α via IKKβ inhibits eNOS expression in endothelial cells [49]. Thus, TNF-α-dependent inhibition of eNOS function may account for the reduction in bioavailability of NO and endothelial dysfunction that characterizes T2D, heart failure. Similarly, in many cell types (adipocytes, liver carcinoma, skeletal muscle cells) TNF-α activates serine/threonine kinases (JNK, p38 kinase, IKKβ) that directly or indirectly increase serine phosphorylation of IRS, leading to the decreased activity of PI3K and the development of insulin resistance [50–52].

TNF-α also stimulates the expression of other inflammatory proteins. One of them, C-reactive protein (CRP), inhibits its endothelium-dependent NO production by uncoupling insulin-activated IR from IRS-PI3K-Akt-eNOS pathway Mechanism of this uncoupling includes phosphorylation of Ser-307 in IRS-1 via activation of the tyrosine kinase Syk-RhoA-JNK signaling axis [53].

To conclude, there is an association between adipocyte-produced proinflammatory cytokines and diabetes, obesity, and hypertension that contributes to the development of insulin resistance, impaired insulin signaling and endothelial dysfunction.

## Adiponectin

Adiponectin (also previously called AdipoQ and ACRP30) is a highly abundant plasma protein (accounting for as much as 0.01% of total plasma protein). This 30 kDa monomeric protein tends to aggregate into polymeric forms. Levels of plasma adiponectin are significantly reduced in obese subjects (compared to nonobese subjects) [54], in patients with T2D (compared to nondiabetic controls) [55], and in patients after acute myocardial infarction [56]. Thus, circulating adiponectin levels are inversely correlated with several cardiovascular risk factors, including high blood pressure and CRP levels [57, 58].

Lower levels of circulating adiponectin have also been found in subjects with MetSyn [59, 60]. The relationship between adiponectin and MetSyn has been further underscored by the identification of specific variants of the adiponectin gene *APM1*. In an analysis of a Japanese population with coronary artery disease (CAD), Ohashi et al. [61] found that the frequency of one *APM1* variant, a missense mutation (I164T) was significantly higher in CAD subjects (2.9%) than in the control (0.8%), with subjects carrying the I164T mutation having significantly lower plasma adiponectin levels than those without the mutation. In addition, CAD subjects containing the I164T mutation had a higher prevalence of the MetSyn phenotype than did patients with the wild-type allele. Several *APM1* polymorphisms or single-nucleotide polymorphisms (SNPs) residing within the promoter region (e.g., G-11377C), exon 2 (e.g., T45G) and intron 2 (e.g., G276T), have been demonstrated to associate with MetSyn phenotypes in various studies with some differences in both phenotype and plasma adiponectin levels noted in different populations [62]. For instance, Hara et al. [63] reported association of the G276 allele with lower plasma adiponectin concentration, especially in the obese Japanese, a finding replicated in independent studies of French, Greek, and Spanish populations [64–66]. On the other hand, in Italians, the T276 allele was associated with underlying lower serum adiponectin concentration [67, 68]. The underlying biological mechanism for the genetic associations may involve allele-specific differential expression of adiponectin gene via reduced transcription or mRNA stability.

While adiponectin has been strongly implicated in cardiovascular and metabolic disorders, including CAD, MetSyn, and diabetes, a cause-effect relationship to any of these disorders remains to be established. While gene studies show a contributing effect, they have also suggested that adiponectin genes and plasma levels can be impacted by other modifiers whether environmental or genetic.

For example, plasma adiponectin levels can be raised by improving insulin sensitivity, as found with significant weight reduction and treatment with insulin sensitizers, including PPAR-γ agonists, such as thiazolidinedione (TZD) suggesting that lower levels of plasma adiponectin can be the result of insulin resistance rather than its cause [69, 70]. It is notable that both insulin and IGF-1 can increase adiponectin synthesis and secretion in white adipose tissue. On the other hand, adiponectin expression and secretion can be suppressed by adipose factors, which are turned on once fat cell mass increases, such as inflammatory cytokines (including TNF-α and IL-6), renin–angiotensin system, β-adrenergic agonists, endothelin-1, glucocorticoid, and increased oxidative stress [71–78].

The multiplicity of actions of adiponectin on the cardiovascular system has been attributed to its diverse effects on signaling pathways (Fig. 16.2) in different cell-types [79]. For instance, in both skeletal and cardiac muscle, adiponectin plays an important role in the regulation of fatty acid and glucose metabolism by promoting glucose uptake and increasing fatty acid oxidation (FAO) rates [80]. The action of adiponectin is mediated by adiponectin receptors AdipoR1 and AdipoR2, both of which are expressed (albeit AdipoR1 more abundantly) in skeletal and cardiac muscle, whereas AdipoR2 is more abundant in liver in humans [81]. The adiponectin-dependent increase in skeletal muscle FAO has been shown to be accompanied by an upregulation of gene expression of the fatty acid transporter, acyl-CoA oxidase (ACO), the mitochondrial uncoupling protein (UCP) 2 and PPAR-α [81, 82]. In addition, phosphorylation and activation of the AMPK appears to be a integral contributory factor in adiponectin-induced FAO in skeletal myocytes [83, 84]. Moreover, adiponectin-induced AMPK activation in skeletal muscle was associated with elevation in the phosphorylation of acetyl-CoA carboxylase (ACC) and decreased levels of malonyl CoA, a competitive inhibitor of carnitine palmitoyl-transferase I and suppressor of mitochondrial FAO [83]. While an adiponectin-dependent increase in FAO has also been reported in cardiac muscle, there is evidence for both an AMPK-dependence and an AMPK-independence in the mechanism [85, 86].

The adiponectin-dependent increase in glucose transport in both skeletal muscle and cardiac muscle primarily results from an enhanced translocation of the glucose transporter GLUT4 to the cell membrane and was also attributable to AMPK activation [83, 84].

The upregulation of glucose uptake in cardiac and skeletal muscle (the latter accounts for the great majority of the

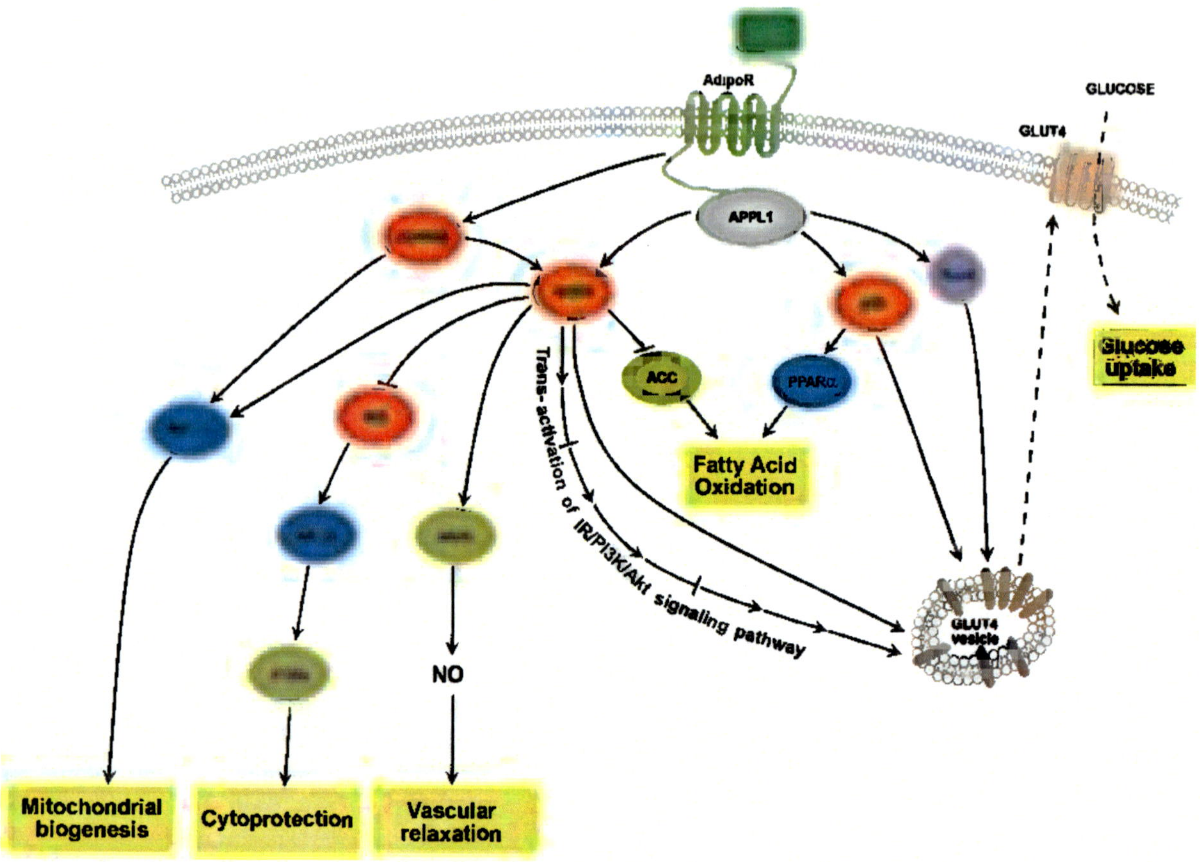


**Fig. 16.2** Adiponectin signal transduction pathways. The binding of adiponectin to its receptor provokes the activation of AMPK and the activation of various signaling molecules. Abbreviations: *ACC* acetyl coenzyme-A carboxylase; *AdipoR* adiponectin receptor; *AMPK* AMP-activated protein kinase; *AMPKK* AMPK kinase; *APPL1* adaptor protein, phosphotyrosine interaction, PH domain and leucine zipper containing 1; *eNOS* endothelial NO synthase; *GLUT* glucose transporter; *IKK* inhibitor of κB kinase; *NF-κB* nuclear factor-κB; *NO* nitric oxide; *p38* p38 mitogen-activated protein kinase; *PDK1* phosphoinositide-dependent protein kinase-1; *PGC-1α* PPAR gamma coactivator 1-alpha; *PI3K* phosphatidylinositol 3-kinase; *PPAR* peroxisome proliferator-activated receptor; *PTEN* phosphatase and tensin homolog; *Rab5* RAS-associated protein

insulin-stimulated glucose disposal) contributes to enhanced insulin sensitivity, glucose tolerance and is likely beneficial in subjects with T2D. Moreover, adiponectin also reduces plasma glucose levels by suppressing gluconeogenesis and glucose production in the liver in part by downregulating hepatic phosphoenolpyruvate carboxylinase (PEPCK) and glucose-6-phosphatase (G6Pase) gene expression, most likely mediated by AMPK phosphorylation [84, 87].

Adiponectin also impacts several signaling pathways leading to atherogenesis. By inhibiting NF-κB activation, adiponectin reduces TNF-α-stimulated monocyte adhesion to endothelial cells, expression of adhesion factors [e.g., intercellular adhesion molecule 1 (ICAM-1), vascular cell adhesion molecule 1 (VCAM-1), and P selectin] and proinflammatory IL 8. Adiponectin also inhibits macrophage-to-foam cell transformation, suppresses both TNF-α expression and scavenger receptors in macrophages and adipose tissues [88]. Kumada et al. [89] have demonstrated that adiponectin selectively increases tissue inhibitor of metalloproteinases (TIMP-1) expression in human monocyte-derived macrophages via the transcriptional induction of the antiinflammatory cytokine IL-10. Adiponectin also impacts both angiogenesis and endothelial function by increasing eNOS expression, activity, and NO production in vascular endothelial cells, dependent on the induction of AMPK signaling [90, 91]. Adiponectin-dependent activation of AMPK signaling pathway has also been linked to the suppression of endothelial cell apoptosis [92] and to the stimulation of human umbilical vein endothelial cells migration and differentiation into capillary structures *in vitro* [91].

Adiponectin also contributes to vascular remodeling by the suppression of smooth muscle cell proliferation and by direct binding to growth factors and interference with their ability to activate VSMC receptor-mediated growth and cellular responses [93, 94].

Transgenic mouse studies have corroborated adiponectin's multiple signaling effects on vascular cells, including endothelial and smooth muscle as well as on macrophages which are likely contributory to the inflammation and endothelial dysfunction implicated in atherogenesis. In a well-established model of atherosclerosis, the ApoE knockout mouse, adiponectin overexpression reduces the formation of aortic atherosclerotic lesions concomitant with the reduction of scavenger receptors, TNF-α, VCAM-1, and can protect against the development of atherosclerosis *in vivo* [95, 96].

Phosphorylation of AMPK induced by adiponectin leads to the inhibition of protein synthesis, and there is growing evidence that this may impact (as well as be an adaptive response to) pathological cardiac hypertrophy. Adiponectin-deficient transgenic mice exhibit increased cardiac hypertrophy in response to either pressure overload or Ang II infusion, and this can be reversed by adenoviral-mediated adiponectin overexpression [97]. Evidences for the involvement of AMPK signaling comes from the observations that AMPK phosphorylation declines in adiponectin-deficient mice and from abrogation of the adiponectin inhibition of cardiac hypertrophy in mouse strains with dominant-negative AMPK. The activation of AMPK inhibits cardiac myocyte protein synthesis, mediated by reduced p70 S6 kinase phosphorylation and increased phosphorylation of eukaryotic elongation factor-2, (eEF-2) [98].

Adiponectin also appears to furnish cardioprotection against myocardial I/R injury by stemming both myocyte apoptosis and TNF-α production. While the abrogation of adiponectin-mediated inhibition of cardiac myocyte apoptosis (but not TNF-α production) in dominant-negative AMPK strains suggested at least a partial involvement of AMPK-dependent signaling in adiponectin's prosurvival actions [99], others have shown that enhanced AMPK activity markedly stimulates FAO during ischemia leading to subsequent intracellular acidosis and cell death putting the role of AMPK signaling in cardiac recovery into question [100, 101]. Moreover, the activation of cyclooxygenase-2 (COX-2) by adiponectin with concomitant increases in prostaglandin E2 (PGE2) synthesis also has been linked to cardioprotective role of adiponectin. Interestingly, pharmacological COX-2 inhibition had no effects on adiponectin-mediated AMPK activation or on the inhibition of myocyte apoptosis, while AMPK inhibition showed no effect on adiponectin-mediated COX-2 induction or on reduced TNF-α production caused by lipopolysaccharides indicating the activation of two independent signaling pathways by adiponectin [99].

## Leptin

The adipokine leptin, product of the *Ob* gene is a 16 kDa protein primarily produced by mature adipocytes and is secreted in plasma. Plasma levels of leptin are strongly correlated with adipose mass in rodents as well as in humans [102]. Glucocorticoids and insulin are potent stimulators of leptin expression, and expression is attenuated by β-adrenergic agonists (via β-adrenergic receptors and cAMP), TZDs (via PPAR-γ), testosterone, thyroid hormone, and fasting. It is well established that leptin (whose expression increases after food intake) inhibits subsequent food intake, reduces body weight and stimulates energy expenditure (initially characterized as a sensor of satiety) and binds to specific receptors in the hypothalamus and central nervous system to maintain energy homeostasis. The subsequent identification of isoforms of the leptin receptor (Ob-Ra, Ob-Rb, and Ob-Re), members of the interleukin-6 cytokine family of receptors, in peripheral tissues, including the heart, VSMCs, and endothelial cells, have led to suggestions of a wide range of actions of this adipokine, including direct effects on the heart and vascular system [103–106]. Moreover, in the heart, increased leptin expression have been recently reported following I/R [107, 108] and leptin levels are increased in the media of cultured neonatal rat cardiomyocytes following the treatment with Ang II or endothelin I suggesting that the heart can serve as a site of leptin production [109].

Leptin after binding to its receptor(s) signals primarily by activating the Janus kinase (JAK), signal transducer and activator of transcription (STAT), IRS, and MAPK pathways [110]. In the well-characterized JAK-STAT pathway, ligand binding promotes Ob receptor oligomerization and binding to JAK2 leading to JAK2 autophosphorylation and to phosphorylation of specific tyrosine residues (985, 1077, and 1138) on the Ob-receptor. Phosphorylation of Tyr-1138 on Ob-Rb receptor triggers recruitment of STAT proteins (primarily STAT3) to the JAK2/Ob-Rb complex (to note: Ob-Rb is the only Ob receptor subtype that contains the STAT-binding site).

Tyr-985 phosphorylation on Ob-Rb leads to the activation of Erk1/2 which belongs to the MAPK family. Erk1/2 activation results in increased expression of c-fos and egr-1, target genes involved in cell proliferatioin and differentiation. In addition, leptin has been reported to promote p38 MAPK phosphorylation in both rat VSMCs and in cardiomyocytes and has been associated with the onset of hypertrophy and apoptotic cell death [107, 111, 112]. NF-κB is a proposed downstream target for both p38- and JNK-involving MAPK signaling pathways giving it an integral role as transcriptional regulator of the proinflammatory cytokines TNF-α and IL-1.

## Ghrelin

Ghrelin is a peptide hormone released from the stomach, stimulates hunger and is considered the counterpart of the leptin. Recent investigations revealed an expression of the receptor to this hormone, GHSR-1a, in many cell types and tissues, including cardiomyocytes, myocardium, vascular endothelium, and monocytes [113, 114]. This raises the possibility that ghrelin may have important physiological actions in peripheral tissues. In 2007, Iantorno et al. [115] reported novel vascular actions of ghrelin which mimic PI3K-dependent actions of insulin to stimulate the production of NO by endothelial cells. Regulatory action of ghrelin in endothelium is realized through a signaling pathway that involves GHSR-1a, PI3K, Akt, and eNOS.

According to a number of clinical observations, circulating levels of ghrelin are low in insulin-resistant conditions [116], and some ghrelin gene polymorphisms are associated with increased incidence of diabetes, impaired glucose tolerance, and hypertension [117]. When patients with the MetSyn were treated to increase levels of ghrelin, their endothelial dysfunction improved as a result of increased NO bioavailability [118].

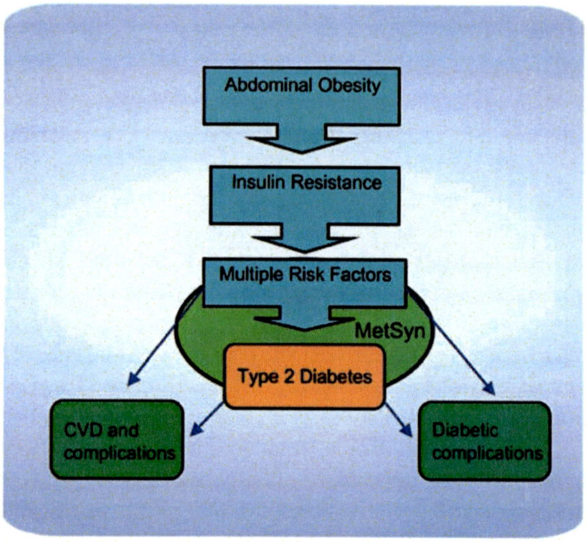


**Fig. 16.3** MetSyn encompasses each stage in the development of risk factors. As the syndrome develops, risks of atherosclerotic cardiovascular disease (CVD) and its complications, including increased cardiac dysrhythmia, heart failure, and thrombosis. With diabetes, other diabetic complications in addition to CVD develop

## Metabolic Syndrome

Term "metabolic syndrome" ("insulin resistance syndrome," "syndrome X") is used to define a constellation of abnormalities that is associated with increased risk for the development of T2D and atherosclerotic vascular disease (e.g., heart disease and stroke) (Fig. 16.3). One of the abnormalities associated with a statistically higher risk of heart disease, hypertension, insulin resistance, and T2D, mellitus is central obesity. One animal model proposed as a model for the development of the MetSyn, includes feeding rats a diet one-third of which is sucrose. The sucrose first elevates blood levels of triglycerides, which induce visceral fat and ultimately result in insulin resistance [119].

There are two common features of this metabolic disorder: adipose tissue dysfunction and elevated levels of tumor necrosis factor-α and adipokines (e.g., adiponectin, resistin). TNF-α has been shown to cause the production of inflammatory cytokines, and the MetSyn is thought to be associated with a chronic low-grade inflammation. In addition, TNF-α triggers cell signaling by the interaction with a TNF-α receptor that may lead to insulin resistance.

It was initially assumed that adipocytes were the predominant source of the elevated adipose tissue TNF-α in obesity. Indeed, obesity in both rodents and humans is associated with increased numbers of apoptotic and necrotic adipocytes in adipose tissue, and one hypothesis is that increased adipocyte death may induce chemoattractant signals that recruit monocytes, and it is likely that adipose tissue macrophages are predominantly responsible for the elevated production of TNF-α during obesity [120].

TNF-α secreted by adipose tissue may function as an autocrine, paracrine, or endocrine factor. Thus, chronic treatment with TNF-α decreases insulin-stimulated glucose uptake in rat skeletal muscle. The molecular basis for the observed impairment in insulin action involves suppression of the expression of many proteins that are required for insulin-stimulated glucose uptake. One of the transcription factors shown to be targeted by TNF-α signaling is PPAR-γ. The detailed mechanism of this influence includes JNK and Erk1/2-catalyzed phosphorylation of serine in the regulatory domain of PPAR-γ which suppresses its activity [121, 122]. In addition, TNF-α downregulates levels of expression of PPAR-γ and another transcription factor, CCAAT/enhancer-binding protein (C/EBPα). It seems very likely that TNF-α-dependent suppression of GLUT4 level occurs via PPAR-γ and C/EBPα because GLUT4 promoter contains response elements for these nuclear factors [123].

Capability of IR and IRS-1 to mediate the effect of insulin on glucose uptake via GLUT4 depends on their tyrosine phosphorylations. There are evidences that TNF-α through its receptor (TNFR1) can inhibit insulin-stimulated tyrosine phosphorylation of both the IR and IRS-1 proteins [124, 125]. In case of IRS-1 it happens because of TNF-α-induced serine phosphorylation of this protein (in many tissues TNF-α activates serine/threonine protein kinases, such as Erk1/2, JNK, PKCθ, and IKKβ which phosphorylate IRS-1) [126–129]. Serine-phosphorylated IRS-1 cannot be tyrosine-phosphorylated by IR and thus stops transducing signal downstream from IR. This results in the impaired insulin-dependent glucose uptake in adipose and nonadipose cells (insulin resistance).

Actions of TNF-α can impact on insulin sensitivity through increased FFA and altered adipokine production. TNF-α reduces FFA oxidation in hepatocytes [130] and skeletal muscle [131] through effects mediated by the induction of protein phosphatase 2C and suppression of AMPK [131]. The reduced rates of FAO are accompanied by increased accumulation of lipids. Hepatic insulin resistance could develop from lipid deposits in the liver. FFA could also be deposited as triglycerides in muscles, resulting in peripheral insulin resistance. Moreover, in experimental animals loading β-cells with fat leads to insulin secretion impairment [132]. Therefore, fat deposition in the wrong organs could provide a "common soil" hypothesis for explaining both insulin resistance and impaired insulin secretion in T2D.

How do elevated levels of plasma FFAs inhibit glucose transport? First, raising FFA concentrations has a negative effect on PI3K, an important component of the insulin-signaling cascade involved in regulating GLUT4 translocation. Second, an elevation in plasma FFA concentration leads to an increase in muscle fatty acyl CoA and bioactive lipid DAG concentrations. Both long-chain fatty acyl CoAs and DAG activate PKCθ, which increases serine phosphorylation with subsequent inhibition of insulin receptor substrate-1 tyrosine phosphorylation [133].

Also TNF-α suppresses the transcription of adiponectin in an adipocytes [134], so levels of this adipokine, which enhances insulin sensitivity, stimulates hepatic glucose usage and reduces FFA levels, are decreased in obese patients.

## Nuclear Receptors

Very important aspect of MetSyn, obesity, diabetes is dyslipidemia, and several nuclear hormone receptors play the pivotal role in altered lipid metabolism.

Nuclear receptors called liver X receptors (LXR) regulate expression of more than a dozen proteins; many of them are parts of the cholesterol and fatty acid metabolic pathways. Of two subtypes of LXR, LXRβ is distributed ubiquitously, whereas LXRα is found predominately in the liver, adipose tissue, and macrophages. Natural LXR agonists are sterol metabolites (several hydroxycholesterols). LXRs are constitutively nuclear, form heterodimer with retinoid X receptor (RXR), and repress transcription of their target genes in the absence of ligand. Binding of ligand causes a conformational change in the LXR that enhances dissociation of corepressors. Thus, ligand binding promotes transcription of genes regulated by LXRs. The activation of LXR is known to have antidiabetic effects through its coordinate regulation of glucose metabolism in both liver and adipose tissue [135]. Same time, LXRα agonists can induce hypertriglyceridemia because they stimulate an expression of several lipogenic regulators in the liver, such as sterol regulatory element-binding protein 1c (SREBP-1c) and fatty acid synthase (FAS) [136, 137]. Fortunately, this side effect of LXR agonists might be improved by concomitant treatment with fibrates (PPAR-α agonists; see below).

One recent hypothesis links obesity-related insulin resistance, and consequently T2D, to a chronic inflammation that involves the accumulation of macrophages in adipose tissue [138, 139]. LXRα agonists would be expected to reduce macrophage-induced inflammation in adipose tissue because LXRα in these cells is shown to regulate cholesterol transporter, ATP-binding cassette transporter (ABCA) 1, responsible for cholesterol efflux [140].

Another key transcriptional regulators in lipid metabolism are PPARs. There are three isoforms in this subfamily of nuclear receptors: PPAR-α, PPAR-β/δ, and PPAR-γ. While PPAR-α and PPAR-β/δ are widely expressed in metabolically active tissues (liver, heart, kidney, skeletal muscle), PPAR-γ is found predominantly in adipose tissue and macrophages. No specific gene targets have been found for PPAR-β/δ. Similar to LXRs, PPARs bind PPAR response elements within the promoter elements of genes as a heterodimers with the RXR subunit. Ligand activation of PPARs increases gene transcription.

There are several drugs and natural substances known to modulate PPAR activity. For example, the TZDs are agonists that activate PPAR-γ and are able to improve the actions of insulin in diabetes [141, 142]. The fibrate drugs (gemfibrozil, fenofibrate) as well as fatty acids and metabolites of the arachadonic acid pathway activate PPAR-α, increase hepatic FAO and therefore reduce triglyceride production [143].

Several investigators have shown that the activation of PPAR-α isoform in isolated cardiomyocytes (PPAR-α agonists, PPAR-α overexpression) induces the expression of many genes involved in fatty acid catabolism [144, 145]. Surprisingly, PPAR-α agonists show little effect of on myocardial PPAR target genes when applied *in vivo*. They have rather unexpected cardiac effects. For example, PPAR-α agonists decrease cardiac FAO in diabetic mice [146]. It is likely that PPAR-α agonists influence cardiac metabolism indirectly via changing the concentration of circulated cardiac substrates, lipids (it is known that PPAR-α agonists inhibit hepatic lipoprotein secretion) [147].

Interestingly, studies on recently developed transgenic mice with cardiac-specific deletion of PPAR-β/δ demonstrate that this isoform is an important regulator of myocardial energy metabolism: animals cardiodefective in PPAR-β/δ have decreased rates of FAO, increased cardiac lipid accumulation, and develop cardiomyopathy and CHF leading to premature death [148].

One important effect of PPAR-γ activation is an increase of glucose uptake in adipose tissue and skeletal muscle. Selective muscle- or adipose-specific PPAR-γ knockout causes severe insulin resistance progressive lipodystrophy in those tissues [149, 150]. Another PPAR-γ target genes are the lipid scavenger receptors SR-B1 and CD36, so overexpression or knockout of PPAR-γ infuences oxidized lipid scavenging [151, 152]. In addition, PPARγ activates lipoprotein lipase, FABPpm, and LXRα in adipocytes and thus promotes fat storage and reduces serum lipid levels.

Detailed mechanism of PPAR-dependent regulation of target gene transcription includes an additional step after agonist-promoted heterodimerization "PPAR-RXR." This step is a recruitment of transcriptional coactivators that are necessary to initiate target gene transcription. The best-characterized coactivator of PPAR-α in the heart is the cardiac-enriched PPAR-γ coactivator-1α (PGC-1α). PGC-1α in the heart induces genes encoding FAO enzymes, stimulates mitochondrial biogenesis and enhances the expression of protein components of the electron transport chain [153, 154]. Transgenic mice with cardiac-specific overexpression of PGC-1α demonstrate mitochondrial ultrastructural abnormalities and dramatic cardiac dysfunction [155]. From the other side, PGC-1α-deficient mouse lines develop cardiomyopathic remodeling especially under an increased workload [156]. Thus, genetic models demonstrate that perturbations in the PGC-1α system could influence cardiomyopathic remodeling.

## Oxidative Stress

There is an activation of the RAAS in cardiometabolic syndrome which leads to elevations in Ang II and aldosterone. Increased Ang II contributes to the development of resistance to the metabolic actions of insulin in cardiovascular as well as skeletal muscle tissue ("insulin resistance") [157, 158], as well as other components of cardiometabolic syndrome, such as hypertension and dyslipidemia.

In the vasculature, inhibitory effects of Ang II include the activation of $AT_1$ receptor, the generation of ROS via activation of the NADPH oxidase enzyme complex, and ROS-related signaling pathways, such as the activation of small molecular-weight G proteins (RhoA, Rac1). A study by Wei et al. [159] has showed that in transgenic rat with increased Ang II levels and increased plasma mineralocorticoids, blood vessels exhibit increased NADPH oxidase activity, ROS levels, and reduced phosphorylation/activation of eNOS by insulin. These abnormalities were markedly improved by *in vivo* treatment of animals with an $AT_1$ receptor blocker or antioxidant (tempol).

Increased ROS also activate NF-κB-dependent redox signaling pathway [160]. NF-κB, in turn, upregulates inflammatory molecules, such as TNF-α [161]. TNF-α activates several serine protein kinases which increase serine phosphorylation of IRS, leading to decreased PI3K-Akt signaling responses, impaired insulin stimulation of eNOS, and vasodilatation [162].

In addition, proinflammatory effects of chronically elevated levels of glucose and fatty acids contribute to endothelial dysfunction and insulin resistance and include increased ROS (see "Mitochondrial Dysfunction" below). Increased ROS induced by hyperglycemia and dyslipidemia further impair insulin signaling and decrease NO bioavailability [163–165].

Another component of RAAS, aldosterone, also influences insulin function. Thus, blocking of aldosterone receptor (mineralocorticoid receptor, MR) with spironolactone improves insulin-stimulated increases in glucose uptake in skeletal muscle, and the effect is based on the reduction in NADPH oxidase activity and attenuation of ROS [166].

## Mitochondrial Dysfunction

Hyperglycemia and diabetes affect cardiac mitochondria function directly. In animals treated with streptozocin to induce diabetes, cardiac mitochondria show pronounced swelling, increased damage, and targeting by lysosomes [167]. Mitochondria from a variety of diabetic animal models show diminished respiratory control as well as increased oxidative stress [168, 169]. These changes in mitochondrial structure and function are reversed with insulin administration.

Insulin resistance occurs in patients with T2D, and associates with central obesity. As a consequence of this abnormality, patients develop hyperglycemia (T2D) and increased cytosolic triglyceride levels in several tissues (obesity). Hyperglycemia and hypertriglyceridemia activate pathways leading to mitochondrial dysfunction and increased production of ROS in mitochondria. Chronic overproduction of mitochondrial ROS damages pancreatic β-cells, increases oxidation of low-density lipoprotein (LDL), affects endothelial cells – and these factors promote atherosclerotic vascular disease.

As we mentioned earlier, hyperlipidemia is accompanied by increased NEFAs. The reaction between palmityol-CoA, an intracellular intermediate of NEFA palmitate, and serine leads to the *de novo* synthesis of the sphingolipid ceramide which has been demonstrated to induce cardiomyocyte apoptosis. Experimental data suggest that ceramide acts on the mitochondria of intact cells probably through nonspecific injury to the mitochondrial membrane lipids, which leads to the release of proapoptotic factors [170]. Interestingly, in addition to participation in *de novo* synthesis of proapoptotic factor (ceramide), palmitate (and other long-chain saturated fatty acids) can induce apoptosis in cultured rat neonatal cardiomyocytes directly [171]. Palmitate-induced myocyte apoptosis may involve opening of the mitochondrial permeability transition pore (MTP) and is associated with an increase in caspase-3 activity, mitochondrial cytochrome *c* release, poly(ADP-ribose) polymerase cleavage (inhibiting DNA repair), DNA laddering, and loss of the mitochondrial membrane potential [170, 172, 173]. Moreover, in contrast to ceramide, palmitate-induced cell death is not dependent on the generation of ROS [174].

Phospholipase isoform $iPLA_2$ which is increased in both diabetes and myocardial ischemia (see subsection "Lipotoxicity") is closely associated with the accumulation of long-chain acylcarnitine and 3-hydroxyacylcarnitine – a potential indication of impaired mitochondrial fatty acid β-oxidation. The acylcarnitine elevation was attenuated by treatment with the $iPLA_2$ inhibitor (BEL) [44]. Other studies have shown that $iPLA_2$ isoform activity markedly increases in the mitochondrial inner membrane during ischaemia. More specifically, increased $iPLA_2$ activity was associated with the inner membrane of mitochondria and resulted in a rapid decrease in phosphatidylcholine and phosphatidylethanolamine glycerophospholipid species. These changes can be partially alleviated by pretreatment of hearts with the BEL which suggests that mitochondria may be a critical locus for the $iPLA_2$ signal transduction associated with myocardial injury [175].

ROS overproduction is one of the major characteristics of dysfunctional mitochondria. Hyperglycemia induces the production of $O_2^{\bullet-}$ in endothelium via activation of electron transport complex (ETC) (inhibitor of ETC Complex II normalizes mitochondrial ROS levels). In addition, superoxide dismutase 2 (SOD2) which quickly dismutates $O_2^{\bullet-}$ in the mitochondrial matrix to hydrogen peroxide ($H_2O_2$), is deficient in dysfunctional mitochondria (overexpression of SOD2 normalizes mitochondrial ROS levels). Also UCPs play a role in ROS generation: these inner mitochondrial membrane anion transporters allow protons to leak back into the mitochondrial matrix, decreasing the mitochondrial membrane potential and ROS generation. Overexpression of UCP-1 normalizes mitochondrial ROS levels [176].

Insulin resistance increases plasma FFAs levels because of enhanced lipolytic activity of adipocytes. Increased oxidation of FFAs by aortic endothelial cells would lead to accelerated production of $O_2^{\bullet-}$. FFA-induced ROS production is mediated by ETC (ROS production was inhibited by overexpression of either UCP-1 or SOD2).

Elevated mitochondrial ROS induce vascular dysfunction and atherosclerosis (Fig. 16.4). One mechanism involves ROS-dependent activation of polyol pathway, which results in sorbitol accumulation and nuclear factor B activation [176]. NF-κB induces the expression of VCAM-1 and monocyte chemoattractant protein-1 in aortic endothelial cells stimulating atherogenesis [177].

In aortic endothelial cells, ROS increase hexosamine pathway activity [178]. The activation of the hexosamine pathway causes the increased glycosylation and subsequent transactivation of transcription factor Sp1, resulting in increased expression of Sp1-dependent genes, such as transforming growth factor-$\beta_1$ (risk factor of atherosclerosis) and plasminogen activator inhibitor-1 (risk factor of ischemic heart disease) [178–180].

In addition, ROS produce DNA strand breaks which activate poly(ADP ribose) polymerase in endothelium. Poly (ADP ribose) polymerase poly(ADP-ribosyl)ates/inhibits glyceraldehyde-3-phosphate dehydrogenase (GAPDH). Inhibition of GAPDH activity increases the entry of upstream glycolytic metabolites into pathways of glucose overuse, including increased flux through advanced AGE and PKC-glucotoxic pathways [181].

Elevated mitochondrial ROS are also responsible for the increased oxidation of LDL in endothelial cultures [182]. Oxidized LDLs is one of the factors that promote atherosclerosis.

Increased production of ROS in mitochondria damages lipids, proteins, and mtDNA, which leads to the mitochondrial dysfunction and further increase of ROS. One important consequence of mitochondrial dysfunction is the opening of MTP. Prolonged opening of MTP leads to matrix swelling and outer mitochondrial membrane rupture which in turn causes the release of proapoptotic molecules within the intermembrane space, leading to cell death via caspase-dependent and caspase-independent mechanisms [183]. This mechanism underlies hyperglycemia-induced apoptosis of human aortic endothelial cells [184]. Decreased expression of genes related to oxidative phosphorylation, in dysfunctional mitochondria is associated with low aerobic capacity observed in cardiovascular disease.

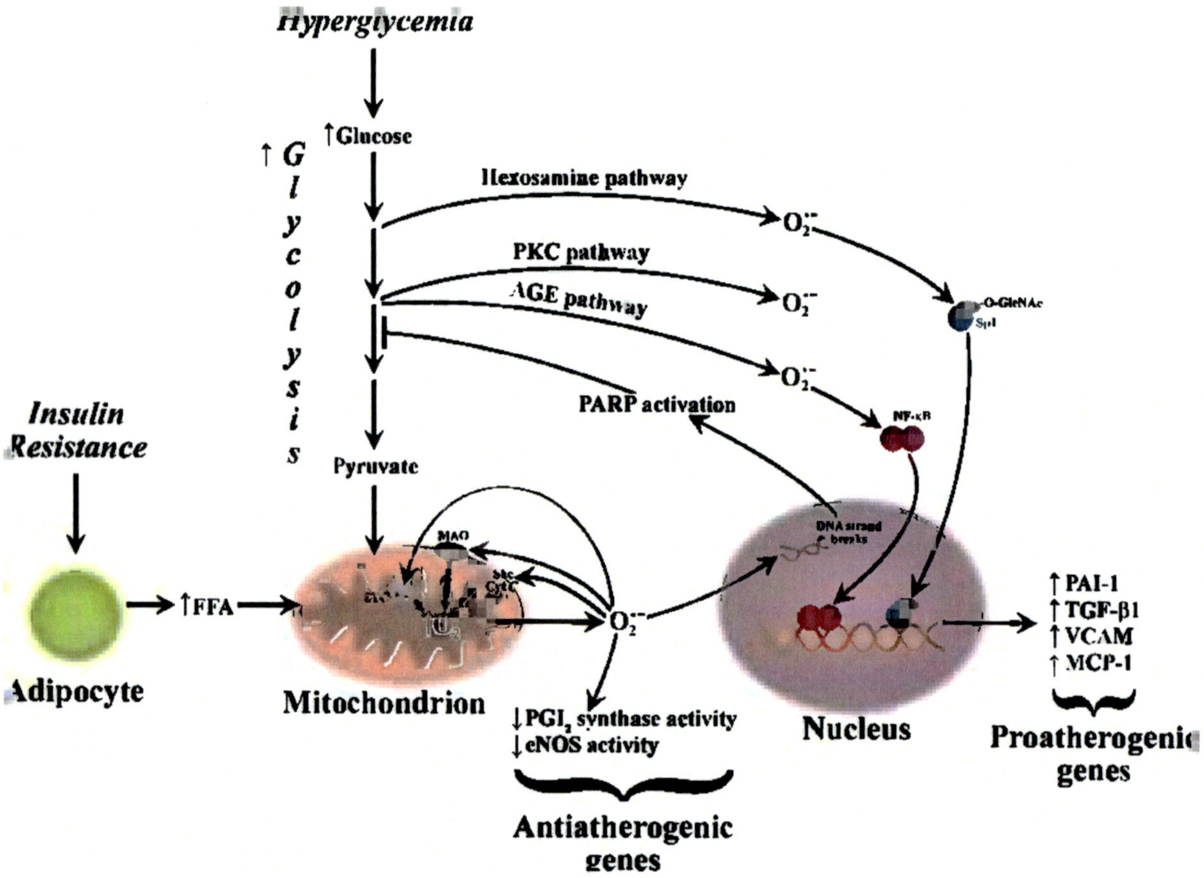


**Fig. 16.4** Activation of atherogenic-signaling pathways via insulin resistance and hyperglycemia-induced overproduction of mitochondrial superoxide products. Major sources of reactive oxygen species ($O_2^{\bullet-}$) in mitochondria are electron transport complex (ETC), monoamine oxidase (MAO), and p66Shc-activated cytochrome C (Cyt C). Overproduction of mitochondrial $O_2^{\bullet-}$ caused by high glucose flux through endothelial cells (hyperglycemia), and/or by increased oxidation of free fatty acids (FFA) by endothelial cells (enhanced lipolytic activity of adipocytes under insulin resistance condition) results in DNA strand breaks and the activation of poly(ADP ribose) polymerase (PARP). PARP disrupts glycolytic pathway by ribosylation/inactivation of glyceraldehyde-3-phosphate dehydrogenase. The upstream metabolites of disrupted glycolytic pathway are processed through hexosamine pathway, PKC pathway, or AGE pathway. Glycosylation of transcription factor Sp1, by the addition of *N*-acetylglucosamine (GlcNAc) results in transactivation of proatherogenic genes: plasminogen activator inhibitor-1 (PAI-1), transforming growth factor β1 (TGF-β1). Activation of transcription factor nuclear factor κB (NF-κB) induces transactivation of proatherogenic genes: vascular cell adhesion molecule-1 (VCAM-1), monocyte chemoattractant protein-1 (MCP-1). In addition, overproduction of mitochondrial $O_2^{\bullet-}$ inactivates antiatherogenic enzymes: eNOS, prostacyclin ($PGI_2$) synthase

## Genetic Basis for Diabetes and Metabolic Syndrome

### *Genes Associated with Diabetes*

The genetic basis for diabetes has been a matter of increasing interest. In several rare forms of diabetes, including neonatal diabetes and maturity onset diabetes, a monogenic etiology has been elucidated. Similar to common forms of hypertension, the common types of diabetes (Type 1 and 2) are polygenic; a large number of genes appear to be involved with extensive modulation by environmental and epigenetic factors.

Recent reports have identified discrete activating mutations in the *KCNJ11* gene, which encodes the Kir6.2 subunit of the sarcolemmal ATP-sensitive potassium channels ($K_{ATP}$) that prevent its closure in the pancreatic β-cells (and affect insulin secretion) as a primary cause of neonatal diabetes mellitus. The mutated $K_{ATP}$ channels do not close in the presence of metabolically generated ATP so the β-cell membrane stays hyperpolarized and insulin secretion does not occur. The degree of $K_{ATP}$ channel overactivity has been shown to correlate with the severity of the diabetic phenotype.

Mutations in channel properties of Kir6.2 that underlie transient neonatal diabetes (I182V) or more severe forms of permanent neonatal diabetes (V59M, Q52R, and I296L) have been identified, which in all cases result in a significant decrease in sensitivity to inhibitory ATP correlating with channel overactivity in intact cells and increasing the $K_{ATP}$ current, which inhibits β-cell electrical activity and insulin secretion [185–188].

The targeted ATP-sensitive potassium channel couples membrane excitability to cellular metabolism and is a critical mediator in the process of glucose-stimulated insulin secretion. These findings have also proven applicable to the studies of T2D. A number of $K_{ATP}$ channel polymorphisms have been described and linked to altered insulin secretion indicating that genes encoding this ion channel could be susceptibility markers for T2D and that genetic variants of $K_{ATP}$ channels may underlie altered β-cell electrical activity, and glucose homeostasis, in addition to increased susceptibility to T2D. In particular, the Kir6.2 E23K polymorphism has been linked to increased susceptibility to T2D in Caucasian populations. As with many of the genetic polymorphisms associated with diabetes (to be discussed), this polymorphism has also been associated with weight gain and obesity, both of which constitute major diabetes risk factors. Mechanistically, it has been proposed that the LCFA accumulation in the plasma and in pancreatic β-cells in obese and Type 2 diabetic patients elicit an enhanced stimulation of $K_{ATP}$ channels containing subunits encoded by the polymorphic Kir6.2 E23K allele [189–191].

In addition, loss-of-function mutations in the genes encoding the two subunits of $K_{ATP}$ channels have been identified which reduce $K_{ATP}$ channel activity and lead to the most common form of congenital hyperinsulinism, resulting in persistent and severe hypoglycemia in the neonatal and infancy period. Moreover, sulfonylureas, which inhibit $K_{ATP}$ channels, can enhance insulin secretion in Type 2 diabetics. This has led to their widespread use in treating patients who were insulin-dependent and provides an important alternative treatment to the use of insulin injections with improved glycemic control [192].

A second type of monogenic subtype of diabetes mellitus, maturity-onset diabetes of the young (MODY), is characterized by an early onset of T2D (usually as children or young adults), including some abnormalities of the β-cell function and an autosomal dominant inheritance with high penetrance [193]. MODY types represent less than 5% of all cases of T2D. Mutations in six genes have been described thus far; these different gene mutations are associated with different clinical forms of the disease. For instance, mutation in the *GCK* gene-encoding glucokinase, a key regulatory glycolytic enzyme of the β-cell, is found in MODY 2 patients. The other mutant *loci* present in MODY 1, 3, 4, 5, and 6, respectively, include defects in specific transcription factors, including the hepatocyte nuclear factors-1α, -4α, and -1β (HNF-1α, HNF-4α, and HNF-1β), the insulin promoter factor-1 (IPF-1) and NeuroD [194–196]. Individuals harboring either of the two most frequently found forms of MODY 2 display a more benign clinical prognosis with an elevated threshold for glucose sensing resulting in mild, regulated hyperglycemia or impaired glucose tolerance with relatively few cardiovascular complications. In contrast, subjects with MODY 3 (with defective HNF-1α) exhibit a much more severe disorder, more typical of T2D and frequently require treatment with sulphonylureas or insulin.

Other monogenic forms of T2D characterized by severe insulin resistance are the result of mutations in genes encoding PPAR-γ (*PPARG*), Akt (*AKT2*), and the insulin receptor (*INSR*) [197–200]. These patients sometimes develop discrete extra-pancreatic phenotypes; for example, lipid abnormalities or a variety of cystic renal diseases.

Efforts to identify genes responsible for more common, polygenic forms of Type 1 diabetes (T1D) and T2D have been less fruitful. Despite intensive research, there is still no definitive genetic test to diagnose T1D or T2D. Data from both animal and clinical studies have revealed multiple overlaps in the genes implicated in both types of diabetes as well as in the pathogenic pathways, including apoptotic remodeling. Moreover, both types of diabetes have been shown to be associated with significant damage to mitochondria in pancreatic β-cells, liver, and heart [201, 202].

In T2D, which accounts for about 85% of all diabetic patients, the body either produces too little insulin, or does not respond well to it. Forms of T2D tend to have a middle/late age of onset and occur with both impaired insulin secretion and insulin resistance. While this type of diabetes might be predicted to be induced by multiple gene defects specifically involved in insulin action and/or insulin secretion, a variety of other genes have been at least partially implicated in its genesis as shown in Table 16.1. Moreover, other genetically influenced traits like obesity and hyperlipidemia are strongly associated with T2D. Interestingly, many of the risk factors implicated in the development of T2D, including weight gain, lack of physical exercise, and increasing age, are associated with an impaired mitochondrial function. Furthermore, recent studies have suggested that mitochondrial bioenergetic dysfunction largely underlies the defects in insulin responsiveness found in skeletal muscle and liver responsible for insulin resistance [203], and for defects in glucose-stimulated insulin secretion by pancreatic β-cells responsible for the progression to hyperglycemia [204–207].

Genes and genetic variants thus far identified in mediating T2D encode calpain 10, PPARγ, Kir6.2, and insulin. In addition, some evidence exists that genes encoding proteins, such as for adiponectin, IRS-1, and some others, may also influence the susceptibility to T2D. The clinical manifestations and course of this disease is fostered by the interaction

**Table 16.1** Polymorphisms associated with diabetes

| Phenotype | Variants (gene) | Polymorphism | Reference |
|---|---|---|---|
| T2D | Methylenetetrahydrofolate reductase (*MTHFR*) | 677C>T | [216] |
| T2D | Transcription factor-activating protein 2β (*TFAP2B*) | Variable # of tandem repeat | [217] |
| T2D | PGC-1α (*PPARGC1A*) | Thr394Thr | [218] |
| | | Gly482Ser | [219] |
| T2D | TNF-α (TNFA) | G-308 A allele | [220, 221] |
| T2D | SUR1/Kir6.2/(*KCNJ11*) | E23K | [222] |
| T2D | Calpain 10 (*CAPN10*) | SNP44 | [223, 224] |
| T2D | Lamin A/C (*LMNA*) | 1908C/T | [225] |
| T2D | IRS-1 (*IRS-1*) | Gly972Arg | [226, 227] |
| T2D/CAD | Adiponectin (*APM1*) | SNP exon 2 (45T/G) | [228] |
| T2D/CAD | Adiponectin (*APM1*) | SNP intron 2 (276G/T) | [229] |
| T2D/CAD | Adiponectin (*APM1*) | Tyr111His | [230] |
| T2D | PPAR-γ (*PPARG*) | Pro12Ala | [231–233] |
| T1D | Lymphoid protein tyrosine phosphatase (*PTPN22*) | 1858T | [213] |
| T1D | Catatonic T-lymphocyte-associated antigen-4 (*CTLA-4*) | A49G | [212, 234, 235] |
| T1D | HLA class II DQ and DR alleles | Several haplotypes in DQb, DQa and DRB1 | [236, 237] |
| T1D | Insulin (*INS*) | Polymorphism 5′ tandem repeats | [238, 239] |

*T2D* Type 2 diabetes; *T1D* Type 1 diabetes; *CAD* Coronary artery disease

of environmental and genetic factors, including frequent polymorphisms of many genes, not just one. These polymorphisms may be localized in the coding or regulatory parts of the genes and are present, although with different frequencies, in both patients as well as healthy individuals. A large number of pathogenic mechanisms for T2D have been proposed, including increased nonesterified fatty acids, inflammatory cytokines, adipokines, and mitochondrial dysfunction for insulin resistance, and glucotoxicity, lipotoxicity, and amyloid formation for β-cell dysfunction [208].

Multifactorial and polygenic T1D is strongly influenced by multiple genes controlling the immune system, within the major histocompatibility complex, primarily HLA-DQ and HLA-DR [209]. Another well-characterized susceptibility locus is the insulin gene, including the variable nucleotide tandem repeat locus within the regulatory region of the gene. This genetic variation affects the expression of insulin in the thymus and may play a role in the modulation of tolerance to this molecule [210]. Moreover, a significant autoimmune component has been identified in a large subset of T1D cases, with measurable autoantibodies a useful diagnostic and prognostic marker [211]. In addition, several predisposition *loci* (Table 16.1), interacting with each other, have significant influence on the susceptibility to T1D. These include polymorphic variants of genes involved in signal transduction, including the cytotoxic T lymphocyte-associated molecule 4 (*CTLA-4*) [212], and the *PTPN22* gene encoding the lymphoid protein tyrosine phosphatase (LYP) [213]. In some populations, genomic variations of vitamin D metabolism and target cell action predispose to T1D. There is increasing epidemiological evidence suggesting that vitamin D deficiency in early life increases the incidence of later onset autoimmune diseases, such as T1D, in genetically predisposed individuals, and that high-dose vitamin D supplementation can be protective against its development in both animals and humans [214, 215]

## Genes Involved in Metabolic Syndrome and Insulin Resistance

Numerous groups are seeking to identify key genes involved as risk factors in MetSyn with many of the common polymorphic genetic variants showing some complicity/association with either the overall syndrome or with some of its central components [240–272], as well as to establish presymptomatic disease biomarkers [273]. Table 16.2 shows information concerning several of the candidate genetic *loci* which have been examined for significant association with MetSyn and its components.

Insight into the understanding of MetSyn has come from recent studies of familial partial lipodystrophy (FPLD) as a potential model of MetSyn. FPLD is a rare monogenic form of insulin resistance with a gradual evolution and marked recapitulation of key clinical and biochemical features of MetSyn [274]. FPLD can be caused by mutations in either *LMNA*, encoding nuclear lamin A/C (subtype FPLD2) [275, 277], or in *PPARG*, encoding (subtype FPLD3), a transcription factor with a key role in adipocyte differentiation and

**Table 16.2** Gene polymorphisms associated with insulin resistance and metabolic syndrome

| Phenotype | Protein (gene) | Polymorphism | Reference |
|---|---|---|---|
| Insulin resistance, FPLD | Lamin A/C (*LMNA*) | 1908C/T, R482Q, R133L, H566H | [240–242] |
| Body fat/insulin sensitivity; MetSyn | Protein tyrosine phosphatase 1β (*PTPN1*) | Pro387Leu Haplotypes | [243–245] |
| Insulin resistance/premature CHD/ obesity/MetSyn | β3-adrenoceptor (*ADRB3*) | Trp64Arg | [246–248] |
| Insulin resistance | PPAR-α (*PPARA*) | Leu162Val | [249, 250] |
| Increased risk of MetSyn | PPAR-γ (*PPARG*) | Pro12Ala; 3 SNPs including P2 -689C>T, Pro12 Ala (C/G) and 1431C>T- act together not independently | [251–253] |
| Improvement in IR after exercise | PPAR-γ (*PPARG*) | Pro12Ala | [254] |
| No significant association with overall MetSyn or with insulin resistance; ↑ BMI with both Ala and T allele | PPAR-γ (*PPARG*) | Pro12Ala C161T | [255] |
| No association with overall MetSyn, IR or BMI but did with other MetSyn components (e.g., ↑ WHR and DBP/HT) | PPAR-γ (*PPARG*) | Pro12Ala | [256] |
| No association with MetSyn | PPAR-δ (*PPARD*) | 14 SNPs | [257] |
| Insulin resistance/MetSyn | $\beta_2$-adrenoceptor (*ADRB2*) | Arg16Gly | [258] |
| Association with MetSyn | PGC-1α (*PPARGC1A*) | Gly482 Ser | [259] |
| Not associated with MetSyn but with some MetSyn components (e.g., ↑ WHR and SBP/HT) | PGC-1α (*PPARGC1A*) | Gly482 Ser | [256, 260] |
| Association with MetSyn | Fatty acid-binding protein 2 gene (*FABP2*) | Ala54Thr | [261–264] |
| No association with MetSyn in CHD patients | Fatty acid-binding protein 2 (*FABP2*) | Ala54Thr | [265] |
| Association with MetSyn | Adiponectin (*APM1*) | I164T | [266] |
| No association with IR or MetSyn | Transcription factor 7-like 2 (*TCF7L2*) | 2 SNPs rs12255372 and rs7903146 | [267, 268] |
| No association with MetSyn | Ghrelin (*ghrelin*) | Leu72Met | [269] |
| Association with BMI + MetSyn | Forkhead box protein C2 (*FOXC2*) | −512C>T | [270] |
| Association with MetSyn | Hepatic nuclear factor-4α (*HNF4A*) | 2 haplotypes containing 5 intronic SNPs | [271] |
| Association with MetSyn | Apolipoprotein C-III (*APOC3*) | C-482T+T-455C | [262, 263, 272] |

*BMI* body mass index; *CHD* coronary heart disease; *FPLD* familial partial lipodystrophy; *MetSyn* metabolic syndrome; *DBP/HT* diastolic blood pressure/hypertension; *SBP/HT* systolic blood pressure/hypertension; *SNPs* single-nucleotide polymorphisms; *WHR* waist:hip ratio

metabolic regulation [278, 279]. Most of the mutations in *LMNA* associated with FPLD2 are missense mutations localized near the 3′ end of the gene proximal to the DNA-binding domain at the C terminus of the protein suggesting that a probable molecular pathogenic mechanism elicited by these mutations involves their interaction with transcription factors or other DNA-binding elements [280]. Interestingly, mutations in lamin A/C also cause cardiomyopathy and Hutchinson–Gilford progeria syndrome (HGPS). Over a dozen mutations in PPAR-γ have been implicated in FPLD3, some acting by a dominant-negative mechanism, others through haploinsufficiency. Dominant-negative mutations in PPAR-γ leading to MetSyn with severe hyperinsulinemia and early onset hypertension have also been reported [e.g., proline-467-leucine (P467L)] [281]. Moreover, recent studies have described several mutations in the DNA-binding domain of human PPAR-γ that lead to lipodystrophy and severe insulin resistance [282]. These mutants lack DNA binding and transcriptional activity but can translocate to the nucleus, interact with PPAR-γ coactivators and have been shown to inhibit coexpressed wild-type receptor. Expression of target genes dependent on PPAR-γ was markedly attenuated.

About 50% of FPLD patients do not have mutations in either *LMNA* or *PPARG* suggesting the potential involvement of either unidentified regulatory mutations, novel causative/pathogenic sequences in these genes or undiscovered genetic *loci* [280].

Given the list of genetic factors that have thus far been implicated from association studies in MetSyn, it is noteworthy that a good proportion are involved in transcription regulation, including the previously discussed laminin A/C,

PPARs (α and γ), PGC-1, the forkhead transcription factor FOXO2, and the hepatic nuclear factor-4α. Alterations in the abundance and activity of transcription factors can lead to complex dysregulation of gene expression, which is pivotal in the generation of insulin resistance-associated and its clustering of coronary risk factors at the cellular or gene regulatory level. Members of the nuclear hormone receptor superfamily – for example, PPARs, PGC-1, RXR-α, and SREBPs have all been implicated in both insulin resistance and MetSyn [283–285]. In addition to their regulation by a host of metabolites and nutrients, these transcription factors are also targets of hormones (like insulin and leptin), growth factors, inflammatory signals, and drugs. Extracellular stimuli are coupled to transcription factors by a variety of signaling pathways, including the MAP kinase cascades. For instance, SREBPs appear to be substrates of MAP kinases and have been proposed to play a contributory role in the development of cellular features belonging to lipid toxicity and MetSyn [100]. Therefore, MetSyn appears to be not only a disease or state of altered glucose tolerance, plasma lipid levels, blood pressure, and body fat distribution, but rather a complex clinical phenomenon of dysregulated gene expression.

In addition, variants of genes encoding protein components of intracellular metabolic signaling pathways, including FABPpm, adiponectin, protein tyrosine phosphatase nonreceptor 1β, and apolipoprotein C and β-adrenergic receptor, have also been implicated or associated with MetSyn.

These findings have suggested an important approach for the clinical management and the treatment of MetSyn [286–288]. Both PPAR-α activators, such as fibric acid class of hypolipidemic drugs and PPAR-γ agonists, including antidiabetic TZDs, have proved to be effective for improving MetSyn. PPAR-α agonists, such as the fibrates, correct dyslipidemia thus decreasing cardiovascular disease risk while PPAR-γ agonists, such as the glitazones, increase insulin sensitivity and decrease plasma glucose levels in patients with diabetes. Moreover, both PPAR-α and PPAR-γ agonists exert antiinflammatory activities in liver, adipose, and vascular tissues.

## Conclusions

Insulin regulates diverse cellular functions, perhaps the most critical being glucose homeostasis and cellular growth. Insulin acts through its cognate receptor to activate at least two major signaling pathways – a PI3K-dependent and a PI3K-independent pathway. As the downstream targets of the Akt pathway specifically involved in insulin-induced growth are more clearly defined, and as the *in vivo* role of the PI3K-independent pathway is better understood, the opportunities for preventing morbidity and mortality in the contexts of diabetic cardiomyopathy and in the treatment of heart failure continue to progress.

Several insulin effects are antiatherogenic. In the presence of metabolic insulin resistance, the resulting compensatory hyperinsulinemia stimulates both the MAP-kinase signaling pathway and prenylation of Ras and Rho proteins and thus becomes proatherogenic.

Obesity, especially abdominal obesity, is related to insulin resistance. The main mediators of insulin resistance are elevated concentrations of FFAs and altered adipokine secretion from "dysfunctional" adipocytes. Adipose tissue-derived TNF-α can contribute to the metabolic complications associated with obesity by altering both adipose tissue function and expandability.

In various cardiometabolic diseases, alterations in AMPK pathway take place. It seems promising for pharmaceutical companies to produce a specific AMPK activator or compounds that target the downstream effectors of AMPK to treat obesity, cardiomyocyte hypertrophy and mitochondrial biogenesis.

Lipid metabolism is strongly regulated by nuclear receptors, such as liver X receptor and PPAR. Pharmacological targeting of these receptors can manage dyslipidemia, other lipid-related diseases as well as associated MetSyn, obesity, and diabetes.

In cardiometabolic syndrome, the activation of the RAAS occurs with subsequent elevation in Ang II and aldosterone. Ang II and aldosterone trigger the formation of ROS which attenuate the insulin-signaling mechanisms in the heart, vasculature, and induce endothelial dysfunction and cardiovascular disease.

Elevated circulating FFAs and hyperglycemia associated with T2D cause mitochondrial dysfunction and regulatory changes. Recent studies implicated mitochondrial dysfunction in the pathophysiology of the adverse cardiac effects of T2D. For instance, muscle cells and adipocytes develop insulin insensitivity and inefficiency in the execution of insulin-stimulated glucose utilization, due to defects in the insulin-signaling pathway, which resulted from excess ROS in the cells with mitochondrial dysfunction. Mitochondria-targeting drugs appear to have potential in the prevention and/or treatment of insulin resistance and T2D.

In several forms of diabetes, a genetic basis for disease has been elucidated. In monogenic types of diabetes, such as neonatal diabetes and maturity onset diabetes, mutations in genes encoding sarcolemmal ion channels, glucokinase, and specific transcription factors account for abnormalities of β-cell function, impaired glucose tolerance, and several other complications. At the same time, multiple genes are implicated in the development of more common polygenic forms of T1D and T2D, including those specifically involved in insulin action, as well as a variety of other genes responsible for apoptotic remodeling, obesity, hyperlipidemia, and autoimmunity; similarly, a large number of genetic *loci* are associated with MetSyn. Overlapping of predisposing genes suggests a close relationship between T2D and MetSyn. Furthermore, several studies have demonstrated that in subjects with MetSyn the risk for progression to T2D is greater than fivefold compared to subjects without the syndrome.

## Summary

- IRS-PI3K-PDK-1-Akt-dependent branch of the insulin-signaling pathway results in many biological actions of insulin, such as vascular relaxation, glucose uptake, glycogen synthesis, and gluconeogenesis.
- Shc-Grb-2-Sos-Ras-Erk1-2-dependent branch of insulin signaling regulates cell growth, mitogenesis, and differentiation.
- An important insulin effect is the activation of endothelial eNOS, leading to the increased production of NO. Also, insulin can directly stimulate the production of another endothelial vasodilator, prostacyclin.
- Insulin-dependent attenuation of VSMC contractility is achieved via IR-PI3K-Akt-NO-cascade and targets MLC. Decreased vasoconstrictor tone is also the result of insulin-dependent decrease of intracellular $Ca^{2+}$.
- Besides its well-known role in glucose transport and metabolism, insulin has a key role in regulating the transport of LCFAs in the myocardium. The regulation of physiological growth of cardiomyocytes by insulin involves the activation of Akt.
- Under the MetSyn conditions of imbalance between PI3K (affected) and MAPK (unaffected)-dependent insulin effects take place which leads to insulin resistance and endothelial dysfunction.
- Under compensatory hyperinsulinemia, induced prenylation of Ras and Rho proteins is related to prohypertensive event – increased proliferatory activity of VSMCs.
- Advanced AGEs accumulate in the diabetic myocardium via the aldose reductase-dependent polyol pathway. They contribute to diabetic vascular complications via RAGE receptors, activating diverse signal transduction cascades and downstream pathways.
- Hyperlipidemia develops during diabetes in the form of increased triglycerides and NEFAs. NEFAs play a central role in altering cellular insulin signaling through several mechanisms leading to insulin resistance and compensatory hyperinsulinemia. In addition, elevated NEFAs can alter several cardiomyocyte functions regulated independently by insulin: they reduce myocardial contractility and induce apoptosis.
- Phospholipid-related signaling pathways, regulated by several phospholipases ($PLA_2$, PLC, and PLD), are central to the pathology of diabetic cardiomyopathy and I/R. Recent studies indicate that $iPLA_2$-related signal transduction associated with myocardial injury takes place in mitochondria.
- TNF-α-dependent inhibition of eNOS function may account for the reduction in bioavailability of NO and endothelial dysfunction that characterizes T2D and heart failure. TNF-α also stimulates the expression of several inflammatory proteins.
- Levels of plasma adiponectin are significantly reduced in obese subjects, in patients with T2D, in patients after acute myocardial infarction.
- In both skeletal and cardiac muscle, adiponectin plays an important role in the regulation of fatty acid and glucose metabolism by promoting glucose uptake and increasing FAO rates.
- The adiponectin-dependent increase in glucose transport in both skeletal muscle and cardiac muscle primarily resulted from an enhanced translocation of the glucose transporter GLUT4 to the cell membrane.
- Adiponectin impacts several signaling pathways leading to atherogenesis: it reduces TNF-α-stimulated monocyte adhesion to endothelial cells, impacts both angiogenesis and endothelial function, and suppresses smooth muscle cell proliferation.
- The adipokine leptin has a wide range of actions on the heart and the vascular system. Leptin after binding to its receptors signals primarily by activating the JAK, STAT, IRS, and MAPK pathways.
- Ghrelin is a peptide hormone released from the stomach. Ghrelin mimics PI3K-dependent effects of insulin to stimulate the production of NO by endothelial cells.
- TNF-α secreted by adipose tissue may function as an autocrine, paracrine, or endocrine factor. It is highly possible that TNF-α-dependent suppression of GLUT4 level occurs via PPAR-γ and C/EBPα. There is evidence that TNFR1 is involved in the development of insulin resistance in adipose and nonadipose cells. In addition, TNF-α reduces FFA oxidation in hepatocytes and skeletal muscle, and elevated levels of plasma FFAs inhibit glucose transport.
- LXRs are activated by sterol metabolites (several hydroxycholesterols). The activation of LXR is known to have antidiabetic effects through its coordinated regulation of glucose metabolism in both liver and adipose tissue.
- Key transcriptional regulators in lipid metabolism are PPARs. The best-characterized coactivator of PPAR-α in the heart is the cardiac-enriched PGC-1α. Genetic models demonstrate that perturbations in the PGC-1α system could influence the cardiomyopathic remodeling.
- The activation of RAAS in cardiometabolic syndrome leads to increases in Ang II and aldosterone. In the vasculature, Ang II generates ROS via activation of the NADPH oxidase.

ROS can reduce the activation of eNOS by insulin. Also ROS upregulate inflammatory molecules, such as TNF-α. Aldosterone also weakens insulin function via activation of NADPH oxidase and ROS increase.

- Ceramide acts on mitochondria of intact cells through nonspecific injury to the mitochondrial membrane lipids, which leads to the release of proapoptotic factors.
- Mitochondria may be a critical locus for the $iPLA_2$ signal transduction associated with myocardial injury.
- ROS overproduction is one of the major characteristics of mitochondria dysfunction. Increased oxidation of FFAs induces ROS production, and this process is mediated by ETC.
- In aortic endothelial cells, ROS increase the expression of Sp1-dependent genes known as risk factors of atherosclerosis and ischemic heart disease. In addition, ROS trigger advanced AGE and PKC-glucotoxic pathways, increase oxidation of LDL and cause the release of proapoptotic molecules within the dysfunctional mitochondria.
- The primary cause of neonatal diabetes mellitus are activating mutations in Kir6.2 subunit of the sarcolemmal $K_{ATP}$ channel, which prevent its closure in the pancreatic β-cells and affect insulin secretion. In addition, loss-of-function mutations in genes encoding the two subunits of $K_{ATP}$ channels, which reduce $K_{ATP}$ channel activity and lead to the most common form of congenital hyperinsulinism, have been identified.
- Mutations in six genes have been associated with different clinical forms of MODY. These genes encode glucokinase and several specific transcription factors.
- T2D might be predicted to be induced by multiple gene defects. Many of the risk factors implicated in the development of T2D are associated with impaired mitochondrial function. Genes and genetic variants, thus far identified in mediating T2D encode calpain 10, PPAR-γ, Kir6.2, and insulin.
- Polygenic T1D is strongly influenced by multiple genes controlling the immune system, within the major histocompatibility complex, primarily HLA-DQ and -DR. In addition, several predispositing *loci* have significant influence on the susceptibility to T1D.
- Many investigators are seeking to identify key genes involved as risk factors in MetSyn, and several of the candidate genetic *loci* showed some significant association with MetSyn.
- Members of the nuclear hormone receptor superfamily (PPARs, PGC-1, RXR-α, and SREBPs) have all been implicated in both insulin resistance and MetSyn. They are not only regulated by a host of metabolites and nutrients, but they are also targets of hormones (insulin, leptin), growth factors, inflammatory signals, and drugs. Thus, MetSyn appears to be a complex clinical phenomenon of dysregulated gene expression.

## References


1. Zeng G, Quon MJ. Insulin-stimulated production of nitric oxide is inhibited by wortmannin. Direct measurement in vascular endothelial cells. J Clin Invest. 1996;98:894–8.
2. Zeng G, Nystrom FH, Ravichandran LV, et al. Roles for insulin receptor, PI3-kinase, and Akt in insulin-signaling pathways related to production of nitric oxide in human vascular endothelial cells. Circulation. 2000;101:1539–45.
3. Montagnani M, Ravichandran LV, Chen H, Esposito DL, Quon MJ. Insulin receptor substrate-1 and phosphoinositide-dependent kinase-1 are required for insulin-stimulated production of nitric oxide in endothelial cells. Mol Endocrinol. 2002;16:1931–42.
4. Trovati M, Massucco P, Mattiello L, et al. Human vascular smooth muscle cells express a constitutive nitric oxide synthase that insulin rapidly activates, thus increasing guanosine 3′:5′-cyclic monophosphate and adenosine 3′:5′-cyclic monophosphate concentrations. Diabetologia. 1999;42:831–9.
5. Sobrevia L, Nadal A, Yudilevich DL, Mann GE. Activation of L-arginine transport (system y+) and nitric oxide synthase by elevated glucose and insulin in human endothelial cells. J Physiol. 1996;490:775–81.
6. Lee JH, Ragolia L. AKT phosphorylation is essential for insulin-induced relaxation of rat vascular smooth muscle cells. Am J Physiol Cell Physiol. 2006;291:C1355–65.
7. Surks HK, Mochizuki N, Kasai Y, et al. Regulation of myosin phosphatase by a specific interaction with cGMP-dependent protein kinase Iα. Science. 1999;286:1583–7.
8. Bolotina VM, Najibi S, Palacino JJ, Pagano PJ, Cohen RA. Nitric oxide directly activates calcium-dependent potassium channels in vascular smooth muscle. Nature. 1994;368:850–3.
9. Kovacic S, Soltys CL, Barr AJ, Shiojima I, Walsh K, Dyck JR. Akt activity negatively regulates phosphorylation of AMP-activated protein kinase in the heart. J Biol Chem. 2003;278:39422–7.
10. Chabowski A, Coort SLM, Calles-Escandon J, et al. Insulin stimulates fatty acid transport by regulating expression of FAT/CD36 but not FABPpm. Am J Physiol Endocrinol Metab. 2004; 287:E781–9.
11. von Lewinski D, Bruns S, Walther S, Kogler H, Pieske B. Insulin causes $[Ca^{2+}]_i$-dependent and $[Ca^{2+}]_i$-independent positive inotropic effects in failing human myocardium. Circulation. 2005;111: 2588–95.
12. McDowell SA, McCall E, Matter WF, Estridge TB, Vlahos CJ. Phosphoinositide 3-kinase regulates excitation-contraction coupling in neonatal cardiomyocytes. Am J Physiol Heart Circ Physiol. 2004;286:H796–805.
13. Rota M, Boni A, Urbanek K, et al. Nuclear targeting of Akt enhances ventricular function and myocyte contractility. Circ Res. 2005;97:1332–41.
14. Walsh K. Akt signaling and growth of the heart. Circulation. 2006;113:2032–4.
15. Samuelsson AM, Bollano E, Mobini R, et al. Hyperinsulinemia: effect on cardiac mass/function, angiotensin II receptor expression, and insulin signaling pathways. Am J Physiol Heart Circ Physiol. 2006;291:H787–96.
16. DeBosch BJ, Muslin AJ. Insulin signaling pathways and cardiac growth. J Mol Cell Cardiol. 2008;44:855–64.
17. Draznin B, Miles P, Kruszynska Y, et al. Effects of insulin on prenylation as a mechanism of potentially detrimental influence of hyperinsulinemia. Endocrinology. 2000;141:1310–6.
18. Golovchenko I, Goalstone ML, Watson P, Brownlee M, Draznin B. Hyperinsulinemia enhances transcriptional activity of nuclear factor-κB induced by angiotensin II, hyperglycemia, and advanced glycosylation end products in vascular smooth muscle cells. Circ Res. 2000;87:746–52.

19. Iwata K, Nishinaka T, Matsuno K, Kakehi T, Katsuyama M, Ibi M, et al. The activity of aldose reductase is elevated in diabetic mouse heart. J Pharmacol Sci. 2007;103:408–16.
20. Cappiello M, Voltarelli M, Cecconi I, Vilardo PG, Dal Monte M, Marini I, et al. Specifically targeted modification of human aldose reductase by physiological disulfides. J Biol Chem. 1996;271: 33539–44.
21. Tan AL, Forbes JM, Cooper ME. AGE, RAGE, and ROS in diabetic nephropathy. Semin Nephrol. 2007;27:130–43.
22. Coughlan MT, Cooper ME, Forbes JM. Renal microvascular injury in diabetes: RAGE and redox signaling. Antioxid Redox Signal. 2007;9:331–42.
23. Goldin A, Beckman JA, Schmidt A, Creager MA. Advanced glycation end products: sparking the development of diabetic vascular injury. Circulation. 2006;114:597–605.
24. Kaneko M, Bucciarelli L, Hwang YC, Lee L, Yan SF, Schmidt AM, et al. Aldose reductase and AGE-RAGE pathways: key players in myocardial ischemic injury. Ann N Y Acad Sci. 2005;1043: 702–9.
25. Hartog JW, Voors AA, Bakker SJ, Smit AJ, van Veldhuisen DJ. Advanced glycation end-products (AGEs) and heart failure: Pathophysiology and clinical implications. Eur J Heart Fail. 2007;9:1146–55.
26. Koyama Y, Takeishi Y, Arimoto T, Niizeki T, Shishido T, Takahashi H, et al. High serum level of pentosidine, an advanced glycation end product (AGE), is a risk factor of patients with heart failure. J Card Fail. 2007;13:199–206.
27. Liu J, Masurekar MR, Vatner DE, Jyothirmayi GN, Regan TJ, Vatner SF, et al. Glycation end-product cross-link breaker reduces collagen and improves cardiac function in aging diabetic heart. Am J Physiol Heart Circ Physiol. 2003;285:H2587–91.
28. Zieman S, Kass D. Advanced glycation end product cross-linking: pathophysiologic role and therapeutic target in cardiovascular disease. Congest Heart Fail. 2004;10:144–9.
29. Bidasee KR, Nallani K, Yu Y, Cocklin RR, Zhang Y, Wang M, et al. Chronic diabetes increases advanced glycation end products on cardiac ryanodine receptors/calcium-release channels. Diabetes. 2003;52:1825–36.
30. Schäfer S, Huber J, Wihler C, Rütten H, Busch AE, Linz W. Impaired left ventricular relaxation in type 2 diabetic rats is related to myocardial accumulation of N(epsilon)-(carboxymethyl) lysine. Eur J Heart Fail. 2006;8:2–6.
31. Cooper ME. Importance of advanced glycation end products in diabetes-associated cardiovascular and renal disease. Am J Hypertens. 2004;17:31S–8.
32. van Heerebeek L, Hamdani N, Handoko ML, Falcao-Pires I, Musters RJ, Kupreishvili K, et al. Diastolic stiffness of the failing diabetic heart: importance of fibrosis, advanced glycation end products, and myocyte resting tension. Circulation. 2008;117:43–51.
33. Kim JK, Kim YJ, Fillmore JJ, et al. Prevention of fat-induced insulin resistance by salicylate. J Clin Invest. 2001;108:437–46.
34. Schwartzbauer G, Robbins J. The tumor suppressor gene PTEN can regulate cardiac hypertrophy and survival. J Biol Chem. 2001;276:35786–93.
35. Liu GX, Hanley PJ, Ray J, Daut J. Long-chain acyl-coenzyme A esters and fatty acids directly link metabolism to $K_{ATP}$ channels in the heart. Circ Res. 2001;88:918–24.
36. Tappia PS. Phospholipid-mediated signaling systems as novel targets for treatment of heart disease. Can J Physiol Pharmacol. 2007;85:25–41.
37. Lamers JM, De Jonge HW, Panagia V, Van Heugten HA. Receptor-mediated signalling pathways acting through hydrolysis of membrane phospholipids in cardiomyocytes. Cardioscience. 1993;4:121–31.
38. Tappia PS, Dent MR, Dhalla NS. Oxidative stress and redox regulation of phospholipase D in myocardial disease. Free Radic Biol Med. 2006;41:349–61.
39. Xu YJ, Panagia V, Shao Q, Wang X, Dhalla NS. Phosphatidic acid increases intracellular free $Ca^{2+}$ and cardiac contractile force. Am J Physiol Heart Circ Physiol. 1996;271:H651–9.
40. Dhalla NS, Xu YJ, Sheu SS, Tappia PS, Panagia V. Phosphatidic acid: a potential signal transducer for cardiac hypertrophy. J Mol Cell Cardiol. 1997;29:2865–71.
41. Pavoine C, Behforouz N, Gauthier C, et al. $\beta_2$-adrenergic signaling in human heart: shift from the cyclic AMP to the arachidonic acid pathway. Mol Pharmacol. 2003;64:1117–25.
42. Engelbrecht AM, Ellis B. Apoptosis is mediated by cytosolic phospholipase $A_2$ during simulated ischaemia/reperfusion-induced injury in neonatal cardiac myocytes. Prostaglandins Leukot Essent Fatty Acids. 2007;77:37–43.
43. Su X, Han X, Mancuso DJ, Abendschein DR, Gross RW. Accumulation of long-chain acylcarnitine and 3-hydroxy acylcarnitine molecular species in diabetic myocardium: identification of alterations in mitochondrial fatty acid processing in diabetic myocardium by shotgun lipidomics. Biochemistry. 2005;44:5234–45.
44. Mancuso DJ, Abendschein DR, Jenkins CM, et al. Cardiac ischemia activates calcium-independent phospholipase $A_2\beta$, precipitating ventricular tachyarrhythmias in transgenic mice: rescue of the lethal electrophysiologic phenotype by mechanism-based inhibition. J Biol Chem. 2003;278:22231–6.
45. Morin CL, Eckel RH, Marcel T, Pagliassotti MJ. High fat diets elevate adipose tissue-derived tumor necrosis factor-α activity. Endocrinology. 1997;138:4665–71.
46. Boden G, She P, Mozzoli M, et al. Free fatty acids produce insulin resistance and activate the proinflammatory nuclear factor-κB pathway in rat liver. Diabetes. 2005;54:3458–65.
47. Ajuwon KM, Spurlock ME. Palmitate activates the NF-κB transcription factor and induces IL-6 and TNFα expression in 3T3-L1 adipocytes. J Nutr. 2005;135:1841–6.
48. Eringa EC, Stehouwer CD, Walburg K, et al. Physiological concentrations of insulin induce endothelin-dependent vasoconstriction of skeletal muscle resistance arteries in the presence of tumor necrosis factor-α dependence on c-Jun N-terminal kinase. Arterioscler Thromb Vasc Biol. 2006;26:274–80.
49. Anderson HD, Rahmutula D, Gardner DG. Tumor necrosis factor-α inhibits endothelial nitric-oxide synthase gene promoter activity in bovine aortic endothelial cells. J Biol Chem. 2004;279:963–9.
50. Nguyen MT, Satoh H, Favelyukis S, et al. JNK and tumor necrosis factor-α mediate free fatty acid-induced insulin resistance in 3T3-L1 adipocytes. J Biol Chem. 2005;280:35361–71.
51. Gao Z, Zuberi A, Quon MJ, Dong Z, Ye J. Aspirin inhibits serine phosphorylation of insulin receptor substrate 1 in tumor necrosis factor-treated cells through targeting multiple serine kinases. J Biol Chem. 2003;278:24944–50.
52. de Alvaro C, Teruel T, Hernandez R, Lorenzo M. Tumor necrosis factor α produces insulin resistance in skeletal muscle by activation of inhibitor κB kinase in a p38 MAPK-dependent manner. J Biol Chem. 2004;279:17070–8.
53. Xu JW, Morita I, Ikeda K, Miki T, Yamori Y. C-reactive protein suppresses insulin signaling in endothelial cells. Role of Syk tyrosine kinase. Mol Endocrinol. 2007;21:564–73.
54. Arita Y, Kihara S, Ouchi N, et al. Paradoxical decrease of an adipose-specific protein, adiponectin, in obesity. Biochem Biophys Res Commun. 1999;257:79–83.
55. Hotta K, Funahashi T, Arita Y, et al. Plasma concentrations of a novel, adipose-specific protein, adiponectin, in type 2 diabetic patients. Arterioscler Thromb Vasc Biol. 2000;20:1595–9.
56. Kojima S, Funahashi T, Sakamoto T, et al. The variation of plasma concentrations of a novel, adipocyte derived protein, adiponectin, in patients with acute myocardial infarction. Heart. 2003;89:667.
57. Iwashima Y, Katsuya T, Ishikawa K, et al. Hypoadiponectinemia is an independent risk factor for hypertension. Hypertension. 2004;43:1318–23.

58. Ouchi N, Kihara S, Funahashi T, et al. Reciprocal association of C-reactive protein with adiponectin in blood stream and adipose tissue. Circulation. 2003;107:671–4.
59. Hulthe J, Hulten LM, Fagerberg B. Low adipocyte-derived plasma protein adiponectin concentrations are associated with the metabolic syndrome and small dense low-density lipoprotein particles: atherosclerosis and insulin resistance study. Metabolism. 2003;52:1612–4.
60. Gilardini L, McTernan PG, Girola A, et al. Adiponectin is a candidate marker of metabolic syndrome in obese children and adolescents. Atherosclerosis. 2006;189:401–7.
61. Ohashi K, Ouchi N, Kihara S, et al. Adiponectin I164T mutation is associated with the metabolic syndrome and coronary artery disease. J Am Coll Cardiol. 2004;43:1195–200.
62. Yang WS, Chuang LM. Human genetics of adiponectin in the metabolic syndrome. J Mol Med. 2006;84:112–21.
63. Hara K, Boutin P, Mori Y, et al. Genetic variation in the gene encoding adiponectin is associated with an increased risk of type 2 diabetes in the Japanese population. Diabetes. 2002;51:536–40.
64. Vasseur F, Helbecque N, Dina C, et al. Single-nucleotide polymorphism haplotypes in the both proximal promoter and exon 3 of the APM1 gene modulate adipocyte-secreted adiponectin hormone levels and contribute to the genetic risk for type 2 diabetes in French Caucasians. Hum Mol Genet. 2002;11:2607–14.
65. González-Sánchez JL, Zabena CA, Martínez-Larrad MT, et al. An SNP in the adiponectin gene is associated with decreased serum adiponectin levels and risk for impaired glucose tolerance. Obes Res. 2005;13:807–12.
66. Xita N, Georgiou I, Chatzikyriakidou A, et al. Effect of adiponectin gene polymorphisms on circulating adiponectin and insulin resistance indexes in women with polycystic ovary syndrome. Clin Chem. 2005;51:416–23.
67. Filippi E, Sentinelli F, Trischitta V, et al. Association of the human adiponectin gene and insulin resistance. Eur J Hum Genet. 2004; 12:199–205.
68. Filippi E, Sentinelli F, Romeo S, et al. The *adiponectin* gene *SNP+276G>T* associates with early-onset coronary artery disease and with lower levels of adiponectin in younger coronary artery disease patients (age ≤50 years). J Mol Med. 2005;83:711–9.
69. Yang WS, Lee WJ, Funahashi T, et al. Weight reduction increases plasma levels of an adipose-derived anti-inflammatory protein, adiponectin. J Clin Endocrinol Metab. 2001;86:3815–9.
70. Maeda N, Takahashi M, Funahashi T, et al. PPARγ ligands increase expression and plasma concentrations of adiponectin, an adipose-derived protein. Diabetes. 2001;50:2094–9.
71. Juan CC, Chuang TY, Chang CL, Huang SW, Ho LT. Endothelin-1 regulates adiponectin gene expression and secretion in 3T3-L1 adipocytes via distinct signaling pathways. Endocrinology. 2007;148:1835–42.
72. Delporte ML, Funahashi T, Takahashi M, Matsuzawa Y, Brichard SM. Pre- and post-translational negative effect of β-adrenoceptor agonists on adiponectin secretion: in vitro and in vivo studies. Biochem J. 2002;367:677–85.
73. Fasshauer M, Klein J, Neumann S, Eszlinger M, Paschke R. Adiponectin gene expression is inhibited by β-adrenergic stimulation via protein kinase A in 3T3-L1 adipocytes. FEBS Lett. 2001;507:142–6.
74. Soares AF, Guichardant M, Cozzone D, et al. Effects of oxidative stress on adiponectin secretion and lactate production in 3T3-L1 adipocytes. Free Radic Biol Med. 2005;38:882–9.
75. Fasshauer M, Kralisch S, Klier M, et al. Adiponectin gene expression and secretion is inhibited by interleukin-6 in 3T3-L1 adipocytes. Biochem Biophys Res Commun. 2003;301:1045–50.
76. Degawa-Yamauchi M, Moss KA, Bovenkerk JE, et al. Regulation of adiponectin expression in human adipocytes: effects of adiposity, glucocorticoids, and tumor necrosis factor α. Obes Res. 2005;13:662–9.
77. Wang B, Jenkins JR, Trayhurn P. Expression and secretion of inflammation-related adipokines by human adipocytes differentiated in culture: integrated response to TNF-α. Am J Physiol Endocrinol Metab. 2005;288:E731–40.
78. Gable DR, Hurel SJ, Humphries SE. Adiponectin and its gene variants as risk factors for insulin resistance, the metabolic syndrome and cardiovascular disease. Atherosclerosis. 2006;188:231–44.
79. Hopkins TA, Ouchi N, Shibata R, Walsh K. Adiponectin actions in the cardiovascular system. Cardiovasc Res. 2007;74:11–8.
80. Karbowska J, Kochan Z. Role of adiponectin in the regulation of carbohydrate and lipid metabolism. J Physiol Pharmacol. 2006;57:103–13.
81. Yamauchi T, Kamon J, Ito Y, et al. Cloning of adiponectin receptors that mediate antidiabetic metabolic effects. Nature. 2003;423:762–9.
82. Yamauchi T, Kamon J, Waki H, et al. The fat-derived hormone adiponectin reverses insulin resistance associated with both lipoatrophy and obesity. Nat Med. 2001;7:941–6.
83. Tomas E, Tsao TS, Saha AK, et al. Enhanced muscle fat oxidation and glucose transport by ACRP30 globular domain: acetyl-CoA carboxylase inhibition and AMP-activated protein kinase activation. Proc Natl Acad Sci USA. 2002;99:16309–13.
84. Yamauchi T, Kamon J, Minokoshi Y, et al. Adiponectin stimulates glucose utilization and fatty-acid oxidation by activating AMP-activated protein kinase. Nat Med. 2002;8:1288–95.
85. Onay-Besikci A, Altarejos JY, Lopaschuk GD. gAd-globular head domain of adiponectin increases fatty acid oxidation in newborn rabbit hearts. J Biol Chem. 2004;279:44320–6.
86. Li L, Wu LL. Effect of AMP-activated protein kinase on cardiovascular protection of adiponectin. Sheng Li Xue Bao. 2007;59:614–8.
87. Combs TP, Berg AH, Obici S, Scherer PE, Rossetti L. Endogenous glucose production is inhibited by the adipose-derived protein Acrp30. J Clin Invest. 2001;108:1875–81.
88. Ouchi N, Kihara S, Funahashi T, Matsuzawa Y, Walsh K. Obesity, adiponectin and vascular inflammatory disease. Curr Opin Lipidol. 2003;14:561–6.
89. Kumada M, Kihara S, Ouchi N, et al. Adiponectin specifically increased tissue inhibitor of metalloproteinase-1 through interleukin-10 expression in human macrophages. Circulation. 2004;109:2046–9.
90. Chen H, Montagnani M, Funahashi T, Shimomura I, Quon MJ. Adiponectin stimulates production of nitric oxide in vascular endothelial cells. J Biol Chem. 2003;278:45021–6.
91. Ouchi N, Kobayashi H, Kihara S, et al. Adiponectin stimulates angiogenesis by promoting cross-talk between AMP-activated protein kinase and Akt signaling in endothelial cells. J Biol Chem. 2004;279:1304–9.
92. Kobayashi H, Ouchi N, Kihara S, et al. Selective suppression of endothelial cell apoptosis by the high molecular weight form of adiponectin. Circ Res. 2004;94:e27–31.
93. Arita Y, Kihara S, Ouchi N, et al. Adipocyte-derived plasma protein adiponectin acts as a platelet-derived growth factor-BB-binding protein and regulates growth factor-induced common postreceptor signal in vascular smooth muscle cell. Circulation. 2002;105:2893–8.
94. Wang Y, Lam KS, Xu JY, et al. Adiponectin inhibits cell proliferation by interacting with several growth factors in an oligomerization-dependent manner. J Biol Chem. 2005;280:18341–7.
95. Okamoto Y, Kihara S, Ouchi N, et al. Adiponectin reduces atherosclerosis in apolipoprotein E-deficient mice. Circulation. 2002;106:2767–70.
96. Yamauchi T, Kamon J, Waki H, et al. Globular adiponectin protected ob/ob mice from diabetes and ApoE-deficient mice from atherosclerosis. J Biol Chem. 2003;278:2461–8.

97. Shibata R, Ouchi N, Ito M, et al. Adiponectin-mediated modulation of hypertrophic signals in the heart. Nat Med. 2004; 10:1384–9.
98. Chan AY, Soltys CL, Young ME, Proud CG, Dyck JR. Activation of AMP-activated protein kinase inhibits protein synthesis associated with hypertrophy in the cardiac myocyte. J Biol Chem. 2004;279:32771–9.
99. Shibata R, Sato K, Pimentel DR, et al. Adiponectin protects against myocardial ischemia-reperfusion injury through AMPK- and COX-2-dependent mechanisms. Nat Med. 2005;11:1096–103.
100. Kudo N, Gillespie JG, Kung L, et al. Characterization of 5′-AMP-activated protein kinase activity in the heart and its role in inhibiting acetyl-CoA carboxylase during reperfusion following ischemia. Biochim Biophys Acta. 1996;1301:67–75.
101. Makinde AO, Gamble J, Lopaschuk GD. Upregulation of 5′-AMP-activated protein kinase is responsible for the increase in myocardial fatty acid oxidation rates following birth in the newborn rabbit. Circ Res. 1997;80:482–9.
102. Considine RV, Sinha MK, Heiman ML, et al. Serum immunoreactive-leptin concentrations in normal-weight and obese humans. N Engl J Med. 1996;334:292–5.
103. Löllmann B, Grüninger S, Stricker-Krongrad A, Chiesi M. Detection and quantification of the leptin receptor splice variants Ob-Ra, b, and, e in different mouse tissues. Biochem Biophys Res Commun. 1997;238:648–52.
104. Bohlen F, Kratzsch J, Mueller M, et al. Leptin inhibits cell growth of human vascular smooth muscle cells. Vascul Pharmacol. 2007;46:67–71.
105. Bouloumié A, Drexler HC, Lafontan M, Busse R. Leptin, the product of Ob gene, promotes angiogenesis. Circ Res. 1998;83:1059–66.
106. Sierra-Honigmann MR, Nath AK, Murakami C, et al. Biological action of leptin as an angiogenic factor. Science. 1998;281:1683–6.
107. Smith CC, Mocanu MM, Davidson SM, Wynne AM, Simpkin JC, Yellon DM. Leptin, the obesity-associated hormone, exhibits direct cardioprotective effects. Br J Pharmacol. 2006;149:5–13.
108. Matsui H, Motooka M, Koike H, et al. Ischemia/reperfusion in rat heart induces leptin and leptin receptor gene expression. Life Sci. 2007;80:672–80.
109. Rajapurohitam V, Javadov S, Purdham DM, Kirshenbaum LA, Karmazyn M. An autocrine role for leptin in mediating the cardiomyocyte hypertrophic effects of angiotensin II and endothelin-1. J Mol Cell Cardiol. 2006;41:265–74.
110. Yang R, Barouch LA. Leptin signaling and obesity: cardiovascular consequences. Circ Res. 2007;101:545–59.
111. Shin HJ, Oh J, Kang SM, et al. Leptin induces hypertrophy via p38 mitogen-activated protein kinase in rat vascular smooth muscle cells. Biochem Biophys Res Commun. 2005;329:18–24.
112. Rajapurohitam V, Gan XT, Kirshenbaum LA, Karmazyn M. The obesity-associated peptide leptin induces hypertrophy in neonatal rat ventricular myocytes. Circ Res. 2003;93:277–9.
113. Gnanapavan S, Kola B, Bustin SA, et al. The tissue distribution of the mRNA of ghrelin and subtypes of its receptor, GHS-R, in humans (Abstract). J Clin Endocrinol Metab. 2002;87:2988.
114. Cao JM, Ong H, Chen C. Effects of ghrelin and synthetic GH secretagogues on the cardiovascular system. Trends Endocrinol Metab. 2006;17:13–8.
115. Iantorno M, Chen H, Kim JA, et al. Ghrelin has novel vascular actions that mimic PI3-kinase-dependent actions of insulin to stimulate production of NO from endothelial cells. Am J Physiol Endocrinol Metab. 2006;292:E756–64.
116. Poykko SM, Kellokoski E, Horkko S, Kauma H, Kesaniemi YA, Ukkola O. Low plasma ghrelin is associated with insulin resistance, hypertension, and the prevalence of type 2 diabetes. Diabetes. 2003;52:2546–53.
117. Mager U, Lindi V, Lindstrom J, et al. Association of the Leu72Met polymorphism of the ghrelin gene with the risk of type 2 diabetes in subjects with impaired glucose tolerance in the Finnish Diabetes Prevention Study. Diabet Med. 2006;23:685–9.
118. Tesauro M, Schinzari F, Iantorno M, et al. Ghrelin improves endothelial function in patients with metabolic syndrome. Circulation. 2005;112:2986–92.
119. Fukuchi S, Hamaguchi K, Seike M, Himeno K, Sakata T, Yoshimatsu H. Role of fatty acid composition in the development of metabolic disorders in sucrose-induced obese rats. Exp Biol Med. 2004;229:486–93.
120. Weisberg S, McCann D, Desai M, Rosenbaum M, Leibel RL, Ferrante Jr AW. Obesity is associated with macrophage accumulation in adipose tissue. J Clin Invest. 2003;112:1796–808.
121. Hu E, Kim JB, Sarraf P, Spiegelman BM. Inhibition of adipogenesis through MAP kinase-mediated phosphorylation of PPARγ. Science. 1996;274:2100–3.
122. Adams M, Reginato MJ, Shao D, Lazar MA, Chatterjee VK. Transcriptional activation by peroxisome proliferator-activated receptor γ is inhibited by phosphorylation at a consensus mitogen-activated protein kinase site. J Biol Chem. 1997;272:5128–32.
123. Jain RG, Phelps KD, Pekala PH. Tumor necrosis factor-α initiated signal transduction in 3T3-L1 adipocytes. J Cell Physiol. 1999;179:58–66.
124. Hotamisligil GS, Peraldi P, Budavari A, Ellis R, White MF, Spiegelman BM. IRS-1-mediated inhibition of insulin receptor tyrosine kinase activity in TNF-α- and obesity-induced insulin resistance. Science. 1996;271:665–8.
125. Zick Y. Ser/Thr phosphorylation of IRS proteins: a molecular basis for insulin resistance. Sci STKE. 2005;2005(pe4):1–3.
126. Engelman JA, Berg AH, Lewis RY, Lisanti MP, Scherer PE. Tumor necrosis factor α-mediated insulin resistance, but not dedifferentiation, is abrogated by MEK1/2 inhibitors in 3T3-L1 adipocytes. Mol Endocrinol. 2000;14:1557–69.
127. Aguirre V, Uchida T, Yenush L, Davis R, White MF. The c-Jun $NH_2$-terminal kinase promotes insulin resistance during association with insulin receptor substrate-1 and phosphorylation of $Ser^{307}$. J Biol Chem. 2000;275:9047–54.
128. Kellerer M, Mushack J, Seffer E, Mischak H, Ullrich A, Häring HU. Protein kinase C isoforms α, δ and θ require insulin receptor substrate-1 to inhibit the tyrosine kinase activity of the insulin receptor in human kidney embryonic cells (HEK 293 cells). Diabetologia. 1998;41:833–8.
129. Gao Z, Hwang D, Bataille F, et al. Serine phosphorylation of insulin receptor substrate 1 by inhibitor κB kinase complex. J Biol Chem. 2002;277:48115–21.
130. Nachiappan V, Curtiss D, Corkey BE, Kilpatrick L. Cytokines inhibit fatty acid oxidation in isolated rat hepatocytes: synergy among TNF, IL-6, and IL-1. Shock. 1994;1:123–9.
131. Steinberg GR, Michell BJ, van Denderen BJ, et al. Tumor necrosis factor α-induced skeletal muscle insulin resistance involves suppression of AMP-kinase signaling. Cell Metab. 2006;4: 465–74.
132. Ioannidis I. The road from obesity to type 2 diabetes. Angiology. 2008;59 Suppl 2:39S–43.
133. Haasch D, Berg C, Clampit JE, et al. PKCθ is a key player in the development of insulin resistance. Biochem Biophys Res Commun. 2006;343:361–8.
134. Degawa-Yamauchi M, Moss KA, Bovenkerk JE, et al. Regulation of adiponectin expression in human adipocytes: effects of adiposity, glucocorticoids, and tumor necrosis factor α. Obes Res. 2005;13:662–9.
135. Laffitte BA, Chao LC, Li J, et al. Activation of liver X receptor improves glucose tolerance through coordinate regulation of glucose metabolism in liver and adipose tissue. Proc Natl Acad Sci USA. 2003;100:5419–24.

136. Peet DJ, Turley SD, Ma W, et al. Cholesterol and bile acid metabolism are impaired in mice lacking the nuclear oxysterol receptor LXRα. Cell. 1998;93:693–704.
137. Schultz JR, Tu H, Luk A, et al. Role of LXRs in control of lipogenesis. Genes Dev. 2000;14:2831–8.
138. Xu H, Barnes GT, Yang Q, et al. Chronic inflammation in fat plays a crucial role in the development of obesity-related insulin resistance. J Clin Invest. 2003;112:1821–30.
139. Weisberg SP, McCann D, Desai M, et al. Obesity is associated with macrophage accumulation in adipose tissue. J Clin Invest. 2003;112:1796–808.
140. Venkateswaran A, Lafitte BA, Joseph SB, et al. Control of cellular cholesterol efflux by the nuclear oxysterol receptor LXRα. Proc Natl Acad Sci USA. 2000;97:12097–102.
141. Lehmann JM, Moore LB, Smith-Oliver TA, et al. An antidiabetic thiazolidinedione is a high affinity ligand for peroxisome proliferator-activated receptor γ. J Biol Chem. 1995;270:12953–6.
142. Willson TM, Cobb JE, Cowan DJ, et al. The structure-activity relationship between peroxisome proliferatoractivated receptor γ agonism and the antihyperglycemic activity of thiazolidinediones. J Med Chem. 1996;39:665–8.
143. Forman BM, Chen J, Evans RM. Hypolipidemic drugs, polyunsaturated fatty acids, and eicosanoids are ligands for peroxisome proliferator-activated receptors α and δ. Proc Natl Acad Sci USA. 1997;94:4312–7.
144. Huss JM, Levy FH, Kelly DP. Hypoxia inhibits the peroxisome proliferator-activated receptor α/retinoid X receptor gene regulatory pathway in cardiac myocytes: a mechanism for $O_2$-dependent modulation of mitochondrial fatty acid oxidation. J Biol Chem. 2001;276:27605–12.
145. Gilde AJ, van der Lee KA, Willemsen PH, et al. Peroxisome proliferator-activated receptor (PPAR) α and PPARβ/δ, but not PPARγ, modulate the expression of genes involved in cardiac lipid metabolism. Circ Res. 2003;92:518–24.
146. Aasum E, Belke DD, Severson DL, et al. Cardiac function and metabolism in Type 2 diabetic mice after treatment with BM 17.0744, a novel PPAR-α activator. Am J Physiol Heart Circ Physiol. 2002;283:H949–57.
147. Dashti N, Ontko JA. Alterations in rat serum lipids and apolipoproteins following clofibrate treatment. Atherosclerosis. 1983;49:255–66.
148. Cheng L, Ding G, Qin Q, et al. Cardiomyocyte-restricted peroxisome proliferator-activated receptor δ deletion perturbs myocardial fatty acid oxidation and leads to cardiomyopathy. Nat Med. 2004;10:1245–50.
149. Hevener A, He W, Barak Y, et al. Muscle-specific PPAR γ deletion causes insulin resistance. Nat Med. 2003;9:1491–7.
150. He W, Barak Y, Hevener A, et al. Adipose-specific peroxisome proliferator-activated receptor γ knockout causes insulin resistance in fat and liver but not in muscle. Proc Natl Acad Sci USA. 2003;100:15712–7.
151. Moore KJ, Rosen ED, Fitzgerald ML, et al. The role of PPAR-γ in macrophage differentiation and cholesterol uptake. Nat Med. 2001;7:41–7.
152. Chawla A, Barak Y, Nagy L, et al. PPAR-γ dependent and independent effects on macrophage gene expression in lipid metabolism and inflammation. Nat Med. 2001;7:48–52.
153. Lehman JJ, Barger PM, Kovacs A, Saffitz JE, Medeiros DM, Kelly DP. Peroxisome proliferator-activated receptor γ coactivator-1 promotes cardiac mitochondrial biogenesis. J Clin Invest. 2000;106:847–56.
154. Huss JM, Torra IP, Staels B, Giguere V, Kelly DP. Estrogen-related receptor α directs peroxisome proliferator-activated receptor α signaling in the transcriptional control of energy metabolism in cardiac and skeletal muscle. Mol Cell Biol. 2004;24:9079–91.
155. Russell LK, Mansfield CM, Lehman JJ, et al. Cardiac-specific induction of the transcriptional coactivator peroxisome proliferator-activated receptor γ coactivator-1α promotes mitochondrial biogenesis and reversible cardiomyopathy in a developmental stage-dependent manner. Circ Res. 2004;94:525–33.
156. Arany Z, He H, Lin J, et al. Transcriptional coactivator PGC-1α controls the energy state and contractile function of cardiac muscle. Cell Metab. 2005;1:259–71.
157. Sowers JR. Insulin resistance and hypertension. Am J Physiol Heart Circ Physiol. 2004;286:H1597–602.
158. Manrique C, Lastra G, Whaley-Connell A, Sowers JR. Hypertension and the cardiometabolic syndrome. J Clin Hypertens. 2005; 7:471–6.
159. Wei Y, Stump CS, Habibi J, et al. NADPH oxidase activation contributes to vascular inflammation, insulin resistance, and remodeling in the transgenic (mRen2) rat. Hypertension. 2007;50: 384–91.
160. Nathan C. Specificity of a third kind: reactive oxygen and nitrogen intermediates in cell signaling. J Clin Invest. 2003;111:769–78.
161. Muller DN, Dechend R, Mervaala EM, et al. NF-κB inhibition ameliorates angiotensin II-induced inflammatory damage in rats. Hypertension. 2000;35:193–201.
162. Kim JA, Yeh DC, Ver M, et al. Phosphorylation of Ser24 in the pleckstrin homology domain of insulin receptor substrate-1 by mouse Pelle-like kinase/interleukin-1 receptor-associated kinase: cross-talk between inflammatory signaling and insulin signaling that may contribute to insulin resistance. J Biol Chem. 2005;280:23173–83.
163. Reusch JE. Diabetes, microvascular complications, and cardiovascular complications: what is it about glucose? J Clin Invest. 2003;112:986–8.
164. Wang XL, Zhang L, Youker K, et al. Free fatty acids inhibit insulin signaling-stimulated endothelial nitric oxide synthase activation through upregulating PTEN or inhibiting Akt kinase. Diabetes. 2006;55:2301–10.
165. Inoguchi T, Li P, Umeda F, et al. High glucose level and free fatty acid stimulate reactive oxygen species production through protein kinase C-dependent activation of NAD(P)H oxidase in cultured vascular cells. Diabetes. 2000;49:1939–45.
166. Lastra G, Manrique CM, Stump CS, et al. Low-dose spironolactone reduces reactive oxygen species generation and improves insulin-stimulated glucose transport in skeletal muscle in TGR(mRen-2)27 rats. Am J Physiol Endocrinol Metab. 2008;295:E110–6.
167. Seager MJ, Singal PK, Orchard R, Pierce GN, Dhalla NS. Cardiac cell damage: a primary myocardial disease in streptozotocin-induced chronic diabetes. Br J Exp Pathol. 1984;65:613–23.
168. Mokhtar N, Lavoie JP, Rousseau-Migneron S, Nadeau A. Physical training reverses defect in mitochondrial energy production in heart of chronically diabetic rats. Diabetes. 1993;42:682–7.
169. Tomita M, Mukae S, Geshi E, Umetsu K, Nakatani M, Katagiri T. Mitochondrial respiratory impairment in streptozotocin induced diabetic rat heart. Jpn Circ J. 1996;60:673–82.
170. Kong JY, Rabkin SW. Mitochondrial effects with ceramide-induced cardiac apoptosis are different from those of palmitate. Arch Biochem Biophys. 2003;412:196–206.
171. Hickson-Bick DL, Buja ML, McMillin JB. Palmitate-mediated alterations in the fatty acid metabolism of rat neonatal cardiac myocytes. J Mol Cell Cardiol. 2000;32:511–9.
172. Sparagna GC, Hickson-Bick DL, Buja LM, McMillin JB. Fatty acid-induced apoptosis in neonatal cardiomyocytes: redox signaling. Antioxid Redox Signal. 2001;3:71–9.
173. Sparagna GC, Hickson-Bick DL, Buja LM, McMillin JB. A metabolic role for mitochondria in palmitate-induced cardiac myocyte apoptosis. Am J Physiol Heart Circ Physiol. 2000;279: H2124–32.

174. Hickson-Bick DL, Sparagna GC, Buja LM, McMillin JB. Palmitate-induced apoptosis in neonatal cardiomyocytes is not dependent on the generation of ROS. Am J Physiol Heart Circ Physiol. 2002;282:H656–64.
175. Williams SD, Gottlieb RA. Inhibition of mitochondrial calcium-independent phospholipase $A_2$ ($iPLA_2$) attenuates mitochondrial phospholipid loss and is cardioprotective. Biochem J. 2002; 362:23–32.
176. Nishikawa T, Edelstein D, Du XL, et al. Normalizing mitochondrial superoxide production blocks three pathways of hyperglycaemic damage. Nature. 2000;404:787–90.
177. Verma S, Li SH, Badiwala MV, et al. Endothelin antagonism and interleukin-6 inhibition attenuate the proatherogenic effects of C-reactive protein. Circulation. 2002;105:1890–6.
178. Du XL, Edelstein D, Rossetti L, et al. Hyperglycemia-induced mitochondrial superoxide overproduction activates the hexosamine pathway and induces plasminogen activator inhibitor-1 expression by increasing Sp1 glycosylation. Proc Natl Acad Sci USA. 2000;97:12222–6.
179. Kohler HP, Grant PJ. Plasminogen-activator inhibitor type 1 and coronary artery disease. N Engl J Med. 2000;342:1792–801.
180. Bobik A. Transforming growth factor-betas and vascular disorders. Arterioscler Thromb Vasc Biol. 2006;26:1712–20.
181. Du X, Matsumura T, Edelstein D, et al. Inhibition of GAPDH activity by poly(ADP-ribose) polymerase activates three major pathways of hyperglycemic damage in endothelial cells. J Clin Invest. 2003;112:1049–57.
182. Mabile L, Meilhac O, Escargueil-Blanc I, et al. Mitochondrial function is involved in LDL oxidation mediated by human cultured endothelial cells. Arterioscler Thromb Vasc Biol. 1997;17:1575–82.
183. Honda HM, Korge P, Weiss JN. Mitochondria and ischemia/reperfusion injury. Ann N Y Acad Sci. 2005;1047:248–58.
184. Recchioni R, Marcheselli F, Moroni F, Pieri C. Apoptosis in human aortic endothelial cells induced by hyperglycemic condition involves mitochondrial depolarization and is prevented by *N*-acetyl-L-cysteine. Metabolism. 2002;51:1384–8.
185. Koster JC, Permutt MA, Nichols CG. Diabetes and Insulin Secretion: The ATP-Sensitive K+ Channel (KATP) Connection. Diabetes. 2005;54:3065–72.
186. Hattersley AT, Ashcroft FM. Activating mutations in Kir6.2 and neonatal diabetes: new clinical syndromes, new scientific insights, and new therapy. Diabetes. 2005;54:2503–13.
187. Gloyn AL, Pearson ER, Antcliff JF, Proks P, Bruining GJ, Slingerland AS, et al. Activating mutations in the gene encoding the ATP-sensitive potassium-channel subunit Kir6.2 and permanent neonatal diabetes. N Engl J Med. 2004;350:1838–49.
188. Sperling MA. Neonatal diabetes mellitus: from understudy to center stage. Curr Opin Pediatr. 2005;17:512–8.
189. Gloyn AL, Weedon MN, Owen KR, Turner MJ, Knight BA, Hitman G, et al. Large-scale association studies of variants in genes encoding the pancreatic beta-cell KATP channel subunits Kir6.2 (KCNJ11) and SUR1 (ABCC8) confirm that the KCNJ11 E23K variant is associated with type 2 diabetes. Diabetes. 2003;52:568–72.
190. Riedel MJ, Steckley DC, Light PE. Current status of the E23K Kir6.2 polymorphism: implications for type-2 diabetes. Hum Genet. 2005;116:133–45.
191. Riedel MJ, Boora P, Steckley D, de Vries G, Light PE. Kir6.2 polymorphisms sensitize beta-cell ATP-sensitive potassium channels to activation by acyl CoAs: a possible cellular mechanism for increased susceptibility to type 2 diabetes? Diabetes. 2003;52:2630–5.
192. Slingerland AS, Hattersley AT. Mutations in the Kir6.2 subunit of the KATP channel and permanent neonatal diabetes: new insights and new treatment. Ann Med. 2005;37:186–95.
193. Malecki MT. Genetics of type 2 diabetes mellitus. Diab Res Clin Pract. 2005;68:S10–21.
194. Gupta RK, Kaestner KH. HNF-4alpha: from MODY to late-onset type 2 diabetes. Trends Mol Med. 2004;10:521–4.
195. Gloyn AL. Glucokinase (GCK) mutations in hyper- and hypoglycemia: maturity-onset diabetes of the young, permanent neonatal diabetes, and hyperinsulinemia of infancy. Hum Mutat. 2003;22:353–62.
196. Mitchell SM, Frayling TM. The role of transcription factors in maturity-onset diabetes of the young. Mol Genet Metab. 2002;77:35–43.
197. George S, Rochford JJ, Wolfrum C, Gray SL, Schinner S, Wilson JC, et al. A family with severe insulin resistance and diabetes due to a mutation in AKT2. Science. 2004;304:1325–8.
198. Hone J, Accili D, al-Gazali LI, Lestringant G, Orban T, Taylor SI. Homozygosity for a new mutation (Ile119->Met) in the insulin receptor gene in five sibs with familial insulin resistance. J Med Genet. 1994;31:715–6.
199. Kusari J, Takata Y, Hatada E, Freidenberg G, Kolterman O, Olefsky JM. Insulin resistance and diabetes due to different mutations in the tyrosine kinase domain of both insulin receptor gene alleles. J Biol Chem. 1991;266:5260–7.
200. Musso C, Cochran E, Moran SA, Skarulis MC, Oral EA, Taylor S, et al. Clinical course of genetic diseases of the insulin receptor (type A and Rabson-Mendenhall syndromes): a 30-year prospective. Medicine. 2004;83:209–22.
201. Shen X, Zheng S, Thongboonkerd V, Xu M, Pierce Jr WM, Klein JB, et al. Cardiac mitochondrial damage and biogenesis in a chronic model of type 1 diabetes. Am J Physiol Endocrinol Metab. 2004;287:E896–905.
202. Ferreira FM, Seica R, Oliveira PJ, Coxito PM, Moreno AJ, Palmeira CM, et al. Diabetes induces metabolic adaptations in rat liver mitochondria: role of coenzyme Q and cardiolipin contents. Biochim Biophys Acta. 2003;1639:113–8.
203. Ritov VB, Menshikova EV, He J, Ferrell RE, Goodpaster BH, Kelley DE. Deficiency of subsarcolemmal mitochondria in obesity and type 2 diabetes. Diabetes. 2005;54:8–14.
204. Petersen KF, Dufour S, Befroy D, Garcia R, Shulman GI. Impaired mitochondrial activity in the insulin-resistant offspring of patients with type 2 diabetes. N Engl J Med. 2004;350:664–71.
205. Silva JP, Kohler M, Graff C, Oldfors A, Magnuson MA, Berggren PO, et al. Impaired insulin secretion and beta-cell loss in tissue-specific knockout mice with mitochondrial diabetes. Nat Genet. 2000;26:336–40.
206. Brownlee M. A radical explanation for glucose-induced beta cell dysfunction. J Clin Invest. 2003;112:1788–90.
207. Lowell BB, Shulman GI. Mitochondrial dysfunction and type 2 diabetes. Science. 2005;307:384–7.
208. Stumvoll M, Goldstein BJ, van Haeften TW. Type 2 diabetes: principles of pathogenesis and therapy. Lancet. 2005;365:1333–46.
209. Malecki MT. Genetics of type 2 diabetes mellitus. Diab Res Clin Pract. 2005;68:S10–21.
210. Kelly MA, Mijovic CH, Barnett AH. Genetics of type 1 diabetes. Best Pract Res Clin Endocrinol Metab. 2001;15:279–91.
211. Achenbach P, Bonifacio E, Ziegler AG. Predicting type 1 diabetes. Curr Diab Rep. 2005;5:98–103.
212. Kavvoura FK, Ioannidis JP. CTLA-4 gene polymorphisms and susceptibility to type 1 diabetes mellitus: a HuGE Review and meta-analysis. Am J Epidemiol. 2005;162:3–16.
213. Bottini N, Musumeci L, Alonso A, Rahmouni S, Nika K, Rostamkhani M, et al. A functional variant of lymphoid tyrosine phosphatase is associated with type I diabetes. Nat Genet. 2004;36:337–8.
214. Mathieu C, Badenhoop K. Vitamin D and type 1 diabetes mellitus: state of the art. Trends Endocrinol Metab. 2005;16:261–6.
215. Luong K, Nguyen LT, Nguyen DN. The role of vitamin D in protecting type 1 diabetes mellitus. Diab Metab Res Rev. 2005; 21:338–46.

216. Pollex RL, Mamakeesick M, Zinman B, Harris SB, Hanley AJ, Hegele RA. Methylenetetrahydrofolate reductase polymorphism 677C>T is associated with peripheral arterial disease in type 2 diabetes. Cardiovasc Diabetol. 2005;4:17.
217. Maeda S, Tsukada S, Kanazawa A, Sekine A, Tsunoda T, Koya D, et al. Genetic variations in the gene encoding TFAP2B are associated with type 2 diabetes mellitus. J Hum Genet. 2005;50: 283–92.
218. Vimaleswaran KS, Radha V, Ghosh S, Majumder PP, Deepa R, Babu HN, et al. Peroxisome proliferator-activated receptor-gamma co-activator-1alpha (PGC-1alpha) gene polymorphisms and their relationship to Type 2 diabetes in Asian Indians. Diabet Med. 2005;22:1516–21.
219. Ek J, Andersen G, Urhammer SA, Gaede PH, Drivsholm T, Borch-Johnsen K, et al. Mutation analysis of peroxisome proliferator-activated receptor-gamma coactivator-1 (PGC-1) and relationships of identified amino acid polymorphisms to Type II diabetes mellitus. Diabetologia. 2001;44:2220–6.
220. Nicaud V, Raoux S, Poirier O, Cambien F, O'Reilly DS, Tiret L. The TNF alpha/G-308A polymorphism influences insulin sensitivity in offspring of patients with coronary heart disease: the European Atherosclerosis Research Study II. Atherosclerosis. 2002;161:317–25.
221. Vendrell J, Fernandez-Real JM, Gutierrez C, Zamora A, Simon I, Bardaji A, et al. A polymorphism in the promoter of the tumor necrosis factor-alpha gene (-308) is associated with coronary heart disease in type 2 diabetic patients. Atherosclerosis. 2003;167:257–64.
222. Florez JC, Burtt N, de Bakker PI, Almgren P, Tuomi T, Holmkvist J, et al. Haplo-type structure and genotype-phenotype correlations of the sulfonylurea receptor and the islet ATP-sensitive potassium channel gene region. Diabetes. 2004;53:1360–8.
223. Hayes MG, Del Bosque-Plata L, Tsuchiya T, Hanis CL, Bell GI, Cox NJ. Patterns of linkage disequilibrium in the type 2 diabetes gene calpain-10. Diabetes. 2005;54:3573–6.
224. Evans JC, Frayling TM, Cassell PG, Saker PJ, Hitman GA, Walker M, et al. Studies of association between the gene for calpain-10 and type 2 diabetes mellitus in the United Kingdom. Am J Hum Genet. 2001;69:544–52.
225. Liang H, Murase Y, Katuta Y, Asano A, Kobayashi J, Mabuchi H. Association of LMNA 1908C/T polymorphism with cerebral vascular disease and diabetic nephropathy in Japanese men with type 2 diabetes. Clin Endocrinol. 2005;63:317–22.
226. Armstrong M, Haldane F, Taylor RW, Humphriss D, Berrish T, Stewart MW, et al. Human insulin receptor substrate-1: variant sequences in familial non-insulin-dependent diabetes mellitus. Diabet Med. 1996;13:133–8.
227. Jellema A, Zeegers MP, Feskens EJ, Dagnelie PC, Mensink RP. Gly972Arg variant in the insulin receptor substrate-1 gene and association with Type 2 diabetes: a meta-analysis of 27 studies. Diabetologia. 2003;46:990–5.
228. Zacharova J, Chiasson JL. STOP-NIDDM Study Group. The common polymorphisms (single nucleotide polymorphism [SNP] +45 and SNP +276) of the adiponectin gene predict the conversion from impaired glucose tolerance to type 2 diabetes: the STOP-NIDDM trial. Diabetes. 2005;54:893–9.
229. Bacci S, Menzaghi C, Ercolino T, Ma X, Rauseo A, Salvemini L, et al. The +276 G/T single nucleotide polymorphism of the adiponectin gene is associated with coronary artery disease in type 2 diabetic patients. Diab Care. 2004;27:2015–20.
230. Ukkola O, Santaniemi M, Rankinen T, Leon AS, Skinner JS, Wilmore JH, et al. Adiponectin polymorphisms, adiposity and insulin metabolism: HERITAGE family study and Oulu diabetic study. Ann Med. 2005;37:141–50.
231. Mori H, Ikegami H, Kawaguchi Y, Seino S, Yokoi N, Takeda J, et al. The Pro12→Ala substitution in PPAR-gamma is associated with resistance to development of diabetes in the general population: possible involvement in impairment of insulin secretion in individuals with type 2 diabetes. Diabetes. 2001;50:891–4.
232. Doney AS, Fischer B, Leese G, Morris AD, Palmer CN. Cardiovascular risk in type 2 diabetes is associated with variation at the PPARG locus: a Go-DARTS study. Arterioscler Thromb Vasc Biol. 2004;24:2403–7.
233. Altshuler D, Hirschhorn JN, Klannemark M, Lindgren CM, Vohl MC, Nemesh J, et al. The common PPARgamma Pro12Ala polymorphism is associated with decreased risk of type 2 diabetes. Nat Genet. 2000;26:76–80.
234. Nistico L, Buzzetti R, Pritchard LE, Van der Auwera B, Giovannini C, Bosi E, et al. The CTLA-4 gene region of chromosome 2q33 is linked to, and associated with, type 1 diabetes. Belgian Diabetes Registry. Hum Mol Genet. 1996;5:1075–80.
235. Van der Auwera BJ, Vandewalle CL, Schuit FC, Winnock F, De Leeuw IH, Van Imschoot S, et al. CTLA-4 gene polymorphism confers susceptibility to insulin-dependent diabetes mellitus (IDDM) independently from age and from other genetic or immune disease markers. The Belgian Diabetes Registry. Clin Exp Immunol. 1997;110:98–103.
236. Redondo MJ, Fain PR, Eisenbarth GS. Genetics of type 1A diabetes. Recent Prog Horm Res. 2001;56:69–89.
237. Erlich HA. HLA class II sequences and genetic susceptibility to insulin dependent diabetes mellitus. Baillières Clin Endocrinol Metab. 1991;5:395–411.
238. Kennedy GC, German MS, Rutter WJ. The minisatellite in the diabetes susceptibility locus IDDM2 regulates insulin transcription. Nat Genet. 1995;9:293–8.
239. Bell GI, Horita S, Karam JH. A polymorphic locus near the human insulin gene is associated with insulin-dependent diabetes mellitus. Diabetes. 1984;33:176–83.
240. Steinle NI, Kazlauskaite R, Imumorin IG, Hsueh WC, Pollin TI, O'Connell JR, et al. Variation in the lamin A/C gene: associations with metabolic syndrome. Arterioscler Thromb Vasc Biol. 2004;24:1708–13.
241. Murase Y, Yagi K, Katsuda Y, Asano A, Koizumi J, Mabuchi H. An LMNA variant is associated with dyslipidemia and insulin resistance in the Japanese. Metabolism. 2002;51:1017–21.
242. Caux F, Dubosclard E, Lascols O, Buendia B, Chazouilleres O, Cohen A, et al. A new clinical condition linked to a novel mutation in lamins A and C with generalized lipoatrophy, insulin-resistant diabetes, disseminated leukomelanodermic papules, liver steatosis, and cardiomyopathy. J Clin Endocrinol Metab. 2003;88: 1006–13.
243. Ukkola O, Rankinen T, Lakka T, Leon AS, Skinner JS, Wilmore JH, et al. Protein tyrosine phosphatase 1B variant associated with fat distribution and insulin metabolism. Obes Res. 2005;13:829–34.
244. Spencer-Jones NJ, Wang X, Snieder H, Spector TD, Carter ND, O'Dell SD. Protein tyrosine phosphatase-1B gene PTPN1: selection of tagging single nucleotide polymorphisms and association with body fat, insulin sensitivity, and the metabolic syndrome in a normal female population. Diabetes. 2005;54:3296–304.
245. Palmer ND, Bento JL, Mychaleckyj JC, Langefeld CD, Campbell JK, Norris JM, Haffner SM, Bergman RN, Bowden DW; insulin resistance atherosclerosis study (IRAS) family study. Association of protein tyrosine phosphatase 1B gene polymorphisms with measures of glucose homeostasis in Hispanic Americans: the insulin resistance atherosclerosis study (IRAS) family study. Diabetes. 2004;53:3013–19
246. Manraj M, Francke S, Hebe A, Ramjuttun US, Froguel P. Genetic and environmental nature of the insulin resistance syndrome in Indo-Mauritian subjects with premature coronary heart disease: contribution of beta3-adrenoreceptor gene polymorphism and beta blockers on triglyceride and HDL concentrations. Diabetologia. 2001;44:115–22.

247. Strazzullo P, Iacone R, Siani A, Cappuccio FP, Russo O, Barba G, et al. Relationship of the Trp64Arg polymorphism of the beta3-adrenoceptor gene to central adiposity and high blood pressure: interaction with age. Cross-sectional and longitudinal findings of the Olivetti Prospective Heart Study. J Hypertens. 2001; 19:399–406.
248. Bracale R, Pasanisi F, Labruna G, Finelli C, Nardelli C, Buono P, et al. Metabolic syndrome and ADRB3 gene polymorphism in severely obese patients from South Italy. Eur J Clin Nutr. 2007;61(10):1213–9.
249. Robitaille J, Brouillette C, Houde A, Lemieux S, Perusse L, Tchernof A, et al. Association between the PPARalpha-L162V polymorphism and components of the metabolic syndrome. J Hum Genet. 2004;49:482–9.
250. Tai ES, Collins D, Robins SJ, O'Connor Jr JJ, Bloomfield HE, Ordovas JM, et al. The L162V polymorphism at the peroxisome proliferator activated receptor alpha locus modulates the risk of cardiovascular events associated with insulin resistance and diabetes mellitus: the Veterans Affairs HDL Intervention Trial (VA-HIT). Atherosclerosis. 2006;187:153–60.
251. Frederiksen L, Brodbaek K, Fenger M, Jorgensen T, Borch-Johnsen K, Madsbad S, et al. Comment: studies of the Pro12Ala polymorphism of the PPAR-gamma gene in the Danish MONICA cohort: homozygosity of the Ala allele confers a decreased risk of the insulin resistance syndrome. J Clin Endocrinol Metab. 2002;87:3989–92.
252. Li S, Chen W, Srinivasan SR, Boerwinkle E, Berenson GS. The Bogalusa Heart Study The peroxisome proliferator-activated receptor-gamma2 gene polymorphism (Pro12Ala) beneficially influences insulin resistance and its tracking from childhood to adulthood: the Bogalusa Heart Study. Diabetes. 2003;52: 1265–9.
253. Meirhaeghe A, Cottel D, Amouyel P, Dallongeville J. Association between peroxisome proliferator-activated receptor gamma haplotypes and the metabolic syndrome in French men and women. Diabetes. 2005;54:3043–8.
254. Kahara T, Takamura T, Hayakawa T, Nagai Y, Yamaguchi H, Katsuki T, et al. PPARgamma gene polymorphism is associated with exercise-mediated changes of insulin resistance in healthy men. Metabolism. 2003;52:209–12.
255. Rhee EJ, Oh KW, Lee WY, Kim SY, Oh ES, Baek KH, et al. Effects of two common polymorphisms of peroxisome proliferator-activated receptor-gamma gene on metabolic syndrome. Arch Med Res. 2006;37:86–94.
256. Sookoian S, Garcia SI, Porto PI, Dieuzeide G, Gonzalez CD, Pirola CJ. Peroxisome proliferator-activated receptor gamma and its coactivator-1 alpha may be associated with features of the metabolic syndrome in adolescents. J Mol Endocrinol. 2005; 35:373–80.
257. Grarup N, Albrechtsen A, Ek J, Borch-Johnsen K, Jorgensen T, Schmitz O, et al. Variation in the peroxisome proliferator-activated receptor delta gene in relation to common metabolic traits in 7,495 middle-aged white people. Diabetologia. 2007;50: 1201–8.
258. Dallongeville J, Helbecque N, Cottel D, Amouyel P, Meirhaeghe A. The Gly16–>Arg16 and Gln27–>Glu27 polymorphisms of beta2-adrenergic receptor are associated with metabolic syndrome in men. J Clin Endocrinol Metab. 2003;88:4862–6.
259. Vohl MC, Houde A, Lebel S, Hould FS, Marceau P. Effects of the peroxisome proliferator-activated receptor-gamma co-activator-1 Gly482Ser variant on features of the metabolic syndrome. Mol Genet Metab. 2005;86:300–6.
260. Ambye L, Rasmussen S, Fenger M, Jorgensen T, Borch-Johnsen K, Madsbad S, et al. Studies of the Gly482Ser polymorphism of the peroxisome proliferator-activated receptor gamma coactivator 1alpha (PGC-1alpha) gene in Danish subjects with the metabolic syndrome. Diab Res Clin Pract. 2005;67:175–9.
261. Boullu-Sanchis S, Lepretre F, Hedelin G, Donnet JP, Schaffer P, Froguel P, et al. Type 2 diabetes mellitus: association study of five candidate genes in an Indian population of Guadeloupe, genetic contribution of FABP2 polymorphism. Diab Metab. 1999;25: 150–6.
262. Guettier JM, Georgopoulos A, Tsai MY, Radha V, Shanthirani S, Deepa R, et al. Polymorphisms in the fatty acid-binding protein 2 and apolipoprotein C-III genes are associated with the metabolic syndrome and dyslipidemia in a South Indian population. J Clin Endocrinol Metab. 2005;90:1705–11.
263. Pollex RL, Hanley AJ, Zinman B, Harris SB, Khan HM, Hegele RA. Metabolic syndrome in aboriginal Canadians: prevalence and genetic associations. Atherosclerosis. 2006;184:121–9.
264. Vimaleswaran KS, Radha V, Mohan V. Thr54 allele carriers of the Ala54Thr variant of FABP2 gene have associations with metabolic syndrome and hyper-triglyceridemia in urban South Indians. Metabolism. 2006;55:1222–6.
265. Erkkila AT, Lindi V, Lehto S, Pyorala K, Laakso M, Uusitupa MI. Variation in the fatty acid binding protein 2 gene is not associated with markers of metabolic syndrome in patients with coronary heart disease. Nutr Metab Cardiovasc Dis. 2002;12:53–9.
266. Ohashi K, Ouchi N, Kihara S, Funahashi T, Nakamura T, Sumitsuji S, et al. Adiponectin I164T mutation is associated with the metabolic syndrome and coronary artery disease. J Am Coll Cardiol. 2004;43:1195–200.
267. Marzi C, Huth C, Kolz M, Grallert H, Meisinger C, Wichmann HE, et al. Variants of the transcription factor 7-like 2 gene (TCF7L2) are strongly associated with type 2 diabetes but not with the metabolic syndrome in the MONICA/KORA surveys. Horm Metab Res. 2007;39:46–52.
268. Melzer D, Murray A, Hurst AJ, Weedon MN, Bandinelli S, Corsi AM, et al. Effects of the diabetes linked TCF7L2 polymorphism in a representative older population. BMC Med. 2006;4:34.
269. Bing C, Ambye L, Fenger M, Jorgensen T, Borch-Johnsen K, Madsbad S, et al. Large-scale studies of the Leu72Met polymorphism of the ghrelin gene in relation to the metabolic syndrome and associated quantitative traits. Diabet Med. 2005;22: 1157–60.
270. Carlsson E, Groop L, Ridderstrale M. Role of the FOXC2–512C>T polymorphism in type 2 diabetes: possible association with the dysmetabolic syndrome. Int J Obes Lond. 2005;29:268–74.
271. Weissglas-Volkov D, Huertas-Vazquez A, Suviolahti E, Lee J, Plaisier C, Canizales-Quinteros S, et al. Common hepatic nuclear factor-4alpha variants are associated with high serum lipid levels and the metabolic syndrome. Diabetes. 2006;55:1970–7.
272. Miller M, Rhyne J, Chen H, Beach V, Ericson R, Luthra K, et al. APOC3 promoter polymorphisms C-482T and T-455C are associated with the metabolic syndrome. Arch Med Res. 2007;38:444–51.
273. Koh KK, Han SH. Quon MJ.Inflammatory markers and the metabolic syndrome: insights from therapeutic interventions. J Am Coll Cardiol. 2005;46:1978–85.
274. Hegele RA. Familial partial lipodystrophy: a monogenic form of the insulin resistance syndrome. Mol Genet Metab. 2000;71:539–44.
275. Caux F, Dubosclard E, Lascols O, Buendia B, Chazouilleres O, Cohen A, et al. A new clinical condition linked to a novel mutation in lamins A and C with generalized lipoatrophy, insulin-resistant diabetes, disseminated leukomelanodermic papules, liver steatosis, and cardiomyopathy. J Clin Endocrinol Metab. 2003;88: 1006–13.
276. Haque WA, Oral EA, Dietz K, Bowcock AM, Agarwal AK, Garg A. Risk factors for diabetes in familial partial lipodystrophy, Dunnigan variety. Diab Care. 2003;26:1350–5.

277. Cao H, Hegele RA. Nuclear lamin A/C R482Q mutation in canadian kindreds with Dunnigan-type familial partial lipodystrophy. Hum Mol Genet. 2000;9:109–12.
278. Hegele RA, Cao H, Frankowski C, Mathews ST, Leff T. PPARG F388L, a transactivation-deficient mutant, in familial partial lipodystrophy. Diabetes. 2002;51:3586–90.
279. Hegele RA, Pollex RL. Genetic and physiological insights into the metabolic syndrome. Am J Physiol Regul Integr Comp Physiol. 2005;289:R663–9.
280. Hegele RA, Joy TR, Al-Attar S, Rutt BK. Lipodystrophies: windows on adipose biology and metabolism. J Lipid Res. 2007;48(7):1433–44.
281. Savage DB, Tan GD, Acerini CL, Jebb SA, Agostini M, Gurnell M, et al. Human metabolic syndrome resulting from dominant-negative mutations in the nuclear receptor peroxisome proliferator-activated receptor-gamma. Diabetes. 2003;52:910–7.
282. Agostini M, Schoenmakers E, Mitchell C, Szatmari I, Savage D, Smith A, et al. Non-DNA binding, dominant-negative, human PPARgamma mutations cause lipodystrophic insulin resistance. Cell Metab. 2006;4:303–11.
283. Kotzka J, Muller-Wieland D. Sterol regulatory element-binding protein (SREBP)-1: gene regulatory target for insulin resistance? Expert Opin Ther Targets. 2004;8:141–9.
284. Koo SH, Satoh H, Herzig S, Lee CH, Hedrick S, Kulkarni R, et al. Montminy M.PGC-1 promotes insulin resistance in liver through PPAR-alpha-dependent induction of TRB-3. Nat Med. 2004; 10:530–4.
285. Shulman AI, Mangelsdorf DJ. Retinoid x receptor heterodimers in the metabolic syndrome. N Engl J Med. 2005;353: 604–15.
286. Berger JP, Akiyama TE, Meinke PT. PPARs: therapeutic targets for metabolic disease. Trends Pharmacol Sci. 2005;26: 244–51.
287. Han SH, Quon MJ, Koh KK. Beneficial vascular and metabolic effects of peroxisome proliferator-activated receptor-alpha activators. Hypertension. 2005;46:1086–92.
288. Chinetti-Gbaguidi G, Fruchart JC, Staels B. Role of the PPAR family of nuclear receptors in the regulation of metabolic and cardiovascular homeostasis: new approaches to therapy. Curr Opin Pharmacol. 2005;5:177–83.

# Chapter 17
# Dysrhythmias/Channelopathies and Signaling Pathways

**Abstract** Cardiac dysrhythmias remain an important cause of morbidity and mortality around the world. Notwithstanding the recent scientific advances in genetic analysis to assign a risk stratification for many of the dysrhythmogenic syndromes, we should take into account the numerous and important epigenetic modifiers/modulator factors, including family history, gender, repolarization abnormalities, and sympathetic tone, each of which may influence disease presentation and severity. Other environmental factors and modulators, such as ethnicity and geographical distribution need to be closely evaluated in assessing causality, and in establishing the specific diagnosis of the rhythm disorder. This can be facilitated not only by monitoring specific cardiac markers, but also by close observation of the pharmacological responses and the electrophysiological phenotypes and analysis of specific signaling pathways affected, i.e., mutations in ankyrin B protein that lead to altered $Ca^{2+}$ signaling in adult cardiomyocytes resulting in premature ventricular beats as well as mutations in other signaling proteins, providing a significant rationale for dysrhythmia. The use of systems pharmacology linking drug targets, disease genes, and signaling protein interaction networks may allow individualization of therapy, which is particularly critical in establishing drug dosages and efficacy in children and aging patients with dysrhythmias and structural heart defects, a population for which pharmacokinetics has proven to be poorly defined and often unpredictable.

**Keywords** Dysrhythmias signaling • Channelopathies • Molecular biology • Ankyrins • Conduction defects

## Introduction

The normal adult human heart beats approximately 100,000 times a day. Regular beating rate (cardiac rhythm) is finely tuned to the organism requirements. Irregularities in cardiac rhythm when the heart rhythm is inappropriately fast or slow constitute cardiac dysrhythmias. These abnormalities can impair cardiac function resulting in derangements, which vary from mild symptoms to severe life-threatening complications.

Despite significant advances in our understanding of the molecular mechanism of many cardiac dysrhythmias, these disorders remain an important cause of morbidity and mortality around the world. In the USA alone, over 300,000 cases of sudden death occur each year due to ventricular dysrhythmias.

As it has been discussed in Chap. 6, ion signaling through cardiac ion channels plays a crucial role in the regulation of cardiac function. The molecular analysis of the inherited cardiac dysrhythmias has been the main driving force behind the identification and characterization of various ion channels in the heart. The underlying cause of dysrhythmias is defective ion channel function leading to abnormal cardiac electrical activity. Considerable progress has been made in understanding the properties and the roles of cardiac ion channels, the genes encoding many of these channels have been cloned and sequenced, and mutations in these genes have been linked to various forms of dysrhythmias. Moreover, silent mutations and functional DNA polymorphisms have also been found to play a significant role in increasing the susceptibility for these disorders, together with nongenetic or environmental factors, such as gender, aging, the presence of cardiac structural defects and ethnicity. This chapter emphasizes the functional consequences of dysrhythmia-associated ion channel mutations and reviews the development of new types of safer and more effective antidysrhythmic therapeutic approaches.

## Inherited Cardiac Dysrhythmias

Inherited or primary dysrhythmias are electrophysiological disorders not associated with structural heart pathology, and include long QT syndrome (LQTS), short QT syndrome (SQTS), catecholamine-induced polymorphic ventricular tachycardia (CPVT), and Brugada syndrome (BrS). The majority of the inherited dysrhythmias with a known genetic basis are caused by mutations in the genes encoding cardiac

J. Marín-García, *Signaling in the Heart*,
DOI 10.1007/978-1-4419-9461-5_17, © Springer Science+Business Media, LLC 2011

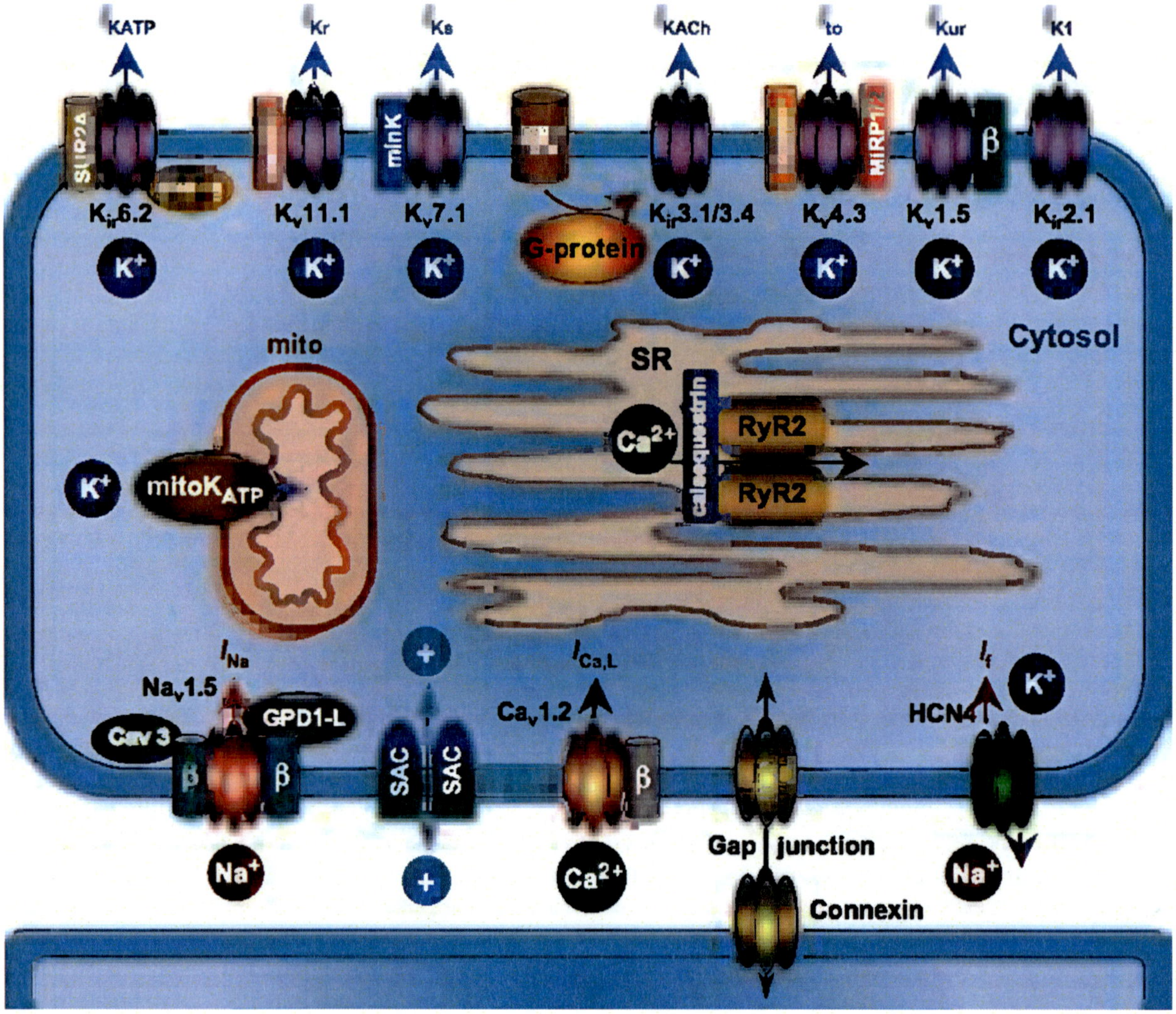


**Fig. 17.1** Molecular basis of dysrhythmias. Dysrhythmias-related cardiac ion channels with associated proteins and corresponding ion currents, which they conduct, are schematically depicted. *Mito* mitochondrion, *SR* sarcoplasmic reticulum, *SAC* stretch activated channel. See text for the details

ion channels subunits (e.g., the pore-forming α-subunits and modulatory β-subunits), as well as the auxiliary associated proteins essential for proper channel functioning. These mutations tend to affect the ion fluxes across the cardiomyocyte membrane resulting in increasing duration of the ventricular action potential (AP) and prolongation of the QT interval. These disorders that can also be referred to as cardiac channelopathies are schematically depicted in Fig. 17.1.

## Long QT Syndrome

LQTS is an inherited dysrhythmia characterized by abnormally prolonged QT interval that can result in ventricular tachycardia (VT), ventricular fibrillation (VF), and sudden death. Diagnostic criteria for the LQTS based on a scoring of clinical parameters as suggested by Schwartz et al. [1] are summarized in Table 17.1.

Similar to many other primary electrical cardiac disorders, LQTS demonstrates a high degree of genetic diversity, 12 *loci* responsible for this disease have so far been reported (Table 17.2).

In 1995, mutations in cardiac voltage-gated $K^+$ channel ($K_v11.1$/hERG) gene *KCNH2* and $Na^+$ channel ($Na_v1.5$) gene *SCN5A* were associated with two LQTS subtypes, LQT2 and LQT3, respectively [2, 3]. Furthermore, it was demonstrated that the mutant $Na^+$ channel exhibited sustained inward current during membrane depolarization, and the persistence of the inward current is responsible for the

**Table 17.1** LQTS diagnostic criteria

| | Points |
|---|---|
| *ECG findings*[a] | |
| QTc[b] | |
| ≥480 ms$^{1/2}$ | 3 |
| 460–470 ms$^{1/2}$ | 2 |
| 450 ms$^{1/2}$ (in males) | 1 |
| *Torsades de pointes*[c] | 2 |
| T-wave alternant | 1 |
| Notched T waves in three leads | 1 |
| Low heart rate for age[d] | 0.5 |
| *Clinical history* | |
| Syncope | |
| With stress | 2 |
| Without stress | 1 |
| Congenital deafness | 0.5 |
| *Family history*[e] | |
| Family member with definitive LQTS[f] | 1 |
| Unexplained SCD of immediate family member below age 30 | 0.5 |

Scoring: ≤1 point, low probability of LQTS; 2–3 points, intermediate probability of LQTS; ≥4 points, high probability of LQTS

[a]In the absence of medications or disorders known to affect these ECG findings

[b]QTc calculated by Bazett's formula where QTc=QT/square root of RR

[c]Mutually exclusive

[d]Resting heart rate below the second percentile for age

[e]The same family member cannot be counted in both categories

[f]LQTS defined by an LQTS score≥4

**Table 17.2** The genetic basis of LQTS

| Type | Locus | Gene | Protein | Function | Frequency |
|---|---|---|---|---|---|
| LQT1 | 11p15.5 | *KCNQ1* | $K_V$7.1 α | $I_{Ks}$ ↓ | 30–35% |
| LQT2 | 7q35 | *KCNH2* | $K_V$11.1 α | $I_{Kr}$ ↓ | 25–30% |
| LQT3 | 3p21 | *SCN5A* | $Na_V$1.5 α | $I_{Na}$ ↑ | 5–10% |
| LQT4 | 4q25 | *ANK2* | Ankyrin-B | $I_{Na,K}$ ↓ $I_{NCX}$ ↓ | 1–2% |
| LQT5 | 21q22.1 | *KCNE1* | minK β | $I_{Ks}$ ↓ | 1% |
| LQT6 | 21q22.1 | *KCNE2* | MiRP1 β | $I_{Kr}$ ↓ | Rare |
| LQT7[a] | 17q23 | *KCNJ2* | $K_{ir}$2.1 α | $I_{K1}$ ↓ | Rare |
| LQT8[b] | 12p13.3 | *CACNA1C* | $Ca_V$ 1.2 a1c | $I_{Ca,L}$ ↑ | Rare |
| LQT9 | 3p25 | *CAV3* | Caveolin-3 | $I_{Na}$ ↑ | Rare |
| LQT10 | 11q23 | *SCN4B* | $Na_V$1.5 b4 | $I_{Na}$ ↑ | Rare |
| LQT11 | 7q21 | *AKAP9* | Yotiao | $I_{Ks}$ ↓ | Rare |
| LQT12 | 20q11.2 | *SNTA1* | α1-syntrophin | $I_{Na}$ ↑ | Rare |

[a]Andersen–Tawil syndrome

[b]Timothy syndrome

abnormally prolonged AP [4]. These breakthrough discoveries were followed by the identification of several mutations responsible for other subgroups of LQTS, including the LQT1, LQT4, LQT5, and LQT6 [5–9].

Mutations within *KCNQ1* and *KCNH2* account for more than 80% of autosomal dominant LQTS, LQT1, and LQT2 (Romano–Ward syndrome), respectively. Mutations of the α-subunit of $K_{ir}$2.1 channel (LQT7) and the β-subunits of $K_V$7.1 (LQT5) and $K_V$11.1 (LQT6) channels represent minor causes of LQTS. Most of these mutations have a dominant negative effect, coassembling with normal subunits leads to impairing their function.

More than 80 *SCN5A* mutations cause the LQT3 variant accounting for 7–10% among the numerous subtypes of LQTS [10, 11] and can be defined as gain-of-function mutations. They cause small increases in inward current during the plateau phase resulting in prolongation of the AP. Patients with LQT3 usually experience cardiac events at rest rather than during exercise or emotional stress [12].

Two LQT-related syndromes have been recently linked to mutations in the cardiac ion channels. LQT7 (Andersen syndrome) is caused by several mutations of *KCNJ2* gene encoding the α-subunit of $K^+$ inward rectifier $K_{ir}$2.1 channel [13]. LQT8 [Timothy syndrome (TS)] results from a missense mutation (G406R) in the *CACN1C* gene encoding the α-subunit of the L-type $Ca^{2+}$ $Ca_V$1.2 channel [14, 15].

Andersen syndrome (also known as Andersen–Tawil syndrome) was first reported in 1971 in a patient with muscle weakness, extrasystoles, and developmental delay [16]. Patients may exhibit more severe cardiac dysrhythmias and a variety of ventricular and bidirectional ventricular tachycardia. Andersen syndrome or LQTS type 7 is inherited as an autosomal dominant entity and exhibits significant phenotypic and genotypic variability, since at least 21 different mutations within the *KCNJ2* gene are associated with the disorder. All these mutations result in the loss of function and dominant-negative suppression of $K_{ir}$2.1 channel function [13, 17]. Intriguely, *KCNJ2* mutations do not account for all cases of this disorder; over 30% of patients with clinically indistinguishable symptoms do not harbor $K_{ir}$2.1 mutations [18]. Given a dominant-negative effect of the *KCNJ2* mutations, it has been suggested that the unexplained cases might be associated with defective auxiliary proteins essential for $K_{ir}$2.1 channel function. Candidates include aberrant scaffolding or adaptor proteins, which anchor the $K_{ir}$2.1 subunits to the specific cell membrane domains, or are involved in its intracellular trafficking. A proteomics approach has recently identified a number of high affinity interactions of cardiac $K_{ir}$2.1 channel with scaffolding proteins, including SAP97 and the Lin7/calcium/calmodulin-dependent serine protein kinase (CASK). Moreover, the $K_{ir}$2.1 channel targeting to the basolateral membrane of polarized epithelial cells and its colocalization with SAP97 and CASK could be abrogated by a mutant form of CASK, which caused the improper channel localization, implicating CASK as a central component of a large multiprotein $K_{ir}$2-containing complex [19, 20]. Importantly, transgenic mice harboring nonfunctional SAP97 and CASK exhibit a similar phenotype to *KCNJ2* knockout mice [21–23]. Early lethality of these transgenic mice complicates the analysis and it remains to be seen whether mutations in $K_{ir}$2.1-interacting domains are involved in Andersen syndrome.

Functional expression analysis of TS, revealed that the G406R mutation in the L-type $Ca^{2+}$ channel encoding gene exhibits a gain-of-function effect producing a maintained inward $Ca^{2+}$ current in multiple cell types and inducing intracellular $Ca^{2+}$ overload. In the heart, this prolonged $Ca^{2+}$ current delays cardiomyocyte repolarization, extends the AP with increased delayed afterdepolarization events leading to an increased risk of lethal dysrhythmia. TS encompasses multiorgan dysfunction, including lethal cardiac dysrhythmias, webbing of fingers and toes, congenital heart disease, immune deficiency, intermittent hypoglycemia, cognitive abnormalities, and autism.

Mutations in the *ANK2* gene encoding the scaffolding protein ankyrin B were initially implicated in the etiology of LQT4. They resulted in a loss of function of ankyrin B and development of fatal cardiac dysrhythmias [7]. Moreover, transgenic mice heterozygous for a null *ANK2* mutation display dysrhythmias similar to humans. The importance of the ubiquitously expressed ankyrin adaptor protein is underscored by the pleiotropic effects of ankyrin B mutations, including the abnormal expression and localization of various ankyrin B-binding proteins, such as the $Na^+/K^+$ ATPase, $Na^+/Ca^{2+}$ exchanger, and inositol-1,4,5-trisphosphate (IP3) receptors. Furthermore, ankyrin B mutations also lead to altered $Ca^{2+}$ signaling in adult cardiomyocytes that results in extrasystoles, providing a significant rationale for dysrhythmias.

Mammals express three ankyrins proteins (ankyrin R, ankyrin B, and ankyrin G), which are encoded by three distinct genes, *ANK1*, *ANK2*, and *ANK3*, respectively. These adaptor proteins interface between various ion channels and transporters and the cytoskeleton. While most vertebrate tissues harbor multiple alternatively spliced forms of each ankyrin, cardiomyocytes are exceptional in expressing a single form of ankyrin B (220 kDa) and one isoform of ankyrin G (190 kDa).

Mohler and Bennett [24] have recently identified a number of loss-of-function mutations in ankyrin B in patients who display cardiac dysrhythmias in association with sinus node dysfunction suggesting a novel class of channelopathies. These patients exhibited severe sinus bradycardia, ventricular and atrial fibrillation, catecholaminergic-induced ventricular tachycardia, syncope, and had an increased risk of sudden cardiac death. Studies in ankyrin B-deficient mice suggested that altered $Ca^{2+}$ homeostasis likely underlies spontaneous $Ca^{2+}$-based depolarization events in response to catecholaminergic stimulation leading eventually to ventricular dysrhythmias.

Recently, a macromolecular complex containing ankyrin B, the $Na^+/K^+$ ATPase, $Na^+/Ca^{2+}$ exchanger, and IP3 receptor localized in cardiomyocyte T-tubules in discrete microdomains has been reported as distinct from classic dihydropyridine receptor/ryanodine receptor "dyads" [25]. The E1425G mutation in the *ANK2* gene abrogates ankyrin B binding to the complex and mediates LQTS. Given low level of ankyrin B and the absence of T-tubule-associated ankyrin B, $Na^+/K^+$ ATPase and the $Na^+/Ca^{2+}$ exchanger in skeletal muscle, the ankyrin B-based complex has been proposed to be a specialized adaptation of cardiomyocytes with a prominent role for cytosolic $Ca^{2+}$ modulation.

The essential role that ankyrins play in the control of heart rhythm is expanding very rapidly. Recently, it has been demonstrated that $K_{ir}6.2$, subunit of the ATP-sensitive $K^+$ channel requires binding to ankyrin B for its proper targeting to the cardiomyocyte membrane [26]. Additionally, a role of ankyrin remodeling in myocardial infarction has been reported [27]. Therefore, based on recent findings and the ubiquitous expression of ankyrins, their dysfunction may be increasingly linked to new classes of human disorders.

Mutations in the caveolin-3 gene (*CAV3*) encoding M-caveolin (Cav3), a principal protein in caveolae, have been identified in patients with LQT9 and have been found to be a rare cause of sudden infant death syndrome (SIDS) in African American children [28, 29]. Scaffolding domain of Cav3 interacts with various ion channels, including the α-subunit of $Na_V1.5$ channel modulating $I_{Na}$ current [30–32]. The *CAV3* mutations associated with LQT9 and SIDS interfere with the interaction between Cav3 and $Na_V1.5$ channel leading to upregulation of $I_{Na}$ current.

The *SCN4B* gene, a member of the *SCNB* gene family, encodes the β1-subunit of the $Na_V1.5$ channel [33]. The β-subunits mediate interaction of the $Na_V1.5$ with ankyrin-B and -G, and modulate its function [34]. A *SCN4B* mutation has recently been identified in a Mexican-mestizo family with LQT10 causing a significant increase in the late $I_{Na}$ current [35].

Upon thorough analysis of 50 LQTS, families with no mutations in other known LQTS-related genes, a single mutation in the A-kinase anchor protein 9 gene (*AKAP9*) has been found (LQT11) [36]. *AKAP9* encodes two proteins, a human homolog of the rat protein AKAP120 and the yotiao protein. The functional $K^+$ channel responsible for the $I_{Ks}$ current consists of the $K_V7.1$ α-subunit, the mink β-subunit, and the yotiao protein. Yotiao interacts with $K_V7.1$, protein kinase A (PKA), and protein phosphatase 1. These interactions are essential for the cAMP-dependent regulation of the $I_{Ks}$ current mediated via PKA-dependent phosphorylation of $K_V7.1$ [37]. The LQT11-causing mutation reduces the binding of yotiao to $K_V7.1$ and the cAMP-dependent phosphorylation of $K_V7.1$ leading to abrogation of the response of the $I_{Ks}$ current to cAMP stimulation [36].

Finally, in 1 out 50 LQTS probands, with no mutations in LQT1–11-related genes, a missense mutation in the α-1 syntrophin gene (*SNTA1*) has been identified in a proband with a very long $QT_c$ interval (LQT12) [38]. α-1 syntrophin is membrane-associated scaffolding protein linking $Na_V1.5$ to the neuronal nitric oxide synthase (nNOS)/plasma membrane $Ca^{2+}$-ATPase (PMCA) complex in the heart [39, 40]. The mutation associated with LQT12 leads to uncoupling of $Ca^{2+}$ transporting, PMCA and the $Na_V1.5$/α-1 syntrophin complex. These defects can result in increased late $I_{Na}$ current similar to the LQT3 phenotype. However, alterations in interactions with other ion channels may also contribute to the phenotype [38].

## Short QT Syndrome

Although the initial emphasis studying inherited dysrhythmias was directed toward LQTS and BrS, recently SQTS has emerged as yet another disorder associated with alterations of ion signaling in the heart. SQTS is a rare heritable channelopathy characterized by an abnormally short QT interval (typically <320 ms) with tall, peaked, narrow-based T waves. This autosomal dominant syndrome can affect infants, children, and young adults. Patients with SQTS display a marked propensity for atrial fibrillation (AF) and increased risk for sudden cardiac death (SCD) from ventricular tachydysrhythmias.

The initial recognition of increased risk of SCD with a short QT interval was reported in 1993 [41], and in 2000 the diagnosis of a new clinical syndrome, SQTS, was proposed [42]. Subsequently, a definitive association between SQTS and familial SCD was reported [43].

The common characteristic in most forms of SQTS is alteration in $K^+$ ion signaling leading to acceleration of repolarization of the AP caused by mutations in the cardiac $K^+$ channels. These gain-of-function mutations accelerate outward potassium currents in phase 2 and 3 of the cardiac AP resulting in shortening of the plateau phase (phase 2), the AP duration, the refractory period, and the overall QT interval (<320 ms). The combination of inhomogeneity during repolarization, short refractory periods and increased dispersion of refractoriness forms a substrate for the development of atrial and ventricular dysrhythmias [44, 45].

Table 17.3 summarizes the presently known mutations responsible for SQTS. This syndrome can be classified into three subtypes caused by gain-of-function mutations in $K^+$ channel genes *KCNH2*, *KCNQ1*, and *KCNJ2* [46–48]. The overlapping subtypes, sometimes also referred to as SQT4 and SQT5, is caused by mutations in the $Ca^{2+}$ channel genes *CACNA1C* and *CACNB2*, respectively, resulting in the loss of function [49]. Patients with SQT4 and SQT5 display in addition to a short QT interval ST segment elevation in the right precordial leads typical of the BrS.

There are currently no well-defined criteria for the clinical diagnosis of SQTS. A short QT interval alone does not mean that the patient has SQTS, the SQTS diagnosis requires that the condition is familial and that syncope or cardiac events exist in the patient history [50]. Typical age at the time of diagnosis is 30 years. There is no gender predominance, as reported in LQTS (females) or BrS (males) [51]. Furthermore, similar to LQTS, there are significant variations in penetrance and expression in patients with SQTS ranging from asymptomatic carriers to patients with AF or ventricular fibrillation or SCD.

The relatively high risk of SCD of patients with SQTS makes the implantation of an implantable cardioverter-defibrillator (ICD) the treatment of choice [52]. However, this treatment is less clear for asymptomatic SQTS patients or affected family members. Pharmacological therapy for SQTS is geared toward prolonging repolarization by blocking outward $K^+$ currents. A few promising treatments of patients with SQT1 (mutation in *KCNH2*) with quinidine/hydroquinidine, disopyramide, and amiodarone combined with beta blockers have been reported [53].

## Brugada Syndrome

BrS is characterized by ST-segment elevation and negative T waves in the right precordial leads (V1–V3) and an increase in SCD resulting from polymorphic ventricular tachycardias [54, 55]. BrS occurs predominantly in male adults and can be influenced by heart rate, autonomic nervous activity, fever, and drugs. Geographically, the incidence of BrS varies, estimated to be from 5 to 50 per 10,000. In endemic areas, such as East/Southeast Asia (particularly Japan and Thailand), BrS is a leading cause of death among young and otherwise healthy adult males [56, 57].

**Table 17.3** The genetic basis of short QT syndrome

| Type | Gene | Locus | Protein | Function | Mechanism |
|---|---|---|---|---|---|
| SQT1 | *KCNH2* | 7q35–7q36 | $K_V11.1$ | α-Subunit $I_{Kr}$ | Gain of function |
| SQT2 | *KCNQ1* | 11p15.5 | $K_V7.1$ | α-Subunit $I_{Ks}$ | Gain of function |
| SQT3 | *KCNJ2* | 17q23.1–17q24.2 | $K_{ir}2.1$ | α-Subunit $I_{K1}$ | Gain of function |
| *Overlap syndromes* | | | | | |
| SQT4 | *CACNA1C* | 12p13.3 | $Ca_V1.2$ | α-Subunit $I_{Ca,L}$ | Loss of function |
| SQT5 | *CACNB2* | 10p12 | $Ca_Vβ2$ | β2-Subunit $I_{Ca,L}$ | Loss of function |

The early repolarization (phase 1) during the AP plays a central role in BrS [58, 59]. A reduced $Na^+$ current enables the outward $I_{to}$ current, the major contributor of the phase 1, to deepen phase 1 notch (Fig. 17.2). This enhancement of the phase 1 delays the $Ca^{2+}$ current ($I_{Ca,L}$) and causes delayed repolarization (phase 2). Further enhancement of the phase 1 notch blocks the rising of the phase 2 dome and abrogates the AP duration [60, 61]. This also affects $Na^+$ channel and $Na^+$/$Ca^{2+}$ exchanger functions [58]. The phase 1 mediates the effects of sex hormones on $Na^+$ channel contributing to the male predominance of BrS. Testosterone induces the outward currents ($I_{Kr}$, $I_{Ks}$, $I_{K1}$) and reduces the inward current ($I_{Ca,L}$) deepening the phase 1 notch, whereas estrogen reduces $I_{to}$ shallowing the phase 1 [62].

Several mutations in the genes encoding $Na^+$, L-type $Ca^{2+}$ and transient outward $K^+$ channels have been identified in patients with BrS (Table 17.4) [63]. Since 1998 when the first mutations in the gene encoding the α-subunit of the $Na^+$ channel (*SCNA5*) were identified in patients with BrS, more than 90 *SCNA5* mutations have been associated with BrS (BS1); they account for 18–30% of clinically diagnosed cases [55, 64–66]. Mutant *SCNA5* as has been noted above can also cause LQT3 [3]. However, unlike the LQT3 mutations causing a gain of function, BrS-associated *SCNA5* mutations invariably cause a loss of the channel function.

The β–subunit of $Na^+$ channel encoded by the *SCN1B* gene and product encoded by the *glycerol-3-phosphate dehydrogenase 1-like* (*GPD1-L*) gene are components of a large multiprotein complex of the functional $Na^+$ channel [67]. Mutations in these genes can also affect $Na^+$ channel function reducing the $I_{Na}$ and causing BrS (BS5 and BS2, respectively; Table 17.4) [68, 69].

Although $Na^+$ channel mutations are the most common, mutations in the genes encoding the α- and β-subunits of the $Ca^{2+}$ channel were identified in some patients with BrS (BS3 and BS4, respectively). These mutations cause loss of the $Ca^{2+}$ channel function, resulting in BrS in combination with short QT intervals and familial SCD [49].

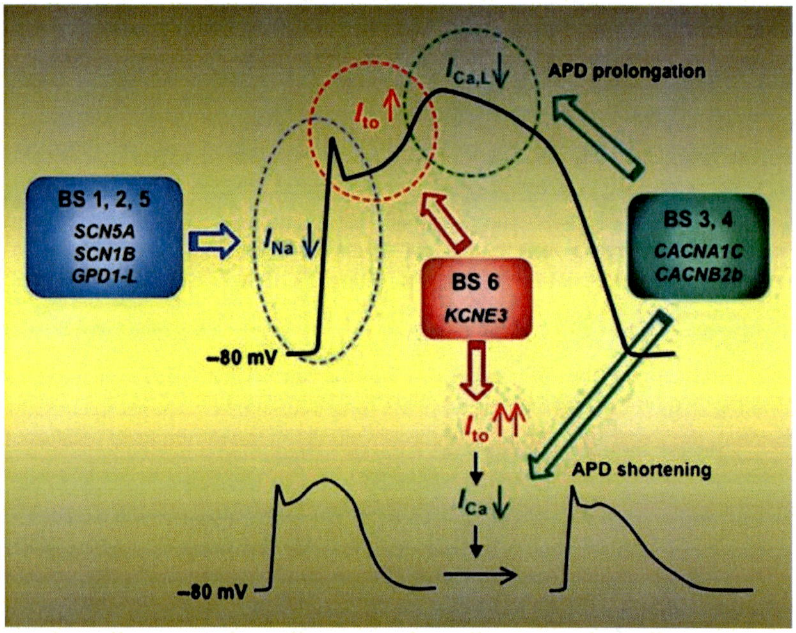


**Fig. 17.2** Ion channel mutations associated with various types of BrS. In BS1, 2 and 5 (SCN5A, GPD1-L, and SCN1B mutations), reduced $I_{Na}$ current induces the transient outward $I_{to}$ current deepening the phase 1 notch. Deep phase 1 mediates delayed activation of $I_{Ca,L}$ current leading to the delayed onset of phase 2 dome and prolongation of the AP duration (APD). Similar delay of $I_{Ca,L}$ is also caused by mutations in the $Ca^{2+}$ channel genes (CACNA1C and CACNB2b) found in BS3 and 4. Excessive upregulation of $I_{to}$ current caused by KCNE3 mutations observed in BS6 can lead to the loss of the phase 2 dome and eventually to the APD shortening [63]

**Table 17.4** The genetic basis of Brugada syndrome

| Type | Locus | Gene | Protein | Affected current | Effect on channel function | Incidence (%) |
|---|---|---|---|---|---|---|
| BS1 | 3p21 | SCN5A | $Na_V1.5$ | $I_{Na}$ | Loss of function | 20 |
| BS2 | 3p24 | GPD1-L | GPD1-L protein | $I_{Na}$ | Loss of function | 1 |
| BS3 | 12p13.3 | CACNA1C | $Ca_V1.2$ | $I_{Ca,L}$ | Loss of function | 7 |
| BS4 | 10p12.33 | CACNB2b | $Ca_V\beta2b$ | $I_{Ca,L}$ | Loss of function | 5 |
| BS5 | 19q13.1 | SCN1B | $Na_V\beta1$ | $I_{Na}$ | Loss of function | 1 |
| BS6 | 11q13–q14 | KCNE3 | MiRP2 | $I_{to}$ | Gain of function | 1 |

*GPD1-L* glycerol-3-phosphate dehydrogenase 1-like gene

A mutation of ankyrin B, which modulates $Ca^{2+}$ and $Na^{+}$ currents, and mutations of the delayed rectifier K channel genes (*KCNH2* and *KCNQ1*) were also reported in patients with BrS [70, 71]. The other myocardial ankyrin isoform, ankyrin G, is directly associated with the principal cardiac voltage-gated $Na_V1.5$ channel, responsible for $I_{Na}$. Loss of the $Na_V1.5$ channel function caused by *SCN5A* mutations can lead to BrS, characterized by right bundle branch block, dysrhythmia and increased the risk of SCD [72]. A *SCN5A* mutation (E1053) in the critical nine amino acid motif required for interaction with ankyrin G has recently been reported in a proband with BrS [73]. These *SCN5A* mutations disrupt the interaction of human $Na_V1.5$ with ankyrin G leading to abnormal $Na_V1.5$ membrane localization and loss of the $Na_V1.5$ channel function. Mutation of *KCNE3* gene encoding the β-subunit of the $K_V4.3$ channel enhances the $I_{to}$ and can also result in phenotypical expression of BrS (BS6; Table 17.4) [74].

The failure to identify gene mutations in most patients with BrS suggests that either unknown mutations or pathophysiological cellular regulations, such as posttranslational modulations may also be responsible for the observed alterations in ion currents and clinical symptoms. Moreover, the adult male dominance of clinical manifestation suggests that gender- and age-related factors play an important role in dysrhythmias in BrS [62, 75].

In contrast to LQTS, only limited analysis of genotype–phenotype correlations has been performed in patients with the BrS phenotype. Patients harboring an *SCN5A* mutation demonstrated longer and progressive conduction delays (PQ, QRS, and HV intervals), frequent occurrences of fragmented QRS complex (f-QRS), and ventricular dysrhythmias of extra right ventricular outflow tract (RVOT) origin [75–77]. On the other hand, about 80% of the patients with BrS without an *SCN5A* mutation had ventricular dysrhythmias of RVOT origin [78]. Patients with a $Ca^{2+}$ channel gene mutation had a short QT interval [49].

We have previously stated that inherited dysrhythmias are primarily electrophysiological disorders not caused by structural cardiac pathology. However, there is evidence that myocarditis and arrhythmogenic right ventricular cardiomyopathy or dysplasia (ARVD) may mimic the BrS phenotype and both may account for the transient development of the ECG abnormalities characteristic for this disorder [79, 80]. On the other hand, it is possible that patients with BrS caused by an *SCN5A* mutation may also have structural heart pathology. It remains to be determined whether the genetic defect contributes to the development of structural heart disease or whether these defects arise independently and together contribute to the clinical manifestations.

Given the fact that BrS may not be fully explained by a single mechanism and many patients with BrS are asymptomatic, thorough genetic screening and genotype–phenotype correlation analysis of this disorder is of paramount importance. Finally, none of the drugs currently available have been fully successful in the treatment of this disease, and only an ICD appears to be an effective alternative. Further research is needed to be able to diagnose all cases of BrS, and to develop new, efficient treatment.

## Catecholaminergic Polymorphic Ventricular Tachycardia

CPVT is an inherited disorder characterized by adrenergically mediated ventricular dysrhythmias leading to syncope, cardiac arrest, and SCD [81]. This disease was originally described in 1978; however, only in 2001 its association with alterations in $Ca^{2+}$ signaling in the heart was identified [81, 82]. Affected individuals develop ventricular dysrhythmias in response to physical exercise or emotional stress while they have normal resting ECGs, making the diagnosis difficult. The mean age of the onset of symptoms is 8 years, approximately 30% of CPVT patients display symptoms before age 10, and 60% have at least 1 syncope episode before age 40. The response of CPVT to medical therapy is poor and the incidence of SCD reaches 30% before age 40 [83].

Mutations in two genes encoding the ryanodine receptor type 2 (*RYR2*) and calsequestrin (*CASQ2*) have been found to be associated with two subtypes of CPVT (Table 17.5) [82, 84, 85].

RyR2 controls $Ca^{2+}$ release from the sarcoplasmic reticulum (SR) into the cytoplasm mediating excitation–contraction coupling. Mutant RyR2 enhances $Ca^{2+}$ release (leakage) that induces extrusion of $Ca^{2+}$ to the extracellular matrix by the $Na^{+}/Ca^{2+}$ exchanger (Fig. 17.3). The $Na^{+}/Ca^{2+}$ exchanger exchanges one $Ca^{2+}$ ion for three $Na^{+}$ ions generating an inward $Na^{+}$ current. This current is responsible for delayed afterdepolarization during phase 4 of the AP, which causes ventricular tachydysrhythmia [86, 87].

Presently, more than 70 different mutations in the *RyR2* gene have been identified in 55% to 60% of patients with CPVT (http://www.fsm.it/cardmoc).

**Table 17.5** Genetic basis of CPVT

| Type | Locus | Gene | Protein | Mechanism | Inheritance |
|---|---|---|---|---|---|
| CPVT1 | 1q42–43 | RYR2 | Ryanodine receptor 2 RyR2 | $Ca^{2+}$ release from the SR | AD |
| CPVT2 | 1p13–21 | CASQ2 | Calsequestrin | | AR |

*AD* autosomal dominant, *AR* autosomal recessive, *SR* sarcoplasmic reticulum

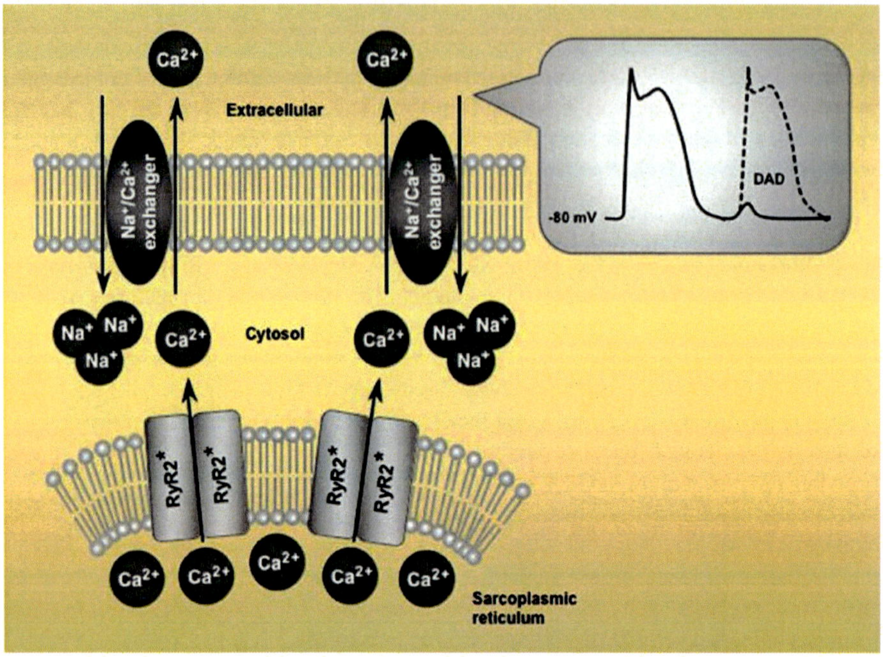


**Fig. 17.3** Molecular defects leading to CPVT. Mutant ryanodine receptor type 2 (RyR2*) conduits the release of $Ca^{2+}$ ions from the sarcoplasmic reticulum into the cytoplasm. This $Ca^{2+}$ leakage induces the $Na^+/Ca^{2+}$ exchanger to efflux $Ca^{2+}$ ions into the extracellular matrix. The $Na^+/Ca^{2+}$ exchanger exchanges one $Ca^{2+}$ ion for three $Na^+$ ions and thereby generates an inward $Na^+$ current, which underlies delayed afterdepolarization (DAD) of the AP [88]

Mutations in *CASQ2* gene encoding the cardiac calsequestin are responsible for the autosomal recessive form of CPVT. Calsequestin is a regulatory $Ca^{2+}$ buffering protein in the SR. Its monomers polymerize in the terminal cisternae of the cardiac SR and regulate $Ca^{2+}$signaling. Calsequestin is not only $Ca^{2+}$ buffering protein but together with junctin and triadin it interacts with RyR2 regulating the RyR2 opening probability. *CASQ2* mutations have been identified in 1–2% of patients with CPVT.

Finally, an autosomal recessive type of CPVT mapping to the chromosome 7p14-22 has recently been identified [89]. However, the affected gene remains to be determined.

Current therapy for this rare disease includes ICD plus β–adrenergic blockers for symptomatic patients and at least β-blockers for asymptomatic carriers.

## Familial Atrial Fibrillation

AF is the most common dysrhythmia in clinical practice affecting approximately 2.3 million patients in the United States and 4.5 million patients in Europe [90]. In the general population, it has a prevalence of approximately 1% that increases steadily with age to ~6% in those 65 years of age, and over 10% by 80 years of age [90, 91]. AF can be defined as a chaotic activation of the atria leading to an irregular ventricular response.

Most AF is associated with other cardiac pathology, such as hypertensive cardiac disease, valvular disease, cardiomyopathy or atherosclerotic cardiovascular disease, however, in 10–30% of patients with AF no obvious structural cause or no previous cardiac pathology is present. This condition is defined as lone AF. Although until very recently the molecular basis of AF has been poorly understood and a genetic basis for this common dysrhythmia has been considered unlikely, the recent analysis of familial AF (FAF) forms has provided definitive insight into the etiology of the disease.

The initial description of FAF was followed by the identification of the first genetic locus for the dysrhythmia on the long arm of chromosome 10 segregated with the affected individuals [92]. Subsequently, Ellinor et al. [93] using linkage and haplotype analysis have mapped a novel *locus* on chromosome 6q14-16 in a large family with autosomal dominant AF. Presently, both genes remain to be identified. In the last years, several *loci* associated with AF have been identified and in some cases the responsible genes have been cloned (Table 17.6).

The identification of a *locus* on chromosome 11p15.5 and *KCNQ1* gene responsible for the disease provided the first association of FAF with alteration of cardiac ion channel function [95]. This gene encodes the pore forming α-subunit of the cardiac $K^+$ channel responsible for the slow repolarizing current $I_{Ks}$, and as we discussed previously its loss-of-function mutation causes LQTS. In contrast to LQTS, a missense mutation (S140G) in a large Chinese family with FAF leads to gain of function in $I_{Ks}$ current explaining the shortening of the AP duration characteristic to AF. Subsequently, mutations in other genes encoding cardiac $K^+$ channels have been reported to be associated with FAF,

**Table 17.6** Genes and loci associated with familial atrial fibrillation (FAF) [94]

| Locus | Inheritance | Gene | Disease |
|---|---|---|---|
| *Familial AF as a monogenic disease* | | | |
| 10q22 | Autosomic dominant | – | |
| 6q14–16 | Autosomic dominant | – | |
| 10p11–q21 | Autosomic dominant | – | |
| 12p12 | Autosomic dominant | – | |
| 5p15 | Autosomic dominant | – | |
| 11p15 | Autosomic dominant | *KCNQ1* | |
| 3p21 | Autosomic dominant | *SCN5A* | |
| 12p13 | Autosomic dominant | *KCNA5* | |
| 11q13–14 | Autosomic dominant | *KCNE3* | |
| 21q22 | Autosomic dominant | *KCNE2* | |
| 17q23 | Autosomic dominant | *KCNJ2* | |
| 1p36–35 | Autosomic dominant | *NPPA* | |
| 5p13 | Autosomic recessive | *NUP155* | |
| *Familial AF associated with other monogenic channelopathies* | | | |
| 11p15 | Autosomic dominant | *KCNQ1* | Short QT |
| 7q35 | Autosomic dominant | *KCNH2* | Short QT |
| 3p21 | Autosomic dominant | *SCN5A* | Long QT |

including gain-of-function mutations in *KCNE2* and *KCNJ2* [96, 97]. Genetic defect identified in *KCNE3* has not lead to the alteration in cardiac electrophysiological characteristics suggesting that it could be a rare polymorphism [98]. Finally, identification of a loss-of-function mutation in *KCNA5* gene encoding $K_V1.5$ channel in patients with AF suggests that not only a shortening but prolongation of the cardiac AP can also cause the disease [99].

The cardiac $Na^+$ channel, $Na_V1.5$, plays a key role in the phase 0 of the AP. Mutations in the *SCN5A* gene encoding this channel are responsible for several primary dysrhythmia syndromes, such as already discussed LQTS and BrS. Mutant *SCN5A* have recently been reported to be associated with FAF [100]. Interestingly, both loss-of-function and gain-of-function mutations can lead to either familial or nonfamilial forms of AF [101]. According to current hypothesis, loss-of-function mutations leading to $I_{Na}$ reduction may induce AF by slowing atrial electrical conduction, whereas gain of function may provoke AF by enhancing spontaneous excitability of atrial myocytes [102].

Two unique families with AF linked to genetic defects in non-ion channels have been described. In one such family, a frame shift mutation in the natriuretic peptide precursor A (*NPPA*) gene has been found [103]. *NPPA* encodes the atrial natriuretic peptide, which modulates ion signaling in cardiomyocytes and can contribute to shortening of the atrial AP and development of AF. In a second family characterized by a neonatal onset with autosomal recessive inheritance, a mutation in *NUP155* has been identified [104]. This gene has been mapped to 5p13 and has been also associated with sudden death in the family. *NUP155* encodes a member of the nuclear pore protein complex, nucleoporin. The mutant nucleoporin displays abnormal localization resulting in reduced nuclear membrane permeability. How this defect in a general intracellular process can lead to the specific AF phenotype observed in this family remains to be determined.

AF has been reported as a concomitant disease with other cardiac disorders including skeletal myopathies, hypertrophic cardiomyopathy, familial amyloidosis as well as channelopathies like LQTS, SQTS, and BrS. The familial forms of AF are rare, the majority of the AF cases are acquired and associated with structural pathologies. Importantly, not all patients with the same cardiac abnormality develop AF suggesting the existence of genetic factors, which predispose some of them to AF.

In conclusion, AF is a highly heterogeneous disorder with multiple factors involved. Although a large number of genes linked to AF have already been reported (Table 17.6) mutations in these genes do not explain all cases of FAF. Thus, our understanding of AF will largely benefit from the identification of disease genes and from the analysis of the alterations in gene expression associated with the familial forms of the disorder.

## Cardiac Conduction Defects

Cardiac conduction disease (CCD) is characterized by progressive defects at the atrial, atrioventricular (AV), and/or ventricular level and is frequently linked to mutations in *SCN5A* gene encoding the $Na_V1.5$ channel. Lev–Lenegre disease is one of the most common forms of CCD. This defect was initially described as an acquired complete AV block with right (RBBB) or left bundle branch block (LBBB) and widening of the QRS complex [105]. In 1999, Schott et al. [106] identified the first autosomal dominant mutation in the *SCN5A* gene that segregated with progressive cardiac conduction defect (PCCD) in a large French family, and a second *SCN5A* mutation which cosegregated in a smaller Dutch family with familial nonprogressive cardiac conduction defect. The mutation led to the abrogation of the cardiac $Na^+$ channel function. Interestingly, none of the affected individuals had LQTS or BrS, although heterozygous *SCN5A* mutations have been associated with LQTS or BrS. Subsequently, several other laboratories have reported loss-of-function mutations in the cardiac *SCN5A* in familial PCCD resulted in the reduced $I_{Na}$ current and enhanced slow inactivation [107–109]. The reason why the same *SCN5A* mutation can lead either to an isolated CCD or to BrS is unclear. Moreover, some patients with the mutant cardiac $Na^+$ channel display combined clinical features of CCD and BrS [110, 111].

More recently, missense mutations in *SCN1B* gene encoding the β1-subunit of the $Na_V1.5$ channel have been identified to be associated with CCD as well as with BrS and AF [68, 112]. In all reported cases, this defect has led to the reduced $I_{Na}$ current suggesting that the β1-subunit is essential for the $Na_V1.5$ channel expression at the cell membrane of atrial and ventricular cardiomyocytes.

Sick sinus node syndrome (SSS) is characterized by sinus bradycardia, sinus arrest, atrial standstill and tachycardia–bradycardia syndrome. In 2003, Benson et al. [113] identified the first loss-of-function mutations in the *SCN5A* gene associated with congenital SSS, which was diagnosed during the first decade of life. Presently, at least 14 *SCN5A* mutations have been reported to be associated with SSS [114]. One mutation (E161K) found in two families contributes to both SSS and BrS phenotypes [115]. Several mutations in *SCN5A* have also been identified in families displaying heterogeneous phenotype characterized by SSS, dilated cardiomyopathy, and atrial or ventricular dysrhythmias [116, 117].

Electrophysiological analysis demonstrated that SSS-associated mutations cause a loss of function in the cardiac $Na^+$ channel leading to reduced $I_{Na}$ current [118]. This defect uncouples electrical activity of the pacemaker cells and the surrounding cells in the sinoatrial (SA) node leading to propagation of electrical activity through the SA node and atria.

As we have previously discussed in Chap. 6, hyperpolarization-activated cyclic nucleotide-gated channel 4 (HCN4) is a major HCN channel isoform expressed in SA node cells. This channel plays a central role in the regulation of heart rhythmicity. Mutations in the *HCN4* gene have recently been identified to be associated with sinus bradycardia and complex cardiac dysrhythmia [119–122]. Interestingly, all affected individuals found so far are heterozygous for the respective mutations in the *HCN4* gene. Based on the analysis of mouse *HCN4* knockout models, disruption of both *HCN4* alleles appears to be embryonic lethal because of the defects in the development cardiac conduction system.

## Sudden Death Infant Syndrome

According to the National Institutes of Health consensus conference sudden infant death can be defined as the "sudden death of infant or young child, which is unexpected by history, and in which a thorough post mortem examination fails to demonstrate an adequate cause of death" [123]. Since SIDS was defined as a syndrome, it is the result more than one disease associated with mutations in several "SIDS-related genes" operating as a polygenic inheritance and predisposing the infant to sudden death, in combination with a number of environmental risk factors [124]. However, some rare cases of SIDS have been reported to be associated with the alterations in ion signaling in the heart caused by mutations in cardiac ion channels. Mutations and polymorphisms in the *KCNQ1* and *KCNH2* genes encoding cardiac $K^+$ channels and in *SCN5A* gene encoding the $Na_V1.5$ channel have been identified to be linked to SIDS [125–127].

## Wolff–Parkinson–White Syndrome

The Wolff–Parkinson–White (WPW) syndrome was first described by Wolf and associates in 1930 [128]. This complex disorder is characterized by paroxysmal atrial tachycardia secondary to reentry via an anomalous pathway, which in the atria becomes a retrograde conduction pathway. In some patients, episodic atrial fibrillation and flutter may occur, as well as VT and VF. Generally, patients with WPW have an abnormal short P-R interval with prolonged QRS duration on their ECG. Sudden death, although infrequent, has been reported. Each case of WPW is highly individual and displays a variety of manifestations.

WPW can present as an inherited (familial) or as an acquired cardiac dysrhythmia. The familial form was first described in a large family in 1995 [129]. The disease exhibited an autosomal dominant pattern and was linked to chromosome 7q3. Subsequently, mutations in the *PRKAG2* gene responsible for WPW have been identified in several different families [130–132]. *PRKAG2* encodes the γ2 regulatory subunit of AMP-activated protein kinase (AMPK).

AMPK is a heterotrimeric enzyme composed of three subunits, α, β, and γ, and expressed in most mammalian tissues, including heart muscle [133, 134]. In the heart, AMPK plays a key role promoting activation of the glycolytic pathway by phosphorylating and activating phosphofructikinase-2, enhancing fatty acid β-oxidation, and ameliorating relative ATP deficiencies [134, 135]. The importance of AMPK increases when stress, such as excessive load or ischemia, is induced under hypoxic conditions.

*PRKAG2* mutations have been found in the conserved nucleotide-sensing regions of the regulatory γ2 subunit of AMPK leading to inappropriate activation of the enzyme under resting conditions. Activation of AMPK results in dramatic glycogen accumulation in the cardiomyocytes leading to cardiac hypertrophy, accessory atrioventricular connections, and degeneration of the physiological conduction system [134]. However, significant phenotypic variability has been found within and between different *PRKAG2* mutations. Given the wide range of AMPK targets, its role in the cardiac hypertrophic process and the effect of AMPK mutations on cardiac function remains to be determined.

Other causes for WPW have also been reported, including mutations in mitochondrial DNA and in 1,4-α-glucosidase [136, 137]. Furthermore, unlike familial WPW syndrome, sporadic WPW seems to be not commonly associated with mutations in the *PRKAG2* gene [138]. It is also unlikely that the polymorphisms in *PRKAG2* identified in some patients predispose to accessory pathway formation, since their incidence has been similar in both affected and normal individuals. Therefore, there are currently many unclarified details with regard to the precise molecular mechanisms responsible for both inherited and sporadic forms of WPW syndrome.

## Acquired Dysrhythmias

In contrast to inherited dysrhythmias linked to mutations in genes encoding cardiac ion channels or their auxiliary proteins, acquired or secondary dysrhythmias are caused by changes in the expression of ion channels or in their gating characteristics. Such changes have been reported for the majority if not all cardiac ion channels. Therefore, we review these diseases based on defects in ion signaling and fluxes rather than on clinical manifestations.

Alterations in $I_{Na}$ and $I_{NaL}$ currents can be associated with acquired dysrhythmias. In chronic tachydysrhythmia, reduced $Na_V1.5$ channel expression is responsible for $I_{Na}$ reduction [139]. In heart failure, $I_{Na}$ current is also reduced, whereas $I_{NaL}$ is enhanced. Reduction of $Na_V1.5$ expression underlies decrease in $I_{Na}$, while phosphorylation of $Na^+$ channels induced by intracellular $Ca^{2+}$ rise leads to increased $I_{NaL}$ [139]. In myocardial infarction, the expression of $Na^+$ channel is reduced and its gating properties are altered in myocytes in the surviving zone of the infracted area. Finally, during myocardial ischemia $I_{NaL}$ current is enhanced suggesting the use of $I_{NaL}$ blockers as a therapy for chronic angina [140]

$I_{to}$ current through the voltage-gated $K^+$ channels is responsible for early repolarization (phase 1) of the cardiac AP. Reduction of $I_{to}$ has been identified in AF, myocardial infarction, and heart failure. In myocardial infarction and probably heart failure, this reduction results from activation of calcineurin, a phosphatase, regulating gene transcription by dephosphorylation of transcription factors [139]. On the other hand, increased $I_{to}$ current has been reported in the hypertrophic phase preceding heart failure. Finally, $I_{to}$ current may be reduced leading to prolongation in QT interval in diabetes, and insulin therapy can induce $K_V4.3$ expression and partially restore $I_{to}$ [141].

$K_V1.5$ is expressed predominantly in the atria is responsible for ultrarapidly activating delayed outward rectifying $I_{Kur}$ current in atrial myocytes. $K_V1.5$ expression is downregulated in the epicardial zone of the infarcted hearts and in AF [139]. Additionally, improper localization of $K_V1.5$ has been reported in the intercalated disks upon myocardial ischemia [141].

Expression of $K_V11.1$ underlying rapidly activating delayed outward rectifying $I_{Kr}$ current is reduced in myocardial infarction but not in AF or heart failure. This causes prolongation in the cardiac AP duration. Interestingly, similar $K_V11.1$ downregulation has been identified in diabetes [142]. On the other hand, upon acute ischemia, $I_{Kr}$ is enhanced resulting in shortening AP duration.

$K_V7.1$ and minK encoded by *KCNQ1* and *KCNE1*, respectively, form $K^+$ channel responsible for slowly activating delayed outward rectifying $I_{Ks}$ current. Reduced $I_{Ks}$ has been demonstrated in atrial, ventricular, and sinoatrial node myocytes during heart failure. Since $I_{Kr}$ has not changed, $I_{Ks}$ downregulation seems to be mainly responsible for prolongation of AP duration under this condition. Moreover, both $K_V7.1$ and minK protein levels are decreased in infarcted border zones during first 2 days postinfarction [139, 141].

$K_{ir}2.1$ channel encoded by *KCNJ2* is responsible for inward rectifying $I_{K1}$ current that stabilizes the resting AP in atrial and ventricular myocytes during phase 4 and contributes to the terminal part of phase 3. In chronic AF, $K_{ir}2.1$ expression is upregulated and $I_{K1}$ is increased accordingly leading to more negative resting potentials and, together with reduced $I_{Ca,L}$, to AP shortening [139]. In animal models for ventricular failure and heart postinfarction, reduction in $I_{K1}$ has been found [141]. Furthermore, acidosis and accumulation of intracellular $Ca^{2+}$, $Mg^{2+}$, and $Na^+$ observed during ischemia may contribute to the inhibition of $I_{K1}$ [143]. It has been suggested that $I_{K1}$ reduction induces spontaneous excitability and eventually triggers dysrhythmia during heart failure or ischemia.

The voltage dependent $Ca_V1.2$ channel is responsible for the L-type inward $I_{Ca,L}$ current contributing mainly to the plateau phase of the cardiac AP. This $Ca^{2+}$ influx induces $Ca^{2+}$ release through ryanodine receptor, RyR2, expressed in the sarcoplasmic reticulum membrane. Defects in $Ca^{2+}$ signaling may induce dysrhythmia and contractile dysfunction.

Reduced $Ca_V1.2$ expression leading to $I_{Ca,L}$ reduction and AP shortening has been found in patients with AF [144]. Interestingly, in heart failure, the expression of $Ca_V1.2$ has also been downregulated, however, $I_{Ca,L}$ has been unchanged because of increased phosphorylation and consequently activation of the channel [139]. $I_{Ca,L}$ current is decreased in the border zone of the infarcted area and during acute ischemia [139, 143]. In the latter case, intracellular accumulation of $Ca^{2+}$ and $Mg^{2+}$ and extracellular acidosis is responsible for this reduction.

The pacemaker or "funny" $I_f$ current enables spontaneous initiation of cardiac electrical activity. HCN4 is a major HCN channel isoform predominantly expressed in sinoatrial node and atrioventricular node myocytes and Purkinje fibers is responsible for $I_f$. Upregulation of HCN4 expression and consequently increased $I_f$ can induce spontaneous excitation of nonpacemaker myocytes resulting in dysrhythmia. Indeed, increased levels of HCN4 and higher $I_f$ amplitude have been identified in atrial and ventricular myocytes of patients with AF and heart failure [139].

During the last decade, lethal ventricular dysrhythmias have been found to be associated with the use of various drugs, most of which have subsequently been withdrawn from the market. Actually, one half of withdrawn drugs mandated by the FDA since 1998 have been attributable to cardiac side effects, mostly dysrhythmias [145]. The implicated drugs include not only antidysrhythmics, but also a number of noncardiac drugs, such as antihistaminics and antipsychotics.

In addition to the type linked to the congenital form of LQTS, acquired *torsades de pointes* (TdP) ventricular tachycardia may be drug induced. Often an aberrantly prolonged AP may lead to TdP [146]. On ECG, TdP presents with short, self-limited bouts of rapid ventricular rhythm together with a changing morphology of the QRS complexes. In the absence of tachycardia, the QT is prolonged (500 ms or more) with the first complex of each run appearing relatively late. The majority of drugs with the potential to induce TdP inhibit the cardiac $K^+$ channel, $K_V11.1$/hERG, leading to delayed repolarization as indicated by LQT. As we previously discussed, rare mutations in *KCNH2* gene encoding $K_V11.1$ have been associated with LQT2, placing this cardiac $K^+$ channel in the intersection between congenital LQTS (cLQTS) and acquired forms of LQTS (aLQTS), i.e., drug-induced TdP [147].

Polymorphisms in the genes associated with LQTS may also contribute to the dysrhythmogenic susceptibility and increase the risk of aLQTS. Analysis of the *KCNE2* gene, encoding the minK-related peptide 1 (MiRP1), a β-subunit of the cardiac $K_V11.1$ channel in patients with aLQTS has been reported [148]. Three individuals with sporadic mutations, and a patient with sulfamethoxazole-associated LQTS, who carried an SNP (found in approximately 1.6% of the general population), have been identified. $K_V11.1$ channel encoded by the mutant gene has displayed decreased $I_{Kr}$ current at baseline and wild-type drug sensitivity, whereas the channel encoded by the SNP-containing allele has carried normal $I_{Kr}$ at baseline but has been inhibited by therapeutic doses of sulfamethoxazole that have not affected the wild-type channel. Importantly, common sequence variation that can provoke life-threatening drug reactions has been clinically silent before drug exposure.

Mutation analysis of five cLQTS genes (*SCN5A, KCNH2, KCNQ1, KCNE1, KCNE2)* in patients with drug-related LQTS with confirmed TdP has suggested that mutations in these genes can explain a minority of aLQTS cases [149]. Moreover, there are a number of nongenetic factors, which may contribute to increased susceptibility to aLQTS/TdP, including female gender, hypokalemia, and other heart diseases. Individual differences in drug metabolism, caused by mutations or functional SNPs in drug-metabolizing enzyme genes, may be an additional significant risk factor for aLQTS, especially if multiple drugs are involved [150].

## Dysrhythmias Associated with Defects in FAO and Mitochondrial Function

Fatty acids determine the fluidity and stability of cellular membranes and have significant impact on cardiac membrane functions, such as transport of ions and substrates, and therefore on the cardiac electrophysiology. Certain defects in FAO have been reported to be associated with cardiac conduction defects and dysrhythmias [151]. Most of these defects have led to the accumulation of long-chain acylcarnitines, an intermediate of fatty acids metabolism. Amphiphilic long-chain acylcarnitines displaying detergent-like properties can significantly modify the electrophysiological function of cardiac membranes, including $Na^+$, $Ca^{2+}$ transport, and impaired gap junction activity. Accumulation of acylcarnitines has been implicated in the genesis of ventricular rhythm disorders during myocardial ischemia [152]. Moreover, patients with cardiomyopathy caused by inborn defects in carnitine translocase often have an increased onset of dysrhythmias [153]. Accordingly, the selective inhibition of carnitine palmitoyltransferase-I prevents the accumulation of potentially toxic acylcarnitines and reduces the risk of electrophysiologic disturbance during hypoxia/ischemia [154].

Research to further define the site of action of these toxic long-chain intermediates within the cardiomyocytes, e.g., at the level of the plasma membrane, mitochondrial matrix, or membrane(s) may be of use in the development of new therapies.

Mitochondria traditionally defined as the cellular powerhouse play a central role providing the energy that the heart needs for pumping blood to oxygenate the tissues of the organism. However, mitochondria perform other important noncanonical physiologic functions, including regulation of ion signaling. Several ion channels have recently been found in the mitochondrial membrane, including $Na^+$ channels, $Na^+/Ca^{2+}$, $Na^+/Mg^{2+}$, and $Na^+/H^+$ exchangers, the voltage-gated $K^+$ channel (mito$K_V$1.3), $Ca^{2+}$-activated $K^+$ channel, the ATP-activated $K^+$ channel (mito$K_{ATP}$), the twin-pore domain TASK-3 channel, and the voltage-dependent anion channel (VDAC) [155, 156]. These channels not only control membrane potential, but also regulate the generation of reactive oxygen species (ROS), mitochondrial respiration, and mitochondrial matrix volume. Unfortunately, the molecular identity of mitochondrial ion channels remains largely unknown.

Most current studies suggest an elevation in intracellular $Na^+$ concentration ($[Na^+]_i$) during cardiac hypertrophy and heart failure. This rise in $[Na^+]_i$ seems to result from increased $Na^+$ influx across cell membrane, but the mitochondrial $Na^+$ channels do not contribute to this process. However, the increased $[Na^+]_i$ during heart failure has been suggested to reduce mitochondrial $[Ca^{2+}]$ that could have significant consequences for mitochondrial metabolism [155]. At present, it is not clear whether $Na^+$ channels or $Na^+/H^+$ exchangers are mainly responsible for the rise in $[Na^+]_i$ during hypertrophy and ischemia. Furthermore, it has been reported that heart failure and hypertrophy are associated with a decrease of the Na/K ATPase expression while expression of $Na^+/Ca^{2+}$ exchangers is increased [157–160].

Ischemia results in immediate alteration of mitochondrial function, including the failure of ATP synthesis and a drop in mitochondrial inner membrane potential ($\Delta\Psi$) accompanying by increase in cytosolic $Ca^{2+}$, phosphate, fatty acids, and ROS originated from the respiratory chain [161]. Opening of the mito$K_{ATP}$ channel stimulates mitochondrial ATP production and energy transfer from mitochondria to the cytosol during high workload conditions; it can also modulate ROS levels [162]. The mito$K_{ATP}$ opening also blocks the onset of the mitochondrial permeability transition, a main cause of necrosis in ischemia–reperfusion injury [163, 164]. Moreover, hypoxia has been reported to stimulate $Ca^{2+}$-activated $K^+$ channel in the inner mitochondrial membrane [165, 166].

The rescue of $\Delta\Psi$ is a key determinant of postischemic functional recovery of the heart. Mitochondrial ROS release causes the collapse of $\Delta\Psi$ and the instability of the AP largely mediated by a mechanism involving a mitochondrial anion channel. It has recently been suggested a new mechanism for postischemic dysrhythmias explaining how mitochondrial dysfunction in the reperfused heart leads to a local fall of electrical currents preventing propagation [167]. These findings also suggest that by stabilizing the mitochondrial membrane potential, the cellular AP can be maintained, blunting the AP shortening during ischemia, and preventing ventricular dysrhythmias during reperfusion.

## Conclusions and Future Directions

In the past two decades, remarkable advancement has been achieved in the understanding of the molecular mechanisms of a number of life-threatening dysrhythmias. Numerous mutations and SNPs in the genes encoding the proteins forming various cardiac ion channels and transporters, associated with inherited or familial forms of dysrhythmias have been identified. The discovery that inherited and acquired electrophysiological heart disorders can be caused by alterations in generic material has opened a new era in cardiology. For example, genotyping single SNPs, using haplotype block analysis by the human genome consortium, has revealed that by the application of modern informatic tools to this vast collection of data we could select from the approximately 10 million polymorphisms present in the human genome, in particular those nonsilent SNPs that actually may have a functional role (by producing sequence changes in specific cardiac proteins) in many dysrhythmogenic syndromes, and other cardiovascular pathologies.

The unexpected complexity of the interplay among multiple proteins and mechanisms leads to appreciation that similar clinical manifestations often result from different genetic defects. For example, various mutations in different genes cause the same LQTS, which account for the 50% of clinically diagnosed cases, while over 30% of patients with clinically indistinguishable symptoms do not harbor mutations in corresponding genes. High heterogeneity and complexity is also present in SQTS, BrS, CPVT, and AF. The failure to identify gene mutations in most patients with dysrhythmias suggests that either unknown mutations or post-translational modulations of the cardiac ion channels may also contribute to the observed alterations in ion currents and clinical symptoms. Moreover, there are significant variations in penetrance and expression in patients with various dysrhythmias ranging from asymptomatic carriers to patients with atrial or ventricular fibrillation or SCD.

Careful screening of several families with dysrhythmia history has identified the high variability of electrophysiological markers. It is apparent that individuals from the same kindred harboring the same disease-associated gene mutation or SNPs may display markedly different phenotypes. Moreover, SNPs are being continually identified in a number of genes, which regulate cardiac excitability. These findings, combined with the increasing recognition that genetic factors can contribute to many common dysrhythmias increase the prospect that screening of the genomic variability in populations may in the future be used to manage patients [168]. Given the fact that most of dysrhythmias may not be fully explained by a single mechanism and many affected individuals are asymptomatic, thorough genetic screening combined with genotype–phenotype correlation analysis of these disorders is of paramount significance. Such high-density genome-wide analysis represents the basis for the development of novel strategies to identify contributors to susceptibility in common cardiac pathologies and most likely lead to discover new and relevant therapeutic targets.

Despite inherited dysrhythmias being rare disorders, insights derived from the studies of the mutated ion channels have tremendously advanced our understanding of heart physiology. Animal models with specific gene modifications has been instrumental in defining precise roles, which cardiac ion channel proteins play under the normal and pathological conditions. The list of these transgenic models is continuously growing providing invaluable information in searching novel therapeutic targets in the cardiac signaling pathways.

Notwithstanding the recent achievements in genetic analysis of dysrhythmogenic syndromes, we must also take into account numerous environmental and epigenetic modulator factors, such as gender, ethnicity, geographical location, age, family history, repolarization, and sympathetic tone abnormalities, influencing both clinical manifestations and disease severity. Moreover, we have to carefully monitor not only specific cardiac markers, but also the individual pharmacological responses and electrophysiological manifestations. Individualizing therapy may be particularly critical in establishing drug dosages and efficacy in children and aging patients with dysrhythmias and structural heart defects, a population for which pharmacokinetics has proven to be poorly defined and often unpredictable.

At present, many fundamental questions still remain to be answered regarding the molecular mechanisms underlying dysrhythmias as well as other cardiac pathologies. To address these questions, emerging novel technologies have been exploited. These approaches include the integrated use of functional genomics and proteomics, bioinformatics, and computational modeling combined with gene and cell transfer therapies and antisense strategies (e.g., ribozymes, antisense oligonucleotides, or RNA interference). The high potential of gene and cell therapies has recently been demonstrated in a number of animal models and in limited clinical studies [169–173]. It is evident that we are still in the very beginning of a new exciting era in cardiology and further intensive studies to establish efficient and safe procedures for widespread clinical application of these novel strategies are required.

## Summary

- Cardiac dysrhythmias remain an important cause of morbidity and mortality around the world. In the USA alone, over 300,000 cases of sudden death occur each year due to ventricular dysrhythmias. Although molecular genetic analysis had shown an unforeseen genetic diversity of the inherited dysrhythmias, making its genetic differentiation rather complex, the evolving technologies appear to offer the promise of fast progress in this important field of cardiac pathology.
- Inherited dysrhythmias are primary electrophysiological disorders not associated with structural heart pathology, and include LQTS, SQTS, CPVT, and BrS. The majority of the inherited dysrhythmias (i.e., channelopathies) with a known genetic basis are caused by mutations in genes encoding cardiac ion channels subunits and/or the auxiliary associated proteins essential for proper channel functioning.
- LQTS is characterized by abnormally prolonged ventricular repolarization, which can result in ventricular tachycardia known also as TdP, ventricular fibrillation, and sudden death. This dysrhythmia may be self limited, but it may also be followed by syncope and sudden death.
- Mutations in genes encoding the cardiac voltage-gated $K^+$, $Na^+$, L-type $Ca^{2+}$ channels, and associated proteins are responsible for different types of LQTS (LQTS1–10). However, extensive phenotypic heterogeneity, even within mutation carriers in the same family, suggests the importance of modifying factors and genes that are yet not known.
- SQTS is a rare autosomal dominant channelopathy characterized by an abnormally short QT interval with tall, peaked, narrow-based T waves. Patients with SQTS (typically infants, children, and young adults) display a marked propensity for AF and increased risk for SCD from ventricular tachydysrhythmias.
- The common characteristic in most forms of SQTS (SQT1-3) is alteration in $K^+$ ion signaling leading to acceleration of repolarization of the AP caused by mutations in the cardiac $K^+$ channels. The overlapping subtypes (SQT4 and SQT5) are caused by mutations in the $Ca^{2+}$ channel genes. Patients with SQT4 and SQT5 display in addition to a short QT interval ST segment elevation in the right precordial leads typical of the BrS.
- BrS is characterized by ST-segment elevation and negative T waves in the right precordial leads (V1–V3) and an increase in SCD resulting from polymorphic ventricular tachycardias. Several mutations in the genes encoding $Na^+$, L-type $Ca^{2+}$, and transient outward $K^+$ channels as well as GPD1-L protein have been identified in patients with different types of BrS (BS1-6). BrS may not be fully explained by a single mechanism and many patients with BrS are asymptomatic.
- CPVT is an inherited disorder characterized by adrenergically mediated ventricular dysrhythmias leading to syncope, cardiac arrest, and SCD. Mutations in two genes encoding two important components of the sarcoplasmic

reticulum essential for excitation–contraction coupling, the *RYR2* and *CASQ2*, have been found to be associated with subtypes of CPVT 1 and 2, respectively.

- AF is the most common dysrhythmia seen in clinical practice affecting approximately 2.3 million patients in the USA. Although until very recently, a genetic basis for this common dysrhythmia has been considered unlikely, the recent analysis of familial AF forms has led to the identification of several *loci* and genes responsible for the disease.
- The familial forms of AF are rare; the majority of the AF cases are acquired and associated with cardiac structural pathologies leading to heart failure.
- CCD is characterized by progressive defects at the atrial, atrioventricular, and/or ventricular level and is frequently linked to mutations in *SCN5A* gene encoding the $Na_V1.5$ channel. Lev–Lenegre disease is one of the most common forms of CCD.
- SSS is characterized by sinus bradycardia, sinus arrest, atrial standstill and tachycardia–bradycardia syndrome. Presently, at least 14 *SCN5A* mutations have been reported to be associated with SSS.
- Mutations in the *HCN4* gene encoding HCN4, a major HCN channel isoform expressed in SA node cells, have recently been identified to be associated with sinus bradycardia and complex cardiac dysrhythmias.
- SIDS is the result of more than one disease associated with mutations in several "SIDS-related genes" predisposing the infant to sudden death, in combination with a number of environmental risk factors. However, some rare cases of SIDS have been reported to be associated with mutations and polymorphisms in the genes encoding cardiac $K^+$ channels and the $Na_V1.5$ channel.
- WPW is complex disorder characterized by an abnormal short P-R interval with prolonged QRS duration on their ECG. Each case of WPW is highly individual and displays a variety of manifestations. Mutations in the *PRKAG2* gene encoding the γ2 regulatory subunit of AMPK have been identified in several different families with WPW.
- Unlike inherited dysrhythmias linked to mutations in genes encoding cardiac ion channels or their auxiliary proteins, acquired or secondary dysrhythmias are caused by changes in expression of ion channels or in their gating characteristics. Such changes have been reported for the majority if not all cardiac ion channels.
- During the last decade, lethal ventricular dysrhythmias have been found to be associated with the use of various drugs, most of which have subsequently been withdrawn from the market. The implicated drugs include not only antidysrhythmics, but also a number of noncardiac drugs, such as antihistaminics and antipsychotics.
- In addition to the inherited LQTS, acquired TdP ventricular tachycardia may be drug induced. Often an aberrantly prolonged AP may lead to TdP. The majority of drugs with the potential to induce TdP inhibit the cardiac $K^+$ channel, $K_V11.1$/hERG, leading to delayed repolarization.
- Besides mutations in LQTS genes, polymorphisms in the genes associated with LQTS may also increase the risk of acquired LQTS. Moreover, there are a number of non-genetic factors, which may contribute to increased susceptibility to acquired LQTS/TdP, including female gender, hypokalemia, and other heart diseases.
- Defects in the mitochondrial ion currents can lead to differential electrical excitability in hearts subjected to ischemia and reperfusion. This may be a new mechanism that could explain postischemic dysrhythmias, in which mitochondrial dysfunction in the reperfused heart leads to a local fall of electrical currents preventing propagation.
- The rescue of mitochondrial inner membrane potential ($\Delta\Psi$) is a key determinant of postischemic functional recovery of the heart.
- Since most of dysrhythmias may not be fully explained by a single mechanism and many affected individuals are asymptomatic, thorough genetic screening combined with genotype–phenotype correlation analysis of these disorders is of paramount significance. This approach represents the basis for the development of novel strategies to identify contributors to susceptibility in common cardiac pathologies and most likely lead to discovery new therapeutic targets.
- Numerous environmental and epigenetic modulator factors (e.g., gender, ethnicity, geographical location, age, family history, repolarization, and sympathetic tone abnormalities) influencing both clinical manifestations and disease severity have to be considered. Moreover, not only specific cardiac markers, but also the individual pharmacological responses and electrophysiological manifestations have to be carefully monitored.
- The constantly growing list of animal models with specific gene modifications has been instrumental in defining the precise roles, which cardiac ion channel proteins play under normal and pathological conditions.
- Transgenic models also provide invaluable information in searching novel therapeutic targets in the cardiac signaling pathways.
- Emerging new technologies, including the integrated use of functional genomics and proteomics, bioinformatics, and computational modeling combined with gene and cell transfer therapies and antisense strategies (e.g., ribozymes, antisense oligonucleotides, or RNA interference) allow the development of highly efficient and individualized therapy of cardiac disease.

## References


1. Schwartz PJ, Moss AJ, Vincent GM, Crampton RS. Diagnostic criteria for the long QT syndrome. An update. Circulation. 1993;88: 782–4.
2. Curran ME, Splawski I, Timothy KW, Vincent GM, Green ED, Keating MT. A molecular basis for cardiac arrhythmia: HERG mutations cause long QT syndrome. Cell. 1995;80:795–803.
3. Wang Q, Shen J, Splawski I, et al. SCN5A mutations associated with an inherited cardiac arrhythmia, long QT syndrome. Cell. 1995;80:805–11.
4. Bennett PB, Yazawa K, Makita N, George Jr AL. Molecular mechanism for an inherited cardiac arrhythmia. Nature. 1995;376:683–5.
5. Wang Q, Curran ME, Splawski I, et al. Positional cloning of a novel potassium channel gene: KVLQT1 mutations cause cardiac arrhythmias. Nat Genet. 1996;12:17–23.
6. Keating MT, Sanguinetti MC. Molecular and cellular mechanisms of cardiac arrhythmias. Cell. 2001;104:569–80.
7. Mohler PJ, Schott JJ, Gramolini AO, et al. Ankyrin-B mutation causes type 4 long-QT cardiac arrhythmia and sudden cardiac death. Nature. 2003;421:634–9.
8. Splawski I, Tristani-Firouzi M, Lehmann MH, Sanguinetti MC, Keating MT. Mutations in the hminK gene cause long QT syndrome and suppress IKs function. Nat Genet. 1997;17:338–40.
9. Abbott GW, Sesti F, Splawski I, et al. MiRP1 forms IKr potassium channels with HERG and is associated with cardiac arrhythmia. Cell. 1999;97:175–87.
10. Splawski I, Shen J, Timothy KW, et al. Spectrum of mutations in long-QT syndrome genes. KVLQT1, HERG, SCN5A, KCNE1, and KCNE2. Circulation. 2000;102:1178–85.
11. Napolitano C, Priori SG, Schwartz PJ, et al. Genetic testing in the long QT syndrome: development and validation of an efficient approach to genotyping in clinical practice. JAMA. 2005;294: 2975–80.
12. Priori SG, Napolitano C, Schwartz PJ, et al. Association of long QT syndrome loci and cardiac events among patients treated with beta-blockers. JAMA. 2004;292:1341–4.
13. Tristani-Firouzi M, Jensen JL, Donaldson MR, et al. Functional and clinical characterization of KCNJ2 mutations associated with LQT7 (Andersen syndrome). J Clin Invest. 2002;110:381–8.
14. Splawski I, Timothy KW, Decher N, et al. Severe arrhythmia disorder caused by cardiac L-type calcium channel mutations. Proc Natl Acad Sci USA. 2005;102:8089–96.
15. Splawski I, Timothy KW, Sharpe LM, et al. Ca(V)1.2 calcium channel dysfunction causes a multisystem disorder including arrhythmia and autism. Cell. 2004;119:19–31.
16. Andersen ED, Krasilnikoff PA, Overvad H. Intermittent muscular weakness, extrasystoles, and multiple developmental anomalies. A new syndrome? Acta Paediatr Scand. 1971;60:559–64.
17. Bendahhou S, Fournier E, Sternberg D, et al. In vivo and in vitro functional characterization of Andersen's syndrome mutations. J Physiol. 2005;565:731–41.
18. Donaldson MR, Yoon G, Fu YH, Ptacek LJ. Andersen-Tawil syndrome: a model of clinical variability, pleiotropy, and genetic heterogeneity. Ann Med. 2004;36 Suppl 1:92–7.
19. Leonoudakis D, Conti LR, Radeke CM, McGuire LM, Vandenberg CA. A multiprotein trafficking complex composed of SAP97, CASK, Veli, and Mint1 is associated with inward rectifier Kir2 potassium channels. J Biol Chem. 2004;279:19051–63.
20. Leonoudakis D, Conti LR, Anderson S, et al. Protein trafficking and anchoring complexes revealed by proteomic analysis of inward rectifier potassium channel (Kir2.x)-associated proteins. J Biol Chem. 2004;279:22331–46.
21. Laverty HG, Wilson JB. Murine CASK is disrupted in a sex-linked cleft palate mouse mutant. Genomics. 1998;53:29–41.
22. Zaritsky JJ, Eckman DM, Wellman GC, Nelson MT, Schwarz TL. Targeted disruption of Kir2.1 and Kir2.2 genes reveals the essential role of the inwardly rectifying K(+) current in K(+)-mediated vasodilation. Circ Res. 2000;87:160–6.
23. Caruana G, Bernstein A. Craniofacial dysmorphogenesis including cleft palate in mice with an insertional mutation in the discs large gene. Mol Cell Biol. 2001;21:1475–83.
24. Mohler PJ, Bennett V. Ankyrin-based cardiac arrhythmias: a new class of channelopathies due to loss of cellular targeting. Curr Opin Cardiol. 2005;20:189–93.
25. Mohler PJ, Davis JQ, Bennett V. Ankyrin-B coordinates the Na/K ATPase, Na/Ca exchanger, and InsP3 receptor in a cardiac T-tubule/ SR microdomain. PLoS Biol. 2005;3:e423.
26. Kline CF, Kurata HT, Hund TJ, et al. Dual role of K ATP channel C-terminal motif in membrane targeting and metabolic regulation. Proc Natl Acad Sci USA. 2009;106:16669–74.
27. Hund TJ, Wright PJ, Dun W, Snyder JS, Boyden PA, Mohler PJ. Regulation of the ankyrin-B-based targeting pathway following myocardial infarction. Cardiovasc Res. 2009;81:742–9.
28. Vatta M, Ackerman MJ, Ye B, et al. Mutant caveolin-3 induces persistent late sodium current and is associated with long-QT syndrome. Circulation. 2006;114:2104–12.
29. Cronk LB, Ye B, Kaku T, et al. Novel mechanism for sudden infant death syndrome: persistent late sodium current secondary to mutations in caveolin-3. Heart Rhythm. 2007;4:161–6.
30. Martinez-Marmol R, Villalonga N, Sole L, et al. Multiple Kv1.5 targeting to membrane surface microdomains. J Cell Physiol. 2008;217:667–73.
31. Ye B, Balijepalli RC, Foell JD, et al. Caveolin-3 associates with and affects the function of hyperpolarization-activated cyclic nucleotide-gated channel 4. Biochemistry. 2008;47:12312–8.
32. Palygin OA, Pettus JM, Shibata EF. Regulation of caveolar cardiac sodium current by a single Gsalpha histidine residue. Am J Physiol Heart Circ Physiol. 2008;294:H1693–9.
33. Meadows LS, Isom LL. Sodium channels as macromolecular complexes: implications for inherited arrhythmia syndromes. Cardiovasc Res. 2005;67:448–58.
34. Yu FH, Westenbroek RE, Silos-Santiago I, et al. Sodium channel beta4, a new disulfide-linked auxiliary subunit with similarity to beta2. J Neurosci. 2003;23:7577–85.
35. Medeiros-Domingo A, Kaku T, Tester DJ, et al. SCN4B-encoded sodium channel beta4 subunit in congenital long-QT syndrome. Circulation. 2007;116:134–42.
36. Chen L, Marquardt ML, Tester DJ, Sampson KJ, Ackerman MJ, Kass RS. Mutation of an A-kinase-anchoring protein causes long-QT syndrome. Proc Natl Acad Sci USA. 2007;104: 20990–5.
37. Kurokawa J, Motoike HK, Rao J, Kass RS. Regulatory actions of the A-kinase anchoring protein Yotiao on a heart potassium channel downstream of PKA phosphorylation. Proc Natl Acad Sci USA. 2004;101:16374–8.
38. Ueda K, Valdivia C, Medeiros-Domingo A, et al. Syntrophin mutation associated with long QT syndrome through activation of the nNOS-SCN5A macromolecular complex. Proc Natl Acad Sci USA. 2008;105:9355–60.
39. Miyagoe-Suzuki Y, Takeda SI. Association of neuronal nitric oxide synthase (nNOS) with alpha1-syntrophin at the sarcolemma. Microsc Res Tech. 2001;55:164–70.
40. Gavillet B, Rougier JS, Domenighetti AA, et al. Cardiac sodium channel Nav1.5 is regulated by a multiprotein complex composed of syntrophins and dystrophin. Circ Res. 2006;99:407–14.
41. Algra A, Tijssen JG, Roelandt JR, Pool J, Lubsen J. QT interval variables from 24 hour electrocardiography and the two year risk of sudden death. Br Heart J. 1993;70:43–8.
42. Gussak I, Brugada P, Brugada J, et al. Idiopathic short QT interval: a new clinical syndrome? Cardiology. 2000;94:99–102.

43. Gaita F, Giustetto C, Bianchi F, et al. Short QT Syndrome: a familial cause of sudden death. Circulation. 2003;108:965–70.
44. Extramiana F, Antzelevitch C. Amplified transmural dispersion of repolarization as the basis for arrhythmogenesis in a canine ventricular-wedge model of short-QT syndrome. Circulation. 2004;110: 3661–6.
45. Schimpf R, Wolpert C, Gaita F, Giustetto C, Borggrefe M. Short QT syndrome. Cardiovasc Res. 2005;67:357–66.
46. Brugada R, Hong K, Dumaine R, et al. Sudden death associated with short-QT syndrome linked to mutations in HERG. Circulation. 2004;109:30–5.
47. Bellocq C, van Ginneken AC, Bezzina CR, et al. Mutation in the KCNQ1 gene leading to the short QT-interval syndrome. Circulation. 2004;109:2394–7.
48. Priori SG, Pandit SV, Rivolta I, et al. A novel form of short QT syndrome (SQT3) is caused by a mutation in the KCNJ2 gene. Circ Res. 2005;96:800–7.
49. Antzelevitch C, Pollevick GD, Cordeiro JM, et al. Loss-of-function mutations in the cardiac calcium channel underlie a new clinical entity characterized by ST-segment elevation, short QT intervals, and sudden cardiac death. Circulation. 2007;115: 442–9.
50. Anttonen O, Junttila MJ, Rissanen H, Reunanen A, Viitasalo M, Huikuri HV. Prevalence and prognostic significance of short QT interval in a middle-aged Finnish population. Circulation. 2007;116:714–20.
51. Giustetto C, Di Monte F, Wolpert C, et al. Short QT syndrome: clinical findings and diagnostic-therapeutic implications. Eur Heart J. 2006;27:2440–7.
52. Schimpf R, Wolpert C, Bianchi F, et al. Congenital short QT syndrome and implantable cardioverter defibrillator treatment: inherent risk for inappropriate shock delivery. J Cardiovasc Electrophysiol. 2003;14:1273–7.
53. Patel U, Pavri BB. Short QT syndrome: a review. Cardiol Rev. 2009;17:300–3.
54. Brugada P, Brugada J. Right bundle branch block, persistent ST segment elevation and sudden cardiac death: a distinct clinical and electrocardiographic syndrome. A multicenter report. J Am Coll Cardiol. 1992;20:1391–6.
55. Antzelevitch C, Brugada P, Borggrefe M, et al. Brugada syndrome: report of the second consensus conference. Heart Rhythm. 2005;2:429–40.
56. Antzelevitch C, Brugada P, Brugada J, et al. Brugada syndrome: a decade of progress. Circ Res. 2002;91:1114–8.
57. Antzelevitch C, Brugada P, Borggrefe M, et al. Brugada syndrome: report of the second consensus conference: endorsed by the Heart Rhythm Society and the European Heart Rhythm Association. Circulation. 2005;111:659–70.
58. Sah R, Ramirez RJ, Oudit GY, et al. Regulation of cardiac excitation-contraction coupling by action potential repolarization: role of the transient outward potassium current (I(to)). J Physiol. 2003;546:5–18.
59. Calloe K, Cordeiro JM, Di Diego JM, et al. A transient outward potassium current activator recapitulates the electrocardiographic manifestations of Brugada syndrome. Cardiovasc Res. 2009;81: 686–94.
60. Fish JM, Antzelevitch C. Role of sodium and calcium channel block in unmasking the Brugada syndrome. Heart Rhythm. 2004;1: 210–7.
61. Morita H, Zipes DP, Morita ST, Wu J. Differences in arrhythmogenicity between the canine right ventricular outflow tract and anteroinferior right ventricle in a model of Brugada syndrome. Heart Rhythm. 2007;4:66–74.
62. Shimizu W, Matsuo K, Kokubo Y, et al. Sex hormone and gender difference – role of testosterone on male predominance in Brugada syndrome. J Cardiovasc Electrophysiol. 2007;18:415–21.
63. Morita H, Zipes DP, Wu J. Brugada syndrome: insights of ST elevation, arrhythmogenicity, and risk stratification from experimental observations. Heart Rhythm. 2009;6:S34–43.
64. Chen Q, Kirsch GE, Zhang D, et al. Genetic basis and molecular mechanism for idiopathic ventricular fibrillation. Nature. 1998;392:293–6.
65. Priori SG, Napolitano C, Gasparini M, et al. Natural history of Brugada syndrome: insights for risk stratification and management. Circulation. 2002;105:1342–7.
66. Tan HL, Bezzina CR, Smits JP, Verkerk AO, Wilde AA. Genetic control of sodium channel function. Cardiovasc Res. 2003;57: 961–73.
67. Abriel H. Roles and regulation of the cardiac sodium channel Na v 1.5: recent insights from experimental studies. Cardiovasc Res. 2007;76:381–9.
68. Watanabe H, Koopmann TT, Le Scouarnec S, et al. Sodium channel beta1 subunit mutations associated with Brugada syndrome and cardiac conduction disease in humans. J Clin Invest. 2008;118: 2260–8.
69. London B, Michalec M, Mehdi H, et al. Mutation in glycerol-3-phosphate dehydrogenase 1 like gene (GPD1-L) decreases cardiac Na+ current and causes inherited arrhythmias. Circulation. 2007;116:2260–8.
70. Mohler PJ, Le Scouarnec S, Denjoy I, et al. Defining the cellular phenotype of "ankyrin-B syndrome" variants: human ANK2 variants associated with clinical phenotypes display a spectrum of activities in cardiomyocytes. Circulation. 2007;115:432–41.
71. Verkerk AO, Wilders R, Schulze-Bahr E, et al. Role of sequence variations in the human ether-a-go-go-related gene (HERG, KCNH2) in the Brugada syndrome. Cardiovasc Res. 2005;68: 441–53.
72. Priori SG, Napolitano C. Genetics of cardiac arrhythmias and sudden cardiac death. Ann N Y Acad Sci. 2004;1015:96–110.
73. Mohler PJ, Rivolta I, Napolitano C, et al. Nav1.5 E1053K mutation causing Brugada syndrome blocks binding to ankyrin-G and expression of Nav1.5 on the surface of cardiomyocytes. Proc Natl Acad Sci USA. 2004;101:17533–8.
74. Delpon E, Cordeiro JM, Nunez L, et al. Functional effects of KCNE3 mutation and its role in the development of Brugada syndrome. Circ Arrhythm Electrophysiol. 2008;1:209–18.
75. Yokokawa M, Noda T, Okamura H, et al. Comparison of long-term follow-up of electrocardiographic features in Brugada syndrome between the SCN5A-positive probands and the SCN5A-negative probands. Am J Cardiol. 2007;100:649–55.
76. Smits JP, Eckardt L, Probst V, et al. Genotype-phenotype relationship in Brugada syndrome: electrocardiographic features differentiate SCN5A-related patients from non-SCN5A-related patients. J Am Coll Cardiol. 2002;40:350–6.
77. Morita H, Kusano KF, Miura D, et al. Fragmented QRS as a marker of conduction abnormality and a predictor of prognosis of Brugada syndrome. Circulation. 2008;118:1697–704.
78. Morita H, Nagase S, Miura D, et al. Differential effects of cardiac sodium channel mutations on initiation of ventricular arrhythmias in patients with Brugada syndrome. Heart Rhythm. 2009;6:487–92.
79. Frustaci A, Priori SG, Pieroni M, et al. Cardiac histological substrate in patients with clinical phenotype of Brugada syndrome. Circulation. 2005;112:3680–7.
80. Saffitz JE. Structural heart disease, SCN5A gene mutations, and Brugada syndrome: a complex menage a trois. Circulation. 2005; 112:3672–4.
81. Leenhardt A, Lucet V, Denjoy I, Grau F, Ngoc DD, Coumel P. Catecholaminergic polymorphic ventricular tachycardia in children. A 7-year follow-up of 21 patients. Circulation. 1995;91:1512–9.
82. Priori SG, Napolitano C, Tiso N, et al. Mutations in the cardiac ryanodine receptor gene (hRyR2) underlie catecholaminergic polymorphic ventricular tachycardia. Circulation. 2001;103:196–200.

83. Priori SG, Napolitano C, Memmi M, et al. Clinical and molecular characterization of patients with catecholaminergic polymorphic ventricular tachycardia. Circulation. 2002;106:69–74.
84. Swan H, Piippo K, Viitasalo M, et al. Arrhythmic disorder mapped to chromosome 1q42-q43 causes malignant polymorphic ventricular tachycardia in structurally normal hearts. J Am Coll Cardiol. 1999;34:2035–42.
85. Lahat H, Pras E, Olender T, et al. A missense mutation in a highly conserved region of CASQ2 is associated with autosomal recessive catecholamine-induced polymorphic ventricular tachycardia in Bedouin families from Israel. Am J Hum Genet. 2001;69: 1378–84.
86. Liu N, Rizzi N, Boveri L, Priori SG. Ryanodine receptor and calsequestrin in arrhythmogenesis: what we have learnt from genetic diseases and transgenic mice. J Mol Cell Cardiol. 2009;46:149–59.
87. Liu N, Ruan Y, Priori SG. Catecholaminergic polymorphic ventricular tachycardia. Prog Cardiovasc Dis. 2008;51:23–30.
88. Amin AS, Tan HL, Wilde AA. Cardiac ion channels in health and disease. Heart Rhythm. 2010;7:117–26.
89. Bhuiyan ZA, Hamdan MA, Shamsi ET, et al. A novel early onset lethal form of catecholaminergic polymorphic ventricular tachycardia maps to chromosome 7p14-p22. J Cardiovasc Electrophysiol. 2007;18:1060–6.
90. Fuster V, Ryden LE, Cannom DS, et al. ACC/AHA/ESC 2006 Guidelines for the Management of Patients with Atrial Fibrillation: a report of the American College of Cardiology/American Heart Association Task Force on Practice Guidelines and the European Society of Cardiology Committee for Practice Guidelines (Writing Committee to Revise the 2001 Guidelines for the Management of Patients With Atrial Fibrillation): developed in collaboration with the European Heart Rhythm Association and the Heart Rhythm Society. Circulation. 2006;114:e257–354.
91. Feinberg WM, Blackshear JL, Laupacis A, Kronmal R, Hart RG. Prevalence, age distribution, and gender of patients with atrial fibrillation. Analysis and implications. Arch Intern Med. 1995;155: 469–73.
92. Brugada R, Tapscott T, Czernuszewicz GZ, et al. Identification of a genetic locus for familial atrial fibrillation. N Engl J Med. 1997;336:905–11.
93. Ellinor PT, Shin JT, Moore RK, Yoerger DM, MacRae CA. Locus for atrial fibrillation maps to chromosome 6q14-16. Circulation. 2003;107:2880–3.
94. Campuzano O, Brugada R. Genetics of familial atrial fibrillation. Europace. 2009;11:1267–71.
95. Chen YH, Xu SJ, Bendahhou S, et al. KCNQ1 gain-of-function mutation in familial atrial fibrillation. Science. 2003;299:251–4.
96. Yang Y, Xia M, Jin Q, et al. Identification of a KCNE2 gain-of-function mutation in patients with familial atrial fibrillation. Am J Hum Genet. 2004;75:899–905.
97. Xia M, Jin Q, Bendahhou S, et al. A Kir2.1 gain-of-function mutation underlies familial atrial fibrillation. Biochem Biophys Res Commun. 2005;332:1012–9.
98. Zhang DF, Liang B, Lin J, Liu B, Zhou QS, Yang YQ. KCNE3 R53H substitution in familial atrial fibrillation. Chin Med J (Engl). 2005;118:1735–8.
99. Olson TM, Alekseev AE, Liu XK, et al. Kv1.5 channelopathy due to KCNA5 loss-of-function mutation causes human atrial fibrillation. Hum Mol Genet. 2006;15:2185–91.
100. Darbar D, Kannankeril PJ, Donahue BS, et al. Cardiac sodium channel (SCN5A) variants associated with atrial fibrillation. Circulation. 2008;117:1927–35.
101. Tsai CT, Lai LP, Hwang JJ, Lin JL, Chiang FT. Molecular genetics of atrial fibrillation. J Am Coll Cardiol. 2008;52:241–50.
102. Li Q, Huang H, Liu G, et al. Gain-of-function mutation of Nav1.5 in atrial fibrillation enhances cellular excitability and lowers the threshold for action potential firing. Biochem Biophys Res Commun. 2009;380:132–7.
103. Hodgson-Zingman DM, Karst ML, Zingman LV, et al. Atrial natriuretic peptide frameshift mutation in familial atrial fibrillation. N Engl J Med. 2008;359:158–65.
104. Zhang X, Chen S, Yoo S, et al. Mutation in nuclear pore component NUP155 leads to atrial fibrillation and early sudden cardiac death. Cell. 2008;135:1017–27.
105. Lev M, Kinare SG, Pick A. The pathogenesis of atrioventricular block in coronary disease. Circulation. 1970;42:409–25.
106. Schott JJ, Alshinawi C, Kyndt F, et al. Cardiac conduction defects associate with mutations in SCN5A. Nat Genet. 1999;23:20–1.
107. Tan HL, Bink-Boelkens MT, Bezzina CR, et al. A sodium-channel mutation causes isolated cardiac conduction disease. Nature. 2001;409:1043–7.
108. Wang DW, Viswanathan PC, Balser JR, George Jr AL, Benson DW. Clinical, genetic, and biophysical characterization of SCN5A mutations associated with atrioventricular conduction block. Circulation. 2002;105:341–6.
109. Bezzina CR, Rook MB, Groenewegen WA, et al. Compound heterozygosity for mutations (W156X and R225W) in SCN5A associated with severe cardiac conduction disturbances and degenerative changes in the conduction system. Circ Res. 2003;92:159–68.
110. Kyndt F, Probst V, Potet F, et al. Novel SCN5A mutation leading either to isolated cardiac conduction defect or Brugada syndrome in a large French family. Circulation. 2001;104:3081–6.
111. Shirai N, Makita N, Sasaki K, et al. A mutant cardiac sodium channel with multiple biophysical defects associated with overlapping clinical features of Brugada syndrome and cardiac conduction disease. Cardiovasc Res. 2002;53:348–54.
112. Watanabe H, Darbar D, Kaiser DW, et al. Mutations in sodium channel beta1- and beta2-subunits associated with atrial fibrillation. Circ Arrhythm Electrophysiol. 2009;2:268–75.
113. Benson DW, Wang DW, Dyment M, et al. Congenital sick sinus syndrome caused by recessive mutations in the cardiac sodium channel gene (SCN5A). J Clin Invest. 2003;112:1019–28.
114. Ruan Y, Liu N, Priori SG. Sodium channel mutations and arrhythmias. Nat Rev Cardiol. 2009;6:337–48.
115. Smits JP, Koopmann TT, Wilders R, et al. A mutation in the human cardiac sodium channel (E161K) contributes to sick sinus syndrome, conduction disease and Brugada syndrome in two families. J Mol Cell Cardiol. 2005;38:969–81.
116. Shi R, Zhang Y, Yang C, et al. The cardiac sodium channel mutation delQKP 1507-1509 is associated with the expanding phenotypic spectrum of LQT3, conduction disorder, dilated cardiomyopathy, and high incidence of youth sudden death. Europace. 2008;10:1329–35.
117. Ge J, Sun A, Paajanen V, et al. Molecular and clinical characterization of a novel SCN5A mutation associated with atrioventricular block and dilated cardiomyopathy. Circ Arrhythm Electrophysiol. 2008;1:83–92.
118. Lei M, Zhang H, Grace AA, Huang CL. SCN5A and sinoatrial node pacemaker function. Cardiovasc Res. 2007;74:356–65.
119. Schulze-Bahr E, Neu A, Friederich P, et al. Pacemaker channel dysfunction in a patient with sinus node disease. J Clin Invest. 2003;111:1537–45.
120. Ueda K, Nakamura K, Hayashi T, et al. Functional characterization of a trafficking-defective HCN4 mutation, D553N, associated with cardiac arrhythmia. J Biol Chem. 2004;279:27194–8.
121. Milanesi R, Baruscotti M, Gnecchi-Ruscone T, DiFrancesco D. Familial sinus bradycardia associated with a mutation in the cardiac pacemaker channel. N Engl J Med. 2006;354:151–7.
122. Nof E, Luria D, Brass D, et al. Point mutation in the HCN4 cardiac ion channel pore affecting synthesis, trafficking, and functional expression is associated with familial asymptomatic sinus bradycardia. Circulation. 2007;116:463–70.

123. Beckwith JB. Discussion of the terminology and definition of the sudden infant death syndrome. In: Bergman AB, Beckwith JB, Ray CG, editors. Proceedings of the second international conference on causes of sudden death in infants. Seattle: University of Washington Press; 1970. p. 14–22.
124. Kinney HC, Thach BT. The sudden infant death syndrome. N Engl J Med. 2009;361:795–805.
125. Schwartz PJ, Priori SG, Dumaine R, et al. A molecular link between the sudden infant death syndrome and the long-QT syndrome. N Engl J Med. 2000;343:262–7.
126. Ackerman MJ, Siu BL, Sturner WQ, et al. Postmortem molecular analysis of SCN5A defects in sudden infant death syndrome. JAMA. 2001;286:2264–9.
127. Lupoglazoff JM, Denjoy I, Villain E, et al. Long QT syndrome in neonates: conduction disorders associated with HERG mutations and sinus bradycardia with KCNQ1 mutations. J Am Coll Cardiol. 2004;43:826–30.
128. Wolff L, Parkinson J. White, P.D. Bundle brunch block with short P-R interval in healthy young people prone to paroxysmal tachycardia. Am Heart J. 1930;5:683–704.
129. MacRae CA, Ghaisas N, Kass S, et al. Familial Hypertrophic cardiomyopathy with Wolff-Parkinson-White syndrome maps to a locus on chromosome 7q3. J Clin Invest. 1995;96:1216–20.
130. Blair E, Redwood C, Ashrafian H, et al. Mutations in the gamma(2) subunit of AMP-activated protein kinase cause familial hypertrophic cardiomyopathy: evidence for the central role of energy compromise in disease pathogenesis. Hum Mol Genet. 2001;10:1215–20.
131. Gollob MH, Green MS, Tang AS, et al. Identification of a gene responsible for familial Wolff-Parkinson-White syndrome. N Engl J Med. 2001;344:1823–31.
132. Arad M, Benson DW, Perez-Atayde AR, et al. Constitutively active AMP kinase mutations cause glycogen storage disease mimicking hypertrophic cardiomyopathy. J Clin Invest. 2002;109: 357–62.
133. Hardie DG. Minireview: the AMP-activated protein kinase cascade: the key sensor of cellular energy status. Endocrinology. 2003;144:5179–83.
134. Arad M, Seidman CE, Seidman JG. AMP-activated protein kinase in the heart: role during health and disease. Circ Res. 2007;100:474–88.
135. Nagata D, Hirata Y. The role of AMP-activated protein kinase in the cardiovascular system. Hypertens Res. 2010;33:22–8.
136. Nikoskelainen EK, Savontaus ML, Huoponen K, Antila K, Hartiala J. Pre-excitation syndrome in Leber's hereditary optic neuropathy. Lancet. 1994;344:857–8.
137. Raben N, Plotz P, Byrne BJ. Acid alpha-glucosidase deficiency (glycogenosis type II, Pompe disease). Curr Mol Med. 2002;2: 145–66.
138. Vaughan CJ, Hom Y, Okin DA, McDermott DA, Lerman BB, Basson CT. Molecular genetic analysis of PRKAG2 in sporadic Wolff-Parkinson-White syndrome. J Cardiovasc Electrophysiol. 2003;14:263–8.
139. Nattel S, Maguy A, Le Bouter S, Yeh YH. Arrhythmogenic ion-channel remodeling in the heart: heart failure, myocardial infarction, and atrial fibrillation. Physiol Rev. 2007;87:425–56.
140. Moss AJ, Zareba W, Schwarz KQ, Rosero S, McNitt S, Robinson JL. Ranolazine shortens repolarization in patients with sustained inward sodium current due to type-3 long-QT syndrome. J Cardiovasc Electrophysiol. 2008;19:1289–93.
141. Tamargo J, Caballero R, Gomez R, Valenzuela C, Delpon E. Pharmacology of cardiac potassium channels. Cardiovasc Res. 2004;62:9–33.
142. Zhang Y, Xiao J, Wang H, et al. Restoring depressed HERG K+ channel function as a mechanism for insulin treatment of abnormal QT prolongation and associated arrhythmias in diabetic rabbits. Am J Physiol Heart Circ Physiol. 2006;291:H1446–55.
143. Mangoni ME, Nargeot J. Genesis and regulation of the heart automaticity. Physiol Rev. 2005;88:919–82.
144. Brette F, Leroy J, Le Guennec JY, Salle L. Ca2+ currents in cardiac myocytes: old story, new insights. Prog Biophys Mol Biol. 2006;91:1–82.
145. Noble D. Unraveling the genetics and mechanisms of cardiac arrhythmia. Proc Natl Acad Sci USA. 2002;99:5755–6.
146. Roden DM. Taking the "idio" out of "idiosyncratic": predicting torsades de pointes. Pacing Clin Electrophysiol. 1998;21:1029–34.
147. Fitzgerald PT, Ackerman MJ. Drug-induced torsades de pointes: the evolving role of pharmacogenetics. Heart Rhythm. 2005;2:S30–7.
148. Sesti F, Abbott GW, Wei J, et al. A common polymorphism associated with antibiotic-induced cardiac arrhythmia. Proc Natl Acad Sci USA. 2000;97:10613–8.
149. Paulussen AD, Gilissen RA, Armstrong M, et al. Genetic variations of KCNQ1, KCNH2, SCN5A, KCNE1, and KCNE2 in drug-induced long QT syndrome patients. J Mol Med. 2004;82:182–8.
150. Aerssens J, Paulussen AD. Pharmacogenomics and acquired long QT syndrome. Pharmacogenomics. 2005;6:259–70.
151. Bonnet D, Martin D, De Pascale L, et al. Arrhythmias and conduction defects as presenting symptoms of fatty acid oxidation disorders in children. Circulation. 1999;100:2248–53.
152. Corr PB, Creer MH, Yamada KA, Saffitz JE, Sobel BE. Prophylaxis of early ventricular fibrillation by inhibition of acylcarnitine accumulation. J Clin Invest. 1989;83:927–36.
153. Stanley CA, Hale DE, Berry GT, Deleeuw S, Boxer J, Bonnefont JP. Brief report: a deficiency of carnitine-acylcarnitine translocase in the inner mitochondrial membrane. N Engl J Med. 1992;327:19–23.
154. Tripp ME. Developmental cardiac metabolism in health and disease. Pediatr Cardiol. 1989;10:150–8.
155. Murphy E, Eisner DA. Regulation of intracellular and mitochondrial sodium in health and disease. Circ Res. 2009;104:292–303.
156. Szewczyk A, Jarmuszkiewicz W, Kunz WS. Mitochondrial potassium channels. IUBMB Life. 2009;61:134–43.
157. Verdonck F, Volders PG, Vos MA, Sipido KR. Increased Na+ concentration and altered Na/K pump activity in hypertrophied canine ventricular cells. Cardiovasc Res. 2003;57:1035–43.
158. Bossuyt J, Ai X, Moorman JR, Pogwizd SM, Bers DM. Expression and phosphorylation of the Na-pump regulatory subunit phospholemman in heart failure. Circ Res. 2005;97:558–65.
159. Sipido KR, Volders PG, Vos MA, Verdonck F. Altered Na/Ca exchange activity in cardiac hypertrophy and heart failure: a new target for therapy? Cardiovasc Res. 2002;53:782–805.
160. Bers DM, Despa S. Cardiac myocytes Ca2+ and Na+ regulation in normal and failing hearts. J Pharmacol Sci. 2006;100:315–22.
161. Turrens JF. Mitochondrial formation of reactive oxygen species. J Physiol. 2003;552:335–44.
162. Facundo HT, Fornazari M, Kowaltowski AJ. Tissue protection mediated by mitochondrial K+ channels. Biochim Biophys Acta. 2006;1762:202–12.
163. Weiss JN, Korge P, Honda HM, Ping P. Role of the mitochondrial permeability transition in myocardial disease. Circ Res. 2003;93:292–301.
164. Di Lisa F, Bernardi P. Mitochondria and ischemia-reperfusion injury of the heart: fixing a hole. Cardiovasc Res. 2006;70:191–9.
165. Gu XQ, Siemen D, Parvez S, et al. Hypoxia increases BK channel activity in the inner mitochondrial membrane. Biochem Biophys Res Commun. 2007;358:311–6.
166. Cheng Y, Gu XQ, Bednarczyk P, Wiedemann FR, Haddad GG, Siemen D. Hypoxia increases activity of the BK-channel in the inner mitochondrial membrane and reduces activity of the permeability transition pore. Cell Physiol Biochem. 2008;22:127–36.
167. Akar FG, Aon MA, Tomaselli GF, O'Rourke B. The mitochondrial origin of postischemic arrhythmias. J Clin Invest. 2005;115: 3527–35.

168. Roden DM. Human genomics and its impact on arrhythmias. Trends Cardiovasc Med. 2004;14:112–6.
169. Xue T, Cho HC, Akar FG, et al. Functional integration of electrically active cardiac derivatives from genetically engineered human embryonic stem cells with quiescent recipient ventricular cardiomyocytes: insights into the development of cell-based pacemakers. Circulation. 2005;111:11–20.
170. Nattel S, Carlsson L. Innovative approaches to anti-arrhythmic drug therapy. Nat Rev Drug Discov. 2006;5:1034–49.
171. Telemaque S, Marsh JD. Modification of cardiovascular ion channels by gene therapy. Expert Rev Cardiovasc Ther. 2009;7: 939–53.
172. Hajjar RJ, Samulski RJ. Heart failure: a silver bullet to treat heart failure. Gene Ther. 2006;13:997.
173. Torella D, Indolfi C, Goldspink DF, Ellison GM. Cardiac stem cell-based myocardial regeneration: towards a translational approach. Cardiovasc Hematol Agents Med Chem. 2008; 6:53–9.

# Chapter 18
# Signaling in Atherosclerosis

**Abstract** Coronary artery disease (CAD), the most common cause of death in the western hemisphere by the year 2020 is becoming the leading cause of morbidity and mortality in the world. Atherosclerosis, the primary cause of CAD, involves a wide spectrum of cell types, organs, and diverse physiological processes which underlie its complex genetic basis.

While atherosclerosis is viewed as a multifactorial, multistep disease, the involvement of chronic inflammation at multiple stages of the disease, from initiation to progression, has led to the suggestion that many risk factors contribute to its pathogenesis by aggravating the underlying inflammatory process. These factors include dyslipidemia, diabetes, and pro- and anticoagulant factors. Significantly, the signaling pathways that are involved in vascular alterations leading to atherogenesis share many of the same components utilized by other vascular remodeling programs, including the development of vascular calcification and angiogenesis. Identification of proatherogenic and also antiatherogenic factors in these signaling pathways assists in the development of better stage-specific biomarkers of the disease and the development of therapeutic intervention and potential therapies.

**Keywords** Atherosclerosis • Signaling pathways • Lipids • Mitochondria • Antioxidant enzymes

## Introduction

Atherosclerosis commonly causes myocardial infarction and ischemic stroke, two main causes of death in developed countries. This chronic progressive disease is characterized by the formation and accumulation of lipid streaks and fibrous elements in large arteries leading to plaque disruption, thrombosis, and myocardial infarction. While atherosclerosis has traditionally been viewed as a lipid-storage disease with great emphasis on the deposition, modification, and cellular uptake of cholesterol and lipoproteins as major factors in its development, recent focus has centered on the involvement of inflammation and endothelial dysfunction in atherogenesis.

Atherosclerosis is increasingly viewed as a multifactorial, multistep disease. The involvement of chronic inflammation of the arterial wall at all stages of the disease, from the initiation to progression and rapture of atherosclerotic plaques, has led to the suggestion that multiple risk factors contribute to its pathogenesis by aggravating the inflammatory process.

It has been increasingly recognized that the vascular endothelium actively regulates vascular tone, vessel growth, lipid breakdown, inflammation, and thrombogenesis, all important factors in atherogenesis. Moreover, it has been amply demonstrated that endothelial dysfunction contributes to the development of atherosclerosis through deregulation of vasoconstriction, monocyte and platelet adhesion, cytokine and growth factor generation, and release and thrombogenesis.

The elucidation of the proatherogenic and antiatherogenic factors in these signaling pathways may assist in the development of better stage-specific biomarkers of the disease and the development of new therapeutic intervention.

## The Role of Lipids

The link between hyperlipidemia and atherosclerosis has been established on the basis of extensive experimental and clinical studies, which dominated the field until the 1970s [1]. Although emerging evidence suggests that multiple factors contribute to atherogenesis a major role of cholesterol in this process is well documented [2–5]. It has been demonstrated that increased low-density lipoprotein (LDL) and decreased high-density lipoprotein (HDL) cholesterol levels in plasma represent one of the most important risk factors for developing atherosclerosis [6–8].

Lipoproteins represent a heterogeneous population of particles differing in size, density, and lipid and apolipoprotein composition. Several subfractions of LDL and HDL have been described based on their ultracentrifuge flotation

J. Marín-García, *Signaling in the Heart*,
DOI 10.1007/978-1-4419-9461-5_18, 

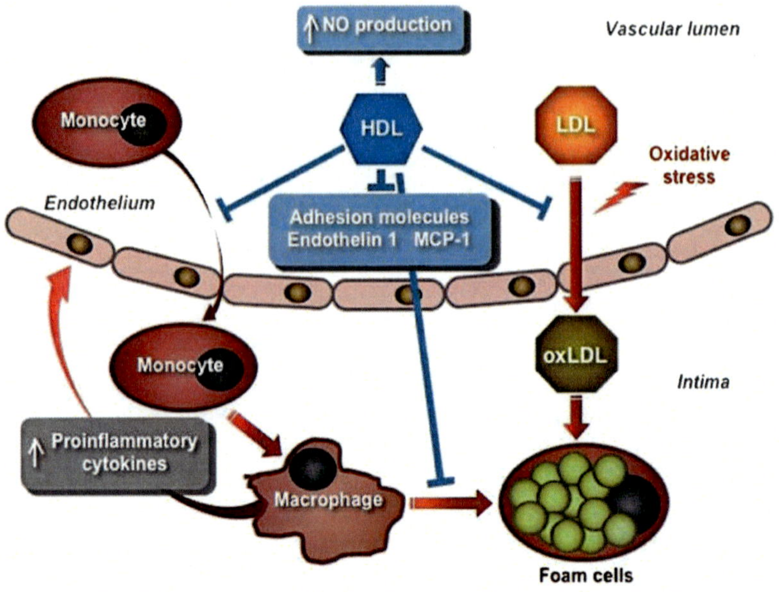


**Fig. 18.1** HDL particles attenuate atherogenesis. HDLs inhibit expression of cellular adhesion molecules (ICAM, VCAM, and CD11b) and monocyte chemoattractant protein-1 (MCP-1) downregulating thereby monocyte adhesion and infiltration between intact endothelial cells into the intima. HDLs interfere with oxidation of LDLs and limit macrophage differentiation into foam cells, characteristic to atherosclerotic plaques. In addition, HDLs activate eNOS resulting in the elevation of antiatherogenic NO levels while downregulate endothelin 1 leading to vascular endothelial relaxation

rates in high-salt solutions and have subsequently been associated to varying degrees with the risk of cardiovascular disease (CVD) [9]. Various metabolic pathways generate LDL and HDL particles of different compositions and sizes.

## LDL Particles

In 1981, Henriksen et al. [10] demonstrated that native LDL particles could be oxidized by endothelial cells (ECs), and oxidized LDLs (oxLDLs) were recognized with higher affinity and taken up much more rapidly by peritoneal macrophages than native LDLs. Various specific plasma membrane receptors, including scavenger receptors A and B [SR-A and SR-B (CD36)] and lectin-like oxLDL receptor, bind oxLDL with high affinity. SR-A and SR-B account for about 90% of the oxLDL uptake by the macrophages [11]. Internalization of oxLDLs, potentially cytotoxic materials, by tissue macrophages may initially play protective role [12]. Increased cholesterol concentration in the macrophages does not downregulate these receptors leading to progressive cholesterol loading in macrophages. Small dense LDL particles appear to be more associated with the development of atherosclerosis than larger LDLs [13].

OxLDLs activate smooth muscle cells and endothelial cells, as well as the expression and presentation of adhesion molecules and secretion of inflammatory mediators leading eventually to the recruitment of leukocytes to the vessel lumen [12, 14]. The resulting local accumulation of inflammatory cells further induces LDL oxidation creating a vicious circle. In the intima, monocytes convert into macrophages able to uptake oxLDLs via their scavenger receptors resulting in the accumulation of cells loaded with lipids. These cells, also called foam cells, are characteristic to atherosclerotic plaques (Fig. 18.1). Although several alternative mechanisms for the formation of foam cells have been suggested, none of them has yet been proven to contribute to this process *in vivo* [12, 15].

Local vascular damage caused by secreted inflammatory mediators and cell apoptosis generate antigenic and thrombogenic debris contributing to atheroma progression. Moreover, intimal macrophages can metabolize oxLDLs and present resultant products as antigens to recruit T cells and activate thereby immune response.

## HDL Particles

In contrast to LDLs, HDLs play a protective role against atherosclerosis. They not only transfer cholesterol from intimal macrophages to the liver, but also improve vascular function and inhibit multiple inflammatory pathways reducing further the risk of atherogenesis [16–19]. Plasma levels of HDLs represent the most potent lipid risk factor for atherosclerosis.

The major protein component of HDL particles is apolipoprotein A-I (apoA-I), which is secreted from the liver and intestines. Additional apolipoproteins, including apoA-V, apoC, apoD, and apoE are involved in HDL formation. Multiple ATP-binding cassette (ABC) transporters, such as ABCA1, ABCG1, and ABCG4, and scavenger receptor class B type I (SR-BI) contribute to transfer cholesterol from peripheral cells to HDLs circulating in the plasma. In HDLs, cholesterol is esterified by lecithin:cholesterol acyltransferase.

HDL particles can be endocytosed by the hepatocytes or they can interact with SR-BI located on the hepatocyte surface transferring cholesterol ester without HDL degradation [20, 21]. In addition, cholesterol ester can be transferred from HDL and LDL particles by cholesterol ester transfer protein (CETP) to triglyceride-rich lipoproteins taken up by the liver. Finally, ABCG5 and ABCG8 transporters excrete cholesterol from the liver into the bile.

Increased levels of HDL particles improve vascular relaxation and infusion of HDLs or apoA-I into patients markedly improves vascular function [22–24]. Major mechanism of this improvement is HDL-mediated stimulation of nitric oxide (NO) synthesis. The reduced availability of NO is an integral feature of endothelial dysfunction that significantly contributes to atherogenesis. Reduced NO availability occurs through a complex combination of its impaired arterial membrane receptor signaling for agonists or physiological stimuli capable of generating NO, reduced NO production by inducible (iNOS) and endothelial (eNOS) NO synthase, abnormal NO release, diffusion to the vascular smooth muscle cell (VSMC) and downstream signaling events, including its interaction with guanylate cyclase, and increased NO destruction by oxygen-free radicals [25]. Moreover, endothelium-mediated vasoconstrictors, adhesion molecules, cytokines, growth factors, and thrombogenic factors, such as endothelin, are increased by oxLDLs. In contrast to LDLs, HDLs stimulate eNOS via the interaction of apoA-I on HDLs with SR-BI on the endothelial cell surface [26]. Accordingly, antibodies against apoA-I or loss of SR-BI inhibit NO production in response to HDLs. Moreover, lysophospholipids from HDL particles can also enhance eNOS activity via the sphingosine-1-phosphate (S1P) receptor and protein kinase Akt activation [27, 28]. Hence, the preservation or restoration of NO-mediated signaling pathways in arteries is a critical target for new interventions aimed at atherosclerosis.

HDL particles inhibit various steps in vascular inflammation (Fig. 18.1). They inhibit the expression of E-selectin, an essential regulator of leukocyte trafficking, and vascular cell adhesion molecule-1 (VCAM-1) and intercellular adhesion molecule-1 (ICAM-1) [29–32]. Furthermore, apoA-I, a major constituent of HDLs, limits macrophage differentiation into foam cells [33].

It has been demonstrated that HDLs possess antioxidant properties attributed mainly to apoA-I. ApoA-I is able to remove oxidation-prone molecules from human LDLs rendering them resistant to oxidation by endothelial cells *in vitro* [34]. HDL particles contain several enzymes able to metabolize oxidized lipids, such as paraoxonase (PON), lipoprotein-associated phospholipase A2 and glutathione peroxidase (GPx) [35]. Moreover, both HDL-associated PON and platelet-activating factor acetylhydrolase inhibit the oxLDL biosynthesis *in vitro* [34, 36]. Accordingly, PON-1 deficient mice are more prone to atherosclerosis than wild-type mice [37, 38]. HDL particles also limit leukocyte trafficking preventing thereby influx of reactive oxygen species (ROS)-producing neutrophils.

Finally, HDL particles contribute to the prevention of arterial and venous thrombosis. Recent clinical studies have demonstrated that increased levels of HDLs decrease the risk of venous thrombosis [39, 40]. Several mechanisms underlying the antithrombotic effect of HDLs have been suggested. They include HDL-mediated activation of endothelial synthesis of antiplatelet prostacyclin and the anticoagulant protein C and S pathways and inhibition of platelet activation and aggregation [41–44].

While HDL particles are usually considered to be antiatherosclerotic factors, the existence of proinflammatory HDLs, also known as dysfunctional HDLs, has been reported (Fig. 18.2). Decrease in antiinflammatory capacity of HDLs is associated with changes in their protein composition. Dysfunctional HDLs from patients with atherosclerosis contain less apoA-I, PON and platelet-activating factor acetylhydrolase and more serum amyloid A compared to HDLs from normal individuals [45]. Functionally, these changes lead to a diminished capacity of dysfunctional HDLs to reduce LDL oxidation, to activate eNOS and to inhibit thrombosis. In addition to decreased apoA-I levels, apoA-I in dysfunctional HDLs is more nitrated and chlorinated than apoA-I in normal HDLs. Importantly, these chemical modifications of apoA-I downregulate the ability of HDLs to

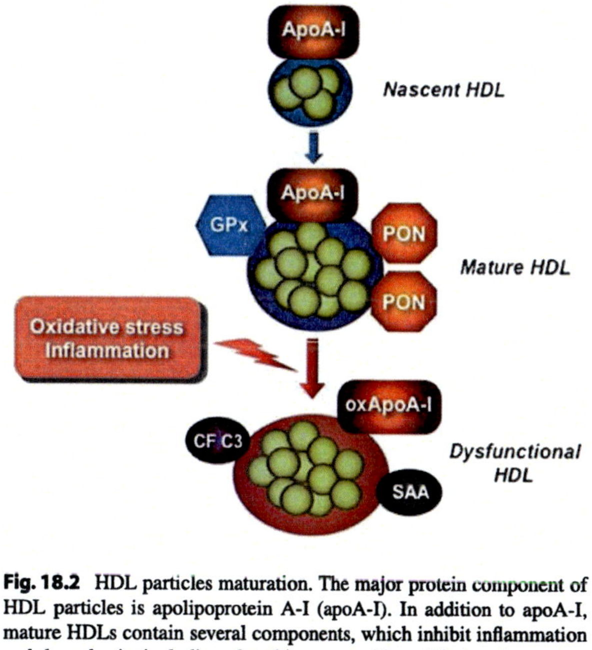


**Fig. 18.2** HDL particles maturation. The major protein component of HDL particles is apolipoprotein A-I (apoA-I). In addition to apoA-I, mature HDLs contain several components, which inhibit inflammation and thrombosis, including glutathione peroxidase (GPx) and paraoxonase (PON). Dysfunctional HDLs contain less apoA-I and antioxidants PON and GPx but more proinflammatory serum amyloid A (SAA) and complement factor C3 (CF C3). In addition to decreased apoA-I levels, apoA-I in dysfunctional HDLs is more oxidized and nitrated than apoA-I in normal HDLs. These changes are associated with significant decrease in antiinflammatory capacity of HDLs

remove cholesterol from the macrophages [45]. ApoA-I and lipids in dysfunctional HDLs can also be oxidized by myeloperoxidase contributing to the conversion of HDLs from antiinflammatory to proinflammatory factors [46].

## Activation of Immune Cells in Atherosclerosis

### *Endothelial Activation, Adhesion Molecules, and Chemokines*

Accumulation of oxLDLs at sites of hemodynamic strain in the arteries initiates a local inflammatory process. The first blood cells, which appear at the sites of endothelial activation, are the platelets [47]. Glycoproteins expressed in the platelet engage the endothelial surface molecules facilitating activation of the endothelial cells, which are resistant to adhesion and aggregation of leukocytes under normal conditions. Activated endothelial cells express various leukocyte adhesion molecules, such as VCAM-1 and P- and E-selectins (Fig. 18.3) [48–51]. Importantly, cholesterol diet induces VCAM-1 expression in endothelial cells in areas prone to atherosclerotic lesions [48]. Moreover, oxLDL-derived metabolites and proinflammatory cytokines interleukin (IL)-1β (IL-1β) or tumor necrosis factor α (TNF-α) induce nuclear factor κ-light-chain-enhancer of activated B cells (NF-κB)-mediated activation of VCAM-1 expression in endothelial cells [52].

VCAM-1 binds preferably the monocytes and T lymphocytes. Monocytes adhered to the activated endothelium are induced by proinflammatory proteins, chemokines,

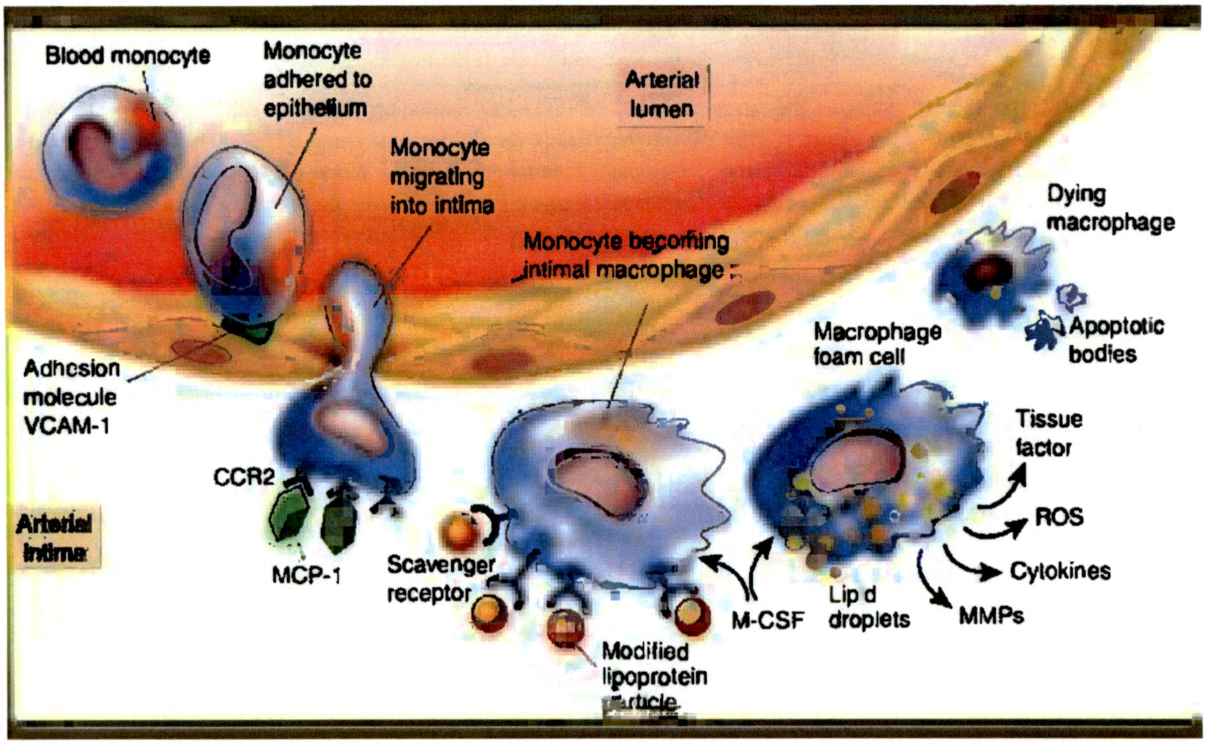


**Fig. 18.3** Roles of mononuclear phagocytes in atherogenesis [53]. Sequence of the major steps in the recruitment of monocytes to the atherosclerotic plaque is schematically shown proceeding from *left to right*. Upon normal conditions, the arterial endothelium is resistant to adhesion and aggregation of leukocytes. However, upon inflammatory activation, the endothelial cells increase the expression of various leukocyte adhesion molecules, such as VCAM-1, and P- and E-selectins. Monocytes adhered to the activated endothelium are induced by proinflammatory proteins, chemokines, to infiltrate between intact endothelial cells into the intima. Binding of monocyte chemoattractant protein-1 (MCP-1) to its receptor CCR2 on the monocyte surface plays a major role in this process. Within the inflamed intima, the recruited monocytes differentiate into macrophages, which induce the expression of scavenger receptors. Scavenger receptors bind and internalize modified, for example by oxidation or glycation, lipid particles. Macrophages loaded with modified lipids and oxLDL-derived cholesterol eventually convert into the foam cells, a hallmark of the atherosclerotic lesion. The macrophages proliferate in the inflamed intima and release various growth factors and cytokines amplifying ROS production and the local vascular inflammation. They also release procoagulant tissue factor and metalloproteinases (MMPs) that degrade the arterial extracellular matrix destabilizing the plaque's fibrous caps and initiating rapture of plaque and thrombus formation. Macrophages can undergo apoptosis and die forming necrotic core characteristic for the advanced atherosclerotic lesions. Reprinted from Libby [53] with kind permission from Nature Publishing Group

to infiltrate into the intima. Within the inflamed intima, the recruited monocytes differentiate into macrophages. This central step in atherogenesis is associated with upregulation of expression of scavenger receptors and toll-like receptors (TLRs) in macrophages [54, 55].

Scavenger receptors internalize oxLDLs and dead cell fragments [55]. Macrophages become loaded with modified lipids and oxLDL-derived cholesterol and eventually convert into characteristic foam cells. TLRs not only bind pathogen-like molecules and oxLDLs, but can also induce a signal cascade resulting in cell activation [54, 56].

The macrophages proliferate in the inflamed intima and release various growth factors and cytokines accelerating the inflammatory process [53, 57–59]. They also release procoagulant tissue factor (TF) and metalloproteinases that contribute to the endothelium damage by degrading the arterial extracellular matrix.

In addition to VCAM-1, monocyte chemoattractant protein-1 (MCP-1) recruits the monocytes into the arterial intima [60, 61]. MCP-1 interacts with the monocyte chemokine receptor CCR-2 facilitating leukocyte infiltration into the intima by diapedesis. Consistently, the loss of MCP-1 or CCR-2 reduces the development of atherosclerosis in apo$E^{-/-}$ mice. In atheroma, multiple chemokines and their receptors, listed in Table 18.1, are expressed contributing to the recruitment of leukocytes and to the vascular inflammatory response during all phases of atherosclerosis [62].

**Table 18.1** Effect of chemokines and their receptors on atherosclerosis (modified from [62])

| Chemokine | Receptor | Effect on lesion size |
|---|---|---|
| CCL2 | CCR2 | ↓ Atherosclerosis [9–11, 14] |
| | | ↓ Neointimal hyperplasia [15] |
| CCL5 | CCR1 | ↑ Atherosclerosis [33] |
| | CCR3 | ↔ Neointimal hyperplasia [36] |
| | CCR5 | ↑ Atherosclerosis [35] |
| CXCL1 | CXCR2 | ↓ Atherosclerosis [41] |
| CXCL8 | | ↑ Neointimal hyperplasia [49] |
| MIF | CXCR2 | ↔ Atherosclerosis [53] |
| | CXCR4 | ↓ Neointimal hyperplasia [60, 61] |
| | | ↓ Atherosclerosis [54] |
| | | ↑ Regression [55] |
| CXCL12 | CXCR4 | ↑ Atherosclerosis [46] |
| | | ↓ Neointimal hyperplasia [63, 70] |
| $CX_3CL1$ | $CX_3CR1$ | ↓ Atherosclerosis [81] |
| CXCL16 | CXCR6 | ↑ Atherosclerosis [101] |
| | | ↓ Atherosclerosis [100] |
| CXCL19 | CXCR7 | ↓ Regression [104] |
| CXCL20 | | |
| CXCL9 | CXCR3 | ↓ Atherosclerosis [105] |
| CXCL10 | | ↓ Atherosclerosis [106, 107] |
| CXCL11 | | ↓ Atherosclerosis [112] |
| CXCL4 | CXCR3B | ↓ Atherosclerosis [119] |

↓, reduction; ↑, increase; ↔, no change

## Heterogeneity of Monocytes

Monocytes involved in the inflammatory response during atherosclerosis represent a heterogeneous population. $CD14^{+}CD16^{+}$ ($Ly6C^{low}CCR2^{-}CX_3CR1^{hi}$ in mice) monocytes, also known as resident monocytes, play mainly surveillance role in homeostasis, whereas $CD14^{hi}CD16^{-}$ ($Ly6C^{hi}CCR2^{+}CX_3CR1^{low}$ in mice) monocytes or inflammatory monocytes enter more efficiently the sites of inflammation [63, 64]. In hypercholesterolemic mice, the numbers of the $Ly6C^{hi}$ monocytes is significantly elevated while the numbers of $Ly6C^{low}$ show no change. $Ly6C^{hi}$ monocytes employ CCR5 and $CX_3CR1$ receptors, in addition to CCR2, to infiltrate into the plaques. Low amounts of $Ly6C^{low}$ monocytes can also be found in plaques; they use CCR5 receptors more frequently than $CX_3CR1$ receptors. $Ly6C^{hi}$ monocytes produce higher amounts of proinflammatory cytokines, myeloperoxidase, proteinases, and P-selectin glycoprotein ligand-1 compared to $Ly6C^{low}$ monocytes [59].

Both resident and inflammatory monocyte subsets are able to differentiate into $CD11c^{+}$ dendritic cells, however, within plaques $Ly6C^{low}$ monocytes are more predisposed to converting to $CD11c^{+}$ cells [64]. Contribution of these dendritic cells in the development of atherosclerosis remains to be determined.

## T-Cell Activation in Vascular Inflammation

In addition to cells of innate immunity, T lymphocytes (T cells) are also involved in vascular inflammatory response during atherosclerosis. These cells of the adaptive immune response represent a smaller population of the leukocytes in atherosclerotic plaques. T cells found in plaques are predominantly $CD4^{+}$ cells [65]. Natural killer T cells represent a minor T cell subpopulation found mostly in early atherosclerotic lesions and their activation in apo$E^{-/-}$ mice induces atherosclerosis [66]. $CD8^{+}$ T cells also present in plaques recognize viral antigens presented by the major histocompatibility complex (MHC) class I [65]. Their activation can lead to the arterial cell death contributing to the development atherosclerosis in apo$E^{-/-}$ mice [67].

T cells infiltrate to plaques via interactions involving various chemoattractant molecules, such as CXCL9, CXCL10, and CXCL11, and their common receptor CXCR3 and RANTES [68–71]. Within atherosclerotic plaques, T cells recognize various antigens presented by the MHC class II and assume different programs of activation to differentiate into type 1 T-helper ($T_H1$) and type 2 T-helper ($T_H2$) cells [58, 72].

$T_H1$ cells produce cytokine interferon-γ (IFN-γ), which induces the classical activation of macrophages, whereas

$T_H2$ cells generate cytokines IL-4 and IL-13 inducing alternative macrophage activation. IFN-γ dominates in the development of atherosclerosis triggering the secretion of proinflammatory mediators, including various proteases, NO, and proinflammatory cytokines TNF and IL-1 and leading to local inflammation and arterial wall damage [73, 74]. In addition, IFN-γ inhibits proliferation of endothelial and smooth muscle cells and production of collagen contributing thereby to fragility of atherosclerotic plaques [75–77]. Decreased atherosclerosis in IFN-γ deficient or IFN-γ receptor-deficient mice support role of IFN-γ as an important proinflammatory mediator [78, 79].

Alternative macrophage activation induced by IL-4 and IL-13 can attenuate some proinflammatory effects of IFN-γ [74]. However, their antiinflammatory effect appears to be less pronounced than IL-10 or transforming growth factor β (TGF-β), and the precise role of IL-4 and IL-13 in atherogenesis remains to be determined. In addition, the $T_H2$-induced pathway can induce elastolytic enzymes contributing thereby to the formation of aneurysms [80].

Complex relationships between T cells and macrophages are mostly regulated by the antigens presented to T cells by the MHC class II on the surface of the antigen-presenting cells. T cells isolated from atherosclerotic lesions recognize oxLDLs, heat shock protein 60 (HSP60) as well as proteins of *Chlamydia pneumoniae*, a microorganism found in plaques [81, 82]. However, contribution of various antigens to atherogenesis is not clear.

Thus, the activation of T cells triggers a potent cytokine cascade involving a variety of downstream inflammatory mediators. Some of them, such as IL-6 and C-reactive proteins, are present in the peripheral circulation explaining how the activation of a limited number of immune cells can induce not only local, but also systemic inflammatory response [72].

## Plaque Rapture

The atherosclerotic lesion form a subendothelial scar-like structure called the fibrous cap that is generated by the intimal collagen-secreting myofibroblasts. Activated macrophages, T cells and mast cells generate a variety of inflammatory cytokines, proteases, radicals, coagulation, and vasoactive factors. They destabilize fibrous caps and initiate rapture of plaque and thrombus formation [76, 77, 83]. Physical fracture of the thinning fibrous cap causes thrombotic occlusion of an artery leading frequently to the sudden onset of myocardial infarction and strokes [84–86].

IFN-γ produced by $T_H1$ cells inhibits the VSMC synthesis of interstitial collagens, the main stabilizing component of the fibrous caps [76, 77]. Within plaques, activated macrophages express proteases that can degrade collagens contributing to plaque rapture [87, 88]. The activation of T cells within the intima induces the expression of proinflammatory cytokine, the CD40 ligand (CD40L or CD154). Ligation of the macrophage CD40 by CD154 induces the generation of matrix metalloproteinases (MMPs) MMP1, MMP8, and MMP13, which can degrade the plaque matrix [89]. Another type of proteases, cysteine proteases, induced by certain cytokines appears to be also involved in plaque rapture [90]. Moreover, ligation of CD40 expressed by macrophages induces expression of the potent procoagulant TF [83, 91]. TF triggers the coagulation cascade boosting thrombogenic potential of the lipid core of the atherosclerotic plaques and leading eventually to plaque erosion and thrombus formation [84].

## Macrophage Death in Atherosclerosis

Macrophage death is a characteristic feature of advanced atherosclerotic lesions although apoptotic macrophages have been observed during all stages of the disease [92, 93]. In early atherosclerotic lesions (before the formation of necrotic core), accelerated macrophage apoptosis has been associated with significant decrease in lesion area and plaque progression [94, 95]. The phagocytic clearance of apoptotic cells – efferocytosis (from the Latin "*effero*," to take to the grave) – results in fewer inflammatory and postapoptotic necrotic cells in early lesions. Moreover, phagocytosis of apoptotic cells attenuates the production of proinflammatory cytokines IL-1β, IL-8, IL-12, IL-23, IL-27, and TNF-α while upregulates the expression of antiinflammatory cytokine IL-10 [96–98]. Thus, at early stage of the disease macrophage death plays a beneficial role by decreasing lesion cellularity and inflammatory response.

Advanced atherosclerotic lesions contain apoptotic cells originated from various cell types, such as macrophages, T lymphocytes, and VSMCs [99]. In advanced lesions, apoptotic cells, mainly macrophages, uncleared by efferocytosis, accumulate and surround the forming necrotic core [100]. Necrotic cores are composed mainly of postapoptotic macrophage debris, a source of inflammatory stimuli, inducing inflammatory response, and damaging surrounding cells. These detrimental processes contribute to rapture of fibrous caps, plaque erosion, and eventually to thrombogenesis [101].

Emerging evidence suggests that endoplasmic reticulum (ER) stress promotes macrophage death and advanced plaque necrosis. ER stress induces the unfolded protein response, which upregulates the expression of the proapoptotic *CEBP-homologous protein* (CHOP, also known as Gadd153) [102]. CHOP-mediated release of $Ca^{2+}$ from the ER has been suggested to be its major apoptotic effect [103, 104].

Expression of CHOP and other ER stress proteins correlate with advanced atherosclerotic plaque progression [105]. Given a protective role of ER stress, prolonged activation and high levels of ER stress would be required to trigger apoptosis. According to the current concept, physiological levels of ER stress combined with a second signal initiate apoptotic response in advanced atherosclerotic lesions [105, 106]. Activation of pattern recognition receptors (PRRs), including scavenger receptors and TLRs, by various molecules with pathogen-associated molecular patterns (PAMPs), such as modified lipids, pathogens, foreign antigens, and endogenous proteins, represents an example of second hit needed to trigger apoptosis. Accumulation of LDL cholesterol by macrophages leads to excess accumulation of cholesterol in the ER membranes, which induces ER stress, while the second signal is binding of molecules with PAMPs by macrophage PRRs. Combination of these signals suppresses protective signaling of ER stress and amplifies apoptotic response [107, 108].

Multiple ligands, such as oxLDLs, oxidized phosphatidylcholine, β-amyloid, and advanced glycation end-products, can engage macrophage SR-A and CD36. Combinatorial activation of SR-A and TLR4 by SR-A ligands or CD36 and TLR2 by oxidized phospholipids initiates apoptosis in ER-stressed macrophages and contributes to detrimental processes in advanced atherosclerotic lesions. While participation of SR-A and CD36 in macrophage apoptosis and plaque necrosis is well documented, involvement of TLR2 and TLR4 in these processes *in vivo* requires further investigation.

Growing evidence suggests that defects in efferocytosis in advanced atherosclerotic lesions are a main driving force of necrotic core formation and plaque vulnerability [106]. However, the molecular mechanisms of defective efferocytosis in advanced atheroma are poorly understood.

## Rho Kinases as Mediators of Atherosclerosis

In the past decade, growing evidence suggests an important preventive role of statins and Rho kinases (ROCKs) against the development atherosclerosis. Statins inhibit the generation of isoprenoid intermediates of the cholesterol biosynthesis modifying composition of the lipid core of the atherosclerotic plaques and reducing their size and stability [109]. Moreover, statins appear to be able to downregulate the number of inflammatory cells in plaques. Isoprenoid intermediates serve also as lipid attachments for the post-translational modifications of various signaling molecules, including three main classes of Rho GTPase – Rac, Rho, and Cdc42 [110]. The Rho downstream effectors ROCKs play an essential role in this signaling pathway and contribute to the pleiotropic effects of statins in atherogenesis.

### *ROCKs*

ROCKs are serine/threonine protein kinases with a molecular mass of ~160 kDa homologous to dystrophia myotonica protein kinase (DMPK), DMPK-related cell division control protein 42 (Cdc42)-binding kinases (MRCKs), and citron kinase [111]. In mammalian cells, two ROCK isoforms are expressed encoded by two distinct genes: ROCK1 (ROCK I, P160-ROCK or ROKβ) and ROCK2 (ROCK II or ROKα) [112]. They share a 65% overall homology with the highest similarity in their kinase domains (92% identity). ROCKs are composed of an amino-terminal kinase domain, followed by a potential coiled-coil region containing Rho-binding domain (RBD), and a carboxy-terminal pleckstrin homology (PH) domain with an internal cysteine-rich domain (CRD). This carboxy-terminal domain serves as an autoinhibitory region, which decreases the kinase activity [113]. The binding of GTP-bound RhoA to the RBD of ROCKs represses autoinhibition of the carboxyl terminus and stimulates the kinase activity. ROCK activation also results from the arachidonic acid binding to the PH domain or from cleavage of the carboxyl terminus by caspase-3 [114–116]. The coiled-coil region appears to be involved in the interactions with other α-helical proteins while the PH domain participates in protein subcellular localization.

ROCK1 and ROCK2 are ubiquitously expressed with the highest ROCK1 mRNA levels in kidney, liver, lung, spleen, and testis while ROCK2 mRNA is predominantly expressed in the skeletal muscles, heart, and brain [112]. Both ROCK isoforms are expressed in vascular smooth and heart muscles. ROCK2 is predominantly localized in the cytoplasm and partially translocates to cell plasma membrane upon the treatment of cells with Rho-activating factors [117–119]. Subcellular localization of ROCK1 is less clear, however, it might colocalize with centrosomes [120].

Engagement of G protein-coupled receptors (GPCRs) and tyrosine kinase receptors recruits and activates Rho guanine nucleotide exchange factors (GEFs), which activate Rho (Fig. 18.4). Activated GTP-bound Rho is able to bind to the RBD of ROCKs inducing an open conformation of ROCKs associated with the kinase activity [117, 121]. In addition to the Rho-dependent activation, ROCKs can be activated by some lipids, especially by arachidonic acid, which bind to its PH domain [122]. Moreover, ROCK1 can be activated by the caspase-3-mediated cleavage of its caboxy-terminal region during apoptosis [115, 116, 123]. Other small GTPases, including GEM and Rad, appear to bind to and inhibit the ROCK activity [124].

In response to Rho activation by S1P and lysophosphatidic acid (LPA), ROCKs phosphorylate numerous cellular targets mediating a variety of cellular responses.

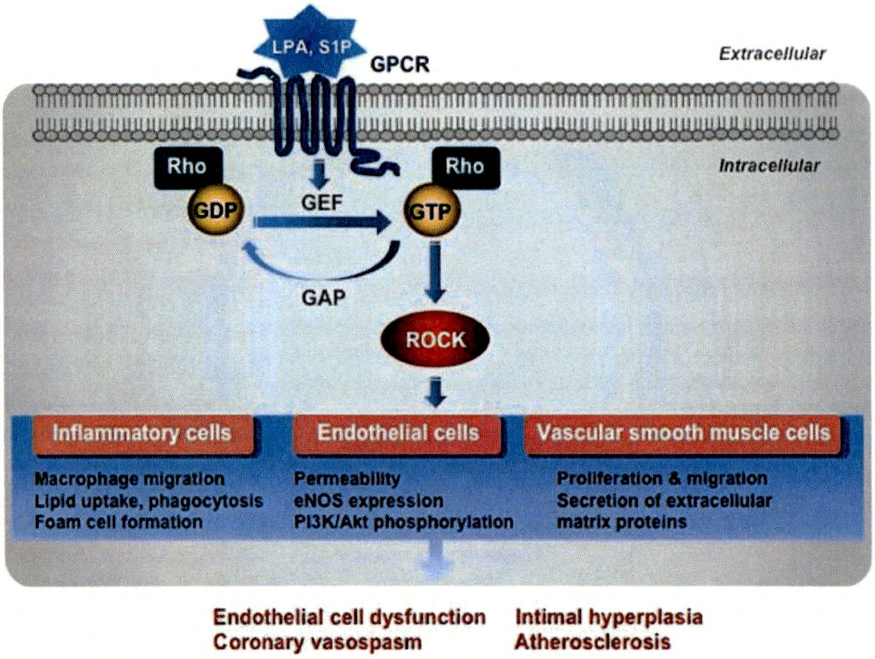


**Fig. 18.4** Rho/ROCK signaling in atherosclerosis. In response to lysophosphatidic acid (LPA) or sphingosine-1-phosphate (S1P) engagement of GPCRs, guanine nucleotide exchange factors (GEF) are activated and promote Rho activation. Activated GTP-bound Rho stimulates ROCKs to phosphorylate numerous downstream targets modulating various processes in inflammatory, endothelial, and vascular smooth muscle cells. Activation of Rho/ROCK signaling pathway eventually contributes to the development of vascular inflammation and atherosclerosis

Among the ROCK substrates are myosin light chain (MLC), myosin-binding subunit of MLC phosphatase (MLCP), LIM protein kinases 1 and 2 and ezrin–radixin–moesin (ERM) proteins, adducin and troponin [111, 125]. Thus, ROCKs are involved in the regulation of actin cytoskeleton assembly and cell contractility, migration, and invasion.

## Statins

Statins have emerged as important cholesterol-lowering factors used in preventive therapy of coronary artery diseases. They efficiently inhibit the 3-hydroxymethylglutaryl coenzyme A (HMG-CoA) reductase by binding to the enzyme's active site and blocking its substrate-product transition state [126, 127]. In addition to the lipid-lowering ability, statins can inhibit synthesis of important isoprenoid intermediates of cholesterol biosynthesis, including farnesyl pyrophosphate and geranylgeranyl pyrophosphate [128]. These intermediates represent lipid attachments to various proteins, such as nuclear lamins, Rac, Rap, Ras, and Rho [110]. Isoprenylation plays an important role for intracellular trafficking and function of small GTPases [129]. Statins inhibit Rho targeting to the cell membrane preventing thereby the Rho-mediated activation of ROCKs. Growing evidence suggests that the pleiotropic effects of statins are mediated by the RhoA/ROCK signaling pathways [130–132].

## ROCKs in Atherogenesis

Rho/ROCK signaling pathways are involved in pathogenesis of atherosclerosis controlling actin cytoskeleton organization, vascular smooth muscle contraction, cell adhesion and motility, and gene expression. Pharmaceutical agents that inhibit ROCKs have been shown to prevent vascular inflammation and atherosclerosis. In animal models, inhibition of ROCK activity results in significant reduction and even regression of atherosclerotic lesions [133, 134].

Rho/ROCK pathway plays a regulatory role in T cell homeostasis and the ROCK inhibitor Y-27632 efficiently downregulates concanavalin A (ConA)-induced T cell proliferation and activation [135–137]. Moreover, Y-27632 has recently been shown to inhibit phosphorylation of ROCK downstream target proteins from ERM family in the atherosclerotic plaques [138]. A recent study has demonstrated that macrophage ROCK1 contributes to the development of atherosclerosis [139]. Indeed, macrophages from ROCK1$^{-/-}$ mice transplanted into atherosclerosis-prone LDL receptor (LDLR)$^{-/-}$ mice possess altered chemotaxis to MCP-1, reduced lipid uptake, and foam cell formation.

In vascular endothelial cells, the activation of Rho/ROCK signaling pathway contributes to oxLDL-induced endothelial cell contractility and to the modulation of endothelial fibrinolytic activity [140, 141]. Rho/ROCKs play a role in VSMC proliferation and migration, angiotensin II (Ang II)-induced MCP-1, and plasminogen activator inhibitor expression [142, 143]. The activation of Rho/ROCK pathways in inflammatory cells and fibroblasts induces the transcriptional activity of several proatherosclerotic factors, including AP-1, NF-κB, and serum response factor [144–146]. ROCK1 is involved in recruitment and infiltration of inflammatory cells to the arterial wall and is the critical mediator of neointima formation while the role of ROCK2 in atherogenesis remains to be determined [147].

As have been noticed above, the antiinflammatory effects of statins are mediated by the inhibition of the Rho/ROCK signaling pathway. Recent study has shown that statin-mediated inhibition of ROCK results in significant improvement of endothelial function in patients with atherosclerosis but not in healthy individuals [148]. Interestingly, hydrophobic statin inhibits ROCKs and improves flow-mediated dilation more efficiently than hydrophilic statin [149]. Statins also induce endothelial eNOS expression through Rho/ROCK inhibition contributing to the regulation of endothelial function [130, 150]. In addition, Rho/ROCK inhibition results in phosphatidylinositol 3-kinase (PI3K)-Akt-mediated phosphorylation and the activation of eNOS [151, 152]. In animal models, administration of ROCK inhibitors fasudil and Y-27632 protected against coronary and cerebral vasospasm confirming the therapeutic potential of ROCK modulators [153, 154].

Thus, the Rho/ROCK signaling pathway is involved in numerous aspects of cardiovascular physiology and pathophysiology, including atherosclerosis, coronary vasospasm, and myocardial ischemia. Targeting of ROCKs with specific pharmaceutical drugs represents a promising therapeutic strategy in the prevention and treatment of cardiovascular disease.

## Oxidative Stress

Oxidative stress (OS) is one of the major causes of the development of inflammation and atherosclerosis (Fig. 18.5) [155, 156]. Vascular endothelial cells express various enzymes, including nicotinamide adenine dinucleotide phosphate (NADPH) oxidase, eNOS, myeloperoxidase, xanthine oxidase, and lipoxygenase, which contribute to the generation of ROS. ROS comprise various molecules, which affect a wide range of cellular functions. ROS are key mediators of signaling pathways that underlie vascular inflammation in atherogenesis, from the formation of fatty streak, through atherogenic lesion progression, and subsequently to plaque rupture.

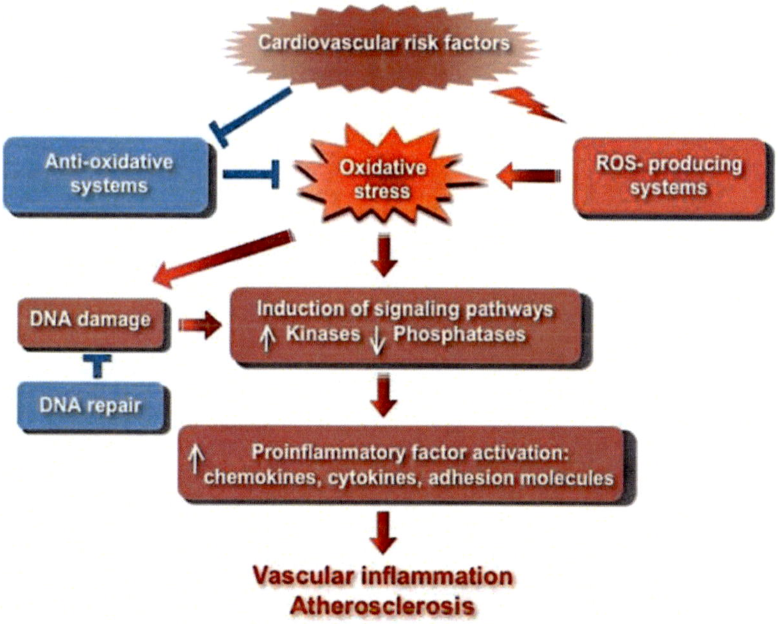


**Fig. 18.5** Role of vascular oxidative stress in the development of atherosclerosis. Cardiovascular risk factors induce ROS-generating systems and inhibit antioxidative systems. The resulting oxidative stress (OS) activates various signaling pathways via activation of protein kinases and inhibition of protein phosphatases. This leads to the activation of multiple proinflammatory factors (chemokines, cytokines, and adhesion molecules) that promote atherogenesis through a number of different mechanisms, including the activation of redox-sensitive transcription factors (which stimulate the expression of proinflammatory genes) and the activation of signaling cascades. OS also promotes nuclear and mitochondrial DNA damage which can be repaired by DNA repair enzyme systems. OS-induced signaling cascades result in the development of vascular inflammation and atherosclerosis

In vascular cells, ROS-generating enzymes are activated in response to inflammation or endothelial cell injury [157, 158]. In addition, enhanced ROS production from dysfunctional mitochondrial electron transport chain (ETC) plays a central role in atherosclerosis and other vascular diseases. Physical forces, such as oscillatory shear at sites where atherosclerosis develops, can also contribute to vascular superoxide ($O_2^{\bullet-}$) production [155]. OS results from an imbalance between the activities of ROS-generating enzyme systems and antioxidant enzyme systems, such as superoxide dismutase (SOD), glutathione peroxidase, catalase, heme oxygenase (HO), and thioredoxin (Trx) peroxidase, in favor of the former (Fig. 18.5).

## *Nitric Oxide Is a Protector Against Atherosclerosis*

NO, a central player in the endothelium-mediated protection against atherosclerosis, is synthesized by eNOS. NO is not only the best characterized relaxing factor blocking abnormal constriction of the arteries, but it also prevents the platelet aggregation, the expression of adhesion molecules in the endothelial cells and the production of vasoconstrictor, endothelin-1 [159–161]. Various products released by aggregating platelets, and thrombin activate eNOS to enhance generation and the release of NO. Interruption of this protective pathway results in the initiation of the inflammatory processes leading eventually to atherosclerosis [161].

Two major mediators secreted by aggregating platelets, which induce eNOS activity to elevate NO production, are ADP and serotonin (5-hydroxytryptamine, 5-HT). While ADP plays a minor role in the activation of eNOS engaging $P_2Y$ purinoceptors, serotonin is a major mediator acting through $5\text{-HT}_{1D}$ receptors on the surface of the endothelial cells (Fig. 18.6). Both, thrombin receptor (protease-activated receptor [PAR]) and serotonergic receptors mediate activating signal to eNOS via $G_i$ proteins, whereas the $P_2Y$ purinoceptors and bradykinin receptors activate NO production through $G_q$ proteins. Thus, under physiological conditions the endothelial cells in response to platelet aggregation upregulate NO production and release. Released NO induces the relaxation of the underlaying smooth muscles leading to increasing blood flow and preventing the development of coagulation. Under pathological conditions, if the endothelial cells are dysfunctional and damaged, unable to release relaxing signal, the aggregating platelets can reach the vascular smooth muscles and induce their contraction by secreting vasoconstrictors [161].

Mammalian NOS family includes three isoforms, neuronal NOS (nNOS; NOS I), iNOS (NOS II), and eNOS (NOS III). Although all isoforms catalyze the oxidation of the terminal guanidine group of L-arginine to generate NO and L-citrulline, they possess distinctive enzymatic and regulatory characteristics [159, 162, 163]. NOS isoforms are encoded by distinct genes and share 50–60% sequence identity.

The NOS N terminus is homologous in all isoforms and the C terminus shares significant homology with cytochrome P450 reductases. The N terminus binds tetrahydrobiopterin ($BH_4$) and heme while the active site located near heme binds

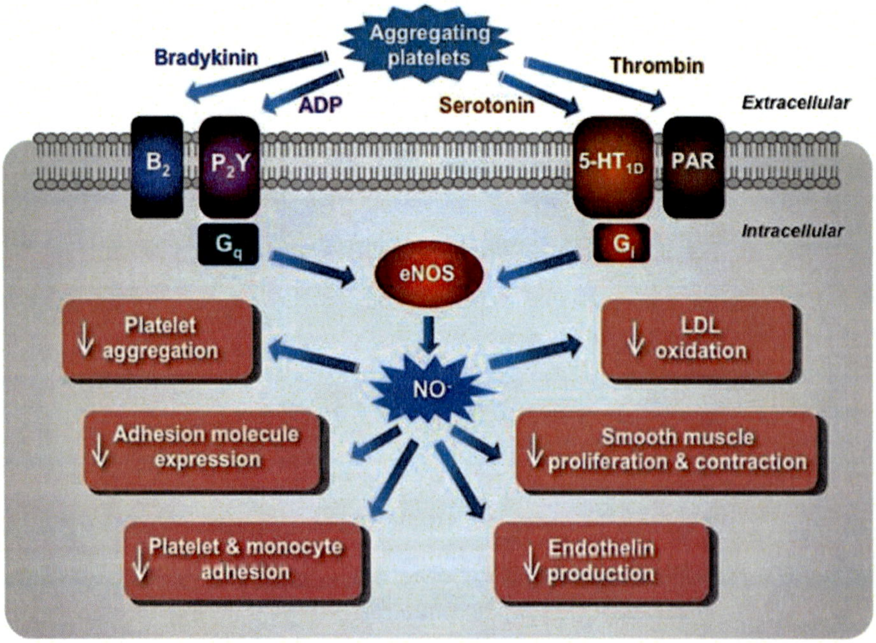


**Fig. 18.6** Schematic representation of eNOS activation. GPCR-mediated signaling induces eNOS activity to elevate NO production. Both thrombin receptor (protease-activated receptor, PAR) and serotonergic receptors ($5\text{-HT}_{1D}$) mediate activating signal to eNOS via $G_i$ proteins, whereas the $P_2Y$ purinoceptors and bradykinin receptors ($B_2$) activate NO production through $G_q$ proteins. Increased NO levels attenuate a variety of proinflammatory processes

L-arginine. $BH_4$ is an essential cofactor for NOS. Upon OS, $BH_4$ can be oxidized to generate 7,8-dihydropterin ($BH_2$) and biopterin. It has been suggested that redox status of $BH_4$ and $BH_2$ may be clinically relevant; however, the precise mechanisms and role of this process remain to be determined [164–166]. The C terminus binds NADPH, flavin mononucleotide and flavin adenine dinucleotide (FMN and FAD) cofactors. During synthesis of NO, NOS transfers electrons within its reductase domain from NADPH to FAD and then to FMN. Dimerization of NOS is required for its maximal enzymatic activity [167].

eNOS activity is regulated by the availability of substrate and cofactors. Under physiological conditions, the intracellular concentrations of L-arginine and $BH_4$ are relatively high to stimulate NO synthesis and to decrease $O_2^{\bullet-}$ production. However, OS and CVD are associated with decreased L-arginine concentration due to the activation of arginase and arginine decarboxylase and with reduced $BH_4$ levels [168, 169]. These pathological conditions lead to the NOS uncoupling, i.e., switch to the generation of $O_2^{\bullet-}$ and $H_2O_2$ resulting in reduced NO availability and eventually in endothelial dysfunction (Fig. 18.7).

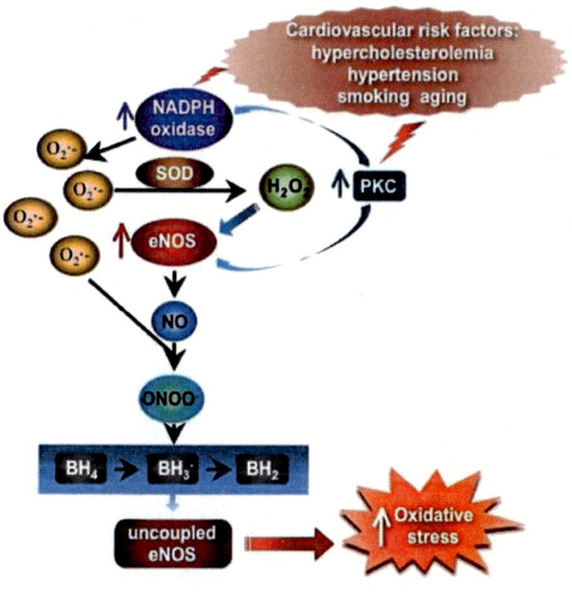


**Fig. 18.7** Cardiovascular risk factors promote eNOS uncoupling. Cardiovascular diseases upregulate NADPH oxidases to produce elevated levels of superoxide ($O_2^{\bullet-}$). Superoxide dismutase (SOD) converts $O_2^{\bullet-}$ into $H_2O_2$ that induces eNOS expression. Protein kinase C (PKC) activation can also contribute to eNOS induction. The products of NADPH oxidases and eNOS, $O_2^{\bullet-}$ and NO, rapidly recombine to generate peroxynitrite ($ONOO^-$). $ONOO^-$ can oxidize the essential cofactor of eNOS, (6R-)5,6,7,8-tetrahydrobiopterin ($BH_4$) to trihydrobiopterin radical ($BH_3^{\bullet}$) and then to the quinonoid 6,7-[8H]-H2-biopterin ($BH_2$). These pathological conditions lead to the eNOS uncoupling, i.e., switch to the generation $O_2^{\bullet-}$ and $H_2O_2$ resulting in reduced NO availability, vascular oxidative stress and eventually in endothelial dysfunction. The increased eNOS expression aggravates the situation

In addition, eNOS is highly regulated by various posttranslational modifications. eNOS can be phosphorylated at several serine, threonine, and tyrosine residues by several protein kinases, including Akt, protein kinase A (PKA), AMP-activated protein kinase (AMPK), cGMP-dependent protein kinase, and $Ca^{2+}$/calmodulin-dependent protein kinase II (CaM kinase II) [159, 160, 163]. Phosphorylation at Ser617 (sequence numbering corresponds to the most characterized bovine eNOS), Ser635, Ser1177, and Tyr83 stimulates eNOS activity, whereas phosphorylation at Ser116, Thr497, and Tyr659 inhibits the enzyme. Various protein phosphatases, such as protein phosphatase 1 and 2A, and calcineurin, are involved in dephosphorylation of eNOS contributing to the regulation of eNOS signaling. However, functional consequences of these phosphorylation events as well as the different extracellular stimuli activating distinct kinase pathways are not completely understood.

The irreversible myristoylation of the N-terminal glycine of eNOS followed by palmitoylation at Cys15 and Cys26 target eNOS to the plasma membrane microdomain called caveolae. N-myristoyltransferase catalyzes N-myristoylation of eNOS while palmitoyltransferase DHHC-21 may contribute to palmitoylation of the enzyme. Prolonged agonist activation of eNOS results in its depalmitoylation by acyl protein thioesterase-1 and intracellular translocation inhibiting eNOS signaling pathway [163, 170].

Reversible S-nitrosylation of eNOS inhibits the enzyme activity while denitrosylation in response to agonist treatment of endothelial cells leads to its temporal activation [171]. In addition to the modulation of eNOS enzymatic activity, S-nitrosylation regulates the subcellular localization of eNOS targeting it to the plasma membrane.

Finally, eNOS is involved in interactions with multiple proteins, including caveolin-1, heat shock protein Hsp90, soluble guanylate cyclase, adapter protein Gab1, protein tyrosine phosphatase SHP2, motor protein dynamin-2, and two novel proteins, the *eNOS interacting protein* (NOSIP) and the *eNOS trafficking inducer* protein (NOSTRIN) [172, 173]. These protein–protein interactions affect enzymatic activity and subcellular localization of eNOS and add complexity to the regulation of the production of NO, an important protector against inflammation and atherosclerosis.

## NADPH Oxidase

Membrane-associated NADPH oxidase is the principal source of ROS in vascular endothelial cells, which plays a central role in the development of atherosclerosis. It belongs to the Nox family of NADPH oxidases, which catalyze

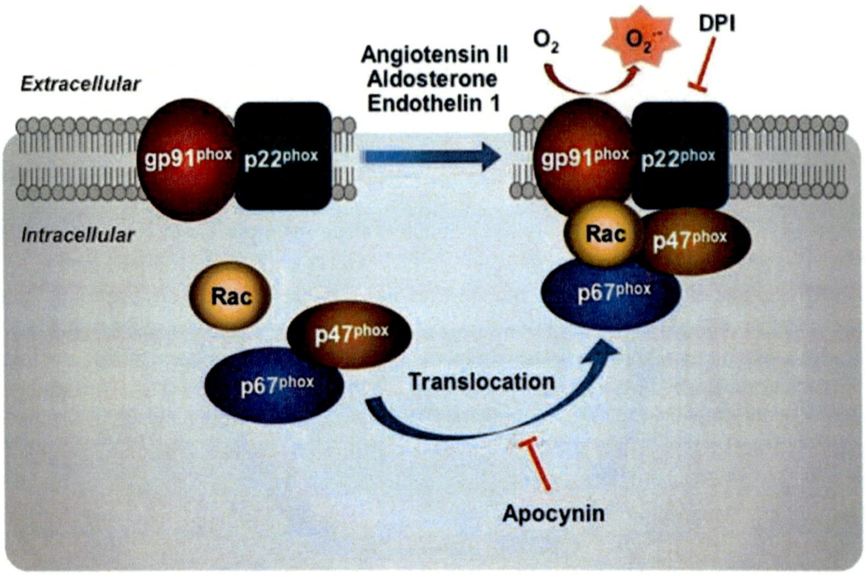


**Fig. 18.8** Schematic representation of NADPH oxidase structure and activation. The prototypical NADPH oxidase contains four subunits: the membrane-associated subunits gp91$^{phox}$ and p22$^{phox}$ and cytosolic subunits p47$^{phox}$ and p67$^{phox}$. Two cytosolic subunits and Rac in response to stimulation by angiotensin II, aldosterone, or endothelin 1 translocate to the plasma membrane to assemble catalytically active NAPDH oxidase complex. Apocynin inhibits NADPH oxidase by blocking translocation of cytosolic subunits and the active complex formation, while diphenyleneiodonium (DPI) is a nonspecific flavoprotein inhibitor of NADPH oxidase

transfer of electrons across cellular membranes generating ROS [174–177]. They function as electrogenic enzymes using NADPH as a source of electrons and molecular oxygen as an acceptor.

The prototypical NADPH oxidase, expressed in phagocytic cells, contains four subunits: the membrane-associated gp91$^{phox}$ (Nox2) and p22$^{phox}$ and cytosolic p47$^{phox}$ and p67$^{phox}$. Two cytosolic subunits in response to stimulation translocate to the plasma membrane to assemble catalytically active NAPDH oxidase complex [178, 179] (Fig. 18.8). Seven Nox isoforms are expressed in various tissues. Nox1, Nox2, Nox4, and Nox5 represent isoforms expressed in the vascular tissue.

NADPH oxidase can be activated by vasoactive hormones (Ang II, aldosterone, endothelin 1), growth factors (platelet-derived growth factor, TGF-β), mechanical stimuli (shear stress and stretch) and cardiovascular risk factors (oxidized lipids, high blood pressure, cigarette smoke) [180, 181]. Vascular NADPH oxidase activation by Ang II is the best characterized and involves receptor tyrosine kinases, protein kinase C (PKC), tyrosine kinase c-Src, and phospholipase D [180, 182].

Genetic disruption of NADPH oxidase subunits in mice has confirmed an essential role of the enzyme in the development of inflammation and atherosclerosis. Loss of p47$^{phox}$ has led to decreased blood pressure response to Ang II and reduced atherogenesis in apoE$^{-/-}$ mice [183, 184]. Similarly, mice lacking Nox1 exhibit reduced vascular ROS production and blood pressure responses to Ang II while Nox1 overexpression potentiates Ang II-induced ROS production and hypertension [185, 186].

Thus, NADPH oxidase-produced ROS affect a variety of processes contributing to the development of vascular endothelial dysfunction, inflammation, and eventually to atherosclerosis (Fig. 18.9).

## *Xantine Oxidase*

Xantine oxidase, a product of proteolysis of xantine dehydrogenase, catalyzes transfer of electrons to molecular oxygen generating the ROS superoxide ($O_2^{\bullet-}$) and hydrogen superoxide. Endothelial cells express xantine dehydrogenase, which can be converted into xantine oxidase. Interestingly, the expression of xantine dehydrogenase is redox sensitive and depends on the endothelial NADPH oxidase [187]. Another source of xantine oxidase is the liver. It has been shown that elevated levels of cholesterol induce its release from the liver into circulating plasma, where it can associate with endothelial glycosaminoglycans [188].

Increased xantine oxidase activity has been associated with hyperuricemia and elevated vascular $O_2^{\bullet-}$ production contributing to endothelial dysfunction [189, 190]. Elevated xantine oxidase activity has also been implicated in human hypertension and administration of xantine oxidase inhibitor allopurinol normalizes blood pressure in patients with hypertension [191, 192].

Although xantine oxidase inhibitor oxypurinol inhibits vascular $O_2^{\bullet-}$ production and improves endothelium function, the importance of xantine oxidase for the development of

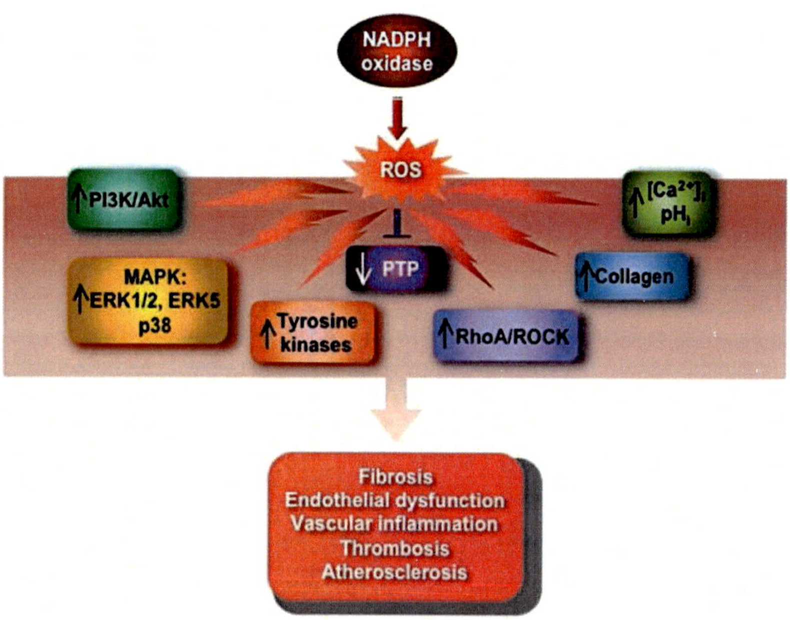


**Fig. 18.9** NADPH oxidase is the principal source of ROS in vascular cells. Activation of NADPH oxidase results in elevated ROS production, which in turn modulates various downstream targets, including inhibition of protein tyrosine phosphatases (PTP) and stimulation of protein kinases, such as MAPKs (ERK1/2, ERK5, and p38), tyrosine kinases, ROCKs, and PI3K/Akt. In addition, increased intracellular ROS production also induces an increase in intracellular pH ($pH_i$) and in intracellular free $Ca^{2+}$ concentration ($[Ca^{2+}]_i$) as well as upregulation of collagen expression. All these changes contribute to the development of fibrosis, endothelial dysfunction, vascular inflammation, thrombosis, and atherosclerosis

endothelial dysfunction and atherosclerosis remains uncertain. Clinical data regarding the use of xantine oxidase inhibitors to improve endothelial function in diabetic and hypercholesterolemic patients are also controversial [193–195].

## Mitochondrial Oxidative Dysfunction

Growing evidence suggests an essential role of mitochondria in the vascular inflammatory process leading to atherosclerotic lesions [196–198]. Under physiologic conditions, NADPH oxidase and mitochondria represent the main sources of ROS in most mammalian cells [199, 200]. Elevated mitochondrial ROS generation is an early step in the development of atherosclerosis.

Approximately 1% of the $O_2$ utilized by mitochondria is reduced by a single electron to form $O_2^{\bullet-}$. In mitochondrial ETC, $O_2^{\bullet-}$ production is catalyzed by the NADH dehydrogenase (complex I) and by the ubiquinol-cytochrome *c* reductase (complex III) [201]. Mitochondrial $O_2^{\bullet-}$ production depends also on the activity of manganese superoxide dismutase 2 (MnSOD or SOD2) localized in the mitochondrial matrix. Superoxide is quickly converted into hydrogen peroxide ($H_2O_2$) by metalloenzymes, SOD2 or copper/zinc superoxide dismutase 1 (Cu/Zn SOD or SOD1) [202]. In mitochondria, $H_2O_2$ is mostly reduced to water by GPx or catalase, the later is present in mitochondria from the heart but not from other tissues [203, 204]. It also can be transformed in highly reactive hydroxyl radical in the presence of reduced transition metals [205]. For the reduction of $H_2O_2$, GPx uses reduced glutathione (GSH), transported from cytosol into mitochondria, as the hydrogen donor. Oxidized glutathione resulted from this reaction can be reduced to GSH by glutathione reductase.

Mitochondria can generate NO, which controls mitochondrial oxygen consumption by the reversible inhibition of cytochrome *c* oxidase [206–208]. NO can react with $O_2^{\bullet-}$ producing an oxidant, peroxynitrite ($ONOO_2^-$), which is able to nitrate irreversibly various proteins, inhibit enzymes, induce DNA lesions and mitochondrial dysfunction [209, 210]. On the other hand, the inhibition of complex I activity by forming S-nitrothiols reduces mitochondrial ROS production [211].

However, mitochondria are not only sources, but they are also targets of ROS. ROS inhibit multiple mitochondrial enzymes, including complexes I, II, and III, aconitase, α-ketoglutarate dehydrogenase, and pyruvate dehydrogenase [212]. Oxidative inactivation of adenine nucleotide translocase (ANT) and mitochondrial DNA (mtDNA) polymerase γ result in the inhibition of oxidative phosphorylation (OxPhos) affecting all cellular ATP-dependent processes [213, 214]. SOD2 is inactivated by nitration of Tyr34 located in its active cite [215]. Moreover, ROS induce the oxidation of cardiolipin, a phospholipid component of the mitochondrial inner membrane, which plays an essential role in mitochondrial bioenergetics [216–218]. Oxidized cardiolipin inhibits the activity of complex I and induces the release of cytochrome *c*.

Mitochondrial ROS production is controlled by multiple factors, including concentration of oxygen, ETC efficacy, availability of electron donors NADH and $FADH_2$, and activity of cytokines and uncoupling proteins (UCPs) [196, 219, 220].

Cholesterol and oxLDLs cause mitochondrial damage and dysfunction [196]. By triggering mitochondrial-mediated apoptosis in various cells, oxLDLs contributed to the development of inflammation and atherosclerosis (Fig. 18.10). In vascular cells, oxLDLs induce two distinct apoptotic pathways [221]. The first pathway involves the activation of the cysteine protease calpain, opening of the permeability transition pore leading to the release of the proapoptotic proteins tBid (truncated version of Bcl-2 family member Bid) and cytochrome *c* with subsequent activation of caspase-3. The second pathway is caspase-independent and mediated by the release of apoptosis-inducing factor.

Free cholesterol accumulated in the macrophages in advanced atherosclerotic lesions, induces the decrease of mitochondrial membrane potential, the release of cytochrome *c*, the activation of caspase-9 and increase in the levels of the proapoptotic Bax [222]. Similarly, oxLDLs also cause mitochondria-mediated apoptosis leading to macrophage lysis while scavengers of peroxide radicals prevent this process [223]. Loss of SOD2 induces mtDNA damage and accelerates the development of atherosclerosis in apoE$^{-/-}$ mice [224]. mtDNA damage and protein nitration are significantly elevated upon hypercholesterolemia demonstrating that atherosclerotic risk factor promotes mitochondrial damage (Fig. 18.10) [225]. Consistently, the copy number of mtDNA in leukocytes of patients with hyperlipidemia is lower than in healthy subject [226].

Thus, OS causes progressive mitochondrial damage and dysfunction. These pathological alterations of mitochondrial function trigger multiple signaling pathways playing critical roles in the onset and progression of atherosclerosis.

## Antioxidant Enzymes

SOD, GPx, catalase, heme oxygenase (HO) and thioredoxin (Trx) peroxidase are the major antioxidant enzymes protecting against vascular OS [227].

SOD catalyzes the conversion of $O_2^{\bullet-}$ into $O_2$ and $H_2O_2$. Three isoforms of SOD are expressed in humans, cytosolic SOD1, mitochondrial SOD2, and extracellular SOD3. SOD1 deficient mice are viable and develop normally and exhibit no marked phenotype under normal conditions, but they are more susceptible to cardiac ischemia-reperfusion injury. The

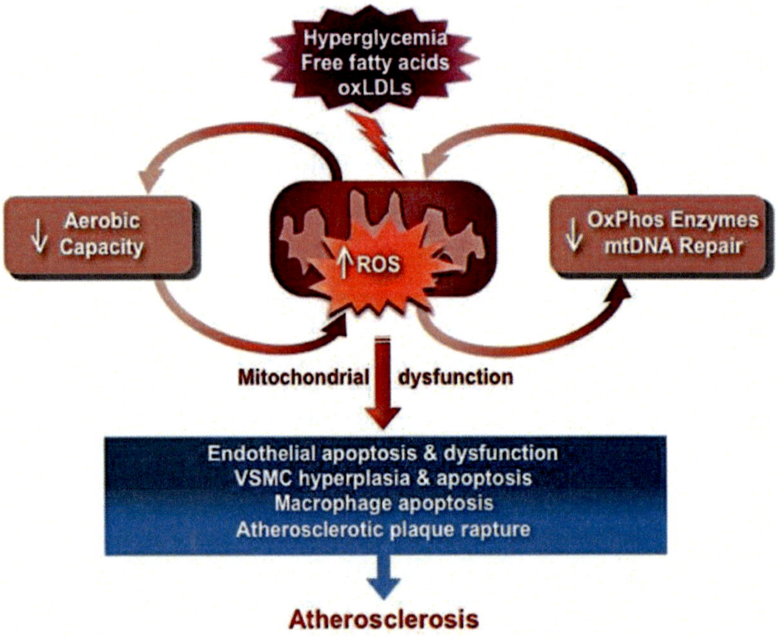


**Fig. 18.10** Role of mitochondrial dysfunction in atherogenesis. Cardiovascular risk factors, such as hyperglycemia, elevated levels of free fatty acids and oxLDLs, induce mitochondria-mediated ROS generation and mitochondrial dysfunction and damage. Mitochondrial dysfunction is associated with the inhibition of oxidative phosphorylation (OxPhos) enzymes leading to decrease in mitochondrial aerobic capacity. Mitochondria-mediated ROS generation induces mtDNA damage and inhibits mtDNA repair contributing further to mitochondrial dysfunction. Elevated ROS production leads to apoptosis of endothelial cells, VSMCs and macrophages, and highly contributes to the progression of atherosclerotic lesions, plaque rapture, and atherosclerosis

role of SOD2 in mitochondrial function has been discussed above. SOD3 decreases $O_2^{\bullet-}$ production and maintains NO levels in the cardiovascular system and its inactivation contributes to the development of hypertension [228].

Several GPx isoforms are known, among them GPx1 is the predominant isozyme expressed in mammalian tissues. GPx1-deficient mice develop normally but are more susceptible to cardiac ischemia-reperfusion injury [229]. Moreover, the loss of GPx1 facilitates atherosclerotic progression in apoE$^{-/-}$ mice [230]. GPx4 knockout mice are early embryonically lethal.

Catalase catalyzes the conversion $H_2O_2$ into $H_2O$ and $O_2$. Catalase knockout mice are viable and exhibit no marked abnormalities [231]. Biological significance of this enzyme remains to be determined.

HO catalyzes the rate-limiting first step in the degradation of the prooxidative heme, producing equimolar amounts of biliverdin, carbon monoxide (CO), and $Fe^{2+}$. OS-induced HO activation is an adaptive response contributing to the protection against vascular remodeling and atherosclerosis [232]. Formed biliverdin is subsequently converted into bilirubin, a scavenger for free radicals and inhibitor of NADPH oxidase, the principal source of ROS in vascular system [233]. Moreover, a decrease in heme amount due to its breakdown by HO1 limits heme availability for the assembly of functional NADPH oxidase reducing thereby ROS production [234]. In addition, CO produced by the HO catalysis possesses antiinflammatory and vasodilator characteristics [235]. Animal genetic models of HO1 overexpression or its deficiency confirm the antiatherosclerotic role of the enzyme [236].

Trx is expressed in vascular endothelial and VSMCs. Its ROS-scavenging capabilities can be attributed to Trx peroxidase activity [237]. Trx reduces the oxidized Trx peroxidase and resultant reduced Trx peroxidase is able to scavenge $H_2O_2$ and $ONOO^-$.

## Phosphatidylinositol 3-Kinase (PI3K) Signaling

As we have already discussed in previous chapter (Chap. 7), PI3K plays important and complex roles in lipid signaling, regulating various processes in the cardiovascular system under physiological and pathological conditions. PI3K signaling controls multiple processes contributing to the development of atherosclerosis and inflammation. Since PI3Kγ is highly expressed in the hematopoietic cells, this isoform plays major role in inflammation associated with atherosclerosis. PI3Kγ signaling is activated by oxLDLs and other proinflammatory lipids, Ang II, and various chemokines, including CCL-2/MCP-1, CCL-3/MIP-1α, and IL-8 [238]. Several other proatherogenic stimuli, such as IFN-γ, TGF-β, and TNF-α, can also induce PI3K signaling pathway through engagement of various receptors [239–241].

The PI3Kγ-specific inhibitor AS605240 has significantly reduced lesion size in a mouse model of early and advance stage atherosclerosis [242]. Treatment with the inhibitor has resulted in a significant decrease in Akt phosphorylation within atherosclerotic plaques suggesting an important role for this effector of PI3K in atheroprotection. In agreement, double apoE-Akt1 deficient mice develop more severe atherosclerotic lesions in the aorta and coronary vessels than apoE deficient mice [243]. Importantly, the loss of Akt1 has been associated with downregulation of eNOS phosphorylation and increased inflammatory signaling.

## TNF in Atherogenesis

Cytokines act as messengers in the immune system via binding specific receptor complexes, which induce in turn signal transduction cascades to modulate the expression of multiple genes, and thereby mediate a wide range of regulatory and effector functions. Homotrimeric cytokine TNF interacts with two specific receptors, TNFRI and TNFRII, to control various cellular processes under physiological and pathological conditions [244–246]. TNF is synthesized in the endoplasmic reticulum and trafficked to the plasma membrane to convert into a functional membrane-bound protein (Fig. 18.11). Membrane-bound cleaving enzyme, TNF-α converting enzyme (TACE), can generate a soluble form of TNF (sTNF).

TNF plays a crucial regulatory role in lipid metabolism and inflammatory response through leukocyte activation and maturation as well as cytokine, chemokine, and bioactive intermediate production and release. As one of the major proinflammatory cytokine, TNF also induces vascular endothelial cells to produce various adhesion molecules and proinflammatory cytokines, chemokines, and their receptors contributing thereby to the recruitment of activated lymphocytes to the atherosclerotic lesions (Fig. 18.12).

### *TNF and Lipid Metabolism*

Growing evidence suggests a close relationship between TNF and lipid metabolism. In animal models, a cholesterol-rich diet increases significantly serum TNF-α levels [247]. TNF-α concentrations correlate positively with the plasma levels of proatherosclerotic very low density lipoproteins (VLDLs) in hyperlipidemic patients, while negatively with the levels of HDL particles [248]. Patients with dyslipidemia exhibit abnormal TNF-α levels that can be normalized by administration of antiatherosclerotic agents, such as statins and peroxisome proliferator-activated receptor (PPAR)-α inhibitors [249]. Administration of anti-TNF-α monoclonal antibody to patients

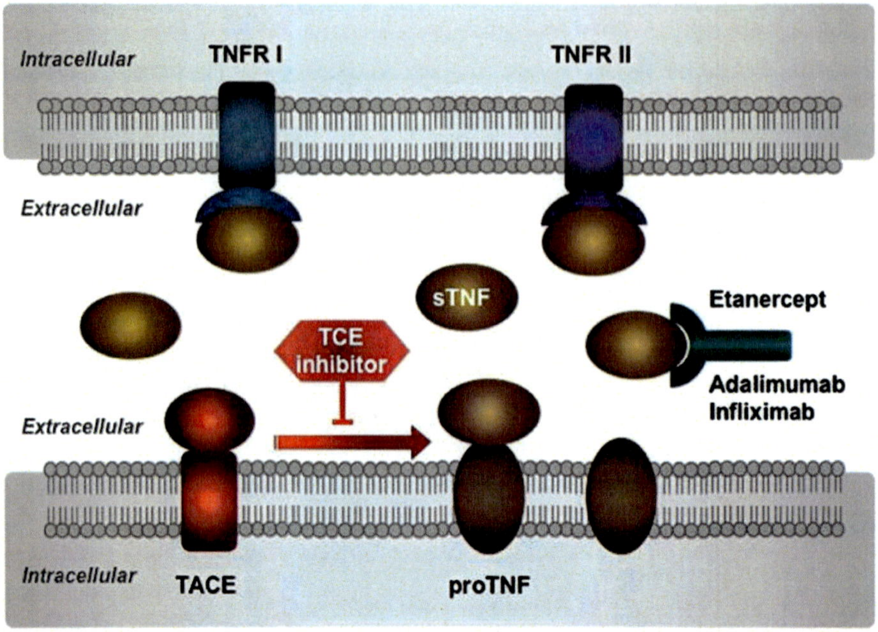


**Fig. 18.11** Membrane-bound and soluble forms of TNF. TNF synthesized in the endoplasmic reticulum is recruited to the plasma membrane to convert into a functional membrane-associated protein. Membrane-bound cleaving enzyme, TNF-α converting enzyme (TACE), can generate a soluble form of TNF (sTNF). TNF can modulate various cellular processes through interactions with its specific receptors TNFRI and TNFRII. Anti-TNF drugs, adalimumab, infliximab, and etanercept, bind proTNF or sTNF and prevent their interactions with the receptors blocking downstream cellular-signaling cascades

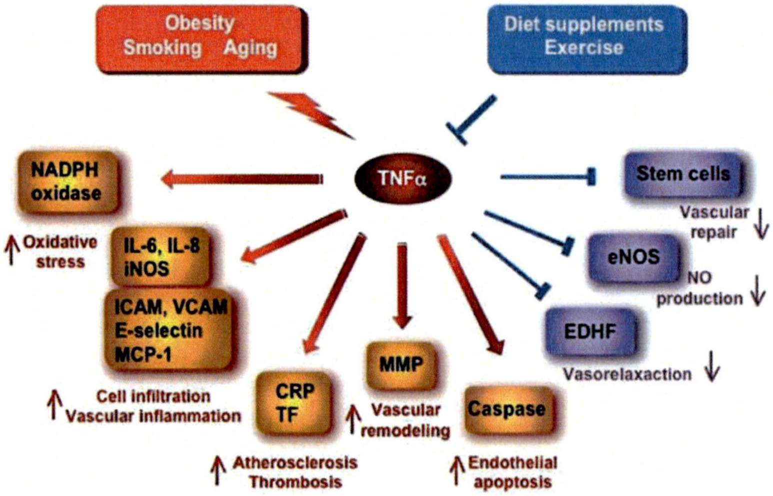


**Fig. 18.12** Proatherosclerotic role of TNF-α signaling. Risk factors, such as obesity, smoking, physical inactivity, and aging (*red box*), converge on TNF-α to induce proinflammatory and proatherosclerotic responses. These factors trigger TNF-α-mediated upregulation of NADPH oxidase, various inflammatory cytokines and chemokines (IL-6, IL-8), adhesion and chemoattractant molecules [ICAM, VCAM, monocyte chemoattractant protein-1 (MCP-1)], E-selectin, matrix metalloproteinases (MMPs), and other proinflammatory and proapoptotic proteins (C-reactive protein (CRP), procoagulant tissue factor (TF), caspase). Activated TNF-α downregulates vascular repair and inhibits NO bioavailability and vasorelaxation. Dietary supplements and regular exercise (*blue box*) attenuate TNF-α-mediated proinflammatory signaling protecting against the development of cardiovascular diseases

with active rheumatoid arthritis (RA) induces increased levels of HDL particles and decreases the atherosclerotic index [250, 251]. In animal models, administration of TNF-α leads to a significant increase in serum triglyceride levels whereas inhibition of TNF-α attenuates the lipopolysaccharide (LPS)-induced increase in serum triglycerides [252, 253]. The molecular mechanisms of the effect of TNF-α on lipid metabolism are complex and remain to be determined.

## TNF Effects on Endothelial Dysfunction

As a potent pro-inflammatory trigger, TNF-α downregulates eNOS expression and activity in endothelial cells while induces expression of another isoform, iNOS [254–256]. The direct inhibitory effect of TNF-α on eNOS expression and NO production has been demonstrated in various vasculatures [257–259]. In addition to eNOS, argininosuccinate synthase (ASS) also regulates NO production in endothelial cells [260–263]. ASS catalyzes the rate-limiting step of the citrulline/NO cycle in arginine regeneration, while eNOS synthesizes NO through the conversion of L-arginine into L-citrulline. It has been demonstrated that TNF-α significantly inhibits ASS expression in aortic endothelial cells downregulating thereby NO production [256]. Interestingly, the citrulline/NO cycle and eNOS activity appear to be regulated in coordinated fashion [264]. However, TNF-α decreases the bioavailability of NO by not only inhibiting its eNOS-mediated production but also by inducing its removal [259].

It has recently been reported that advance glycation end product (AGE) with its receptor RAGE, and NF-κB signaling pathway are involved in upregulation of vascular TNF-α [259]. The elevated TNF-α levels induce ROS production resulting eventually in endothelial dysfunction. The increased expression of TNF-α observed upon myocardial ischemia-reperfusion injury has been demonstrated to activate xanthine oxidase-mediated $O_2^{\bullet-}$ production leading to coronary endothelial dysfunction [265].

### *TNF and ROS Generation*

The generation of ROS induces NF-κB-mediated transcription of various cytokine genes, which initiates a cytokine signaling cascade leading to upregulation of TNF [266, 267]. As it has previously been discussed, $O_2^{\bullet-}$ is the precursor to many ROS including $H_2O_2$ and $ONOO^-$. In endothelial cells and neutrophils, TNF-α contributes to $O_2^{\bullet-}$ production through activation of NADPH oxidase, xantine oxidase, ceramide-activated protein kinase and NOS [199, 268–270]. TNF-α-mediated activation of NADPH oxidase, a principal source of $O_2^{\bullet-}$, involves upregulation of the membrane-associated $p22^{phox}$ as well as the cytosolic $p47^{phox}$ and $p67^{phox}$ subunits [271, 272]. This complex signaling cascade involves multiple downstream pathways, such as mitogen-activated protein kinases (MAPKs), Janus kinase (JAK)/signal transducer and activator of transcription (STAT), PI3K, cdc42/rac, p21ras, and NF-κB [273].

TNF-α triggers signaling through the IκB kinase (IKK) complex leading to the release of NF-κB from the inhibitor of NF-κB (IκBα) and its translocation to the nucleus (see also Chap. 19, Fig. 19.3). In the nucleus, NF-κB regulates the expression of various genes involved in inflammatory and OS response [274–276]. In diabetic mouse model, NF-κB induces the expression of endothelial TNF-α, IL-6, MCP-1, and several adhesion molecules accentuating OS and resulting in endothelial dysfunction [246, 277]. Furthermore, AGE/RAGE signaling cascade through TNF-α triggers $O_2^{\bullet-}$ overproduction leading to eNOS uncoupling and eventually to vascular inflammation and atherosclerosis [246, 278].

Aging represents an independent risk factor for cardiovascular diseases and aging-related alteration of TNF-α signaling plays an essential role in proinflammatory shift in vascular function [279]. Serum levels of TNF-α are increased in the elderly [280]. Upregulation of TNF-α signaling pathway found in arteries of aged rats appears to induce apoptosis of endothelial cells contributing to endothelial dysfunction and atherogenesis in the elderly [281, 282]. Abnormal TNF-α expression is associated with cardiovascular aging and pharmacological TNF blockade may exert antiaging vasculoprotective effects.

Finally, it has been suggested that regular exercise suppresses several proinflammatory factors, including TNF-α, IL-6, and C-reactive protein (CRP), while induces antiinflammatory factors such as IL-4, IL-10, TGF-β, and adiponectin [283–286]. The antiinflammatory effects of exercise represent the mechanism of its protective role against the development of cardiovascular disease.

In summary, TNF is one of key regulators of vascular homeostasis playing a pivotal role in endothelial dysfunction, vascular inflammation, and atherosclerosis. Various risk factors, including obesity, smoking and aging, converge on TNF to initiate OS, vascular remodeling and inflammation, apoptosis, and thrombosis leading to vascular damage. Clinical studies suggest the potential benefits of anti-TNF therapy for ameliorating cardiovascular disorders and vascular complications in various diseases.

## Antiinflammatory Factors

### *TGF-β Signaling Pathway*

T-cell-mediated proinflammatory activation is under inhibitory control by two antiinflammatory cytokines, IL-10 and TGF-β. Various cells, including endothelial cells, smooth muscle cells, platelets, macrophages, and regulatory T ($T_{REG}$) cells, are able to produce these antiinflammatory cytokines, however, the $T_{REG}$ cells ($CD4^+CD25^+FoxP3$) represent their primary source. The $T_{REG}$ cell depletion aggravates atherogenesis while their transplantation into atherosclerosis-susceptible mice inhibits the disease [287]. Blockage of IL-10 by gene targeting or by pharmacologic inhibition accelerates the development of atherosclerosis and coronary thrombosis in hypercholesterolemic mice [288–290].

TGF-β exerts potent antiinflammatory actions on all cell types that contribute to atherogenesis, including endothelial cells, VSMCs, macrophages, and lymphocytes [291]. TGF-β represents a small family composed of three members

TGF-β, -β2, and -β3 in mammalian cells [292]. TGF-β-like proteins include activins, inhibins, growth/differentiation factors, and bone morphogenic proteins (BMPs). TGF-β exists as a latent complex bound to extracellular matrix proteins, such as collagen type IV, fibronectin, and thrombospondin, and can be activated by several proteases, including plasmin, translutaminase, and thrombospondin [293].

Growing evidence suggests that TGF-β is a key regulator of vascular repair playing two important roles in inflammation resolution [292, 294]. First, TGF-β attenuates inflammatory processes especially those mediated by $T_{REG}$ cells and phagocytes [295, 296]. Second, it induces collagen production by intimal fibroblasts and initiates thereby a protective scar-like structure formation stabilizing atherosclerotic plaque [297]. Alterations in the scar response in intimal VSMCs contribute to thinning of fibrous caps leading eventually to plaque rapture [298]. Inhibition of TGF-β signaling by anti-TGF-β antibodies or by decoy soluble TGF-β receptor in $apoE^{-/-}$ mice accelerate vulnerable plaque formation with enlarged necrotic cores and thin fibrous caps [299, 300]. Similarly, disruption of TGF-β signaling pathway in T cells results in significant increase of atherogenesis [301]. TGF-β overexpression in atherosclerosis-susceptible mice stabilizes atherosclerotic lesions and decreases the development of atherosclerosis [297]. Moreover, significant downregulation of TGF-β has been found in patients with atherosclerosis.

B cells can also contribute to the prevention of vascular inflammation and atherosclerosis through the production of specific antibodies against plaque antigens. These antibodies can recognize components of oxLDL particles and apoptotic cell membranes contributing to their elimination and inhibition of atherosclerotic lesion formation [302, 303].

## *Peroxisome Proliferator-Activated Receptors*

Recently, another group of antiatherosclerotic factors known as PPARs has been discovered to be mediators of pharmacologic agents, which reduce T cell activation attenuating vascular inflammation and atherosclerosis [304]. PPARs belong to the nuclear hormone-receptor family and consist of three related members: PPARα, PPARβ/δ, and PPARγ encoded by the distinct genes *PPARA, PPARD*, and *PPARG*, respectively [305]. They function as nuclear transcription factors regulating in addition to peroxisome proliferation a variety of cellular functions, including glucose metabolism, insulin sensitivity, fatty acid oxidation, cytokines production, and cardiovascular protection.

Various natural ligands can bind and activate PPARs. Ligand-activated PPARs form heterodimers with the retinoid X receptor (RXR), then recognize and bind a DNA sequence upstream of the PPAR target genes, which are called the peroxisome proliferator response element (PPRE). Recruitment of coactivators or corepressors results eventually in the activation or repression of PPAR target gene expression [306–308]. Moreover, PPARs can interact physically with other transcription factors, such as NF-κB, AP-1, Smad-3, and STATs causing the so-called transrepression of certain genes. The phosphorylation of PPARs by various protein kinases, including p38 MAPK, extracellular-signal-regulated kinase (Erk), "breakpoint claster region" protein kinase (Bcr) and PKA, modulates their basal and ligand-dependent transcriptional activity and their pleiotropic biological functions [309–313].

Natural PPAR ligands can be divided into major groups, fatty acids and phospholipids. Both oxidized lipids and phospholipids and prostaglandins also bind and activate PPAR signaling pathway [304]. Unsaturated fatty acids, eicosanoid derivatives and fibrates are also able to stimulate PPARs.

PPARγ is predominantly expressed in adipose tissue and is involved in the regulation of adipocyte differentiation and function as well as in lipid metabolism. PPARγ is also expressed in T cells, macrophages, endothelial cells, and VSMCs and plays an essential role in cardiovascular physiology and pathophysiology [304]. It has been found that the expression of PPARγ is significantly elevated upon the formation of the atherosclerotic lesions [314, 315]. Ligand-activated PPARγ appears to attenuate inflammatory response in vascular endothelial cells. In addition, agonists of both PPARγ and PPARα attenuate the production of proinflammatory IFN-γ, IL-2, and TNF by T cells [316].

PPARα, like PPARγ, is also expressed in cardiovascular cells and in differentiated macrophages [317–320]. PPARα appears to play a protective role against the development of atherosclerosis in various cells [304]. PPARα agonists modulate macrophage cholesterol homeostasis and inhibit inflammatory activation of VSMC [317]. Moreover, PPARα contributes to the prevention of atherosclerotic thrombosis by inhibiting tissue factor (TF) expression and activity in macrophages [320].

PPARδ is ubiquitously expressed, including the cardiovascular system where it plays important regulatory role. Ligand-activated PPARδ stimulates endothelial expression and the release of vascular endothelial growth factor (VEGF), increasing thereby endothelial cell proliferation and angiogenesis [321]. Moreover, it activates the expression of several antioxidant genes, such as SOD1, catalase, and thioredoxin, thus downregulating OS [322]. In endothelial cells, PPARδ can also attenuate $H_2O_2$-induced apoptosis through the activation of 14-3-3ε protein and PI3K-Akt signaling pathway [323–325]. Although PPARδ overexpression leads to elevated inflammatory response via macrophage activation, PPARδ agonists inhibit the expression of inflammatory genes IFN-γ, MCP-1, TNF-α, ICAM-1, and VCAM-1 in the atherosclerotic plaques [326]. In VSMCs, the activation of the

PPARδ/MCP-1/TGF-β induces antiinflammatory signaling mechanisms [327]. These experimental data support a protective role of PPARδ against atherosclerosis.

Thus, growing evidence suggests an antiinflammatory and atheroprotective role for PPARs, however, conflicting experimental data exist and the administration of PPAR agonists (e.g., thiazolidinedione and fibrates) has so far been disappointing in the clinical setting.

## *Liver X Receptor Signaling*

Liver X receptor (LXR) family of transcription factors function at the intersection of lipid metabolism, innate immunity, and inflammation playing a critical role in inflammation resolution [328]. LXRs, initially identified as orphan receptors, have subsequently been found to be transcriptional activators functioning as cholesterol sensors [329]. They upregulate expressions of various target genes controlling conversion of cholesterol to bile acid, reverse cholesterol transport, and interstitial cholesterol absorption.

Two LXRs are encoded by distinct genes: LXRα is expressed predominantly in the liver, macrophages, kidney, and adipose and adrenal tissues while LXRβ is expressed ubiquitously. While the loss of either LXRα or LXRβ has little effect, deletion of both LXR isoforms leads to decrease in serum triglyceride and HDL levels and to increase in LDLs associated with significant accumulation of lipids and foam cells in the subendothelium of the aortic root [330]. Recent studies suggest that LXRα appears to play a more prominent role in maintaining cholesterol homeostasis upon cholesterol overload [331, 332].

Uptake and accumulation of oxLDLs by macrophages through their scavenger receptors results in the formation of fatty steaks and leads eventually to atherosclerotic lesions. Macrophage LXRs antagonize this process by upregulating the ABC transporters (ABCA1 and ABCG1) mediating increased reverse cholesterol transport (Fig. 18.13) [333]. Transplantation of bone marrow deficient for LXRα and LXRβ into atherosclerosis-prone $LDLR^{-/-}$ mice results in significant acceleration of atherogenesis while macrophage-specific LXRα overexpression reduces the disease [334, 335].

Macrophage LXRs can also modulate innate immunity. It has been demonstrated that the activation of LXRs inhibits LPS-induced expression of several proinflammatory factors, such as IL-6, NOS, and cyclooxygenase-2 (COX-2), in macrophages from wild-type but not from LXR-deficient mice [328]. Antiinflammatory effect of LXRs is mediated through inhibition of NF-κB signaling pathway [336]. It has been shown that LXRs inhibit expression of the NF-κB target gene, encoding matrix metalloproteinase-9 [328, 337]. Matrix metalloproteinase-9 is located within atherosclerotic lesions

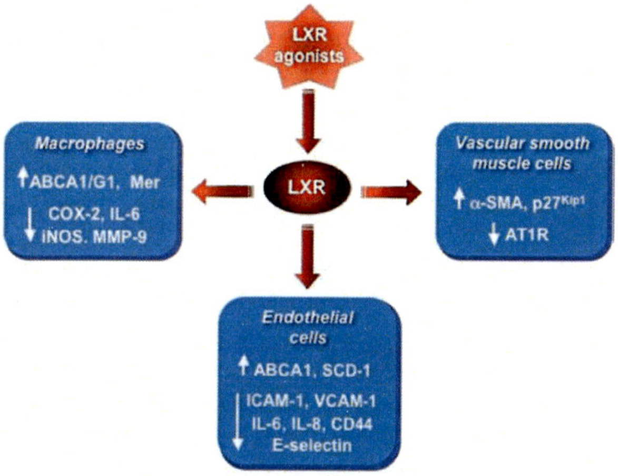


**Fig. 18.13** Effects of LXR signaling on cells of arterial wall. Major cell types involved in inflammatory response in arterial wall and corresponding cellular downstream targets of LXR signaling are schematically shown. See text for details

and contributes to the degradation of extracellular matrix leading eventually to plaque rapture.

Phagocytosis of apoptotic cells activates LXR-signaling pathway leading to induction of apoptotic cell receptor Mer [338]. Importantly, activated LXRs promote apoptotic cell clearance within atherosclerotic lesions attenuating thereby the inflammatory response. Consistently, $LXR^{-/-}$ macrophages lose their characteristic ability to upregulate Mer expression and phagocytosis of apoptotic cells; they are also unable to inhibit induction of proinflammatory IL-1β and MCP-1 [338].

Bacterial and viral pathogens initiate signaling through TLRs. It has been demonstrated that the activation of TLRs attenuates LXR signaling leading to decreased ABC transporter-mediated cholesterol efflux [339]. Moreover, *Chlamydia pneumonia*-induced acceleration of atherosclerosis can be inhibited by TLR2/4/myeloid differentiation primary response gene 88 (MyD88) deficiency but increased by the loss of LXRα [340]. This LXR-mediated antiinflammatory interference with the TLR pathway has been better studied in macrophages.

Endothelial cells express predominantly LXRβ, and LXR agonists induce antiadhesive and antiinflammatory effects in the endothelium (Fig. 18.13). LXR agonists have been reported to downregulate various proinflammatory factors, including CD44, E-selectin, ICAM-1, VCAM-1, IL-6, and IL-8 in artery and vein endothelial cells [341]. Activation of LXR pathway upregulates ABCA1 transporter in endothelial cells while oxLDLs downregulate ABCA1 expression by inhibiting LXR [342–344]. Furthermore, a recent study suggests that areas of healthy arteries of high blood flow are characterized by increased endothelial LXR expression mediating antiatherosclerotic effects, while regions of low turbulent flow exhibit decreased LXR expression [345]. In healthy arteries, LXR signaling upregulates stearoyl-coenzyme A desaturase-1 (SCD-1), which can also be induced by laminar shear stress [346, 347]. SCD-1 is the rate limiting enzyme involved in the conversion of saturated fatty acids to monounsaturated fatty acids. Importantly, downregulation of SCD-1 is associated with apoptosis, lipotoxicity, and inflammation leading to endothelial dysfunction.

VSMCs play a central role in vascular contractility. They also contribute to stabilization of the atherosclerotic plaques through fibrous cap formation. LXRβ and lower levels of LXRα have been identified in VSMCs to be involved in the regulation of their contractility, proliferation, calcification, and apoptosis [348–350]. Ang II type 1 receptor (AT1R) appears to be involved in these processes and its expression is attenuated by LXR agonists (Fig. 18.13) [351]. Moreover, the activation of LXR signaling leads to upregulation of α-smooth muscle actin (α-SMA) and the cell cycle regulator $p27^{Kip1}$, suggesting induced differentiation [352]. However, the role of these events in atherogenesis remains to be determined. Also, while the effect of LXR ligands on VSMC calcification has been established, it is not clear if this contributes to the prevention of plaque rapture [353, 354].

Thus, a highly controlled balance between inflammatory and antiinflammatory signals governs the development of atherosclerosis. Multiple signaling pathways in all cell types involved in atherogenesis contribute to regulation of this balance. Metabolic factors, such as inflammatory cytokines TNF and IL-6 as well as cytokines produced by the adipose tissue, leptin, adiponectin, and resistin also contribute to this process [355, 356].

## Angiogenesis and Atherosclerosis

Growing attention has been focused on the link between angiogenesis and the development of atherosclerosis. Increased neovascularization of the intima and the atherosclerotic plaques is thought to arise from the adventitial *vasa vasorum*, leading to abnormal vascular development. The resulting network of immature vessels is a viable source of invading inflammatory cells that can contribute to plaque instability leading to plaque rupture, intravascular thrombosis, and tissue infarction. This pathological angiogenesis can be triggered by inflammatory cells (e.g., monocytes), which produce and secrete growth factors and cytokines.

VEGF, a potent angiogenesis factor, stimulates proliferation, migration, and tube formation of ECs, primarily through the receptor tyrosine kinase VEGF receptor type 2 (VEGFR2/Flk1/KDR). VEGF binding triggers receptor dimerization, kinase activation, and autophosphorylation of specific tyrosine residues within VEGFR2, which results in the activation of downstream mediators, including PKC, PI3K/Akt, Erk1/2, p38 MAPK, phospholipase C, and eNOS. Furthermore, similar to epidermal growth factor (EGF), cytokines, shear stress, hypoxia, and GPCR agonists, VEGF stimulates endothelial ROS production via the activation of Nox2 subunit of NADPH oxidase [357, 358]. VEGFR2-mediated ROS generation has been implicated in the modulation of endothelial cell migration, proliferation, and postnatal angiogenesis linked to atherogenesis.

A key role of VEGF/VEGFR2 signaling in intimal neovascularization in atherosclerosis has been suggested by findings demonstrating that VEGF-A is expressed primarily in endothelial cells and macrophages associated with the atherosclerotic plaques in human coronary arteries [359]. Importantly, VEGF-A expression is significantly upregulated during the course of atherosclerosis [360]. Emerging evidence suggests that VEGF-A signaling accelerates atherogenesis through the stimulation of *vasa vasorum* growth that can recruit inflammatory cells, such as circulating monocytes and macrophages, to the atherosclerotic lesions [357].

## Conclusions and Future Directions

According to the current concept, atherosclerosis is a chronic inflammatory disease elicited by deregulated balance of lipid accumulation and immune response leading to the formation of lipid-rich plaques in the large arteries. Multiple cell types are involved in the development of the disease. Therefore, it is not surprising that all major signaling pathways contribute to the development of atherosclerosis. Importantly, the signaling cascades that are involved in pathological vascular alterations leading to atherosclerosis share multiple components and mechanisms with pathways, which regulate normal cellular functions.

It is well established that inflammation is involved in all stages of atherosclerosis from initiation and progression to advanced lesion formation and thrombotic complications. Substantial experimental and clinical observations support the use of the inflammatory status as a guide for the prevention and treatment of atherosclerosis. Potent antiinflammatory and immunosuppressant agents can provide promising treatment for acute coronary artery disease (CAD) [84]. However, for prolonged treatment of atherosclerosis a more specific approach is required.

Statins are attractive cellular factors displaying pleiotropic effects that could be used in preventive therapy for atherosclerosis. Remarkably, in addition to their lipid-lowering properties, statins possess potent antiinflammatory capability mediated through the RhoA/ROCK signaling and other signaling pathways involved in the control of cell division and antigen presentation. Moreover, by lowering cholesterol levels in cell membranes, statins inhibit the clustering of T cell receptors attenuating immune activation [361]. Two clinical trials have demonstrated beneficial antiinflammatory effects of statin therapy in patients with atherosclerosis [362, 363]. Importantly, improved clinical outcome after statin therapy has been independent of the reduction in plasma cholesterol concentrations.

Inhibitors of eicosanoid synthesis, such as the cyclooxygenase-2 inhibitor rofecoxib, represent another example of promising antiinflammatory drugs. However, given the high complexity of eicosanoid biology and the ability of these compounds to inhibit enzymes involved in the production of antithrombotic eicosanoids, further studies are required prior to the use of this type of antiinflammatory agents in patients with atherosclerosis [364].

In the past decade, impressive advances in understanding the critical role of the cytokine system in the development of atherosclerosis have been made and several drugs have been suggested to modulate cytokine pathways and improve clinical outcomes. However, clinical trials of inhibitors of cytokines and their receptors have revealed that therapeutic targeting of the cytokine system in atherosclerosis present a greater challenge than initially expected [365]. The high complexity and redundancy of the cytokine system combined with the multifaceted nature of inflammatory response could explain the relatively slow progress in this field [366]. In this regard, several drugs able to modulate cytokine effectors, such as p38 MAPK, c-Jun N-terminal kinase (JNK), JAK1-3, and spleen tyrosine kinase (Syk), are in clinical development for therapeutic treatment of inflammatory diseases [245].

TNF signaling pathway plays a central role in the development of atherosclerosis. Promising clinical data demonstrate beneficial effects of TNF blockade in attenuating inflammation and improving vascular function. Three TNF antagonists, infliximab, adalimumab, and etanercept, have been demonstrated to be effective in the treatment of rheumatoid arthritis and are currently licensed for therapy of this chronic inflammatory disease [367–369].

Like the reduction in LDL levels does not account for all benefits of statins, beneficial effects of modulation of Ang II signaling extend far beyond blood pressure lowering. Ang II, an important vasoactive peptide of the renin–angiotensin system, acts as proinflammatory mediator inducing MCP-1 and VCAM-1 expression by vascular endothelial cells, and IL-6 production by VSMCs. The antiinflammatory potential of this agent has to be further investigated in clinical settings for therapeutic treatment against atherosclerosis.

Combined pharmacological inhibition of chemokine receptors CCR2, CX3CR1, and CCR5 has been effective in the treatment of atherosclerosis in mouse model [370]. Since the CCR5-antagonist maraviroc is licensed in the USA and Europe, this compound is a reasonable candidate for treating atherosclerosis. However, prolonged treatment desirable for antiatherogenic therapy may have side effects and requires further attentive investigation.

Modified chemokines represent a novel type of therapeutic agents [371–374]. They possess altered properties but retain the ability to inhibit pathological immune responses locally at the sites of inflammation without affecting physiological immune functions. The possible side effects of this recently emerged strategy remain to be determined.

Defective inflammation resolution in advanced atheroma represents a critical step in the development of atherosclerosis [106]. Although due to recent advances in the field, we have a rather complete list of mediators and effectors involved in inflammation resolution, the precise molecular mechanisms of their action remain to be determined. Efferocytosis, a key step in the resolution of inflammation, removes efficiently apoptotic inflammatory cells in early atherosclerotic lesions. However, in advanced lesions efferocytosis becomes inefficient, and highly elevated apoptosis is associated with plaque necrosis [101, 375]. The molecular mechanism underlying this dramatic change in efferocytosis efficiency is poorly understood.

While the major inflammation resolution cells within atherosclerotic lesions are macrophages and dendritic cells, it would also be important to evaluate the roles of neutrophils and mast cells in normal and pathological inflammation resolution. Recent studies have reported the participation of these two cell types in plaque progression and atherosclerosis [376, 377].

In conclusion, our current knowledge about the molecular events involved in atherosclerosis has provided exciting insights into its pathogenesis. Unraveling the mechanisms underlying these pathological processes offers new opportunity for prediction and prevention of atherosclerosis, and undoubtedly leads to the development of new, more efficient therapeutic strategies to treat this life-threatening disease.

## Summary

- Atherosclerosis is increasingly viewed as a multifactorial, multistep disease. The involvement of chronic inflammation of the arterial wall at all stages of the disease, from the initiation to progression and rapture of atherosclerotic plaques, has led to the suggestion that multiple risk factors contribute to its pathogenesis by aggravating the inflammatory process.
- Although emerging evidence suggests that multiple factors contribute to atherogenesis, a major role of cholesterol in this process is well documented. It has been demonstrated that increased LDL and decreased HDL cholesterol levels in plasma represent one of the most important risk factors for developing atherosclerosis.

Various metabolic pathways generate LDL and HDL particles of different compositions and sizes.

- Native LDL particles can be oxidized by endothelial cells, and oxLDLs are recognized with higher affinity and taken up much more rapidly by peritoneal macrophages than native LDLs. Various specific plasma membrane receptors, including SR-A and SR-B (CD36) and lectin-like oxLDL receptor, bind oxLDL with high affinity. OxLDLs activate smooth muscle and endothelial cells, as well as the expression and presentation of adhesion molecules and secretion of inflammatory mediators, leading eventually to the recruitment of leukocytes to the vessel lumen and contributing to atheroma progression.
- In contrast to LDLs, HDLs play a protective role against atherosclerosis. They not only transfer cholesterol from intimal macrophages to the liver, but also improve vascular function and inhibit multiple inflammatory pathways, reducing further the risk of atherogenesis. Plasma levels of HDLs represent the most potent lipid risk factor for atherosclerosis.
- HDLs inhibit the expression of E-selectin, an essential regulator of leukocyte trafficking, and adhesion molecules VCAM-1 and ICAM-1. ApoA-I, a major constituent of HDLs, limits macrophage differentiation into foam cells and removes oxidation-prone molecules from LDLs, rendering them resistant to oxidation. HDLs also limit leukocyte trafficking preventing thereby influx of ROS-producing neutrophils. Finally, HDLs contribute to the prevention of arterial and venous thrombosis.
- Accumulation of oxLDLs at sites of hemodynamic strain in the arteries initiates a local inflammatory process. Activated endothelial cells express various leukocyte adhesion molecules, such as VCAM-1 and P- and E-selectins. Moreover, oxLDL-derived metabolites and proinflammatory cytokines IL-1β or TNF-α induce NF-κB-mediated the activation of VCAM-1 expression in endothelial cells.
- Monocytes adhered to the activated endothelium are induced by proinflammatory proteins, chemokines, to infiltrate into the intima. Within the inflamed intima, the recruited monocytes differentiate into macrophages. Macrophages internalize oxLDLs via scavenger receptors, become loaded with modified lipids and oxLDL-derived cholesterol and eventually convert into characteristic foam cells.
- The macrophages proliferate in the inflamed intima and release various growth factors and cytokines accelerating the inflammatory process. In atheroma, multiple chemokines and their receptors are expressed contributing to the recruitment of leukocytes and to the vascular inflammatory response during all phases of atherosclerosis.
- In addition to cells of innate immunity, cells of the adaptive immune response, T lymphocytes, are also involved in vascular inflammation during atherosclerosis. T cells found in plaques are predominantly $CD4^+$ cells. T cells infiltrate to plaques via interactions involving various chemoattractant molecules, such as CXCL9, CXCL10, and CXCL11 and their common receptor CXCR3 and RANTES.
- Within plaques, T cells recognize various antigens presented by the MHC class II and assume different programs of activation to differentiate into $T_H1$ and $T_H2$ cells. $T_H1$ cells produce cytokine IFN-γ, which induces the classical activation of macrophages, whereas $T_H2$ cells generate cytokines IL-4 and IL-13 inducing alternative macrophage activation. IFN-γ dominates in the development of atherosclerosis triggering the secretion of proinflammatory mediators, including various proteases, NO, and proinflammatory cytokines TNF and IL-1 and leading to local inflammation and arterial wall damage.
- Activated macrophages, T cells, and mast cells generate a variety of inflammatory cytokines, proteases, radicals, coagulation, and vasoactive factors. They destabilize fibrous caps and initiate rapture of plaque and thrombus formation. Physical fracture of the thinning fibrous cap causes thrombotic occlusion of an artery leading frequently to the sudden onset of myocardial infarction and strokes.
- Macrophage death is a characteristic feature of advanced atherosclerotic lesions. The phagocytic clearance of apoptotic cells – efferocytosis – results in fewer inflammatory and postapoptotic necrotic cells and attenuates the production of proinflammatory cytokines in early lesions. In advanced lesions, however, apoptotic cells uncleared by efferocytosis accumulate and surround the forming necrotic core contributing to the rapture of fibrous caps, plaque erosion, and eventually to thrombogenesis.
- ER stress promotes macrophage death and advanced plaque necrosis. ER stress induces the unfolded protein response, which upregulates the expression of the proapoptotic protein CHOP (Gadd153). Prolonged activation and high levels of ER stress are required to trigger apoptosis. According to the current concept, physiological levels of ER stress combined with a second signal initiate apoptotic response in advanced atherosclerotic lesions.
- Accumulation of LDL cholesterol by macrophages leads to excess accumulation of cholesterol in the ER membranes, which induces ER stress, while the second signal is binding of molecules with PAMPs by macrophage PRRs. Combination of these signals suppresses protective signaling of ER stress and amplifies apoptotic response.
- Statins and ROCKs play an important preventive role against the development atherosclerosis. In response to Rho activation by S1P and LPA, ROCKs phosphorylate numerous cellular targets, such as MLC, myosin-binding subunit of MLCP, LIM protein kinases 1 and 2, and ERM proteins, adducin and troponin, mediating a variety of cellular responses.

- Statins are important cholesterol-lowering factors used in the preventive therapy of coronary artery diseases. In addition, statins inhibit Rho targeting to the cell membrane preventing thereby the Rho-mediated activation of ROCKs affecting multiple downstream targets.
- Rho-ROCK-signaling pathways are involved in pathogenesis of atherosclerosis controlling actin cytoskeleton organization, vascular smooth muscle contraction, cell adhesion, motility, and gene expression. Pharmaceutical agents that inhibit ROCKs have been shown to prevent vascular inflammation and atherosclerosis.
- OS is one of the major causes of the development of inflammation and atherosclerosis. Vascular endothelial cells express various enzymes, including NADPH oxidase, eNOS, myeloperoxidase, xanthine oxidase, and lipoxygenase, which contribute to the generation of ROS. ROS are key mediators of signaling pathways that underlie vascular inflammation in atherogenesis, from the formation of fatty streak, through atherogenic lesion progression, and subsequently to plaque rupture.
- NO, synthesized by eNOS, is a central player in the endothelium-mediated protection against atherosclerosis. NO is not only the best characterized relaxing factor blocking abnormal constriction of the arteries, but it also prevents the platelet aggregation, the expression of adhesion molecules in the endothelial cells and the production of vasoconstrictor, endothelin-1. Under physiological conditions, the endothelial cells in response to platelet aggregation upregulate NO production and release. Released NO induces the relaxation of the underlaying smooth muscles leading to increasing blood flow and preventing the development of coagulation. If the endothelial cells are dysfunctional, unable to release relaxing signal, the aggregating platelets can reach the vascular smooth muscles and induce their contraction by secreting vasoconstrictors.
- eNOS activity is regulated by the availability of substrate and cofactors. Under physiological conditions, the intracellular concentrations of L-arginine and $BH_4$ are relatively high to stimulate NO synthesis and to decrease $O_2^{\bullet-}$ production. However, OS and CVD are associated with decreased L-arginine and $BH_4$ levels, which lead to the NOS uncoupling resulting in reduced NO availability and eventually in endothelial dysfunction.
- Membrane-associated NADPH oxidase is the principal source of ROS in vascular endothelial cells, which plays a central role in the development of atherosclerosis. NADPH oxidase can be activated by vasoactive hormones (Ang II, aldosterone, endothelin 1), growth factors (platelet-derived growth factor, TGF-β), mechanical stimuli (shear stress and stretch), and cardiovascular risk factors (oxidized lipids, high blood pressure, cigarette smoke).
- Xantine oxidase catalyzes transfer of electrons to molecular oxygen generating the ROS superoxide ($O_2^{\bullet-}$) and hydrogen superoxide. Although xantine oxidase inhibitor oxypurinol inhibits vascular $O_2^{\bullet-}$ production and improves endothelium function, the importance of xantine oxidase for the development of endothelial dysfunction and atherosclerosis remains uncertain.
- Mitochondria play an essential role in the vascular inflammatory process leading to atherosclerotic lesions. Under physiologic conditions, NADPH oxidase and mitochondria represent the main sources of ROS in most mammalian cells. Elevated mitochondrial ROS generation is an early step in the development of atherosclerosis. However, mitochondria are not only sources, but they are also targets of ROS. ROS inhibit multiple mitochondrial enzymes, including complexes I, II, and III, aconitase, α-ketoglutarate dehydrogenase, and pyruvate dehydrogenase.
- Cholesterol and oxLDLs cause mitochondrial damage and dysfunction. oxLDLs trigger mitochondrial-mediated apoptosis in various cells leading to macrophage lysis while scavengers of peroxide radicals prevent this process. OS causes progressive mitochondrial damage and dysfunction playing a central role in the onset and progression of inflammation and atherosclerosis.
- SOD, GPx, catalase, HO, and Trx peroxidase are the major antioxidant enzymes protecting against vascular OS. SOD catalyzes the conversion of $O_2^{\bullet-}$ into $O_2$ and $H_2O_2$ and maintains NO levels in the cardiovascular system. Its inactivation contributes to the development of hypertension. Among several GPx isoforms, GPx1 is the predominant isozyme expressed in mammalian tissues. Loss of GPx1 facilitates atherosclerotic progression in apo$E^{-/-}$ mice. HO catalyzes the degradation of the prooxidative heme producing equimolar amounts of biliverdin, CO and $Fe^{2+}$. OS-induced HO activation is an adaptive response contributing to the protection against vascular remodeling and atherosclerosis. Trx is expressed in vascular endothelial cells and VSMCs. Its ROS-scavenging capabilities can be attributed to Trx peroxidase activity.
- PI3Kγ, highly expressed in the hematopoietic cells, plays major role in inflammation associated with atherosclerosis. PI3Kγ signaling is activated by oxLDLs and other proinflammatory lipids, Ang II and various chemokines, including CCL-2/MCP-1, CCL-3/MIP-1α, and IL-8. Several other proatherogenic stimuli, such as IFN-γ, TGF-β, and TNF-α, can also induce PI3K-signaling pathway through the engagement of various receptors. The PI3Kγ-specific inhibitor AS605240 has significantly reduced lesion size in a mouse model of early and advanced stage atherosclerosis.
- TNF plays a crucial role in lipid metabolism and inflammatory response through leukocyte activation and

maturation as well as cytokine, chemokine, and bioactive intermediate production and release. TNF also induces vascular endothelial cells to produce various adhesion molecules and proinflammatory cytokines, chemokines, and their receptors contributing thereby to the recruitment of activated lymphocytes to the atherosclerotic lesions. Various risk factors, including obesity, smoking, and aging, converge on TNF to initiate OS, vascular remodeling and inflammation, apoptosis and thrombosis leading to vascular damage.

- TGF-β exerts potent antiinflammatory actions on all cell types contributed to the atherogenesis, including endothelial cells, VSMCs, macrophages, and lymphocytes. First, TGF-β attenuates inflammatory processes especially those mediated by $T_{REG}$ ($CD4^{+}CD25^{+}$) cells and phagocytes. Second, it induces collagen production by intimal fibroblasts and initiates thereby a protective scar-like structure formation stabilizing atherosclerotic plaque. Disruption of TGF-β -signaling pathway in T cells results in significant increase in atherogenesis.
- Recently, another group of antiatherosclerotic factors, known as PPARs, has been discovered to be mediators of pharmacologic agents, which reduce T cell activation, attenuating vascular inflammation, and atherosclerosis. Various ligands can bind and activate PPARs to form heterodimers with the RXR, which then recognize and bind a DNA sequence upstream of the PPAR target genes.
- LXR family of transcription factors function at the intersection of lipid metabolism, innate immunity, and inflammation playing a critical role in inflammation resolution. They upregulate expressions of various target genes controlling the conversion of cholesterol to bile acid, reverse cholesterol transport, and interstitial cholesterol absorption. Macrophage LXRs antagonize uptake and accumulation oxLDLs by upregulating the ABC transporters, and also modulate innate immunity. Activated LXRs promote apoptotic cell clearance within atherosclerotic lesions attenuating thereby inflammatory response.
- Growing evidence suggests the link between angiogenesis and the development of atherosclerosis. Increased neovascularization of the intima and the atherosclerotic plaques is a viable source of invading inflammatory cells that can contribute to plaque rupture, thrombosis, and tissue infarction. VEGF-VEGFR2 signaling plays a key role in intimal neovascularization in atherosclerosis. Furthermore, VEGF stimulates endothelial ROS production via the activation of NADPH oxidase. Importantly, VEGF-A expression is significantly upregulated in atherosclerosis.
- Current knowledge about the molecular events involved in atherosclerosis has provided exciting insights into its pathogenesis. Unraveling the mechanisms underlying these pathological processes offers new opportunity for prediction and prevention of atherosclerosis, and undoubtedly leads to the development of new, more efficient therapeutic strategies to treat this life-threatening disease.

## References

1. Ross R, Harker L. Hyperlipidemia and atherosclerosis. Science. 1976;193:1094–100.
2. Neaton JD, Wentworth D. Serum cholesterol, blood pressure, cigarette smoking, and death from coronary heart disease. Overall findings and differences by age for 316,099 white men. Multiple Risk Factor Intervention Trial Research Group. Arch Intern Med. 1992;152:56–64.
3. Pearson TA, LaCroix AZ, Mead LA, Liang KY. The prediction of midlife coronary heart disease and hypertension in young adults: the Johns Hopkins multiple risk equations. Am J Prev Med. 1990;6:23–8.
4. Steinberg D. Thematic review series: the pathogenesis of atherosclerosis. An interpretive history of the cholesterol controversy, part V: the discovery of the statins and the end of the controversy. J Lipid Res. 2006;47:1339–51.
5. Steinberg D. The pathogenesis of atherosclerosis. An interpretive history of the cholesterol controversy, part IV: the 1984 coronary primary prevention trial ends it – almost. J Lipid Res. 2006;47: 1–14.
6. Gordon T, Castelli WP, Hjortland MC, Kannel WB, Dawber TR. High density lipoprotein as a protective factor against coronary heart disease. The Framingham Study. Am J Med. 1977;62:707–14.
7. Soloff LA. Intermediate lipoproteins, atherosclerosis, and Gofman. Circulation. 1998;97:708.
8. Grundy SM, Cleeman JI, Daniels SR, et al. Diagnosis and management of the metabolic syndrome: an American Heart Association/National Heart, Lung, and Blood Institute Scientific Statement. Circulation. 2005;112:2735–52.
9. Krauss RM. Lipoprotein subfractions and cardiovascular disease risk. Curr Opin Lipidol. 2010;21:305–11.
10. Henriksen T, Mahoney EM, Steinberg D. Enhanced macrophage degradation of low density lipoprotein previously incubated with cultured endothelial cells: recognition by receptors for acetylated low density lipoproteins. Proc Natl Acad Sci USA. 1981;78: 6499–503.
11. Kunjathoor VV, Febbraio M, Podrez EA, et al. Scavenger receptors class A-I/II and CD36 are the principal receptors responsible for the uptake of modified low density lipoprotein leading to lipid loading in macrophages. J Biol Chem. 2002;277:49982–8.
12. Steinberg D. The LDL modification hypothesis of atherogenesis: an update. J Lipid Res. 2009;50(Suppl):S376–81.
13. Berneis KK, Krauss RM. Metabolic origins and clinical significance of LDL heterogeneity. J Lipid Res. 2002;43:1363–79.
14. Witztum JL, Steinberg D. Role of oxidized low density lipoprotein in atherogenesis. J Clin Invest. 1991;88:1785–92.
15. Steinberg D, Witztum JL. Lipoproteins and atherogenesis. Current concepts. JAMA. 1990;264:3047–52.
16. Simons K, Ikonen E. How cells handle cholesterol. Science. 2000;290:1721–6.
17. Shaul PW, Mineo C. HDL action on the vascular wall: is the answer NO? J Clin Invest. 2004;113:509–13.
18. Haas MJ, Mooradian AD. Regulation of high-density lipoprotein by inflammatory cytokines: establishing links between immune dysfunction and cardiovascular disease. Diabetes Metab Res Rev. 2010;26:90–9.

19. Lowenstein CJ, Cameron SJ. High-density lipoprotein metabolism and endothelial function. Curr Opin Endocrinol Diabetes Obes. 2010;17:166–70.
20. Glass C, Pittman RC, Weinstein DB, Steinberg D. Dissociation of tissue uptake of cholesterol ester from that of apoprotein A-I of rat plasma high density lipoprotein: selective delivery of cholesterol ester to liver, adrenal, and gonad. Proc Natl Acad Sci USA. 1983;80:5435–9.
21. Acton S, Rigotti A, Landschulz KT, Xu S, Hobbs HH, Krieger M. Identification of scavenger receptor SR-BI as a high density lipoprotein receptor. Science. 1996;271:518–20.
22. Zeiher AM, Schachlinger V, Hohnloser SH, Saurbier B, Just H. Coronary atherosclerotic wall thickening and vascular reactivity in humans. Elevated high-density lipoprotein levels ameliorate abnormal vasoconstriction in early atherosclerosis. Circulation. 1994;89:2525–32.
23. Spieker LE, Sudano I, Hurlimann D, et al. High-density lipoprotein restores endothelial function in hypercholesterolemic men. Circulation. 2002;105:1399–402.
24. Bisoendial RJ, Hovingh GK, Levels JH, et al. Restoration of endothelial function by increasing high-density lipoprotein in subjects with isolated low high-density lipoprotein. Circulation. 2003;107:2944–8.
25. Napoli C, Ignarro LJ. Nitric oxide and atherosclerosis. Nitric Oxide. 2001;5:88–97.
26. Yuhanna IS, Zhu Y, Cox BE, et al. High-density lipoprotein binding to scavenger receptor-BI activates endothelial nitric oxide synthase. Nat Med. 2001;7:853–7.
27. Igarashi J, Thatte HS, Prabhakar P, Golan DE, Michel T. Calcium-independent activation of endothelial nitric oxide synthase by ceramide. Proc Natl Acad Sci USA. 1999;96:12583–8.
28. Nofer JR, van der Giet M, Tolle M, et al. HDL induces NO-dependent vasorelaxation via the lysophospholipid receptor S1P3. J Clin Invest. 2004;113:569–81.
29. Cockerill GW, Rye KA, Gamble JR, Vadas MA, Barter PJ. High-density lipoproteins inhibit cytokine-induced expression of endothelial cell adhesion molecules. Arterioscler Thromb Vasc Biol. 1995;15:1987–94.
30. Ashby DT, Rye KA, Clay MA, Vadas MA, Gamble JR, Barter PJ. Factors influencing the ability of HDL to inhibit expression of vascular cell adhesion molecule-1 in endothelial cells. Arterioscler Thromb Vasc Biol. 1998;18:1450–5.
31. Cockerill GW, Saklatvala J, Ridley SH, et al. High-density lipoproteins differentially modulate cytokine-induced expression of E-selectin and cyclooxygenase-2. Arterioscler Thromb Vasc Biol. 1999;19:910–7.
32. Cockerill GW, Huehns TY, Weerasinghe A, et al. Elevation of plasma high-density lipoprotein concentration reduces interleukin-1-induced expression of E-selectin in an in vivo model of acute inflammation. Circulation. 2001;103:108–12.
33. Dansky HM, Charlton SA, Barlow CB, et al. Apo A-I inhibits foam cell formation in Apo E-deficient mice after monocyte adherence to endothelium. J Clin Invest. 1999;104:31–9.
34. Navab M, Hama SY, Cooke CJ, et al. Normal high density lipoprotein inhibits three steps in the formation of mildly oxidized low density lipoprotein: step 1. J Lipid Res. 2000;41:1481–94.
35. Shih DM, Gu L, Hama S, et al. Genetic-dietary regulation of serum paraoxonase expression and its role in atherogenesis in a mouse model. J Clin Invest. 1996;97:1630–9.
36. Navab M, Reddy ST, Van Lenten BJ, Anantharamaiah GM, Fogelman AM. The role of dysfunctional HDL in atherosclerosis. J Lipid Res. 2009;50(Suppl):S145–9.
37. Shih DM, Gu L, Xia YR, et al. Mice lacking serum paraoxonase are susceptible to organophosphate toxicity and atherosclerosis. Nature. 1998;394:284–7.
38. Tward A, Xia YR, Wang XP, et al. Decreased atherosclerotic lesion formation in human serum paraoxonase transgenic mice. Circulation. 2002;106:484–90.
39. Deguchi H, Pecheniuk NM, Elias DJ, Averell PM, Griffin JH. High-density lipoprotein deficiency and dyslipoproteinemia associated with venous thrombosis in men. Circulation. 2005;112: 893–9.
40. Eichinger S, Pecheniuk NM, Hron G, et al. High-density lipoprotein and the risk of recurrent venous thromboembolism. Circulation. 2007;115:1609–14.
41. Spector AA, Scanu AM, Kaduce TL, Figard PH, Fless GM, Czervionke RL. Effect of human plasma lipoproteins on prostacyclin production by cultured endothelial cells. J Lipid Res. 1985;26: 288–97.
42. Symons JD. Longitudinal and cross-sectional studies of the relationship between 6-keto PGF1 alpha and high density lipoproteins. Prostaglandins Leukot Essent Fatty Acids. 1990;39:159–65.
43. Griffin JH, Kojima K, Banka CL, Curtiss LK, Fernandez JA. High-density lipoprotein enhancement of anticoagulant activities of plasma protein S and activated protein C. J Clin Invest. 1999;103: 219–27.
44. Naqvi TZ, Shah PK, Ivey PA, et al. Evidence that high-density lipoprotein cholesterol is an independent predictor of acute platelet-dependent thrombus formation. Am J Cardiol. 1999;84: 1011–7.
45. Zheng L, Nukuna B, Brennan ML, et al. Apolipoprotein A-I is a selective target for myeloperoxidase-catalyzed oxidation and functional impairment in subjects with cardiovascular disease. J Clin Invest. 2004;114:529–41.
46. Undurti A, Huang Y, Lupica JA, Smith JD, DiDonato JA, Hazen SL. Modification of high density lipoprotein by myeloperoxidase generates a pro-inflammatory particle. J Biol Chem. 2009;284: 30825–35.
47. Massberg S, Brand K, Gruner S, et al. A critical role of platelet adhesion in the initiation of atherosclerotic lesion formation. J Exp Med. 2002;196:887–96.
48. Cybulsky MI, Gimbrone Jr MA. Endothelial expression of a mononuclear leukocyte adhesion molecule during atherogenesis. Science. 1991;251:788–91.
49. Cybulsky MI, Iiyama K, Li H, et al. A major role for VCAM-1, but not ICAM-1, in early atherosclerosis. J Clin Invest. 2001;107: 1255–62.
50. Eriksson EE, Xie X, Werr J, Thoren P, Lindbom L. Importance of primary capture and L-selectin-dependent secondary capture in leukocyte accumulation in inflammation and atherosclerosis in vivo. J Exp Med. 2001;194:205–18.
51. Johnson RC, Chapman SM, Dong ZM, et al. Absence of P-selectin delays fatty streak formation in mice. J Clin Invest. 1997;99: 1037–43.
52. Collins T, Cybulsky MI. NF-kappaB: pivotal mediator or innocent bystander in atherogenesis? J Clin Invest. 2001;107:255–64.
53. Libby P. Inflammation in atherosclerosis. Nature. 2002;420: 868–74.
54. Janeway Jr CA, Medzhitov R. Innate immune recognition. Annu Rev Immunol. 2002;20:197–216.
55. Peiser L, Mukhopadhyay S, Gordon S. Scavenger receptors in innate immunity. Curr Opin Immunol. 2002;14:123–8.
56. Miller YI, Chang MK, Binder CJ, Shaw PX, Witztum JL. Oxidized low density lipoprotein and innate immune receptors. Curr Opin Lipidol. 2003;14:437–45.
57. Mestas J, Ley K. Monocyte-endothelial cell interactions in the development of atherosclerosis. Trends Cardiovasc Med. 2008;18: 228–32.
58. Rocha VZ, Libby P. Obesity, inflammation, and atherosclerosis. Nat Rev Cardiol. 2009;6:399–409.

59. Libby P, Okamoto Y, Rocha VZ, Folco E. Inflammation in atherosclerosis: transition from theory to practice. Circ J. 2010;74:213–20.
60. Boring L, Gosling J, Cleary M, Charo IF. Decreased lesion formation in CCR2$^{-/-}$ mice reveals a role for chemokines in the initiation of atherosclerosis. Nature. 1998;394:894–7.
61. Gu L, Okada Y, Clinton SK, et al. Absence of monocyte chemoattractant protein-1 reduces atherosclerosis in low density lipoprotein receptor-deficient mice. Mol Cell. 1998;2:275–81.
62. Zernecke A, Weber C. Chemokines in the vascular inflammatory response of atherosclerosis. Cardiovasc Res. 2010;86:192–201.
63. Swirski FK, Libby P, Aikawa E, et al. Ly-6Chi monocytes dominate hypercholesterolemia-associated monocytosis and give rise to macrophages in atheromata. J Clin Invest. 2007;117:195–205.
64. Tacke F, Alvarez D, Kaplan TJ, et al. Monocyte subsets differentially employ CCR2, CCR5, and CX3CR1 to accumulate within atherosclerotic plaques. J Clin Invest. 2007;117:185–94.
65. Hansson GK. Immune mechanisms in atherosclerosis. Arterioscler Thromb Vasc Biol. 2001;21:1876–90.
66. Tupin E, Nicoletti A, Elhage R, et al. CD1d-dependent activation of NKT cells aggravates atherosclerosis. J Exp Med. 2004;199:417–22.
67. Ludewig B, Freigang S, Jaggi M, et al. Linking immune-mediated arterial inflammation and cholesterol-induced atherosclerosis in a transgenic mouse model. Proc Natl Acad Sci USA. 2000;97:12752–7.
68. Mach F, Sauty A, Iarossi AS, et al. Differential expression of three T lymphocyte-activating CXC chemokines by human atheroma-associated cells. J Clin Invest. 1999;104:1041–50.
69. Heller EA, Liu E, Tager AM, et al. Chemokine CXCL10 promotes atherogenesis by modulating the local balance of effector and regulatory T cells. Circulation. 2006;113:2301–12.
70. van Wanrooij EJ, de Jager SC, van Es T, et al. CXCR3 antagonist NBI-74330 attenuates atherosclerotic plaque formation in LDL receptor-deficient mice. Arterioscler Thromb Vasc Biol. 2008;28:251–7.
71. Veillard NR, Kwak B, Pelli G, et al. Antagonism of RANTES receptors reduces atherosclerotic plaque formation in mice. Circ Res. 2004;94:253–61.
72. Hansson GK. Inflammation, atherosclerosis, and coronary artery disease. N Engl J Med. 2005;352:1685–95.
73. Frostegard J, Ulfgren AK, Nyberg P, et al. Cytokine expression in advanced human atherosclerotic plaques: dominance of pro-inflammatory (Th1) and macrophage-stimulating cytokines. Atherosclerosis. 1999;145:33–43.
74. Gordon S. Alternative activation of macrophages. Nat Rev Immunol. 2003;3:23–35.
75. Friesel R, Komoriya A, Maciag T. Inhibition of endothelial cell proliferation by gamma-interferon. J Cell Biol. 1987;104:689–96.
76. Hansson GK, Hellstrand M, Rymo L, Rubbia L, Gabbiani G. Interferon gamma inhibits both proliferation and expression of differentiation-specific alpha-smooth muscle actin in arterial smooth muscle cells. J Exp Med. 1989;170:1595–608.
77. Amento EP, Ehsani N, Palmer H, Libby P. Cytokines and growth factors positively and negatively regulate interstitial collagen gene expression in human vascular smooth muscle cells. Arterioscler Thromb. 1991;11:1223–30.
78. Gupta S, Pablo AM, Jiang X, Wang N, Tall AR, Schindler C. IFN-gamma potentiates atherosclerosis in ApoE knock-out mice. J Clin Invest. 1997;99:2752–61.
79. Buono C, Come CE, Stavrakis G, Maguire GF, Connelly PW, Lichtman AH. Influence of interferon-gamma on the extent and phenotype of diet-induced atherosclerosis in the LDLR-deficient mouse. Arterioscler Thromb Vasc Biol. 2003;23:454–60.
80. Shimizu K, Shichiri M, Libby P, Lee RT, Mitchell RN. Th2-predominant inflammation and blockade of IFN-gamma signaling induce aneurysms in allografted aortas. J Clin Invest. 2004;114:300–8.
81. Stemme S, Faber B, Holm J, Wiklund O, Witztum JL, Hansson GK. T lymphocytes from human atherosclerotic plaques recognize oxidized low density lipoprotein. Proc Natl Acad Sci USA. 1995;92:3893–7.
82. Kol A, Sukhova GK, Lichtman AH, Libby P. Chlamydial heat shock protein 60 localizes in human atheroma and regulates macrophage tumor necrosis factor-alpha and matrix metalloproteinase expression. Circulation. 1998;98:300–7.
83. Mach F, Schonbeck U, Bonnefoy JY, Pober JS, Libby P. Activation of monocyte/macrophage functions related to acute atheroma complication by ligation of CD40: induction of collagenase, stromelysin, and tissue factor. Circulation. 1997;96:396–9.
84. Libby P, Aikawa M. Stabilization of atherosclerotic plaques: new mechanisms and clinical targets. Nat Med. 2002;8:1257–62.
85. Kume T, Okura H, Yamada R, et al. Frequency and spatial distribution of thin-cap fibroatheroma assessed by 3-vessel intravascular ultrasound and optical coherence tomography: an ex vivo validation and an initial in vivo feasibility study. Circ J. 2009;73:1086–91.
86. Kashiwagi M, Tanaka A, Kitabata H, et al. Relationship between coronary arterial remodeling, fibrous cap thickness and high-sensitivity C-reactive protein levels in patients with acute coronary syndrome. Circ J. 2009;73:1291–5.
87. Henney AM, Wakeley PR, Davies MJ, et al. Localization of stromelysin gene expression in atherosclerotic plaques by in situ hybridization. Proc Natl Acad Sci USA. 1991;88:8154–8.
88. Galis ZS, Sukhova GK, Lark MW, Libby P. Increased expression of matrix metalloproteinases and matrix degrading activity in vulnerable regions of human atherosclerotic plaques. J Clin Invest. 1994;94:2493–503.
89. Dollery CM, Libby P. Atherosclerosis and proteinase activation. Cardiovasc Res. 2006;69:625–35.
90. Liu J, Sukhova GK, Sun JS, Xu WH, Libby P, Shi GP. Lysosomal cysteine proteases in atherosclerosis. Arterioscler Thromb Vasc Biol. 2004;24:1359–66.
91. Mach F, Schonbeck U, Sukhova GK, et al. Functional CD40 ligand is expressed on human vascular endothelial cells, smooth muscle cells, and macrophages: implications for CD40-CD40 ligand signaling in atherosclerosis. Proc Natl Acad Sci USA. 1997;94:1931–6.
92. Kockx MM, De Meyer GR, Muhring J, Jacob W, Bult H, Herman AG. Apoptosis and related proteins in different stages of human atherosclerotic plaques. Circulation. 1998;97:2307–15.
93. Kockx MM, Herman AG. Apoptosis in atherosclerosis: beneficial or detrimental? Cardiovasc Res. 2000;45:736–46.
94. Arai S, Shelton JM, Chen M, et al. A role for the apoptosis inhibitory factor AIM/Spalpha/Api6 in atherosclerosis development. Cell Metab. 2005;1:201–13.
95. Liu J, Thewke DP, Su YR, Linton MF, Fazio S, Sinensky MS. Reduced macrophage apoptosis is associated with accelerated atherosclerosis in low-density lipoprotein receptor-null mice. Arterioscler Thromb Vasc Biol. 2005;25:174–9.
96. Fadok VA, Bratton DL, Konowal A, Freed PW, Westcott JY, Henson PM. Macrophages that have ingested apoptotic cells in vitro inhibit proinflammatory cytokine production through autocrine/paracrine mechanisms involving TGF-beta, PGE2, and PAF. J Clin Invest. 1998;101:890–8.
97. Savill J, Fadok V. Corpse clearance defines the meaning of cell death. Nature. 2000;407:784–8.
98. Chung EY, Kim SJ, Ma XJ. Regulation of cytokine production during phagocytosis of apoptotic cells. Cell Res. 2006;16:154–61.
99. Bjorkerud S, Bjorkerud B. Apoptosis is abundant in human atherosclerotic lesions, especially in inflammatory cells (macrophages and T cells), and may contribute to the accumulation of gruel and plaque instability. Am J Pathol. 1996;149:367–80.

100. Ball RY, Stowers EC, Burton JH, Cary NR, Skepper JN, Mitchinson MJ. Evidence that the death of macrophage foam cells contributes to the lipid core of atheroma. Atherosclerosis. 1995;114:45–54.
101. Tabas I. Consequences and therapeutic implications of macrophage apoptosis in atherosclerosis: the importance of lesion stage and phagocytic efficiency. Arterioscler Thromb Vasc Biol. 2005;25: 2255–64.
102. Zinszner H, Kuroda M, Wang X, et al. CHOP is implicated in programmed cell death in response to impaired function of the endoplasmic reticulum. Genes Dev. 1998;12:982–95.
103. Li G, Mongillo M, Chin KT, et al. Role of ERO1-alpha-mediated stimulation of inositol 1,4,5-triphosphate receptor activity in endoplasmic reticulum stress-induced apoptosis. J Cell Biol. 2009;186: 783–92.
104. Timmins JM, Ozcan L, Seimon TA, et al. Calcium/calmodulin-dependent protein kinase II links ER stress with Fas and mitochondrial apoptosis pathways. J Clin Invest. 2009;119:2925–41.
105. Seimon T, Tabas I. Mechanisms and consequences of macrophage apoptosis in atherosclerosis. J Lipid Res. 2009;50(Suppl):S382–7.
106. Tabas I. Macrophage death and defective inflammation resolution in atherosclerosis. Nat Rev Immunol. 2010;10:36–46.
107. Devries-Seimon T, Li Y, Yao PM, et al. Cholesterol-induced macrophage apoptosis requires ER stress pathways and engagement of the type A scavenger receptor. J Cell Biol. 2005;171:61–73.
108. Seimon TA, Obstfeld A, Moore KJ, Golenbock DT, Tabas I. Combinatorial pattern recognition receptor signaling alters the balance of life and death in macrophages. Proc Natl Acad Sci USA. 2006;103:19794–9.
109. Fukumoto Y, Libby P, Rabkin E, et al. Statins alter smooth muscle cell accumulation and collagen content in established atheroma of watanabe heritable hyperlipidemic rabbits. Circulation. 2001;103:993–9.
110. Van Aelst L, D'Souza-Schorey C. Rho GTPases and signaling networks. Genes Dev. 1997;11:2295–322.
111. Riento K, Ridley AJ. Rocks: multifunctional kinases in cell behaviour. Nat Rev Mol Cell Biol. 2003;4:446–56.
112. Nakagawa O, Fujisawa K, Ishizaki T, Saito Y, Nakao K, Narumiya S. ROCK-I and ROCK-II, two isoforms of Rho-associated coiled-coil forming protein serine/threonine kinase in mice. FEBS Lett. 1996;392:189–93.
113. Amano M, Chihara K, Nakamura N, Kaneko T, Matsuura Y, Kaibuchi K. The COOH terminus of Rho-kinase negatively regulates rho-kinase activity. J Biol Chem. 1999;274:32418–24.
114. Feng J, Ito M, Ichikawa K, et al. Inhibitory phosphorylation site for Rho-associated kinase on smooth muscle myosin phosphatase. J Biol Chem. 1999;274:37385–90.
115. Coleman ML, Sahai EA, Yeo M, Bosch M, Dewar A, Olson MF. Membrane blebbing during apoptosis results from caspase-mediated activation of ROCK I. Nat Cell Biol. 2001;3:339–45.
116. Sebbagh M, Renvoize C, Hamelin J, Riche N, Bertoglio J, Breard J. Caspase-3-mediated cleavage of ROCK I induces MLC phosphorylation and apoptotic membrane blebbing. Nat Cell Biol. 2001;3:346–52.
117. Leung T, Manser E, Tan L, Lim L. A novel serine/threonine kinase binding the Ras-related RhoA GTPase which translocates the kinase to peripheral membranes. J Biol Chem. 1995;270:29051–4.
118. Sin WC, Chen XQ, Leung T, Lim L. RhoA-binding kinase alpha translocation is facilitated by the collapse of the vimentin intermediate filament network. Mol Cell Biol. 1998;18:6325–39.
119. Royal I, Lamarche-Vane N, Lamorte L, Kaibuchi K, Park M. Activation of cdc42, rac, PAK, and rho-kinase in response to hepatocyte growth factor differentially regulates epithelial cell colony spreading and dissociation. Mol Biol Cell. 2000;11:1709–25.
120. Chevrier V, Piel M, Collomb N, et al. The Rho-associated protein kinase p160ROCK is required for centrosome positioning. J Cell Biol. 2002;157:807–17.
121. Ishizaki T, Maekawa M, Fujisawa K, et al. The small GTP-binding protein Rho binds to and activates a 160 kDa Ser/Thr protein kinase homologous to myotonic dystrophy kinase. EMBO J. 1996;15:1885–93.
122. Feng J, Ito M, Kureishi Y, et al. Rho-associated kinase of chicken gizzard smooth muscle. J Biol Chem. 1999;274:3744–52.
123. Ishizaki T, Naito M, Fujisawa K, et al. p160ROCK, a Rho-associated coiled-coil forming protein kinase, works downstream of Rho and induces focal adhesions. FEBS Lett. 1997;404:118–24.
124. Ward Y, Yap SF, Ravichandran V, et al. The GTP binding proteins Gem and Rad are negative regulators of the Rho-Rho kinase pathway. J Cell Biol. 2002;157:291–302.
125. Dong M, Yan BP, Yu CM. Current status of rho-associated kinases (ROCKs) in coronary atherosclerosis and vasospasm. Cardiovasc Hematol Agents Med Chem. 2009;7:322–30.
126. Alberts AW, Chen J, Kuron G, et al. Mevinolin: a highly potent competitive inhibitor of hydroxymethylglutaryl-coenzyme A reductase and a cholesterol-lowering agent. Proc Natl Acad Sci USA. 1980;77:3957–61.
127. Istvan ES, Deisenhofer J. Structural mechanism for statin inhibition of HMG-CoA reductase. Science. 2001;292:1160–4.
128. Goldstein JL, Brown MS. Regulation of the mevalonate pathway. Nature. 1990;343:425–30.
129. Hall A. G proteins and small GTPases: distant relatives keep in touch. Science. 1998;280:2074–5.
130. Laufs U, La Fata V, Plutzky J, Liao JK. Upregulation of endothelial nitric oxide synthase by HMG CoA reductase inhibitors. Circulation. 1998;97:1129–35.
131. Laufs U, Marra D, Node K, Liao JK. 3-Hydroxy-3-methylglutaryl-CoA reductase inhibitors attenuate vascular smooth muscle proliferation by preventing rho GTPase-induced down-regulation of p27(Kip1). J Biol Chem. 1999;274:21926–31.
132. Mundy G, Garrett R, Harris S, et al. Stimulation of bone formation in vitro and in rodents by statins. Science. 1999;286:1946–9.
133. Shimokawa H, Morishige K, Miyata K, et al. Long-term inhibition of Rho-kinase induces a regression of arteriosclerotic coronary lesions in a porcine model in vivo. Cardiovasc Res. 2001;51:169–77.
134. Mallat Z, Gojova A, Sauzeau V, et al. Rho-associated protein kinase contributes to early atherosclerotic lesion formation in mice. Circ Res. 2003;93:884–8.
135. Woodside DG, Wooten DK, McIntyre BW. Adenosine diphosphate (ADP)-ribosylation of the guanosine triphosphatase (GTPase) rho in resting peripheral blood human T lymphocytes results in pseudopodial extension and the inhibition of T cell activation. J Exp Med. 1998;188:1211–21.
136. Vicente-Manzanares M, Rey M, Perez-Martinez M, et al. The RhoA effector mDia is induced during T cell activation and regulates actin polymerization and cell migration in T lymphocytes. J Immunol. 2003;171:1023–34.
137. Tharaux PL, Bukoski RC, Rocha PN, et al. Rho kinase promotes alloimmune responses by regulating the proliferation and structure of T cells. J Immunol. 2003;171:96–105.
138. Rekhter M, Chandrasekhar K, Gifford-Moore D, et al. Immunohistochemical analysis of target proteins of Rho-kinase in a mouse model of accelerated atherosclerosis. Exp Clin Cardiol. 2007;12:169–74.
139. Wang HW, Liu PY, Oyama N, et al. Deficiency of ROCK1 in bone marrow-derived cells protects against atherosclerosis in $LDLR^{-/-}$ mice. FASEB J. 2008;22:3561–70.
140. Essig M, Nguyen G, Prie D, Escoubet B, Sraer JD, Friedlander G. 3-Hydroxy-3-methylglutaryl coenzyme A reductase inhibitors increase fibrinolytic activity in rat aortic endothelial cells. Role of geranylgeranylation and Rho proteins. Circ Res. 1998;83:683–90.
141. Essler M, Retzer M, Bauer M, Heemskerk JW, Aepfelbacher M, Siess W. Mildly oxidized low density lipoprotein induces contraction of human endothelial cells through activation of Rho/Rho

kinase and inhibition of myosin light chain phosphatase. J Biol Chem. 1999;274:30361–4.
142. Funakoshi Y, Ichiki T, Shimokawa H, et al. Rho-kinase mediates angiotensin II-induced monocyte chemoattractant protein-1 expression in rat vascular smooth muscle cells. Hypertension. 2001;38:100–4.
143. Takeda K, Ichiki T, Tokunou T, et al. Critical role of Rho-kinase and MEK/ERK pathways for angiotensin II-induced plasminogen activator inhibitor type-1 gene expression. Arterioscler Thromb Vasc Biol. 2001;21:868–73.
144. Perona R, Montaner S, Saniger L, Sanchez-Perez I, Bravo R, Lacal JC. Activation of the nuclear factor-kappaB by Rho, CDC42, and Rac-1 proteins. Genes Dev. 1997;11:463–75.
145. Montaner S, Perona R, Saniger L, Lacal JC. Multiple signalling pathways lead to the activation of the nuclear factor kappaB by the Rho family of GTPases. J Biol Chem. 1998;273:12779–85.
146. Hajra L, Evans AI, Chen M, Hyduk SJ, Collins T, Cybulsky MI. The NF-kappa B signal transduction pathway in aortic endothelial cells is primed for activation in regions predisposed to atherosclerotic lesion formation. Proc Natl Acad Sci USA. 2000;97:9052–7.
147. Noma K, Rikitake Y, Oyama N, et al. ROCK1 mediates leukocyte recruitment and neointima formation following vascular injury. J Clin Invest. 2008;118:1632–44.
148. Nohria A, Prsic A, Liu PY, et al. Statins inhibit Rho kinase activity in patients with atherosclerosis. Atherosclerosis. 2009;205:517–21.
149. Rawlings R, Nohria A, Liu PY, et al. Comparison of effects of rosuvastatin (10 mg) versus atorvastatin (40 mg) on rho kinase activity in caucasian men with a previous atherosclerotic event. Am J Cardiol. 2009;103:437–41.
150. Laufs U, Liao JK. Post-transcriptional regulation of endothelial nitric oxide synthase mRNA stability by Rho GTPase. J Biol Chem. 1998;273:24266–71.
151. Wolfrum S, Grimm M, Heidbreder M, et al. Acute reduction of myocardial infarct size by a hydroxymethyl glutaryl coenzyme A reductase inhibitor is mediated by endothelial nitric oxide synthase. J Cardiovasc Pharmacol. 2003;41:474–80.
152. Wolfrum S, Dendorfer A, Rikitake Y, et al. Inhibition of Rho-kinase leads to rapid activation of phosphatidylinositol 3-kinase/protein kinase Akt and cardiovascular protection. Arterioscler Thromb Vasc Biol. 2004;24:1842–7.
153. Sato M, Tani E, Fujikawa H, Kaibuchi K. Involvement of Rho-kinase-mediated phosphorylation of myosin light chain in enhancement of cerebral vasospasm. Circ Res. 2000;87:195–200.
154. Sato S, Ikegaki I, Asano T, Shimokawa H. Antiischemic properties of fasudil in experimental models of vasospastic angina. Jpn J Pharmacol. 2001;87:34–40.
155. Harrison D, Griendling KK, Landmesser U, Hornig B, Drexler H. Role of oxidative stress in atherosclerosis. Am J Cardiol. 2003;91:7A–11.
156. Kondo T, Hirose M, Kageyama K. Roles of oxidative stress and redox regulation in atherosclerosis. J Atheroscler Thromb. 2009; 16:532–8.
157. Vasquez-Vivar J, Kalyanaraman B, Martasek P, et al. Superoxide generation by endothelial nitric oxide synthase: the influence of cofactors. Proc Natl Acad Sci USA. 1998;95:9220–5.
158. Ozaki M, Kawashima S, Yamashita T, et al. Overexpression of endothelial nitric oxide synthase accelerates atherosclerotic lesion formation in apoE-deficient mice. J Clin Invest. 2002;110:331–40.
159. Dudzinski DM, Igarashi J, Greif D, Michel T. The regulation and pharmacology of endothelial nitric oxide synthase. Annu Rev Pharmacol Toxicol. 2006;46:235–76.
160. Dudzinski DM, Michel T. Life history of eNOS: partners and pathways. Cardiovasc Res. 2007;75:247–60.
161. Vanhoutte PM, Shimokawa H, Tang EH, Feletou M. Endothelial dysfunction and vascular disease. Acta Physiol (Oxf). 2009; 196:193–222.
162. Forstermann U, Boissel JP, Kleinert H. Expressional control of the 'constitutive' isoforms of nitric oxide synthase (NOS I and NOS III). FASEB J. 1998;12:773–90.
163. Shaul PW. Regulation of endothelial nitric oxide synthase: location, location, location. Annu Rev Physiol. 2002;64:749–74.
164. Martinez-Hervas S, Fandos M, Real JT, et al. Insulin resistance and oxidative stress in familial combined hyperlipidemia. Atherosclerosis. 2008;199:384–9.
165. Crabtree MJ, Tatham AL, Al-Wakeel Y, et al. Quantitative regulation of intracellular endothelial nitric-oxide synthase (eNOS) coupling by both tetrahydrobiopterin-eNOS stoichiometry and biopterin redox status: insights from cells with tet-regulated GTP cyclohydrolase I expression. J Biol Chem. 2009;284:1136–44.
166. Crabtree MJ, Tatham AL, Hale AB, Alp NJ, Channon KM. Critical role for tetrahydrobiopterin recycling by dihydrofolate reductase in regulation of endothelial nitric-oxide synthase coupling: relative importance of the de novo biopterin synthesis versus salvage pathways. J Biol Chem. 2009;284:28128–36.
167. Balligand JL, Feron O, Dessy C. eNOS activation by physical forces: from short-term regulation of contraction to chronic remodeling of cardiovascular tissues. Physiol Rev. 2009;89:481–534.
168. Morris Jr SM. Recent advances in arginine metabolism: roles and regulation of the arginases. Br J Pharmacol. 2009;157:922–30.
169. Channon KM. Tetrahydrobiopterin: regulator of endothelial nitric oxide synthase in vascular disease. Trends Cardiovasc Med. 2004;14:323–7.
170. Feron O, Balligand JL. Caveolins and the regulation of endothelial nitric oxide synthase in the heart. Cardiovasc Res. 2006;69: 788–97.
171. Erwin PA, Mitchell DA, Sartoretto J, Marletta MA, Michel T. Subcellular targeting and differential S-nitrosylation of endothelial nitric-oxide synthase. J Biol Chem. 2006;281:151–7.
172. Michel T, Vanhoutte PM. Cellular signaling and NO production. Pflugers Arch. 2010;459:807–16.
173. Fleming I. Molecular mechanisms underlying the activation of eNOS. Pflugers Arch. 2010;459:793–806.
174. Bedard K, Krause KH. The NOX family of ROS-generating NADPH oxidases: physiology and pathophysiology. Physiol Rev. 2007;87:245–313.
175. Brandes RP, Schroder K. Differential vascular functions of Nox family NADPH oxidases. Curr Opin Lipidol. 2008;19:513–8.
176. Nauseef WM. Biological roles for the NOX family NADPH oxidases. J Biol Chem. 2008;283:16961–5.
177. Leto TL, Morand S, Hurt D, Ueyama T. Targeting and regulation of reactive oxygen species generation by Nox family NADPH oxidases. Antioxid Redox Signal. 2009;11:2607–19.
178. Chanock SJ, el Benna J, Smith RM, Babior BM. The respiratory burst oxidase. J Biol Chem. 1994;269:24519–22.
179. Cross AR, Erickson RW, Curnutte JT. The mechanism of activation of NADPH oxidase in the cell-free system: the activation process is primarily catalytic and not through the formation of a stoichiometric complex. Biochem J. 1999;341(Pt 2):251–5.
180. Lassegue B, Clempus RE. Vascular NAD(P)H oxidases: specific features, expression, and regulation. Am J Physiol Regul Integr Comp Physiol. 2003;285:R277–97.
181. Orosz Z, Csiszar A, Labinskyy N, et al. Cigarette smoke-induced proinflammatory alterations in the endothelial phenotype: role of NAD(P)H oxidase activation. Am J Physiol Heart Circ Physiol. 2007;292:H130–9.
182. Touyz RM, Chen X, Tabet F, et al. Expression of a functionally active gp91phox-containing neutrophil-type NAD(P)H oxidase in smooth muscle cells from human resistance arteries: regulation by angiotensin II. Circ Res. 2002;90:1205–13.
183. Barry-Lane PA, Patterson C, van der Merwe M, et al. p47phox is required for atherosclerotic lesion progression in ApoE(−/−) mice. J Clin Invest. 2001;108:1513–22.

184. Landmesser U, Cai H, Dikalov S, et al. Role of p47(phox) in vascular oxidative stress and hypertension caused by angiotensin II. Hypertension. 2002;40:511–5.
185. Matsuno K, Yamada H, Iwata K, et al. Nox1 is involved in angiotensin II-mediated hypertension: a study in Nox1-deficient mice. Circulation. 2005;112:2677–85.
186. Dikalova A, Clempus R, Lassegue B, et al. Nox1 overexpression potentiates angiotensin II-induced hypertension and vascular smooth muscle hypertrophy in transgenic mice. Circulation. 2005;112:2668–76.
187. McNally JS, Davis ME, Giddens DP, et al. Role of xanthine oxidoreductase and NAD(P)H oxidase in endothelial superoxide production in response to oscillatory shear stress. Am J Physiol Heart Circ Physiol. 2003;285:H2290–7.
188. White CR, Darley-Usmar V, Berrington WR, et al. Circulating plasma xanthine oxidase contributes to vascular dysfunction in hypercholesterolemic rabbits. Proc Natl Acad Sci USA. 1996;93:8745–9.
189. Viel EC, Benkirane K, Javeshghani D, Touyz RM, Schiffrin EL. Xanthine oxidase and mitochondria contribute to vascular superoxide anion generation in DOCA-salt hypertensive rats. Am J Physiol Heart Circ Physiol. 2008;295:H281–8.
190. Spiekermann S, Schenk K, Hoeper MM. Increased xanthine oxidase activity in idiopathic pulmonary arterial hypertension. Eur Respir J. 2009;34:276.
191. Kanbay M, Ozkara A, Selcoki Y, et al. Effect of treatment of hyperuricemia with allopurinol on blood pressure, creatinine clearence, and proteinuria in patients with normal renal functions. Int Urol Nephrol. 2007;39:1227–33.
192. Feig DI, Soletsky B, Johnson RJ. Effect of allopurinol on blood pressure of adolescents with newly diagnosed essential hypertension: a randomized trial. JAMA. 2008;300:924–32.
193. Cardillo C, Kilcoyne CM, Cannon 3rd RO, Quyyumi AA, Panza JA. Xanthine oxidase inhibition with oxypurinol improves endothelial vasodilator function in hypercholesterolemic but not in hypertensive patients. Hypertension. 1997;30:57–63.
194. O'Driscoll JG, Green DJ, Rankin JM, Taylor RR. Nitric oxide-dependent endothelial function is unaffected by allopurinol in hypercholesterolaemic subjects. Clin Exp Pharmacol Physiol. 1999;26:779–83.
195. Butler R, Morris AD, Belch JJ, Hill A, Struthers AD. Allopurinol normalizes endothelial dysfunction in type 2 diabetics with mild hypertension. Hypertension. 2000;35:746–51.
196. Ballinger SW. Mitochondrial dysfunction in cardiovascular disease. Free Radic Biol Med. 2005;38:1278–95.
197. Madamanchi NR, Runge MS. Mitochondrial dysfunction in atherosclerosis. Circ Res. 2007;100:460–73.
198. Puddu P, Puddu GM, Cravero E, De Pascalis S, Muscari A. The emerging role of cardiovascular risk factor-induced mitochondrial dysfunction in atherogenesis. J Biomed Sci. 2009;16:112.
199. Sorescu D, Griendling KK. Reactive oxygen species, mitochondria, and NAD(P)H oxidases in the development and progression of heart failure. Congest Heart Fail. 2002;8:132–40.
200. Zhang DX, Gutterman DD. Mitochondrial reactive oxygen species-mediated signaling in endothelial cells. Am J Physiol Heart Circ Physiol. 2007;292:H2023–31.
201. Turrens JF, Alexandre A, Lehninger AL. Ubisemiquinone is the electron donor for superoxide formation by complex III of heart mitochondria. Arch Biochem Biophys. 1985;237:408–14.
202. Okado-Matsumoto A, Fridovich I. Subcellular distribution of superoxide dismutases (SOD) in rat liver: Cu, Zn-SOD in mitochondria. J Biol Chem. 2001;276:38388–93.
203. Chance B, Sies H, Boveris A. Hydroperoxide metabolism in mammalian organs. Physiol Rev. 1979;59:527–605.
204. Green K, Brand MD, Murphy MP. Prevention of mitochondrial oxidative damage as a therapeutic strategy in diabetes. Diabetes. 2004;53 Suppl 1:S110–8.
205. Ide T, Tsutsui H, Kinugawa S, et al. Direct evidence for increased hydroxyl radicals originating from superoxide in the failing myocardium. Circ Res. 2000;86:152–7.
206. Cleeter MW, Cooper JM, Darley-Usmar VM, Moncada S, Schapira AH. Reversible inhibition of cytochrome c oxidase, the terminal enzyme of the mitochondrial respiratory chain, by nitric oxide. Implications for neurodegenerative diseases. FEBS Lett. 1994; 345:50–4.
207. Ghafourifar P, Richter C. Nitric oxide synthase activity in mitochondria. FEBS Lett. 1997;418:291–6.
208. Giulivi C, Poderoso JJ, Boveris A. Production of nitric oxide by mitochondria. J Biol Chem. 1998;273:11038–43.
209. Ballinger SW, Patterson C, Yan CN, et al. Hydrogen peroxide- and peroxynitrite-induced mitochondrial DNA damage and dysfunction in vascular endothelial and smooth muscle cells. Circ Res. 2000;86:960–6.
210. Radi R, Cassina A, Hodara R. Nitric oxide and peroxynitrite interactions with mitochondria. Biol Chem. 2002;383:401–9.
211. Dahm CC, Moore K, Murphy MP. Persistent S-nitrosation of complex I and other mitochondrial membrane proteins by S-nitrosothiols but not nitric oxide or peroxynitrite: implications for the interaction of nitric oxide with mitochondria. J Biol Chem. 2006;281: 10056–65.
212. Zeevalk GD, Bernard LP, Song C, Gluck M, Ehrhart J. Mitochondrial inhibition and oxidative stress: reciprocating players in neurodegeneration. Antioxid Redox Signal. 2005;7: 1117–39.
213. Yan LJ, Sohal RS. Mitochondrial adenine nucleotide translocase is modified oxidatively during aging. Proc Natl Acad Sci USA. 1998;95:12896–901.
214. Graziewicz MA, Day BJ, Copeland WC. The mitochondrial DNA polymerase as a target of oxidative damage. Nucleic Acids Res. 2002;30:2817–24.
215. MacMillan-Crow LA, Crow JP, Thompson JA. Peroxynitrite-mediated inactivation of manganese superoxide dismutase involves nitration and oxidation of critical tyrosine residues. Biochemistry. 1998;37:1613–22.
216. Schlame M, Rua D, Greenberg ML. The biosynthesis and functional role of cardiolipin. Prog Lipid Res. 2000;39:257–88.
217. Nomura K, Imai H, Koumura T, Kobayashi T, Nakagawa Y. Mitochondrial phospholipid hydroperoxide glutathione peroxidase inhibits the release of cytochrome c from mitochondria by suppressing the peroxidation of cardiolipin in hypoglycaemia-induced apoptosis. Biochem J. 2000;351:183–93.
218. Paradies G, Petrosillo G, Pistolese M, Di Venosa N, Federici A, Ruggiero FM. Decrease in mitochondrial complex I activity in ischemic/reperfused rat heart: involvement of reactive oxygen species and cardiolipin. Circ Res. 2004;94:53–9.
219. Droge W. Free radicals in the physiological control of cell function. Physiol Rev. 2002;82:47–95.
220. Turrens JF. Mitochondrial formation of reactive oxygen species. J Physiol. 2003;552:335–44.
221. Vindis C, Elbaz M, Escargueil-Blanc I, et al. Two distinct calcium-dependent mitochondrial pathways are involved in oxidized LDL-induced apoptosis. Arterioscler Thromb Vasc Biol. 2005;25: 639–45.
222. Yao PM, Tabas I. Free cholesterol loading of macrophages is associated with widespread mitochondrial dysfunction and activation of the mitochondrial apoptosis pathway. J Biol Chem. 2001;276: 42468–76.
223. Asmis R, Begley JG. Oxidized LDL promotes peroxide-mediated mitochondrial dysfunction and cell death in human macrophages: a caspase-3-independent pathway. Circ Res. 2003;92:e20–9.
224. Ohashi M, Runge MS, Faraci FM, Heistad DD. MnSOD deficiency increases endothelial dysfunction in ApoE-deficient mice. Arterioscler Thromb Vasc Biol. 2006;26:2331–6.

225. Knight-Lozano CA, Young CG, Burow DL, et al. Cigarette smoke exposure and hypercholesterolemia increase mitochondrial damage in cardiovascular tissues. Circulation. 2002;105:849–54.
226. Liu CS, Kuo CL, Cheng WL, Huang CS, Lee CF, Wei YH. Alteration of the copy number of mitochondrial DNA in leukocytes of patients with hyperlipidemia. Ann N Y Acad Sci. 2005;1042:70–5.
227. Forstermann U. Nitric oxide and oxidative stress in vascular disease. Pflugers Arch. 2010;459:923–39.
228. Jung O, Marklund SL, Xia N, Busse R, Brandes RP. Inactivation of extracellular superoxide dismutase contributes to the development of high-volume hypertension. Arterioscler Thromb Vasc Biol. 2007;27:470–7.
229. Yoshida T, Maulik N, Engelman RM, et al. Glutathione peroxidase knockout mice are susceptible to myocardial ischemia reperfusion injury. Circulation. 1997;96:II216–20.
230. Torzewski M, Ochsenhirt V, Kleschyov AL, et al. Deficiency of glutathione peroxidase-1 accelerates the progression of atherosclerosis in apolipoprotein E-deficient mice. Arterioscler Thromb Vasc Biol. 2007;27:850–7.
231. Ho YS, Xiong Y, Ma W, Spector A, Ho DS. Mice lacking catalase develop normally but show differential sensitivity to oxidant tissue injury. J Biol Chem. 2004;279:32804–12.
232. Stocker R, Perrella MA. Heme oxygenase-1: a novel drug target for atherosclerotic diseases? Circulation. 2006;114:2178–89.
233. Jiang F, Roberts SJ, Datla S, Dusting GJ. NO modulates NADPH oxidase function via heme oxygenase-1 in human endothelial cells. Hypertension. 2006;48:950–7.
234. Taille C, El-Benna J, Lanone S, et al. Induction of heme oxygenase-1 inhibits NAD(P)H oxidase activity by down-regulating cytochrome b558 expression via the reduction of heme availability. J Biol Chem. 2004;279:28681–8.
235. Morita T. Heme oxygenase and atherosclerosis. Arterioscler Thromb Vasc Biol. 2005;25:1786–95.
236. Hoekstra KA, Godin DV, Cheng KM. Protective role of heme oxygenase in the blood vessel wall during atherogenesis. Biochem Cell Biol. 2004;82:351–9.
237. Yamawaki H, Haendeler J, Berk BC. Thioredoxin: a key regulator of cardiovascular homeostasis. Circ Res. 2003;93:1029–33.
238. Chang JD, Sukhova GK, Libby P, et al. Deletion of the phosphoinositide 3-kinase p110gamma gene attenuates murine atherosclerosis. Proc Natl Acad Sci USA. 2007;104:8077–82.
239. Harvey EJ, Li N, Ramji DP. Critical role for casein kinase 2 and phosphoinositide-3-kinase in the interferon-gamma-induced expression of monocyte chemoattractant protein-1 and other key genes implicated in atherosclerosis. Arterioscler Thromb Vasc Biol. 2007;27:806–12.
240. Wang SE, Xiang B, Guix M, et al. Transforming growth factor beta engages TACE and ErbB3 to activate phosphatidylinositol-3 kinase/Akt in ErbB2-overexpressing breast cancer and desensitizes cells to trastuzumab. Mol Cell Biol. 2008;28:5605–20.
241. Pincheira R, Castro AF, Ozes ON, Idumalla PS, Donner DB. Type 1 TNF receptor forms a complex with and uses Jak2 and c-Src to selectively engage signaling pathways that regulate transcription factor activity. J Immunol. 2008;181:1288–98.
242. Fougerat A, Gayral S, Gourdy P, et al. Genetic and pharmacological targeting of phosphoinositide 3-kinase-gamma reduces atherosclerosis and favors plaque stability by modulating inflammatory processes. Circulation. 2008;117:1310–7.
243. Fernandez-Hernando C, Ackah E, Yu J, et al. Loss of Akt1 leads to severe atherosclerosis and occlusive coronary artery disease. Cell Metab. 2007;6:446–57.
244. Feldmann M, Brennan FM, Maini RN. Role of cytokines in rheumatoid arthritis. Annu Rev Immunol. 1996;14:397–440.
245. McKellar GE, McCarey DW, Sattar N, McInnes IB. Role for TNF in atherosclerosis? Lessons from autoimmune disease. Nat Rev Cardiol. 2009;6:410–7.
246. Zhang H, Park Y, Wu J, et al. Role of TNF-alpha in vascular dysfunction. Clin Sci (Lond). 2009;116:219–30.
247. Zhao SP, Wu ZH, Wu J, Hong SC, Deng P. Effect of atorvastatin on tumor necrosis factor alpha serum concentration and mRNA expression of adipose in hypercholesterolemic rabbits. J Cardiovasc Pharmacol. 2005;46:185–9.
248. Jovinge S, Hamsten A, Tornvall P, et al. Evidence for a role of tumor necrosis factor alpha in disturbances of triglyceride and glucose metabolism predisposing to coronary heart disease. Metabolism. 1998;47:113–8.
249. Okopien B, Krysiak R, Kowalski J, et al. Monocyte release of tumor necrosis factor-alpha and interleukin-1beta in primary type IIa and IIb dyslipidemic patients treated with statins or fibrates. J Cardiovasc Pharmacol. 2005;46:377–86.
250. Popa C, Netea MG, Radstake T, et al. Influence of anti-tumour necrosis factor therapy on cardiovascular risk factors in patients with active rheumatoid arthritis. Ann Rheum Dis. 2005;64:303–5.
251. Spanakis E, Sidiropoulos P, Papadakis J, et al. Modest but sustained increase of serum high density lipoprotein cholesterol levels in patients with inflammatory arthritides treated with infliximab. J Rheumatol. 2006;33:2440–6.
252. Memon RA, Grunfeld C, Moser AH, Feingold KR. Tumor necrosis factor mediates the effects of endotoxin on cholesterol and triglyceride metabolism in mice. Endocrinology. 1993;132:2246–53.
253. Feingold KR, Marshall M, Gulli R, Moser AH, Grunfeld C. Effect of endotoxin and cytokines on lipoprotein lipase activity in mice. Arterioscler Thromb. 1994;14:1866–72.
254. MacNaul KL, Hutchinson NI. Differential expression of iNOS and cNOS mRNA in human vascular smooth muscle cells and endothelial cells under normal and inflammatory conditions. Biochem Biophys Res Commun. 1993;196:1330–4.
255. Zhang J, Patel JM, Li YD, Block ER. Proinflammatory cytokines downregulate gene expression and activity of constitutive nitric oxide synthase in porcine pulmonary artery endothelial cells. Res Commun Mol Pathol Pharmacol. 1997;96:71–87.
256. Goodwin BL, Pendleton LC, Levy MM, Solomonson LP, Eichler DC. Tumor necrosis factor-alpha reduces argininosuccinate synthase expression and nitric oxide production in aortic endothelial cells. Am J Physiol Heart Circ Physiol. 2007;293:H1115–21.
257. De Palma C, Meacci E, Perrotta C, Bruni P, Clementi E. Endothelial nitric oxide synthase activation by tumor necrosis factor alpha through neutral sphingomyelinase 2, sphingosine kinase 1, and sphingosine 1 phosphate receptors: a novel pathway relevant to the pathophysiology of endothelium. Arterioscler Thromb Vasc Biol. 2006;26:99–105.
258. Picchi A, Gao X, Belmadani S, et al. Tumor necrosis factor-alpha induces endothelial dysfunction in the prediabetic metabolic syndrome. Circ Res. 2006;99:69–77.
259. Gao X, Belmadani S, Picchi A, et al. Tumor necrosis factor-alpha induces endothelial dysfunction in Lepr(db) mice. Circulation. 2007;115:245–54.
260. Xie L, Hattori Y, Tume N, Gross SS. The preferred source of arginine for high-output nitric oxide synthesis in blood vessels. Semin Perinatol. 2000;24:42–5.
261. Hattori Y, Campbell EB, Gross SS. Argininosuccinate synthetase mRNA and activity are induced by immunostimulants in vascular smooth muscle. Role in the regeneration or arginine for nitric oxide synthesis. J Biol Chem. 1994;269:9405–8.
262. Husson A, Brasse-Lagnel C, Fairand A, Renouf S, Lavoinne A. Argininosuccinate synthetase from the urea cycle to the citrulline-NO cycle. Eur J Biochem. 2003;270:1887–99.
263. Goodwin BL, Solomonson LP, Eichler DC. Argininosuccinate synthase expression is required to maintain nitric oxide production and cell viability in aortic endothelial cells. J Biol Chem. 2004;279:18353–60.
264. Oyadomari S, Gotoh T, Aoyagi K, Araki E, Shichiri M, Mori M. Coinduction of endothelial nitric oxide synthase and arginine

recycling enzymes in aorta of diabetic rats. Nitric Oxide. 2001;5: 252–60.
265. Zhang C, Xu X, Potter BJ, et al. TNF-alpha contributes to endothelial dysfunction in ischemia/reperfusion injury. Arterioscler Thromb Vasc Biol. 2006;26:475–80.
266. De Martin R, Hoeth M, Hofer-Warbinek R, Schmid JA. The transcription factor NF-kappa B and the regulation of vascular cell function. Arterioscler Thromb Vasc Biol. 2000;20:E83–8.
267. Ye J, Wang L, Zhang X, Tantishaiyakul V, Rojanasakul Y. Inhibition of TNF-alpha gene expression and bioactivity by site-specific transcription factor-binding oligonucleotides. Am J Physiol Lung Cell Mol Physiol. 2003;284:L386–94.
268. Downey JM, Omar B, Ooiwa H, McCord J. Superoxide dismutase therapy for myocardial ischemia. Free Radic Res Commun. 1991;12–13(Pt 2):703–20.
269. Pritchard Jr KA, Groszek L, Smalley DM, et al. Native low-density lipoprotein increases endothelial cell nitric oxide synthase generation of superoxide anion. Circ Res. 1995;77:510–8.
270. Cai H, Harrison DG. Endothelial dysfunction in cardiovascular diseases: the role of oxidant stress. Circ Res. 2000;87:840–4.
271. Guzik TJ, West NE, Black E, et al. Vascular superoxide production by NAD(P)H oxidase: association with endothelial dysfunction and clinical risk factors. Circ Res. 2000;86:E85–90.
272. Guzik TJ, Mussa S, Gastaldi D, et al. Mechanisms of increased vascular superoxide production in human diabetes mellitus: role of NAD(P)H oxidase and endothelial nitric oxide synthase. Circulation. 2002;105:1656–62.
273. Feng L, Matsumoto C, Schwartz A, Schmidt AM, Stern DM, Pile-Spellman J. Chronic vascular inflammation in patients with type 2 diabetes: endothelial biopsy and RT-PCR analysis. Diab Care. 2005;28:379–84.
274. Rimbach G, Valacchi G, Canali R, Virgili F. Macrophages stimulated with IFN-gamma activate NF-kappa B and induce MCP-1 gene expression in primary human endothelial cells. Mol Cell Biol Res Commun. 2000;3:238–42.
275. Kumar A, Takada Y, Boriek AM, Aggarwal BB. Nuclear factor-kappaB: its role in health and disease. J Mol Med. 2004;82:434–48.
276. dela Paz NG, Simeonidis S, Leo C, Rose DW, Collins T. Regulation of NF-kappaB-dependent gene expression by the POU domain transcription factor Oct-1. J Biol Chem. 2007;282:8424–34.
277. Russo G, Leopold JA, Loscalzo J. Vasoactive substances: nitric oxide and endothelial dysfunction in atherosclerosis. Vascul Pharmacol. 2002;38:259–69.
278. Zou W, Amcheslavsky A, Takeshita S, Drissi H, Bar-Shavit Z. TNF-alpha expression is transcriptionally regulated by RANK ligand. J Cell Physiol. 2005;202:371–8.
279. Csiszar A, Labinskyy N, Smith K, Rivera A, Orosz Z, Ungvari Z. Vasculoprotective effects of anti-tumor necrosis factor-alpha treatment in aging. Am J Pathol. 2007;170:388–98.
280. Bruunsgaard H, Skinhoj P, Pedersen AN, Schroll M, Pedersen BK. Ageing, tumour necrosis factor-alpha (TNF-alpha) and atherosclerosis. Clin Exp Immunol. 2000;121:255–60.
281. Belmin J, Bernard C, Corman B, Merval R, Esposito B, Tedgui A. Increased production of tumor necrosis factor and interleukin-6 by arterial wall of aged rats. Am J Physiol. 1995;268:H2288–93.
282. Csiszar A, Ungvari Z, Koller A, Edwards JG, Kaley G. Proinflammatory phenotype of coronary arteries promotes endothelial apoptosis in aging. Physiol Genomics. 2004;17:21–30.
283. Das UN. Anti-inflammatory nature of exercise. Nutrition. 2004;20:323–6.
284. Keller C, Keller P, Giralt M, Hidalgo J, Pedersen BK. Exercise normalises overexpression of TNF-alpha in knockout mice. Biochem Biophys Res Commun. 2004;321:179–82.
285. Bruunsgaard H. Physical activity and modulation of systemic low-level inflammation. J Leukoc Biol. 2005;78:819–35.
286. de Lemos ET, Reis F, Baptista S, et al. Exercise training is associated with improved levels of C-reactive protein and adiponectin in ZDF (type 2) diabetic rats. Med Sci Monit. 2007;13: BR168–74.
287. Ait-Oufella H, Salomon BL, Potteaux S, et al. Natural regulatory T cells control the development of atherosclerosis in mice. Nat Med. 2006;12:178–80.
288. Mallat Z, Besnard S, Duriez M, et al. Protective role of interleukin-10 in atherosclerosis. Circ Res. 1999;85:e17–24.
289. Pinderski LJ, Fischbein MP, Subbanagounder G, et al. Overexpression of interleukin-10 by activated T lymphocytes inhibits atherosclerosis in LDL receptor-deficient Mice by altering lymphocyte and macrophage phenotypes. Circ Res. 2002;90: 1064–71.
290. Caligiuri G, Rudling M, Ollivier V, et al. Interleukin-10 deficiency increases atherosclerosis, thrombosis, and low-density lipoproteins in apolipoprotein E knockout mice. Mol Med. 2003;9:10–7.
291. Saltis J, Agrotis A, Bobik A. Regulation and interactions of transforming growth factor-beta with cardiovascular cells: implications for development and disease. Clin Exp Pharmacol Physiol. 1996;23:193–200.
292. McCaffrey TA. TGF-beta signaling in atherosclerosis and restenosis. Front Biosci (Schol Ed). 2009;1:236–45.
293. Falcone DJ, McCaffrey TA, Haimovitz-Friedman A, Vergilio JA, Nicholson AC. Macrophage and foam cell release of matrix-bound growth factors. Role of plasminogen activation. J Biol Chem. 1993;268:11951–8.
294. Grainger DJ. TGF-beta and atherosclerosis in man. Cardiovasc Res. 2007;74:213–22.
295. Huynh ML, Fadok VA, Henson PM. Phosphatidylserine-dependent ingestion of apoptotic cells promotes TGF-beta1 secretion and the resolution of inflammation. J Clin Invest. 2002;109:41–50.
296. Wahl SM, Swisher J, McCartney-Francis N, Chen W. TGF-beta: the perpetrator of immune suppression by regulatory T cells and suicidal T cells. J Leukoc Biol. 2004;76:15–24.
297. Frutkin AD, Otsuka G, Stempien-Otero A, et al. TGF-[beta]1 limits plaque growth, stabilizes plaque structure, and prevents aortic dilation in apolipoprotein E-null mice. Arterioscler Thromb Vasc Biol. 2009;29:1251–7.
298. Virmani R, Burke AP, Kolodgie FD, Farb A. Vulnerable plaque: the pathology of unstable coronary lesions. J Interv Cardiol. 2002;15:439–46.
299. Mallat Z, Gojova A, Marchiol-Fournigault C, et al. Inhibition of transforming growth factor-beta signaling accelerates atherosclerosis and induces an unstable plaque phenotype in mice. Circ Res. 2001;89:930–4.
300. Lutgens E, Gijbels M, Smook M, et al. Transforming growth factor-beta mediates balance between inflammation and fibrosis during plaque progression. Arterioscler Thromb Vasc Biol. 2002;22:975–82.
301. Robertson AK, Rudling M, Zhou X, Gorelik L, Flavell RA, Hansson GK. Disruption of TGF-beta signaling in T cells accelerates atherosclerosis. J Clin Invest. 2003;112:1342–50.
302. Caligiuri G, Nicoletti A, Poirier B, Hansson GK. Protective immunity against atherosclerosis carried by B cells of hypercholesterolemic mice. J Clin Invest. 2002;109:745–53.
303. Binder CJ, Horkko S, Dewan A, et al. Pneumococcal vaccination decreases atherosclerotic lesion formation: molecular mimicry between *Streptococcus pneumoniae* and oxidized LDL. Nat Med. 2003;9:736–43.
304. Hamblin M, Chang L, Fan Y, Zhang J, Chen YE. PPARs and the cardiovascular system. Antioxid Redox Signal. 2009;11:1415–52.
305. Dreyer C, Krey G, Keller H, Givel F, Helftenbein G, Wahli W. Control of the peroxisomal beta-oxidation pathway by a novel family of nuclear hormone receptors. Cell. 1992;68:879–87.
306. Marcus SL, Miyata KS, Zhang B, Subramani S, Rachubinski RA, Capone JP. Diverse peroxisome proliferator-activated receptors bind to the peroxisome proliferator-responsive elements of the rat hydratase/dehydrogenase and fatty acyl-CoA oxidase genes but

differentially induce expression. Proc Natl Acad Sci USA. 1993;90:5723–7.
307. Desvergne B, Wahli W. Peroxisome proliferator-activated receptors: nuclear control of metabolism. Endocr Rev. 1999;20:649–88.
308. Robyr D, Wolffe AP, Wahli W. Nuclear hormone receptor coregulators in action: diversity for shared tasks. Mol Endocrinol. 2000;14:329–47.
309. Adams M, Reginato MJ, Shao D, Lazar MA, Chatterjee VK. Transcriptional activation by peroxisome proliferator-activated receptor gamma is inhibited by phosphorylation at a consensus mitogen-activated protein kinase site. J Biol Chem. 1997;272: 5128–32.
310. Camp HS, Tafuri SR. Regulation of peroxisome proliferator-activated receptor gamma activity by mitogen-activated protein kinase. J Biol Chem. 1997;272:10811–6.
311. Barger PM, Browning AC, Garner AN, Kelly DP. p38 mitogen-activated protein kinase activates peroxisome proliferator-activated receptor alpha: a potential role in the cardiac metabolic stress response. J Biol Chem. 2001;276:44495–501.
312. Alexis JD, Wang N, Che W, et al. Bcr kinase activation by angiotensin II inhibits peroxisome-proliferator-activated receptor gamma transcriptional activity in vascular smooth muscle cells. Circ Res. 2009;104:69–78.
313. Lazennec G, Canaple L, Saugy D, Wahli W. Activation of peroxisome proliferator-activated receptors (PPARs) by their ligands and protein kinase A activators. Mol Endocrinol. 2000;14:1962–75.
314. Marx N, Sukhova G, Murphy C, Libby P, Plutzky J. Macrophages in human atheroma contain PPARgamma: differentiation-dependent peroxisomal proliferator-activated receptor gamma(PPARgamma) expression and reduction of MMP-9 activity through PPARgamma activation in mononuclear phagocytes in vitro. Am J Pathol. 1998;153:17–23.
315. Ricote M, Huang J, Fajas L, et al. Expression of the peroxisome proliferator-activated receptor gamma (PPARgamma) in human atherosclerosis and regulation in macrophages by colony stimulating factors and oxidized low density lipoprotein. Proc Natl Acad Sci USA. 1998;95:7614–9.
316. Marx N, Kehrle B, Kohlhammer K, et al. PPAR activators as anti-inflammatory mediators in human T lymphocytes: implications for atherosclerosis and transplantation-associated arteriosclerosis. Circ Res. 2002;90:703–10.
317. Staels B, Koenig W, Habib A, et al. Activation of human aortic smooth-muscle cells is inhibited by PPARalpha but not by PPARgamma activators. Nature. 1998;393:790–3.
318. Chinetti G, Gbaguidi FG, Griglio S, et al. CLA-1/SR-BI is expressed in atherosclerotic lesion macrophages and regulated by activators of peroxisome proliferator-activated receptors. Circulation. 2000;101:2411–7.
319. Marx N, Sukhova GK, Collins T, Libby P, Plutzky J. PPARalpha activators inhibit cytokine-induced vascular cell adhesion molecule-1 expression in human endothelial cells. Circulation. 1999;99:3125–31.
320. Marx N, Mackman N, Schonbeck U, et al. PPARalpha activators inhibit tissue factor expression and activity in human monocytes. Circulation. 2001;103:213–9.
321. Piqueras L, Reynolds AR, Hodivala-Dilke KM, et al. Activation of PPARbeta/delta induces endothelial cell proliferation and angiogenesis. Arterioscler Thromb Vasc Biol. 2007;27:63–9.
322. Fan Y, Wang Y, Tang Z, et al. Suppression of pro-inflammatory adhesion molecules by PPAR-delta in human vascular endothelial cells. Arterioscler Thromb Vasc Biol. 2008;28:315–21.
323. Liou JY, Lee S, Ghelani D, Matijevic-Aleksic N, Wu KK. Protection of endothelial survival by peroxisome proliferator-activated receptor-delta mediated 14-3-3 upregulation. Arterioscler Thromb Vasc Biol. 2006;26:1481–7.
324. Brunelli L, Cieslik KA, Alcorn JL, Vatta M, Baldini A. Peroxisome proliferator-activated receptor-delta upregulates 14-3-3 epsilon in human endothelial cells via CCAAT/enhancer binding protein-beta. Circ Res. 2007;100:e59–71.
325. Han JK, Lee HS, Yang HM, et al. Peroxisome proliferator-activated receptor-delta agonist enhances vasculogenesis by regulating endothelial progenitor cells through genomic and nongenomic activations of the phosphatidylinositol 3-kinase/Akt pathway. Circulation. 2008;118:1021–33.
326. Li AC, Binder CJ, Gutierrez A, et al. Differential inhibition of macrophage foam-cell formation and atherosclerosis in mice by PPARalpha, beta/delta, and gamma. J Clin Invest. 2004;114: 1564–76.
327. Kim HJ, Ham SA, Kim SU, et al. Transforming growth factor-beta1 is a molecular target for the peroxisome proliferator-activated receptor delta. Circ Res. 2008;102:193–200.
328. Joseph SB, Castrillo A, Laffitte BA, Mangelsdorf DJ, Tontonoz P. Reciprocal regulation of inflammation and lipid metabolism by liver X receptors. Nat Med. 2003;9:213–9.
329. Calkin AC, Tontonoz P. Liver x receptor signaling pathways and atherosclerosis. Arterioscler Thromb Vasc Biol. 2010;30:1513–8.
330. Schuster GU, Parini P, Wang L, et al. Accumulation of foam cells in liver X receptor-deficient mice. Circulation. 2002;106: 1147–53.
331. Bradley MN, Hong C, Chen M, et al. Ligand activation of LXR beta reverses atherosclerosis and cellular cholesterol overload in mice lacking LXR alpha and apoE. J Clin Invest. 2007;117:2337–46.
332. Bischoff ED, Daige CL, Petrowski M, et al. Non-redundant roles for LXRalpha and LXRbeta in atherosclerosis susceptibility in low density lipoprotein receptor knockout mice. J Lipid Res. 2010;51:900–6.
333. Venkateswaran A, Laffitte BA, Joseph SB, et al. Control of cellular cholesterol efflux by the nuclear oxysterol receptor LXR alpha. Proc Natl Acad Sci USA. 2000;97:12097–102.
334. Tangirala RK, Bischoff ED, Joseph SB, et al. Identification of macrophage liver X receptors as inhibitors of atherosclerosis. Proc Natl Acad Sci USA. 2002;99:11896–901.
335. Teupser D, Kretzschmar D, Tennert C, et al. Effect of macrophage overexpression of murine liver X receptor-alpha (LXR-alpha) on atherosclerosis in LDL-receptor deficient mice. Arterioscler Thromb Vasc Biol. 2008;28:2009–15.
336. Ghisletti S, Huang W, Jepsen K, et al. Cooperative NCoR/SMRT interactions establish a corepressor-based strategy for integration of inflammatory and anti-inflammatory signaling pathways. Genes Dev. 2009;23:681–93.
337. Castrillo A, Joseph SB, Marathe C, Mangelsdorf DJ, Tontonoz P. Liver X receptor-dependent repression of matrix metalloproteinase-9 expression in macrophages. J Biol Chem. 2003;278:10443–9.
338. A-Gonzalez N, Bensinger SJ, Hong C, et al. Apoptotic cells promote their own clearance and immune tolerance through activation of the nuclear receptor LXR. Immunity. 2009;31:245–58.
339. Castrillo A, Joseph SB, Vaidya SA, et al. Crosstalk between LXR and toll-like receptor signaling mediates bacterial and viral antagonism of cholesterol metabolism. Mol Cell. 2003;12:805–16.
340. Naiki Y, Sorrentino R, Wong MH, et al. TLR/MyD88 and liver X receptor alpha signaling pathways reciprocally control *Chlamydia pneumoniae*-induced acceleration of atherosclerosis. J Immunol. 2008;181:7176–85.
341. Morello F, Saglio E, Noghero A, et al. LXR-activating oxysterols induce the expression of inflammatory markers in endothelial cells through LXR-independent mechanisms. Atherosclerosis. 2009; 207:38–44.
342. Liao H, Langmann T, Schmitz G, Zhu Y. Native LDL upregulation of ATP-binding cassette transporter 1 in human vascular endothelial cells. Arterioscler Thromb Vasc Biol. 2002;22:127–32.
343. Norata GD, Ongari M, Uboldi P, Pellegatta F, Catapano AL. Liver X receptor and retinoic X receptor agonists modulate the expression

of genes involved in lipid metabolism in human endothelial cells. Int J Mol Med. 2005;16:717–22.
344. Zhu Y, Liao H, Xie X, et al. Oxidized LDL downregulates ATP-binding cassette transporter-1 in human vascular endothelial cells via inhibiting liver X receptor (LXR). Cardiovasc Res. 2005;68: 425–32.
345. Zhu M, Fu Y, Hou Y, et al. Laminar shear stress regulates liver X receptor in vascular endothelial cells. Arterioscler Thromb Vasc Biol. 2008;28:527–33.
346. Sun Y, Hao M, Luo Y, et al. Stearoyl-CoA desaturase inhibits ATP-binding cassette transporter A1-mediated cholesterol efflux and modulates membrane domain structure. J Biol Chem. 2003;278:5813–20.
347. Qin X, Tian J, Zhang P, et al. Laminar shear stress up-regulates the expression of stearoyl-CoA desaturase-1 in vascular endothelial cells. Cardiovasc Res. 2007;74:506–14.
348. Blaschke F, Leppanen O, Takata Y, et al. Liver X receptor agonists suppress vascular smooth muscle cell proliferation and inhibit neointima formation in balloon-injured rat carotid arteries. Circ Res. 2004;95:e110–23.
349. Leik CE, Carson NL, Hennan JK, et al. GW3965, a synthetic liver X receptor (LXR) agonist, reduces angiotensin II-mediated pressor responses in Sprague-Dawley rats. Br J Pharmacol. 2007;151:450–6.
350. Hsu JJ, Lu J, Huang MS, et al. T0901317, an LXR agonist, augments PKA-induced vascular cell calcification. FEBS Lett. 2009;583:1344–8.
351. Imayama I, Ichiki T, Patton D, et al. Liver X receptor activator downregulates angiotensin II type 1 receptor expression through dephosphorylation of Sp1. Hypertension. 2008;51:1631–6.
352. Kang MA, Jeoung NH, Kim JY, et al. Up-regulation of skeletal muscle LIM protein 1 gene by 25-hydroxycholesterol may mediate morphological changes of rat aortic smooth muscle cells. Life Sci. 2007;80:460–7.
353. Watson KE, Bostrom K, Ravindranath R, Lam T, Norton B, Demer LL. TGF-beta 1 and 25-hydroxycholesterol stimulate osteoblast-like vascular cells to calcify. J Clin Invest. 1994;93:2106–13.
354. Saito E, Wachi H, Sato F, Seyama Y. 7-ketocholesterol, a major oxysterol, promotes pi-induced vascular calcification in cultured smooth muscle cells. J Atheroscler Thromb. 2008;15:130–7.
355. Arner P. The adipocyte in insulin resistance: key molecules and the impact of the thiazolidinediones. Trends Endocrinol Metab. 2003;14:137–45.
356. Yudkin JS, Juhan-Vague I, Hawe E, et al. Low-grade inflammation may play a role in the etiology of the metabolic syndrome in patients with coronary heart disease: the HIFMECH study. Metabolism. 2004;53:852–7.
357. Matsumoto T, Mugishima H. Signal transduction via vascular endothelial growth factor (VEGF) receptors and their roles in atherogenesis. J Atheroscler Thromb. 2006;13:130–5.
358. Ushio-Fukai M. VEGF signaling through NADPH oxidase-derived ROS. Antioxid Redox Signal. 2007;9:731–9.
359. Inoue M, Itoh H, Ueda M, et al. Vascular endothelial growth factor (VEGF) expression in human coronary atherosclerotic lesions: possible pathophysiological significance of VEGF in progression of atherosclerosis. Circulation. 1998;98:2108–16.
360. Ramos MA, Kuzuya M, Esaki T, et al. Induction of macrophage VEGF in response to oxidized LDL and VEGF accumulation in human atherosclerotic lesions. Arterioscler Thromb Vasc Biol. 1998;18:1188–96.
361. Ehrenstein MR, Jury EC, Mauri C. Statins for atherosclerosis – as good as it gets? N Engl J Med. 2005;352:73–5.
362. Nissen SE, Tuzcu EM, Schoenhagen P, et al. Statin therapy, LDL cholesterol, C-reactive protein, and coronary artery disease. N Engl J Med. 2005;352:29–38.
363. Ridker PM, Cannon CP, Morrow D, et al. C-reactive protein levels and outcomes after statin therapy. N Engl J Med. 2005;352:20–8.
364. Bresalier RS, Sandler RS, Quan H, et al. Cardiovascular events associated with rofecoxib in a colorectal adenoma chemoprevention trial. N Engl J Med. 2005;352:1092–102.
365. Koenen RR, Weber C. Therapeutic targeting of chemokine interactions in atherosclerosis. Nat Rev Drug Discov. 2010;9:141–53.
366. Horuk R. Chemokine receptor antagonists: overcoming developmental hurdles. Nat Rev Drug Discov. 2009;8:23–33.
367. Bathon JM, Martin RW, Fleischmann RM, et al. A comparison of etanercept and methotrexate in patients with early rheumatoid arthritis. N Engl J Med. 2000;343:1586–93.
368. Feldmann M, Maini RN. Anti-TNF alpha therapy of rheumatoid arthritis: what have we learned? Annu Rev Immunol. 2001;19: 163–96.
369. Weinblatt ME, Keystone EC, Furst DE, et al. Adalimumab, a fully human anti-tumor necrosis factor alpha monoclonal antibody, for the treatment of rheumatoid arthritis in patients taking concomitant methotrexate: the ARMADA trial. Arthritis Rheum. 2003;48:35–45.
370. Combadiere C, Potteaux S, Rodero M, et al. Combined inhibition of CCL2, CX3CR1, and CCR5 abrogates Ly6C(hi) and Ly6C(lo) monocytosis and almost abolishes atherosclerosis in hypercholesterolemic mice. Circulation. 2008;117:1649–57.
371. Martin AP, Canasto-Chibuque C, Shang L, Rollins BJ, Lira SA. The chemokine decoy receptor M3 blocks CC chemokine ligand 2 and CXC chemokine ligand 13 function in vivo. J Immunol. 2006;177:7296–302.
372. Potzinger H, Geretti E, Brandner B, et al. Developing chemokine mutants with improved proteoglycan affinity and knocked-out GPCR activity as anti-inflammatory recombinant drugs. Biochem Soc Trans. 2006;34:435–7.
373. Schuksz M, Fuster MM, Brown JR, et al. Surfen, a small molecule antagonist of heparan sulfate. Proc Natl Acad Sci USA. 2008;105:13075–80.
374. Deruaz M, Frauenschuh A, Alessandri AL, et al. Ticks produce highly selective chemokine binding proteins with antiinflammatory activity. J Exp Med. 2008;205:2019–31.
375. Schrijvers DM, De Meyer GR, Herman AG, Martinet W. Phagocytosis in atherosclerosis: molecular mechanisms and implications for plaque progression and stability. Cardiovasc Res. 2007;73:470–80.
376. Zernecke A, Bot I, Djalali-Talab Y, et al. Protective role of CXC receptor 4/CXC ligand 12 unveils the importance of neutrophils in atherosclerosis. Circ Res. 2008;102:209–17.
377. Liu J, Divoux A, Sun J, et al. Genetic deficiency and pharmacological stabilization of mast cells reduce diet-induced obesity and diabetes in mice. Nat Med. 2009;15:940–5.

# Part VII
# Targeting Signaling in Cardiovascular Medicine

# Chapter 19
# Stem Cells Signaling Pathways in the Heart

**Abstract** Limited capacity of regeneration and proliferation of human cardiomyocytes can prevent neither the scar formation that follows myocardial infarction (MI) nor the loss of cardiac function occurring in patients with cardiomyopathy and heart failure (HF). The replacement and regeneration of functional cardiac muscle is an important objective that could be achieved either by the stimulation of autologous resident cardiomyocytes or by the transplantation of stem cells to directly repopulate these tissues, a viable therapeutic approach for repairing the injured myocardium. Regardless of the right type of stem cells and the technique of their delivery in the right place at right time, better understanding of the molecular mechanisms of multiple signaling pathways contributing to the regulation of essential functions of transplanted or endogenous stem and progenitor cells is required for conceiving improved therapy of injured heart. Although the key signaling molecules represent attractive targets for pharmaceutical or genetic modulations to induce and facilitate cardiac regeneration, involvement of many of them in various signaling pathways responding to different stimuli and leading to different outcomes further complicates their clinical use. Unleashing the novel powerful tools of modern cell biology, genetics and pharmacology combined with functional genomics and proteomics will undoubtedly provide significant impact on the future strategy for cell-based cardiac repair. In this chapter, we review the cell types used in myocardial transplantation and present information concerning signaling pathways, preclinical findings, and potential clinical application of myocardial cell engineering.

**Keywords** Stem cells • Embryonic cells • Cardiac progenitor cells • Cell delivery • Signaling pathways

## Introduction

In developed countries, cardiovascular disease (CVD) remains a leading cause of morbidity and mortality. In the USA, approximately 2.5 million deaths each year (one in every 2.8 death) are resulted from CVD [1, 2]. Two main CVDs, myocardial infarction and heart failure, affect more than eight and five million people in the USA, respectively. Despite current state-of-the-art interventional and pharmacological therapies for CVD, these grim statistics demonstrate the limits of therapeutic efficiency of these approaches and underscore the great need for new and efficacious strategies.

Progenitor and stem cell-based therapy has emerged as a potential new therapeutic option for heart diseases. During the past decade, growing experimental evidence in animal models and in small clinical trials suggests that stem cell transplantation into the infarcted myocardium beneficially remodels and repairs injured tissue and improves cardiac function. Limitations of low survival of the engrafted cells, difficulties of their integration into the host myocardium, and modest and transient improvement of cardiac function induce extensive efforts to understand and eventually to modify signaling pathways involved in cardiac infarction with regard to stem cells. In this chapter, we discuss the different stem cell types that have being studied, cell delivery techniques, and the main signaling pathways associated with these cells emphasizing the potential modifications of these pathways aiming to repair the infracted and dysfunctional hearts.

One of the major barriers to the development of cell-based therapy for myocardial repair is the ability to generate a sufficient number of cells for therapeutic use. According to the rough estimation, a typical myocardial infarct results in approximately $1–2 \times 10^9$ cardiomyocyte loss [3]. Even though not all lost cells have to be replaced to improve cardiac function, a cell replacement procedure has to generate and deliver hundreds of millions of cells capable to form cardiomyocytes for a single patient. Moreover, although cardiomyocyte replacement is a primary aim, other cell types, including cardiac fibroblasts, smooth muscle, and endothelial cells, play essential roles in cardiac function and their replacement is also required. Given current cell delivery protocols suffer from substantial inefficiency with 50–90% cell loss, it will be necessary to introduce at least $10^{10}$ to $10^{11}$ cells [4, 5]. Therefore, the scale of heart cell-based therapy represents a considerable, often underestimated, challenge.

J. Marín-García, *Signaling in the Heart*,
DOI 10.1007/978-1-4419-9461-5_19, 

## Stem Cell Types

Stem cells, regardless of their origin and type, are undifferentiated cells characterized by the potential for self-renewal and by the ability to differentiate into various cell types known as pluripotency. Various cells, differing significantly in regards to their origin, structure, and function as well as to their ability to differentiate into one or more cell types, have been considered as candidates for myocardial cell-based repair therapy (Table 19.1).

### *Embryonic Stem Cells*

The embryonic stem cells (ESCs), the prototypical stem cells, are pluripotent cells capable to differentiate into a variety cell types, including cardiomyocytes [7]. Human ESCs were first isolated in 1998 by Thomson and associates from the inner cell mass of preimplantation embryo [8]. During the last decade, protocols to differentiate human ESCs into cardiomyocytes have been developed and successful isolations of cardiomyocytes from differentiating ESCs have been reported [9]. The ESC-derived cardiomyocytes exhibit all major molecular, structural, mechanical, and electrophysiological characteristics of primary cardiomyocytes. However, the ESC-derived cardiomyocyte features resemble fetal, rather than adult, cardiomyocytes [7, 10]. Transplantation experiments with the ESC-derived cardiomyocytes in animal models have demonstrated that grafted cells survive and mature for at least 12 weeks and can couple electrically with the host cardiomyocytes [11]. Improvement in cardiac function have been reported 4 weeks after transplantation; however, no beneficial effects have been detected 12 weeks after transplantation [12, 13].

**Table 19.1** Major cell types with potentials for cardiac cell therapy [6]

| Cell type | Source | Advantages | Limitations |
|---|---|---|---|
| Cardiac stem cells | Allogenic fetal, neonatal, or adult heart | 1. Recognition of myocardial growth factors and recruitment to myocardium are likely faster and more efficient than other cell types<br>2. *In vivo* electrical coupling of transplanted cells to existing myocardium has been demonstrated | 1. Poor cell growth in vitro<br>2. Transplanted cells are very sensitive to ischemic insult and apoptotic cell death<br>3. Availability from either fetal (F), neonatal (N), or adult sources is low at present; likely immune rejection; F and N cells pose ethical difficulties |
| Skeletal myoblast | Autologous skeletal muscle biopsy | 1. Cells proliferate in vitro (allowing for autologous transplant)<br>2. Ischemia resistant<br>3. Transplanted myoblasts can differentiate into slow-twitch myocytes (similar to cardiomyocytes) enabling cellular cardiomyoplasty<br>4. Reduces progressive ventricular dilatation and improves cardiac function<br>5. Can use adult cells | 1. Likely do not develop new cardiomyocytes *in vivo*<br>2. Electrical coupling to surrounding myocardial cells is unclear (may cause dysrhythmias)<br>3. Long-term stability of differentiated phenotype unknown |
| Adult bone marrow stem cells | Autologous bone marrow stromal cells (mesenchymal); bone marrow (endothelial progenitor cells) | 1. Pluripotent stem cells can develop into cardiomyocytes<br>2. Stem cells are easy to isolate and grow well in culture<br>3. Neovascularization can occur at the site of myocardial scar reducing ischemia<br>4. Transdifferentiation of cells into cardiomyocyte *in vivo* has been shown<br>5. Can be derived from autologous source; no immune-suppression treatment<br>6. Can improve myocardial contractile function | 1. New program of cell differentiation is required<br>2. Efficiency of the differentiation into adult cardiomyocytes appears limited<br>3. Signaling, stability, and regulation of differentiation unknown |
| Embryonic stem cells | Allogenic blastocyst (inner mass) | 1. Easy propagation and well-defined cardiomyocyte differentiation process<br>2. *In vivo* electrical coupling of transplanted cells to existing myocardial cells<br>3. Pluripotent cells | 1. Potential for tumor formation and immune rejection (allogenic)<br>2. Incomplete response to physiological stimuli<br>3. Legal and ethical issues<br>4. Donor availability |

ESCs have several potential disadvantages, which has to be overcome before clinical translation. First, their defining capability for pluripotent proliferation increases risk of teratoma formation after transplantation into various tissues including the heart [14]. Immunological mismatch of ESCs due to their allogenic source would require immunosuppression to prevent immune rejection after transplantation [15]. Finally, the therapeutic use of ESCs has raised the ethical debates although progress has recently been made to circumvent some of the central issues.

## Bone Marrow-Derived Cells

The bone marrow represents a typical source of adult stem cells containing two main cell subsets, hematopoietic stem cells (HSCs), and mesenchymal stem cells (MSCs). Bone marrow cells can be isolated from the peripheral circulation or by bone marrow aspiration.

HSCs characterized by the expression of CD34 and CD113 cell-surface antigens have been identified in the peripheral blood and bone marrow [16]. Primitive $CD34^{-}$ HSCs may also exist [17, 18]. HSCs demonstrate high self-renewal ability and serve as precursors of blood cell lineages. They appear to be capable to differentiate into cardiomyocytes or generate chimeras via fusion with donor cardiomyocytes after injection [19, 22].

MSCs characterized by the absence of HSC markers have been identified in bone marrow and adipose tissues [23]. They express CD44, CD90, CD105, CD106, and CD166 cell-surface markers, while they are negative for hematopoietic (CD14, CD34, and CD45) and endothelial (VEGFR2, CD34, and CD105) markers and are capable to differentiate into several nonhematopoietic cell types [24, 25]. Like HSCs, MSCs can transdifferentiate into cardiomyocytes and fuse with host cells affecting functional recovery after myocardial infarction [22, 26, 27]. MSCs can be isolated and expanded in culture easier compared to HSCs. Furthermore, they do not express MHC class II, do not induce local inflammatory responses, and therefore avoid allogenic rejection [28]. These features of MSCs make them attractive candidates for cell-based therapy for myocardial disease.

Presently, mainly autologous adult bone marrow-derived progenitor cells have been used in human clinical studies. Initial reports demonstrated a robust transdifferentiation of transplanted bone marrow cells into cardiomyocytes have boosted significant enthusiasm in the field [29, 30]. However, subsequent studies have demonstrated that the injected bone marrow cells fuse with the recipient cardiomyocytes rather than transdifferentiate into cardiomyocytes [20, 21]. Furthermore, although clinical trials using bone marrow-derived progenitor cells have proved that the procedure is safe, their outcomes have been conflicting. Some trials have demonstrated modest beneficial effects attributed to a poorly understood paracrine effects, while others have failed to show significant differences compared to the control groups [3, 31]. Nevertheless, current consensus is that no myocardial regeneration has been documented in clinical trials using autologus bone marrow cells, and modest improvement in cardiac function probably results from angiogenesis induced by the growth factor secretion by the transplanted cells [32].

## Skeletal Myoblasts

The hypothesis to use contractile cells to repair the failing heart was initiated in 1989 [33]. Initial studies have demonstrated that human skeletal myoblasts co-cultured *in vitro* with rat cardiomyocytes have adopted cardiac features and their transplantation into failing heart in animal models have improved cardiac function and survival [34]. However, despite reported beneficial effect on left ventricular function, subsequent studies have provided little evidence that injected skeletal myoblasts transdifferentiate into cardiomyocytes [35, 36]. Postmortem pathology has failed to demonstrate integration of transfected skeletal myoblasts into recipient myocardium [37]. Moreover, the reported modest contractile benefit has not been sustained and has been associated with elevated ventricular dysrhythmias [38–40].

## Resident Cardiac Progenitor Cells

Traditionally, the myocardium has been considered as a terminally differentiated tissue. Its poor regeneration capacity was thought to result from the absence of endogenous cardiac progenitor cells and from the inability of mature cardiomyocytes to turn over by dividing. However, emerging evidence suggests that the mammalian heart contains a population of resident cardiac stem cells (CSCs) and cardiomyocytes are able to divide, albeit at very low level. Several laboratories using various cell markers, such as c-$kit^{+}$, Sca-$1^{+}$, $Isl1^{+}$, and different isolation methods have identified clusters of resident CSCs in the heart [41–47]. Currently, it remains to be determined to what extent these cell populations described by different laboratories overlap. They also could represent the same cell type at different stages of differentiation.

Although in the embryonic mammalian heart several endogenous CSC populations are present which, are able to differentiate into cardiomyocytes, smooth muscle, and endothelial cells, their numbers rapidly decline with age.

However, CSCs can be isolated from the normal human myocardium using a minimally invasive biopsy procedure. Isolated human c-kit+ CSCs are able to expand and differentiate *in vitro* predominantly into cardiomyocytes and, to a lesser extent, into smooth muscle and endothelial cells. Importantly, these cells, injected in the infracted myocardium of immunodeficient or immunosuppressed rodents, integrate functionally with rodent myocardium and contribute to the improvement of cardiac function [48, 49]. Although these findings point to the possibility of using CSCs in cell-based therapy for the diseased heart, no clinical data using CSCs are available yet and many critical questions about these cells remain unanswered.

### Induced Pluripotent Stem Cells

Recently, a revolutionary approach to generate ESC-like cell population from differentiated cells, such as adult fibroblasts, has been suggested. This strategy relies on the reprogramming of adult fully differentiated cells into so-called induced pluripotent stem cells (iPSCs) using ectopic expression of specific genes responsible for pluripotency [50–52]. Human iPSCs were initially generated from human skin fibroblasts using retroviral introduction of four transgenes encoding transcription factors Oct4, Sox2, c-myc, and Kif4 [53, 54]. Subsequently, this technique has been successfully applied to other cell types obtained from patients [55–58]. It has been shown that both Oct4 and Sox2 are essential for the reprogramming of somatic cells into multipotent embryonic-like cells, whereas oncogenes may be dispensable, and iPCS can be created in the absence of c-myc transduction [59–61]. Moreover, recent studies have demonstrated that iPSCs can be directed to differentiate into cardiovascular and hematopoietic progenitor cells [62–65].

The original protocol for the generation of iPSCs has been based on the use of virus-mediated delivery of reprogramming factors. However, viral integration into the genome of cells undergoing the reprogramming poses the danger of induced mutagenesis leading to malignant transformation. To address this important issue and improve safety of the technique, it has recently been developed virus-free delivery systems including direct delivery of reprogramming proteins [66–69]. However, it remains to be determined if iPSCs produced by these novel methods are similar to iPSCs created using virus-mediated delivery.

It also worth to mention that generation of "disease- and patient-specific" iPSCs will provide valuable cellular models for understanding the mechanisms of various diseases [70]. Since iPSCs are derived from adult somatic cells they overcome the ethical issues raised by the use of human embryos. Moreover, the use of iPSCs can circumvent the immunocompatibility problems as the starting cell material can be obtained from the same patient.

## Cell Delivery Techniques

Effective cell-based therapy to repair injured tissue requires targeted delivery of chosen cells to an area of the damaged tissue. The heart as a constantly contracting organ changing its geometry creates additional difficulties to transplant cells into localized heart region and immobilize them there. Experimental animal models have demonstrated that only approximately 15% of cells transplanted to infracted myocardium stay in the heart [71]. Therefore, it is difficult to predict an adequate cell dose and the size of the graft after transplantation.

Although several routes of cell delivery including intravenous, intracoronary, direct epicardial injection, catheter-based transendocardial injection, and transvenous injection into coronary veins have been described, none has yet emerged as the preferred approach [72, 73]. Intracoronary infusion is considered as a safe procedure and is one of the most common cell delivery methods in clinical studies. The major drawbacks of this technique include an elevated incidence of coronary events and low efficiency as a significant amount of transplanted cells fails to home in the infracted myocardium [74, 75]. Transendocardial injection delivers implanted cells to the border zone of the infracted myocardium by an injection catheter guided by the mapping system and therefore improves the efficiency of cell therapy. This delivery technique is currently under clinical trials.

## Stem Cell Signaling Pathways

Common characteristics of stem cells essential for their use in cell-based therapy are their ability for self-renewal and their capability to differentiate into different cell types. Commitment to developmental cell lineages represents a complex stepwise process orchestrated by intricate signaling networks. Transplanted stem cells or resident CSCs are exposed to numerous signals from the microenvironment, which regulates their survival, proliferation, and differentiation. These signaling pathways include Wnt, nuclear factor-κB (NF-κB), stromal cell-derived factor-1 (SDF-1)/C-X-C chemokine receptor type 4 (CXCR4), mitogen-activated protein kinase (MAPK), and phosphatidylinositol 3-kinase (PI3K)-Akt pathways. Understanding the molecular mechanisms of these signaling pathways will improve the clinical outcome of cell-based therapy for heart failure.

## Wnt Signaling

Wnt proteins are a large family of secreted signaling proteins, which modulate various developmental processes including embryonic induction, cell proliferation, survival, migration, and the specification of cell fate [76, 77]. Traditionally, cardiovascular development has been thought to be mediated by noncanonical, Jun-N-terminal kinase (JNK)-dependent, Wnt signaling, whereas cardiac specification has been downregulated by canonical, β-catenin-dependent, Wnt signaling. More recent model has reconsidered the role of canonical Wnt signaling; it activates initial specification while inhibits later cardiomyocyte differentiation. Major players involved in cardiac Wnt signaling are listed in Table 19.2 [78].

Nineteen mammalian Wnt proteins have currently been identified suggesting the existence of a highly complex signaling network. Wnt proteins display several common characteristics including a signal sequence essential for their secretion, several highly charged amino acids residues, specific location of 22 cysteine residues, and multiple glycosylation sites. In addition to glycosylation, Wnts are also modified by the palmitate attachment to the first conserved cysteine and to a serine residue in the middle of the molecule [79]. These lipid modifications explain hydrophobic properties of Wnt proteins and are essential for their targeting to the cell membrane although the exact function of Wnt palmitoylation remains to be determined.

Studies on *Drosophila*, *Caenorhabditis elegans*, and mammalian tissue culture have demonstrated that the transmembrane protein Wntless/Evi, located in the plasma membrane and in the Golgi, controls transport and secretion of Wnt proteins from producing cells [80–82]. It has further been proposed that the retromer complex including Vps35 protein is implicated in Wnt transport in a form of lipoprotein particles [83–85].

Frizzled (Fzd) receptors with seven transmembrane domains bind various Wnts through the cysteine-rich domain (CRD) localized on their extracellular region [86–88]. Canonical Wnt signaling requires association of Fzd proteins with the lipoprotein receptor related proteins 5 and 6 (Lrp5 and -6) [89, 90]. Following Wnt binding, intracellular domain of Fzd receptor interacts directly with the dishevelled (Dvl) proteins to transduce a Wnt signal [91]. Dvl inactivates a multiprotein complex composed of the scaffolding protein axin, adenomatosis polyposis coli (APC), and glycogen synthase kinase 3β (Gsk3β) [92, 93]. Since this complex mediates phosphorylation of β-catenin that targets it for degradation, canonical Wnt signaling results in β-catenin stabilization and its accumulation in the nucleus. β-Catenin binds LEF/TCF family of transcription factors and stimulates transcription of specific genes (Fig. 19.1).

Noncanonical Wnt signaling is mediated via the $Ca^{2+}$-protein kinase C (PKC) or RhoA-JNK pathways [94] (Fig. 19.1). $Ca^{2+}$-PKC pathway is initiated by the Wnt-dependent induction of Fzd receptors leading to G protein-dependent activation of intracellular $Ca^{2+}$ release. Increased $Ca^{2+}$ levels activate in turn several $Ca^{2+}$-dependent protein kinases including PKC and calmodulin-dependent protein kinase II. Rho-JNK

**Table 19.2** Role of various Wnt signaling components in cardiovascular development

| Gene | Phenotype | Pathway | Species |
|---|---|---|---|
| β-Catenin | Required for postnatal hypertrophic growth (mouse), acts upstream of FGF signaling to regulate SHF progenitors (mouse), essential for cardiac valve development (zebrafish) | Canonical | Mouse and zebrafish |
| Wnt ligands | | | |
| *Wnt2a* | Placental vascular failure | Canonical | Mouse |
| *Wnt5a* | Cardiac outflow tract septation, endothelial proliferation, and vessel branching *in vitro* | Noncanonical | Mouse |
| *Wnt7b* | Loss of pulmonary vascular smooth muscle integrity and hemorrhage | Canonical | Mouse |
| *Wnt9a* | Epicardial/coronary development | Unknown | Mouse |
| *Wnt11* | Cardiac specification (frogs and zebrafish), cardiac outflow tract development (mouse) | Noncanonical | Mouse, frogs, and zebrafish |
| Frizzled receptors | | | |
| *Fzd4* | Retinal vascular development | Unknown | Mouse |
| *Fzd5* | Placental vascular development | Unknown | Mouse |
| Dishevelled | | | |
| *Dvl1* | Myofibroblast wound healing | Unknown | Mouse |
| *Dvl2* | OFT development | Unknown | Mouse |
| *Apc* | Cardiac valve development | Canonical | Zebrafish |
| *Gsk3β* | Positively regulates cardiomyocyte proliferation, required for postnatal hypertrophic growth | Canonical | Mouse |
| *Vangl2* | OFT development | Noncanonical | Mouse |

*Apc* adenomatous polypopsis coli, *Gsk3β* glycogen synthase kinase 3β, *OFT* outflow tract, *FGF* fibroblast growth factor, *SHF* second heart field

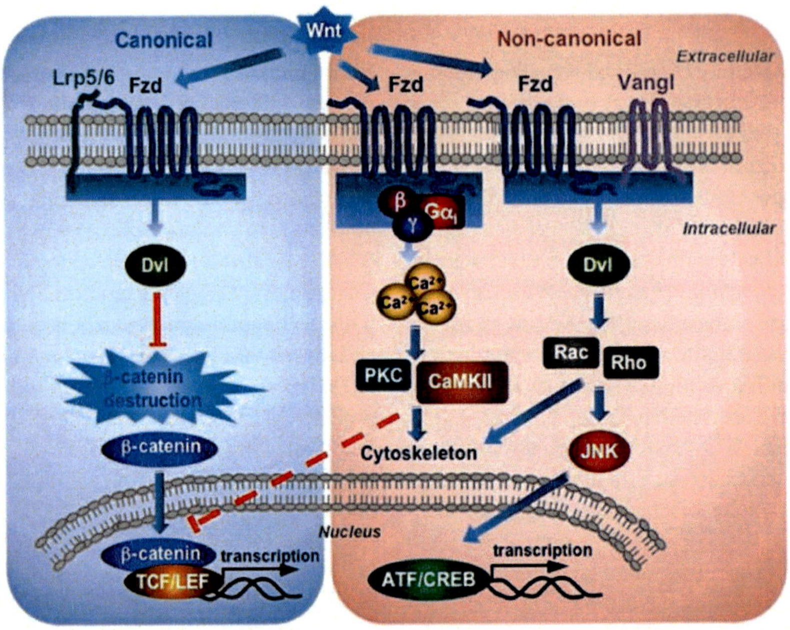


**Fig. 19.1** Canonical and noncanonical Wnt signaling pathways. In canonical Wnt signaling, Wnt ligands engage the Fzd-Lrp5/6 co-receptor complex to activate effector protein dishelled (Dvl) leading to the inhibition of β-catenin degradation. β-Catenin stabilization results in its accumulation in the nucleus and in the activation of TCF/LEF-dependent transcription. Noncanonical Wnt signaling involves the $Ca^{2+}$/PKC- and Rho/JNK-mediated pathways. In the former, Fzd receptor activation results in G protein-mediated release of $Ca^{2+}$ leading to the activation of PKC and calmodulin-dependent protein kinase II (CaMKII). This contributes to rearrangements of cytoskeleton and to the inhibition of β-catenin-dependent canonical signaling via poorly understood mechanism. In the latter, Wnt engages Fzd and Vangl receptors to activate Dvl-mediated Rho/Rac and JNK signaling resulting in the stimulation of the ATF/CREB-dependent transcription

pathway operates through Dvl-dependent activation of Rho family GTPases, RhoA and Rac, and their downstream effectors such as Rho-associated kinase (ROCK) and JNK, leading eventually to stimulation of activating transcription factor (ATF)/cAMP response element-binding (CREB)-mediated transcription.

According to the current paradigm, Wnt proteins can initiate either canonical or noncanonical Wnt signaling pathway depending on the Wnt–Fzd combinations and intracellular downstream effectors [95, 96]. However, the mechanism determining the choice is poorly understood.

In addition to Fzd receptors, Wnt proteins can interact with other receptors containing Wnt-binding domains (Fig. 19.2). Ryk/Derailed type receptors contain a Wnt inhibitory factor (WIF) ligand-binding domain and a tyrosine kinase motif [97]. Ryk/Derailed receptors are involved in Wnt3a-mediated canonical Wnt signaling in mammalian cells [98]. Another alternative Wnt receptor with the cysteine-rich Wnt-binding domain is the single-pass tyrosine kinase Ror2, which promotes the Wnt5a-mediated inhibition of β-catenin-TCF signaling [95]. Intracellular signaling pathways initiated by these alternative Wnt receptors and their relation to known Wnt signaling cascades remain to be determined.

Initial cardiac specification depends on specific expression pattern of components of canonical Wnt signaling cascade. Canonical Wnt1 and Wnt3a expressed in the neural plate and dorsal neural tube downregulate cardiac specification in the posterior-medial mesoderm, and activation of canonical Wnt signaling in the anterior mesoderm abrogates the expression of early cardiac *Nkx2-5* and *Gata4* genes, whereas the downregulation of canonical Wnt signaling induces heart formation from posterior mesoderm [99].

In addition to specification, canonical Wnt signaling governs cardiac progenitor cell proliferation and differentiation. β-Catenin mediates this pathway via binding directly and thereby activating transcription from the *Isl1* and *Fgf10* promoters [100, 101]. Moreover, the *Fgf3*, *Fgf16*, and *Fgf20* genes, encoding components of the fibroblast growth factor (FGF) pathway, can also be regulated by β-catenin [100]. Interestingly, it has recently been reported that Wnt-FGF signaling is essential for zebrafish fin regeneration suggesting a general tissue regeneration potential of this pathway [102].

Recent studies of mouse ESCs suggest a biphasic role for canonical Wnt signaling in heart development [103, 104]. Wnt2a and most likely Wnt2b appears to activate canonical Wnt signaling positively regulating early cardiac

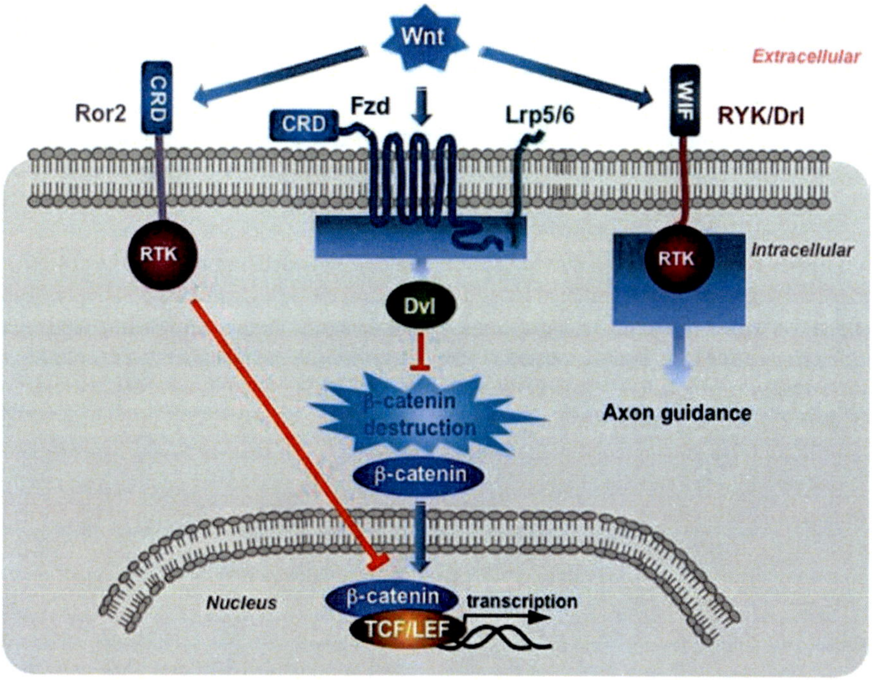


**Fig. 19.2** Wnt proteins can activate various receptors. In addition to Fzd receptors, Wnt ligands interact with the Ror2 tyrosine kinases through the cysteine-rich domain (CRD) antagonizing β-catenin-dependent canonical Wnt signaling. Wnt can also bind to the RYK/Derailed (Drl) receptors through a Wnt inhibitory factor (WIF) binding domain contributing to axon guidance in *Drosophila*. Both Ror2 and RYK/Drl contain an intracellular tyrosine kinase motif (RTK)

gene expression but inhibiting cardiac differentiation at later stages.

Noncanonical Wnt signaling also plays an important role in heart development. Wnt11 activates noncanonical Wnt signaling inducing cardiogenesis without activation of expression of mesodermal markers [105, 106]. The phenotypes of recently generated mice deficient in noncanonical Wnt5a and Wnt11 suggest an essential role for this signaling pathway in later stages of the outflow tract development [107, 108]. Treatment of differentiating mouse ESCs with Wnt11 also induces cardiac progenitor cell development [104]. Interestingly, *Wnt11* expression in the outflow tract is controlled by canonical Wnt signaling suggesting intertalk between canonical and noncanonical Wnt signaling pathways [101, 109]. It appears that both $Ca^{2+}$-PKC and RhoA-JNK downstream pathways mediate the Wnt11-activated cardiogenesis [106].

Bmp ligands are the members of the transforming growth factor-β (TGF-β) family of signaling proteins. Bmp-mediated pathway has been shown to be essential for cardiac mesoderm differentiation [110–112]. Studies of mouse models suggest that Bmp4 and Bmp7 are involved in Wnt-β-catenin and FGF signaling and are required for cardiac progenitor cell development [113–116]. While inducing differentiation, Bmp can downregulate progenitor proliferation playing a role of molecular switch from cardiac progenitor cell expansion to differentiation mode [117, 118].

## Nuclear Factor-κB Signaling

The NF-κB signaling controls a variety of cellular responses and has been linked to multiple physiological and pathological processes in the myocardium. Although the NF-κB pathway is conserved in mammals, it has initially been studied in the immune cells. The NF-κB proteins represent a family of structurally related transcription factors consisting of five members including NF-κB1 (p50), NF-κB2 (p52), c-Rel, RelA (p65), and RelB [119, 120]. c-Rel, RelA (p65), and RelB contain the Rel homology domain (RHD) at their N-terminus required for dimerization and DNA binding and a C-terminal transcription activation domain. NF-κB1 (p50) and NF-κB2 (p52), synthesized as precursor proteins, p105 and p100, respectively, contain an N-terminal RHD similar to Rel proteins and a C-terminal ankyrin repeat domain.

Precursor p105 and p100 are proteolytically processed at the C-terminus to produce mature p50 and p52, which contain the N-terminal RHD while lack the C-terminal transcription activation domain [120].

Different NF-κB proteins form dimers, most common p50/p65 heterodimers found in all cells, to increase affinity of NF-κB to DNA. Major cellular fraction of NF-κB dimers is localized in the cytoplasm in an inactive state due to binding to the inhibitors of NF-κB (IκBs). In response to various stimuli, the canonical NF-κB signaling is initiated by phosphorylation of IκBs at two conserved serine residues, Ser-32 and Ser-36, and/or a tyrosine residue, Tyr-42, located at the N-terminal signal response domain (SRD). This initial step is catalyzed by the IκB kinase (IKK) complex composed of a regulatory subunit, IKKγ (also known as NF-κB essential modifier, NEMO) and two kinase catalytic subunits, IKKα and IKKβ. Phosphorylated IκB is targeted for polyubiquitination and subsequent rapid degradation by the 26S proteosome. The degradation of the inhibitory IκB releases NF-κB dimers and translocate them into the nucleus where they activate transcription of multiple specific genes [121] (Fig. 19.3).

Canonical NF-κB signaling pathway depends on IKKβ and IKKγ, while the alternative pathway involves IKKα and NF-κB inducing kinase (NIK) resulting in phosphorylation and processing of p100 precursor and formation of p52/RelB heterodimers [121] (Fig. 19.3). According to the current paradigm, activation of the canonical or alternative NF-κB signaling pathway depends on the type of stimuli.

Activated NF-κB signaling is transient and can be downregulated by a feedback loop. NF-κB induces IκBα expression, newly synthesized IκBα translocates to the nucleus, and binds to NF-κB dimers displacing them from DNA. The IκB/NF-κB complex translocates from the nucleus to the cytoplasm as an inactive complex since the association with IκB masks the nuclear localization domain in NF-κB [120, 121].

Emerging evidence suggests that the NF-κB signaling is an essential mediator of cardiomyocyte responses to various stimuli, including cytokines, lipopolysaccharide (LPS), oxidative stress, cardiac remodeling, and ischemia reperfusion [122–125]. Recent study of stem cells has demonstrated that the NF-κB signaling is involved in regulation of proliferation and differentiation [126]. In human MSCs, tumor necrosis factor-α (TNF-α), LPS, or hypoxia induce the NF-κB-mediated increased production of growth factors, vascular endothelial growth factor (VEGF), FGF2, and hepatocyte growth factor (HGF) [127]. Accordingly, inhibition of the NF-κB pathway results in significant downregulation of the production of these growth factors. Interestingly, although TNF-α, LPS, or hypoxia induce also extracellular-signal-regulated kinase (Erk) and JNK activation, their inhibition has no significant effect on MSC production of VEGF, FGF2, and HGF. Therefore, the NF-κB-mediated upregulation of growth factor production may play an important role in stem cell-stimulated angiogenesis and cardiac protection. Moreover, pharmacological stimulation of the NF-κB signaling may be a promising strategy to enhance paracrine effects before stem cell transplantation.

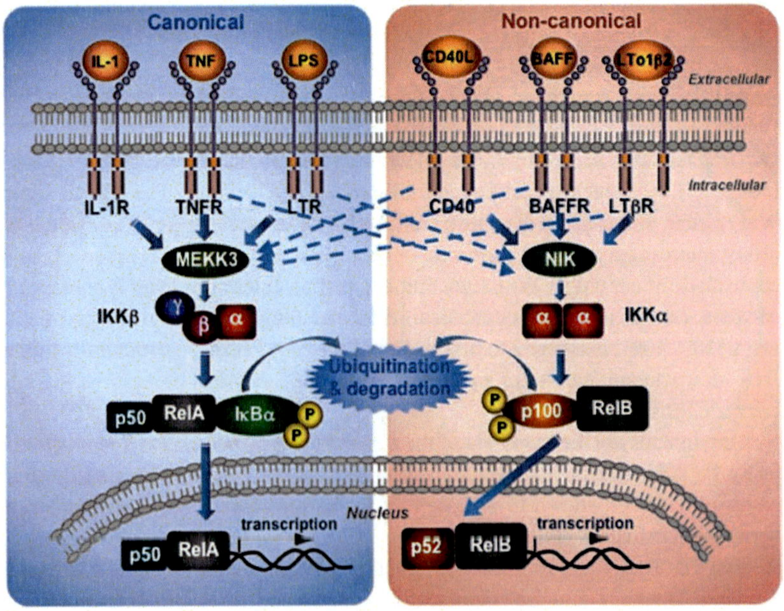


**Fig. 19.3** Canonical and noncanonical NF-κB signaling pathways. Major receptors, intracellular mediators, and effectors are schematically shown. See text for further details

## *SDF-1/CXCR4 Signaling*

Various strategies of stem cell-based therapy for cardiac repair relay on efficient homing and engrafting of transplanted cells in the myocardium. Therefore, understanding of the mechanisms that regulate progenitor cell migration and mobilization is critically important for the development of efficient stem cell therapy to treat cardiovascular disorders.

The SDF-1 signaling through its corresponding receptor CXCR4 is an important regulator of HSC trafficking and homing [128]. In preclinical study, it has been demonstrated that intramyocardial delivery of recombinant SDF-1 results in recruitment of progenitor cells to the heart and attenuation of ischemic cardiomyopathy [129]. Accordingly, inhibition of the SDF-1 degradation in the ischemic heart improves cardiac function and survival after acute myocardial infarction suggesting a cardioprotective effect of SDF-1 signaling [130, 131].

SDF-1 also known as CXCL12 was initially identified as a pre-B-cell stimulating factor [132]. It is a member of a family of structurally related low-molecular-mass proteins that are characterized by chemostatic activity and play essential roles in cell trafficking, homing, and differentiation. Mammalian SDF-1s are highly homologous; the human and mouse SDF-1α differ only in a single amino acid substitution Valin 18-Isoleucin.

Similar to other chemokines, the SDF-1 is composed of three antiparallel β-sheets and an overlying α-helix. The N-terminal region of SDF-1 is required for receptor binding and activation and cleavage of the N-terminus by multiple peptidases abrogates SDF-1 activity [133–135]. There is an additional receptor binding and activation site within the RFFESH sequence.

SDF-1 is expressed in various organs including the heart, kidney, liver, and spleen. Unlike the inflammatory-induced expression of other chemokines, expression of SDF-1 is constitutive suggesting essential roles in organ homeostasis and development. In agreement with such fundamental function, loss of SDF-1 or its receptor CXCR4 results in late embryonic or early perinatal lethality caused by defects in hematopoesis, cardio-, and angiogenesis although the reason for the mortality is not entirely clear [136–139]. SDF-1 and CXCR4 are expressed in various stem cells including HSCs, ESCs, primordial stem cells, and cardiopoietic progenitor cells [140–145] (Table 19.3). Based on these findings, it has been suggested that CXCR4 may be used as a marker for various type of stem cells.

SDF-1 binds and activates two related G protein-coupled receptors (GPCRs), CXCR4 and CXCR7. Binding of SDF1 to the receptors results in the formation of CXCR4/CXCR4 homodimers or CXCR4/CXCR7 heterodimers [146, 147]. In the heart, SDF-1 does not interact with CXCR7, while in endothelial cells SDF-1 induces the formation of CXCR4/CXCR7 heterodimers [148]. Engagement of CXCR4 induces several $G_i$ protein-mediated signal transduction pathways including activation of various kinases (PKC, PI3K-Akt, paxilin, Erk1/2), phospholipase C (PLC), and $Ca^{2+}$ efflux [149] (Fig. 19.4).

**Table 19.3** CXCR4 expression in stem and progenitor cells

| Cell type | Sources | Cell markers |
|---|---|---|
| Embryonic stem cells | Blastocyst | SSEA1, Oct-4, Nanog, and rex-1 |
| Primordial germ cells | Embryo | Alkaline phosphatase and Oct-4 |
| Very small embryonic-like stem cells | Bone marrow | Mouse: $CXCR4^+/Sca\text{-}1^+/lin^-/CD45^-$<br>Human: $CXCR4^+/CD34^+/AC133^+/CD45^-$ |
| Haematopoietic stem cells | Bone marrow | Mouse: c-kit$^+$, sca-1$^+$, and Lin$^-$<br>Human: CD34$^+$ |
| Haematopoietic stem cells | Human cord blood | CD34 |
| Endothelial progenitor cells | Peripheral blood | Mouse: CD31$^+$, CD34$^+$, and Flk-1$^+$<br>Human: CD34$^+$, CD133$^+$, and KDR$^+$ |
| Cardiac stem cells | Heart | Lin$^-$ and c-kit$^+$ |
| Muscle satellite cells | Skeletal muscle | CD29$^+$, CD34$^+$, and Pax7$^+$ |
| Neural progenitor cells | Brain | Nestin |

Upon SDF-1 binding to CXCR4, the heterotrimeric $G_i$ protein dissociates into α and β/γ subunits. Pertussis toxin abrogates SDF-1/CXCR4-mediated cell migration confirming essential role of SDF-1 signaling through $G_i$ in this process. β/γ subunit activates two major downstream mediators, PI3K and PLC. PLC hydrolyzes phosphatidylinositol 4,5-bisphosphate ($PIP_2$) into inositol 1,4,5-trisphosphate ($IP_3$) and diacylglycerol (DAG) inducing release of $Ca^{2+}$ from intracellular stores. This in turn results in PKC activation and atypical PKCζ may play an essential role in SDF-1-mediated stem cell trafficking [150]. Activated PI3K and PKC contribute to phosphorylation of tyrosine in multiple focal adhesion proteins including focal adhesion kinase (FAK), proline-rich kinase-2 (PYK-2), and paxilin and Crk-associated substrate (p130/Cas) [151]. These proteins are involved in the formation of focal adhesions and rearrangement of actin cytoskeleton in progenitor cells. In addition to phosphorylation of focal adhesion proteins, induction of PI3K leads to phosphorylation and activation of Akt pathway playing an important role in cell proliferation and survival. Moreover, SDF-1 signaling through CXCR4 can also induce JAK/signal transducer and activator of transcription (STAT) and NF-κB pathways. Internalization of CXCR4 represents a mechanism

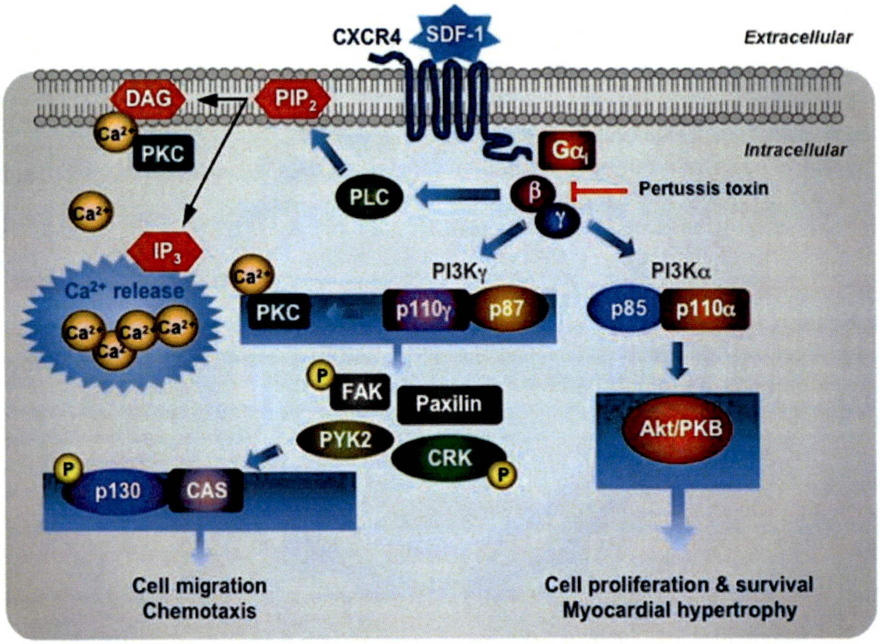


**Fig. 19.4** SDF-1 signaling through CXCR4 receptor. SDF-1 binding to the GPCR CXCR4 induces dissociation of the heterotrimeric G protein into $G\alpha_i$ and heterodimeric β/γ subunits. β/γ Subunits activate PI3Kα /Akt and PI3Kγ pathways. They also stimulate phospholipase C (PLC) catalyzing hydrolysis of $PIP_2$ into $IP_3$ and DAG, and thereby inducing the release of $Ca^{2+}$ from intracellular stores and PKC activation. Pertussis toxin inhibits β/γ subunit-mediated effects. PI3Kγ and PKC can phosphorylate and activate focal adhesion proteins FAK, PYK-2, CRK, paxilin, and p130/Cas

of desensitization of the receptor to SDF-1 signals. It is controlled by CXCR4 phosphorylation catalyzed by PKC and GPCR kinases, which induces binding of β-arrestin to the C-terminus of CXCR4 and subsequent endocytosis of this complex [152].

In early embryonic development, SDF-1 signaling governs the migration of $CXCR4^+$ primordial stem cells to the sites of neovascularization. Importantly, high levels of CXCR4 expression are also colocalized in these areas [153]. Postnatally, the SDF-1/CXCR4 signaling plays an important role during myocardial ischemia. After acute myocardial infarction, SDF-1 expression is upregulated by hypoxia-inducible factor 1 (HIF-1) [154]. Leakage of SDF-1 from damaged cardiomyocytes recruits bone marrow progenitor cells to ischemic myocardium resulting in enhanced neovascularization and decreased apoptosis of cardiomyocytes [129, 130].

Thus, emerging evidence supports a fundamental role for the SDF-1/CXCR4 signaling pathway in progenitor cell trafficking and mobilization. Given expression of CXCR4 in various stem and progenitor cells, including endothelial, hematopoietic, and endogenous CSCs, upregulation of its ligand SDF-1 represents a promising strategy to target stem cells to the sites of myocardial injury.

## Mitogen-Activated Protein Kinases

The MAPKs are a large family of serine/threonine kinases which can be subdivided into six groups based on their structural similarity: Erks 1 and 2; Erk3; Erk5; Erk7; p38MAPKs (p38α, β, γ, and δ); and JNK/stress-activated protein kinases (SAPKs) 1, 2, and 3. They are ubiquitously expressed and regulate diverse cellular functions ranging from proliferation and differentiation to migration and apoptosis [155, 156]. The most characterized MAPKs are Erk1/2, p38α, β, γ, and δ, and JNK1, 2, and 3.

Activation of MAPKs requires dual phosphorylation on threonine (Thr) and tyrosine (Tyr) residues within a conserved motif (Thr-*X*-Tyr) specific for each kinase group. Specific motifs for three major groups are Thr-Glu-Tyr for Erk1/2, Thr-Gly-Tyr for the p38MAPKs, and Thr-Pro-Tyr for JNK/SAPKs [157]. Phosphorylation of these sites on MAPKs induces a conformational change enhancing substrate accessibility and catalytic activity. Phosphorylation of each group of MAPKs is mediated by a distinct kinase cascade. MAP kinase kinase kinases (MAP3Ks, MKKKs, or MEKKs) phosphorylate and activate dual-specificity MAP kinase kinases (MAP2Ks, MKKs, or MEKs) that in turn phosphorylate MAPKs (Fig. 19.5). These cascades of

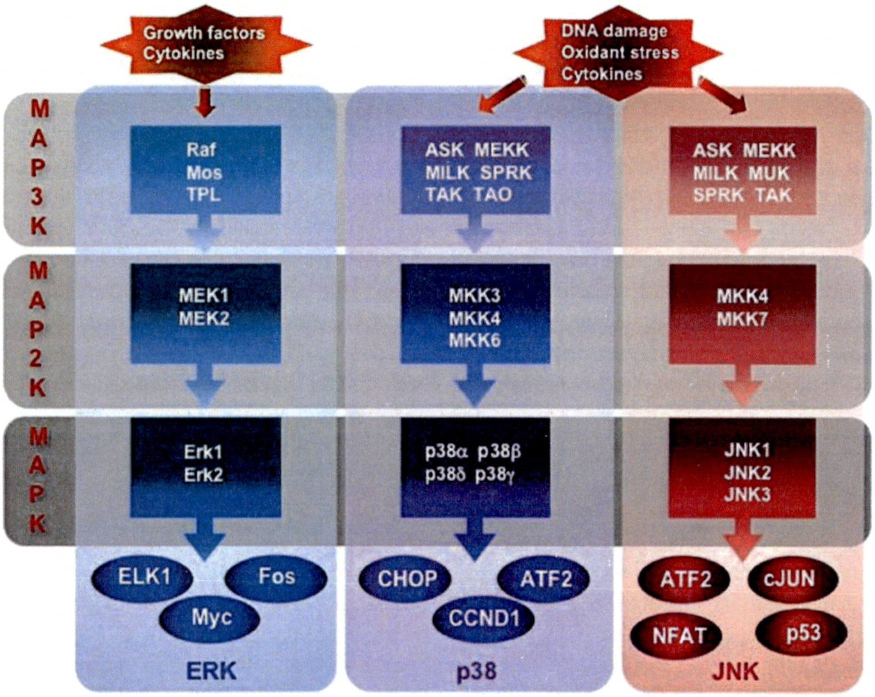


**Fig. 19.5** Scheme of the major MAPK signaling cascades. Erk1/2, p38, and JNK MAPKs represent the major and best characterized groups of MAPKs. They are activated in response to various extracellular stimuli. Phosphorylation of each group of MAPKs is mediated by a distinct kinase cascade. MAP3Ks or MEKKs phosphorylate and activate dual specificity of MAP2Ks or MEKs that in turn phosphorylate MAPKs. Activated MAPKs can phosphorylate a variety of target substrates modulating their activities

phosphorylation events eventually result in activation of a variety of downstream pathways including various MAPK-activated protein serine/threonine kinases (RSKs, MSKs, and MNKs) and transcriptional regulators (c-Myc, c-Jun, Elk1, Fos, and many others).

A variety of extracellular stimuli activates the MAPK signaling cascades. The Erk1/2 pathway is predominantly induced by mitogens including the serum and growth factors and is involved in regulation of cell-cycle progression and cell proliferation and survival. The p38 and JNK signaling pathways are activated in response to stress stimuli such as radiation and other DNA-damaging agents and cytokines [158]. However, despite specificity of MAPK response, a single signal often induces multiple MAPK groups.

Interaction of MAPKs with their MAP2Ks and downstream targets through docking sites (D-motifs) determines MAPK substrate specificity [159, 160]. Moreover, various scaffold proteins such as MEK partner 1 (MP1), JNK-interacting protein (JIP1), 2, and 3, paxilin, oncostatin-M (OSM), and β-arrestins are involved in the assembly of specific MAPK complexes facilitating their activation and modulating signaling outcomes [161].

Finally, specific phosphoprotein phosphatases dephosphorylate MAPKs determining the duration of kinase activation [156, 162]. They can be subdivided into three groups: phosphatases capable to dephosphorylate phosphoserine/threonine, phosphotyrosine, or both types of residues (DSP). DSPs that dephosphorylate MAPKs, also called MKPs, can inactivate single or several MAPK subgroups [162]. MKP-3 is capable to dephosphorylate Erk1/2 while MKP-1 dephosphorylates Erk, JNK, and p38.

Erk1 ($p44^{MAPK}$) and Erk2 ($p42^{MAPK}$) encoded by the *MAPK3* and *MAPK1* genes, respectively, were originally described as protein kinases that phosphorylate myelin basic protein and microtubule-associated protein 2 (MAP-2) in response of growth factor stimulation [163]. They are expressed in most eukaryotic cells, share more than 80% amino acid sequence similarity, and are activated by growth factors, cytokines, serum, and ligands for GPCRs [156, 164]. The upstream MAP3Ks associated with the Erk1/2 signaling pathway are the Raf isoforms (A-, B-, and C-Raf/Raf-1), Mos kinase, and tumor progression locus-2 kinase (Tpl-2) (Fig. 19.5). MEK1 and 2 are MAP2Ks specific for Erk1/2. After signal activation, Erk1/2 phosphorylate various cytoplasmic and nuclear proteins including p90 ribosomal protein S6 kinase (p90 RSK), mitogen- and stress-activated protein kinase (MSK), MAP kinase-interacting kinase (MNK), and transcription factors Elk-1, c-Fos, and c-Myc [156]. Activated Erk1/2 can translocate into the nucleus to facilitate phosphorylation of transcription factors and other nuclear proteins affecting gene expression. Protein phosphatase 2A (PP2A) dephosphorylates several components of the Erk1/2 cascade regulating signal activation.

An essential role of the Erk1/2 signaling cascade in cell proliferation and survival has been well documented [165, 166]. Erk1 knockout mice are viable displaying no significant developmental abnormalities, whereas loss of Erk2 leads to embryonic lethality caused by defects in placental development [167, 168].

p38 was initially identified as a tyrosine-phosphorylated protein in macrophages stimulated with LPS [169]. The four p38 isoforms, p38α, β, γ, and δ, are encoded by four distinct *MAPK* genes, *MAPK14*, *11*, *12*, and *13*, respectively. p38 MAPKs share only approximately 60% sequence identity, suggesting rather diverse cellular functions. p38α and β are ubiquitously expressed while expression of p38γ and δ is more tissue-restricted [170].

Several upstream MAP3Ks, including apoptosis signal-regulating kinases (ASKs), MEKKs, mixed lineage kinases (MLKs), SH3 domain-containing proline-rich kinase (SPRK), TGF-β-activated kinases (TAKs), and "thousand and one amino acid" kinases (TAOs), and at least three MAP2Ks, such as MKK3, 4, and 6, participate in the p38 MAPK cascade (Fig. 19.5). The p38 pathway is activated in response to cellular stress including UV radiation, hypoxia, osmotic shock, and pro-inflammatory cytokines [156]. Activated p38 MAPKs phosphorylate a wide range of proteins including protein kinases p90 RSK, MAPKAP2 and 3, various heat shock proteins (HSPs), and transcription factors ATF-1 and 2, C/EBPβ, STAT1, and 3.

Few scaffold proteins have been identified for p38 signaling pathway. Upon osmotic stress, a specific scaffold for p38, OSM, binds to MEKK3, MEK3, and p38 and recruits p38 with associated kinases to membrane raffles in proximity to the actin cytoskeleton facilitating thereby cellular adaptation to osmotic stress [171]. Moreover, p38 shares the JIP and JNK/stress-activated protein kinase associated protein-1 (JSAP) scaffolds with JNK pathway. JIP2 binds to MLK3 (MAP3K), MEK3 (MAP2K), and p38α or γ enhancing p38 activation and signaling [172, 173]. Finally, several DSPs such as MKP1, 5, 7, and 8 as well as PP2A and B can dephosphorylate and inactivate p38 representing a feedback mechanism for this signaling pathway [156, 174, 175].

Only p38α knockout mice are embryonically lethal as a result of severe defects in placental development and in erythropoiesis while deficiency in other p38 isoforms shows no obvious phenotypes suggesting redundancy between various isoforms [176–178]. Moreover, p38 appears to be involved in phosphorylation of the ETS-related transcription factor TEL essential for postnatal hematopoiesis for all cell lineages [179].

JNK/SAPKs were originally described as a UV-stimulated protein kinase, which could bind and phosphorylate the N-terminal activation domain of the transcription factor c-Jun [180]. Members of JNK/SAPK family JNK1 (SAPKβ), 2 (SAPKα), and 3 (SAPKγ) are encoded by three distinct genes *MAPK8*, *9*, and *10*, respectively. Furthermore, there are four splice variants of each JNK1 and 2 and two splice variants of JNK3 increasing total number of JNK isoforms at least to 10 [181]. Amino acid sequences of JNKs are greater than 85% identical. JNK1 and 2 are expressed ubiquitously while JNK3 is expressed predominantly in the brain, testis, and heart [182].

Upstream MAP3Ks in the JNK pathway include MEKKs (MEKK1, 2, and 3), MLKs (MLK1, 2, and 3; dual-leucine zipper kinase, DLK), Tpl-2, ASKs, TAOs, and TAK1 [183]. Major MAP2Ks that mediate signaling in this pathway are MEK4 (MKK4) and 7 (MKK7). Under certain conditions, they can be also involved in activation of the p38 signaling adding complexity to the MAPK signaling. Multiple stress stimuli including heat shock, oxidant stress, ionizing radiation and other DNA-damaging agents, protein synthesis inhibitors as well as pro-inflammatory cytokines, and growth factors can activate the JNK signaling pathway. In response to these stimuli, JNKs phosphorylate number of transcription factors ATF-2, c-Jun, c-Myc, Elk-1, nuclear factor of activated T cells (NFAT) and p53, MAPs and members of the pro-apoptotic Bcl2 family, Bid, Bax, and Bim [183, 184].

Scaffolding proteins JIP1 and 2 significantly stimulate the MLK-mediated JNK activation. JIP3 functions as a scaffold between MLK3, MEK7, and JNK, while JIP4 (JLP) interacts with both JNK and p38, their upstream kinases MEK4 and MEKK3, and downstream targets Max and c-Myc [156, 185]. Multidomain scaffolding protein axin also stimulates JNK signaling tethering it and MEKK1 and 4 [186].

Several JNK phosphoserine/threonine phosphatases belong to the MKP family, DSP2, MKP1, 2, 4, 5, and 6, some of them also dephosphorylate p38α and β [187]. However, much less is presently known about protein phosphatases, which dephosphorylate upstream MAP3Ks and MAP2Ks involved in the JNK pathway.

Among the JNKs, mice deficient in JNK1 or JNK2 are viable displaying defects in T cell activation, whereas double knockout of JNK1/2 is embryonic lethal because of defects in specific apoptosis during early brain development [188, 189]. Loss of JNK3 expressed predominantly in the brain results in a higher resistance of the hippocampus neurons to apoptosis [190]. Moreover, MKK4 or 7 knockout mice are early embryonic lethal supporting essential roles for the JNK signaling in control of embryonic development [184].

Over the past decade, extensive work has focused on the characterization of the role of MAPK signaling cascades in CVD. Studies on transgenic mice suggest that the Raf-MKK1/2-Erk1/2 pathway stimulates cardiomyocyte growth and survival, whereas the MKKK-MKK3/6-p38α and MKKK-MKK4/7-JNK1/2 cascades inhibit the growth of cardiomyocytes [191, 192]. It has been suggested that negative effect of p38α and JNK1/2 signaling on cardiac hypertrophy is mediated through inactivation of NFAT transcription factors [193]. The opposite roles of Erk1/2 vs. p38 and JNK1/2 signaling in the development of cardiac hypertrophy have been supported by experiments with specific pharmacological inhibitors. Based on this model, activation of Erk1/2 and/or inhibition of p38 and JNK pathways may be promising therapeutic strategy for patients with dilated cardiomyopathy.

The role of MAPK signaling in pathological cardiac remodeling after myocardial infarction (MI) has also been extensively studied. In rodent experimental models, all three major MAPK cascades are activated within minutes after MI [194, 195]. Interestingly, p38α activation has been maintained for several weeks after the initial insult [196–198]. One potential mechanism for the p38-mediated pathological cardiac remodeling appears to be inhibition of the antiapoptotic Bcl-$x_L$ and Bcl-2 [195, 199]. Ability of p38α to inhibit cardiomyocyte proliferation could also contribute to this effect [200]. Cardiac MKKK involved in activation of both p38α and JNK1/2 signaling cascades is ASK1. In ASK knockout mice, JNK1/2 activation has been significantly reduced in the infarction border zone while p38α activation has not been changed suggesting a role for ASK1-JNK1/2 pathway in pathological cardiac remodeling [201]. The role of the Erk1/2 signaling in cardiac remodeling after MI is less well understood.

Finally, MAPK cascades play essential roles in atherosclerosis and vascular restenosis. JNK2 and probably p38α activation but not Erk1/2 appears to be required for macrophage-mediated atherosclerotic lesion formation. Moreover, JNK2 and p38α can be involved in regulation of the uptake of oxidized low-density lipoprotein (LDL) by macrophages [202, 203]. Activation of all three major MAPKs, Erk1/2, JNK1/2, and p38α, promotes neointima formation and vascular smooth muscle cell proliferation after vascular injury [204–206].

Recent studies have addressed an important role of MAPK signaling cascades in proliferation, maintenance, and commitment of progenitor cells. MEK1-Erk1/2 signaling appears to contribute to the regulation of early myeloid commitment of HSCs [207]. In addition to inhibition of various antiproliferative genes, the Raf-MEK-Erk pathway contributes to the regulation of expression of apoptosis regulatory factors, such as Bad, Bim, Mcl-1, and caspase 9 [166]. Moreover, cytokine-induced MEK-Erk activation leads to change in expression of a variety of cell-cycle modulating genes and controls thereby the balance between proliferation and apoptosis of hematopoietic progenitor cells [208]. Interestingly, ES cells do not require Erk signaling for proliferation and self-renewal but they become dependent on this pathway upon lineage commitment [209, 210].

The p38 signaling is activated in HSCs from mice deficient in a large stress-induced "ataxia-telangiectasia mutated" protein kinase (ATM) involved in cell-cycle checkpoint and in control of oxidant levels [211]. Activation of p38 in *Atm*$^{-/-}$ HSCs is associated with defects in self-renewal and maintenance of these cells, whereas inhibition of p38 can rescue HSC self-renewal capacity [212]. Moreover, p38 activation is induced by erythropoietin and is required for the initiation of erythroid differentiation [213, 214]. The p38 cascade can be also activated by myelosuppressive cytokines: interferons (IFNs)-α and -β, TGF-β, and TNF-α and can contribute to the regulation of myeloid differentiation [215–217]. Although emerging evidence suggests that JNK signaling is also involved in the regulation of erythroid and myeloid development, mechanistically its role in these processes is less well defined.

In human MSCs, all three major MAPK signaling cascades are activated in response to TNF-α, LPS, or hypoxia exposure. However, only activation of p38 MAPK but not Erk or JNK has been associated with increased production and release of paracrine growth factors such as VEGF, FGF2, HGF, and insulin-like growth factor 1 (IGF-1) [127, 218]. Since stem cells appear to mediate cardiac repair via paracrine effects, delineating the molecular mechanism of these signaling cascades will provide novel therapeutic strategy for the treatment of CVD.

## PI3K-Akt-mTOR Signaling

As it has previously been discussed in Chap. 7, the family of PI3Ks phosphorylates the 3′-OH group of inositol lipids to generate a second messenger phosphatidylinositol (3,4,5)-trisphosphate ($PIP_3$). PI3K family comprises three functional classes, I, II, and III, based on their protein structure, substrate specificity, associated regulatory subunits, and activation mechanisms [219–221]. Class I PI3Ks are the best studied enzymes and three members of class I isoforms, PI3Kα, -β, and -γ, are expressed in the heart, among them PI3Kα and PI3Kγ are the best characterized.

Heterodimeric PI3Kα (p110α/p85) is activated by receptor tyrosine kinases (RTKs), including epidermal growth factor (EGF) receptor, IGF-1 receptor, and platelet-derived growth factor (PDGF) receptor. PI3Kγ heterodimer (p110γ/p87) is activated by GPCRs via interaction with the $G_{\beta\gamma}$ subunits (Fig. 19.6).

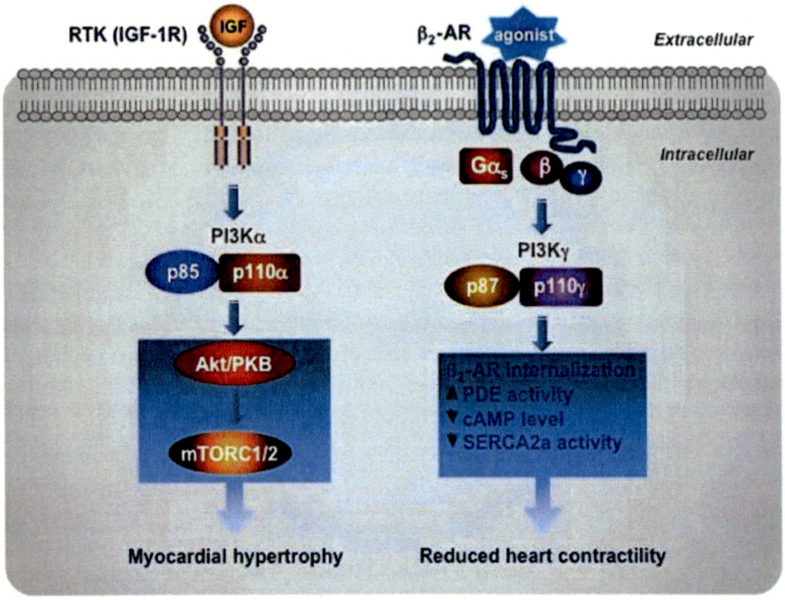


**Fig. 19.6** Schematic representation of PI3Kα and PI3Kγ signaling. Engagement of receptor tyrosine kinases (RTKs, e.g., IGF-1R) activates PI3Kα via interaction of RTKs with its adaptor subunit (p85). Activated PI3Kα stimulates Akt/PKB and mTORC1/2 resulting in myocardial hypertrophy under physiologic conditions, without affecting contractility. $\beta_2$-AR signaling activates PI3Kγ via $G_{\beta\gamma}$ subunits leading to the upregulation of PDE activity and $\beta_2$-AR desensitization and internalization. This results in the downregulation of cAMP levels and SERCA2a activity and eventually in the reduced heart contractility

Serine/threonine protein kinase Akt, also known as protein kinase B (PKB), is a central mediator of PI3K signaling. There are three highly conserved Akt isoforms: Akt1/α, Akt2/β, and Akt3/γ [222]. In response to the RTK engagement, PI3K is activated to produce a second messenger $PIP_3$, which in turn recruits PH domain containing proteins, such as Akt and phosphoinositide-dependent protein kinase-1 (PDK1), to the plasma membrane. Akt recruitment leads to a conformational change enabling its phosphorylation on Thr308 by PDK1 and on Ser473 by the mammalian target of rapamycin (mTOR)-Rictor complex 2 (mTORC2) [223, 224]. Activated Akt phosphorylates a wide variety of substrates, including GSK3β, endothelial nitric oxide synthase, $p21^{Cip1/WAF1}$, $p27^{Kip1}$, Raf, FOXO family of Forkhead transcription factors, Bad, caspase-9, mdm2, and many others [225, 226]. This explains a regulatory role of PI3K-Akt signaling pathway in multiple cardiac processes ranging from cardiomyocyte metabolism, growth and proliferation to apoptosis, contractility and coronary angiogenesis. Given an essential role of the Akt-mediated signaling for progenitor cell survival, PI3K-Akt pathway has received a particular attention as a strategy to improve stem cell survival upon transplantation into the heart.

mTOR belongs to the PI3K-related kinase family; however, it functions exclusively as a serine/threonine protein kinase [227]. mTOR forms two multiprotein complexes, mTOR complex 1 (mTORC1) and mTORC2. mTORC1 contains mTOR, Raptor, mLST8, PRAS40, FKBP38, and Deptor proteins and is inhibited by the rapamycin. mTORC2 is composed of mTOR, Rictor, mLST8, Sin1, Protor, and Deptor proteins and is rapamycin insensitive.

In response to growth factors, Akt activates mTORC1 through inactivation of the tuberous sclerosis complex, TSC1/TSC2 complex, an upstream inhibitor of mTORC1. In addition, Akt also phosphorylates and inactivates PRAS40, which acts as an inhibitor of mTORC1 [228]. Activated mTORC1 phosphorylates eukaryotic translation initiation factor 4E-binding protein 1 (4E-BP1) and p70 ribosomal protein S6 kinase (p70S6K). Thus, mTORC1 controls various steps of protein biosynthesis regulating cell proliferation and survival as well as angiogenesis [229].

Activation of mTORC2 depends on PI3K and TSC1/TSC2 complex; however, its mechanism is poorly understood [230]. mTORC2 phosphorylates PKCα and PKCε and appears to contribute to the regulation of cytoskeleton organization through control of actin polymerization [230, 231].

Several lipid and protein phosphatases function as negative regulators of PI3K-Akt-mTOR signaling pathway. A dual-specificity, lipid-protein phosphatase PTEN is the major phosphatase responsible for the hydrolysis of $PIP_3$ and $PIP_2$ [232, 233]. Two Src homology domain-containing inositol phosphatases, SHIP-1 and SHIP-2, can also dephosphorylate $PIP_3$ to $PIP_2$ [234]. Finally, PP2A and PH domain leucine-rich repeat protein phosphatases 1 and 2 (PHLPP1 and -2) deactivate Akt by dephosphorylating its Thr308 and both Thr308 and Ser473 residues, respectively [235, 236].

As we have previously discussed, PI3Kα- and PI3Kγ-mediated signaling pathways play different roles in the myocardium (see Chap. 7). PI3Kα signaling regulates myocardial hypertrophy, apoptosis, maintenance of cardiac structure and function under mainly physiological (e.g., exercise training) conditions, and reduces cardiac fibrosis.

In contrast, activation of PI3Kγ pathway leads to reduced cardiomyocyte contractility through downregulation of beta-adrenoceptor (β-AR) signaling, cAMP production, and protein kinase A (PKA) and sarco/endoplasmic reticulum $Ca^{2+}$-ATPase (SERCA2a) activities and therefore has a detrimental effect on cardiac function.

Emerging evidence suggests that a highly complex PI3K-Akt-mTor signaling network play specific roles in control of self-renewal, maintenance, proliferation, survival, and differentiation of stem cells. Most adult HSCs are maintained in a quiescent state, which appears to be one of the main mechanisms to maintain stem cell function [237]. Emerging evidence suggests that PI3K-Akt-mTor signaling is involved in the regulation of HSC maintenance. Loss of PTEN in mouse HSCs leads to transient increase in their number followed by a depletion of the HSC population [238]. Importantly, rapamycin rescues this phenotype suggesting that mTORC1 activation contributes to the regulation of HSC maintenance [239]. Members of the FOXO family of transcription factors, FOXO1, 3, and 4, which are downstream targets of Akt, are also required for the long-term HSC maintenance. HSCs from FOXO-deficient animals are characterized by increased exit from quiescence and elevated apoptosis leading to a reduction of the HSC pool [240]. Moreover, it has recently been reported that SHIP is not an intrinsic requirement for HSC function but it is rather required for the BM milieu determining the repopulation ability of HSCs [241].

PI3K-Akt pathway is also involved in control lineage commitment, a highly regulated process, by which a multipotent stem cell becomes restricted to a specific progenitor cell lineage. Indeed, inhibition of PI3K-Akt signaling or expression of a dominant negative Akt leads to upregulation of C/EBPα phosphorylation, which is a central transcription factor controlling lineage choices during myelopoiesis [242, 243]. Furthermore, erythropoietin (EPO)- and stem cell factor (SCF)-dependent activation of PI3K-Akt-mTOR signaling appears to play a regulatory role in the balance between proliferation and maturation of erythroid progenitors [244]. It has recently been reported that PI3K-Akt-mTOR signaling cascades contribute to the regulation of megacaryocyte differentiation through inhibition of apoptosis and mTORC1/2 induction [245, 246].

MSCs overexpressing Akt injected into the perinfarct border zone in rat hearts prevent cardiac remodeling and restore performance of infracted hearts compared to MSCs without Akt overexpression [247]. In addition, MSCs overexpressing Akt may be able to differentiate into cardiomyocytes 2 weeks after transplantation. Media from hypoxia-treated MSCs overexpressing Akt reduce infarction size and improve cardiac function supporting a paracrine mechanism for MSC-mediated cardioprotection [248]. Finally, overexpression of various growth factors and cytokines secreted by MSCs can further improve their cardioprotective potential and represents a promising strategy for treating acute and chronic heart disease [249–251].

## Conclusions and Future Directions

The field of cardiac cell therapy has demonstrated exciting developments since the first studies in 2001 reported the possibility of cardiac repair in animal models using BSCs followed by multiple clinical trials. Multiple stem cell types and their derivatives including ESCs, bone marrow-derived stem cells, endogenous cardiac progenitor cells, and iPSCs have been examined in transplantation studies. Among those cell sources, autologous stem cells represent the preferable cell type due to their immunocompatibility and decreased risk of tumorigenesis. Normal somatic cells genetically modified into embryonic-like iPSCs may be optimal for cell-based therapy purposes. However, the long-term effects of iPSC-derived cardiomyocytes need to be critically evaluated. It is also possible that different types of stem cells may be preferable for therapeutical treatment of different myocardial pathologies.

Traditional view on the adult heart as a terminally differentiated organ has recently been challenged by outstanding discoveries that adult cardiomyocytes are able to renew themselves, albeit at a very low level, and that endogenous CSCs are present in the adult murine and human hearts. However, their origin and ability to differentiate into the different lineages of the heart remain more obscure. Continued elucidation of the molecular mechanisms involved in the regulation of their self-renewal, survival, and differentiation is required prior the use of resident CSCs in cell-replacing therapy for cardiac disease.

Various cell delivery strategies via infusion or injection have been tested; however, their efficiency as well as retention of implanted cells remain lower than hoped. The timing of cell delivery is also presently poorly understood. The challenge of poor retention of implanted cells may be overcome by engineered cardiac tissue *ex vivo* to provide artificial scaffold for the regenerating myocardium. Breakthrough creation of a prototype cardiac tissue has recently been reported [252, 253].

Further development of safe, sensitive, and effective cell tracking systems is critical for understanding of the fate of transplanted cells as well as their contribution to myocardial repair. Currently available labeling techniques allow cell tracking only short periods after transplantation as labeling agents would be diluted upon cell division or degraded over time. Introduction of novel metabolic or genetic markers combined with positron emission tomography and magnetic resonance imaging can improve cell tracking *in vivo*.

More importantly, regardless of the right type of stem cells and the technique of their delivery in the right place at right time, better understanding of the molecular mechanisms of signaling pathways contributing to the regulation of essential functions of transplanted or endogenous stem and progenitor cells is required for conceiving improved therapy of injured heart. Multiple signaling cascades converge to control progenitor cell recruitment, proliferation, survival, and differentiation in response to myocardial injury. Many of the signaling pathways such as Wnt, MAPK, and PI3K-Akt-mTOR can play very different roles in response to different extracellular stimuli. These key signaling molecules represent attractive targets for pharmaceutical or genetic modulations to induce and facilitate cardiac regeneration. However, involvement of many of them in various signaling pathways responding to different stimuli and leading to different outcomes further complicates their clinical use. Unleashing the novel powerful tools of modern cell biology, genetics and pharmacology combined with functional genomics and proteomics will undoubtedly provide a significant impact on the future strategy for cell-based cardiac repair.

## Summary

- During the past decade, progenitor and stem cell-based therapy has emerged as a potential new therapeutic option for heart diseases. Growing experimental evidence in animal models and in small clinical trials suggests that stem cell transplantation into the infarcted myocardium beneficially remodels and repairs injured tissue and improves cardiac function.
- Stem cells, regardless of their origin and type, are undifferentiated cells characterized by the potential for self-renewal and by the ability to differentiate into various cell types known as pluripotency. Various cells, differing significantly in regards to their origin, structure, function, and ability to differentiate into one or more cell types, have been considered as candidates for myocardial cell-based repair therapy.
- ESCs, the prototypical stem cells, are pluripotent cells capable to differentiate into a variety of cell types, including cardiomyocytes. During the last decade, protocols to differentiate human ESCs into cardiomyocytes have been developed and successful isolations of cardiomyocytes from differentiating ESCs have been reported.
- The bone marrow represents a typical source of adult stem cells containing two main cell subsets, HSCs and MSCs. HSCs characterized by the expression of CD34 and CD113 cell-surface antigens demonstrate high self-renewal ability and serve as precursors of blood cell lineages. MSCs express CD44, CD90, CD105, CD106, and CD166 cell-surface markers, while they are negative for hematopoietic and endothelial markers, and are capable to differentiate into several nonhematopoietic cell types. Both HSCs and MSCs appear to be capable to transdifferentiate into cardiomyocytes and fuse with host cells affecting functional recovery after myocardial infarction.
- Initial studies have demonstrated that human skeletal myoblasts co-cultured *in vitro* with rat cardiomyocytes have adopted cardiac features and their transplantation into failing heart in animal models have improved cardiac function and survival. However, subsequent work has provided little evidence that injected skeletal myoblasts transdifferentiate into cardiomyocytes.
- Traditionally, the myocardium has been considered as a terminally differentiated tissue with poor regeneration capacity. However, the mammalian heart contains resident CSCs and cardiomyocytes are able to divide, albeit at very low level. CSCs, expressing c-kit$^+$, Sca-1$^+$, Isl1$^+$, can be isolated from the normal human myocardium, expanded, and differentiated *in vitro* predominantly into cardiomyocytes, smooth muscle, and endothelial cells. Importantly, these cells, injected in the infracted myocardium of immunodeficient rodents, appear to integrate functionally with the myocardium and improve cardiac function.
- Recently, a revolutionary approach to reprogram fully differentiated cells into ESC-like iPSCs, using ectopic expression of specific genes responsible for pluripotency, has been suggested. Although the original protocol for the generation of iPSCs has been based on the use of virus-mediated delivery of reprogramming factors, virus-free delivery systems has recently been proposed.
- Effective cell-based therapy to repair injured tissue requires targeted delivery of chosen cells to an area of the damaged tissue. The heart as a constantly contracting organ creates additional difficulties to transplant cells into localized heart region and immobilize them there. Although several routes of cell delivery have been described, none has yet emerged as the preferred approach.
- Common characteristics of stem cells essential for their use in cell-based therapy are their ability for self-renewal and differentiation into different cell types. Multiple signaling cascades in transplanted stem cells or resident CSCs are involved in the regulation of their proliferation, survival, and differentiation. These signaling pathways include Wnt, NF-κB, SDF-1/CXCR4, MAPK, and PI3K-Akt-mTOR pathways.
- Wnt proteins are a large family of secreted signaling proteins, which modulate various developmental processes. Traditionally, cardiovascular development has been thought to be mediated by noncanonical, JNK-dependent, Wnt signaling, whereas cardiac specification has been downregulated by canonical, β-catenin-dependent, Wnt signaling.

Recent studies have demonstrated essential roles of FGF and Bmp signaling pathways as downstream mediators of Wnt signaling in cardiac progenitor cell development.

- Initial cardiac specification depends on specific expression of components of canonical Wnt signaling cascade. In addition to specification, canonical Wnt signaling governs cardiac progenitor cell proliferation and differentiation. Noncanonical Wnt signaling also plays an important role in heart development. Wnt11 activates noncanonical Wnt signaling inducing cardiogenesis without activation of expression of mesodermal markers.
- The NF-κB signaling controls a variety of cellular responses and has been linked to multiple physiological and pathological processes in the myocardium. Canonical NF-κB signaling pathway depends on IKKβ and IKKγ, while the alternative pathway involves IKKα and NIK. The NF-κB signaling is an essential mediator of cardiomyocyte responses to various stimuli, including cytokines, LPS, oxidative stress, cardiac remodeling, and ischemia reperfusion. In human MSCs, TNF-α, LPS, or hypoxia induce the NF-κB-mediated increased production of growth factors and may play an important role in stem cell-stimulated angiogenesis and cardiac protection.
- SDF-1 and its receptor CXCR4 are expressed in various stem cells including HSCs, ESCs, primordial stem cells, and CSCs. It has recently been demonstrated that intramyocardial delivery of recombinant SDF-1 results in recruitment of progenitor cells to the heart and attenuation ischemic cardiomyopathy. Accordingly, inhibition of the SDF-1 degradation in the ischemic heart improves cardiac function and survival. Thus, the SDF-1/CXCR4 signaling pathway plays a fundamental role in stem and progenitor cell trafficking and mobilization.
- MAPKs are a large family of serine/threonine kinases, which regulate diverse cellular functions ranging from proliferation and differentiation to migration and apoptosis. The most characterized MAPKs are Erk1/2, p38α, β, γ, and δ, and JNK1, 2, and 3. Activation of MAPKs requires dual phosphorylation on threonine and tyrosine residues leading to a conformational change that enhances substrate accessibility and catalytic activity. Phosphorylation of each group of MAPKs is mediated by a distinct kinase cascade. MAP3Ks or MEKKs phosphorylate and activate dual-specificity MAP2Ks or MEKs that in turn phosphorylate MAPKs. These phosphorylation cascades eventually result in activation of multiple downstream pathways including various MAPK-activated protein serine/threonine kinases and transcriptional regulators.
- Over the past decade, extensive work has focused on the characterization of the role of MAPK signaling cascades in CVD. Studies on transgenic mice suggest that the Raf-MKK1/2- Erk1/2 pathway stimulates cardiomyocyte growth and survival, whereas the MKKK-MKK3/6-p38 and MKKK-MKK4/7-JNK1/2 cascades inhibit the growth of cardiomyocytes. In rodent experimental models after MI, all three major MAPK cascades are activated within minutes and play roles in pathological cardiac remodeling.
- MEK1-Erk1/2 signaling appears to contribute to the regulation of early myeloid commitment of HSCs. Moreover, cytokine-induced MEK-Erk activation leads to change in expression of a variety of cell-cycle modulating genes and controls thereby the balance between proliferation and apoptosis of HSCs. Interestingly, ES cells appear to not require Erk signaling for proliferation and self-renewal but they become dependent on this pathway upon lineage commitment.
- The p38 signaling is activated in *Atm*$^{-/-}$ HSCs and associated with defects in self-renewal and maintenance of these cells, whereas inhibition of p38 can rescue HSC self-renewal capacity. Moreover, p38 activation induced by erythropoietin is required for the initiation of erythroid differentiation. Although emerging evidence suggests that JNK signaling is also involved in the regulation of erythroid and myeloid development, its role in these processes is less well defined.
- PI3Kα- and PI3Kγ-mediated signaling play different roles in the heart. PI3Kα pathway regulates myocardial hypertrophy, apoptosis, maintenance of cardiac structure, and function under mainly physiological conditions. In contrast, activation of PI3Kγ pathway leads to reduced cardiomyocyte contractility through downregulation of β-AR signaling and therefore has a detrimental effect on cardiac function. Emerging evidence suggests that PI3K-Akt-mTOR signaling is involved in the regulation of HSC maintenance as well as in the control of lineage commitment via the Akt-mediated C/EBPα phosphorylation. Finally, MSCs overexpressing Akt injected into the perinfarct border zone in rat hearts prevent cardiac remodeling and restore performance of infracted hearts compared to MSCs without Akt overexpression.
- Regardless of the right type of stem cells and the technique of their delivery in the right place at right time, better understanding of the molecular mechanisms of multiple signaling pathways contributing to the regulation of essential functions of transplanted or endogenous stem and progenitor cells is required for conceiving improved therapy of injured heart. Although the key signaling molecules represent attractive targets for pharmaceutical or genetic modulations to induce and facilitate cardiac regeneration, involvement of many of them in various signaling pathways responding to different stimuli and leading to different outcomes further complicates their clinical use. Unleashing the novel powerful tools of

modern cell biology, genetics and pharmacology combined with functional genomics and proteomics will undoubtedly provide significant impact on the future strategy for cell-based cardiac repair.

## References


1. Rosamond W, Flegal K, Friday G, et al. Heart disease and stroke statistics – 2007 update: a report from the American Heart Association Statistics Committee and Stroke Statistics Subcommittee. Circulation. 2007;115:e69–171.
2. Taylor DA, Zenovich AG. Cardiovascular cell therapy and endogenous repair. Diabetes Obes Metab. 2008;10 Suppl 4:5–15.
3. Laflamme MA, Murry CE. Regenerating the heart. Nat Biotechnol. 2005;23:845–56.
4. Hofmann M, Wollert KC, Meyer GP, et al. Monitoring of bone marrow cell homing into the infarcted human myocardium. Circulation. 2005;111:2198–202.
5. Hou D, Youssef EA, Brinton TJ, et al. Radiolabeled cell distribution after intramyocardial, intracoronary, and interstitial retrograde coronary venous delivery: implications for current clinical trials. Circulation. 2005;112:I150–6.
6. Marin-Garcia M. Heart failure: bench to bedside. New York: Springer; 2010.
7. Kehat I, Kenyagin-Karsenti D, Snir M, et al. Human embryonic stem cells can differentiate into myocytes with structural and functional properties of cardiomyocytes. J Clin Invest. 2001;108:407–14.
8. Thomson JA, Itskovitz-Eldor J, Shapiro SS, et al. Embryonic stem cell lines derived from human blastocysts. Science. 1998;282: 1145–7.
9. Passier R, Denning C, Mummery C. Cardiomyocytes from human embryonic stem cells. Handb Exp Pharmacol. 2006;101–22.
10. Mummery C, Ward-van Oostwaard D, Doevendans P, et al. Differentiation of human embryonic stem cells to cardiomyocytes: role of coculture with visceral endoderm-like cells. Circulation. 2003;107:2733–40.
11. Passier R, van Laake LW, Mummery CL. Stem-cell-based therapy and lessons from the heart. Nature. 2008;453:322–9.
12. Laflamme MA, Chen KY, Naumova AV, et al. Cardiomyocytes derived from human embryonic stem cells in pro-survival factors enhance function of infarcted rat hearts. Nat Biotechnol. 2007;25:1015–24.
13. van Laake LW, Passier R, Monshouwer-Kloots J, et al. Human embryonic stem cell-derived cardiomyocytes survive and mature in the mouse heart and transiently improve function after myocardial infarction. Stem Cell Res. 2007;1:9–24.
14. Liu YP, Dovzhenko OV, Garthwaite MA, et al. Maintenance of pluripotency in human embryonic stem cells stably over-expressing enhanced green fluorescent protein. Stem Cells Dev. 2004;13: 636–45.
15. Saric T, Frenzel LP, Hescheler J. Immunological barriers to embryonic stem cell-derived therapies. Cells Tissues Organs. 2008;188: 78–90.
16. Yin AH, Miraglia S, Zanjani ED, et al. AC133, a novel marker for human hematopoietic stem and progenitor cells. Blood. 1997;90: 5002–12.
17. Goodell MA, Rosenzweig M, Kim H, et al. Dye efflux studies suggest that hematopoietic stem cells expressing low or undetectable levels of CD34 antigen exist in multiple species. Nat Med. 1997;3:1337–45.
18. Sato T, Laver JH, Ogawa M. Reversible expression of CD34 by murine hematopoietic stem cells. Blood. 1999;94:2548–54.
19. Eisenberg LM, Burns L, Eisenberg CA. Hematopoietic cells from bone marrow have the potential to differentiate into cardiomyocytes in vitro. Anat Rec A Discov Mol Cell Evol Biol. 2003;274:870–82.
20. Alvarez-Dolado M, Pardal R, Garcia-Verdugo JM, et al. Fusion of bone-marrow-derived cells with Purkinje neurons, cardiomyocytes and hepatocytes. Nature. 2003;425:968–73.
21. Nygren JM, Jovinge S, Breitbach M, et al. Bone marrow-derived hematopoietic cells generate cardiomyocytes at a low frequency through cell fusion, but not transdifferentiation. Nat Med. 2004;10:494–501.
22. Zhang N, Mustin D, Reardon W, et al. Blood-borne stem cells differentiate into vascular and cardiac lineages during normal development. Stem Cells Dev. 2006;15:17–28.
23. Tomita S, Li RK, Weisel RD, et al. Autologous transplantation of bone marrow cells improves damaged heart function. Circulation. 1999;100:II247–56.
24. Haider H, Ashraf M. Bone marrow stem cell transplantation for cardiac repair. Am J Physiol Heart Circ Physiol. 2005;288: H2557–67.
25. Pittenger MF, Mackay AM, Beck SC, et al. Multilineage potential of adult human mesenchymal stem cells. Science. 1999;284:143–7.
26. Kawada H, Fujita J, Kinjo K, et al. Nonhematopoietic mesenchymal stem cells can be mobilized and differentiate into cardiomyocytes after myocardial infarction. Blood. 2004;104:3581–7.
27. Rota M, Kajstura J, Hosoda T, et al. Bone marrow cells adopt the cardiomyogenic fate in vivo. Proc Natl Acad Sci USA. 2007;104: 17783–8.
28. Ryan JM, Barry FP, Murphy JM, Mahon BP. Mesenchymal stem cells avoid allogeneic rejection. J Inflamm. 2005;2:8.
29. Orlic D, Kajstura J, Chimenti S, et al. Bone marrow cells regenerate infarcted myocardium. Nature. 2001;410:701–5.
30. Jackson KA, Majka SM, Wang H, et al. Regeneration of ischemic cardiac muscle and vascular endothelium by adult stem cells. J Clin Invest. 2001;107:1395–402.
31. Segers VF, Lee RT. Stem-cell therapy for cardiac disease. Nature. 2008;451:937–42.
32. Korf-Klingebiel M, Kempf T, Sauer T, et al. Bone marrow cells are a rich source of growth factors and cytokines: implications for cell therapy trials after myocardial infarction. Eur Heart J. 2008;29:2851–8.
33. Kao RL, Rizzo C, Magovern GJ. Satellite cells for myocardial regeneration [abstract]. Physiologist. 1989;32:220.
34. Taylor DA, Atkins BZ, Hungspreugs P, et al. Regenerating functional myocardium: improved performance after skeletal myoblast transplantation. Nat Med. 1998;4:929–33.
35. Siminiak T, Fiszer D, Jerzykowska O, et al. Percutaneous transcoronary-venous transplantation of autologous skeletal myoblasts in the treatment of post-infarction myocardial contractility impairment: the POZNAN trial. Eur Heart J. 2005;26:1188–95.
36. Menasche P. Skeletal myoblasts as a therapeutic agent. Prog Cardiovasc Dis. 2007;50:7–17.
37. Hagege AA, Carrion C, Menasche P, et al. Viability and differentiation of autologous skeletal myoblast grafts in ischaemic cardiomyopathy. Lancet. 2003;361:491–2.
38. Smits PC, van Geuns RJ, Poldermans D, et al. Catheter-based intramyocardial injection of autologous skeletal myoblasts as a primary treatment of ischemic heart failure: clinical experience with six-month follow-up. J Am Coll Cardiol. 2003;42:2063–9.
39. Menasche P, Alfieri O, Janssens S, et al. The Myoblast Autologous Grafting in Ischemic Cardiomyopathy (MAGIC) trial: first randomized placebo-controlled study of myoblast transplantation. Circulation. 2008;117:1189–200.
40. Fernandes S, Amirault JC, Lande G, et al. Autologous myoblast transplantation after myocardial infarction increases the inducibility of ventricular arrhythmias. Cardiovasc Res. 2006;69:348–58.

41. Oh H, Bradfute SB, Gallardo TD, et al. Cardiac progenitor cells from adult myocardium: homing, differentiation, and fusion after infarction. Proc Natl Acad Sci USA. 2003;100:12313–8.
42. Beltrami AP, Barlucchi L, Torella D, et al. Adult cardiac stem cells are multipotent and support myocardial regeneration. Cell. 2003;114:763–76.
43. Messina E, De Angelis L, Frati G, et al. Isolation and expansion of adult cardiac stem cells from human and murine heart. Circ Res. 2004;95:911–21.
44. Laugwitz KL, Moretti A, Lam J, et al. Postnatal isl1+ cardioblasts enter fully differentiated cardiomyocyte lineages. Nature. 2005;433:647–53.
45. Pfister O, Mouquet F, Jain M, et al. CD31- but Not CD31+ cardiac side population cells exhibit functional cardiomyogenic differentiation. Circ Res. 2005;97:52–61.
46. Smith RR, Barile L, Messina E, Marban E. Stem cells in the heart: what's the buzz all about? Part 2: Arrhythmic risks and clinical studies. Heart Rhythm. 2008;5:880–7.
47. Smith RR, Barile L, Messina E, Marban E. Stem cells in the heart: what's the buzz all about? – Part 1: Preclinical considerations. Heart Rhythm. 2008;5:749–57.
48. Smith RR, Barile L, Cho HC, et al. Regenerative potential of cardiosphere-derived cells expanded from percutaneous endomyocardial biopsy specimens. Circulation. 2007;115:896–908.
49. Bearzi C, Rota M, Hosoda T, et al. Human cardiac stem cells. Proc Natl Acad Sci USA. 2007;104:14068–73.
50. Takahashi K, Yamanaka S. Induction of pluripotent stem cells from mouse embryonic and adult fibroblast cultures by defined factors. Cell. 2006;126:663–76.
51. Yamanaka S. Strategies and new developments in the generation of patient-specific pluripotent stem cells. Cell Stem Cell. 2007;1:39–49.
52. Jaenisch R, Young R. Stem cells, the molecular circuitry of pluripotency and nuclear reprogramming. Cell. 2008;132:567–82.
53. Takahashi K, Tanabe K, Ohnuki M, et al. Induction of pluripotent stem cells from adult human fibroblasts by defined factors. Cell. 2007;131:861–72.
54. Yu J, Vodyanik MA, Smuga-Otto K, et al. Induced pluripotent stem cell lines derived from human somatic cells. Science. 2007;318:1917–20.
55. Dimos JT, Rodolfa KT, Niakan KK, et al. Induced pluripotent stem cells generated from patients with ALS can be differentiated into motor neurons. Science. 2008;321:1218–21.
56. Ebert AD, Yu J, Rose Jr FF, et al. Induced pluripotent stem cells from a spinal muscular atrophy patient. Nature. 2009;457: 277–80.
57. Hotta A, Cheung AY, Farra N, et al. Isolation of human iPS cells using EOS lentiviral vectors to select for pluripotency. Nat Methods. 2009;6:370–6.
58. Maehr R, Chen S, Snitow M, et al. Generation of pluripotent stem cells from patients with type 1 diabetes. Proc Natl Acad Sci USA. 2009;106:15768–73.
59. Chambers I, Smith A. Self-renewal of teratocarcinoma and embryonic stem cells. Oncogene. 2004;23:7150–60.
60. Nakagawa M, Koyanagi M, Tanabe K, et al. Generation of induced pluripotent stem cells without Myc from mouse and human fibroblasts. Nat Biotechnol. 2008;26:101–6.
61. Wernig M, Meissner A, Cassady JP, Jaenisch R. c-Myc is dispensable for direct reprogramming of mouse fibroblasts. Cell Stem Cell. 2008;2:10–2.
62. Hanna J, Wernig M, Markoulaki S, et al. Treatment of sickle cell anemia mouse model with iPS cells generated from autologous skin. Science. 2007;318:1920–3.
63. Narazaki G, Uosaki H, Teranishi M, et al. Directed and systematic differentiation of cardiovascular cells from mouse induced pluripotent stem cells. Circulation. 2008;118:498–506.
64. Schenke-Layland K, Rhodes KE, Angelis E, et al. Reprogrammed mouse fibroblasts differentiate into cells of the cardiovascular and hematopoietic lineages. Stem Cells. 2008;26:1537–46.
65. Kuzmenkin A, Liang H, Xu G, et al. Functional characterization of cardiomyocytes derived from murine induced pluripotent stem cells in vitro. FASEB J. 2009;23:4168–80.
66. Kim D, Kim CH, Moon JI, et al. Generation of human induced pluripotent stem cells by direct delivery of reprogramming proteins. Cell Stem Cell. 2009;4:472–6.
67. Kaji K, Norrby K, Paca A, Mileikovsky M, Mohseni P, Woltjen K. Virus-free induction of pluripotency and subsequent excision of reprogramming factors. Nature. 2009;458:771–5.
68. Woltjen K, Michael IP, Mohseni P, et al. piggyBac transposition reprograms fibroblasts to induced pluripotent stem cells. Nature. 2009;458:766–70.
69. Yu J, Hu K, Smuga-Otto K, et al. Human induced pluripotent stem cells free of vector and transgene sequences. Science. 2009;324: 797–801.
70. Saha K, Jaenisch R. Technical challenges in using human induced pluripotent stem cells to model disease. Cell Stem Cell. 2009;5: 584–95.
71. Muller-Ehmsen J, Whittaker P, Kloner RA, et al. Survival and development of neonatal rat cardiomyocytes transplanted into adult myocardium. J Mol Cell Cardiol. 2002;34:107–16.
72. Perin EC, Lopez J. Methods of stem cell delivery in cardiac diseases. Nat Clin Pract Cardiovasc Med. 2006;3 Suppl 1:S110–3.
73. Beeres SL, Atsma DE, van Ramshorst J, Schalij MJ, Bax JJ. Cell therapy for ischaemic heart disease. Heart. 2008;94:1214–26.
74. Freyman T, Polin G, Osman H, et al. A quantitative, randomized study evaluating three methods of mesenchymal stem cell delivery following myocardial infarction. Eur Heart J. 2006;27:1114–22.
75. Kurpisz M, Czepczynski R, Grygielska B, et al. Bone marrow stem cell imaging after intracoronary administration. Int J Cardiol. 2007;121:194–5.
76. Logan CY, Nusse R. The Wnt signaling pathway in development and disease. Annu Rev Cell Dev Biol. 2004;20:781–810.
77. Nusse R. Wnt signaling and stem cell control. Cell Res. 2008;18:523–7.
78. Cohen ED, Tian Y, Morrisey EE. Wnt signaling: an essential regulator of cardiovascular differentiation, morphogenesis and progenitor self-renewal. Development. 2008;135:789–98.
79. Takada R, Satomi Y, Kurata T, et al. Monounsaturated fatty acid modification of Wnt protein: its role in Wnt secretion. Dev Cell. 2006;11:791–801.
80. Ching W, Nusse R. A dedicated Wnt secretion factor. Cell. 2006;125:432–3.
81. Banziger C, Soldini D, Schutt C, Zipperlen P, Hausmann G, Basler K. Wntless, a conserved membrane protein dedicated to the secretion of Wnt proteins from signaling cells. Cell. 2006;125:509–22.
82. Bartscherer K, Pelte N, Ingelfinger D, Boutros M. Secretion of Wnt ligands requires Evi, a conserved transmembrane protein. Cell. 2006;125:523–33.
83. Verges M, Luton F, Gruber C, et al. The mammalian retromer regulates transcytosis of the polymeric immunoglobulin receptor. Nat Cell Biol. 2004;6:763–9.
84. Panakova D, Sprong H, Marois E, Thiele C, Eaton S. Lipoprotein particles are required for Hedgehog and Wingless signalling. Nature. 2005;435:58–65.
85. Coudreuse DY, Roel G, Betist MC, Destree O, Korswagen HC. Wnt gradient formation requires retromer function in Wnt-producing cells. Science. 2006;312:921–4.
86. Bhanot P, Brink M, Samos CH, et al. A new member of the frizzled family from *Drosophila* functions as a Wingless receptor. Nature. 1996;382:225–30.
87. Hsieh JC, Rattner A, Smallwood PM, Nathans J. Biochemical characterization of Wnt-frizzled interactions using a soluble,

biologically active vertebrate Wnt protein. Proc Natl Acad Sci USA. 1999;96:3546–51.
88. Wu CH, Nusse R. Ligand receptor interactions in the Wnt signaling pathway in *Drosophila*. J Biol Chem. 2002;277:41762–9.
89. Wehrli M, Dougan ST, Caldwell K, et al. arrow encodes an LDL-receptor-related protein essential for Wingless signalling. Nature. 2000;407:527–30.
90. Li Y, Bu G. LRP5/6 in Wnt signaling and tumorigenesis. Future Oncol. 2005;1:673–81.
91. Chen W, ten Berge D, Brown J, et al. Dishevelled 2 recruits beta-arrestin 2 to mediate Wnt5A-stimulated endocytosis of Frizzled 4. Science. 2003;301:1391–4.
92. Mao J, Wang J, Liu B, et al. Low-density lipoprotein receptor-related protein-5 binds to Axin and regulates the canonical Wnt signaling pathway. Mol Cell. 2001;7:801–9.
93. Zeng X, Tamai K, Doble B, et al. A dual-kinase mechanism for Wnt co-receptor phosphorylation and activation. Nature. 2005;438:873–7.
94. Veeman MT, Axelrod JD, Moon RT. A second canon. Functions and mechanisms of beta-catenin-independent Wnt signaling. Dev Cell. 2003;5:367–77.
95. Mikels AJ, Nusse R. Purified Wnt5a protein activates or inhibits beta-catenin-TCF signaling depending on receptor context. PLoS Biol. 2006;4:e115.
96. Tu X, Joeng KS, Nakayama KI, et al. Noncanonical Wnt signaling through G protein-linked PKCdelta activation promotes bone formation. Dev Cell. 2007;12:113–27.
97. Kroiher M, Miller MA, Steele RE. Deceiving appearances: signaling by "dead" and "fractured" receptor protein-tyrosine kinases. Bioessays. 2001;23:69–76.
98. Lu W, Yamamoto V, Ortega B, Baltimore D. Mammalian Ryk is a Wnt coreceptor required for stimulation of neurite outgrowth. Cell. 2004;119:97–108.
99. Marvin MJ, Di Rocco G, Gardiner A, Bush SM, Lassar AB. Inhibition of Wnt activity induces heart formation from posterior mesoderm. Genes Dev. 2001;15:316–27.
100. Cohen ED, Wang Z, Lepore JJ, et al. Wnt/beta-catenin signaling promotes expansion of Isl-1-positive cardiac progenitor cells through regulation of FGF signaling. J Clin Invest. 2007;117:1794–804.
101. Lin L, Cui L, Zhou W, et al. Beta-catenin directly regulates Islet1 expression in cardiovascular progenitors and is required for multiple aspects of cardiogenesis. Proc Natl Acad Sci USA. 2007;104:9313–8.
102. Stoick-Cooper CL, Weidinger G, Riehle KJ, et al. Distinct Wnt signaling pathways have opposing roles in appendage regeneration. Development. 2007;134:479–89.
103. Naito AT, Shiojima I, Akazawa H, et al. Developmental stage-specific biphasic roles of Wnt/beta-catenin signaling in cardiomyogenesis and hematopoiesis. Proc Natl Acad Sci USA. 2006; 103:19812–7.
104. Ueno S, Weidinger G, Osugi T, et al. Biphasic role for Wnt/beta-catenin signaling in cardiac specification in zebrafish and embryonic stem cells. Proc Natl Acad Sci USA. 2007;104:9685–90.
105. Schneider VA, Mercola M. Wnt antagonism initiates cardiogenesis in *Xenopus laevis*. Genes Dev. 2001;15:304–15.
106. Pandur P, Lasche M, Eisenberg LM, Kuhl M. Wnt-11 activation of a non-canonical Wnt signalling pathway is required for cardiogenesis. Nature. 2002;418:636–41.
107. Zhou W, Lin L, Majumdar A, et al. Modulation of morphogenesis by noncanonical Wnt signaling requires ATF/CREB family-mediated transcriptional activation of TGFbeta2. Nat Genet. 2007;39:1225–34.
108. Schleiffarth JR, Person AD, Martinsen BJ, et al. Wnt5a is required for cardiac outflow tract septation in mice. Pediatr Res. 2007;61: 386–91.
109. Park EJ, Ogden LA, Talbot A, et al. Required, tissue-specific roles for Fgf8 in outflow tract formation and remodeling. Development. 2006;133:2419–33.
110. Schultheiss TM, Burch JB, Lassar AB. A role for bone morphogenetic proteins in the induction of cardiac myogenesis. Genes Dev. 1997;11:451–62.
111. Shi Y, Katsev S, Cai C, Evans S. BMP signaling is required for heart formation in vertebrates. Dev Biol. 2000;224:226–37.
112. Brand T. Heart development: molecular insights into cardiac specification and early morphogenesis. Dev Biol. 2003;258:1–19.
113. Liu W, Selever J, Wang D, et al. Bmp4 signaling is required for outflow-tract septation and branchial-arch artery remodeling. Proc Natl Acad Sci USA. 2004;101:4489–94.
114. Tirosh-Finkel L, Elhanany H, Rinon A, Tzahor E. Mesoderm progenitor cells of common origin contribute to the head musculature and the cardiac outflow tract. Development. 2006;133:1943–53.
115. Ai D, Fu X, Wang J, et al. Canonical Wnt signaling functions in second heart field to promote right ventricular growth. Proc Natl Acad Sci USA. 2007;104:9319–24.
116. Park EJ, Watanabe Y, Smyth G, et al. An FGF autocrine loop initiated in second heart field mesoderm regulates morphogenesis at the arterial pole of the heart. Development. 2008;135:3599–610.
117. Prall OW, Menon MK, Solloway MJ, et al. An Nkx2-5/Bmp2/Smad1 negative feedback loop controls heart progenitor specification and proliferation. Cell. 2007;128:947–59.
118. Klaus A, Saga Y, Taketo MM, Tzahor E, Birchmeier W. Distinct roles of Wnt/beta-catenin and Bmp signaling during early cardiogenesis. Proc Natl Acad Sci USA. 2007;104:18531–6.
119. Ghosh S, May MJ, Kopp EB. NF-kappa B and Rel proteins: evolutionarily conserved mediators of immune responses. Annu Rev Immunol. 1998;16:225–60.
120. Hayden MS, Ghosh S. Signaling to NF-kappaB. Genes Dev. 2004;18: 2195–224.
121. Hayden MS, Ghosh S. Shared principles in NF-kappaB signaling. Cell. 2008;132:344–62.
122. Sasaki H, Galang N, Maulik N. Redox regulation of NF-kappaB and AP-1 in ischemic reperfused heart. Antioxid Redox Signal. 1999;1:317–24.
123. Craig R, Wagner M, McCardle T, Craig AG, Glembotski CC. The cytoprotective effects of the glycoprotein 130 receptor-coupled cytokine, cardiotrophin-1, require activation of NF-kappa B. J Biol Chem. 2001;276:37621–9.
124. Wright G, Singh IS, Hasday JD, et al. Endotoxin stress-response in cardiomyocytes: NF-kappaB activation and tumor necrosis factor-alpha expression. Am J Physiol Heart Circ Physiol. 2002;282: H872–9.
125. Hall G, Singh IS, Hester L, Hasday JD, Rogers TB. Inhibitor-kappaB kinase-beta regulates LPS-induced TNF-alpha production in cardiac myocytes through modulation of NF-kappaB p65 subunit phosphorylation. Am J Physiol Heart Circ Physiol. 2005;289:H2103–11.
126. Widera D, Mikenberg I, Kaltschmidt B, Kaltschmidt C. Potential role of NF-kappaB in adult neural stem cells: the underrated steersman? Int J Dev Neurosci. 2006;24:91–102.
127. Crisostomo PR, Wang Y, Markel TA, Wang M, Lahm T, Meldrum DR. Human mesenchymal stem cells stimulated by TNF-alpha, LPS, or hypoxia produce growth factors by an NF kappa B- but not JNK-dependent mechanism. Am J Physiol Cell Physiol. 2008;294: C675–82.
128. Broxmeyer HE. Chemokines in hematopoiesis. Curr Opin Hematol. 2008;15:49–58.
129. Askari AT, Unzek S, Popovic ZB, et al. Effect of stromal-cell-derived factor 1 on stem-cell homing and tissue regeneration in ischaemic cardiomyopathy. Lancet. 2003;362:697–703.
130. Zaruba MM, Theiss HD, Vallaster M, et al. Synergy between CD26/DPP-IV inhibition and G-CSF improves cardiac function after acute myocardial infarction. Cell Stem Cell. 2009;4:313–23.

131. Zaruba MM, Franz WM. Role of the SDF-1-CXCR4 axis in stem cell-based therapies for ischemic cardiomyopathy. Expert Opin Biol Ther. 2010;10:321–35.
132. Shirozu M, Nakano T, Inazawa J, et al. Structure and chromosomal localization of the human stromal cell-derived factor 1 (SDF1) gene. Genomics. 1995;28:495–500.
133. Crump MP, Gong JH, Loetscher P, et al. Solution structure and basis for functional activity of stromal cell-derived factor-1; dissociation of CXCR4 activation from binding and inhibition of HIV-1. EMBO J. 1997;16:6996–7007.
134. Proost P, Struyf S, Schols D, et al. Processing by CD26/dipeptidyl-peptidase IV reduces the chemotactic and anti-HIV-1 activity of stromal-cell-derived factor-1alpha. FEBS Lett. 1998;432: 73–6.
135. Delgado MB, Clark-Lewis I, Loetscher P, et al. Rapid inactivation of stromal cell-derived factor-1 by cathepsin G associated with lymphocytes. Eur J Immunol. 2001;31:699–707.
136. Nagasawa T, Hirota S, Tachibana K, et al. Defects of B-cell lymphopoiesis and bone-marrow myelopoiesis in mice lacking the CXC chemokine PBSF/SDF-1. Nature. 1996;382:635–8.
137. Tachibana K, Hirota S, Iizasa H, et al. The chemokine receptor CXCR4 is essential for vascularization of the gastrointestinal tract. Nature. 1998;393:591–4.
138. Zou YR, Kottmann AH, Kuroda M, Taniuchi I, Littman DR. Function of the chemokine receptor CXCR4 in haematopoiesis and in cerebellar development. Nature. 1998;393:595–9.
139. Ara T, Tokoyoda K, Okamoto R, Koni PA, Nagasawa T. The role of CXCL12 in the organ-specific process of artery formation. Blood. 2005;105:3155–61.
140. Christopherson 2nd KW, Cooper S, Broxmeyer HE. Cell surface peptidase CD26/DPPIV mediates G-CSF mobilization of mouse progenitor cells. Blood. 2003;101:4680–6.
141. Dabusti M, Lanza F, Campioni D, et al. CXCR-4 expression on bone marrow CD34+ cells prior to mobilization can predict mobilization adequacy in patients with hematologic malignancies. J Hematother Stem Cell Res. 2003;12:425–34.
142. Ara T, Nakamura Y, Egawa T, et al. Impaired colonization of the gonads by primordial germ cells in mice lacking a chemokine, stromal cell-derived factor-1 (SDF-1). Proc Natl Acad Sci USA. 2003;100:5319–23.
143. Kucia M, Reca R, Miekus K, et al. Trafficking of normal stem cells and metastasis of cancer stem cells involve similar mechanisms: pivotal role of the SDF-1-CXCR4 axis. Stem Cells. 2005;23:879–94.
144. Guo Y, Hangoc G, Bian H, Pelus LM, Broxmeyer HE. SDF-1/CXCL12 enhances survival and chemotaxis of murine embryonic stem cells and production of primitive and definitive hematopoietic progenitor cells. Stem Cells. 2005;23:1324–32.
145. Nelson TJ, Faustino RS, Chiriac A, Crespo-Diaz R, Behfar A, Terzic A. CXCR4+/FLK-1+ biomarkers select a cardiopoietic lineage from embryonic stem cells. Stem Cells. 2008;26:1464–73.
146. Vila-Coro AJ, Rodriguez-Frade JM, Martin De Ana A, Moreno-Ortiz MC, Martinez AC, Mellado M. The chemokine SDF-1alpha triggers CXCR4 receptor dimerization and activates the JAK/STAT pathway. FASEB J. 1999;13:1699–710.
147. Murphy PM, Baggiolini M, Charo IF, et al. International union of pharmacology. XXII. Nomenclature for chemokine receptors. Pharmacol Rev. 2000;52:145–76.
148. Sierro F, Biben C, Martinez-Munoz L, et al. Disrupted cardiac development but normal hematopoiesis in mice deficient in the second CXCL12/SDF-1 receptor, CXCR7. Proc Natl Acad Sci USA. 2007;104:14759–64.
149. Ratajczak MZ, Zuba-Surma E, Kucia M, Reca R, Wojakowski W, Ratajczak J. The pleiotropic effects of the SDF-1-CXCR4 axis in organogenesis, regeneration and tumorigenesis. Leukemia. 2006;20:1915–24.
150. Petit I, Goichberg P, Spiegel A, et al. Atypical PKC-zeta regulates SDF-1-mediated migration and development of human CD34+ progenitor cells. J Clin Invest. 2005;115:168–76.
151. Wang JF, Park IW, Groopman JE. Stromal cell-derived factor-1alpha stimulates tyrosine phosphorylation of multiple focal adhesion proteins and induces migration of hematopoietic progenitor cells: roles of phosphoinositide-3 kinase and protein kinase C. Blood. 2000;95:2505–13.
152. Orsini MJ, Parent JL, Mundell SJ, Marchese A, Benovic JL. Trafficking of the HIV coreceptor CXCR4. Role of arrestins and identification of residues in the c-terminal tail that mediate receptor internalization. J Biol Chem. 1999;274:31076–86.
153. McGrath KE, Koniski AD, Maltby KM, McGann JK, Palis J. Embryonic expression and function of the chemokine SDF-1 and its receptor, CXCR4. Dev Biol. 1999;213:442–56.
154. Ceradini DJ, Kulkarni AR, Callaghan MJ, et al. Progenitor cell trafficking is regulated by hypoxic gradients through HIF-1 induction of SDF-1. Nat Med. 2004;10:858–64.
155. Pearson G, Robinson F, Beers Gibson T, et al. Mitogen-activated protein (MAP) kinase pathways: regulation and physiological functions. Endocr Rev. 2001;22:153–83.
156. Raman M, Chen W, Cobb MH. Differential regulation and properties of MAPKs. Oncogene. 2007;26:3100–12.
157. Anderson NG, Maller JL, Tonks NK, Sturgill TW. Requirement for integration of signals from two distinct phosphorylation pathways for activation of MAP kinase. Nature. 1990;343:651–3.
158. Roux PP, Blenis J. ERK and p38 MAPK-activated protein kinases: a family of protein kinases with diverse biological functions. Microbiol Mol Biol Rev. 2004;68:320–44.
159. Biondi RM, Nebreda AR. Signalling specificity of Ser/Thr protein kinases through docking-site-mediated interactions. Biochem J. 2003;372:1–13.
160. Tanoue T, Nishida E. Molecular recognitions in the MAP kinase cascades. Cell Signal. 2003;15:455–62.
161. Kolch W. Coordinating ERK/MAPK signalling through scaffolds and inhibitors. Nat Rev Mol Cell Biol. 2005;6:827–37.
162. Owens DM, Keyse SM. Differential regulation of MAP kinase signalling by dual-specificity protein phosphatases. Oncogene. 2007;26:3203–13.
163. Boulton TG, Cobb MH. Identification of multiple extracellular signal-regulated kinases (ERKs) with antipeptide antibodies. Cell Regul. 1991;2:357–71.
164. Yoon S, Seger R. The extracellular signal-regulated kinase: multiple substrates regulate diverse cellular functions. Growth Factors. 2006;24:21–44.
165. Chambard JC, Lefloch R, Pouyssegur J, Lenormand P. ERK implication in cell cycle regulation. Biochim Biophys Acta. 2007;1773:1299–310.
166. McCubrey JA, Steelman LS, Chappell WH, et al. Roles of the Raf/MEK/ERK pathway in cell growth, malignant transformation and drug resistance. Biochim Biophys Acta. 2007;1773:1263–84.
167. Pages G, Guerin S, Grall D, et al. Defective thymocyte maturation in p44 MAP kinase (Erk 1) knockout mice. Science. 1999;286:1374–7.
168. Hatano N, Mori Y, Oh-hora M, et al. Essential role for ERK2 mitogen-activated protein kinase in placental development. Genes Cells. 2003;8:847–56.
169. Chen Z, Gibson TB, Robinson F, et al. MAP kinases. Chem Rev. 2001;101:2449–76.
170. Zarubin T, Han J. Activation and signaling of the p38 MAP kinase pathway. Cell Res. 2005;15:11–8.
171. Uhlik MT, Abell AN, Johnson NL, et al. Rac-MEKK3-MKK3 scaffolding for p38 MAPK activation during hyperosmotic shock. Nat Cell Biol. 2003;5:1104–10.
172. Buchsbaum RJ, Connolly BA, Feig LA. Interaction of Rac exchange factors Tiam1 and Ras-GRF1 with a scaffold for the p38

mitogen-activated protein kinase cascade. Mol Cell Biol. 2002;22:4073–85.
173. Schoorlemmer J, Goldfarb M. Fibroblast growth factor homologous factors and the islet brain-2 scaffold protein regulate activation of a stress-activated protein kinase. J Biol Chem. 2002;277:49111–9.
174. Salojin KV, Owusu IB, Millerchip KA, Potter M, Platt KA, Oravecz T. Essential role of MAPK phosphatase-1 in the negative control of innate immune responses. J Immunol. 2006;176:1899–907.
175. Hu JH, Chen T, Zhuang ZH, et al. Feedback control of MKP-1 expression by p38. Cell Signal. 2007;19:393–400.
176. Tamura K, Sudo T, Senftleben U, Dadak AM, Johnson R, Karin M. Requirement for p38alpha in erythropoietin expression: a role for stress kinases in erythropoiesis. Cell. 2000;102:221–31.
177. Beardmore VA, Hinton HJ, Eftychi C, et al. Generation and characterization of p38beta (MAPK11) gene-targeted mice. Mol Cell Biol. 2005;25:10454–64.
178. Aouadi M, Binetruy B, Caron L, Le Marchand-Brustel Y, Bost F. Role of MAPKs in development and differentiation: lessons from knockout mice. Biochimie. 2006;88:1091–8.
179. Arai H, Maki K, Waga K, et al. Functional regulation of TEL by p38-induced phosphorylation. Biochem Biophys Res Commun. 2002;299:116–25.
180. Derijard B, Hibi M, Wu IH, et al. JNK1: a protein kinase stimulated by UV light and Ha-Ras that binds and phosphorylates the c-Jun activation domain. Cell. 1994;76:1025–37.
181. Davis RJ. Signal transduction by the c-Jun N-terminal kinase. Biochem Soc Symp. 1999;64:1–12.
182. Davis RJ. Signal transduction by the JNK group of MAP kinases. Cell. 2000;103:239–52.
183. Kyriakis JM, Avruch J. Mammalian mitogen-activated protein kinase signal transduction pathways activated by stress and inflammation. Physiol Rev. 2001;81:807–69.
184. Wang X, Destrument A, Tournier C. Physiological roles of MKK4 and MKK7: insights from animal models. Biochim Biophys Acta. 2007;1773:1349–57.
185. Lee CM, Onesime D, Reddy CD, Dhanasekaran N, Reddy EP. JLP: a scaffolding protein that tethers JNK/p38MAPK signaling modules and transcription factors. Proc Natl Acad Sci USA. 2002; 99:14189–94.
186. Luo W, Ng WW, Jin LH, Ye Z, Han J, Lin SC. Axin utilizes distinct regions for competitive MEKK1 and MEKK4 binding and JNK activation. J Biol Chem. 2003;278:37451–8.
187. Alonso A, Sasin J, Bottini N, et al. Protein tyrosine phosphatases in the human genome. Cell. 2004;117:699–711.
188. Kuan CY, Yang DD, Samanta Roy DR, Davis RJ, Davis RJ, Rakic P, et al. The Jnk1 and Jnk2 protein kinases are required for regional specific apoptosis during early brain development. Neuron. 1999;22:667–76.
189. Sabapathy K, Jochum W, Hochedlinger K, Chang L, Karin M, Wagner EF. Defective neural tube morphogenesis and altered apoptosis in the absence of both JNK1 and JNK2. Mech Dev. 1999;89:115–24.
190. Yang DD, Kuan CY, Whitmarsh AJ, et al. Absence of excitotoxicity-induced apoptosis in the hippocampus of mice lacking the Jnk3 gene. Nature. 1997;389:865–70.
191. Heineke J, Molkentin JD. Regulation of cardiac hypertrophy by intracellular signalling pathways. Nat Rev Mol Cell Biol. 2006;7:589–600.
192. Wang Y. Mitogen-activated protein kinases in heart development and diseases. Circulation. 2007;116:1413–23.
193. Braz JC, Bueno OF, Liang Q, et al. Targeted inhibition of p38 MAPK promotes hypertrophic cardiomyopathy through upregulation of calcineurin-NFAT signaling. J Clin Invest. 2003;111:1475–86.
194. Yoshida K, Yoshiyama M, Omura T, et al. Activation of mitogen-activated protein kinases in the non-ischemic myocardium of an acute myocardial infarction in rats. Jpn Circ J. 2001;65:808–14.
195. Ren J, Zhang S, Kovacs A, Wang Y, Muslin AJ. Role of p38alpha MAPK in cardiac apoptosis and remodeling after myocardial infarction. J Mol Cell Cardiol. 2005;38:617–23.
196. Qin F, Liang MC, Liang CS. Progressive left ventricular remodeling, myocyte apoptosis, and protein signaling cascades after myocardial infarction in rabbits. Biochim Biophys Acta. 2005;1740:499–513.
197. Matsumoto-Ida M, Takimoto Y, Aoyama T, Akao M, Takeda T, Kita T. Activation of TGF-beta1-TAK1-p38 MAPK pathway in spared cardiomyocytes is involved in left ventricular remodeling after myocardial infarction in rats. Am J Physiol Heart Circ Physiol. 2006;290:H709–15.
198. Tenhunen O, Soini Y, Ilves M, et al. p38 Kinase rescues failing myocardium after myocardial infarction: evidence for angiogenic and anti-apoptotic mechanisms. FASEB J. 2006;20:1907–9.
199. Kaiser RA, Bueno OF, Lips DJ, et al. Targeted inhibition of p38 mitogen-activated protein kinase antagonizes cardiac injury and cell death following ischemia-reperfusion in vivo. J Biol Chem. 2004;279:15524–30.
200. Engel FB, Schebesta M, Duong MT, et al. p38 MAP kinase inhibition enables proliferation of adult mammalian cardiomyocytes. Genes Dev. 2005;19:1175–87.
201. Yamaguchi O, Higuchi Y, Hirotani S, et al. Targeted deletion of apoptosis signal-regulating kinase 1 attenuates left ventricular remodeling. Proc Natl Acad Sci USA. 2003;100:15883–8.
202. Zhao M, Liu Y, Wang X, New L, Han J, Brunk UT. Activation of the p38 MAP kinase pathway is required for foam cell formation from macrophages exposed to oxidized LDL. APMIS. 2002;110:458–68.
203. Rahaman SO, Lennon DJ, Febbraio M, Podrez EA, Hazen SL, Silverstein RL. A CD36-dependent signaling cascade is necessary for macrophage foam cell formation. Cell Metab. 2006;4:211–21.
204. Gennaro G, Menard C, Michaud SE, Deblois D, Rivard A. Inhibition of vascular smooth muscle cell proliferation and neointimal formation in injured arteries by a novel, oral mitogen-activated protein kinase/extracellular signal-regulated kinase inhibitor. Circulation. 2004;110:3367–71.
205. Izumi Y, Kim S, Namba M, et al. Gene transfer of dominant-negative mutants of extracellular signal-regulated kinase and c-Jun NH2-terminal kinase prevents neointimal formation in balloon-injured rat artery. Circ Res. 2001;88:1120–6.
206. Ohashi N, Matsumori A, Furukawa Y, et al. Role of p38 mitogen-activated protein kinase in neointimal hyperplasia after vascular injury. Arterioscler Thromb Vasc Biol. 2000;20:2521–6.
207. Hsu CL, Kikuchi K, Kondo M. Activation of mitogen-activated protein kinase kinase (MEK)/extracellular signal regulated kinase (ERK) signaling pathway is involved in myeloid lineage commitment. Blood. 2007;110:1420–8.
208. Geest CR, Buitenhuis M, Groot Koerkamp MJ, Holstege FC, Vellenga E, Coffer PJ. Tight control of MEK-ERK activation is essential in regulating proliferation, survival, and cytokine production of CD34+-derived neutrophil progenitors. Blood. 2009;114:3402–12.
209. Kunath T, Saba-El-Leil MK, Almousailleakh M, Wray J, Meloche S, Smith A. FGF stimulation of the Erk1/2 signalling cascade triggers transition of pluripotent embryonic stem cells from self-renewal to lineage commitment. Development. 2007;134: 2895–902.
210. Binetruy B, Heasley L, Bost F, Caron L, Aouadi M. Concise review: regulation of embryonic stem cell lineage commitment by mitogen-activated protein kinases. Stem Cells. 2007;25:1090–5.
211. Ito K, Hirao A, Arai F, et al. Reactive oxygen species act through p38 MAPK to limit the lifespan of hematopoietic stem cells. Nat Med. 2006;12:446–51.
212. Jang YY, Sharkis SJ. A low level of reactive oxygen species selects for primitive hematopoietic stem cells that may reside in the low-oxygenic niche. Blood. 2007;110:3056–63.

213. Nagata Y, Moriguchi T, Nishida E, Todokoro K. Activation of p38 MAP kinase pathway by erythropoietin and interleukin-3. Blood. 1997;90:929–34.
214. Uddin S, Ah-Kang J, Ulaszek J, Mahmud D, Wickrema A. Differentiation stage-specific activation of p38 mitogen-activated protein kinase isoforms in primary human erythroid cells. Proc Natl Acad Sci USA. 2004;101:147–52.
215. Verma A, Deb DK, Sassano A, et al. Activation of the p38 mitogen-activated protein kinase mediates the suppressive effects of type I interferons and transforming growth factor-beta on normal hematopoiesis. J Biol Chem. 2002;277:7726–35.
216. Katsoulidis E, Li Y, Yoon P, et al. Role of the p38 mitogen-activated protein kinase pathway in cytokine-mediated hematopoietic suppression in myelodysplastic syndromes. Cancer Res. 2005;65:9029–37.
217. Navas TA, Mohindru M, Estes M, et al. Inhibition of overactivated p38 MAPK can restore hematopoiesis in myelodysplastic syndrome progenitors. Blood. 2006;108:4170–7.
218. Wang M, Crisostomo PR, Herring C, Meldrum KK, Meldrum DR. Human progenitor cells from bone marrow or adipose tissue produce VEGF, HGF, and IGF-I in response to TNF by a p38 MAPK-dependent mechanism. Am J Physiol Regul Integr Comp Physiol. 2006;291:R880–4.
219. Stephens LR, Jackson TR, Hawkins PT. Agonist-stimulated synthesis of phosphatidylinositol(3,4,5)-trisphosphate: a new intracellular signalling system? Biochim Biophys Acta. 1993;1179:27–75.
220. Katso R, Okkenhaug K, Ahmadi K, White S, Timms J, Waterfield MD. Cellular function of phosphoinositide 3-kinases: implications for development, homeostasis, and cancer. Annu Rev Cell Dev Biol. 2001;17:615–75.
221. Foster FM, Traer CJ, Abraham SM, Fry MJ. The phosphoinositide (PI) 3-kinase family. J Cell Sci. 2003;116:3037–40.
222. Brazil DP, Yang ZZ, Hemmings BA. Advances in protein kinase B signalling: AKTion on multiple fronts. Trends Biochem Sci. 2004;29:233–42.
223. Sarbassov DD, Guertin DA, Ali SM, Sabatini DM. Phosphorylation and regulation of Akt/PKB by the rictor-mTOR complex. Science. 2005;307:1098–101.
224. Guertin DA, Stevens DM, Thoreen CC, et al. Ablation in mice of the mTORC components raptor, rictor, or mLST8 reveals that mTORC2 is required for signaling to Akt-FOXO and PKCalpha, but not S6K1. Dev Cell. 2006;11:859–71.
225. Manning BD, Cantley LC. AKT/PKB signaling: navigating downstream. Cell. 2007;129:1261–74.
226. Franke TF. PI3K/Akt: getting it right matters. Oncogene. 2008;27:6473–88.
227. Memmott RM, Dennis PA. Akt-dependent and -independent mechanisms of mTOR regulation in cancer. Cell Signal. 2009;21:656–64.
228. Wang L, Harris TE, Lawrence Jr JC. Regulation of proline-rich Akt substrate of 40 kDa (PRAS40) function by mammalian target of rapamycin complex 1 (mTORC1)-mediated phosphorylation. J Biol Chem. 2008;283:15619–27.
229. Dunlop EA, Tee AR. Mammalian target of rapamycin complex 1: signalling inputs, substrates and feedback mechanisms. Cell Signal. 2009;21:827–35.
230. Huang J, Manning BD. A complex interplay between Akt, TSC2 and the two mTOR complexes. Biochem Soc Trans. 2009;37:217–22.
231. Jacinto E, Loewith R, Schmidt A, et al. Mammalian TOR complex 2 controls the actin cytoskeleton and is rapamycin insensitive. Nat Cell Biol. 2004;6:1122–8.
232. Keniry M, Parsons R. The role of PTEN signaling perturbations in cancer and in targeted therapy. Oncogene. 2008;27:5477–85.
233. Stiles BL. Phosphatase and tensin homologue deleted on chromosome 10: extending its PTENtacles. Int J Biochem Cell Biol. 2009;41:757–61.
234. Kalesnikoff J, Sly LM, Hughes MR, et al. The role of SHIP in cytokine-induced signaling. Rev Physiol Biochem Pharmacol. 2003;149:87–103.
235. Eichhorn PJ, Creyghton MP, Bernards R. Protein phosphatase 2A regulatory subunits and cancer. Biochim Biophys Acta. 2009;1795:1–15.
236. Brognard J, Newton AC. PHLiPPing the switch on Akt and protein kinase C signaling. Trends Endocrinol Metab. 2008;19:223–30.
237. Orford KW, Scadden DT. Deconstructing stem cell self-renewal: genetic insights into cell-cycle regulation. Nat Rev Genet. 2008;9:115–28.
238. Zhang J, Grindley JC, Yin T, et al. PTEN maintains haematopoietic stem cells and acts in lineage choice and leukaemia prevention. Nature. 2006;441:518–22.
239. Yilmaz OH, Valdez R, Theisen BK, et al. Pten dependence distinguishes haematopoietic stem cells from leukaemia-initiating cells. Nature. 2006;441:475–82.
240. Tothova Z, Kollipara R, Huntly BJ, et al. FoxOs are critical mediators of hematopoietic stem cell resistance to physiologic oxidative stress. Cell. 2007;128:325–39.
241. Hazen AL, Smith MJ, Desponts C, Winter O, Moser K, Kerr WG. SHIP is required for a functional hematopoietic stem cell niche. Blood. 2009;113:2924–33.
242. Friedman AD. Transcriptional control of granulocyte and monocyte development. Oncogene. 2007;26:6816–28.
243. Buitenhuis M, Verhagen LP, van Deutekom HW, et al. Protein kinase B (c-akt) regulates hematopoietic lineage choice decisions during myelopoiesis. Blood. 2008;111:112–21.
244. Grech G, Blazquez-Domingo M, Kolbus A, et al. Igbp1 is part of a positive feedback loop in stem cell factor-dependent, selective mRNA translation initiation inhibiting erythroid differentiation. Blood. 2008;112:2750–60.
245. Kozuma Y, Kojima H, Yuki S, Suzuki H, Nagasawa T. Continuous expression of Bcl-xL protein during megakaryopoiesis is post-translationally regulated by thrombopoietin-mediated Akt activation, which prevents the cleavage of Bcl-xL. J Thromb Haemost. 2007;5:1274–82.
246. Fuhler GM, Tyl MR, Olthof SG, Lyndsay Drayer A, Blom N, Vellenga E. Distinct roles of the mTOR components Rictor and Raptor in MO7e megakaryocytic cells. Eur J Haematol. 2009;83:235–45.
247. Mangi AA, Noiseux N, Kong D, et al. Mesenchymal stem cells modified with Akt prevent remodeling and restore performance of infarcted hearts. Nat Med. 2003;9:1195–201.
248. Gnecchi M, He H, Noiseux N, et al. Evidence supporting paracrine hypothesis for Akt-modified mesenchymal stem cell-mediated cardiac protection and functional improvement. FASEB J. 2006;20:661–9.
249. Yau TM, Kim C, Li G, Zhang Y, Weisel RD, Li RK. Maximizing ventricular function with multimodal cell-based gene therapy. Circulation. 2005;112:I123–8.
250. Spiegelstein D, Kim C, Zhang Y, et al. Combined transmyocardial revascularization and cell-based angiogenic gene therapy increases transplanted cell survival. Am J Physiol Heart Circ Physiol. 2007;293:H3311–6.
251. Zhang D, Fan GC, Zhou X, et al. Over-expression of CXCR4 on mesenchymal stem cells augments myoangiogenesis in the infarcted myocardium. J Mol Cell Cardiol. 2008;44:281–92.
252. Furuta A, Miyoshi S, Itabashi Y, et al. Pulsatile cardiac tissue grafts using a novel three-dimensional cell sheet manipulation technique functionally integrates with the host heart, in vivo. Circ Res. 2006;98:705–12.
253. Domian IJ, Chiravuri M, van der Meer P, et al. Generation of functional ventricular heart muscle from mouse ventricular progenitor cells. Science. 2009;326:426–9.

# Chapter 20
# Cardioprotection and Signaling Pathways

**Abstract** Cardioprotection (CP) is a way to guard the heart from damage secondary to different insults, including ischemia, ischemia/reperfusion (I/R), and chemical, metabolic, and physical stressors. Ischemic preconditioning (IPC), by single or multiple brief periods of ischemia, protects the heart against a more prolonged ischemic insult. Although our understanding of CP has increased substantially within the last few years, the precise signaling pathways involved and the mechanisms mediating the role of mitochondria in CP remain to be established. Better understanding of the cellular, molecular, and biochemical events occurring in CP may allow the development of new interventions to improve the outcome of patients with myocardial diseases.

**Keywords** Cardioprotection • Myocardial ischemia • Ischemic preconditioning • Signaling • Kinases

## Introduction

Myocardial ischemia has a large number of effects on cardiovascular physiology. Its lethality most likely stems from its marked perturbation of metabolism ultimately depriving the cardiomyocyte of the bioenergy needed to provide pumping energy and electrical signaling. Early ischemic damage shares with other physiological stresses (e.g., heat-shock) a significant perturbation of the mitochondrial phenotype, including increased organelle swelling, and uncoupling of respiration and oxidative phosphorylation (OXPHOS). Myocardial ischemia also results in the critical depletion of the mitochondrial inner membrane phospholipid cardiolipin (CL), involved in cytochrome *c* insertion, retention, and electron transport function [1]. Decrease in cardiolipin content accompanied by persistent decrements in the content of cytochrome *c* and oxidation through cytochrome *c* oxidase has been considered a potential mechanism of additional myocyte injury during reperfusion [2]. Moreover, the onset of ATP depletion and subsequent cell de-energization found in sustained ischemia results in both necrotic cell death and in signaling apoptosis (programmed cell death) [3, 4].

It is worth noting that most of the available experience with CP has been obtained from studies in young and middle-aged animal and cells. Thus, application of this experience to the aging or senescent human heart may not be relevant since the aging heart has a decreased capacity to tolerate and respond to various forms of stress, and the likelihood of myocardial ischemia and cardiac dysfunction significantly increases. Future studies will address this important variable.

In this chapter, a discussion on the cellular, molecular, and biochemical mechanisms operating in IPC-CP and recommendations for future studies and potential therapeutic interventions is provided.

## Reperfusion and Cardioprotection

Until recently, reperfusion of the ischemic myocardium was the only method available to restore the cellular functions affected by ischemia, including the prevention of cell death (by necrosis or apoptosis). However, reperfusion may result in significant myocardial damage, including myocardial stunning, dysrhythmias, and myocyte cell death; the functional recovery of the heart may occur only after a period of abnormal cardiac contractility that may last hours or days. In an apparent contradiction, functional mitochondria can exacerbate ischemic damage, especially at the onset of reperfusion. During reperfusion, a pronounced increase in fatty acid influx and unbalanced fatty acid β-oxidation (FAO) occur resulting in an excess of acetyl-CoA that saturates the transluminal coronary angioplasty (TCA) cycle at the expense of glucose and pyruvate oxidation that eventually is inhibited. Increased mitochondrial OXPHOS causes reactive oxygene species (ROS) accumulation with associated elevated lipid peroxidation; this results in lower cardiolipin levels in the inner membrane with consequent effect on respiratory enzyme activities.

J. Marín-García, *Signaling in the Heart*,
DOI 10.1007/978-1-4419-9461-5_20, 

It has been suggested that reperfusion injury entails apoptotic cell death, in contrast to ischemic injury which primarily causes necrotic cell death [5]. By the way, since ischemic cell death (in a number of models) is accompanied by swelling, the term oncosis (proposed by von Reckling-hausen in 1910) is often used instead of necrosis [6]. Oncosis (or unprogrammed cell death) leads to necrosis with karyolysis in contrast to apoptosis that leads to necrosis with karyorhexis and cell shrinkage [6]. In general, myocardial ischemic injury relates to both major modes of cell death, oncosis, and apoptosis; these mechanisms of cell death may occur together in ischemic myocytes, with oncotic mechanisms and changes in morphology dominating the end stage of irreversible injury [7]. Timely reperfusion produces significant effects on ischemic myocardium, including a component of reperfusion injury and a greater amount of salvage of myocardium [7]. Interestingly, preconditioning significantly reduces DNA fragmentation and apoptotic myocyte death associated with ischemia–reperfusion; however, the potential mechanisms underlying this effect remain unclear [8].

## Ischemic Preconditioning

Ischemic preconditioning (IPC), by single or multiple brief periods of ischemia, protects the heart against a prolonged ischemic insult (index ischemia). Chemical, metabolic, and even physical stressors are also able to generate cardioprotective reactions that share aspects of the IPC model and in these reactions mitochondria are key factors. Besides regulating the cellular bioenergetic supply in the form of adenosine triphosphate (ATP) the organelle plays many other critical roles (see later section on "Mitochondrial End Effectors of IPC Cardioprotection").

Two distinct pathways of IPC cardioprotection (CP) have been shown: (1) Acute protection that result from brief periods of ischemia, applied 1–2 h before a longer ischemic insult, occurs within a few minutes after the initial stimulus, lasting for 2–3 h and is considered the classical preconditioning model. This early phase of IPC protects against myocardial infarction (MI), but not myocardial stunning; (2) Delayed preconditioning, often referred to as a second window of protection (SWOP), appears about 12–24 h after the preconditioning event and lasts several days [9]. The delayed preconditioning pathway is generally recognized as having greater clinical relevance, with a longer protective phase, and greater effectiveness against both MI and stunning.

First, demonstrated by Murry in a canine model [10], IPC has subsequently been successfully carried out in other animal studies. A variety of interventions and cardioprotective drugs have been tested with these animal models (Table 20.1) in an effort to identify the mechanisms involved in CP.

**Table 20.1** Pharmacological approaches to cardioprotection

| Class | Specific drug/chemical |
|---|---|
| Potassium channel openers | Nicorandil |
| | Diazoxide |
| | Pinacidil |
| | Cromakalin |
| | Hydrogen sulfide |
| | Bromoenol lactone (BEL) |
| Potassium channel blockers | Glibenclamide |
| | 5-HD |
| Inhibitors of the MTP pore | Cyclosporin A |
| Receptor-mediated signaling pathways/ligands | Adenosine |
| | Opioids |
| | Bradykinin |
| | Acetylcholine |
| | Endothelin |
| | PKC-TIP |
| | Tyrosine kinases |
| Sphingolipid/ceramide signaling | Ceramide |
| | Sphingosine |
| | Chelerythrine (Pan–PKC inhibitors) |
| Mitochondrial-generated ROS and lipid peroxidation | Idebenone |
| | FCCP |
| | DNP |
| | Coenzyme Q |
| | Quercetin |
| | Carvedilol |
| Other mitochondria-mediated products | MnSOD induction |
| | Mitochondrial $Ca^{2+}$ levels |
| | Lipophilics |
| | Estrogen |
| Other pharmacological agents | Erythropoietin (EPO) |
| | Statin (avorstatin) |
| | Glucose-insulin-potassium (GIK) |
| | Insulin |
| | Nitroglycerin |
| | Pyruvate |
| | Glucagon-like peptide |
| | Sidenafil |
| | Monophosphoryl lipid A |
| | Amobarbital |

*DNP* 2,4-Dinitrophenol, *FCCP* Carbonyl cyanide p-trifluoromethoxyphenylhydrazone

IPC-like effects have also been observed in cultured cells exposed to hypoxia and metabolic inhibition, and can reverse defects in both the functioning and the structure of mitochondria in a manner similar to animal studies. While IPC has been demonstrated in every specie in which it has been tested, a number of conflicting observations suggest that significance of interspecific differences may underlie the cellular and molecular mechanisms of CP.

The duration of the IPC, the index ischemia and the subsequent reperfusion are critical determinants of both the signaling pathways involved as well as the cardioprotective outcome. The transient preconditioning state has been described for periods ranging from one cycle of ischemia/reperfusion (I/R) of 1.25 min to five cycles of 5 min-ischemia/5 min-reperfusion [11]. When the duration of the index ischemia was extended beyond 3 h (with no intermediary reperfusion), CP was abolished. This suggests that reperfusion after a damaging (but not prolonged) index ischemia is required for CP [10]. These observations also strongly suggest that IPC delays the lethal effects, i.e., cell death, of ischemia rather than completely preventing it.

## Cellular and Molecular Events in IPC

While making the heart resistant to infarction, IPC triggers changes in the physiology of the heart, including signal transduction pathways that carry the signal for CP and subsequently converge on one or more end-effectors. Triggers exert their action prior to the index ischemia. After the index ischemia has been initiated, mediators convey the signal to the end-effectors which actually promote CP during the lethal ischemic insult (index ischemia) and/or the subsequent reperfusion period. There is evidence that somewhere within the signal transduction pathways between the trigger signal and the end-effector resides a memory element that is set by the preconditioning protocol and which keeps the myocardium in a preconditioned state.

### *Triggering Early IPC*

Adenosine and its G protein coupled receptors represent an important triggering stimulus, as well as a locus for feedback control of IPC [12]. Adenosine is important trigger of IPC, and one of three autacoids (adenosine, opioids, and bradykinin) released by the ischemic tissue. Adenosine exerts effects through adenosine receptor subtypes: A(1)AR, A(2A)AR, A(2B)AR, and A(3)AR; all are expressed in myocardial and vascular cells and couple to G proteins to trigger a range of responses (sometimes beneficial) [13].

Although our understanding of the role that adenosinergic pathways play in CP is still evolving, some investigators believe that the A(2) subtypes are important factors in the mediation of CP [13]. Others suggest that mainly the A(1) and to some extent A(3) receptors participate in the intracellular signaling that triggers CP [14]. During myocardial ischemia, adenosine is generated at high levels from ATP metabolism.

IPC-induced CP is abrogated by the treatment with adenosine receptor antagonists suggesting that adenosine produced during IPC, acting on cell-surface receptors is a critical event in triggering IPC [15]. Subsequent observations extended these findings after the discovery that intravenous administration of adenosine A(1) receptor agonists could be substituted for the IPC regimen resulting in the same CP, an early example of pharmacological preconditioning [16, 17]. Other triggering ligands have been identified, including opioids, bradykinin, norepinephrine, and endothelin, which the heart can release during a brief ischemic period. They bind specific G protein coupled receptors, and elicit IPC-mediated CP. Interestingly, unlike bradykinin and opioids, adenosine is not dependent on the opening of mitochondrial ATP-sensitive potassium channels (mito$K_{ATP}$) or the release of ROS, but rather activates phospholipase C (PLC) and/or protein kinase C (PKC) directly. As it will be discussed later, another signaling cascade during reperfusion involves activated PKC which initiates binding to and activation of an A(2) adenosine receptor. Cohen and Downey [14] have reported that the adenosine receptor is the A(2B) and although this is the low-affinity receptor, its interaction with PKC increases and its affinity makes it responsive to the accumulated tissue adenosine. A(2B) agonists, but not adenosine or A(1) agonists, infused during reperfusion can initiate this second signaling cascade mimicking the protection of preconditioning. Moreover, A(2B) receptors are critical for the protection of postconditioning. Taken together, adenosine is both an important trigger and a mediator of CP. At least three surface receptors acting in parallel can trigger preconditioning [14]; receptors that are thought to represent parallel and redundant pathways [18]. Whether these multiple receptor pathways act in concert or synergistically may have significant clinical impact on the development of more effective therapeutic approach, e.g., using smaller drug dosages thereby avoiding potential toxicity and adverse side-effects. Ligand occupation of these specific receptors triggers different signaling pathways, and binding of specific ligands to their membrane-bound receptors can lead to the subsequent activation of PLC and production of diacylglycerol (DAG) which in turn stimulates the pivotal signal transducer PKC.

Evidence has implicated tyrosine kinases as signaling cardioprotective responses in conjunction with nonadenosine receptors, as well as the involvement of sphingolipid signaling pathway in IPC-mediated CP [19, 20]. As a consequence of the receptor redundancy, blockade of a single receptor type often does not completely block CP, but rather raises the ischemic threshold required for protection [21].

ROS and nitric oxide (NO) are triggering elements in the early IPC pathway [22]. Treatment with a free radical scavenger (e.g., N-2-mercaptopropionylglycine or MPG) can raise the threshold of preconditioning abolishing CP from a single cycle of IPC but not from a more extensive four cycle regimen [23]. Moreover, in isolated rabbit hearts, exposure for 5 min to a free radical generator (i.e., xanthine oxidase) prior to index ischemia and reperfusion, could trigger a preconditioned state based on its subsequent cardioprotective effect [24]. There is evidence that ROS directly activate protective kinases [25–28], and trigger the activation of the mito$K_{ATP}$ channel as a cardioprotective mediator [22], resulting in further generation of mitochondrial-produced ROS; however, the precise source of the triggered ROS has not yet been established.

While the role of endogenous NO in IPC is not clear, there is evidence that exogenous NO (from administration of an NO donor) has a role in CP [29–31]. Although the involvement of NO as an early IPC triggering event remains controversial, some observations suggest that ROS and NO are both required for preconditioning protection [22]. Also, exogenous NO (derived from perfusion with the NO donor S-nitroso-N-acetylpenicillamine [32]) administered to isolated rabbit hearts, prior to I/R, showed CP similar to that with IPC [33]. CP is PKC dependent; however, no evidence for endogenous NO generation or triggering was found. Other experimental studies in either rat or swine model have shown that endogenous NO synthesis does not play a contributory role to early IPC [34, 35]. Findings from the endothelial nitric oxide synthase (eNOS) knock-out mice revealed that the loss of endogenous NO had no obvious effect on the development of IPC cardioprotection, with higher levels of IPC. Lower levels of IPC in the eNOS null-mice (that were effective in the development of CP in the wild-type mice) resulted in significant attenuation of CP [36]. Taken together, these findings suggest that eNOS or endogenous NO are not likely required for robust early IPC, although NO may contribute to early IPC by lowering the ischemic threshold for protection. Gathered data after interruption of the cardioprotective effect of IPC by L-Nitro-Arginine Methyl Ester (L-NAME) suggested a dual role for NO in the heart; (1) a negative role in the nonadapted myocardium subjected to ischemia–reperfusion, and (2) a positive one, due to its involvement in the protection triggered by short-term cardiac adaptation by preconditioning [37].

## Mediators of Early IPC

A broad array of signal transducing elements has been identified as mediators within the cardioprotective pathway. Several intracellular channels have been implicated as transducers located at both the cell membrane (e.g., sarcolemmal $K_{ATP}$ (sarc$K_{ATP}$) channels) and within the mitochondrial membranes (e.g., mito$K_{ATP}$ channels) [32]. In acute CP, a primary mechanism by which the activation of the trigger/mediator signaling pathways leads to the activation of the end effectors appears to involve posttranslational modification, such as the phosphorylation mediated by kinases. Despite a wealth of information concerning the involvement of these cardioprotective transducing elements, the sequence of events in the intracellular cascades remains unclear. A diagrammatic representation of the major signaling components in early (or acute) IPC pathway is shown in Fig. 20.1.

## The Phosphatidylinositol 3-Kinase Pathway

One of the earliest mediators of ischemic CP (triggering a variety of other mediating events) is phosphatidylinositol 3-kinase (PI3K) [38]. The activation of the PI3K pathway is cardioprotective in a number of experimental preconditioning models. Moreover, IPC leads to the activation of protein kinase B (also called Akt) and kinase p70S6k downstream of PI3K, consistent with a prominent role for PI3K in IPC [39, 40]. In an isolated nonworking rat heart model, transient preischemic exposure to nanomolar concentrations of a β1-adrenoceptor agonist was protective against I/R, and PI3K, PKC, and protein kinase A (PKA) appear to be involved in the trigger phase of PC [41].

Inhibitors of PI3K, such as wortmannin and LY294002, attenuate the CP afforded by IPC [39, 40]. In addition, in isolated perfused rat heart, a standard IPC stimulus induced Akt activation and Akt-mediated phosphorylation of its downstream target glycogen synthase kinase 3β (GSK3β) immediately before the index ischemic period [40]. To the best of our knowledge no specific inhibitor is readily available to block Akt in IPC, and then it is unclear whether the activation of Akt, which has been shown to accompany IPC, is in fact required for IPC. Parenthetically, it has been recently found that global ischemia in intact rat brain triggered expression and the activation of the Akt inhibitor carboxyl-terminal modulator protein (CTMP) in vulnerable hippocampal neurons and that CTMP bound and extinguished Akt activity and was essential to ischemia-induced neuronal death [42].

PI3K activation occurs after the triggering ligands (e.g., adenosine, bradykinin, or opioids) bind the G protein coupled receptors and initiates a signaling cascade involving PI3K activation. Krieg et al. [43] have reported that the activation of the G protein coupled receptor in response to acetylcholine-induced IPC leads to the transactivation of the epidermal growth factor receptor (EGFR) and associated tyrosine kinases which then activate the PI3K-Akt (see Fig. 20.1).

## Protein Kinase C

Studies in a variety of preconditioning models have shown that PKC activation is essential for IPC. Blockade of PKC with PKC inhibitors eliminated the CP resulting from IPC [44, 45]. Furthermore, the activation of PKC by phorbol esters can mimic IPC in the isolated rabbit heart [45].

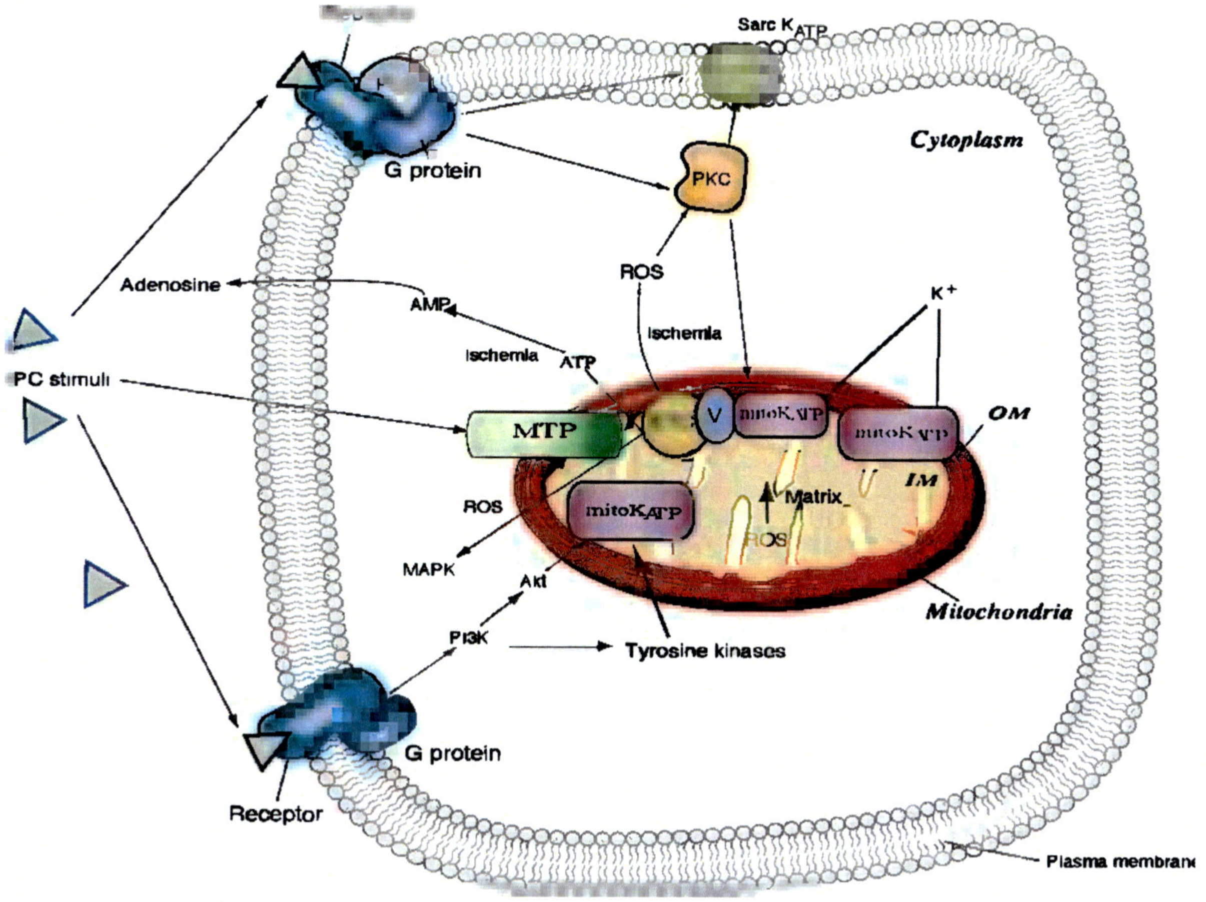


**Fig. 20.1** Schematic representation depicting cellular events occurring in early ischemic preconditioning. Adenosine, produced by the breakdown of adenine nucleotides during ischemia, leads to the activation of phospholipase C (PLC) which in turn activates and translocates protein kinase C (PKC) to target mitochondrial membranes modulating the opening of mitochondrial $K_{ATP}$ and to a lesser extent, sarcolemmal $K_{ATP}$ channels. Other agonists of G protein coupled receptors also stimulate parallel signaling pathways with either PKC-dependent or tyrosine kinase-dependent and phosphatidylinositol 3-kinase (PI3K) cascades. Multiple mitochondrial events are integral to cardioprotection, including increased mitochondrial ROS generation, modulation of permeability transition pore (MTP pore) opening, and mitochondrial ETC activity, albeit the temporal sequence of these events is presently undetermined. Mito$K_{ATP}$ channel opening also provides positive feedback by altering components, such as ROS, mitogen-activated protein kinase (MAPK), or PKC activation

A serine/threonine kinase activated by lipid cofactors, derived from the breakdown of membrane lipids by PLC, PKC has multiple isoforms in the heart, each of them with similar substrate specificity. PKC isoforms achieve specificity by their physical translocation to specific docking sites located on specific intracellular organelles. The activation of the PKC isoforms at those sites results in further mediation of the cardioprotective signal, although the precise intracellular targets have not yet been identified. Experimental studies have shown potential contributory roles in IPC for PKCδ and PKCα isoforms and a primary role for myocardial PKCε. While data is available in support of a relationship between PKCδ and delayed IPC [46, 47], a positive role for PKCδ in early IPC appears questionable [48–50]. Limited observational data are available regarding the translocation of PKCα to sarcolemmal membrane in IPC [51]. In contrast, in early (acute) IPC a positive role for PKCε has been documented in several studies [49, 50, 52, 53]. Targeted disruption of the PKCε gene in isolated perfused mouse hearts abolished the cardioprotective infarct size reduction resulting from IPC [54]. Specific activation and intracellular translocation of PKCε during IPC has been demonstrated. Interestingly, PKCε translocates to the mitochondria, where it interacts with mitogen-activated protein kinase (MAPK) in modular functional signaling complexes. Furthermore, PKCε-MAPK complex formation can lead to the activation of mitochondrial signaling extracellular signal-regulated kinases (Erks) and subsequently to the phosphorylation of the proapoptotic Bad protein which could reduce or delay cell death [55]. Other potential mitochondrial targets for PKCε that may be affected in preconditioning, both as mediators and end-effectors of the pathway, include the mitochondrial permeability transition pore (MTP) and the mito$K_{ATP}$ channel. A direct protein–protein interaction between PKCε and several components of the MTP [e.g., voltage-dependent anion channel (VDAC), adenine nucleotide translocator (ANT), and hexokinase II] has been reported by Baines and associates [56]. This interaction may inhibit the pathological opening of the pore (including $Ca^{2+}$-induced opening and subsequent mitochondrial swelling) contributing to PKCε-induced CP. While such a direct physical relationship between PKCε and the mito$K_{ATP}$ channel has not yet been validated, several observations suggested that myocardial PKCε is involved in IPC upstream of mito$K_{ATP}$ channels [57, 58].

PKC can be activated by PI3K [38, 40]. Wortmannin, the PI3K inhibitor was shown to block the PC-induced translocation of PKC. The precise role of PI3K in the activation of PKC is not known, but phosphoinositide products have been reported to activate several isoforms of PKC [59]. Okada et al. [60] have reported that an interplay between PI3K and PKC activation may be involved in bringing about cardioprotective memory.

## Tyrosine Kinases and MAP Kinases

The activation of a tyrosine kinase pathway appears to be recruited in acute IPC as an early mediating event. This pathway has been reported by some investigators to be downstream of the PKC pathway [61] while others have assigned it as a parallel pathway [62, 63]. There is also evidence that subfamilies of the MAPKs, including p42/p44 Erk1/2, the 52–54 kDa c-JUN kinase, and the 38 kDa p38 MAPK, may be involved in IPC. The role of Erk1/2 as a potential signaling component of IPC is controversial, with some observations showing that classic IPC results in Erk1/2 activation that occurs before the index ischemic period [64], while other studies report no changes in Erk1/2 phosphorylation in response to an IPC [39]. Most evidence supports a downstream role for p38 MAPK, which we will discuss later.

## IPC, ATP-Sensitive Potassium Channels and Potassium Channel Openers

Another essential component of the early cardioprotective pathway in IPC is the ATP-sensitive potassium channel ($K_{ATP}$). This channel opens when the intracellular levels of ATP decline as occurs in ischemia. Sarcolemmal $K_{ATP}$ (sarc$K_{ATP}$) channels were initially thought to be implicated in IPC but actually the focus has shifted to the mitochondrial $K_{ATP}$ (mito $K_{ATP}$) channels. Data about IPC and the $K_{ATP}$ channels has also been derived from the use of potassium channel openers (KCOs), such as nicorandil, diazoxide, pinacidil, and cromakalin, and also from the use of $K_{ATP}$ channel blockers (e.g., glibenclamide and 5-hydroxydecanoic acid [5-HD]), which diminish the beneficial effects on cardiac tissue [65] of short ischemic events [66]. Garlid et al. [67] reported that the KCO diazoxide was 1,000–2,000 times more potent in opening the mito$K_{ATP}$ channel than the sarc$K_{ATP}$ channel. The negative effect of 5-HD that abolishes diazoxide's protective effect entirely, further suggested that mito$K_{ATP}$ mediates IPC, since 5-HD is highly selective in interacting with the mito$K_{ATP}$ as compared to the sarc$K_{ATP}$.

While 5-HD treatment has shown to modulate mitochondrial respiration and FAO independently of its interaction with the mito$K_{ATP}$ channel [68], further experimental data support the view that both mitochondrial and sarc$K_{ATP}$ channels have contributory roles in CP [69, 70]. Under specific conditions, diazoxide and 5-HD can modulate sarc$K_{ATP}$ channel opening [69], whereas mito$K_{ATP}$ can bind molecules previously thought to interact solely with sarc$K_{ATP}$ channels, and in some species (e.g., dogs) both the mitochondrial and sarc$K_{ATP}$ channels must be blocked to entirely abolish IPC cardioprotection [70]. Another ion channel previously thought to be exclusive of the myocardial sarcolemmal membrane (i.e., the calcium-activated $K^+$ channel) has been located on the mitochondrial inner membrane and found to be functionally cardioprotective against MI [70].

Biochemical structural analysis of the proteins comprising the $K_{ATP}$ channels has been carried out with functional channels isolated from both cardiac sarcolemmal and mitochondrial membranes [71, 72]. The channel activities of these preparations derived from either heart or brain have been reconstituted in proteoliposomes and shown to be regulated by the same ligands as *in vivo*. While the sarc$K_{ATP}$ channels are heteromultimeric complexes of sulfonylurea receptors (SUR) and potassium inward rectifier (Kir) gene products [73], the precise molecular composition of the mito$K_{ATP}$ channel remained to be identified [74]. Using immunoblot analysis, there was no evidence for the presence of the known sulfonylurea receptors (SUR1 or SUR2) in mitochondria [75]. SUR1 subunits are strongly expressed at the sarcolemmal surface of ventricular myocytes (but not in the coronary vasculature), whereas SUR2 protein has been predominantly located in cardiomyocytes and coronary vessels (mostly in smaller vessels). Immunocytochemistry of isolated ventricular myocytes shows colocalization of Kir6.2 and SUR2 proteins in a striated sarcomeric pattern, suggesting t-tubular expression of these proteins [76].

While there is evidence for the presence of Kir6.1 and Kir6.2 in cardiac mitochondria [77], gene knock-out studies demonstrated no discernible functional role for these proteins in the mito$K_{ATP}$ channel [74]. Upon reconstitution into proteoliposomes and lipid bilayers a multiprotein complex can be purified from the mitochondrial inner membrane that functions as a mito$K_{ATP}$ channel, fully sensitive to channel activators and blockers [78]. This complex contains five mitochondrial proteins, including ANT, the mitochondrial ATP-binding cassette protein 1 (mABC1), the phosphate carrier (PIC), ATP synthase, and succinate dehydrogenase (SDH). However, the pore-forming component of the mito$K_{ATP}$ channel remains undetermined as do the potential identity of other proteins within this complex. The identity of major bioenergetic proteins (ATP synthase and SDH) within the mito$K_{ATP}$ channel and their overlap with structural components of the MTP (e.g., ANT) have important ramifications in the mechanism of CP and its relatedness to both bioenergetic function and apoptosis regulation. However, the lack of a defined structure for the mito$K_{ATP}$ channel has led to question both its significance and even its existence.

A number of the pharmacological treatments used to promote CP may have compromised "specificity" (mainly when used at high dosage) impacting on more than a single cellular event, thereby complicating the interpretation of their action. Some observations have shown that diazoxide and 5-HD have distinctive effects on mitochondria, independent of the mito$K_{ATP}$ channel targets [79–82], making the evaluation of upstream and downstream events difficult. For example, respiratory complex II (SDH) activity appears to be directly inhibited by some KCOs (e.g., diazoxide) [83] which may make difficult the interpretation of experiments using diazoxide to elicit mito$K_{ATP}$ channel opening (a standard procedure in this field). This finding, however, takes on added significance with the suggestion that SDH may be a primary component of the mito$K_{ATP}$ channel [78]. This is also consistent with the uncovering that specific SDH inhibitors, such as 3-nitropropionic acid (3-NPA) can result in increased $K^+$ transport by the mito$K_{ATP}$ channel and can mimic the CP provided by IPC [78, 84]. Moreover, $K_{ATP}$ channel blockers, which entirely abolish the cardioprotective effect of diazoxide, reverse the effect of complex II inhibitors.

## Mitochondrial Events in IPC

Treatment with KCOs causes a reduction in mitochondrial $Ca^{2+}$ influx [85, 86]. Korge et al. [85] have carried out experiments with isolated mitochondria which are anoxic and "de-energized" simulating the ischemic state; under these conditions, a mild depolarization of the mitochondrial membrane potential occurs that prevents the uptake into the matrix of extramitochondrial $Ca^{2+}$ (preventing mitochondrial $Ca^{2+}$ overload which could ensue during ischemia or reperfusion), and also inhibits the induction of the MTP. If isolated mitochondria are previously loaded with $Ca^{2+}$ and subsequently depolarized (with diazoxide), the MTP is opened [87, 88]. Moreover, the opening of MTP occurs in the first minutes of reperfusion following ischemia [89], in association with elevated levels of mitochondrial $Ca^{2+}$, increased oxidative stress (OS) and increased matrix pH [90]. The opening of the MTP allows $H_2O$ and solutes to enter the mitochondria, increasing matrix volume and rupturing the outer membrane and might result in cell death by either apoptosis or necrosis. Inhibition of the MTP opening during reperfusion by cyclosporin A (CsA) has been found to be cardioprotective [91].

IPC not only protects against necrotic cell death but blunts the progression of myocyte apoptosis. Apoptosis occurring as a result of I/R, is reperfusion-triggered [92]; this may be because reperfusion rapidly restores intracellular ATP levels, which are required to allow the progression of the apoptotic pathway. Reduction of apoptosis by IPC is elicited by the inhibition of inflammatory cell activation and by altering the expression of apoptogenic proteins (e.g., Bax) and PKC activity. Early hallmark events of apoptosis occur at the mitochondrial inner membrane and include opening of the MTP and the release of cytochrome *c* from mitochondria; upon CP, cytochrome *c* release is inhibited along with MTP opening, and the $Ca^{2+}$ flux into mitochondria is stemmed. Interestingly, the IPC-modulation of myocardial apoptotic events has been documented in early but not delayed IPC [92].

The connection of mitochondrial respiratory activity and ATP levels to IPC and to the opening of mitochondrial $K_{ATP}$ channels is considered to be critical in the regulation of myocardial bioenergetics and in the survival of the ischemic cardiomyocyte. A modest but significant level of uncoupling of respiration and OXPHOS occurs with IPC [93]. Under certain circumstances, uncoupling of OXPHOS allows fine-tuning of the cellular bioenergetic efficiency based on the cell needs, and it has been considered a primary *raison d'etre* for mitochondrial uncoupling proteins. The importance of uncoupling is supported by observations where treatment with uncouplers (e.g. 2,4-dinitrophenol) mimics IPC [48] [91]. Moreover, it has been demonstrated in myocytes, that a critical determinant in the opening of both sarc$K_{ATP}$ and mito$K_{ATP}$ channels is the cytoplasmic ATP pool, produced mainly by mitochondrial OXPHOS. Either uncoupling of mitochondrial OXPHOS or inhibition of cyclooxygenase (COX) results in the opening of cardiac $K_{ATP}$ channels, even in the presence of glucose [94]. Modulation of COX activity by NO should be expected to directly affect the opening of $K_{ATP}$ channels and it may contribute to the effect of NO on IPC [95]. In addition, increased myocardial ATP levels have been shown in rat hearts treated with IPC and with diazoxide (in comparison to levels of untreated animals) [96]. IPC and KCO treatment inhibits cardiac mitochondrial ATPase and reduce ATP depletion during the early stages of sustained ischemia, and this effect persists during the critical early phase of reperfusion [97]. The inhibition of ATP hydrolysis allows the conservation of high-energy phosphates, improve the energy state of the heart during ischemia and contribute to postischemic recovery. Similarly, in isolated mitochondria subjected to anoxia, diazoxide if present during the anoxic insult preserves ADP-stimulated respiration and ATP levels, relative to untreated anoxic-stressed mitochondria [98]. However, mitochondrial ATP depletion and decreased mitochondrial ATPase activity are improbably responsible for the cardioprotective effects of IPC; support for this comes from experiments showing that ATP depletion can remain unchanged and even be enhanced during IPC cardioprotection [99, 100].

The effect of ATP differential compartmentation within the cardiomyocyte may also contribute to CP since subsarcolemmal ATP would be more readily available for membrane transporters and ion pumps needed for cardiomyocyte function [93].

It is generally accepted that mito$K_{ATP}$ channels close when mitochondrial ATP level is high, and open under conditions of low ATP levels (responding to ATP/ADP ratio, as well as to localized changes in adenosine content derived from adenine nucleotide metabolism). Adenine nucleotide translocator, VDAC, and creatine kinase play important roles in the control and cellular distribution of mitochondrial ATP and also as components of the MTP [101]. The opening of mito$K_{ATP}$ channels, either by KCOs or by ischemic preconditioning, modulates the MTP opening; although it is not clear whether the MTP is an upstream target of the cardioprotective stimuli or a downstream end-effector [55]. Moreover, the role of $K_{ATP}$ channel opening in mediating VDAC activity and mitochondrial outer membrane permeability, to regulate both the efflux and influx of mitochondrial ATP, may be critical in CP although it remains to be clarified. On this point, it is noteworthy that VDAC is a pivotal control site for triggering myocardial apoptosis, with high affinity for both the proapoptotic and the antiapoptotic proteins of the Bcl-2 family and kinases, such as PKCε [55]. Another important aspect of respiratory and OXPHOS regulation is the critical role played by ROS generation (a by-product of OXPHOS), which as discussed below represents a critical downstream mediator of preconditioning. ROS induces a significant decrease in mitochondrial NADH-supported respiration, which can be reproduced by pharmacological inhibition of the activity of complex I, a primary site of mitochondrial ROS generation [102].

IPC and treatment with KCOs significantly increase cardiomyocyte mitochondrial volume, probably due to the influx of $K^+$ [103]. In isolated mitochondria, $K^+$-induced mitochondrial swelling was not necessary for the increased ROS production mediated by diazoxide, suggesting that diazoxide may modulate ROS by a mechanism independent of the $K^+$ channel opening [98]. Related to mito$K_{ATP}$ channel opening is the preservation of normal low mitochondrial outer membrane permeability to nucleotides and cytochrome *c*, beneficial effects abrogated by the mito$K_{ATP}$ channel inhibitor 5-HD [104]. The opening of mito$K_{ATP}$ channels during ischemia may contribute to the tight structure of the intermembrane space needed to preserve low outer membrane permeability to adenine nucleotides during ischemia. Thus, mito$K_{ATP}$ channel opening functions as both a signaling trigger and a mediator/effector of the CP pathway [105].

## Cardioprotection

Several conflicting findings have been noted in CP: (1) Collected experimental data frequently do not translate among different models or species; (2) There are conflicts in regard to the common definition of cellular events by using a variety of pharmaceutical reagents (e.g., inhibitors of specific protein kinases, ROS scavengers and mimetics, channel openers and inhibitors, metabolic uncouplers, etc.). In addition, signaling protein kinases have multiple isoforms, some with demonstrably opposing effects on CP, as well as markedly different affinities for the inhibitors used [106]. Moreover, some of the pharmacological treatments used, in particular at high dosage, compromises "specificity," by impacting on more than a single cellular event, complicating the interpretation. Both diazoxide and 5-HD have distinctive effects on mitochondria, independent of targeting mito$K_{ATP}$ channel [79–82], making the pathway evaluation of upstream and downstream events rather difficult. For example, respiratory complex II (SDH) activity appears to be directly inhibited by some KCOs, e.g., diazoxide [83], which may complicate the interpretation of experiments utilizing diazoxide to elicit mito$K_{ATP}$ channel opening (a standard procedure in this field). This finding, however, takes on added significance with the discovery that SDH is a component of the mito$K_{ATP}$ channel [78]. This is also consistent with the finding that specific complex II inhibitors, such as 3-NPA can result in increased $K^+$ transport by the mito$K_{ATP}$ channel, and can mimic the CP provided by ischemic preconditioning (IPC) [84]. Moreover, $K_{ATP}$ channel blockers which entirely abolish the cardioprotective effect of diazoxide, also reverse the effect of complex II inhibitors [78]. The use of anesthesia to generate CP in some animal models has been put into question. Presumably, through targeting of mito$K_{ATP}$ channels, several anesthetics have marked effect on CP [107], as well as on several mitochondrial metabolic processes, including the reduction of mitochondrial complex I activity [108] and mitochondrial membrane potential [66]. Conflicting results have also emerged when comparing studies of isolated mitochondria from cultured cells, organs, or whole animal. Another confounding variable that may affect CP studies is gender, as suggested by a report showing that testosterone can induce cytoprotection by specifically activating the cardiac mito$K_{ATP}$ channel in rat cardiomyocytes [109]. However, more information is needed to confirm these findings in *in vivo* heart and to also assess their clinical implications.

### *ROS and CP*

ROS is involved in CP. For example, diazoxide-mediated CP results in ROS production, and blocking the production of ROS by antioxidants blunts CP [110]. Moreover, it has been established that mitochondrial ROS production occurs downstream and as a consequence of mito$K_{ATP}$ channel opening [111]. A further link between the signaling pathways, ROS generation and mitochondrial bioenergetic function has been confirmed by the discovery that the second messenger, ceramide directly targets complex III activity, which is a pivotal site of mitochondrial ROS generation [111].

Interestingly, ROS function in the CP pathway(s) as a downstream intracellular signal or second messenger which can lead to further protein kinase and G protein activation, the activation of the nuclear poly (ADP-ribose) polymerase (PARP) or direct modulation of the mito$K_{ATP}$ channel [112–114]; however, the increasing ROS accumulation that occurs in ischemic cardiomyocytes in response to reperfusion is reduced either by diazoxide or by hypoxic preconditioning [114, 115]. Furthermore, ROS levels are reduced in preconditioned tissues following index ischemia at the time of reperfusion, which is considered to be a consequence of preconditioning preventing MTP opening [116].

These findings characterized ROS as dual-sided in CP, both in signaling or triggering preconditioning and mediating ischemic cell death (which CP can stem). While it is accepted that the initial signaling mechanism of ROS generation lies downstream of mito$K_{ATP}$ channel opening, further research is necessary to precisely delineate its temporal location in the series of events that comprise the cardioprotective pathway, and to also outline how ROS generation occurs as a consequence of $K_{ATP}$ channel opening [117].

## Early and Late IPC Pathways

Previously, we have described the two distinct pathways of IPC cardioprotection, the acute and the delayed preconditioning, or SWOP [109]. Delayed preconditioning has generally been recognized as having greater clinical relevance since the protective phase is longer and appears to be effective against both MI and stunning.

While delayed preconditioning involves a similar spectrum of stimuli (e.g., adenosine, opioids), involvement of their receptors and transducers (e.g., the mito$K_{ATP}$ channels, protein kinases) as well as pathophysiological mediators (e.g., ROS) utilized by the acute IPC pathway, there appears to be a greater complexity in its regulation because a number of effectors of delayed CP need to be synthesized entirely de novo, a response involving the coordinated regulation of multiple genes expression.

Earlier, it was thought that the acute preconditioning pathway did not involve protein synthesis since blockade at either the transcription or translational levels had no effect on IPC. However, there is evidence that actinomycin D-transcriptional inhibition abolished IPC-induced CP, with marked decreases in the activation of several MAP kinases and subsequent transcription and phosphorylation of specific transcription factors [118]. Unlike the acute protection pathway, the delayed IPC pathway appears to be more dependent on the involvement of NO and its synthase, heat-shock proteins (e.g., HSP27, HSP72), mitochondrial superoxide dismutase (MnSOD), and cytokine induction [119–123] (Fig. 20.2) than the acute protection pathway, although NO involvement in concert with the acute protection pathway has been recently reported [22]. The ability of NO to act in concert with ROS signaling (generated in the CP after cell exposure to KCOs and other preconditioning stimuli), may be a critical factor in targeting the mitochondrial respiratory chain and modulating mito$K_{ATP}$ channels to provide protection against ischemia [124]. This is supported by data showing that: (1) inhibitors of NOS abrogate delayed IPC-mediated protection [122]; (2) NO donors can mimic IPC in conferring CP [124] and (3) free radical scavengers can reduce NOS activation, NO generation, and the induction of cardioprotective signaling [9].

As in early preconditioning, delayed preconditioning also promotes CP by increasing cardiomyocyte bioenergetic capacity in response to ischemic and anoxic insult. The modulation of mitochondrial respiration in response to delayed preconditioning has been shown to occur primarily as a consequence of the induction of electron transport chain (ETC) and ANT transcription, promoted by the ROS-dependent upregulation of nuclear transcription factors NRF-1 and PGC-1α.

Mitochondria play a central role in IPC and CP. Besides the contribution of the mito$K_{ATP}$ channels, mitochondrial-generated ROS and ETC induction in the events of both types of preconditioning, a mitochondrial-specific antioxidant response has also been demonstrated, mainly in the events of delayed preconditioning [120]. The time course of MnSOD induction, a mitochondrial-specific protective response to ROS mediated damage, correlates well with the appearance of ischemic tolerance. In contrast, the activities of other cellular antioxidants [i.e., catalase and cytosolic superoxide dismutase (CuSOD)] were not affected in a similar way. Induction of MnSOD in delayed IPC is further supported by observations that treatment with antisense oligonucleotides to MnSOD, immediately after IPC, completely abolished both IPC cardioprotective effects and MnSOD induction [120]. Likely, MnSOD induction occurs downstream of ROS generation and ROS-mediated the activation of specific transcription factors (e.g., NF-κB), and subsequent induction of cytokines (e.g., TNF-α and IL-1β) [20, 125]. Interestingly, the myocyte intracellular pathways affected by TNF-α

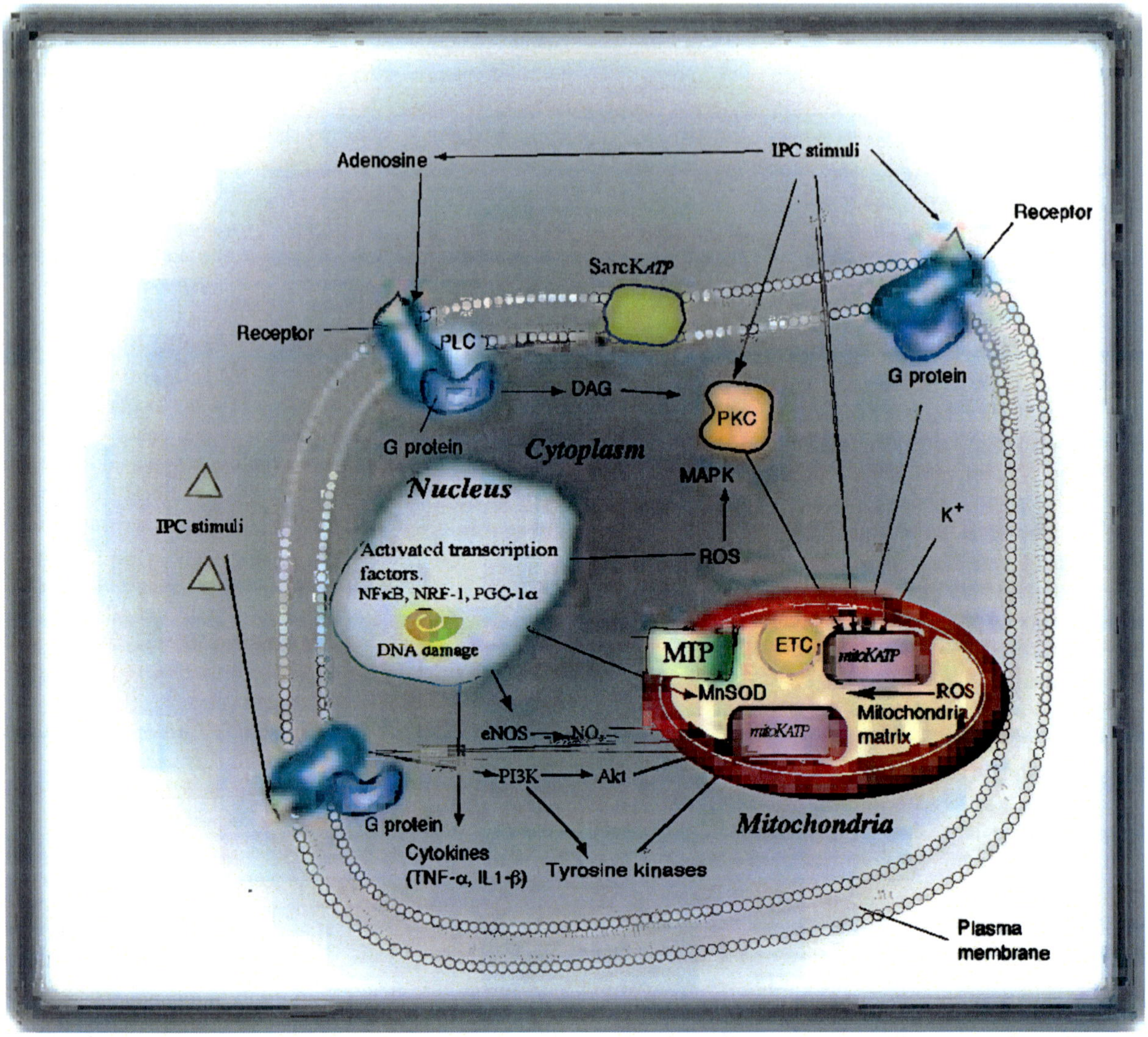


**Fig. 20.2** Schematic representation depicting cellular events occurring in delayed ischemic preconditioning. The contributions of mito$K_{ATP}$ channels and the parallel signal transduction pathways utilizing a variety of agonists, G protein-coupled receptors and protein kinases are depicted as found with early preconditioning model. Also, a protein kinase-mediated activation (with ROS involvement) of specific nuclear transcription factors (e.g., NF-κB, NRF-1, PGC-1α) is involved, leading to increased gene expression and synthesis of protective proteins, enabling sustained CP. These include: mitochondrial MnSOD, ETC, eNOS and several stress-activated heat-shock proteins (HSPs) as well as activation of cytokines (including TNF-α and IL-1β). These proteins provide several levels of CP by acting at a variety of cellular sites, including mitochondria

include the rapid activation of the sphingolipid-ceramide signaling pathway generating endogenous second-messengers ceramide and sphingosine [20].

Thus, in the delayed IPC pathway mitochondria response to relatively short-term metabolic stresses contributes to a dynamic intracellular cross talk and signaling between multiple cellular compartments, enabling the cardiomyocyte to adopt a stress-tolerant state. Moreover, in addition to IPC, several physical stresses can serve to promote MnSOD-associated delayed cardioprotective induction, including whole body hyperthermia and exercise [120, 125].

## Potential Applications of CP to Clinical Medicine

Understanding the molecular and biochemical mechanism of CP offers great potential in the clinical treatment of myocardial diseases. However, unequivocal data and more refined techniques for precise evaluation of the molecular events occurring in CP are needed, mainly prior to the treatment with drugs and regimens whose specificity of action, and overall metabolic and physiological role (as well as potential benefit) are not clear. New research may be instrumental in advancing our understanding of the temporal pathway of mitochondrial-mediated CP, and may allow the development of novel and more effective drugs with increased specificity of action, as well as acquisition of critical information necessary to know when to apply CP. For instance, the addition of specific cardioprotective agents at the immediate onset of reperfusion has been found to significantly reduce infarct size [126–128]. Exploration of alternative methods to improve mitochondrial cardioprotective responses also seems warranted. For example, the finding that MTP opening during reperfusion can be modulated by either mitochondrial $Ca^{2+}$ levels or OS suggests that regulation of mitochondrial $Ca^{2+}$ and/or ROS levels prior to reperfusion might be effective in exerting CP. Moreover, the use of gene therapy in animal models of myocardial ischemia [129, 130], has shown promising results with the upregulation of cardioprotective genes [e.g., inducible NOS (iNOS) and heme oxygenase]. Notwithstanding focused research is needed to understand the role that mitochondria play in the protection against the cell damage that may occur with transient ischemia and reperfusion, such as hypercontracture and swelling that finally may lead to sarcoplasmic rupture and cell death.

We must keep in mind that most of the IPC events studied so far took place primarily in the healthy heart of experimental animals; thus, it may be possible that PC may behave differently and may be less effective in human cardiac dysfunction secondary to heart failure/cardiomyopathies. While the application of PC thus far in human cardiovascular diseases (CVD) has been rather limited, there is supportive evidence that CP may be a useful adjunct in the management of CVD. Several clinical studies have shown that the IPC-mediated CP [131], or the careful use of stress or pharmacological stimuli (e.g., exercise, nitroglycerin, and adenosine) can be of benefit in patients with angina, and in those undergoing coronary angioplasty or open heart surgery [132–134]. Furthermore, the screening of potential PC-mimetic candidates in patients undergoing angioplasty appeared feasible, as well as evaluation of PC-based strategies in clinical trials of heart transplant and bypass surgery. Nevertheless, definitive evidence of their beneficial effects needs to be established prior to use in aged and diseased heart [135].

## *Mitochondrial End-Effectors of IPC Cardioprotection*

End targets for the CP pathways include: metabolic effects (i.e., an improved cardiomyocyte energy balance), modulation of mito$K_{ATP}$ channel, MTP opening, $Na^+/H^+$ exchanger (NHE), osmotic swelling, decreased cytoskeletal fragility, changes in gap junction function, reduced levels of TNF-α and ROS, and most importantly, changes in apoptotic remodeling [32]. End-effector(s) of IPC cardioprotection must be involved in the reduction of cell death (either apoptotic or necrotic and likely both) which is paramount to its success. Since necrotic cell death (and not apoptosis) is characterized by a loss of plasma membrane integrity, the end-effector(s) of IPC must address and correct this defect. Significantly, overexpression of antiapoptotic factors, such as Bcl-2, has been shown to inhibit necrosis [136].

A number of the potential candidates for end-effectors are associated with mitochondria. This is not surprising since mitochondria are the primary organelles involved in ATP production, have been implicated in maintaining cell integrity, and mitochondrial events are pivotally linked to the initiation of apoptosis. A key participant in the trigger and mediation of cardioprotective responses, mitochondria, houses and regulates critical early events in the apoptotic pathway. Mitochondrial plasticity enables its functioning both as a target and as a player in myocardial signal transduction events, in the generation of ROS in response to a variety of cellular insults, and providing an appropriate antioxidant response (Table 20.2). Evidence suggests an important role for several mitochondrial proteins as end effectors; the mito$K_{ATP}$ channel, apoptotic proteins, such as Bad and the MTP. A mitochondrial fraction containing five coimmunoprecipitating proteins (SDH, ANT, ATP synthase, inorganic phosphate carrier, and ABC1) has been shown to display mitochondrial ATP-sensitive $K^+$-channel activity, and the mABC1 protein plays a major role in cellular protection against OS. Thus, ABC1 may be an important target in CP [137, 138]. Because of their common mitochondrial location, it is possible that the mechanisms involved in the regulation of these different mitochondrial proteins are related.

As a potential unifying mechanism, modulation of the activity of mito$K_{ATP}$ channels could reduce cell death in several ways. These include ROS control, inhibition of mitochondrial $Ca^{2+}$ uptake, and the regulation of mitochondrial volume that could alter mitochondrial permeability by VDAC or the MTP. Gathered evidence from experimental ischemia studies in rat cardiomyocytes, reperfusion in rabbit myocytes,

**Table 20.2** Mitochondrial functional plasticity

| Regulatable mitochondrial functions |
|---|
| Respiration and oxidative phosphorylation |
| Ion ($K^+$, $Ca^{2+}$) transport and storage |
| ATP production |
| Cytochrome *c* release |
| Redox state |
| ROS generation |
| Permeability transition pore opening |
| Antioxidant response (e.g., MnSOD, glutathione) |
| Role in apoptotic progression |
| Responsive to $K_{ATP}$ channel openers (e.g., diazoxide, nicorandil) |
| Responsive to $Na^+/H^+$ exchanger inhibitor cariporide |

perfused rat hearts and in isolated cardiac mitochondria, that $K_{ATP}$ channel openers, such as diazoxide make mitochondria resistant to $Ca^{2+}$ overload, by reducing $Ca^{2+}$ influx and activating $Ca^{2+}$ release [86, 139–141]. Thus, the mito$K_{ATP}$ channel may provide cardioprotective effects on end-effector by acting against myocardial necrosis and stunning through alleviation of OS and $Ca^{2+}$ overload.

In IPC, MTP and the mito$K_{ATP}$ channel, by their close physical proximity as well as by their response to KCOs, seem to keep a close relationship, although a direct interaction between them has not been established [138]. Experimental studies have shown that changes in mitochondrial $Ca^{2+}$ may inhibit the MTP opening. In isolated mitochondria, diazoxide prevents opening of the MTP induced by $Ca^{2+}$ addition and cytochrome *c* loss [85]; also, the activation of PKC mimicked the effect of diazoxide on the MTP and was blocked by 5-HD.

The NHE inhibitor cariporide attenuates mitochondrial $Ca^{2+}$ overload and MTP opening by increasing the time taken to induce the MTP, and decrease ouabain or phenylarsine oxide (PAO)-induced hypercontracture. Observations by Toda et al. [142] indicated that cariporide attenuates mitochondrial $Ca^{2+}$ overload and that these effects might potentially contribute to the mechanisms of CP afforded by NHE inhibitors. In addition, they found that cariporide and cyclosporine A inhibit tetramethylrhodamine methyl ester (TMRM)-induced MTP opening and PAO-induced hypercontracture, but diazoxide did not. MTP opening allows water, foreign substances and solutes to enter the mitochondria, increasing matrix volume, disrupting mitochondrial function, rupturing the outer membrane, and may result in cell death by either apoptosis or necrosis. Inhibiting MTP opening during reperfusion using CsA, which has a high affinity for binding the MTP component cyclophilin-D (CYP-D), is cardioprotective [91], although insulin has been found to be superior to CsA in ischemia and reoxygenation protection of human myocardium [143]. While the role of the MTP in IPC has not been directly demonstrated, CP afforded from either IPC, diazoxide or an adenosine agonist treatment could be blocked by pretreatment with atractyloside, an opener of the MTP [144, 145].

VDAC, an essential component of the MTP is affected by events at the mito$K_{ATP}$ channel and interacts with some of the Bcl-2 proapoptotic family of proteins (such as Bax, Bad, Bak, and Bok among others). Release of cytochrome *c* showed the same dependence on Bax and VDAC [146, 147], and both can associate after an apoptotic stimulus; this interaction is inhibited by Bcl-xL *in vitro*. Purified VDAC and recombinant Bax act together to allow permeability of cytochrome *c* and sucrose through a lipid membrane with no apparent changes in channel conductance. Moreover, Bax induces the loss of mitochondrial membrane potential in yeast mitochondria that can be inhibited by Bcl-xL, and this is dependent on the presence of yeast VDAC [148]. Interestingly, induction of apoptosis by inhibition of the antiapoptotic Bcl-2 proteins requires covert levels of direct activators of Bax and Bak at the outer mitochondrial membrane [149].

Moreover, hearts deficient in cyclophilin-D [CYP-D (–/–) mice], another critical component of the MTP, are resistant to the infarct-limiting effects of ischemic and pharmacological preconditioning and postconditioning, validating the essential role that MTP plays in CP [150].

The hypothesis that the activation of the mito$K_{ATP}$ channel induces protection via modulation of VDAC is of interest considering that overexpression of Bcl-2 modulates CP via inhibition of VDAC [151]. The cardiac-specific overexpression of Bcl-2 has been shown to reduce myocyte death after I/R, reducing the rate of decline in ATP during ischemia and attenuated ischemia-mediated acidification [152]. Similar to the effects of cardiac Bcl-2 overexpression, reduction of VDAC activity might explain many of the intracellular changes associated with preconditioning, such as the attenuated rate of decline in ATP, reduced cardiomyocyte acidification, and reduced cell death. Diazoxide addition has also been reported to reduce acidification during ischemia [153], and perhaps this effect is mediated via $K_{ATP}$-dependent closure of VDAC [104]. Furthermore, VDAC interaction with proapoptotic proteins potentially contributes to IPC cardioprotection. There is evidence that early apoptotic progression involves proapoptotic family members, such as Bax, targeting the outer mitochondrial membrane, and forming a large conductance channel that promotes cytochrome *c* release from the mitochondria [147, 154]. The precise mechanism by which Bax mediates the release of cytochrome *c* is not well understood, nor the association of channel-forming Bax with other mitochondrial membrane proteins, such as VDAC, or the larger MTP complex.

Bad is another proapoptotic protein that binds and sequesters Bcl-2, in such a way that Bcl-2 can no longer inhibit apoptosis. Phosphorylation of Bad releases it, thereby freeing Bcl-2 to inhibit apoptosis. Earlier, we pointed out that a mitochondrial-localized signaling complex containing PKCε and MAPK has been isolated and characterized; this signaling complex is associated with increased mitochondrial phosphorylation of Bad [55]. Similarly, PKCε forms a functional signaling complex with components of the MTP, including VDAC, ANT, and hexokinase [56].

IPC not only protects against necrotic cell death occurring with I/R but blunts the progression of apoptosis, a process that is increased with aging. It has been proposed that cardiomyocyte apoptosis occurring during I/R, is primarily reperfusion-triggered [155]. Reperfusion rapidly restores intracellular ATP levels required for apoptosis. Altering the expression of apoptogenic proteins (e.g., Bax) and PKC activity by IPC can also reduce apoptosis. Early hallmark events of apoptosis occur at the mitochondrial inner membrane, including MTP opening and cytochrome *c* release from mitochondria (see Chap. 9, Fig. 20.2); upon CP, cytochrome *c* release is inhibited along with MTP opening and mitochondrial $Ca^{2+}$ flux stemmed. Interestingly, modulation of the myocardial apoptotic events has been documented in early but not in delayed IPC [92].

## Other Targets in IPC

Besides mitochondria, there are numerous other nonmitochondrial targets in cardiomyocyte IPC pathways. These include cytoskeletal proteins and membrane proteins (e.g., $Na^+/H^+$ exchangers) that may be plausible targets for CP because they might assist in maintaining cell integrity. Probably, the same signaling kinases that target mitochondrial proteins can also modify nonmitochondrial proteins and modulate other aspects of cardiomyocyte metabolism and function. IPC has been reported to alter connexins (a major component of gap junctions) [156–158], $Ca^{2+}$ handling by sarcoplasmic reticulum [156]. Opening of mito$K_{ATP}$ channels has also been shown to activate p38 MAPK [156, 157] and MAPK has been suggested to act as a downstream mediator in IPC signaling because its involvement in the phosphorylation and redistribution of heat shock proteins (e.g., HSP27); impacting on the integrity and stability of myocardial cytoskeleton [159].

## Gene Expression in Early IPC

Although it is known that delayed CP involves upregulation of gene expression with *de novo* protein synthesis, involvement of gene expression in early or acute preconditioning has generally been excluded. The time for preconditioning appeared to be too short and the preconditioned state, following a preconditioning protocol, too transient (1–2 h) for new mRNA and protein synthesis to fully kick in to provide early CP. Earlier observations showed that protein synthesis inhibition with either actinomycin D or cycloheximide did not block the protection of preconditioning, precluding gene expression as a possible mechanism of IPC memory [159]. However, higher concentrations of cycloheximide (but not actinomycin D) could abrogate preconditioning protection suggesting that a *de novo* protein synthesized from a preexisting transcript might be involved [160]. Other studies have found that actinomycin D-transcriptional inhibition abolished early IPC-induced CP with marked reductions in the activation of several MAP kinases and subsequent transcription and phosphorylation of specific transcription factors [118].

Myocardial gene expression profile by microarray has been used in preconditioned tissues to confirm these findings, and large-scale changes in gene expression with classic IPC have been found [161, 162]. Genes with abnormal myocardial expression secondary to ischemia and/or preconditioning include encoding proteins participating in stress responses, apoptosis, protein degradation, metabolic enzymes, regulatory proteins, as well as a number of genes with unknown cellular functions [161]. Genes exhibiting altered expression with preconditioning include, oligoadenylate synthase, chaperonin subunit ε, a cGMP phosphodiesterase (PDE9A1), a secretory carrier membrane protein, an amino acid transporter, and a protease 28 subunit have not been previously implicated in CP, suggesting that these are new end-effectors of IPC. Interestingly, using anesthetics specific patterns of myocardial gene expression (with some shared elements) were found in preconditioning in comparison to IPC [162]. Support to the view that transcription is likely required for both classical ischemic and pharmacological preconditioning became available when both types of preconditioning were abolished in isolated mice hearts with defective transcription factor, signal transducer, and activator of transcription-3 (STAT-3) [163].

## *Second Window of Protection/ Delayed Preconditioning*

Depending on the species and end-point used for CP, delayed preconditioning, or SWOP lasts for up to 72 h. In delayed PC, phosphorylation of transcription factors, initiating the synthesis of late appearing effector proteins that promote cell survival during subsequent ischemia, may be a crucial event. Gene expression is tailored in a regulated fashion to induce new proteins that promote cell repair and to protect against subsequent I/R insult. The nature of the preconditioning stimulus determines the activation of a variety of transcription factors, regulating a large number of target genes. Among transcription factors, NF-κB and AP-1 have consistently been activated in delayed IPC; this activation has a specific time curve, and is slightly influenced by the number of PC cycles [164]. IPC and NO donors are able to induce NF-κB activation; on the other hand, direct inhibition of NF-κB abrogates delayed IPC cardioprotection [165].

Several genes are upregulated in both delayed ischemic and heat-shock preconditioning. Fauchon et al. [166] using representational difference analysis of cDNA (cDNA, RDA) identified upregulated genes in rabbit myocardium *in vivo* by ischemic preconditioning following posttreatment reperfusion for 2 h, 4 h, and 6 h. Three novel genes and six genes with known function where identified, including the TGF-β receptor interacting protein 1, the α isoform of the A subunit of protein phosphatase 2 (PP2) and the cap-binding protein NCBP1. These genes were induced in rabbit myocardium *in vivo* by both ischemia and heat shock, consistent with a fundamental role in the development of delayed adaptation. The role of PP2 in modulating the mitogen-activated protein kinase pathway and promoting cell survival was consistent with the reperfusion injury salvage kinase pathway mediating the protective effects of IPC. Expression of Trip1 and Ncbp1 also implicates TGF-β signaling pathways and RNA processing and transport in delayed adaptation to stress in the myocardium.

Recently, using a multicycle ischemic preconditioning protocol, Lochner et al. [167] find no evidence for α-1 adrenergic cascade or PKC activation in preconditioning. In contrast, during a multicycle preconditioning protocol they found cyclic increases in tissue cyclic nucleotides cAMP and cGMP, suggesting specific roles for the β-adrenergic signaling pathway and nitric oxide as triggers of CP. This was also confirmed by findings that (1) administration of the β-adrenergic agonist, isoproterenol, or the NO donors SNAP or SNP before sustained ischemia elicited CP, similar to ischemic preconditioning; (2) β-adrenergic blockade or NOS inhibition during an ischemic preconditioning protocol abolished protection. In addition, effectors downstream of cAMP, such as p38 MAPK and CREB, were involved in the triggering process. Evaluation of intracellular signaling during sustained ischemia and reperfusion showed that ischemic preconditioned-induced CP, compared to nonpreconditioned hearts, was associated with a significant reduction in tissue cAMP, attenuation of p38 MAPK activation, increased tissue cGMP levels, and HSP27 activation. The role of the stress kinase p38 MAPK was further investigated using the inhibitor SB203580. Taken together, these findings suggest that injury by necrosis and apoptosis shared the activation of p38 MAPK as a common signal transduction pathway and that pharmacological targeting of this kinase has the potential to manipulate both these processes during I/R injury.

Unlike the acute protection pathway, the delayed IPC pathway appears to be more dependent on the involvement of NO and NOS, heat-shock proteins (e.g., HSP27, HSP72), MnSOD, and cytokines induction (Fig. 20.3). During delayed IPC, NO exerts a profound influence on the development of the protective preconditioned phenotype [9]. In the initial phase of delayed IPC, endothelial NOS-derived NO serves as the initiator of a molecular cascade culminating in the subsequent activation of inducible NOS, which then confers protection [168]. Therefore, eNOS is activated within an early timeframe and iNOS is activated later, and both events appear to be absolute requirements for the generation of delayed IPC-mediated CP. The ability of NO to act in concert with ROS signaling (generated after cell exposure to KCOs and other preconditioning stimuli), may be a critical factor in targeting the mitochondrial respiratory chain and modulating

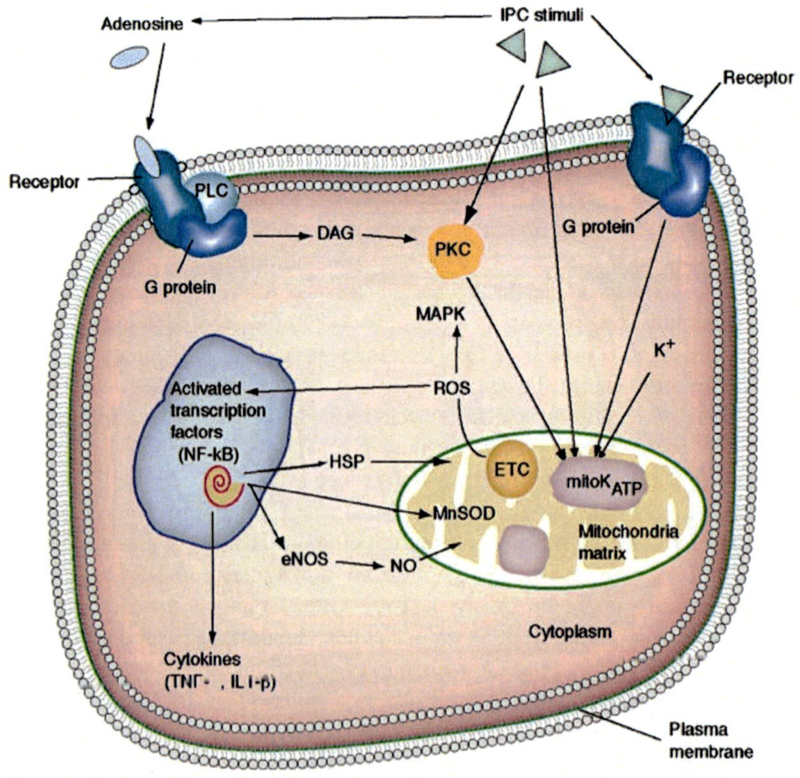


**Fig. 20.3** Schematic representation depicting further cellular events occurring in delayed IPC (see text)

mito$K_{ATP}$ channels to provide protection against ischemia [124]. This is further supported by data showing that: (a) inhibitors of NOS abrogate delayed IPC-mediated protection [146]; (b) NO donors can mimic IPC in conferring CP [124], and (c) free radical scavengers can reduce NOS activation, NO generation and the induction of cardioprotective signaling [9]. Experimental data have provided genetic evidence that COX-2 is an obligatory downstream effector of iNOS-dependent CP and that NF-κB is a critical link between iNOS and COX-2. Thus, iNOS imparts its protective effects, at least in part, by recruiting NF-κB, leading to COX-2 upregulation. On the other hand, COX-2 does not play an important cardioprotective role under basal conditions (when iNOS is not upregulated) [169].

Finally, delayed preconditioning, as in early preconditioning, also promotes CP by augmenting the cardiomyocyte bioenergetic capacity in response to ischemic and anoxic insult. The modulation of mitochondrial respiration in response to delayed preconditioning primarily occurs as a consequence of the induction of ETC and ANT transcription, promoted by the ROS-dependent upregulation of nuclear transcription factors NRF-1 and PGC-1α [170].

## Postconditioning and Cardioprotection

Postconditioning is a series of brief mechanical interruptions of reperfusion following a specific prescribed algorithm applied at the very onset of reperfusion. This algorithm lasts only from 1 to 3 min, depending on species. Unlike preconditioning, which exerts its effects both during the index ischemia and in the reperfusion stage, postconditioning appears to exert its effects during reperfusion alone. Postconditioning modifies the early phase of reperfusion by reducing the oxidant burden and consequent oxidant-induced injury, by attenuating the local inflammatory response to reperfusion, and by engaging end-effectors and signaling pathways implicated in ischemic and pharmacologic preconditioning. As with IPC, the postconditioning phenomenon has observed in both large and small animal *in vivo* models, as well as in *ex vivo* and cell culture models. A comparison between these two models or protocols of cardioprotection is shown in Fig. 20.4.

Postconditioning sets in motion triggers and signals that are functionally related to reduced cell death, in part by promoting the upregulation of survival kinases known to attenuate the pathogenesis of apoptosis and possibly necrosis. Adenosine has been implicated in the CP of postconditioning, as has eNOS, NO, and guanylyl cyclase, opening of $K_{ATP}$ channels and closing of MTP [127, 171, 172]. Although we do not know each of the potential pathways involved in CP by postconditioning, this has been associated with the activation of intracellular survival pathways, such as Erk1/2 and PI3K- Akt pathways [173]. Many of the pathways involved in postconditioning have also been identified in IPC, although some may not be involved in preconditioning (Erk1/2). In contrast to preconditioning, which

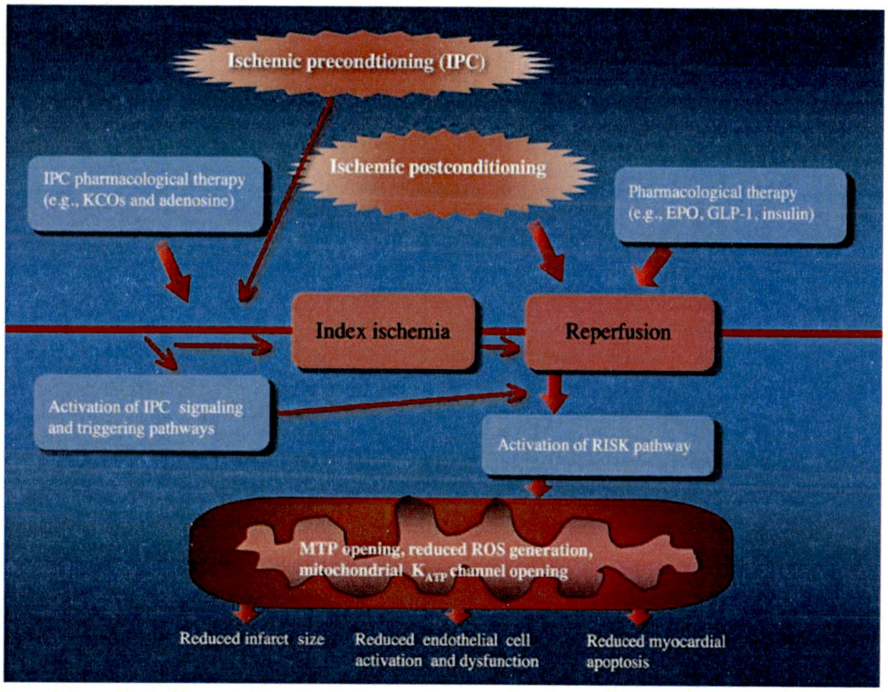


**Fig. 20.4** Flow-chart comparing postconditioning and preconditioning protocols

**Table 20.3** Comparison of IPC and postconditioning

| | IPC | Postconditioning |
|---|---|---|
| Time of intervention | Before onset of ischemia | After onset of ischemia at start of reperfusion |
| *Physiological phenotype* | | |
| Reduced infarct size | + | + |
| Reduced apoptosis | + | + |
| Reduced tissue edema | + | + |
| Reduced endothelial dysfunction | + | + |
| *Signaling components* | | |
| Adenosine-mediated | + | + |
| NO, eNOS, and guanylyl cyclase | + | + |
| Opening of mitoK$_{ATP}$ channels | + | + |
| Closing of the MTP pore | + | + |
| Activation of PI3K-Akt survival pathway | + | + |
| Activation of Erk1/2 | + | + |
| Opioid mediated | + | ? |
| Reduced expression of leukocyte adhesion molecules and cytokines | + | + |
| Reduced lipid peroxidation | + | + |
| Reduced ROS generation coincident with reduced cell death | + | + |
| Genomic response (mostly with delayed IPC) | + | – |
| Pharmacological agents for clinical application | Adenosine, nicorandil | Insulin, GLP-1, erythropoietin, atorvastatin |

*GLP-1* glucagon-like peptide-1

requires a foreknowledge of the ischemic event, postconditioning can be applied at the onset of reperfusion, like at the point of clinical service, i.e., angioplasty, stenting, cardiac surgery, and transplantation [174].

A comparison of key signaling components utilized in preconditioning and postconditioning is shown in Table 20.3.

## Remote Conditioning

Remote, preconditioning-induced protection was initially carried out using a canine model. Brief occlusions of the circumflex (Cx) artery protected not only the Cx vascular bed but preconditioned as well the left anterior descending (LAD) vascular bed [175]. Later studies demonstrated that brief episodes of ischemia in peripheral sites (i.e., kidney, hind-limb, and brain) can protect the heart against infarction [176–178]. Remote preconditioning has been mainly reported in animal species, including rats, mice, rabbits, and pigs, conferring CP in most of them (usually assessed by limitation of MI size) to a similar extent as classical IPC. Evidence documenting remote preconditioning in humans although still limited is increasing. Remote ischemic preconditioning using transient upper limb ischemia was employed by Hausenloy et al. [179] in a group of 57 patients undergoing coronary artery bypass graft surgery. They found significantly reduced patients overall serum troponin-T release at 6, 12, 24, and 48 h after surgery. Recently, Bøtker et al. [180] have reported a study of 333 consecutive adult patients with suspected first acute MI. Patients were randomly assigned in a 1:1 ratio by computerized block randomization to receive primary percutaneous coronary intervention with ($n$=166 patients) or without ($n$=167) remote conditioning (intermittent arm ischemia through four cycles of 5-min inflation and 5-min deflation of a blood-pressure cuff). Remote conditioning was provided during transport to hospital and primary percutaneous coronary intervention in hospital. The primary endpoint was myocardial salvage index at 30 days after primary percutaneous coronary intervention, measured by myocardial perfusion imaging. In the 212 patients that met the criteria for inclusion and have completed follow-up, the median salvage index was 0·75 in the remote conditioning group versus 0·55 in the control group with a mean salvage index of 0·69 versus 0·57 (0·26), with mean difference of 0·12. Taken together, these data suggest that remote ischemic conditioning before hospital admission increases myocardial salvage, and has a favorable safety profile. Nevertheless, more trials are needed to assess the effect of remote conditioning on clinical outcomes.

As with classical IPC, both early and delayed (or SWOP) components have been described in remote preconditioning [11]. Opioids [13], adenosine [181], and stimulation of myocardial adenosine receptors [178], PKCε activation [182], ROS involvement [183], as well as the activation of the mitoK$_{ATP}$ channel [184] plays a role in the signaling pathway of remote preconditioning. Remote preconditioning modified myocardial gene expression by upregulating cardioprotective genes, include genes protecting against OS (e.g., Hadhsc, Prdx4, and Fabp4) and cytoprotection (Hsp73), and suppressing genes like proinflammatory genes (e.g., Egr-1, Dusp 1 and 6) that may be involved in I/R injury pathogenesis [185]. These findings support earlier observations showing that remote IPC (a brief forearm ischemia) provided a stimulus for mediating inflammatory gene transcription in circulating human leukocytes [186].

Notwithstanding the mechanisms of remote conditioning are less well defined than classical IPC; including, the temporal sequence of events mediating CP and delineation of the elusive identity of the diffusible protective factor(s), which appears to be released from the heart and other organs to initiate protection at remote sites. It is also unclear whether this factor is a humoral protein, a neurogenic factor or both [11]. Attempts to identify a specific protein factor by proteomic analysis were unsuccessful [187].

## Application in Human

The concept that CP mediated by ischemic and pharmacological conditioning occurs in humans has been well documented. Human ventricular cardiomyocytes exhibit delayed CP 24 h after a short period of simulated ischemia in addition to classic preconditioning [188].

Human myocardial tissue can be preconditioned, as suggested by patients undergoing procedures that involve brief periods of ischemia, such as coronary angioplasty (suggestive of IPC). They experienced amelioration of the severity of chest pain, decreased ST-segment elevation, decreased lactate production, and decreased release of myocardial markers, such as creatine kinase-MB (CK-MB) with subsequent balloon inflations, constituting a myocardial CP [11]. Given that many individuals experience brief episodes of ischemia before an acute MI, it is possible that preinfarct angina has the potential to precondition the myocardium, thereby reducing infarct size and improving survival. A retrospective analysis of the Thrombolysis in Myocardial Infarction-4 study (TIMI-4) [189] revealed that the presence of preinfarct angina was associated with smaller infarct size, based on creatine kinase release, improved left ventricular function with reduced incidence of congestive heart failure (CHF) and shock, and reduced mortality, findings confirmed in other studies. Also, older patients (> or = 60 years of age) in the TIMI-4 study appeared to benefit from a history of angina prior to acute MI suggesting that this may lead to successful preconditioning in the older heart [190]. The time interval between the last episode of angina and the index MI is a critical factor as suggested by findings that prodromal angina is only protective if it occurs within 24–72 h of MI, a time course remarkably similar to that of the delayed IPC in animal models [191, 192].

The clinical use of IPC has also been applied in the setting of coronary artery bypass grafting (CABG) to protect the myocardium from global ischemia. Vahlhaus et al. [193] have investigated whether IPC is associated with phosphorylation of p38 MAPK (characteristic for early preconditioning) or with increased protein expression of HSP-72 (characteristic for delayed preconditioning) at the time of CABG in patients. They found that in the human left ventricular myocardium there is an SWOP lasting for at least 48 h, while at that time the early phase of preconditioning has already gone. Interestingly, IPC by unstable angina reduced the release of CK-MB following CABG and stimulated left ventricular HSP-72 protein expression. It has also been suggested that preceding unstable angina and/or IPC-activated genes regulated by NF-κB or hypoxia-inducible factor 1α (HIF-1α) in parallel to improved cardiac function. Czibik et al. [194] have noted that expression of the NF-κB-regulated genes increased after cardioplegia and reperfusion, but this was not influenced by IPC in stable patients. TNF and iNOS expression were found reduced after IPC in patients with unstable angina, but there were no significant differences in gene expression between stable and unstable patients before cardioplegia and IPC. NF-κB translocation during reperfusion was reduced in stable preconditioned and unstable control patients compared to stable controls. Heat shock protein-72 expression increased after preconditioning of unstable patients. Taken together heart function improved by IPC in both stable and unstable patients; however, unstable angina per se had no effect. Moreover, NF-κB-regulated genes were influenced by IPC, but hypoxia-inducible genes were not. IPC by unstable angina reduced the release of CK-MB following CABG and stimulates left ventricular HSP-72 protein expression [193]. Also, the use of pharmacological preconditioning using adenosine A1 receptor agonists was found to confer some myocardial protection in CABG but not as much as obtained with cross-clamping IPC [195].

### *Clinical Trials*

Assessment of the effect of IPC in human has been initiated in studies involving cardiac surgery, coronary angioplasty, and treadmill exercise. Unfortunately, in most circumstances clinical application of IPC is not desirable or feasible because repeated cross-clamping of the aorta poses potential risks, such as atheroembolism from calcifications, and also because IPC can prolong the length of surgery. Protecting the heart of coronary artery disease (CAD) patients most likely requires a pharmacological approach using drugs which either enhance or may mimic IPC. The *I*mpact *O*f *N*icorandil in *A*ngina (IONA) trial have provided evidence of the clinical utility of pharmacological preconditioning by its demonstration that chronic administration of the KCO nicorandil, significantly improved the cardiovascular prognosis in CAD patients [196].

A limiting factor in the efficacious use of either ischemic or pharmacological preconditioning pertains to the temporal boundaries of its protection; it is required that they be applied prior to the ischemic insult, which in most clinical situations [other than CABG or percutaneous transluminal coronary angioplasty (PTCA)] is difficult to predict. In this respect, the use of pharmacological postconditioning (administered during reperfusion) may circumvent the unpredictable nature of the ischemic insult and may be clinically more applicable. Preclinical studies have shown robust CP with the application of agents, such as insulin, erythropoietin, and glucagon-like peptide 1 during the first few minutes of reperfusion, as acute pharmacological activators of myocardial survival kinases. However, at present there is only limited clinical data concerning pharmacological preconditioning.

The first Acute Myocardial Infarction Study of Adenosine (AMISTAD) trial [197] was designed to prospectively test the hypothesis that adenosine, as an adjunct to thrombolysis would reduce MI size, as measured by SPECT imaging. The benefit of this trial was limited to patients with anterior infarction, where relative reduction in infarct size was 67%, with little evidence of benefit for infarcts located elsewhere. In fact the incidence of adverse events was higher in the adenosine group.

In the follow-up AMISTAD-II trial [198], the effect of intravenous adenosine on clinical outcomes and infarct size in ST-segment elevation myocardial infarction (STEMI) patients undergoing reperfusion therapy was assessed. Only a weak trend toward improvement in clinical outcomes was achieved, but the difference in primary endpoints (new onset CHF or death from any cause) between the pooled adenosine and placebo groups was not statistically significant. However, a significant relationship between infarct size and occurrence of primary endpoints was seen, with infarct size reduction with the 70-μg/kg/min adenosine infusion, a finding that correlated with fewer adverse clinical events. On the other hand, Kloner et al. [199] went beyond the primary goals of the AMISTAD-II to determine whether the efficacy of adenosine *versus* placebo was dependent on the timing of reperfusion therapy in the AMISTAD-II. They found that 3 h adenosine infusion administered as an adjunct to reperfusion therapy within the first 3.17 h onset of evolving anterior ST-segment elevation acute myocardial infarction (AMI) enhanced early and late survival, and reduced the composite clinical endpoint of death or CHF at 6 months. A critical caveat in this study was the post hoc character of this study that cannot replace a well-designed prospective clinical trial.

## Conclusions and Future Perspective

Despite significant progress in the medical and surgical management of CAD, acute ischemia and MI remain major public health problems with very high accumulative mortality and morbidity. Understanding the cellular, molecular, and biochemical mechanisms of CP are essential to develop new methodologies and therapies to improve the outcome of so many patients. These include the signals, transducing mediators, and effectors in the temporal pathway(s) of CP that may be instrumental for the development of novel pharmaceutical agents with increased efficacy and specificity.

Protecting the heart of CAD patients likely requires a pharmacological approach by using drugs which either enhance or mimic IPC. Also, a limiting factor in the efficacious use of either ischemic or pharmacological preconditioning relates to the temporal boundaries of its protection since it is required that they be applied prior to the ischemic insult, which in most clinical situations (other than CABG or PTCA) is difficult to predict. Toward this goal, the use of pharmacological postconditioning (administered during reperfusion) may circumvent the unpredictable nature of the ischemic insult and may be more clinically applicable. Preclinical studies have shown significant CP with the application of agents, such as insulin, statins, erythropoietin, sildenafil, and glucagon-like peptide 1 during the first few minutes of reperfusion, as acute pharmacological activators of myocardial survival kinases. In the *A*torvastatin for *R*eduction of *MY*ocardial *D*amage during *A*ngioplasty (ARMYDA), pretreatment with atorvastatin significantly reduced procedural myocardial injury in individuals with chronic stable angina undergoing elective coronary interventions [200].

At present, the availability of clinical data following pharmacological preconditioning is limited. Also, information regarding if the CP period can be extended beyond 72 h of delayed preconditioning without losing effect is needed, and if a combination of various pharmacological triggers may improve CP. Prior to clinical trials with drugs and regimens whose specificity of action, including their overall metabolic and physiological role and potential benefits are not absolutely clear, unequivocal data and more advances techniques to evaluate the molecular events of CP are needed. Data accumulated thus far from animal models show that ischemic and pharmacological preconditioning, and postconditioning may offer a desirable alternative and/or adjunct to the currently available drug armamentarium and surgical interventions for CAD. The use of pharmacological postconditioning may provide the most attractive approach, but again large-scale trials are needed to test their efficacy.

Presently, a number of questions need to be answered prior to the routine use of these interventions as therapy:

1. Because most IPC events have been studied primarily in the healthy heart of experimental young and middle-aged animals, are these cardioprotective interventions applicable to human myocardial ischemia, including the aging heart?
2. Besides myocardial ischemia, could we achieve IPC-CP in patients with primary cardiomyopathies and cardiomyopathy of aging?

3. Can preconditioning or CP be successfully applied in the treatment of acute MI, after heart transplant and following bypass surgery for CAD?
4. Can the time of CP (whether IPC, pharmacological, or postconditioning) be extended even beyond the 72 h period of delayed preconditioning without losing effect.
5. Can combinations of various pharmacological triggers improve CP?

Finally, the discovery that the MTP opens during reperfusion and that it can be modulated (by either mitochondrial $Ca^{2+}$ levels or OS) suggests that controlled management of mitochondrial $Ca^{2+}$ and/or ROS levels prior to reperfusion might be effective in providing CP. Intensive and focused research effort *is* needed to implement mitochondria function modulation as a protective therapy against cell death and cardiac dysfunction (both in the adult and aging human heart). Since previous and most of ongoing research in CP has been carried out in animal models and cultured cells, better understanding of the cellular and molecular events involved in IPC and CP in human is critical to successfully treat these devasting cardiovascular diseases.

## Summary

- Despite significant progress in the medical and surgical management of CAD, acute ischemia, and MI remain major public health problems with very high accumulative mortality and morbidity.
- CP shields the heart from damage secondary to insults, such as ischemia.
- Protecting the heart of CAD patients likely requires a pharmacological approach by using drugs which either enhance or mimic IPC.
- Ischemic preconditioning, by using single or multiple brief periods of ischemia, protects the heart against prolonged ischemic insult (index ischemia).
- Prospective interventions and new drugs should be evaluated not only in young and middle-aged animals, but also in the aging or senescent heart, a population in which CP is most relevant.
- CP interventions await to be successfully applied in humans.
- The myocardial signaling pathways of IPC, including numerous triggering and mediating components that are increasingly being identified with a number of the components showing redundancy (i.e., parallel pathways).
- Early and delayed stages of CP have differences in their signaling components as well as in their effects on cardiac phenotype.
- Pharmacological preconditioning can be effectively used to mimic IPC and appears to be more applicable to clinical settings.
- The use of erythropoietin, statins, sidenafil, nitroglycerin, potassium channel openers, adenosine receptor agonists, and insulin has provided CP against I/R injury in preclinical models.
- The aging heart may have a diminished capacity to tolerate and respond to various forms of stresses, and the likelihood of myocardial ischemia and cardiac dysfunction increases.
- Prospective interventions and new drugs should be evaluated not only in young and middle-aged animals, but also in the aging or senescent heart, a population in which CP is most relevant.
- Cardioprotective interventions await to be successfully applied in humans.
- In the ARMYDA , pretreatment with atorvastatin significantly reduced procedural myocardial injury in individuals with chronic stable angina undergoing elective coronary interventions.
- Prior to clinical trials with drugs and regimens whose specificity of action, including their overall metabolic and physiological role and potential benefits, are not absolutely clear, unequivocal data, and more advanced techniques to evaluate the molecular events of CP are needed.

## References


1. Paradies G, Petrosillo G, Pistolese M, Di Venosa N, Serena D, Ruggiero FM. Lipid peroxidation and alterations to oxidative metabolism in mitochondria isolated from rat heart subjected to ischemia and reperfusion. Free Radic Biol Med. 1999;27: 42–50.
2. Lesnefsky EJ, Chen Q, Slabe TJ, et al. Ischemia, rather than reperfusion, inhibits respiration through cytochrome oxidase in the isolated, perfused rabbit heart: role of cardiolipin. Am J Physiol Heart Circ Physiol. 2004;287:H258–67.
3. Eefting F, Rensing B, Wigman J, et al. Role of apoptosis in reperfusion injury. Cardiovasc Res. 2004;61:414–26.
4. Honda HM, Korge P, Weiss JN. Mitochondria and ischemia/reperfusion injury. Ann N Y Acad Sci. 2005;1047:248–58.
5. Gottlieb RA, Burleson KO, Kloner RA, Babior BM, Engler RL. Reperfusion injury induces apoptosis in rabbit cardiomyocytes. J Clin Invest. 1994;94:1621–8.
6. Majno G, Joris I. Apoptosis, oncosis, and necrosis. An overview of cell death. Am J Pathol. 1995;146:3–15.
7. Buja LM. Myocardial ischemia and reperfusion injury. Cardiovasc Pathol. 2005;14:170–5.
8. Iliodromitis EK, Lazou A, Kremastinos DT. Ischemic preconditioning: protection against myocardial necrosis and apoptosis. Vasc Health Risk Manag. 2007;3:629–37.
9. Bolli R. The late phase of preconditioning. Circ Res. 2000;87: 972–83.
10. Murry CE, Jennings RB, Reimer KA. Preconditioning with ischemia: a delay of lethal cell injury in ischemic myocardium. Circulation. 1986;74:1124–36.
11. Riksen NP, Smits P, Rongen GA. Ischaemic preconditioning: from molecular characterisation to clinical application–part I. Neth J Med. 2004;62:353–63.

12. Mubagwa K, Flameng W. Adenosine, adenosine receptors and myocardial protection: an updated overview. Cardiovasc Res. 2001; 52:25–39.
13. Peart JN, Headrick JP. Adenosinergic cardioprotection: multiple receptors, multiple pathways. Pharmacol Ther. 2007;114:208–21.
14. Cohen MV, Downey JM. Adenosine: trigger and mediator of cardioprotection. Basic Res Cardiol. 2008;103:203–15.
15. Liu GS, Thornton J, Van Winkle DM, Stanley AW, Olsson RA, Downey JM. Protection against infarction afforded by preconditioning is mediated by A1 adenosine receptors in rabbit heart. Circulation. 1991;84:350–6.
16. Thornton JD, Liu GS, Olsson RA, Downey JM. Intravenous pretreatment with A1-selective adenosine analogues protects the heart against infarction. Circulation. 1992;85:659–65.
17. Fremes SE, Zhang J, Furukawa RD, Mickle DA, Weisel RD. Adenosine pretreatment for prolonged cardiac storage. An evaluation with St. Thomas' Hospital and University of Wisconsin solutions. J Thorac Cardiovasc Surg. 1995;110:293–301.
18. Cohen MV, Baines CP, Downey JM. Ischemic preconditioning: from adenosine receptor to KATP channel. Annu Rev Physiol. 2000;62:79–109.
19. Qin Q, Downey JM, Cohen MV. Acetylcholine but not adenosine triggers preconditioning through PI3-kinase and a tyrosine kinase. Am J Physiol Heart Circ Physiol. 2003;284:H727–34.
20. Lecour S, Smith RM, Woodward B, Opie LH, Rochette L, Sack MN. Identification of a novel role for sphingolipid signaling in TNF alpha and ischemic preconditioning mediated cardioprotection. J Mol Cell Cardiol. 2002;34:509–18.
21. Goto M, Liu Y, Yang XM, Ardell JL, Cohen MV, Downey JM. Role of bradykinin in protection of ischemic preconditioning in rabbit hearts. Circ Res. 1995;77:611–21.
22. Lebuffe G, Schumacker PT, Shao ZH, Anderson T, Iwase H, Vanden Hoek TL. ROS and NO trigger early preconditioning: relationship to mitochondrial KATP channel. Am J Physiol Heart Circ Physiol. 2003;284:H299–308.
23. Baines CP, Goto M, Downey JM. Oxygen radicals released during ischemic preconditioning contribute to cardioprotection in the rabbit myocardium. J Mol Cell Cardiol. 1997;29:207–16.
24. Tritto I, D'Andrea D, Eramo N, et al. Oxygen radicals can induce preconditioning in rabbit hearts. Circ Res. 1997;80:743–8.
25. Gopalakrishna R, Anderson WB. $Ca^{2+}$- and phospholipid-independent activation of protein kinase C by selective oxidative modification of the regulatory domain. Proc Natl Acad Sci USA. 1989;86:6758–62.
26. von Ruecker AA, Han-Jeon BG, Wild M, Bidlingmaier F. Protein kinase C involvement in lipid peroxidation and cell membrane damage induced by oxygen-based radicals in hepatocytes. Biochem Biophys Res Commun. 1989;163:836–42.
27. Oh YI, Kim JH, Kang CW. Oxidative stress in MCF-7 cells is involved in the effects of retinoic acid-induced activation of protein kinase C-delta on insulin-like growth factor-I secretion and synthesis. Growth Horm IGF Res. 2010;20:101–9.
28. Kabir AM, Clark JE, Tanno M, et al. Cardioprotection initiated by reactive oxygen species is dependent on activation of PKCepsilon. Am J Physiol Heart Circ Physiol. 2006;291:H1893–9.
29. Qin Q, Yang XM, Cui L, et al. Exogenous NO triggers preconditioning via a cGMP- and mitoKATP-dependent mechanism. Am J Physiol Heart Circ Physiol. 2004;287:H712–8.
30. Lefer DJ, Nakanishi K, Johnston WE, Vinten-Johansen J. Antineutrophil and myocardial protecting actions of a novel nitric oxide donor after acute myocardial ischemia and reperfusion of dogs. Circulation. 1993;88:2337–50.
31. Lochner A, Marais E, Genade S, Moolman JA. Nitric oxide: a trigger for classic preconditioning? Am J Physiol Heart Circ Physiol. 2000;279:H2752–65.
32. Yellon DM, Downey JM. Preconditioning the myocardium: from cellular physiology to clinical cardiology. Physiol Rev. 2003;83:1113–51.
33. Nakano A, Liu GS, Heusch G, Downey JM, Cohen MV. Exogenous nitric oxide can trigger a preconditioned state through a free radical mechanism, but endogenous nitric oxide is not a trigger of classical ischemic preconditioning. J Mol Cell Cardiol. 2000;32:1159–67.
34. Bell RM, Yellon DM. The contribution of endothelial nitric oxide synthase to early ischaemic preconditioning: the lowering of the preconditioning threshold. An investigation in eNOS knockout mice. Cardiovasc Res. 2001;52:274–80.
35. Post H, Schulz R, Behrends M, Gres P, Umschlag C, Heusch G. No involvement of endogenous nitric oxide in classical ischemic preconditioning in swine. J Mol Cell Cardiol. 2000;32:725–33.
36. Weselcouch EO, Baird AJ, Sleph P, Grover GJ. Inhibition of nitric oxide synthesis does not affect ischemic preconditioning in isolated perfused rat hearts. Am J Physiol. 1995;268:H242–9.
37. Andelova E, Bartekova M, Pancza D, Styk J, Ravingerova T. The role of NO in ischemia/reperfusion injury in isolated rat heart. Gen Physiol Biophys. 2005;24:411–26.
38. Murphy E. Primary and secondary signaling pathways in early preconditioning that converge on the mitochondria to produce cardioprotection. Circ Res. 2004;94:7–16.
39. Mocanu MM, Bell RM, Yellon DM. PI3 kinase and not p42/p44 appears to be implicated in the protection conferred by ischemic preconditioning. J Mol Cell Cardiol. 2002;34:661–8.
40. Tong H, Chen W, Steenbergen C, Murphy E. Ischemic preconditioning activates phosphatidylinositol-3-kinase upstream of protein kinase C. Circ Res. 2000;87:309–15.
41. Robinet A, Hoizey G, Millart H. PI 3-kinase, protein kinase C, and protein kinase A are involved in the trigger phase of beta1-adrenergic preconditioning. Cardiovasc Res. 2005;66:530–42.
42. Miyawaki T, Ofengeim D, Noh KM, et al. The endogenous inhibitor of Akt, CTMP, is critical to ischemia-induced neuronal death. Nat Neurosci. 2009;12:618–26.
43. Krieg T, Qin Q, McIntosh EC, Cohen MV, Downey JM. ACh and adenosine activate PI3-kinase in rabbit hearts through transactivation of receptor tyrosine kinases. Am J Physiol Heart Circ Physiol. 2002;283:H2322–30.
44. Mitchell MB, Meng X, Ao L, Brown JM, Harken AH, Banerjee A. Preconditioning of isolated rat heart is mediated by protein kinase C. Circ Res. 1995;76:73–81.
45. Ytrehus K, Liu Y, Downey JM. Preconditioning protects ischemic rabbit heart by protein kinase C activation. Am J Physiol. 1994;266: H1145–52.
46. Kudo M, Wang Y, Xu M, Ayub A, Ashraf M. Adenosine A(1) receptor mediates late preconditioning via activation of PKC-delta signaling pathway. Am J Physiol Heart Circ Physiol. 2002;283:H296–301.
47. Zhao TC, Kukreja RC. Protein kinase C-delta mediates adenosine A3 receptor-induced delayed cardioprotection in mouse. Am J Physiol Heart Circ Physiol. 2003;285:H434–41.
48. Fryer RM, Hsu AK, Wang Y, Henry M, Eells J, Gross GJ. PKC-delta inhibition does not block preconditioning-induced preservation in mitochondrial ATP synthesis and infarct size reduction in rats. Basic Res Cardiol. 2002;97:47–54.
49. Liu H, McPherson BC, Yao Z. Preconditioning attenuates apoptosis and necrosis: role of protein kinase C epsilon and -delta isoforms. Am J Physiol Heart Circ Physiol. 2001;281:H404–10.
50. Gray MO, Zhou HZ, Schafhalter-Zoppoth I, Zhu P, Mochly-Rosen D, Messing RO. Preservation of base-line hemodynamic function and loss of inducible cardioprotection in adult mice lacking protein kinase C epsilon. J Biol Chem. 2004;279:3596–604.
51. Wang Y, Ashraf M. Activation of alpha1-adrenergic receptor during $Ca^{2+}$ pre-conditioning elicits strong protection against $Ca^{2+}$ overload injury via protein kinase C signaling pathway. J Mol Cell Cardiol. 1998;30:2123–35.
52. Liu GS, Cohen MV, Mochly-Rosen D, Downey JM. Protein kinase C-epsilon is responsible for the protection of preconditioning in rabbit cardiomyocytes. J Mol Cell Cardiol. 1999;31:1937–48.

53. Ping P, Zhang J, Qiu Y, et al. Ischemic preconditioning induces selective translocation of protein kinase C isoforms epsilon and eta in the heart of conscious rabbits without subcellular redistribution of total protein kinase C activity. Circ Res. 1997;81:404–14.
54. Saurin AT, Pennington DJ, Raat NJ, Latchman DS, Owen MJ, Marber MS. Targeted disruption of the protein kinase C epsilon gene abolishes the infarct size reduction that follows ischaemic preconditioning of isolated buffer-perfused mouse hearts. Cardiovasc Res. 2002;55:672–80.
55. Baines CP, Zhang J, Wang GW, et al. Mitochondrial PKCepsilon and MAPK form signaling modules in the murine heart: enhanced mitochondrial PKCepsilon-MAPK interactions and differential MAPK activation in PKCepsilon-induced cardioprotection. Circ Res. 2002;90:390–7.
56. Baines CP, Song CX, Zheng YT, et al. Protein kinase Cepsilon interacts with and inhibits the permeability transition pore in cardiac mitochondria. Circ Res. 2003;92:873–80.
57. Hassouna A, Matata BM, Galinanes M. PKC-epsilon is upstream and PKC-alpha is downstream of mitoKATP channels in the signal transduction pathway of ischemic preconditioning of human myocardium. Am J Physiol Cell Physiol. 2004;287:C1418–25.
58. Ohnuma Y, Miura T, Miki T, et al. Opening of mitochondrial K(ATP) channel occurs downstream of PKC-epsilon activation in the mechanism of preconditioning. Am J Physiol Heart Circ Physiol. 2002;283:H440–7.
59. Toker A, Meyer M, Reddy KK, et al. Activation of protein kinase C family members by the novel polyphosphoinositides PtdIns-3,4-P2 and PtdIns-3,4,5-P3. J Biol Chem. 1994;269:32358–67.
60. Okada T, Otani H, Wu Y, et al. Integrated pharmacological preconditioning and memory of cardioprotection: role of protein kinase C and phosphatidylinositol 3-kinase. Am J Physiol Heart Circ Physiol. 2005;289:H761–7.
61. Baines CP, Wang L, Cohen MV, Downey JM. Protein tyrosine kinase is downstream of protein kinase C for ischemic preconditioning's anti-infarct effect in the rabbit heart. J Mol Cell Cardiol. 1998;30:383–92.
62. Fryer RM, Schultz JE, Hsu AK, Gross GJ. Importance of PKC and tyrosine kinase in single or multiple cycles of preconditioning in rat hearts. Am J Physiol. 1999;276:H1229–35.
63. Vahlhaus C, Schulz R, Post H, Rose J, Heusch G. Prevention of ischemic preconditioning only by combined inhibition of protein kinase C and protein tyrosine kinase in pigs. J Mol Cell Cardiol. 1998;30:197–209.
64. Fryer RM, Pratt PF, Hsu AK, Gross GJ. Differential activation of extracellular signal regulated kinase isoforms in preconditioning and opioid-induced cardioprotection. J Pharmacol Exp Ther. 2001;296:642–9.
65. O'Rourke B. Myocardial K(ATP) channels in preconditioning. Circ Res. 2000;87:845–55.
66. Szewczyk A, Wojtczak L. Mitochondria as a pharmacological target. Pharmacol Rev. 2002;54:101–27.
67. Garlid KD, Paucek P, Yarov-Yarovoy V, et al. Cardioprotective effect of diazoxide and its interaction with mitochondrial ATP-sensitive $K^+$ channels. Possible mechanism of cardioprotection. Circ Res. 1997;81:1072–82.
68. Hanley PJ, Mickel M, Loffler M, Brandt U, Daut J. K(ATP) channel-independent targets of diazoxide and 5-hydroxydecanoate in the heart. J Physiol. 2002;542:735–41.
69. D'Hahan N, Moreau C, Prost AL, et al. Pharmacological plasticity of cardiac ATP-sensitive potassium channels toward diazoxide revealed by ADP. Proc Natl Acad Sci USA. 1999;96:12162–7.
70. Sanada S, Kitakaze M, Asanuma H, et al. Role of mitochondrial and sarcolemmal K(ATP) channels in ischemic preconditioning of the canine heart. Am J Physiol Heart Circ Physiol. 2001;280:H256–63.
71. Paucek P, Mironova G, Mahdi F, Beavis AD, Woldegiorgis G, Garlid KD. Reconstitution and partial purification of the glibenclamide-sensitive, ATP-dependent $K^+$ channel from rat liver and beef heart mitochondria. J Biol Chem. 1992;267:26062–9.
72. Bajgar R, Seetharaman S, Kowaltowski AJ, Garlid KD, Paucek P. Identification and properties of a novel intracellular (mitochondrial) ATP-sensitive potassium channel in brain. J Biol Chem. 2001;276: 33369–74.
73. Babenko AP, Gonzalez G, Aguilar-Bryan L, Bryan J. Reconstituted human cardiac KATP channels: functional identity with the native channels from the sarcolemma of human ventricular cells. Circ Res. 1998;83:1132–43.
74. Seharaseyon J, Ohler A, Sasaki N, et al. Molecular composition of mitochondrial ATP-sensitive potassium channels probed by viral Kir gene transfer. J Mol Cell Cardiol. 2000;32:1923–30.
75. Lacza Z, Snipes JA, Kis B, Szabo C, Grover G, Busija DW. Investigation of the subunit composition and the pharmacology of the mitochondrial ATP-dependent $K^+$ channel in the brain. Brain Res. 2003;994:27–36.
76. Morrissey A, Rosner E, Lanning J, et al. Immunolocalization of KATP channel subunits in mouse and rat cardiac myocytes and the coronary vasculature. BMC Physiol. 2005;5:1.
77. Lacza Z, Snipes JA, Miller AW, Szabo C, Grover G, Busija DW. Heart mitochondria contain functional ATP-dependent $K^+$ channels. J Mol Cell Cardiol. 2003;35:1339–47.
78. Ardehali H, Chen Z, Ko Y, Mejia-Alvarez R, Marban E. Multiprotein complex containing succinate dehydrogenase confers mitochondrial ATP-sensitive $K^+$ channel activity. Proc Natl Acad Sci USA. 2004;101:11880–5.
79. Hanley PJ, Gopalan KV, Lareau RA, Srivastava DK, von Meltzer M, Daut J. Beta-oxidation of 5-hydroxydecanoate, a putative blocker of mitochondrial ATP-sensitive potassium channels. J Physiol. 2003; 547:387–93.
80. Lim KH, Javadov SA, Das M, Clarke SJ, Suleiman MS, Halestrap AP. The effects of ischaemic preconditioning, diazoxide and 5-hydroxydecanoate on rat heart mitochondrial volume and respiration. J Physiol. 2002;545:961–74.
81. Das M, Parker JE, Halestrap AP. Matrix volume measurements challenge the existence of diazoxide/glibencamide-sensitive KATP channels in rat mitochondria. J Physiol. 2003;547: 893–902.
82. Javadov SA, Clarke S, Das M, Griffiths EJ, Lim KH, Halestrap AP. Ischaemic preconditioning inhibits opening of mitochondrial permeability transition pores in the reperfused rat heart. J Physiol. 2003;549:513–24.
83. Grimmsmann T, Rustenbeck I. Direct effects of diazoxide on mitochondria in pancreatic B-cells and on isolated liver mitochondria. Br J Pharmacol. 1998;123:781–8.
84. Akao M, O'Rourke B, Kusuoka H, Teshima Y, Jones SP, Marban E. Differential actions of cardioprotective agents on the mitochondrial death pathway. Circ Res. 2003;92:195–202.
85. Korge P, Honda HM, Weiss JN. Protection of cardiac mitochondria by diazoxide and protein kinase C: implications for ischemic preconditioning. Proc Natl Acad Sci USA. 2002;99:3312–7.
86. Murata M, Akao M, O'Rourke B, Marban E. Mitochondrial ATP-sensitive potassium channels attenuate matrix Ca(2+) overload during simulated ischemia and reperfusion: possible mechanism of cardioprotection. Circ Res. 2001;89:891–8.
87. Katoh H, Nishigaki N, Hayashi H. Diazoxide opens the mitochondrial permeability transition pore and alters $Ca^{2+}$ transients in rat ventricular myocytes. Circulation. 2002;105:2666–71.
88. Xu M, Wang Y, Ayub A, Ashraf M. Mitochondrial K(ATP) channel activation reduces anoxic injury by restoring mitochondrial membrane potential. Am J Physiol Heart Circ Physiol. 2001;281: H1295–303.
89. Griffiths EJ, Halestrap AP. Mitochondrial non-specific pores remain closed during cardiac ischaemia, but open upon reperfusion. Biochem J. 1995;307(Pt 1):93–8.

90. Crompton M, Costi A. Kinetic evidence for a heart mitochondrial pore activated by $Ca^{2+}$, inorganic phosphate and oxidative stress. A potential mechanism for mitochondrial dysfunction during cellular $Ca^{2+}$ overload. Eur J Biochem. 1988;178:489–501.
91. Minners J, van den Bos EJ, Yellon DM, Schwalb H, Opie LH, Sack MN. Dinitrophenol, cyclosporin A, and trimetazidine modulate preconditioning in the isolated rat heart: support for a mitochondrial role in cardioprotection. Cardiovasc Res. 2000;47: 68–73.
92. Zhao ZQ, Vinten-Johansen J. Myocardial apoptosis and ischemic preconditioning. Cardiovasc Res. 2002;55:438–55.
93. Opie LH, Sack MN. Metabolic plasticity and the promotion of cardiac protection in ischemia and ischemic preconditioning. J Mol Cell Cardiol. 2002;34:1077–89.
94. Knopp A, Thierfelder S, Doepner B, Benndorf K. Mitochondria are the main ATP source for a cytosolic pool controlling the activity of ATP-sensitive K(+) channels in mouse cardiac myocytes. Cardiovasc Res. 2001;52:236–45.
95. Trochu JN, Bouhour JB, Kaley G, Hintze TH. Role of endothelium-derived nitric oxide in the regulation of cardiac oxygen metabolism: implications in health and disease. Circ Res. 2000;87:1108–17.
96. Miura T, Liu Y, Kita H, Ogawa T, Shimamoto K. Roles of mitochondrial ATP-sensitive K channels and PKC in anti-infarct tolerance afforded by adenosine A1 receptor activation. J Am Coll Cardiol. 2000;35:238–45.
97. Bosetti F, Yu G, Zucchi R, Ronca-Testoni S, Solaini G. Myocardial ischemic preconditioning and mitochondrial F1F0-ATPase activity. Mol Cell Biochem. 2000;215:31–7.
98. Ozcan C, Holmuhamedov EL, Jahangir A, Terzic A. Diazoxide protects mitochondria from anoxic injury: implications for myopreservation. J Thorac Cardiovasc Surg. 2001;121:298–306.
99. Green DW, Murray HN, Sleph PG, et al. Preconditioning in rat hearts is independent of mitochondrial F1F0 ATPase inhibition. Am J Physiol. 1998;274:H90–7.
100. Asimakis GK, Inners-McBride K, Medellin G, Conti VR. Ischemic preconditioning attenuates acidosis and postischemic dysfunction in isolated rat heart. Am J Physiol. 1992;263:H887–94.
101. Laclau MN, Boudina S, Thambo JB, et al. Cardioprotection by ischemic preconditioning preserves mitochondrial function and functional coupling between adenine nucleotide translocase and creatine kinase. J Mol Cell Cardiol. 2001;33:947–56.
102. da Silva MM, Sartori A, Belisle E, Kowaltowski AJ. Ischemic preconditioning inhibits mitochondrial respiration, increases H2O2 release, and enhances $K^+$ transport. Am J Physiol Heart Circ Physiol. 2003;285:H154–62.
103. Garlid KD. Opening mitochondrial K(ATP) in the heart – what happens, and what does not happen. Basic Res Cardiol. 2000;95:275–9.
104. Dos Santos P, Kowaltowski AJ, Laclau MN, et al. Mechanisms by which opening the mitochondrial ATP- sensitive K(+) channel protects the ischemic heart. Am J Physiol Heart Circ Physiol. 2002;283:H284–95.
105. Oldenburg O, Cohen MV, Yellon DM, Downey JM. Mitochondrial K(ATP) channels: role in cardioprotection. Cardiovasc Res. 2002;55:429–37.
106. Ping P, Murphy E. Role of p38 mitogen-activated protein kinases in preconditioning: a detrimental factor or a protective kinase? Circ Res. 2000;86:921–2.
107. Zaugg M, Lucchinetti E, Spahn DR, Pasch T, Garcia C, Schaub MC. Differential effects of anesthetics on mitochondrial K(ATP) channel activity and cardiomyocyte protection. Anesthesiology. 2002;97:15–23.
108. Hanley PJ, Ray J, Brandt U, Daut J. Halothane, isoflurane and sevoflurane inhibit NADH: ubiquinone oxidoreductase (complex I) of cardiac mitochondria. J Physiol. 2002;544:687–93.
109. Er F, Michels G, Gassanov N, Rivero F, Hoppe UC. Testosterone induces cytoprotection by activating ATP-sensitive $K^+$ channels in the cardiac mitochondrial inner membrane. Circulation. 2004;110: 3100–7.
110. Pain T, Yang XM, Critz SD, et al. Opening of mitochondrial K(ATP) channels triggers the preconditioned state by generating free radicals. Circ Res. 2000;87:460–6.
111. Gudz TI, Tserng KY, Hoppel CL. Direct inhibition of mitochondrial respiratory chain complex III by cell-permeable ceramide. J Biol Chem. 1997;272:24154–8.
112. Krenz M, Oldenburg O, Wimpee H, et al. Opening of ATP-sensitive potassium channels causes generation of free radicals in vascular smooth muscle cells. Basic Res Cardiol. 2002;97:365–73.
113. Cohen MV, Yang XM, Liu GS, Heusch G, Downey JM. Acetylcholine, bradykinin, opioids, and phenylephrine, but not adenosine, trigger preconditioning by generating free radicals and opening mitochondrial K(ATP) channels. Circ Res. 2001;89: 273–8.
114. Halmosi R, Berente Z, Osz E, Toth K, Literati-Nagy P, Sumegi B. Effect of poly(ADP-ribose) polymerase inhibitors on the ischemia-reperfusion-induced oxidative cell damage and mitochondrial metabolism in Langendorff heart perfusion system. Mol Pharmacol. 2001;59:1497–505.
115. Ozcan C, Bienengraeber M, Dzeja PP, Terzic A. Potassium channel openers protect cardiac mitochondria by attenuating oxidant stress at reoxygenation. Am J Physiol Heart Circ Physiol. 2002;282:H531–9.
116. Hausenloy DJ, Maddock HL, Baxter GF, Yellon DM. Inhibiting mitochondrial permeability transition pore opening: a new paradigm for myocardial preconditioning? Cardiovasc Res. 2002;55: 534–43.
117. Yue Y, Qin Q, Cohen MV, Downey JM, Critz SD. The relative order of mK(ATP) channels, free radicals and p38 MAPK in preconditioning's protective pathway in rat heart. Cardiovasc Res. 2002;55:681–9.
118. Strohm C, Barancik M, von Bruehl M, et al. Transcription inhibitor actinomycin-D abolishes the cardioprotective effect of ischemic reconditioning. Cardiovasc Res. 2002;55:602–18.
119. Zhao TC, Kukreja RC. Late preconditioning elicited by activation of adenosine A(3) receptor in heart: role of NF-kappa B, iNOS and mitochondrial K(ATP) channel. J Mol Cell Cardiol. 2002;34: 263–77.
120. Hoshida S, Yamashita N, Otsu K, Hori M. The importance of manganese superoxide dismutase in delayed preconditioning: involvement of reactive oxygen species and cytokines. Cardiovasc Res. 2002;55:495–505.
121. Wang Y, Kudo M, Xu M, Ayub A, Ashraf M. Mitochondrial K(ATP) channel as an end effector of cardioprotection during late preconditioning: triggering role of nitric oxide. J Mol Cell Cardiol. 2001;33:2037–46.
122. Ockaili R, Emani VR, Okubo S, Brown M, Krottapalli K, Kukreja RC. Opening of mitochondrial KATP channel induces early and delayed cardioprotective effect: role of nitric oxide. Am J Physiol. 1999;277:H2425–34.
123. Latchman DS. Heat shock proteins and cardiac protection. Cardiovasc Res. 2001;51:637–46.
124. Rakhit RD, Mojet MH, Marber MS, Duchen MR. Mitochondria as targets for nitric oxide-induced protection during simulated ischemia and reoxygenation in isolated neonatal cardiomyocytes. Circulation. 2001;103:2617–23.
125. Yamashita N, Hoshida S, Otsu K, Taniguchi N, Kuzuya T, Hori M. The involvement of cytokines in the second window of ischaemic preconditioning. Br J Pharmacol. 2000;131:415–22.
126. Inagaki K, Chen L, Ikeno F, et al. Inhibition of delta-protein kinase C protects against reperfusion injury of the ischemic heart in vivo. Circulation. 2003;108:2304–7.

127. Zhao ZQ, Corvera JS, Halkos ME, et al. Inhibition of myocardial injury by ischemic postconditioning during reperfusion: comparison with ischemic preconditioning. Am J Physiol Heart Circ Physiol. 2003;285:H579–88.
128. Xu Z, Jiao Z, Cohen MV, Downey JM. Protection from AMP 579 can be added to that from either cariporide or ischemic preconditioning in ischemic rabbit heart. J Cardiovasc Pharmacol. 2002;40:510–8.
129. Li Q, Guo Y, Xuan YT, et al. Gene therapy with inducible nitric oxide synthase protects against myocardial infarction via a cyclooxygenase-2-dependent mechanism. Circ Res. 2003;92: 741–8.
130. Melo LG, Agrawal R, Zhang L, et al. Gene therapy strategy for long-term myocardial protection using adeno-associated virus-mediated delivery of heme oxygenase gene. Circulation. 2002;105: 602–7.
131. Leesar MA, Stoddard MF, Xuan YT, Tang XL, Bolli R. Nonelectrocardiographic evidence that both ischemic preconditioning and adenosine preconditioning exist in humans. J Am Coll Cardiol. 2003;42:437–45.
132. Leesar MA, Stoddard MF, Dawn B, Jasti VG, Masden R, Bolli R. Delayed preconditioning-mimetic action of nitroglycerin in patients undergoing coronary angioplasty. Circulation. 2001;103: 2935–41.
133. Crisafulli A, Melis F, Tocco F, et al. Exercise-induced and nitroglycerin-induced myocardial preconditioning improves hemodynamics in patients with angina. Am J Physiol Heart Circ Physiol. 2004;287:H235–42.
134. Mikhail P, Verma S, Fedak PW, Weisel RD, Li RK. Does ischemic preconditioning afford clinically relevant cardioprotection? Am J Cardiovasc Drugs. 2003;3:1–11.
135. Kloner RA, Speakman MT, Przyklenk K. Ischemic preconditioning: a plea for rationally targeted clinical trials. Cardiovasc Res. 2002;55:526–33.
136. Chen Z, Chua CC, Ho YS, Hamdy RC, Chua BH. Overexpression of Bcl-2 attenuates apoptosis and protects against myocardial I/R injury in transgenic mice. Am J Physiol Heart Circ Physiol. 2001;280:H2313–20.
137. Ardehali H, O'Rourke B, Marban E. Cardioprotective role of the mitochondrial ATP-binding cassette protein 1. Circ Res. 2005;97: 740–2.
138. Weiss JN, Korge P, Honda HM, Ping P. Role of the mitochondrial permeability transition in myocardial disease. Circ Res. 2003;93: 292–301.
139. Ishida H, Hirota Y, Genka C, Nakazawa H, Nakaya H, Sato T. Opening of mitochondrial K(ATP) channels attenuates the ouabain-induced calcium overload in mitochondria. Circ Res. 2001;89: 856–8.
140. Eaton M, Hernandez LA, Schaefer S. Ischemic preconditioning and diazoxide limit mitochondrial Ca overload during ischemia/reperfusion: role of reactive oxygen species. Exp Clin Cardiol. 2005;10:96–103.
141. Holmuhamedov EL, Wang L, Terzic A. ATP-sensitive $K^+$ channel openers prevent $Ca^{2+}$ overload in rat cardiac mitochondria. J Physiol. 1999;519(Pt 2):347–60.
142. Toda T, Kadono T, Hoshiai M, et al. $Na^+/H^+$ exchanger inhibitor cariporide attenuates the mitochondrial $Ca^{2+}$ overload and PTP opening. Am J Physiol Heart Circ Physiol. 2007;293: H3517–23.
143. Schneider A, Ad N, Izhar U, Khaliulin I, Borman JB, Schwalb H. Protection of myocardium by cyclosporin A and insulin: in vitro simulated ischemia study in human myocardium. Ann Thorac Surg. 2003;76:1240–5.
144. Hausenloy D, Wynne A, Duchen M, Yellon D. Transient mitochondrial permeability transition pore opening mediates preconditioning-induced protection. Circulation. 2004;109:1714–7.
145. Hausenloy DJ, Yellon DM, Mani-Babu S, Duchen MR. Preconditioning protects by inhibiting the mitochondrial permeability transition. Am J Physiol Heart Circ Physiol. 2004;287: H841–9.
146. Shimizu S, Narita M, Tsujimoto Y. Bcl-2 family proteins regulate the release of apoptogenic cytochrome c by the mitochondrial channel VDAC. Nature. 1999;399:483–7.
147. Shimizu S, Ide T, Yanagida T, Tsujimoto Y. Electrophysiological study of a novel large pore formed by Bax and the voltage-dependent anion channel that is permeable to cytochrome c. J Biol Chem. 2000;275:12321–5.
148. Green DR SD, McStay, G. et al. Voltage-dependent anion channel (VDAC) to Bax (Bax). Science Signaling 2009. http://stke.sciencemag.org/cgi/cm/stkecm;CMR_16290.
149. Chipuk JE, Fisher JC, Dillon CP, Kriwacki RW, Kuwana T, Green DR. Mechanism of apoptosis induction by inhibition of the anti-apoptotic BCL-2 proteins. Proc Natl Acad Sci USA. 2008;105: 20327–32.
150. Lim SY, Davidson SM, Hausenloy DJ, Yellon DM. Preconditioning and postconditioning: the essential role of the mitochondrial permeability transition pore. Cardiovasc Res. 2007;75:530–5.
151. Imahashi K, Schneider MD, Steenbergen C, Murphy E. Transgenic expression of Bcl-2 modulates energy metabolism, prevents cytosolic acidification during ischemia, and reduces ischemia/reperfusion injury. Circ Res. 2004;95:734–41.
152. Chatterjee S, Stewart AS, Bish LT, et al. Viral gene transfer of the antiapoptotic factor Bcl-2 protects against chronic postischemic heart failure. Circulation. 2002;106:I212–7.
153. Forbes RA, Steenbergen C, Murphy E. Diazoxide-induced cardioprotection requires signaling through a redox-sensitive mechanism. Circ Res. 2001;88:802–9.
154. Cheng EH, Wei MC, Weiler S, et al. BCL-2, BCL-X(L) sequester BH3 domain-only molecules preventing BAX- and BAK-mediated mitochondrial apoptosis. Mol Cell. 2001;8:705–11.
155. Zhao ZQ, Nakamura M, Wang NP, et al. Reperfusion induces myocardial apoptotic cell death. Cardiovasc Res. 2000;45: 651–60.
156. Schwanke U, Konietzka I, Duschin A, Li X, Schulz R, Heusch G. No ischemic preconditioning in heterozygous connexin43-deficient mice. Am J Physiol Heart Circ Physiol. 2002;283:H1740–2.
157. Jain SK, Schuessler RB, Saffitz JE. Mechanisms of delayed electrical uncoupling induced by ischemic preconditioning. Circ Res. 2003;92:1138–44.
158. Totzeck A, Boengler K, van de Sand A, et al. No impact of protein phosphatases on connexin 43 phosphorylation in ischemic preconditioning. Am J Physiol Heart Circ Physiol. 2008;295:H2106–12.
159. Schulz R, Cohen MV, Behrends M, Downey JM, Heusch G. Signal transduction of ischemic preconditioning. Cardiovasc Res. 2001; 52:181–98.
160. Matsuyama N, Leavens JE, McKinnon D, Gaudette GR, Aksehirli TO, Krukenkamp IB. Ischemic but not pharmacological preconditioning requires protein synthesis. Circulation. 2000;102: III312–8.
161. Onody A, Zvara A, Hackler Jr L, Vigh L, Ferdinandy P, Puskas LG. Effect of classic preconditioning on the gene expression pattern of rat hearts: a DNA microarray study. FEBS Lett. 2003;536:35–40.
162. Sergeev P, da Silva R, Lucchinetti E, et al. Trigger-dependent gene expression profiles in cardiac preconditioning: evidence for distinct genetic programs in ischemic and anesthetic preconditioning. Anesthesiology. 2004;100:474–88.
163. Smith RM, Suleman N, Lacerda L, et al. Genetic depletion of cardiac myocyte STAT-3 abolishes classical preconditioning. Cardiovasc Res. 2004;63:611–6.
164. Jancso G, Lantos J, Borsiczky B, Szanto Z, Roth E. Dynamism of NF-kappaB and AP-1 activation in the signal transduction

of ischaemic myocardial preconditioning. Eur Surg Res. 2004;36: 129–35.
165. Xuan YT, Tang XL, Banerjee S, et al. Nuclear factor-kappaB plays an essential role in the late phase of ischemic preconditioning in conscious rabbits. Circ Res. 1999;84:1095–109.
166. Fauchon MA, Pell TJ, Baxter GF, et al. Representational difference analysis of cDNA identifies novel genes expressed following preconditioning of the heart. Exp Mol Med. 2005;37:311–22.
167. Lochner A, Marais E, Genade S, Huisamen B, du Toit EF, Moolman JA. Protection of the ischaemic heart: investigations into the phenomenon of ischaemic preconditioning. Cardiovasc J Afr. 2009;20:43–51.
168. Bolli R, Dawn B, Tang XL, et al. The nitric oxide hypothesis of late preconditioning. Basic Res Cardiol. 1998;93:325–38.
169. Li Q, Guo Y, Tan W, et al. Cardioprotection afforded by inducible nitric oxide synthase gene therapy is mediated by cyclooxygenase-2 via a nuclear factor-kappaB dependent pathway. Circulation. 2007;116:1577–84.
170. McLeod CJ, Jeyabalan AP, Minners JO, Clevenger R, Hoyt Jr RF, Sack MN. Delayed ischemic preconditioning activates nuclear-encoded electron-transfer-chain gene expression in parallel with enhanced postanoxic mitochondrial respiratory recovery. Circulation. 2004;110:534–9.
171. Argaud L, Gateau-Roesch O, Raisky O, Loufouat J, Robert D, Ovize M. Postconditioning inhibits mitochondrial permeability transition. Circulation. 2005;111:194–7.
172. Kin H, Zatta AJ, Lofye MT, et al. Postconditioning reduces infarct size via adenosine receptor activation by endogenous adenosine. Cardiovasc Res. 2005;67:124–33.
173. Hausenloy DJ, Tsang A, Yellon DM. The reperfusion injury salvage kinase pathway: a common target for both ischemic preconditioning and postconditioning. Trends Cardiovasc Med. 2005;15: 69–75.
174. Yellon DM, Hausenloy DJ. Realizing the clinical potential of ischemic preconditioning and postconditioning. Nat Clin Pract Cardiovasc Med. 2005;2:568–75.
175. Przyklenk K, Bauer B, Ovize M, Kloner RA, Whittaker P. Regional ischemic 'preconditioning' protects remote virgin myocardium from subsequent sustained coronary occlusion. Circulation. 1993; 87:893–9.
176. Pell TJ, Baxter GF, Yellon DM, Drew GM. Renal ischemia preconditions myocardium: role of adenosine receptors and ATP-sensitive potassium channels. Am J Physiol. 1998;275:H1542–7.
177. Tokuno S, Hinokiyama K, Tokuno K, Lowbeer C, Hansson LO, Valen G. Spontaneous ischemic events in the brain and heart adapt the hearts of severely atherosclerotic mice to ischemia. Arterioscler Thromb Vasc Biol. 2002;22:995–1001.
178. Schulte G, Sommerschild H, Yang J, et al. Adenosine A receptors are necessary for protection of the murine heart by remote, delayed adaptation to ischaemia. Acta Physiol Scand. 2004;182:133–43.
179. Hausenloy DJ, Mwamure PK, Venugopal V, et al. Effect of remote ischaemic preconditioning on myocardial injury in patients undergoing coronary artery bypass graft surgery: a randomised controlled trial. Lancet. 2007;370:575–9.
180. Botker HE, Kharbanda R, Schmidt MR, et al. Remote ischaemic conditioning before hospital admission, as a complement to angioplasty, and effect on myocardial salvage in patients with acute myocardial infarction: a randomised trial. Lancet. 2010;375:727–4.
181. Takaoka A, Nakae I, Mitsunami K, et al. Renal ischemia/reperfusion remotely improves myocardial energy metabolism during myocardial ischemia via adenosine receptors in rabbits: effects of "remote preconditioning". J Am Coll Cardiol. 1999;33: 556–64.
182. Wolfrum S, Schneider K, Heidbreder M, Nienstedt J, Dominiak P, Dendorfer A. Remote preconditioning protects the heart by activating myocardial PKCepsilon-isoform. Cardiovasc Res. 2002;55:583–9.
183. Weinbrenner C, Schulze F, Sarvary L, Strasser RH. Remote preconditioning by infrarenal aortic occlusion is operative via delta1-opioid receptors and free radicals in vivo in the rat heart. Cardiovasc Res. 2004;61:591–9.
184. Kristiansen SB, Henning O, Kharbanda RK, et al. Remote preconditioning reduces ischemic injury in the explanted heart by a KATP channel-dependent mechanism. Am J Physiol Heart Circ Physiol. 2005;288:H1252–6.
185. Konstantinov IE, Arab S, Li J, et al. The remote ischemic preconditioning stimulus modifies gene expression in mouse myocardium. J Thorac Cardiovasc Surg. 2005;130:1326–32.
186. Konstantinov IE, Arab S, Kharbanda RK, et al. The remote ischemic preconditioning stimulus modifies inflammatory gene expression in humans. Physiol Genomics. 2004;19:143–50.
187. Lang SC, Elsasser A, Scheler C, et al. Myocardial preconditioning and remote renal preconditioning – identifying a protective factor using proteomic methods? Basic Res Cardiol. 2006;101:149–58.
188. Arstall MA, Zhao YZ, Hornberger L, et al. Human ventricular myocytes in vitro exhibit both early and delayed preconditioning responses to simulated ischemia. J Mol Cell Cardiol. 1998;30: 1019–25.
189. Kloner RA, Shook T, Przyklenk K, et al. Previous angina alters in-hospital outcome in TIMI 4. A clinical correlate to preconditioning? Circulation. 1995;91:37–45.
190. Kloner RA, Shook T, Cannon CP, Przyklenk K. Ischemic preconditioning: implications for the geriatric heart. Am J Geriatr Cardiol. 2001;10:145–8. quiz 149–151.
191. Kloner RA, Shook T, Antman EM, et al. Prospective temporal analysis of the onset of preinfarction angina versus outcome: an ancillary study in TIMI-9B. Circulation. 1998;97:1042–5.
192. Yamagishi H, Akioka K, Hirata K, et al. Effects of preinfarction angina on myocardial injury in patients with acute myocardial infarction: a study with resting 123I-BMIPP and 201T1 myocardial SPECT. J Nucl Med. 2000;41:830–6.
193. Vahlhaus C, Neumann J, Luss H, et al. Ischemic preconditioning by unstable angina reduces the release of CK-MB following CABG and stimulates left ventricular HSP-72 protein expression. J Card Surg. 2005;20:412–9.
194. Czibik G, Wu Z, Berne GP, et al. Human adaptation to ischemia by preconditioning or unstable angina: involvement of nuclear factor kappa B, but not hypoxia-inducible factor 1 alpha in the heart. Eur J Cardiothorac Surg. 2008;34:976–84.
195. Teoh LK, Grant R, Hulf JA, Pugsley WB, Yellon DM. The effect of preconditioning (ischemic and pharmacological) on myocardial necrosis following coronary artery bypass graft surgery. Cardiovasc Res. 2002;53:175–80.
196. Argaud L, Ovize M. How to use the paradigm of ischemic preconditioning to protect the heart? Med Sci (Paris). 2004;20:521–5.
197. Mahaffey KW, Puma JA, Barbagelata NA, et al. Adenosine as an adjunct to thrombolytic therapy for acute myocardial infarction: results of a multicenter, randomized, placebo-controlled trial: the Acute Myocardial Infarction STudy of ADenosine (AMISTAD) trial. J Am Coll Cardiol. 1999;34:1711–20.
198. Ross AM, Gibbons RJ, Stone GW, Kloner RA, Alexander RW. A randomized, double-blinded, placebo-controlled multicenter trial of adenosine as an adjunct to reperfusion in the treatment of acute myocardial infarction (AMISTAD-II). J Am Coll Cardiol. 2005; 45:1775–80.
199. Kloner RA, Forman MB, Gibbons RJ, Ross AM, Alexander RW, Stone GW. Impact of time to therapy and reperfusion modality on the efficacy of adenosine in acute myocardial infarction: the AMISTAD-2 trial. Eur Heart J. 2006;27:2400–5.
200. Pasceri V, Patti G, Nusca A, Pristipino C, Richichi G, Di Sciascio G. Randomized trial of atorvastatin for reduction of myocardial damage during coronary intervention: results from the ARMYDA (Atorvastatin for Reduction of MYocardial Damage during Angioplasty) study. Circulation. 2004;110:674–8.

# Chapter 21
# Targeting Signaling Pathways

**Abstract** A driving force for the assessment of cardiovascular signaling is that it can enhance our understanding of cardiac abnormalities that are known to result from signaling dysfunction. Defective signaling mechanisms are central to the pathogenesis of heart failure (HF), atherosclerosis, and hypertension, and the main goal of the ongoing increasing research in the field of signal transduction pathways is to identify specific targets that may allow the development of new treatment modalities for these diseases. For example, in HF, dysregulation of cardiac $\beta_1$-adrenoceptor signaling and transduction are key features in the disease progression. Drugs interfering with this pathway, traditionally, target membrane receptors (β-adrenoceptor blockers and the inhibitors of the renin–angiotensin–aldosterone axis), although the effector systems of this signaling cascade are also interesting targets. Moreover, development of isoform-selective stimulator and/or inhibitor compounds for adenylyl cyclase, which functions as a signaling catalyst, could lead to organ-specific pharmacotherapeutics to treat HF. Also, $Ca^{2+}$ signaling plays a central role in cardiac cell function, and disturbed $Ca^{2+}$-handling and $Ca^{2+}$-dependent signaling are hallmarks of HF.

In this chapter, we discuss how targeting major components of cardiovascular signaling pathways may enable the development and characterization of reagents with high specificity to heart signaling pathways, and the arrival of new technologies for the inhibition of stress signaling (e.g., specific kinase inhibitors) in cardiac and vascular cells.

**Keywords** Targeting cardiovascular signaling • Antioxidants • Metabolic signaling • Inflammation • Prosurvival pathways

## Introduction

Defective signaling mechanisms are central to the pathogenesis of cardiovascular diseases, including heart failure, atherosclerosis, and hypertension. Previously, in Chap. 18, we have presented a comprehensive discussion on the several signaling systems that contribute to the pathophysiology of atherosclerosis. It will suffice to briefly mention here those systems dealing with: (1) Chronic inflammation, where the complexity of the inflammatory response makes it very challenging to develop effective and targeted therapies. (2) Apoptosis, which is associated with loss of cells following myocardial infarction (MI), atherosclerotic plaque instability, and congestive heart failure. Understanding the molecular mechanisms that regulate apoptosis and the mediators that either prevent or trigger cell death provides a great opportunity for new strategies targeting apoptotic/prosurvival pathways. (3) Redox signaling – reactive oxygen species (ROS) at low levels induce subtle changes in intracellular signaling pathways, but at high levels cause cell damage. The imbalance between ROS and cellular antioxidant defenses will produce oxidative stress (OS). Again, OS is involved in the pathophysiology of hypertension, atherosclerosis, and HF. Potential therapies may specifically target enzymes important in redox signaling (rather than the nonspecific antioxidant approaches that have, to date, been disappointing in clinical trials).

Important risk factors for the development of cardiovascular diseases (CVDs) are obesity-linked diseases such as diabetes mellitus and atherosclerosis, and one of their impacts on the heart is significant changes in metabolic fuel supply and handling. Metabolic abnormalities associated with diabetes are coordinated by adiponectin and nuclear receptors, such as peroxisome proliferator-activated receptors (PPARs); thus, their metabolic modulation has emerged as a new therapeutic strategy for the treatment of HF. This strategy is aimed to optimize myocardial energy utilization, via shifting the substrate utilization from free fatty acids to glucose, and focus the attention on nuclear receptors that function as regulators of energy homeostasis (such as PPARs).

Also, disorders of glucose metabolism are important risk factors to develop cardiovascular diseases. Under hyperglycemic conditions (diabetes mellitus, hypertension, oxidative stress, and aging), reducing sugars nonenzymaticaly modify the proteins leading to formation of advanced glycation end products (AGEs). AGEs and the receptor for AGEs (RAGE) form a signaling pathway, which triggers a series of cellular

J. Marín-García, *Signaling in the Heart*,
DOI 10.1007/978-1-4419-9461-5_21, 

signaling events, including the activation of nuclear factor kappa-light-chain-enhancer of activated B cells (NF-κB), leading to the production of proinflammatory cytokines, and inflammation. Recent studies in rodent models of atherosclerosis demonstrated beneficial effects of RAGE antagonism and genetic deletion, and breaking AGEs appears to be a promising tool in the therapy of cardiovascular insults related to diabetes, hypertension, and aging.

In this chapter, we present currently available information and new perspectives on targeting specific signaling pathways as therapies for a number of cardiovascular infirmities.

## Reactive Oxygen Species Generation, Effects: Antioxidant Response

Reactive oxygen species are derived from many sources including mitochondria, xanthine oxidase, uncoupled nitric oxide (NO) synthases (NOS), and NADPH oxidases. The latter enzymes are important in redox signaling and play role in the pathophysiology of hypertension, atherosclerosis, and HF.

In atherosclerosis, level of ROS increases as a result of activation of ROS-producing enzymes in vascular endothelial cells, such as NADPH oxidase, xanthine oxidase, lipoxygenases, NOS, and dysfunctional mitochondrial respiratory chain. NADPH oxidase-deficient mice show significantly less experimentally induced atherosclerosis compared with normal animals, which implicates this enzyme in the development of the disease [1]. NADPH oxidase and ROS production are increased in vascular cells by a number of agonists associated with atherosclerosis such as angiotensin II, thrombin, platelet-derived growth factor (PDGF), and tumor necrosis factor alpha (TNF-α) [2–5]. Extensively produced ROS is the source for the oxidative modification of low-density lipoproteins. Oxidized low-density lipoprotein (ox-LDL) damages/induces the chronic activation of the vascular endothelium and components of the immune system. One of the mechanisms by which ROS affects atherogenesis is the activation of mitogenic signaling pathways in smooth muscle cells (SMCs) [6]. In addition, upregulation of inducible NOS (iNOS) in macrophages results in elevated NO, which in turn contributes to OS and tissue damage through formation of peroxynitrite [7].

Increasing evidence implicates the redox-sensitive pathways in the development of cardiac hypertrophy. The mechanism involves ROS-mediated activation of various mitogen-activated protein kinases and the transcription factor NF-κB. Key sources of these ROS include NADPH oxidases and nitric oxide synthases. For example, angiotensin II-induced hypertrophy of cardiomyocytes depends on Nox2 isoform of catalytic subunit of NADPH oxidase, and genetically modified mice lacking this enzyme showed the inhibition of angiotensin II-dependent left ventricle (LV) hypertrophy (LVH) [8, 9]. Interestingly, development of pressure overload-induced LV hypertrophy involves another NADPH oxidase, Nox4, and probably uncoupled NOS [10, 11], and in animal models antioxidant *N*-acetylcysteine attenuates this type of LVH [10].

Nox- and xanthine oxidase-derived ROS are also responsible for several other alterations that accompany or lead to congestive heart failure. They increase expression of profibrotic growth factors and genes, activate NF-κB, and thus stimulate interstitial cardiac fibrosis [9, 12]; contribute to myocardial contractile dysfunction via inactivation of nitric oxide [13, 14]; and potentiate adverse LV remodeling after MI [15].

New observations emphasize the role of mitochondria-originated ROS in cardiac cell injury and the contribution of monoamine oxidase (MAO) and adaptor protein $p66^{Shc}$ to both mitochondrial ROS formation and cardiac cell injury [16].

These data suggest that specifically targeting certain ROS-dependent pathways may be an attractive therapeutic alternative. The successful effects of some drugs that are already in use support this idea. Thus, angiotensin I-converting enzyme (ACE) inhibitors and angiotensin II receptor blockers which inhibit NADPH oxidase, as well as statins (inhibit Rac-dependent activation of NADPH oxidase), are shown to be protective in cardiac hypertrophy and remodeling after MI [17].

## Metabolic Signaling Targets

A number of metabolic disorders are high risk factors for heart disease. Among them is obesity, or increased mass of adipose tissue, which increases the risk for hypertension, atherosclerosis, and other CVDs. Thus, understanding the molecular mechanisms of obesity and obesity-linked diseases would facilitate the development of antidiabetic and anti-atherogenic drugs.

One of the key signaling pathways involved in adipocyte differentiation and hypertrophy is adiponectin/adiponectin receptor pathway. Adiponectin, a protein hormone that is secreted from adipose tissue, modulates a number of metabolic processes, including glucose regulation and fatty acid catabolism. Clinical observations indicate that plasma adiponectin levels are low in obese individuals and hypoadiponectinemia is associated with diabetes, coronary artery disease, and hypertension [18–21]. In agreement with clinical findings, adiponectin-deficient mice showed insulin resistance and an atherogenic phenotype. A number of experimental studies have shown that adiponectin plays a protective role in the development of insulin resistance and exerts beneficial effects on the heart under pathological conditions such as hypertension, CVD, and acute ischemia [22–24].

The inhibitory effects of adiponectin on inflammatory and hypertrophic responses, and the stimulation of endothelial cell responses are mainly attributed to modulation of signaling molecules. So far, two adiponectin receptors have been identified: AdipoR1 and AdipoR2 [25], and these receptors affect the downstream targets. One of the adiponectin-dependent pathways involves AMP-activated protein kinase (AMPK), an important cellular metabolic rate control point. For example, through AMPK-dependent phosphorylation of endothelial NO synthase (eNOS), adiponectin stimulates NO production in endothelial cells, and this improves angiogenesis and endothelial function under ischemic conditions [26]. AMPK signaling is also involved in prosurvival (under hypoxia–reoxygenation conditions, adiponectin via AMPK suppressed apoptosis of cultured cardiomyocytes) [23] and in the antihypertrophy [27] effects of adiponectin.

Independent from above is the anti-inflammatory effect of adiponectin in the heart from acute ischemia. In this situation, adiponectin stimulates expression/activity of cyclooxygenase-2 (COX-2) via sphingosine-1-phosphate receptor, and COX-2 in turn activates synthesis of the anti-inflammatory factor prostaglandin $E_2$ [23, 28].

Adiponectin appears to be a promising therapeutic target for CVDs. For example, thiazolidinediones are shown to improve insulin resistance and have beneficial effects on CVD via the activation of peroxisome proliferator-activated receptor γ (PPAR-γ), which controls the synthesis/secretion of adiponectin by adipocyte [29]. Thus, nutritional approaches that influence adiponectin production could be useful in the management of CVDs.

Besides obesity, diabetes is another metabolic defect that may participate in the development of CVD. One aspect of diabetes-caused changes in the heart is related to alterations in the available sources of metabolic fuel, in the lack of insulin (or resistance to its action). In healthy adult heart, lipids are the main source of energy, and their oxidation produces 60–80% of total cardiac ATP (20–40% of ATP is synthesized with the use of glucose as an energy source). In diabetes, more than 90% of total ATP is generated by the "burning" of fatty acids, whereas the capability of the heart to use glucose as a metabolic fuel is greatly diminished. These metabolic abnormalities are associated with the functioning of a regulatory signaling system, called peroxisome proliferator-activated receptor(s) (PPAR). PPARs are ligand-activated nuclear transcription factors that belong to the nuclear receptor superfamily. Three known isoforms of PPAR (α, δ, and γ) regulate glucose and lipid redistribution: PPAR-α is involved in fatty acid oxidation (FAO); PPAR-γ is mainly expressed in adipose tissue and controls adipocyte differentiation and insulin sensitivity; and PPAR-δ is ubiquitously expressed, but its function is not clear.

PPARs regulate the expression of a large number of genes involved in carbohydrate and lipid metabolism. Their activation depends on/and is regulated by ligand binding, cofactor recruitment, and phosphorylation. The main endogenous PPAR ligands include fatty acids, derivatives of fatty acids (products of lipooxygenase- and cyclooxygenase-catalyzed reactions). PPAR-α controls the expression of a wide range of proteins involved in both transport and β-oxidation of free fatty acids. When PPAR-α is absent (PPAR-α null mice) or is not active enough (diabetes and hyperlipidemia), β-oxidation of lipids is not effective, and they accumulate in liver, heart, and kidney (tissues with a high rate of lipid oxidation), which may ultimately lead to lipotoxicity and death (Fig. 21.1).

PPAR-γ is a regulator of genes expression involved in fatty acid uptake and lipogenesis as well as glucose transportation in adipocytes. Activation of PPAR-γ results in the stimulation of adipogenesis and the formation of insulin-sensitive adipocytes.

## PPAR Isoform-Specific Agonists

PPAR-α agonists (fibrates) through the activation of PPAR-α induce hepatic lipid uptake and increase lipoprotein lysis (due to β-oxidation of fatty acids) and can be used to treat disorders with elevated levels of plasma lipids (dyslipidemias).

PPAR-γ agonists (thiazolidinediones) improve insulin sensitivity in diabetics, as they restore insulin-sensitive adipocytes and improve the overall glucose homeostasis (increase glucose transport in adipocytes and decrease glucose formation). On the other hand, PPARs impact on cardiovascular (patho)physiology is not only via their known effects on carbohydrate and lipid metabolism, but also by their influence on the physiology of all the major components of the cardiovascular system. Thus, PPARs play a beneficial anti-inflammatory role in atherosclerosis because they inhibit in macrophages the activity of transcription factors NF-κB, STAT, and activator protein-1, which upregulate the expression of proinflammatory cytokines [30, 31]. Moreover, PPARs downregulate cytokine-induced genes [vascular cell adhesion protein-1 (VCAM-1), tissue factor] and monocytic chemotactic protein-1 in vascular endothelial cells [32–34], and also inhibit vascular smooth muscle cell proliferation and migration [35, 36].

Interestingly, although predominantly expressed in adipocytes, PPAR-γ expression has been found also in rat heart and is cardioprotective under MI/reperfusion conditions. Increasing number of observations have demonstrated that in animal models of ischaemia–reperfusion, agonists of PPAR-γ significantly reduce infarct size and improve myocardial function [37–39], whereas the specific PPAR-γ antagonist, GW9662, significantly increased infarct size following

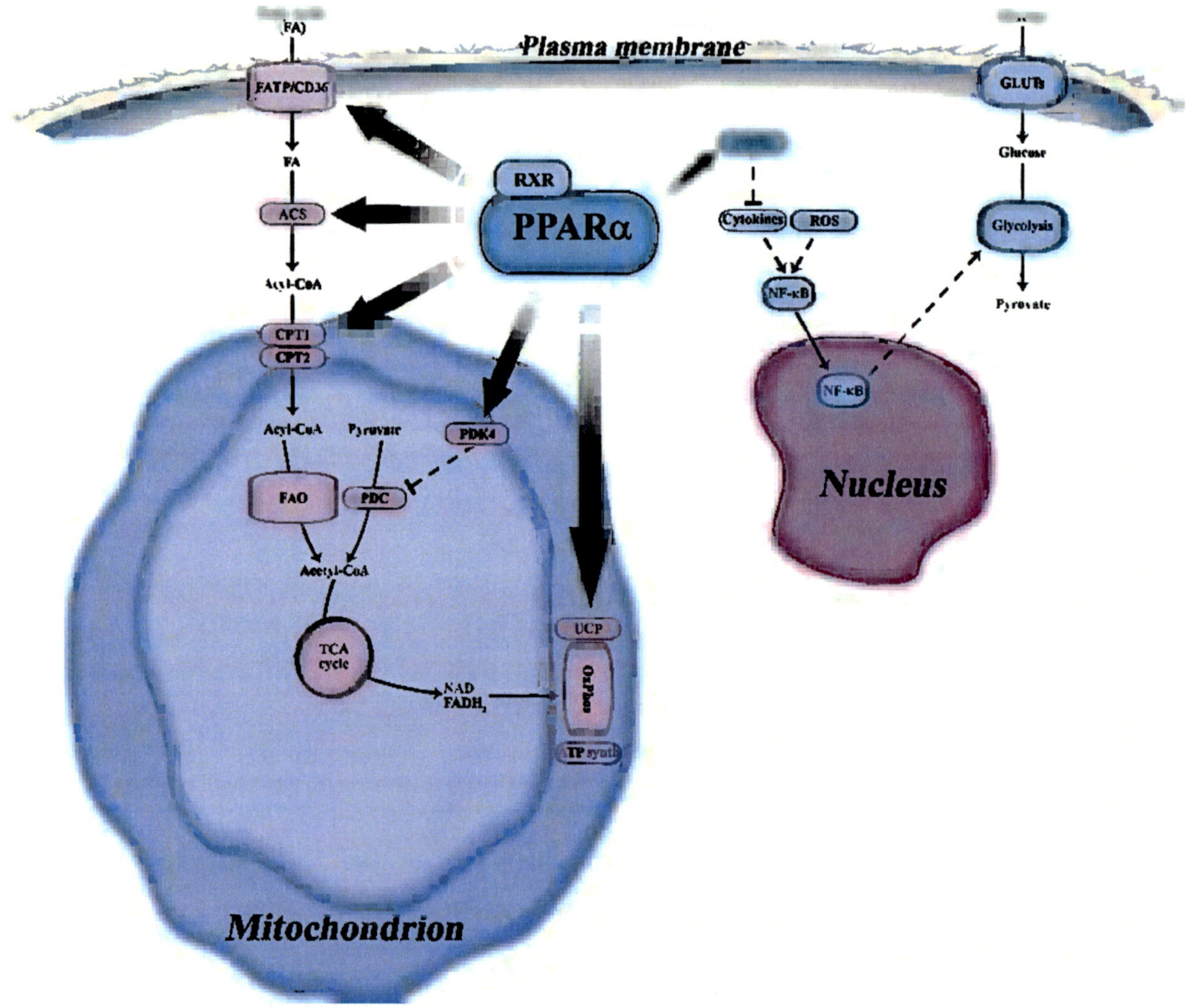


**Fig. 21.1** PPAR-α modulates lipid and glucose metabolism in cardiac cell. PPAR-α-dependent regulations are indicated by *thick black arrows*; other activatory/inhibitory events are indicated by *dashed lines*. Abbreviations: *ACS* acyl-CoA synthetase, *AMPK* AMP-activated protein kinase, *ATP synth* ATP synthase, *CD36* fatty acid translocase, *CPT* carnitine palmitoyltransferase, *FA* fatty acid, *FAO* fatty acid oxidation, *FATP* fatty acid transporter protein, *GLUTs* glucose transporters, *NF-κB* nuclear factor-κB, *OxPhos* oxidative phosphorylation, *PDC* pyruvate dehydrogenase complex, *PDK* pyruvate dehydrogenase kinase, *PPAR* peroxisome proliferator-activated receptor, *ROS* reactive oxygen species, *RXR* retinoid X receptor, *TCA cycle* tricarboxylic acid cycle, *UCP* uncoupling protein

ischaemia–reperfusion [40]. The cardioprotective effects of PPAR-γ seem to be realized via regulation of a number of signaling cascades: inhibition of NF-κB [41], reduction of leukocytes infiltration [42], and inhibition of apoptosis [37].

The role of PPARs in the development of hypertrophy is still a subject of debate (e.g., whether PPAR activity changes in hypertrophy is adaptive or relates to pathological myocardial remodeling). Several observations suggest that downregulation of the major cardiac isoform of PPAR (PPAR-α) in hypertrophy is adaptive because it leads to decrease expression of PPAR-α-controlled enzymes involved in fatty acid oxidation (Fig. 21.1), and hypertrophied myocardium shifts to glucose oxidation metabolism to easy the oxygen consumption burden. In agreement with these observations, pharmacological reactivation of PPAR-α with agonists, or overexpression of PPAR-α leads to a depression of cardiac efficiency and cardiac dysfunction in animal models of cardiac hypertrophy [43, 44].

On the other hand, several recent studies have shown that the downregulation of PPAR-α may be maladaptive, for example, during the decompensated stage of cardiac hypertrophy (see Chap. 15). Thus, PPAR-α agonists fenofibrate and Wy 14,643 inhibit the development of myocardial fibrosis [via inhibition of endothelin-1 (ET-1)-mediated fibroblast proliferation] and attenuate the development of myocardial hypertrophy (through inhibition of ET-1- and NF-κB-induced hypertrophy of cardiomyocytes) [45–47]. Based on these findings, it appears that pharmacological reactivation of PPAR-α may attenuate the progression of hypertrophy and associated contractile dysfunction.

## Targeting Advanced Glycation End Products

Under hyperglycemic OS or inflammatory conditions, reducing sugars (i.e., glucose or fructose) interact nonenzymatically with proteins, lipids, and polynucleotides. Glycation (nonenzymatic glycosylation) and oxidation of them, and subsequent slow complex reactions form a heterogeneous group of adducts called "advanced glycation end products" [48]. Examples of AGEs include pentosidine, pyrraline, imidazolone, and carboxymethyllysine. Accumulation of AGEs increases under physiological aging or diabetes. AGEs are able to bind to a 35 kDa-membrane receptor, the receptor for advanced glycation endproducts. RAGE is as a pattern recognition receptor and belongs to the immunoglobulin super family. Binding of AGE to RAGE results in ROS generation and subsequent activation of the redox-sensitive NF-κB, leading to the production of proinflammatory cytokines, and causing inflammation. The key step that activates NF-κB is the stimulus-dependent, residue-specific phosporylation of NF-κB inhibitor, IκB. The kinase complex that is responsible for the specific phosphorylation of IκB is the IκB kinase (IKK) signalsome [49]. This phosphorylation leads to ubiquitination and proteasomal degradation of the inhibitor, release and activation of NF-κB.

AGEs and their receptor are implicated in normal aging of the arterial wall, as well as in many age-related chronic diseases, including hyperglycemia, diabetes, and CVD. A number of AGE/RAGE-induced changes affect vascular endothelium. Thus, AGE/RAGE stimulate OS leading to the production of peroxynitrite instead of NO. Peroxynitrite is a proatherosclerosis factor damaging endothelial cells and activating platelets [50]. In addition, AGE/RAGE in vascular endothelial cells stimulate NF-κB-dependent expression of a variety of atherosclerosis-related genes (intercellular adhesion molecule-1, vascular cell adhesion molecule-1, etc.) [51], resulting in severe inflammation [52].

Interestingly, ligand-activated RAGE can also function as an adhesion receptor facilitating the early steps of atherogenesis – leukocyte infiltration through the vascular endothelium and adhesion to endothelial cells. In RAGE-deficient mice, inflammation-induced leukocyte recruitment to the endothelium is significantly impaired, but can be restored by vascular endothelium-specific expression of RAGE. RAGE-mediated leukocyte adhesion to endothelial cell is based on a direct interaction between endothelial RAGE and leukocyte integrin [53, 54]. Taking into account the important role of leukocyte migration in atherogenesis and arterial aging, further studies on mechanisms of RAGE-mediated cell migration are warranted.

The inhibition of AGE formation or AGE–RAGE interaction, downregulation of RAGE expression or blockade of the RAGE downstream signaling may be promising therapeutic targets in the prevention of various life-threatening cardiovascular disorders. For example, AGE cross-linkage breaking compounds, such as alagebrium, have been shown to reduce RAGE expression, improve endothelial function in patients with hypertension, and have an anti-inflammatory effect [55, 56]. Yet another approach to neutralize AGEs is the treatment of inflammatory diseases with C-terminally truncated RAGE (sRAGE). This fragment of RAGE is fully capable of ligand (i.e., AGE) binding, but is devoid of signaling function. In animal models, sRAGE suppressed the formation of atherosclerotic lesions, while it stabilized established atherosclerotic lesions [57, 58]. These observations suggest that sRAGE can be used in the treatment of metabolic and inflammatory diseases, as well as an antiaging reagent.

## Inflammation Control

As previously discussed in Chap. 18, atherosclerosis, which underlies CVD, is an inflammatory disease, and inflammatory events and signaling pathways play a central role in the recruitment of inflammatory cells to the artery walls, with elevated levels of LDL cholesterol, as well as in the formation of atherosclerotic plaques. Furthermore, hypertension and other risk factors (smoking, diabetes, etc.) lead to dysfunction of vascular endothelium. This is a start-point of a series of inflammatory receptor-mediated processes such as leukocyte recruitment to the damaged endothelium, adhesion of leukocytes, and penetration through vessel wall to interstitium. Dysfunctional endothelial cells release inflammatory chemokines, which serve as chemoattractants for monocytes: monocytes express G protein-coupled receptors for chemokines [CC chemokine receptors (CCR), CXC chemokine receptors (CXCR)], so they detect gradients of chemokines and migrate towards the chemotactic substance [59]. For instance, macrophage chemo-attractant protein-1 (MCP-1) has a significant role in the recruitment of monocytes to the endothelium: MCP-1 knockout animals do not develop atherosclerosis [60]. The MCP-1 receptor (CCR-2) knockouts also show reduced atherosclerosis [61]. Deficiency in macrophage migration inhibitory factor (MIF) has also been shown to be atheroprotective in mouse model of atherosclerosis [62, 63]. Another factor, allograft inflammatory factor-1 (AIF-1), is also tightly associated with monocyte activation and migration to the areas of forming atherosclerotic lesions and acts via activation of

Akt, p44/42 Erk, and p38 kinase [64]. Thus, activation of pathway phosphatidylinositol 3-kinase (PI3K)-Akt-NF-κB common for several monocyte-activating factors (MIF and AIF-1) indicates that it can be considered as a target when designing therapeutic strategies.

In addition to chemokines, damaged vascular endothelial cells express cell adhesion molecules which are also critical in monocyte recruitment in immune-mediated and hematological diseases – mice deficient in adhesion molecules are protected against atherosclerosis when fed an atherogenic diet [65, 66]. Upregulation of adhesion molecules/receptors on the surface of dysfunctional endothelial cells include E- and P-selectins. These proteins are receptors for carbohydrate ligands on leukocyte surfaces. In addition, endothelial cells express a number of ligands, which interact with integrins present in leukocytes: intracellular adhesion molecule (ICAM), vascular cell adhesion molecule 1, and endothelial leukocyte adhesion molecule (ELAM) [67]. Upregulation of VCAM-1 and ICAM-1, which can be induced by cytokines [TNF-α and interleukin (IL)-1], includes activation of PI3K-Src-NF-κB cascade [65] and occurs as a result of increased gene transcription. While VCAM-1 is an endothelial ligand for integrin "Very Late Antigen-4" (VLA-4), ICAM-1 is a ligand for integrin "Lymphocyte function-associated antigen 1" (LFA-1), both found on leukocytes. As a result of protein–protein interactions between endothelial cells and recruited leukocytes, the latter cells first adhere to the endothelial cells and then move between them to vessel adventitia. Signaling pathways that regulate expression of ICAM, VCAM-1 may be potential targets for treating atherosclerosis, which is characterized by upregulation of cell adhesion molecules.

Within vessel adventitia, leukocytes bind to extracellular matrix proteins via expressed integrins and CD44 (receptor for hyaluronic acid, collagens, and matrix metalloproteinases), and interact with other cells including T cells and vascular smooth muscle cells via various receptors and cytokines released by them. This leads to an enhanced inflammatory response. As monocytes/macrophages are central to the initiation and progression of atherosclerosis, they may be appropriate targets for designing anti-inflammatory drugs.

## Apoptotic and Prosurvival (Proliferative) Pathways

In response to chronic increases in hemodynamic pressure and volume overload (e.g., long-standing hypertension and postmyocardial injury), the heart initiates an adaptive response of compensatory hypertrophy that leads to cardiac enlargement and maintenance of normal cardiac function. However, there can be a transition point, where biomechanical stimuli subsequently activate pathways that lead to cardiac dysfunction, myocyte loss, replacement fibrosis, and overt congestive heart failure. Several studies have documented activation of the proapoptotic signaling pathway(s) as a critical point in the transition between compensatory cardiac hypertrophy and heart failure [68, 69]. Progression from hypertrophy to apoptosis *in vitro* and *in vivo* was coincident with activation of p38 and Jun kinases [69]. Thus, inhibition of cardiac myocytes apoptosis is clearly a valid target for the development of new therapeutic agents for HF.

The actual combination of functioning signaling cascades in the hypertrophied myocardium is rather complex and includes, for instance, the prosurvival PI3K-Akt-pathways. Activation of Akt can be accomplished by several prosurvival signaling pathways, including insulin [70], insulin-like growth factor 1 (IGF-1) [71], gp130 cytokine [72], estrogen [73], neuregulin-1 [74], and agonists of $G_q$ protein coupled receptors [75], all of them activating PI3K with subsequent stimulation of Akt. There are many downstream targets of Akt involved in the prosurvival effect of this kinase in cardiomyocytes. Two of them, B-cell lymphoma 2 (Bcl-2) protein-associated death promoter (BAD) and glycogen synthase kinase-3β (GSK-3β), belong to apoptotic pathway. Activated Akt phosphorylates/inhibits these proteins [76].

Contribution of GSK-3β to cardiomyocyte survival/apoptosis is based on the involvement of this kinase in the regulation of the mitochondrial permeant transition pore (MTP). While under normal physiological conditions, MTP is closed, a variety of causes (including ischemia/reperfusion) lead to its opening, and this results in mitochondrial dysfunction, destabilization of cardiomyocyte homeostasis, and apoptosis. GSK-3β via phosphorylation of several components of the MTP (voltage-dependent anion channel and p53 protein) and indirect interactions with another components of the MTP (cyclophilin-D and adenine nucleotide translocator) increases the sensitivity of the MTP to $Ca^{2+}$, a trigger of channel opening [77, 78]. A variety of prosurvival stimuli activate PI3K/Akt-, Erk-, and PKC, which phosphorylate/inactivate GSK-3β. As a result, the opening threshold of MTP is elevated, and it stays closed. Considering the essential role that GSK-3β plays in cardiomyocyte death, targeting this protein kinase seems a promising strategy cardioprotection in coronary artery disease and HF.

In addition, Akt is involved in a cascade of phosphorylations resulting in the activation of NF-κB: Akt directly phosphorylates the inhibitor of κB kinase, which allows IKK to phosphorylate the inhibitory subunit of NF-κB. Once phosphorylated, the inhibitory subunit dissociates from NF-κB, and NF-κB translocates to the nucleus, where it upregulates transcriptionally antiapoptotic target genes [79].

Besides Akt, active PI3K in cardiomyocytes triggers at least two other documented prosurvival kinases. One of them is serum- and glucocorticoid-responsive kinase-1 (SGK1). According to Aoyama et al. [80], IGF-1 via PI3K-SGK1 protects cardiac cells and inhibits their apoptosis. Also, downstream of PI3K, although independent of Akt, is the integrin-linked kinase (ILK). ILK contributes to cardiomyocyte survival/preservation of cardiac function in mice after coronary ligation [81].

In summary, potential pharmacological targets include:

1. PI3K-Akt pathway. Overexpression of Akt reduced apoptosis and infarct size after ischemic injury [82]. Similarly, overexpression of Akt or PI3K inhibits *in vitro* cardiomyocyte apoptosis caused by hypoxia [83]. However, it is pertinent to note that high level and/or chronic overexpression of Akt may be maladaptive [84].
2. IKK-NF-κB pathway. Overexpression of IKK is sufficient to activate NF-κB and prevent cardiomyocyte death and dysfunction following hypoxic stress [85].
3. Inhibition of caspases, which are core components of apoptotic pathways, has a promising therapeutic potential in MI. For example, irreversible caspase peptide inhibitor benzyloxycarbonyl-Val-Ala-Asp-fluoromethylketone (z-VAD-fmk) improved MI-caused apoptosis of myocytes in an animal model [86]. Several other peptide and peptidomimetic caspase inhibitors designed by Idun pharmaceuticals, Maxim pharmaceuticals, and Pfizer have shown to reduce heart muscle damage and prevent apoptosis in animal models [87].
4. Apoptosis signal-regulating kinase 1 (ASK1) induces apoptosis in cardiomyocytes via activation of p38 and Jun kinases (see above). Hikoso et al. [88] transcoronary transferred dominant-negative mutant of ASK1 efficiently prevented HF progression in cardiomyopathic hamsters. These investigators concluded that viral vector-mediated gene therapy can be applied to target ASK1 to treat HF patients.

Cell apoptosis and proliferation are very important components of the pathophysiology of another common cardiac disease, atherosclerosis. They contribute to the inflammatory response of damaged vessel, plaque vulnerability and destabilization, and thrombus formation.

Atherosclerotic plaques are known to contain different types of cells (endothelial cells, SMCs, lymphocytes and macrophages), and all of them undergo apoptosis with varying outcome. However, the significance of apoptotic cell death in atherosclerosis remains unclear. Apoptosis of macrophages and T lymphocytes may be beneficial as it could attenuate the inflammatory response by removal of inflammatory cells from the plaque. At the same time, apoptosis of SMCs decreases the synthesis of collagen fibers, weakens the fibrous cap, and thus can be harmful.

One of the signaling factors associated with apoptosis and the expression of apoptosis-related proteins in both early and advanced atherosclerotic lesions are oxidative forms of low-density lipoproteins, accumulating in the subendothelial matrix. Oxidized LDL-induced apoptosis of endothelial and SMCs is mediated by receptors of the TNF receptor family, including Fas (APO-1/CD95) and TNF receptors I and II [89]. Observations that support oxLDL-dependent induction of Fas/caspase-apoptotic pathway include upregulation of Fas ligand (FasL), downregulation of caspase-8-related FLICE-inhibitory protein (FLIP), and activation of class I and class II caspases [89, 90]. In addition, in the vicinity of atherosclerotic plaque, oxLDLs upregulate p53 in macrophages and endothelial cells, which contribute to apoptosis [89, 91]. Reduction of FasL (for example, by antiFasL antibodies) may be therapeutically beneficial as it was shown to reduce apoptosis caused by oxLDLs [89].

It has been also brought to our attention that the response of macrophages to oxidized LDL (proliferation or death) depends on the oxidation degree, the exposure time, and the concentration of oxLDL: lightly oxidized LDL at relatively low concentrations induced proliferation of macrophages [92]. Also, the sensitivity of macrophages to oxLDL dramatically changes after they start expressing scavenger receptors and convert into foam cells uploaded with oxLDL. Moreover, expression of macrophages scavenger receptors A (MSR-As) stimulates expression of prosurvival signaling proteins in these cells, such as Bcl-2, proto-oncogene serine/threonine-protein kinase (PIM-1), and p21 [93]. Interestingly, SMCs in atherosclerotic plaques are heterogeneous in response to Fas-dependent apoptosis: some of them are relatively apoptosis-resistant because they have changed levels of signaling proteins downstream to the Fas receptor [less Fas ligand (FasL) and caspases, but more FLIP] [94].

## β-Adrenergic Pathways and Calcium Signaling

Heart failure is characterized by activation of the sympathetic nervous system (SNS) and subsequent stimulation of cardiac β-adrenergic receptors (β-ARs) in an attempt to compensate for reduced cardiac output. Prolonged SNS activation harmfully affects myocardial excitation–contraction coupling, disturbs $Ca^{2+}$ handling and enhances apoptotic pathways, playing a central role in the progression of chronic heart failure [95, 96].

Heart failure alters the components of the β-AR signaling pathway: decreases the density of $\beta_1$-ARs, uncouples $\beta_1$-AR from $G_s$ protein, and increases the level of $G_i$ protein [97]. In addition, β-AR kinase 1 and phosphoinositide 3-kinase are recruited to the ligand-chronically activated $\beta_1$-AR [98]. Prolonged exposure to norepinephrine is cardiotoxic: it stimulates apoptosis of cardiomyocytes via the adrenergic and the ROS/caspase signaling pathways and produces damaging reactive catecholamine intermediates [99, 100].

The β-AR blockers (β-blockers) form the mainstay of current medical management of chronic heart failure. Chronic β-blocker therapy inhibits catecholamine cardiotoxic effects, upregulates $\beta_1$-ARs, attenuates proapoptotic factors release, and improves myocardial performance [101].

As an alternative to the β-AR blockade, direct targeting of cardiac adenylyl cyclase isoform (AC5) might be an effective therapy in the treatment of cardiovascular diseases [102]. Recently, it has been demonstrated that AC5 deletion results in desensitization after long-term catecholamine stress, and protects against cardiomyocyte apoptosis [103]. On the other hand, agents that activate AC5 might mimic selective cardiac specific β-AR agonist and could be used for acute HF treatment [104].

Some of the detrimental effects of chronic $\beta_1$-AR stimulation are the result of $Ca^{2+}$/calmodulin-dependent protein kinase II (CaMKII) activation. Inhibition of CaMKII protects against apoptosis during excessive $\beta_1$-AR stimulation [105] and prevents the chronic catecholamine stimulation-caused progression of structural heart disease [106]. The proapoptotic actions of CaMKII are related to its ability to cause diastolic $Ca^{2+}$ leak from the sarcoplasmic reticulum (SR) as a result of hyperphosphorylation of phospholamban and ryanodine receptors (RyRs) [107, 108]. This markedly increased diastolic $Ca^{2+}$ could result in increased $Ca^{2+}$ uptake by mitochondria and cardiomyocyte apoptosis via mitochondrial death pathway [109]. Since inactivation of CaMKII has a beneficial impact on cardiac function under pathological stress, CaMKII is a promising potential target for HF treatment.

Another β-AR-regulated protein kinase, protein kinase A (PKA), is also involved in RyR hyperphosphorylation/diastolic SR $Ca^{2+}$ leak during HF-induced hyperfunction of the neurohormonal system [110].

Advances in transgenic and gene therapy methodologies may allow novel strategies to target functional alterations of the β-AR system and $Ca^{2+}$-handling in HF. For example, the delivery of the $\beta_2$-AR gene to failing rat hearts (pressure-overload hypertrophy model), resulted in amelioration of LV function [111]. Recently, *in vivo* gene therapy targeted to inhibit β-AR kinase 1 [112], cardiospecific overexpression of AC6 [113], and restoration of $Ca^{2+}$ binding/sensing protein S100A1 expression [114, 115] in animal models lead to enhanced cardiac function and delayed the progression of HF.

## Conclusions

Increasing evidence from studies in animal models and humans is implicating oxidative stress in the development of cardiovascular disease. This also strongly suggests that OS plays an important pathophysiological role in several aspects of the failing heart phenotype. A better understanding of the ROS-dependent signal transduction mechanisms, their localization, and the integration of both ROS-dependent transcriptional and signaling pathways may provide the basis for the development of new therapeutic strategies for CVD and HF.

A decreased capacity for glucose oxidation, as observed in diabetes, is linked to cardiac dysfunction. The PPAR family of nuclear receptors is involved in the regulation of glucose utilization as well as many other aspects of metabolic homeostasis. As suggested by preliminary studies, metabolic modulation strategies targeting PPAR-regulated pathways appears to have significant therapeutic potential. However, better understanding of the role of PPARs in the cardiovascular system is required to delineate the precise mechanisms by which PPARs may modify cellular CVD processes and to identify new therapeutic targets.

Formation of AGEs and its pathophysiological effects contribute to the cardiovascular disorders associated with aging, diabetes, and hypertension. The pivotal role of AGE–RAGE system in the pathophysiological processes suggests that this signaling network is an attractive target for therapeutic intervention. Pharmacological approaches aimed at abolishing the negative effects of AGEs involve agents that can prevent AGEs formation, break existing cross-links, or block their receptors.

Several lines of evidence suggest that apoptosis and alterations in apoptotic signaling contribute to the pathogenesis of HF and atherosclerosis. These observations prompted therapeutic targeting of the apoptotic pathways in heart diseases. However, several questions still remain unanswered. For instance, long-term use of caspase inhibitors could be toxic because it is not known whether caspases have physiological roles beyond that of apoptosis. Another open question is whether preservation of cell survival means preservation of their function.

In the normal and diseased human heart, both β-adrenergic signaling and $Ca^{2+}$ handling in cardiomyocytes are critical. Abnormal β-AR signaling and $Ca^{2+}$ cycling appear to be promising targets to either prevent or treat chronic heart failure. An alternative to the classical β-adrenergic-antagonist therapy against chronic heart failure, selective inhibition of members of the β-AR signaling pathway, including adenylyl cyclase isoforms, CaMKII, and β-AR kinase offer the opportunity to heart-specific manipulation of cAMP.

## Summary

- In atherosclerosis, level of ROS increases as a result of activation of ROS-producing enzymes in vascular endothelial cells. ROS is the source for the oxidative modification of low-density lipoproteins. Oxidized low-density lipoprotein damages/induces the chronic activation of the vascular endothelium and components of the immune system.

- Redox-sensitive pathways are implicated in the development of cardiac hypertrophy. The mechanism involves ROS-mediated activation of various mitogen-activated protein kinases and the transcription factor NF-κB.
- Recent findings emphasize role of mitochondria-originated ROS in cardiac cell injury.
- Specific targeting of certain ROS-dependent pathways after myocardial infarction has been shown to be protective against cardiac hypertrophy and remodeling.
- Adiponectin is a protein hormone that modulates glucose regulation and fatty acid catabolism. Adiponectin plays a protective role in the development of insulin resistance and has proved to be beneficial in CVDs.
- Inhibitory effects of adiponectin on inflammatory and hypertrophic responses involve receptor-dependent activation of AMP-activated protein kinase and cyclooxygenase-2.
- Diabetes-caused heart alterations are related to the changes in available sources of metabolic fuel. Peroxisome proliferator-activated receptors (PPARs) are ligand-activated nuclear transcription factors that regulate glucose and lipid redistribution, and are associated with cardiac metabolic abnormalities.
- PPAR isoform-specific agonists (fibrates and thiazolidinediones) can be used in the treatment of dyslipidemias and improve glucose homeostasis.
- PPARs play a beneficial anti-inflammatory role in atherosclerosis because they downregulate macrophage proinflammatory cytokines, endothelial adhesion proteins, and migratory activity of SMCs.
- While PPARs are cardioprotective in myocardial infarction/reperfusion injury, their role in other cardiac diseases is still a subject of debate.
- Advanced glycation end products (AGEs) are adducts of nonenzymatic modification of proteins, lipids, and polynucleotides with reducing sugars. They progressively accumulate under hyperglycemic, OS or inflammatory conditions, as well as with physiological aging. AGEs, via their receptor (RAGE), cause inflammation.
- C-terminally truncated RAGE (sRAGE) can be used for the treatment of metabolic and inflammatory diseases, as well as an antiaging reagent.
- Dysfunctional endothelial cells release inflammatory chemokines (MCP-1, MIF and AIF-1), which serve as chemoattractants for monocytes. Migration of monocytes leads to development of atherosclerotic lesions.
- In addition to chemokines, damaged endothelial cells express cell adhesion molecules, which are also critical in monocyte recruitment in immune-mediated and hematological diseases: E- and P-selectins, ICAM, and VCAM-1.
- PI3K-Akt-NF-κB- and PI3K-Src-NF-κB-signaling pathways regulate expression of inflammatory chemokines and adhesion molecules/receptors, and may be potential targets for treating atherosclerosis.
- Several studies have documented activation of the proapoptotic signaling pathway(s) as a critical point in the transition between compensatory cardiac hypertrophy and HF. Thus, inhibition of apoptosis is a valid target for the development of therapeutic agents for HF.
- Prosurvival PI3K-Akt-pathway also functions in the hypertrophied myocardium. There are many downstream targets of Akt involved in the prosurvival effect of this kinase in cardiomyocytes: BAD, GSK-3β, and NF-κB.
- Potential antiapoptotic pharmacological targets include PI3K-Akt pathway, IKK-NF-κB pathway, caspases, and ASK1.
- Apoptosis and proliferation play an important role in the pathophysiology of atherosclerosis. Oxidized LDLs induce apoptosis of endothelial and SMCs via Fas-caspase-apoptotic pathway. Response of macrophages to oxidized LDL (proliferation or death) depends on the oxidation degree, the exposure time and the concentration of oxidized LDL.
- HF alters the components of the β-AR signaling pathway: decreases density of $\beta_1$-ARs and uncouples $\beta_1$-AR from $G_s$ protein. During HF, prolonged exposure to norepinephrine is cardiotoxic: it stimulates apoptosis of cardiomyocytes. Chronic β-blocker therapy inhibits the catecholamine cardiotoxic effects, upregulates $\beta_1$-ARs, attenuates proapoptotic systems, and improves myocardial performance.
- As an alternative to the β-AR blockade, a direct targeting of cardiac adenylyl cyclase isoform (AC5) might be effective in the treatment of CVDs.
- CaMKII hyperactivation is a proapoptotic factor: CaMKII disturbs $Ca^{2+}$ handling that triggers cardiomyocyte apoptosis via mitochondrial death pathway. Targeting CaMKII is a promising modality of HF treatment.
- Gene therapy strategies to target functional alterations of the β-AR system and $Ca^{2+}$-handling in HF appears to be promising.

## References


1. Barry-Lane PA, Patterson C, van der Merwe M, et al. p47*phox* is required for atherosclerotic lesion progression in *ApoE–/–* mice. J Clin Invest. 2001;108:1513–22.
2. Ushio-Fukai M, Zafari AM, Fukui T, Ishizaka N, Griendling KK. p22phox is a critical component of the superoxide-generating NADH/NADPH oxidase system and regulates angiotensin

II-induced hypertrophy in vascular smooth muscle cells. J Biol Chem. 1996;271:23317–21.
3. Patterson C, Ruef J, Madamanchi NR, Barry-Lane P, Hu Z, Horaist C, et al. Stimulation of vascular cell NAD(P)H oxidase by thrombin; evidence that p47phox may participate in forming this oxidase in vitro and in vivo. J Biol Chem. 1999;274:19814–22.
4. Sundaresan M, Yu ZX, Ferrans VJ, Irani K, Finkel T. Requirement for generation of $H_2O_2$ for platelet-derived growth factor signal transduction. Science. 1995;270:296–9.
5. De Keulenaer GW, Alexander RW, Ushio-Fukai M, Ishizaka N, Griendling KK. Tumor necrosis factor α activates a p22phox-based NADH oxidase in vascular smooth muscle cells. Biochem J. 1998;329:653–7.
6. Vendrov AE, Hakim ZS, Madamanchi NR, Rojas M, Madamanchi C, Runge MS. Atherosclerosis is attenuated by limiting superoxide generation in both macrophages and vessel wall cells. Arterioscler Thromb Vasc Biol. 2007;27:2714–21.
7. Esaki T, Hayashi T, Muto E, et al. Expression of inducible nitric oxide synthase and Fas/Fas ligand correlates with the incidence of apoptotic cell death in atheromatous plaques of human coronary arteries. Nitric Oxide. 2000;4:561–71.
8. Nakagami H, Takemoto M, Liao JK. NADPH oxidase-derived superoxide anion mediates angiotensin II-induced cardiac hypertrophy. J Mol Cell Cardiol. 2003;35:851–9.
9. Bendall JK, Cave AC, Heymes C, et al. Pivotal role of a gp91phox-containing NADPH oxidase in angiotensin II-induced cardiac hypertrophy in mice. Circulation. 2002;105:293–6.
10. Byrne JA, Grieve DJ, Bendall JK, et al. Contrasting roles of NADPH oxidase isoforms in pressure-overload versus angiotensin II-induced cardiac hypertrophy. Circ Res. 2003;93:802–5.
11. Takimoto E, Champion HC, Li M, et al. Oxidant stress from nitric oxide synthase-3 uncoupling stimulates cardiac pathologic remodeling from chronic pressure load. J Clin Invest. 2005;115: 1221–31.
12. Johar S, Cave AC, Narayanapanicker A, et al. Aldosterone mediates angiotensin II-induced interstitial cardiac fibrosis via a Nox2-containing NADPH oxidase. FASEB J. 2006;20:1546–8.
13. MacCarthy PA, Grieve DJ, Li JM, et al. Impaired endothelial regulation of ventricular relaxation in cardiac hypertrophy: role of reactive oxygen species and NADPH oxidase. Circulation. 2001;104:2967–74.
14. Berry CE, Hare JM. Xanthine oxidoreductase and cardiovascular disease: molecular mechanisms and pathophysiological implications. J Physiol. 2004;555:589–606.
15. Engberding N, Spiekermann S, Schaefer A, et al. Allopurinol attenuates left ventricular remodeling and dysfunction after experimental myocardial infarction: a new action for an old drug? Circulation. 2004;110:2175–9.
16. Di Lisa F, Kaludercic N, Carpi A, Menabo R, Giorgio M. Mitochondrial pathways for ROS formation and myocardialinjury: the relevance of p66$^{Shc}$ and monoamine oxidase. Basic Res Cardiol. 2009;104:131–9.
17. Seddon M, Looi YH, Shah AM. Oxidative stress and redox signalling in cardiac hypertrophy and heart failure. Heart. 2007;93:903–7.
18. Arita Y, Kihara S, Ouchi N, et al. Paradoxical decrease of an adipose-specific protein, adiponectin, in obesity. Biochem Biophys Res Commun. 1999;257:79–83.
19. Spranger J, Kroke A, Mohlig M, et al. Adiponectin and protection against type 2 diabetes mellitus. Lancet. 2003;361:226–8.
20. Komura N, Kihara S, Sonoda M, et al. Clinical significance of high-molecular weight form of adiponectin in male patients with coronary artery disease. Circ J. 2008;72:23–8.
21. Iwashima Y, Katsuya T, Ishikawa K, et al. Hypoadiponectinemia is an independent risk factor for hypertension. Hypertension. 2004;43:1318–23.
22. Pischon T, Girman CJ, Hotamisligil GS, Rifai N, Hu FB, Rimm EB. Plasma adiponectin levels and risk of myocardial infarction in men. JAMA. 2004;291:1730–7.
23. Shibata R, Sato K, Pimentel DR, et al. Adiponectin protects against myocardial ischemia-reperfusion injury through AMPK- and COX-2-dependent mechanisms. Nat Med. 2005;11:1096–103.
24. Tao L, Gao E, Jiao X, et al. Adiponectin cardioprotection after myocardial ischemia/reperfusion involves the reduction of oxidative/nitrative stress. Circulation. 2007;115:1408–16.
25. Yamauchi T, Kamon J, Ito Y, et al. Cloning of adiponectin receptors that mediate antidiabetic metabolic effects. Nature. 2003;423:762–9.
26. Cheng KK, Lam KS, Wang Y, et al. Adiponectin-induced endothelial nitric oxide synthase activation and nitric oxide production are mediated by APPL1 in endothelial cells. Diabetes. 2007;56:1387–94.
27. Fujioka D, Kawabata KI, Saito Y, et al. Role of adiponectin receptors in endothelin-induced cellular hypertrophy in cultured cardiomyocytes and their expression in infarcted heart. Am J Physiol Heart Circ Physiol. 2006;290:H2409–16.
28. Ikeda Y, Ohashi K, Shibata R, et al. Cyclooxygenase-2 induction by adiponectin is regulated by a sphingosine kinase-1 dependent mechanism in cardiac myocytes. FEBS Lett. 2008;582:1147–50.
29. Miyazaki Y, Mahankali A, Wajcberg E, Bajaj M, Mandarino LJ, DeFronzo RA. Effect of pioglitazone on circulating adipocytokine levels and insulin sensitivity in type 2 diabetic patients. J Clin Endocrinol Metab. 2004;89:4312–9.
30. Ricote M, Li AC, Willson TM, Kelly CJ, Glass CK. The peroxisome proliferator-activated receptor-gamma is a negative regulator of macrophage activation. Nature. 1998;391:79–82.
31. Jiang C, Ting AT, Seed B. PPAR-γ agonists inhibit production of monocyte inflammatory cytokines. Nature. 1998;391:82–6.
32. Marx N, Sukhova GK, Collins T, Libby P, Plutzky J. PPARα activators inhibit cytokine-induced vascular cell adhesion molecule-1 expression in human endothelial cells. Circulation. 1999;99:3125–31.
33. Marx N, Mach F, Sauty A, et al. Peroxisome proliferator-activated receptor-gamma activators inhibit IFN-gamma-induced expression of the T cell-active CXC chemokines IP-10, Mig, and I-TAC in human endothelial cells. J Immunol. 2000;164:6503–8.
34. Bruemmer D, Blaschke F, Law RE. New targets for PPARγ in the vessel wall: implications for restenosis. Int J Obes (Lond). 2005;29 Suppl 1:S26–30.
35. Delerive P, De BK, Besnard S, et al. Peroxisome proliferator-activated receptor alpha negatively regulates the vascular inflammatory gene response by negative cross-talk with transcription factors NF-κB and AP-1. J Biol Chem. 1999;274:32048–54.
36. Law RE, Goetze S, Xi XP, et al. Expression and function of PPARγ in rat and human vascular smooth muscle cells. Circulation. 2000;101:1311–8.
37. Liu HR, Tao L, Gao E, et al. Antiapoptotic effects of rosiglitazone in hypercholesterolemic rabbits subjected to myocardial ischemia and reperfusion. Cardiovasc Res. 2004;62:135–44.
38. Gonon AT, Bulhak A, Labruto F, Sjoquist PO, Pernow J. Cardioprotection mediated by rosiglitazone, a peroxisome proliferator-activated receptor gamma ligand, in relation to nitric oxide. Basic Res Cardiol. 2007;102:80–9.
39. Zingarelli B, Hake PW, Mangeshkar P, et al. Diverse cardioprotective signaling mechanisms of peroxisome proliferator-activated receptor-gamma ligands, 15-deoxy-δ12,14-prostaglandin J2 and ciglitazone, in reperfusion injury: role of nuclear factor-κB, heat shock factor 1, and Akt. Shock. 2007;28:554–63.
40. Sivarajah A, McDonald MC, Thiemermann C. The cardioprotective effects of preconditioning with endotoxin, but not ischemia, are abolished by a peroxisome proliferator-activated receptor-gamma antagonist. J Pharmacol Exp Ther. 2005;313:896–901.

41. Wayman NS, Hattori Y, McDonald MC, et al. Ligands of the peroxisome proliferator-activated receptors (PPAR-γ and PPAR-α) reduce myocardial infarct size. FASEB J. 2002;16:1027–40.
42. Ito H, Nakano A, Kinoshita M, Matsumori A. Pioglitazone, a peroxisome proliferator-activated receptor-gamma agonist, attenuates myocardial ischemia/reperfusion injury in a rat model. Lab Invest. 2003;83:1715–21.
43. Young ME, Laws FA, Goodwin GW, Taegtmeyer H. Reactivation of peroxisome proliferator-activated receptor α is associated with contractile dysfunction in hypertrophied rat heart. J Biol Chem. 2001;276:44390–5.
44. Finck BN, Lehman JJ, Leone TC, et al. The cardiac phenotype induced by PPARα overexpression mimics that caused by diabetes mellitus. J Clin Invest. 2002;109:121–30.
45. Ogata T, Miyauchi T, Sakai S, Irukayama-Tomobe Y, Goto K, Yamaguchi I. Stimulation of peroxisome-proliferator-activated receptor alpha (PPAR α) attenuates cardiac fibrosis and endothelin-1 production in pressure-overloaded rat hearts. Clin Sci (Lond). 2002;103 Suppl 48:284S–8.
46. Irukayama-Tomobe Y, Miyauchi T, Sakai S, et al. Endothelin-1-induced cardiac hypertrophy is inhibited by activation of peroxisome proliferator-activated receptor-α partly via blockade of c-Jun $NH_2$-terminal kinase pathway. Circulation. 2004;109:904–10.
47. Smeets PJ, Teunissen BE, Planavila A, et al. Inflammatory pathways are activated during cardiomyocyte hypertrophy and attenuated by peroxisome proliferator-activated receptors PPARα and PPARδ. J Biol Chem. 2008;283:29109–18.
48. Bucala R, Cerami A. Advanced glycosylation: chemistry, biology, and implications for diabetes and aging. Adv Pharmacol. 1992;23:1–34.
49. Hacker H, Karin M. Regulation and function of IKK and IKK-related kinases. Sci STKE. 2006;2006:re13.
50. Bucala R, Tracey KJ, Cerami A. Advanced glycosylation products quench nitric oxide and mediate defective endothelium-dependent vasodilatation in experimental diabetes. J Clin Invest. 1991;87: 432–8.
51. Vlassara H, Fuh H, Donnelly T, Cybulsky M. Advanced glycation endproducts promote adhesion molecule (VCAM-1, ICAM-1) expression and atheroma formation in normal rabbits. Mol Med. 1995;1:447–56.
52. Chavakis T, Bierhaus A, Nawroth PP. RAGE (receptor for advanced glycation end products): a central player in the inflammatory response. Microbes Infect. 2004;6:1219–25.
53. Chavakis T, Bierhaus A, Al-Fakhri N, et al. The pattern recognition receptor (RAGE) is a counterreceptor for leukocyte integrins: a novel pathway for inflammatory cell recruitment. J Exp Med. 2003;198:1507–15.
54. Orlova VV, Choi EY, Xie C, et al. A novel pathway of HMGB1-mediated inflammatory cell recruitment that requires Mac-1-integrin. EMBO J. 2007;26:1129–39.
55. Vaitkevicius PV, Lane M, Spurgeon H, et al. A cross-link breaker has sustained effects on arterial and ventricular properties in older rhesus monkeys. Proc Natl Acad Sci USA. 2001;98:1171–5.
56. Zieman SJ, Melenovsky V, Clattenburg L, et al. Advanced glycation endproduct crosslink breaker (alagebrium) improves endothelial function in patients with isolated systolic hypertension. J Hypertens. 2007;25:577–83.
57. Park L, Raman KG, Lee KJ, et al. Suppression of accelerated diabetic atherosclerosis by the soluble receptor for advanced glycation endproducts. Nat Med. 1998;4:1025–31.
58. Bucciarelli LG, Wendt T, Qu W, et al. RAGE blockade stabilizes established atherosclerosis in diabetic apolipoprotein E-null mice. Circulation. 2002;106:2827–35.
59. Kraaijeveld AO, de Jager SC, van Berkel TJ, Biessen EA, Jukema JW. Chemokines and atherosclerotic plaque progression: towards therapeutic targeting? Curr Pharm Des. 2007;13:1039–52.
60. Gu L, Okada Y, Clinton SK, et al. Absence of monocyte chemoattractant protein-1 reduces atherosclerosis in low density lipoprotein receptor-deficient mice. Mol Cell. 1998;2:275–81.
61. Boring L, Gosling J, Cleary M, Charo IF. Decreased lesion formation in CCR2–/– mice reveals a role for chemokines in the initiation of atherosclerosis. Nature. 1998;394:894–7.
62. Pan JH, Sukhova GK, Yang JT, et al. Macrophage migration inhibitory factor deficiency impairs atherosclerosis in low-density lipoprotein receptor-deficient mice. Circulation. 2004;109:3149–53.
63. Bernhagen J, Krohn R, Lue H, et al. MIF is a noncognate ligand of CXC chemokine receptors in inflammatory and atherogenic cell recruitment. Nat Med. 2007;13:587–96.
64. Tian Y, Kelemen SE, Autieri MV. Inhibition of AIF-1 expression by constitutive siRNA expression reduces macrophage migration, proliferation, and signal transduction initiated by atherogenic stimuli. Am J Physiol Cell Physiol. 2006;290:C1083–91.
65. Amin MA, Haas CS, Zhu K, et al. Migration inhibitory factor up-regulates vascular cell adhesion molecule-1 and intercellular adhesion molecule-1 via Src, PI3 kinase, and NFκB. Blood. 2006;107:2252–61.
66. Nageh MF, Sandberg ET, Marotti KR, et al. Deficiency of inflammatory cell adhesion molecules protects against atherosclerosis in mice. Arterioscler Thromb Vasc Biol. 1997;17:1517–20.
67. Galkina E, Ley K. Vascular adhesion molecules in atherosclerosis. Arterioscler Thromb Vasc Biol. 2007;27:2292–301.
68. Hirota H, Chen J, Betz UAK, et al. Loss of a gp130 cardiac muscle cell survival pathway is a critical event in the onset of heart failure during biomechanical stress. Cell. 1999;97:189–98.
69. Adams JW, Sakata Y, Davis MG, et al. Enhanced Gαq signaling: a common pathway mediates cardiac hypertrophy and apoptotic heart failure. Proc Natl Acad Sci USA. 1998;95:10140–5.
70. Baines CP, Wang L, Cohen MV, Downey JM. Myocardial protection by insulin is dependent on phosphatidylinositol 3-kinase but not protein kinase C or KATP channels in the isolated rabbit heart. Basic Res Cardiol. 1999;94:188–98.
71. Li B, Setoguchi M, Wang X, et al. Insulin-like growth factor-1 attenuates the detrimental impact of nonocclusive coronary artery constriction on the heart. Circ Res. 1999;84:1007–19.
72. Oh H, Fujio Y, Kunisada K, et al. Activation of phosphatidylinositol 3-kinase through glycoprotein 130 induces protein kinase B and p70 S6 kinase phosphorylation in cardiac myocytes. J Biol Chem. 1998;273:9703–10.
73. Patten RD, Pourati I, Aronovitz MJ, et al. 17β-estradiol reduces cardiomyocyte apoptosis *in vivo* and *in vitro* via activation of phospho-inositide-3 kinase/Akt signaling. Circ Res. 2004;95:692–9.
74. Fukazawa R, Miller TA, Kuramochi Y, et al. Neuregulin-1 protects ventricular myocytes from anthracycline-induced apoptosis via erbB4-dependent activation of PI3-kinase/Akt. J Mol Cell Cardiol. 2003;35:1473–9.
75. Naga Prasad SV, Esposito G, Mao L, Koch WJ, Rockman HA. Gβγ-dependent phosphoinositide 3-kinase activation in hearts with *in vivo* pressure overload hypertrophy. J Biol Chem. 2000;275:4693–8.
76. Yin H, Chao L, Chao J. Kallikrein/kinin protects against myocardial apoptosis after ischemia/reperfusion via Akt-GSK-3 and Akt-Bad-14-3-3 signaling pathways. J Biol Chem. 2005;280:8022–30.
77. Pastorino JG, Hoek JB, Shulga N. Activation of glycogen synthase kinase 3β disrupts the binding of hexokinase II to mitochondria by phosphorylating voltage-dependent anion channel and potentiates chemotherapy-induced cytotoxicity. Cancer Res. 2005;65:10545–54.
78. Venkatapuram S, Wang C, Krolikowski JG, et al. Inhibition of apoptotic protein p53 lowers the threshold of isoflurane-induced cardioprotection during early reperfusion in rabbits. Anesth Analg. 2006;103:1400–5.

79. Romashkova JA, Makarov SS. NF-κB is a target of AKT in anti-apoptotic PDGF signaling. Nature. 1999;401:86–90.
80. Aoyama T, Matsui T, Novikov M, Park J, Hemmings B, Rosenzweig A. Serum and glucocorticoid-responsive kinase-1 (SGK1) regulates cardiomyocyte survival and hypertrophic response. Circulation. 2005;111:1652–9.
81. Bock-Marquette I, Saxena A, White MD, Dimaio JM, Srivastava D. Thymosin β4 activates integrin-linked kinase and promotes cardiac cell migration, survival and cardiac repair. Nature. 2004;432: 466–72.
82. Matsui T, Tao J, del Monte F, et al. Akt activation preserves cardiac function and prevents injury after transient cardiac ischemia *in vivo*. Circulation. 2001;104:330–5.
83. Matsui T, Li L, del Monte F, et al. Adenoviral gene transfer of activated PI3-kinase and Akt inhibits apoptosis of hypoxic cardiomyocytes *in vitro*. Circulation. 1999;100:2373–9.
84. Matsui T, Li L, Wu JC, et al. Phenotypic spectrum caused by transgenic overexpression of activated Akt in the heart. J Biol Chem. 2002;277:22896–901.
85. Regula KM, Baetz D, Kirshenbaum LA. Nuclear factor-κB represses hypoxia-induced mitochondrial defects and cell death of ventricular myocytes. Circulation. 2004;110:3795–802.
86. Yaoita H, Ogawa K, Maehara K, Maruyama Y. Attenuation of ischemia/reperfusion injury in rats by a caspase inhibitor. Circulation. 1998;97:276–81.
87. Fischer U, Schulze-Osthoff K. New approaches and therapeutics targeting apoptosis in disease. Pharmacol Rev. 2005;57:87–215.
88. Hikoso S, Ikeda Y, Yamaguchi O, et al. Progression of heart failure was suppressed by inhibition of apoptosis signal-regulating kinase 1 via transcoronary gene transfer. J Am Coll Cardiol. 2007; 50:453–62.
89. Napoli C, Quehenberger O, de Nigris F, et al. Mildly oxidized low density lipoprotein activates multiple apoptotic signaling pathways in human coronary cells. FASEB J. 2000;14:1996–2007.
90. Sata M, Walsh K. Endothelial cell apoptosis induced by oxidized LDL is associated with the downregulation of the cellular caspase inhibitor FLIP. J Biol Chem. 1998;273:33103–6.
91. Kinscherf R, Claus R, Wagner M, et al. Apoptosis caused by oxidized LDL is manganese superoxide dismutase and p53 dependent. FASEB J. 1998;12:461–7.
92. Han CY, Pak YK. Oxidation-dependent effects of oxidized LDL: proliferation or cell death. Exp Mol Med. 1999;31:165–73.
93. Shiffman D, Mikita T, Tai JTN, et al. Large scale gene expression analysis of cholesterol-loaded macrophages. J Biol Chem. 2000;275:37324–32.
94. Chan S-W, Hegyi L, Scott S, et al. Sensitivity to Fas-mediated apoptosis is determined below receptor level in human vascular smooth muscle cells. Circ Res. 2000;86:1038–46.
95. Piacentino 3rd V, Weber CR, Chen X, et al. Cellular basis of abnormal calcium transients of failing human ventricular myocytes. Circ Res. 2003;92:651–8.
96. Olivetti G, Abbi R, Quaini F, et al. Apoptosis in the failing human heart. N Engl J Med. 1997;336:1131–41.
97. Bristow MR, Ginsburg R, Minobe W, et al. Decreased catecholamine sensitivity and β-adrenergic-receptor density in failing human hearts. N Engl J Med. 1982;307:205–11.
98. Perrino C, Naga Prasad SV, Patel M, Wolf MJ, Rockman HA. Targeted inhibition of β-adrenergic receptor kinase-1-associated phosphoinositide-3 kinase activity preserves β-adrenergic receptor signaling and prolongs survival in heart failure induced by calsequestrin overexpression. J Am Coll Cardiol. 2005;45:1862–70.
99. Fu YC, Chi CS, Yin SC, Hwang B, Chiu YT, Hsu SL. Norepinephrine induces apoptosis in neonatal rat cardiomyocytes through a reactive oxygen species-TNF α-caspase signaling pathway. Cardiovasc Res. 2004;62:558–67.
100. Neri M, Cerretani D, Fiaschi AI, et al. Correlation between cardiac oxidative stress and myocardial pathology due to acute and chronic norepinephrine administration in rats. J Cell Mol Med. 2007;11:156–70.
101. Adamson PB, Gilbert EM. Reducing the risk of sudden death in heart failure with beta-blockers. J Card Fail. 2006;12:734–46.
102. Iwatsubo K, Minamisawa S, Tsunematsu T, et al. Direct inhibition of type 5 adenylyl cyclase prevents myocardial apoptosis without functional deterioration. J Biol Chem. 2004;279:40938–45.
103. Okumura S, Vatner DE, Kurotani R, et al. (2007) Disruption of type 5 adenylyl cyclase enhances desensitization of cyclic adenosine monophosphate signal and increases Akt signal with chronic catecholamine stress. Circulation. 2007;116:1776–83.
104. Sugiyama A, Satoh Y, Takahara A, Nakamura Y, Hashimoto K. Cardiovascular and adenylate cyclase stimulating effects of colforsin daropate, a water-soluble forskolin derivative, compared with those of isoproterenol, dopamine and dobutamine. Circ J. 2002;66:1150–4.
105. Yang Y, Zhu WZ, Joiner ML, et al. Calmodulin kinase II inhibition protects against myocardial cell apoptosis in vivo. Am J Physiol Heart Circ Physiol. 2006;291:H3065–75.
106. Zhang R, Khoo MS, Wu Y, et al. Calmodulin kinase II inhibition protects against structural heart disease. Nat Med. 2005;11:409–17.
107. Couchonnal LF, Anderson ME. The role of calmodulin kinase II in myocardial physiology and disease. Physiology. 2008;23:151–9.
108. Kohlhaas M, Zhang T, Seidler T, et al. Increased sarcoplasmic reticulum calcium leak but unaltered contractility by acute CaMKII overexpression in isolated rabbit cardiac myocytes. Circ Res. 2006;98:235–44.
109. Zhu W, Woo AY, Yang D, Cheng H, Crow MT, Xiao RP. Activation of CaMKIIδC is a common intermediate of diverse death stimuli-induced heart muscle cell apoptosis. J Biol Chem. 2007;282:10833–9.
110. Marx SO, Reiken S, Hisamatsu Y, et al. PKA phosphorylation dissociates FKBP12.6 from the calcium release channel (ryanodine receptor): defective regulation in failing hearts. Cell. 2000;101:365–76.
111. Kawahira Y, Sawa Y, Nishimura M, et al. In vivo transfer of a β2-adrenergic receptor gene into the pressure-overloaded rat heart enhances cardiac response to β-adrenergic agonist. Circulation. 1998;98:II262–7.
112. Shah AS, White DC, Emani S, et al. In vivo ventricular gene delivery of a β-adrenergic receptor kinase inhibitor to the failing heart reverses cardiac dysfunction. Circulation. 2001;103:1311–6.
113. Lai NC, Roth DM, Gao MH, et al. Intracoronary adenovirus encoding adenylyl cyclase VI increases left ventricular function in heart failure. Circulation. 2004;110:330–6.
114. Most P, Pleger ST, Volkers M, et al. Cardiac adenoviral S100A1 gene delivery rescues failing myocardium. J Clin Invest. 2004;114:1550–63.
115. Pleger ST, Remppis A, Heidt B, et al. S100A1 gene therapy preserves in vivo cardiac function after myocardial infarction. Mol Ther. 2005;12:1120–9.

# Part VIII
# Looking to the Future

# Chapter 22
# Signaling and the Frontiers Ahead

**Abstract** Signal transduction is a process by which cells convert one kind of signal or stimulus into another. Because of its complexity, the study of cell signaling is of necessity multidisciplinary, using tools from biochemistry, cell biology, structural biology, bioinformatics, and computational biology, together with *in vivo* and *in vitro* studies, including cells culture and whole animal models, to address a wide range of questions in cardiovascular medicine; indeed, to probe into some of the core processes that define and regulate life itself. As discussed previously, a number of cardiovascular pathological conditions are associated with defects in the complex structures formed by signaling pathways; however, increasing knowledge of the genetic and molecular changes that occur in cardiovascular diseases, in general, in the complex field of signal transduction pathways is being deciphered. Nevertheless, fundamental questions still remain unanswered regarding the underlying molecular and biochemical mechanisms involved in signaling, and how this information can be used in improving cardiac diagnosis and treatment. To address these questions, emerging technologies are being recruited, some tested so far in animal models and others are being investigated in clinical trials. Novel approaches include the use of molecular genetics, microarrays, proteomics, and integrated systems biology. In this chapter, a discussion on new frontiers in CV signaling is presented.

**Keywords** Integrative cardiology • Microarray • Genetic biomarkers • Modeling system • Computational biology

## Introduction

In the cardiovascular as in other systems, signal transduction pathways are biochemical processes that control the responses of cells/cardiomyocytes to signals in their internal or external milieu forming complex interconnected networks. The functions of a plurality of cells from plants, vertebrates to mammals, including the human heart, are under the control of similar signaling pathways. Understanding these regulatory systems is central to investigators in diverse disciplines, including neurobiology, immunology, cell biology, and developmental biology – in fields, such as pharmacology and cardiovascular biology [1]. Most of the signal transduction pathways involve the ordered sequences of biochemical reactions inside the cell that are carried out by enzymes and activated by second messengers. These processes are mostly rapid, on the order of milliseconds in the case of ion flux, or minutes for the activation of protein- and lipid-mediated kinase cascades, but in some cases may take hours, and even days (as it happened with gene expression) for completion. The amount of proteins and other molecules participating in the events involving signal transduction increases as the process develops from the initial stimulus, resulting in a "signaling cascade," where relatively small stimulus elicits a large response – this process is known as amplification of the signal.

In this chapter, we discuss current and future new technologies like proteomics, systems biology, bioinformatics, etc., that are also part of future frontiers in cardiovascular signaling. For example, a number of cardiovascular pathological conditions are associated with defects in energy metabolism and in the signaling processes involved. However, using transgenic technology, genetic modifications of key steps in the uptake and metabolism of glucose and fatty acids have enhanced our understanding on how these pathways may influence CVD pathogenesis, and how this knowledge can be used to develop new clinical diagnostic, prognostic markers, and new therapies for cardiovascular diseases.

## Caveats in the Application of Targeting Specific Signaling

A driving force in the assessment of cardiovascular signaling is that it may enhance our understanding of a widening spectrum of cardiac abnormalities known to result from signaling dysfunction, and eventually contributing to the discovery of new treatment modalities. The recognition that cytoprotective

J. Marín-García, *Signaling in the Heart*,
DOI 10.1007/978-1-4419-9461-5_22, © Springer Science+Business Media, LLC 2011

signaling plays an essential role in cardioprotection has generated great interest in the pharmacological manipulation of cardiac metabolism and signaling; however, several important caveats pertain to the application of targeting specific signaling molecules within the clinical setting, such as many signaling molecules participate in multiple pathways. For instance, inhibiting reactive oxygen species (ROS) production might prove helpful in reducing the negative consequences of myocardial ischemia/reperfusion (I/R), but could prove counterproductive in cardioprotection and in oxygen sensing. Moreover, the existence of redundant signaling pathways that trigger cardiac dysfunction also offers challenges to develop new therapeutic interventions. Clearly, understanding of the precise order of intracellular events, their downstream consequences, the overall interrelatedness and regulation of these pathways is necessary for discovery of new therapies. Development and characterization of reagents with high specificity to heart signaling pathways and the arrival of new technologies for the inhibition of stress signaling (e.g., specific kinase inhibitors) in cardiac and vascular cells represent important goals.

## Emerging Concepts in an Integrative Approach to Cardiovascular Signaling

In both physiological and pathological conditions, a broad array of postgenomic approaches, presently available, has shown that an increasing number of genes and their products are critical to cardiomyocyte signaling. While considerable progress has been made in the identification of the downstream targets of these genes and the pathways they constitute, the majority of targets and the complex interrelationships between the signaling pathways involved still remain largely undetermined. Furthermore, a large number of signaling interactions between the nucleus and other cellular organelles and between organelles themselves remain to be identified. For example, understanding of cross talk between mitochondria and plasma membrane receptors, exocytotic events and excitable channels, endoplasmic reticulum (ER), Golgi bodies, peroxisomes, nuclear effects on import/export, regulatory gene expression, signaling cell cycle progression and cardiac-specific and developmentally specific factors involved in myocardial signaling require an increasingly and continuous research effort.

Important information about the nature of these signaling processes can be obtained from the analysis of the multiprotein complexes involved. Specific focus on macromolecular aggregates, including delineation of protein–protein interactions and subproteomic analysis, is of considerable significance in the identification of a number of protein components within tyrosine kinase modules, G protein complexes, ion and mitochondrial permeability transition (MTP) pore channels. Besides, identification of kinases, ligands, and second messengers, an entire panoply of docking and scaffolding proteins, which organize the aggregate structure, and numerous modulators which affect the signaling processes may eventually allow to decipher the three-dimensional architecture of cardiomyocyte signaling. Manipulation of these components within the hearts of transgenic animals and in cardiomyocytes grown *in vitro* is possible using sophisticated techniques of gene transfer and RNAi targeting specific gene expression. On the other hand, reconstitution of functional complexes within artificial membranes has shown less success in signaling studies, underlying the view that the overall context of molecular interactions that constitute cardiomyocyte signaling can be lost in such reductionist schemes. Against this reductionist scheme is the report by Azuaje et al. [2] on an Integrative Pathway-Centric Modeling of Ventricular Dysfunction after Myocardial Infarction. This approach could likely take over traditional reductionist methods relying on gene expression data and identification of individual biomarkers in isolation. These established methodologies present major limitations not only for improving prediction power, but also for model interpretability. The report noted that a diversity of pathway-level perturbations can be profiled in samples of patients with ventricular dysfunction following Myocardial infarction (IM), most of which represent major reductions of gene expression. Signaling pathways implicated in antigen-dependent B-cell activation and the synthesis of leucine was highly altered. By analyzing patient-specific samples encoded with information derived from those significantly perturbed pathways, it is possible to visualize differential prognostic patterns and to perform computational classification of patients with areas under the receiver operating characteristic curve above 0.75. These investigators also showed how the integration of the outcomes generated by different pathway-based analysis models may improve ventricular dysfunction prediction performance. Take together this approach presents an alternative view of key relationships and perturbations that may trigger the development or prevention of ventricular dysfunction after MI.

## Current and Future Technology in Cardiovascular Signaling

In the development of left ventricle (LV) dysfunction and heart failure (HF), cardiac remodeling is a critical step. Structural changes include hypertrophy and subsequent chamber dilatation with progressive impairment in cardiac function. This remodeling process encompasses changes in both the cardiomyocytes and the extracellular matrix (ECM); the latter includes the activation of proteolytic enzymes that lead to degradation and reorganization of collagens. In addition, neurohumoral signals acting through many interrelated signal transduction pathways may lead to pathological cardiac hypertrophy and subsequently progression to HF. Many agonists act through cell surface receptors coupled with G proteins to mobilize intracellular calcium, with consequent activation of downstream kinases and the calcium- and calmodulin-dependent phosphatase calcineurin; also, mitogen-activated protein kinase (MAPK) signaling pathways are interconnected at multiple levels with calcium-dependent kinases and calcineurin [3].

Strategies for the normalization of gene expression in HF with small molecules that control signal transduction pathways directed at transcription factors and associated chromatin-modifying enzymes have been proposed [4]. As previously discussed in Chap. 1, microarray technology has facilitated the way to simultaneously assess the expression of tens of thousands of gene transcripts in a single experiment, providing a resolution and precision of phenotypic characterization not previously possible. Differential gene expression profiling in genetic and multifactorial cardiovascular diseases have been recommended and the data obtained with microarray platforms, used to explore transcriptome alterations in cardiovascular diseases, have been reported [5]. Within the heart, many examples of genetic and protein changes that correlate with dysfunction have been noted, both during normal growth and development and during the progression to congestive HF of diverse etiologies. Also, detailed gene profiling has been performed on structural and functional changes in energy metabolism and intracellular calcium handling. Recently, Uchida et al. [6] reported a simple algorithm named Database-dependent Gene Selection and Analysis (DGSA) able to identify genes with unknown functions but involved in organ development, mainly in the heart. A large number of yet uncharacterized genes were identified that are expressed during heart development. Initial functional characterization of genes by loss-of-function analysis, employing morpholino injections into zebrafish embryos, disclosed severe developmental defects indicating a critical function of selected genes for developmental processes. Thus, DGSA appears to be a versatile and efficient tool for database mining allowing the selection of uncharacterized genes for functional analysis.

At the present time, many clinical cardiologists and investigators believe that myocyte apoptosis is a critical factor in the pathogenesis and pathophysiology of HF, particularly following myocardial infarction [7]. Indeed, in animal models of HF a number of molecular and genetic studies have demonstrated that cardiomyocyte apoptosis play a central role in this process, and if this paradigm were to be confirmed in humans, apoptosis regulation will be a potential and logical target in the treatment of HF. On the other hand, it has been also reported that while the failing human and monkey heart is characterized by significant apoptosis, apoptosis in nonmyocytes cells was eight- to ninefold greater than in myocytes [8]. Nonetheless, and despite of successes in establishing the mechanistic link between apoptosis signaling and HF, understanding of the molecular mechanisms that regulate cell death, specifically in HF remains unclear. Fundamental questions are still unanswered regarding the underlying molecular and biochemical mechanisms triggering these signaling pathways involved in HF, and how this information can be used to improve diagnosis and treatment. In order to address these questions besides currently available modalities, new, emerging technologies are being recruited that eventually may be able to address this conundrum, so far some have been tested in animal models and others are being investigated in clinical trials. The integration of molecular genetics, microarrays, and proteomics under the so-called systems biology, the increasing number of novel cardiac biomarkers for cardiac diagnosis together with gene and cell transplantation therapies represent some of the more current and other forthcoming advances in cardiology.

## *Integrating "Omics" in Cardiomyocyte Signaling*

Two major approaches have been employed in proteomics research, the complete proteomic approach (large platforms) in which the entire proteome-all proteins are characterized, and a more limited proteomic analysis which is either targeted to specific candidate proteins or limited to specific classes of proteins, or subproteomes [9]. A well-thought out argument has been presented regarding the limitation of current proteomic technology to carry out a complete proteomic analysis of plasma proteins [10]. This is mainly due to the high sensitivity required to evaluate in depth the plasma proteome, reflecting both the dynamic range of its constituent-specific proteins, such as regulatory factors and cytokines, simply undetectable by present proteomic technology, as well as the surprising level of posttranslational modifications that most proteins undergo. Proteomic analysis can be used in the identification of novel markers of human cardiac allograft rejection. Besides cardiac allograft rejection, large platforms with hundreds of proteins are also necessary for early detection of HF, for determining its pathogenesis, and for the assessment of HF patient's response to therapy. However, the study of cell signaling on a system-wide level in solid tissue is often not feasible using several techniques, including mass spectrometry (MS), because these techniques generally require a quantity of tissue that usually cannot be obtained from the heart by myocardial biopsy. Interestingly, new miniaturized proteomic technologies are emerging that may circumvent these limitations and offer the ability to monitor protein abundances and posttranslational modification states in a multiplexed and quantitative fashion. Gujral and MacBeath [11] have recently reported emerging technologies that may dissect in detail cellular signaling and advance progress on clinical research. These technologies include nanofluidic proteomic immunoassay (NIA), lysate microarray, and flow cytometry-based Luminex xMAP assay.

Similarly, the potential of proteomics to evaluate global changes in protein expression and their posttranslational state in response to normal or pathological stimuli in the heart has been delayed because its requirement for significant quantity of tissue to be able to perform a comprehensive proteomic characterization. Indeed, this has severely limited its technical application in cardiac diseases. Recently, Gutstein et al. [12] analyzed ongoing efforts to adapt standard methods of tissue sampling (e.g., myocardial biopsy), protein extraction and arraying, with emphasis on those appropriate to smaller tissue samples ranging in size from several microliters down to single cells. The effects of miniaturization on these analyses were highlighted using neuroscience-related examples, as were statistical issues unique to the high-dimensional datasets generated by proteomic experiments.

While currently there are various high-throughput methods for studying cell signaling and regulatory networks, their understanding is not clear. As reported by Huang and Fraenkel [13], the majority of hits identified by transcriptional, proteomic, and genetic assays lie outside of the expected pathways. These investigators have recently reported a technique that uses previously described protein–protein and protein–DNA interactions to determine how these hits are organized into functionally coherent pathways, identifying many components of the cellular response that are not apparent in the original data. This approach was applied simultaneously to phosphoproteomic and transcriptional data for the yeast pheromone response, and was able to identify changes in diverse cellular processes beyond the expected pathways. Similar approaches could be used to mine for unreported data when assessing cardiomyocytes signal transduction pathways, including regulatory factors.

Proteomics by high-resolution two dimensional gel electrophoresis (2-DE) has been looked at as an alternative to high-throughput proteomics. Schwab et al. [14] have study in the murine heart the identification and characterization of protein patterns involved in the maintenance of normal heart physiology at the protein species level, performing an adapted 2-DE/MS approach, as well as identification and characterization of posttranslational modified and truncated protein species using female and male animals of different age, receiving the phytoestrogen genistein. To identify sex and genistein treatment-related effects comparative analyses were performed. Selected 2-DE spots, which exposed varying abundance between animal groups and identified identical proteins, were subject to multi-protease cleavage to generate an elevated sequence coverage enabling the characterization of posttranslational modifications and truncation *loci* via high-resolution MS. Interestingly, while several truncated, phosphorylated, and acetylated species were identified for mitochondrial ATP synthase, malate dehydrogenase and trifunctional enzyme subunit α, confirmation of several of these modifications by manual spectra analysis failed. Nonetheless, it is recommended that these results should be taken with caution for the blind trust in software output. For the regulatory light chain of myosin, the investigators identified an N-terminal processed species that so far has been related to ischemic conditions only. Taken together the information content of protein species separated by high-resolution 2-DE seems to be an alternative to high-throughput proteomics, whose main interest is to enumerate protein names instead of protein species identity. Furthermore, the proteome of different heart structures at three stages of chicken embryonic development have been described by Bon et al. [15] A combination of gel separation, nanoLC separation, and mass spectrometry was used to identify a total of 267 proteins in different tissue structures of the chicken heart. Using spectral counting, as a semiquantitative measure of protein abundance, differences in protein abundance for a number of proteins were detected between different tissues and time points of development. Interestingly, differences in protein levels for the different stages as well as different structures (great arteries, outflow tract and ventricles) were detected in myosin-heavy chain 6, myosin-heavy chain 7, titin, connectin, collagen α-1, and xin. Pathway analysis identified proteins that were related to theoretical protein networks. Myosin-heavy chain 6 and myosin heavy chain 7 were the more abundantly present during cardiovascular development, with myosin-heavy chain 6 highly regulated in a stage and heart tissue-specific manner.

## Microarray and Genetic Biomarkers

Gene expression profiling may also be used to identify a pattern of genes (a molecular signature) that serves as biomarker of relevant clinical parameters in cardiovascular pathologies (e.g., disease presence, progression or response to therapy) [16]. While not as easy or convenient to perform as the screening of circulating biomarkers, the deployment of gene/transcriptome profiling targeted to specific tissues or cell populations can often provide important information about signaling transduction pathways unavailable with other approaches.

## Modeling Systems Approaches to Assess Signaling

A major challenge facing cardiovascular medicine is how to translate the wealth of reductionist detail about molecules, cells, and tissues into a real understanding of how these systems function in health and are perturbed in the disease processes. There is increasing interest in integrating the large comprehensive and ever-growing databases containing genomic, proteomic, biochemical, anatomical, and physiological information that can be searched and retrieved via the Internet to further understand the functioning of the heart in health and disease. Eventually, a model of cardiac function at both the genetic, molecular, and physiological level that may allow predictions about the potential interventions to be made is available. It is anticipated that such simulated modeling would also highlight the gaps in our knowledge and elicit new perspectives on how to fill them [17].

On a limited scale, early work on this type of integrative approach (prior to the acquisition of much of the proteomic/genomic/molecular data) has employed data on ion concentration and metabolite levels during cardiac ischemia to construct a simulated computer model that integrates cardiac energetics with electrophysiological changes (a novel approach to studying myocardial ischemia). This model proved to be informative in predicting the effects of specific therapeutic interventions [18]. Furthermore, this model aided the identification of electrophysiological effects of therapeutic interventions, such as $Na^+$-$H^+$ block, and suggested that it may be an effective strategy to control cardiac dysrhythmias during calcium overload, by regulating sodium–calcium exchange. In addition to myocardial ischemia, other types of cardiovascular disorders that so far have been "modeled" using this type of approach include cardiac contractile disorders [19–21].

The construction of these models, together to the array of data discussed above, requires an extensive and sophisticated mathematical and computer algorithmic treatment [22]. Its undertaking is clearly a multidisciplinary approach involving mathematics, computer skills, molecular and cell biology, genetics, physiology, and anatomy.

Supporting integrative approaches, it had been recently suggested that graph theory provides a useful and powerful tool for the analysis of cellular signaling networks because intracellular components, such as signaling proteins, transcription factors, and genes, are connected by links, representing various types of chemical interactions that result in functional consequences. Thereby, localization of components is often used as a regulatory mechanism to achieve specific effects in response to different receptor signals. Lipshtat et al. [23] described an approach for incorporating spatial distribution into graphs and for the development of mixed graphs, where links are specified by mutual chemical affinity as well as colocalization and suggested that such mixed graphs provide more accurate descriptions of functional cellular networks and their regulatory capabilities, and aid in the development of large-scale predictive models of cellular behavior.

It is important to keep in mind that these models underscore that much critical postgenomic data remains to be obtained, including the identification of gene-environmental factors and protein–protein interactions, which appear to underlie many phenotypic changes and signaling events in the cardiomyocyte. Biomolecular interactions revealed by proteomic information are important for unraveling metabolic and signaling pathways operating in the cell, and in particular in response to disease and injury. The identification of functional interactions between signaling pathways and genetic networks is also of key interest, as they provide a window into regulation of the coordinate expression of functional groups of genes, with a few key pathways switching between alternative cell fates. Also underlying these models is the important recognition that the heart and vascular system are dynamic and more than just electrical circuitry or mechanical pumps, and they have the ability to grow and remodel in response to changing environments, partly determined by genes and partly by their physical environment [24].

Recently, Lengar [25] expressed his opinion on "why we need quantitative dynamic models." The reason behind was that to understand the mechanisms by which interacting components in signal transduction pathways become regulatory systems, it is necessary to have a quantitative understanding of the system. At the cardiomyocyte level, this may allow to know the concentrations of cellular components, such as proteins and the reaction rates for interactions between components. As it has been suggested [25], a mechanistic understanding of regulatory behavior is helpful in developing predictive models of relationships between complex genotypes and variable phenotypes. Also, in support of a more integrative, rather holistic approach to study cardiovascular signaling pathways is the recent report by Alexander et al. [26] stating that a fundamental goal in system biology is to understand how cellular behavior emerges from the interaction in time and space of genetically encoded molecular parts, as well as nongenetically encoded small molecules. As noted, most approaches to molecular network analysis rely to some extent on the assumption that molecular networks are modular – that is, they are separable and can be studied to some degree in isolation. The investigators described recent advances in the analysis of modularity in biological networks, focusing on the increasing realization that a dynamic perspective is essential to grouping molecules into modules and determining their collective function. From this, it is apparent that in the near future a system approach may need to be applied to metabolic events occurring in the mitochondria of the failing heart, as well as to events occurring in other cardiovascular pathologies.

## Bioinformatics/Mathematics/Computational Biology in Signaling

Signaling pathways topologies can be deducted by multiple alterations measurements of specific biochemical species.

Xu et al. [27] have reported that specification of biological decisions by signaling pathways is encoded by interplay between activation dynamics and network topologies. The authors develop a mathematical approach to rank the possible paths through a signaling pathway and to develop hypotheses that can be rationally tested. This approach named Bayesian inference (or statistical inference in which evidence or observations are used to update or to newly infer the probability that a hypothesis may be true) – based modeling (BIBm) explores signaling pathways by which epidermal growth factor (EGF) stimulates extracellular signal-regulated kinase (Erk). The predicted and experimentally validated model shows that both Raf-1 and, unexpectedly the B-Raf, are needed to fully activate Erk in two different cell lines. This methodology evidentially supported pathway topologies even when a limited number of biochemical and kinetic measurements are available.

Formalized rules for protein–protein interactions have been proposed to represent the binding and enzymatic activities of proteins in cellular signaling. Rules can be processed to automatically generate mathematical or computational models for a system, which enables explanatory and predictive insights into the system's behavior [28]. These rules are independent units of a model specification that facilitate model revision. Then, instead of changing a large number of equations or lines of code, as may be required in the case of a conventional mathematical model, protein interaction can be introduced or modified simply by adding or changing a single rule that represents the interaction of interest.

To be useful, it is evident that the increasing amount of information coming from technologies, such as genomics, proteomics, and other molecular research technologies, need to be integrated and standardized to advance the knowledge on integrative signaling pathways. This allows new interpretations and theories to be experimentally and clinically tested. The application of information technology or *Bioinformatics* to the field of molecular cardiovascular biology facilitates the creation and maintenance of database storing information coming from transcriptomics, miRNA and protein level and sequences that allow the profiling and identification of causal genes, protein expression, and signaling pathways from diverse cardiovascular pathologies. The actual process of analyzing and interpreting data is referred to as *computational biology*. Development of this type of database involves not only design issues, but also the development of complex interfaces whereby researchers could both access existing data as well as submit new or revised data. Presently, bioinformatics has evolved in such a way that the most pressing task involves the analysis and interpretation of various types of data, including nucleotide and amino acid sequences, protein domains, and protein structures.

What sets bioinformatics apart from other approaches is its focus on developing and applying computationally intensive techniques (e.g., data mining and machine learning algorithms) to achieve this goal. Major research in this field is being focused on sequence alignment, gene finding, genome assembly, protein structure alignment, protein structure prediction, prediction of gene expression, and protein–protein interactions, and the modeling of evolution. It is evident that to understand how cardiomyocytes structure function and proliferation are altered in cardiovascular diseases, the collected biological data must be combined to provide a comprehensive picture. In doing so, new insights into the etiology, diagnostic and prognostication of diseases, such as HF, as well as for new therapies might be attained.

## Postgenomic Contributions to New and Future Therapeutic Options in Cardiovascular Medicine

Molecular genetic analysis has substantially improved our understanding of the structure and functioning of the heart, both in early development and in aging, and has opened the door to further unraveling the order of molecular/cellular events and the principal molecules involved in both normal and malformed/dysfunctional hearts. Information derived from transgenic models has been instrumental in defining numerous therapeutic targets in signaling transduction pathways, how heart and cardiovascular system respond to stresses and insults, the elucidation of both apoptotic and survival/proliferative pathways affecting cardiomyocyte growth, oxidative stress, hypertrophy, aging and cell death, and in defining metabolic pathways essential for energy transduction necessary for contractile function and electrical excitability. Animal models with specific gene dysfunction have been highly informative in defining the roles that specific contractile and ion channel proteins play in both, the normal and diseased heart. Some of this information has been very important in the design of pharmacological strategies to develop novel targets in the treatment of CVDs, ranging from HF to stroke, MI, and acute and chronic inflammatory diseases [29]. Therapies aimed at metabolic remodeling may be developed to effectively complement the treatment of myocardial ischemia, HF, the more obvious metabolic-based CVDs (i.e., metabolic syndrome), as well as underlying therapies, such as calorie restriction aimed at modulating overall longevity and the aging processes [30–32].

The potential of combining genetic and cell engineering has been demonstrated in a number of animal models and in limited number of clinical studies, and this may be a useful adjunct in repairing the "broken" heart and damaged vessels, and in enhancing the heart's regenerative potential. New developments in stem cells and genomic-related signaling pathways have been previously discussed and the potential for cell-based therapy and tissue reengineering is progressing toward the future of novel and effective developments.

## Conclusions and Future Frontiers

Like in other systems, most of the cardiovascular signaling transduction pathways involve the ordered sequences of biochemical reactions inside the cell that are carried out by enzymes and activated by second messengers. Knowledge of the multiple signals that participate in cardiac pathophysiology and pathogenesis is changing the way that we approach the treatment of cardiac diseases. Also, new insights into the cardiovascular consequences of abnormal gene function and expression is impacting on the development of targeted therapeutic strategies and disease management, and it will likely replace less effective treatment modalities directed solely at rectifying structural cardiac defects and temporal improvement of function.

With the completion of the Human Genome Project, the likely identification of novel genes involved in pathogenesis will serve as an important foundation toward understanding how specific gene defects generate different cardiac disease phenotypes. Methods, such as Bioinformatics, can be employed to search existing databases with the routine use of reverse genetics techniques, allowing subsequent cloning of novel genes/cDNAs of interest followed by the characterization of spatial–temporal patterns of specific gene expression. Moreover, in an integrating approach both transcriptome and proteomic methodologies can be used to further delineate the functions of the gene products, defining their precise role in pathogenesis, elucidating their interaction with other proteins in the subcellular pathways and potentially enabling their application as clinical markers of coronary artery disease and HF.

Further research into the identification of molecular receptors and regulators that control cardiomyocyte proliferation in the human heart is critical. Understanding the molecular basis of cardiomyocyte proliferation could greatly impact future clinical attempts to repair the damaged heart. Comprehensive gene expression profiles of adult, embryonic and postnatal myocytes, as well as generation of myocyte cell culture lines with the capacity to respond to proliferative inducers are essential to understand the mechanisms of cell growth regulation. As previously discussed in Chap. 19, cellular transplantation is an alternative approach with which to augment myocyte number in the diseased, damaged heart. Nonetheless, intense research efforts are still necessary to define the optimal conditions necessary for cardiomyocyte differentiation and proliferation, and for the fully functional integration of stem cells in the myocardium, as well as to assess the ability of transplanted stem cells to repair myocardial alterations occurring in MI and in the failing heart. Research focusing on the basic science of stem cells, including signaling pathways, isolation of factors involved in the acceleration of specific cell-type differentiation, migration, homing, and proliferation *in vitro* as well as *in vivo* and their responses to oxidative stress, apoptosis, and injury may speed their use in clinical transplantation as a new treatment modality for cardiovascular pathologies.

Although many of the available methodologies, including gene profiling, proteomics, genetic analysis, development of relevant animal and cellular models, proteomics and pharmacogenomics are promising, at this time they all are relatively new approaches still in the process to be learned. Hopefully, the use of these methodologies in an integrated system biology approach further fosters our knowledge on the critical interactions between cells and organ systems, their cross talk and the blend of multiple signaling pathways into functioning or unified hole. We concur with Adler [33] that among the most recent and exciting advances in mammalian biology are new therapeutic applications of research that are changing the established boundaries of signaling pathways, such as unexpected mechanisms to prolong life and prevention of aging.

In summary, we are witnessing the transition from the cardiology of the past to a future that will include the study of integrated systems biology, the constructive cycle of computational model building, and experimental verification capable of providing the necessary input to achieve new and exciting discoveries and also finding renew hope.

## Summary

- Signaling transduction pathways involve the ordered sequences of biochemical reactions inside the cell that are carried out by enzymes and activated by second messengers.
- Proteomics and gene profiling techniques are critical techniques that can be used in the identification of novel biomarkers.
- Also proteomic techniques, such as mass spectroscopy and protein microarrays, offer improved methods for rapid and highly accurate screening of multiple markers.
- Mitochondrial-based defects have been identified in a variety of CVDs due to the primary role of mitochondria in generating metabolic energy, as well as its role in oxidative stress signaling and ROS generation, and in the early events of apoptotic cell death.
- The use of gene therapy to target specific genetic defects or supplement deficiencies has an enormous potential

with many possible useful applications; however, we are still awaiting the development of safe, effective, and easily testable vector systems.

- Cardiac targeted therapies may involve the direct modulation of myocardial cells with transfected genes or injected gene products, or the transplantation of cells or groups of cells (cell sheets).
- Cell-based therapies will likely play a critical role in the future since a number of cell-types have shown the ability to be recruited into damaged myocardium with potentially beneficial results. These cells include several types of stem cells (either embryonic or adult), neonatal cardiomyocytes, and skeletal myocytes.
- Both stem cell biology and the cardiac milieu optimal for their homing, integration, and long-term survival need to be better characterized. In addition to cardiac transplantation, both stem cells and a combination of angiogenic factors can be utilized in therapeutic angiogenesis with its potential application in myocardial ischemia.
- A variety of biomaterials have been employed to simulate the cardiac environment, and they have been useful in tissue reengineering of valves (e.g., aortic) and vessel remodeling.
- The application of information technology or *Bioinformatics* to the field of molecular cardiovascular biology allows the creation and maintenance of database storing information coming from transcriptomics, miRNA, and protein level and sequences that allow the profiling and identification of causal genes and protein expression in HF.
- What sets bioinformatics apart from other approaches is its focus on developing and applying computationally intensive techniques (e.g., data mining and machine learning algorithms) to achieve this goal.
- We are witnessing the transition from conventional cardiology practice to the study of systems biology, computational model building, and experimental verification capable of providing the necessary input to achieve new and exciting discoveries.

## References


1. Yaffe MB. Seeing the signaling forest and the trees. Sci Signal. 2008;1:eg5.
2. Azuaje F, Devaux Y, Wagner DR. Integrative pathway-centric modeling of ventricular dysfunction after myocardial infarction. PLoS One. 2010;5:e9661.
3. Sugden PH, Clerk A. "Stress-responsive" mitogen-activated protein kinases (c-Jun N-terminal kinases and p38 mitogen-activated protein kinases) in the myocardium. Circ Res. 1998;83:345–52.
4. McKinsey TA, Olson EN. Toward transcriptional therapies for the failing heart: chemical screens to modulate genes. J Clin Invest. 2005;115:538–46.
5. Nanni L, Romualdi C, Maseri A, Lanfranchi G. Differential gene expression profiling in genetic and multifactorial cardiovascular diseases. J Mol Cell Cardiol. 2006;41:934–48.
6. Uchida S, Schneider A, Wiesnet M, et al. An integrated approach for the systematic identification and characterization of heart-enriched genes with unknown functions. BMC Genomics. 2009;10:100.
7. Mani K, Kitsis RN. Myocyte apoptosis: programming ventricular remodeling. J Am Coll Cardiol. 2003;41:761–4.
8. Park M, Shen YT, Gaussin V, et al. Apoptosis predominates in nonmyocytes in heart failure. Am J Physiol Heart Circ Physiol. 2009;297:H785–91.
9. Stanley BA, Gundry RL, Cotter RJ, Van Eyk JE. Heart disease, clinical proteomics and mass spectrometry. Dis Markers. 2004;20:167–78.
10. Anderson L. Candidate-based proteomics in the search for biomarkers of cardiovascular disease. J Physiol. 2005;563:23–60.
11. Gujral TS, MacBeath G. Emerging miniaturized proteomic technologies to study cell signaling in clinical samples. Sci Signal. 2009;2:pe65.
12. Gutstein HB, Morris JS, Annangudi SP, Sweedler JV. Microproteomics: analysis of protein diversity in small samples. Mass Spectrom Rev. 2008;27:316–30.
13. Huang SS, Fraenkel E. Integrating proteomic, transcriptional, and interactome data reveals hidden components of signaling and regulatory networks. Sci Signal. 2009;2:ra40.
14. Schwab K, Neumann B, Scheler C, Jungblut PR, Theuring F. Adaptation of proteomic techniques for the identification and characterization of protein species from murine heart. Amino Acids. 2011;41:401–14.
15. Bon E, Steegers R, Steegers EA, et al. Proteomic analyses of the developing chicken cardiovascular system. J Proteome Res. 2010;9:268–74.
16. Kittleson MM, Hare JM. Molecular signature analysis: using the myocardial transcriptome as a biomarker in cardiovascular disease. Trends Cardiovasc Med. 2005;15:130–8.
17. Bassingthwaighte JB, Qian H, Li Z. The Cardiome Project. An integrated view of cardiac metabolism and regional mechanical function. Adv Exp Med Biol. 1999;471:541–53.
18. Ch'en F, Clarke K, Vaughan-Jones R, Noble D. Modeling of internal pH, ion concentration, and bioenergetic changes during myocardial ischemia. Adv Exp Med Biol. 1997;430:281–90.
19. Noble D. Modeling the heart – from genes to cells to the whole organ. Science. 2002;295:1678–82.
20. Rudy Y. From genome to physiome: integrative models of cardiac excitation. Ann Biomed Eng. 2000;28:945–50.
21. Hunter P, Smith N, Fernandez J, Tawhai M. Integration from proteins to organs: the IUPS Physiome Project. Mech Ageing Dev. 2005;126:187–92.
22. Crampin EJ, Halstead M, Hunter P, et al. Computational physiology and the Physiome Project. Exp Physiol. 2004;89:1–26.
23. Lipshtat A, Neves SR, Iyengar R. Specification of spatial relationships in directed graphs of cell signaling networks. Ann N Y Acad Sci. 2009;1158:44–56.
24. Noble D. Modelling the heart: insights, failures and progress. Bioessays. 2002;24:1155–63.
25. Lyengar R. Why we need quantitative dynamic models. Sci Signal. 2009;2:eg3.
26. Alexander RP, Kim PM, Emonet T, Gerstein MB. Understanding modularity in molecular networks requires dynamics. Sci Signal. 2009;2:pe44.
27. Xu TR, Vyshemirsky V, Gormand A, et al. Inferring signaling pathway topologies from multiple perturbation measurements of specific biochemical species. Sci Signal. 2010;3:ra20.
28. Hlavacek WS, Faeder JR, Blinov ML, Posner RG, Hucka M, Fontana W. Rules for modeling signal-transduction system. Sci STKE. 2006;2006:re6.

29. Kreuter M, Langer C, Kerkhoff C, Reddanna P, Kania AL, Maddika S, et al. Stroke, myocardial infarction, acute and chronic inflammatory diseases: caspases and other apoptotic molecules as targets for drug development. Arch Immunol Ther Exp (Warsz). 2004;52: 141–55.
30. Roth GS, Lane MA, Ingram DK. Caloric restriction mimetics: the next phase. Ann N Y Acad Sci. 2005;1057:365–71.
31. Ingram DK, Anson RM, de Cabo R, et al. Development of calorie restriction mimetics as a prolongevity strategy. Ann N Y Acad Sci. 2004;1019:412–23.
32. Dirks AJ, Leeuwenburgh C. Caloric restriction in humans: potential pitfalls and health concerns. Mech Ageing Dev. 2006;127:1–7.
33. Adler EM. 2009: signaling breakthroughs of the year. Sci Signal. 2010;3:eg1.

# Glossary

**AA** Arachidonic acid.

**ABC** Adenosine triphosphate-binding cassette.

**AC** Adenylyl cyclase, a membrane-bound enzyme which catalyzes the synthesis of the second messenger cyclic AMP from ATP in conjunction with specific signaling ligand. (e.g., adrenergic), receptors and G proteins.

**ACC** Acetyl CoA carboxylase, an enzyme that synthesizes malonyl-CoA from cytoplasmic and peroxisomal acetyl-CoA.

**ACE** Angiotensin converting enzyme, a central element of the renin–angiotensin system, converts the decapeptide angiotensin I to the potent pressor octapeptide angiotensin II (Ang II), mediating peripheral vascular tone, as well as glomerular filtration in the kidney.

**Acetyl CoA** Small water-soluble molecule that carries acetyl groups linked to coenzyme A (CoA) by a thioester bond.

**ACH** Acetylcholine.

**ACS** Acute coronary syndrome.

**ACS cells** Adult cardiac stem cells.

**Adenovirus** Common vector for gene transfer with high efficiency of transfection *in vivo*, but limited by transient transgene expression and host immunogenic response.

**ADP** Adenosine diphosphate.

**Adrenoceptors** Members of the G protein-coupled receptor superfamily, linking adrenergic signaling from the sympathetic nervous system and the cardiovascular system, with integral roles in the rapid regulation of myocardial function.

**AF** Atrial fibrillation, the most common dysrhythmia seen in clinical cardiology, can be familial with both monogenic and more heterogeneous genetic cases reported.

**AGE** Advanced glycation end-products.

**AHF** Anterior heart field.

**AIF** Apoptosis-inducing factor. Released from mitochondrial intermembrane space in early apoptosis and subsequently involved in nuclear DNA fragmentation.

**AKAP** Specific PKA anchoring proteins; regulators of PKA function and signaling by directing and concentrating PKA at specific subcellular sites.

**Akt** Protein kinase B (PKB). Myocardial Akt phosphorylates a number of downstream targets, including cardioprotective factors involved in glucose and mitochondrial metabolism, apoptosis, and regulators of protein synthesis.

**Alagille syndrome** An autosomal dominant disorder with a wide spectrum of developmental anomalies and commonly presenting with TOF caused by mutations in the *JAG1* gene encoding a notch ligand (jagged1).

**ALCAR** Acetyl-L-carnitine, supplementation with LA appears to improve myocardial bioenergetics and decrease oxidative stress associated with aging.

**ALKs** Activin receptor-like kinase.

**Allele** One of the several alternate forms of a single gene occupying a given *locus* on a chromosome or mtDNA.

**Allotopic expression** Alternative method of mitochondrial gene therapy in which a mitochondrial gene is reengineered for expression from the nucleus and targeting its translation product to the mitochondria.

**AM** Adrenomedullin. G protein-coupled receptor.

**AMHC1** Atrial myosin heavy chain-1.

**AMI** Acute myocardial infarction.

**AMISTAD** Acute myocardial infarction study of adenosine.

**AMPK** AMP-activated protein kinase involved in myocardial metabolic energy sensing.

**Amplification** Generation of many copies of a specific region of DNA.

J. Marín-García, *Signaling in the Heart*,
DOI 10.1007/978-1-4419-9461-5, © Springer Science+Business Media, LLC 2011

**Anderson syndrome** Also known as Anderson–Tawil syndrome or LQT7, an autosomal dominant characterized by a heterogeneous phenotype, including a variety of cardiac dysrhythmias with many patients having mutations in the *KCNJ2* gene coding the $K^+$ channel inward rectifier ($I_{K1}$) Kir2 channel.

**Angiogenesis** Formation of new vessels from preexisting ones, and in particular the sprouting of new capillaries from postcapillary venules.

**ANP** Atrial natriuretic peptide (also ANF).

**ANT** Adenine nucleotide translocator. A mitochondrial inner membrane carrier of ADP and ATP and constituent of the MTP.

**Antimycin A** Specific inhibitor of complex III activity.

**Antisense RNA** RNA complementary to a specific transcript of a gene that can hybridize to the specific RNA and block its function.

**AP** Action potential.

**Apaf-1** Apoptotic protease activating factor 1.

**APC** Anesthetic preconditioning.

**APD** Action potential duration.

**APLA** Antiphospholipid antibodies.

**Apoptosis** Programmed cell death.

**Apoptosome** Cytosolic complex involved in the activation of apoptotic caspases.

**AR** Androgen receptor.

**ARC protein** Apoptosis repressor with a CARD, inhibitor of both the intrinsic and extrinsic apoptosis pathways.

**ARE** Androgen response element, specific regulatory sites within the DNA of target nuclear genes, which bound by AR produce long-term genomic effects of testosterone.

**ARH** Autosomal recessive familial hypercholesterolemia; a rare disorder with a clinical phenotype similar to homozygous FH, but less severe, more variable and responsive to lipid-lowering therapy.

**ARMYDA** Atorvastatin for reduction of myocardial damage during angioplasty.

**Arteriogenesis** Process of maturation and/or *de novo* growth of specifically collateral arteries, which mainly occurs following ischemic vascular disease.

**ARVC** Arrhythmogenic right ventricular cardiomyopathy (same as ARVD).

**ARVD** Arrhythmogenic right ventricular dysplasia; the most common symptoms are ventricular dysrhythmias, heart palpitations, fainting or loss of consciousness (syncope), and sudden death.

**ASK1** Apoptosis signal-regulating kinase 1. **α-SMA** α-smooth muscle actin.

**ASO** Antisense oligonucleotides. These short, synthetic DNA molecules can reduce specific gene expression by acting either directly or as decoys of transcription factors.

**ASS** Argininosuccinate synthase.

**α2-ARs** α2-adrenergic receptors, receptors for endogenous catecholamine agonists (e.g., norepinephrine and epinephrine) that mediate a number of physiological and pharmacological responses, such as changes in blood pressure and heart rate.

**ATP** Adenosine triphosphate.

**Atractyloside** Inhibitor of the adenine nucleotide translocator, and MTP opener.

**Autosomal inheritance pattern** The gene of interest is present on any of the nonsex chromosomes.

**AV** Atrioventricular.

**AVC** Atrioventricular canal.

**AVJ** Atrioventricular junction.

**AV node** Atrioventricular node; a group of specialized cells located between the atria and ventricles that regulate electrical current passing to the ventricles.

**Bacteriophage** A virus that infects bacteria; useful as a vector for gene transfer.

**β-AR** Beta-adrenergic receptor, G protein-coupled receptors containing seven transmembrane domains involved in signaling pathways of diverse cardiovascular functions including blood pressure control and cardiac contractility.

**β-ARK** Beta-adrenergic receptor kinase, a GRK which mediates the desensitization of the β-adrenergic receptor by phosphorylation of agonist-occupied receptors.

**BAT** Brown adipose tissue.

**BDNF** Brain-derived neurotrophic factor.

**BEL** Bromoenol lactone.

**BF** Blood flow.

**BH Domains** Features of proapoptotic proteins, (BH1–4) are essential for homo- and heterocomplex formation, as well as to induce cell death. Proapoptotic homologs can be subdivided into two major subtypes, the multidomain Bax subfamily (e.g., Bax and Bak), which possesses BH1–3 domains, and the BH3-only subfamily (e.g., Bad and Bid).

**bHLH** Basic helix-loop-helix.

**BIBm** Bayesian inference-based modeling.

**Bid** A proapoptotic Bcl-2-related protein, which links the extrinsic and intrinsic apoptotic pathways.

**Bilayer** Arrangement of phospholipids in biological membranes.

**BK** Bradykinin. Vasoactive peptide plays a fundamental role in controlling the functional and structural integrity of blood vessel wall.

**BMDC** Bone marrow-derived cells.

**BMK1** Big mitogen-activated protein kinase1.

**BMP** Bone morphogenetic protein, a class of ligands which bind specific membrane-bound receptors involved in signaling events in early cardiomyocyte differentiation.

**BMPR** Bone morphogenetic protein receptor.

**BNP** Brain natriuretic peptide, hemodynamic marker of neurohumoral and vascular stress.

**bp** Base pairs.

**BRDU** Bromodeoxyuridine, a DNA synthesis inhibitor.

**BrS** Brugada syndrome. A form of idiopathic ventricular fibrillation which in children can be inherited as an autosomal dominant trait with variable penetrance, and can lead to sudden death in healthy young individuals; a subset of cases have *SCN5A* mutations.

**BSS** Bernard–Soulier syndrome, a rare autosomal recessive disorder caused by mutations in various polypeptides in the GpIb/IX/V complex, the principal platelet receptor for vWF.

**BST** Biotin-switch technique. Allows biotinylation of S-nitrosylated proteins.

**CA** Catalase, antioxidant enzyme.

**CABG** Coronary artery bypass grafting.

**CAD** Coronary artery disease. (See ischemic heart disease).

**Calcineurin** Intracellular $Ca^{2+}$-regulated phosphatase implicated as a mediator of cardiac hypertrophy.

**CALCRL** Calcitonin receptor-like.

**CaM** Calmodulin, an intracellular $Ca^{2+}$ sensor that selectively activates downstream signaling pathways in response to local changes in $Ca^{++}$.

**CaMK** $Ca^{2+}$/CaM-dependent protein kinase.

**cAMP** Cyclic AMP; second messenger used extensively in cell signaling. Product of AC.

**CAR** Coxsackievirus–adenovirus receptor; cell-surface receptor specifically interacts with both CVB and adenoviruses.

**CARD** Caspase activation and recruitment domain.

**Cardiolipin** Anionic phospholipid located primarily in mitochondrial inner membrane.

**Cardiomyocyte** A single cell of a heart muscle.

**Carnitine** Carrier molecule involved in the transport of long-chain fatty acids into the mitochondria for β-FAO.

**CASK** Calcium/calmodulin-dependent serine protein kinase.

**Caspases** Intracellular cysteine proteases activated during apoptosis that cleave substrates at their aspartic acid residues.

**Caveolae** Specialized flask-shaped subdomains of the plasma membrane particularly abundant in cardiovascular cells that function both in protein trafficking and signal transduction.

**CBP** CREB-binding protein.

**CCCP** Carbonyl cyanide m-chlorophenyl hydrazone. A potent uncoupler.

**CCS** Cardiac conduction system, a heterogenous complex of cells responsible for establishing and maintaining the rhythmic excitation of the mature heart.

**CDKs** Cyclin-dependent kinases.

**cDNA** Complementary DNA. DNA fragment that is synthesized from the RNA strand by reverse transcriptase. This DNA copy of a mature RNA lacks the introns that are present in the genomic DNA.

**cDNA library** Collection of cDNAs synthesized from the mRNA of an organism cloned into a vector.

**Cell cycle** The period between the release of a cell as one of the progeny of a division and its own subsequent division by *mitosis* into two daughter cells.

**Cell fusion** Fusion of two somatic cells creating a hybrid cell.

**CETP** Cholesteryl ester transfer protein plays role in reverse cholesterol transport with transfer of cholesteryl ester-rich HDL to triglyceride-rich lipoproteins (VLDL).

**CFTR** Cystic fibrosis conductance regulator.

**cGK1** cGMP-dependent protein kinase 1.

**cGMP** Cyclic guanosine monophosphate.

**CGRP** Calcitonin gene-related peptide, cardioprotective agent.

**Chagas disease** Infectious disease that often results in myocarditis and/or dilated cardiomyopathy endemic to South and Central America caused by *Trypanosoma cruzi* (*T. cruzi*), a flagellated protozoan parasite.

**CHAMP** Cardiac helicase activated by MEF2 protein.

**Chaperone** Protein that assists in the proper folding and assembly into larger complexes of unfolded or misfolded proteins.

**Char syndrome** Autosomal dominant syndrome with PDA, facial dysmorphology, and fifth-finger malformation due to mutations in *TFAP2B*, the gene encoding a transcription factor AP2β, which is highly expressed in neural crest cells.

**CHARGE syndrome** Acronym for a wide range of developmental abnormalities, including colobomata, heart defects, choanal atresia, retardation of growth and development, genital and ear abnormalities.

**CHD** Congenital heart defect.

**Cholesterol 7-Hydroxylase deficiency** Deficiency in the first enzyme in the pathway of cholesterol catabolism and bile synthesis; leads to high levels of LDL cholesterol and hypercholesterolemia.

**Chromatin** The complex of DNA and histone and nonhistone proteins found in the nucleus of a eukaryotic cell that constitutes the chromosomes.

**Chromosome** A very long, continuous piece of DNA, which hold together many genes, regulatory elements and other intervening nucleotide sequences.

**CICR** $Ca^{2+}$-induced $Ca^{2+}$-release.

***Cis*-acting elements** Regulatory DNA sequences that affect the expression of genes only on the molecule of DNA where they reside; not protein encoding.

**CK** Creatine kinase. Both mitochondrial and cytosolic isoforms of this enzyme that catalyze the reversible phosphorylation of creatine by ATP to form the high-energy compound phosphocreatine.

**CL** Cardiolipin, anionic phospholipid present in mitochondrial membranes. Deficiency in Barth syndrome. Involved in stabilization of ETC complexes and in apoptosis.

**CNG** Cyclic nucleotide-gated.

**CNS** Central nervous system.

**Codon** A three-nucleotide sequence in mRNA specifying a unique amino acid.

**Complementary DNA** A DNA copy of a RNA template.

**Complex I** NADH-ubiquinone oxidoreductase.

**Complex II** Succinate CoQ oxidoreductase.

**Complex III** CoQ-cytochrome *c* oxidoreductase.

**Complex IV** Cytochrome *c* oxidase.

**Complex V** Oligomycin-sensitive ATP synthase. Also termed F0-F1 ATPase.

**Connexins** A group of transmembrane proteins that form gap junctions between cells.

**Consomic rats** Strains produced by repeated backcrossing of a whole chromosome onto an inbred strain.

**CoQ** Coenzyme Q (also ubiquinone). Electron carrier and antioxidant.

**Cosmid vector** A type of plasmid constructed by the insertion of *cos* sequences from Phage lambda.

**COT curve** Expresses the ratio of the concentration of denatured single-stranded DNA to the initial concentration of DNA (Co) times time or Cot.

**COX** Cytochrome *c* oxidase (complex IV).

**CP** Cardioprotection.

**CPCs** Cardiac progenitor cells.

**CpG islands** GC-rich regions of DNA often found in promoter regions.

**CPT-I** Carnitine palmitoyltransferase I.

**CPT-II** Carnitine palmitoyltransferase II.

**CPVT** Catecholaminergic polymorphic ventricular tachycardia; characterized by syncope, seizures or SD, in response to exercise or emotional stress, and affecting mainly young children with morphologically normal hearts. Mutations in *RyR2* (autosomal dominant) or in *CASQ2* encoding calsequestrin 2 (autosomal recessive) have been found leading to CPVT.

**CR** Caloric restriction, a restricted dietary regimen which has been shown to increase lifespan in a number of organisms, including mammals and may have anti-aging effects in the heart.

**CREB** cAMP-response element-binding protein.

**Cristae** Folding of inner mitochondrial membrane to enlarge the surface area.

**CRP** C-reactive protein; a significant marker of inflammation and atherosclerotic progression. Serum CRP levels are predictive of future cardiovascular events.

**CsA** Cyclosporin A. An inhibitor of the MTP pore opening.

**CSCs** Cardiac stem cells.

**CT-1** Cardiotrophin, an interleukin 6-related cytokine, has been shown to promote both the survival and proliferation of cultured neonatal cardiomyocytes.

**cTnI** Cardiac troponin I, widely used marker of myocardial ischemia and necrosis.

**cTnT** Cardiac troponin T, widely used marker of myocardial ischemia and necrosis.

**Cu-ZnSOD** Copper-zinc superoxide dismutase.

**CVB** Coxsackievirus B viruses, commonly identified infectious agents (particularly CVB3) causing human viral myocarditis.

**CVD** Cardiovascular disease.

**Cx** Connexin.

**CYP** Cytochrome P450 monooxygenase.

**CyP-D** Cyclophilin D. CsA-binding mitochondrial matrix protein; component of the MTP.

**Cytochrome** A family of proteins that contain heme as a prosthetic group involved in electron transfer and identifiable by their absorption spectra.

**Cytochrome c** A mitochondrial protein involve in ETC at complex IV. Its release from the mitochondrial into the cytosol is a trigger of caspase activation and early myocardial apoptosis.

**DAD** Delayed afterdepolarization.

**DAF** Decay accelerating factor.

**DAG** Diacylglycerol, second messenger produced by phospholipase C (see PLC).

**DCA** Dichloroacetate. By inhibiting PDH kinase, DCA stimulates PDH, promoting aerobic oxidation and reducing lactic acidosis.

**DCFH-DA** 2′,7′-dichlorodihydrofluorescein diacetate, membrane-permeable fluorometric indicator of ROS levels in the cytosol.

**DCM** Dilated cardiomyopathy.

**Delayed preconditioning** Often referred to as a second window of protection, this pathway appears about 12–24 h after the preconditioning event and lasts several days.

**DES** Drug-eluting stent.

**Desmoglein** A transmembrane desmosomal protein; mutations lead to ARVD.

**Desmoplakin** Desmosomal protein; mutations lead to either dominant or recessive ARVD.

**Desmosome** Highly organized intercellular junctions that provide mechanical integrity to tissues by anchoring intermediate filaments to sites of strong adhesion.

**Dexrazoxane** Antioxidant that prevents site Fe-based oxidative damage by chelating free iron; provides clinical cardioprotection against doxorubicin-induced oxidative damage.

**DGGE** Denaturing gradient gel electrophoresis, mutation screening technique.

**DGS** DiGeorge or velocardiofacial syndrome, characterized by an assortment of craniofacial defects, conotruncal heart abnormalities (PTA, TOF, and VSD), aortic arch defects and hypoplasia; most commonly caused by large-scaled deletions at chromosome 22 q11.

**DGSA** Database-dependent gene selection and analysis.

**DHF** 2′,7′-dichlorofluorescin, membrane-permeable fluorometric indicator of ROS levels (particularly in mitochondrial matrix).

**DHT** Dihydrotestosterone.

**DIC** Disseminated intravascular coagulation.

**Differential display** Technique used to identify genes that are differentially expressed; RNA from the samples being compared is reverse transcribed, and the cDNA is further amplified using random primers. Genes that are differentially expressed in the chosen samples can be identified by electrophoresis.

**DISC** Death-inducing signaling complex, a multiprotein complex involved in the extrinsic apoptotic pathway triggered by the binding of specific ligands to the death receptor.

**D-loop** Noncoding regulatory region of mtDNA involved in controlling its replication and transcription.

**DNA footprinting** A technique that detects DNA–protein interaction knowing that a protein bound to DNA may protect that DNA from enzymatic cleavage.

**DNA primary structure** The specific nucleotide sequence from the beginning to the end of the molecule.

**DOR** δ-opiod receptor.

**DORV** Double outlet right ventricle, a rare group of cardiac anomalies characterized by both great arteries (pulmonary and aorta) arising primarily from the right ventricle. Patients with DORV have been reported with mutations in connexin (Cx43) or with deletions in chromosome 22q11.

**Double helix** A molecular model of DNA made of two complementary strands of the bases guanine, adenine, thymine, and cytosine, covalently linked through phosphodiester bonds. Each strand forms a helix, and the two helices are held together through hydrogen bonds, ionic forces, hydrophobic interactions, and van der Waals forces.

**Doxorubicin** Also called adriamycin. Used to treat leukemia but also causes extensive mitochondrial defects, myocardial apoptosis and induces cardiomyopathy.

**DQAsomes** Liposome-like vesicles formed in aqueous medium with a dicationic amphiphile dequalinium used as a mitochondrial-specific delivery system for gene therapy.

**DSCR1** Down syndrome critical region 1 alias modulatory calcineurin interacting protein1 (MCIP1).

**dsRNA** Double-stranded RNA.

**EAD** Early afterdepolarization.

**EAM** Experimental autoimmune myocarditis.

**Early preconditioning** Also called acute or classic preconditioning, results from brief periods of ischemia applied 1–2 h before the index ischemia, occurs within a few minutes after the initial stimulus, and lasts for 2–3 h.

**EB** Embryoid bodies, aggregations of embryonic stem cells which can differentiate spontaneously *in vitro* to a variety of cell types, including cardiomyocytes.

**EC** Endocardial cushion.

**ECD** Extracellular domain.

**ECE-1** Endothelin-Converting Endopeptidase-1.

**ECM** Extracellular matrix.

**EDCFs** Endothelium-derived contracting factor.

**ED-FRAP** Enzyme-dependent fluorescence recovery after photobleaching used to measure the dehydrogenase activities associated with mitochondrial NADH generation.

**Ees** End-systolic elastance, a major determinant of systolic function and ventricular–arterial interaction.

**EGF** Epidermal growth factor. A low-molecular weight (6 kDa) polypeptide first purified from the mouse submandibular gland, but since then it has been found in many human tissues.

**EGFR** Epidermal growth factor receptor.

**ELAM** Endothelial leukocyte adhesion molecule.

**Electroporation** Method to transfect cells with either exogenous genes or proteins using electrical field.

**EMT** Endothelial–mesenchymal transdifferentiation.

**Endonucleases** Enzymes that cleave DNA at specific sites.

**Enhancer** A regulatory, short region of DNA that can dramatically enhance specific gene expression, often lying outside the promoter.

**eNOS** Endothelial nitric oxide synthase.

**EPC** Endothelial progenitor cell.

**EPDCs** Epicardium-derived cell.

**Epigenetic** Acquired and reversible modification of genetic material (e.g., methylation).

**Episomes** Genetic elements with extrachromosomal location (e.g., plasmids).

**Epitope** Part of a foreign organism or its proteins that is recognized by the immune system and targeted by antibodies, cytotoxic T cells or both.

**EPR** Epiregulin.

**ER** Endoplasmic reticulum. A cytosolic compartment, where lipids and membrane-bound proteins are synthesized.

**ErbB** Family of receptor tyrosine kinases, which mediate cell proliferation, migration, differentiation, adhesion, and apoptosis in numerous cell types, binding to a wide variety of ligands (e.g., EGF, neuregulins, TGF, and HB-EGF) and contribute to regulating endocardial cushion remodeling and valve formation.

**ERE** Estrogen response element, specific regulatory site within the DNA of target nuclear genes, which bound by ERs, produce long-term genomic effects of testosterone.

**ERK** Extracellular-regulated kinase.

**ERM** Ezrin–radixin–moesin proteins. A ROCK substrate.

**ERs** Estrogen receptors.

**ES** Embryonic stem cell.

**ESI** Electrospray ionization.

**Essential hypertension** The most common form of hypertension for which there is no recognized primary cause.

**EST** Expressed sequence tag. A sequence fragment of a transcribed protein-coding or non-protein coding DNA.

**ET** Endothelin, signaling peptides (ET-1, ET-2, and ET-3) modulate contractile function and growth stimulation of cardiomyocytes by binding specific G protein-coupled receptors and triggering downstream signaling (e.g., DAG, IP3).

**ETC** Electron transport chain. A series of complexes in the mitochondrial inner membrane to conduct electrons from the oxidation of NADH and succinate to oxygen.

**Exon** Segment of a gene that remains after the splicing of the primary RNA transcript and contains the coding sequences as well as 5′ and 3′ untranslated regions.

**Exonucleases** Enzymes that cleave nucleotides one at a time from an end of a polynucleotide chain.

**Expression vector** A vector that contains elements necessary for high-level and accurate transcription and translation of an inserted cDNA in a particular host or tissue.

**FA** Friedreich ataxia. An autosomal-dominant neuromuscular disorder with frequent HCM.

**FACS** Fluorescence-activated cell sorting, which can evaluate and isolate specific cell-types and cell-cycle stages.

**FAD** Flavin adenine dinucleotide. Common coenzyme of dehydrogenases. In the ETC, FAD is covalently linked to SDH.

**FADD** Fas-associated via death domain, adaptor protein recruiting procaspase into the apoptotic-promoting complex DISC.

**$FADH_2$** Flavin adenine dinucleotide (reduced form).

**FAF** Familial atrial fibrillation.

**Familial ligand-defective apoB** A common monogenic dominant hypercholesterolemic disorder resulting from mutation and deficiency of apolipoprotein B, the major protein of LDL, which reduces its affinity and binding to LDLR.

**FAO** Fatty acid oxidation.

**FAS** Fatty acid synthase.

**FasL** Fas ligand, death ligand in extrinsic apoptotic pathway.

**FCHL** Familial combined hyperlipidemia, characterized by elevated levels of plasma triglycerides, LDL and VLDL-cholesterol or both is the most common discrete hyperlipidemia and a common cause of premature atherosclerosis.

**FDA** Fluorescein diacetate, a fluorochrome used to discriminate between necrosis and apoptosis in intact cardiomyocytes.

**FDCM** Familial-dilated cardiomyopathy.

**FDG-PET** Fluor-Deoxy-Glucose-Positron-Emission-Tomography, used in the evaluation of levels of myocardial bioenergetic substrates, such as glucose.

**FFA** Free fatty acid.

**FGF** Fibroblast growth factor.

**FGF-2** Fibroblast growth factor-2.

**FH** Familial hypercholesterolemia; an autosomal-dominant disorder characterized by elevated cholesterol, and premature CAD, is the result of mutations that affect the LDLR.

**FHCM** Familial hypertrophic cardiomyopathy.

**FHF** First heart field.

**Fish eye disease** A rare autosomal-dominant disorder effecting HDL levels due to a deficiency in lecithin: cholesterol acyltransferase (LCAT).

**FLIP** Flice-inhibitory protein.

**FnBP** Fibronectin-binding proteins.

**FoxO** Forkhead transcription factor.

**FPLD** Familial partial lipodystrophy.

**f-QRS** Fragmented QRS complex.

**Frameshift mutation** Changes in the reading frame resulting from either an insertion or deletion of nucleotides.

**FRDA** Gene for frataxin, a mitochondrial-localized protein. Mutations in FRDA are responsible for FA.

**FRET** Fluorescence resonance energy transfer, technique useful in monitoring the fluctuations (and localization) of molecules (e.g., cyclic AMP) in living cells using fluorophores.

**FRTA** Free radical theory of aging.

**FTICR** Fourier transform ion cyclotron resonance.

**Functional genomics** A branch of molecular biology that makes use of the enormous amount of data produced by genome sequencing to delineate genome function.

**GATA** Family of zinc finger-containing transcription factors which contribute to the activation of the cardiac-specific gene program involved in cardiac cell differentiation.

**GC** Guanylyl cyclase.

**GDF** Growth differentiation factor.

**GEF** Guanine nucleotide-exchange factor.

**Gel shift** Differential mobility during gel electrophoresis; is used to gauge whether a specific DNA fragment (containing the regulatory motif of interest) is bound by an extract of nuclear proteins. The specific protein–DNA interaction is detectable by retardation in the mobility of the DNA fragment that is successfully bound.

**Gene product** The protein, tRNA, or rRNA encoded by a gene.

**Gene transfection** Introduction of DNA into eukaryotic cells.

**Genetic code** Correspondence between nucleotide triplets (codons) and specific amino acids in proteins.

**Genome** Total genetic information carried by a cell or an organism.

**Genomic library** Collection of DNA fragments (each inserted into a vector molecule) representative of the entire genome.

**Genotype** Genetic constitution of a cell or an organism.

**GFP** Green fluorescent protein. Useful marker for imaging localized proteins.

**GH** Growth hormone.

**GIK** Glucose, insulin, and potassium. Applied as a metabolic "cocktail" to provide beneficial preconditioning effects to injured myocardium.

**GK rat** Goto-Kakizaki rat, experimental model of type 2 diabetes.

**GLP-1** Glucagon-like peptide 1, cardioprotective agent.

**GLUT** Glucose transporter.

**Glycolysis** Cytosolic-located metabolic pathway present in all cells catalyzing the anaerobic conversion of glucose to pyruvate.

**Glycosylation** An enzyme-directed site-specific process, resulting in the addition of carbohydrate residues to proteins and lipids.

**Gordon's syndrome** Also pseudohypoaldosteronism Type II (PHA type II), a rare monogenic Mendelian trait characterized by familial hypertension with increased renal salt reabsorption, and impaired $K^+$ and $H^+$ excretion, and low renin activity found to be due to mutations in the WNK gene.

**GPCR** G protein-coupled receptor.

**GPD1-L** Glycerol-3-phosphate dehydrogenase 1-like. The β-subunit of $Na^+$ channel encoded by the *SCN1B* gene. The product encoded by the glycerol-3-phosphate dehydrogenase 1-like gene is a component of a large multiprotein complex of the functional $Na^+$ channel.

**G protein** A heterotrimeric membrane-associated GTP-binding protein involved in cell-signaling pathways; activated by specific hormone or ligand binding to GPCR.

**GPx** Glutathione peroxidase. An antioxidant enzyme with both mitochondrial and cytosolic isoforms.

**Grb-2** Growth factor receptor-bound protein-2.

**GRK** G protein-regulated kinase.

**GSH** Reduced glutathione.

**GSK-3B** Glycogen synthase kinase 3B, negative regulator of cardiac hypertrophy and of both normal and pathologic stress-induced growth.

**GSSG** Glutathione disulfide.

**GTP** Guanosine triphosphate.

**HA** Hyaluronic acid, a glycosaminoglycan composed of alternating glucuronic acid and *N*-acetylglucosamine residues, present in the ECM, to expand the extracellular space, regulate ligand availability and direct remodeling events in the cardiac jelly.

**HAGR** Human aging genomic resources.

**HAND proteins** Basic helix-loop-helix transcription factors (e.g., hand, eland) that play critical roles in early cardiac development.

**HAT** Histone acetyltransferase.

**HB-EGF** Heparin-binding epidermal growth factor.

**HCM** Hypertrophic cardiomyopathy.

**HCN** Hyperpolarization-activated cyclic nucleotide-gated channel.

**5-HD** 5-Hydroxydecanoic acid, selective mito$K_{ATP}$ channel blocker.

**HDs** Homology domains. All catalytic subunits of class I PI3Ks are 110-kDa proteins and share common structural features: they contain four HDs, including the catalytic domain (HR1), PIK domain (HR2), C2 domain (HR3), and Ras-binding domain (RBD or HR4).

**HDAC** Histone deacetylase. A class of enzymes that influence transcription by selectively removing acetyl groups from an ε-*N*-acetyl lysine amino acid on histone proteins.

**HDL** High-density lipoprotein.

**Helicase** Enzyme that separates the strands of DNA.

**HETE** Hydroxyeicosatetraenoic acid.

**Heterochromatin** Condensed regions of chromosomes containing less active genes.

**Heteroplasmy** Presence of more than 1 genotype in a cell.

**HF** Heart failure.

**H-FABP** Heart-type fatty acid-binding protein, an intracellular-binding protein, with potential clinical utility as an indicator of cardiac ischemia and necrosis.

**HGP** Human genome project.

**HiCM** Histiocytoid cardiomyopathy.

**HIF** Hypoxia-inducible factor.

**Histones** Chief proteins of chromatin acting as spools around which DNA winds. They play a role in regulation of gene expression.

**HLHS** Hypoplastic left heart syndrome, a heterogeneous group of developmental abnormalities in which there is a small or absent left ventricle with hypoplastic mitral and aortic valves rendering the left ventricle nonfunctional with systemic outflow obstruction appears to have a genetic component in its etiology.

**HNE** 4-Hydroxynonenal. A major product of endogenous lipid peroxidation.

**HNF** Hepatocyte nuclear factor.

**$H_2O_2$** Hydrogen peroxide; a form of ROS and marker of oxidative stress.

**HO-1** Heme oxygenase, antioxidant enzyme with cardioprotective function.

**Homocysteine** A reactive amino acid intermediate in methionine metabolism whose adverse effects include endothelial dysfunction with associated platelet activation and thrombus formation and accumulation of vascular atherosclerotic lesions.

**Homoplasmy** Presence of a single genotype in a cell.

**HOP** Homeodomain only protein, a small divergent protein that lacks certain conserved residues required for DNA binding, initiates gene expression early in cardiogenesis and is involved in control of cardiac growth during embryogenesis and early prenatal development and acts by transcription factor recruitment and chromatin remodeling.

**HOS** Holt-Oram syndrome, a rare inherited disease characterized mainly by abnormalities of the upper limbs and CHD resulting from mutations and haploinsufficiency of *TBX5*.

**HPETE** Hydroperoxyeicosatetraenoic acid.

**HRE** Hypoxia response element.

**HRT** Hormone replacement therapy.

**HSC** Hematopoietic stem cell.

**HSP** Heat-shock protein. A family of chaperones involved in protein folding.

**Htl** Heartless. The Heartless fibroblast growth factor receptor is required for the differentiation of a variety of mesodermal tissues in the *Drosophila* embryo.

**Hybridization** Binding of nucleic acid sequences through complementary base pairing. The hybridization rate is influenced by temperature, G-C composition, extent of homology, and length of the sequences involved.

**Hydrophobic** Lipophilic. Insoluble in water.

**IAA** Interrupted aortic arch, an extremely rare CHD defined as the loss of luminal continuity between the ascending and descending aorta caused by a variety of genetic defects.

**IAP** Inhibitor of apoptosis protein.

**IC** Inner curvature. Looping leads to the appropriate juxtaposition of heart regions and the formation of an inner curvature, and a trabeculated outer curvature, the future ventricular myocardium.

**ICAM-1** Inter-Cellular adhesion molecule 1. Cell surface adhesion molecule also known as cluster of differentiation 54 (CD54).

**ICD** Implantable cardioverter-defibrillator, implantation effective for treatment of short QT.

**IFM** Interfibrillar mitochondria.

**IFN-γ** Interferon-γ.

**IGF-1** Insulin-like growth factor, stimulates proliferative cardiomyocyte pathways and cell growth.

**IKK** κB kinase.

**ILK** Integrin-linked kinase.

**IMA** Ischemia-modified albumin, indicator of early myocardial ischemia and ACS.

**IMAC** Anion channel of the inner mitochondrial membrane.

**Infective endocarditis** Microbial infection of the endocardial surface of the heart, which commonly involves the heart valves.

***In situ* hybridization** Technique using DNA probes to localize specific transcripts within the cell in conjunction with microscopy.

**Integral membrane protein** Protein with at least one transmembrane segment requiring detergent for solubilization.

**Integrins** Class of transmembrane, cell-surface receptor molecules that constitute part of the link between the extracellular matrix and the cardiomyocyte cytoskeleton and which act as signaling molecules and transducers of mechanical force.

**Intermembrane space** Space between inner and outer membranes of mitochondria.

**Intron** A segment of a nuclear gene that is transcribed into the primary RNA transcript but is excised during RNA splicing and not present in the mature transcript.

**Ion channels** Multisubunit transmembrane protein complexes that perform the task of mediating selective flow of millions of ions per second across cell membranes, and are the fundamental functional units of biological excitability.

**IONA** Impact of nicorandil in angina trial have provided evidence of the clinical utility of pharmacological preconditioning by its demonstration that chronic administration of the KCO nicorandil, significantly improved the cardiovascular prognosis in CAD patients.

**Ionophore** Small hydrophobic molecule that promotes the transfer of specific ions through the membrane bilayer.

**IP3** Inositol trisphosphate, second messenger produced by phospholipase C (see PLC).

**IPC** Ischemic preconditioning.

**I/R** Ischemia/reperfusion.

**Iron-sulfur center** Nonheme iron ions complexed with cysteine chains and inorganic sulfide atoms making a protein capable of conducting electrons in electron transport or redox reactions.

**Iroquois homeobox genes** Family of homeodomain-containing transcription factor genes, implicated in cardiac chamber-specific gene expression.

**IRS-1** Insulin receptor substrate-1.

**Ischemic heart disease** Also called coronary artery disease (CAD) and coronary heart disease (CHD), this condition is caused by narrowing of the coronary arteries, thereby causing decreased blood supply to the heart.

**Isoforms** Related forms of the same protein generated by alternative splicing, transcriptional starts or encoded by entirely different genes.

**Isoschizomers** Pairs of restriction enzymes specific to the same DNA recognition sequence.

**JC-1** Fluorometric dye used for measuring/imaging mitochondrial membrane potential.

**JAK** Janus kinase.

**KAP** Kinase adaptor protein.

**Karyotype** A snap-shot of the number of chromosomes in the normal diploid cell, as well as their size distribution.

**Kawasaki disease** An acute self-limited vasculitis of infancy and early childhood has undetermined etiology and is the leading cause of acquired pediatric heart disease in US and Japan.

**KCOs** Potassium channel openers (e.g., nicorandil, diazoxide, and pinacidil) can mediate cardioprotection.

**KIR** Potassium inward rectifier.

**KLOTHO** A single-pass transmembrane protein that function in signaling pathways that suppress aging and which has β-glucuronidase activity.

**Knock-out mutation** A null mutation in a gene, abolishing its function (usually in transgenic mouse); allows evaluation of its phenotypic role.

**Krebs cycle** Central metabolic pathway of aerobic respiration occurring in the mitochondrial matrix; involves oxidation of acetyl groups derived from pyruvate to $CO_2$, NADH, and $H_2O$. The NADH from this cycle is a central substrate in the OXPHOS pathway. Also termed TCA or citric acid cycle.

**KSS** Kearns–Sayre syndrome. A mitochondrial neuropathy characterized by ptosis, ophthalmoplegia, and retinopathy with frequent cardiac conduction defects and cardiomyopathy.

**LA** Lipoic acid, a potent thiol antioxidant and mitochondrial metabolite, appears to increase low molecular weight antioxidants, decreasing age-associated oxidative damage.

**LAD** Left anterior descending.

**LAP** Latency associated peptide.

**LBBB** Left bundle branch block.

**LBD** Ligand-binding domain.

**LCAD** Long-chain acyl CoA dehydrogenase involved in FAO.

**LCFA** Long-chain fatty acid.

**LCHAD** Long-chain 3-hydroxylacyl-CoA dehydrogenase.

**LDL** Low-density lipoprotein, a cholesteryl ester-rich particle (containing only apoB100) whose plasma levels are elevated in several monogenic disorders of lipoprotein metabolism and lead to atherosclerosis.

**LDLR** LDL receptor, cell-surface receptor in liver or peripheral tissues responsible for LDL removal from blood; defective LDLR results in FH.

**LEOPARD Syndrome** Autosomal dominant syndrome characterized by multiple lentigines and cafe-au-lait spots, cardiac conduction abnormalities, obstructive cardiomyopathy, pulmonary stenosis, retardation of growth, and deafness, allelic to NS and due to mutations in *PTPN11*, a gene encoding tyrosine phosphatase SHP-2.

**LFA-1** Lymphocyte function-associated antigen 1.

**LHON** Leber hereditary optical neuropathy.

**Liddle's syndrome** An autosomal dominant monogenic form of hypertension with both hypokalemia and increased sodium reabsorption due to specific defects in either the β or γ subunit of the epithelial sodium channel (*ENaC*) causing gain-of-function of channel activation.

**Ligand** Any molecule that binds to a specific site on a protein or a receptor molecule.

**Ligase** Enzyme that joins together two molecules in an energy-dependent process; ligases are involved in DNA replication and repair. Extensively used in recombinant DNA.

**Liposomes** Lipid spheres with a fraction of aqueous fluid in the center used as vectors for gene transfection with plasmid DNA or oligonucleotides.

**LMP1** Latent membrane protein 1.

**L-NAME** L-nitro-arginine methyl ester.

**LOX** Lypoxigenase.

**LQT** Long QT syndrome; prolongation of the QT interval is a significant cause of syncope and SCD in children; delayed or prolonged repolarization of the cardiac myocyte can be acquired (e.g., drugs) or congenital (e.g., mutations in specific ion channels).

**LR** Left versus right identity in the developing heart.

**LTA** Lymphotoxin-alpha, an inflammation-mediating cytokine implicated in coronary artery plaque formation; polymorphic gene variants associated with MI.

**LTs** Leukotrienes. Like prostaglandins, they are biologically active lipids that play essential roles in the regulation of myocardial bioenergetics, contractility and various signaling pathways.

**Luciferase** ATP-dependent photoprotein; luciferase is used to fluorometrically assess the specific organelle ATP levels.

**LV** Left ventricle.

**LVAD** Left-ventricular assist device.

**LVH** Left ventricular hypertrophy.

**LTQ** Linear trap quadrupole.

**LXR** Liver X receptor. Nuclear receptors that regulate expression of more than a dozen proteins, many of them are parts of the cholesterol and fatty acid metabolic pathways.

**MAC** Mitochondrial apoptosis-induced channels.

**mAChR** Muscarinic acetylcholine receptor.

**MALDI** Matrix-assisted laser desorption/ionization. Method of ionization of proteins, peptides, sugars, and large organic molecules are ionized using for mass spectrometry.

**MAPC** Multipotent adult precursor cell.

**MAPK** Mitogen-activated protein kinases. A family of conserved serine/threonine protein kinases activated as a result of a wide range of signals involved in cell proliferation and differentiation; includes JNK and ERK.

**Marfan syndrome** Autosomal dominant connective tissue disorder presents with skeletal, ocular, and cardiovascular abnormalities, including high neonatal mortality due to polyvalvular involvement, aortic root, fatal aortic dissection or aortic insufficiency and severe CHF.

**Matrix** Space enclosed by the mitochondrial inner membrane.

**MAVS** Mitochondrial antiviral signaling.

**MBs** Molecular beacons. Hairpin-forming oligonucleotides labeled at one end with a quencher, and at the other end with a fluorescent reporter dye.

**MCAD** Medium-chain acyl-CoA dehydrogenase, FAO enzyme.

**MCD** Malonyl-CoA decarboxylase, an enzyme involved in regulation of malonyl CoA turnover.

**MCM** Mitochondrial cardiomyopathy.

**MCP-1** Monocyte chemoattractant protein-1.

**MCU** Mitochondrial $Ca^{2+}$ uniporter.

**MDA-5** Melanoma differentiation-associated gene 5.

**MEF2C** Myocyte enhancer factor 2C.

**MELAS** Mitochondrial encephalomyopathy with lactic acidosis and strokelike episodes.

**Membrane potential** Combination of proton and ion gradients across the membrane making the inside negative relative to the outside.

**MERRF** Mitochondrial cytopathy including myotonus, epilepsy, and ragged-red fibers.

**MetSyn** Metabolic syndrome.

**MHC** Myosin heavy chain.

**MI** Myocardial infarction.

**Microarray** A range of oligonucleotides immobilized onto a surface (chip) that can be hybridized to determine quantitative transcript expression or mutation detection.

**MIF** Migration inhibitory factor.

**Mineralocorticoid-induced hypertension** A monogenic autosomal dominant form of an early-onset hypertension, markedly exacerbated during pregnancy due to mutations in the *MR* hormone-binding domain.

**Minisatellites** Repetitive and variable DNA sequences, generally GC-rich, ranging in length from 10 to over 100 bp.

**MiRNAs** MicroRNAs. Regulators of cell proliferation.

**Mirtrons** MicroRNAs that are located in the introns of the mRNA-encoding host genes.

**MIS** Müllerian-inhibiting substance.

**Missense mutation** Mutation that causes substitution of one amino acid for another.

**$MitoK_{ATP}$ channel** ATP-sensitive mitochondrial inner membrane channel; activation of **$MitoK_{ATP}$** channel has been implicated as a central signaling event (both as trigger and end effector) in IPC and other cardioprotection pathways.

**MitoQ** Synthetic ubiquinone analog which can be selectively targeted to mitochondria; used to provide antioxidant cardioprotection.

**Mito VitE** Synthetic analog of vitamin E which can reduce mitochondrial lipid peroxidation and protein damage and accumulate after oral administration at therapeutic concentrations within the cardiac tissue.

**MKKK** MAPK kinase kinase.

**MLA** Monophosphoryl lipid A, a nontoxic derivative of the endotoxin pharmacophore lipid A, cardioprotective agent.

**MLC** Myosin light chain.

**MLCP** Myosin light chain phosphatase.

**MLP** Muscle LIM protein, localized in the cardiomyocyte cytoskeleton, a positive regulator of myogenic differentiation.

**MMPs** Metalloproteinases, enzymes involved in extracellular matrix remodeling.

**MNCX** Mitochondrial $Na^+/Ca^{2+}$ exchanger.

**Mn-SOD** Manganese-superoxide dismutase.

**Mobile carrier** Small molecule shuttling electrons between complexes in the ETC.

**Modifier gene** A gene that modifies a trait encoded by another gene.

**MODY** Maturity-onset diabetes of the young.

**MOMP** Mitochondrial outer-membrane permeabilization, an apoptotic event in part mediated by binding of proapoptotic proteins (e.g., Bad, Bax, and Bid) to mitochondria.

**Motif homology searching** Search for similarity patterns between proteins, which can prove highly informative about the structural and functional properties of the encoded protein.

**MPG** N-2-mercaptopropionylglycine, a free radical scavenger.

**MPO** Myeloperoxidase, indicator of pathological inflammation and for the risk of ACS.

**MR** Mineralocorticoid receptor.

**mRNA** Messenger RNA. Specifies the amino acid sequence of a protein; translated into protein on ribosomes. Transcript of RNA polymerase II.

**MRTF** Myocardin-related transcription factor.

**MSC** Mesenchymal stem cell.

**MSR-A** Macrophages scavenger receptor A.

**MT** Metallothionein. An inducible antioxidant metal-binding protein with cardioprotective properties.

**mtDNA** Mitochondrial DNA.

**mTOR** Mammalian target of rapamycin.

**MTP** Mitochondrial permeability transition pore. A non-specific megachannel in the mitochondrial inner membrane.

**Mutation** Change occurring in the genetic material (usually DNA or RNA).

**MVH** Midventricular hypertrophy. An unusual pattern of hypertrophy in which papillary muscle hypertrophy leads to obstruction of the midventricular cavity.

**MVP** Mitral valve prolapse.

**NADH** Nicotinamide adenine dinucleotide (reduced form).

**NAG** *N*-acetylglucosamine.

**NCC** Neural crest cell.

**NCOA1** Nuclear receptor coactivator 1. Polymorphism in this receptor is associated with diastolic blood pressure in women.

**NCX** $Na^+$-$Ca^{2+}$ exchanger.

**ND1** One of seven ND subunits in mtDNA encoding complex I.

**NE** Norepinephrine.

**NEFAs** Nonesterified fatty acid.

**NER** Nucleotide excision repair.

**NET-1** NE tranporter-1.

**Neuregulin** Family of endocardial-expressed peptide growth factors acting as potent ligands for tyrosine kinase receptors (ErbBs), and shown to promote growth and differentiation of embryonic cardiomyocytes, epithelial and glial cells.

**NFAT** Nuclear factor of activated T cells, a family of transcription factors controlled by the $Ca^{2+}$-regulated phosphatase, calcineurin.

**NFATc** A member of the NFAT family, exclusively expressed on the endocardium which plays a prominent role in the morphogenesis of the semilunar valves and septa.

**NF-κB** Nuclear Factor-Kappa B. Family of transcription factors involved in the control of a number of normal cellular and organismal processes, including immune and inflammatory responses, developmental processes, cellular growth, and apoptosis.

**NHERF** $Na^+/H^+$-exchanger regulatory factor.

**NIK** NF-κB-inducing kinase.

**NIP-1** Neuropilin-interacting protein-1.

**NKX2.5** An NK-class homeodomain-containing transcription factor homologous to *Drosophila melanogaster* *tinman* gene; is crucial in cardiac differentiation processes, including the establishment or maintenance of a ventricular gene expression program. While not essential for the specification of heart cell lineage, or for heart tube formation, it is required for the completion of the looping of the heart.

**nNOS** Neuronal NOS.

**NO** Nitric oxide; vasodilator.

**Nonmendelian inheritance** Cytoplasmic inheritance due to genes located in mitochondria.

**Nonsense codon** Any one of three triplets (UAG, UGA, or UAA) that cause termination of protein synthesis (same as stop codon).

**Nonsense mutation** Change in DNA specifying replacement of an amino acid codon by a nonsense codon.

**Northern blot** Molecular biology technique by which RNAs separated by electrophoresis are transferred and immobilized for the detection of specific transcripts by hybridization with labeled probe.

**NOS** Nitric oxide synthase.

**NOSIP** eNOS-interacting protein.

**NOSTRIN** eNOS trafficking inducer protein.

**Notch** A receptor family mediating an evolutionarily conserved signaling pathway involved in cell fate specification. Mutations in Notch signaling regulator have been implicated in aortic valve disease.

**NPD** Niemann–Pick disease. Rare recessively inherited lysosomal disorder characterized by multiorgan abnormalities resulting from lysosomal sphingomyelinase accumulation.

**NPPA** Natriuretic peptide precursor A gene.

**NPY** Neuropeptide Y.

**NRF-1 and NRF-2** Nuclear respiratory factors. Transcription factors that modulate expression of nuclear-DNA-encoded mitochondrial proteins.

**NRG-1** Neuroregulin-1. This protein is one of the four proteins in the neuregulin family, which plays key roles in both trabeculae formation and the differentiation of CCS cells.

**NRPs** Neuropilins. Transmembrane glycoproteins, which do not transduce signals themselves, but mediate functional responses as a result of complex formation with VEGF receptors.

**NS** Noonan syndrome. An autosomal-dominant disorder characterized by craniofacial dysmorphia, short stature, and cardiovascular defects, including pulmonary valvular stenosis, HCM and ASD, allelic to LS and due to mutations in *PTPN11*, a gene encoding tyrosine phosphatase SHP-2.

**Nt** Nucleotide, the basic unit of DNA composed; of a purine or pyrimidine base, a sugar (deoxyribose), and a phosphate group.

**NTG** Nitroglycerin.

**NT-proBNP** N-terminal pro-brain natriuretic peptide, indicator of hemodynamic stress and for the risk of congestive HF.

**Nuclease** Enzyme that catalyzes the degradation of DNA (DNAse) or RNA (RNAse); specific nucleases have been identified that target either the 5′ or 3′ ends of DNA (exonuclease) or that can digest nucleic acids from internal sites (endonucleases).

**Null mutation** Ablation or knock-out of a gene.

**NVPs** Nodal vesicular parcels. Leftward movement of sheathed lipidic particles.

**N-WASP** Neural Wiskott–Aldrich syndrome protein.

**OH** Origin of replication for mtDNA, heavy strand.

**OL** Origin of replication for mtDNA, light strand.

**Oligomycin** Specific inhibitor of mitochondrial ATP synthase and OXPHOS.

**Oligonucleotide** Short polymer of DNA or RNA that is usually synthetic in origin.

**ORF** Open reading frame. It contains a start codon and a stop codon.

**ORI** Origin of replication. Unique DNA sequence at which DNA replication is initiated from this point replication may proceed either bidirectionally or unidirectionally.

**OS** Oxidative stress.

**OxLDL** Oxidized LDL, a primary substrate for macrophage activation; involved in atherosclerosis progression.

**OXPHOS** Oxidative phosphorylation. A process in mitochondria in which ATP formation is driven by electron transfer from NADH and $FADH_2$ to molecular oxygen and by the generation of a pH gradient and chemiosmotic coupling.

**PAA** Pharyngeal arch arterie.

**PAAP** Platelet aggregation-associated protein.

**PAGE** Polyacrylamide gel electrophoresis.

**PAI-1** Plasminogen activator inhibitor-1, a principal regulator of fibrinolysis.

**PAR** Protease-activated receptors; PARs are G proteins coupled transmembrane receptors are activated by extracellular proteolytic cleavage by serine proteases, such as thrombin and trypsin.

**Paraoxonase** Antioxidant enzyme.

**PARP** Poly (ADP-ribose) polymerase.

**PCR** Polymerase chain reaction. An amplification of DNA fragments using a thermostable DNA polymerase and paired oligonucleotide primers subjected to repeated reactions with thermal cycling.

**PDA** Patent ductus arteriosus, a relatively common CHD that results when the ductus arteriosus, a muscular artery connecting the pulmonary artery to the descending aorta fails to remodel and close after birth, resulting in a left-to-right shunt.

**PDE** Phophodiesterase.

**PDGF** Platelet-derived growth factor.

**PDH** Pyruvate dehydrogenase (Also PDC and PDHC).

**PDK-1** Phosphoinositide-dependent kinase 1, enzyme downstream of PIP3 production in the PI3K pathway which becomes activated in part by its translocation to the plasma membrane, in proximity to its substrates which include Akt (PKB).

**PE** Proepicardium.

**Penetrance** The proportion of individuals with a specific genotype expressing the related phenotype.

**PEPCK** Phosphoenolpyruvate carboxykinase.

**Peptide** Short polymer of amino acids that can be produced synthetically.

**Peripheral membrane protein** Protein associated with membrane via protein–protein interactions; solubilized by changes in pH or salt.

**Peroxisome** Small organelle that uses oxygen to oxidize organic molecules, including fatty acids and contain enzymes that generate and degrade hydrogen peroxide ($H_2O_2$) (e.g., catalase).

**pFOX** Partial fatty acid oxidation.

**PG** Prostaglandin.

**PGC-1α** Peroxisome proliferator-activated receptor γ coactivator. Transcriptional regulator of mitochondrial bioenergetic and biogenesis operative during physiological transitions.

**PH** Pleckstrin homology.

**Phage display** Technique involving fusing proteins with a bacteriophage coat protein resulting in the display of the fused protein on the exterior surface of the phage, while the DNA encoding the fusion protein resides within; this permits the selection of proteins (and their physically attached DNAs) for specific-binding characteristics (e.g., antibody, DNA or specific ligand-binding) or functional features (e.g., enzymatic assay) by an *in vitro* screening process.

**Pharmaceutical preconditioning** A large variety of drugs, including the targeted use of volatile anesthetics, potassium channel openers, nitric oxide donors, and modulators of downstream pathways, including erythropoietin, statins, insulin, and pyruvate, mimicking ischemic preconditioning and provide cardioprotection when either substituted for the preconditioning period or applied at reperfusion.

**Pharmacogenetics** Study of the role of inheritance in interindividual variation in drug response.

**Pharmacogenomics** A branch of pharmaceutics dealing with the influence of genetic changes on drug response by correlating gene expression or SNPs with the drug's effect.

**Phenotype** Observable physical characteristics of a cell or organism resulting from the interaction of its genetic constitution (genotype) with its environment.

**Phospholamban** Negative regulator of SERCA.

**Phosphorothioate-modified antisense oligonucleotides** Substitution of sulfur for one of the oxygens in the phosphate backbone; renders the oligonucleotide more stable to nuclease degradation.

**PI3K** Phosphatidylinositol 3-kinase.

**PI3KR2** p85-β regulatory subunit of PI3K.

**PIP3** Phosphatidylinositol 3,4,5-triphosphate, product of PI3K activity.

**PKA** Protein kinase A. Activated by cAMP.

**PKB** Protein kinase B; also called *Akt*.

**PKC** Protein kinase C.

**Plakoglobin** Cytoplasmic protein of catenin family, a key component of desmosomes and adherens junctions. Mutant forms can promote ARVD and Naxos disease.

**Plakophilin** Desmosomal protein; mutational variants cause ARVD.

**Plasmid** A relatively autonomous replicating nonchromosomal DNA molecule primarily found in bacteria, which can be used as a vector for transferring recombinant genes to cells or tissues.

**$PLA_2$** Phospholipase $A_2$.

**PLB** Phospholamban. Transmembrane protein which is involved in the β-adrenergic regulation of cardiac cells.

**PLC** Phospholipase C, a potent effector enzyme catalyzing the hydrolysis of inositol phospholipids and production of second messengers, such as IP3 and DAG in response to agonists (e.g., acetylcholine) binding to membrane bound GPCRs. This signaling promotes a downstream increase in intracellular $Ca^{2+}$ levels, and PKC activation, which modulate myocardial contraction.

**Pleiotropic mutation** A single mutation with multiple (often unrelated) effects on organism.

**PMAC** Phosphate Metal Affinity Chromatography. Allows any phosphorylated protein to be enriched from cellular extracts.

**PMCA** Plasma membrane $Ca^{2+}$ ATPase.

**PNA** Peptide nucleic acid; an alternative delivery system for nucleic acids to mitochondria.

**Polγ** Nuclear-encoded catalytic subunit of mtDNA polymerase γ.

**Polyadenylation** Addition of a sequence of polyadenylic acid (poly A residues) to the 3′ end of most messenger RNAs after their transcription.

**Polygenic** A large number of genes each contributing a small amount to the phenotype.

**Porin** Pore-forming protein in the outer mitochondrial membrane (see VDAC).

**Positional cloning** A technique used to identify polymorphisms that flank the mapped allele, as well as to map and to clone mutant alleles that are not tagged for rapid cloning.

**Postconditioning** A series of brief interruptions of reperfusion applied at the very onset of reperfusion; can reduce infarct size and apoptosis and provide cardioprotection.

**Posttranslational modification** Postsynthetic modification of proteins by glycosylation, phosphorylation, proteolytic cleavage, or other covalent changes involving side chains or termini.

**p70s6k** 70-kDa ribosomal protein S6 kinase. It plays a key role in translational control of cell proliferation in response to growth factors in mammalian cells.

**PPAR** Peroxisome proliferator-activated receptor. Nuclear receptor transcription factors that function as transcriptional regulators in a variety of tissues, including the heart.

**PPH** Primary pulmonary hypertension, also called pulmonary arterial hypertension, (PAH). A rare autosomal dominant disease with incomplete penetrance characterized by distinctive changes in pulmonary arterioles that lead to increased pulmonary artery pressures, right ventricular failure, and death which can be acquired or congenital arising most commonly from mutations in BMPR2, ALK1, or ENG (encoding *endoglin*).

**PPI** Protein–protein interaction.

**Pre-RC** Pre-replication complex. It is formed on DNA replication origins by the assembly of several factors: origin recognition complex, Cdc6, Cdt1, and minichromosome maintenance (MCM) 2–7 helicase complex.

**Primer** Short nucleotide sequence that is paired with one strand of DNA and provides a free 3′-OH end at which a DNA polymerase starts the synthesis of a nascent chain.

**Promoter** Noncoding regulatory region of DNA sequence upstream of the gene-coding sequence involved in the binding of RNA polymerase to initiate transcription.

**Protein kinase** Enzyme that transfers the terminal phosphate group of ATP to a specific amino acid of a target protein.

**Proteome** Entire complement of proteins contained within the eukaryotic cell.

**Pseudogenes** Nonfunctional genes that are likely relics of evolution with strong homology to functional gene.

**PTA** Persistent truncus arteriosus.

**PTB** Phosphotyrosine binding.

**PTM** Posttranslational modification.

**PTP** Protein tyrosine phosphatase.

**PTPN** Protein tyrosine phosphatase nonreceptor.

**PTX** Pertussis toxin.

**PUFA** Polyunsaturated fatty acid.

**QTL** Quantitative trait *loci* for blood pressure regulation; have been identified in rat studies by genome-wide scanning and linkage studies in the rat.

**RA** Retinoic acid; plays role in cardiomyocyte differentiation.

**RAAS** Renin–angiotensin–aldosterone system.

**RACK** Receptor for activated C kinase.

**RAGE** Receptor for advanced glycation endproducts.

**RALDH2** Retinaldehyde dehydrogenase 2. Enzyme that catalyzes the second oxidative step in RA biosynthesis; its loss of function creates a severe embryonic RA deficiency.

**RaM** Rapid mode of uptake.

**RAMP2** Receptor activity-modifying protein 2.

**Ras** A small G protein (see small G proteins).

**Ras-GTP** Activated GTP-bound Ras.

**Rb** Retinoblastoma susceptibility protein.

**RBBB** Right bundle branch block.

**RCM** Restrictive cardiomyopathy.

**Reading frame** Contiguous and non-overlapping set of three nucleotide codons in DNA or RNA used to predict amino acid sequence.

**Real-Time PCR** Quantitative PCR technique employs simultaneous DNA amplification and quantification often using fluorescent dyes that intercalate with double-stranded DNA, and modified DNA oligonucleotide probes which fluoresce when hybridized with a complementary DNA.

**Recombinant DNA** An artificial DNA sequence resulting from the combination of two DNA sequences in a plasmid.

**Redox reactions** Oxidation–reduction reactions in which there is a transfer of electrons from an electron donor (the reducing agent) to an electron acceptor (oxidizing agent).

**Remote conditioning** Preconditioning, which is not confined to one organ, but also limits infarct size in remote, non-preconditioned organs.

**Reporter gene** A gene that is attached to another gene or regulatory element to be identified in cell culture, animals, or plants.

**Restenosis** The reclosing or renarrowing of an artery after an interventional procedure, such as angioplasty or stent placement.

**Restriction endonucleases** Endonucleases that recognize a specific sequence in a DNA molecule (usually palindromic) and cleave the DNA at or near that site.

**RFLP** Restriction fragment-length polymorphism. A variation in the length of restriction fragments due to the presence or absence of a restriction site.

**RGS proteins** Regulators of G protein-signaling proteins, a family of proteins that accelerate intrinsic GTP hydrolysis on $\alpha$ subunits of trimeric G proteins and play crucial roles in the physiological regulation of G protein-mediated cell signaling.

**Rhod-2AM** Fluorescent calcium indicator used to assess $Ca^{2+}$ uptake, localization and levels in cardiomyocytes.

**Rhodamine 123** A fluorescent dye used to stain mitochondria in living cells.

**Rho-GEF** Rho-guanine nucleotide exchange factor.

**Ribosome** A factory-like organelle that builds proteins from a set of genetic instructions. Composed of rRNA and ribosomal proteins, it translates mRNA into a polypeptide chain.

**Ribozyme** RNA molecule with endonucleolytic activity, which can be used to selectively target specific gene expression.

**RIG-I** Retinoic acid-induced protein I.

**RISC** RNA-induced silencing.

**RNAi** RNA interference, use of a specific double-stranded RNA (dsRNA) construct to silence specific gene expression posttranscriptionally.

**RNA polymerase** Enzyme responsible for transcribing DNA as template into RNA.

**RNS** Reactive nitrogen species.

**ROK** Rho-associated kinase.

**ROS** Reactive oxygen species, including superoxide, hydroxyl radicals, and hydrogen peroxide.

**Rotenone** Specific inhibitor of complex I activity.

**rRNA** Ribosomal RNA. A central component of the ribosome.

**RSK** Ribosomal S6 kinase.

**R-Smad** Receptor-associated Smad protein.

**RTK** Receptor tyrosine kinase; this large family of proteins includes receptors for many growth factors and insulin; ligand-binding results in dimerization and phosphorylation of downstream-signaling targets as well as autophosphorylation.

**RT-PCR** Reverse transcription (RT) of RNA to DNA with the enzyme reverse transcriptase can be combined with traditional PCR to allow the amplification and determination of the abundance of specific RNA.

**RV** Right ventricle.

**RXFP** Relaxin family peptide.

**RXR** Retinoid X receptor. On binding 9-*cis* retinoic acid, RXR acts as a heterodimer and as a repressor or activator of specific gene transcription, playing a key role in cardiac development and physiological gene expression.

**Ryanodine receptor** Major SR $Ca^{2+}$ release channel in cardiac muscle; mutations in the cardiac isoform encoded by *RyR2* result in ARVD and CVPT.

**SAGE** Serial analysis of gene expression. Quantitative analysis of RNA transcripts by using short sequence tags to generate a characteristic expression profile.

**SAME** Syndrome of apparent mineralocorticoid excess. A rare, autosomal recessive form of hypertension presenting in infancy or childhood as an acquired syndrome resulting from the inhibition of 11 $\beta$-hydroxysteroid dehydrogenase, an enzyme responsible for converting cortisol to cortisone or from mutations in the renal-specific isoform gene for 11 β-hydroxysteroid dehydrogenase (*11-β HSD*).

**SA node** The sinoatrial node is a group of specialized cells in the top of the right atrium which produces electrical impulses (a relatively simple action potential) that travel down to eventually reach the ventricular muscle causing the heart to contract; serves as the "natural" pacemaker of the heart.

**Sca-1** Stem cell antigen-1.

**SCAD** Short-chain acyl CoA dehydrogenase, enzyme involved in FAO.

**SCD** Sudden cardiac death.

**SCD-1** Stearoyl-coenzyme A desaturase-1.

**SD** Sudden death.

**SDDS** Sudden dysrhythmia death syndrome.

**SDH** Succinate dehydrogenase. A TCA cycle enzyme associated with respiratory complex II.

**SDS** Sodium dodecyl sulfate. An ionic detergent used for the solubilization, denaturation of proteins, and their size separation in PAGE.

**SELE** E-selectin gene. In Japanese patients, a Ser128Arg polymorphism has been found in this gene in association with myocardial infarction.

**SELP** P-selectin gene. Associated with increased risk for myocardial infarction.

**Septal defect** A hole in the wall of the heart separating the atria (ASD) or in the wall separating the ventricles (VSD).

**SERCA** Sarcoplasmic reticulum $Ca^{2+}$-ATPase. There are three major isoforms which are variably expressed in different muscle types.

**SERM** Selective estrogen receptor modulator.

**SfA** Sanglifehrin A. Mediates inhibition of MTP opening.

**SGK1** Serum and glucocorticoid-responsive kinase-1.

**SHF** Second heart field. Provided a potential explanation for the observations that many genes and transgenes are expressed in the RV and OFT, but not in the left ventricle (LV) or atria, and that there are several mutations in mice and diseases in humans that selectively impact the RV.

**SHR** Spontaneously hypertensive rat.

**Sick sinus syndrome** The failure of the sinus node to regulate the heart's rhythm.

**Signal sequence** Amino acid sequence for targeting proteins into specific organelles (e.g., mitochondria, nucleus) often but not invariably located at N-terminus.

**Silent mutation** Mutation that alters a particular codon, but not the amino acid and does not affect protein or phenotype.

**SINES** Single interspersed element. Family of repeated sequence present in the human genome.

**SI/R** Simulated ischemia/reperfusion.

**siRNA** Small interfering RNA. Sometimes known as short interfering RNAs, they are a class of 20–25 nucleotide-long RNA molecules that interfere with the expression of genes.

**Sitosterolemia** A rare autosomal recessive hypercholesterolemic disorder caused by mutations in either sterolin-1 or -2, members of the ABC transporter family.

**SK** Sphingosine kinase.

**Smac/Diablo** Mitochondrial intermembrane protein released into the cytosol during early apoptosis stimulating caspase activation.

**Smads** A group of cytoplasmic-signaling proteins that upon phosphorylation in response to BMP, TGF-β, and BMPR signaling, translocate to the nucleus and directly regulate gene transcription.

**Small G proteins** Superfamily of guanine nucleotide-binding proteins, including Ras, Rho, Rab, Ran, and ADP ribosylation factor(s), which act as molecular switches to regulate cardiac myocyte hypertrophy and survival associated with cell growth and division, cytoskeletal events, vesicular transport, and myofibrillar apparatus. As with heterotrimeric G proteins, they are activated by exchange of GDP to GTP, and inactivated by return to a GDP-bound state but not mediated by agonist-occupied receptors rather primarily mediated by activation of guanine nucleotide exchange factors (GEFs).

**SMase** Sphingomyelinase.

**SMC** Smooth muscle cell.

**Snail/Slug factors** A group of zinc-finger transcription factors that primarily act as transcriptional repressors.

**SNIP1** Smad nuclear-interacting protein 1.

**S-nitrosylation** A ubiquitous posttranslational modification involving the covalent attachment of NO to cysteine thiol moieties on targeted proteins.

**SNP** Single nucleotide polymorphism.

**SNX27** Sorting nexin 27. Interaction with this protein sequesters the 5-$HT_{4(a)}$ receptor in early endosomes.

**SOCS** Suppressor of cytokine signaling.

**SOD** Superoxide dismutase. An antioxidant ROS-scavenging enzyme with both mitochondrial (MnSOD) and cytosolic (CuSOD) isoforms.

**Southern blot** Detection of separated restriction fragments after size separation on agarose gels, transfer to membranes and hybridization with labeled gene probes.

**SP cells** Side population cells; rare groups of multipotent progenitor cells capable of proliferation and differentiation.

**SPECT** Single-photon emission computed tomography used to assess myocardial metabolism and screen for CAD.

**SPIs** Serine protease inhibitors, termed serpins; key regulators of numerous cardiovascular pathways that initiate inflammation, coagulation, angiogenesis, apoptosis, extracellular matrix composition, and complement activation responses.

**Splicing** Reaction in the nucleus in which introns are removed from primary nuclear RNA and exons joined to generate mRNA.

**SPRED-1** Sprouty-related protein-1. This protein localizes in a lipid raft/caveola and inhibits Erk activation together with caveolin-1.

**SQTS** Short QT syndrome. A disorder associated with alterations of ion signaling in the heart. This autosomal-dominant syndrome can affect infants, children, and young adults. Patients display a marked propensity for atrial fibrillation and increased risk for sudden cardiac death from ventricular tachydysrhythmias.

**SR** Sarcoplasmic reticulum. A network of internal membranes in muscle-cell cytosol that contains high $Ca^{2+}$ concentration, which is released on excitation.

**SREBP-1c** Sterol regulatory element-binding protein 1c.

**SRF** Serum response factor, a transcription factor, required for the formation of vertebrate mesoderm leading to the origin of the cardiovascular system.

**SSCP** Single strand conformation polymorphism; technique of mutation detection.

**SSM** Subsarcolemmal mitochondria.

**SSS** Sick sinus node syndrome.

**STAT** Signal transducer and activator of transcription.

**Statins** HMG-CoA reductase inhibitors used to treat patients with elevated plasma LDL.

**STEMI** ST-segment elevation myocardial infarction.

**Sticky ends** Results from restriction endonuclease enzymes making a staggered cut across the two strands of DNA with one of the two molecules containing a short single-stranded unique overhang; useful in recombinant DNA engineering.

**Stop codon** See nonsense codon.

**SUMO** Small ubiquitin-like modifier.

**Supravalvular aortic stenosis** Discrete narrowing of the ascending aorta resulting from mutations in the gene encoding a component of the extracellular matrix (i.e., elastin).

**SUR** Sulfonylurea receptor.

**SVT** Supraventricular tachycardia.

**T3** Triiodothyronine.

**TACE** Tumor necrosis factor-α-converting enzyme.

**TAG** Intracellular triacylglycerol.

**Taq polymerase** Thermostable DNA polymerase isolated from the bacterium *Thermus aquaticus* used extensively in PCR.

**T box genes** Family of highly conserved transcription factors involved in early embryonic cell fate decisions, regulatory development of extraembryonic structures, embryonic patterning, and many aspects of organogenesis.

**TCA cycle** Tricarboxylic acid cycle (see Krebs cycle).

**TCE** Trichloroethylene, a halogenated hydrocarbon, associated with increased CHD incidence in animal studies and in pregnant women exposed to TCE-contaminated water.

**TC-PTP** T-cell protein tyrosine phosphatase.

**T2D** Type 2 diabetes.

**TD** Tangier disease, a rare monogenic autosomal codominant atherosclerotic disease characterized by the absence of HDL and very low plasma levels of apoA. It is caused by mutations in the ATP-binding cassette transporter gene (*ABCA1*).

**TdP** *Torsade de pointes*, a polymorphic ventricular tachycardia that can be followed by syncope and SD; this can be acquired by exercise (swimming) or congenital (any of the LQTs).

**Telomerase** An enzyme that recognizes the G-rich strand, and elongates it using an RNA template that is a component of the enzyme itself.

**Telomere** Special structure containing tandem repeats of a short G-rich sequence present at the end of a chromosome.

**TERC** Telomerase RNA component.

**TERT** Telomerase reverse transcriptase catalytic subunit.

**TF** Tissue factor.

**TFAM** Mitochondrial transcription factor A. Also known as mtTFA.

**TFP** Mitochondrial trifunctional protein. A multienzyme complex of the β-oxidation cycle.

**TGA** Transposition of the great arteries, the most common cyanotic neonatal lesion has shown in several cases a monogenic inheritance pattern.

**TGCE** Temperature gradient capillary electrophoresis, a sensitive technique coupling heteroduplex analysis with capillary electrophoresis; used to efficiently scan an entire coding region to identify a wide spectrum of mutations.

**TGF** Transforming growth factor.

**TH** Thyroid hormone (also thyroxin), a stimulus of cardiac hypertrophic growth, and myocardial mitochondrial biogenesis.

**Thrombospondins** Family of extracellular matrix glycoproteins with a role in platelet adhesion, modulation of vascular injury, coagulation, and angiogenesis and MI.

**TIM** Protein complex in mitochondrial inner membrane required for protein import.

**TIMP** Tissue inhibitor of the metalloproteinase.

**TIRFM** Total internal reflection fluorescent microscopy. In this technique, fluorescent proteins are selectively excited by evanescent wave of light, which penetrates to very short depth into the sample.

**Titin** Large polypeptide, anchored in the Z-disc spanning the sarcomere; contributes to sarcomere organization, myofibrillar elasticity and myofibrillar cell signaling.

**TLR** Toll-like receptor; involved in the innate-immunity signaling response of the macrophage, including pattern-recognition of pathogens and oxidized LDL, leukocyte recruitment and production of local inflammation and downstream signaling in atherosclerotic progression.

**TNF-α** Tumor necrosis factor α.

**TOF** Tetralogy of Fallot, most common form of complex CHD featuring VSD, obstructed right ventricular outflow, aortic dextroposition, and right ventricular hypertrophy.

**TOM** Protein complex in mitochondrial outer membrane required for protein import.

**Topoisomerase** Enzyme that changes the supercoiling of DNA.

**TP** Thromboxane receptor. Stimulation of G protein-coupled receptor leads to smooth muscle contraction.

**tPA** Tissue-type plasminogen activator, a primary regulator of fibrinolysis.

**TR** Thyroid hormone receptor, mediates both nuclear genomic effects of TH (largely as a transcription factor) as well as nongenomic effects of TH.

***Trans*-acting elements** Regulatory elements that mediate specific gene expression. They are not located within or near the gene (e.g., proteins that bind and regulate specific promoters).

**Transcript** Specific RNA product of DNA transcription.

**Transcription factor** Protein required for the initiation of transcription by RNA polymerase at specific sites and functioning as a regulatory factor in gene expression.

**Transcriptome** Comprehensive transcript analysis for expression profiling.

**Transgenesis** Introduction of an exogenous gene – called a transgene – into a living organism so that the organism exhibits a new property and transmits that property to its offspring.

**Transgenic animal** Animal that has stably incorporated one or more genes from another cell or organism.

**Translation** Synthesis of protein from the mRNA template at the ribosome.

**Transposons** Sequences of DNA that can move/transpose around to different locations within the genome of a cell.

**TRCP** Transient receptor potential channels.

**TREs** Thyroid hormone-responsive elements.

**TRF2** Telomere repeat-binding factor, telomere-associated protein critical for the control of telomere structure and function.

**Triplet Repeat Syndromes** Inherited neuromuscular disorders caused by expanded repeats of trinucleotide sequences within specific genes including Friedreich ataxia (FA) and myotonic muscular dystrophy (MMD).

**TRK** Tropomyosin receptor kinase.

**tRNA** Transfer RNA. A small RNA molecule used in protein synthesis as an adaptor between mRNA and amino acids.

**TS** Timothy syndrome (also termed LQT8), a multiorgan disease that includes lethal cardiac dysrhythmias, webbing of fingers and toes, congenital heart disease, immune deficiency, intermittent hypoglycemia, cognitive abnormalities, and autism. It is due to a missense mutation (G406R) in the *CACNA1c* gene, encoding the L-type $Ca^{2+}$ α-subunit Cav1.2.

**TTP** Thrombotic thrombocytopenic purpura, an autosomal recessive relapsing form of severe thrombotic microangiopathy characterized by marked thrombocytopenia, systemic platelet aggregation, erythrocyte fragmentation, and organ ischemia and caused by mutations in the *ADAMTS13* gene.

**TUNEL** dUTP nick end labeling.

**Two-dimensional electrophoresis** Technique for separating proteins based on their size and charge differences.

**Two-hybrid system** Method to detect proteins that interact with each other using yeast gene expression.

**TX** Thromboxane.

**TZD** Thiazolidinedion; agonists that activate PPAR-γ and are able to improve the actions of insulin in diabetes.

**Ub** Ubiquitin. Many intracellular processes are regulated through an enzymatic conjugation of cellular proteins' lysines to the C-terminal glycine residue of conserved 76 amino acid protein ubiquitin, a process named ubiquitination.

**UCP** Uncoupling protein.

**Uncoupler** Protein or other molecule capable of uncoupling electron transport from oxidative phosphorylation.

**uPA** Urokinase-type plasminogen activator.

**Vasculogenesis** The *de novo* formation of the first primitive vascular plexus and postnatal vascularization.

**VASP** Vasodilator-stimulated phosphoprotein that may be involved in PKG I regulation of electrical coupling.

**VCAM-1** Vascular cell adhesion molecule.

**VDAC** Voltage-dependent anion channel in mitochondrial outer membrane (see porin).

**VDR** Vitamin D receptor, involved in signaling cardiac morphogenesis.

**VDRE** Vitamin D response element.

**VEGF** Vascular endothelial growth factor.

**Versican** An ECM-localized chondroitin sulfate proteoglycan, which binds HA; expressed in the pathways of neural crest cell migration and in prechondrogenic regions and has been associated with valvulogenesis in the developing heart.

**VGCC** Voltage-gated calcium channel.

**VIP** Vasoactive intestinal peptide.

**VLA-4** Very late antigen-4.

**VLCAD** Very long-chain acyl CoA-dehydrogenase; enzyme involved in mitochondrial β-oxidation of fatty acids.

**VLDL** Very low density lipoprotein, a triglyceride-rich lipoprotein containing apoB100 which progressively become enriched in cholesteryl ester (CE) as a result of CE transfer from HDL and is converted by lipolysis to LDL and/or taken up as VLDL remnants by the liver.

**VMHC1** Ventricle myosin heavy chain-1.

**VSD** Ventricular septal defect.

**VSMC** Vascular smooth muscle cell.

**VWF** Von Willebrand Factor. Blood glycoprotein implicated in hemostasis.

**Western blot** Immunochemical detection of proteins immobilized on a filter after size separation by PAGE.

**Wild-type** The common genotype or phenotype of a given organism occurring in nature.

**Williams syndrome** A rare autosomal-dominant disorder characterized by supravalvular aortic stenosis and stenosis of systemic and/or pulmonary arteries and has been associated with a large-scale deletion in chromosome 7 (including the elastin gene).

**WPW** Wolff–Parkinson–White syndrome presents with hypertrophic cardiomyopathy, ventricular preexcitation, conduction defects, and accumulation of cardiac glycogen.

**X-linked inheritance pattern** The presence of the gene of interest on the X-chromosome.

**XIAP** X-linked inhibitor of apoptosis.

**XO** Xanthine oxidase, cytosolic enzyme involved in purine metabolism, involved in myocardial ROS production (e.g., superoxide radicals) particularly after I/R injury.

**YAC** Yeast artificial chromosomes, shuttle vectors designed to contain yeast chromosomal elements (i.e., centromeric and telomeric sequences) which segregate as chromosomes and allow the incorporation of very large inserts of heterologous DNA.

**Y2H** Two-hybrid screening system. A technique used to discover protein–protein interactions and protein–DNA interactions.

**Z-discs** Cardiomyocytes component positioned at the junction between the cytoskeleton and the myofilaments, providing a physical connection between the sarcomere, nucleus, membrane, and sarcoplasmic reticulum (SR) with a role in cardiac contraction and signaling.

# Index

**D**

**I**



**J**



**K**



**L**